MOORE'S
Essential
Clinical
Anatomy

SIXTH EDITION

MOORE'S
Essential
Clinical
Anatomy

SIXTH EDITION

Anne M. R. Agur, BSc (OT), MSc, PhD
Professor, Division of Anatomy, Department of Surgery, Faculty of Medicine
Division of Physical Medicine and Rehabilitation, Department of Medicine
Department of Physical Therapy, Department of Occupational Science &
   Occupational Therapy
Division of Biomedical Communications, Institute of Communication, Culture,
   Information, and Technology
Institute of Medical Science
Rehabilitation Sciences Institute
Graduate Department of Dentistry
University of Toronto
Toronto, Ontario, Canada

Arthur F. Dalley II, PhD, FAAA
Professor Emeritus and Research Professor, Department of Cell and
   Developmental Biology
Adjunct Professor, Department of Orthopaedic Surgery and Rehabilitation
Vanderbilt University School of Medicine
Adjunct Professor for Anatomy
Belmont University School of Physical Therapy
Nashville, Tennessee

Founding coauthor (with Anne M. R. Agur) and coauthor for first to fifth editions:

Keith L. Moore, MSc, PhD, Hon. DSc, FIAC, FRSM, FAAA
Professor Emeritus, Division of Anatomy
Department of Surgery
Former Chair of Anatomy
Associate Dean for Basic Medical Sciences
Faculty of Medicine, University of Toronto
Toronto, Ontario, Canada

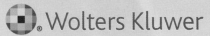 Wolters Kluwer

Philadelphia • Baltimore • New York • London
Buenos Aires • Hong Kong • Sydney • Tokyo

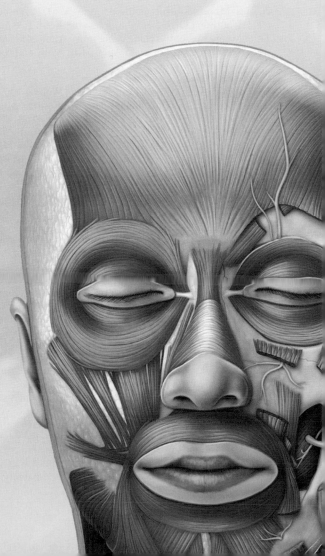

Acquisitions Editor: Crystal Taylor
Development Editor: Kathleen H. Scogna (freelance); Andrea Vosburgh (in-house)
Editorial Coordinator: Lindsay Ries
Marketing Manager: Jason Oberacker
Production Project Manager: Joan Sinclair
Design Coordinator: Stephen Druding
Art Director: Jennifer Clements
Artists: Imagineeringart.com, lead artist Natalie Intven, MSc, BMC; Dragonfly Media Group, lead artist Rob Duckwall
Manufacturing Coordinator: Margie Orzech
Prepress Vendor: Absolute Service, Inc.

Sixth Edition

9 8 7 6 5 4 3 2 1

Printed in China

Not authorised for sale in United States, Canada, Australia, New Zealand, Puerto Rico, and U.S. Virgin Islands.

Library of Congress Cataloging-in-Publication Data

Names: Moore, Keith L., author. | Agur, A. M. R., author. | Dalley, Arthur
F., II, author. | Digest of (work): Moore, Keith L. Clinically oriented anatomy.
Title: Moore's essential clinical anatomy / Anne M.R. Agur, Authur F. Dalley
II ; founding coauthor (with Anne M.R. Agur) and coauthor for first to
fifth editions, Keith L. Moore.
Other titles: Essential clinical anatomy
Description: Sixth edition. | Philadelphia : Wolters Kluwer, [2019] |
Preceded by Essential clinical anatomy / Keith L. Moore, Anne M.R. Agur,
Arthur F. Dalley II. Fifth edition. [2015]. | Keith L. Moore's name
appears first in the previous editions. | Parent text: Clinically oriented
anatomy / Keith L. Moore, Arthur F. Dalley, Anne M.R. Agur. Eighth
edition. [2018]. | Includes bibliographical references and index.
Identifiers: LCCN 2018048208 | ISBN 9781496369659 (paperback)
Subjects: | MESH: Anatomy | Handbooks
Classification: LCC QM23.2 | NLM QS 39 | DDC 612—dc23 LC record available
at https://lccn.loc.gov/2018048208

shop.lww.com

RRS1811

To my husband, Enno, and my family, Kristina, Erik, and Amy, for their support and encouragement.
—AMRA

To Muriel,
my bride, best friend, counselor, and mother of our sons,
and to our family—
Tristan, Lana, Elijah, Finley, Sawyer, and Dashiell; Denver; Skyler, Sara, and Dawson—with love
and great appreciation for their support, understanding, good humor,
and, most of all, patience.
—AFD

In Loving Memory of Marion
My best friend, wife, colleague, mother of our five children, and grandmother of our nine
grandchildren for her love, unconditional support, and understanding.
Wonderful memories keep you in our hearts and minds.
—KLM

And with sincere appreciation for the anatomical donors without whom our studies would not be
possible and for the support and patience of their families.

Anne M.R. Agur, BSc (OT), MSc, PhD

Arthur F. Dalley II, PhD, FAAA

**Keith L. Moore, MSc, PhD,
Hon. DSc, FIAC, FRSM, FAAA**

Twenty-two years have passed since the first edition of *Essential Clinical Anatomy* was published. The main aim of the sixth edition is to provide a compact yet thorough textbook of clinical anatomy for students and practitioners in the health care professions and related disciplines. With each edition, we strive to make the book even more student friendly. The basic approach that underlies this textbook is to

- provide a basic text of human clinical anatomy for use in current health sciences curricula.
- present an appropriate amount of clinically relevant anatomical material in a readable and interesting form.
- place emphasis on clinical anatomy that is important for practice.
- provide a concise clinically oriented anatomical overview for clinical courses in subsequent years.
- serve as a rapid review when preparing for examinations, particularly those prepared by the National Board of Medical Examiners.
- offer enough information for those wishing to refresh their knowledge of clinical anatomy.

This edition has been thoroughly revised, keeping in mind the many invaluable comments received from students, colleagues, and reviewers. Key features include the following:

- The art program continues to undergo revision and refinement with each edition. All of the illustrations are full color, highlight important facts, and show anatomy in relation to clinical medicine and surgery. A great effort has been made to further improve clarity of labeling and to place illustrations on the pages being viewed as the illustrations are cited in the text.
- New overview illustrations of the sensory and motor innervation of the upper and lower limbs facilitate integration.
- A description of the structure and function of the enteric nervous system and its unique role in the innervation of the digestive tract has been added that highlights important new information about this system's structure and function.
- New surface anatomy photographs of clinical procedures and their relevant anatomy emphasize the importance of knowledge of clinical anatomy.
- More illustrated clinical correlations, known as "clinical blue boxes," have been included to help students

understand the practical value of anatomy. In response to our readers' suggestions, the clinical boxes have been grouped. They are also classified by the following icons to indicate the type of clinical information covered:

Anatomical variations. These blue boxes feature anatomical variations that may be encountered in the dissection lab or in practice, emphasizing the clinical importance of awareness of such variations.

Life cycle. These blue boxes emphasize prenatal developmental factors that affect postnatal anatomy and anatomical phenomena specifically associated with stages of life—childhood, adolescence, adult, and advanced age.

Trauma icon. The effect of traumatic events—such as fractures of bones or dislocations of joints—on normal anatomy and the clinical manifestations and dysfunction resulting from such injuries are featured in these blue boxes.

Diagnostic procedures icon. Anatomical features and observations that play a role in physical diagnosis are targeted in these blue boxes.

Surgical procedures. These blue boxes address such topics as the anatomical basis of surgical procedures, such as the planning of incisions and the anatomical basis of regional anesthesia.

Pathology. The effect of disease on normal anatomy such as cancer of the breast, and anatomical structures or principles involved in the confinement or dissemination of disease within the body are the types of topics covered in these blue boxes.

- Surface anatomy is integrated into the discussion of each region to demonstrate the relationship between anatomy and physical examination, diagnosis, and clinical procedures.
- Medical images of radiographic, computed tomography (CT), magnetic resonance imaging (MRI), and ultrasonographic studies have been included, often with correlative illustrations. Current diagnostic imaging techniques demonstrate anatomy as it is often viewed clinically.

- Student resources, including case studies accompanied by clinico-anatomical problems and USMLE-style multiple-choice questions, are available to students online at http://thePoint.lww.com/MooreECA6e, providing a convenient and comprehensive means of self-testing and review.
- Instructors may contact their sales representative through http://thePoint.lww.com/MooreECA6e for information about accessing the instructor resources, including images, for use in their teaching and course materials.

The terminology adheres to the *Terminologia Anatomica* (1998) approved by the International Federation of Associations of Anatomists (IFAA). The official English equivalent terms are used throughout the present edition. When new terms are introduced, however, the Latin forms as used in Europe, Asia, and other parts of the world appear in parentheses. The roots and derivation of terms are included to help students understand the meaning of the terminology. Eponyms, although not endorsed by the IFAA, appear in parentheses to assist students during their clinical studies.

The parent of this book, *Clinically Oriented Anatomy* (*COA*), is recommended as a resource for more detailed descriptions of human anatomy and its relationship and importance to medicine and surgery. *Moore's Essential Clinical Anatomy*, in addition to its own unique illustrations and manuscript, has utilized materials from *Clinically Oriented Anatomy* and *Grant's Atlas of Anatomy*.

We again welcome your comments and suggestions for improvements in future editions.

Anne M. R. Agur
University of Toronto
Faculty of Medicine

Arthur F. Dalley II
Vanderbilt University
School of Medicine

Acknowledgments

We wish to thank the following colleagues who were invited by the publisher to assist with the development of this sixth edition.

REVIEWERS FOR THE SIXTH EDITION

Keiichi Akita, MD, PhD
Professor and Chair
Department of Clinical Anatomy
Tokyo Medical and Dental University
Tokyo, Japan

Quentin A. Fogg, BSc (Hons), PhD, FRCPS (Glasg)
Associate Professor, Clinical Anatomy
Department of Anatomy and Neuroscience
The University of Melbourne
Victoria, Australia

Chelsea M. Lohman-Bonfiglio, PhD, ATC, CSCS
Associate Professor and Director of Curriculum and
    Instruction, Clinical Anatomy
Department of Interdisciplinary Health Sciences
Arizona School of Health Sciences
A.T. Still University
Mesa, Arizona

Geoffroy Noel, PhD
Associate Professor and Director, Division of
    Anatomical Sciences
Department of Anatomy and Cell Biology
McGill University
Montreal, Quebec, Canada

Bassam Nyaeme, MD
Faculty, Basic Health Sciences
British Columbia Institute of Technology
Burnaby, British Columbia, Canada

Monica Oblinger, PhD
Professor, Vice Chair and Associate Vice President for
    Research Compliance
Department of Cell Biology and Anatomy
Chicago Medical School, School of Graduate and
    Postdoctoral Studies
Chicago, Illinois

Rebecca L. Pratt, PhD
Professor of Anatomy
Department of Foundational Medical Studies
Oakland University William Beaumont School of Medicine
Rochester, Michigan

Hanan Dawood Yassa, MD
Assistant Professor and Head of Anatomy and
    Embryology Department
Department of Anatomy and Embryology
Beni Suef University
Beni Suef, Beni Suef Governorate, Egypt

REVIEWERS FOR THE FIFTH EDITION

Kacie Bhushan
Nova Southeastern University
Fort Lauderdale, Florida

Leonard J. Cleary, PhD
Professor
The University of Texas Health Science Center Medical School
Houston, Texas

Alan Crandall, MS
Idaho State University
Pocatello, Idaho

Bertha Escobar-Poni, MD
Loma Linda University
Loma Linda, California

Thomas Gillingwater, PhD
Professor of Neuroanatomy
University of Edinburgh
Edinburgh, United Kingdom

William Huber, PhD
Professor
St. Louis Community College at Forest Park
St. Louis, Missouri

Lorraine Jadeski, PhD
Associate Professor
University of Guelph
Ontario, Canada

Marta Lopez, LM, CPM, RMA
Program Coordinator/Professor
Medical Assisting Program
Miami Dade College
Miami, Florida

Yogesh Malam
University College London
London, United Kingdom

Volodymyr Mavrych, MD, PhD, DSc
Professor
St. Matthew's University
West Bay, Cayman Islands

Karen McLaren

Monica Oblinger, MS, PhD
Professor
Rosalind Franklin University of Medicine and Science
North Chicago, Illinois

Onyekwere Onwumere, MA, MPhil
Adjunct Faculty
The College of New Rochelle
New Rochelle, New York

Simon Parson, BSc, PhD
Professor
University of Edinburgh
Edinburgh, United Kingdom

Gaurav Patel
Windsor University School of Medicine
Cayon, Saint Kitts

Ryan Splittgerber, PhD
Associate Professor
Department of Surgery Administration
Vanderbilt University School of Medicine
Nashville, Tennessee

Christy Tomkins-Lane, PhD
Assistant Professor
Mount Royal University
Calgary, Alberta, Canada

Victor Emmanuel Usen
Medical University of Lublin
Lublin, Poland

Edward Wolfe, DC
Instructor
Central Piedmont Community College
Charlotte, North Carolina

Andrzej Zeglen
Lincoln Memorial University-DeBusk College of
   Osteopathic Medicine
Harrogate, Tennessee

In addition to reviewers, many people, some of them unknowingly, helped us by discussing parts of the manuscript and/or providing constructive criticism of the text and illustrations in the present and previous editions:

- *Dr. Peter H. Abrahams*, Emeritus Professor of Clinical Anatomy, Warwick Medical School, Coventry, United Kingdom
- *Dr. Edna Becker*, Associate Professor of Medical Imaging, University of Toronto Faculty of Medicine, Toronto, Ontario, Canada
- *Dr. Robert T. Binhammer*, Emeritus Professor of Genetics, Cell Biology and Anatomy, University of Nebraska Medical Center, Omaha, Nebraska
- *Dr. Stephen W. Carmichael*, Professor Emeritus, Mayo Medical School, Rochester, Minnesota
- *Dr. James D. Collins*, Professor Emeritus of Radiological Sciences, University of California, Los Angeles School of Medicine/Center for Health Sciences, Los Angeles, California
- *Dr. Raymond F. Gasser*, Emeritus Professor of Cell Biology and Anatomy and Adjunct Professor of Obstetrics and Gynecology, Louisiana State University School of Medicine, New Orleans, Louisiana
- *Dr. Douglas J. Gould*, Professor of Neuroscience and Chair, Department of Foundational Medical Studies, Oakland University William Beaumont School of Medicine, Rochester, Michigan
- *Dr. Daniel O. Graney*, Professor of Biological Structure, University of Washington School of Medicine, Seattle, Washington
- *Dr. David G. Greathouse*, Director of Clinical Electrophysiology Services, Texas Physical Therapy Specialists, New Braunfels, Texas
- *Dr. Masoom Haider*, Associate Professor of Medical Imaging, University of Toronto Faculty of Medicine, Toronto, Ontario, Canada
- *Dr. John S. Halle*, Professor and former Chair, Belmont University School of Physical Therapy, Nashville, Tennessee
- *Dr. June A. Harris*, Professor of Anatomy, Faculty of Medicine, Memorial University of Newfoundland Health Sciences Centre, St. John's, Newfoundland and Labrador, Canada
- *Dr. Walter Kucharczyk*, Professor and Neuroradiologist Senior Scientist, Department of Medical Resonance Imaging, University Health Network, Toronto, Ontario, Canada
- *Dr. Randy J. Kulesza, Jr.*, Professor of Anatomy and Assistant Dean for Medical Education, Lake Erie College of Osteopathic Medicine, Erie, Pennsylvania
- *Dr. Nirusha Lachman*, Professor of Anatomy, Mayo Medical School, Rochester, Minnesota
- *Dr. H. Wayne Lambert*, Associate Professor, Department of Neurobiology and Anatomy, West Virginia University School of Medicine, Morgantown, West Virginia
- *Dr. Lillian Nanney*, Professor Emeritus of Plastic Surgery, Vanderbilt University School of Medicine, Nashville, Tennessee

- *Dr. Todd R. Olson*, Professor Emeritus of Anatomy and Structural Biology, Albert Einstein College of Medicine, Bronx, New York
- *Dr. Wojciech Pawlina*, Professor and Chair of Anatomy, Mayo Medical School, Rochester, Minnesota
- *Dr. T. V. N. Persaud*, Professor Emeritus of Human Anatomy and Cell Science, Faculties of Medicine and Dentistry, University of Manitoba, Winnipeg, Manitoba, Canada. Professor of Anatomy and Embryology, St. George's University, Granada, West Indies
- *Dr. Cathleen C. Pettepher*, Professor of Cancer Biology and Assistant Dean for Assessment, Vanderbilt University School of Medicine, Nashville, Tennessee
- *Dr. Thomas H. Quinn*, Professor of Biomedical Sciences, Creighton University School of Medicine, Omaha, Nebraska
- *Dr. Tatsuo Sato*, Professor and Head (retired), Second Department of Anatomy, Tokyo Medical and Dental University, Faculty of Medicine, Tokyo, Japan
- *Dr. Carol Scott-Conner*, Professor Emeritus, Department of Surgery, University of Iowa, Roy J. and Lucille A. Carver College of Medicine, Iowa City, Iowa
- *Dr. Ryan Splittgerber*, Associate Professor, Department of Surgery Administration, Vanderbilt University School of Medicine, Nashville, Tennessee.
- *Dr. Joel A. Vilensky*, Professor of Anatomy, Indiana University School of Medicine, Indianapolis, Indiana
- *Dr. Edward C. Weber*, Diagnostic Radiologist, The Imaging Center, Fort Wayne, Indiana
- *Dr. David G. Whitlock*, Professor Emeritus of Anatomy, University of Colorado Medical School, Denver, Colorado

Art plays a major role in facilitating learning, especially in anatomy. We extend our sincere gratitude and appreciation for the skills, and talents, of our medical illustrator, Jennifer Clements from Wolters Kluwer for this edition. We also thank Kam Yu, who prepared the illustrations for the first edition. We continue to benefit from the extensive surface anatomy project photographed by E. Anne Raynor, Senior Photographer, Vanderbilt Medical Art Group, under the direction of authors Arthur F. Dalley II and Anne M. R. Agur, with the support of Wolters Kluwer.

We wish to thank Dr. Edward C. Weber and Dr. Joel A. Vilensky for their review of clinical material, contributions to the Clinical Box features, and Medical Imaging photos.

Without the expertise and dedication of Kathleen Scogna, freelance developmental editor, this book would not have been possible. Our appreciation and thanks are extended to the editorial and production teams at Wolters Kluwer Health who provided their expertise in the development of this edition: Crystal Taylor, Senior Acquisitions Editor; Andrea Vosburgh, Development Editor; Lindsay Ries, Editorial Coordinator; Jennifer Clements, Art Director; and Joan Sinclair, Production Coordinator. We also thank Harold Medina of Absolute Service, Inc. Finally, thanks to the Sales Division at Wolters Kluwer Health, which has played a key role in the success of this book.

Anne M. R. Agur
Arthur F. Dalley II

Contents

4 THORAX 183

5 ABDOMEN 253

6 PELVIS AND PERINEUM 339

9 NECK 595

10 REVIEW OF CRANIAL NERVES 641

Clinical Boxes

4 THORAX 183

5 ABDOMEN 253

9 NECK 595

10 REVIEW OF CRANIAL NERVES 641

Figure Credits

All sources are published by Lippincott Williams & Wilkins/ Wolters Kluwer unless otherwise noted.

 OVERVIEW AND BASIC CONCEPTS 1

Figure 1.15B–E Cormack DH. *Essential Histology*. 2nd ed. 2001; Plates 11.1, 11.2, 11.3, and 11.4.

Figure 1.32 Courtesy of Dr. E.L. Lansdown, Professor of Medical Imaging, University of Toronto, Ontario, Canada.

Figure 1.33B,C Wicke L. *Atlas of Radiologic Anatomy*. 6th ed. Taylor AN, trans-ed. 1998. [Wicke L. *Roentgen-Anatomie Normalbefunde*. 5th ed. Munich, Germany: Urban & Schwarzenberg; 1995.]

Figure 1.34B,C Wicke L. *Atlas of Radiologic Anatomy*. 6th ed. Taylor AN, trans-ed. 1998. [Wicke L. *Roentgen-Anatomie Normalbefunde*. 5th ed. Munich, Germany: Urban & Schwarzenberg; 1995.]

Figure 1.35A Wicke L. *Atlas of Radiologic Anatomy*. 6th ed. Taylor AN, trans-ed. 1998. [Wicke L. *Roentgen-Anatomie Normalbefunde*. 5th ed. Munich, Germany: Urban & Schwarzenberg; 1995.], **B** Dean D, Herbener TE. *Cross-sectional Human Anatomy*. 2000.E54

Figure 1.36 Knight L. *Medical Terminology: An Illustrated Guide Canadian Edition*. 2nd ed. 2013; Fig. 17-18C.

Figure B1.1 Courtesy of Dr. D. Armstrong, University of Toronto, Ontario, Canada.

Figure B1.2 Based on Willis MC. *Medical Terminology: A Programmed Learning Approach to the Language of Health Care*. 2002, p. 198.

Figure B1.3 Reprinted with permission from *Roche Lexikon Medizin*, 4th ed. Munich, Germany: Urban & Schwarzenberg; 1998.

 BACK 45

Figure 2.1C Based on Nathwani B, Olson TR. *A.D.A.M. Student Atlas of Anatomy*. Baltimore: Williams & Wilkins, 1997.

Figure 2.3C Courtesy of Dr. Joel A. Vilensky, Indiana University School of Medicine, Fort Wayne, Indiana, and Dr. Edward C. Weber, The Imaging Center, Fort Wayne, Indiana.

Figure 2.4C Courtesy of Dr. D. Salonen, University of Toronto, Ontario, Canada.

Figure 2.4E Courtesy of Dr. D. Armstrong, University of Toronto, Ontario, Canada.

Figure 2.5D Becker RF, Wilson JW, Gehweiler JA. *Anatomical Basis of Medical Practice*. 1974.

Figure 2.6C,E Courtesy of Dr. J. Heslin, University of Toronto, Ontario, Canada.

Figure 2.6D Becker RF, Wilson JW, Gehweiler JA. *Anatomical Basis of Medical Practice*. 1974.

Figure 2.22B–E Based on Nathwani B, Olson TR. *A.D.A.M. Student Atlas of Anatomy*. Baltimore: Williams & Wilkins, 1997.

Figure 2.26B,C Wicke L. *Atlas of Radiologic Anatomy*. 6th ed. Taylor AN, trans-ed. 1998. [Wicke L. *Roentgen-Anatomie Normalbefunde*. 5th ed. Munich, Germany: Urban & Schwarzenberg; 1995.]

Figure 2.27A,B Courtesy of the Visible Human Project, National Library of Medicine, Visible Man, 1715; **C** Courtesy of Dr. D. Salonen, University of Toronto, Ontario, Canada; **D** Courtesy of Dr. D. Armstrong, University of Toronto, Ontario, Canada.

Figure B2.3 Moore KL, Persaud TVN, Torchia MG. *The Developing Human: Clinically Oriented Embryology*. 10th ed. Philadelphia, PA: Elsevier/Saunders; 2016.

Figure B2.4B Clark CR. *The Cervical Spine*. 3rd ed. 1998.

Figure B2.5A Image reproduced with permission from Zubin I. *Spondylolisthesis Imaging*. Medscape Drugs and Diseases; 2018. https://emedicine.medscape.com/article/396016-overview.

Figure B2.7 Yochum TR, Rowe LJ. *Yochum and Rowe's Essentials of Skeletal Radiology*. 3rd ed. 2004; Figs. 14-3A, 14-1C, and 14-5.

Figure B2.8 C Choi SJ, Song JS, Kim C, et al. The use of magnetic resonance imaging to predict the clinical outcome of non-surgical treatment for lumbar intervertebral disc herniation. *Korean J Radiol*. 2007;8:156–163:5a.

Figure B2.10 Courtesy of Organ LW, Papadopoulos P, Pérez J. *Radiofrequency Neurotomy of Lumbar Medial Branch*. Diros/Owl Monographs; 2013. https://dirostech.com/techniques-procedures/#!

Figure B2.13 Modified from Finneson BE. *Low Back Pain*. 2nd ed. 1980:302.

Figure B2.14 Modified from White AA, Panjabi MM. *Clinical Biomechanics of the Spine*. 1978:331.

Figure B2.15 Bickley LS. *Bates' Guide to Physical Examination and History Taking*. 12th ed. 2017; Fig. 17.63.

 UPPER LIMB 91

Figure 3.9 Courtesy Dr. E. Becker, University of Toronto, Ontario, Canada.

Figure 3.11A Modified from Tank PW, Gest TR. *Lippincott Williams & Wilkins Atlas of Anatomy*. 2008; Plate 2.53.

Figure 3.13 Central image from Tank PW, Gest TR. *Lippincott Williams & Wilkins Atlas of Anatomy*. 2008; Plate 2.46. Brachial, radial, and ulnar pulse photos from Bickley LS. *Bates' Guide to Physical Examination and History Taking*. 12th ed. 2017; Figs. 4.8, 9.30, and 12.26.

Figure 3.17 Modified from Tank PW, Gest TR. *Lippincott Williams & Wilkins Atlas of Anatomy*. 2008; Plates 2.47A, 2.48, 2.49, and 2.50.

Figure 3.18B–E Adapted with permission from David Pounds (author/illustrator), from Clay JH, Pounds DM. *Basic Clinical Massage Therapy: Integrating Anatomy and Treatment*. 2nd ed. 2008; Figs. 4.1, 4.4, 4.9, and 4.49.

Figure 3.21D Adapted with permission from David Pounds (author/illustrator), from Clay JH, Pounds DM. *Basic Clinical Massage Therapy: Integrating Anatomy and Treatment*. 2nd ed. 2008; Fig. 4.31.

Figure 3.27 Modified from Tank PW, Gest TR. *Lippincott Williams & Wilkins Atlas of Anatomy*. 2008; Plate 2.14.

Figure 3.30 Adapted with permission from David Pounds (author/illustrator), from Clay JH, Pounds DM. *Basic Clinical Massage Therapy: Integrating Anatomy and Treatment*. 2nd ed. 2008; Figs. 5.3, 5.4, and 5.10.

Figure 3.31D Based on Hoppenfeld S, de Boer P. *Surgical Exposures in Orthopaedics*. 3rd ed. Philadelphia: Lippincott Williams & Wilkins, 2003; Fig. 2.27.

Figure 3.56C Modified from Hamill J, Knutzen KM, Derrick TR. *Biomechanical Basis of Human Movement*. 4th ed. 2015; Fig. 5.8.

Figure 3.58A Courtesy of Dr. E. Lansdown, University of Toronto, Ontario, Canada.

Figure 3.59A,B Courtesy of Dr. E. Becker, University of Toronto, Ontario, Canada.

Figure 3.62C Courtesy of Dr. J. Heslin, University of Toronto, Ontario, Canada.

Figure 3.65A–C Dean D, Herbener TE. *Cross-sectional Human Anatomy*. 2000; Plates 7.2, 7.5, and 7.8.

Figure 3.66A Courtesy of Dr. W. Kucharczyk, University of Toronto, Ontario, Canada.

Figure 3.66B,C Lee JKT, Sagel SS, Stanley, RJ, et al. *Computed Body Tomography with MRI Correlation*. 4th ed. 2006; Fig. 22.13A,C.

Figure B3.2B Based on Hoppenfeld S, de Boer P. *Surgical Exposures in Orthopaedics*. 3rd ed. Philadelphia: Lippincott Williams & Wilkins, 2003; Fig. 2.27.

Figure B3.5 Rowland LP. *Merritt's Textbook of Neurology*. 9th ed. 1995.

Figure B3.7 Anderson MK, Hall SJ, Martin M. *Foundations of Athletic Training*. 3rd ed. 1995.

Figure B3.8 Bickley LS. *Bates' Guide to Physical Examination and History Taking*. 10th ed. 2009:697.

Figure B3.13 Modified from Salter RB. *Textbook of Disorders and Injuries of the Musculoskeletal System*. 3rd ed. 1999; Fig. 17-1 (colorized).

Figure B3.14 Modified from Werner R. *A Massage Therapist's Guide to Pathology*. 6th ed. 2015; Fig 3.33.

Figure B3.21 Modified from Salter RB. *Textbook of Disorders and Injuries of the Musculoskeletal System*. 3rd ed. 1999; Fig. 11-65 (colorized).

Figure B3.24A,B Yochum TR, Rowe LJ. *Yochum and Rowe's Essentials of Skeletal Radiology*. 3rd ed. 2004; Fig. 9.192A,B.

Figure B3.25 Redrawn from Anderson MK. *Fundamentals of Sports Injury Management*. 2nd ed. 2002.

 THORAX   183

Figure 4.8B Courtesy of Dr. Joel A. Vilensky, Indiana University School of Medicine, Fort Wayne, Indiana, and Dr. Edward C. Weber, The Imaging Center, Fort Wayne, Indiana.

Figure 4.20A Courtesy of Dr. Joel A. Vilensky, Indiana University School of Medicine, Fort Wayne, Indiana, and Dr. Edward C. Weber, The Imaging Center, Fort Wayne, Indiana.

Figure 4.27A Courtesy of Dr. Joel A. Vilensky, Indiana University School of Medicine, Fort Wayne, Indiana, and Dr. Edward C. Weber, The Imaging Center, Fort Wayne, Indiana.

Figure 4.50A,B Courtesy of I. Morrow, University of Manitoba, Canada.

Figure 4.50C Courtesy of I. Verschuur, Joint Department of Medical Imaging, UHN/Mount Sinai Hospital, Toronto, Canada.

Figure 4.51A–C Courtesy of I. Verschuur, Joint Department of Medical Imaging, UHN/Mount Sinai Hospital, Toronto, Canada.

Figure B4.4A,C Based on Bickley LS. *Bates' Guide to Physical Examination and History Taking*. 10th ed. 2009; Table 10-2, p. 414.

Figure B4.4B Left: Evans RJ, Evans MK, Brown YMR. *Canadian Maternity, Newborn & Women's Health Nursing*. 2nd ed. 2015; Fig. 2.8.

Right: Hatfield NT, Kincheloe CA. *Introductory Maternity & Pediatric Nursing*. 4th ed. 2018; Fig. 4.1C.

Figure B4.7B Daffner RH, Hartman MS. *Clinical Radiology: The Essentials*. 4th ed. 2014.

Figure B4.10 *Stedman's Medical Dictionary*. 27th ed. 2000 (Artist: Neil O. Hardy, Westport, CT); Photographs of bronchus, carina, and trachea from Feinsilver SH, Fein A. *Textbook of Bronchoscopy*. 1995; Photograph of bronchoscopy procedure courtesy of Temple University Hospital, Philadelphia.

Figure B4.12 Dean D, Herbener TE. *Cross-sectional Human Anatomy*. 2000.

Figure B4.13 Based on *Stedman's Medical Dictionary*. 27th ed. 2000 (Artist: Neil O. Hardy, Westport, CT).

Figure B4.15 Based on figures provided by the Anatomical Chart Company.

Figure B4.17 Based on *Stedman's Medical Dictionary*. 27th ed. 2000 (Artist: Neil O. Hardy, Westport, CT).

Figure B4.18 Feigenbaum H, Armstrong WF, Ryan T. *Feigenbaum's Echocardiography*. 5th ed. 2005:116.

Figure SA4.5B,F Bickley LS. *Bates' Guide to Physical Examination and History Taking*. 12th ed. 2017:322.

Figure SA4.5C *Stedman's Medical Dictionary*. 28th ed. 2006 (Artist: Neil Hardy).

Figure SA4.5D Bickley LS. *Bates' Guide to Physical Examination and History Taking*. 11th ed. 2013:309.

Figure SA4.7B Modified from Bickley LS. *Bates' Guide to Physical Examination and History Taking*. 10th ed. 2009:330.

5 ABDOMEN 253

Figure 5.4B–E Adapted with permission from David Pounds (author/illustrator), from Clay JH, Pounds DM. *Basic Clinical Massage Therapy: Integrating Anatomy and Treatment.* 2nd ed. 2008; Plate 7-3.

Figure 5.19A Based on *Stedman's Medical Dictionary.* 27th ed. 2000 (Artist: Neil O. Hardy, Westport, CT).

Figure 5.21C Courtesy of Dr. E.L. Lansdown, Professor of Medical Imaging, University of Toronto, Ontario, Canada.

Figure 5.28A Based on *Stedman's Medical Dictionary.* 27th ed. 2000 (Artist: Neil O. Hardy, Westport, CT).

Figure 5.28C,D Based on Sauerland EK. *Grant's Dissector.* 12th ed. 1999.

Figure 5.39B,C Reprinted with permission from Karaliotas C, Broelsch C, Habib N. *Liver and Biliary Tract Surgery: Embryological Anatomy to 3D-Imaging and Transplant Innovations.* Vienna, Austria: Springer; 2007: Fig. 2.13, p. 28. Copyright 2007.

Figure 5.41A,C Courtesy of Dr. G.B. Haber, University of Toronto, Ontario, Canada.

Figure 5.49 Photo courtesy of Dr. Joel A. Vilensky, Indiana University School of Medicine, Fort Wayne, Indiana, and Dr. Edward C. Weber, The Imaging Center, Fort Wayne, Indiana.

Figure 5.58B Adapted with permission from David Pounds (author/illustrator), from Clay JH, Pounds DM. *Basic Clinical Massage Therapy: Integrating Anatomy and Treatment.* 2nd ed. 2008; Fig. 4-64.

Figure 5.69A–F Courtesy of A.M. Arenson, University of Toronto, Ontario, Canada.

Figure 5.70A–C part II. Courtesy of Tom White, Department of Radiology. The Health Sciences Center, University of Tennessee, Memphis, Tennessee.

Figure 5.71A, C, & D Courtesy of Dr. M.A. Haider, University of Toronto, Toronto, Canada.

Figure 5.72A Courtesy of M. Asch, University of Toronto, Ontario, Canada.

Figure 5.72B Dean D, Herbener TE. *Cross-sectional Human Anatomy.* 2000.

Figure 5.72C Courtesy of Dr. C.S. Ho, University of Toronto, Ontario, Canada.

Figure B5.5 Based on Tank PW, Gest TR. *Lippincott Williams & Wilkins Atlas of Anatomy.* 2008; Plate 5.11B,C.

Figure B5.8 Linn-Watson T. *Radiographic Pathology.* 2nd ed. 2014; Fig. 4.9.

Figure B5.9 Mitros FA. *Atlas of Gastrointestinal Pathology.* New York, NY: Gower Medical; 1998: Fig. 5.46.

Figure B5.10A Scott-Conner CE, Dawson DL. Essential Operative Techniques and Anatomy. 4th ed. 2013; **B** Mitros FA. *Atlas of Gastrointestinal Pathology.* New York, NY: Gower Medical; 1998: Fig. 10.42.

Figure B5.11 Courtesy of Dr. Joel A. Vilensky, Indiana University School of Medicine, Fort Wayne, Indiana, and Dr. Edward C. Weber, The Imaging Center, Fort Wayne, Indiana.

Figure B5.12 Mitros FA. *Atlas of Gastrointestinal Pathology.* New York, NY: Gower Medical; 1998: Fig. 1.10.

Figure B5.12 Inset *Stedman's Medical Dictionary.* 28th ed. 2006.

Figure B5.13 Bickley LS. *Bates' Guide to Physical Examination and History Taking.* 10th ed. 2009:429.

Figure B5.14B Based on Eckert P, Haring R, Satter P, et al. *Fibrinklebung, Indikation und Anwendung.* München, Germany: Urban & Schwarzenberg; 1986.

Figure SA5.2B Based on Basmajian JV, Slonecker CE. *Grant's Method of Anatomy.* 11th ed. 1989; Fig. 12.30.

Figure SA5.3C *Stedman's Medical Dictionary.* 27th ed. 2000 (Artist: Neil O. Hardy, Westport, CT).

Figure SA5.4 Based on Bickley LS. *Bates' Guide to Physical Examination and History Taking.* 10th ed. 2009:440.

6 PELVIS AND PERINEUM 339

Figure 6.5D Courtesy of Dr. E.L. Lansdown, University of Toronto, Ontario, Canada.

Figure 6.8E Based on DeLancey JO. Structural support of the urethra as it relates to stress urinary incontinence: the hammock hypothesis. *Am J Obstet Gynecol.* 1994;170:1713–1720.

Figure 6.20B Modified from Detton AJ. *Grant's Dissector.* 16th ed. 2017; Fig. 5.37.

Figure 6.27A Left: Based on Dauber W. *Pocket Atlas of Human Anatomy.* 5th rev ed. New York, NY: Thieme; 2007:195. **B** Courtesy of Dr. A.M. Arenson, University of Toronto, Toronto, Ontario, Canada (ultrasound image).

Figure 6.42 Based on Clemente CD. *Anatomy: A Regional Atlas of the Human Body.* 5th ed. 2006; Fig. 272.1.

Figure 6.59A–D Courtesy of M.A. Heider, University of Toronto, Ontario, Canada.

Figure 6.60A–E Courtesy of M.A. Heider, University of Toronto, Ontario, Canada.

Figure 6.61A Beckmann CR. *Obstetrics and Gynecology.* 5th ed. 2006.

Figure 6.61B,C Courtesy of A.M. Arenson, University of Toronto, Ontario, Canada.

Figure 6.61D Daffner RH. *Clinical Radiology: The Essentials.* 2nd ed. 1999.

Figure 6.61E Erkonen WE, Smith WL. *Radiology 101: The Basics and Fundamentals of Imaging.* 3rd ed. 2010.

Figure 6.61F Daffner RH. *Clinical Radiology: The Essentials.* 2nd ed. 1999.

Figure B6.2 Hartwig W. *Fundamental Anatomy.* 2008:176.

Figure B6.4A Based on *Stedman's Medical Dictionary.* 27th ed. 2000.

Figure B6.6A,B Based on *Stedman's Medical Dictionary.* 27th ed. 2000.

Figure B6.7 Based on Tank PW, Gest TR. *Lippincott Williams and Wilkins Atlas of Anatomy.* 2008; Plate 6.19A.

Figure B6.8 Based on Fuller J, Schaller-Ayers J. *Health Assessment: A Nursing Approach.* 2nd ed. 1994; Fig. B3.11 (Artist: Larry Ward, Salt Lake City, UT).

Figure B6.10A Illustration based on *Stedman's Medical Dictionary.* 27th ed. 2000; **B** Laparoscopic photograph: With permission from Bristow RE, Johns Hopkins School of Medicine, Baltimore, MD.

7 LOWER LIMB 409

Figure 7.11D Modified from Egol KA, Bazylewicz SC. *The Ortho-paedic Manual: From the Office to the OR.* 2018.

Figure 7.11E,F Bickley LS. *Bates' Guide to Physical Examination and History Taking.* 12th ed. 2017; Figs. 12-19 and 12-23.

Figure 7.12D Based on Melloni JL. *Melloni's Illustrated Review of Human Anatomy: By Structures—Arteries, Bones, Muscles, Nerves, Veins.* 1988.

Figure 7.14A–F Modified from Tank PW, Gest TR. *Lippincott Williams & Wilkins Atlas of Anatomy.* 2008; Plates 3.63, 3.64, 3.65C, 3.66A–C, and 3.67A,B.

Figure 7.15B,C Adapted with permission from David Pounds (author/illustrator), from Clay JH, Pounds DM. *Basic Clinical Massage Therapy: Integrating Anatomy and Treatment.* 2nd ed. 2008; Plate 9.2.

Figure 7.16B–G Adapted with permission from David Pounds (author/illustrator), from Clay JH, Pounds DM. *Basic Clinical Massage Therapy: Integrating Anatomy and Treatment.* 2nd ed. 2008; Figs. 9.24–9.28.

Figure 7.22C–F Adapted with permission from David Pounds (author/illustrator), from Clay JH, Pounds DM. *Basic Clinical Massage Therapy: Integrating Anatomy and Treatment.* 2nd ed. 2008; Figs. 8.16–8.18 and Plate 9.5.

Figure 7.25F–H Adapted with permission from David Pounds (author/illustrator), from Clay JH, Pounds DM. *Basic Clinical Massage Therapy: Integrating Anatomy and Treatment.* 2nd ed. 2008; Figs. 9.12–9.14.

Figure 7.30D–F Adapted with permission from David Pounds (author/illustrator), Clay JH, Pounds DM. *Basic Clinical Massage Therapy: Integrating Anatomy and Treatment.* 2nd ed. 2008; Figs. 10.10, 10.14, and 10.16.

Figure 7.32B,C Adapted with permission from David Pounds (author/illustrator), from Clay JH, Pounds DM. *Basic Clinical Massage Therapy: Integrating Anatomy and Treatment.* 2nd ed. 2008; Plate 10.3.

Figure 7.33B–D Adapted with permission from David Pounds (author/illustrator), from Clay JH, Pounds DM. *Basic Clinical Massage Therapy: Integrating Anatomy and Treatment.* 2nd ed. 2008; Plate 10.4, Figs. 10.22 and 10.29.

Figure 7.35D,E Adapted with permission from David Pounds (author/illustrator), from Clay JH, Pounds DM. *Basic Clinical Massage Therapy: Integrating Anatomy and Treatment.* 2nd ed. 2008; Fig. 10.30.

Figure 7.41 Adapted with permission from David Pounds (author/illustrator), from Clay JH, Pounds DM. *Basic Clinical Massage Therapy: Integrating Anatomy and Treatment.* 2nd ed. 2008; Fig. 10.41.

Figure 7.42C–G Adapted with permission from David Pounds (author/illustrator), from Clay JH, Pounds DM. *Basic Clinical Massage Therapy: Integrating Anatomy and Treatment.* 2nd ed. 2008; Plates 10.5 and 10.6.

Figure 7.45 Based on Rose J, Gamble JG. *Human Walking.* 2nd ed. 1994.

Figure 7.46A Adapted with permission from David Pounds (author/illustrator), fromClay JH, Pounds DM. *Basic Clinical Massage Therapy: Integrating Anatomy and Treatment.* 2nd ed. 2008; Plate 9.1.

Figure 7.47C Based on Kapandji, IA. *The Physiology of the Joints. Volume 2: Lower Limb.* 5th ed. Edinburgh, United Kingdom: Churchill Livingstone; 1987.

Figure 7.50B,D Courtesy of Dr. P. Bobechko, University of Toronto, Ontario, Canada.

Figure 7.51B Courtesy of Dr. D. Salonen, University of Toronto, Ontario, Canada.

Figure 7.53D Courtesy of Dr. D. Salonen, University of Toronto, Ontario, Canada.

Figure 7.57A Adapted with permission from David Pounds (author/illustrator), from Clay JH, Pounds DM. *Basic Clinical Massage Therapy: Integrating Anatomy and Treatment.* 2nd ed. 2008; Plate 10.1.

Figure 7.57B Wicke L. *Atlas of Radiologic Anatomy.* 6th ed. Taylor AN, trans-ed. 1998. [Wicke L. *Roentgen-Anatomie Normalbefunde.* 5th ed. Munich, Germany: Urban & Schwarzenberg; 1995.]

Figure 7.57C,D Courtesy of Dr. P. Bobechko and Dr. E. Becker, Department of Medical Imaging, University of Toronto, Ontario, Canada.

Figure 7.61A Radiograph courtesy of Dr. W. Kucharczyk, University of Toronto, Ontario, Canada.

Figure 7.63C,D Courtesy of Dr. D. Salonen, University of Toronto, Ontario, Canada.

Figure 7.64D–F Courtesy of Dr. D. Salonen, University of Toronto, Ontario, Canada.

Figure B7.3B Yochum TR, Rowe LJ. *Yochum and Rowe's Essentials of Skeletal Radiology.* 3rd ed. 2004.

Figure B7.4 From Joshi A. *Osgood-Schlatter disease imaging, updated* Apr 17, 2017. https://emedicine.medscape.com/article/411842 -overview. © eMedicine.com, 2017.

Figure B7.6A Reprinted with permission from *Roche Lexikon Medizin.* 4th ed. Munich, Germany: Urban & Schwarzenberg; 1998.

Figure B7.6B–D *Stedman's Medical Dictionary.* 28th ed. 2006 (Artist: Neil O. Hardy, Westport, CT).

Figure B7.13 and B7.14 Bickley LS. *Bates' Guide to Physical Examination and History Taking.* 10th ed. 2009:485.

Figure B7.15 and B7.16 Bickley LS. *Bates' Guide to Physical Examination and History Taking.* 8th ed. 2003; unn0336-016-065, unn0336-016-068.

Figure B7.17A Willis MC. *Medical Terminology: A Programmed Learning Approach to the Language of Health Care.* 2002.

Figure B7.17B Daffner RH. *Clinical Radiology: The Essentials.* 2nd ed. 1999.

Figure B7.19A–C Modified from Palastanga NP, Field DG, Soames R. *Anatomy and Human Movement.* 4th ed. Oxford, United Kingdom: Butterworth-Heinemann; 2002.

Figure B7.19D,F *Stedman's Medical Dictionary.* 27th ed. 2000.

Figure B7.19E Daffner RH. *Clinical Radiology: The Essentials.* 2nd ed. 1999.

Figure B7.20 *Stedman's Medical Dictionary.* 27th ed. 2000.

Figure B7.22A *Stedman's Medical Dictionary.* 27th ed. 2000.

Figure B7.23 Berg D, Worzala K. *Atlas of Adult Physical Diagnosis.* 2006; Fig. 13.6.

8 HEAD 499

Figure 8.8B Based on Tank PW, Gest TR. *Lippincott Williams & Wilkins Atlas of Anatomy.* 2008; Plate 7.60B.

Figure 8.15A,B Tank PW, Gest TR. *Lippincott Williams & Wilkins Atlas of Anatomy.* 2008; Plate 7.29.

Figure 8.19 Based on Tank PW, Gest TR. *Lippincott Williams & Wilkins Atlas of Anatomy.* 2008; Plate 7.73.

Figure 8.20 Based on Tank PW, Gest TR. *Lippincott Williams & Wilkins Atlas of Anatomy.* 2008; Plate 7.74.

Figure 8.24E Courtesy of Dr. W. Kucharczyk, University of Toronto, Ontario, Canada.

Figure 8.25A Tank PW, Gest TR. *Lippincott Williams & Wilkins Atlas of Anatomy.* 2008; Plate 7.78.

Figure 8.28A Based on Melloni JL. *Melloni's Illustrated Review of Human Anatomy: By Structures—Arteries, Bones, Muscles, Nerves, Veins.* 1988:149.

Figure 8.28B Based on Van de Graaff K. *Human Anatomy.* 4th ed. Dubuque, IA: WC Brown; 1995: Fig. 15.18.

Figure 8.29A Welch Allyn, Inc., Skaneateles Falls, NY.

Figure 8.29C–D Courtesy of J. Spilkin, OD, University Optometric Clinic, Toronto, Ontario, Canada.

Figure 8.30 Based on Van de Graaff K. *Human Anatomy.* 4th ed. Dubuque, IA: WC Brown; 1995: Fig. 15.17.

Figure 8.33A,B Based on Melloni JL. *Melloni's Illustrated Review of Human Anatomy: By Structures—Arteries, Bones, Muscles, Nerves, Veins.* 1988:141, 143.

Figure 8.33D Courtesy of Dr. W. Kucharczyk, University of Toronto, Ontario, Canada.

Figure 8.35B–E Based on Girard L. *Anatomy of the Human Eye. II. The Extra-ocular Muscles.* Houston, TX: Teaching Films, Inc.; n.d.

Figure 8.37A Based on Melloni JL. *Melloni's Illustrated Review of Human Anatomy: By Structures—Arteries, Bones, Muscles, Nerves, Veins.* 1988:189.

Figure 8.41A–C Adapted with permission from David Pounds (author/illustrator), from Clay JH, Pounds DM. *Basic Clinical Massage Therapy: Integrating Anatomy and Treatment.* 2nd ed. 2008; Figs. 3.15, 3.16, and 3.19.

Figure 8.46D,E Langland OE, Langlais RP, Preece JW. *Principles of Dental Imaging.* 2002; Fig. 11.32A,B.

Figure 8.51B Courtesy of Dr. M.J. Phatoah, University of Toronto, Ontario, Canada.

Figure 8.57 Courtesy of Dr. B. Liebgott, University of Toronto, Ontario, Canada.

Figure 8.58A Based on Tank PW, Gest TR. *Lippincott Williams & Wilkins Atlas of Anatomy.* 2008; Plate 7.40A.

Figure 8.58C Based on Tank PW, Gest TR. *Lippincott Williams & Wilkins Atlas of Anatomy.* 2008; Plate 7.38C.

Figure 8.62B Based on Paff GH. *Anatomy of the Head and Neck.* Philadelphia, PA: W.B. Saunders Co.; 1973; Figs. 238–240.

Figure 8.64A,B Based on Paff GH. *Anatomy of the Head and Neck.* Philadelphia, PA: W.B. Saunders Co.; 1973; Figs. 238–240.

Figure 8.64D,E Based on Hall-Craggs ECB. *Anatomy as a Basis for Clinical Medicine.* 2nd ed. Baltimore, MD: Urban & Schwarzenberg; 1990; Fig. 9.100.

Figure 8.68B Courtesy of Dr. E. Becker, University of Toronto, Ontario, Canada.

Figure 8.68C Courtesy of Dr. D. Armstrong, University of Toronto, Ontario, Canada.

Figure 8.72A,B Based on Tank PW, Gest TR. *Lippincott Williams & Wilkins Atlas of Anatomy.* 2008; Plate 7.66B,C.

Figure 8.79 Based on Seeley RR, Stephens TR, Tate P. *Anatomy and Physiology.* 6th ed. New York, NY: McGraw-Hill; 2003: Fig. 15.28.

Figure 8.80A Courtesy of Dr. E. Becker, University of Toronto, Ontario, Canada.

Figure 8.80B,C Courtesy of Dr. D. Armstrong, University of Toronto, Ontario, Canada.

Figure 8.81A Courtesy of Dr. W. Kucharczyk, University of Toronto, Ontario, Canada.

Figure 8.81B Courtesy of Dr. D. Armstrong, University of Toronto, Ontario, Canada.

Figure 8.81C–F Photos courtesy of the Visible Human Project, National Library of Medicine, Visible Man 1107 & 1168.

Figure B8.1A Courtesy of Trauma.org.

Figure B8.3 © Visuals Unlimited, Hollis, New Hampshire.

Figure B8.4B Courtesy of Dr. Joel A. Vilensky, Indiana University School of Medicine, Fort Wayne, Indiana, and Dr. Edward C. Weber, The Imaging Center, Fort Wayne, Indiana.

Figure B8.5 Skin Cancer Foundation.

Figure B8.7 Photo courtesy of Welch Allyn, Inc., Skaneateles Falls, NY.

Figure B8.8 Cohen BJ. *Medical Terminology.* 4th ed. 2003.

Figure B8.10 Mann IC. *The Development of the Human Eye.* New York, NY: Grune & Stratton; 1974.

Figure B8.13 Courtesy of Dr. Joseph B. Jacobs, NYU Medical Center, New York, NY.

Figure B8.14 Hall-Craggs ECB. *Anatomy as a Basis for Clinical Medicine.* 3rd ed. 1995.

Figure B8.15 Bechara Y. Ghorayeb, MD, Houston, TX.

9 NECK 595

Figure 9.2 Based on Tank PW, Gest TR. *Lippincott Williams & Wilkins Atlas of Anatomy.* 2008; Plate 7.100,B.

Figure 9.3A Adapted with permission from David Pounds (author/illustrator), from Clay JH, Pounds DM. *Basic Clinical Massage Therapy: Integrating Anatomy and Treatment.* 2nd ed. 2008; Fig. 3.28.

Figure 9.16B Courtesy of Dr. D. Salonen, University of Toronto, Ontario, Canada.

Figure 9.22A Based on Tank PW, Gest TR. *Lippincott Williams & Wilkins Atlas of Anatomy.* 2008; Plate 7.10.

Figure 9.23B Based on Liebgott B. *The Anatomical Basis of Dentistry.* Philadelphia, PA: Saunders; 1982: Fig 9.22.

Figure 9.24B Based on Tank PW, Gest TR. *Lippincott Williams & Wilkins Atlas of Anatomy.* 2008; Plate 7.21.

Figure 9.27 Courtesy of Dr. J. Heslin, University of Toronto, Ontario, Canada.

Figure 9.28A Courtesy of Dr. M. Keller, University of Toronto, Ontario, Canada.

Figure 9.28B Courtesy of Dr. Walter Kucharczyk, University of Toronto, Ontario, Canada.

Figure 9.29A Courtesy of I. Veschuur, UHN/Mount Sinai Hospital, Toronto, Ontario, Canada.

Figure 9.29B Reproduced from Lee H, Yi HA, Baloh RW. Sudden bilateral simultaneous deafness with vertigo as a sole manifestation of vertebrobasilar insufficiency. *J Neurol Neurosurg Psychiatry*. 2003;74:540. Copyright 2003. With permission from BMJ Publishing Group Ltd.

Figure 9.30 Siemens Medical Solutions USA, Inc.

Figure B9.1 Printed with permission from Akron Children's Hospital, Akron, Ohio.

Figure B9.5 Klima G. *Schilddrüsen-Sonographie*. München, Germany: Urban & Schwarzenberg; 1989.

Figure B9.6 and B9.8 Rohen JW, Yokochi C, Lutjen-Drecoll E. *Color Atlas of Anatomy: A Photographic Study of the Human Body*. 5th ed. 2003.

10 REVIEW OF CRANIAL NERVES 641

Figure 10.9A Based on Melloni, JL. *Melloni's Illustrated Review of Human Anatomy: By Structures—Arteries, Bones, Muscles, Nerves, Veins*. 1988.

Figure B10.6 Left: Bickely LS. *Bates' Guide to Physical Examination and History Taking*. 12th ed. 2017; Fig. 17-15; **Right:** Weber JR, Kelley JH. *Health Assessment in Nursing*. 4th ed. 2018; Fig. 27-14. © B. Proud.

Figure B10.7 Modified from Campbell, WW. *DeJong's The Neurologic Examination*. 7th ed. 2013; Fig. 20.3.

Note: Credits for figures based on illustrations from **Grant's Atlas of Anatomy and Clinically Oriented Anatomy** *are available at* http://thepoint.lww.com.

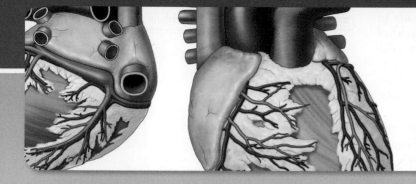

Overview and Basic Concepts

CLINICAL BOX KEY

Anatomical Variations

Diagnostic Procedures

Life Cycle

Surgical Procedures

Trauma

Pathology

Essential Clinical Anatomy relates the structure and function of the body to what is commonly required in the general practice of medicine, dentistry, and the allied health sciences. Because the number of details in anatomy overwhelms many beginning students, *Essential Clinical Anatomy* simplifies, correlates, and integrates the information so that it is easier to understand. The *clinical correlation boxes* (blue boxes) and *clinical case studies* (http://thePoint.lww.com) illustrate the clinical applications of anatomy. The *surface anatomy boxes* (orange boxes) provide an understanding of what lies under the skin, and the *medical imaging techniques* (green boxes), included throughout and at the end of each chapter, illustrate how anatomy is visualized clinically.

APPROACHES TO STUDYING ANATOMY

There are three main approaches to studying human gross anatomy: regional, systemic, and clinical (applied). In this introductory chapter, the systemic approach is used; in subsequent chapters, the clinical and regional approaches are used.

Regional anatomy is based on the organization of the body into parts: head, neck, trunk (further subdivided into thorax, abdomen, pelvis/perineum, and back), and paired upper and lower limbs. Emphasis is placed on the relationships of various systemic structures (e.g., muscles, nerves, and arteries) within the region (Fig. 1.1). Each region is not an isolated part and must be put into the context of adjacent regions and of the body as a whole. Surface anatomy is an essential part of the regional approach, providing a knowledge of what structures are visible and/or palpable (perceptible to touch) in the living body at rest and in action. The physical examination of patients is the clinical extension of surface anatomy. In people with stab wounds, for example, the health care worker must be able to visualize the deep structures that might be injured.

Systemic anatomy is an approach to anatomical study organized by *organ systems* that work together to carry out complex functions. None of the organ systems functions in isolation. For example, much of the skeletal, articular, and muscular systems constitute the *locomotor system*. And although the structures directly responsible for locomotion are the muscles, bones, joints, and ligaments, other systems are involved as well. The arteries and veins of the circulatory system supply oxygen to them and remove waste from them, and the nerves of the nervous system stimulate them to act. Brief descriptions of the systems of the body and their fields of study (in parentheses) follow:

- *Integumentary system* (dermatology) consists of the skin (integument) and its appendages, such as the hair and nails.

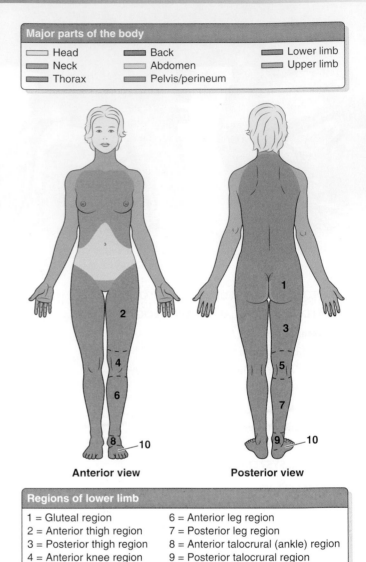

Major parts of the body

☐ Head		☐ Back		☐ Lower limb
☐ Neck		☐ Abdomen		☐ Upper limb
☐ Thorax		☐ Pelvis/perineum		

Anterior view   Posterior view

Regions of lower limb

1 = Gluteal region	6 = Anterior leg region
2 = Anterior thigh region	7 = Posterior leg region
3 = Posterior thigh region	8 = Anterior talocrural (ankle) region
4 = Anterior knee region	9 = Posterior talocrural region
5 = Posterior knee region	10 = Foot region

FIGURE 1.1. **Anatomical position and regions of body.**

The skin, an extensive sensory organ, forms a protective covering for the body.

- *Skeletal system* (osteology, orthopedics) consists of bones and cartilage. It provides support for the body and protects vital organs. The muscular system acts on the skeletal system to produce movements.
- *Articular system* (arthrology) consists of joints and their associated ligaments. It connects the bony parts of the skeletal system and provides the sites at which movements occur.
- *Muscular system* (myology) consists of muscles that act (contract) to move or position parts of the body (e.g., the bones that articulate at joints).
- *Nervous system* (neurology) consists of the *central nervous system* (brain and spinal cord) and the *peripheral nervous system* (nerves and ganglia, together with

their motor and sensory endings). The nervous system controls and coordinates the functions of the organ systems.

- *Circulatory system* (angiology) consists of the cardiovascular and lymphatic systems, which function in parallel to distribute fluids within the body.
  - *Cardiovascular system* (cardiology) consists of the heart and blood vessels that propel and conduct blood through the body.
  - *Lymphoid system* consists of a network of lymphatic vessels that withdraws excess tissue fluid (lymph) from the body's interstitial (intercellular) fluid compartment, filters it through lymph nodes, and returns it to the bloodstream.
- *Digestive* or *alimentary system* (gastroenterology) consists of the organs and glands associated with the ingestion, mastication (chewing), deglutition (swallowing), digestion and absorption of food, and the elimination of feces (solid wastes) after the nutrients have been absorbed.
- *Respiratory system* (pulmonology) consists of the air passages and lungs that supply oxygen and eliminate carbon dioxide. The control of airflow through the system produces tone, which is further modified into speech.
- *Urinary system* (urology) consists of the kidneys, ureters, urinary bladder, and urethra, which filter blood and subsequently produce, transport, store, and intermittently excrete liquid waste (urine).
- *Reproductive system* (obstetrics and gynecology for females, andrology for males) consists of the gonads (ovaries and testes) that produce oocytes (eggs) and sperms and the other genital organs concerned with reproduction.
- *Endocrine system* (endocrinology) consists of discrete ductless glands (e.g., thyroid gland) as well as cells of the intestine and blood vessel walls and specialized nerve endings that secrete hormones. Hormones are distributed by the cardiovascular system to reach receptor organs in all parts of the body. These glands influence metabolism and coordinate and regulate other processes (e.g., the menstrual cycle).

Clinical (applied) anatomy emphasizes aspects of the structure and function of the body important in the practice of medicine, dentistry, and the allied health sciences. It encompasses both the regional and the systemic approaches to studying anatomy and stresses clinical application.

ANATOMICOMEDICAL TERMINOLOGY

Anatomy has an international vocabulary that is the foundation of medical terminology. This nomenclature enables precise communication among health professionals worldwide as well as among scholars in basic and applied health sciences.

Although *eponyms* (names of structures derived from the names of people) are not used in official anatomical terminology, those commonly used by clinicians appear in parentheses throughout this book to aid students in their clinical years. Similarly, formerly used terms appear in parentheses on first mention—for example, internal thoracic artery (internal mammary artery). The terminology in this book conforms to the *Terminologia Anatomica: International Anatomical Terminology* (Federative Committee on Anatomical Terminology, 1998).

Anatomical Position

All anatomical descriptions are expressed in relation to the anatomical position (Fig. 1.1) to ensure that the descriptions are not ambiguous. The anatomical position refers to people—regardless of the actual position they may be in—as if they were standing erect, with their

- Head, eyes (gaze), and toes directed anteriorly (forward).
- Upper limbs by the sides with the palms facing anteriorly.
- Lower limbs close together with the feet parallel and the toes directed anteriorly.

Anatomical Planes and Sections

Anatomical descriptions relating to sectional anatomy and planar medical imaging (e.g., CT or MRI—see "Medical Imaging" at the end of this chapter) are based on conceptual planes that intersect the body in the anatomical position (Fig. 1.2). There are unlimited sagittal, frontal, transverse, and oblique planes, but there is only one median plane.

- **Median (median sagittal) plane** is the vertical plane passing longitudinally through the center of the body, dividing it into right and left halves.
- **Sagittal planes** are vertical planes passing through the body *parallel to the median plane*. It is helpful to give a point of reference to indicate the position of a specific plane—for example, a sagittal plane through the midpoint of the clavicle. A plane parallel to and near the median plane may be referred to as a *paramedian plane*.
- **Frontal (coronal) planes** are vertical planes passing through the body *at right angles to the median plane*, dividing it into anterior (front) and posterior (back) portions—for example, a frontal plane through the heads of the mandible.
- **Transverse planes** are planes passing through the body *at right angles to the median and frontal planes*. A transverse plane divides the body into superior (upper) and inferior (lower) parts—for example, a transverse plane through the umbilicus. Radiologists refer to transverse planes as *transaxial planes* or simply *axial planes*.
- **Oblique planes or sections** are planes or sections that do not align with the preceding planes.

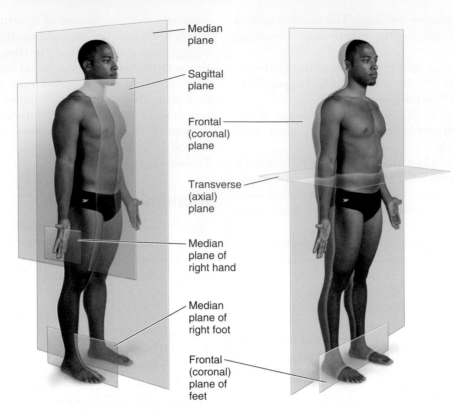

FIGURE 1.2. Planes of body.

Terms of Relationship and Comparison

Various adjectives, arranged as pairs of opposites, describe the relationship of parts of the body in the anatomical position and compare the position of two structures relative to each other. These pairs of adjectives are explained and illustrated in Figure 1.3. For example, the eyes are superior to the nose, whereas the nose is inferior to the eyes.

Combined terms describe intermediate positional arrangements:

- **Inferomedial** means nearer to the feet and closer to the median plane—for example, the anterior parts of the ribs run inferomedially.
- **Superolateral** means nearer to the head and farther from the median plane.

Proximal and **distal** are directional terms used when describing positions—for example, whether structures are nearer to the trunk or point of origin (i.e., proximal). **Dorsum** refers to the superior or dorsal (back) surface of any part that protrudes anteriorly from the body, such as the *dorsum of the foot*, *hand*, *penis*, or *tongue*. It is easier to understand why these surfaces are considered dorsal if one thinks of a quadrupedal plantigrade animal that walks on its soles, such as a dog. The **sole (plantar surface)** indicates the inferior aspect or bottom of the foot, much of which is in contact with the ground when standing barefoot. The **palm (palmar surface)** refers to the flat anterior aspect of the hand, excluding the five digits, and is the opposite of the dorsum of the hand.

Terms of Laterality

Paired structures having right and left members (e.g., the kidneys) are **bilateral**, whereas those occurring on one side only (e.g., the spleen) are **unilateral**. **Ipsilateral** means occurring on the same side of the body; the right thumb and right great toe are ipsilateral, for example. **Contralateral** means occurring on the opposite side of the body; the right hand is contralateral to the left hand.

Terms of Movement

Various terms describe movements of the limbs and other parts of the body (Fig. 1.4). Although most movements take place at joints where two or more bones or cartilages articulate with one another, several nonskeletal structures exhibit movement (e.g., tongue, lips, and eyelids). Movements taking place at joints are described relative to the axes around which the part of the body moves and the plane in which the movement takes place—for example, flexion and extension of the shoulder take place in the sagittal plane around a frontal (coronal) axis.

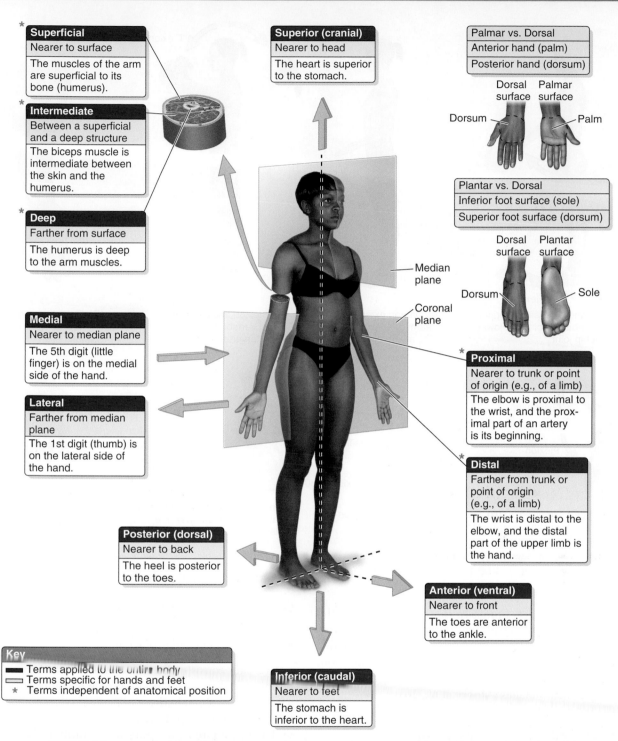

Superficial

Nearer to surface

The muscles of the arm are superficial to its bone (humerus).

Intermediate

Between a superficial and a deep structure

The biceps muscle is intermediate between the skin and the humerus.

Deep

Farther from surface

The humerus is deep to the arm muscles.

Medial

Nearer to median plane

The 5th digit (little finger) is on the medial side of the hand.

Lateral

Farther from median plane

The 1st digit (thumb) is on the lateral side of the hand.

Superior (cranial)

Nearer to head

The heart is superior to the stomach.

Median plane

Coronal plane

Posterior (dorsal)

Nearer to back

The heel is posterior to the toes.

Palmar vs. Dorsal

Anterior hand (palm)

Posterior hand (dorsum)

Dorsal surface   Palmar surface

Dorsum   Palm

Plantar vs. Dorsal

Inferior foot surface (sole)

Superior foot surface (dorsum)

Dorsal surface   Plantar surface

Dorsum   Sole

Proximal

Nearer to trunk or point of origin (e.g., of a limb)

The elbow is proximal to the wrist, and the proximal part of an artery is its beginning.

Distal

Farther from trunk or point of origin (e.g., of a limb)

The wrist is distal to the elbow, and the distal part of the upper limb is the hand.

Anterior (ventral)

Nearer to front

The toes are anterior to the ankle.

Key

Terms applied to the entire body

Terms specific for hands and feet

* Terms independent of anatomical position

Inferior (caudal)

Nearer to feet

The stomach is inferior to the heart.

FIGURE 1.3. Terms of relationship and comparison. These terms describe the position of one structure to another.

Anatomical Variations

Although anatomy books describe the structure of the body observed in most people (i.e., the most common pattern), the structure of individuals and even the right and left sides of the same individual may vary considerably in the details. Students are often frustrated because the bodies they are examining or dissecting do not conform to the atlas or textbook they are using. Students should expect anatomical variations when dissecting or studying prosected specimens. The bones of the skeleton vary not only in their basic shape but also in the details of surface structure. There is also a wide variation in the size, shape, and form of the

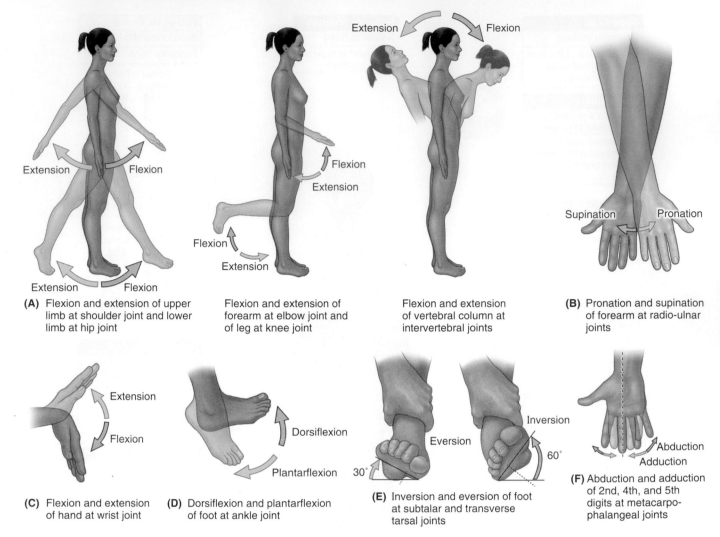

(A) Flexion and extension of upper limb at shoulder joint and lower limb at hip joint

Flexion and extension of forearm at elbow joint and of leg at knee joint

Flexion and extension of vertebral column at intervertebral joints

(B) Pronation and supination of forearm at radio-ulnar joints

(C) Flexion and extension of hand at wrist joint

(D) Dorsiflexion and plantarflexion of foot at ankle joint

(E) Inversion and eversion of foot at subtalar and transverse tarsal joints

(F) Abduction and adduction of 2nd, 4th, and 5th digits at metacarpophalangeal joints

FIGURE 1.4. Terms of movement. These terms describe movements of the limbs and other parts of the body; most movement takes place at joints where two or more bones or cartilages articulate with each other. *(continued)*

attachment of muscles. Similarly, there is variation in the method of division of vessels and nerves, and the greatest variation occurs in veins. Apart from racial and sexual differences, humans exhibit considerable genetic variation. Approximately 3% of newborns show one or more significant congenital anomalies (Moore et al., 2016).

INTEGUMENTARY SYSTEM

The skin, the largest organ of the body, is readily accessible and is one of the best indicators of general health (Swartz, 2014). *The skin serves the following functions*:

- *Protection* for the body from environmental effects, such as abrasions and harmful substances
- *Containment* of the tissues, organs, and vital substances of the body, preventing dehydration
- *Heat regulation* through sweat glands, blood vessels, and fat deposits

- *Sensation* (e.g., pain) by way of superficial nerves and their sensory endings
- *Synthesis and storage* of vitamin D

The skin consists of a superficial cellular layer, the epidermis, which creates a tough protective outer surface, and a basal (deep) regenerative and pigmented connective tissue layer, the dermis (Fig. 1.5A).

The **epidermis** is a keratinized stratified (layered) epithelium with a tough outer surface composed of keratin (a fibrous protein). The outer layer of the epidermis is continuously "shed" or rubbed away with replacement of new cells from the basal layer. This process renews the epidermis of the entire body every 25–45 days. The epidermis is avascular (no blood vessels or lymphatics) and is nourished by the vessels in the underlying dermis. The skin is supplied by afferent nerve endings that are sensitive to touch, irritation (pain), and temperature. Most nerve terminals are in the dermis, but a few penetrate the epidermis.

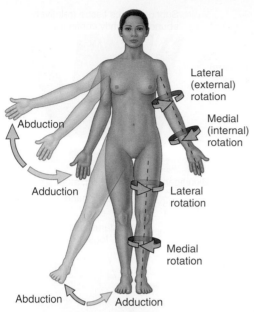

Lateral (external) rotation

Medial (internal) rotation

Abduction

Adduction

Lateral rotation

Medial rotation

Abduction

Adduction

Abduction

Abduction Adduction

(G) Abduction and adduction of right limbs and rotation of left limbs at glenohumeral and hip joints

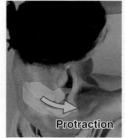

Circumduction

(H) Circumduction (circular movement) of lower limb at hip joint

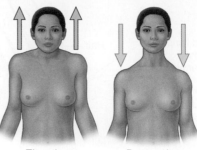

Elevation   Depression

(I) Elevation and depression of shoulders (scapula and clavicle)

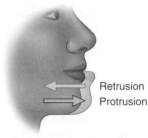

Retraction

Protraction

(K) Protraction and retraction of scapula on thoracic wall

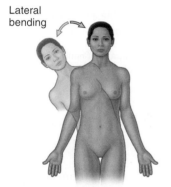

Lateral bending

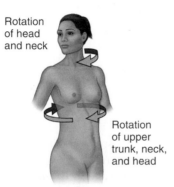

Rotation of head and neck

Rotation of upper trunk, neck, and head

(J) Lateral bending (lateral flexion) of trunk and rotation of upper trunk, neck, and head

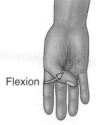

Retrusion

Protrusion

(L) Protrusion and retrusion of mandible (jaw) at temporomandibular joints

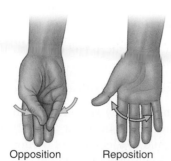

Opposition   Reposition

(M) Opposition and reposition of thumb and little finger at carpometacarpal joint of thumb combined with flexion at metacarpophalangeal joints

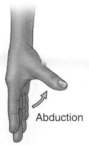

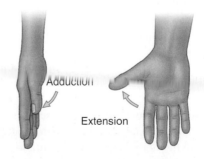

Adduction

Abduction

Extension

Flexion

(N) The thumb is rotated 90 degrees relative to other structures. Abduction and adduction at metacarpophalangeal joint occurs in a sagittal plane; flexion and extension at metacarpophalangeal and interphalangeal joints occurs in frontal planes, opposite to these movements at other joints.

FIGURE 1.4. Terms of movement. *(continued)*

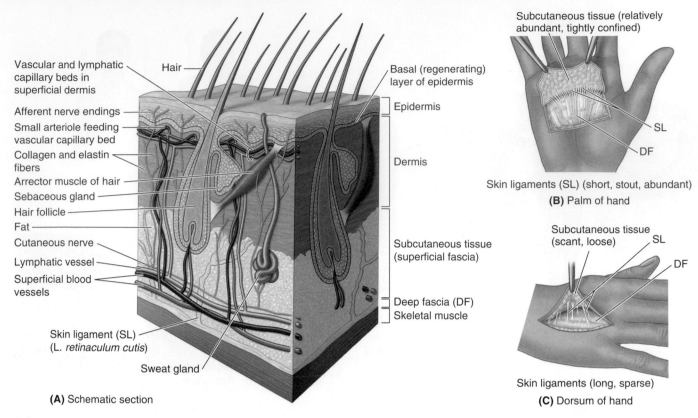

FIGURE 1.5. Structure of skin and subcutaneous tissue. A. Skin and some of its specialized structures. **B.** Skin ligaments of palm of hand. The skin of the palm, like that of the sole of the foot, is firmly attached to the underlying deep fascia. **C.** Skin ligaments of dorsum of hand. The long, relatively sparse skin ligaments allow the mobility of the skin in this region.

The **dermis** is formed by a dense layer of interlacing *collagen* and *elastic fibers*. These fibers provide skin tone and account for the strength and toughness of the skin. The primary direction of collagen fibers determines the characteristic tension lines (cleavage lines) and wrinkle lines in the skin. The deep layer of the dermis contains hair follicles, with their associated smooth arrector (L. *arrector pili*) muscles and sebaceous glands. Contraction of the **arrector muscles** erects the hairs (causing goose bumps), thereby compressing the sebaceous glands and helping them secrete their oily product onto the skin. Other integumentary structures include the hair, nails, mammary glands, and the enamel of teeth.

The **subcutaneous tissue** (superficial fascia) is composed of loose connective tissue and fat. Located between the dermis and underlying deep fascia, the subcutaneous tissue contains the deepest parts of the sweat glands, the blood and lymphatic vessels, and cutaneous nerves. The subcutaneous tissue provides for most of the body's fat storage, so its thickness varies greatly depending on the person's nutritional state. **Skin ligaments** (L. *retinacula cutis*), consisting of numerous small fibrous bands, extend through the subcutaneous tissue and attach the deep surface of the dermis to the underlying deep fascia (Fig. 1.5B,C). The length and density of these ligaments determine the mobility of the skin over deep structures.

The **deep fascia** is a dense, organized connective tissue layer, devoid of fat, that envelops most of the body deep to the skin and subcutaneous tissue. Extensions from its internal surface

- Invest deeper structures, such as individual muscles and neurovascular bundles (**investing fascia**)
- Divide muscles into groups or compartments (**intermuscular septa**)
- Lie between the musculoskeletal walls and the serous membranes lining body cavities (**subserous fascia**)

The deep fascia also forms (1) **retinacula**, which hold tendons in place during joint movement, and (2) **bursae** (closed sacs containing fluid), which prevent friction and enable structures to move freely over one another.

In living people, **fascial planes** (interfascial and intrafascial) are potential spaces between adjacent fascias or fascia-lined structures. During surgical procedures, surgeons take advantage of these planes, separating structures to create actual spaces that allow access to deeper structures. These planes are often fused in embalmed cadavers.

CLINICAL BOX

SKIN INCISIONS AND WOUNDS

Tension Lines

 Tension lines (cleavage lines) keep the skin taut yet allow for creasing with movement. Lacerations or surgical incisions that parallel the tension lines usually heal well with little scarring because there is minimal disruption of the collagen fibers. An incision or laceration across tension lines disrupts a greater number of collagen fibers, causing the wound to gape and possibly heal with excessive (keloid) scarring. Surgeons make their incisions parallel with the tension lines when other considerations (e.g., adequate exposure, avoiding nerves) are not of greater importance.

Stretch Marks in Skin

 The collagen and elastic fibers in the dermis form a tough, flexible meshwork of tissue. The skin can distend considerably when the abdomen enlarges, as during pregnancy, for example. However, if stretched too far or too rapidly, it can result in damage to the collagen fibers in the dermis. Bands of thin wrinkled skin, initially red, become purple and later white. Stretch marks appear on the abdomen, buttocks, thighs, and breasts during pregnancy. These marks also form in obese individuals. Stretch marks generally fade (but never disappear completely) after pregnancy and weight loss.

Burns

 Burns are tissue injuries caused by thermal, electrical, radioactive, or chemical agents.

- In *superficial burns*, the damage is limited to the superficial part of the epidermis.
- In *partial thickness burns*, the damage extends through the epidermis into the superficial part of the dermis. However, except for their most superficial parts, the sweat glands and hair follicles are not damaged and can provide the source of replacement cells for the basal layer of the epidermis.
- In *full-thickness burns*, the entire epidermis and dermis, and perhaps underlying muscle, are damaged. A minor degree of healing may occur at the edges, but the open ulcerated portions require skin grafting.

The extent of the burn (percent of total body surface affected) is generally more significant than the degree (severity of depth) in estimating its effect on the well-being of the victim.

SKELETAL SYSTEM

The skeleton of the body is composed of bones and cartilages and has two main parts (Fig. 1.6):

- The **axial skeleton** consists of the bones of the head (cranium or skull), neck (cervical vertebrae), and trunk (ribs, sternum, vertebrae, and sacrum).
- The **appendicular skeleton** consists of the bones of the limbs, including those forming the pectoral (shoulder) and pelvic girdles.

Cartilage

Cartilage is a resilient, semirigid, avascular type of connective tissue that forms parts of the skeleton where more flexibility is necessary (e.g., the costal cartilages that attach the ribs to the sternum). The articulating surfaces of bones participating in a synovial joint are capped with **articular cartilage**, which provides smooth, low-friction gliding surfaces for free movement of the articulating bones (e.g., blue areas of the humerus in Fig. 1.6). Cartilage is avascular, and therefore, its cells obtain oxygen and nutrients by diffusion. The proportion of bone and cartilage in the skeleton changes as the body grows; the younger a person is, the greater the contribution of cartilage. The bones of a newborn infant are soft and flexible because they are mostly composed of cartilage.

Bone

Bone, a living tissue, is a highly specialized, hard form of connective tissue that makes up most of the skeleton and is the chief supporting tissue of the body. Bones serve the following functions:

- Protection for vital structures
- Support for the body and its vital cavities
- The mechanical basis for movement
- Storage for salts (e.g., calcium)
- A continuous supply of new blood cells (produced by the marrow in the medullary cavity of many bones)

There are two types of bone: **compact bone** and **spongy** (trabecular or cancellous) **bone**. The differences between these types of bone depend on the relative amount of solid matter and the number and size of the spaces they contain (Fig. 1.7). All bones have a superficial thin layer of compact bone around a central mass of spongy bone, except where the latter is replaced by a **medullary (marrow) cavity**. Within this cavity of adult bones and between the spicules of spongy

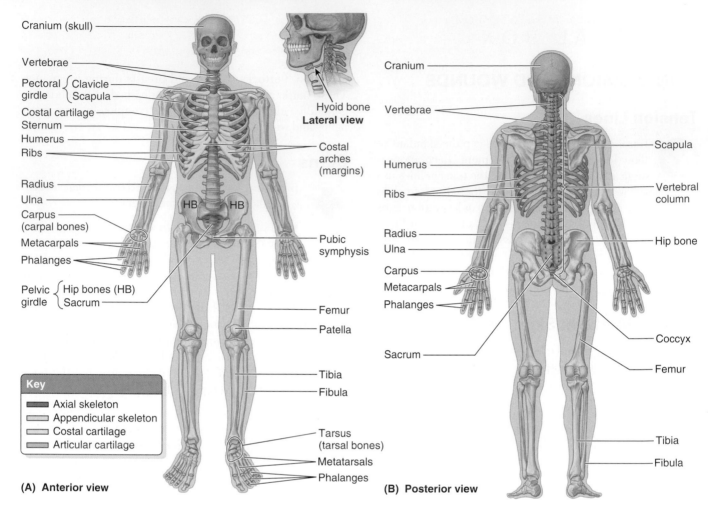

Cranium (skull)

Vertebrae

Pectoral { Clavicle
girdle { Scapula

Costal cartilage

Sternum

Humerus

Ribs

Radius

Ulna

Carpus
(carpal bones)

Metacarpals

Phalanges

Pelvic { Hip bones (HB)
girdle { Sacrum

HB   HB

Hyoid bone
Lateral view

Costal
arches
(margins)

Pubic
symphysis

Femur

Patella

Tibia

Fibula

Tarsus
(tarsal bones)

Metatarsals

Phalanges

Key
- Axial skeleton
- Appendicular skeleton
- Costal cartilage
- Articular cartilage

(A) Anterior view

Cranium

Vertebrae

Humerus

Ribs

Radius

Ulna

Carpus

Metacarpals

Phalanges

Sacrum

Scapula

Vertebral
column

Hip bone

Coccyx

Femur

Tibia

Fibula

(B) Posterior view

FIGURE 1.6. Skeletal system.

bone, blood cells and platelets are formed. The architecture of spongy and compact bone varies according to function.

Compact bone provides strength for weight bearing. In long bones, designed for rigidity and attachment of muscles and ligaments, the amount of compact bone is greatest near the middle of the shaft (body) of the bone, where it is liable to buckle. Living bones have some elasticity (flexibility) and great rigidity (hardness).

The fibrous connective tissue covering that surrounds bone is called **periosteum** (see Fig. 1.10); the tissue surrounding cartilage elements, excluding articular cartilage, is called **perichondrium**. The periosteum and perichondrium help nourish the tissue, are capable of laying down more cartilage or bone (particularly during fracture healing), and provide an interface for attachment of tendons and ligaments.

CLINICAL BOX

BONE DYNAMICS

Heterotopic Bone

Bone sometimes forms in soft tissues where it is not normally present. Horse riders often develop heterotopic bone in their thighs or buttocks (*rider's bones*), probably because of chronic muscle strain resulting

in small hemorrhagic (bloody) areas that undergo calcification and eventual ossification.

Bone Adaptation

Bones are living organs that hurt when injured, bleed when fractured, remodel in relationship to stress placed on them, and change with age. Like other organs, bones have blood vessels, lymphatic vessels, and nerves, and they may become diseased. Unused bones, such as in a paralyzed or immobilized limb, *atrophy* (decrease in size). Bone may be absorbed, which occurs in the mandible after teeth are extracted. Bones undergo *hypertrophy* (enlarge) when they have increased weight to support for a long period.

Bone Trauma and Repair

Trauma to a bone may *fracture* (break) it. For a fracture to heal properly, the broken ends must be brought together, approximating their normal position (*reduction of fracture*). During bone healing, the surrounding *fibroblasts* (connective tissue cells) proliferate and secrete collagen that forms a *collar of callus* to hold the bones together. Remodeling of bone occurs in the fracture area, and the callus calcifies. Eventually, the callus is resorbed and replaced by bone.

Degeneration—Osteoporosis

 As people age, both the organic and inorganic components of bone decrease, often resulting in *osteoporosis*, an abnormal reduction in the quantity of bone, or atrophy of skeletal tissue. The bones become brittle, lose their elasticity, and fracture easily.

CLASSIFICATION OF BONES

Bones are classified according to their shape (Fig. 1.6):

- **Long bones** are tubular structures (e.g., humerus in the arm, phalanges in the fingers).
- **Short bones** are cuboidal and are found only in the ankle (tarsus) and wrist (carpus).
- **Flat bones** usually serve protective functions (e.g., those of the cranium protect the brain).
- **Irregular bones**, such as those in the face, have various shapes other than long, short, or flat.

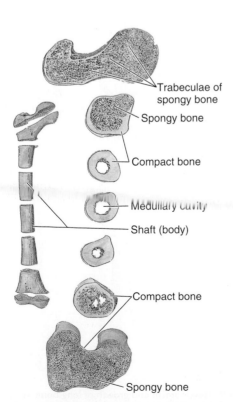

FIGURE 1.7. Transverse sections of femur (thigh bone). Observe the trabeculae (tension and pressure lines) related to the weight-bearing function of this bone.

Labels: Trabeculae of spongy bone; Spongy bone; Compact bone; Medullary cavity; Shaft (body); Compact bone; Spongy bone

- **Sesamoid bones** (e.g., patella, or kneecap) develop in certain tendons. These bones protect the tendons from excessive wear and often change the angle of the tendons as they pass to their attachments.

BONE MARKINGS

Bone markings appear wherever tendons, ligaments, and fascia are attached or where arteries lie adjacent to or enter bones. Other formations occur in relation to the passage of a tendon (often to direct the tendon or improve its leverage) or to control the type of movement occurring at a joint. *Some markings and features of bones are as follows* (Fig. 1.8):

- **Condyle**: rounded articular area (e.g., condyles of the femur)
- **Crest**: ridge of bone (e.g., iliac crest)
- **Epicondyle**: eminence superior to a condyle (e.g., epicondyles of the humerus)
- **Facet**: smooth, flat area, usually covered with cartilage, where a bone articulates with another bone (e.g., articular facets of a vertebra)
- **Foramen**: passage through a bone (e.g., obturator foramen)
- **Fossa**: hollow or depressed area (e.g., infraspinous fossa of the scapula)
- **Line (linea)**: linear elevation (e.g., soleal line of the tibia)
- **Malleolus**: rounded prominence (e.g., lateral malleolus of the fibula)
- **Notch**: indentation at the edge of a bone (e.g., greater sciatic notch in the posterior border of the hip bone)
- **Process**: projecting spine-like part (e.g., spinous process of a vertebra)
- **Protuberance**: projection of bone (e.g., external occipital protuberance of the cranium)
- **Spine**: thorn-like process (e.g., spine of the scapula)
- **Trochanter**: large, blunt elevation (e.g., greater trochanter of the femur)
- **Tubercle**: small, raised eminence (e.g., greater tubercle of the humerus)
- **Tuberosity**: large, rounded elevation (e.g., ischial tuberosity of the hip bone)

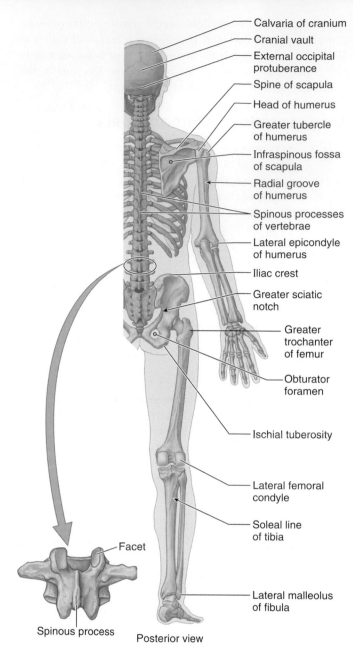

FIGURE 1.8. **Bony markings and formations.**

BONE DEVELOPMENT

All bones are derived from **mesenchyme** (embryonic connective tissue) by one of two different processes: intramembranous ossification (directly from mesenchyme) and endochondral ossification (from cartilage derived from mesenchyme). The histology of a bone is the same either way.

- In **intramembranous ossification** (membranous bone formation), mesenchymal models of bone form during the embryonic period, and direct ossification of the mesenchyme begins in the fetal period.

- In **endochondral ossification** (cartilaginous bone formation), cartilage models of bones form from mesenchyme during the fetal period, and bone subsequently replaces most of the cartilage.

The following brief description of endochondral ossification explains how long bones grow. The mesenchymal cells condense and differentiate into *chondroblasts*, dividing cells in growing cartilage tissue, thereby forming a *cartilaginous bone model* (Fig. 1.9A). In the midregion of the bone model, the cartilage *calcifies* and *periosteal capillaries* (capillaries from the fibrous sheath surrounding the model) grow into the calcified cartilage of the bone model and supply its interior. These blood vessels, together with associated *osteogeni* (bone-forming) cells, form a **periosteal bud**.

The capillaries initiate the **primary ossification center**, so named because the bone tissue it forms replaces most of the cartilage in the shaft of the bone model. The shaft of a bone ossified from a primary ossification center is the **diaphysis**, which grows as the bone develops.

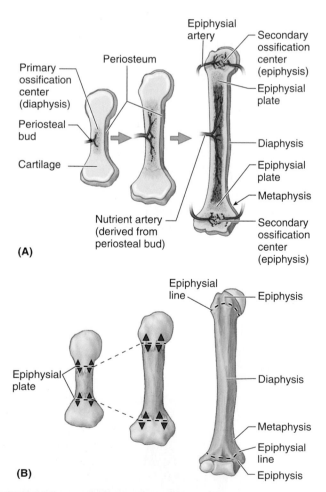

FIGURE 1.9. **Development and growth of long bone. A.** Formation of primary and secondary centers of ossification. **B.** Growth in the length of the bone occurs on both sides of the epiphysial plates (*arrowheads*).

Most **secondary ossification centers** appear in other parts of the developing bone after birth; the parts ossified from these centers are **epiphyses**. **Epiphysial arteries** grow into the developing cavities with associated osteogenic cells. The flared part of the diaphysis nearest to the epiphysis is the **metaphysis** (Fig. 1.9B). For growth to continue, the bone formed from the primary center in the diaphysis does not fuse with that formed from the secondary centers in the epiphyses until the bone reaches its adult size. Thus, during growth of a long bone, cartilaginous **epiphysial plates** intervene between the diaphysis and the epiphyses. These growth plates are eventually replaced by bone at each of its two sides, diaphysial and epiphysial. When this occurs, bone growth ceases, and the diaphysis fuses with the epiphyses. The seam formed during this process (*synostosis*) is dense and appears in radiographs as an **epiphysial line** (Fig. 1.10). The epiphysial fusion of bones occurs progressively from puberty to maturity.

VASCULATURE AND INNERVATION OF BONES

Bones are richly supplied with blood vessels (Fig. 1.10). The arterial supply is from the following vessels:

- **Nutrient arteries** (one or more per bone) that arise outside the periosteum, pass through the shaft of a long bone via **nutrient foramina**, and split in the medullary cavity into longitudinal branches. These vessels supply the bone marrow, spongy bone, and deeper portions of the compact bone.
- Small branches from the **periosteal arteries** of the periosteum supply most of the compact bone. Consequently, if the periosteum is removed, the bone will die.
- **Metaphysial** and **epiphysial arteries** supply the ends of the bones. These vessels arise mainly from the arteries that supply the joints.

Veins accompany arteries through the *nutrient foramina*. Many large veins leave through foramina near the articular ends of the bones. Lymphatic vessels are abundant in the periosteum.

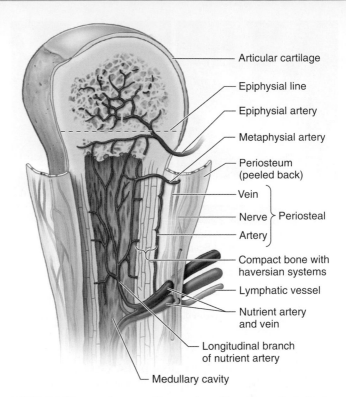

FIGURE 1.10. **Vasculature and innervation of long bone.** The bulk of compact bone is composed of haversian systems (osteons). The haversian canal in the system houses one or two small blood vessels for nourishing the osteocytes (bone cells).

Nerves accompany the blood vessels supplying bones. The periosteum is richly supplied with sensory nerves—**periosteal nerves**—that carry pain fibers. The periosteum is especially sensitive to tearing or tension, which explains the acute pain from bone fractures. Bone itself is relatively sparsely supplied with sensory endings. Within bones, *vasomotor nerves* cause constriction or dilation of blood vessels, regulating blood flow through the bone marrow.

CLINICAL BOX

Accessory Bones

 Accessory (supernumerary) bones develop when additional ossification centers appear and form extra bones. Many bones develop from several centers of ossification, and the separate parts normally fuse. Sometimes, one of these centers fails to fuse with the main bone, giving the appearance of an extra bone; however, careful study shows that the apparent extra bone is a missing part of the main bone. Accessory bones are common in the foot and calvarium (cranial vault [Fig. 1.8]).

Assessment of Bone Age

Knowledge of the sites where ossification centers occur, the times of their appearance, the rate at which they grow, and the times of fusion (*synostosis*) of the sites is used to determine the age of a person in clinical medicine, forensic science, and anthropology (Fig. B1.1). The main criteria for determining bone age are (1) the appearance of calcified material in the diaphysis

(*Continued on next page*)

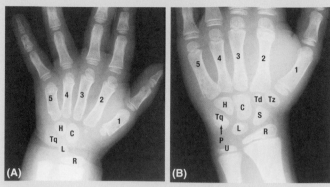

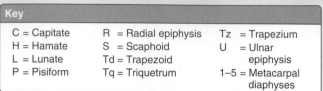

Key

C = Capitate	R = Radial epiphysis	Tz = Trapezium
H = Hamate	S = Scaphoid	U = Ulnar
L = Lunate	Td = Trapezoid	epiphysis
P = Pisiform	Tq = Triquetrum	1–5 = Metacarpal diaphyses

FIGURE B1.1. Anteroposterior view, right hand of (A) a 2.5-year-old and (B) an 11-year-old.

and/or epiphyses and (2) the disappearance of the dark line representing the epiphysial plate (absence of this line indicates epiphysial fusion has occurred; fusion occurs at specific times for each epiphysis). The fusion of epiphyses with the diaphysis occurs 1–2 years earlier in girls than in boys. *Bone age* (achieved level of skeletal development) during the growing years can be determined by radiographic study of the ossification centers of the hand.

Displacement and Separation of Epiphyses

 An injury that causes a fracture in an adult usually causes the displacement of an epiphysis in a child. Without knowledge of bone growth and the appearance of bones in radiographic and other diagnostic images at various ages, a displaced epiphysial plate could be mistaken for a fracture, and separation of an epiphysis could be interpreted as a displaced piece of fractured bone. Bone is smoothly curved on each side of the epiphysial plate, whereas fractures leave sharp, often uneven edges of bone.

Avascular Necrosis

Loss of blood supply to an epiphysis or other parts of a bone results in death of bone tissue, or *avascular necrosis* (G. *nekrosis*, deadness). After every fracture, small areas of adjacent bone undergo necrosis. In some fractures, avascular necrosis of a large fragment of bone may occur.

Degenerative Joint Disease

Synovial joints are well designed to withstand wear, but heavy use over many years can cause degenerative changes. Beginning early in adult life and progressing slowly thereafter, aging of articular cartilage occurs on the ends of the articulating bones, particularly those of the hip, knee, vertebral column, and hands. These irreversible degenerative changes in joints result in the articular cartilage becoming less effective as a shock absorber and low-friction surface. As a result, the articulation becomes vulnerable to the repeated impacts and friction that occur during joint movements (e.g., during running). In some people, these changes cause considerable pain. *Degenerative joint disease*, or *osteoarthritis* (osteoarthrosis), is often accompanied by stiffness, discomfort, and pain. Osteoarthritis is common in older people and usually affects joints that support the weight of their bodies (e.g., hips and knees).

Joints

A **joint** is an articulation, or the place of union or junction, between two or more rigid components (bones, cartilages, or even parts of the same bone). Joints exhibit a variety of forms and functions. Some joints have no movement, others allow only slight movement, and some are freely movable, such as the glenohumeral (shoulder) joint.

CLASSIFICATION OF JOINTS

The three types of joints (fibrous, cartilaginous, and synovial) are classified according to the manner or type of material by which the articulating bones are united (Table 1.1):

- The articulating bones of **fibrous joints** are united by fibrous tissue. The amount of movement occurring at a fibrous joint depends in most cases on the length of the fibers uniting the articular bones. A **syndesmosis** type of fibrous joint unites the bones with a sheet of fibrous tissue, either a ligament or fibrous membrane. Consequently, this type of joint is partially movable. A **gomphosis** (*dentoalveolar syndesmosis*) is a type of fibrous joint found in teeth. A peg-like process fits into a socket (*alveolus*) to stabilize the tooth and provides proprioceptive information from short periodontal ligaments about how hard we are chewing or clenching our teeth.

- The articulating structures of **cartilaginous joints** are united by hyaline cartilage (*primary cartilaginous joints* [*synchondroses*]) or fibrocartilage (*secondary cartilaginous joints* [*symphyses*]). **Synchondroses** permit growth of the length of the bone and allow slight bending during early life until the epiphysial plate converts to bone and

TABLE 1.1. TYPES OF JOINTS

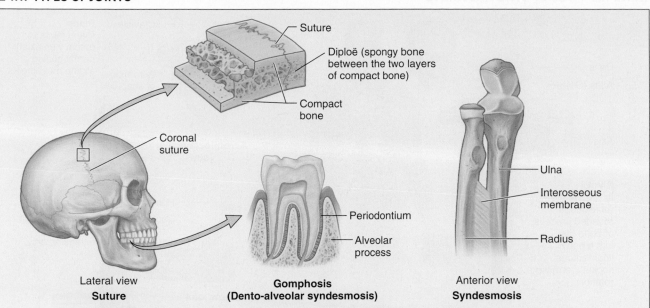

Suture

Diploë (spongy bone between the two layers of compact bone)

Compact bone

Coronal suture

Ulna

Interosseous membrane

Radius

Periodontium

Alveolar process

Lateral view
Suture

Gomphosis
(Dento-alveolar syndesmosis)

Anterior view
Syndesmosis

In **fibrous joints,** articulating bones are joined by fibrous tissue. Sutures of the cranium are fibrous joints in which bones are close together and united by fibrous tissue, often interlocking along a wavy line. Flat bones consist of two plates of compact bone separated by spongy bone and marrow (diploë). In a **syndesmosis joint,** the bones are joined by an interosseous ligament or a sheet of fibrous tissue (e.g., the interosseous membrane joining the forearm bones). In a **gomphosis joint,** a peg-like process fits into a socket (e.g., the articulation between the root of the tooth and the alveolar process). Fibrous tissue, the periodontium, anchors the tooth in the socket.

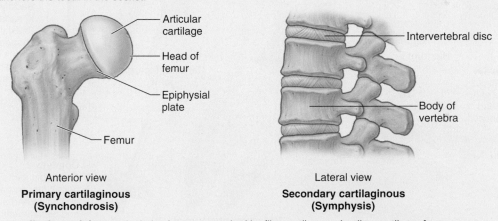

Articular cartilage

Head of femur

Epiphysial plate

Femur

Intervertebral disc

Body of vertebra

Anterior view
Primary cartilaginous
(Synchondrosis)

Lateral view
Secondary cartilaginous
(Symphysis)

In **cartilaginous joints,** articulating bones are united by fibrocartilage or hyaline cartilage. In a **synchondrosis,** such as that in a developing long bone, the bony epiphysis and body are joined by an epiphysial plate (hyaline cartilage). In a **symphysis,** the binding tissue is a fibrocartilaginous disc (e.g., between two vertebrae).

In a **synovial joint** (articulation), the two bones are separated by the characteristic joint cavity (containing synovial fluid) but are joined by an articular capsule (fibrous capsule lined with synovial membrane). The bearing surfaces of the bones are covered with articular cartilage. Synovial joints are functionally the most common and important type of joint. They provide free movement between the bones they join and are typical of nearly all joints of the limbs.

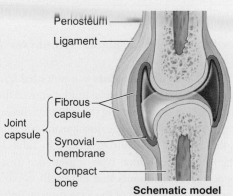

Periosteum

Ligament

Joint capsule

Fibrous capsule

Synovial membrane

Compact bone

Schematic model

TABLE 1.2. TYPES OF SYNOVIAL JOINTS

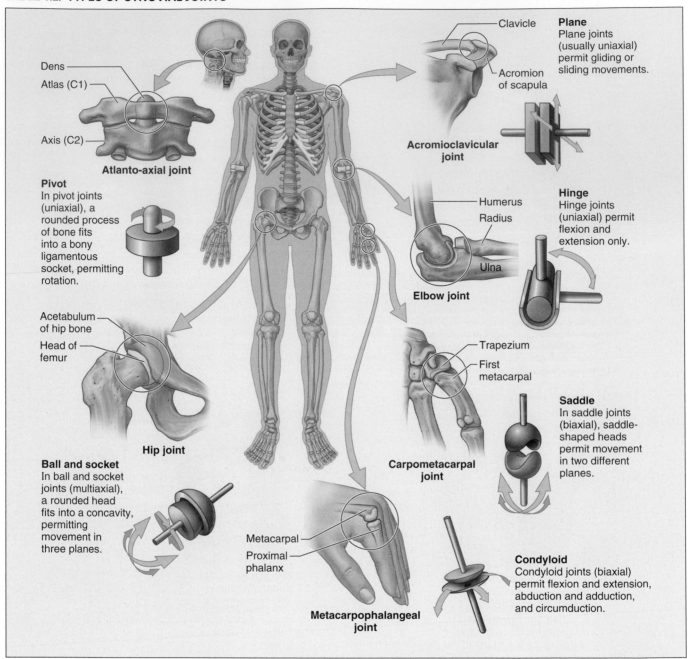

the epiphyses fuse with the diaphysis. **Symphyses** are strong, slightly mobile joints.

- The articular cavity of **synovial joints** is a potential space that contains a small amount of synovial fluid. Synovial fluid serves the dual function of nourishing the articular cartilage and lubricating the joint surfaces. The distinguishing features of a synovial joint are illustrated and described in Table 1.1. Synovial joints, the most common type of joint, are usually reinforced by accessory ligaments that either are separate (extrinsic) or are a thickened part of the joint capsule (intrinsic). Some synovial joints have other distinguishing features, such as fibrocartilaginous *articular discs* or *menisci*, which

are present when the articulating surfaces of the bones are incongruous. The six major types of synovial joints are classified according to the shape of the articulating surfaces and/or the type of movement they permit (Table 1.2).

VASCULATURE AND INNERVATION OF JOINTS

Joints receive blood from *articular arteries* that arise from vessels around the joint. The arteries often *anastomose* (communicate) to form networks (*peri-articular arterial anastomoses*), which ensure a continuous blood supply to a joint throughout its range of movement. *Articular veins* are communicating veins that accompany the arteries (L. *venae comitantes*) and, like the arteries, are located in the joint capsule, mostly in the synovial membrane.

Joints have a rich nerve supply; the nerve endings are numerous in the joint capsule. In the distal parts of limbs, the *articular nerves* are branches of the cutaneous nerves supplying the overlying skin. Otherwise, most articular nerves are branches of nerves that supply the muscles that cross and therefore move the joint. The Hilton Law states that the nerves supplying a joint also supply the muscles moving the joint and the skin covering their attachments.

Pain fibers are numerous in the fibrous layer of the joint capsule and associated ligaments; the synovial membrane is relatively insensitive. Joints transmit a sensation called *proprioception*, information that provides an awareness of movement and position of the parts of the body.

MUSCULAR SYSTEM

Muscle cells, often called *muscle fibers* because they are long and narrow when relaxed, are specialized contractile cells organized into tissues that move body parts or temporarily alter the shape of internal organs. The associated connective tissue conveys nerve fibers and capillaries to the muscle fibers as it binds them into bundles or fascicles.

There are three types of muscle fibers (Table 1.3): (1) **skeletal striated muscle**, which moves bones and other structures (e.g., the eyes); (2) **cardiac striated muscle**, which forms most of the walls of the heart and adjacent parts of the great vessels; and (3) **smooth muscle**, which forms part of the walls of most vessels and hollow organs, moves substances through viscera such as the intestine, and controls movement through blood vessels.

Skeletal Muscles

All skeletal muscles have a fleshy contractile portion (one or more *heads* or *bellies*) composed of skeletal striated muscle and a noncontractile portion composed mainly of collagen bundles: *tendons* (*rounded*) and *aponeuroses* (*flat sheets*).

When referring to the length of a muscle, both the belly and the tendons are included. Most skeletal muscles are attached directly or indirectly through tendons and aponeuroses to bones, cartilages, ligaments, or fascia or

TABLE 1.3. TYPES OF MUSCLE

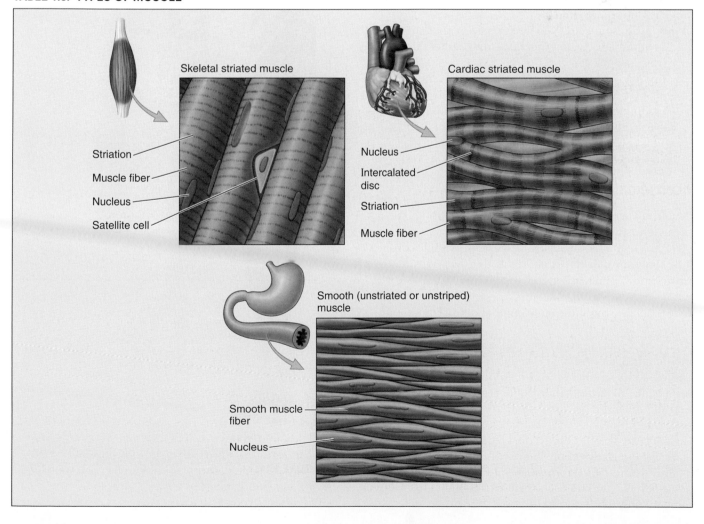

Skeletal striated muscle

- Striation
- Muscle fiber
- Nucleus
- Satellite cell

Cardiac striated muscle

- Nucleus
- Intercalated disc
- Striation
- Muscle fiber

Smooth (unstriated or unstriped) muscle

- Smooth muscle fiber
- Nucleus

to some combination of these structures; however, some muscles are attached to organs (e.g., the eyeball), to skin (e.g., facial muscles), and to mucous membranes (e.g., intrinsic tongue muscles). Muscles are organs of movement, but they also provide static support, give form to the body, and provide heat. Figure 1.11 identifies some of the superficial muscles; the deep muscles are identified when each region is studied.

Most muscles are named on the basis of their function or the bones to which they are attached. The abductor digiti minimi, for example, abducts the little finger. The sternocleidomastoid (L. *kleidos*, bolt) attaches inferiorly to the sternum and clavicle and superiorly to the mastoid process of the temporal bone of the cranium. Other muscles are named on the basis of their shape (G. *deltoid*, triangle), position (medial, lateral, anterior, or posterior), length (*brevis*, short; *longus*, long), relative size (maximus, minimus), or number of heads or bellies (biceps, triceps, digastric). Muscles also may be classified according to their shape and architecture (Fig. 1.12), as in the following examples:

- **Pennate muscles** are feather-like in the arrangement of their fascicles (fiber bundles): unipennate, bipennate, or multipennate (L. *pennatus*, feather).
- **Fusiform muscles** are spindle-shaped (round, thick belly, and tapered ends).
- In **parallel muscles**, the fascicles lie parallel to the long axis of the muscle; flat muscles with parallel fibers often have aponeuroses.
- **Convergent muscles** have a broad attachment from which the fascicles converge to a single tendon.
- **Circular muscles** surround a body opening or orifice, constricting it when contracted.
- **Digastric muscles** feature two bellies in series, sharing a common intermediate tendon.

CONTRACTION OF MUSCLES

When muscles contract, the fibers shorten to about 70% of their resting length. Muscles with a long, parallel fascicle arrangement shorten the most, providing considerable range of movement at a joint, but are not powerful. Muscle power increases as the total number of muscle cells increases. Therefore, the shorter, wide pennate muscles that "pack in" the most fiber bundles shorten less but are the most powerful.

When a muscle contracts and shortens, one of its attachments usually remains fixed and the other one moves. Attachments of muscles are commonly described as the origin and insertion; the *origin* is usually the proximal end of the muscle, which remains fixed during muscular contraction; the *insertion* is usually the distal end of the muscle, which is movable. However, some muscles can act in both directions under different circumstances. Therefore, the terms *proximal* and *distal* or *medial* and *lateral* are used in this book when describing most muscle attachments.

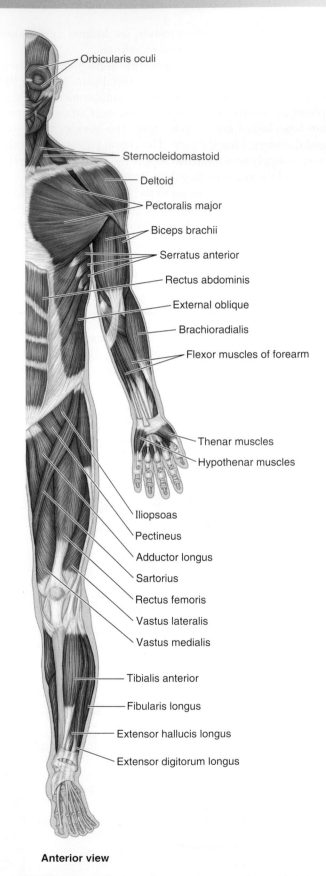

Orbicularis oculi
Sternocleidomastoid
Deltoid
Pectoralis major
Biceps brachii
Serratus anterior
Rectus abdominis
External oblique
Brachioradialis
Flexor muscles of forearm
Thenar muscles
Hypothenar muscles
Iliopsoas
Pectineus
Adductor longus
Sartorius
Rectus femoris
Vastus lateralis
Vastus medialis
Tibialis anterior
Fibularis longus
Extensor hallucis longus
Extensor digitorum longus

Anterior view

FIGURE 1.11. Skeletal muscles. Some larger muscles are labeled.

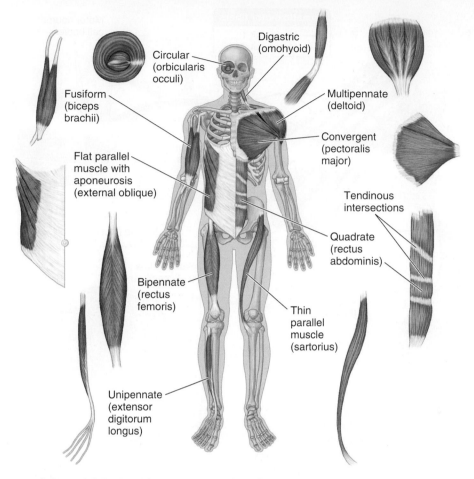

FIGURE 1.12. Architecture and shape of skeletal muscles. Various types of muscles are shown whose shapes depend on the arrangement of fiber bundles.

Skeletal muscle can undergo contraction in three ways:

1. **Reflexive contraction** is automatic and not voluntarily controlled (even though it may involve "voluntary" skeletal muscle)—for example, respiratory movements of the diaphragm. Muscle stretch evokes reflexive contraction produced by tapping a tendon with a reflex hammer.
2. **Tonic contraction** is a slight contraction (**muscle tone**) that does not produce movement or active resistance but gives the muscle firmness, assisting the stability of joints and the maintenance of posture.
3. There are two principal types of **phasic contraction**. In **isometric contractions**, the muscle length remains the same—no movement occurs but muscle tension is increased above tonic levels (e.g., the deltoid holds the arm steady in abduction). In **isotonic contractions**, the muscle changes length to produce movement. There are two forms of isotonic contraction: **concentric contraction**, in which movement occurs owing to muscle shortening (e.g., the deltoid muscle shortens to raise the arm into abduction), and **eccentric contraction**, in which there is progressive relaxation of a contracted muscle (controlled lengthening [e.g., the deltoid lengthens, allowing gravity to lower the arm to the adducted position]).

The *structural unit* of a muscle is a **muscle fiber** (Fig. 1.13). Connective tissue covering individual muscle fibers is called **endomysium**, a group of fibers (fiber bundles) is invested by **perimysium**, and the entire muscle is surrounded by **epimysium**. The *functional unit* of a muscle, consisting of a motor neuron and the muscle fibers it controls, is a **motor unit**. When a motor neuron in the spinal cord is stimulated, it initiates an impulse that causes all the muscle fibers supplied by that motor unit to contract simultaneously. The number of muscle fibers in a motor unit varies from one to several hundred according to the size and function of the muscle. Large motor units, in which one neuron supplies several hundred muscle fibers, are found in the large trunk and thigh muscles. In the small eye and hand muscles, where precision movements are required, the motor units contain only a few muscle fibers.

Muscles serve specific functions in moving and positioning the body. The same muscle may act as a prime mover, antagonist, synergist, or fixator under specific conditions. The functions are described as follows:

- A **prime mover** or **agonist** is the main muscle responsible for producing a specific movement of the body (e.g., concentric contraction).

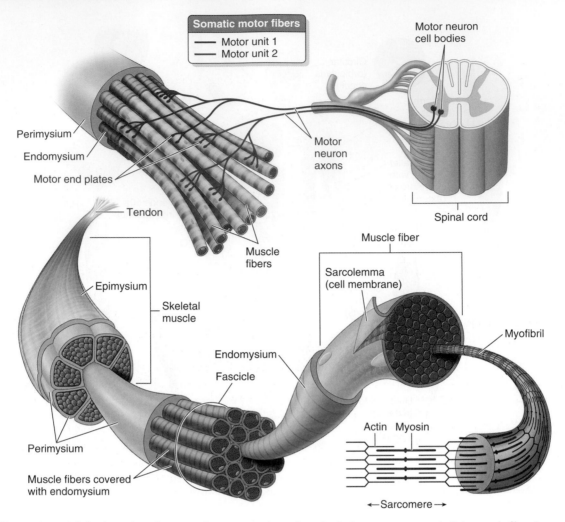

FIGURE 1.13. **Structure of skeletal muscle and motor unit.** A motor unit consists of a single motor neuron and all the muscle fibers innervated by it. Actin (thin) and myosin (thick) filaments are contractile elements (myofibrils) in the muscle fibers.

- **Fixators** steady the proximal parts of a limb while movements are occurring in distal parts.
- A **synergist** complements the action of prime movers—for example, by preventing movement of the intervening joint when a prime mover passes over more than one joint.
- An **antagonist** is a muscle that opposes the action of a prime mover. As a prime mover contracts, the antagonist progressively relaxes, producing a smooth movement.

Cardiac Striated Muscle

Cardiac striated muscle forms the muscular wall of the heart—the **myocardium** (Table 1.3). Some cardiac muscle is also present in the walls of the aorta, pulmonary vein, and superior vena cava (Fig. 1.14). Cardiac muscle contractions are not under voluntary control. Rhythmic contractions are generated intrinsically by *pacemaker nodes* composed of special cardiac muscle fibers, the rate of which is influenced by the autonomic nervous system (discussed later in this chapter). Contractile stimuli are largely propagated myogenically (from muscle fiber to muscle fiber) rather than by direct nerve stimulation.

Smooth Muscle

Smooth muscle, named for the absence of microscopic striations, forms a large part of the middle coat or layer (tunica media) of the walls of most blood vessels and the muscular part of the wall of the digestive tract and ducts (Fig. 1.15A and Table 1.3). Smooth muscle is also found in skin (*arrector muscles* associated with hair follicles [Fig. 1.5A]) and in the eyeball (to control lens thickness and pupil size). Like cardiac muscle, smooth muscle is innervated by the autonomic nervous system (Table 1.3); hence, it is an *involuntary muscle* that can undergo partial contraction for long periods. This is important in regulating the size of the lumen of tubular structures. In the walls of the digestive tract, uterine tubes, and ureters, the smooth muscle cells undergo synchronized rhythmic contractions (peristaltic waves). This process (**peristalsis**) propels the contents along these tubular structures under control of the enteric nervous system.

CLINICAL BOX

Muscle Testing

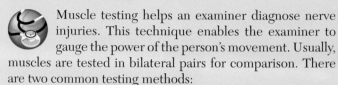

Muscle testing helps an examiner diagnose nerve injuries. This technique enables the examiner to gauge the power of the person's movement. Usually, muscles are tested in bilateral pairs for comparison. There are two common testing methods:

1. The person performs movements that resist those produced by the examiner (active). When testing flexion of the forearm, the examiner asks the person to flex his or her forearm while the examiner resists the effort.
2. The examiner performs movements against resistance produced by the person. For example, the person keeps the forearm flexed while the examiner attempts to extend it.

Electromyography

The electrical stimulation of muscles through electromyography (EMG) is another method for testing muscle action. The examiner places surface electrodes over a muscle and asks the person to perform certain movements. The examiner then amplifies and records the differences in electrical action potentials of the muscles. A normal resting muscle shows only a baseline activity (tonus), which disappears only during sleep, during paralysis, and when under anesthesia. Contracting muscles demonstrate variable peaks of phasic activity.

EMG makes it possible to analyze the activity of an individual muscle during different movements. EMG may also be part of the treatment program for restoring the action of muscles.

Muscular Atrophy

Wasting of the muscular tissue (atrophy) of a limb, for example, may result from a primary disorder of the muscle or from a lesion of a nerve. Muscle atrophy may also be caused by prolonged immobilization of a limb, such as with a cast or sling.

Compensatory Hypertrophy and Myocardial Infarction

In *compensatory hypertrophy*, the myocardium responds to increasing demands by increasing the size of its fibers (cells). When cardiac muscle fibers are damaged during a heart attack, the tissue becomes necrotic (dies) and the fibrous scar tissue that develops forms a *myocardial infarct* (MI), an area of *myocardial necrosis* (pathological death of myocardial tissue). Smooth muscle cells also undergo compensatory hypertrophy in response to increased demands. During pregnancy, the smooth muscle cells in the wall of the uterus increase not only in size (*hypertrophy*) but also in number (*hyperplasia*).

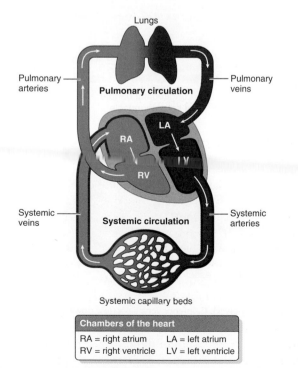

FIGURE 1.14. Schema of cardiovascular system. The continuous circuit consists of two loops: the pulmonary and systemic circulations, served by separate halves of the heart.

Labels in figure:
Lungs
Pulmonary arteries
Pulmonary veins
Pulmonary circulation
LA
RA
LV
RV
Systemic circulation
Systemic veins
Systemic arteries
Systemic capillary beds

Chambers of the heart
RA = right atrium   LA = left atrium
RV = right ventricle   LV = left ventricle

CARDIOVASCULAR SYSTEM

The **circulatory system** transports fluids throughout the body; it consists of the cardiovascular and lymphatic systems. The heart and blood vessels form the blood transportation network, the **cardiovascular system** (Fig. 1.14). The heart pumps blood through the body's vast system of vessels. The blood carries nutrients, oxygen, and waste products to and from cells.

The **heart** consists of two muscular pumps that, although adjacently located, act in a series, dividing the cardiovascular system into two circulations. In the *pulmonary circulation*, the right heart propels low-oxygen blood returned to it into the lungs, where carbon dioxide is exchanged for oxygen. In the *systemic circulation*, oxygen-rich blood returned to the left heart is pumped to the remainder of the body, exchanging oxygen and nutrients for carbon dioxide.

There are three types of blood vessels: *arteries*, *veins*, and *capillaries* (Fig. 1.15). Blood under high pressure leaves the heart and is distributed to the body by a branching system of thicker-walled (more muscular) arteries. The final distributing vessels, *arterioles*, deliver oxygenated blood to capillaries. Minute but numerous thin-walled capillaries form a *capillary bed*, where the interchange of

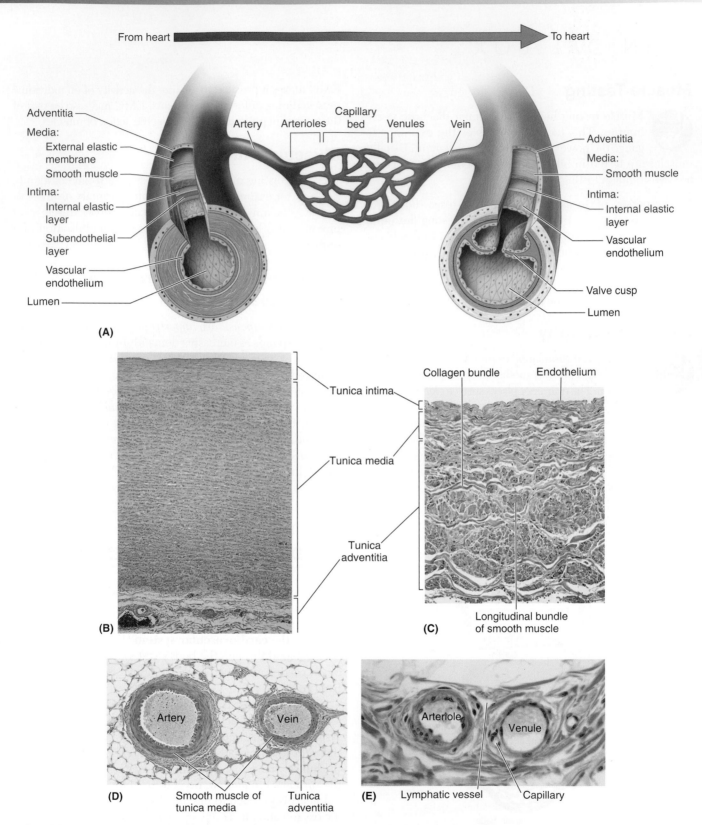

FIGURE 1.15. Structures of arteries and veins. A. Overview. **B.** Aorta, an elastic artery (low power). **C.** Inferior vena cava (low power). **D.** Muscular artery and vein (low power). **E.** Arteriole and venule (high power).

oxygen, nutrients, waste products, and other substances with the extracellular fluid occurs (Fig. 1.15A). Blood from the capillary bed passes into *venules*, which resemble wide capillaries. Venules drain into small veins that open into larger veins. The largest veins, the superior vena cava (SVC) and inferior vena cava (IVC), return poorly oxygenated blood to the heart.

Most vessels of the circulatory system have three tunics or coats: **tunica intima**, the thin endothelial lining of vessels; **tunica media**, the middle smooth muscle layer; and **tunica adventitia**, the outer connective tissue coat.

Arteries

Arteries carry blood away from the heart and distribute it to the body (Fig. 1.16A). Blood passes from the heart through arteries of ever-decreasing caliber. The different types of arteries are distinguished from each other on the basis of overall size, relative amounts of elastic tissue or muscle in the tunica media, and the thickness of the wall relative to the lumen (Fig. 1.15A). Artery size and type is a continuum—that is, there is a gradual change in morphological characteristics from one type to another.

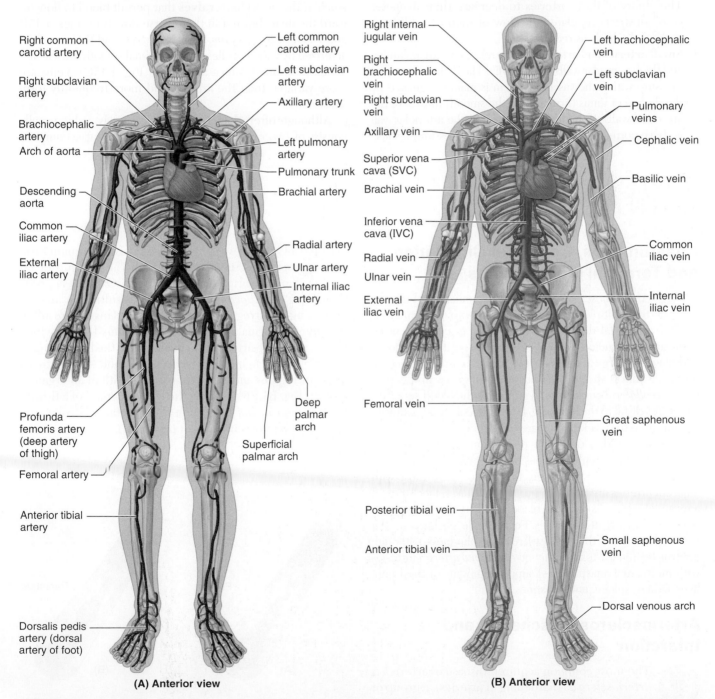

(A) Anterior view

(B) Anterior view

FIGURE 1.16. Systemic portion of cardiovascular system. A. Principal arteries. **B.** Principal veins. Superficial veins are shown in the left limbs; deep veins are shown in the right limbs.

There are three types of arteries:

- **Large elastic arteries** (conducting arteries) have many elastic layers in their walls; examples are the aorta and its branches from the arch of the aorta (Fig. 1.15B). The maintenance of blood pressure in the arterial system between contractions of the heart results from the elasticity of these arteries. This quality allows them to expand when the heart contracts and to return to normal between cardiac contractions.
- **Medium muscular arteries** (distributing arteries) have walls that consist mainly of smooth muscle circularly arranged; one example is the femoral artery (Fig. 1.15D). The ability of these arteries to decrease their diameter (vasoconstrict) regulates the flow of blood to different parts of the body as required.
- **Small arteries** and **arterioles** have relatively narrow lumina and thick muscular walls (Fig. 1.15E). The degree of arterial pressure within the vascular system is mainly regulated by the degree of tonus (firmness) in the smooth muscle of the arteriolar walls. If the tonus of muscle in the arteriolar wall is above normal, *hypertension* (high blood pressure) results.

Veins

Veins return poorly oxygenated blood to the heart from the capillary beds (Fig. 1.16B). The large pulmonary veins are atypical in that they carry well-oxygenated blood from the lungs to the heart. Because of the lower blood pressure in the venous system, the walls of veins are thinner than those of their companion arteries (Fig. 1.15A,C–E). The smallest veins, **venules**, unite to form larger veins that usually form *venous plexuses*, such as the **dorsal venous arch** of the foot (Fig. 1.16B). **Medium veins** in the limbs, where the flow of blood is opposed by the pull of gravity, and other locations (such as the neck) have **valves** that permit blood to flow toward the heart but not in the reverse direction (Figs. 1.15B and 1.17A). **Large veins**, such as the SVC and IVC, are characterized by wide bundles of longitudinal smooth muscle and a well-developed tunica adventitia (Fig. 1.15C). Veins are more variable than the arteries and more frequently form anastomoses.

Although often depicted as single vessels, veins usually consist of two or more vessels. The veins that accompany

CLINICAL BOX

Anastomoses, Collateral Circulation, and Terminal (End) Arteries

Anastomoses (communicating connections) between the multiple branches of an artery provide numerous potential detours for blood flow in case the usual pathway is obstructed by compression, the position of a joint, pathology, or surgical ligation. If a main channel is occluded, the smaller alternate channels can usually increase in size, providing a *collateral circulation* that ensures the blood supply to structures distal to the blockage. However, collateral pathways require time to develop; they are usually insufficient to compensate for sudden occlusion or ligation. There are areas where collateral circulation does not exist or is inadequate to replace the main vessel. Arteries that do not anastomose with adjacent arteries are true *terminal (end) arteries*. Occlusion of a terminal artery disrupts the blood supply to the structure or segment of an organ it supplies. For example, occlusion of the terminal arteries of the retina will result in blindness. Although not true terminal arteries, *functional terminal arteries* (arteries with ineffectual anastomoses) supply segments of the brain, liver, kidney, spleen, and intestines.

Arteriosclerosis: Ischemia and Infarction

The most common acquired disease of arteries is *arteriosclerosis* (hardening of arteries), a group of diseases characterized by thickening and loss of elasticity of arterial walls. *Atherosclerosis*, a common form of arteriosclerosis, is associated with the buildup of fat (mainly cholesterol) in the arterial walls. Calcium deposits then form an *atheromatous plaque*, resulting in arterial narrowing and irregularity (Fig. B1.2A). This may result in *thrombosis* (formation of a local *thrombus* [clot]), which may occlude the artery or be flushed into the bloodstream, resulting in *ischemia* (reduction of blood supply to an organ or region) and *infarction* (local death of an organ or tissue) (Fig. B1.2B). Among the consequences of a thrombus are *myocardial infarction* (heart attack), stroke, and *gangrene* (necrosis in parts of the limbs).

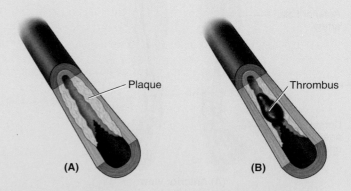

(A) Plaque

(B) Thrombus

FIGURE B1.2. Atheromatous plaque and thrombus.

CLINICAL BOX

Varicose Veins

When the walls of veins lose their elasticity or deep fascia becomes incompetent in sustaining the musculovenous pump, the veins become weak and dilate under the pressure of supporting a column of blood against gravity. This results in *varicose veins*, abnormally swollen, twisted veins, most often seen in the legs (Fig. B1.3).

Varicose veins have a caliber greater than normal, and their valve cusps do not meet or have been destroyed by inflammation. These veins have *incompetent valves*; thus, the column of blood ascending toward the heart is unbroken, placing increased pressure on the weakened walls of the veins and exacerbating their varicosities.

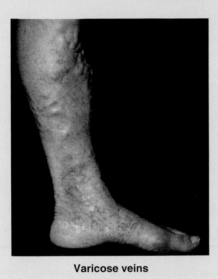

Varicose veins

FIGURE B1.3. Varicose veins.

deep arteries (accompanying veins) surround them in a branching network (Fig. 1.17B) and occupy a relatively unyielding *vascular sheath* with the artery they accompany. As a result, they are stretched and flattened as the artery expands during contraction of the heart, which assists in driving the venous blood toward the heart. The outward expansion of the bellies of contracting skeletal muscles in the legs, for example, compresses the veins, "milking" the blood superiorly toward the heart; this is known as the *musculovenous pump* (Fig. 1.17A).

Capillaries

Capillaries are simple endothelial tubes connecting the arterial and venous sides of the circulation. They are

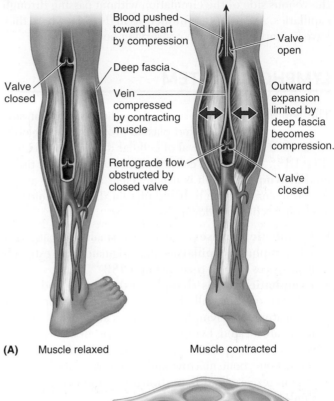

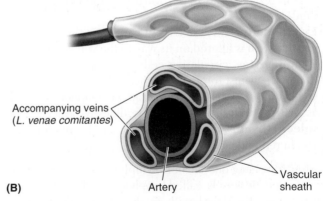

FIGURE 1.17. Veins. A. The musculovenous pump. Muscular contractions in the limbs function with the venous valves to move blood toward the heart. The outward expansion of the bellies of contracting muscles is limited by deep fascia and becomes a compressive force, propelling the blood against gravity. **B.** Accompanying veins (*L. venae comitantes*).

generally arranged in networks (**capillary beds**) between the arterioles and venules (Figs. 1.14 and 1.15A). The blood flowing through capillaries is brought to them by arterioles and carried away from them by venules. As the hydrostatic pressure in the arterioles forces blood through the capillary bed, oxygen, nutrients, and other cellular materials are exchanged with the surrounding tissue. In some regions, such as in the fingers, there are direct connections between the small arteries and veins proximal to the capillary beds they supply and drain. The sites of such communications—**arteriovenous anastomoses (AV shunts)**—permit blood to pass directly from the arterial to the venous side of the circulation without passing through capillaries. AV shunts are numerous in the skin, where they have an important role in conserving body heat.

LYMPHOID SYSTEM

The lymphatic system provides for the drainage of surplus tissue fluid and leaked plasma proteins to the bloodstream and for the removal of cellular debris and infection (Fig. 1.18). This system collects surplus extracellular tissue fluid as **lymph**. Lymph is usually clear and watery and is similar in composition to blood plasma. The lymphoid system consists of the following structures:

- **Lymphatic plexuses**, networks of small lymphatic vessels; **lymphatic capillaries**, that originate in the extracellular spaces of most tissues (Fig. 1.18B)
- **Lymphatic vessels (lymphatics)**, a nearly body-wide network of thin-walled vessels with abundant *valves* originating from lymphatic plexuses along which lymph nodes are located. Lymphatic vessels occur almost everywhere blood capillaries are found, except, for example, teeth, bone, bone marrow, and the entire central nervous system (excess fluid here drains into the cerebrospinal fluid).
- **Lymph nodes**, small masses of lymphatic tissue through which lymph is filtered on its way to the venous system
- **Lymphocytes**, circulating cells of the immune system that react against foreign materials
- **Lymphoid organs**, sites that produce lymphocytes, such as that found in the walls of the digestive tract; in the **spleen**, **thymus**, and lymph nodes; and in **myeloid tissue** in red bone marrow

After traversing one or more lymph nodes, lymph enters larger lymphatic vessels, called lymphatic trunks, which unite to form either the right lymphatic duct or the thoracic duct (Fig. 1.18A):

- The **right lymphatic duct** drains lymph from the body's right upper quadrant (right side of head, neck, and thorax and the entire right upper limb). The duct ends in the angular junction of the right subclavian and internal jugular veins, called the **right venous angle**.
- The **thoracic duct** drains lymph from the remainder of the body. This duct begins in the abdomen as a dilatation, the **cisterna chyli**, and ascends through the thorax and enters the junction of the left internal jugular and left subclavian veins, called the **left venous angle**.

Superficial lymphatic vessels in the skin and subcutaneous tissue eventually drain into a *deep lymphatic vessel*. The deep vessels accompany the major blood vessels.

Additional functions of the lymphatic system include the following:

- *Absorption and transport of dietary fat*, in which special lymphatic capillaries (lacteals) receive all absorbed fat (chyle) from the intestine and convey it through the thoracic duct to the venous system
- *Formation of a defense mechanism for the body*. When foreign protein drains from an infected area, antibodies specific to the protein are produced by immunologically competent cells and/or lymphocytes and dispatched to the infected area.

NERVOUS SYSTEM

The nervous system enables the body to react to continuous changes in its external and internal environments. It controls and integrates various activities of the body, such as circulation and respiration. For descriptive purposes, the human nervous system is divided as follows:

- Structurally into the *central nervous system* (CNS), made up of the brain and spinal cord, and the *peripheral nervous system*, consisting of nerve fibers and cell bodies outside the CNS
- Functionally into the sensory (afferent) nervous system, which carries information to the CNS, and the motor (efferent) nervous system, which carries stimulatory impulses from the CNS to effector organs, such as muscles or glands. Both the sensory and motor systems have somatic and visceral components. The somatic (voluntary) motor system supplies skeletal muscle, and the visceral (involuntary) motor system (also known as the autonomic nervous system) supplies smooth muscle, glands, and the conducting system of the heart. The somatic sensory nervous system carries sensation (e.g., touch and pain) from

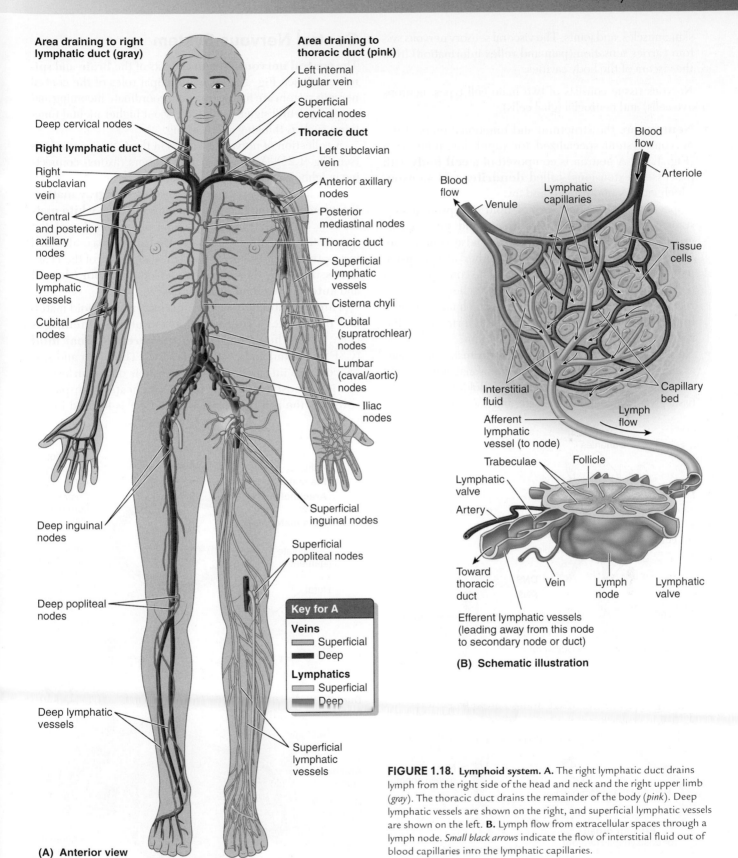

Area draining to right lymphatic duct (gray)

Deep cervical nodes

Right lymphatic duct

Right subclavian vein

Central and posterior axillary nodes

Deep lymphatic vessels

Cubital nodes

Deep inguinal nodes

Deep popliteal nodes

Deep lymphatic vessels

Area draining to thoracic duct (pink)

Left internal jugular vein

Superficial cervical nodes

Thoracic duct

Left subclavian vein

Anterior axillary nodes

Posterior mediastinal nodes

Thoracic duct

Superficial lymphatic vessels

Cisterna chyli

Cubital (supratrochlear) nodes

Lumbar (caval/aortic) nodes

Iliac nodes

Superficial inguinal nodes

Superficial popliteal nodes

Superficial lymphatic vessels

Key for A

Veins
Superficial
Deep

Lymphatics
Superficial
Deep

(A) Anterior view

Blood flow

Arteriole

Blood flow

Venule

Lymphatic capillaries

Tissue cells

Interstitial fluid

Capillary bed

Afferent lymphatic vessel (to node)

Lymph flow

Trabeculae

Follicle

Lymphatic valve

Artery

Toward thoracic duct

Vein

Lymph node

Lymphatic valve

Efferent lymphatic vessels (leading away from this node to secondary node or duct)

(B) Schematic illustration

FIGURE 1.18. Lymphoid system. A. The right lymphatic duct drains lymph from the right side of the head and neck and the right upper limb (*gray*). The thoracic duct drains the remainder of the body (*pink*). Deep lymphatic vessels are shown on the right, and superficial lymphatic vessels are shown on the left. **B.** Lymph flow from extracellular spaces through a lymph node. *Small black arrows* indicate the flow of interstitial fluid out of blood capillaries into the lymphatic capillaries.

skin, muscles, and joints. The visceral sensory system carries sensation (pain and reflex information) from the viscera of the body cavities.

Nervous tissue consists of two main cell types: neurons (nerve cells) and neuroglia (glial cells).

- **Neurons** are the structural and functional units of the nervous system specialized for rapid communication (Fig. 1.19). A neuron is composed of a **cell body** with processes (extensions) called **dendrites** and an **axon**, which carries impulses to and away from the cell body, respectively. **Myelin**, layers of lipid and protein substances, forms a **myelin sheath** around some axons, greatly increasing the velocity of impulse conduction. Neurons communicate with each other at **synapses**, points of contact between neurons. The communication occurs by means of *neurotransmitters*, chemical agents released or secreted by one neuron, which may excite or inhibit another neuron, continuing or terminating the relay of impulses or the response to them.
- **Neuroglia** (glial cells or glia) are approximately five times as abundant as neurons and are nonneuronal, nonexcitable cells that form a major component (scaffolding) of nervous tissue. Neuroglia support, insulate, and nourish the neurons.

Central Nervous System

The **central nervous system** consists of the **brain** and **spinal cord** (see Fig. 1.21). The principal roles of the central nervous system are to integrate and coordinate incoming and outgoing neural signals and to carry out higher mental functions, such as thinking and learning.

A collection of nerve cell bodies in the CNS is a **nucleus** (see Fig. 1.21). A bundle of nerve fibers (axons) connecting neighboring or distant nuclei of the CNS is a **tract**. The nerve cell bodies lie within and constitute the **gray matter**; the interconnecting fiber tract systems form the **white matter** (Fig. 1.20). In transverse sections of the spinal cord, the gray matter appears roughly as an H-shaped area embedded in a matrix of white matter. The struts (supports) of the H are **horns**; therefore, there are right and left **posterior (dorsal)** and **anterior (ventral) gray horns**.

Three membranous layers—pia mater, arachnoid mater, and dura mater—collectively constitute the **meninges** (Fig. 1.20). The meninges and the **cerebrospinal fluid** (CSF) surround and protect the CNS. The brain and spinal cord are intimately covered on their outer surface by the innermost meningeal layer, a delicate, transparent covering, the **pia mater** (pia). The CSF is located in the

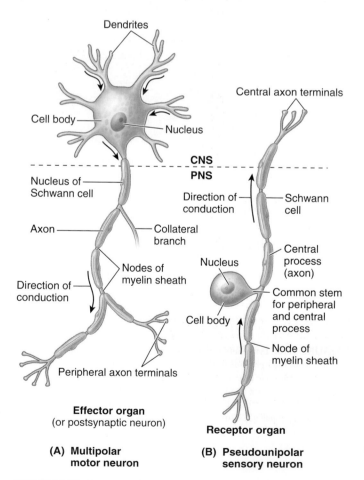

FIGURE 1.19. **Structure of a motor neuron.** Parts of a motor neuron are demonstrated.

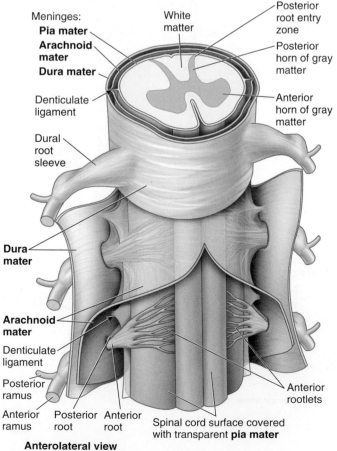

FIGURE 1.20. **Spinal cord and meninges.**

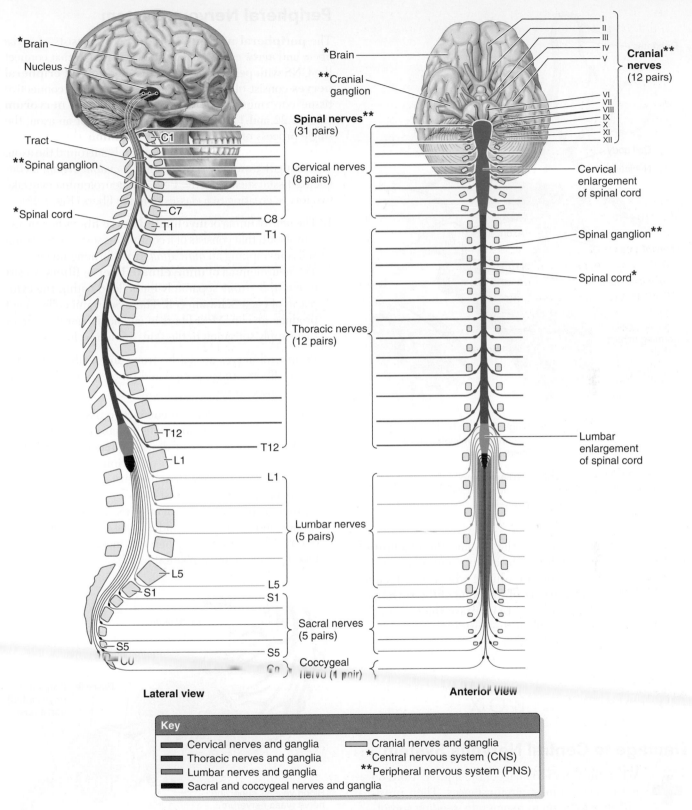

FIGURE 1.21. Basic organization of nervous system.

Lateral view

Anterior View

*Brain

Nucleus

Tract

**Spinal ganglion

*Spinal cord

C1

C7

C8

T1

T1

T12

T12

L1

L1

L5

S1

S5

Co

Spinal nerves
(31 pairs)

Cervical nerves
(8 pairs)

Thoracic nerves
(12 pairs)

Lumbar nerves
(5 pairs)

Sacral nerves
(5 pairs)

Coccygeal
nerve (1 pair)

*Brain

**Cranial
ganglion

I
II
III
IV
V

VI
VII
VIII
IX
X
XI
XII

Cranial
nerves
(12 pairs)

Cervical
enlargement
of spinal cord

Spinal ganglion**

Spinal cord*

Lumbar
enlargement
of spinal cord

Key

Cervical nerves and ganglia

Thoracic nerves and ganglia

Lumbar nerves and ganglia

Sacral and coccygeal nerves and ganglia

Cranial nerves and ganglia

*Central nervous system (CNS)

**Peripheral nervous system (PNS)

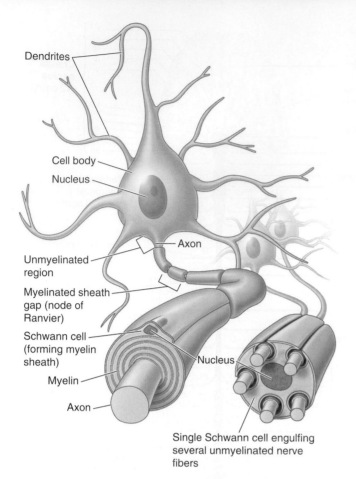

FIGURE 1.22. **Myelinated and unmyelinated nerves.** The myelin sheath gaps (nodes of Ranvier) are intervals in the myelin sheath (i.e., where short lengths of the axon are not covered by myelin).

subarachnoid space between the pia and the **arachnoid mater** (arachnoid). External to the pia and arachnoid is the thick, tough **dura mater** (dura), which is intimately related to the internal aspect of the bone of the surrounding neurocranium (braincase). In contrast, the dura of the spinal cord is separated from the vertebral column by a fat-filled space, the *epidural space*.

CLINICAL BOX

Damage to Central Nervous System

When the CNS is damaged, the injured axons do not recover in most circumstances. Their proximal stumps begin to regenerate, sending sprouts into the area of the lesion; however, growth is blocked by astrocyte (a type of glial cell) proliferation at the site of injury. As a result, permanent disability follows destruction of a tract in the CNS.

Peripheral Nervous System

The **peripheral nervous system** (PNS) consists of *nerve fibers* and *nerve cell bodies* outside of the CNS that connect the CNS with peripheral structures (Fig. 1.19). **Peripheral nerves** consist of bundles of nerve fibers; their connective tissue coverings; and blood vessels, the **vasa nervorum** (Figs. 1.22 and 1.23). A *nerve fiber* consists of an axon, the single process of a neuron; its **neurolemma**, the cell membranes of Schwann cells that immediately surround the axon, separating it from other axons; and its *endoneurium*, a connective tissue sheath. In the PNS, the neurolemma may take two forms, creating two classes of nerve fibers (Fig. 1.22):

1. The neurolemma of **myelinated nerve fibers** have a myelin sheath that consists of a continuous series of Schwann cells enwrapping an *individual axon*, forming myelin.
2. The neurolemma of **unmyelinated nerve fibers** consist of *multiple axons* separately embedded within the cytoplasm of each Schwann cell. These Schwann cells do not produce myelin. Most fibers in cutaneous nerves (nerves that supply sensation to the skin) are unmyelinated.

Peripheral nerves are fairly strong and resilient because the nerve fibers are supported and protected by three connective tissue coverings (Fig. 1.23):

1. **Endoneurium**, a delicate connective tissue sheath that surrounds the neurolemma cells and axons
2. **Perineurium**, a layer of dense connective tissue that encloses a fascicle (bundle) of peripheral nerve fibers, providing an effective barrier against penetration of the nerve fibers by foreign substances
3. **Epineurium**, a thick connective tissue sheath that surrounds and encloses a bundle of fascicles, forming the outermost covering of the nerve; it includes fatty tissues, blood vessels, and lymphatics.

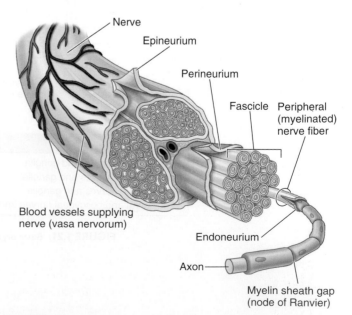

FIGURE 1.23. **Arrangement and ensheathment of peripheral nerve fibers.**

A peripheral nerve is much like a telephone cable. The axons are the individual wires insulated by the neurolemma and endoneurium; the insulated wires are bundled by the perineurium; and the bundles are surrounded in turn by the epineurium, forming the outer wrapping of the "cable."

A collection of nerve cell bodies outside the CNS is a **ganglion** (Fig. 1.21). There are both motor (autonomic) and sensory ganglia.

Peripheral nerves are either cranial or spinal nerves. Of the 12 pairs of **cranial nerves (CNs)**, 11 pairs arise from the brain, and 1 pair (CN XI) arises from the superior part of the spinal cord. All CNs exit the cranial cavity through foramina in the cranium (G. *kranion*, skull). All 31 pairs of **spinal nerves**—8 cervical (C), 12 thoracic (T), 5 lumbar (L), 5 sacral (S), and 1 coccygeal (Co)—arise from the spinal cord and exit through intervertebral foramina in the vertebral column (see Fig. 1.21).

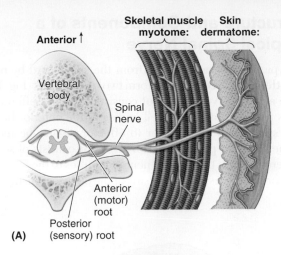

CLINICAL BOX

Peripheral Nerve Degeneration

When peripheral nerves are crushed or severed, their axons degenerate distal to the lesion because they depend on their cell bodies for survival. A *crushing nerve injury* damages or kills the axons distal to the injury site; however, the nerve cell bodies usually survive and the connective tissue coverings of the nerve are intact. No surgical repair is needed for this type of nerve injury because the intact connective tissue sheaths guide the growing axons to their destinations. Surgical intervention is necessary if the nerve is cut because the regeneration of axons requires apposition of the cut ends by sutures through the epineurium. The individual fascicles (bundles of nerve fibers) are realigned as accurately as possible. Compromising a nerve's blood supply for a long period produces *ischemia* by compression of the vasa nervorum (Fig. 1.23), which can also cause nerve degeneration. Prolonged ischemia of a nerve may result in damage no less severe than that produced by crushing or even cutting the nerve.

Somatic Nervous System

The **somatic nervous system** is composed of somatic parts of the CNS and PNS and provides general sensory and motor innervation to all parts of the body (G. *soma*), except the viscera in the body cavities, smooth muscle, and glands. The *somatic (general) sensory fibers* transmit sensations of touch, pain, temperature, and position from sensory receptors (Fig. 1.24). The *somatic motor fibers* stimulate skeletal (voluntary) muscle exclusively, evoking voluntary and reflexive movement by causing its contraction.

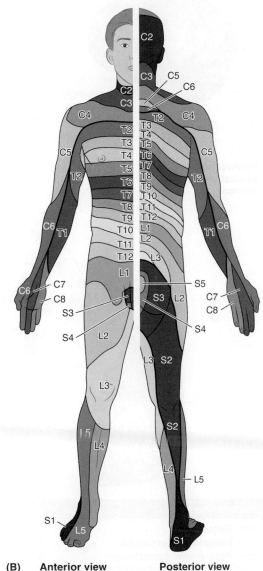

FIGURE 1.24. Dermatomes and myotomes. A. Schematic representation of a dermatome (the unilateral area of skin) and a myotome (the unilateral portion of skeletal muscle) receiving innervation from a single spinal nerve. **B.** Dermatome map. This map is based on the studies of Foerster, 1993 and reflects both anatomical (actual) distribution or segmental innervation and clinical experience.

Structure and Components of a Typical Spinal Nerve

A typical spinal nerve arises from the spinal cord by **nerve rootlets**, which converge to form two **nerve roots** (Fig. 1.20). The **anterior (ventral) root** consists of motor (efferent) fibers passing from nerve cell bodies in the anterior horn of the spinal cord gray matter to effector organs located peripherally. The **posterior (dorsal) root** consists of sensory (afferent) fibers that convey neural impulses to the CNS from sensory receptors in various parts of the body (e.g., in the skin).

The posterior root carries general sensory fibers to the posterior horn of the spinal cord. The anterior and posterior roots unite at the intervertebral foramen to form a spinal nerve, which immediately divides into two **rami** (branches): a posterior ramus and an anterior ramus (Fig. 1.25). As branches of a mixed spinal nerve, the anterior and posterior rami also carry both motor and sensory nerves, as do all their branches:

- The **posterior rami** supply nerve fibers to synovial joints of the vertebral column, deep muscles of the back, and the overlying skin.

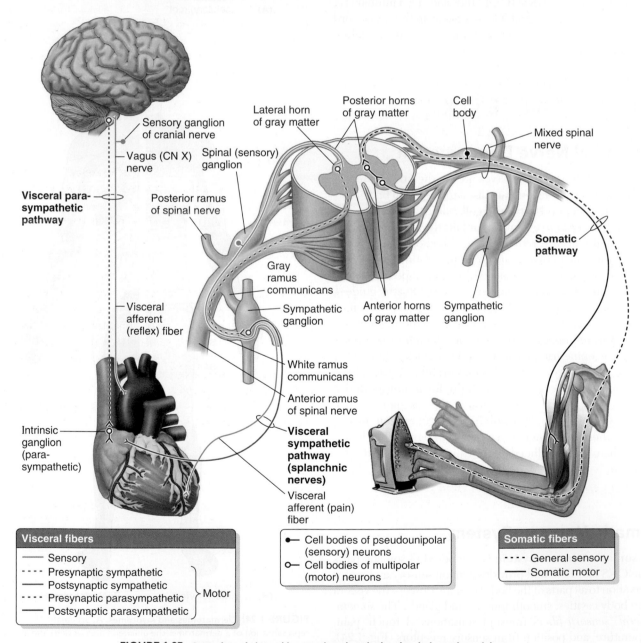

FIGURE 1.25. Somatic and visceral innervation via spinal, splanchnic, and cranial nerves.

- The **anterior rami** supply nerve fibers to the much larger remaining area, consisting of anterior and lateral regions of the trunk and the upper and lower limbs arising from them.

A typical spinal nerve includes the following components:

- **Somatic sensory fibers and motor fibers**
  - *General sensory* (*general somatic afferent*) *fibers* transmit sensations from the body to the CNS; they may be *exteroceptive sensations* (pain, temperature, touch, and pressure) from the skin (Fig. 1.25 right) or pain and *proprioceptive sensations* from muscles, tendons, and joints. Proprioceptive sensations are subconscious sensations that convey information on joint position and the tension of tendons and muscles, providing information on how the body and limbs are oriented in space, independent of visual input. The unilateral area of skin innervated by the general sensory fibers of a single spinal nerve is called a **dermatome** (Fig. 1.24A). From clinical studies of lesions of the posterior roots or spinal nerves, *dermatome maps* have been devised that indicate the typical pattern of innervation of the skin by specific spinal nerves (Fig. 1.24B). However, a lesion of a single posterior root or spinal nerve would rarely result in numbness over the area demarcated for that nerve in these maps because the general sensory fibers conveyed by adjacent spinal nerves overlap as they are distributed to the skin, providing a type of double coverage. Clinicians need to understand the dermatomal innervation of the skin so they can determine, using sensory testing (e.g., with a pin), whether a particular spinal nerve/spinal cord segment is functioning normally.

    Somatic motor (*general somatic efferent*) fibers transmit impulses to skeletal (voluntary) muscles (Fig. 1.25 right). The unilateral muscle mass receiving innervation from the somatic motor fibers conveyed by a single spinal nerve is a **myotome** (Fig. 1.24A). Each skeletal muscle is usually innervated by the somatic motor fibers of several spinal nerves; therefore, the muscle myotome will consist of several segments. The muscle myotomes have been grouped by joint movement to facilitate clinical testing—for example, muscles that flex the glenohumeral (shoulder) joint are innervated primarily by the C5 spinal nerve, and muscles that extend the knee joint are innervated by the L3 and L4 spinal nerves.
- **Visceral motor fibers** of the sympathetic part of the autonomic nervous system (explained in the following section) are conveyed by all branches of all spinal nerves to the smooth muscle of blood vessels and to sweat glands and arrector pili muscles of the skin. (Visceral motor fibers of the parasympathetic part of the autonomic nervous system and visceral afferent fibers have very limited association with spinal nerves.)
- Connective tissue coverings (Fig. 1.23)
- **Vasa nervorum**, blood vessels supplying the nerves

Autonomic Nervous System

The autonomic nervous system (ANS), or *visceral motor system*, consists of **visceral efferent (motor) fibers** that stimulate smooth (involuntary) muscle in the walls of blood vessels and organs, modified cardiac muscle (the intrinsic stimulating and conducting tissue of the heart), and glands (Table 1.4). However, the visceral efferent fibers of the ANS serving viscera of the body cavities are accompanied by **visceral afferent (sensory) fibers**. As the afferent component of autonomic reflexes and the conductors of pain impulses from internal organs, these visceral afferent fibers also regulate visceral functions (Fig. 1.25 left).

VISCERAL MOTOR INNERVATION

The efferent nerve fibers and ganglia of the ANS are organized into two systems or divisions:

1. **Sympathetic (thoracolumbar) division**. In general, the effects of sympathetic stimulation are *catabolic* (preparing the body for "flight or fight").
2. **Parasympathetic (craniosacral) division**. In general, the effects of parasympathetic stimulation are *anabolic* (promoting normal function and conserving energy).

Although both sympathetic and parasympathetic systems often innervate the same structures, they have different (usually contrasting) but coordinated effects (Table 1.4). Conduction of impulses from the CNS to the effector organ involves a series of two neurons in both sympathetic and parasympathetic systems. The cell body of the **presynaptic (preganglionic) neuron** (*first neuron*) is located in the gray matter of the CNS. Its fiber (axon) synapses on the cell body of a **postsynaptic (postganglionic) neuron**, the *second neuron* in the series (Fig. 1.25 left). The cell bodies of such second neurons are located outside the CNS in autonomic ganglia, with the postsynaptic fibers terminating on the effector organ (smooth muscle, modified cardiac muscle, or glands). A functional distinction of pharmacological importance in medical practice is that the postsynaptic neurons of the two systems generally liberate different neurotransmitter substances: *norepinephrine by the sympathetic division* (except in the case of sweat glands) and *acetylcholine by the parasympathetic division*. The anatomical distinction between the sympathetic and the parasympathetic motor divisions of the ANS is based primarily on (1) the location of the presynaptic cell bodies and (2) which nerves conduct the presynaptic fibers from the CNS. These differences are discussed in more detail later in this chapter.

TABLE 1.4. FUNCTIONS OF AUTONOMIC NERVOUS SYSTEM

Organ, Tract, or System		Effect of Sympathetic Stimulation	Effect of Parasympathetic Stimulation
Eyes	Pupil	Dilates pupil (admits more light for increased acuity at a distance)	Constricts pupil (protects retina from excessively bright light)
	Ciliary body	No effect (does not innervate)	Contracts ciliary muscle, allowing lens to thicken for near vision (accommodation)
Skin	Arrector muscle of hair	Causes hairs to stand on end (gooseflesh or goose bumps)	No effect (does not innervate)[a]
	Peripheral blood vessels	Vasoconstricts (blanching of skin and lips; turning fingertips blue)	
	Sweat glands	Promotes sweating[b]	
Other glands	Lacrimal glands	Slightly decreases secretion (a result of vasoconstriction)[c]	Secretomotor (promotes secretion)
	Salivary glands	Secretion decreases, becomes thicker, more viscus (as above).[c]	Secretomotor (promotes abundant, watery secretion)
Heart		Increases rate and strength of contraction; inhibits effect of parasympathetic system on coronary vessels, allowing them to dilate[c]	Decreases rate and strength of contraction (conserving energy); constricts coronary vessels in relation to reduced demand
Lungs		Inhibits effect of parasympathetic system, resulting in bronchodilation and reduced secretion, allowing for maximum air exchange	Constricts bronchi (conserving energy) and promotes bronchial secretion
Digestive tract		Inhibits peristalsis and constricts blood vessels to digestive tract so blood is available to skeletal muscle; maintains tonus of internal anal sphincter to aid fecal continence at rest	Stimulates peristalsis and secretion of digestive juices; stimulates peristalsis of rectum and inhibits internal anal sphincter causing defecation
Liver and gallbladder		Promotes breakdown of glycogen to glucose (for increased energy)	Promotes building/conservation of glycogen; increases secretion of bile
Urinary tract		Vasoconstriction of renal vessels slows urine formation; internal sphincter of bladder contracted to maintain urinary continence.	Contracts detrusor muscle of bladder wall, causing urination; inhibits contraction of internal sphincter of bladder in males
Genital system		Causes ejaculation and vasoconstriction, resulting in remission of erection	Produces engorgement (erection) of erectile tissues of external genitals
Suprarenal medulla		Release of adrenaline into blood	No effect (does not innervate)

[a]The parasympathetic system is restricted in its distribution to the head, neck, and body cavities (except for erectile tissues of genitalia); otherwise, parasympathetic fibers are never found in the body wall and limbs. Sympathetic fibers, by comparison, are distributed to all vascularized portions of the body.
[b]With the exception of the sweat glands, glandular secretion is parasympathetically stimulated.
[c]With the exception of the coronary arteries, vasoconstriction is sympathetically stimulated; the effects of sympathetic stimulation on glands (other than sweat glands) are the indirect effects of vasoconstriction.

Sympathetic Visceral Motor Innervation

The cell bodies of *presynaptic* neurons of the sympathetic division of the ANS are located in the **intermediolateral cell columns** (IMLs) or nuclei of the spinal cord (Fig. 1.26). The paired (right and left) IMLs are a part of the gray matter, extending between the 1st thoracic (T1) and the 2nd or 3rd lumbar (L2 or L3) segments of the spinal cord. In horizontal sections of this part of the spinal cord, the IMLs appear as small **lateral horns** of the H-shaped gray matter, looking somewhat like an extension of the crossbar of the H between the posterior and the anterior horns of gray matter. The cell bodies of *postsynaptic* neurons of the sympathetic nervous system occur in two locations, the paravertebral and prevertebral ganglia (Figs. 1.27 and 1.28):

- **Paravertebral ganglia** are linked to form right and left *sympathetic trunks (chains)* on each side of the vertebral column that extend essentially the length of this column. The superior paravertebral ganglion—the **superior cervical ganglion** of each sympathetic trunk—lies at the base of the cranium. A **ganglion impar** often forms inferiorly, where the two trunks unite at the level of the coccyx.

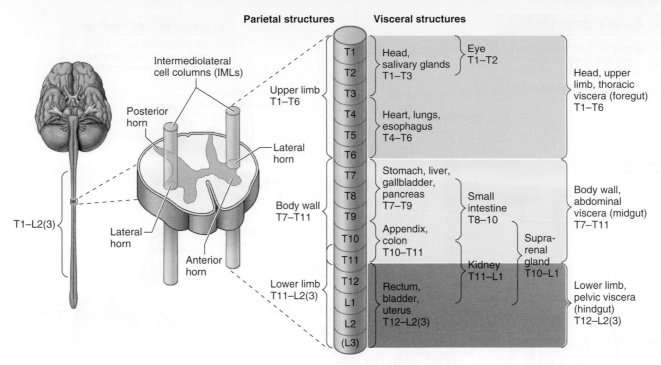

Parietal structures   **Visceral structures**

T1 — Head, salivary glands T1–T3 — Eye T1–T2

Head, upper limb, thoracic viscera (foregut) T1–T6

Upper limb T1–T6 — T2, T3

T4, T5 — Heart, lungs, esophagus T4–T6

T6

T7 — Stomach, liver, gallbladder, pancreas T7–T9

Body wall T7–T11 — T8, T9 — Small intestine T8–10

Body wall, abdominal viscera (midgut) T7–T11

T10, T11 — Appendix, colon T10–T11 — Supra-renal gland T10–L1

Kidney T11–L1

Lower limb T11–L2(3) — T12, L1, L2 — Rectum, bladder, uterus T12–L2(3)

Lower limb, pelvic viscera (hindgut) T12–L2(3)

(L3)

T1–L2(3)

Posterior horn
Intermediolateral cell columns (IMLs)
Lateral horn
Lateral horn
Anterior horn

FIGURE 1.26. Intermediolateral cell columns.

- **Prevertebral ganglia** are in the plexuses that surround the origins of the main branches of the abdominal aorta (for which they are named), such as the large **celiac ganglia** that surround the origin of the celiac trunk (a major vessel arising from the aorta) and the aortic, hypogastric, and pelvic plexuses that descend from them.

Because they are motor fibers, the axons of presynaptic neurons leave the spinal cord through anterior roots, then and enter the anterior rami of spinal nerves T1 through L2 or L3 (Figs. 1.26 and 1.28). Almost immediately after entering the rami, all the presynaptic sympathetic fibers leave the anterior rami of these spinal nerves and pass to the *sympathetic trunks* through **white rami communicantes**. Within the sympathetic trunks, presynaptic fibers follow one of four possible courses: (1) ascend, or (2) descend in the sympathetic trunk to synapse with a postsynaptic neuron of a higher or lower paravertebral ganglion, or (3) enter and synapse immediately with a postsynaptic neuron of the paravertebral ganglion at that level, or (4) pass through the sympathetic trunk without synapsing, continuing on within an abdominopelvic splanchnic nerve (innervates abdominopelvic viscera) to reach the prevertebral ganglia (Figs. 1.28 and 1.29).

Presynaptic sympathetic fibers that provide autonomic innervation within the head, neck, body wall, limbs, and thoracic cavity follow one of the first three courses, synapsing within the paravertebral ganglia.

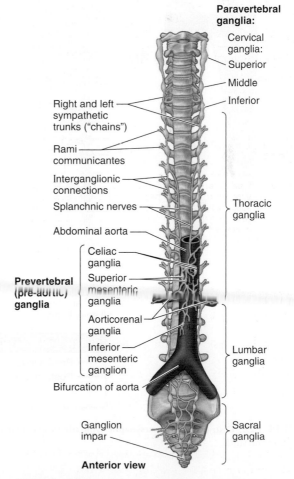

Paravertebral ganglia:
Cervical ganglia:
Superior
Middle
Inferior

Right and left sympathetic trunks ("chains")
Rami communicantes
Interganglionic connections
Splanchnic nerves
Abdominal aorta

Thoracic ganglia

Prevertebral (pre-aortic) ganglia
Celiac ganglia
Superior mesenteric ganglia
Aorticorenal ganglia
Inferior mesenteric ganglion
Bifurcation of aorta
Ganglion impar

Lumbar ganglia

Sacral ganglia

Anterior view

FIGURE 1.27. Ganglia of sympathetic nervous system.

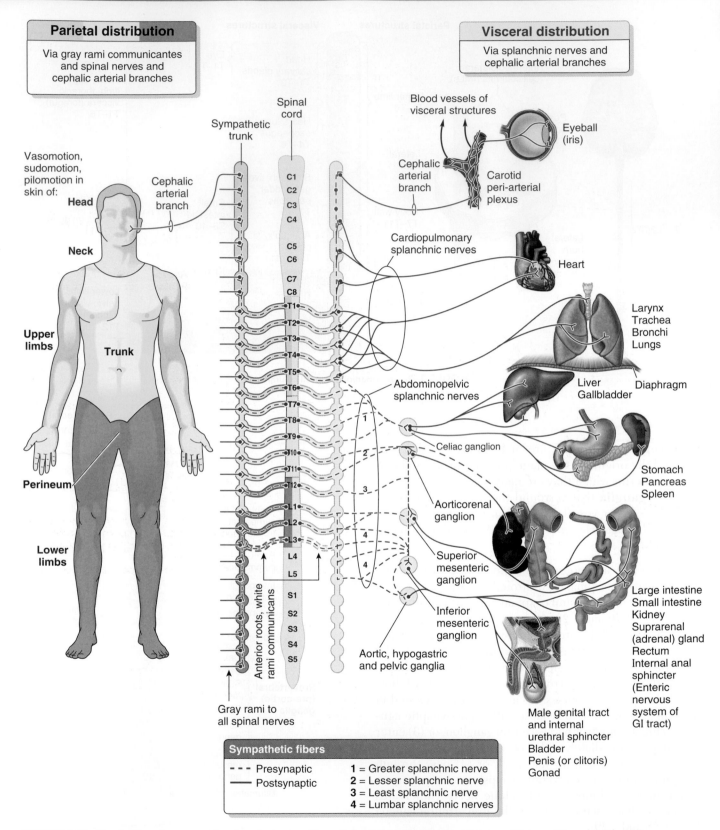

Parietal distribution

Via gray rami communicantes and spinal nerves and cephalic arterial branches

Visceral distribution

Via splanchnic nerves and cephalic arterial branches

Spinal cord

Sympathetic trunk

Vasomotion, sudomotion, pilomotion in skin of:

Cephalic arterial branch

Head

Neck

Upper limbs

Trunk

Perineum

Lower limbs

Anterior roots, white rami communicans

Gray rami to all spinal nerves

Blood vessels of visceral structures

Cephalic arterial branch

Carotid peri-arterial plexus

Eyeball (iris)

Cardiopulmonary splanchnic nerves

Heart

Larynx Trachea Bronchi Lungs

Abdominopelvic splanchnic nerves

Liver Gallbladder

Diaphragm

Celiac ganglion

Stomach Pancreas Spleen

Aorticorenal ganglion

Superior mesenteric ganglion

Inferior mesenteric ganglion

Aortic, hypogastric and pelvic ganglia

Large intestine Small intestine Kidney Suprarenal (adrenal) gland Rectum Internal anal sphincter (Enteric nervous system of GI tract)

Male genital tract and internal urethral sphincter Bladder Penis (or clitoris) Gonad

Sympathetic fibers

- - - Presynaptic
——— Postsynaptic

1 = Greater splanchnic nerve
2 = Lesser splanchnic nerve
3 = Least splanchnic nerve
4 = Lumbar splanchnic nerves

FIGURE 1.28. Distribution of postsynaptic sympathetic nerve fibers. Splanchnic nerves: Greater (1), Lesser (2), Least (3), and Lumbar (4). *GI,* gastrointestinal.

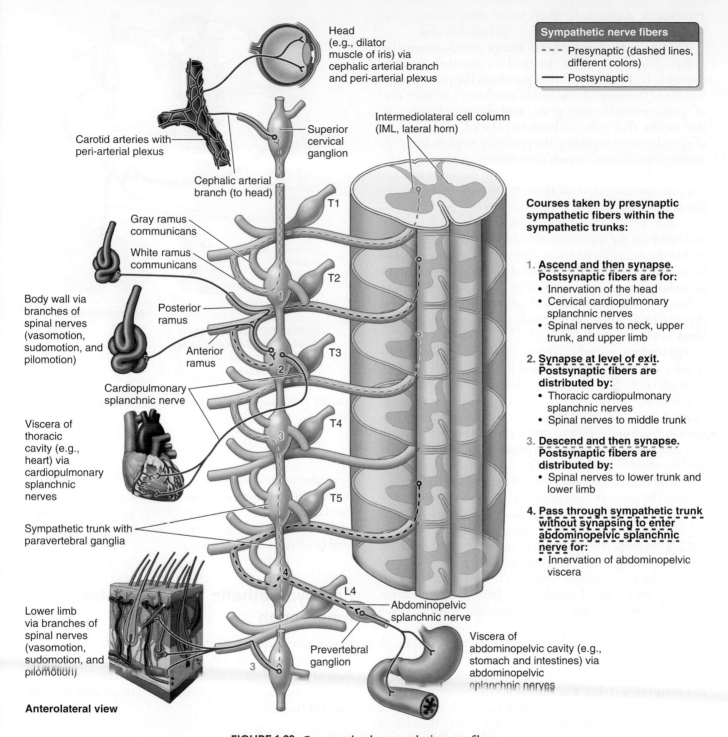

Head
(e.g., dilator
muscle of iris) via
cephalic arterial branch
and peri-arterial plexus

Sympathetic nerve fibers
- - - Presynaptic (dashed lines, different colors)
——— Postsynaptic

Intermediolateral cell column
(IML, lateral horn)

Superior
cervical
ganglion

Carotid arteries with
peri-arterial plexus

Cephalic arterial
branch (to head)

T1

Gray ramus
communicans

White ramus
communicans

T2

Body wall via
branches of
spinal nerves
(vasomotion,
sudomotion, and
pilomotion)

Posterior
ramus

Anterior
ramus

T3

Cardiopulmonary
splanchnic nerve

T4

Viscera of
thoracic
cavity (e.g.,
heart) via
cardiopulmonary
splanchnic
nerves

Sympathetic trunk with
paravertebral ganglia

T5

Lower limb
via branches of
spinal nerves
(vasomotion,
sudomotion, and
pilomotion)

L4

Abdominopelvic
splanchnic nerve

Prevertebral
ganglion

Viscera of
abdominopelvic cavity (e.g.,
stomach and intestines) via
abdominopelvic
splanchnic nerves

Anterolateral view

**Courses taken by presynaptic
sympathetic fibers within the
sympathetic trunks:**

1. **Ascend and then synapse.**
   Postsynaptic fibers are for:
   - Innervation of the head
   - Cervical cardiopulmonary
     splanchnic nerves
   - Spinal nerves to neck, upper
     trunk, and upper limb

2. **Synapse at level of exit.**
   Postsynaptic fibers are
   distributed by:
   - Thoracic cardiopulmonary
     splanchnic nerves
   - Spinal nerves to middle trunk

3. **Descend and then synapse.**
   Postsynaptic fibers are
   distributed by:
   - Spinal nerves to lower trunk and
     lower limb

4. **Pass through sympathetic trunk
   without synapsing to enter
   abdominopelvic splanchnic
   nerve for:**
   - Innervation of abdominopelvic
     viscera

FIGURE 1.29. Courses taken by sympathetic motor fibers.

Presynaptic sympathetic fibers innervating viscera within the abdominopelvic cavity follow the fourth course.

Postsynaptic sympathetic fibers greatly outnumber presynaptic fibers. Those destined for distribution within the neck, body wall, and limbs pass from the paravertebral ganglia of the sympathetic trunks to adjacent anterior rami of spinal nerves through **gray rami communicantes**. By this means, they enter all branches of each of the 31 pairs of spinal nerves, including the posterior rami, to stimulate contraction of blood vessels (*vasomotion*) and the arrector muscles of hair (*pilomotion*, resulting in goose bumps) and to cause sweating (*sudomotion*). Postsynaptic sympathetic fibers that perform these functions in the head (plus innervation of the dilator muscle of the iris) all have their cell bodies in the superior cervical ganglion at the superior end of the sympathetic trunk. They pass from the ganglion by means of a **cephalic arterial branch** to form **periarterial plexuses** of nerves (Figs. 1.28 and 1.29), which follow branches of the carotid arteries, or they may pass directly to nearby CNs to reach their destination in the head.

Splanchnic nerves convey visceral efferent (autonomic) and afferent fibers to and from viscera of the body cavities (Figs. 1.27 to 1.29). Postsynaptic sympathetic fibers destined for viscera of the thoracic cavity (e.g., heart, lungs, and esophagus) pass through *cardiopulmonary splanchnic nerves* to enter the cardiac, pulmonary, and esophageal plexuses. The presynaptic sympathetic fibers involved in innervation of viscera of the abdominopelvic cavity (e.g., the stomach, intestines, and pelvic organs) pass to the prevertebral ganglia through *abdominopelvic splanchnic nerves* (the greater, lesser, least, and lumbar splanchnic nerves). All presynaptic sympathetic fibers of the abdominopelvic splanchnic nerves, except those involved in innervating the suprarenal (adrenal) glands, synapse in the prevertebral ganglia. The postsynaptic fibers from the prevertebral ganglia form peri-arterial plexuses, which follow branches of the abdominal aorta to reach their destination.

Some presynaptic sympathetic fibers that pass through the prevertebral (aorticorenal) ganglia without synapsing terminate directly on cells in the medulla of the **suprarenal gland** (Fig. 1.30). The suprarenal medullary cells function as a special type of postsynaptic neuron that, instead of releasing their neurotransmitter substance onto the cells of a specific effector organ, release it into the bloodstream to circulate throughout the body, producing a widespread sympathetic response. Thus, the sympathetic innervation of this gland is exceptional.

As described earlier, postsynaptic sympathetic fibers are components of virtually all branches of all spinal nerves. By this means and via peri-arterial plexuses, they extend to and innervate all the body's blood vessels (the sympathetic system's primary function) as well as sweat glands, arrector muscles of hairs, and visceral structures. Thus, the

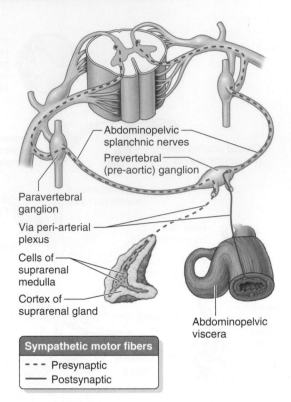

FIGURE 1.30. Sympathetic supply to medulla of suprarenal (adrenal) gland.

- Abdominopelvic splanchnic nerves
- Prevertebral (pre-aortic) ganglion
- Paravertebral ganglion
- Via peri-arterial plexus
- Cells of suprarenal medulla
- Cortex of suprarenal gland
- Abdominopelvic viscera

Sympathetic motor fibers
- - - Presynaptic
—— Postsynaptic

sympathetic nervous system reaches virtually all parts of the body, with the rare exception of avascular tissues, such as cartilage and nails. Presynaptic fibers are relatively short, whereas postsynaptic fibers are relatively long, having to extend to all parts of the body.

Parasympathetic Visceral Motor Innervation

Presynaptic parasympathetic neuron cell bodies are located in two sites within the CNS—a cranial and sacral site—and their fibers exit by two routes (Fig. 1.31). This accounts for the alternate name of the parasympathetic (craniosacral) division of the ANS.

- Cranial site: From the gray matter of the brainstem, fibers exit the CNS within CN III, CN VII, CN IX, and CN X; these fibers constitute the **cranial parasympathetic outflow**.
- Sacral site: From the gray matter of the sacral segments of the spinal cord (S2–S4), the fibers exit the CNS through the anterior roots of spinal nerves S2–S4 and the pelvic splanchnic nerves that arise from their anterior rami; these fibers constitute the **sacral parasympathetic outflow**.

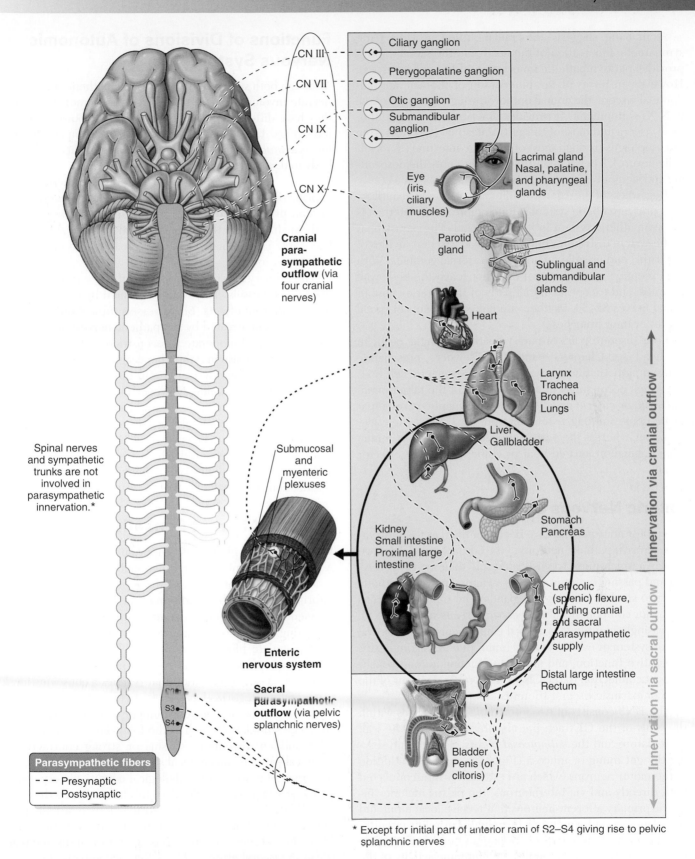

Ciliary ganglion

Pterygopalatine ganglion

Otic ganglion
Submandibular ganglion

CN III

CN VII

CN IX

CN X

Cranial para-sympathetic outflow (via four cranial nerves)

Lacrimal gland
Nasal, palatine, and pharyngeal glands

Eye (iris, ciliary muscles)

Parotid gland

Sublingual and submandibular glands

Heart

Larynx
Trachea
Bronchi
Lungs

Liver
Gallbladder

Stomach
Pancreas

Left colic (splenic) flexure, dividing cranial and sacral parasympathetic supply

Distal large intestine
Rectum

Bladder
Penis (or clitoris)

Spinal nerves and sympathetic trunks are not involved in parasympathetic innervation.*

Submucosal and myenteric plexuses

Kidney
Small intestine
Proximal large intestine

Enteric nervous system

Sacral parasympathetic outflow (via pelvic splanchnic nerves)

S3
S4

Innervation via cranial outflow →

← Innervation via sacral outflow

Parasympathetic fibers	
- - -	Presynaptic
——	Postsynaptic

* Except for initial part of anterior rami of S2–S4 giving rise to pelvic splanchnic nerves

FIGURE 1.31. Distribution of parasympathetic nerve fibers.

As its name suggests, the cranial outflow provides parasympathetic innervation of the head, and the sacral outflow provides parasympathetic innervation of the pelvic viscera. However, in terms of the innervation of thoracic and abdominal viscera, the cranial outflow through the vagus nerve (CN X) is dominant. It provides innervation to all the thoracic viscera and most of the gastrointestinal (GI) tract from the esophagus through most of the large intestine (to its left colic flexure). The sacral outflow supplies only the descending and sigmoid colon and rectum.

Regardless of the extensive influence of its cranial outflow, the parasympathetic system is much more restricted than is the sympathetic system in its distribution. The parasympathetic system distributes only to the head, visceral cavities of the trunk, and erectile tissues of the external genitalia. With the exception of the latter, it does not reach the body wall or limbs, and except for initial parts of the anterior rami of spinal nerves S2–S4, its fibers are not components of spinal nerves or their branches.

Four discrete pairs of parasympathetic ganglia occur in the head (see Chapters 8 and 10). Elsewhere, presynaptic parasympathetic fibers synapse with postsynaptic cell bodies, which occur singly in or on the wall of the target organ (*intrinsic* or *enteric ganglia*). Most presynaptic parasympathetic fibers are long, extending from the CNS to the effector organ, whereas the postsynaptic fibers are short, running from a ganglion located near or embedded in the effector organ.

Enteric Nervous System

The motor neurons that have been identified as the postsynaptic parasympathetic neurons of the GI tract are now known to play a much more sophisticated role than merely receiving and passing on input from presynaptic parasympathetic fibers to smooth muscles and glands. These motor neurons are major components of the *enteric nervous system* (ENS), increasingly identified as a third component of the visceral motor system or even a "second brain" due to its complexity, integrative function, and ability to function autonomously, without connection to the CNS via the other divisions of the ANS or extrinsic visceral afferents.

The ENS consists of two interconnected plexuses within the walls of the GI tract: the *myenteric plexus* of the wall musculature and the *submucosal plexus*, deep to and serving the gut lining or mucosa (Fig. 1.31, inset). In addition to the motor neurons, which are extensively interconnected both directly and via interneurons, the plexus includes intrinsic primary afferent neurons that receive local input and stimulate the motor neurons, forming local reflex circuitry that intrinsically integrates exocrine and endocrine secretion, vasomotion, micro-motility, and immune activity of the gut. This local activity is only modulated by the input from the extrinsic parasympathetic and sympathetic fibers. More detailed information about the enteric nervous system is provided in Chapter 5.

Functions of Divisions of Autonomic Nervous System

Although both sympathetic and parasympathetic systems innervate involuntary (and often affect the same) structures, they have different, usually contrasting yet coordinated, effects (Figs. 1.28 and 1.31). In general, the sympathetic system is a *catabolic* (energy-expending) *system* that enables the body to deal with stresses, such as when preparing the body for the fight-or-flight response. The parasympathetic system is primarily a *homeostatic* or *anabolic* (energy-conserving) *system*, promoting the quiet and orderly processes of the body, such as those that allow the body to feed and assimilate. Table 1.4 summarizes the specific functions of the ANS and its divisions.

The primary function of the sympathetic system is to regulate blood vessels. This is accomplished by several means having different effects. Blood vessels throughout the body are tonically innervated by sympathetic nerves, maintaining a resting state of moderate vasoconstriction. In most vascular beds, an increase in sympathetic signals causes increased vasoconstriction, and a decrease in the rate of sympathetic signals allows vasodilation. However, in certain regions of the body, sympathetic signals are vasodilatory (i.e., sympathetic transmitter substances inhibit active vasoconstriction, allowing the blood vessels to be passively dilated by the blood pressure). In the coronary vessels, the vessels of skeletal muscles, and the external genitalia, sympathetic stimulation results in vasodilation (Wilson-Pauwels et al., 2010).

Visceral Afferent Sensation

Visceral afferent fibers have important relationships to the ANS, both anatomically and functionally. We are usually unaware of the sensory input of these fibers, which provides information about the condition of the body's internal environment. This information is integrated in the CNS, often triggering visceral or somatic reflexes or both. Visceral reflexes regulate blood pressure and chemistry by altering such functions as heart and respiratory rates and vascular resistance. Visceral sensation that reaches a conscious level is generally categorized as pain that is usually poorly localized and may be perceived as hunger or nausea. However, adequate stimulation, such as the following, may elicit true pain: sudden distention, spasms or strong contractions, chemical irritants, mechanical stimulation (especially when the organ is active), and pathological conditions (especially *ischemia*—inadequate blood supply) that lower the normal thresholds of stimulation. Normal activity usually produces no sensation but may do so when there is ischemia. Most visceral reflex (unconscious) sensation and some pain travel in visceral afferent fibers that accompany the parasympathetic fibers retrograde. Most visceral pain impulses (from the heart and most organs of the peritoneal cavity) travel centrally along visceral afferent fibers accompanying sympathetic fibers.

MEDICAL IMAGING

Body Systems

Familiarity with imaging techniques commonly used in clinical settings enables one to recognize abnormalities such as congenital anomalies, tumors, and fractures. The introduction of contrast media allows the study of various luminal or vascular organs and potential or actual spaces, such as the digestive or alimentary system, blood vessels, kidneys, synovial cavities, and subarachnoid space. This section consists of short descriptions of the principles of some of the commonly used diagnostic imaging techniques:

- Conventional radiography (ordinary X-ray images)
- Computerized tomography (CT)
- Ultrasonography (US)
- Magnetic resonance imaging (MRI)
- Positron emission tomography (PET)

Conventional Radiography

In a radiological examination, a highly penetrating beam of X-rays transilluminates the patient, showing tissues of differing densities of mass within the body as images of differing densities of light and dark on the X-ray film (Fig. 1.32). A tissue or organ that is relatively dense in mass, such as compact bone in a rib, absorbs more X-rays than does a less dense tissue, such as spongy (cancellous) bone (see Box 1.1, "Basic Principles of X-Ray Image Formation"). Consequently, a dense tissue or organ produces a relatively transparent area on the X-ray film because relatively fewer X-rays reach the emulsion in the film. Therefore, relatively fewer grains of silver are developed at this area when the film is processed. A very dense substance is *radiopaque*, whereas a substance of less density is *radiolucent*.

Many of the same principles that apply to making a shadow apply to conventional radiography. Radiographs are made with the part of the patient's body being studied close to the

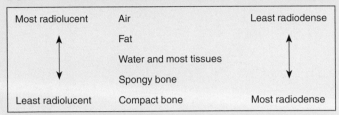

BOX 1.1. BASIC PRINCIPLES OF X-RAY IMAGE FORMATION

Most radiolucent	Air	Least radiodense
	Fat	
↕	Water and most tissues	↕
	Spongy bone	
Least radiolucent	Compact bone	Most radiodense

X-ray film or detector to maximize the clarity of the image and minimize magnification artifacts. In basic radiological nomenclature, *postero-anterior (PA) projection* refers to a radiograph in which the X-rays traverse the patient from posterior (P) to anterior (A); the X-ray tube is posterior to the patient, and the X-ray film or detector is anterior. A radiograph using *anteroposterior (AP) projection* radiography is the opposite. Both PA and AP projection radiographs are viewed as if you and the patient are facing each other (the patient's right side is opposite your left); this is referred to as an *anteroposterior view*. Thus, the standard chest X-ray, taken to examine the heart and lungs, is an AP view of a PA projection. For lateral radiographs, radiopaque letters (R or L) are used to indicate the side placed closest to the film or detector, and the image is viewed from the same direction that the beam was projected.

The introduction of contrast media (radiopaque fluids such as iodine compounds or barium) allows the study of various luminal or vascular organs and potential or actual spaces—such as the digestive tract, blood vessels, kidneys, synovial cavities, and the subarachnoid space—that are not visible in plain films. Most radiological examinations are performed in at least two projections at right angles to each other. Because each radiograph presents a two-dimensional (2-D) representation of a three-dimensional (3-D) structure, structures sequentially penetrated by the X-ray beam overlap each other. Thus, more than one view is usually necessary to detect and localize an abnormality accurately.

Computerized Tomography

CT shows images of the body that resemble transverse anatomical sections (Fig. 1.33). A beam of X-rays is passed through the body as the X-ray tube and detector rotate around the axis of the body. The amount of radiation absorbed by each different type of tissue of the chosen body plane varies with the amount of fat, bone, and water in each element. A computer compiles and generates images as 2-D slices and total 3-D reconstructions.

Ultrasonography

US is a technique that allows visualization of superficial or deep structures in the body by recording pulses of ultrasonic waves reflecting off the tissues (Fig. 1.34). The images can

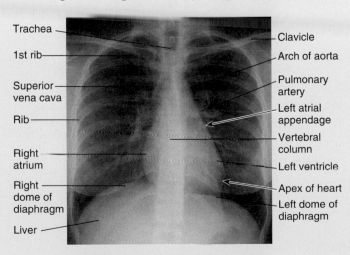

Trachea
1st rib
Superior vena cava
Rib
Right atrium
Right dome of diaphragm
Liver
Clavicle
Arch of aorta
Pulmonary artery
Left atrial appendage
Vertebral column
Left ventricle
Apex of heart
Left dome of diaphragm

FIGURE 1.32. Radiograph of thorax (chest).

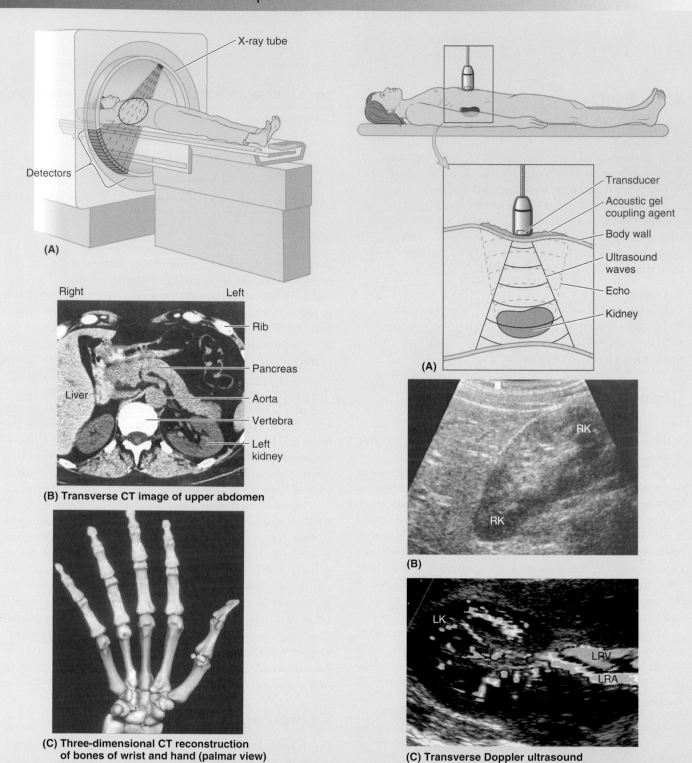

(A)

(B) Transverse CT image of upper abdomen

Right — Left

Liver

Rib
Pancreas
Aorta
Vertebra
Left kidney

(C) Three-dimensional CT reconstruction of bones of wrist and hand (palmar view)

Transducer
Acoustic gel coupling agent
Body wall
Ultrasound waves
Echo
Kidney

(A)

(B)

RK
RK

(C) Transverse Doppler ultrasound

LK
LRV
LRA

FIGURE 1.33. Computerized tomography. A. The X-ray tube rotates around the person in the computerized tomography (CT) scanner and sends a fan-shaped beam of X-rays through the person's body from a variety of angles. X-ray detectors on the opposite side of the person's body measure the amount of radiation that passes through a transverse section of the person. **B** and **C.** A computer reconstructs the CT images. Transverse scans are oriented so they appear the way an examiner would view the section when standing at the foot of the bed and looking toward a supine person's head.

FIGURE 1.34. Ultrasonography. A. The image results from the echo of ultrasound waves from structures of different densities. **B.** A longitudinal image of a right kidney (*RK*) is displayed. **C.** Doppler US shows blood flow to and away from the kidney. *LK*, left kidney; *LRA*, left renal artery; *LRV*, left renal vein.

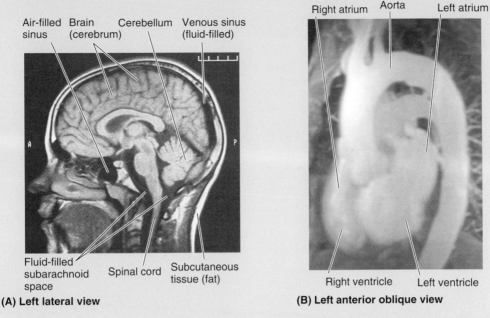

Air-filled sinus Brain (cerebrum) Cerebellum Venous sinus (fluid-filled)

Fluid-filled subarachnoid space Spinal cord Subcutaneous tissue (fat)

(A) Left lateral view

Right atrium Aorta Left atrium

Right ventricle Left ventricle

(B) Left anterior oblique view

FIGURE 1.35. Magnetic resonance imaging. A. Sagittal magnetic resonance imaging (MRI) study of the head and upper neck. **B.** Magnetic resonance angiogram of heart and great vessels.

be viewed in real time to demonstrate the motion of structures and flow within blood vessels (Doppler US) and then recorded as single images or as a movie. Because US is noninvasive and does not use radiation, it is the standard method of evaluating the growth and development of the embryo and fetus.

Magnetic Resonance Imaging

MRI shows images of the body similar to those produced by CT, but they are better for tissue differentiation (Fig. 1.35). Using MRI, the clinician is able to reconstruct the tissues in *any plane*, even arbitrary oblique planes. The person is placed in a scanner with a strong magnetic field, and the body is pulsed with radio waves. Signals subsequently emitted from the patient's tissues are stored in a computer and may be reconstructed in 2-D or 3-D images. The appearance of tissues on the generated images can be varied by controlling how radiofrequency pulses are sent and received. Scanners can be gated or paced to visualize moving structures, such as the heart and blood flow, in real time.

Positron Emission Tomography

PET scanning uses cyclotron-produced isotopes of extremely short half-life that emit positrons. PET scanning is used to evaluate the physiological functions of organs such as the brain on a dynamic basis. Areas of increased brain activity will show selective uptake of the injected isotope (Fig. 1.36).

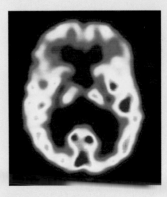

FIGURE 1.36. Positron emission tomography. Transverse scan showing regions of brain activity.

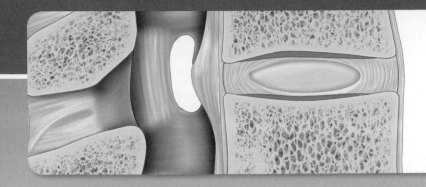

2

Back

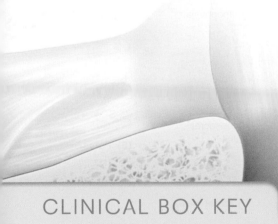

CLINICAL BOX KEY

Anatomical
Variations

Diagnostic
Procedures

Life Cycle

Surgical
Procedures

Trauma

Pathology

The back, the posterior aspect of the trunk inferior to the neck and superior to the gluteal region (buttocks), is the region of the body to which the head, neck, and limbs are attached. Because of their close association with the trunk, the back of the neck and the posterior and deep cervical muscles and vertebrae are described in this chapter. The back consists of the following:

- Skin
- Subcutaneous tissue
- Deep fascia
- Muscles (a superficial layer, concerned with positioning and moving the upper limbs, and deeper layers, concerned with posture, moving, or maintaining the position of the axial skeleton)
- Ligaments
- Vertebral column
- Ribs (in the thoracic region)
- Spinal cord and meninges (membranes covering the spinal cord)
- Various segmental nerves and vessels

VERTEBRAL COLUMN

The **vertebral column** (spine), extending from the cranium (skull) to the apex of the coccyx, forms the skeleton of the neck and back and is the main part of the axial skeleton (articulated bones of the cranium, vertebral column, ribs, and sternum). The vertebral column protects the spinal cord and spinal nerves, supports the weight of the body superior to the level of the pelvis, provides a partly rigid and flexible axis for the body and a pivot for the head, and plays an important role in posture and locomotion.

The adult vertebral column typically consists of 33 vertebrae arranged in five regions: 7 cervical, 12 thoracic, 5 lumbar, 5 sacral, and 4 coccygeal (Fig. 2.1A–D). The **lumbosacral angle** is located at the junction of the lumbar region of the vertebral column and sacrum. Significant motion occurs between only the superior 25 vertebrae. The 5 sacral vertebrae (segments) are fused in adults to form the sacrum, and the 4 coccygeal vertebrae (segments) are fused to form the coccyx. The vertebrae gradually become larger as the vertebral column descends to the sacrum and then become progressively smaller toward the apex of the coccyx. These structural differences are related to the fact that the successive vertebrae bear increasing amounts of the body's weight. The vertebrae reach maximum size immediately superior to the sacrum, which transfers the weight to the pelvic girdle at the sacro-iliac joints. The presacral vertebral column is flexible because it consists of **vertebrae** joined together by semirigid **intervertebral (IV) discs**. The 25 cervical, thoracic, lumbar, and first sacral vertebrae also articulate at synovial *zygapophysial joints*, which facilitate and control the vertebral column's flexibility.

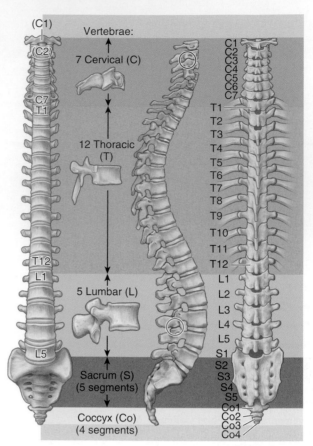

(A) Anterior view   **(B) Right lateral view**   **(C) Posterior view with vertebral ends of ribs**

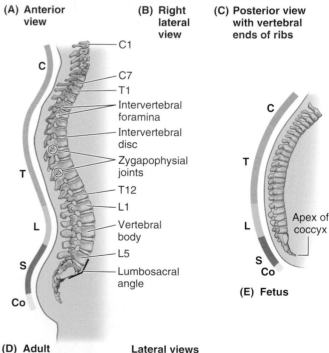

(D) Adult   **Lateral views**

FIGURE 2.1. Vertebral column and curvatures. A–C. Regions of adult vertebral column. Zygapophysial (facet) joints representative of each region are *circled*. **D.** Curvatures of adult vertebral column. **E.** Curvatures of fetal vertebral column. Cervical + thoracic + lumbar = presacral vertebral column.

The vertebral bodies contribute approximately three quarters of the height of the presacral vertebral column, and the fibrocartilage of IV discs contributes approximately one quarter. The shape and strength of the vertebrae and IV discs, ligaments, and muscles provide stability to the vertebral column.

Curvatures of Vertebral Column

The vertebral column in adults has four curvatures: cervical, thoracic, lumbar, and sacral (Fig. 2.1D). The **thoracic** and **sacral** (pelvic) **curvatures (kyphoses)** are concave anteriorly, whereas the **cervical** and **lumbar cur-**vatures (lordoses) are concave posteriorly. The thoracic and sacral curvatures are **primary curvatures** developing during the fetal period (Fig. 2.1E). Primary curvatures are retained throughout life as a consequence of differences in height between the anterior and the posterior parts of the vertebrae. The cervical and lumbar curvatures are **secondary curvatures**, which begin to appear in the cervical region during the fetal period but do not become obvious until infancy. Secondary curvatures are maintained primarily by differences in thickness between the anterior and the posterior parts of the IV discs. The **cervical curvature** becomes prominent when an infant begins to hold his or her head erect.

Curvatures of Vertebral Column

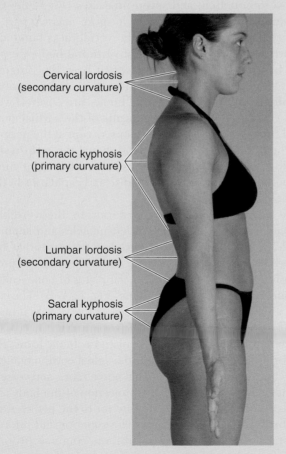

Cervical lordosis (secondary curvature)

Thoracic kyphosis (primary curvature)

Lumbar lordosis (secondary curvature)

Sacral kyphosis (primary curvature)

FIGURE SA2.1. Lateral view of normal curvatures of back.

When the posterior surface of the trunk is observed, especially in a lateral view, the normal curvatures of the vertebral column are apparent.

Abnormal Curvatures of Vertebral Column

 Abnormal curvatures in some people result from developmental anomalies and in others from pathological processes such as *osteoporosis*. Osteoporosis is characterized by a net demineralization of bones and results from a disruption of the normal balance of calcium deposition and resorption. The bones become weakened and brittle and are subject to fracture. Vertebral body osteoporosis occurs in all vertebrae but is most common in thoracic vertebrae and is an especially common finding in postmenopausal women.

Excessive thoracic kyphosis (clinically shortened to **kyphosis**) is characterized by an abnormal increase in the thoracic curvature; the vertebral column curves posteriorly (Fig. B2.1A,B). This abnormality can result from erosion of the anterior part of one or more vertebrae. Progressive erosion and collapse of vertebrae results in an overall loss of height. *Dowager hump* is a colloquial name for excess thoracic kyphosis in older women resulting from osteoporosis; however, kyphosis occurs in geriatric people of both sexes.

Excessive lumbar lordosis (clinically shortened to **lordosis**) is characterized by an anterior rotation of the pelvis, producing an abnormal increase in the lumbar curvature; the vertebral column becomes more convex anteriorly (Fig. B2.1A,C). This *abnormal extension deformity* may be associated with weakened trunk musculature, especially of the anterolateral abdominal wall. To compensate for alterations to their normal line of gravity, women develop a temporary lordosis during late pregnancy.

(*Continued on next page*)

Scoliosis (curved back) is characterized by an abnormal lateral curvature that is accompanied by rotation of the vertebrae (Fig. B2.1D,E). The spinous processes turn toward the cavity of the abnormal curvature. Scoliosis is the most common deformity of the vertebral column in pubertal girls (aged 12–15 years). Asymmetric weakness of the intrinsic back muscles (*myopathic scoliosis*), failure of half of a vertebra to develop (*hemivertebra*), and a difference in the length of the lower limbs are causes of scoliosis.

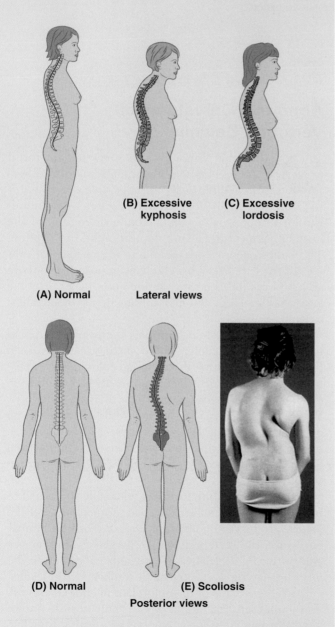

(B) Excessive kyphosis

(C) Excessive lordosis

(A) Normal   **Lateral views**

(D) Normal   **(E) Scoliosis**

Posterior views

FIGURE B2.1. Normal and abnormal curvatures of vertebral column.

The lumbar curvature becomes apparent when an infant begins to walk and assumes the upright posture. This curvature, generally more pronounced in females, ends at the lumbosacral angle, formed at the junction of the L5 vertebra with the sacrum. The **sacral curvature** of females is reduced so that the coccyx protrudes less into the pelvic outlet (birth canal).

The curvatures provide additional flexibility (shock-absorbing resilience) to the vertebral column, augmenting that provided by the IV discs. Although the flexibility provided by the IV disc is passive and limited primarily by the zygapophysial (facet) joints and longitudinal ligaments, that provided by the curvatures is actively resisted by the contraction of muscle groups antagonistic to the movement.

Structure and Function of Vertebrae

Vertebrae vary in size and other characteristics from one region of the vertebral column to another and to a lesser degree within each region. A *typical vertebra* consists of a vertebral body, vertebral arch, and seven processes (Fig. 2.2A–C). The **vertebral body** (the anterior, more massive part of the vertebra) gives strength to the vertebral column and supports body weight. The size of vertebral bodies, especially from T4 inferiorly, increases to bear the progressively greater body weight. In life, most of the superior and inferior surfaces of vertebral bodies are covered with hyaline cartilage, which are remnants of the cartilaginous model from which the bone develops, except at the periphery, where there is a ring of smooth bone, the **epiphysial rim** (Fig. 2.2A). The cartilaginous remnants permit some diffusion of fluid between the IV disc and capillaries in the vertebral body.

The **vertebral arch** lies posterior to the vertebral body and is formed by right and left pedicles and laminae (Fig. 2.2C). The **pedicles** are short, stout processes that join the vertebral arch to the vertebral body. The pedicles project posteriorly to meet two broad, flat plates of bone, called **laminae**, which unite in the midline (Fig. 2.2A–B). The vertebral arch and the posterior surface of the vertebral body form the walls of the **vertebral foramen**. The succession of vertebral foramina in the articulated column forms the **vertebral canal**, which contains the spinal cord, meninges (protective membranes), fat, spinal nerve roots, and vessels. The indentations formed by the projection of the body and articular processes superior and inferior to the pedicles are **vertebral notches** (Fig. 2.2B). The superior and inferior vertebral notches of adjacent vertebrae combine to form the **IV foramina**, which give passage to spinal nerve roots and accompanying vessels and contain the spinal ganglia (Fig. 2.2D).

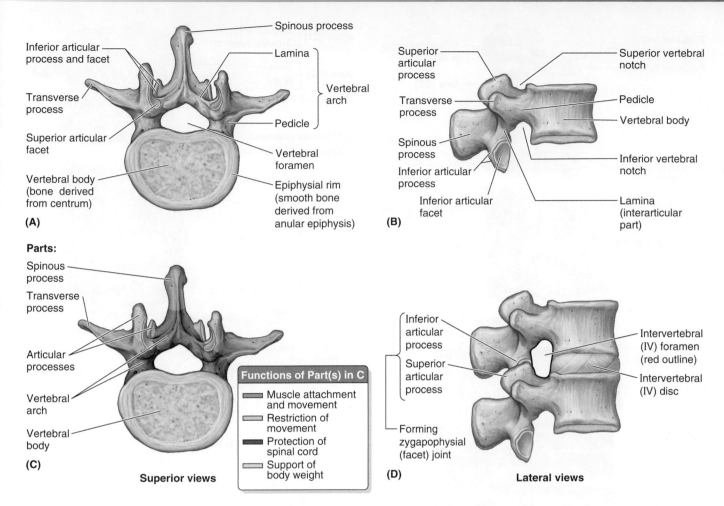

FIGURE 2.2. **Typical vertebra, represented by second lumbar vertebra. A and B.** Bony features. **C.** Functional components. **D.** Formation of IV foramen.

Seven processes arise from the vertebral arch of a typical vertebra (Fig. 2.2):

- One median **spinous process** projects posteriorly (and usually inferiorly) from the vertebral arch at the junction of the laminae.
- Two **transverse processes** project posterolaterally from the junctions of the pedicles and laminae.
- Four **articular processes**—two **superior** and two **inferior**—also arise from the junctions of the pedicles and laminae, each bearing an **articular surface (facet)**.

The spinous process and two transverse processes project from the vertebral arch and provide attachments for deep back muscles, serving as levers in moving the vertebrae (Fig. 2.2C). The four articular processes are in apposition with corresponding processes of vertebrae superior and inferior to them, forming *zygapophysial (facet) joints* (Fig. 2.2D). The direction of the articular facets on the articular processes determines the types of movements

permitted and restricted between adjacent vertebrae of each region. The interlocking of the articular processes also assists in keeping adjacent vertebrae aligned, particularly preventing one vertebra from slipping anteriorly on the vertebra below.

Regional Characteristics of Vertebrae

Each of the 33 vertebrae is unique. However, most of them demonstrate characteristic features identifying them as belonging to one of the five regions of the vertebral column (e.g., cervical vertebrae are characterized by the presence of foramina in their transverse processes). In each region, the articular facets are oriented in a characteristic direction that determines the type of movement permitted in aggregate for the region. Regional variations in the size and shape of the vertebral canal accommodate the varying thickness of the spinal cord. The main regional characteristics of vertebrae are summarized in Tables 2.1 through 2.3 and Figures 2.3 through 2.7.

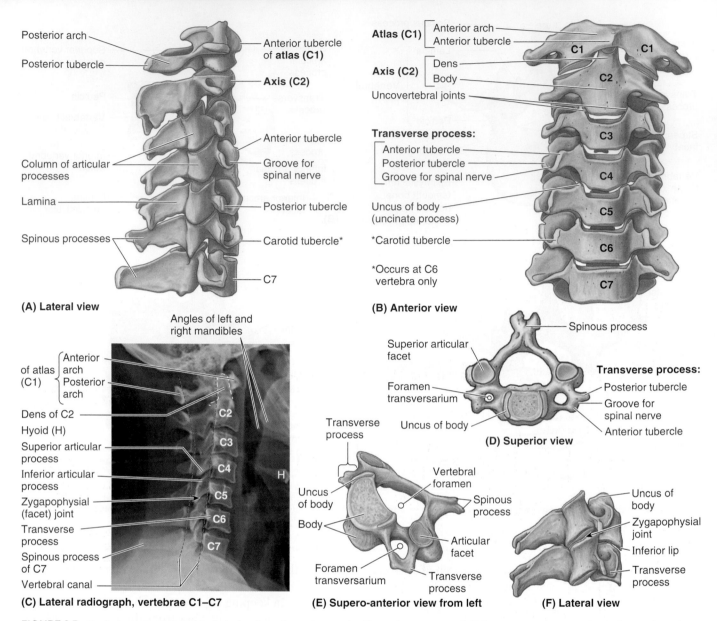

(A) Lateral view

(B) Anterior view

(C) Lateral radiograph, vertebrae C1–C7

(D) Superior view

(E) Supero-anterior view from left

(F) Lateral view

FIGURE 2.3. **Cervical vertebrae. A and B.** Articulated vertebrae. **C.** Lateral radiograph (compare with **[A]**). **D–F.** Bony features of typical cervical vertebrae.

TABLE 2.1. CERVICAL VERTEBRAE

Part (Typical Vertebrae)	Distinctive Characteristics
Body	Small and wider from side to side than anteroposteriorly; superior surface is concave between adjacent (uncinate) processes; inferior surface is convex.
Vertebral foramen	Large and triangular
Transverse processes	Foramina transversaria; small or absent in C7; vertebral arteries and accompanying venous and sympathetic plexuses pass through foramina (except C7, which transmits only small accessory vertebral veins); anterior and posterior tubercles
Articular processes	Superior facets directed superoposteriorly; inferior facets directed infero-anteriorly
Spinous process	C3–C5 short and bifid[a] (split in two parts); process of C6 is long but that of C7 is longer (C7 is called vertebra prominens)

[a]Less common in black individuals.

(continued)

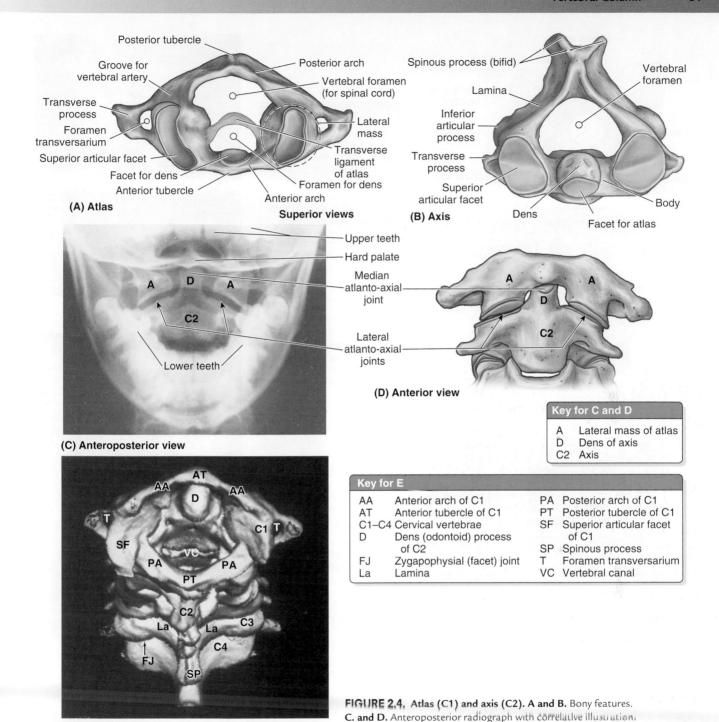

FIGURE 2.4. Atlas (C1) and axis (C2). **A and B.** Bony features. **C. and D.** Anteroposterior radiograph with correlative illustration. **E.** Three-dimensional reconstructed CT image.

Key for C and D

A	Lateral mass of atlas
D	Dens of axis
C2	Axis

Key for E

AA	Anterior arch of C1	PA	Posterior arch of C1
AT	Anterior tubercle of C1	PT	Posterior tubercle of C1
C1–C4	Cervical vertebrae	SF	Superior articular facet of C1
D	Dens (odontoid) process of C2	SP	Spinous process
FJ	Zygapophysial (facet) joint	T	Foramen transversarium
La	Lamina	VC	Vertebral canal

TABLE 2.1. CERVICAL VERTEBRAE *(continued)*

Part (Typical Vertebrae)	Distinctive Characteristics
Atlas (C1)	• Ring-like; somewhat kidney-shaped when viewed superiorly or inferiorly • No spinous process or body; consists of two lateral masses connected by anterior and posterior arches • Concave superior articular facets form atlanto-occipital joints with the occipital condyles; flat inferior facets meet with the C2 vertebra to form lateral atlanto-axial joints
Axis (C2)	• Strongest cervical vertebra • Distinguishing feature is the dens, which projects superiorly from its body and provides a pivot around which the atlas turns and carries the cranium. • Articulates anteriorly with the anterior arch of the atlas and posteriorly with the transverse ligament of the atlas

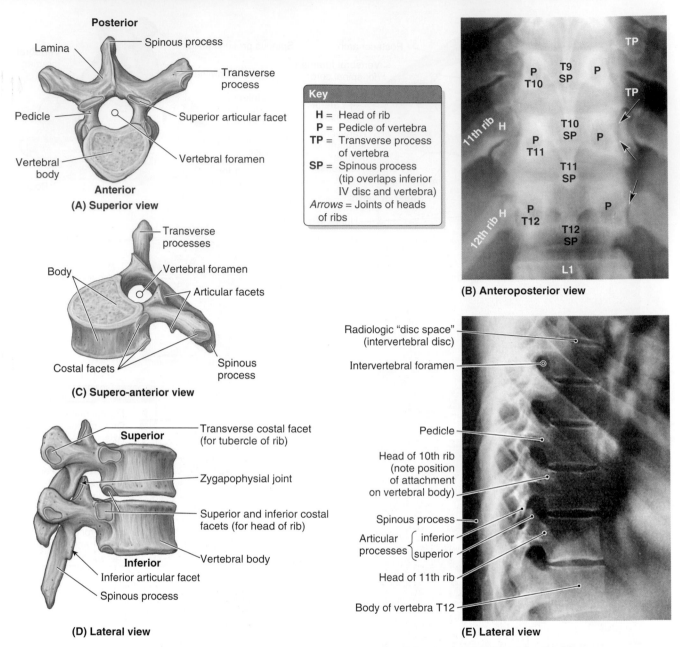

FIGURE 2.5. Thoracic vertebrae. Thoracic vertebrae (T1–T12) form the posterior part of the skeleton of the thorax and articulate with the ribs. **A.** Bony features of typical vertebra. **B.** Anteroposterior radiograph. **C.** Bony features of typical vertebra. **D.** Articulated vertebrae. **E.** Lateral radiograph. The apparent space between the vertebral bodies in radiographs is the site of the radiolucent IV disc.

TABLE 2.2. THORACIC VERTEBRAE

Part	Distinctive Characteristics
Body	Heart-shaped; bears one or two bilateral costal facets for articulation with head of rib (*H*)
Vertebral foramen	Circular and smaller than those in cervical and lumbar regions
Transverse process (*TP*)	Long and strong; extends posterolaterally; length diminishes from T1–T12; those of T1–T10 have transverse costal facets for articulation with tubercle of rib.
Articular processes	Superior articular facets directed posteriorly and slightly laterally; inferior articular facets directed anteriorly and slightly medially
Spinous process (*SP*)	Long; slopes postero-inferiorly, overlapping subadjacent vertebral body (sometimes completely)

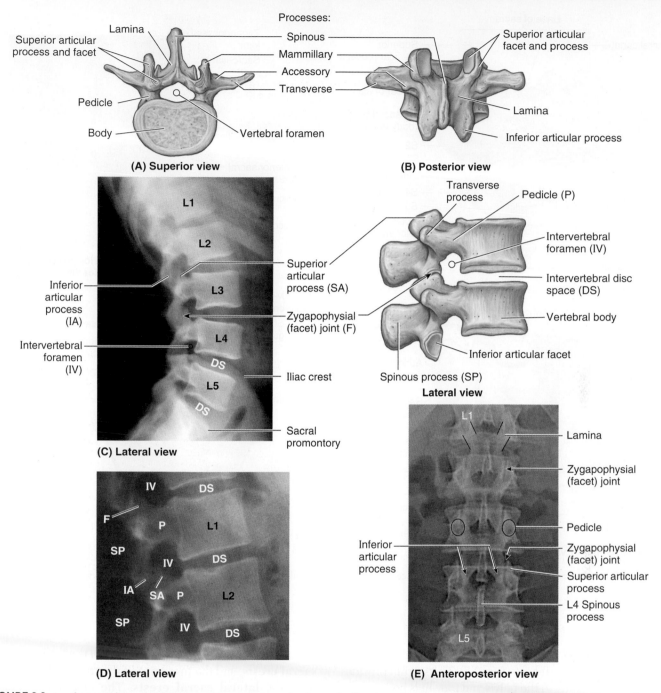

FIGURE 2.6. Lumbar vertebrae. A and B. Bony features. **C.** Lateral radiograph of lumbar spine. **D.** Lateral radiograph of L1–L2 region. Letters refer to structures labeled in **C. E.** Anteroposterior radiograph.

TABLE 2.3. LUMBAR VERTEBRAE

Part	Distinctive Characteristics
Body	Massive; kidney-shaped when viewed superiorly; larger and heavier than those of other regions
Vertebral foramen	Triangular; larger than in thoracic vertebrae and smaller than in cervical vertebrae
Transverse processes	Long and slender; accessory process on posterior surface of base of each process
Articular processes	Superior articular facets directed posteromedially (or medially); inferior articular facets directed anterolaterally (or laterally); mammillary process on posterior surface of each superior articular process
Spinous process	Short and sturdy; hatchet-shaped

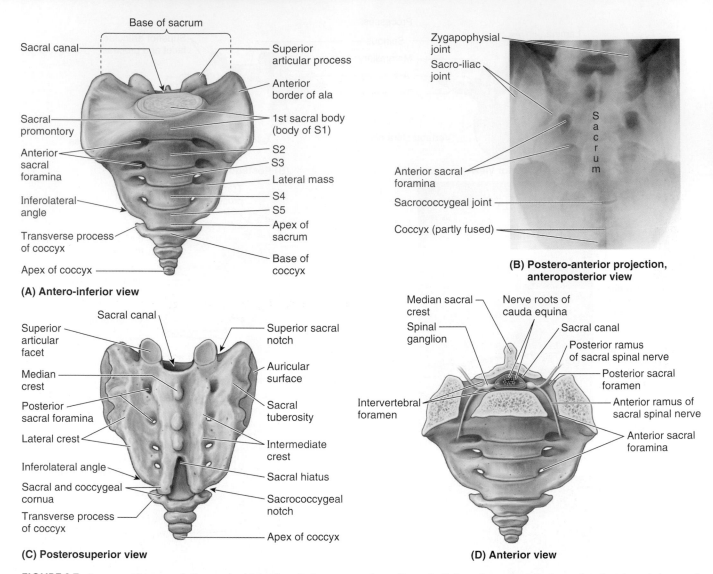

FIGURE 2.7. Sacrum and coccyx. A. Base and pelvic surface. **B.** Postero-anterior radiograph. **C.** Posterior surface. **D.** Coronal section through 1st sacral foramina.

The large, wedge-shaped **sacrum** in adults is composed of five fused sacral vertebrae (Fig. 2.7). The sacrum provides strength and stability to the pelvis and transmits body weight to the pelvic girdle through the **sacro-iliac joints.** The **base of the sacrum** is formed by the superior surface of the S1 vertebra. Its superior articular processes articulate with the inferior articular processes of the L5 vertebra. The projecting anterior edge of the body of the first sacral vertebra is the **sacral promontory.** On the pelvic and dorsal surfaces are four pairs of sacral foramina for the exit of the rami of the first four sacral nerves and the accompanying vessels. The pelvic surface of the sacrum is smooth and concave. The four transverse lines indicate where fusion of the sacral vertebrae occurred. The posterior surface of the sacrum is rough and convex. The fused spinous processes form the **median sacral crest.** The fused articular processes form the **intermediate sacral crests,** and the fused tips of the transverse processes form the **lateral sacral crests.** The inverted U-shaped **sacral hiatus** results from the absence of the laminae and spinous processes of the S4 and S5 vertebrae. The hiatus leads into the sacral canal, the inferior end of the vertebral canal. The **sacral cornua** (L. *horns*), representing the inferior articular processes of the S5 vertebra, project inferiorly on each side of the sacral hiatus and are a helpful guide to its location. The lateral surface of the sacrum has an ear-shaped (auricular) articular surface that participates in the sacro-iliac joint. The four vertebrae of the tapering **coccyx** are remnants of the skeleton of the embryonic tail-like caudal eminence. The distal three vertebrae fuse during middle life to form the coccyx, a beak-like bone that articulates with the sacrum.

CLINICAL BOX

Spina Bifida

The most common congenital anomaly of the vertebral column is **spina bifida occulta**, in which the laminae (embryonic neural arches) of L5 and/or S1 fail to develop normally and fuse. This bony defect, present in up to 24% of people, is concealed by skin, but its location is often indicated by a tuft of hair. Most people with spina bifida occulta have no back problems (Moore et al., 2016). In severe types of the anomaly, such as **spina bifida cystica**, one or more vertebral arches may almost completely fail to develop (Fig. B2.2). Spina bifida cystica is associated with herniation of the meninges (*meningocele*) and/or the spinal cord (*meningomyelocele*). Usually, neurological symptoms are present in severe cases of meningomyelocele (e.g., paralysis of limbs and disturbances in bladder and bowel control).

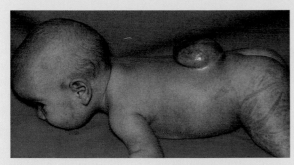

Infant with spina bifida cystica

FIGURE B2.2. Spina bifida cystica with meningomyelocele.

Laminectomy

A **laminectomy** is the surgical excision of one or more spinous processes and their supporting laminae (*1* in Fig. B2.3). The term is also commonly used to denote the removal of most of the vertebral arch by transecting the pedicles (*2* in Fig. B2.3). Laminectomies provide access to the vertebral canal to relieve pressure on the spinal cord or nerve roots, commonly caused by a tumor, herniated IV disc, or bony hypertrophy (excess growth).

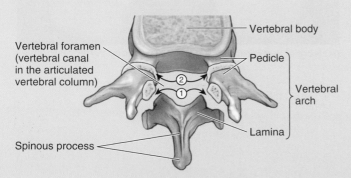

Vertebral body

Vertebral foramen
(vertebral canal
in the articulated
vertebral column)

Pedicle

Vertebral
arch

Lamina

Spinous process

FIGURE B2.3. Laminectomy.

Fractures of Vertebrae

Fractures and fracture–dislocations of the vertebral column usually result from sudden forceful flexion, as in an automobile accident. Typically, the injury is a crush or compression fracture of the body of one or more vertebrae. If violent anterior movement of the vertebra occurs in addition to compression, a vertebra may be displaced anteriorly on the vertebra inferior to it. Usually, this movement dislocates and fractures the articular facets between the two vertebrae and ruptures the interspinous ligaments. Irreparable injuries to the spinal cord accompany most severe flexion injuries of the vertebral column.

Fracture and Dislocation of Atlas

Vertical forces (e.g., striking the bottom of a pool in a diving accident) compressing the lateral masses between the occipital condyles and the axis drive them apart, fracturing one or both of the anterior or posterior arches (Fig. B2.4A). If the force is sufficient, rupture of the transverse ligament that links them will also occur (Fig. B2.4B). The resulting Jefferson or burst fracture in itself does not necessarily result in spinal cord injury because the dimensions of the bony ring actually increase. Spinal cord injury is more likely, however, if the transverse ligament has also been ruptured.

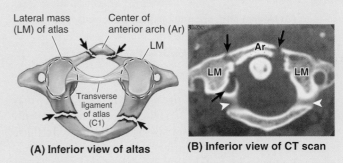

Lateral mass
(LM) of atlas

Center of
anterior arch (Ar)

LM

Ar

LM   LM

Transverse
ligament
of atlas
(C1)

(A) Inferior view of altas   **(B) Inferior view of CT scan**

FIGURE B2.4. Jefferson (burst) fracture of atlas. *Red arrows* indicate fractures.

Dislocation of Vertebrae

The bodies of the cervical vertebrae can be dislocated in neck injuries with less force than is required to fracture them. Because of the large vertebral canal in the cervical region, slight dislocation can occur without damaging the spinal cord; however, severe dislocations may injure the spinal cord. If the dislocation does not result in "facet jumping" with locking of the displaced articular processes, the cervical vertebrae may self-reduce ("slip back into place") so that a radiograph may not indicate that the cord has been injured. Magnetic resonance imaging (MRI) may reveal the resulting soft tissue damage.

(*Continued on next page*)

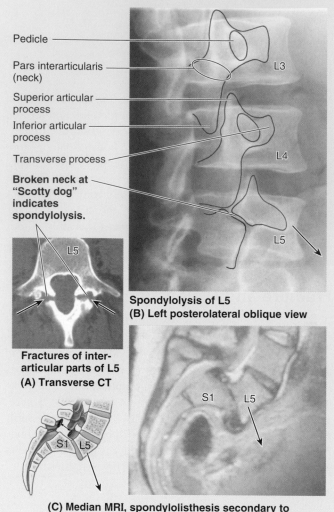

Pedicle

Pars interarticularis (neck)

Superior articular process

Inferior articular process

Transverse process

Broken neck at "Scotty dog" indicates spondylolysis.

L3

L4

L5

Spondylolysis of L5
(B) Left posterolateral oblique view

L5

Fractures of inter-articular parts of L5
(A) Transverse CT

S1 L5

S1 L5

(C) Median MRI, spondylolisthesis secondary to spondylolysis of L5

FIGURE B2.5. Spondylolysis and spondylolisthesis.

Severe *hyperextension of the neck* (whiplash injury) may occur during rear-end motor vehicle collisions, especially when the head restraint is too low or too far back. In these types of hyperextension injuries, the anterior longitudinal ligament is severely stretched and may be torn.

Dislocation of vertebrae in the thoracic and lumbar regions is uncommon because of the interlocking of their articular processes; however, owing to the abrupt transition from the relatively inflexible thoracic region to the much more mobile lumbar region, T11 and T12 are the most commonly fractured noncervical vertebrae.

Fractures of the interarticular parts of the vertebral laminae of L5 (*spondylolysis of L5* [Fig. B2.5A,B]) may result in forward displacement of the L5 vertebral body relative to the sacrum (*spondylolisthesis* [Fig. B2.5C]). Spondylolysis of L5,

or susceptibility to it, probably results from a failure of the centrum of L5 to unite adequately with the neural arches during development. Spondylolisthesis at the L5–S1 articulation may result in pressure on the spinal nerves of the cauda equina as they pass into the superior part of the sacrum, causing back and lower limb pain (Fig. B2.5C). The intrusion of the L5 body into the pelvic inlet reduces the anteroposterior diameter of the pelvic inlet.

Severe hyperextension is most likely to injure the posterior parts of the vertebrae—the vertebral arches and their processes. Severe hyperextension of the neck (e.g., as occurs in diving injuries) may pinch the posterior arch of C1 vertebra between the occipital bone and the C2 vertebra. In these cases, the C1 vertebra usually breaks at one or both grooves for the vertebral arteries. The anterior longitudinal ligament and adjacent anulus fibrosus of the C2–C3 IV disc may also rupture. If this occurs, the cranium, C1, and C2 are separated from the rest of the axial skeleton, and the spinal cord is usually severed. Individuals with this injury seldom survive.

Lumbar Spinal Stenosis

Lumbar spinal stenosis describes a stenotic (narrow) vertebral foramen in one or more lumbar vertebrae (Fig. B2.6). Stenosis of a lumbar vertebral foramen alone may cause compression of one or more of the spinal nerve roots occupying the vertebral canal. Surgical treatment may consist of decompressive laminectomy. Lumbar spinal stenosis may be a hereditary anomaly that can make a person more vulnerable to age-related degenerative changes such as IV disc protrusion. When IV disc protrusion occurs in a patient with spinal stenosis, it further compromises the size of the vertebral canal, as does arthritic proliferation and ligamentous degeneration. It should also be noted that lumbar spinal nerves increase in size as the vertebral column descends, but the IV foramina decrease in size.

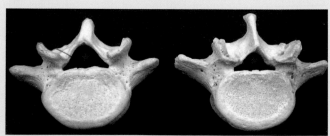

(A) Normal vertebral foramen **(B) Stenotic vertebral foramen**
Superior views

FIGURE B2.6. Spinal stenosis.

Vertebral Body Osteoporosis

Vertebral body osteoporosis is a common metabolic bone disease that is often detected during routine radiographic studies. Osteoporosis results from a net demineralization of the bones caused by a disruption of the normal balance of calcium deposition and resorption. As a result, the quality of bone is reduced and atrophy of skeletal tissue occurs. Although osteoporosis affects the entire skeleton, the most affected areas are the neck of the femur, the bodies of vertebrae, the metacarpals (bones of the hand), and the radius. These bones become weakened and brittle and are subject to fracture.

Radiographs taken during early to moderate osteoporosis demonstrate demineralization, which is evident as diminished radiodensity of the trabecular (spongy) bone of the vertebral bodies, causing the thinned cortical bone to appear relatively prominent (Fig. B2.7A,B). Osteoporosis especially affects the horizontal trabeculae of the trabecular bone of the vertebral body (see Fig. B2.9A). Consequently, vertical striping may become apparent, reflecting the loss of the horizontal supporting trabeculae and thickening of the vertical struts (Fig. B2.7A). Radiographs in later stages may reveal vertebral collapse (compression fractures) and increased thoracic kyphosis (Fig. B2.7C). Vertebral body osteoporosis occurs in all vertebrae but is most common in thoracic vertebrae and is an especially common finding in postmenopausal females.

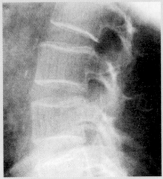

(A) Left lateral view

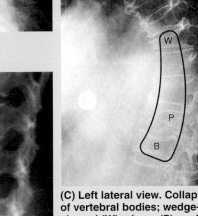

(C) Left lateral view. Collapse of vertebral bodies; wedge-shaped (W), planar (P), and biconcave (B).

(B) Left lateral view

FIGURE B2.7. Vertebral body osteoporosis.

SURFACE ANATOMY

Vertebral Column

Spinous processes can be observed in the upper back when the back is flexed (Fig. SA2.2A,B), but most of the spinous processes can be palpated, even in obese individuals, because the fat is typically more sparse in the midline. Although the spinous process of C7 is usually the most superior process that is visible (hence, the name *vertebra prominens*), that of T1 may be the most prominent. The spinous processes of C2–C6 may be palpated in the nuchal groove between the neck muscles; the C3–C5 spinous processes are deeply placed, separated from the surface by the nuchal ligament, making them harder to palpate. C1 has no spinous process. The transverse processes of the C1, C6, and C7 vertebrae are also palpable. Those of C1 can be palpated by deep pressure postero-inferior to the tips of the mastoid processes of the temporal bones (bony prominences posterior to the ears).

When the neck and back are flexed, the spinous processes of upper thoracic vertebrae may be observed and palpated counting from superior to inferior starting at the C7 spinous process. The tips of the thoracic spinous processes do not indicate the level of the corresponding vertebral bodies because they overlap (lie at the level of) the vertebra below. The transverse processes of the thoracic vertebrae can usually be palpated on each side of the spinous processes in the thoracic region; in thin individuals, the ribs can be palpated from tubercle to angle, at least in the lower back (inferior to the scapula).

The spinous processes of the lumbar vertebrae are large and easy to observe when the trunk is flexed (Fig. SA2.2B) and can be palpated in the **posterior median furrow** (Fig. SA2.2C) when erect. A horizontal line joining the highest points of the iliac crests passes through the tip of the L4

(Continued on next page)

spinous process and the L4–L5 IV disc. This is a useful landmark when performing lumbar puncture to obtain a sample of cerebrospinal fluid (CSF) (see the "Lumbar Spinal Puncture" clinical box discussed later in this chapter). The transverse processes are covered with thick muscles and may or may not be palpable.

The S2 spinous process lies at the middle of a line drawn between the posterior superior iliac spines, indicated by the skin dimples formed by the attachment of skin and deep fascia to these spines (Fig. SA2.2C,D). This level indicates the inferior extent of the subarachnoid space (lumbar cistern).

The *median sacral crest* can be palpated in the midline inferior to the L5 spinous process (Fig. 2.7). The *sacral hiatus* can be palpated at the inferior end of the sacrum in the superior part of the *intergluteal (natal) cleft* between the buttocks. Clinically, the coccyx is examined with a gloved finger in the anal canal and its apex (tip) can be palpated approximately 2.5 cm posterosuperior to the anus. The **sacral triangle** is formed by the lines joining the posterior superior iliac spines and the superior part of the intergluteal cleft. The sacral triangle outlining the sacrum is a common area of pain resulting from low back sprains.

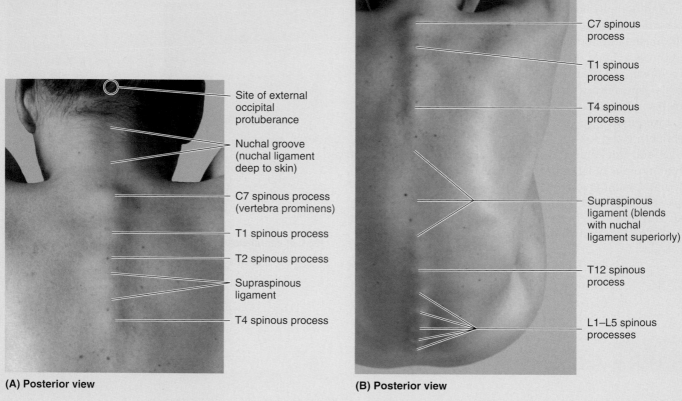

(A) **Posterior view**

Site of external occipital protuberance

Nuchal groove (nuchal ligament deep to skin)

C7 spinous process (vertebra prominens)

T1 spinous process

T2 spinous process

Supraspinous ligament

T4 spinous process

(B) **Posterior view**

C7 spinous process

T1 spinous process

T4 spinous process

Supraspinous ligament (blends with nuchal ligament superiorly)

T12 spinous process

L1–L5 spinous processes

FIGURE SA2.2. **A and B. Neck and back flexed with scapulae protracted.** *(continued)*

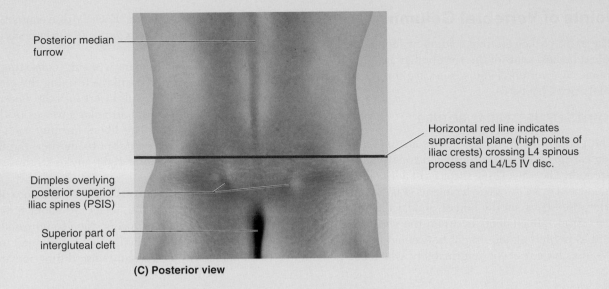

Posterior median furrow

Horizontal red line indicates supracristal plane (high points of iliac crests) crossing L4 spinous process and L4/L5 IV disc.

Dimples overlying posterior superior iliac spines (PSIS)

Superior part of intergluteal cleft

(C) Posterior view

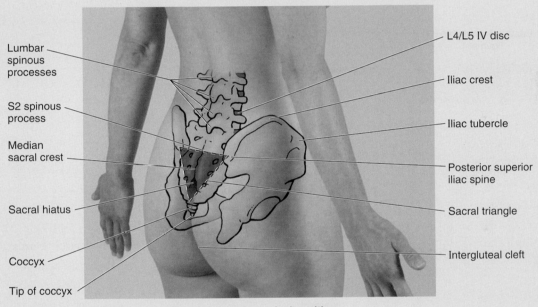

Lumbar spinous processes

S2 spinous process

Median sacral crest

Sacral hiatus

Coccyx

Tip of coccyx

L4/L5 IV disc

Iliac crest

Iliac tubercle

Posterior superior iliac spine

Sacral triangle

Intergluteal cleft

(D) Right posterolateral view, anatomical position

FIGURE SA2.2. *(continued)* **C and D.**

Joints of Vertebral Column

The joints of the vertebral column include the joints of vertebral bodies, joints of the vertebral arches, craniovertebral joints, costovertebral joints (see Chapter 4), and sacro-iliac joints (see Chapter 6).

JOINTS OF VERTEBRAL BODIES

The joints of the vertebral bodies are *symphyses (secondary cartilaginous joints)* designed for weight bearing and strength. The articulating surfaces of adjacent vertebrae are connected by *IV discs* and ligaments (Fig. 2.8). The IV discs, interposed between the bodies of adjacent vertebrae, provide strong attachments between the vertebral bodies. As well as permitting movement between adjacent vertebrae, the discs have resilient deformability, which allows them to serve as shock absorbers. Each IV disc consists of an *anulus fibrosus*, an outer fibrous part, and a gelatinous central mass, the *nucleus pulposus*.

The **anulus fibrosus** is a ring consisting of concentric lamellae of fibrocartilage forming the circumference of the IV disc. The anuli insert into the smooth, rounded *epiphysial rims* on the articular surfaces of the vertebral bodies (Fig. 2.8C). The fibers forming each lamella run obliquely from one vertebra to another; the fibers of one lamella typically run at right angles to those of adjacent ones.

The **nucleus pulposus** is the central core of the IV disc (Fig. 2.8). At birth, the nuclei are about 85% water. The pulpy nuclei become broader when compressed and thinner when tensed or stretched. Compression and tension occur simultaneously in the same disc during movement of the

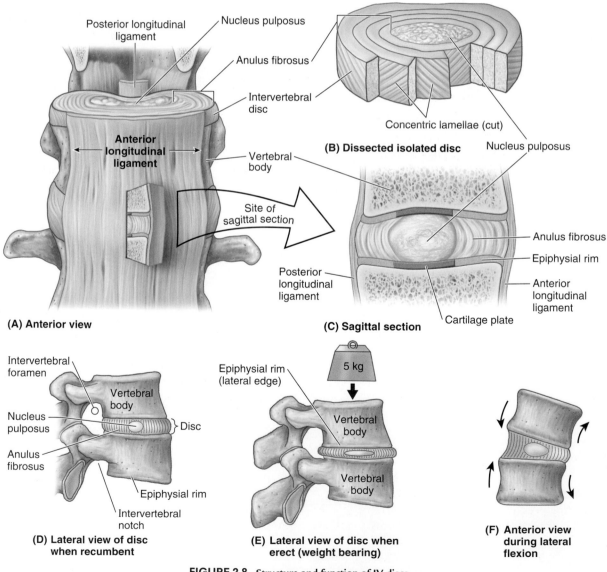

FIGURE 2.8. Structure and function of IV discs.

vertebral column (e.g., anterior and lateral flexion, extension, rotation); the turgid nucleus acts as a semifluid fulcrum (Fig. 2.8D–F). The nuclei pulposi dehydrate with age and lose elastin and proteoglycans while gaining collagen, eventually becoming dry and granular. As a result, the IV discs lose their turgor, becoming thinner, stiffer, and more resistant to deformation. As this occurs, the anulus assumes a greater share of the vertical load and the associated stresses and strains.

The lamellae of the anulus thicken with age and often develop fissures and cavities. Because the lamellae are thinner and less numerous posteriorly, the nucleus pulposus is not centered in the disc but is more posteriorly placed (Fig. 2.8C). The nucleus pulposus is avascular. It receives its nourishment by diffusion from blood vessels at the periphery of the anulus fibrosus and vertebral body.

There is no IV disc between the C1 (atlas) and C2 (axis) vertebrae. The most inferior functional disc is between the L5 and S1 vertebrae. The discs vary in thickness in different regions. They are thicker in the cervical and lumbar regions and thinnest in the superior thoracic region. Their relative thickness is related to the range of movement, and their varying shapes largely produce the secondary curvatures of the vertebral column being thicker anteriorly in the cervical and lumbar regions. Their thickness is most uniform in the thoracic region.

Uncovertebral "joints" (of Luschka) are located between the uncus of the bodies (uncinate processes) of the C3–C6 vertebrae and the beveled inferolateral surfaces of the vertebral bodies superior to them (Fig. 2.9). The joints are at the lateral and posterolateral margins of the IV discs. The articulating surfaces of these joint-like structures are covered with cartilage and enclose a cavity filled with fluid. They are considered to be synovial joints by some; others consider them to be degenerative spaces (fissures) in the discs occupied by extracellular fluid. The uncovertebral joints are frequent sites of spur formation (projecting processes of bone) that may cause neck pain.

The **anterior longitudinal ligament** is a strong, broad fibrous band that covers and connects the anterolateral aspects of the vertebral bodies and IV discs (Figs. 2.8A and 2.10A). The ligament extends from the pelvic surface of the sacrum to the anterior tubercle of the C1 vertebra (atlas) and the occipital bone anterior to the foramen magnum. The anterior longitudinal ligament maintains the stability of the IV joints and limits extension of the vertebral column.

The **posterior longitudinal ligament** is a much narrower, somewhat weaker band compared to the anterior longitudinal ligament. The ligament runs within the vertebral canal along the posterior aspect of the vertebral bodies (Fig. 2.10A,C). It is attached mainly to the IV discs and less so to the posterior edges of the vertebral bodies from C2 (axis) to the sacrum. The posterior longitudinal ligament helps prevent hyperflexion of the vertebral column and

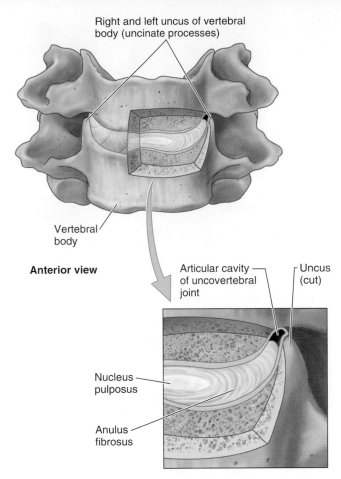

FIGURE 2.9. Uncovertebral joints. These joints are at the posterolateral margin of the cervical IV discs.

posterior herniation of the IV discs. It is well innervated with nociceptive (pain) nerve endings.

JOINTS OF VERTEBRAL ARCHES

The joints of the vertebral arches are the **zygapophysial joints** (facet joints) (see Fig. 2.6). These articulations are synovial, plane joints between the superior and the inferior articular processes (G. *zygapophyses*) of adjacent vertebrae. Each joint is surrounded by a thin, loose **joint (articular) capsule**, which is attached to the margins of the articular surfaces of the articular processes of adjacent vertebrae (Fig. 2.10C). Accessory ligaments unite the laminae, transverse processes, and spinous processes and help stabilize the joints. The zygapophysial joints permit gliding movements between the articular processes; the shape and disposition of the articular surfaces determine the type of movement possible. The zygapophysial joints are innervated by articular branches that arise from the medial branches of the posterior rami of spinal nerves (Fig. 2.11). Each posterior ramus supplies two adjacent joints; therefore, each joint is supplied by two adjacent spinal nerves.

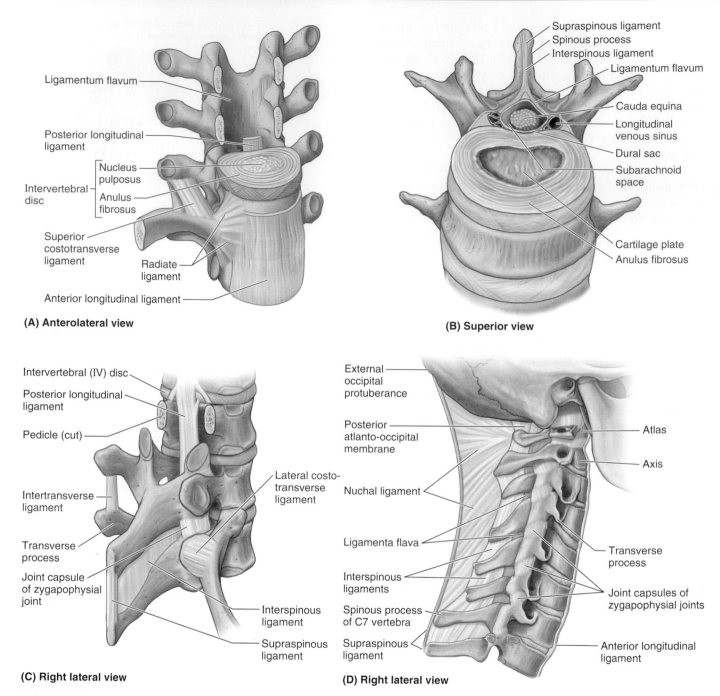

(A) Anterolateral view

(B) Superior view

(C) Right lateral view

(D) Right lateral view

FIGURE 2.10. Joints and ligaments of vertebral column. A. The pedicles of the superior vertebrae have been sawn through, and their bodies have been removed. A rib and its costovertebral joint and associated ligaments are also shown. **B.** In this transverse section of an IV disc, the nucleus pulposus has been removed to show the hyaline cartilage plate covering the superior surface of the vertebral body. **C.** The vertebral arch of the superior vertebra has been removed to show the posterior longitudinal ligament. **D.** Ligaments of the cervical region.

ACCESSORY LIGAMENTS OF INTERVERTEBRAL JOINTS

The laminae of adjacent vertebral arches are joined by broad, pale, yellow elastic fibrous tissue called the **ligamenta flava** (L. *flavus*, yellow), which extend almost vertically from the lamina above to the lamina below (Fig. 2.10A). The ligaments bind the laminae of the adjoining vertebrae together, forming alternating sections of the posterior wall of the vertebral canal. The ligamenta flava resist separation of the vertebral laminae by arresting abrupt flexion of the vertebral column and thereby preventing injury to the IV discs.

The strong elastic ligamenta flava help preserve posture and assist with straightening the column after flexing. Adjacent spinous processes are united by weak, almost membranous **interspinous ligaments** and strong fibrous **supraspinous ligaments** (Fig. 2.10B,C). The supraspinous

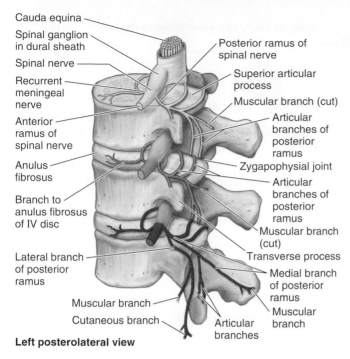

Left posterolateral view

FIGURE 2.11. Innervation of zygapophysial joints.

ligament merges superiorly with the **nuchal ligament** (L. *ligamentum nuchae*), the strong median ligament of the neck (Fig. 2.10D). The nuchal ligament is composed of thickened fibro-elastic tissue extending from the external occipital protuberance and posterior border of the foramen magnum to the spinous processes of the cervical vertebrae. Because of the shortness of the C3–C5 spinous processes, the nuchal ligament substitutes for bone in providing muscular attachments.

CRANIOVERTEBRAL JOINTS

The craniovertebral joints include the atlanto-occipital joints, between the atlas (C1 vertebra) and the occipital bone of the cranium, and the atlanto-axial joints, between the C1 and the C2 vertebrae. *Atlanto*, a Greek prefix, refers to the atlas and is derived from Atlas, the Titan who bore the celestial sphere on his shoulders much as vertebra C1 supports the cranium. These craniovertebral articulations are synovial joints that have no IV discs. Their design allows a wider range of movement than in the rest of the vertebral column.

ATLANTO-OCCIPITAL JOINTS

The atlanto-occipital joints, between the lateral masses of C1 (atlas) and the occipital condyles (Fig. 2.12C), permit nodding of the head, such as the neck flexion and extension that occurs when indicating approval (the "yes" movement—see Table 2.9 figures). The main movement is flexion, with a little lateral flexion (sideways tilting of the head) and some rotation. These joints also permit sideways tilting of the head. The atlanto-occipital joints are synovial joints of the

condyloid type and have thin, loose joint capsules. The cranium and C1 are also connected by **anterior** and **posterior atlanto-occipital membranes** that extend from the anterior and posterior arches of C1 to the anterior and posterior margins of the foramen magnum (Fig. 2.12B). The anterior and posterior atlanto-occipital membranes help prevent excessive movement of these joints.

ATLANTO-AXIAL JOINTS

There are three atlanto-axial articulations: two (right and left) **lateral atlanto-axial joints** between the lateral masses of C1 and the superior facets of C2 (Fig. 2.12C) and one **median atlanto-axial joint** between the dens of C2 and the anterior arch and transverse ligament of the atlas (Fig. 2.12A,B). The median atlanto-axial joint is a pivot joint, whereas the lateral atlanto-axial joints are plane-type synovial joints. Movement at all three atlanto-axial joints permits the head to be turned from side to side, as occurs when rotating the head to indicate disapproval (the "no" movement). During this movement, the cranium and C1 vertebra rotate on the C2 vertebra as a unit. During rotation of the head, the dens of C2 is the pivot, which is held in a socket formed anteriorly by the anterior arch of the atlas and posteriorly by the transverse ligament of the atlas (see the figure for Table 2.10).

The **transverse ligament of the atlas** is a strong band extending between the tubercles on the medial aspects of the lateral masses of the C1 vertebrae (Fig. 2.12A). Vertically oriented but much weaker superior and inferior **longitudinal bands** pass from the transverse ligament to the occipital bone superiorly and to the body of C2 inferiorly. Together, the transverse ligament and the longitudinal bands form the **cruciate ligament** (formerly the cruciform ligament), so named because of its resemblance to a cross (Fig. 2.12C).

Stout **alar ligaments** extend from the sides of the dens to the lateral margins of the foramen magnum. These short, rounded cords attach the cranium to the C2 vertebra and serve as check ligaments, preventing excessive rotation at the joints.

The **tectorial membrane** is the strong superior continuation of the posterior longitudinal ligament across the median atlanto-axial joint through the foramen magnum to the central floor of the cranial cavity. It runs from the body of C2 to the internal surface of the occipital bone and covers the alar ligaments and transverse ligaments of the atlas (Fig. 2.12B,C).

Movements of Vertebral Column

Movements of the vertebral column include flexion, extension, lateral flexion, and rotation (Fig. 2.13). The range of movement of the vertebral column varies according to the region and the individual. The normal range of movement possible in healthy young adults is typically reduced by 50% during advanced age. The mobility of the column results

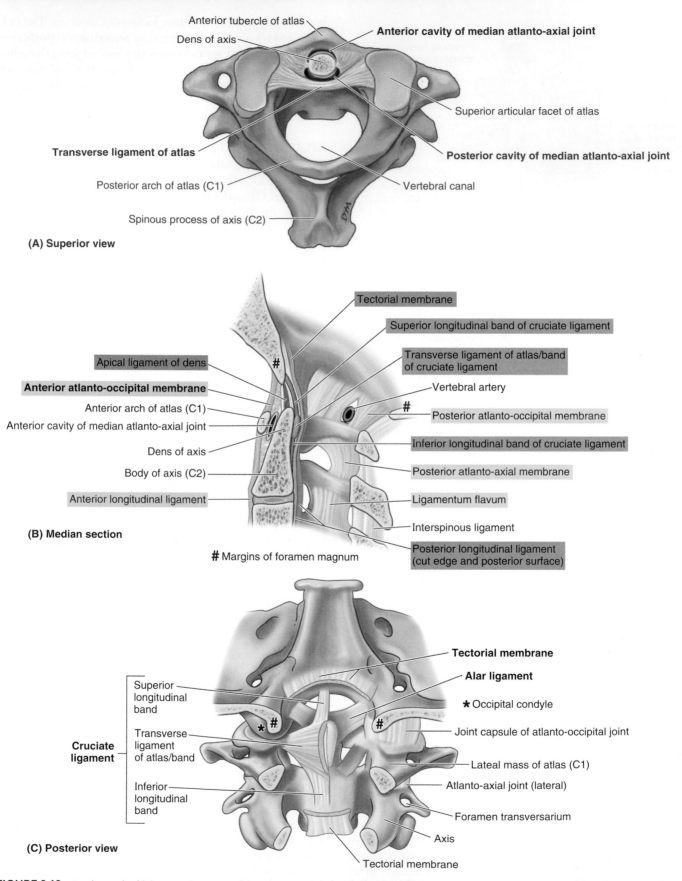

Anterior tubercle of atlas
Dens of axis
Anterior cavity of median atlanto-axial joint
Superior articular facet of atlas
Posterior cavity of median atlanto-axial joint
Transverse ligament of atlas
Posterior arch of atlas (C1)
Vertebral canal
Spinous process of axis (C2)

(A) Superior view

Tectorial membrane
Superior longitudinal band of cruciate ligament
Transverse ligament of atlas/band of cruciate ligament
Apical ligament of dens
Anterior atlanto-occipital membrane
Vertebral artery
Anterior arch of atlas (C1)
Posterior atlanto-occipital membrane
Anterior cavity of median atlanto-axial joint
Dens of axis
Inferior longitudinal band of cruciate ligament
Body of axis (C2)
Posterior atlanto-axial membrane
Anterior longitudinal ligament
Ligamentum flavum
Interspinous ligament
(B) Median section
Posterior longitudinal ligament (cut edge and posterior surface)

Margins of foramen magnum

Tectorial membrane
Alar ligament
Superior longitudinal band
* Occipital condyle
Transverse ligament of atlas/band
Joint capsule of atlanto-occipital joint
Cruciate ligament
Lateral mass of atlas (C1)
Inferior longitudinal band
Atlanto-axial joint (lateral)
Foramen transversarium
Axis
(C) Posterior view
Tectorial membrane

FIGURE 2.12. Craniovertebral joints. A. Ligaments of the atlanto-occipital and atlanto-axial joints. The large vertebral foramen of the atlas (C1 vertebra) is divided into two foramina by the transverse ligament of atlas. The larger posterior foramen is for the spinal cord, and the smaller anterior foramen is for the dens of the axis (C2 vertebra). **B.** The hemisected craniovertebral region shows the median joints and membranous continuities of the ligamenta flava and longitudinal ligaments in the craniovertebral region. **C.** Bands of cruciate ligament.

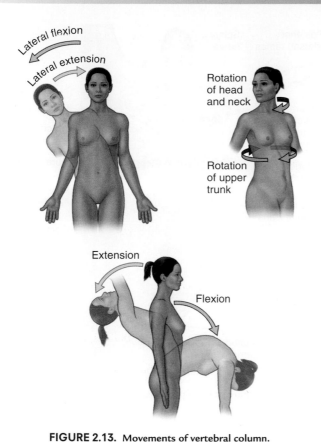

FIGURE 2.13. Movements of vertebral column.

primarily from the compressibility and elasticity of the IV discs. The range of movement of the vertebral column is limited by the following factors:

- Thickness, elasticity, and compressibility of the IV discs
- Shape and orientation of the articular facets
- Tension of the joint capsules of the above joints
- Resistance of the back muscles and ligaments (such as the ligamenta flava and the posterior longitudinal ligament)
- Attachment to the thoracic (rib) cage
- Bulk of the surrounding tissues

The back muscles producing movements of the vertebral column are discussed subsequently; however, the movements are not produced exclusively by the back muscles. They are assisted by gravity and the action of the anterolateral abdominal muscles (e.g., rectus abdominis and oblique muscles; see Table 2.8). Movements between adjacent vertebrae occur at the resilient IV discs and at the zygapophysial joints.

The orientation of the latter joints permits some movements and restricts others. Although movements between adjacent vertebrae are relatively small, especially in the thoracic region, the summation of all the small movements produces a considerable range of movement of the vertebral column as a whole (e.g., when flexing to touch the toes). Movements of the vertebral column are freer in the cervical and lumbar regions than in the thoracic region.

Flexion, extension, lateral flexion, and rotation of the neck are especially free because the

- IV discs, although thin relative to most other discs, are thick relative to the small size of the vertebral bodies at this level.
- articular surfaces of the zygapophysial joints are relatively large and the joint planes are almost horizontal.
- joint capsules of the zygapophysial joints are loose.
- neck is relatively slender (with less surrounding soft tissue bulk).

Flexion of the vertebral column is greatest in the cervical region. The sagittally oriented joint planes of the lumbar region are conducive to flexion and extension. Extension of the vertebral column is most marked in the lumbar region and usually is more extensive than flexion; however, the interlocking articular processes here prevent rotation. The lumbar region, like the cervical region, has large IV discs (the largest ones occur here) relative to the size of the vertebral bodies. Lateral flexion of the vertebral column is greatest in the cervical and lumbar regions.

The thoracic region, in contrast, has IV discs that are thin relative to the size of the vertebral bodies. Relative stability is also conferred on this part of the vertebral column through its connection to the sternum by the ribs and costal cartilages. The joint planes here lie on an arc that is centered on the vertebral body (see Fig. 2.5A), permitting rotation in the thoracic region. This rotation of the upper trunk, in combination with the rotation permitted in the cervical region and that at the atlanto-axial joints, enables the torsion of the axial skeleton that occurs as one looks back over the shoulder (see part E of Table 2.8). However, flexion is limited in the thoracic region, including lateral flexion.

Vasculature of Vertebral Column

Vertebrae are supplied by *periosteal* and *equatorial branches* of the major cervical and segmental arteries and their spinal branches. *Spinal branches* supplying the vertebrae are branches of the (Fig. 2.14)

- vertebral and ascending cervical arteries in the neck
- posterior intercostal arteries in the thoracic region
- subcostal and lumbar arteries in the abdomen
- iliolumbar and lateral and medial sacral arteries in the pelvis

Periosteal and **equatorial branches** arise from these arteries as they cross the external (anterolateral) surfaces of the vertebrae. **Spinal branches** enter the IV foramina and divide into **anterior** and **posterior vertebral canal branches** that pass to the vertebral body and vertebral arch, respectively, and give rise to ascending and descending branches that anastomose with spinal canal branches of adjacent levels. Anterior vertebral canal branches send nutrient arteries into the vertebral bodies. The spinal branches continue as terminal *radicular arteries* distributed to the posterior and anterior roots of the spinal nerves and their coverings or as *segmental medullary arteries* that continue to the spinal cord.

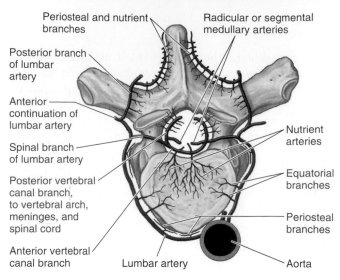

FIGURE 2.14. Blood supply of vertebrae.

Spinal veins form venous plexuses along the vertebral column both inside (**internal vertebral epidural venous plexus**) and outside (**external vertebral venous plexus**) the vertebral canal (Fig. 2.15). The large, tortuous **basivertebral veins** form within the vertebral bodies and emerge from foramina on the surfaces of the vertebral bodies (mostly the posterior aspect) and drain into the external and especially the internal vertebral venous plexuses. The **IV veins** receive veins from the spinal cord and vertebral venous plexuses as they accompany the spinal nerves through the IV foramina to drain into the *vertebral veins* of the neck and *segmental veins* of the trunk.

Innervation of Vertebral Column

Other than the zygapophysial joints (innervated by articular branches of the medial branches of the posterior rami),

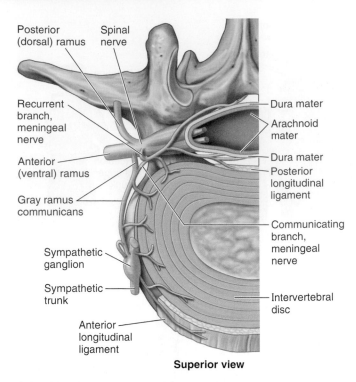

Superior view

FIGURE 2.16. Innervation of periosteum and ligaments of vertebral column and meninges.

the vertebral column is innervated by **meningeal branches of the spinal nerves** (Fig. 2.16). Recurrent branches of the meningeal nerves run back through the IV foramen, but some branches remain outside the canal. The branches outside the canal supply the anuli fibrosi and anterior longitudinal ligament; recurrent branches supply the periosteum, ligamenta flava, anuli fibrosi posteriorly, posterior longitudinal ligament, spinal dura mater, and blood vessels within the vertebral canal.

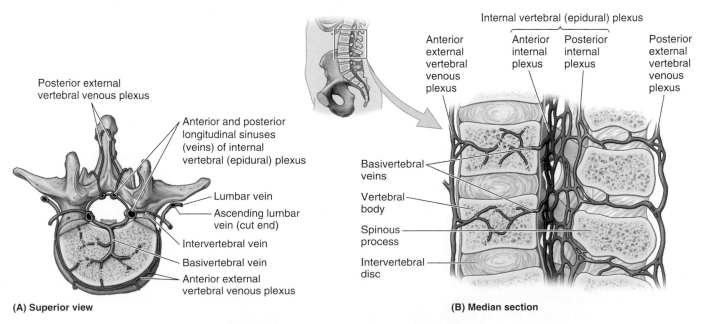

(A) Superior view

(B) Median section

FIGURE 2.15. Venous drainage of vertebral column.

CLINICAL BOX

Herniation of Nucleus Pulposus

Herniation or protrusion of the gelatinous nucleus pulposus into or through the anulus fibrosus is a well-recognized cause of low back and lower limb pain. If degeneration of the posterior longitudinal ligament and wearing of the anulus fibrosus has occurred, the nucleus pulposus may herniate into the vertebral canal and compress the spinal cord or nerve roots of spinal nerves in the cauda equina (Fig. B2.8). Herniations usually occur posterolaterally, where the anulus is relatively thin and does not receive support from the posterior or anterior longitudinal ligaments. A posterolateral herniation is more likely to be symptomatic because of the proximity of the spinal nerve roots.

The *localized back pain* of a herniated disc results from pressure on the longitudinal ligaments and periphery of the anulus fibrosus and from local inflammation resulting from chemical irritation by substances from the ruptured nucleus pulposus. *Chronic pain* resulting from the spinal nerve roots being compressed by the herniated disc is referred to the area (dermatome) supplied by that nerve. Posterolateral herniation is most common in the lumbar region; approximately 95% of protrusions occur at the L4–L5 or L5–S1 levels. In older patients, the nerve roots are more likely being compressed by increased ossification (osteophytes) of the IV foramen as they exit. *Sciatica*, pain in the lower back and hip and radiating down the back of the thigh into the leg, is often caused by a herniated lumbar IV disc or osteophytes that compress the L5 or S1 component of the sciatic nerve. The spinal nerve roots descend to the IV foramen and join to form the spinal nerve. The spinal nerve that exits a given IV foramen passes through the superior half of the foramen and thus lies above and is not affected by a herniating disc at that level. However, the nerve roots passing to the IV foramen immediately and farther below pass directly across the area of herniation (i.e., herniation of the L4–L5 disc affects the L5 nerve root) (Fig. B2.8D).

Symptom-producing IV disc protrusions occur in the cervical region almost as often as in the lumbar region. In the cervical region, the IV discs are centrally placed and extend to the anterior border of the IV foramen. Therefore, a herniating cervical disc compresses the spinal nerve exiting at that level. Recall, however, that cervical spinal nerves exit superior to the vertebra of the same number. Cervical disc protrusions result in pain in the neck, shoulder, arm, and hand.

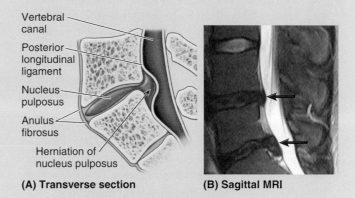

(A) Transverse section

Vertebral canal
Posterior longitudinal ligament
Nucleus pulposus
Anulus fibrosus
Herniation of nucleus pulposus

(B) Sagittal MRI

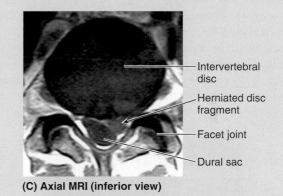

Intervertebral disc
Herniated disc fragment
Facet joint
Dural sac

(C) Axial MRI (inferior view)

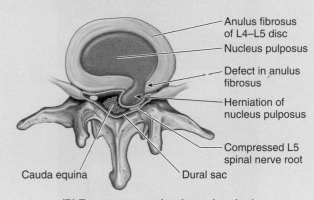

Anulus fibrosus of L4–L5 disc
Nucleus pulposus
Defect in anulus fibrosus
Herniation of nucleus pulposus
Compressed L5 spinal nerve root
Cauda equina   Dural sac

(D) Transverse section (superior view)

FIGURE B2.8. Herniation of nucleus pulposus.

Rupture of Transverse Ligament of Atlas

When the transverse ligament of the atlas ruptures, the dens is set free, resulting in *atlanto-axial subluxation* or incomplete dislocation of the median atlanto-axial joint. When complete dislocation occurs, the dens may be driven into the upper cervical region of the spinal cord, causing *quadriplegia* (paralysis of all four limbs), or into the medulla of the brainstem, causing death.

Rupture of Alar Ligaments

The alar ligaments are weaker than the transverse ligament of the atlas. Consequently, combined flexion and rotation of the head may tear one or both alar ligaments. Rupture of an alar ligament results in an increase of approximately 30% in the range of movement to the opposite side.

(*Continued on next page*)

Aging of Vertebrae and Intervertebral Discs

During middle and older age, there is an overall decrease in bone density and strength, particularly centrally within the vertebral body, that results in the superior and inferior surfaces of the vertebrae becoming increasingly concave (Fig. B2.9A). The nuclei pulposi dehydrate and lose elastin and proteoglycans while gaining collagen. As a result, the IV discs lose their turgor, becoming stiffer and more resistant to deformation. The lamellae of the anulus thicken and often develop fissures and cavities. Although the margins of adjacent vertebral bodies approach more closely as the superior and inferior surfaces of the body become concave, it has been shown that the IV discs increase in size with age. Not only do the IV discs become increasingly convex but also, between the ages of 20 and 70 years, their diameter increases (Bogduk, 1997). Aging of the IV discs, combined with the changing shape of the vertebrae, results in an increase in compressive forces at the periphery of the vertebral bodies where the discs attach. In response, *osteophytes* (bony spurs) commonly develop around the margins of the vertebral bodies (Fig. B2.9B).

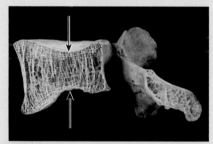

(A) Medial view of right half of lumbar vertebra

(B) Left anterior superior oblique view   ★ osteophytes

FIGURE B2.9. Effects of aging on vertebrae.

Injury and Disease of Zygapophysial Joints

When the zygapophysial joints are injured or develop osteophytes during aging (osteoarthritis), the related spinal nerves are often affected. This causes pain along the distribution pattern of the dermatomes and

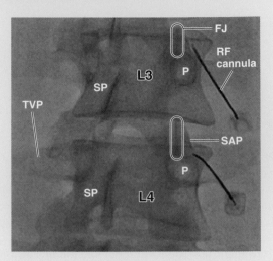

FIGURE B2.10. **Radiofrequency neurotomy of medial branch nerves (L3, L4).** *FJ*, facet joint; *P*, pedicle; *RF*, radiofrequency; *SAP*, superior articular process; *SP*, spinous process; *TVP*, transverse process.

spasm in the muscles derived from the associated myotomes (a myotome consists of all the muscles or parts of muscles receiving innervation from one spinal nerve). *Denervation of lumbar zygapophysial joints* is a procedure that may be used for treatment of back pain caused by disease of these joints. The nerves are sectioned near the joints or are destroyed by radiofrequency *percutaneous rhizolysis* (root dissolution) (Fig. B2.10). The denervation process is directed at the articular branches of two adjacent posterior rami of the spinal nerves because each joint receives innervation from both the nerve exiting that level and the superjacent nerve.

Back Pain

Back pain in general, and lower back pain in particular, is an immense health problem. In terms of health factors causing lost workdays, backache is second only to headache.

Five categories of structures receive innervation in the back and can be sources of pain:

- Fibroskeletal structures: periosteum, ligaments, and anuli fibrosi of IV discs
- Meninges: coverings of the spinal cord
- Synovial joints: capsules of the zygapophysial joints
- Muscles: intrinsic muscles of the back
- Nervous tissue: spinal nerves or nerve roots exiting the IV foramina

Of these, the first two are innervated by (recurrent) meningeal branches of the spinal nerves and the next two are innervated by posterior rami (articular and muscular branches). Pain from nervous tissue—that is, caused by compression or irritation of spinal nerves or nerve roots—is typically *referred pain*, perceived as coming from the cutaneous or subcutaneous area (dermatome) supplied by that nerve, but it may be accompanied by localized pain.

Localized *lower back pain (LBP)* (pain perceived as coming from the back) is generally muscular, joint, or fibroskeletal pain. *Muscular pain* is usually related to reflexive cramping (spasms) producing *ischemia*, often secondarily as a result of *guarding* (contraction of muscles in anticipation of pain). *Zygapophysial joint pain* is generally associated with aging (osteoarthritis) or disease (rheumatoid arthritis) of the joints. Pain from vertebral fractures and dislocations is no different than that from other bones and joints: The sharp pain following a fracture is mostly periosteal in origin, whereas pain from dislocations is ligamentous. The acute localized pain associated with an IV disc herniation emanates from the disrupted posterolateral anulus fibrosis and impingement on the posterior longitudinal ligament. Pain in all of these latter instances is conveyed initially by the meningeal branches of the spinal nerves.

SPINAL CORD AND MENINGES

The spinal cord, spinal meninges, spinal nerve roots, and neurovascular structures that supply them are in the vertebral canal (Fig. 2.17). The **spinal cord**, the major reflex center and conduction pathway between the body and the brain, is a cylindrical structure that is slightly flattened anteriorly and posteriorly. It is protected by the vertebrae and their associated ligaments and muscles, the spinal meninges, and the CSF. The spinal cord begins as a continuation of the **medulla oblongata** (commonly called the medulla), the caudal part of the brainstem. In the newborn, the inferior end of the spinal cord usually is opposite the IV disc between the L2 and the L3 vertebrae. In adults, the spinal cord usually ends opposite the IV disc between the L1 and the L2 vertebrae; however, its tapering end, the **conus medullaris**, may terminate as high as T12 or as low as L3. Thus, the spinal cord occupies only the superior two thirds of the vertebral canal. The spinal cord is enlarged in two regions for innervation of the limbs:

- The **cervical enlargement** extends from the C4 through the T1 segments of the spinal cord, and most of the anterior rami of the spinal nerves arising from it form the *brachial plexus of nerves*, which innervates the upper limbs (see Chapter 3).
- The **lumbosacral (lumbar) enlargement** extends from the L1 through the S3 segments of the spinal cord, and the anterior rami of the spinal nerves arising from it contribute to the *lumbar* and *sacral plexuses of nerves*, which innervate the lower limbs (see Chapter 7). The spinal nerve roots arising from the lumbosacral enlargement and conus medullaris form the **cauda equina**, the bundle of spinal nerve roots running inferior to the spinal cord through the *lumbar cistern* (subarachnoid space).

Structure of Spinal Nerves

A total of 31 pairs of spinal nerves are attached to the spinal cord: 8 cervical, 12 thoracic, 5 lumbar, 5 sacral, and 1 coccygeal (Fig. 2.17A). Multiple rootlets attach to the posterior and anterior surfaces of the spinal cord and converge to form **posterior** and **anterior roots of the spinal nerves** (Fig. 2.18A,B). The part of the spinal cord to which the rootlets of one bilateral pair of roots attach is a **segment of the spinal cord**. The posterior roots of the spinal nerves contain afferent (or sensory) fibers from skin, subcutaneous and deep tissues, and, often, viscera. The anterior roots of spinal nerves contain efferent (motor) fibers to skeletal muscle, and many contain presynaptic autonomic fibers. The cell bodies of somatic axons contributing to the anterior roots are in the **anterior horns of gray matter** of the spinal cord (Fig. 2.18C), whereas the cell bodies of axons making up the posterior roots are outside the spinal cord in the **spinal ganglia** (posterior root ganglia) at the distal ends of the posterior roots. The posterior and anterior nerve roots unite at their points of exit from the vertebral canal to form a **spinal nerve**. The C1 nerves lack posterior roots in 50% of people, and the coccygeal nerve (Co1) may be absent. Each spinal nerve divides almost immediately into a **posterior (dorsal) ramus** and **anterior (ventral) ramus** (Fig. 2.18A). The posterior rami supply the zygapophysial joints, deep muscles of the back, and overlying skin; the anterior rami supply the muscles, joints, and skin of the limbs and the remainder of the trunk.

In adults, the spinal cord is shorter than the vertebral column; hence, there is a progressive obliquity of the spinal nerve roots as the cord descends (Fig. 2.17). Because of the increasing distance between the spinal cord segments and the corresponding vertebrae, the length of the nerve roots increases progressively as the inferior end of the vertebral column is approached. The lumbar and sacral nerve rootlets are the longest. They descend until they reach the IV foramina of exit in the lumbar and sacral regions of the vertebral column, respectively. The bundle of spinal nerve roots in the **lumbar cistern** of the subarachnoid space caudal to the termination of the spinal cord resembles a horse's tail, hence its name *cauda equina* (L. horse tail) (Figs. 2.17B and 2.18C).

The inferior end of the spinal cord has a conical shape and tapers into the *conus medullaris*. From its inferior end, the **filum terminale internum** descends among the spinal nerve roots in the cauda equina. It consists primarily of pia mater, but its proximal end also includes vestiges of neural tissue, connective tissue, and neuroglial tissue (nonneuronal cellular elements of the nervous system). The filum terminale takes on layers of arachnoid and dura mater as it penetrates the inferior end of the dural sac becoming the **filum**

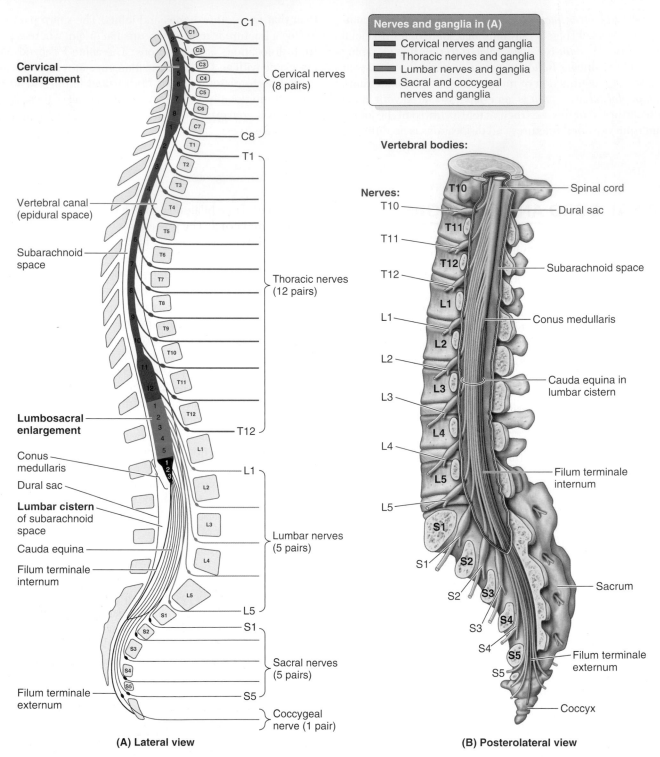

Nerves and ganglia in (A)

▬	Cervical nerves and ganglia
▬	Thoracic nerves and ganglia
▬	Lumbar nerves and ganglia
▬	Sacral and coccygeal nerves and ganglia

Cervical enlargement

Cervical nerves (8 pairs)

Vertebral canal (epidural space)

Subarachnoid space

Thoracic nerves (12 pairs)

Lumbosacral enlargement

Conus medullaris

Dural sac

Lumbar cistern of subarachnoid space

Cauda equina

Filum terminale internum

Lumbar nerves (5 pairs)

Filum terminale externum

Sacral nerves (5 pairs)

Coccygeal nerve (1 pair)

(A) Lateral view

Vertebral bodies:

Nerves:

Spinal cord

Dural sac

Subarachnoid space

Conus medullaris

Cauda equina in lumbar cistern

Filum terminale internum

Sacrum

Filum terminale externum

Coccyx

(B) Posterolateral view

FIGURE 2.17. Relationship of vertebral column, spinal cord, and spinal nerves. Note the relation of the spinal cord segments and spinal nerves to the vertebral column.

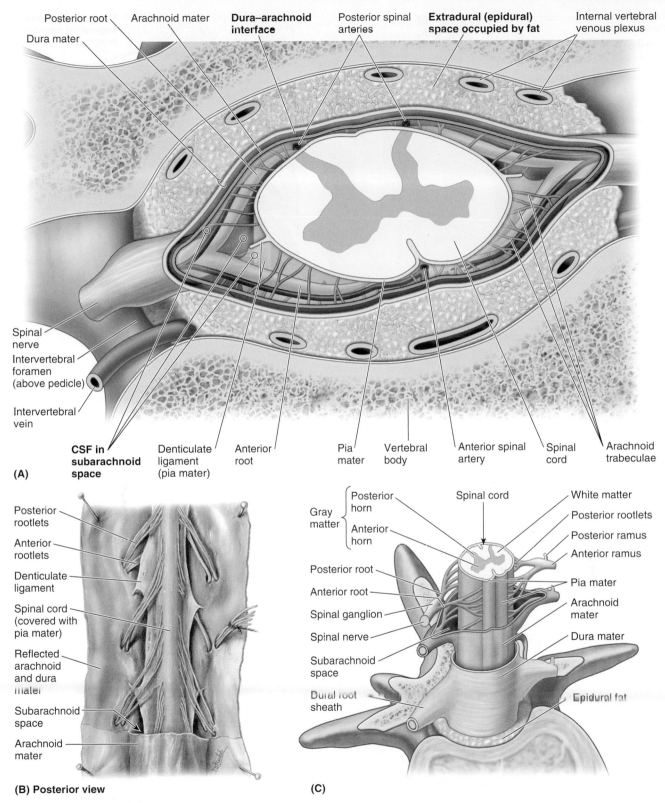

FIGURE 2.18. Spinal cord and spinal meninges. A. Cross section of spinal cord within its meninges. **B.** The meninges have been cut and spread out. The pia mater covers the spinal cord and projects laterally as the denticulate ligament. **C.** Spinal cord, spinal nerves, and spinal meninges. The term "mater" is often omitted, referring simply to "dura," "arachnoid," and "pia."

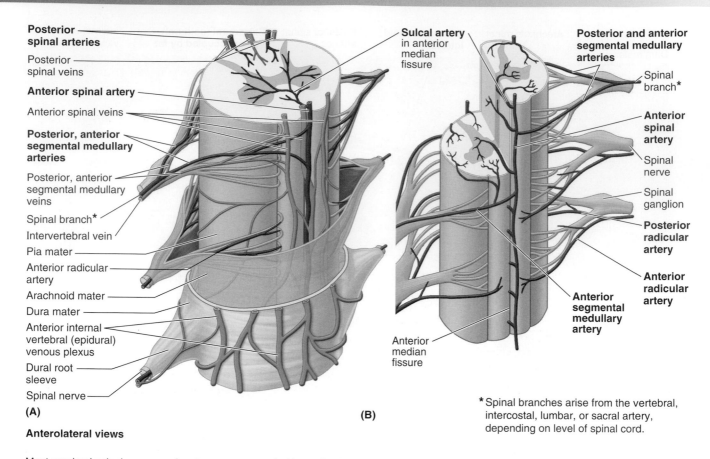

(A)

Anterolateral views

(B)

*Spinal branches arise from the vertebral, intercostal, lumbar, or sacral artery, depending on level of spinal cord.

Most proximal spinal nerves and roots are accompanied by **radicular arteries**, which do not reach the posterior, anterior, or spinal arteries. **Segmental medullary arteries** occur irregularly *in the place of* radicular arteries—they are really just larger vessels that make it all the way to the spinal arteries.

FIGURE 2.19. Spinal cord in situ: vasculature and meninges with associated spaces.

terminale externum that passes through the sacral hiatus to attach ultimately to the coccyx posteriorly. The filum terminale serves as an anchor for the inferior ends of the spinal cord and dural sac.

Spinal Meninges and Cerebrospinal Fluid

Collectively, the dura mater, arachnoid mater, and pia mater surrounding the spinal cord form the **spinal meninges**. These membranes and CSF surround, support, and protect the spinal cord and the spinal nerve roots, including those in the cauda equina.

The **spinal dura mater**, composed of tough, fibrous, and some elastic tissue, is the outermost covering membrane of the spinal cord (Fig. 2.18). The spinal dura mater is separated from the vertebrae by the **extradural (epidural) space** (Fig. 2.19 and Table 2.4). The dura forms the **spinal dural sac**, a long tubular sheath within the vertebral canal (Fig. 2.17). The spinal dural sac adheres to the margin of the foramen magnum of the cranium, where it is continuous with the cranial dura mater. The spinal dural sac is pierced by the

TABLE 2.4. SPACES ASSOCIATED WITH SPINAL MENINGES

Space	Location	Contents
Extradural (epidural)	Between wall of vertebral canal and dura mater	Epidural fat (fatty matrix); internal vertebral venous plexuses; each pair of posterior and anterior roots as they extend to their exit from the vertebral canal at the IV foramina
Subarachnoid	Between arachnoid and pia mater	CSF; arachnoid trabeculae; radicular, segmental medullary, and spinal arteries; veins

spinal nerves and is anchored inferiorly to the coccyx by the *filum terminale externum*. The spinal dura extends into the IV foramina and along the posterior and anterior nerve roots distal to the spinal ganglia to form **dural root sheaths**, or thecal sleeves (Fig. 2.18A). These sheaths blend with the epineurium (outer connective tissue covering of spinal nerves) that adheres to the periosteum lining the IV foramina.

The **spinal arachnoid mater** is a delicate, avascular membrane composed of fibrous and elastic tissue that lines the dural sac and the dural root sheaths. It encloses the CSF-filled subarachnoid space containing the spinal cord, spinal nerve roots, and spinal ganglia (Fig. 2.18B,C). The arachnoid mater is not attached to the dura but is pressed against the inner surface of the dura by the pressure of the CSF. In a *lumbar spinal puncture*, the needle traverses the dura and arachnoid mater simultaneously. Their apposition is the **dura–arachnoid interface**, often erroneously referred to as the "subdural space" (Fig. 2.19). No actual space occurs naturally at this site; it is rather a weak cell layer (Haines, 2013). Bleeding into this layer creates a pathological space at the dura–arachnoid junction in which a *subdural hematoma* is formed. In the cadaver, because of the absence of CSF, the arachnoid falls away from the internal surface of the dura and lies loosely on the spinal cord. In life, the arachnoid mater is separated from the pia mater on the surface of the spinal cord by the *subarachnoid space* containing CSF (Figs. 2.18 and 2.19 and Table 2.4). Delicate strands of connective tissue, the **arachnoid trabeculae**, span the subarachnoid space connecting the arachnoid and pia (Fig. 2.18C).

The **spinal pia mater**, the innermost covering membrane of the spinal cord, consists of flattened cells with long, equally flattened processes that closely follow all the surface features of the spinal cord (Fig. 2.18B,C). The pia mater also directly covers the roots of the spinal nerves and spinal blood vessels. Inferior to the conus medullaris, the pia continues as the filum terminale.

The spinal cord is suspended in the dural sac by the filum terminale and especially by the right and left sawtooth **denticulate ligaments** (L. *denticulus*, small tooth), which run longitudinally along each side of the spinal cord. These ligaments consist of a fibrous sheet of pia mater extending midway between the posterior and the anterior nerve roots. Between 20 and 22 of these processes, shaped much like sharks' teeth, attach to the internal surface of the arachnoid-lined dural sac. The superior processes (uppermost part) of the right and left denticulate ligament attach to the cranial dura mater immediately superior to the foramen magnum. The inferior process extends from the conus medullaris passing between the T12 and the L1 nerve roots.

SUBARACHNOID SPACE

The subarachnoid space lies between the arachnoid mater and the pia mater and is filled with CSF (Figs. 2.17B, 2.18C, and 2.19 and Table 2.4). The enlargement of the subarachnoid space in the dural sac, caudal to the conus medullaris, and containing CSF and the cauda equina, is the lumbar cistern (Fig. 2.17B).

Vasculature of Spinal Cord and Spinal Nerve Roots

The arteries supplying the spinal cord are branches of the vertebral, ascending cervical, deep cervical, intercostal, lumbar, and lateral sacral arteries (Figs. 2.19 and 2.20). Three longitudinal arteries supply the spinal cord: an **anterior spinal artery**, formed by the union of branches of vertebral arteries, and paired **posterior spinal arteries**, each of which is a branch of either the vertebral artery or the posterior inferior cerebellar artery.

The spinal arteries run longitudinally from the medulla of the brainstem to the conus medullaris of the spinal cord. By themselves, the anterior and posterior spinal arteries supply only the short superior part of the spinal cord. The circulation to much of the spinal cord depends on spinal branches of ascending cervical, deep cervical, vertebral, posterior intercostal, and lumbar arteries that enter the vertebral canal through the IV foramina. The **anterior** and **posterior segmental medullary arteries** are derived from spinal branches and supply the spinal cord by joining anterior and posterior spinal arteries. These arteries are chiefly located where the need for a good blood supply to the spinal cord is greatest: the cervical and lumbosacral enlargements. The **great anterior segmental medullary artery** (of Adamkiewicz) reinforces the circulation to two thirds of the spinal cord, including the lumbosacral enlargement. It is much larger than the other segmental medullary arteries and usually arises on the left side at low thoracic or upper lumbar levels.

Posterior and anterior roots of the spinal nerves and their coverings are supplied by **posterior** and **anterior radicular arteries**, which run along the nerve roots. These vessels do not reach the posterior or anterior spinal arteries. Segmental medullary arteries occur irregularly in the place of radicular arteries; they are larger vessels that supply blood to the spinal arteries.

The three **anterior** and three **posterior spinal veins** are arranged longitudinally; they communicate freely with each other and are drained by up to 12 **anterior** and **posterior medullary** and **radicular veins**. The veins draining the spinal cord join the internal vertebral venous plexus in the epidural space (Fig. 2.15). This venous plexus passes superiorly through the foramen magnum to communicate with the dural venous sinuses and veins in the cranium (see Chapter 8). The internal vertebral plexus also communicates with the external vertebral venous plexus on the external surface of the vertebrae.

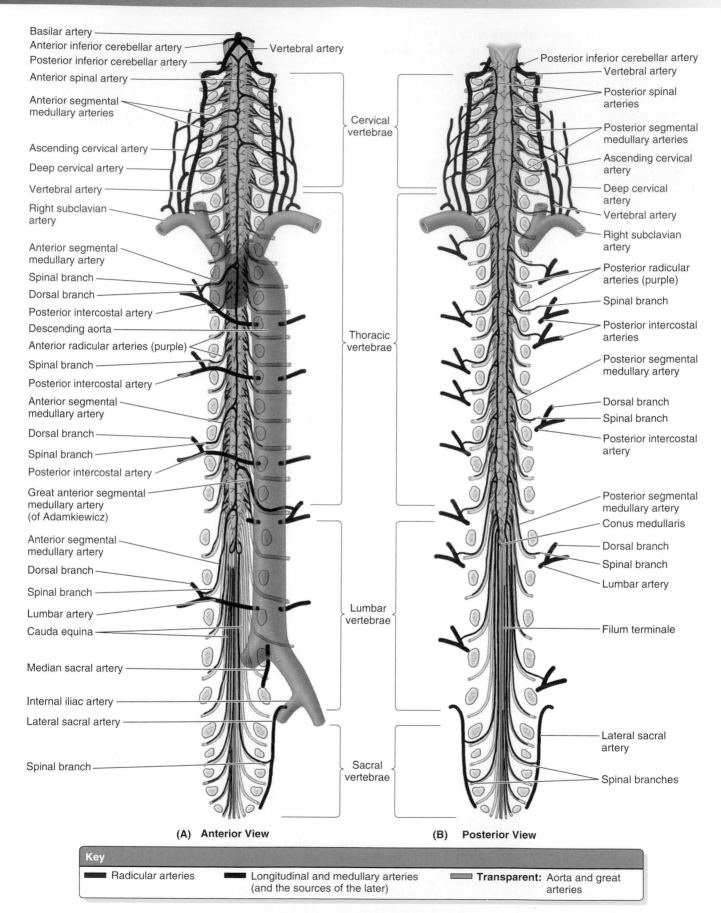

Basilar artery
Anterior inferior cerebellar artery
Posterior inferior cerebellar artery
Anterior spinal artery
Anterior segmental medullary arteries
Ascending cervical artery
Deep cervical artery
Vertebral artery
Right subclavian artery
Anterior segmental medullary artery
Spinal branch
Dorsal branch
Posterior intercostal artery
Descending aorta
Anterior radicular arteries (purple)
Spinal branch
Posterior intercostal artery
Anterior segmental medullary artery
Dorsal branch
Spinal branch
Posterior intercostal artery
Great anterior segmental medullary artery (of Adamkiewicz)
Anterior segmental medullary artery
Dorsal branch
Spinal branch
Lumbar artery
Cauda equina
Median sacral artery
Internal iliac artery
Lateral sacral artery
Spinal branch

Vertebral artery

Cervical vertebrae
Thoracic vertebrae
Lumbar vertebrae
Sacral vertebrae

Posterior inferior cerebellar artery
Vertebral artery
Posterior spinal arteries
Posterior segmental medullary arteries
Ascending cervical artery
Deep cervical artery
Vertebral artery
Right subclavian artery
Posterior radicular arteries (purple)
Spinal branch
Posterior intercostal arteries
Posterior segmental medullary artery
Dorsal branch
Spinal branch
Posterior intercostal artery
Posterior segmental medullary artery
Conus medullaris
Dorsal branch
Spinal branch
Lumbar artery
Filum terminale
Lateral sacral artery
Spinal branches

(A)   **Anterior View**   (B)   **Posterior View**

Key		
Radicular arteries	Longitudinal and medullary arteries (and the sources of the later)	**Transparent:** Aorta and great arteries

FIGURE 2.20. Arterial supply of spinal cord.

CLINICAL BOX

Ischemia of Spinal Cord

The segmental reinforcements of blood supply from the segmental medullary arteries are important in supplying blood to the anterior and posterior spinal arteries. Fractures, dislocations, and fracture–dislocations may interfere with the blood supply to the spinal cord from the spinal and medullary arteries. Deficiency of blood supply (ischemia) of the spinal cord affects its function and can lead to muscle weakness and paralysis.

The spinal cord may also sustain circulatory impairment if the segmental medullary arteries, particularly the great anterior segmental medullary artery (of Adamkiewicz), are narrowed by *obstructive arterial disease*. Sometimes, the aorta is purposely occluded ("cross-clamped") during surgery. Patients undergoing such surgeries, and those with ruptured aneurysms of the aorta or occlusion of the great anterior segmental medullary artery, may lose all sensation and voluntary movement inferior to the level of impaired blood supply to the spinal cord (*paraplegia*). This is secondary to death of neurons in the part of the spinal cord supplied by the anterior spinal artery.

When systemic blood pressure drops severely for 3–6 minutes, blood flow from the segmental medullary arteries to the anterior spinal artery supplying the midthoracic region of the spinal cord may be reduced or stopped. These patients may also lose sensation and voluntary movement in the areas supplied by the affected level of the spinal cord.

Alternative Circulation Pathways

The *vertebral venous plexuses* are important because blood may return from the pelvis or abdomen through these plexuses and reach the heart via the superior vena cava when the inferior vena cava is obstructed. These veins also can provide a route for metastasis of cancer cells to the vertebrae or the brain from an abdominal or pelvic tumor (e.g., prostate cancer).

Lumbar Spinal Puncture

To obtain a sample of CSF from the lumbar cistern, a lumbar puncture needle, fitted with a stylet, is inserted into the subarachnoid space. *Lumbar spinal puncture* (spinal tap) is performed with the patient leaning forward or lying on the side with the back flexed. Flexion of the vertebral column facilitates insertion of the needle by spreading the laminae and spinous processes apart, stretching the ligament flava (Fig. B2.11). Under aseptic conditions, the needle is inserted in the midline between the spinous processes of the L3 and L4 (or the L4 and L5) vertebrae. At these levels in adults, there is reduced danger of damaging the spinal cord.

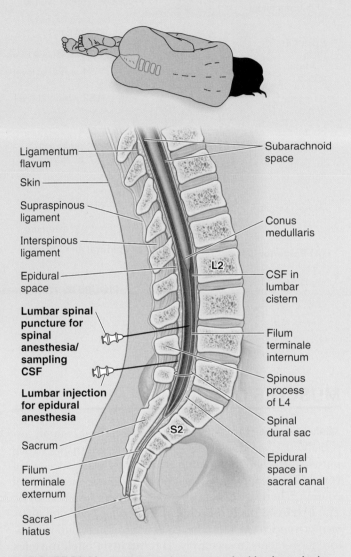

Labels on figure:
- Ligamentum flavum
- Skin
- Supraspinous ligament
- Interspinous ligament
- Epidural space
- **Lumbar spinal puncture for spinal anesthesia/ sampling CSF**
- **Lumbar injection for epidural anesthesia**
- Sacrum
- Filum terminale externum
- Sacral hiatus
- Subarachnoid space
- Conus medullaris
- L2
- CSF in lumbar cistern
- Filum terminale internum
- Spinous process of L4
- Spinal dural sac
- Epidural space in sacral canal
- S2

FIGURE B2.11. Lumbar spinal puncture and epidural anesthesia.

Epidural Anesthesia (Blocks)

An anesthetic agent can be injected into the lumbar extradural (epidural) space using the position described for lumbar spinal puncture. The anesthetic has a direct effect on the spinal nerve roots of the cauda equina after they exit from the dural sac (Fig. B2.12). The patient loses sensation inferior to the level of the block.

An anesthetic agent can also be injected into the extradural space in the sacral canal through the sacral hiatus (*caudal epidural anesthesia*) or through the posterior sacral foramina (*trans-sacral epidural anesthesia*) (Fig. B2.12). The distance the agent ascends (and hence the number of nerves affected) depends on the amount injected and the position assumed by the patient.

(*Continued on next page*)

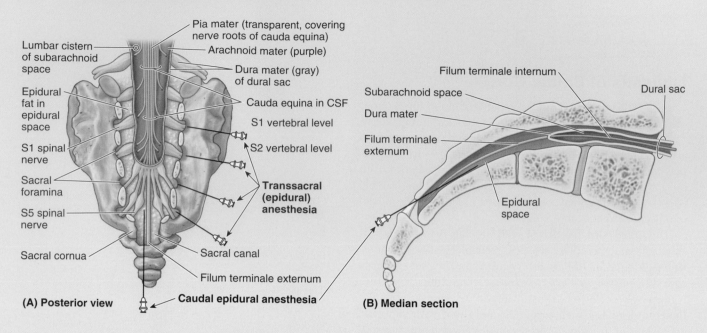

(A) Posterior view   **Caudal epidural anesthesia**   **(B) Median section**

FIGURE B2.12. Trans-sacral and caudal anesthesia.

MUSCLES OF BACK

Most body weight is anterior to the vertebral column, especially in obese people. For this reason, the many strong muscles attached to the spinous and transverse processes of vertebrae are necessary to support and move the vertebral column. There are two major groups of muscles in the back. The **extrinsic back muscles** include *superficial* and *intermediate muscles* that produce and control limb and respiratory movements, respectively. The **intrinsic back muscles** include *deep muscles* that specifically act on the vertebral column, producing its movements and maintaining posture.

Extrinsic Back Muscles

The **superficial extrinsic back muscles** (trapezius, latissimus dorsi, levator scapulae, and rhomboids) connect the upper limbs to the trunk (see Chapter 3). These muscles, although located in the back region, for the most part, receive their nerve supply from the anterior rami of cervical nerves and act on the upper limb. The trapezius receives its motor fibers from a cranial nerve, the spinal accessory nerve (CN XI). The **intermediate extrinsic back muscles** (serratus posterior superior and inferior) are thin muscles and are commonly designated superficial respiratory muscles but are more likely proprioceptive rather than motor in function. They are described with muscles of the thoracic wall (see Chapter 4).

Intrinsic Back Muscles

The intrinsic back muscles (*muscles of back proper*, deep back muscles) are innervated by the posterior rami of spinal nerves and act to maintain posture and control movements of the vertebral column. These muscles, extending from the pelvis to the cranium, are enclosed by deep fascia that attaches medially to the nuchal ligament, the tips of the spinous processes of the vertebrae, the supraspinous ligament, and the median crest of the sacrum. The fascia attaches laterally to the cervical and lumbar transverse processes and to the angles of the ribs. The thoracic and lumbar parts of the deep fascia constitute the **thoracolumbar fascia** (Fig. 2.21). The deep back muscles are grouped into superficial, intermediate, and deep layers according to their relationship to the surface (Table 2.5).

SUPERFICIAL LAYER OF INTRINSIC BACK MUSCLES

The **splenius muscles** (L. *musculi splenii*) are thick and flat and lie on the lateral and posterior aspects of the neck, covering the vertical muscles somewhat like a bandage, which explains their name (L. *splenion*, bandage). The splenii arise from the midline and extend superolaterally to the cervical vertebrae (**splenius cervicis**) and cranium (**splenius capitis**). These muscles cover the deep neck muscles (Fig. 2.22B and Table 2.5).

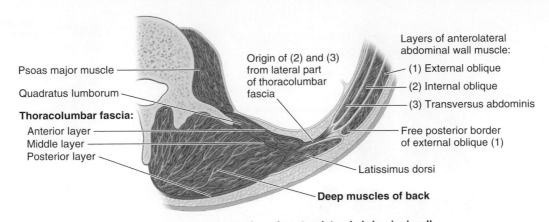

Inferior view of transverse section of posterolateral abdominal wall

FIGURE 2.21. Transverse section of the intrinsic back muscles and layers of thoracolumbar fascia.

INTERMEDIATE LAYER OF INTRINSIC BACK MUSCLES

The **erector spinae muscles** (sacrospinalis) lie in a "groove" on each side of the vertebral column between the spinous processes and the angles of the ribs (Fig. 2.22). The massive **erector spinae**, the chief extensor of the vertebral column, divides into three muscle columns:

- Iliocostalis: lateral column
- Longissimus: intermediate column
- Spinalis: medial column

Each column is divided regionally into three parts according to its superior attachments (e.g., iliocostalis lumborum, iliocostalis thoracis, and iliocostalis cervicis). The common origin of the three erector spinae columns is through a broad tendon that attaches inferiorly to the posterior part of the iliac crest, the posterior aspect of the sacrum, the sacro-iliac ligaments, and the sacral and inferior lumbar spinous processes (Fig. 2.22). Although the muscle columns are generally identified as isolated muscles, each column is actually composed of many overlapping shorter fibers—a design that provides stability, localized action, and segmental vascular and neural supply. The attachments, nerve supply, and actions of the erector spinae are described in Table 2.5.

TABLE 2.5. SUPERFICIAL AND INTERMEDIATE LAYERS OF INTRINSIC BACK MUSCLES

Muscle	Origin	Insertion	Nerve Supply	Main Action(s)
Superficial layer of intrinsic back muscles				
Splenius	Arises from nuchal ligament and spinous processes of C7–T6 vertebrae	*Splenius capitis:* Fibers run superolaterally to mastoid process of temporal bone and lateral third of superior nuchal line of occipital bone. *Splenius cervicis:* tubercles of transverse processes of C1–C3 or C4 vertebrae	Posterior rami of spinal nerves	*Acting alone:* laterally flexes neck and rotates head to side of active muscles *Acting together:* extend head and neck
Intermediate layer of intrinsic back muscles (erector spinae)				
Iliocostalis Longissimus Spinalis	Arise by broad tendon from posterior part of iliac crest, posterior surface of sacrum, sacro-iliac ligaments, sacral and inferior lumbar spinous processes, and supraspinous ligament	*Iliocostalis* (lumborum, thoracis, and cervicis): Fibers run superiorly to angles of lower ribs and cervical transverse processes. *Longissimus* (thoracis, cervicis, and capitis): Fibers run superiorly to ribs between tubercles and angles to transverse processes in thoracic and cervical regions and to mastoid process of temporal bone. *Spinalis* (thoracis, cervicis, and capitis): Fibers run superiorly to spinous processes in upper thoracic region and to cranium.	Posterior rami of spinal nerves	*Acting bilaterally:* extend vertebral column and head; as back is flexed, control movement by gradually lengthening their fibers *Acting unilaterally:* laterally flex vertebral column

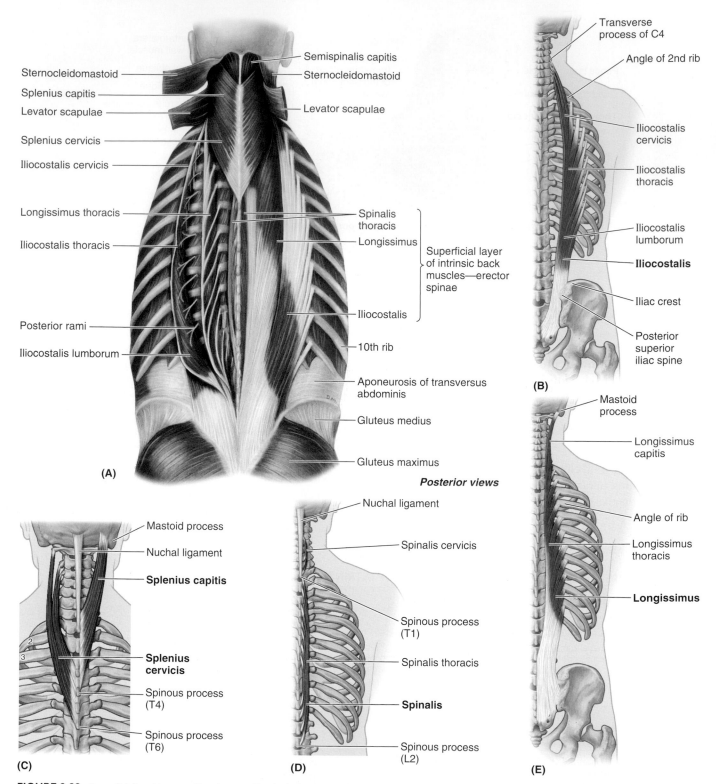

(A)

Posterior views

(B)

(C)

(D)

(E)

FIGURE 2.22. **Superficial and intermediate layers of intrinsic back muscles. A.** Overview. **B.** Iliocostalis. **C.** Splenius capitis and splenius cervicis. **D.** Spinalis. **E.** Longissimus.

SURFACE ANATOMY

Back Muscles

In the midline of the erect back, there is a *posterior median furrow* that overlies the tips of the spinous processes of the vertebrae (Fig. SA2.3). The furrow is continuous superiorly with the nuchal groove in the neck and ends in the flattened triangular area covering the sacrum superior to the intergluteal cleft. The erector spinae muscles produce prominent vertical bulges on each side of the furrow. When the upper limbs are elevated, the scapulae move laterally on the thoracic wall, making the rhomboid and teres major muscles visible. The superficially located trapezius (*D*, descending [superior] part; *T*, transverse [middle] part; *A*, ascending [inferior] part) and latissimus dorsi muscles connecting the upper limbs to the vertebral column are also clearly visible in lean individuals or when the muscles are well developed. Note the dimples indicating the site of the posterior superior iliac spines.

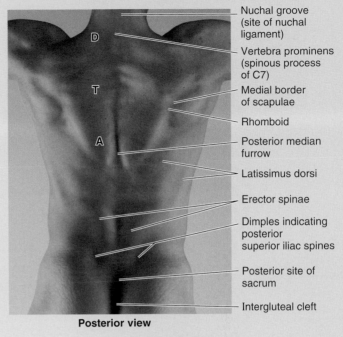

Nuchal groove (site of nuchal ligament)

Vertebra prominens (spinous process of C7)

Medial border of scapulae

Rhomboid

Posterior median furrow

Latissimus dorsi

Erector spinae

Dimples indicating posterior superior iliac spines

Posterior site of sacrum

Intergluteal cleft

Posterior view

FIGURE SA2.3.

DEEP LAYER OF INTRINSIC BACK MUSCLES

Deep to the erector spinae muscles is an obliquely disposed group of muscles—the **transversospinales muscle group**, which is composed of the semispinalis, multifidus, and rotatores. These muscles originate from transverse processes of vertebrae and pass to spinous processes of more superior vertebrae. They occupy the "gutter" between the transverse and spinous processes (Fig. 2.23 and Table 2.6).

- The semispinalis is superficial, spanning four to six segments.
- The multifidus is deeper, spanning two to four segments.
- The rotatores are deepest, spanning one to two segments.

The **semispinalis**, as its name indicates, arises from approximately half of the vertebral column. It is divided into three parts according to the vertebral level of its superior attachments: semispinalis capitis, semispinalis cervicis, and semispinalis thoracis.

The **semispinalis capitis** is responsible for the longitudinal bulge on each side in the back of the neck near the median plane. It ascends from the cervical and thoracic transverse processes to the occipital bone.

The **semispinalis thoracis and cervicis** pass superomedially from the transverse processes to the thoracic and cervical spinous processes of more superior vertebrae.

The **multifidus** consists of short, triangular muscular bundles that are thickest in the lumbar region. Each bundle passes obliquely, superiorly, and medially and attaches along the whole length of the spinous process of the adjacent superior vertebra.

The **rotatores**—best developed in the thoracic region—are the deepest of the three layers of transversospinales muscles. They arise from the transverse process of one vertebra and insert into the root of the spinous processes of the next one or two vertebrae superiorly.

The **interspinales**, **intertransversarii**, and **levatores costarum** are the smallest of the deep back muscles. The interspinales and intertransversarii muscles connect spinous and transverse processes, respectively.

MUSCLES PRODUCING MOVEMENTS OF INTERVERTEBRAL JOINTS

The principal muscles producing movements of the cervical, thoracic, and lumbar IV joints and structures limiting these movements are summarized in Tables 2.7 and 2.8. The back muscles are relatively inactive in the stand-easy position. It is actually the interaction of anterior (abdominal) and posterior (back) muscles that provides the stability and produces motion of the axial skeleton.

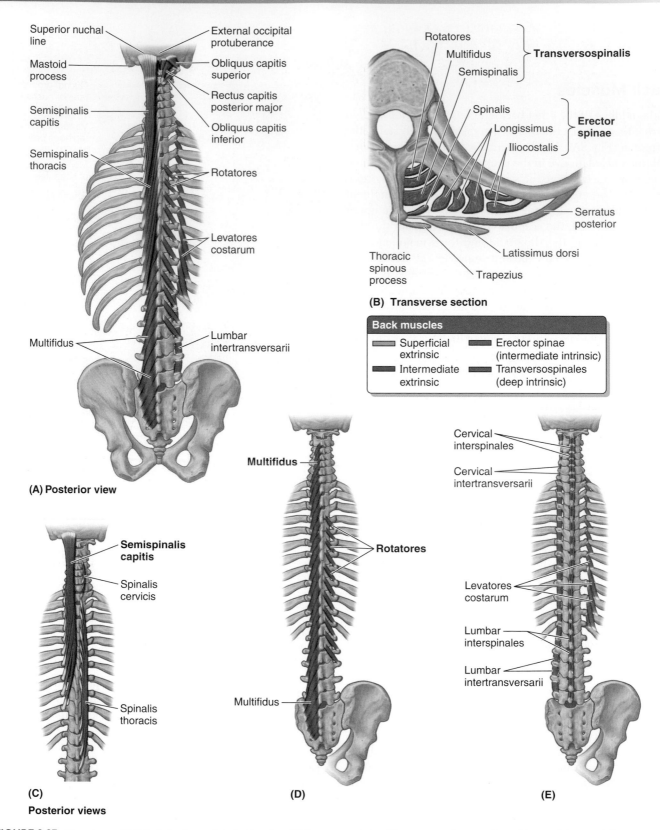

(A) Posterior view

(B) Transverse section

Back muscles
- Superficial extrinsic
- Intermediate extrinsic
- Erector spinae (intermediate intrinsic)
- Transversospinales (deep intrinsic)

(C) Posterior views

(D)

(E)

FIGURE 2.23. Deep layer of intrinsic back muscles. A. Overview. **B.** Transverse section. The erector spinae consists of three columns and the transversospinales consists of three layers: semispinalis **(C)**, multifidus **(D)**, and rotatores **(A and D)**. **E.** Interspinales, intertransversarii, and levatores costarum.

TABLE 2.6. DEEP LAYERS OF INTRINSIC BACK MUSCLES

Muscle	Origin	Insertion	Nerve Supply	Main Action(s)
Deep layer of intrinsic back muscles (transversospinales)				
Semispinalis (thoracis, cervicis, and capitis)	Arises from transverse processes of C4–T10 vertebrae	Fibers run superomedially to occipital bone and spinous processes in upper thoracic and cervical regions, spanning four to six segments.	Posterior rami of spinal nerves	Extends head and thoracic and cervical regions of vertebral column and rotates them contralaterally
Multifidus	Arises from posterior sacrum, posterior superior iliac spine of ilium, aponeurosis of erector spinae, sacro-iliac ligaments, mammillary processes of lumbar vertebrae, transverse processes of thoracic vertebrae, and articular processes of C4–C7	Thickest in lumbar region, fibers pass obliquely superomedially to entire length of spinous processes of vertebrae located two to four segments superior to origin		Unilateral contraction rotates to contralateral side; stabilizes vertebrae during local movements of vertebral column
Rotatores (brevis and longus)	Arise from transverse processes of vertebrae; are best developed in thoracic region	Fibers pass superomedially to attach to junction of lamina and transverse process or spinous process of vertebra immediately (brevis) or two segments (longus) superior to vertebra of origin.		May function as organs of proprioception; possibly stabilize vertebrae and assist with local extension and rotatory movements of vertebral column
Minor deep layer of intrinsic back muscles				
Interspinales	Superior surfaces of spinous processes of cervical and lumbar vertebrae	Inferior surfaces of spinous processes of vertebrae superior to vertebrae of origin	Posterior rami of spinal nerves	Aid in extension and rotation of vertebral column
Intertransversarii	Transverse processes of cervical and lumbar vertebrae	Transverse processes of adjacent vertebrae	Posterior and anterior rami of spinal nerves[a]	Aid in lateral flexion of vertebral column; acting bilaterally, stabilize vertebral column
Levatores costarum	Tips of transverse processes of C7 and T1–T11 vertebrae	Pass inferolaterally and insert on rib between its tubercle and angle	Posterior rami of C8–T11 spinal nerves	Elevate ribs, assisting respiration; assist with lateral flexion of vertebral column

[a]Most back muscles are innervated by posterior rami of the spinal nerves, but a few are innervated by anterior rami.

Smaller muscles generally have higher densities of *muscle spindles* (sensors of *proprioception*—the sense of one's position—that are interdigitated among the muscle's fibers) than do large muscles. It has been presumed that this is because small muscles are used for the most precise movements, such as fine postural movements or manipulation, and therefore require more proprioceptive feedback. The movements described for small muscles are assumed from the location of their attachments, from the direction of the muscle fibers, and from activity measured by *electromyography*. Muscles such as the rotatores, however, are so small and are placed in positions of such relatively poor mechanical advantage that their ability to produce the movements described is somewhat questionable. It has been proposed that the smaller muscles of small–large muscle pairs function more as "kinesiological monitors" (organs of proprioception) and that the larger muscles are the producers of motion.

Suboccipital Region

The **suboccipital region**—superior part of the back of the neck—is the triangular area (*suboccipital triangle*) inferior to the occipital region of the head, including the posterior aspects of the C1 and C2 vertebrae.

The **suboccipital triangle** lies deep to the trapezius and semispinalis capitis muscles (Fig. 2.24). The four small muscles in the suboccipital region—rectus capitis posterior major and minor and obliquus capitis superior and inferior—are innervated by the posterior ramus of C1, the **suboccipital nerve**. The nerve emerges as the vertebral artery courses deeply between the occipital bone and the atlas (C1 vertebra) within the suboccipital triangle.

The suboccipital muscles are mainly postural muscles, but they act on the head—directly or indirectly—as indicated by *capitis* in their name.

- **Rectus capitis posterior major** arises from the spinous process of the C2 vertebra and inserts into the lateral part of the inferior nuchal line of the occipital bone.
- **Rectus capitis posterior minor** arises from the posterior tubercle on the posterior arch of the C1 vertebra and inserts into the medial third of the inferior nuchal line.
- **Obliquus capitis inferior** arises from the spinous process of the C2 vertebra and inserts into the transverse process of the C1 vertebra. The name of this muscle is

TABLE 2.7. PRINCIPAL MUSCLES PRODUCING MOVEMENT OF CERVICAL INTERVERTEBRAL JOINTS

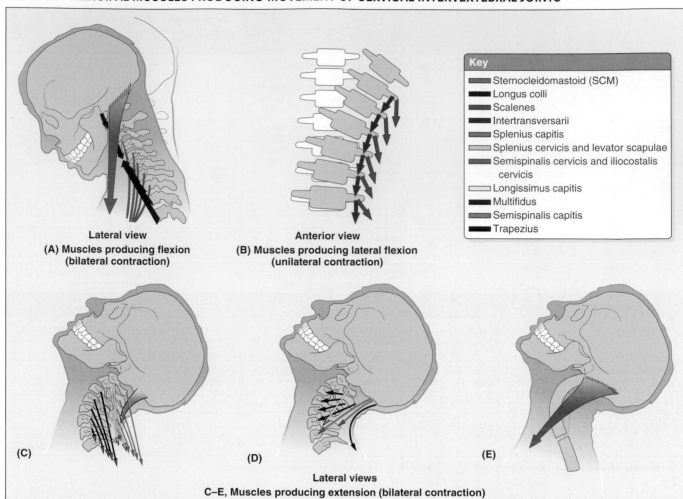

Lateral view
(A) Muscles producing flexion
(bilateral contraction)

Anterior view
(B) Muscles producing lateral flexion
(unilateral contraction)

Key
- Sternocleidomastoid (SCM)
- Longus colli
- Scalenes
- Intertransversarii
- Splenius capitis
- Splenius cervicis and levator scapulae
- Semispinalis cervicis and iliocostalis cervicis
- Longissimus capitis
- Multifidus
- Semispinalis capitis
- Trapezius

(C)   **(D)**   **(E)**

Lateral views
C–E, Muscles producing extension (bilateral contraction)

Movement	Flexion	Extension	Lateral Bending	Rotation
Principle muscles producing movement	Bilateral action of • Longus colli • Scalene • Sternocleidomastoid	Bilateral action of deep neck muscles • Semispinalis cervicis and iliocostalis cervicis • Splenius cervicis and levator scapulae • Splenius capitis • Multifidus • Longissimus capitis • Semispinalis capitis • Trapezius	Unilateral action of • Iliocostalis cervicis • Longissimus capitis and cervicis • Splenius capitis • Splenius cervicis • Intertransversarii and scalenes	Ipsilateral action of • Rotatores • Semispinalis capitis and cervicis • Multifidus • Splenius cervicis Contralateral action of • Sternocleidomastoid
Structures limiting or opposing movement	• Ligaments: posterior atlanto-axial, posterior longitudinal, flavum, tectorial membrane • Posterior neck muscles • Anulus fibrosus (tension posteriorly)	• Ligaments: anterior longitudinal, anterior atlanto-axial • Anterior neck muscles • Anulus fibrosus (tension anteriorly) • Spinous processes (contact between adjacent processes)	• Ligaments: Alar ligament tension limits movement to contralateral side. • Anulus fibrosus (tension anteriorly) • Zygapophysial (facet) joints	• Ligaments: Alar ligament tension limits movement to ipsilateral side. • Anulus fibrosus

TABLE 2.8. PRINCIPAL MUSCLES PRODUCING MOVEMENTS OF THORACIC AND LUMBAR INTERVERTEBRAL JOINTS

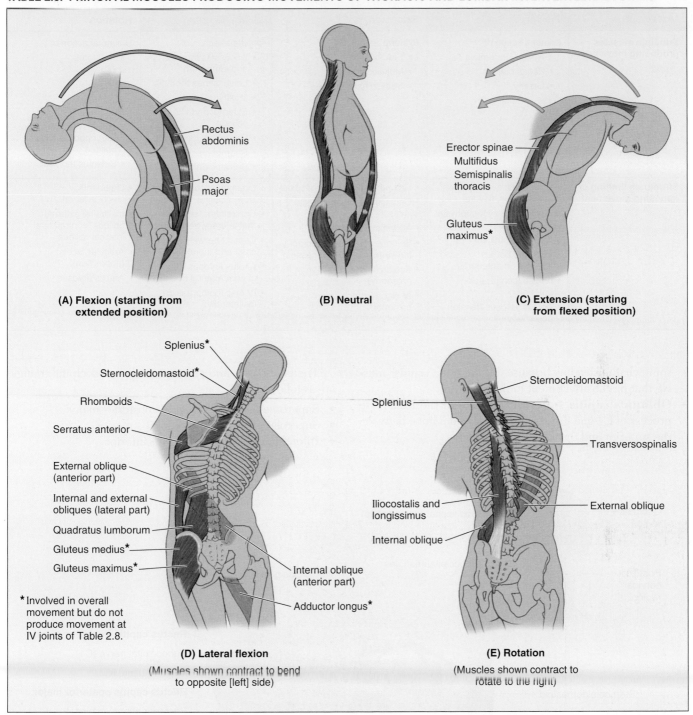

Rectus abdominis

Psoas major

(A) Flexion (starting from extended position)

(B) Neutral

Erector spinae
Multifidus
Semispinalis thoracis

Gluteus maximus*

(C) Extension (starting from flexed position)

Splenius*

Sternocleidomastoid*

Rhomboids

Serratus anterior

External oblique (anterior part)

Internal and external obliques (lateral part)

Quadratus lumborum

Gluteus medius*

Gluteus maximus*

Internal oblique (anterior part)

Adductor longus*

*Involved in overall movement but do not produce movement at IV joints of Table 2.8.

(D) Lateral flexion
(Muscles shown contract to bend to opposite [left] side)

Sternocleidomastoid

Splenius

Transversospinalis

Iliocostalis and longissimus

Internal oblique

External oblique

(E) Rotation
(Muscles shown contract to rotate to the right)

(continued)

TABLE 2.8. STRUCTURES AFFECTING MOVEMENTS OF THORACIC AND LUMBAR INTERVERTEBRAL JOINTS (continued)

Movement	Flexion	Extension	Lateral Bending	Rotation
Principle muscles producing movement	Bilateral action of • Rectus abdominis • Psoas major • Gravity	Bilateral action of • Erector spinae • Multifidus • Semispinalis thoracis	Unilateral action of • Iliocostalis thoracis and lumborum • Longissimus thoracis • Multifidus • External and internal oblique • Quadratus lumborum • Rhomboids • Serratus anterior	Unilateral action of • Rotatores • Multifidus • Iliocostalis • Longissimus • External oblique acting synchronously with opposite internal oblique • Splenius thoracis
Structures limiting or opposing movement	• Ligaments: supraspinous, interspinous, flavum • Capsules of zygapophysial (facet) joints • Extensor muscles • Vertebral bodies (apposition anteriorly) • IV disc (compression anteriorly) • Anulus fibrosus (tension posteriorly)	• Ligaments: anterior longitudinal • Capsules of zygapophysial joints • Abdominal muscles • Spinous processes (contact between adjacent processes) • Anulus fibrosus (tension anteriorly) • IV discs (compression posteriorly)	• Ligaments: contralateral side • Contralateral muscles that laterally bend trunk • Contact between iliac crest and thorax • Anulus fibrosus (tension of contralateral fibers) • IV disc (compression ipsilaterally)	• Ligaments: costovertebral • Ipsilateral external oblique, contralateral internal oblique • Articular facets (apposition) • Anulus fibrosus

somewhat misleading because it is the only "capitis" muscle that has no attachment to the cranium.

• **Obliquus capitis superior** arises from the transverse process of C1 and inserts into the occipital bone between the superior and the inferior nuchal lines.

The boundaries and contents of the suboccipital triangle are as follows:

• Superomedially, rectus capitis posterior major
• Superolaterally, obliquus capitis superior
• Inferolaterally, obliquus capitis inferior

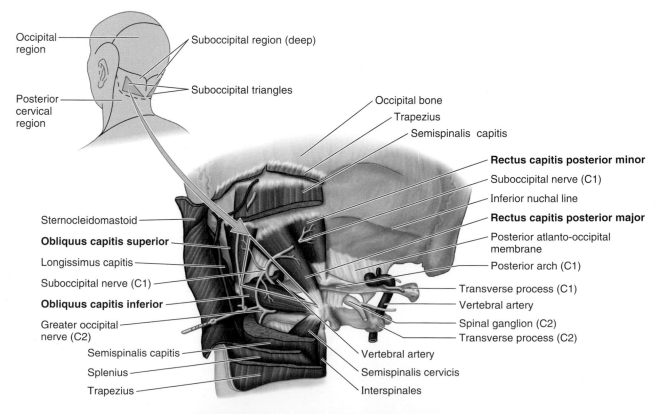

FIGURE 2.24. Suboccipital muscles and suboccipital triangle.

TABLE 2.9. PRINCIPAL MUSCLES PRODUCING MOVEMENT OF ATLANTO-OCCIPITAL JOINTS

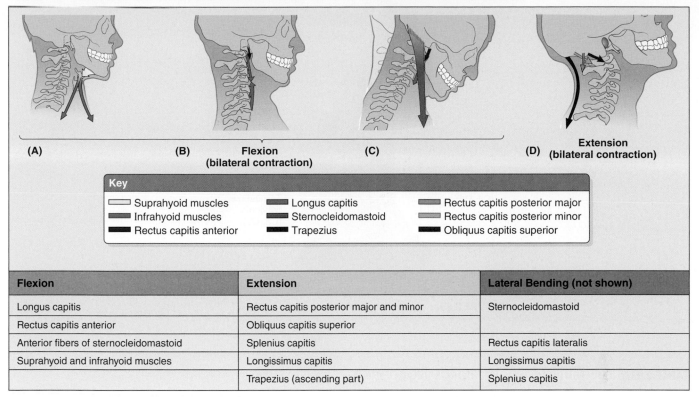

(A)　(B)　**Flexion
(bilateral contraction)**　(C)　(D)　**Extension
(bilateral contraction)**

Key

▢ Suprahyoid muscles	▬ Longus capitis	▬ Rectus capitis posterior major
▬ Infrahyoid muscles	▬ Sternocleidomastoid	▬ Rectus capitis posterior minor
▬ Rectus capitis anterior	▬ Trapezius	▬ Obliquus capitis superior

Flexion	Extension	Lateral Bending (not shown)
Longus capitis	Rectus capitis posterior major and minor	Sternocleidomastoid
Rectus capitis anterior	Obliquus capitis superior	
Anterior fibers of sternocleidomastoid	Splenius capitis	Rectus capitis lateralis
Suprahyoid and infrahyoid muscles	Longissimus capitis	Longissimus capitis
	Trapezius (ascending part)	Splenius capitis

- Floor, posterior atlanto-occipital membrane and posterior arch of C1
- Roof, semispinalis capitis
- Contents, *vertebral artery* and *suboccipital nerve* (C1)

The actions of the suboccipital group of muscles are to extend the head on C1 and rotate the head and the C1 on C2 vertebrae. The principal muscles producing movements of the craniovertebral joints are summarized in Tables 2.9 and 2.10. The motor innervation of the muscles and the cutaneous innervation of the posterior aspect of the head and neck are summarized in Figure 2.25 and Table 2.11.

**TABLE 2.10.　PRINCIPAL MUSCLES PRODUCING
MOVEMENT OF ATLANTO-AXIAL JOINTS**

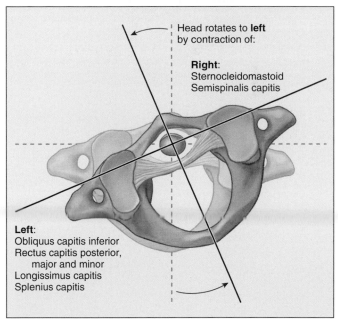

Head rotates to **left**
by contraction of:

Right:
Sternocleidomastoid
Semispinalis capitis

Left:
Obliquus capitis inferior
Rectus capitis posterior,
　major and minor
Longissimus capitis
Splenius capitis

Rotation is the specialized movement at these joints. Movement of one joint involves the other.

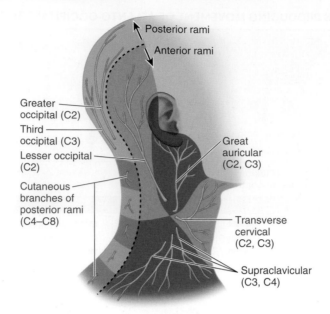

FIGURE 2.25. Sensory innervation of suboccipital region and head.

TABLE 2.11. NERVE SUPPLY OF POSTERIOR ASPECT OF HEAD AND NECK

Nerve	Origin	Course	Distribution
Suboccipital	Posterior ramus of C1 spinal nerve	Runs between cranium and C1 vertebra to reach suboccipital triangle	Muscles of suboccipital triangle
Greater occipital	Posterior ramus of C2 spinal nerve	Emerges inferior to obliquus capitis inferior and ascends to posterior scalp	Skin over neck and occipital bone
Lesser occipital	Anterior rami of spinal nerves C2–C3	Pass directly to skin	Skin of superior posterolateral neck and scalp posterior to ear
Posterior rami, nerves C3–C7	Posterior rami of spinal nerves C3–C7	Pass segmentally to muscles and skin	Intrinsic muscles of back and overlying skin adjacent to vertebral column

CLINICAL BOX

Back Sprains and Strains

Back sprain is an injury in which only ligamentous tissue, or the attachment of ligament to bone, is involved without dislocation or fracture. It results from excessively strong contractions related to movements of the vertebral column, such as excessive extension or rotation.

Back strain involves some degree of stretching or microscopic tearing of muscle fibers. The muscles usually involved are those producing movements of the lumbar IV joints, especially the erector spinae. If the weight is not properly balanced on the vertebral column, strain is exerted on the muscles. This is the most common cause of low back pain.

Back spasms can be the result of muscle or ligament injury, for example, after performing an activity or movement, often sudden, that puts excessive stress on the back, or other pathologies including a herniated/ruptured disc or arthritis (Fig. B2.13). Thus, a common cause of back spasm is heavy lifting. If the back musculature is weak, the likelihood of injury increases. Weak abdominal muscles will also contribute to injury as they also help support the back. As a protective mechanism, the back muscles go into spasm in response to inflammation following an injury. A spasm is a sudden involuntary contraction of one or more muscle groups. Spasms result in cramps, pain, and interference with function, producing involuntary movement and distortion of the vertebral column.

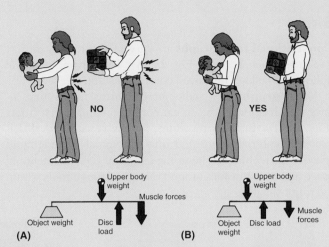

FIGURE B2.14. Load on IV discs created by proper and improper lifting techniques. A. Ergonomics of improper lifting technique. **B.** Ergonomics of proper lifting technique. In **A**, the body weight is a greater distance away from the disc center than in **B**. The load on the discs is dependent on the weight of the object, upper body weight, forces of back muscles, and their respective lever arms relative to the center of the disc. The lever balances below each figure demonstrate that smaller muscle forces and disc loads are present when the object is carried close to the body, that is, closer to the IV disc.

Using the back as a lever when lifting or holding heavy objects puts an enormous strain on the vertebral column and its ligaments and muscles. These strains can be minimized if the lifter crouches, holds the back as straight as possible, and uses the muscles of the buttocks and lower limbs to assist with the lifting. Loads should be carried as close to the trunk as possible (Fig. B2.14).

Straight Leg Test

The *straight leg test*, also called Lasègue test/sign, is performed to determine if a patient with low back pain has a herniated IV disc. The clinician passively flexes the patient's hip with the knee in full extension (Fig. B2.15). This maneuver will cause traction on the nerve roots forming the sciatic nerve, and in the case of a herniated disc will reproduce the pain.

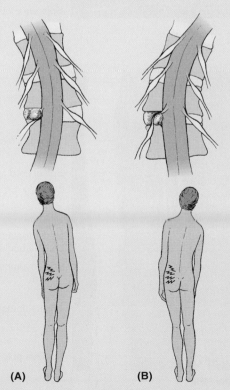

FIGURE B2.13. Muscle spasm following protrusion of an IV disc on the left side. Protrusions are shown passing lateral (**A**) and medial (**B**) to the nerve root. Leaning in a direction that compresses the nerve against the protrusion increases pain as shown; leaning in the opposite direction (not shown) reduces nerve compression, relieving pain.

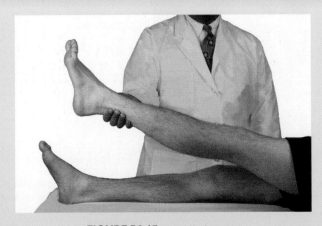

FIGURE B2.15. Straight leg test.

MEDICAL IMAGING

Back

Conventional radiographs are very good for high-contrast structures such as bone (Fig. 2.26A). The advent of digital radiography allows improved contrast resolution.

Myelography is a radiopaque contrast study that allows visualization of the spinal cord and spinal nerve roots (Fig. 2.26B). In this procedure, largely replaced by MRI, contrast material is injected into the spinal subarachnoid space. This technique shows the extent of the subarachnoid space and its extensions around the spinal nerve roots within the dural sheaths.

Computerized tomography (CT) differentiates between the white and the gray matter of the brain and spinal cord. It has also improved the radiological assessment of fractures of the vertebral column, particularly in determining the degree of compression of the spinal cord. The dense vertebrae attenuate much of the X-ray beam and therefore appear white on the scans (Figs. 2.26B and 2.27B). The IV discs have a higher density than the surrounding adipose tissue in the extradural space and the CSF in the subarachnoid space. Three-dimensional reconstruction of CT images is shown in Figure 2.27D.

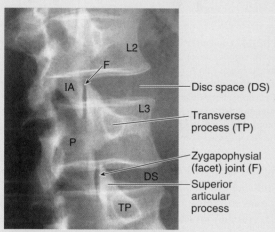

L2
F
IA
Disc space (DS)
L3
Transverse process (TP)
P
Zygapophysial (facet) joint (F)
DS
Superior articular process
TP

(A) Oblique radiograph: P, pedicle; IA, inferior articular process

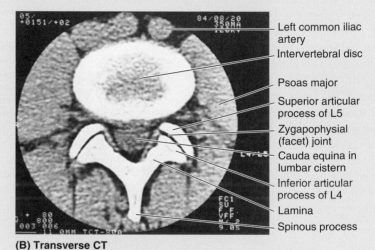

Left common iliac artery
Intervertebral disc
Psoas major
Superior articular process of L5
Zygapophysial (facet) joint
Cauda equina in lumbar cistern
Inferior articular process of L4
Lamina
Spinous process

(B) Transverse CT

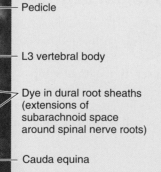

Pedicle
L3 vertebral body
Dye in dural root sheaths (extensions of subarachnoid space around spinal nerve roots)
Cauda equina
Lumbar cistern within dural sac

(C) Anteroposterior myelogram

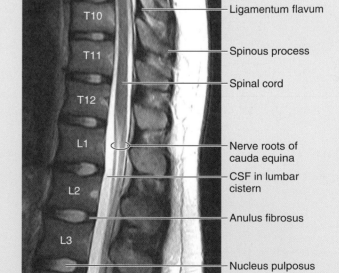

Dural sac
Ligamentum flavum
T10
T11
Spinous process
T12
Spinal cord
L1
Nerve roots of cauda equina
CSF in lumbar cistern
L2
Anulus fibrosus
L3
Nucleus pulposus

(D) Sagittal MRI

FIGURE 2.26. Imaging of the vertebral column. A. Oblique radiograph of lumbar spine. **B.** Transverse (axial) CT scan of L4–L5 IV disc. **C.** Myelogram of lumbar region. **D.** Sagittal MRI scan of vertebral column.

Magnetic resonance imaging (MRI), like CT, is a computer-assisted imaging procedure, but X-rays are not used as with CT. MRI produces extremely good images of the vertebral column, spinal cord, and CSF (Fig. 2.26D). MRI clearly demonstrates the components of IV discs and shows their relationship to the vertebral bodies and longitudinal ligaments. Herniations of the nucleus pulposus and its relationship to the spinal nerve roots also are well defined. MRI is the imaging procedure of choice for evaluating IV disc disorders.

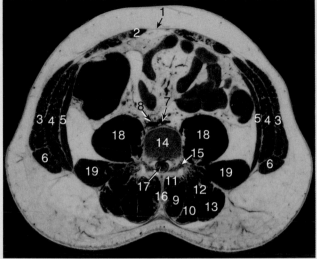

(A) Transverse anatomical section, inferior view

(B) Transverse CT

Key for A and B			
1 Linea alba	6 Latissimus dorsi	11 Multifidus	16 Spinous process
2 Rectus abdominis	7 Descending aorta	12 Rotatores	17 Cauda equina
3 External oblique	8 Inferior vena cava	13 Iliocostalis	18 Psoas major
4 Internal oblique	9 Spinalis	14 4th lumbar vertebra	19 Quadratus lumborum
5 Transversus abdominis	10 Longissimus	15 Transverse process	

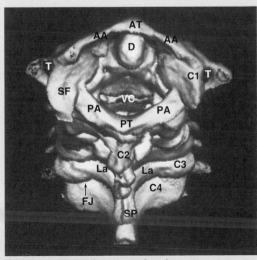

(C) Coronal MRI

(D) CT reconstruction, posterior view

Key for C and D			
AA	Anterior arch of C1	Lu Lungs	SP Spinous process
AT	Anterior tubercle of C1	MP Mastoid process	St Sternocleidomastoid
C1–T1	Vertebrae	PA Posterior arch of C1	T Foramen transversarium
D	Dens (odontoid process) of C2	PT Posterior tubercle of C1	VA Vertebral artery
FJ	Zygapophysial (facet) joint	Sc Scalenes	VC Vertebral canal
La	Lamina	SF Superior articular facet of C1	

FIGURE 2.27. CT imaging. A. Transverse section of cadaveric specimen at L4 vertebra. **B.** Transverse (axial) CT scan at L4 vertebra. **C.** Coronal MRI scan of cervical region. **D.** Three-dimensional reconstructed CT image of cervical spine.

thePoint® Go to http://thePoint.lww.com for helpful study tools, including USMLE-style questions, case studies, images, and more!

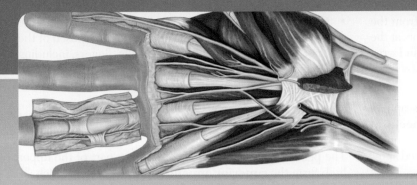

Upper Limb

3

CLINICAL BOX KEY

Anatomical Variations

Diagnostic Procedures

Life Cycle

Surgical Procedures

Trauma

Pathology

The upper limb is characterized by its mobility and ability to grasp, strike, and perform fine motor skills (*manipulation*). These characteristics are especially marked in the hand. Efficiency of hand function results in a large part from the ability to place it in the proper position by movements at the scapulothoracic, glenohumeral, elbow, radio-ulnar, and wrist joints. The upper limb consists of four segments, which are further subdivided into regions (Figs. 3.1 and 3.2):

- **Shoulder**, which includes the deltoid, pectoral, scapular, and lateral part of lateral cervical region. The **pectoral (shoulder) girdle** is a bony ring, incomplete posteriorly, formed by the scapulae and clavicles and completed anteriorly by the manubrium of the sternum.
- **Arm** (L. *brachium*) is between the shoulder and the elbow and is centered around the humerus. It consists of the anterior and posterior regions of the arm.

- **Forearm** (L. *antebrachium*) is between the elbow and the wrist and contains the ulna and radius. It consists of the anterior and posterior regions of the forearm.
- **Hand** (L. *manus*) is distal to the forearm and contains the carpus, metacarpus, and phalanges. It is composed of the wrist, palm, dorsum of hand, and digits (fingers, including the opposable thumb) and is richly supplied with sensory endings for touch, pain, and temperature.

BONES OF UPPER LIMB

The pectoral girdle and bones of the free part of the upper limb form the **superior appendicular skeleton**, which articulates with the axial skeleton only at the sternoclavicular joint, allowing great mobility (Fig. 3.3). The pectoral girdle is supported, stabilized, and propelled by axio-appendicular

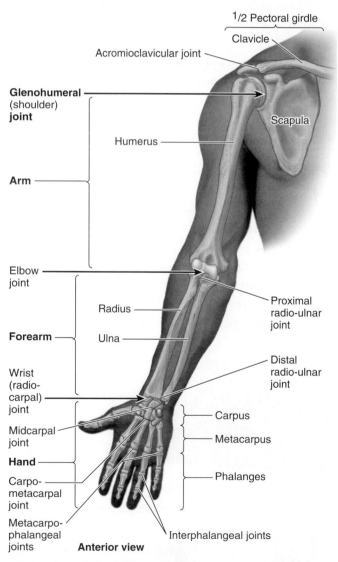

FIGURE 3.1. **Segments and bones of upper limb.** The upper limb is divided into four main segments: shoulder, arm, forearm, and hand.

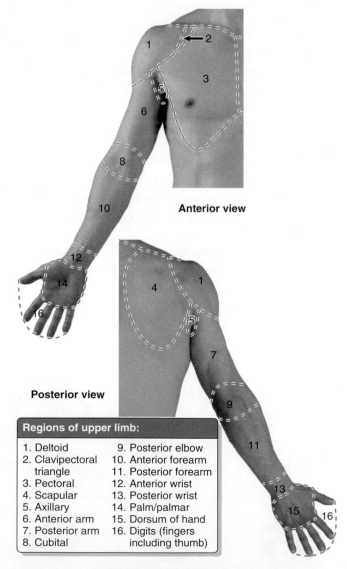

Regions of upper limb:

1. Deltoid	9. Posterior elbow
2. Clavipectoral triangle	10. Anterior forearm
3. Pectoral	11. Posterior forearm
4. Scapular	12. Anterior wrist
5. Axillary	13. Posterior wrist
6. Anterior arm	14. Palm/palmar
7. Posterior arm	15. Dorsum of hand
8. Cubital	16. Digits (fingers including thumb)

FIGURE 3.2. **Regions of upper limb.**

muscles, which attach to the ribs, sternum, and vertebrae of the *axial skeleton*.

Clavicle

The **clavicle** (collar bone) connects the upper limb to the trunk. Its **sternal end** articulates with the **manubrium of the sternum** at the *sternoclavicular* (SC) *joint*. Its **acromial end** articulates with the *acromion of the scapula* at the *acromioclavicular* (AC) *joint* (Figs. 3.3 and 3.4). The medial two thirds of the shaft of the clavicle are convex anteriorly, whereas the lateral third is flattened and concave anteriorly. These curvatures increase the resilience of the clavicle and give it the appearance of an elongated capital S. The clavicle

- Serves as a pivoting strut (rigid support) from which the scapula and free limb are suspended, keeping the free limb lateral to the thorax so that the arm has maximum freedom of motion; fixing the strut in position, especially after its elevation, enables elevation of the ribs for deep inspiration
- Forms one of the boundaries of the *cervico-axillary canal* (passageway between neck and arm), affording protection to the neurovascular bundle supplying the upper limb
- Transmits shocks (traumatic impacts) from the upper limb to the axial skeleton

Although designated as a long bone, the clavicle has no medullary (marrow) cavity. It consists of spongy (trabecular) bone with a shell of compact bone.

Scapula

The **scapula** (shoulder blade) is a triangular flat bone that lies on the posterolateral aspect of the thorax, overlying the 2nd through 7th ribs (Figs. 3.3 and 3.4). The convex **posterior surface** of the scapula is unevenly divided by the **spine of the scapula** into a small **supraspinous fossa** and a much larger **infraspinous fossa**. The concave **costal surface** of the scapula has a large **subscapular fossa**. The triangular **body of the scapula** is thin and translucent superior and inferior to the scapular spine.

The scapula has **medial** (axillary), **lateral** (vertebral), and **superior borders** and **superior** and **inferior angles**. The lateral border of scapula is the thickest part of the bone, which, superiorly, includes the **head of the scapula** where the glenoid cavity is located. The **neck of the scapula** is just medial to the head (Fig. 3.4B). The superior border of the scapula is marked near the junction of its medial two thirds and lateral third by the **suprascapular notch**.

The spine of the scapula continues laterally, expanding to form the **acromion**, the subcutaneous point of the shoulder that articulates with the acromial end of the clavicle (Fig. 3.3C).

(Continued on page 96)

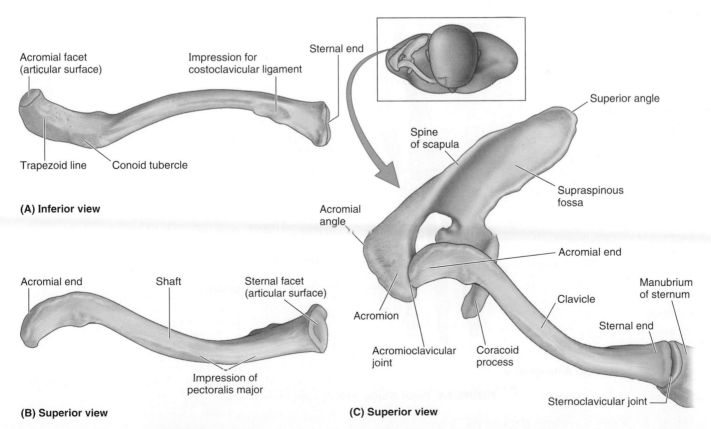

FIGURE 3.3. Clavicle. A. Inferior surface. **B.** Superior surface. **C.** Articulations of clavicle.

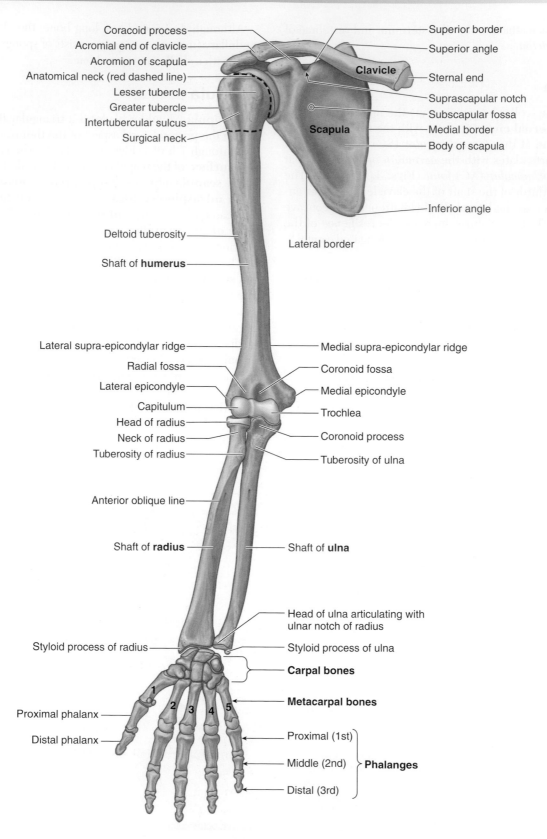

Coracoid process
Acromial end of clavicle
Acromion of scapula
Anatomical neck (red dashed line)
Lesser tubercle
Greater tubercle
Intertubercular sulcus
Surgical neck

Superior border
Superior angle
Clavicle
Sternal end
Suprascapular notch
Subscapular fossa
Medial border
Body of scapula

Scapula

Inferior angle

Deltoid tuberosity

Shaft of **humerus**

Lateral border

Lateral supra-epicondylar ridge
Radial fossa
Lateral epicondyle
Capitulum
Head of radius
Neck of radius
Tuberosity of radius

Medial supra-epicondylar ridge
Coronoid fossa
Medial epicondyle
Trochlea
Coronoid process
Tuberosity of ulna

Anterior oblique line

Shaft of **radius**

Shaft of **ulna**

Head of ulna articulating with
ulnar notch of radius

Styloid process of radius

Styloid process of ulna

Carpal bones

1 2 3 4 5

Metacarpal bones

Proximal phalanx

Distal phalanx

Proximal (1st)
Middle (2nd)   **Phalanges**
Distal (3rd)

(A) Anterior view

FIGURE 3.4. **Bones of upper limb. A.** Anterior view. *(continued)*

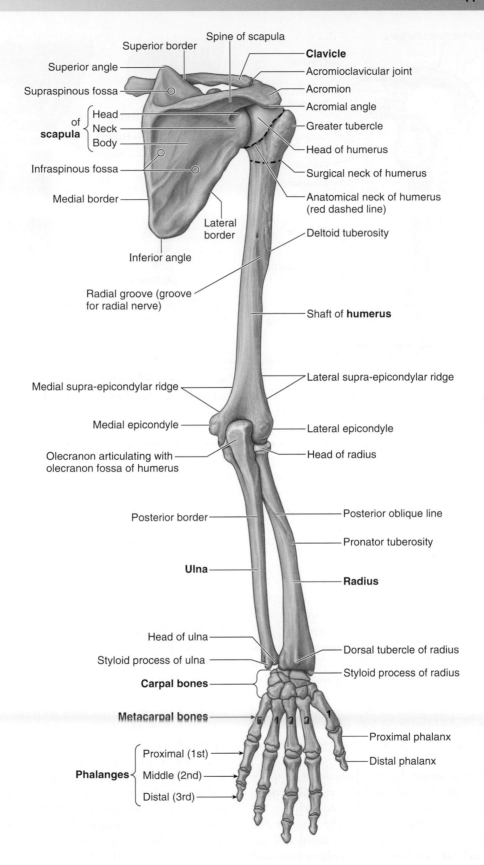

(B) Posterior view

FIGURE 3.4. **Bones of upper limb.** *(continued)* B. Posterior view.

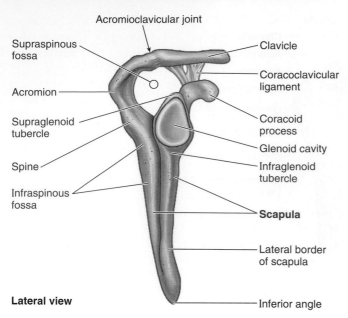

FIGURE 3.5. Right scapula.

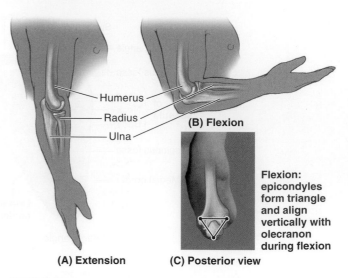

FIGURE 3.6. Bones of right elbow during extension and flexion of elbow joint. **A.** Elbow extended. **B.** Elbow flexed. **C.** Triangulation of epicondyles and olecranon in full flexion.

Superolaterally, the lateral surface of the head of the scapula has a **glenoid cavity**, which articulates with the head of the humerus at the glenohumeral (shoulder) joint (Fig. 3.5). The glenoid (G. socket) cavity is a shallow, concave, oval fossa, which is directed anterolaterally and slightly superiorly and is considerably smaller than the head of the humerus for which it serves as a socket. The beak-like **coracoid process** is superior to the glenoid cavity and projects anterolaterally.

Humerus

The **humerus** (arm bone), the largest bone in the upper limb, articulates with the scapula at the glenohumeral joint and the radius and ulna at the elbow joint (Fig. 3.4). Proximally, the ball-shaped **head of the humerus** articulates with the glenoid cavity of the scapula. The **intertubercular sulcus** (bicipital groove) of the proximal end of the humerus separates the lesser tubercle from the greater tubercle. Just distal to the humeral head, the **anatomical neck of the humerus** separates the head from the tubercles. Distal to the tubercles is the narrow **surgical neck of the humerus**.

The **shaft of the humerus** has two prominent features: the **deltoid tuberosity** laterally and the **radial groove** (groove for radial nerve, spiral groove) posteriorly for the radial nerve and profunda brachii artery. The inferior end of the humeral shaft widens as the sharp medial and lateral **supra-epicondylar** (supracondylar) **ridges** form and then end distally in the prominent **medial epicondyle** and **lateral epicondyle**.

The distal end of the humerus, including the trochlea, capitulum, olecranon, coronoid, and radial fossae, makes up the **condyle of the humerus**. It has two articular surfaces: a lateral **capitulum** (L. little head) for articulation with the head of the radius and a medial **trochlea** (L. pulley) for articulation with the trochlear notch of the ulna. Superior to the trochlea anteriorly is the **coronoid fossa**, which receives the coronoid process of the ulna during full flexion of the elbow (Figs. 3.4A and 3.6). Posteriorly, the **olecranon fossa** accommodates the olecranon of the ulna during extension of the elbow. Superior to the capitulum anteriorly, the shallow **radial fossa** accommodates the edge of the head of the radius when the elbow is fully flexed.

Ulna and Radius

The **ulna**, the stabilizing bone of the forearm, is the medial and longer of the two forearm bones (Fig. 3.4). Its proximal end has two prominent projections—the **olecranon** posteriorly and the **coronoid process** anteriorly; they form the walls of the **trochlear notch**. The trochlear notch of the ulna articulates with the trochlea of the humerus. Inferior to the coronoid process is the **tuberosity of the ulna**. On the lateral side of the coronoid process is a smooth, rounded concavity, the **radial notch**, which articulates with the head of radius (Fig. 3.7A). Distal to the radial notch is a prominent ridge, the **supinator crest**, and between it and the distal part of the coronoid process is a concavity, the **supinator fossa**. Proximally, the **shaft of the ulna** is thick, but it tapers, diminishing in diameter distally. At its narrow distal end is the rounded **head of ulna** with the small, conical **ulnar styloid process** (Fig. 3.4). The ulna does not articulate directly with the carpal bones. It is separated from the carpals by a fibrocartilaginous articular disc.

Olecranon — Trochlear notch
Ulna — Coronoid process
— Radial notch
— Tuberosity of ulna
Supinator crest — Supinator fossa
Shaft (body) — Interosseous border

(A) Lateral view, proximal end of ulna

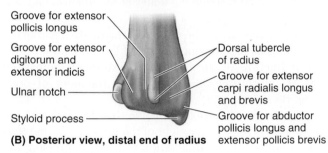

Groove for extensor pollicis longus
Groove for extensor digitorum and extensor indicis — Dorsal tubercle of radius
Ulnar notch — Groove for extensor carpi radialis longus and brevis
Styloid process — Groove for abductor pollicis longus and extensor pollicis brevis

(B) Posterior view, distal end of radius

FIGURE 3.7. Ulna and radius. A. Proximal end of ulna. **B.** Distal end of radius.

The **radius** is the lateral and shorter of the two forearm bones. Its proximal end consists of a cylindrical head, a short neck, and a projection from the medial surface, the **radial tuberosity** (Fig. 3.4A). Proximally, the smooth superior aspect of the **head of the radius** is concave for articulation with the capitulum of humerus. The head also articulates medially with the radial notch of ulna (Fig. 3.7A). The **neck of the radius** is the narrow part between the head and the radial tuberosity. The radial tuberosity demarcates the proximal end (head and neck) from the shaft. The **shaft of the radius** has a lateral convexity and gradually enlarges as it passes distally. The medial aspect of the distal end of the radius forms a concavity, the **ulnar notch**, which accommodates the head of the ulna (Fig. 3.7B). Its lateral aspect terminates distally as the **radial styloid process**. The radial styloid process is larger than the ulnar styloid process and extends farther distally. This relationship is clinically important when the ulna and/or radius is fractured (see Fig. B3.3). The **dorsal tubercle of the radius** lies between two of the shallow grooves for passage of the tendons of forearm muscles and serves as a trochlea (pulley) for the tendon of the long extensor of the thumb.

Bones of Hand

The **wrist**, or **carpus**, is composed of eight **carpal bones** (carpals) arranged in proximal and distal rows of four (Figs. 3.8 and 3.9). These small bones give flexibility to the wrist. The carpus is markedly convex from side to side posteriorly and concave anteriorly. Augmenting movement at the wrist, the two rows of carpals glide on each other; each carpal bone also glides on those adjacent to it. The proximal surfaces of the proximal row of carpals articulate with the inferior end of the radius and the articular disc of the wrist joint. The distal surfaces of these bones articulate with the distal row of carpals.

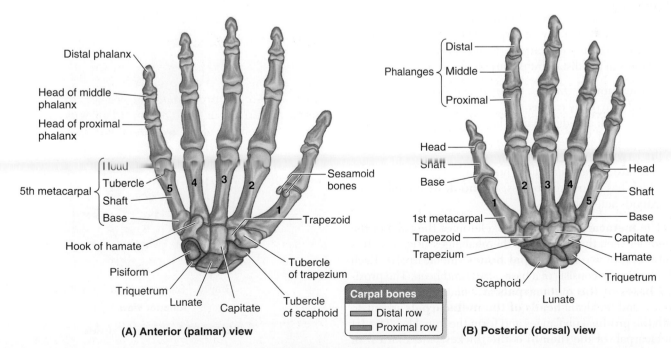

Distal phalanx
Head of middle phalanx
Head of proximal phalanx
5th metacarpal { Head, Tubercle, Shaft, Base }
Hook of hamate
Pisiform
Triquetrum
Lunate
Capitate
Sesamoid bones
Trapezoid
Tubercle of trapezium
Tubercle of scaphoid

(A) Anterior (palmar) view

Phalanges { Distal, Middle, Proximal }
Head
Shaft
Base
Head
Shaft
Base
1st metacarpal
Trapezoid
Trapezium
Scaphoid
Lunate
Capitate
Hamate
Triquetrum

Carpal bones
▭ Distal row
▭ Proximal row

(B) Posterior (dorsal) view

FIGURE 3.8. Bones of hand.

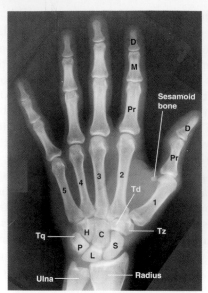

FIGURE 3.9. Radiograph of right hand.

From lateral to medial, the four bones in the proximal row of carpals are as follows:

- **Scaphoid** (G. *skaphé*, skiff, boat): a boat-shaped bone that has a prominent **scaphoid tubercle**
- **Lunate** (L. *luna*, moon): a moon-shaped bone that is broader anteriorly than posteriorly
- **Triquetrum** (L. *triquetrus*, three-cornered): a pyramidal bone on the medial aspect of the carpus
- **Pisiform** (L. *pisum*, pea): a small, pea-shaped bone that lies on the palmar surface of the triquetrum

The proximal surfaces of the distal row of carpals articulate with the proximal row of carpals, and their distal surfaces articulate with the metacarpals. From lateral to medial, the four bones in the distal row of carpals are as follows:

- **Trapezium** (G. *trapeze*, table): a four-sided bone on the lateral side of the carpus
- **Trapezoid:** a wedge-shaped bone
- **Capitate** (L. *caput*, head): the head-shaped bone that is the largest bone in the carpus
- **Hamate** (L. *hamulus*, little hook): a wedge-shaped bone, which has a hooked process, the **hook of hamate**, that extends anteriorly

The **metacarpus** forms the skeleton of the palm of the hand between the carpus and the phalanges (Fig. 3.9). It is composed of five **metacarpal bones** (metacarpals). Each of these bones consists of a base, shaft, and head. The proximal **bases of the metacarpals** articulate with the carpal bones, and the distal **heads of the metacarpals** articulate with the proximal phalanges and form the knuckles. The 1st metacarpal (of the thumb) is the thickest and shortest of these bones.

Each digit has three **phalanges** (**proximal**, **middle**, and **distal**) except for the first (thumb), which has only two (**proximal** and **distal**). Each phalanx has a **base** proximally, a **shaft** (**body**), and a **head** distally. The distal phalanges are flattened and expanded at their distal ends, which underlie the nail beds.

CLINICAL BOX

Fracture of Clavicle

The clavicle is commonly fractured, often by an indirect force transmitted from an outstretched hand through the bones of the forearm and arm to the shoulder during a fall. A fracture may also result from a fall directly on the shoulder. The weakest part of the clavicle is at the junction of its middle and lateral thirds. After fracture of the clavicle, the sternocleidomastoid (SCM) muscle elevates the medial fragment of bone (Fig. B3.1).

The trapezius muscle is unable to hold up the lateral fragment owing to the weight of the upper limb, and thus the shoulder drops. In addition to being depressed, the

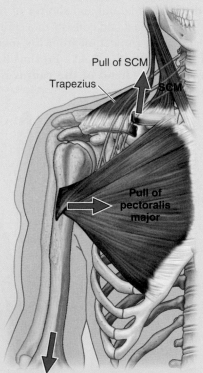

FIGURE B3.1. Fracture of clavicle.

(*Continued on next page*)

lateral fragment of the clavicle may be pulled medially by muscles that normally adduct the arm at the shoulder joint, such as the pectoralis major. Overriding of the bone fragments shortens the clavicle.

Ossification of Clavicle

 The clavicle is the first long bone to ossify (via *intramembranous ossification*), beginning during the fifth and sixth embryonic weeks from medial and lateral primary ossification centers that are close together in the shaft of the clavicle. The ends of the clavicle later pass through a cartilaginous phase (*endochondral ossification*); the cartilages form growth zones similar to those of other long bones.

A secondary ossification center appears at the sternal end and forms a scale-like epiphysis that begins to fuse with the shaft (diaphysis) between 18 and 25 years of age; it is completely fused to it between 25 and 31 years of age. This is the last of the epiphyses of long bones to fuse. An even smaller scale-like epiphysis may be present at the acromial end of the clavicle; it must not be mistaken for a fracture.

Sometimes, fusion of the two ossification centers of the clavicle fails to occur; as a result, a bony defect forms between the lateral and the medial thirds of the clavicle. Awareness of this possible birth defect should prevent diagnosis of a fracture in an otherwise normal clavicle. When doubt exists, both clavicles are radiographed because this defect is usually bilateral.

Fracture of Scapula

Fracture of the scapula is usually the result of severe trauma, as occurs in pedestrian–vehicle accidents. Usually, there are also fractured ribs. Most fractures require little treatment because the scapula is covered on both sides by muscles. Most fractures involve the protruding subcutaneous acromion.

Fractures of Humerus

Fractures of the surgical neck of the humerus are especially common in elderly people with *osteoporosis* (Fig. B3.2A). Even a low-energy fall on the hand, with the force being transmitted up the forearm bones of the extended limb, may result in a fracture. *Transverse fractures of the shaft of humerus* frequently result from a direct blow to the arm. Fracture of the distal part of the humerus, near the supra-epicondylar ridges, is a *supra-epicondylar* (supracondylar) *fracture*. Because nerves are in contact with the humerus, they may be injured when the associated part of the humerus is fractured: surgical neck, axillary nerve; radial groove, radial nerve; distal humerus, median nerve; and medial epicondyle, ulnar nerve (Fig. B3.2B).

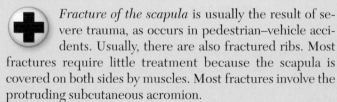

Comminuted   **Transverse**

Spiral   **Greenstick**

Compound   **Oblique**

(A) Types of fracture

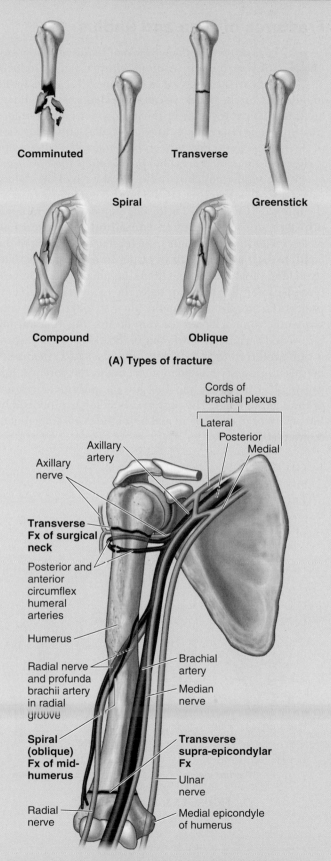

(B) Neurovascular structures related to humeral fractures (Fx)

FIGURE B3.2. Fractures of humerus.

(*Continued on next page*)

Fractures of Ulna and Radius

Fractures of both the ulna and radius are the result of severe injury. A direct injury usually produces transverse fractures at the same level, often in the middle third of the bones. Because the shafts of these bones are firmly bound together by the interosseous membrane, a fracture of one bone is likely to be associated with dislocation of the nearest joint. *Fracture of the distal end or the radius is the most common fracture in people older than 50 years of age.* A complete fracture of the distal 2 cm of the radius, called a **Colles fracture**, is the most common fracture of the forearm (Fig. B3.3). The distal fragment of the radius is displaced dorsally and often **comminuted** (broken into pieces). The fracture results from forced dorsiflexion of the hand, usually as the result of trying to ease a fall by out-stretching the upper limb. Often, the ulnar styloid process is **avulsed** (broken off). Normally, the radial styloid process projects farther distally than the ulnar styloid process; consequently, when a Colles fracture occurs, this relationship is reversed because of shortening of the radius. This fracture is often referred to as a *dinner fork (silver fork) deformity* because a posterior angulation occurs in the forearm just proximal to the wrist and the normal anterior curvature of the relaxed hand. The posterior bending is produced by the posterior displacement and tilt of the distal fragment of the radius.

Fractures of Hand

Fractures of the hand may involve one or more of the 27 bones of the wrist, palm, and digits. *Fracture of the scaphoid* often results from a fall on the palm with the hand abducted (Fig. B3.4). The fracture occurs across the narrow part ("waist") of the scaphoid. Pain occurs primarily on the lateral side of the wrist, especially during dorsiflexion and abduction of the hand. Initial radiographs of the wrist may not reveal a fracture, but radiographs taken 10–14 days later may reveal a fracture because bone resorption has occurred. Owing to the poor blood supply to the proximal part of the scaphoid, union of the fractured parts may take several months. *Avascular necrosis of the proximal fragment of the scaphoid* (pathological death of bone resulting from poor blood supply) may occur and produce *degenerative joint disease of the wrist*.

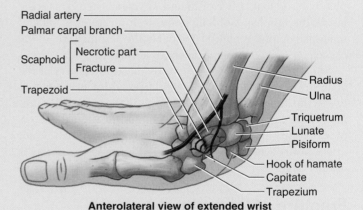

Radial artery
Palmar carpal branch
Scaphoid — Necrotic part
Scaphoid — Fracture
Trapezoid
Radius
Ulna
Triquetrum
Lunate
Pisiform
Hook of hamate
Capitate
Trapezium

Anterolateral view of extended wrist

FIGURE B3.4. Fracture of scaphoid.

Fracture of the hamate may result in nonunion of the fractured bony parts because of the traction produced by the attached muscles. Because the ulnar nerve is close to the hook of the hamate, the nerve may be injured by this fracture, causing decreased grip strength of the hand. The ulnar artery may also be damaged when the hamate is fractured.

Severe crushing injuries of the hand may produce multiple metacarpal fractures, resulting in instability of the hand. Similar injuries of the distal phalanges are common (e.g., when a finger is caught in a car door). A *fracture of a distal phalanx* is usually comminuted, and a painful hematoma (collection of blood) develops. *Fractures of the proximal and middle phalanges* are usually the result of crushing or hyperextension injuries.

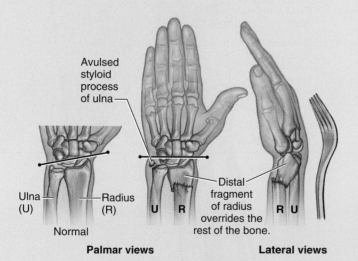

Avulsed styloid process of ulna

Ulna (U)   Radius (R)

Normal

Distal fragment of radius overrides the rest of the bone.

U R   R U

Palmar views   **Lateral views**

FIGURE B3.3. Colles fracture.

SURFACE ANATOMY

Upper Limb Bones

Most bones of the upper limb offer a palpable segment or surface, enabling the skilled examiner to discern abnormalities owing to trauma or malformation (Fig. SA3.1A). The clavicle is subcutaneous and can be palpated throughout its length. Its sternal end projects superior to the manubrium of the sternum. Between the elevated sternal ends of the clavicles is the **jugular notch** (suprasternal notch). The **acromial end of the clavicle** often rises higher than the acromion, forming a palpable elevation at the acromioclavicular joint.

The acromial end can be palpated 2–3 cm medial to the lateral border of the acromion, particularly when the arm is alternately flexed and extended (Fig. SA3.1A).

The **coracoid process of scapula** can be felt deeply at the lateral end of the clavicle in the clavipectoral (deltopectoral) triangle (Fig. SA3.1B). The **acromion of the scapula** is felt easily and is often visible. The lateral and posterior borders of the acromion meet to form the **acromial angle** (Fig. SA3.1A). Inferior to the acromion, the *deltoid muscle* forms the rounded curve of the shoulder.

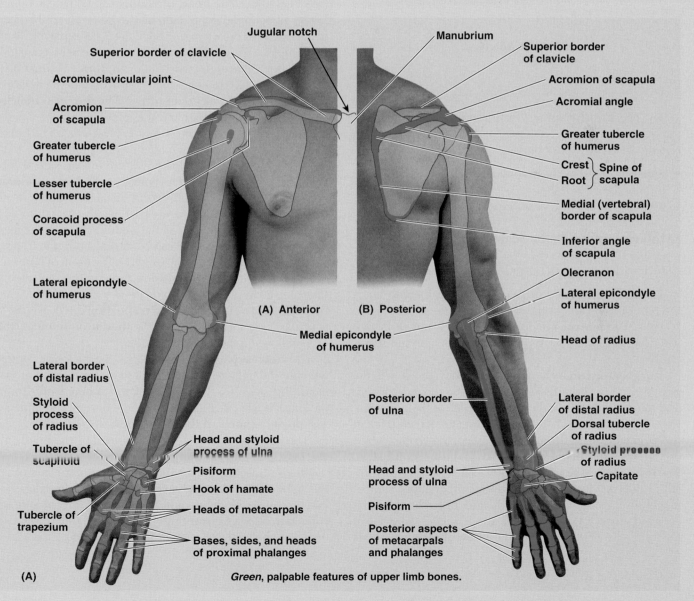

Green, palpable features of upper limb bones.

FIGURE SA3.1A. Surface projection and palpable features of bones of upper limb. *(continued)*

(Continued on next page)

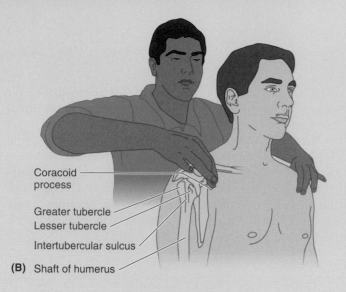

Coracoid
process

Greater tubercle
Lesser tubercle

Intertubercular sulcus

(B) Shaft of humerus

FIGURE SA3.1B.

The **crest of the spine of the scapula** is subcutaneous throughout and can be easily palpated. When the upper limb is in the anatomical position, the

- Superior angle of the scapula (not palpable) lies at the level of the T2 vertebra.
- Medial end of the root of the scapular spine is opposite the spinous process of the T3 vertebra.
- **Inferior angle of the scapula** (easily felt and often visible) lies at the level of the T7 vertebra, near the inferior border of the 7th rib and 7th intercostal space.

The **medial border of scapula** is palpable inferior to the root of the spine of the scapula as it crosses the 3rd to 7th ribs. The **lateral border of scapula** is not easily palpated because it is covered by the teres major and minor muscles.

The **greater tubercle of humerus** may be felt with the person's arm by the side on deep palpation through the deltoid muscle, inferior to the lateral border of the acromion. In this position, the tubercle is the most lateral bony point of the shoulder. When the arm is abducted, the greater tubercle is pulled beneath the acromion and is no longer palpable. The **lesser tubercle of the humerus** may be felt with difficulty by deep palpation through the anterior deltoid, approximately 1 cm laterally and slightly inferior to the tip of the coracoid process. Rotation of the arm facilitates palpation of this tubercle. The location of the **intertubercular sulcus**

or groove, between the greater and the lesser tubercles, is identifiable during flexion and extension of the elbow joint by palpating in an upward direction along the tendon of the long head of the biceps brachii as it moves through the intertubercular sulcus. The **shaft of humerus** may be felt with varying distinctness through the muscles surrounding it. The **medial and lateral epicondyles of the humerus** are palpated on the medial and lateral aspects of the elbow region.

The **olecranon** and **posterior border of the ulna** can be palpated easily. When the elbow joint is extended, observe that the tip of the olecranon and the humeral epicondyles lie in a straight line. When the elbow is flexed, the olecranon forms the apex of an approximately equilateral triangle, of which the epicondyles form the angles at its base (see Fig. 3.6C). The **head of radius** can be palpated and felt to rotate in the depression on the posterolateral aspect of the extended elbow, just distal to the lateral epicondyle of the humerus. The **radial styloid process** can be palpated on the lateral side of the wrist in the **anatomical snuff box** (see Fig. SA3.4C); it is larger and approximately 1 cm more distal than the ulnar styloid process. The **dorsal tubercle of radius** is easily felt around the middle of the dorsal aspect of the distal end of the radius (Fig. SA3.1C). The **head of ulna** forms a rounded subcutaneous prominence that can be easily seen and palpated on the medial side of the dorsal aspect of the wrist. The pointed subcutaneous **ulnar styloid process** may be felt slightly distal to the ulnar head when the hand is supinated.

The **pisiform** can be felt on the anterior aspect of the medial border of the wrist and can be moved from side to side when the hand is relaxed (Fig. SA3.1D). The **hook of hamate** can be palpated on deep pressure over the medial side of the palm, about 2 cm distal and lateral to the pisiform. The **tubercles of the scaphoid and trapezium** can be palpated at the base and medial aspect of the **thenar eminence** (ball of thumb) when the hand is extended.

The **metacarpals**, although overlain by the long extensor tendons of the digits, can be palpated on the dorsum of the hand (Fig. SA3.1C). The **heads of the metacarpals** form the knuckles; the 3rd metacarpal head is the most prominent. The **dorsal aspects of the phalanges** can be palpated easily. The knuckles of the fingers are formed by the **heads of the proximal and middle phalanges**.

When measuring upper limb length, or segments of it, the acromial angle, lateral epicondyle of the humerus, styloid process of the radius, and tip of the 3rd finger are most commonly used as measuring points, with the limb relaxed (dangling) but with the palm directed anteriorly.

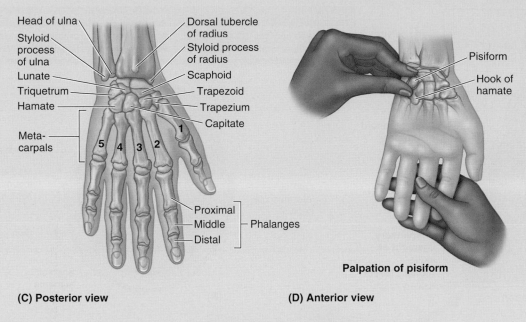

Head of ulna

Styloid process of ulna

Lunate

Triquetrum

Hamate

Meta-carpals

5 4 3 2

Dorsal tubercle of radius

Styloid process of radius

Scaphoid

Trapezoid

Trapezium

Capitate

1

Proximal
Middle — Phalanges
Distal

(C) Posterior view

Pisiform

Hook of hamate

Palpation of pisiform

(D) Anterior view

FIGURE SA3.1C AND D.

FASCIA, VESSELS, AND NERVES OF UPPER LIMB

Subcutaneous Tissue and Fascia

Deep to the skin is subcutaneous tissue (superficial fascia) containing fat and deep fascia surrounding the muscles. If no structure (muscle or tendon, for example) intervenes between the skin and the bone, the deep fascia usually attaches to bone.

The **pectoral fascia** invests the pectoralis major and is continuous inferiorly with the fascia of the anterior abdominal wall. The pectoral fascia leaves the lateral border of the pectoralis major and becomes the **axillary fascia** (Fig. 3.10A,B), which forms the floor of the axilla. Deep to the pectoral fascia and the pectoralis major, another fascial layer, the **clavipectoral fascia**, descends from the clavicle, enclosing the subclavius and then the pectoralis minor, becoming continuous inferiorly with the axillary fascia. The part of the clavipectoral fascia between the pectoralis minor and the subclavius, the **costocoracoid membrane**, is pierced by the lateral pectoral nerve, which primarily supplies the pectoralis major. The part of the clavipectoral fascia inferior to the pectoralis minor, the **suspensory ligament of axilla** (Fig. 3.10A), supports the axillary fascia and pulls it and the skin inferior to it upward during abduction of the arm, forming the axillary fossa.

The scapulohumeral muscles that cover the scapula and form the bulk of the shoulder are also ensheathed by deep fascia. The **deltoid fascia** invests the deltoid and is continuous with the pectoral fascia anteriorly and the dense infraspinous fascia posteriorly (Fig. 3.10A,B). The muscles that cover the anterior and posterior surfaces of the scapula are covered superficially by strong and opaque deep fascia, which is attached to the margins of the scapula. This arrangement creates osseofibrous *subscapular*, *supraspinous*, and *infraspinous compartments*.

The **brachial fascia**, a sheath of deep fascia, encloses the arm like a snug sleeve; it is continuous superiorly with the deltoid, pectoral, axillary, and infraspinous fasciae. The brachial fascia is attached inferiorly to the epicondyles of the humerus and the olecranon of the ulna and is continuous with the antebrachial fascia, the deep fascia of the forearm. Two intermuscular septa, the **medial** and **lateral intermuscular septa**, extend from the deep surface of the brachial fascia and attach to the central shaft and medial and lateral supraepicondylar ridges of the humerus. These septa divide the arm into **anterior (flexor)** and **posterior (extensor) fascial compartments**, each of which contains muscles serving similar functions and sharing common innervation (Fig. 3.10B).

In the forearm, similar fascial compartments are surrounded by the **antebrachial fascia** and separated by the *interosseous membrane* connecting the radius and ulna (Fig. 3.10C). The antebrachial fascia thickens posteriorly over the distal ends of the radius and ulna to form a transverse band, the **extensor retinaculum**, which holds the extensor tendons in position (Fig. 3.10D). The antebrachial fascia also forms an anterior thickening, which is continuous with the extensor retinaculum but is officially unnamed; some authors identify it as the *palmar carpal ligament*. Immediately distal, but at a deeper level to the latter, the

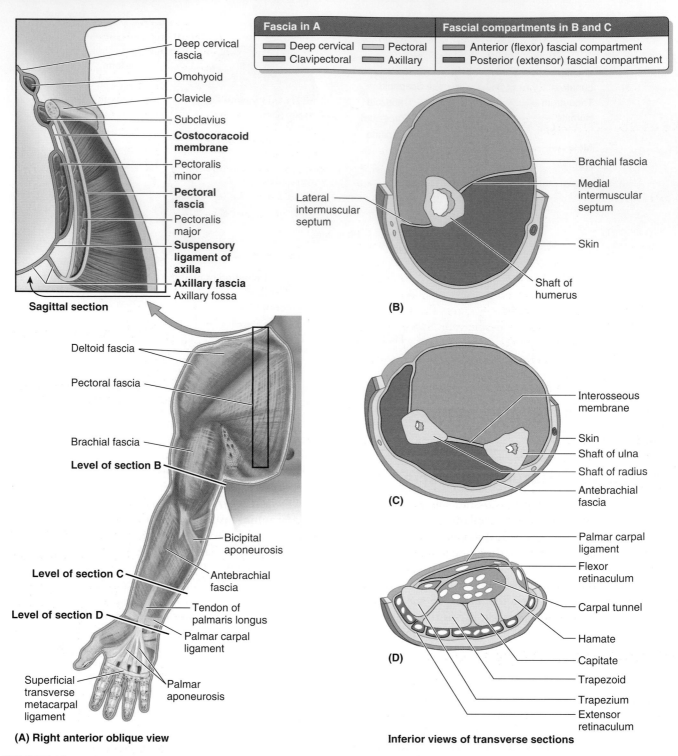

Fascia in A		Fascial compartments in B and C	
Deep cervical	Pectoral	Anterior (flexor) fascial compartment	
Clavipectoral	Axillary	Posterior (extensor) fascial compartment	

Deep cervical fascia
Omohyoid
Clavicle
Subclavius
Costocoracoid membrane
Pectoralis minor
Pectoral fascia
Pectoralis major
Suspensory ligament of axilla
Axillary fascia
Axillary fossa

Sagittal section

Brachial fascia
Medial intermuscular septum
Skin
Shaft of humerus
Lateral intermuscular septum

(B)

Deltoid fascia
Pectoral fascia
Brachial fascia
Level of section B
Bicipital aponeurosis
Level of section C
Antebrachial fascia
Level of section D
Tendon of palmaris longus
Palmar carpal ligament
Superficial transverse metacarpal ligament
Palmar aponeurosis

Interosseous membrane
Skin
Shaft of ulna
Shaft of radius
Antebrachial fascia

(C)

Palmar carpal ligament
Flexor retinaculum
Carpal tunnel
Hamate
Capitate
Trapezoid
Trapezium
Extensor retinaculum

(D)

(A) Right anterior oblique view

Inferior views of transverse sections

FIGURE 3.10. Fascia and compartments of upper limb. A. Fascia. **B.** Fascial compartments of arm. **C.** Fascial compartments of forearm. **D.** Flexor retinaculum and carpal tunnel.

antebrachial fascia is also continued as the **flexor retinaculum** (transverse carpal ligament). This fibrous band extends between the anterior prominences of the outer carpal bones and converts the anterior concavity of the carpus into the *carpal tunnel* through which the flexor tendons and median nerve pass.

The deep fascia of the upper limb continues beyond the extensor and flexor retinacula as the **palmar fascia**. The central part of the palmar fascia, the **palmar aponeurosis**, is thick, tendinous, and triangular. The aponeurosis forms four distinct thickenings that radiate to the bases of the fingers and become continuous with the fibrous tendon sheaths of the digits (Fig. 3.10A). The bands are traversed distally by the **superficial transverse metacarpal ligament**, which forms the base of the palmar aponeurosis. Strong *skin ligaments* extend from the palmar aponeurosis to the skin, holding the palmar skin close to the aponeurosis.

Venous Drainage of Upper Limb

The main **superficial veins of the upper limb**, the cephalic and basilic veins, originate in the subcutaneous tissue on the dorsum of the hand from the **dorsal venous network** (Fig. 3.11). **Perforating veins** form communications between the superficial and the deep veins.

The **cephalic vein** (G. *kephalé*, head) ascends in the subcutaneous tissue from the lateral aspect of the dorsal venous network, proceeding along the lateral border of the wrist and the anterolateral surface of the forearm and arm. Anterior to the elbow, the cephalic vein communicates with the **median cubital vein**, which passes obliquely across the anterior aspect of the elbow and joins the basilic vein. Superiorly, the cephalic vein passes between the deltoid and the pectoralis major muscles and enters the *clavipectoral triangle*, where it pierces the costocoracoid membrane, part of the clavipectoral fascia, and joins the terminal part of the axillary vein.

The **basilic vein** ascends in the subcutaneous tissue from the medial end of the dorsal venous network along the medial side of the forearm and inferior part of the arm. It then passes deeply near the junction of the middle and inferior thirds of the arm, piercing the brachial fascia and running superiorly parallel to the brachial artery, where it merges with the accompanying veins (L. *venae comitantes*) of the brachial artery to form the axillary vein (Fig. 3.11A). The **median antebrachial vein** (median vein of forearm) ascends in the middle of the anterior aspect of the forearm.

Deep veins lie internal to the deep fascia and usually occur as paired, continually interanastomosing, accompanying veins that travel with and bear the same name as the major arteries of the upper limb (Fig. 3.12).

Arterial Supply of Upper Limb

The axillary artery is the main blood supply to the upper limb (Fig. 3.13). The axillary artery is the continuation of the subclavian artery distal to the lateral border of the 1st rib. The pulsations of the axillary artery can be palpated in the axilla. The axillary artery becomes the brachial artery at the inferior border of the teres major. The brachial artery ends in the cubital fossa opposite the neck of the radius, where it divides into the radial and ulnar arteries. The brachial artery is relatively superficial and palpable throughout its course. At first, it lies medial to the humerus, where its pulsations are palpable in the medial bicipital groove. Then, it passes anterior to the medial supra-epicondylar ridge and trochlea of the humerus. As it passes inferolaterally, the pulsations of the brachial artery can be palpated in the cubital fossa. A major branch of the brachial artery is the **profunda brachii**

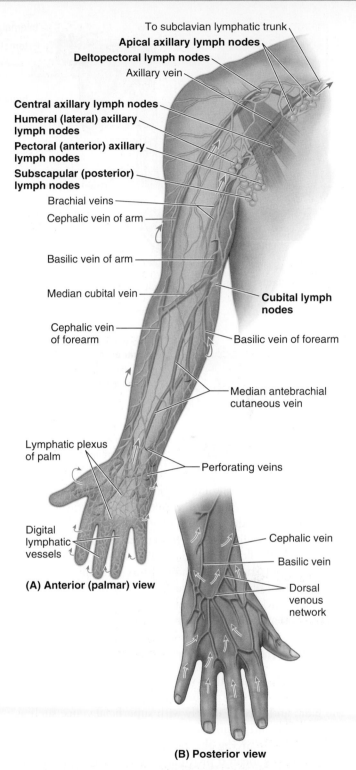

FIGURE 3.11. Superficial venous and lymphatic drainage of upper limb. A. Anterior view of upper limb. **B.** Posterior view of distal forearm and hand. *Green arrows,* superficial lymphatic drainage to lymph nodes.

artery (deep artery of arm) that travels posterior to the humerus in the radial groove and helps form the peri-articular cubital anastomosis of the elbow region. The **ulnar artery** descends through the anterior compartment of the forearm. Pulsations of the ulnar artery can be palpated on the lateral

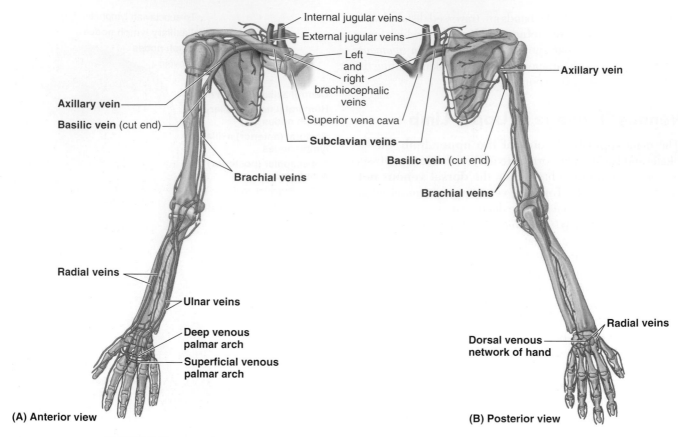

FIGURE 3.12. Deep veins of upper limb. Deep veins bear the same name as the arteries they accompany.

side of the flexor carpi ulnaris tendon, where it lies anterior to the ulnar head. The **radial artery** courses through the forearm laterally deep to the brachioradialis. It leaves the forearm by coursing around the lateral aspect of the wrist and crossing the floor of the anatomical snuff box to reach the hand. The pulsation of the radial artery is usually measured on the distal radius. The ulnar and radial arteries and their branches provide all the blood to the hand primarily via the **superficial** and **deep palmar arches**.

Lymphatic Drainage of Upper Limb

Superficial lymphatic vessels arise from **lymphatic plexuses** in the skin of the fingers, palm, and dorsum of the hand and ascend mostly with superficial veins, such as the cephalic and basilic veins (Fig. 3.11). Some lymphatic vessels accompanying the basilic vein enter the **cubital lymph nodes** located proximal to the medial epicondyle. Efferent vessels from these nodes ascend in the arm and terminate in the **humeral (lateral) axillary lymph nodes**. Most lymphatic vessels accompanying the cephalic vein cross the proximal part of the arm and anterior aspect of the shoulder to enter the **apical axillary lymph nodes**. Some vessels enter the more superficial **deltopectoral lymph nodes**.

Deep lymphatic vessels, less numerous than superficial vessels, accompany the major deep veins and terminate in the humeral (lateral) axillary lymph nodes (Fig. 3.11).

Cutaneous and Motor Innervation of Upper Limb

CUTANEOUS INNERVATION

Cutaneous nerves in the subcutaneous tissue supply the skin of the upper limb (Fig. 3.14). The area of skin supplied by cutaneous branches from a single spinal nerve is a *dermatome* (Fig. 3.15). The dermatomes of the limb follow a general pattern that is easy to understand if one notes that developmentally, the limbs grow as lateral protrusions of the trunk, with the 1st digit (thumb or great toe) located on the cranial side. Thus, the lateral surface of the upper limb is more cranial than the medial surface. There are two dermatome maps in common use. One corresponds to the concepts of limb development (Keegan & Garrett, 1948), and the other is based on clinical findings and is generally preferred by neurologists (Foerster, 1933). Both maps are approximations, delineating dermatomes as distinct zones when actually there is much overlap between adjacent dermatomes and much variation. In both maps, observe the orderly progression of the segmental innervation (dermatomes) of the various cutaneous areas around the limb (Fig. 3.15).

Most cutaneous nerves of the upper limb are multisegmental peripheral nerves derived from the **brachial plexus**, a major nerve network formed by the anterior rami of the C5–T1 spinal nerves. The cutaneous nerves to the shoulder are derived from the **cervical plexus**, a nerve network consisting of a series of nerve loops formed between adjacent anterior rami

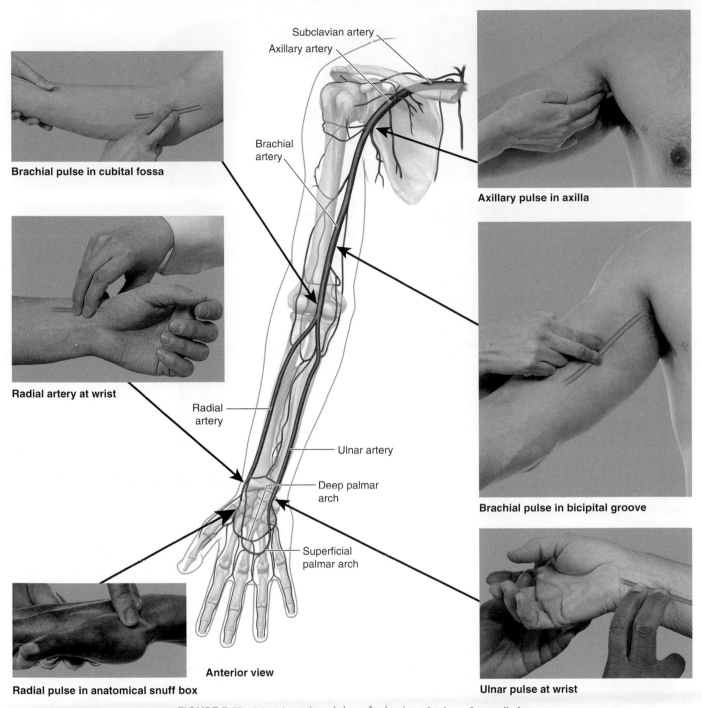

Brachial pulse in cubital fossa

Radial artery at wrist

Radial pulse in anatomical snuff box

Subclavian artery
Axillary artery

Brachial artery

Radial artery

Ulnar artery

Deep palmar arch

Superficial palmar arch

Anterior view

Axillary pulse in axilla

Brachial pulse in bicipital groove

Ulnar pulse at wrist

FIGURE 3.13. Arterial supply and sites of palpation of pulses of upper limb.

of the first four cervical nerves. The cervical plexus lies deep to the SCM on the lateral aspect of the neck. The cutaneous nerves of the arm and forearm (and the spinal cord segments from which they are derived) are as follows (Fig. 3.14):

- **Supraclavicular nerves** (C3, C4) pass anterior to the clavicle, immediately deep to the platysma, and supply the skin over the clavicle and the superolateral aspect of the pectoralis major.
- **Posterior cutaneous nerve of the arm** (C5–C8), a branch of the *radial nerve*, supplies the skin on the posterior surface of the arm.

- **Posterior cutaneous nerve of the forearm** (C5–C8), also a branch of the *radial nerve*, supplies the skin on the posterior surface of the forearm.
- **Superior lateral cutaneous nerve of the arm** (C5, C6), a terminal branch of the *axillary nerve*, emerges from beneath the posterior margin of the deltoid to supply the skin over the lower part of this muscle and on the lateral side of the midarm.
- **Inferior lateral cutaneous nerve of the arm** (C5, C6), a branch of the *radial nerve*, supplies the skin over the inferolateral aspect of the arm; it is frequently

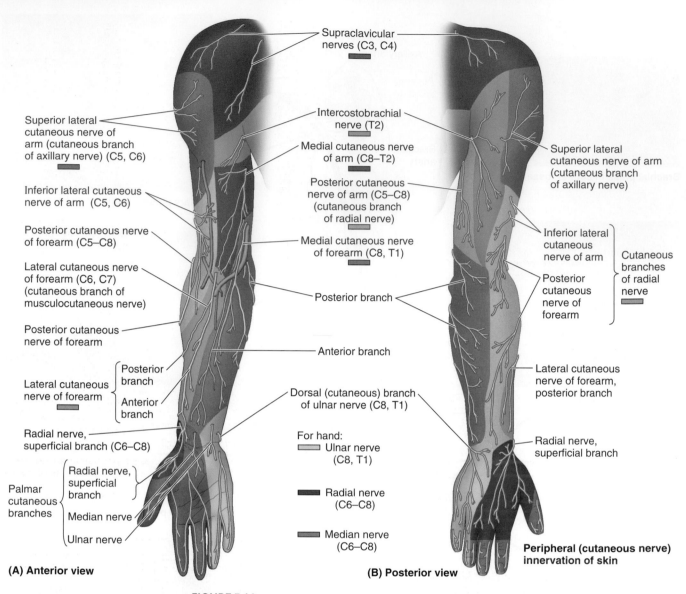

Supraclavicular nerves (C3, C4)

Superior lateral cutaneous nerve of arm (cutaneous branch of axillary nerve) (C5, C6)

Inferior lateral cutaneous nerve of arm (C5, C6)

Posterior cutaneous nerve of forearm (C5–C8)

Lateral cutaneous nerve of forearm (C6, C7) (cutaneous branch of musculocutaneous nerve)

Posterior cutaneous nerve of forearm

Lateral cutaneous nerve of forearm
Posterior branch
Anterior branch

Radial nerve, superficial branch (C6–C8)

Palmar cutaneous branches
Radial nerve, superficial branch
Median nerve
Ulnar nerve

Intercostobrachial nerve (T2)

Medial cutaneous nerve of arm (C8–T2)

Posterior cutaneous nerve of arm (C5–C8) (cutaneous branch of radial nerve)

Medial cutaneous nerve of forearm (C8, T1)

Posterior branch

Anterior branch

Dorsal (cutaneous) branch of ulnar nerve (C8, T1)

For hand:
Ulnar nerve (C8, T1)

Radial nerve (C6–C8)

Median nerve (C6–C8)

Superior lateral cutaneous nerve of arm (cutaneous branch of axillary nerve)

Inferior lateral cutaneous nerve of arm
Posterior cutaneous nerve of forearm
Cutaneous branches of radial nerve

Lateral cutaneous nerve of forearm, posterior branch

Radial nerve, superficial branch

Peripheral (cutaneous nerve) innervation of skin

(A) Anterior view

(B) Posterior view

FIGURE 3.14. Peripheral (cutaneous) innervation of upper limb.

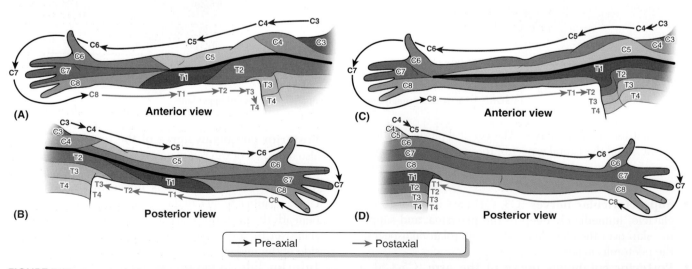

(A) Anterior view

(B) Posterior view

(C) Anterior view

(D) Posterior view

→ Pre-axial   → Postaxial

FIGURE 3.15. Segmental (dermatomal) innervation. A and B. The pattern of segmental innervation proposed by Foerster (1933). **C and D.** The pattern of segmental innervation proposed by Keegan and Garrett (1948).

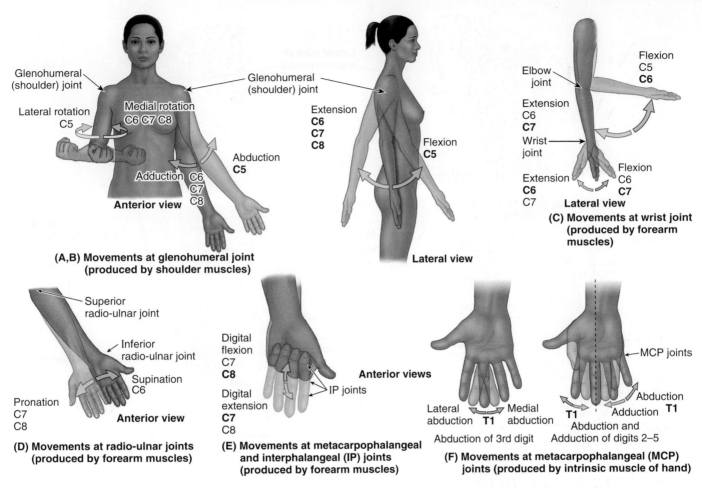

FIGURE 3.16. Segmental innervation of movements of the upper limb. A–F. Most movements involve portions of multiple myotomes.

a branch of the posterior cutaneous nerve of the forearm.

- **Lateral cutaneous nerve of the forearm** (C6, C7), the terminal branch of the *musculocutaneous nerve*, supplies the skin on the lateral side of the forearm.
- **Medial cutaneous nerve of the arm** (C8–T2) arises from the *medial cord of the brachial plexus*, often uniting in the axilla with the lateral cutaneous branch of the 2nd intercostal nerve. It supplies the skin on the medial side of the arm.
- **Intercostobrachial nerve** (T2), a lateral cutaneous branch of the *2nd intercostal nerve*, also contributes to the innervation of the skin on the medial surface of the arm.
- **Medial cutaneous nerve of the forearm** (C8, T1) arises from the *medial cord of the brachial plexus* and supplies the skin on the anterior and medial surfaces of the forearm.

MOTOR INNERVATION

The unilateral embryological muscle mass receiving innervation from a single spinal cord segment or spinal nerve comprises a *myotome*. Upper limb muscles usually receive motor

TABLE 3.1. UPPER LIMB MYOTOMES

Spinal Nerve	Myotome	Spinal Nerve	Myotome
C5	Shoulder abduction	C8	Wrist flexion, finger flexion
C5, C6	Elbow flexion	T1	Intrinsic hand muscles
C7	Elbow extension, wrist extension, finger extension		

fibers from several spinal cord segments via multisegmental peripheral nerves. Thus, most muscles include more than one myotome, and most often, multiple spinal cord segments are involved in producing the movements. The muscle myotomes are grouped by joint movement to facilitate clinical testing (Fig. 3.16 and Table 3.1).

The same mixed multisegmental peripheral nerves that convey sensory fibers to the skin, muscles, and joints of the upper limb convey somatic motor (general somatic efferent) fibers to the upper limb muscles. The somatic (peripheral) motor and sensory innervation of the upper limb is summarized in Figure 3.17.

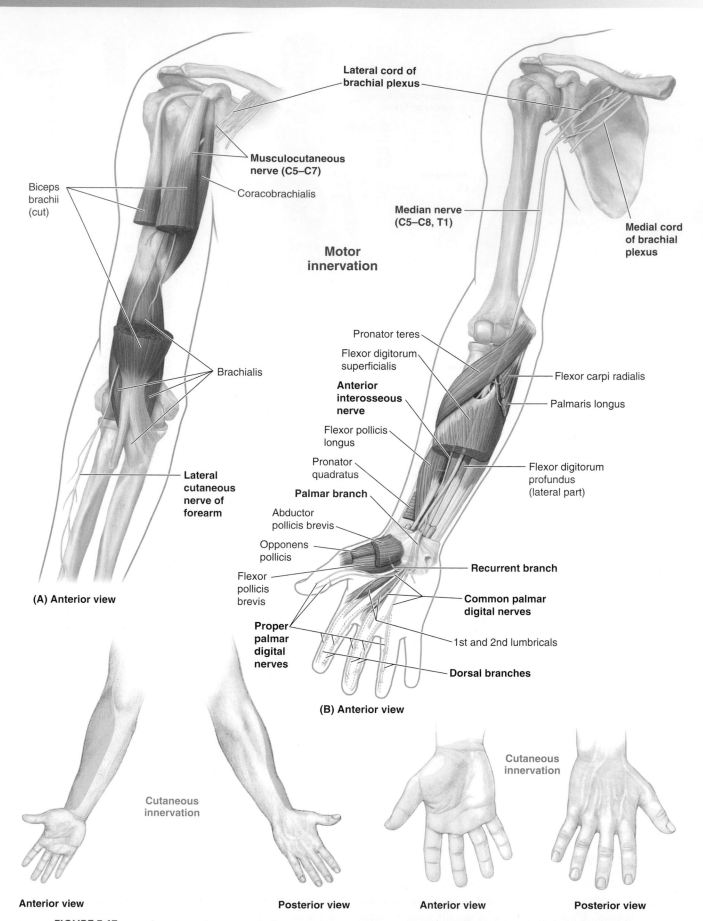

Lateral cord of brachial plexus

Musculocutaneous nerve (C5–C7)

Coracobrachialis

Biceps brachii (cut)

Median nerve (C5–C8, T1)

Medial cord of brachial plexus

Motor innervation

Brachialis

Pronator teres

Flexor digitorum superficialis

Anterior interosseous nerve

Flexor carpi radialis

Palmaris longus

Flexor pollicis longus

Pronator quadratus

Flexor digitorum profundus (lateral part)

Lateral cutaneous nerve of forearm

Palmar branch

Abductor pollicis brevis

Opponens pollicis

Flexor pollicis brevis

Recurrent branch

Common palmar digital nerves

Proper palmar digital nerves

1st and 2nd lumbricals

Dorsal branches

(A) Anterior view

(B) Anterior view

Cutaneous innervation

Cutaneous innervation

Anterior view

Posterior view

Anterior view

Posterior view

FIGURE 3.17. Overview of peripheral nerves innervating the upper limb. A. Musculocutaneous nerve. **B.** Median nerve. *(continued)*

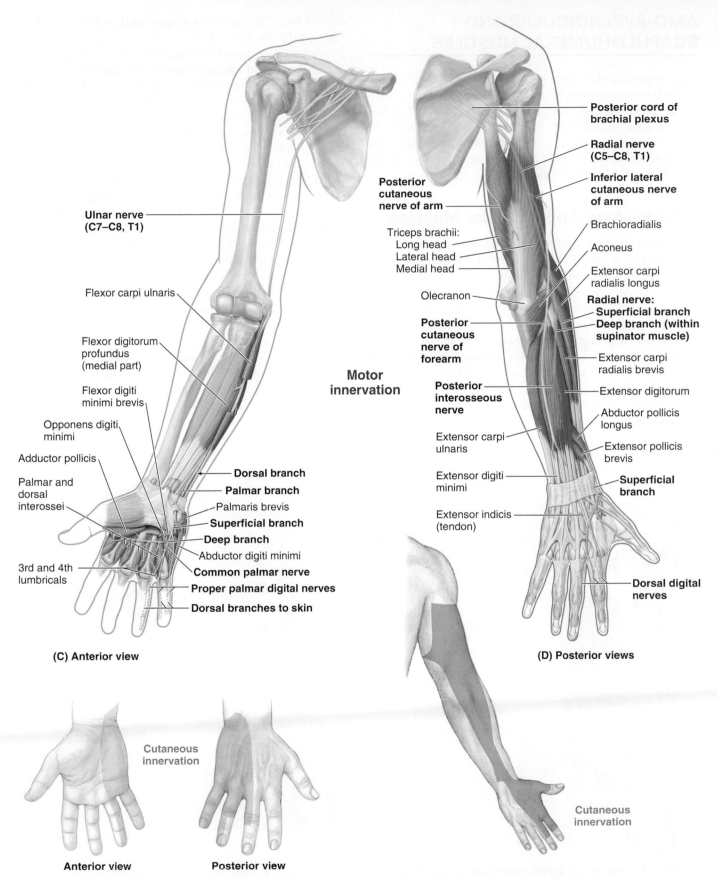

Motor innervation

Ulnar nerve (C7–C8, T1)

Flexor carpi ulnaris

Flexor digitorum profundus (medial part)

Flexor digiti minimi brevis

Opponens digiti minimi

Adductor pollicis

Palmar and dorsal interossei

3rd and 4th lumbricals

Dorsal branch
Palmar branch
Palmaris brevis
Superficial branch
Deep branch
Abductor digiti minimi
Common palmar nerve
Proper palmar digital nerves
Dorsal branches to skin

(C) Anterior view

Posterior cord of brachial plexus
Radial nerve (C5–C8, T1)
Inferior lateral cutaneous nerve of arm
Brachioradialis
Aconeus
Extensor carpi radialis longus
Radial nerve:
Superficial branch
Deep branch (within supinator muscle)
Extensor carpi radialis brevis
Extensor digitorum
Abductor pollicis longus
Extensor pollicis brevis
Superficial branch

Posterior cutaneous nerve of arm

Triceps brachii:
Long head
Lateral head
Medial head

Olecranon

Posterior cutaneous nerve of forearm

Posterior interosseous nerve

Extensor carpi ulnaris

Extensor digiti minimi

Extensor indicis (tendon)

Dorsal digital nerves

(D) Posterior views

Cutaneous innervation

Anterior view   Posterior view

Cutaneous innervation

FIGURE 3.17. Overview of peripheral nerves innervating the upper limb. *(continued)* **C.** Ulnar nerve. **D.** Radial nerve.

AXIO-APPENDICULAR AND SCAPULOHUMERAL MUSCLES

Axio-appendicular muscles (*extrinsic shoulder muscles*) attach the superior appendicular skeleton of the upper limb to the axial skeleton (vertebral column); most act at the physiological *scapulothoracic joint*, moving the scapula on the chest wall. Scapulohumeral muscles (intrinsic shoulder muscles) attach the scapula to the humerus and act at the glenohumeral (shoulder) joint.

Anterior Axio-Appendicular Muscles

Four **anterior axio-appendicular (thoraco-appendicular or pectoral) muscles** move the pectoral girdle: pectoralis major, pectoralis minor, subclavius, and serratus anterior (Fig. 3.18). Attachments of these muscles are illustrated in Figure 3.19. The attachments, nerve supply, and main actions of these muscles are summarized in Table 3.2.

The fan-shaped **pectoralis major** covers the superior part of the thorax. It has **clavicular** and **sternocostal heads** (Fig. 3.18B). The sternocostal head is much larger, and its lateral border forms most of the anterior wall of the axilla, with its inferior border forming the *anterior axillary fold* (see "Axilla" later in this chapter). The pectoralis major and adjacent

deltoid form the narrow **deltopectoral groove**, in which the cephalic vein runs. However, the muscles diverge slightly from each other superiorly and, along with the clavicle, form the **clavipectoral (deltopectoral) triangle** (Fig. 3.18A).

The triangular **pectoralis minor** lies in the anterior wall of the axilla (Fig. 3.18E), where it is almost completely covered by the pectoralis major. The pectoralis minor stabilizes the scapula and is used when stretching the upper limb forward to touch an object that is just out of reach. With the coracoid process, the pectoralis minor forms a "bridge" under which vessels and nerves pass to the arm. Thus, the pectoralis minor is a useful anatomical and surgical landmark for structures in the axilla.

The **subclavius** lies almost horizontally when the arm is in the anatomical position (Fig. 3.18D). This small, round muscle is located inferior to the clavicle and affords some protection to the subclavian vessels and the superior trunk of the brachial plexus if the clavicle fractures.

The **serratus anterior** overlies the lateral part of the thorax and forms the medial wall of the axilla (Fig. 3.18C). This broad sheet of thick muscle was given its name because of the sawtooth appearance of its fleshy slips or digitations (L. *serratus*, a saw). By keeping the scapula closely applied to the thoracic wall, the serratus anterior anchors this bone, enabling other muscles to use it as a fixed bone for movements of the humerus.

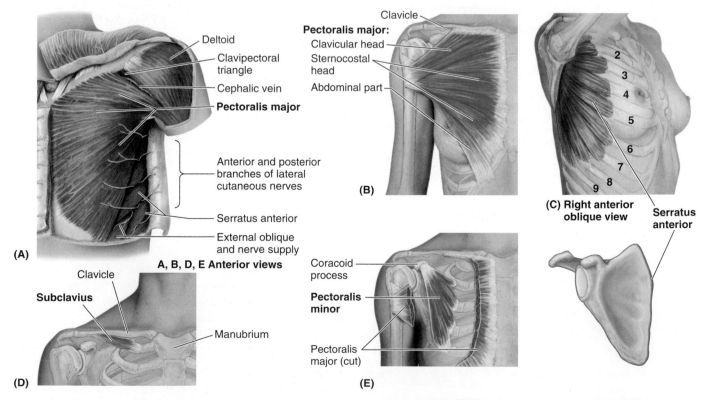

FIGURE 3.18. Anterior axio-appendicular muscles. A. Superficial dissection of pectoral region. **B.** Pectoralis major. **C.** Serratus anterior. *Inset*, scapular attachment of serratus anterior (*blue*). **D.** Subclavius. **E.** Pectoralis minor.

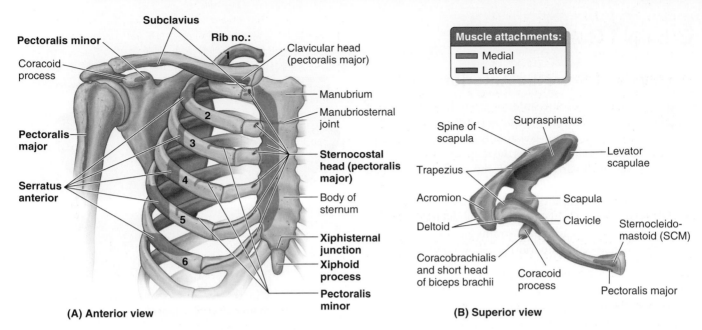

FIGURE 3.19. Attachments of anterior axio-appendicular muscles.

(A) Anterior view

(B) Superior view

TABLE 3.2. ANTERIOR AXIO-APPENDICULAR MUSCLES

Muscle	Proximal Attachment	Distal Attachment	Innervation[a]	Main Action(s)
Pectoralis major	*Clavicular head*: anterior surface of medial half of clavicle *Sternocostal head*: anterior surface of sternum, superior six costal cartilages, aponeurosis of external oblique muscle	Lateral lip of intertubercular sulcus (groove) of humerus	Lateral and medial pectoral nerves, clavicular head (C5, **C6**), sternocostal head (**C7, C8**, T1)	Adducts and medially rotates shoulder joint, draws scapula anteriorly and inferiorly Acting alone, clavicular head flexes shoulder joint and sternocostal head and extends it from the flexed position.
Pectoralis minor	3rd–5th ribs near their costal cartilages	Medial border and superior surface of coracoid process of scapula	Medial pectoral nerve (C8, T1)	Stabilizes scapula by drawing inferiorly and anteriorly against thoracic wall
Subclavius	Junction of 1st rib and its costal cartilage	Inferior surface of middle third of clavicle	Subclavian nerve (**C5**, C6)	Anchors and depresses clavicle
Serratus anterior	External surfaces of lateral parts of 1st–8th ribs	Anterior surface of medial border of scapula	Long thoracic nerve (C5, **C6, C7**)	Protracts scapula and holds it against thoracic wall; rotates scapula

[a]The spinal cord segmental innervation is indicated (e.g., "C5, C6" means that the nerves supplying the deltoid are derived from the 5th and 6th cervical segments of the spinal cord). Numbers in boldface (**C5**) indicate the main segmental innervation. Damage to one or more of the listed spinal cord segments or to the motor nerve roots arising from them results in paralysis of the muscles concerned.

Posterior Axio-Appendicular Muscles

The **posterior axio-appendicular muscles** (also described as *superficial extrinsic back muscles* in Chapter 2) and their attachments are illustrated in Figure 3.20; the attachments, nerve supply, and main actions of these muscles are summarized in Table 3.3. These muscles are divided into two groups:

- *Superficial posterior axio-appendicular muscles*: trapezius and latissimus dorsi
- *Deep posterior axio-appendicular muscles*: levator scapulae and rhomboids

SUPERFICIAL POSTERIOR AXIO-APPENDICULAR MUSCLES

The **trapezius** provides a direct attachment of the pectoral girdle to the trunk. This large triangular muscle covers the posterior aspect of the neck and the superior half of the trunk. The trapezius attaches the pectoral girdle to the cranium and vertebral column and assists in suspending the upper limb. The fibers of the trapezius are divided into three parts that have different actions at the scapulothoracic joint between the scapula and the thoracic wall:

- **Descending (superior) part** elevates the scapula.
- **Middle part** retracts the scapula (i.e., pulls it posteriorly).
- **Ascending (inferior) part** depresses the scapula and lowers the shoulder.

CLINICAL BOX

Paralysis of Serratus Anterior

When the serratus anterior is paralyzed because of *injury to the long thoracic nerve*, the medial border of the scapula moves laterally and posteriorly away from the thoracic wall. This gives the scapula the appearance of a wing. When the arm is raised, the medial border and inferior angle of the scapula pull markedly away from the posterior thoracic wall, a deformation known as a *winged scapula* (Fig. B3.5). The arm cannot be abducted above the horizontal position because the serratus anterior is unable to rotate the glenoid cavity superiorly to allow complete abduction of the limb.

Venipuncture

Because of the prominence and accessibility of the superficial veins, they are commonly used for *venipuncture* (to draw blood or inject a solution). By applying a tourniquet to the arm, the venous return is occluded, and the veins distend and usually are visible and/or palpable. Once a vein is punctured, the tourniquet is removed so that when the needle is removed, the vein will not bleed extensively. The median cubital vein is commonly used

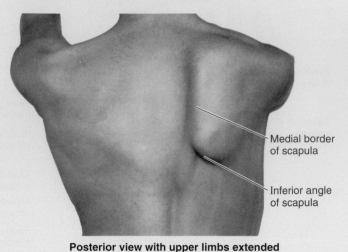

Medial border of scapula

Inferior angle of scapula

Posterior view with upper limbs extended

FIGURE B3.5. Winged scapula.

for venipuncture. The veins forming the *dorsal venous network* and the cephalic and basilic veins are commonly used for long-term introduction of fluids (*intravenous feeding*). The cubital veins are also a site for the introduction of cardiac catheters.

The descending (superior) and ascending (inferior) parts of trapezius act together in rotating the scapula on the thoracic wall. The trapezius also braces the shoulders by pulling the scapulae posteriorly and superiorly, fixing them in position with tonic contraction; consequently, weakness of this muscle causes drooping of the shoulders.

The **latissimus dorsi** is a large, fan-shaped muscle that covers a wide area of the back (Fig. 3.20A,C). It passes from the trunk to the humerus and acts directly on the glenohumeral (shoulder) joint and indirectly on the pectoral girdle (scapulothoracic joint). In conjunction with the pectoralis major, the latissimus dorsi raises the trunk to the arm, which occurs when the limb is fixed and the body moves, as when performing chin-ups (hoisting oneself so the chin touches an overhead bar) or climbing a tree. These movements are also used when the trunk is fixed and the limb moves, as when chopping wood, paddling a canoe, and swimming.

DEEP POSTERIOR AXIO-APPENDICULAR MUSCLES

The superior third of the **levator scapulae** lie deep to the SCM; the inferior third is deep to the trapezius. True to its name, the levator scapulae act with the superior part of

trapezius to elevate the scapula. With the rhomboids and pectoralis minor, the levator scapulae rotate the scapula, depressing the glenoid cavity. Acting bilaterally, they extend the neck; acting unilaterally, the muscle may contribute to lateral flexion of the neck.

The two **rhomboids** (**major** and **minor**) lie deep to the trapezius and form parallel bands that pass inferolaterally from the vertebrae to the medial border of the scapula (Fig. 3.20A,B and Table 3.3). The rhomboids retract and rotate the scapula, depressing the glenoid cavity. They also assist the serratus anterior in holding the scapula against the thoracic wall and fixing the scapula during movements of the upper limb.

Scapulohumeral Muscles

The six **scapulohumeral muscles** (the deltoid, teres major, supraspinatus, infraspinatus, subscapularis, and teres minor) and their attachments are illustrated in Figure 3.21; the attachments, nerve supply, and main actions of these muscles are summarized in Table 3.4).

The **deltoid** is a thick, powerful muscle forming the rounded contour of the shoulder. The muscle is divided into *clavicular (anterior), acromial (middle),* and *spinal (posterior)*

(Continued on page 117)

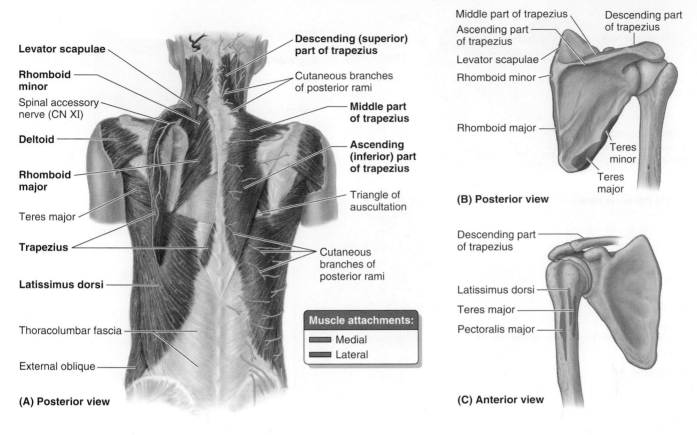

FIGURE 3.20. Posterior axio-appendicular muscles. **A.** Overview. **B and C.** Bony attachments.

TABLE 3.3. POSTERIOR AXIO-APPENDICULAR MUSCLES

Muscle	Medial Attachment	Lateral Attachment	Innervation[a]	Main Action(s)
Superficial posterior thoraco-appendicular (extrinsic shoulder) muscles				
Trapezius	Medial third of superior nuchal line, external occipital protuberance, nuchal ligament, spinous processes of C7–T12 vertebrae	Lateral third of clavicle, acromion and spine of scapula	Spinal accessory nerve (CN XI; motor fibers) and C3, C4 spinal nerves (pain and proprioceptive fibers)	*Descending* (*superior*) *part* elevates, *ascending* (*inferior*) *part* depresses, and *middle part* (or all parts together) retracts scapula; descending and ascending parts act together to rotate glenoid cavity superiorly.
Latissimus dorsi	Spinous processes of inferior six thoracic vertebrae, thoracolumbar fascia, iliac crest, and inferior three or four ribs	Floor of intertubercular sulcus (groove) of humerus	Thoracodorsal nerve (**C6**, **C7**, C8)	Extends, adducts, and medially rotates shoulder joint; raises body toward arms during climbing
Deep posterior thoraco-appendicular (extrinsic shoulder) muscles				
Levator scapulae	Posterior tubercles of transverse processes of C1–C4 vertebrae	Medial border of scapula superior to root of spine	Dorsal scapular (C5) and cervical (C3, C4) spinal nerves	Elevates scapula and rotates glenoid cavity inferiorly
Rhomboid minor and major	*Minor*: nuchal ligament; spinous processes of C7 and T1 vertebrae *Major*: spinous processes of T2–T5 vertebrae	*Minor*: triangular area at medial end of scapular spine *Major*: medial border of scapula from level of spine to inferior angle	Dorsal scapular nerve (C4, **C5**)	Retracts scapula and rotates glenoid cavity inferiorly; fix scapula to thoracic wall

[a]The spinal cord segmental innervation is indicated (e.g., "**C6**, **C7**, C8" means that the nerves supplying the latissimus dorsi are derived from the 6th through 8th cervical segments of the spinal cord). Numbers in boldface (**C6**, **C7**) indicate the main segmental innervation. Damage to one or more of the listed spinal cord segments or to the motor nerve roots arising from them results in paralysis of the muscles concerned.

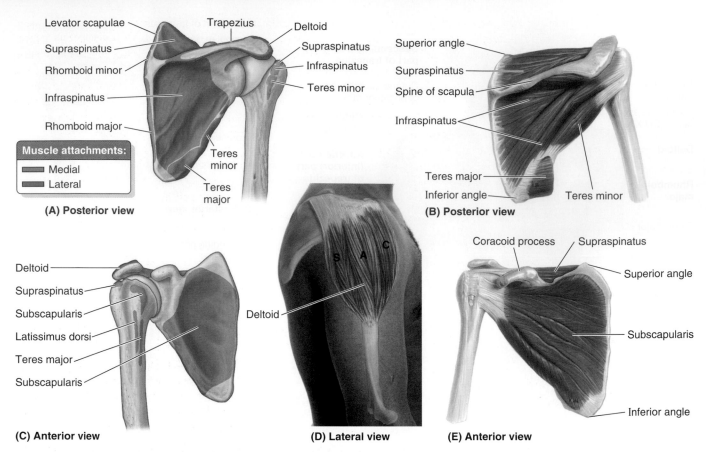

FIGURE 3.21. Scapulohumeral muscles. A and C. Bony attachments. **B.** Supraspinatus, infraspinatus, and teres minor. **D.** Deltoid muscle. *A,* acromial part; *C,* clavicular part; *S,* spinal part. **E.** Subscapularis

TABLE 3.4. SCAPULOHUMERAL (INTRINSIC SHOULDER) MUSCLES

Muscle	Proximal Attachment	Distal Attachment	Innervation[a]	Main Action(s)
Deltoid	Lateral third of clavicle; acromion and spine of scapula	Deltoid tuberosity of humerus	Axillary nerve (**C5**, C6)	Clavicular (anterior) part flexes and medially rotates shoulder joint; acromial (middle) part abducts shoulder joint; spinal (posterior) part extends and laterally rotates shoulder joint.
Supraspinatus[b]	Supraspinous fossa of scapula	Superior facet	Suprascapular nerve (C4, **C5**, C6)	Initiates and assists deltoid in abduction of shoulder joint and acts with other rotator cuff muscles[b]
Infraspinatus[b]	Infraspinous fossa of scapula	Middle facet	of greater tubercle of humerus	Suprascapular nerve (**C5**, C6)	Laterally rotate shoulder joint; help hold humeral head in glenoid cavity of scapula
Teres minor[b]	Middle part of lateral border of scapula	Inferior facet	Axillary nerve (**C5**, C6)	
Teres major	Inferior part of lateral border of scapula and posterior surface of inferior angle of scapula	Medial lip of intertubercular sulcus of humerus	Lower subscapular nerve (C5, **C6**)	Adducts and medially rotates shoulder joint
Subscapularis[b]	Subscapular fossa (most of anterior surface of scapula)	Lesser tubercle of humerus	Upper and lower subscapular nerves (C5, **C6**, C7)	Medially rotates and adducts shoulder joint; helps hold humeral head in glenoid cavity

[a]The spinal cord segmental innervation is indicated (e.g., "**C5**, C6" means that the nerves supplying the deltoid are derived from the 5th and 6th cervical segments of the spinal cord). Numbers in boldface (**C5**) indicate the main segmental innervation. Damage to one or more of the listed spinal cord segments or to the motor nerve roots arising from them results in paralysis of the muscles concerned.

[b]Collectively, the supraspinatus, infraspinatus, teres minor, and subscapularis muscles are referred to as the rotator cuff, or SITS, muscles. Their primary function during all movements of the glenohumeral (shoulder) joint is to hold the humeral head in the glenoid cavity of the scapula.

parts that can act separately or as a whole. When all three parts contract simultaneously, the shoulder joint is abducted. The clavicular and spinal parts act like guy ropes to steady the arm as it is abducted. When the shoulder joint is fully adducted, the line of pull of the deltoid coincides with the axis of the humerus; thus, it pulls directly upward on the bone and cannot initiate or produce abduction. The deltoid is, however, able to act as a shunt muscle, resisting inferior displacement of the head of the humerus from the glenoid cavity. From the fully adducted position, abduction must be initiated by the supraspinatus or by leaning to the side, allowing gravity to initiate the movement. The deltoid becomes fully effective as an abductor after the initial 15 degrees of abduction.

The **teres major** is a thick rounded muscle that lies on the inferolateral third of the scapula. It adducts and medially rotates the arm, but along with the deltoid and rotator cuff muscles, it is an important stabilizer of the humeral head in the glenoid cavity during movement.

Four of the scapulohumeral muscles (intrinsic shoulder muscles)—*S*upraspinatus, *I*nfraspinatus, *T*eres minor, and *S*ubscapularis (referred to as **SITS** muscles)—are called **rotator cuff muscles** because they form a **musculotendinous** **rotator cuff** around the glenohumeral joint. All except the supraspinatus are rotators of the humerus. The supraspinatus, besides being part of the rotator cuff, initiates and assists the deltoid in the first 15 degrees of abduction of the arm. The tendons of the SITS or rotator cuff muscles blend with the joint capsule of the glenohumeral joint, reinforcing it as the musculotendinous rotator cuff, which protects the joint and gives it stability. Tonic contraction of these muscles holds the relatively large head of the humerus firmly against the small and shallow glenoid cavity during arm movements. Bursae around the glenohumeral (shoulder) joint, between the tendons of the rotator cuff muscles and the fibrous layer of the joint capsule, reduce friction on the tendons passing over the bones or other areas of resistance.

AXILLA

The **axilla** is the pyramidal compartment inferior to the glenohumeral joint and superior to the skin and axillary fascia at the junction of the arm and thorax (Fig. 3.22).

The shape and size of the axilla vary depending on the position of the arm; it almost disappears when the shoulder joint is fully abducted. The axilla provides a passageway for vessels and nerves going to and from the upper limb. The axilla has an apex, base, and four walls, three of which are muscular:

- The *apex of the axilla* is the **cervico-axillary canal**, the passageway between the neck and the axilla. It is bounded by the 1st rib, clavicle, and superior edge of the scapula. The arteries, veins, lymphatics, and nerves traverse this superior opening to pass to or from the upper limb.
- The *base of the axilla* is formed by the concave skin, subcutaneous tissue, and axillary (deep) fascia extending from the arm to the thoracic wall forming the **axillary fossa** (armpit).
- The *anterior wall of the axilla* is formed by the pectoralis major and minor and the pectoral and clavipectoral fascia associated with them. The **anterior axillary fold** is the inferiormost part of the anterior wall.
- The *posterior wall of the axilla* is formed chiefly by the scapula and subscapularis on its anterior surface and inferiorly by the teres major and latissimus dorsi. The **posterior axillary fold** is the inferiormost part of the posterior wall that may be grasped.
- The *medial wall of the axilla* is formed by the thoracic wall and the overlying serratus anterior.
- The *lateral wall of the axilla* is the narrow bony wall formed by the *intertubercular sulcus* of the humerus.

The axilla contains the axillary artery and its branches, axillary vein and its tributaries, nerves of the cords and branches of the brachial plexus, lymphatic vessels, and several groups of *axillary lymph nodes* all embedded in *axillary fat*. Proximally, the neurovascular structures are ensheathed in a sleeve-like extension of the cervical prevertebral fascia, the **axillary sheath** (see Fig. 3.28B).

CLINICAL BOX

Injury to Axillary Nerve

Atrophy of the deltoid occurs when the axillary nerve (C5 and C6) is severely damaged (e.g., as might occur when the surgical neck of the humerus is fractured). As the deltoid atrophies unilaterally, the rounded contour of the shoulder disappears, resulting in visible asymmetry of the shoulder outlines. This gives the shoulder a flattened appearance and produces a slight hollow inferior to the acromion. A loss of sensation may occur over the lateral side of the proximal part of the arm, the area supplied by the superior lateral cutaneous nerve of the arm. To test the deltoid (or the function of the axillary nerve), the arm is abducted, against resistance, starting from approximately 15 degrees

Rotator Cuff Injuries and Supraspinatus

Injury or disease may damage the rotator cuff, producing instability of the glenohumeral joint. Rupture or tear of the supraspinatus tendon is the most common injury of the rotator cuff. *Degenerative tendinitis of the rotator cuff* is common, especially in older people. These syndromes are discussed in detail later in this chapter in the Clinical Box on p. 167

SURFACE ANATOMY

Pectoral and Scapular Regions (Anterior and Posterior Axio-Appendicular and Scapulohumeral Muscles)

The large vessels and nerves to the upper limb pass posterior to the convexity in the clavicle. The **clavipectoral (deltopectoral) triangle** is the slightly depressed area just inferior to the lateral part of the clavicle (Fig. SA3.2A). The clavipectoral triangle is bounded by the clavicle superiorly, the deltoid laterally, and the clavicular head of pectoralis major medially. When the arm is abducted and then adducted against resistance, the two heads of the **pectoralis major** are visible and palpable. As this muscle extends from the thoracic wall to the arm, it forms the **anterior axillary fold**. Digitations of the **serratus anterior** appear inferolateral to the pectoralis major. The **coracoid process of the scapula** is covered by the anterior part of deltoid; however, the tip of the process can be felt on deep palpation in the clavipectoral triangle.

The **deltoid** forms the contour of the shoulder (Fig. SA3.2B); as its name indicates, it is shaped like the inverted Greek letter delta.

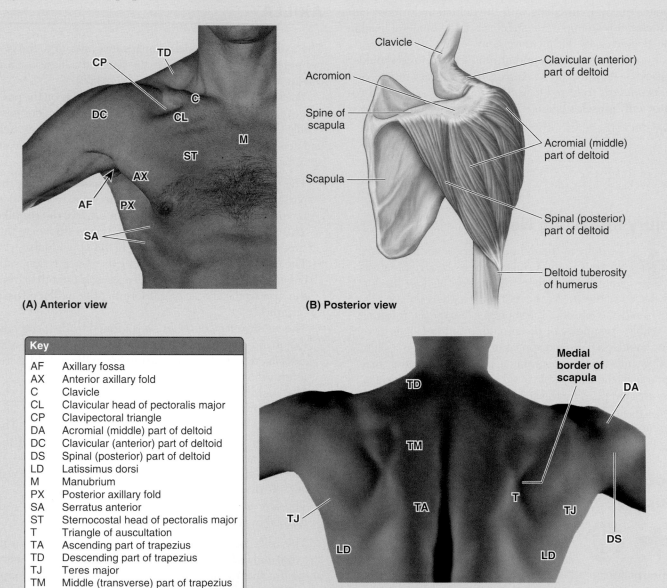

Key	
AF	Axillary fossa
AX	Anterior axillary fold
C	Clavicle
CL	Clavicular head of pectoralis major
CP	Clavipectoral triangle
DA	Acromial (middle) part of deltoid
DC	Clavicular (anterior) part of deltoid
DS	Spinal (posterior) part of deltoid
LD	Latissimus dorsi
M	Manubrium
PX	Posterior axillary fold
SA	Serratus anterior
ST	Sternocostal head of pectoralis major
T	Triangle of auscultation
TA	Ascending part of trapezius
TD	Descending part of trapezius
TJ	Teres major
TM	Middle (transverse) part of trapezius

(A) Anterior view

(B) Posterior view

(C) Posterior view

FIGURE SA3.2. Surface anatomy of pectoral and scapular regions.

(*Continued on next page*)

The superior border of the **latissimus dorsi** and a part of the **rhomboid major** are overlapped by the trapezius (Fig. SA3.2C). The area formed by the superior border of latissimus dorsi, the medial border of the scapula, and the inferolateral border of the trapezius is called the **triangle of auscultation**. This gap in the thick back musculature is a good place to auscultate the posterior segments of the lungs with a stethoscope. When the scapulae are drawn anteriorly by folding the arms across the thorax and the trunk is flexed, the triangle of auscultation enlarges. The **teres major** forms a raised oval area on the inferolateral third of the posterior aspect of the scapula when the arm is adducted against resistance. The **posterior axillary fold** is formed by the teres major and the tendon of the latissimus dorsi. Between the anterior and posterior axillary folds lies the **axillary fossa** (Fig. SA3.2A).

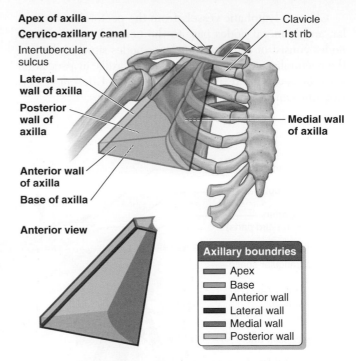

FIGURE 3.22. Location and boundaries of axilla.

Axillary Artery and Vein

The **axillary artery** begins at the lateral border of the 1st rib as the continuation of the subclavian artery and ends at the inferior border of the teres major (Fig. 3.23 and Table 3.5). It passes posterior to the pectoralis minor into the arm and becomes the brachial artery when it passes distal to the inferior border of the teres major. For descriptive purposes, the axillary artery is divided into three parts relative to the pectoralis minor (the part number also indicates the number of its branches):

- The **first part of the axillary artery** is located between the lateral border of the 1st rib and the medial border of the pectoralis minor; it is enclosed in the *axillary sheath* and has one branch: the *superior thoracic artery*.
- The **second part of the axillary artery** lies posterior to the pectoralis minor and has two branches: the *thoracoacromial artery* and *lateral thoracic artery*, which pass medial and lateral to the muscle, respectively.
- The **third part of the axillary artery** extends from the lateral border of the pectoralis minor to the inferior border of the teres major and has three branches. The *subscapular artery* is the largest branch of the axillary artery. Opposite the origin of this artery, the *anterior circumflex humeral artery* and *posterior circumflex humeral artery* arise.

The **axillary vein** lies initially (distally) on the anteromedial side of the axillary artery, with its terminal part antero-inferior to the artery (Fig. 3.24; also see Fig. 3.28A). This large vein is formed by the union of the *accompanying brachial veins* and the basilic vein at the inferior border of the teres major (see Fig. 3.11A). The axillary vein ends at the lateral border of the 1st rib, where it becomes the **subclavian vein** (Fig. 3.24). The veins of the axilla are more abundant than the arteries, are highly variable, and frequently anastomose.

Axillary Lymph Nodes

Many lymph nodes are found in the axillary fat. There are five principal groups of axillary lymph nodes: pectoral, subscapular, humeral, central, and apical (Figs. 3.24 and 3.25).

The **pectoral (anterior) nodes** consist of three to five nodes that lie along the medial wall of the axilla, around the lateral thoracic vein and inferior border of the pectoralis minor. The pectoral nodes receive lymph mainly from the anterior thoracic wall, including most of the breast (see Chapter 4).

The **subscapular (posterior) nodes** consist of six or seven nodes that lie along the posterior axillary fold and subscapular blood vessels. These nodes receive lymph from the posterior aspect of the thoracic wall and scapular region.

The **humeral (lateral) nodes** consist of four to six nodes that lie along the lateral wall of the axilla, medial and posterior to the axillary vein. These humeral nodes receive nearly all the lymph from the upper limb, except that carried by lymphatic vessels accompanying the cephalic vein, which primarily drain to the apical axillary and infraclavicular nodes (Figs. 3.24 and 3.25).

Efferent lymphatic vessels from the pectoral, subscapular, and humeral nodes pass to the **central nodes**. These nodes consist of three or four large nodes situated deep to the pectoralis minor near the base of the axilla, in association with the second part of the axillary artery. Efferent vessels from the central nodes pass to the apical nodes.

The **apical nodes** are located at the apex of the axilla along the medial side of the axillary vein and the first part of the axillary artery. These nodes receive lymph from all other groups of axillary nodes as well as from lymphatics accompanying the proximal cephalic vein. Efferent vessels from the apical nodes traverse the *cervico-axillary*

(Continued on page 123)

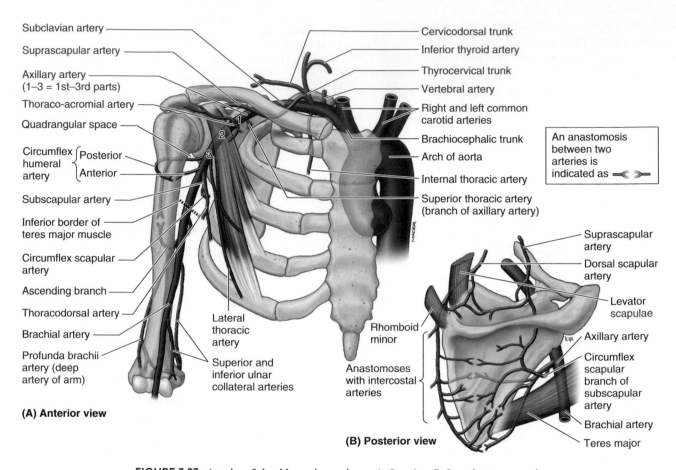

FIGURE 3.23. Arteries of shoulder region and arm. A. Overview. **B.** Scapular anastomosis.

TABLE 3.5. ARTERIES OF PROXIMAL UPPER LIMB (SHOULDER REGION AND ARM)

Artery	Origin	Course
Internal thoracic	Inferior surface of first part ⎫ Subclavian artery	Descends, inclining anteromedially, posterior to sternal end of clavicle and 1st costal cartilage; enters thorax to descend in parasternal plane; gives rise to perforating branches, anterior intercostal, musculophrenic, and superior epigastric arteries
Thyrocervical trunk	Anterior surface of first part ⎭	Ascends as a short trunk often giving rise to two branches: inferior thyroid artery and cervicodorsal trunk. Arising from the cervicodorsal trunk are the suprascapular and dorsal scapular arteries (may also arise directly from thyrocervical trunk).
Suprascapular	Thyrocervical (or as direct branch of subclavian artery)	Passes inferolaterally crossing anterior scalene muscle, phrenic nerve, subclavian artery, and brachial plexus, running laterally posterior and parallel to clavicle; next passes over transverse scapular ligament to supraspinous fossa; then lateral to scapular spine (deep to acromion) to infraspinous fossa on posterior surface of scapula

(continued)

TABLE 3.5. ARTERIES OF PROXIMAL UPPER LIMB (SHOULDER REGION AND ARM) *(continued)*

Artery	Origin		Course
Superior thoracic	First part		Runs anteromedially along superior border of pectoralis minor and then passes between it and pectoralis major to thoracic wall; helps supply 1st and 2nd intercostal spaces and superior part of serratus anterior
Thoraco-acromial	Second part	Axillary artery	Curls around superomedial border of pectoralis minor; pierces costocoracoid membrane (clavipectoral fascia); divides into four branches: pectoral, deltoid, acromial, and clavicular
Lateral thoracic			Descends along axillary border of pectoralis minor; follows it onto thoracic wall, supplying lateral aspect of breast
Circumflex humeral (anterior and posterior)	Third part		Encircle surgical neck of humerus, anastomosing with each other laterally; larger posterior branch traverses quadrangular space.
Subscapular			Descends from level of inferior border of subscapularis along lateral border of scapula, dividing within 2–3 cm into terminal branches, the circumflex scapular and thoracodorsal arteries
Circumflex scapular	Subscapular artery		Curves around lateral border of scapula to enter infraspinous fossa, anastomosing with suprascapular artery
Thoracodorsal			Continues course of subscapular artery, descending with thoracodorsal nerve to enter apex of latissimus dorsi
Profunda brachii artery	Near its origin	Brachial artery	Accompanies radial nerve along radial groove of humerus, supplying posterior compartment of arm and participating in peri-articular cubital arterial anastomoses around elbow joint
Superior ulnar collateral	Near middle of arm		Accompanies ulnar nerve to posterior aspect of elbow; anastomoses with posterior ulnar recurrent artery
Inferior ulnar collateral	Superior to medial epicondyle of humerus		Passes anterior to medial epicondyle of humerus to anastomose with anterior ulnar recurrent artery

CLINICAL BOX

Compression of Axillary Artery

Compression of the third part of the axillary artery against the humerus may be necessary when profuse bleeding occurs. If compression is required at a more proximal site, the axillary artery can be compressed at its origin at the lateral border of the 1st rib by exerting downward pressure in the angle between the clavicle and the attachment of the SCM. See also the Clinical Box on thoracic outlet syndrome.

Arterial Anastomoses Around Scapula

Many *arterial anastomoses* (communications between arteries) occur around the scapula (Fig. 3.23). Several arteries join to form networks on the anterior and posterior surfaces of the scapula: the dorsal scapular, suprascapular, and subscapular (via the circumflex scapular branch). The importance of the *collateral circulation* made possible by these anastomoses becomes apparent when ligation of a lacerated subclavian or axillary artery is necessary. For example, the axillary artery may have to be ligated between the 1st rib and subscapular artery; in other cases, *vascular stenosis* (narrowing) of the axillary artery may result from an atherosclerotic lesion that causes reduced blood flow. In either case, the direction of blood flow in the subscapular artery is reversed, enabling blood to reach the third part of the axillary artery. Note that the subscapular artery receives blood through several anastomoses with the suprascapular artery, transverse cervical artery, and intercostal arteries. *Slow occlusion of an artery* (e.g., resulting from disease) often enables sufficient collateral circulation to develop, preventing *ischemia* (deficiency of blood). Sudden occlusion usually does not allow sufficient time for adequate collateral circulation to develop; as a result, ischemia of the upper limb occurs. *Abrupt surgical ligation of the axillary artery* between the origins of the subscapular and the profunda brachii artery will cut off the blood supply to the arm because the collateral circulation is inadequate.

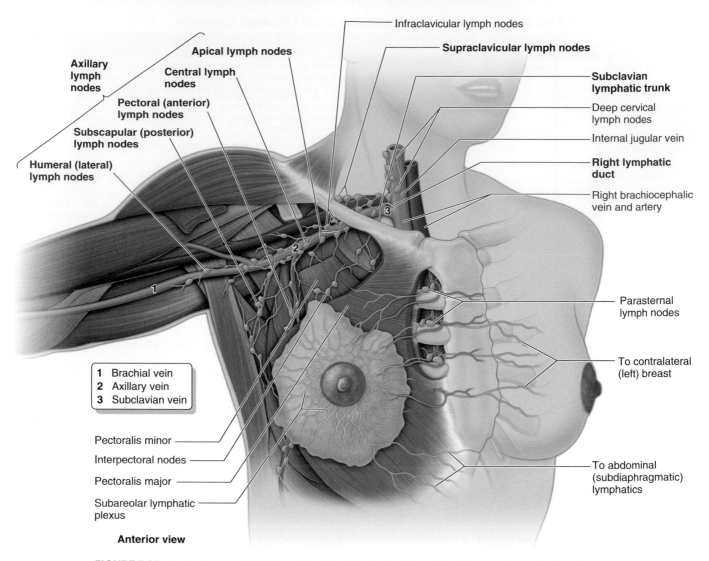

Infraclavicular lymph nodes

Supraclavicular lymph nodes

Apical lymph nodes

Axillary lymph nodes

Central lymph nodes

Subclavian lymphatic trunk

Pectoral (anterior) lymph nodes

Deep cervical lymph nodes

Internal jugular vein

Subscapular (posterior) lymph nodes

Right lymphatic duct

Humeral (lateral) lymph nodes

Right brachiocephalic vein and artery

1 Brachial vein
2 Axillary vein
3 Subclavian vein

Parasternal lymph nodes

To contralateral (left) breast

Pectoralis minor
Interpectoral nodes
Pectoralis major
Subareolar lymphatic plexus

To abdominal (subdiaphragmatic) lymphatics

Anterior view

FIGURE 3.24. Proximal upper limb veins, axillary lymph nodes, and lymphatic drainage of upper limb and breast.

CLINICAL BOX

Injury to Axillary Vein

Wounds in the axilla often involve the axillary vein because of its large size and exposed position. When the arm is fully abducted, the axillary vein overlaps the axillary artery anteriorly. A wound in the proximal part of the vein is particularly dangerous not only because of profuse bleeding but also because of the risk of air entering the vein and producing *air emboli* (air bubbles) in the blood.

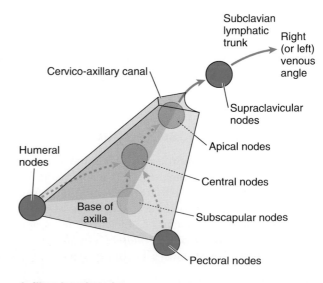

Subclavian lymphatic trunk

Right (or left) venous angle

Cervico-axillary canal

Supraclavicular nodes

Humeral nodes

Apical nodes

Central nodes

Base of axilla

Subscapular nodes

Pectoral nodes

Axillary lymph nodes

FIGURE 3.25. Location and drainage pattern of axillary lymph nodes, schematic illustration.

canal and unite to form the **subclavian lymphatic trunk**, although some vessels may drain en route through the **clavicular (infraclavicular and supraclavicular) nodes**. The subclavian lymphatic trunk may be joined by the jugular and bronchomediastinal trunks on the right side to form the **right lymphatic duct**, or it may enter the **right venous angle** independently (Fig. 3.24). On the left side, the subclavian trunk most commonly joins the **thoracic duct**.

Brachial Plexus

The brachial plexus is a major network of nerves supplying the upper limb. It begins in the lateral cervical region (posterior triangle) and extends into the axilla. The brachial plexus is formed by the union of the anterior rami of the C5–T1 nerves, which constitute the **roots of brachial plexus** (Fig. 3.26 and Table 3.6). The roots usually pass through the gap between the anterior and middle scalene muscles with the subclavian artery. The sympathetic fibers carried by each root of the plexus are received from gray rami of the middle and inferior cervical ganglia as the roots pass between the scalene muscles (see Chapter 9, Neck).

In the inferior part of the neck, the roots of the brachial plexus unite to form three trunks (Figs. 3.27 and 3.28):

- A **superior trunk**, from the union of the C5 and C6 roots
- A **middle trunk**, which is a continuation of the C7 root
- An **inferior trunk**, from the union of the C8 and T1 roots

Each trunk of the brachial plexus divides into anterior and posterior divisions as the plexus passes through the *cervico-axillary canal* posterior to the clavicle. **Anterior divisions**

(Continued on page 128)

CLINICAL BOX

Enlargement of Axillary Lymph Nodes

An infection in the upper limb can cause the axillary nodes to enlarge and become tender and inflamed, a condition called *lymphangitis* (inflammation of lymphatic vessels). The humeral group of nodes is usually the first ones to be involved. Lymphangitis is characterized by warm, red streaks in the skin of the limb. Infections in the pectoral region and breast, including the superior part of the abdomen, can also produce enlargement of the axillary nodes. These nodes are also the most common site of metastases (spread) of cancer of the breast.

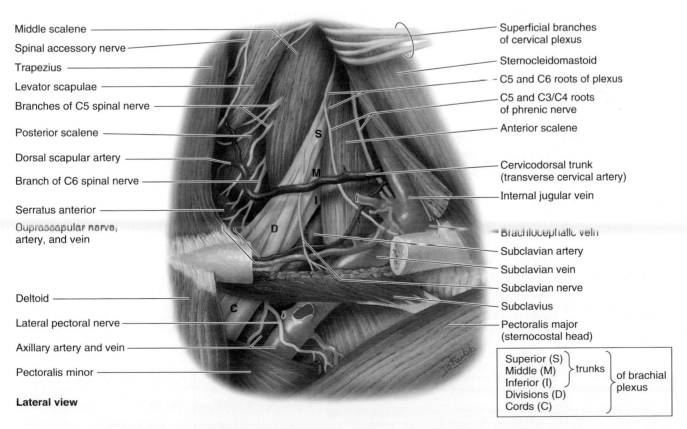

Middle scalene
Spinal accessory nerve
Trapezius
Levator scapulae
Branches of C5 spinal nerve
Posterior scalene
Dorsal scapular artery
Branch of C6 spinal nerve
Serratus anterior
Suprascapular nerve, artery, and vein
Deltoid
Lateral pectoral nerve
Axillary artery and vein
Pectoralis minor

Lateral view

Superficial branches of cervical plexus
Sternocleidomastoid
C5 and C6 roots of plexus
C5 and C3/C4 roots of phrenic nerve
Anterior scalene
Cervicodorsal trunk (transverse cervical artery)
Internal jugular vein
Brachiocephalic vein
Subclavian artery
Subclavian vein
Subclavian nerve
Subclavius
Pectoralis major (sternocostal head)

Superior (S)		
Middle (M)	trunks	of brachial plexus
Inferior (I)		
Divisions (D)		
Cords (C)		

FIGURE 3.26. Brachial plexus and subclavian vessels in lateral cervical region (posterior triangle of neck).

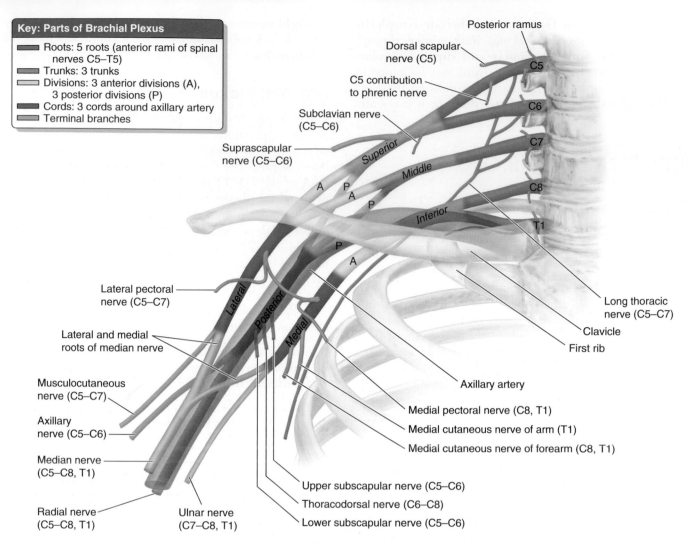

Key: Parts of Brachial Plexus

- Roots: 5 roots (anterior rami of spinal nerves C5–T5)
- Trunks: 3 trunks
- Divisions: 3 anterior divisions (A), 3 posterior divisions (P)
- Cords: 3 cords around axillary artery
- Terminal branches

Posterior ramus

Dorsal scapular nerve (C5)

C5 contribution to phrenic nerve

Subclavian nerve (C5–C6)

Suprascapular nerve (C5–C6)

Superior

Middle

Inferior

C5
C6
C7
C8
T1

Long thoracic nerve (C5–C7)

Clavicle

First rib

Lateral pectoral nerve (C5–C7)

Lateral and medial roots of median nerve

Musculocutaneous nerve (C5–C7)

Axillary nerve (C5–C6)

Median nerve (C5–C8, T1)

Radial nerve (C5–C8, T1)

Ulnar nerve (C7–C8, T1)

Axillary artery

Medial pectoral nerve (C8, T1)

Medial cutaneous nerve of arm (T1)

Medial cutaneous nerve of forearm (C8, T1)

Upper subscapular nerve (C5–C6)

Thoracodorsal nerve (C6–C8)

Lower subscapular nerve (C5–C6)

FIGURE 3.27. Brachial plexus.

TABLE 3.6. BRACHIAL PLEXUS AND NERVES OF UPPER LIMB

Nerve	Origin[a]	Course	Structures Innervated
Supraclavicular branches			
Dorsal scapular	Posterior aspect of anterior ramus of **C5** with a frequent contribution from C4	Pierces middle scalene; descends deep to levator scapulae and rhomboids	Rhomboids; occasionally supplies levator scapulae
Long thoracic	Posterior aspect of anterior rami of **C5**, **C6**, C7	Superior two rami pierce middle scalene; passes through cervico-axillary canal, descending posterior to C8 and T1 anterior rami; runs inferiorly on superficial surface of serratus anterior	Serratus anterior
Suprascapular	Superior trunk, receiving fibers from **C5**, C6, and often C4	Passes laterally across lateral cervical region (posterior triangle of neck), superior to brachial plexus; then through scapular notch deep to transverse scapular ligament	Supraspinatus and infraspinatus muscles; glenohumeral (shoulder) joint
Subclavian nerve (nerve to subclavius)	Superior trunk, receiving fibers from C5, **C6**, and often C4	Descends posterior to clavicle and anterior to brachial plexus and subclavian artery; often giving an *accessory root to phrenic nerve*	Subclavius and sternoclavicular joint (accessory phrenic root innervates diaphragm)

(continued)

TABLE 3.6. BRACHIAL PLEXUS AND NERVES OF UPPER LIMB (*continued*)

Nerve	Origin[a]	Course	Structures Innervated
Infraclavicular branches			
Lateral pectoral	Side branch of lateral cord, receiving fibers from C5, **C6**, C7	Pierces costocoracoid membrane to reach deep surface of pectoral muscles; a *communicating branch to the medial pectoral nerve* passes anterior to axillary artery and vein.	Primarily pectoralis major, but some lateral pectoral nerve fibers pass to pectoralis minor via branch to medial pectoral nerve
Musculocutaneous	Terminal branch of lateral cord, receiving fibers from C5–C7	Exits axilla by piercing coracobrachialis; descends between biceps brachii and brachialis, supplying both; continues as *lateral cutaneous nerve of forearm*	Muscles of anterior compartment of arm (coracobrachialis, biceps brachii, and brachialis); skin of lateral aspect of forearm
Median	*Lateral root of median nerve* is a terminal branch of lateral cord (C6, C7 fibers); *medial root of median nerve* is a terminal branch of medial cord (C8, T1 fibers).	Lateral and medial roots merge to form median nerve lateral to axillary artery; descends through arm adjacent to brachial artery, with nerve gradually crossing anterior to artery to lie medial to artery in cubital fossa	Muscles of anterior forearm compartment (except for flexor carpi ulnaris and ulnar half of flexor digitorum profundus); five intrinsic muscles in thenar half of palm and palmar skin
Medial pectoral		Passes between axillary artery and vein and then pierces pectoralis minor and enters deep surface of pectoralis major; although it is called *medial* for its origin from medial cord, it lies lateral to lateral pectoral nerve.	Pectoralis minor and sternocostal part of pectoralis major
Medial cutaneous nerve of arm	Side branches of medial cord, receiving fibers from C8, T1	Smallest nerve of plexus; runs along medial side of axillary and brachial veins; communicates with *intercostobrachial nerve*	Skin of medial side of arm, as far distally as medial epicondyle of humerus and olecranon of ulna
Median cutaneous nerve of forearm		Initially runs with ulnar nerve (with which it may be confused) but pierces deep fascia with basilic vein and enters subcutaneous tissue, dividing into anterior and posterior branches	Skin of medial side of forearm, as far distally as wrist
Ulnar	Larger terminal branch of medial cord, receiving fibers from C8, T1, and often C7	Descends medial arm, passes posterior to medial epicondyle of humerus, then descends ulnar aspect of forearm to hand	Flexor carpi ulnaris and ulnar half of flexor digitorum profundus (forearm); most intrinsic muscles of hand; skin of hand medial to axial line of digit 4
Upper subscapular	Side branch of posterior cord, receiving fibers from **C5**	Passes posteriorly, entering subscapularis directly	Superior portion of subscapularis
Lower subscapular	Side branch of posterior cord, receiving fibers from **C6**	Passes inferolaterally, deep to subscapular artery and vein	Inferior portion of subscapularis and teres major
Thoracodorsal	Side branch of posterior cord, receiving fibers from C6, **C7**, C8	Arises between upper and lower subscapular nerves and runs inferolaterally along posterior axillary wall to apical part of latissimus dorsi	Latissimus dorsi
Axillary	Terminal branch of posterior cord, receiving fibers from **C5**, C6	Exits axillary fossa posteriorly, passing through quadrangular space[b] with posterior circumflex humeral artery; gives rise to *superior lateral brachial cutaneous nerve*; then winds around surgical neck of humerus deep to deltoid	Glenohumeral (shoulder) joint, teres minor and deltoid muscles, skin of superolateral arm (over inferior part of deltoid)
Radial	Larger terminal branch of posterior cord (largest branch of plexus), receiving fibers from C5–T1	Exits axillary fossa posterior to axillary artery; passes posterior to humerus in radial groove with profunda brachii artery between lateral and medial heads of triceps; perforates lateral intermuscular septum; enters cubital fossa, dividing into *superficial* (cutaneous) and *deep* (motor) *branches*	All muscles of posterior compartments of arm and forearm; skin of posterior and inferolateral arm, posterior forearm, and dorsum of hand lateral to axial line of digit 4

[a]Boldface **C5** indicates primary component of the nerve.
[b]Bounded superiorly by the subscapularis, head of humerus, and teres minor; inferiorly by the teres major; medially by the long head of the triceps; and laterally by the coracobrachialis and surgical neck of the humerus.

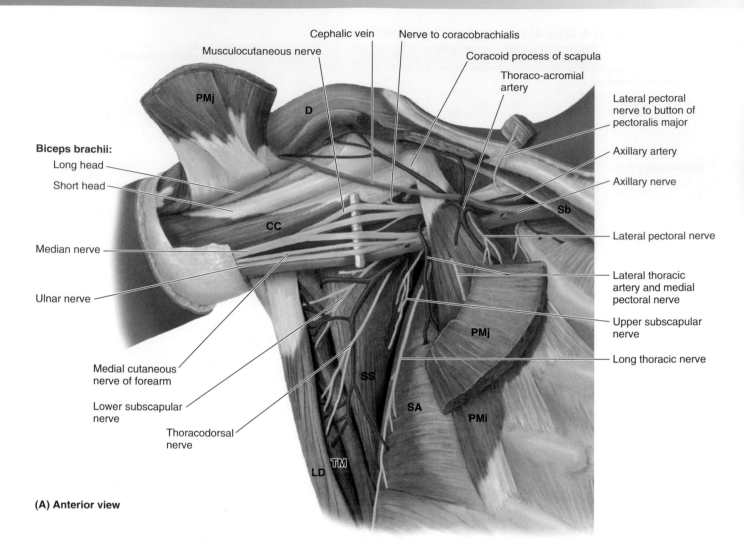

(A) Anterior view

Cephalic vein

Musculocutaneous nerve

Nerve to coracobrachialis

Coracoid process of scapula

Thoraco-acromial artery

Lateral pectoral nerve to button of pectoralis major

Axillary artery

Axillary nerve

Lateral pectoral nerve

Lateral thoracic artery and medial pectoral nerve

Upper subscapular nerve

Long thoracic nerve

Biceps brachii:
Long head
Short head

Median nerve

Ulnar nerve

Medial cutaneous nerve of forearm

Lower subscapular nerve

Thoracodorsal nerve

PMj

D

CC

Sb

PMj

SS

SA

PMi

LD TM

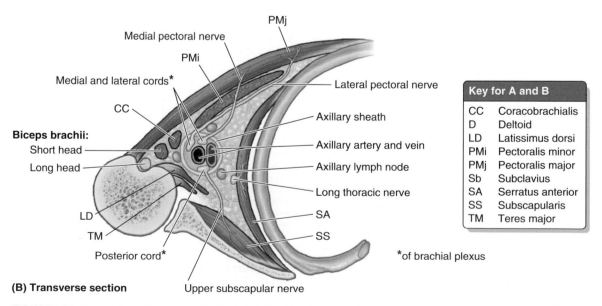

(B) Transverse section

PMj

Medial pectoral nerve

PMi

Medial and lateral cords*

CC

Biceps brachii:
Short head
Long head

LD

TM

Posterior cord*

Upper subscapular nerve

Lateral pectoral nerve

Axillary sheath

Axillary artery and vein

Axillary lymph node

Long thoracic nerve

SA

SS

*of brachial plexus

Key for A and B	
CC	Coracobrachialis
D	Deltoid
LD	Latissimus dorsi
PMi	Pectoralis minor
PMj	Pectoralis major
Sb	Subclavius
SA	Serratus anterior
SS	Subscapularis
TM	Teres major

FIGURE 3.28. Boundaries and contents of axilla. A. Relationship of nerves and vessels to pectoralis minor. **B.** Contents of axilla, transverse section. *(continued)*

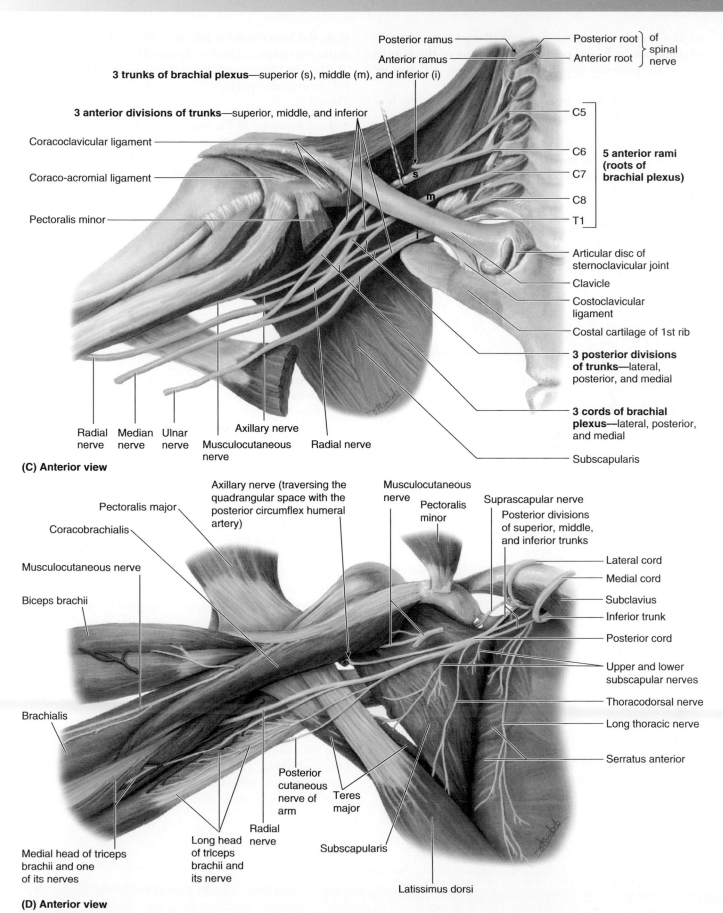

Posterior ramus — Posterior root ⎫ of
Anterior ramus — Anterior root ⎬ spinal
                                nerve

3 trunks of brachial plexus—superior (s), middle (m), and inferior (i)

3 anterior divisions of trunks—superior, middle, and inferior

Coracoclavicular ligament

Coraco-acromial ligament

Pectoralis minor

C5
C6   5 anterior rami
C7   (roots of
C8   brachial plexus)
T1

Articular disc of
sternoclavicular joint

Clavicle

Costoclavicular
ligament

Costal cartilage of 1st rib

**3 posterior divisions
of trunks**—lateral,
posterior, and medial

**3 cords of brachial
plexus**—lateral, posterior,
and medial

Subscapularis

Radial   Median   Ulnar
nerve   nerve   nerve   Axillary nerve
         Musculocutaneous   Radial nerve
         nerve

(C) Anterior view

Axillary nerve (traversing the   Musculocutaneous
quadrangular space with the   nerve
posterior circumflex humeral   Pectoralis   Suprascapular nerve
artery)   minor
                                                        Posterior divisions
Pectoralis major   of superior, middle,
                                                        and inferior trunks
Coracobrachialis
                                                        Lateral cord
Musculocutaneous nerve   Medial cord

Biceps brachii   Subclavius
                                                        Inferior trunk
                                                        Posterior cord

                                                        Upper and lower
                                                        subscapular nerves

                                                        Thoracodorsal nerve

                                                        Long thoracic nerve

Brachialis   Serratus anterior

                Posterior
                cutaneous
                nerve of   Teres
                arm   major
                   Radial
Medial head of triceps   nerve
brachii and one   Long head   Subscapularis
of its nerves   of triceps
                  brachii and
                  its nerve   Latissimus dorsi

(D) Anterior view

FIGURE 3.28 Boundaries and contents of axilla. *(continued)* **C.** Formation of brachial plexus. **D.** Posterior wall of axilla with posterior cord of brachial plexus and its branches.

of the trunks supply the *anterior (flexor) compartments* of the upper limb, and **posterior divisions of the trunks** supply the *posterior (extensor) compartments* of the upper limb.

The divisions of the trunks form three cords of the brachial plexus within the axilla (Figs. 3.27 and 3.28C):

- Anterior divisions of the superior and middle trunks unite to form the **lateral cord**.
- The anterior division of the inferior trunk continues as the **medial cord**.
- Posterior divisions of all three trunks unite to form the **posterior cord**.

The cords of the brachial plexus are named for their position in relation to the second part of the axillary artery (e.g., the lateral cord is lateral to the axillary artery, most easily seen when the limb is abducted).

The brachial plexus is divided into **supraclavicular** and **infraclavicular parts** by the clavicle (Figs. 3.27 and 3.28; Table 3.6):

- Four *branches of the supraclavicular part of the plexus* arise from the roots (anterior rami) and trunks of the plexus (dorsal scapular nerve, long thoracic nerve, nerve to the subclavius, and suprascapular nerve) and are approachable through the neck. *Muscular branches* arise from the anterior rami of C5–T1 to supply the scalene and longus colli muscles.
- *Branches of the infraclavicular part of the plexus* arise from the cords of the brachial plexus and are approachable through the axilla.

CLINICAL BOX

Variations of Brachial Plexus

Variations in the brachial plexus formation are common. In addition to the five anterior rami (C5–T1) that form the roots of the plexus, small contributions may be made by the anterior rami of C4 or T2. When the superiormost root (anterior ramus) of the plexus is C4 and the inferiormost root is C8, it is called a *prefixed brachial plexus*. Alternatively, when the superior root is C6 and the inferior root is T2, it is a *postfixed brachial plexus*. In the latter type, the inferior trunk of the plexus may be compressed by the 1st rib, producing neurovascular symptoms in the upper limb. Variations also may occur in the formation of trunks, divisions, and cords; in the origin and/or combination of branches; and in the relationship to the axillary artery and scalene muscles.

Brachial Plexus Injuries

Injuries to the brachial plexus affect movements and cutaneous sensations in the upper limb. Disease, stretching, and wounds in the lateral cervical region (posterior triangle of the neck) or in the axilla may produce brachial plexus injuries (see Chapter 9). Signs and symptoms depend on which part of the plexus is involved. Injuries to the brachial plexus result in loss of muscular movement (*paralysis*) and loss of cutaneous sensation (*anesthesia*). In *complete paralysis*, no movement is detectable. In *incomplete paralysis*, not all muscles are paralyzed; therefore, the person can move, but the movements are weak compared to those on the uninjured side.

Injuries to superior parts of the brachial plexus (C5 and C6; Klumpke paralysis) usually result from an excessive increase in the angle between the neck and the shoulder. These injuries can occur in a person who is thrown from a motorcycle or a horse and lands on the shoulder in a way that widely separates the neck and shoulder (Fig. B3.6A). When thrown, the person's shoulder often hits something (e.g., a tree or the ground) and stops, but the head and trunk continue to move. This stretches or ruptures superior parts of the brachial plexus or **avulses** (tears) the roots of the plexus from the spinal cord. Injury to the superior trunk is apparent by the characteristic position of the limb ("waiter's tip position") in which the limb hangs by the side in medial rotation (Fig. B3.6B). *Upper brachial plexus injuries* can also occur in a newborn when excessive stretching of the neck occurs during delivery (Fig. B3.6C). As a result of injuries to the superior parts of the brachial plexus (*Erb-Duchenne palsy*), paralysis of the muscles of the shoulder and arm supplied by C5–C6 occurs. The usual clinical appearance is an upper limb with an adducted shoulder, medially rotated arm, and extended elbow. The lateral aspect of the upper limb also experiences loss of sensation. *Chronic microtrauma* to the superior trunk of the brachial plexus from carrying a heavy

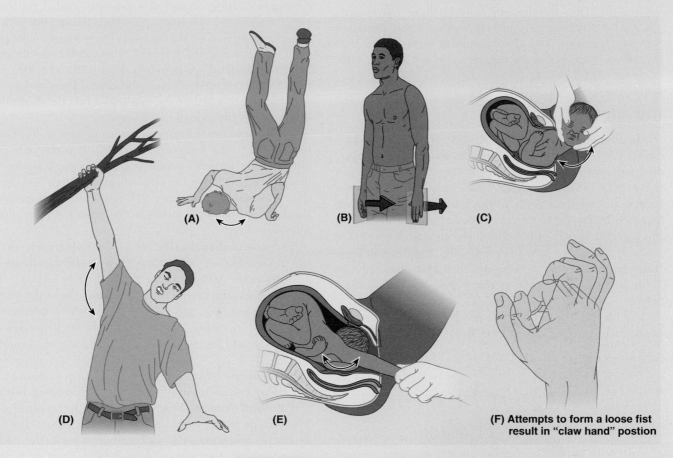

(F) Attempts to form a loose fist result in "claw hand" postion

FIGURE B3.6. Brachial plexus injuries.

backpack can produce motor and sensory deficits in the distribution of the musculocutaneous and radial nerves.

Injuries to inferior parts of the brachial plexus (Klumpke paralysis) are much less common. These injuries may occur when the upper limb is suddenly pulled superiorly—for example, when a person grasps something to break a fall or when a baby's limb is pulled excessively during delivery (Fig. B3.6D,E). These events injure the inferior trunk of the plexus (C8 and T1) and may avulse the roots of the spinal nerves from the spinal cord. The short muscles of the hand are affected and a *claw hand* results (Fig. B3.6F).

Brachial Plexus Block

Injection of an anesthetic solution into or immediately surrounding the axillary sheath interrupts nerve impulses and produces anesthesia of the structures supplied by the branches of the cords of the plexus. Combined with an occlusive tourniquet technique to retain the anesthetic agent, this procedure enables surgeons to operate on the upper limb without using a general anesthetic. The brachial plexus can be anesthetized using a number of approaches, such as interscalene, supraclavicular, and axillary.

ARM

The arm extends from the shoulder to the elbow. Two types of movement occur between the arm and the forearm at the elbow joint: flexion–extension and pronation–supination. The muscles performing these movements are clearly divided into anterior (*flexor*) and posterior (*extensor*) groups. The chief action of both groups is at the elbow joint, but some muscles also act at the glenohumeral joint.

Muscles of Arm

Of the four arm muscles, three flexors (biceps brachii, brachialis, and coracobrachialis) are in the anterior (flexor) compartment and are supplied by the musculocutaneous nerve (Figs. 3.28A and 3.29). One three-headed extensor muscle (triceps brachii) is in the posterior compartment, supplied by the radial nerve. A small triangular muscle on the posterior aspect of the elbow, the anconeus, covers the posterior aspect of the ulna proximally. Figure 3.30 illustrates the muscles of the arm and their bony attachments; Table 3.7 lists their attachments, nerve supply, and main actions.

The **biceps brachii** has two heads (*bi*, two + L. *caput*, head): a **long head** and a **short head**. A broad band, the

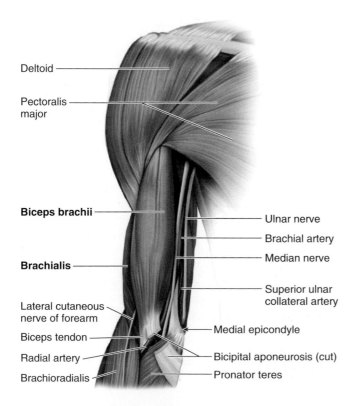

Deltoid

Pectoralis major

Biceps brachii

Brachialis

Lateral cutaneous nerve of forearm

Biceps tendon

Radial artery

Brachioradialis

Ulnar nerve

Brachial artery

Median nerve

Superior ulnar collateral artery

Medial epicondyle

Bicipital aponeurosis (cut)

Pronator teres

FIGURE 3.29. Muscles, arteries, and nerves of anterior arm.

transverse humeral ligament, passes from the lesser to the greater tubercle of the humerus and converts the intertubercular groove into a canal for the tendon of the long head of the biceps. When the elbow is extended, the biceps is a simple flexor of the elbow joint; however, as the elbow flexion approaches 90 degrees and more power is needed, the biceps with the forearm in supination produces flexion, but with the forearm in pronation, the biceps is the primary (most powerful) supinator of the forearm. A triangular membranous band, the **bicipital aponeurosis** (Figs. 3.29 and 3.30A), runs from the biceps tendon across the cubital fossa and merges with the antebrachial (deep) fascia covering the flexor muscles in the medial side of the forearm.

The **brachialis**, a flattened fusiform muscle, lies posterior (deep) to the biceps (Fig. 3.30A). It is the only pure elbow flexor muscle, producing the greatest amount of flexion force. It flexes the elbow in all positions and during slow and quick movements. When the elbow is extended slowly, the brachialis steadies the movement by slowly relaxing.

The **coracobrachialis**, an elongated muscle in the superomedial part of the arm, is a useful landmark for locating other structures in the arm (Fig. 3.30A). The musculocutaneous nerve pierces it, and the distal part of its attachment indicates the location of the nutrient foramen of the humerus. The coracobrachialis helps flex and adduct the arm and stabilize the glenohumeral (shoulder) joint.

The **triceps brachii** is a large fusiform muscle in the posterior compartment of the arm that has **long**, **lateral**, and **medial heads** (Figs. 3.30B and 3.31; Table 3.7). The triceps is the chief extensor of the elbow. Because its long head crosses the glenohumeral joint, the triceps helps stabilize the adducted joint by serving as a shunt muscle, resisting inferior displacement of the head of the humerus along with the deltoid and coracobrachialis. Just proximal to the distal attachment of the triceps is a friction-reducing *subtendinous olecranon bursa*, between the triceps tendon and the olecranon. The **anconeus** muscle assists the triceps extend the elbow joint and may abduct the ulna during pronation of the forearm (Fig. 3.30B and Table 3.7).

Arteries and Veins of Arm

The **brachial artery** provides the main arterial supply to the arm and is the continuation of the axillary artery (Figs. 3.29, 3.31D, and 3.32; Table 3.5). It begins at the inferior border of the teres major and ends in the cubital fossa opposite the neck of the radius under cover of the bicipital aponeurosis, where it divides into the radial and ulnar arteries. The brachial artery, relatively superficial and palpable throughout its course, lies anterior to the triceps and brachialis. Initially, it is medial to the humerus, deep to the

(Continued on page 133)

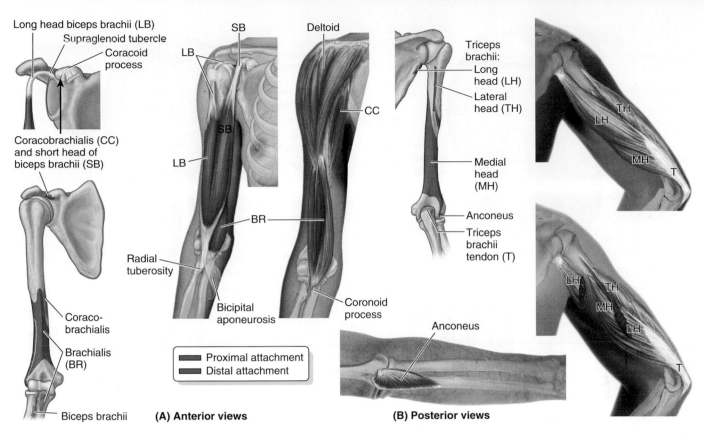

FIGURE 3.30. Muscles of arm and bony attachments. A. Muscles of anterior compartment. **B.** Muscles of posterior compartment.

TABLE 3.7. MUSCLES OF ARM

Muscle	Proximal Attachment	Distal Attachment	Innervation[a]	Main Action(s)
Biceps brachii	Short head: tip of coracoid process of scapula Long head: supraglenoid tubercle of scapula	Tuberosity of radius and fascia of forearm via bicipital aponeurosis	Musculocutaneous nerve[b] (C5, **C6**)	Supinates forearm and, when it is supinated, flexes elbow joint; flexes shoulder joint; short head resists dislocation of shoulder.
Brachialis	Distal half of anterior surface of humerus	Coronoid process and tuberosity of ulna		Flexes elbow joint in all positions
Coracobrachialis	Tip of coracoid process of scapula	Middle third of medial surface of humerus	Musculocutaneous nerve (C5, **C6**, C7)	Helps flex and adduct shoulder joint; resists dislocation of shoulder
Triceps brachii	Long head: infraglenoid tubercle of scapula Lateral head: posterior surface of humerus, superior to radial groove Medial head: posterior surface of humerus, inferior to radial groove	Proximal end of olecranon of ulna and fascia of forearm	Radial nerve (C6, **C7, C8**)	Chief extensor of elbow joint; long head extends shoulder joint and resists dislocation of humerus (especially important during abduction).
Anconeus	Lateral epicondyle of humerus	Lateral surface of olecranon and superior part of posterior surface of ulna	Radial nerve (C7, C8, T1)	Assists triceps in extending elbow joint; stabilizes elbow joint; abducts ulna during pronation

[a]The spinal cord segmental innervation is indicated (e.g., "C5, **C6**" means that the nerves supplying the biceps brachii are derived from the 5th and 6th cervical segments of the spinal cord). Numbers in boldface (**C6**) indicate the main segmental innervation. Damage to one or more of the listed spinal cord segments or to the motor nerve roots arising from them results in paralysis of the muscles concerned.
[b]Some of the lateral part of the brachialis is innervated by a branch of the radial nerve.

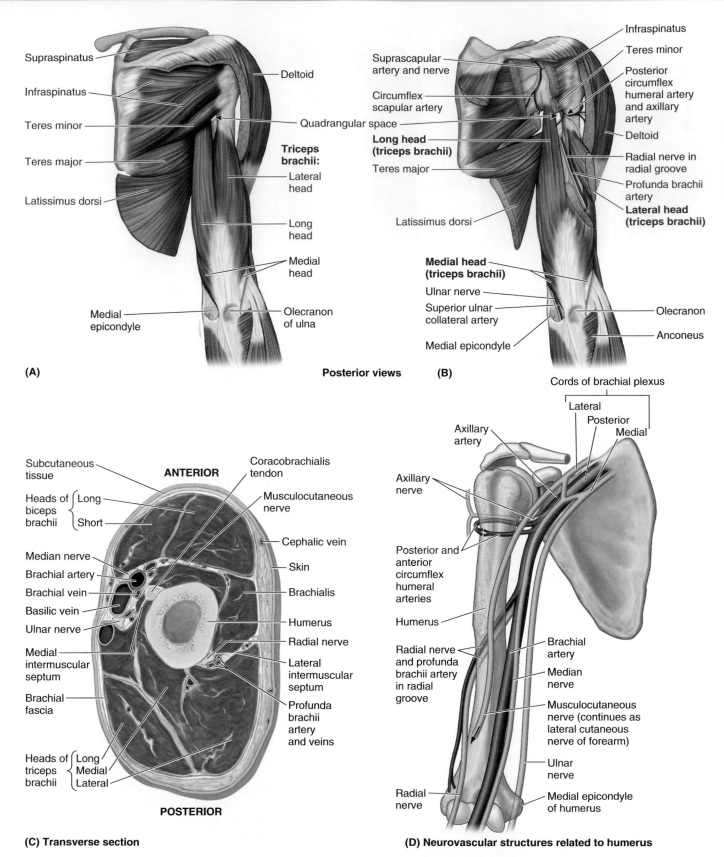

(A)   **Posterior views**   **(B)**

(C) Transverse section

(D) Neurovascular structures related to humerus

FIGURE 3.31. Muscles, arteries, and nerves of posterior arm. A. Superficial dissection. **B.** Deep dissection. **C.** Transverse section. **D.** Relationship of arteries and nerves to humerus.

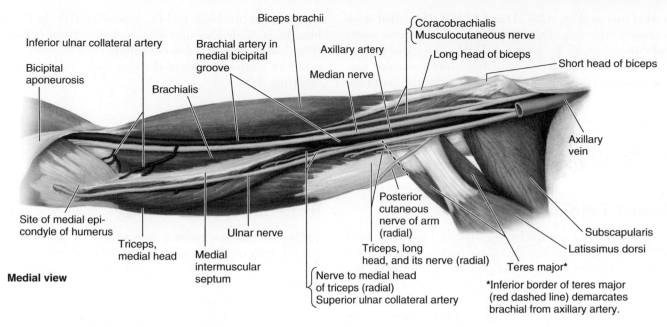

FIGURE 3.32. Muscles and neurovascular structures of arm.

medial bicipital groove. It then runs anterior to the medial supra-epicondylar ridge and trochlea of the humerus. During its inferolateral course, the brachial artery accompanies the median nerve, which crosses anterior to the artery. During its course through the arm, the brachial artery gives rise to unnamed *muscular branches* and the *humeral nutrient artery*, which arise from its lateral aspect. The main named branches of the brachial artery that arise from its medial aspect are the profunda brachii artery (deep artery of arm) (Fig. 3.31B,D) and the **superior** and **inferior ulnar collateral arteries** (Fig. 3.32). The latter vessels help form the peri-articular **cubital anastomosis** of the elbow region (Table 3.5; see also Fig. 3.43).

Two sets of *veins of the arm*, superficial and deep, anastomose freely with each other. The two main **superficial veins of the arm**, the *cephalic* and *basilic veins*, are described earlier (Figs. 3.11 and 3.28A). Paired deep veins, collectively constituting the **brachial vein,** accompany the brachial artery (see Fig. 3.12). The brachial vein begins at the elbow by union of the *accompanying veins of the ulnar and radial arteries* and ends by merging with the basilic vein to form the axillary vein. Both superficial and deep veins have valves, but the deep veins have more.

Nerves of Arm

Four main nerves pass through the arm: median, ulnar, musculocutaneous, and radial (Figs. 3.28, 3.29, 3.31, and 3.32; Table 3.6). The **median nerve** in the arm is formed in the axilla by the union of medial and lateral roots from the medial and lateral cords of the brachial plexus, respectively (see Fig. 3.28A,C). The nerve runs distally in the arm, initially on the lateral side of the brachial artery until it reaches the middle of the arm, where it crosses to the medial side and contacts the brachialis (Fig. 3.32). The median nerve then descends into the cubital fossa, where it lies deep to the bicipital aponeurosis and median cubital vein. The median and ulnar nerves supply no branches to the arm; however, they supply articular branches to the elbow joint.

The **ulnar nerve** in the arm arises from the medial cord of the brachial plexus, conveying fibers mainly from the C8 and T1 nerves (see Fig. 3.28C). It passes distally, anterior to the insertion of teres major and to the long head of triceps, on the medial side of the brachial artery. Around the middle of the arm, it pierces the medial intermuscular septum with the superior ulnar collateral artery and descends between the septum and the medial head of triceps. The ulnar nerve passes posterior to the medial epicondyle of the humerus to enter the forearm (see Figs. 3.26 and 3.33).

The **musculocutaneous nerve** arises from the lateral cord of the brachial plexus, pierces the coracobrachialis, and then continues distally between the brachialis and the biceps (see Fig. 3.28A,C). After supplying all three muscles of the anterior compartment of the arm, the nerve emerges lateral to the biceps brachii as the *lateral cutaneous nerve of the forearm* (Fig. 3.29).

The **radial nerve** enters the arm posterior to the brachial artery, medial to the humerus, and anterior to the long head of triceps (Figs. 3.28C,D and 3.31D). The radial nerve descends inferolaterally with the profunda brachii artery and curves around the humeral shaft in the radial groove. The radial nerve pierces the lateral intermuscular septum and continues inferiorly in the anterior compartment between the brachialis and the brachioradialis. In the cubital fossa, it divides into *deep* and *superficial branches* (Fig. 3.33B). The radial nerve supplies the muscles in the posterior compartments of the arm and forearm and the overlying skin.

CLINICAL BOX

Biceps Tendinitis

The tendon of the long head of the biceps, enclosed by a synovial sheath, moves back and forth in the intertubercular sulcus (groove) of the humerus. Wear and tear of this mechanism can cause shoulder pain. Inflammation of the tendon (*biceps tendinitis*) usually is the result of repetitive microtrauma in sports involving throwing (e.g., baseball).

Rupture of Tendon of Long Head of Biceps

Rupture of the tendon of the long head of the biceps usually results from wear and tear of an inflamed tendon (*biceps tendinitis*). Normally, the tendon is torn from its attachment to the supraglenoid tubercle of the scapula. The rupture is commonly dramatic and is associated with a snap or pop. The detached muscle belly forms a ball near the center of the distal part of the anterior aspect of the arm (Popeye deformity) (Fig. B3.7).

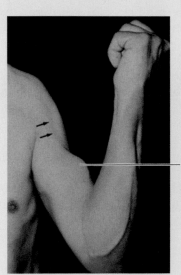

Arrows indicate site of adhered distal fragment of ruptured tendon of long head of biceps brachii.

Distally displaced belly of long head of biceps brachii

FIGURE B3.7. Rupture of biceps tendon.

Bicipital Myotatic Reflex

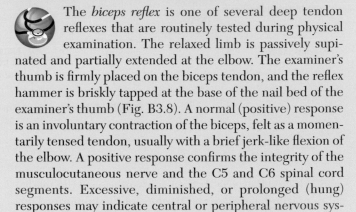

The *biceps reflex* is one of several deep tendon reflexes that are routinely tested during physical examination. The relaxed limb is passively supinated and partially extended at the elbow. The examiner's thumb is firmly placed on the biceps tendon, and the reflex hammer is briskly tapped at the base of the nail bed of the examiner's thumb (Fig. B3.8). A normal (positive) response is an involuntary contraction of the biceps, felt as a momentarily tensed tendon, usually with a brief jerk-like flexion of the elbow. A positive response confirms the integrity of the musculocutaneous nerve and the C5 and C6 spinal cord segments. Excessive, diminished, or prolonged (hung) responses may indicate central or peripheral nervous system disease.

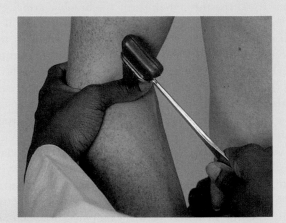

FIGURE B3.8. Method of eliciting biceps reflex.

Injury to Musculocutaneous Nerve

Injury to the musculocutaneous nerve in the axilla is usually inflicted by a weapon such as a knife. A musculocutaneous nerve injury results in *paralysis of the coracobrachialis, biceps, and brachialis*; consequently, flexion of the elbow and supination of the forearm are greatly weakened. Loss of sensation may occur on the lateral surface of the forearm supplied by the lateral cutaneous nerve of the forearm.

Injury to Radial Nerve

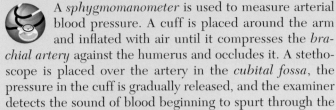

Injury to the radial nerve superior to the origin of its branches to the triceps brachii results in *paralysis of the triceps*, *brachioradialis*, *supinator, and extensor muscles of the wrist and fingers*. Loss of sensation occurs in areas of skin supplied by this nerve. When the radial nerve is injured in the radial groove, the triceps is usually not completely paralyzed but only weakened because only the medial head is affected; however, the muscles in the posterior compartment of the forearm that are supplied by more distal branches of the radial nerve are paralyzed. The characteristic clinical sign of radial nerve injury is *wrist-drop* (inability to extend the wrist and fingers at the metacarpophalangeal joints) (Fig. B3.9). Instead, the wrist remains in the flexed position because of unopposed tonus of the flexor muscles and gravity.

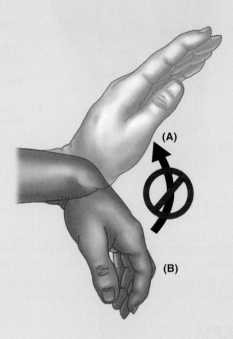

FIGURE B3.9. Wrist-drop.

Occlusion or Laceration of Brachial Artery

Although collateral pathways confer some protection against gradual, temporary, and partial occlusion, sudden complete *occlusion or laceration of the brachial artery* creates a surgical emergency because paralysis of muscles results from ischemia within a few hours. After this, fibrous scar tissue develops and causes the involved muscles to shorten permanently, producing a flexion deformity—*ischemic compartment syndrome* (Volkmann ischemic contracture). Flexion of the fingers and sometimes the wrist results in loss of hand power.

Measuring Blood Pressure

A *sphygmomanometer* is used to measure arterial blood pressure. A cuff is placed around the arm and inflated with air until it compresses the *brachial artery* against the humerus and occludes it. A stethoscope is placed over the artery in the *cubital fossa*, the pressure in the cuff is gradually released, and the examiner detects the sound of blood beginning to spurt through the artery. The first audible spurt indicates *systolic blood pressure*. As the pressure is completely released, the point at which the pulse can no longer be heard indicates *diastolic blood pressure*.

Compression of Brachial Artery

The best place to compress the brachial artery to control hemorrhage is near the middle of the arm. The biceps must be pushed laterally to detect pulsations of the artery (Fig. B3.10). Because the arterial anastomoses around the elbow provide a functionally and surgically important collateral circulation, the brachial artery may be clamped distal to the inferior ulnar collateral artery without producing tissue damage. The anatomical basis for this is that the ulnar and radial arteries still receive sufficient blood through the anastomoses. Ischemia of the elbow and forearm results from clamping the brachial artery proximal to the deep artery of the arm for an extended period.

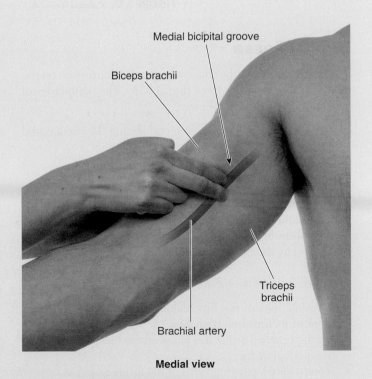

Medial bicipital groove

Biceps brachii

Triceps brachii

Brachial artery

Medial view

FIGURE B3.10. Compression of brachial artery.

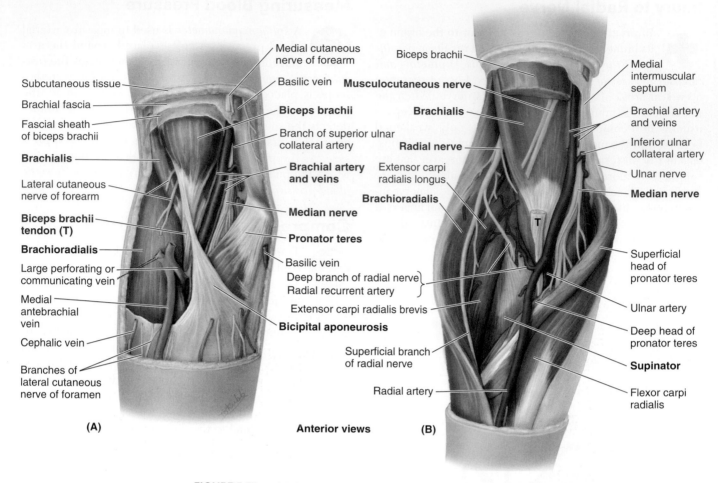

FIGURE 3.33. **Cubital fossa. A.** Superficial dissection. **B.** Deep dissection.

Cubital Fossa

The **cubital fossa** is the shallow triangular depression on the anterior surface of the elbow (Fig. 3.33A). The boundaries of the cubital fossa are as follows:

- Superiorly, an imaginary line connecting the medial and lateral epicondyles
- Medially, the pronator teres
- Laterally, the brachioradialis

The *floor of the cubital fossa* is formed by the brachialis and supinator muscles. The *roof of the cubital fossa* is formed by the continuity of brachial and antebrachial (deep) fascia, reinforced by the bicipital aponeurosis, subcutaneous tissue, and skin.

The *contents of the cubital fossa* are include the (Fig. 3.33B)

- Terminal part of the brachial artery and the commencement of its terminal branches, the radial and ulnar arteries; the brachial artery lies between the biceps tendon and the median nerve.
- (Deep) accompanying veins of the arteries
- Biceps brachii tendon
- Median nerve
- Radial nerve, dividing into superficial and deep branches

In the subcutaneous tissue overlying the cubital fossa are the *median cubital vein* (see Fig. 3.11A), lying anterior to the brachial artery, and the *medial and lateral cutaneous nerves of the forearm*, related to the basilic and cephalic veins (Fig. 3.33A).

FOREARM

The **forearm** is between the elbow and the wrist and contains two bones, the *radius* and *ulna*, which are joined by an interosseous membrane (Fig. 3.34). The role of forearm movement, occurring at the elbow and radio-ulnar joints, is to assist the shoulder in the application of force and in controlling the placement of the hand in space.

Muscles of Forearm

The tendons of the forearm muscles pass through the distal part of the forearm and continue into the wrist, hand, and fingers. The flexors and pronators of the forearm are in the anterior compartment and are served mainly by the *median nerve*; the one and a half exceptions are innervated by the *ulnar nerve*. The extensors and supinators of the forearm are in the posterior compartment and are all innervated by the *radial nerve* (Fig. 3.34).

SURFACE ANATOMY

Arm and Cubital Fossa

The **borders of the deltoid** are visible when the arm is abducted against resistance. The **distal attachment of the deltoid** can be palpated on the lateral surface of the humerus. The **three heads of the triceps** form a bulge on the posterior aspect of the arm and are identifiable when the forearm is extended from the flexed position against resistance (Fig. SA3.3A). The **triceps tendon** may be felt as it descends along the posterior aspect of the arm to the olecranon. The **belly of the biceps brachii** forms a bulge on the anterior aspect of the arm; it becomes more prominent when the elbow is flexed and supinated against resistance (Fig. SA3.3B). Medial and lateral **bicipital grooves** separate the bulges formed by the biceps and triceps. The cephalic vein runs superiorly in the lateral bicipital groove, and the basilic vein ascends in the medial bicipital groove. The **biceps tendon** can be palpated in the cubital fossa, immediately lateral to the midline. The proximal part of the **bicipital aponeurosis** can be palpated where it passes obliquely over the brachial artery and median nerve. The **brachial artery** may be felt pulsating deep to the medial bicipital groove.

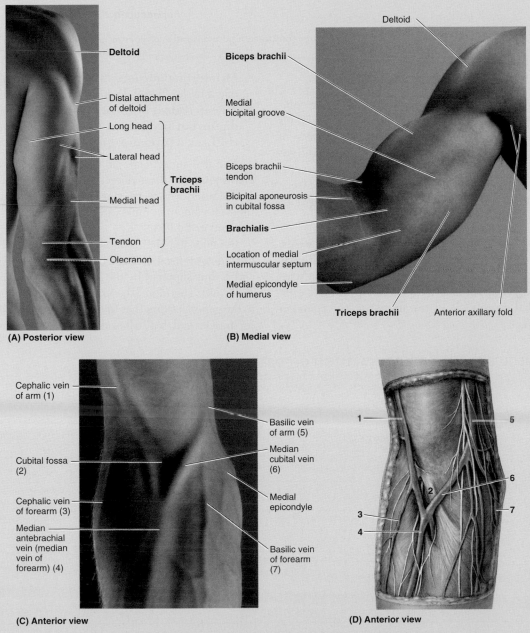

(A) Posterior view

(B) Medial view

(C) Anterior view

(D) Anterior view

FIGURE SA3.3. Surface anatomy of arm and cubital fossa.

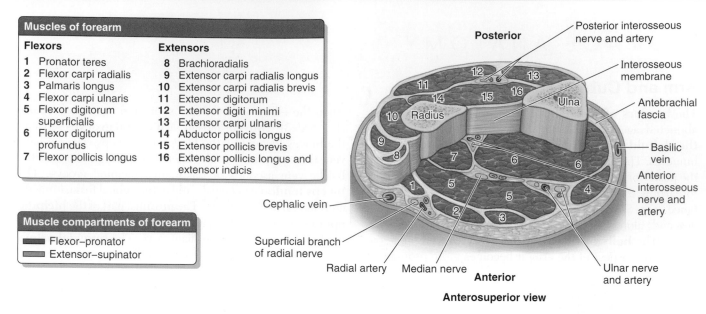

FIGURE 3.34. Stepped transverse section (mid forearm) demonstrating compartments of forearm.

FLEXOR–PRONATOR MUSCLES OF FOREARM

The **flexor–pronator muscles** are in the anterior compartment of the forearm (Figs. 3.34 and 3.35). The tendons of most flexor muscles pass across the anterior surface of the wrist and are held in place by the **palmar carpal ligament** (see Fig. 3.10) and the *flexor retinaculum (transverse carpal ligament)*, thickenings of the antebrachial fascia. The flexor muscles are arranged in three layers or groups (Fig. 3.35):

- A **superficial (first) layer or group** of four muscles: **pronator teres**, **flexor carpi radialis (FCR)**, **palmaris longus**, and **flexor carpi ulnaris (FCU)**. These muscles are all attached proximally by a *common flexor tendon* to the medial epicondyle of the humerus, the *common flexor origin*.

- An **intermediate (second) layer or group**, consisting of one muscle: **flexor digitorum superficialis (FDS)**
- A **deep (third) layer or group** of three muscles: **flexor digitorum profundus (FDP)**, **flexor pollicis longus (FPL)**, and **pronator quadratus**

The five superficial and intermediate muscles cross the elbow joint; the three deep muscles do not. The attachments of the anterior forearm muscles are illustrated in Fig. 3.36; their attachments, innervations, and main actions are summarized in Table 3.8.

Functionally, the *brachioradialis* is a flexor of the elbow joint, but it is located in the extensor (posterior) compartment and is thus supplied by the radial nerve (see Fig. 3.37A and Table 3.9). Therefore, the brachioradialis is a major exception to the generalization that the radial nerve supplies only extensor muscles and that all flexors are in the anterior compartment.

TABLE 3.8. MUSCLES OF ANTERIOR COMPARTMENT OF FOREARM

Muscle	Proximal Attachment	Distal Attachment	Innervation[a]	Main Action
Superficial (first) layer				

Pronator teres (PT)
   Ulnar head
   Humeral head
	Coronoid process of ulna	Middle of convexity of lateral surface of radius	Median nerve (C6, **C7**)	Pronates forearm and flexes elbow joint

Flexor carpi radialis (FCR)	Medial epicondyle of humerus	Base of 2nd (3rd) metacarpal		Flexes and abducts hand at wrist
Palmaris longus		Distal half of flexor retinaculum, palmar aponeurosis	Median nerve (C7, C8)	Flexes hand (at wrist) and tenses palmar aponeurosis

Flexor carpi ulnaris (FCU):
   Humeral head
   Ulnar head
	Olecranon and posterior border of ulna (via aponeurosis)	Pisiform, hook of hamate, 5th metacarpal	Ulnar nerve (C7, **C8**)	Flexes and adducts hand at wrist

(continued)

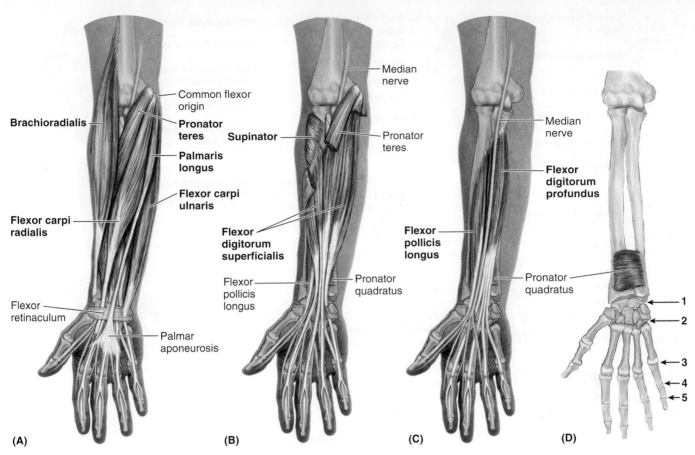

FIGURE 3.35. Muscles of anterior compartment of forearm. **A.** First layer. **B.** Second layer. **C.** Third layer. **D.** Fourth layer. *1*, Wrist joint; *2*, Carpometacarpal joint; *3*, Metacarpophalangeal joint; *4*, Proximal interphalangeal joint; *5*, Distal interphalangeal joint.

TABLE 3.8. MUSCLES OF ANTERIOR COMPARTMENT OF FOREARM *(continued)*

Muscle	Proximal Attachment	Distal Attachment	Innervation[a]	Main Action
Intermediate (second) layer				
Flexor digitorum superficialis (FDS)	*Humero-ulnar head*: medial epicondyle of humerus and coronoid process of ulna *Radial head*: oblique line of radius	Shafts (bodies) of middle phalanges of medial four digits	Median nerve (C7, **C8**, T1)	Flexes wrist joint, carpometacarpal joints Flexes proximal interphalangeal joints of middle four digits; acting more strongly, it also flexes proximal phalanges at metacarpophalangeal joints.
Deep (third) layer				
Flexor digitorum profundus (FDP)	Proximal three quarters of medial and anterior surfaces of ulna and interosseous membrane	Bases of distal phalanges of 2nd, 3rd, 4th, and 5th digits	*Lateral part (to digits 2 and 3)*: median nerve (**C8**, T1) (anterior interosseous branch) *Medial part (to digits 4 and 5)*: ulnar nerve (C8, **T1**)	Flexes wrist joint, carpometacarpal joints Flexes distal interphalangeal joints of digits 2, 3, 4, and 5; assists with wrist flexion
Flexor pollicis longus (FPL)	Anterior surface of radius and adjacent interosseous membrane	Base of distal phalanx of thumb	Anterior interosseous nerve, from median nerve (**C8**, T1)	Flexes wrist joint, carpometacarpal joints Flexes metacarpophalangeal and interphalangeal joints of thumb
Pronator quadratus	Distal quarter of anterior surface of ulna	Distal quarter of anterior surface of radius		Pronates forearm; deep fibers bind radius and ulna together.

[a]The spinal cord segmental innervation is indicated (e.g., "C6, C7" means that the nerves supplying the pronator teres are derived from the 6th and 7th cervical segments of the spinal cord). Numbers in boldface (**C7**) indicate the main segmental innervation. Damage to one or more of the listed spinal cord segments or to the motor nerve roots arising from them results in paralysis of the muscles concerned.

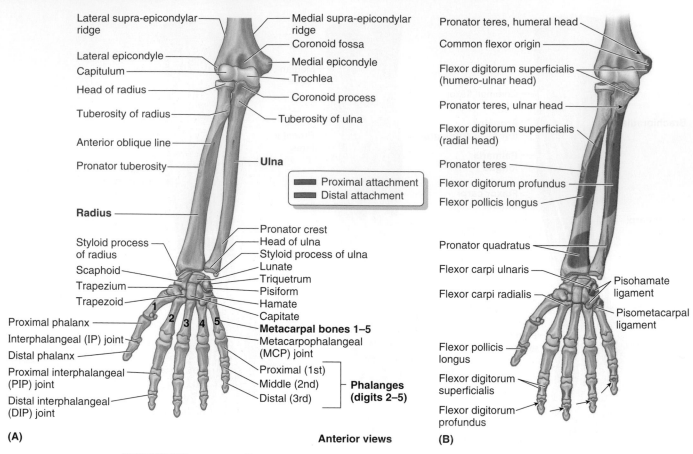

FIGURE 3.36. Features of bones and attachments of muscles of anterior compartment of forearm.

The **long flexors of the digits** (FDS and FDP) also flex the metacarpophalangeal and wrist joints.

The FDP flexes the fingers in slow action; this action is reinforced by the FDS when speed and flexion against resistance are required. When the wrist is flexed at the same time that the metacarpophalangeal and interphalangeal joints are flexed, the long flexor muscles of the fingers are operating over a shortened distance between attachments, and the action resulting from their contraction is consequently weaker. Extending the wrist increases their operating distance, and thus their contraction is more efficient in producing a strong grip. Tendons of the long flexors of the digits pass through the distal part of the forearm, wrist, and palm and continue to the medial four fingers. The FDS flexes the middle phalanges; the FDP flexes the distal phalanges (see Fig. 3.40C).

The pronator quadratus is the prime mover for pronation (Fig. 3.35D). It initiates pronation and is assisted by the *pronator teres* when more speed and power are needed. The pronator quadratus also helps the interosseous membrane hold the radius and ulna together, particularly when upward thrusts are transmitted through the wrist (e.g., during a fall on the hand).

EXTENSOR MUSCLES OF FOREARM

The extensor muscles are in the posterior (extensor–supinator) compartment of the forearm (Figs. 3.34 and 3.37), and all are innervated by branches of the radial nerve (see Fig. 3.17). These muscles may be organized into three functional groups:

- Muscles that extend and abduct or adduct the hand at the wrist joint: extensor carpi radialis longus (ECRL), extensor carpi radialis brevis (ECRB), and extensor carpi ulnaris (ECU)
- Muscles that extend the medial four digits: extensor digitorum, extensor indicis, and extensor digiti minimi (EDM)
- Muscles that extend or abduct the thumb: abductor pollicis longus (APL), extensor pollicis brevis (EPB), and extensor pollicis longus (EPL).

The extensor tendons are held in place in the wrist region by the extensor retinaculum, which prevents bowstringing of the tendons when the hand is extended at the wrist joint. As the tendons pass over the dorsum of the wrist, they are covered with **synovial tendon sheaths**, which reduce friction for the extensor tendons as they traverse the osseofibrous tunnels formed by the attachment of the extensor retinaculum to the distal radius and ulna (Fig. 3.38).

The extensor muscles are organized anatomically into superficial and deep layers. Four *superficial extensors* (ECRB, extensor digitorum, EDM, and ECU) are attached proximally by a *common extensor tendon* to the lateral epicondyle (Figs. 3.37A and 3.39; Table 3.9).

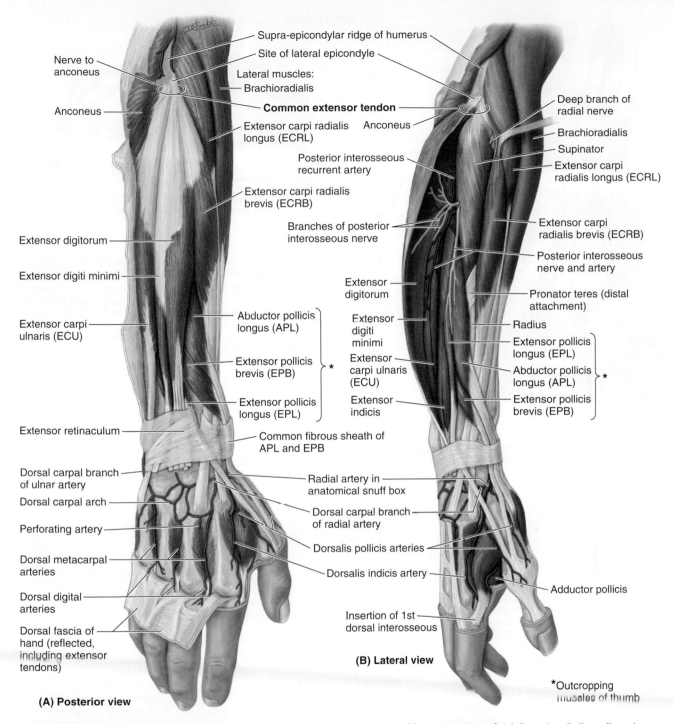

Nerve to anconeus

Anconeus

Extensor digitorum

Extensor digiti minimi

Extensor carpi ulnaris (ECU)

Extensor retinaculum

Dorsal carpal branch of ulnar artery

Dorsal carpal arch

Perforating artery

Dorsal metacarpal arteries

Dorsal digital arteries

Dorsal fascia of hand (reflected, including extensor tendons)

(A) Posterior view

Supra-epicondylar ridge of humerus

Site of lateral epicondyle

Lateral muscles:
Brachioradialis

Common extensor tendon

Extensor carpi radialis longus (ECRL)

Anconeus

Posterior interosseous recurrent artery

Extensor carpi radialis brevis (ECRB)

Branches of posterior interosseous nerve

Extensor digitorum

Extensor digiti minimi

Extensor carpi ulnaris (ECU)

Extensor indicis

Abductor pollicis longus (APL)

Extensor pollicis brevis (EPB)

Extensor pollicis longus (EPL)

Common fibrous sheath of APL and EPB

Radial artery in anatomical snuff box

Dorsal carpal branch of radial artery

Dorsalis pollicis arteries

Dorsalis indicis artery

Insertion of 1st dorsal interosseous

(B) Lateral view

Deep branch of radial nerve

Brachioradialis

Supinator

Extensor carpi radialis longus (ECRL)

Extensor carpi radialis brevis (ECRB)

Posterior interosseous nerve and artery

Pronator teres (distal attachment)

Radius

Extensor pollicis longus (EPL)

Abductor pollicis longus (APL)

Extensor pollicis brevis (EPB)

Adductor pollicis

*Outcropping muscles of thumb

FIGURE 3.37. **Muscles and neurovascular structures of posterior compartment of forearm. A.** Superficial dissection. **B.** Deep dissection.

The proximal attachment of the other two superficial extensors (brachioradialis and ECRL) is to the lateral supra-epicondylar ridge of the humerus and the adjacent lateral intermuscular septum (Fig. 3.39B). The four flat tendons of the extensor digitorum pass deep to the extensor retinaculum to the medial four fingers (Fig. 3.38). The common tendons of the index and little fingers are joined on their medial sides near the knuckles by the respective tendons of the extensor indicis and EDM (extensors of index and little fingers, respectively). The extensor indicis tendon joins the tendons of extensor digitorum to pass deep to the extensor retinaculum through the **tendinous sheath of extensor digitorum and extensor indicis** (common extensor synovial sheath). On the dorsum of the hand, the tendons of extensor digitorum spread out as they run toward the fingers. Adjacent tendons are linked proximal to the metacarpophalangeal

(Continued on page 144)

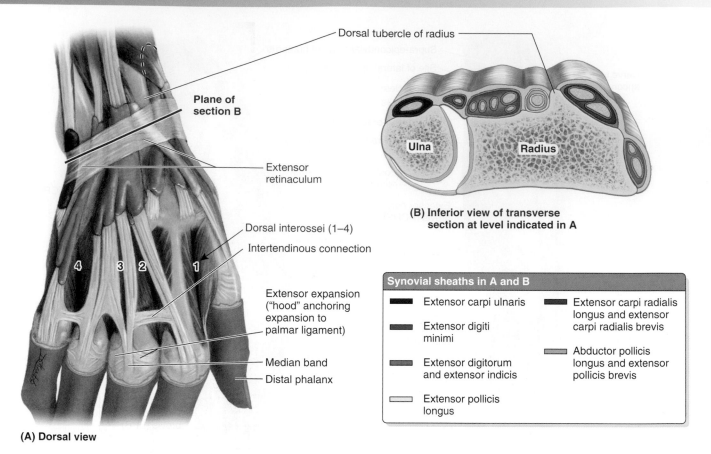

- Dorsal tubercle of radius

Plane of section B

- Extensor retinaculum

(B) **Inferior view of transverse section at level indicated in A**

Ulna   Radius

- Dorsal interossei (1–4)
- Intertendinous connection

4   3   2   1

- Extensor expansion ("hood" anchoring expansion to palmar ligament)
- Median band
- Distal phalanx

Synovial sheaths in A and B	
▬ Extensor carpi ulnaris	▬ Extensor carpi radialis longus and extensor carpi radialis brevis
▬ Extensor digiti minimi	▬ Abductor pollicis longus and extensor pollicis brevis
▬ Extensor digitorum and extensor indicis	
▢ Extensor pollicis longus	

(A) **Dorsal view**

FIGURE 3.38. **Synovial sheaths of extensor tendons on distal forearm and dorsum of hand. A.** Illustration with color-coded synovial sheaths. **B.** Transverse section through distal end of radius and ulna to show extensor tendons in their synovial sheaths.

TABLE 3.9. MUSCLES OF POSTERIOR COMPARTMENT OF FOREARM

Muscle	Proximal Attachment	Distal Attachment	Innervation[a]	Main Action
Superficial layer				
Brachioradialis	Proximal two thirds of lateral supra-epicondylar ridge of humerus	Lateral surface of distal end of radius proximal to styloid process	Radial nerve (C5, **C6**, C7)	Relatively weak flexion of elbow joint, maximal when forearm is in midpronated position
Extensor carpi radialis longus	Lateral supra-epicondylar ridge of humerus	Dorsal aspect of base of 2nd metacarpal	Radial nerve (C6, C7)	Extends and abduct wrist joint; extends carpometacarpal joints (extensor carpi radialis brevis active during fist clenching)
Extensor carpi radialis brevis	Lateral epicondyle of humerus (common extensor origin)	Dorsal aspect of base of 3rd metacarpal	Deep branch of radial nerve (**C7**, C8)	
Extensor digitorum		Extensor expansions of medial four fingers	Posterior interosseous nerve (**C7**, C8), continuation of deep branch of radial nerve	Extends wrist joint, carpometacarpal joints Extends medial four fingers primarily at metacarpophalangeal joints, secondarily at interphalangeal joints
Extensor digiti minimi		Extensor expansion of 5th finger		Extends wrist joint, carpometacarpal joints Extends 5th finger primarily at metacarpophalangeal joint, secondarily at interphalangeal joint
Extensor carpi ulnaris	Lateral epicondyle of humerus; posterior border of ulna via a shared aponeurosis	Dorsal aspect of base of 5th metacarpal		Extends and adducts wrist joint, carpometacarpal joints (also active during fist clenching)
Deep Layer				
Supinator	Lateral epicondyle of humerus, radial collateral and anular ligaments, supinator fossa, crest of ulna	Lateral, posterior, and anterior surfaces of proximal third of radius	Deep branch of radial nerve (C7, **C8**)	Supinates forearm; rotates radius to turn palm anteriorly or superiorly (if elbow is flexed)

(continued)

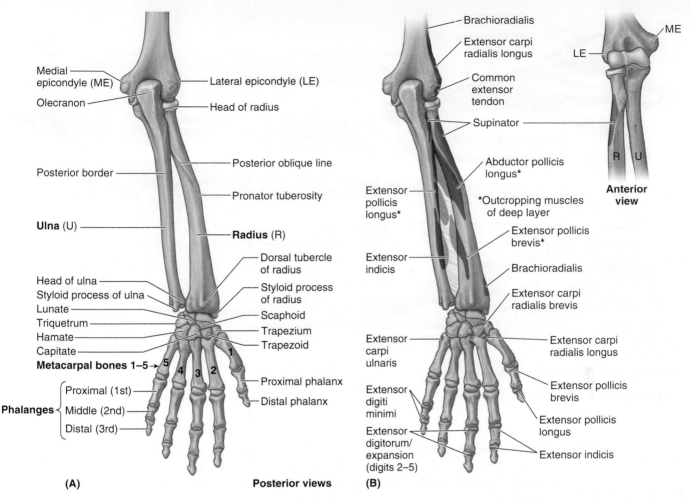

FIGURE 3.39. Features of bones and attachments of muscles of posterior compartment of forearm.

TABLE 3.9. MUSCLES OF POSTERIOR COMPARTMENT OF FOREARM *(continued)*

Muscle	Proximal Attachment	Distal Attachment	Innervation[a]	Main Action
"Outcropping" muscles of deep layer				
Abductor pollicis longus	Posterior surface of proximal halves of ulna, radius, and interosseous membrane	Base of 1st metacarpal	Posterior interosseous nerve (C7, **C8**), continuation of deep branch of radial nerve	Extends wrist joint, carpometacarpal joints Abducts thumb and extends it at carpometacarpal joint
Extensor pollicis longus	Posterior surface of middle third of ulna and interosseous membrane	Dorsal aspect of base of distal phalanx of thumb		Extends wrist joint, carpometacarpal joints Extends distal phalanx of thumb at interphalangeal joint; extends metacarpophalangeal and carpometacarpal joints
Extensor pollicis brevis	Posterior surface of distal third of radius and interosseous membrane	Dorsal aspect of base of proximal phalanx of thumb		Extends wrist joint, carpometacarpal joints Extends proximal phalanx of thumb at metacarpophalangeal joint; extends carpometacarpal joint
Extensor indicis	Posterior surface of distal third of ulna and interosseous membrane	Extensor expansion of 2nd finger		Extends wrist joint, carpometacarpal joints Extends 2nd finger (enabling its independent extension); helps extend hand at wrist

[a]The spinal cord segmental innervation is indicated (e.g., "C7, C8" means that the nerves supplying the extensor carpi radialis brevis are derived from the 7th and 8th cervical segments of the spinal cord). Numbers in boldface (**C7**) indicate the main segmental innervation. Damage to one or more of the listed spinal cord segments or to the motor nerve roots arising from them results in paralysis of the muscles concerned.

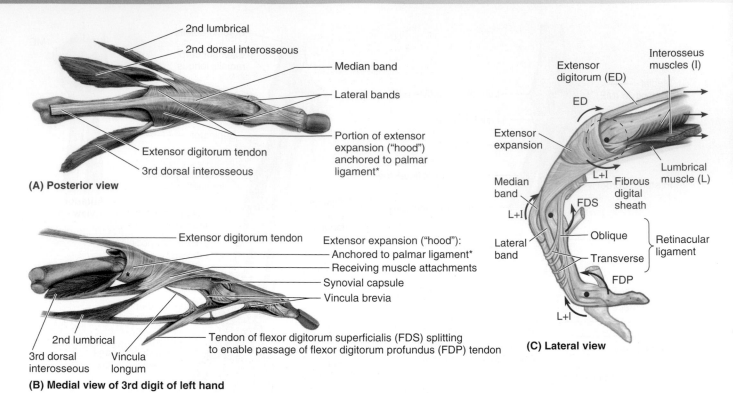

(A) Posterior view

(B) Medial view of 3rd digit of left hand

(C) Lateral view

FIGURE 3.40. Extensor expansion and vincula. A and B. Parts of extensor expansion. The vincula are fibrous bands that convey small vessels to the tendons. **C.** Retinacular ligaments.

joints by three oblique **intertendinous connections** that restrict independent extension of the fingers (Fig. 3.38A). Consequently, normally no finger can remain fully flexed as the other ones are fully extended.

On the distal ends of the metacarpals and along the phalanges, the four tendons of extensor digitorum flatten to form **extensor expansions** (Figs. 3.38 and 3.40). Each extensor expansion (dorsal expansion or "hood") is a triangular tendinous aponeurosis that wraps around the dorsum and sides of a head of the metacarpal and base of the proximal phalanx. The visor-like "hood" of the extensor expansion over the head of the metacarpal is anchored on each side to the **palmar ligament** (a thickened portion of the fibrous layer of the joint capsule of the metacarpophalangeal joints). In forming the extensor expansion, each extensor digitorum tendon divides into a **median band**, which passes to the base of the middle phalanx, and two **lateral bands**, which pass to the base of the distal phalanx. The tendons of the interosseous and lumbrical muscles of the hand join the lateral bands of the extensor expansion (Fig. 3.40).

The **retinacular ligament** is a delicate fibrous band that runs from the proximal phalanx and fibrous digital sheath obliquely across the middle phalanx and two interphalangeal joints (Fig. 3.40C). During flexion of the distal interphalangeal joint, the retinacular ligament becomes taut. The taut retinacular ligament pulls the proximal interphalangeal joint into flexion. Similarly, on extending the proximal joint, the distal joint is pulled by the retinacular ligament into nearly complete extension.

The deep extensor muscles of forearm (APL, EPB, and **extensor pollicis longus**) act on the thumb. The **extensor indicis** confers independence to the index finger in that it may act alone or together with the extensor digitorum to extend the index finger (Figs. 3.37 and 3.39; Table 3.9). The three muscles acting on the thumb (APL, EPB, and EPL) are deep to the superficial extensors and emerge ("crop out") from a furrow in the lateral part of the forearm that divides the extensors. Because of this characteristic, they are referred to as *outcropping muscles*. The tendons of the APL and EPB bound the triangular anatomical snuff box laterally, and the tendon of the EPL bounds it medially (Fig. 3.37B). The snuff box is visible as a hollow on the lateral aspect of the wrist when the thumb is extended fully; this draws the APL, EPB, and EPL tendons up and produces a concavity between them. Observe that the

- *radial artery* lies on the floor of the snuff box.
- *radial styloid process* can be palpated proximally, and the base of the 1st metacarpal can be palpated distally in the snuff box.
- *scaphoid and trapezium* can be felt in the floor of the snuff box between the radial styloid process and the 1st metacarpal.

Nerves of Forearm

The major **nerves of the forearm** are the median, ulnar, and radial (Figs. 3.41 and 3.42). Although the radial nerve appears

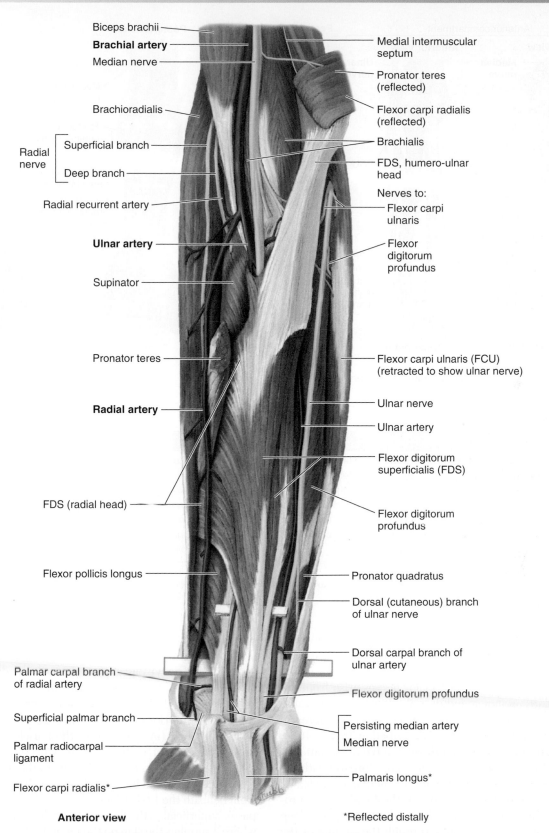

Biceps brachii

Brachial artery

Median nerve

Brachioradialis

Radial nerve
— Superficial branch
— Deep branch

Radial recurrent artery

Ulnar artery

Supinator

Pronator teres

Radial artery

FDS (radial head)

Flexor pollicis longus

Palmar carpal branch
of radial artery

Superficial palmar branch

Palmar radiocarpal
ligament

Flexor carpi radialis*

Medial intermuscular
septum

Pronator teres
(reflected)

Flexor carpi radialis
(reflected)

Brachialis

FDS, humero-ulnar
head

Nerves to:
— Flexor carpi
ulnaris
— Flexor
digitorum
profundus

Flexor carpi ulnaris (FCU)
(retracted to show ulnar nerve)

Ulnar nerve

Ulnar artery

Flexor digitorum
superficialis (FDS)

Flexor digitorum
profundus

Pronator quadratus

Dorsal (cutaneous) branch
of ulnar nerve

Dorsal carpal branch of
ulnar artery

Flexor digitorum profundus

Persisting median artery
Median nerve

Palmaris longus*

Anterior view

*Reflected distally

FIGURE 3.41. Muscles, vessels, and nerves of anterior aspect of forearm.

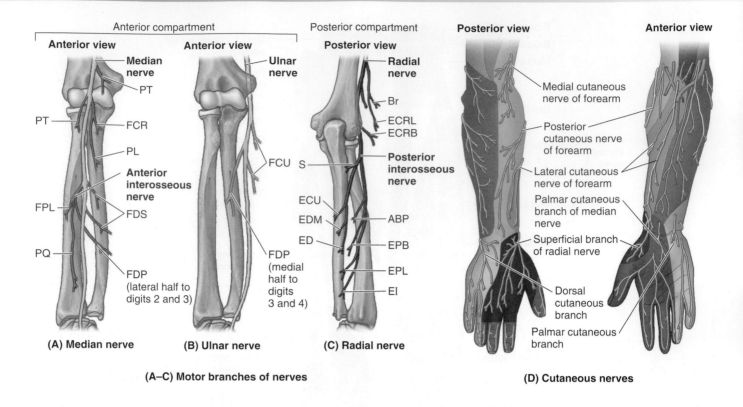

Key

ABP	Abductor pollicis longus	EDM	Extensor digiti minimi	FDS/FDP	Flexor digitorum superficialis and profundus	For hand:	
Br	Brachioradialis	EI	Extensor indicis	FPL	Flexor pollicis longus		Ulnar nerve (C8, T1)
ECRL/ECRB	Extensor carpi radialis longus and brevis	EPL/EPB	Extensor pollicis longus and brevis	PQ	Pronator quadratus		Radial nerve (C6–C8)
ECU	Extensor carpi ulnaris	FCR	Flexor carpi radialis	PL	Palmaris longus		Median nerve (C6–C8)
ED	Extensor digitorum	FCU	Flexor carpi ulnaris	PT	Pronator teres		
				S	Supinator		

FIGURE 3.42. Nerves of forearm. A–C. Motor innervation. **D.** Cutaneous innervation.

in the cubital region, it soon enters the posterior compartment of the forearm. Besides the cutaneous branches, there are only two nerves of the anterior aspect of the forearm: the median and ulnar nerves. The major nerves arise from the brachial plexus as shown and described in Figure 3.27 and Table 3.6, and their courses and distributions in the forearm are illustrated in Figure 3.42 and described in Table 3.10.

The **median nerve** is the principal nerve of the anterior compartment of the forearm. It enters the forearm with the brachial artery and lies medial to it. The median nerve leaves the cubital fossa by passing between the heads of the pronator teres, giving branches to them, and then passes deep to the FDS, continuing distally through the middle of the forearm, between the FDS and the FDP (Fig. 3.41). Near the wrist, the median nerve becomes superficial by passing between the tendons of the FDS and FCR deep to the palmaris longus tendon. The **anterior interosseous nerve** is its major branch (Fig. 3.42). Articular and muscular

branches and a palmar cutaneous branch are also derived from the median nerve.

The **ulnar nerve** passes posterior to the medial epicondyle of the humerus and enters the forearm by passing between the heads of the FCU (Fig. 3.41), giving branches to them. It then passes inferiorly between the FCU and the FDP, supplying the ulnar (medial) part of the muscle that sends tendons to digits 4 and 5. The ulnar nerve becomes superficial at the wrist, running on the medial side of the ulnar artery and the lateral side of the FCU tendon. The ulnar nerve emerges from beneath the FCU tendon just proximal to the wrist and passes superficial to the flexor retinaculum to enter the hand, where it supplies the skin on the medial side of the hand. The branches of the ulnar nerve in the forearm (articular, muscular, and palmar and dorsal cutaneous branches) are described in Table 3.10.

The **radial nerve** leaves the posterior compartment of the arm to cross the anterior aspect of the lateral epicondyle

TABLE 3.10. NERVES OF FOREARM

Nerve	Origin	Course in Forearm
Median	By union of lateral root of median nerve (C6, C7, from lateral cord of brachial plexus) with medial root (C8, T1) from medial cord	Enters cubital fossa medial to brachial artery; exits by passing between heads of pronator teres; descends in fascial plane between flexors digitorum superficialis and profundus; runs deep to palmaris longus tendon as it approaches flexor retinaculum to traverse carpal tunnel
Anterior interosseous	Median nerve in distal part of cubital fossa	Descends on anterior aspect of interosseous membrane with artery of same name, between FDP and FPL, to pass deep to pronator quadratus
Palmar cutaneous branch of median nerve	Median nerve of middle to distal forearm, proximal to flexor retinaculum	Passes superficial to flexor retinaculum to reach skin of central palm
Ulnar	Larger terminal branch of medial cord of brachial plexus (C8, T1, often receives fibers from C7)	Enters forearm by passing between heads of flexor carpi ulnaris, after passing posterior to medial epicondyle of humerus; descends forearm between FCU and FDP; becomes superficial in distal forearm
Palmar cutaneous branch of ulnar nerve	Ulnar nerve near middle of forearm	Descends anterior to ulnar artery; perforates deep fascia in distal forearm; runs in subcutaneous tissue to palmar skin medial to axis of 4th digit
Dorsal cutaneous branch of ulnar nerve	Ulnar nerve in distal half of forearm	Passes postero-inferiorly between ulna and flexor carpi ulnaris; enters subcutaneous tissue to supply skin of dorsum medial to axis of 4th digit
Radial	Larger terminal branch of posterior cord of brachial plexus (C5–T1)	Enters cubital fossa between brachioradialis and brachialis; anterior to lateral epicondyle divides into terminal superficial and deep branches.
Posterior cutaneous nerve of forearm	Radial nerve, as it traverses radial groove of posterior humerus	Perforates lateral head of triceps; descends along lateral side of arm and posterior aspect of forearm to wrist
Superficial branch of radial nerve	Sensory terminal branch of radial nerve, in cubital fossa	Descends between pronator teres and brachioradialis, emerging from latter to arborize over anatomical snuff box and supply skin of dorsum lateral to axis of 4th finger
Deep branch of radial/ posterior interosseous nerve	Motor terminal branch of radial nerve, in cubital fossa	Deep branch exits cubital fossa winding around neck of radius, penetrating and supplying supinator; emerges in posterior compartment of forearm as posterior interosseous nerve; descends on membrane with artery of same name.
Lateral cutaneous nerve of forearm	Continuation of musculocutaneous nerve distal to muscular branches	Emerges lateral to biceps brachii on brachialis, running initially with cephalic vein; descends along lateral border of forearm to wrist
Medial cutaneous nerve of forearm	Medial cord of brachial plexus, receiving C8 and T1 fibers	Perforates deep fascia of arm with basilic vein proximal to cubital fossa; descends medial aspect of forearm in subcutaneous tissue to wrist

FCU, flexor carpi ulnaris; *FDP*, flexor digitorum profundus; *FPL*, flexor pollicis longus.

of the humerus. In the cubital region, the radial nerve divides into deep and superficial branches (see Fig. 3.33B). The *deep branch of radial nerve* arises anterior to the lateral epicondyle and pierces the supinator. The deep branch winds around the lateral aspect of the neck of the radius and enters the posterior (extensor–pronator) compartment of the forearm, where it continues as the *posterior interosseous nerve* (Fig 3.42C and Table 3.10). The superficial branch of the radial nerve is a cutaneous and articular nerve that descends in the forearm under cover of the brachioradialis (Fig. 3.41). The *superficial branch of the radial nerve* (sensory or cutaneous) emerges in the distal part of the forearm and crosses the roof of the anatomical snuff box. It is distributed to skin on the dorsum of the hand and to a number of joints in the hand (Fig 3.42D).

Arteries and Veins of Forearm

The *brachial artery* ends in the distal part of the cubital fossa opposite the neck of the radius by dividing into the ulnar and radial arteries, the main arteries of the forearm (Fig. 3.41). The branches of the ulnar and radial arteries are illustrated in Figure 3.43 and described in Table 3.11.

The ulnar artery descends through the anterior (flexor–pronator) compartment of the forearm, deep to the pronator teres. Pulsations of the ulnar artery can be palpated on the lateral side of the FCU tendon, where it lies anterior to the ulnar head (Fig. 6.41). The ulnar nerve is on the medial side of the ulnar artery. When the brachioradialis is pulled laterally, the entire length of the artery is visible until the distal part of the forearm. The radial artery leaves the forearm by winding around the lateral aspect of the wrist and crossing the floor of the anatomical snuff box to reach the hand (Figs. 6.37 and 3.43). The pulsation of the radial artery is usually measured on the distal radius between the tendons of FCR and APL (Fig. 6.41).

There are superficial and deep veins in the forearm: *superficial* ascends in the subcutaneous tissue; *deep veins* accompany the deep arteries (e.g., radial and ulnar).

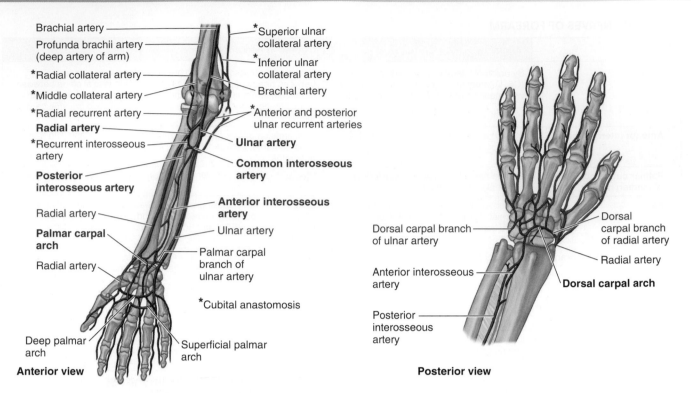

FIGURE 3.43. Arteries of forearm and hand.

TABLE 3.11. ARTERIES OF FOREARM AND WRIST

Artery	Origin	Course in Forearm
Ulnar	As larger terminal branch of brachial artery in cubital fossa	Descends inferomedially and then directly inferiorly deep to superficial pronator teres, palmaris longus, and flexor digitorum superficialis to reach medial side of forearm; passes superficial to flexor retinaculum at wrist in ulnar (Guyon) canal to enter hand
Anterior ulnar recurrent artery	Ulnar artery just distal to elbow joint	Passes superiorly between brachialis and pronator teres, supplying both; then anastomoses with inferior ulnar collateral artery anterior to medial epicondyle
Posterior ulnar recurrent artery	Ulnar artery distal to anterior ulnar recurrent artery	Passes superiorly, posterior to medial epicondyle and deep to tendon of flexor carpi ulnaris; then anastomoses with superior ulnar collateral artery
Common interosseous	Ulnar artery in cubital fossa, distal to bifurcation of brachial artery	Passes laterally and deeply, terminating by dividing into anterior and posterior interosseous arteries
Anterior interosseous	As terminal branches of common interosseous artery, between radius and ulna	Passes distally on anterior aspect of interosseous membrane to proximal border of pronator quadratus; pierces membrane and continues distally to join dorsal carpal arch on posterior aspect of interosseous membrane
Posterior interosseous		Passes to posterior aspect of interosseous membrane, giving rise to recurrent interosseous artery; runs distally between superficial and deep extensor muscles, supplying both
Recurrent interosseous	Posterior interosseous artery, between radius and ulna	Passes superiorly, posterior to proximal radio-ulnar joint, to anastomose with middle collateral artery (from deep artery of arm)
Palmar carpal branch	Ulnar artery in distal forearm	Runs across anterior aspect of wrist, deep to tendons of flexor digitorum profundus, to anastomose with the palmar carpal branch of the radial artery, forming palmar carpal arch
Dorsal carpal branch	Ulnar artery, proximal to pisiform	Passes across dorsal surface of wrist, deep to extensor tendons, to anastomose with dorsal carpal branch of radial artery, forming dorsal carpal arch
Radial	As smaller terminal branch of brachial artery in cubital fossa	Runs inferolaterally under cover of brachioradialis; lies lateral to flexor carpi radialis tendon in distal forearm; winds around lateral aspect of radius and crosses floor of anatomical snuff box to pierce 1st dorsal interosseous muscle
Radial recurrent	Lateral side of radial artery, just distal to brachial artery bifurcation	Ascends between brachioradialis and brachialis, supplying both (and elbow joint); then, anastomoses with radial collateral artery (from profunda brachii artery)
Palmar carpal branch	Distal radial artery near distal border of pronator quadratus	Runs across anterior wrist deep to flexor tendons to anastomose with the palmar carpal branch of ulnar artery to form palmar carpal arch
Dorsal carpal branch	Distal radial artery in proximal part of snuff box	Runs medially across wrist deep to pollicis and extensor radialis tendons, anastomoses with ulnar dorsal carpal branch forming dorsal carpal arch

CLINICAL BOX

Muscle Testing of Flexor Digitorum Superficialis and Flexor Digitorum Profundus

To test the FDS, one finger is flexed at the proximal interphalangeal joint against resistance, and the other three fingers are held in an extended position to inactivate the FDP (Fig. B3.11A). *To test the FDP*, the proximal interphalangeal joint is held in the extended position while the person attempts to flex the distal interphalangeal joint (Fig. B3.11B).

(A) Flexor digitorum superficialis (FDS) muscle test

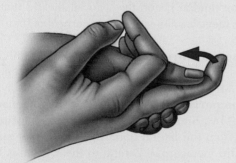

(B) Flexor digitorum profundus (FDP) muscle test

FIGURE B3.11. Muscle testing of FDS and FDP.

Elbow Tendinitis or Lateral Epicondylitis

Elbow tendinitis (tennis elbow) is a painful musculoskeletal condition that may follow repetitive use of the superficial extensor muscles of the forearm. Pain is felt over the lateral epicondyle and radiates down the posterior surface of the forearm. People with elbow tendinitis often feel pain when they open a door or lift a glass. Repeated forceful flexion and extension of the wrist strain the attachment of the common extensor tendon, producing inflammation of the periosteum of the lateral epicondyle (*lateral epicondylitis*). Associated tears of the common extensor tendon, which may be surgically repaired, are visible on magnetic resonance imaging (MRI).

Synovial Cyst of Wrist

Sometimes, a nontender cystic swelling appears on the hand, most commonly on the dorsum of the wrist (Fig. B3.12). The thin-walled cyst contains clear mucinous fluid. Clinically, this type of swelling is called a "ganglion" (G. swelling or knot). These synovial cysts are close to and often communicate with the synovial sheaths. The distal attachment of the ECRB tendon is a common site for such a cyst. A cystic swelling of the common flexor synovial sheath on the anterior aspect of the wrist can enlarge enough to produce compression of the median nerve by narrowing the carpal tunnel (*carpal tunnel syndrome*).

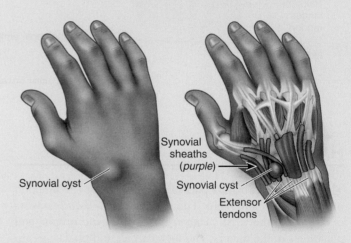

Synovial cyst

Synovial sheaths (*purple*)

Synovial cyst

Extensor tendons

FIGURE B3.12. Synovial cyst of wrist.

Mallet or Baseball Finger

Sudden severe tension on a long extensor tendon may avulse part of its attachment to the phalanx. The most common result of this injury is *mallet* or *baseball finger*. This deformity results from the distal interphalangeal joint suddenly being forced into extreme flexion (hyperflexion) when the tendon is attempting to extend the distal phalanx—for example, when a baseball is miscaught (hyperflexing it) or the finger is jammed into a base pad. These actions avulse the attachment of the tendon from the base of the distal phalanx. As a result, the person is unable to extend the distal interphalangeal joint (Fig. B3.13).

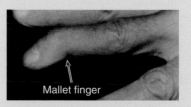

Mallet finger

FIGURE B3.13. Mallet finger.

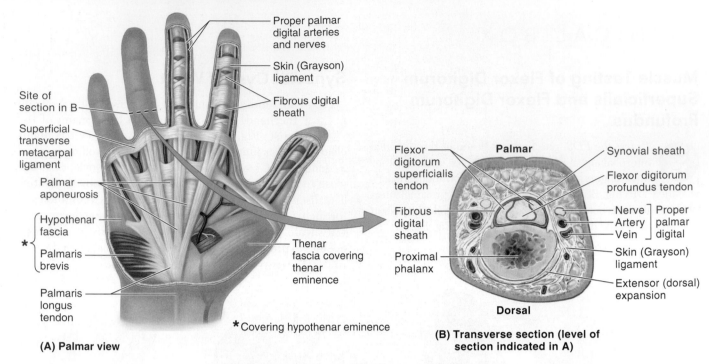

FIGURE 3.44. Palmar fascia and fibrous digital sheaths.

HAND

The wrist, the proximal part of the hand, is at the junction of the forearm and hand. The *skeleton of the hand* consists of *carpals* in the wrist, *metacarpals* in the hand proper, and *phalanges* in the fingers. The metacarpals and phalanges are numbered from 1 to 5, beginning with the thumb and ending with the little finger. The palmar aspect of the hand features a central concavity that separates two eminences: a lateral more prominent **thenar eminence** at the base of the thumb and a medial, smaller **hypothenar eminence** proximal to the base of the 5th finger (Fig. 3.44A).

Fascia of Palm

The **fascia of the palm** is continuous with the antebrachial fascia and the fascia of the dorsum of the hand. This fascia is thin over the thenar and hypothenar eminences, but it is thick centrally where it forms the fibrous palmar aponeurosis and in the fingers where it forms the digital sheaths (Fig. 3.44). The palmar aponeurosis, a strong, well-defined part of the deep fascia of the palm, covers the soft tissues and overlies the long flexor tendons. The proximal end or apex of the triangular palmar aponeurosis is continuous with the flexor retinaculum and the palmaris longus tendon. Distal to the apex, the palmar aponeurosis forms four longitudinal digital bands that radiate from the apex and attach distally to the bases of the proximal phalanges, where they become

continuous with the fibrous digital sheaths (Fig. 3.44). The **fibrous digital sheaths** are ligamentous tubes that enclose the flexor tendon(s) and the synovial sheaths that surround them as they pass along the palmar aspect of their respective digit.

A **medial fibrous septum** extends deeply from the medial border of the palmar aponeurosis to the 5th metacarpal. Medial to this septum is the medial or **hypothenar compartment** containing the hypothenar muscles (Fig. 3.45A). Similarly, a **lateral fibrous septum** extends deeply from the lateral border of the palmar aponeurosis to the 3rd metacarpal. Lateral to the septum is the lateral or **thenar compartment** containing the thenar muscles. Between the hypothenar and the thenar compartments is the **central compartment** containing the flexor tendons and their sheaths, the lumbrical muscles, the superficial palmar arterial arch, and the digital vessels and nerves (Fig. 3.45A). The deepest muscular plane of the palm is the **adductor compartment** containing the adductor pollicis. Between the flexor tendons and the fascia covering the deep palmar muscles are two potential spaces: the **thenar space** and the **midpalmar space** (Fig. 3.45). These spaces are bounded by fibrous septa passing from the edges of the palmar aponeurosis to the metacarpals. Between the two spaces is the especially strong lateral fibrous septum, which is attached to the 3rd metacarpal. The midpalmar space is continuous with the anterior compartment of the forearm via the carpal tunnel.

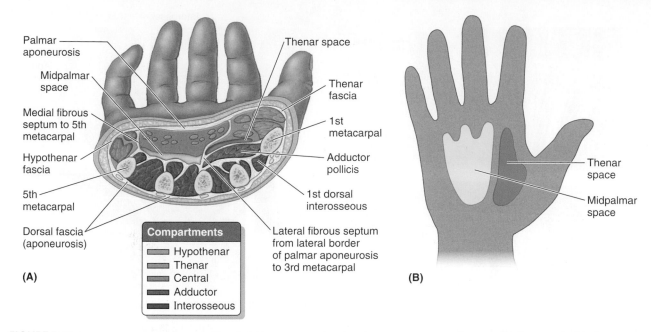

FIGURE 3.45. Compartments and spaces of hand. A. Transverse section showing compartments and spaces. **B.** Thenar and midpalmar spaces.

Muscles of Hand

The intrinsic muscles of the hand are located in five compartments (Fig. 3.45A; see also Table 3.12):

- Thenar muscles in the *thenar compartment*: abductor pollicis brevis, flexor pollicis brevis, and opponens pollicis
- Hypothenar muscles in the *hypothenar compartment*: abductor digiti minimi, flexor digiti minimi brevis, and opponens digiti minimi
- Adductor pollicis in the *adductor compartment*
- The short muscles of the hand, the lumbricals, in the *central compartment* with the long flexor tendons
- The interossei in separate *interosseous compartments* between the metacarpals

THENAR MUSCLES

The **thenar muscles** form the *thenar eminence* on the radial aspect of the palm (Figs. 3.44 and 3.46). Normal movement of the thumb is important for the precise activities of the hand. The high degree of freedom of movements of the thumb results from the 1st metacarpal being independent, with mobile joints at both ends. Several muscles are required to control its freedom of movement (Figs. 3.46 and 3.47):

- *Abduction*: **APL** and **abductor pollicis brevis** (APB)
- *Adduction*: **adductor pollicis** (AD) and 1st dorsal interosseous
- *Extension*: EPL, EPB, and APL
- *Flexion*: **FPL** and **flexor pollicis brevis** (FPB)
- *Opposition*: **opponens pollicis**

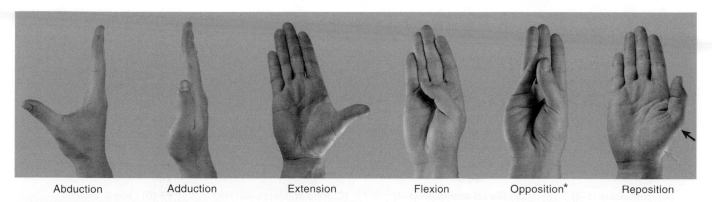

Abduction   Adduction   Extension   Flexion   Opposition*   Reposition

FIGURE 3.46. Movements of thumb. *Red arrow*, location of carpometacarpal joint of thumb; *green arrow*, location of metacarpophalangeal joint of thumb.

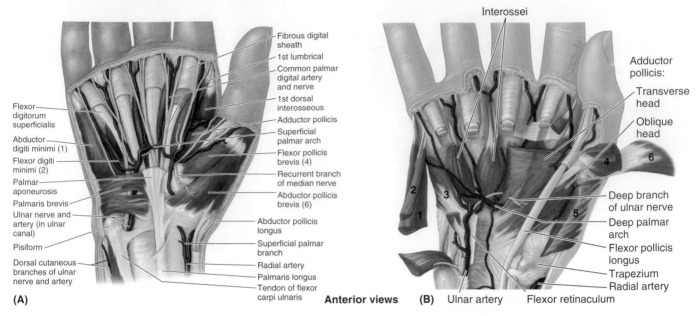

FIGURE 3.47. Palm of hand. A. Superficial dissection showing the muscles, superficial palmar arch and the distribution of median and ulnar nerves. **B.** Deep dissection showing opponens pollicis. *3*, deep palmar arch and deep branch of ulnar nerve; *5*, opponens digiti minimi.

Opposition occurs at the carpometacarpal joint of the thumb (Fig. 3.46). The complex movement of opposition begins with the thumb in the extended position and initially involves abduction and medial rotation of the 1st metacarpal ("cupping" of the palm) produced by the action of the opponens pollicis and then flexion at the metacarpophalangeal joint (Fig. 3.47). The reinforcing action of the AD and FPL increases the pressure that the opposed thumb can exert on the fingertips.

HYPOTHENAR MUSCLES

The **hypothenar muscles** (abductor digiti minimi, flexor digiti minimi brevis, and opponens digiti minimi) are in the hypothenar compartment and produce the *hypothenar eminence* on the medial side of the palm (Fig. 3.47). The **palmaris brevis** is a small muscle in the subcutaneous tissue of the hypothenar

eminence; it is not in the hypothenar compartment. It wrinkles the skin of the hypothenar eminence, deepening the hollow of the palm, thereby aiding the palmar grip. The palmaris brevis covers and protects the ulnar nerve and artery. It is attached proximally to the medial border of the palmar aponeurosis and to the skin on the medial border of the hand.

The attachments, innervations, and main actions of the thenar and hypothenar muscles are illustrated in Figures 3.47, 3.48, and 3.49 and summarized in Table 3.12.

SHORT MUSCLES OF HAND

The **short hand muscles** are the lumbricals and interossei (Figs. 3.47, 3.48, 3.49; Table 3.12). The four slender **lumbrical muscles** were named because of their worm-like appearance (L. *lumbricus*, earthworm). The four **dorsal interosseous**

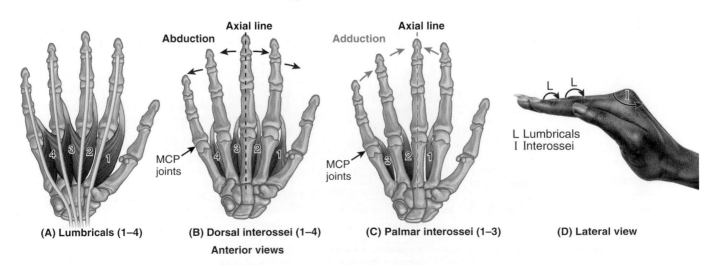

(A) Lumbricals (1–4)   **(B) Dorsal interossei (1–4)**   **(C) Palmar interossei (1–3)**   **(D) Lateral view**

Anterior views

FIGURE 3.48. Lumbricals and palmar and dorsal interossei. A. Lumbricals. **B.** Dorsal interossei. **C.** Palmar interossei. **D.** Z-movement. Metacarpophalangeal (*MCP*) joints are flexed and interphalangeal joints are extended.

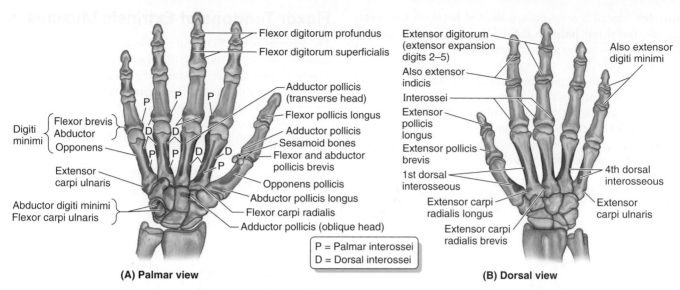

FIGURE 3.49. Attachments of hand muscles.

TABLE 3.12. INTRINSIC MUSCLES OF HAND

Muscle	Proximal Attachment	Distal Attachment	Innervation[a]	Main Action
Thenar muscles				
Opponens pollicis	Flexor retinaculum and tubercles of scaphoid and trapezium	Lateral side of 1st metacarpal	Recurrent branch of median nerve (**C8**, T1)	To oppose thumb, it draws 1st metacarpal medially to center of palm and rotates it medially.
Abductor pollicis brevis		Lateral side of base of proximal phalanx of thumb		Abducts thumb; helps oppose it
Flexor pollicis brevis: Superficial head Deep head				Flexes thumb
Adductor pollicis: Oblique head	Bases of 2nd and 3rd metacarpals, capitate, adjacent carpals	Medial side of base of proximal phalanx of thumb	Deep branch of ulnar nerve (**C8**, T1)	Adducts thumb toward lateral border of palm
Transverse head	Anterior surface of shaft of 3rd metacarpal			
Hypothenar muscles				
Abductor digiti minimi	Pisiform	Medial side of base of proximal phalanx of 5th finger	Deep branch of ulnar nerve (**C8**, T1)	Abducts 5th finger; assists in flexion of its proximal phalanx
Flexor digiti minimi brevis	Hook of hamate and flexor retinaculum			Flexes proximal phalanx of 5th finger
Opponens digiti minimi		Medial border of 5th metacarpal		Draws 5th metacarpal anterior and rotates it, bringing 5th finger into opposition with thumb
Short muscles				
Lumbricals				
1 and 2	Lateral two tendons of flexor digitorum profundus (as unipennate muscles)	Lateral sides of extensor expansions of 2nd–5th fingers	Median nerve (C8, **T1**)	Flex metacarpophalangeal joints; extend interphalangeal joints of 2nd–5th fingers
3 and 4	Medial three tendons of flexor digitorum profundus (as bipennate muscles)			
Dorsal interossei, 1–4	Adjacent sides of two metacarpals (as bipennate muscles)	Bases of proximal phalanges; extensor expansions of 2nd–4th fingers	Deep branch of ulnar nerve (C8, **T1**)	Abduct 2nd–4th fingers from axial line; act with lumbricals in flexing metacarpophalangeal joints and extending interphalangeal joints
Palmar interossei, 1–3	Palmar surfaces of 2nd, 4th, and 5th metacarpals (as unipennate muscles)	Bases of proximal phalanges; extensor expansions of 2nd, 4th, and 5th fingers		Adduct 2nd, 4th, and 5th fingers toward axial line; assist lumbricals in flexing metacarpophalangeal joints and extending interphalangeal joints

[a]The spinal cord segmental innervation is indicated (e.g., "**C8**, T1" means that the nerves supplying the opponens pollicis are derived from the 8th cervical segment and 1st thoracic segment of the spinal cord). Numbers in boldface (**C8**) indicate the main segmental innervation. Damage to one or more of the listed spinal cord segments or to the motor nerve roots arising from them results in paralysis of the muscles concerned.

muscles (dorsal interossei) are located between the metacarpals; the three **palmar interosseous muscles** (palmar interossei) are on the palmar surfaces of the 2nd, 4th, and 5th metacarpals (Fig. 3.48A–C). The four dorsal interossei abduct the fingers at the metacarpophalangeal (MCP) joints, and the three palmar interossei adduct them. As a mnemonic device, use the following acronyms: **d**orsal **ab**duct (**DAB**) and **p**almar **ad**duct (**PAD**). Acting together, the dorsal and palmar interossei and lumbricals produce flexion at the metacarpophalangeal joints and extension of the interphalangeal joints (Z-movement—Fig. 3.48D). This occurs because of their attachment to the lateral bands of the extensor expansions (see Fig. 3.40).

Flexor Tendons of Extrinsic Muscles

The tendons of the FDS and FDP enter the **common flexor sheath** (ulnar bursa) deep to the flexor retinaculum (Fig. 3.50). The tendons enter the central compartment of the hand and fan out to enter the respective **digital synovial sheaths**. The common flexor and digital sheaths enable the tendons to slide freely past each other during movements of the fingers. Near the base of the proximal phalanx, the tendon of the FDS splits and surrounds the tendon of the FDP (Fig. 3.50B). The halves of the FDS tendon are attached to the margins of the anterior aspect of the shaft of the middle phalanx. The tendon of the

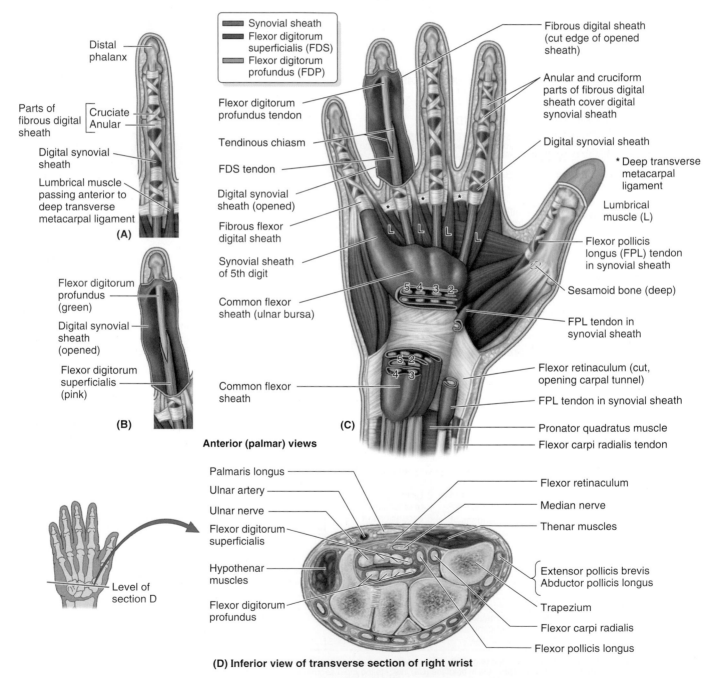

FIGURE 3.50. Synovial and fibrous digital sheaths of long flexor tendons of hand. A. Parts of fibrous digital sheath. **B.** Digital synovial sheath opened. **C.** Dissection of common flexor sheath and synovial sheaths of digits 1–5 (*purple*). **D.** Transverse section of wrist showing carpal tunnel and its contents.

FDP, after passing through the split in the FDS tendon, the *tendinous chiasm*, passes distally to attach to the anterior aspect of the base of the distal phalanx (Fig. 3.50A,B).

The fibrous digital sheaths are strong ligamentous tunnels containing the flexor tendons and their synovial sheaths (Figs. 3.50 and 3.51). The sheaths extend from the heads of the metacarpals to the bases of the distal phalanges. These sheaths prevent the tendons from pulling away from the digits (bow-stringing). The fibrous digital sheaths attach to the bones to form **osseofibrous tunnels** through which the tendons pass to reach the digits. The **anular** and **cruciform parts (ligaments) of the fibrous sheath** (often referred to clinically as "pulleys") are thickened reinforcements of these sheaths. The long flexor tendons are supplied by small blood vessels that pass to them within synovial folds (*vincula*) from the periosteum of the phalanges (Fig. 3.51).

The tendon of FPL passes deep to the flexor retinaculum to the thumb within its own synovial sheath (Fig. 3.50C,D). At the head of the metacarpal, the tendon runs between two *sesamoid bones*—one in the combined tendon of the FPB and APB and the other in the tendon of the AD (Fig. 3.49).

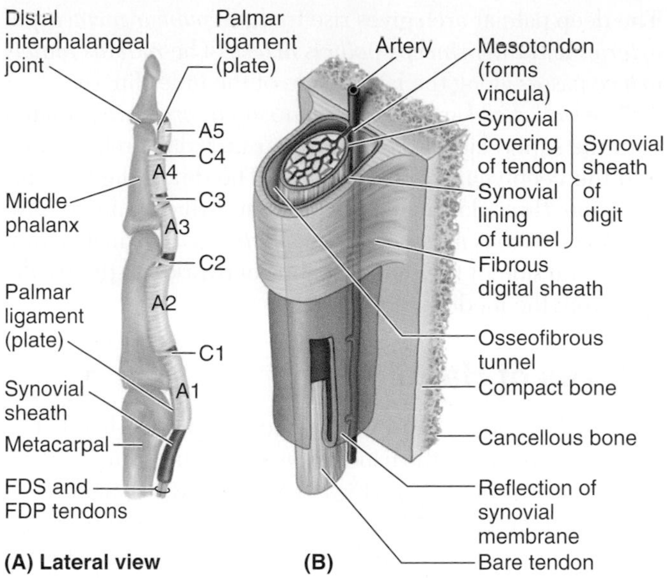

Arteries and Veins of Hand

The ulnar and radial arteries and their branches provide all the blood to the hand (Figs. 3.47 and 3.52). The *ulnar artery* enters the hand anterior to the flexor retinaculum between the pisiform and the hook of hamate via the *ulnar canal* (Guyon canal). The ulnar artery lies lateral to the ulnar nerve. It gives rise to the deep (palmar) branch and then continues superficial to the long flexor tendons, where it is the main contributor to the **superficial palmar arch** (see Fig. 3.47A). The *superficial palmar arch* gives rise to three **common palmar digital arteries** that anastomose with **palmar metacarpal arteries** from the deep palmar arch. Each common

palmar digital artery divides into a pair of **proper palmar digital arteries** that run along the adjacent sides of the 2nd to 4th fingers. The radial artery curves dorsally around the scaphoid and trapezium in the floor of the *anatomical snuff box* (see Fig. 3.37A,B) and enters the palm by passing between the heads of the 1st dorsal interosseous muscle. It then turns medially and passes between the heads of the AD (see Fig. 3.47B). The radial artery ends by anastomosing with the deep branch of the ulnar artery to form the deep palmar arch (Figs. 3.47B and 3.52). This arch, formed mainly by the radial artery, lies across the metacarpals just distal to their bases.

FIGURE 3.51. Fibrous digital sheaths of digits. **A.** Anular and cruciate parts ("pulleys"). **B.** Structure of an osseofibrous tunnel of a finger. *FDS,* flexor digitorum superficialis; *FDP,* flexor digitorum profundus.

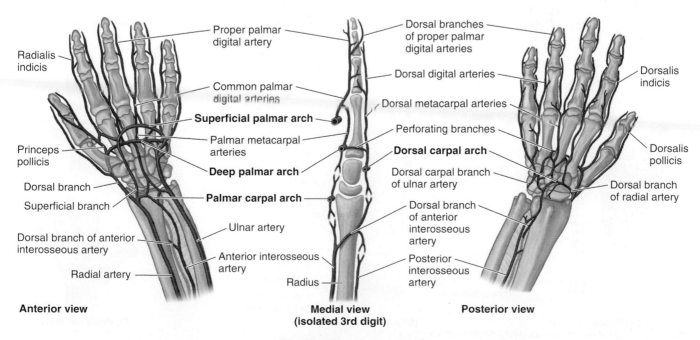

FIGURE 3.52. Arteries of hand.

The deep palmar arch gives rise to three *palmar metacarpal arteries* and the *princeps pollicis artery*. The *radialis indicis artery* passes along the lateral side of the index finger.

The *superficial* and *deep palmar venous arches*, associated with the superficial and deep palmar (arterial) arches, drain into the deep veins of the forearm. The dorsal digital veins drain into three dorsal metacarpal veins, which unite to form the *dorsal venous network*. The *cephalic vein* originates from the lateral side of the dorsal venous network and the *basilic vein* from the medial side.

Nerves of Hand

The median, ulnar, and radial nerves supply the hand. The median nerve enters the hand through the carpal tunnel, deep to the flexor retinaculum (Figs. 3.50D and 3.53), along with the tendons of the FDS, FDP, and FPL. The **carpal tunnel** is the passageway deep to the flexor retinaculum between the tubercles of the scaphoid and the trapezium bones on the lateral side and the pisiform and the hook of hamate on the medial side (Fig. 3.53). Distal to the carpal tunnel, the median nerve supplies two and a half thenar muscles and the 1st and 2nd lumbricals (Table 3.12). It also sends sensory fibers to the skin on the lateral palmar surface, the sides of the first three digits, the lateral half of the 4th digit, and the dorsum of the distal halves of these digits (Fig. 3.54). Note, however, that the *palmar cutaneous branch of the median nerve*, which supplies the central palm, arises proximal to the flexor retinaculum and passes superficial to it (i.e., it does not pass through the carpal tunnel).

The ulnar nerve leaves the forearm by emerging from deep to the tendon of the FCU (see Fig. 3.47A). It continues distally to the wrist via the *ulnar (Guyon) canal*. Here, the ulnar nerve is bound by fascia to the anterior surface of the flexor retinaculum. It then passes alongside the lateral border of the pisiform; the ulnar artery is on its lateral side. Just proximal to the wrist, the ulnar nerve gives off a *palmar cutaneous branch* that passes superficial to the flexor retinaculum and

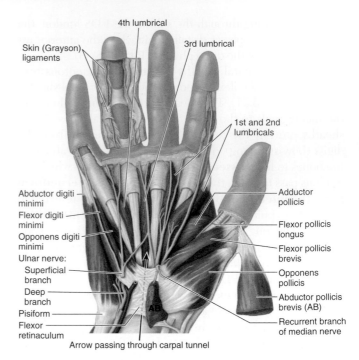

FIGURE 3.53. **Muscles and nerves of hand.**

palmar aponeurosis; it supplies skin on the medial side of the palm (Fig. 3.54B). The ulnar nerve also gives off a *dorsal cutaneous branch*, which supplies the medial half of the dorsum of the hand, the 5th finger, and the medial half of the 4th finger (Fig. 3.54A). The ulnar nerve ends at the distal border of the flexor retinaculum by dividing into superficial and deep branches (Fig. 3.53). The *superficial branch of the ulnar nerve* supplies cutaneous branches to the anterior surfaces of the medial one and a half fingers. The *deep branch of the ulnar nerve* supplies the hypothenar muscles, the medial two lumbricals, the AD, the deep head of FPB, and all the interossei (Fig. 3.47B and Table 3.12). The deep branch also supplies several joints (wrist, intercarpal, carpometacarpal,

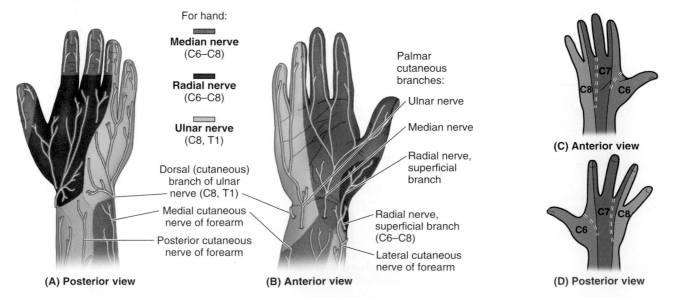

FIGURE 3.54. **Cutaneous innervation of hand. A and B.** Distribution of peripheral cutaneous nerves. **C. and D.** Segmental (dermatomal) innervation.

and intermetacarpal). The ulnar nerve is referred to as the *nerve of fine movements* because it innervates muscles that are concerned with intricate hand movements.

The radial nerve supplies no hand muscles. Its terminal branches, superficial and deep, arise in the cubital fossa (see Fig. 3.33B). The *superficial branch of the radial nerve* is entirely sensory (Fig. 3.54A,B). It pierces the deep fascia near the dorsum of the wrist to supply the skin and fascia over the lateral two thirds of the dorsum of the hand, the dorsum of the thumb, and the proximal parts of the lateral one and a half digits.

CLINICAL BOX

Dupuytren Contracture of Palmar Fascia

 Dupuytren contracture is a disease of the palmar fascia resulting in progressive shortening, thickening, and fibrosis of the palmar fascia and palmar aponeurosis. The fibrous degeneration of the longitudinal digital bands of the aponeurosis on the medial side of the hand pulls the 4th and 5th fingers into partial flexion at the metacarpophalangeal and proximal interphalangeal joints (Fig. B3.14).

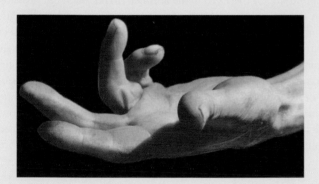

FIGURE B3.14. Dupuytren contracture.

The contracture is frequently bilateral. Treatment of the contracture usually involves surgical excision of the fibrotic parts of the palmar fascia to free the fingers.

Tenosynovitis

Injuries such as puncture of a digit by a rusty nail can cause infection of the digital synovial sheaths.

When inflammation of the tendon and synovial sheath (*tenosynovitis*) occurs, the digit swells and movement becomes painful. Because the tendons of the 2nd through 4th digits nearly always have separate synovial sheaths, the infection usually is confined to the infected digit. If the infection is untreated, however, the proximal ends of these sheaths may rupture, allowing the infection to spread to the midpalmar space (see Fig. 3.45A,B). Because the synovial sheath of the little finger is usually continuous with the common flexor sheath, tenosynovitis in this digit may spread to the common sheath and thus through the carpal tunnel to the forearm. How far an infection spreads from the digits depends on variations in their connections with the common flexor sheath.

The tendons of the APL and EPB are in the same tendinous sheath on the dorsum of the wrist. Excessive friction of these tendons results in fibrous thickening of the sheath and stenosis of the osseofibrous tunnel, *de Quervain tenosynovitis*. This condition causes pain in the wrist that radiates proximally to the forearm and distally to the thumb.

If the tendons of the FDS and FDP enlarge (forming a nodule) proximal to the tunnel, the person is unable to extend the finger. When the finger is extended passively, a snap is audible. This condition is called *stenosing tenosynovitis* (trigger finger or snapping finger) (Fig. B3.15).

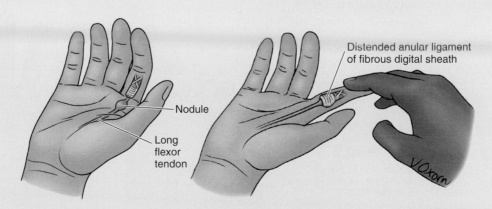

Nodule

Long flexor tendon

Distended anular ligament of fibrous digital sheath

Stenosing tenosynovitis (trigger finger)

FIGURE B3.15. Trigger finger.

(Continued on next page)

Carpal Tunnel Syndrome

 Carpal tunnel syndrome results from any lesion that significantly reduces the size of the carpal tunnel or, more commonly, increases the size of some of the structures (or their coverings) that pass through it (e.g., inflammation of the synovial sheaths). The median nerve is the most sensitive structure in the carpal tunnel and, therefore, it is the most affected (see Fig. 3.50C). The median nerve has two terminal sensory branches that supply the skin of the hand; hence, **paresthesia** (tingling), **hypesthesia** (diminished sensation), or **anesthesia** (absence of tactile sensation) may occur in the lateral three and a half digits. Recall, however, that the *palmar cutaneous branch of the median nerve* arises proximal to and does not pass through the carpal tunnel; thus, sensation in the central palm remains unaffected. This nerve also has one terminal motor branch, the recurrent branch, which innervates the three thenar muscles.

Wasting of the thenar eminence and progressive loss of coordination and strength in the thumb (owing to weakness of the APB and opponens pollicis) may occur if the cause of the compression is not alleviated. Individuals with carpal tunnel syndrome are unable to oppose the thumb (Fig. B3.16). To relieve the compression and resulting symptoms, partial or complete surgical division of the flexor retinaculum, a procedure called **carpal tunnel release**, may be necessary. The incision for carpal tunnel release is made toward the medial side of the wrist and flexor retinaculum to avoid possible injury to the recurrent branch of the median nerve.

Trauma to Median Nerve

Lesions of the median nerve usually occur in two places: the forearm and wrist. The most common site is where the nerve passes through the carpal tunnel. Laceration of the wrist often causes median nerve injury because this nerve is relatively close to the surface. This results in paralysis and wasting of the thenar muscles and the first two lumbrical muscles. Hence, opposition of the thumb is not possible, and fine movements of the 2nd and 3rd digits are impaired. Sensation is also lost over the thumb and adjacent two and a half digits.

Median nerve injury resulting from a perforating wound in the elbow region results in loss of flexion of the proximal and distal interphalangeal joints of the 2nd and 3rd digits. The ability to flex the metacarpophalangeal joints of these digits is also affected because digital branches of the median nerve supply the 1st and 2nd lumbricals. This results in a deformity in which thumb movements are limited to flexion and extension of the thumb in the plane of the palm. This condition is caused by the inability to oppose and by limited abduction of the thumb (Fig. B3.16).

Ulnar Nerve Injury

Ulnar nerve injury usually occurs in one of four places: (1) posterior to the medial epicondyle of the humerus (most common), (2) in the cubital tunnel formed by the tendinous arch connecting the humeral and ulnar heads of the FCU, (3) at the wrist, and (4) in the hand. Ulnar nerve injury occurring at the elbow, wrist, or hand may result in extensive motor and sensory loss to the hand. An injury to the nerve in the distal part of the forearm denervates most intrinsic hand muscles. The power of wrist adduction is impaired, and when an attempt is made to flex the wrist joint, the hand is drawn to the lateral side by the FCR in the absence of the "balance" provided by the FCU. After ulnar nerve injury, the person has difficulty making a fist because, in the absence of opposition, the metacarpophalangeal joints become hyperextended, and he or she cannot flex the 4th and 5th fingers at the distal interphalangeal joints when trying to make a fist. Furthermore, the person cannot extend the interphalangeal joints when trying to straighten the fingers. This characteristic appearance of the hand is known as *claw hand* (Fig. B3.17A). This deformity results from atrophy of the interosseous muscles of the hand. The claw is produced by the unopposed action of the extensors and FDP.

Compression of the ulnar nerve also may occur at the wrist where it passes between the pisiform and the hook of hamate. The depression between these bones is converted by the pisohamate ligament into an osseofibrous ulnar tunnel (Guyon tunnel). **Ulnar canal syndrome** is manifest by hypoesthesia in the medial one and one half fingers (Fig. B3.17B)

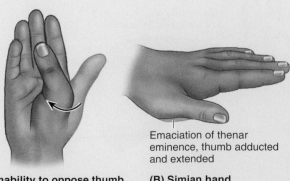

(A) **Inability to oppose thumb** (movement occurs at carpometacarpal joint)

(B) **Simian hand**

Emaciation of thenar eminence, thumb adducted and extended

FIGURE B3.16. Median nerve injury.

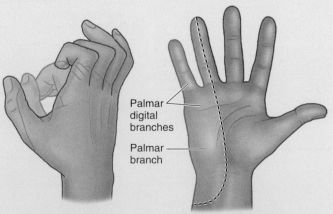

Palmar digital branches

Palmar branch

(A) **Claw hand** Unable to make loose fist

(B) **Sensory distribution of ulnar nerve**

FIGURE B3.17. Ulnar nerve injury.

and weakness of the intrinsic hand muscles. Clawing of the 4th and 5th fingers may occur, but in contrast to proximal ulnar nerve injury, their ability to flex is unaffected, and there is no radial deviation of the hand.

Radial Nerve Injury

 Although the radial nerve supplies no muscles in the hand, *radial nerve injury* in the arm by a fracture of the humeral shaft can produce serious disability of the hand. This injury is proximal to the branches to the extensors of the wrist, so wrist-drop is the primary clinical manifestation. The hand is flexed at the wrist and lies flaccid, and the digits also remain in the flexed position at the metacarpophalangeal joints. The extent of anesthesia is minimal, even in serious radial nerve injuries, and usually is confined to a small area on the lateral part of the dorsum of the hand. Severance of the deep branch results in an inability to extend the thumb and the metacarpophalangeal joints of the other digits. Loss of sensation does not occur because the deep branch is entirely muscular and articular in distribution.

Laceration of Palmar Arches

 Bleeding is usually profuse when the palmar (arterial) arches are lacerated. It may not be sufficient to ligate (tie off) only one forearm artery when the arches are lacerated because these vessels usually have numerous communications in the forearm and hand and bleed from both ends. To obtain a bloodless surgical operating field for treating complicated hand injuries, it may be necessary to compress the brachial artery and its branches proximal to the elbow (e.g., using a pneumatic tourniquet). This procedure prevents blood from reaching the ulnar and radial arteries through the anastomoses around the elbow.

Palmar Wounds and Surgical Incisions

The location of superficial and deep palmar arches should be kept in mind when examining wounds of the palm and when making palmar incisions (see Fig. 3.47). Furthermore, it is important to know that the superficial palmar arch is at the same level as the distal extremity of the common flexor sheath. Incisions or wounds along the medial surface of the thenar eminence may injure the recurrent branch of the median nerve to the thenar muscles.

Ischemia of Digits

Intermittent bilateral attacks of *ischemia of the digits*, marked by cyanosis and often accompanied by paresthesia and pain, are characteristically brought on by cold and emotional stimuli. The condition may result from an anatomical abnormality or an underlying disease. When the cause of the condition is idiopathic (unknown) or primary, it is called *Raynaud syndrome* (disease).

The arteries of the upper limb are innervated by sympathetic nerves. Postsynaptic fibers from the sympathetic ganglia enter nerves that form the brachial plexus and are distributed to the digital arteries through branches arising from the plexus. When treating ischemia resulting from Raynaud syndrome, it may be necessary to perform a cervicodorsal *presynaptic sympathectomy* (excision of a segment of a sympathetic nerve) to dilate the digital arteries.

SURFACE ANATOMY

Forearm and Hand

The **cubital fossa**, the triangular hollow area on the anterior surface of the elbow, is bounded medially by the prominence formed by the flexor–pronator group of muscles that are attached to the medial epicondyle. To estimate the position of these muscles, put your thumb posterior to your medial epicondyle and place your fingers on your forearm as shown in Figure SA3.4A.

A common place for measuring the radial pulse rate is where the radial artery lies on the anterior surface of the distal end of the radius, lateral to the FCR tendon (Fig. SA3.4B). Here, the artery can be felt pulsating between the tendons of the FCR and the APL and where it can be compressed against the radius. The **tendons of the FCR and palmaris longus** can be palpated anterior to the wrist. These tendons are a little lateral to the middle of the wrist and are usually observed by flexing the closed fist against resistance. The **tendon of the palmaris longus** serves as a guide to the median nerve, which lies deep to it. The **FCU tendon** can be palpated as it crosses the anterior aspect of the wrist near the medial side and inserts into the pisiform. The FCU tendon serves as a guide to the ulnar nerve and artery. The **tendons of the FDS** can be palpated as the digits are alternately flexed and extended (Fig SA3.4B).

(*Continued on next page*)

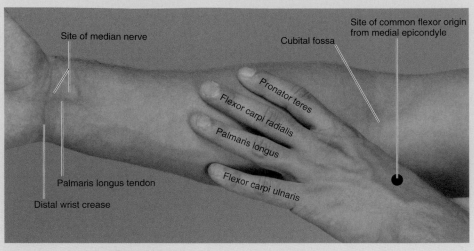

Site of median nerve

Cubital fossa

Site of common flexor origin from medial epicondyle

Pronator teres

Flexor carpi radialis

Palmaris longus

Flexor carpi ulnaris

Palmaris longus tendon

Distal wrist crease

(A) Anterior view of supinated forearm

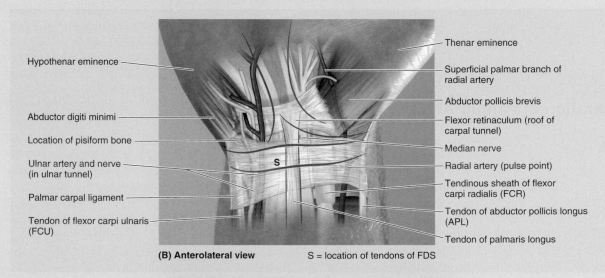

Hypothenar eminence

Thenar eminence

Superficial palmar branch of radial artery

Abductor pollicis brevis

Abductor digiti minimi

Flexor retinaculum (roof of carpal tunnel)

Location of pisiform bone

Median nerve

Ulnar artery and nerve (in ulnar tunnel)

Radial artery (pulse point)

Palmar carpal ligament

Tendinous sheath of flexor carpi radialis (FCR)

Tendon of flexor carpi ulnaris (FCU)

Tendon of abductor pollicis longus (APL)

Tendon of palmaris longus

S

(B) Anterolateral view   S = location of tendons of FDS

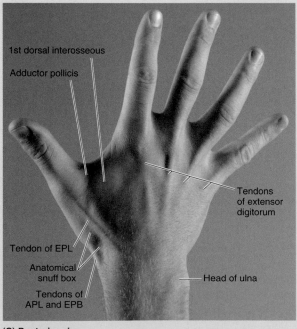

1st dorsal interosseous

Adductor pollicis

Tendons of extensor digitorum

Tendon of EPL

Anatomical snuff box

Head of ulna

Tendons of APL and EPB

(C) Posterior view

FIGURE SA3.4. Surface anatomy of forearm and hand.

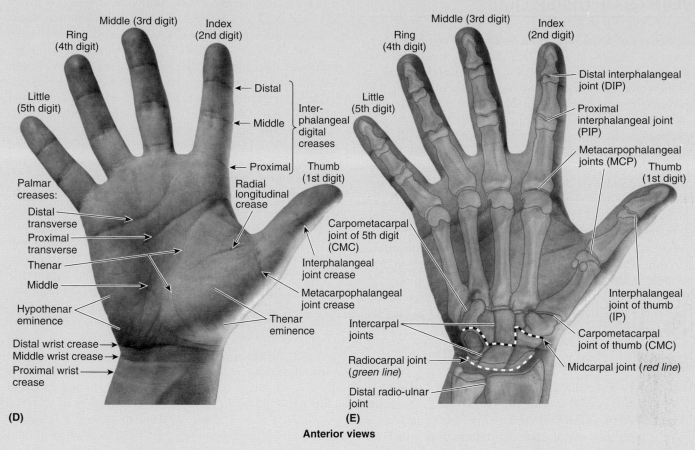

FIGURE SA3.4. *(Continued)*

The **tendons of the APL and EPB** indicate the lateral (anterior) boundary of the anatomical snuff box, and the **tendon of the EPL** indicates the medial (posterior) boundary of the box (Fig. SA3.4C). The **radial artery** crosses the floor of the snuff box, where its pulsations may be felt. The scaphoid and, less distinctly, the trapezium are palpable in the floor of the snuff box.

If the dorsum of the hand is examined with the wrist extended against resistance and the digits abducted, the **tendons of the extensor digitorum** to the fingers stand out (Fig. SA3.4C). These tendons are not visible far beyond the knuckles because they flatten here to form the extensor expansions of the fingers. Under the loose subcutaneous tissue and extensor tendons, the **metacarpals** can be palpated. The knuckles that become visible when a fist is made are produced by the heads of the metacarpals.

The palmar skin presents several more or less constant *flexion creases* where the skin is firmly bound to the deep fascia (Fig. SA3.4D):

- **Wrist creases**: Proximal, **middle**, **distal**. The *distal wrist crease* indicates the proximal border of the flexor retinaculum.
- **Palmar creases**: **Radial longitudinal crease** (the "life line" of palmistry), proximal and distal transverse palmar creases
- **Transverse digital flexion creases**: The **proximal digital crease** is located at the root of the digit, approximately 2 cm distal to the metacarpophalangeal joint. The proximal digital crease of the thumb crosses obliquely, proximal to the 1st metacarpophalangeal joint. The **middle digital crease** lies over the proximal interphalangeal joint, and the **distal digital crease** lies proximal to the distal interphalangeal joint. The thumb, having two phalanges, has only two flexion creases.

JOINTS OF UPPER LIMB

Movement of the pectoral girdle involves the sternoclavicular, acromioclavicular, and glenohumeral joints, usually all moving simultaneously (see Fig. 3.55). Functional defects in any of these joints impair movements of the pectoral girdle. Mobility of the scapula is essential for the freedom of movement of the upper limb. When testing *the range of motion of the pectoral girdle*, both scapulothoracic (movement of the scapula on the thoracic wall) and glenohumeral movements must be considered. Although the initial 30 degrees may occur without scapular motion, in the overall movement of fully elevating the arm, the movement occurs in a 2:1 ratio. For every 3 degrees of elevation, approximately 2 degrees occur at the glenohumeral joint and 1 degree at the scapulothoracic joint (see Fig. 3.56C). This is known as **scapulohumeral rhythm**. The important movements of the pectoral girdle are scapular movements: elevation and depression, protraction (lateral or forward movement of the scapula), and retraction (medial or backward movement of the scapula) and rotation of the scapula.

Sternoclavicular Joint

The **sternoclavicular (SC) joint** is a synovial articulation between the sternal end of the clavicle and the manubrium of the sternum and the 1st costal cartilage. The SC joint is a saddle type of joint but functions as a ball-and-socket joint (see Fig. 3.55). The SC joint is divided into two compartments by an **articular disc**. The disc is firmly attached to the *anterior* and *posterior SC ligaments*, thickenings of the fibrous layer of the joint capsule as well as to the *interclavicular ligament*. The great strength of the SC joint is a consequence of these attachments. Thus, although the articular disc serves as a shock absorber of forces transmitted along the clavicle from the upper limb, dislocation of the clavicle is unusual, whereas fracture of the clavicle is common. The SC joint, the only articulation between the upper limb and the axial skeleton, can be readily palpated because the sternal end of the clavicle lies superior to the manubrium of the sternum.

The **joint capsule** surrounds the SC joint, including the epiphysis at the sternal end of the clavicle. The *fibrous layer of the capsule* is attached to the margins of the articular surfaces, including the periphery of the articular disc. A *synovial membrane* lines the internal surfaces of the fibrous layer of the capsule. **Anterior** and **posterior SC ligaments** reinforce the joint capsule anteriorly and posteriorly. The **interclavicular ligament** strengthens the capsule superiorly (see Fig. 3.55). It extends from the sternal end of one clavicle to the sternal end of the other clavicle; it is also attached to the superior border of the manubrium. The **costoclavicular ligament** anchors the inferior surface of the sternal end of the clavicle to the 1st rib and its costal cartilage, limiting elevation of the pectoral girdle.

Although the SC joint is extremely strong, it is significantly mobile to allow movements of the pectoral girdle and upper limb. During full elevation of the limb, the clavicle is

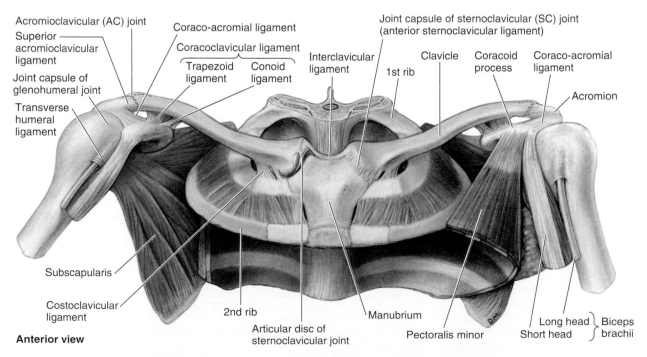

FIGURE 3.55. Joints of pectoral girdle and associated tendons and ligaments.

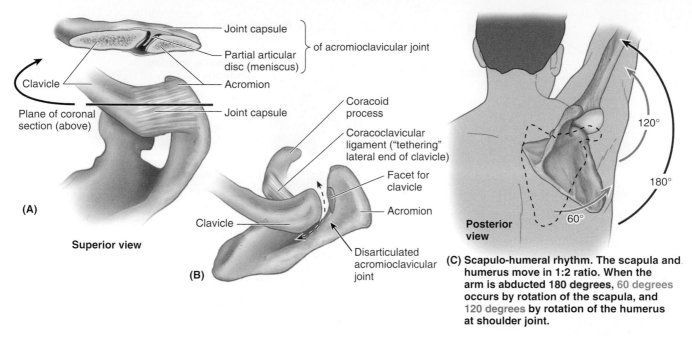

FIGURE 3.56. Acromioclavicular and scapulothoracic joints. A. Joint capsule and partial articular disc. **B.** Coracoclavicular ligament and articular facets. **C.** Rotation of scapula at the physiological scapulothoracic joint.

raised to approximately a 60-degree angle. The SC joint can also be moved anteriorly or posteriorly over a range up to 25–30 degrees.

The SC joint is supplied by internal thoracic and suprascapular arteries (Table 3.4). Branches of the medial supraclavicular and the subclavian nerves supply the SC joint.

Acromioclavicular Joint

The **acromioclavicular (AC) joint** is a plane synovial articulation (Figs. 3.55 and 3.56AB). It is located 2–3 cm from the "point" of the shoulder formed by the lateral part of the acromion of the scapula. The acromial end of the clavicle articulates with the acromion. The articular surfaces, covered with fibrocartilage, are separated by an incomplete wedge-shaped *articular disc*.

The sleeve-like, relatively loose *fibrous layer of the joint capsule* is attached to the margins of the articular surfaces. A *synovial membrane* lines the internal surface of the fibrous layer of the capsule. Although relatively weak, the joint capsule is strengthened superiorly by fibers of the trapezius.

The superior **AC ligament**, a fibrous band extending from the acromion to the clavicle, strengthens the AC joint superiorly (Fig. 3.55). Most of its strength comes from the coracoclavicular ligament. It maintains its integrity and prevents the acromion from being driven under the clavicle even when the AC joint is separated. The strong, extra-articular

coracoclavicular ligament (subdivided into conoid and trapezoid ligaments) is located several centimeters from the AC joint, which anchors the clavicle to the coracoid process of the scapula (Figs. 3.55 and 3.56B). The apex of the vertical **conoid ligament** is attached to the root of the *coracoid process*. Its wide attachment (base) is to the *conoid tubercle* on the inferior surface of the clavicle (see Fig. 3.3A,B). The nearly horizontal **trapezoid ligament** is attached to the superior surface of the coracoid process and extends laterally and posteriorly to the trapezoid line on the inferior surface of the clavicle. In addition to augmenting the AC joint, the coracoclavicular ligament provides the means by which the scapula and free limb are (passively) suspended from the clavicle.

The acromion of the scapula rotates on the acromial end of the clavicle. These movements are associated with motion at the physiological scapulothoracic joint. The axio-appendicular muscles that attach to and move the scapula cause the acromion to move on the clavicle (Fig. 3.57). Factors limiting scapular movements are listed in Table 3.13. The AC joint is supplied by the suprascapular and thoracoacromial arteries. Supraclavicular, lateral pectoral, and axillary nerves supply the joint.

Glenohumeral Joint

The **glenohumeral (shoulder) joint** is a ball-and-socket, synovial joint that permits a wide range of movement; however, its mobility makes the joint relatively unstable.

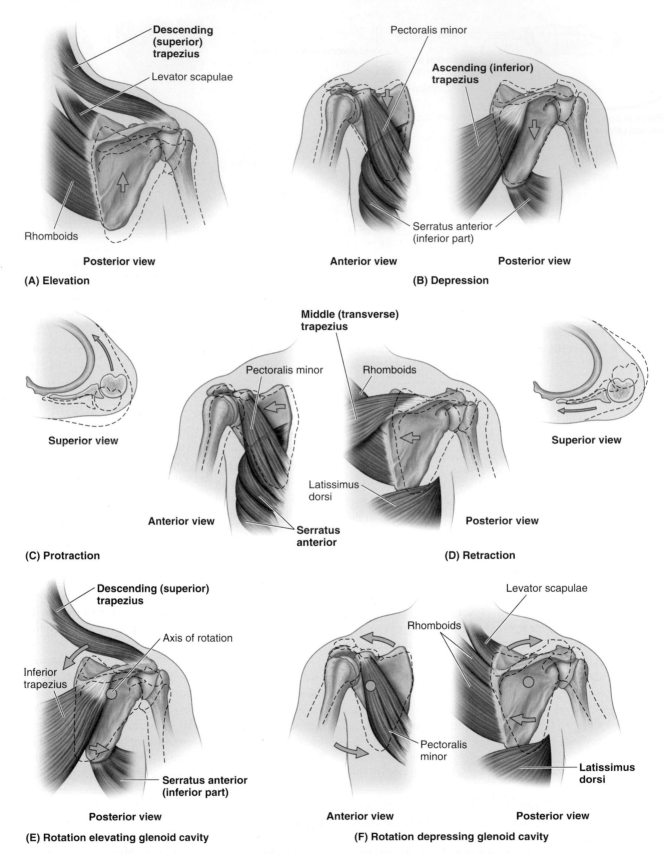

(A) Elevation

(B) Depression

(C) Protraction

(D) Retraction

(E) Rotation elevating glenoid cavity

(F) Rotation depressing glenoid cavity

FIGURE 3.57. Scapular movements. Scapula moves on the thoracic wall at the conceptual "scapulothoracic joint." *Dotted lines,* the starting position of each movement.

TABLE 3.13. STRUCTURES LIMITING MOVEMENTS OF SHOULDER REGION

Movement	Joint(s)	Limiting Structures (Tension)
Flexion (0–180 degrees)	Sternoclavicular Acromioclavicular Glenohumeral Scapulothoracic	*Ligaments*: posterior part of coracohumeral, trapezoid, and posterior part of joint capsule of glenohumeral joint *Muscles*: rhomboids, levator scapulae, extensor and external rotator muscles, rotator muscles of glenohumeral joint
Abduction (0–180 degrees)	Sternoclavicular Acromioclavicular Glenohumeral Scapulothoracic	*Ligaments*: middle and inferior glenohumeral, trapezoid, and inferior part of joint capsule of glenohumeral joint *Muscles*: rhomboids, levator scapulae, adductor muscles of glenohumeral joint *Bony apposition* between greater tubercle of humerus and superior part of glenoid cavity/labrum or lateral aspect of acromion
Extension		*Ligaments*: anterior part of coracohumeral and anterior part of joint capsule of glenohumeral joint *Muscles*: clavicular head of pectoralis major
Medial (internal) rotation	Glenohumeral	*Ligaments*: posterior glenohumeral joint capsule *Muscles*: infraspinatus and teres minor
Lateral (external) rotation		*Ligaments*: glenohumeral, coracohumeral, anterior glenohumeral joint capsule *Muscles*: latissimus dorsi, teres major, pectoralis major, subscapularis

Modified from Clarkson HM. *Musculoskeletal Assessment: Joint Motion and Muscle Testing*. 3rd ed. Baltimore, MD: Lippincott Williams & Wilkins; 2012.

ARTICULATION AND JOINT CAPSULE OF GLENOHUMERAL JOINT

The large spherical *humeral head* articulates with the relatively small and shallow *glenoid cavity* of the scapula, which is deepened slightly by the ring-like, fibrocartilaginous **glenoid labrum** (L. lip). Both articular surfaces are covered with hyaline cartilage (Fig. 3.58A–C). The glenoid cavity accepts little more than a third of the humeral head, which is held in the cavity by the tonus of the *musculotendinous rotator cuff* (supraspinatus, infraspinatus, teres minor, and subscapularis).

The loose *fibrous layer of the joint capsule* surrounds the glenohumeral joint and is attached medially to the margin of the glenoid cavity and laterally to the anatomical neck of the humerus. Superiorly, the fibrous layer encloses the proximal attachment of the long head of biceps brachii to the supraglenoid tubercle of the scapula within the joint. The inferior part of the joint capsule, the only part not reinforced by the rotator cuff muscles, is its weakest area. Here, the capsule is particularly lax and lies in folds when the arm is adducted; however, it becomes taut when the arm is abducted (Fig. 3.58B,D).

The *synovial membrane* lines the internal surface of the fibrous capsule and reflects from it onto the humerus as far as the articular margin of its head (see Fig. 3.54B). The synovial membrane also forms a tubular sheath for the tendon of the long head of the biceps brachii. Anteriorly, there is a communication between the *subscapular bursa* and the synovial cavity of the joint (see Fig. 3.54C).

LIGAMENTS OF GLENOHUMERAL JOINT

The **glenohumeral ligaments**, evident only on the internal aspect of the capsule, strengthen the anterior aspect of the capsule (Fig. 3.58C,D). The **coracohumeral ligament**, a strong band that passes from the base of the coracoid process to the anterior aspect of the greater tubercle, strengthens the capsule superiorly (Fig. 3.58D). The glenohumeral ligaments are intrinsic ligaments that are part of the fibrous layer of the capsule. The **transverse humeral ligament** is a broad, fibrous band that runs from the greater to the lesser tubercle, bridging over the intertubercular sulcus (groove) and converting the sulcus into a canal for the tendon of the long head of biceps brachii and its synovial sheath.

The **coraco-acromial arch** is an extrinsic, protective structure formed by the smooth inferior aspect of the *acromion* and *coracoid process* of the scapula, with the **coraco-acromial ligament** spanning between them (Figs. 3.55 and 3.58D). The coraco-acromial arch overlies the head of the humerus, preventing its superior displacement from the glenoid cavity. The arch is so strong that a forceful superior thrust of the humerus will not fracture it; the shaft of the humerus or clavicle fractures first.

MOVEMENTS OF GLENOHUMERAL JOINT

The glenohumeral joint has more freedom of movement than any other joint in the body. This freedom results from the laxity of its joint capsule and the configuration of the spherical humeral head and shallow glenoid cavity. The glenohumeral joint allows movements around the three axes and permits flexion–extension, abduction–adduction, rotation (medial and lateral) of the humerus, and circumduction. Table 3.13 lists structures that limit movements of the glenohumeral joint. Lateral rotation of the humerus increases the range of abduction. When the arm is abducted without rotation, the greater tubercle contacts the *coraco-acromial arch*, preventing further abduction. If the arm is then laterally rotated 180 degrees, the tubercles are rotated posteriorly, and more articular surface becomes available to continue elevation. Stiffening or fixation of the joints of the pectoral girdle (*ankylosis*) results in a much more restricted range of movement, even if the glenohumeral joint is normal.

The muscles moving the joint are the *axio-appendicular muscles*, which may act indirectly on the joint (i.e., act on the pectoral girdle), and the *scapulohumeral muscles*, which act

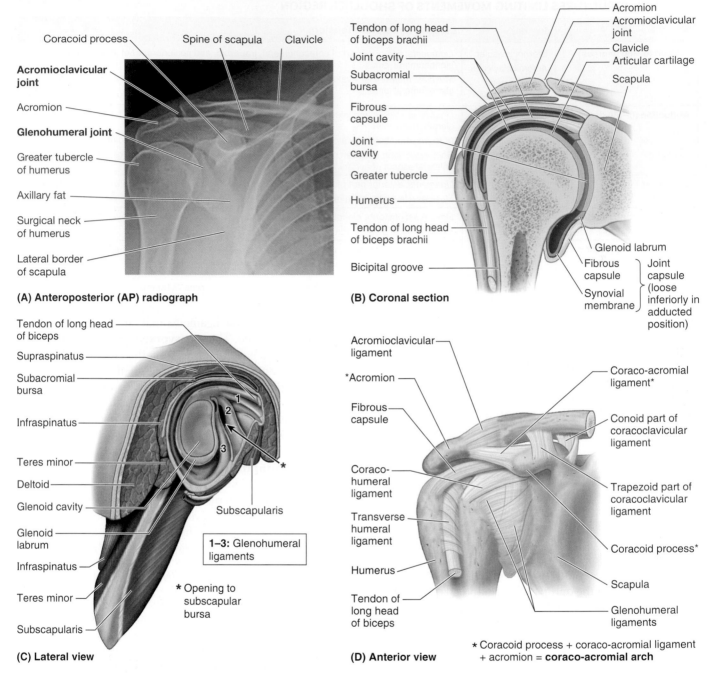

Coracoid process Spine of scapula Clavicle

Acromioclavicular joint

Acromion

Glenohumeral joint

Greater tubercle of humerus

Axillary fat

Surgical neck of humerus

Lateral border of scapula

(A) Anteroposterior (AP) radiograph

Tendon of long head of biceps brachii
Joint cavity
Subacromial bursa
Fibrous capsule
Joint cavity
Greater tubercle
Humerus
Tendon of long head of biceps brachii
Bicipital groove

Acromion
Acromioclavicular joint
Clavicle
Articular cartilage
Scapula

Glenoid labrum
Fibrous capsule
Synovial membrane
Joint capsule (loose inferiorly in adducted position)

(B) Coronal section

Tendon of long head of biceps
Supraspinatus
Subacromial bursa
Infraspinatus
Teres minor
Deltoid
Glenoid cavity
Glenoid labrum
Infraspinatus
Teres minor
Subscapularis

1
2
3
*

Subscapularis

1–3: Glenohumeral ligaments

* Opening to subscapular bursa

(C) Lateral view

Acromioclavicular ligament
*Acromion
Fibrous capsule
Coraco-humeral ligament
Transverse humeral ligament
Humerus
Tendon of long head of biceps

Coraco-acromial ligament*
Conoid part of coracoclavicular ligament
Trapezoid part of coracoclavicular ligament
Coracoid process*
Scapula
Glenohumeral ligaments

(D) Anterior view

* Coracoid process + coraco-acromial ligament + acromion = **coraco-acromial arch**

FIGURE 3.58. Glenohumeral and acromioclavicular joints. A. Radiograph. **B.** Coronal section of glenohumeral joint. **C.** Lateral view of glenoid cavity and related structures following disarticulation of humerus. **D.** Ligaments.

directly on the joint (Tables 3.2 to 3.4). Other muscles serve the glenohumeral joint as *shunt muscles*, acting to resist dislocation without producing movement at the joint, or maintain the head of the humerus in the glenoid cavity. For example, when the arms are at one's side, the deltoid functions as a shunt muscle.

BLOOD SUPPLY AND INNERVATION OF GLENOHUMERAL JOINT

The glenohumeral joint is supplied by the *anterior* and *posterior circumflex humeral arteries* and branches of the *suprascapular*

artery (Table 3.5). The *suprascapular, axillary,* and *lateral pectoral nerves* supply the glenohumeral joint (Table 3.6).

BURSAE AROUND GLENOHUMERAL JOINT

Several **bursae** containing capillary films of *synovial fluid* are located near the joint where tendons rub against bone, ligaments, or other tendons and where skin moves over a bony prominence. Some bursae communicate with the joint cavity; hence, opening a bursa may mean entering the cavity of the joint.

The **subacromial bursa**, sometimes referred to as the *subdeltoid bursa* (Fig. 3.58B,C), is located between the acromion, coraco-acromial ligament, and deltoid superiorly and the supraspinatus tendon and joint capsule of the glenohumeral joint inferiorly. Thus, it facilitates movement of the supraspinatus tendon under the coraco-acromial arch and of the deltoid over the joint capsule and the greater tubercle of the humerus.

The **subscapular bursa** is located between the tendon of the subscapularis and the neck of the scapula. This bursa protects the tendon where it passes inferior to the root of the coracoid process and over the neck of the scapula. It usually communicates with the cavity of the glenohumeral joint through an opening in the fibrous layer of the joint capsule (Fig. 3.58C).

CLINICAL BOX

Rotator Cuff Injuries

The musculotendinous rotator cuff is commonly injured during repetitive use of the upper limb above the horizontal (e.g., during throwing and racquet sports, swimming, and weight lifting). Recurrent inflammation of the rotator cuff, especially the relatively avascular area of the supraspinatus tendon, is a common cause of shoulder pain and results in tears of the rotator cuff (Fig. B3.18). Repetitive use of the rotator cuff muscles (e.g., by baseball pitchers) may allow the humeral head and rotator cuff to impinge on the coraco-acromial arch, producing irritation of the arch and inflammation of the rotator cuff. As a result, *degenerative tendinitis of the rotator cuff* develops. Attrition of the supraspinatus tendon also occurs. Because the supraspinatus muscle is no longer functional with a complete tear of the rotator cuff, the person cannot initiate abduction of the upper limb. If the arm is passively abducted 15 degrees or more, the person can usually initiate abduction by leaning or using their hip, then maintain or continue the abduction using the deltoid.

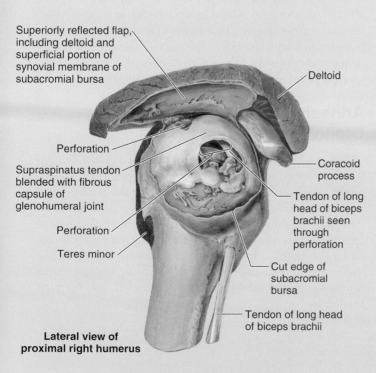

Superiorly reflected flap, including deltoid and superficial portion of synovial membrane of subacromial bursa

Perforation

Supraspinatus tendon blended with fibrous capsule of glenohumeral joint

Perforation

Teres minor

Deltoid

Coracoid process

Tendon of long head of biceps brachii seen through perforation

Cut edge of subacromial bursa

Tendon of long head of biceps brachii

Lateral view of proximal right humerus

FIGURE B3.18. Rotator cuff injury.

Dislocation of Acromioclavicular Joint

Although its extrinsic (coracoclavicular) ligament is strong, the AC joint itself is weak and easily injured by a direct blow. In contact sports such as football, soccer, and hockey, it is not uncommon for *dislocation of the AC joint* to result from a hard fall on the shoulder or on the outstretched upper limb (Fig. B3.19). Dislocation of the AC joint also can occur when a hockey player is, for example, driven violently into the boards. An AC dislocation, often called a "shoulder separation," is severe when both the AC and the coracoclavicular ligaments are torn. When the coracoclavicular ligament tears, the shoulder separates from the clavicle and falls because of the weight of the upper limb.

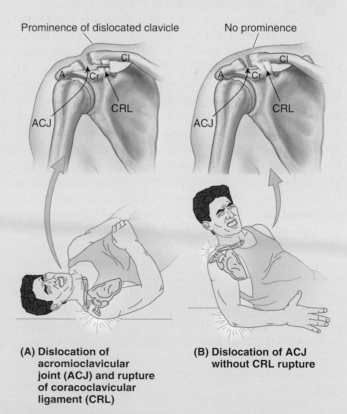

Prominence of dislocated clavicle

No prominence

Cl

A   Cr

CRL

ACJ

Cl

A   Cr

CRL

ACJ

(A) Dislocation of acromioclavicular joint (ACJ) and rupture of coracoclavicular ligament (CRL)

(B) Dislocation of ACJ without CRL rupture

FIGURE B3.19. Dislocation of acromioclavicular joint. *A*, acromion; *Cl*, clavicle; *Cr*, coracoid process.

(*Continued on next page*)

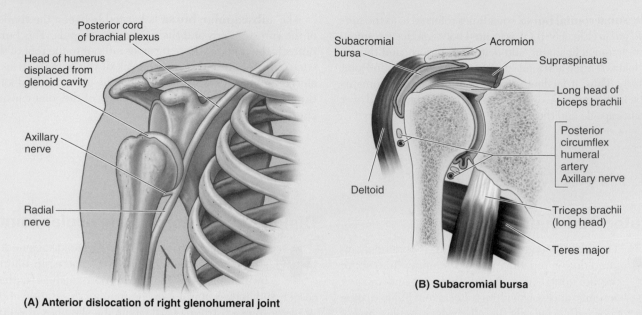

Posterior cord
of brachial plexus

Head of humerus
displaced from
glenoid cavity

Axillary
nerve

Radial
nerve

(A) Anterior dislocation of right glenohumeral joint

Subacromial
bursa — Acromion

Supraspinatus

Long head of
biceps brachii

Posterior
circumflex
humeral
artery
Axillary nerve

Deltoid

Triceps brachii
(long head)

Teres major

(B) Subacromial bursa

FIGURE B3.20. Dislocation of glenohumeral joint.

Dislocation of the AC joint makes the acromion more prominent, and the clavicle may move superior to the acromion.

Dislocation of Glenohumeral Joint

Because of its freedom of movement and instability, the glenohumeral joint is commonly dislocated by direct or indirect injury. Most dislocations of the humeral head occur in the downward (inferior) direction but are described clinically as anterior or (more rarely) posterior dislocations, indicating whether the humeral head has descended anterior or posterior to the infraglenoid tubercle and the long head of triceps. *Anterior dislocation of the glenohumeral joint* occurs most often in young adults (Fig. B3.20A), particularly athletes. It is usually caused by excessive extension and lateral rotation of the humerus. The head of the humerus is driven infero-anteriorly, and the fibrous layer of the joint capsule and glenoid labrum may be stripped from the anterior aspect of the glenoid cavity. A hard blow to the humerus when the glenohumeral joint is fully abducted tilts the head of the humerus inferiorly onto the inferior weak part of the joint capsule. This may tear the capsule and dislocate the joint so that the humeral head comes to lie inferior to the glenoid cavity and anterior to the infraglenoid tubercle. Subsequently, the strong flexor and adductor muscles of the glenohumeral joint usually pull the humeral head anterosuperiorly into a subcoracoid position. Unable to use the arm, the person commonly supports it with the other hand. The axillary nerve may be injured when the glenohumeral joint dislocates because of its close relation to the inferior part of the capsule of this joint (Fig. B3.20B).

Calcific Supraspinatus Tendinitis

Inflammation and calcification of the subacromial bursa result in pain, tenderness, and limitation of movement of the glenohumeral joint. This condition is also known as *calcific scapulohumeral bursitis*. Deposition of calcium in the supraspinatus tendon may irritate the overlying subacromial bursa, producing an inflammatory reaction, *subacromial bursitis*. As long as the glenohumeral joint is adducted, no pain usually results because in this position, the painful lesion is away from the inferior surface of the acromion. In most people, the pain occurs during 50–130 degrees of abduction (*painful arc syndrome*) because during this arc, the supraspinatus tendon is in intimate contact with the inferior surface of the acromion. The pain usually develops in males 50 years of age and older after unusual or excessive use of the glenohumeral joint.

Adhesive Capsulitis of Glenohumeral Joint

Adhesive fibrosis and scarring between the inflamed capsule of the glenohumeral joint, rotator cuff, subacromial bursa, and deltoid usually cause *adhesive capsulitis* ("frozen shoulder"). A person with this condition has difficulty abducting the arm but can obtain an apparent abduction of up to 45 degrees by elevating and rotating the scapula. Injuries that may initiate this condition include glenohumeral dislocations, calcific supraspinatus tendinitis, partial tearing of the rotator cuff, and bicipital tendinitis.

Elbow Joint

The **elbow joint**, a hinge type of synovial joint, is located 2–3 cm inferior to the humeral epicondyles.

ARTICULATION AND JOINT CAPSULE OF ELBOW JOINT

The spool-shaped *trochlea* and spheroidal *capitulum* of the humerus articulate with the *trochlear notch* of the ulna and the slightly concave superior aspect of the *head of radius*, respectively; therefore, there are *humero-ulnar* and *humero-radial articulations* (Fig. 3.59A,B).

The *fibrous layer of the joint capsule* surrounding the joint is attached to the humerus at the margins of the lateral and medial ends of the articular surfaces of the capitulum and trochlea. Anteriorly and posteriorly, it is carried superiorly, proximal to the coronoid and olecranon fossae (Fig. 3.59A). The *synovial membrane* lines the internal surface of the

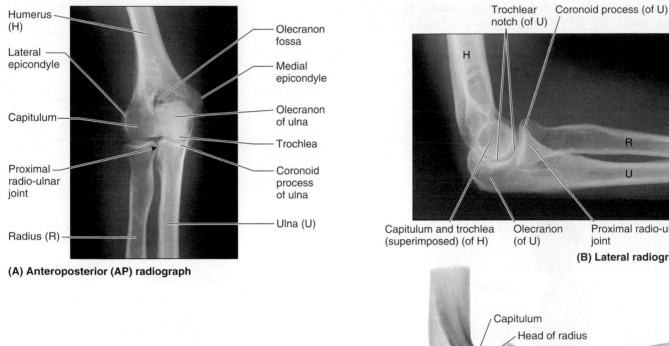

(A) Anteroposterior (AP) radiograph

(B) Lateral radiograph

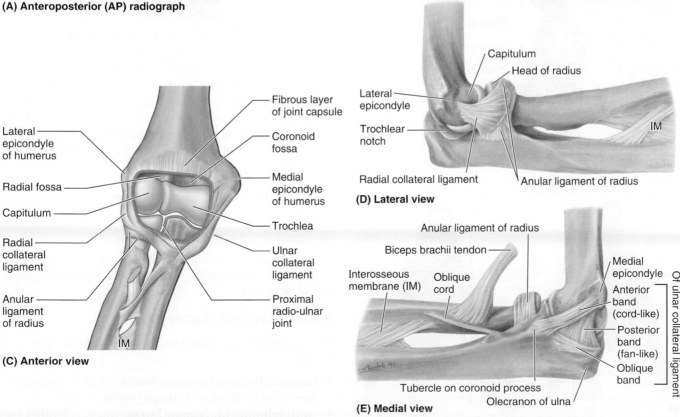

(C) Anterior view

(D) Lateral view

(E) Medial view

FIGURE 3.59. **Elbow and proximal radio-ulnar joints. A.** Anteroposterior radiograph. **B.** Lateral radiograph. **C.** Articulating surfaces. The thin anterior aspect of the joint capsule has been removed. **D.** Lateral ligaments. **E.** Medial ligaments.

fibrous layer of the joint capsule and the intracapsular non-articular parts of the humerus. It is continuous inferiorly with the synovial membrane of the proximal radio-ulnar joint. The joint capsule is weak anteriorly and posteriorly but is strengthened on each side by ligaments.

LIGAMENTS OF ELBOW JOINT

The **collateral ligaments of the elbow joint** are strong triangular bands that are medial and lateral thickenings of the fibrous layer of the joint capsule. The lateral, fan-like **radial collateral ligament** extends from the lateral epicondyle of the humerus and blends distally with the **anular ligament of the radius** (Fig. 3.59D). This ligament encircles and holds the head of the radius in the radial notch of the ulna, forming the proximal radio-ulnar joint and permitting pronation and supination of the forearm. The medial, triangular **ulnar collateral ligament** extends from the medial epicondyle of the humerus to the coronoid process and olecranon of the ulna. It consists of three bands: (1) the *anterior cord-like band* is the strongest, (2) the *posterior fan-like band* is the weakest, and (3) the slender *oblique band* deepens the socket for the trochlea of the humerus (Fig. 3.59E).

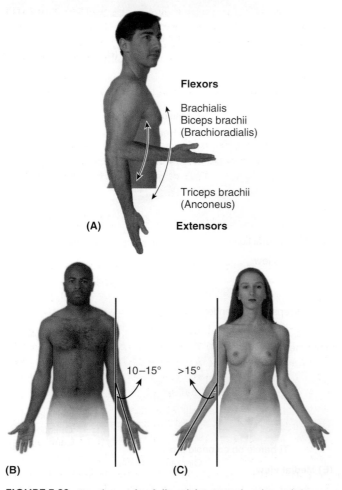

(A)   Flexors

Brachialis
Biceps brachii
(Brachioradialis)

Triceps brachii
(Anconeus)

Extensors

(B)   10–15°

(C)   >15°

FIGURE 3.60. Carrying angle of elbow joint. Note that the angle is greater in the woman.

TABLE 3.14. STRUCTURES LIMITING MOVEMENTS OF ELBOW AND RADIO-ULNAR JOINTS

Joint(s)	Movement	Limiting Structures (Tension)
Humero-ulnar Humeroradial	**Extension**	*Muscles:* flexor muscles of elbow *Joint capsule:* anteriorly *Bony apposition* between olecranon of ulna and olecranon fossa of humerus
Humero-ulnar Humeroradial	**Flexion**	*Muscle:* triceps brachii *Joint capsule:* posteriorly *Soft tissue apposition* between anterior forearm and arm *Bony apposition* between head of radius and radial fossa of humerus
Humeroradial Proximal radio-ulnar Distal radio-ulnar	**Pronation**	*Muscles:* supinator, biceps brachii *Ligaments:* posterior inferior radio-ulnar, interosseous membrane *Bony apposition* of the radius on ulna
Humeroradial Proximal radio-ulnar Distal radio-ulnar	**Supination**	*Muscles:* pronator teres, pronator quadratus *Ligaments:* anterior inferior radio-ulnar, interosseous membrane

Modified from Clarkson HM. *Musculoskeletal Assessment: Joint Motion and Muscle Testing.* 3rd ed. Baltimore, MD: Lippincott Williams & Wilkins; 2012.

MOVEMENTS OF ELBOW JOINT

Flexion and extension occur at the elbow joint. The long axis of the fully extended ulna makes an angle of approximately 170 degrees with the long axis of the humerus. This angle is called the **carrying angle** and is named for the way the forearm angles away from the body when something is carried, such as a pail of water (Fig. 3.60). The obliquity of the angle is more pronounced in women than in men. Table 3.14 lists structures limiting movements of the elbow joint.

BLOOD SUPPLY AND INNERVATION OF ELBOW JOINT

The arteries supplying the elbow are derived from the anastomosis of arteries around the elbow joint (see Fig. 3.43A). The elbow joint is supplied by the musculocutaneous, radial, and ulnar nerves.

BURSAE AROUND ELBOW JOINT

The clinically important bursae are the (Figs. 3.61 and 3.62B)

- **intratendinous olecranon bursa**, which is sometimes present in the tendon of triceps brachii
- **subtendinous olecranon bursa**, which is located between the olecranon and the triceps tendon, just proximal to its attachment to the olecranon

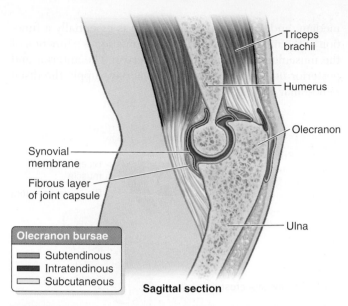

Synovial membrane

Fibrous layer of joint capsule

Triceps brachii

Humerus

Olecranon

Ulna

Olecranon bursae
- Subtendinous
- Intratendinous
- Subcutaneous

Sagittal section

FIGURE 3.61. Coronal section through humero-ulnar articulation of elbow joint showing relationships of bursae.

- **subcutaneous olecranon bursa**, which is located in the subcutaneous connective tissue over the olecranon

The *bicipitoradial bursa* (biceps bursa) separates the biceps tendon from the anterior part of the radial tuberosity.

Proximal Radio-Ulnar Joint

The **proximal (superior) radio-ulnar joint** is a pivot type of synovial joint that allows movement of the head of the radius on the ulna (Figs. 3.59A–C and 3.62).

ARTICULATION AND JOINT CAPSULE OF PROXIMAL RADIO-ULNAR JOINT

The head of the radius articulates with the radial notch of the ulna. The radial head is held in place by the *anular ligament of the radius*. The *fibrous layer of the joint capsule* encloses the joint and is continuous with that of the elbow joint. The *synovial membrane* lines the internal surface of the fibrous layer and nonarticulating aspects of the bones. The synovial membrane is an inferior prolongation of the synovial membrane of the elbow joint (Fig. 3.59C).

LIGAMENTS OF PROXIMAL RADIO-ULNAR JOINT

The **anular ligament** attaches to the ulna, anterior and posterior to the radial notch, which forms a collar that, with the radial notch, forms a ring that completely encircles the head of the radius (Fig. 3.62A). The deep surface of the anular ligament is lined with synovial membrane, which continues distally as a **sacciform recess of the proximal radio-ulnar joint** on the neck of the radius. This arrangement allows the radius to rotate within the anular ligament without binding, stretching, or tearing of the synovial membrane.

Distal Radio-Ulnar Joint

The **distal (inferior) radio-ulnar joint** is a pivot type of synovial joint. The radius moves around the relatively fixed distal end of the ulna (Fig. 3.62).

ARTICULATION AND JOINT CAPSULE OF DISTAL RADIO-ULNAR JOINT

The rounded head of the ulna articulates with the ulnar notch on the medial side of the distal end of the radius. A fibrocartilaginous **articular disc of the distal radio-ulnar joint** binds the ends of the ulna and radius together and is the main uniting structure of the joint (see Fig. 3.63). The base of the disc attaches to the medial edge of the ulnar notch of the radius, and its apex is attached to the lateral side of the base of the styloid process of ulna. The proximal surface of this triangular disc articulates with the distal aspect of the head of the ulna. Hence, the joint cavity is L-shaped in a coronal section, with the vertical bar of the *L* between the radius and the ulna and the horizontal bar between the ulna and the articular disc. The articular disc separates the cavity of the distal radio-ulnar joint from the cavity of the wrist joint.

The *fibrous layer of the joint capsule* encloses the joint but is deficient superiorly. The *synovial membrane* extends superiorly between the radius and the ulna to form the **sacciform recess of the distal radio-ulnar joint** (see Fig. 3.63C). This redundancy of the synovial membrane accommodates the twisting of the capsule that occurs when the distal end of the radius travels around the relatively fixed distal end of the ulna during pronation and supination of the forearm.

LIGAMENTS OF DISTAL RADIO-ULNAR JOINT

Dorsal and anterior inferior radio-ulnar ligaments strengthen the fibrous layer of the joint capsule. These relatively weak transverse bands extend from the radius to the ulna across the anterior and posterior surfaces of the joint.

MOVEMENTS OF PROXIMAL AND DISTAL RADIO-ULNAR JOINTS

During pronation and supination of the forearm, the head of the radius rotates within the cup-shaped anular ligament, and the distal end of the radius rotates around the head of the ulna (Fig. 3.62C,D). **Supination** turns the palm anteriorly, or superiorly when the forearm is flexed. **Pronation** turns the palm posteriorly, or inferiorly when the forearm is flexed. During pronation and supination, it is the radius that rotates. Table 3.14 lists the structures that limit movements of the proximal and distal radio-ulnar joint.

Supination is produced by the supinator (when resistance is absent) and by the biceps brachii (when resistance is present), with some assistance from the EPL and ECRL. *Pronation* is produced by the pronator quadratus (primarily) and pronator teres (secondarily), with some assistance from the FCR, palmaris longus, and brachioradialis (when the forearm is in the midpronated position).

ARTERIES AND NERVES OF PROXIMAL AND DISTAL RADIO-ULNAR JOINTS

The proximal radio-ulnar joint is supplied by the radial portion of the peri-articular cubital **anastomosis** of the elbow joint (see Fig. 3.43). It is innervated by the musculocutaneous, median, and radial nerves. Pronation is essentially a function of the median nerve, whereas supination is a function of the musculocutaneous and radial nerves. The anterior and posterior *interosseous arteries* and *nerves* supply the distal radio-ulnar joint.

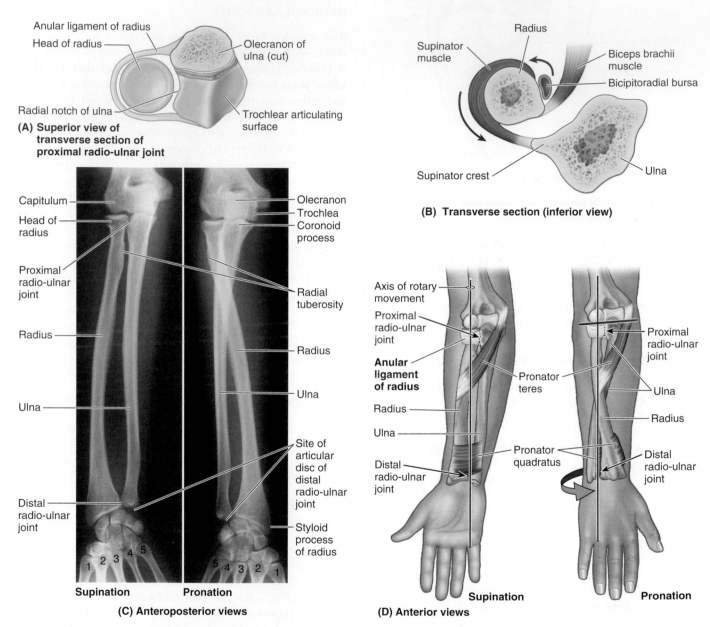

FIGURE 3.62. Proximal and distal radio-ulnar joints. A. Proximal radio-ulnar joint. The head of the radius rotates in the "socket" formed by the anular ligament. **B.** Actions of supinator and biceps brachii in producing supination are shown. **C.** Radiograph. **D.** Position of radius and ulna in supination and pronation.

CLINICAL BOX

Bursitis of Elbow

The *subcutaneous olecranon bursa* is exposed to injury during falls on the elbow and to infection from abrasions of the skin covering the olecranon. Repeated excessive pressure and friction produces a friction *subcutaneous olecranon bursitis* (e.g., "student's elbow") (Fig. B3.21). *Subtendinous olecranon bursitis* results from excessive friction between the triceps tendon and the olecranon—for example, resulting from repeated flexion–extension of the forearm as occurs during certain assembly-line jobs. The pain is severe during flexion of the forearm because of pressure exerted on the inflamed subtendinous olecranon bursa by the triceps tendon.

Avulsion of Medial Epicondyle

Avulsion of the medial epicondyle in children can result from a fall that causes severe abduction of the extended elbow. The resulting traction on the ulnar collateral ligament pulls the medial epicondyle distally. The anatomical basis of *avulsion of the medial epicondyle* is that the epiphysis for the medial epicondyle may not fuse with the distal end of the humerus until up to age 20 years. *Traction injury of the ulnar nerve* is a complication of the abduction type of avulsion of the medial epicondyle.

Ulnar Collateral Ligament Reconstruction

Rupture, tearing, and stretching of the ulnar collateral ligament are increasingly common injuries related to athletic throwing (primarily baseball pitching and also football passing, javelin throwing, and playing water polo). *Reconstruction of the ulnar collateral ligament*, commonly known as a "Tommy John procedure" (named after the first pitcher to undergo the surgery), involves an autologous transplant of a long tendon from the contralateral forearm or leg (e.g., the palmaris longus or plantaris tendon). A 10- to 15-cm length of tendon is passed through holes drilled through the medial epicondyle of the humerus and the lateral aspect of the coronoid process of the ulna (Fig. B3.22).

Dislocation of Elbow Joint

Posterior dislocation of the elbow joint may occur when children fall on their hands with their elbows flexed. Dislocations of the elbow may result from hyperextension or a blow that drives the ulna posteriorly or posterolaterally. The distal end of the humerus is driven through the weak anterior part of the fibrous layer of the joint capsule as the radius and ulna dislocate posteriorly. Injury to the ulnar nerve may also occur.

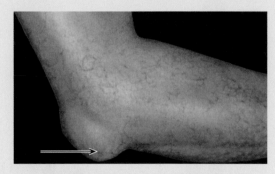

FIGURE B3.21. Subcutaneous olecranon bursitis.

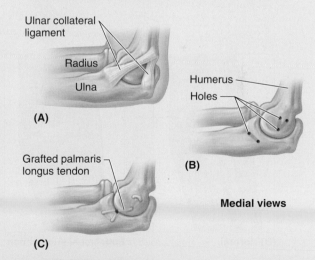

FIGURE B3.22. Ulnar collateral ligament reconstruction.

(*Continued on next page*)

Subluxation and Dislocation of Radial Head

Preschool children, particularly girls, are vulnerable to transient *subluxation* (incomplete temporary dislocation) *of the head of the radius* ("pulled elbow"). The history of these cases is typical. The child is suddenly lifted (jerked) by the upper limb when the forearm is pronated (Fig. B3.23). The child may cry out and refuse to use the limb, which is protected by holding it with the elbow flexed and the forearm pronated. The sudden pulling of the upper limb tears the distal attachment of the anular ligament, where it is loosely attached to the neck of the radius. The radial head then moves distally, partially out of the anular ligament. The proximal part of the torn ligament may become trapped between the head of the radius and the capitulum of the humerus. The source of pain is the pinched anular ligament. The treatment of subluxation consists of supination of the child's forearm while the elbow is flexed. The tear in the anular ligament soon heals when the limb is placed in a sling for about 2 weeks.

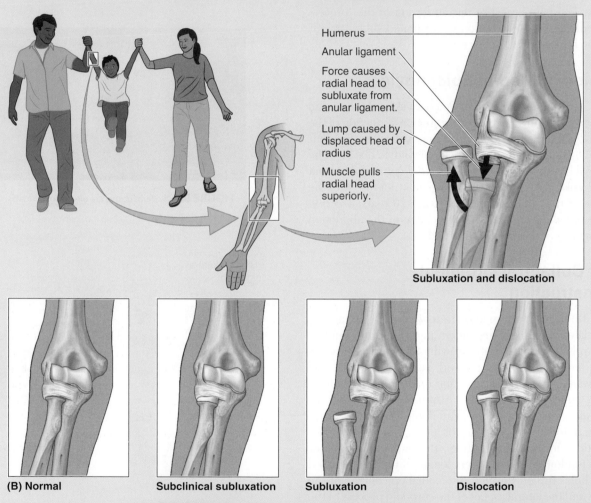

Humerus
Anular ligament
Force causes radial head to subluxate from anular ligament.
Lump caused by displaced head of radius
Muscle pulls radial head superiorly.

Subluxation and dislocation

(B) Normal   **Subclinical subluxation**   **Subluxation**   **Dislocation**

FIGURE B3.23. Subluxation and dislocation of radial head.

TABLE 3.15. WRIST AND CARPAL JOINTS

Joint	Type	Articulation	Joint Capsule	Ligaments	Movements	Nerve Supply
Wrist (radiocarpal)	Condyloid synovial joint	Distal end of radius and articular disc with proximal row of carpal bones (except pisiform)	Fibrous layer of joint capsule surrounds joint and attaches to distal ends of radius and ulna and proximal row of carpal bones; lined by synovial membrane.	Anterior and posterior ligaments strengthen fibrous capsule; ulnar collateral ligament attaches to styloid process of ulna and triquetrum; radial collateral ligament attaches to styloid process of radius and scaphoid.	Flexion–extension, abduction–adduction, circumduction	Anterior interosseous branch of median nerve, posterior interosseous branch of radial nerve, and dorsal and deep branches of ulnar nerve
Carpal (intercarpal)	Plane synovial joint	Between carpal bones of proximal row; joints between carpal bones of distal row *Midcarpal joint*: synovial joint between proximal and distal rows of carpal bones *Pisiform joint*: synovial joint between pisiform and triquetrum	Fibrous layer of joint capsule surrounds joints; lined by synovial membrane; pisiform joint is separate from other carpal joints.	Carpal bones united by anterior, posterior, and interosseous ligaments	Small amount of gliding movement possible; flexion and abduction of hand occur at midcarpal joint.	
Carpometacarpal (CMC) and intermetacarpal (IM)	Plane synovial joints, except for CMC joint of thumb (saddle-shaped synovial joint)	Carpals and metacarpals with each other; CMC joint of thumb between trapezium and base of 1st metacarpal	Fibrous layer of joint capsule surrounds joints; lined on internal surface by synovial membrane.	Bones united by anterior, posterior, and interosseous ligaments	Flexion–extension and abduction–adduction of CMC joint of 1st digit; almost no movement at 2nd and 3rd digits; 4th digit slightly mobile; 5th digit very mobile	

Joints of Hand

The movements that take place at the carpal and digital joints and the structures limiting these movements are summarized in Tables 3.15 to 3.18.

The wrist (carpus), the proximal segment of the hand, is a complex of eight carpal bones. The carpus articulates proximally with the forearm at the wrist joint and distally with the five metacarpals (Fig. 3.63). The joints formed by the carpus include the *wrist (radiocarpal joint)* and the *intercarpal, carpometacarpal*, and *intermetacarpal joints*.

Augmenting movement at the wrist joint, the two rows of carpals glide on each other; in addition, each bone glides on those adjacent to it.

Each digit has three phalanges except the thumb, which has two. The proximal phalanges articulate with the metacarpal bones at the metacarpophalangeal joints. The joint between the proximal and the middle phalanx is the proximal interphalangeal joint, and that between the middle and the distal phalanx is the distal interphalangeal joint (Figs. 3.63 and 3.64). The thumb has one interphalangeal joint.

TABLE 3.16. STRUCTURES LIMITING MOVEMENTS OF WRIST AND CARPAL JOINTS

Movement	Limiting Structures (Tension)
Flexion	*Ligaments*: posterior radiocarpal and posterior part of joint capsule
Extension	*Ligaments*: anterior radiocarpal and anterior part of joint capsule *Bony apposition* between radius and carpal bones
Abduction	*Ligaments*: ulnar collateral ligament and medial part of joint capsule *Bony apposition* between styloid process of radius and scaphoid
Adduction	*Ligaments*: radial collateral and lateral part of joint capsule

Modified from Clarkson HM. *Musculoskeletal Assessment: Joint Motion and Muscle Testing*. 3rd ed. Baltimore, MD: Lippincott Williams & Wilkins; 2012.

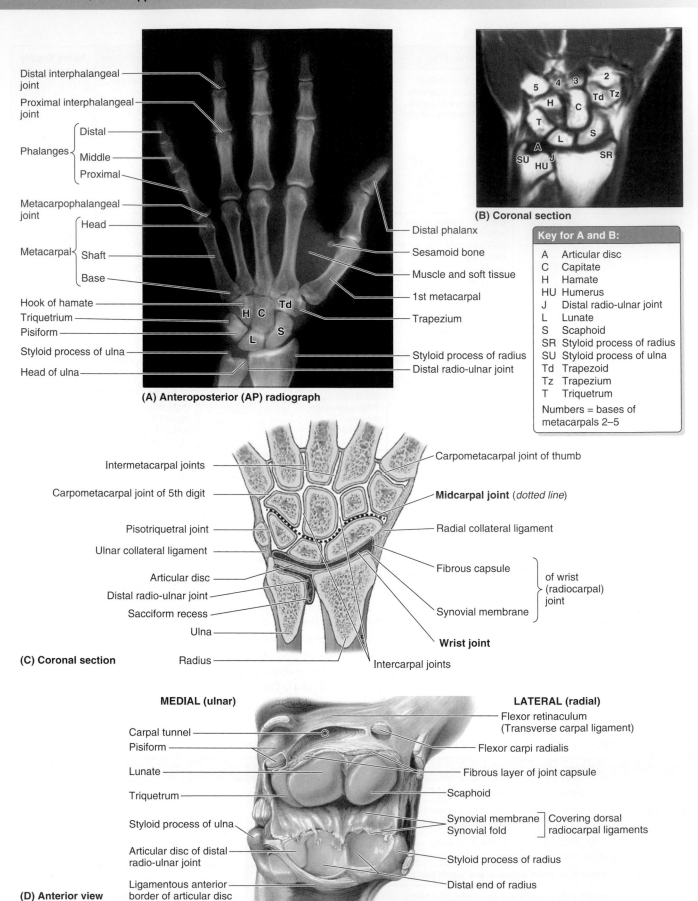

Distal interphalangeal joint

Proximal interphalangeal joint

Phalanges
- Distal
- Middle
- Proximal

Metacarpophalangeal joint

Metacarpal
- Head
- Shaft
- Base

Hook of hamate
Triquetrium
Pisiform
Styloid process of ulna
Head of ulna

Distal phalanx
Sesamoid bone
Muscle and soft tissue
1st metacarpal
Trapezium
Styloid process of radius
Distal radio-ulnar joint

(A) Anteroposterior (AP) radiograph

(B) Coronal section

Key for A and B:

A Articular disc
C Capitate
H Hamate
HU Humerus
J Distal radio-ulnar joint
L Lunate
S Scaphoid
SR Styloid process of radius
SU Styloid process of ulna
Td Trapezoid
Tz Trapezium
T Triquetrum
Numbers = bases of metacarpals 2–5

Intermetacarpal joints
Carpometacarpal joint of 5th digit
Pisotriquetral joint
Ulnar collateral ligament
Articular disc
Distal radio-ulnar joint
Sacciform recess
Ulna
Radius

Carpometacarpal joint of thumb
Midcarpal joint (*dotted line*)
Radial collateral ligament
Fibrous capsule
Synovial membrane
of wrist (radiocarpal) joint
Wrist joint
Intercarpal joints

(C) Coronal section

MEDIAL (ulnar)

Carpal tunnel
Pisiform
Lunate
Triquetrum
Styloid process of ulna
Articular disc of distal radio-ulnar joint
Ligamentous anterior border of articular disc

LATERAL (radial)

Flexor retinaculum (Transverse carpal ligament)
Flexor carpi radialis
Fibrous layer of joint capsule
Scaphoid
Synovial membrane
Synovial fold
Covering dorsal radiocarpal ligaments
Styloid process of radius
Distal end of radius

(D) Anterior view

FIGURE 3.63. Wrist and hand joints. A. Radiograph. **B.** Coronal MRI of wrist. **C.** Coronal section of distal radio-ulnar, wrist, and carpal joints. **D.** Dissection. The wrist joint is opened anteriorly, with the dorsal radiocarpal ligaments acting as a hinge.

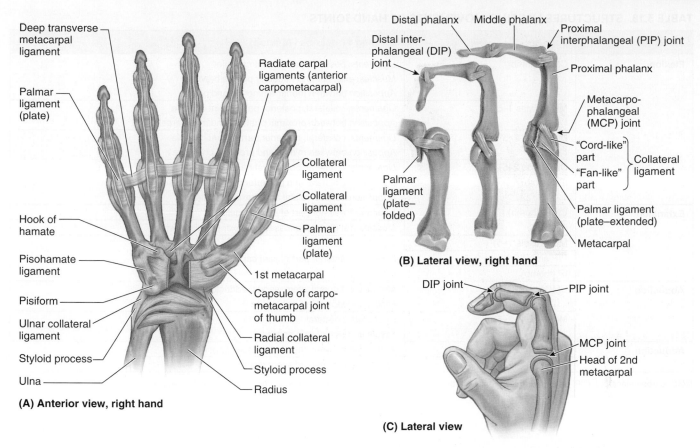

FIGURE 3.64. Joints of hand. A. Palmar ligaments. **B.** Metacarpophalangeal (MCP) and interphalangeal (IP) joints. The palmar ligaments (plates) are modifications of the anterior aspect of the MCP and IP joint capsules. **C.** Joints of digit.

TABLE 3.17. METACARPOPHALANGEAL AND INTERPHALANGEAL JOINTS

Joint	Type	Articulation	Joint Capsule	Ligaments	Movements	Nerve Supply
Metacarpophalangeal (MCP)	Condyloid synovial joints	Heads of metacarpals with base of proximal phalanges	Fibrous layer of joint capsule encloses each joint; lined on internal surface by synovial membrane	Strong palmar ligaments attached to phalanges and metacarpals; deep transverse metacarpal ligaments unite 2nd–5th joints holding heads of metacarpals together; collateral ligaments pass from heads of metacarpals to bases of phalanges.	Flexion–extension, abduction–adduction, and circumduction of 2nd–5th digits; flexion–extension of thumb occurs but abduction–adduction is limited.	Digital nerves arising from ulnar and median nerves
Interphalangeal (IP)	Hinge synovial joints	Heads of phalanges with bases of more distally located phalanges	Fibrous capsule encloses each joint; lined on internal surface by synovial membrane	Similar to metacarpophalangeal joints, except they unite phalanges	Flexion–extension	Digital nerves arising from ulnar and median nerves

TABLE 3.18. STRUCTURES LIMITING MOVEMENTS OF HAND JOINTS

Movement	Joint(s)	Limiting Structures (Tension)
Flexion	CMC (thumb)	*Ligaments*: posterior part of joint capsule *Muscles*: extensor and abductor pollicis brevis *Apposition* between thenar eminence and palm
	MCP (digits 1–5)	*Ligaments*: collateral, posterior part of joint capsule *Apposition* between proximal phalanx and metacarpal
	PIP (digits 2–5)	*Ligaments*: collateral, posterior part of joint capsule *Apposition* between middle and proximal phalanges
	DIP (digits 2–5)	*Ligaments*: collateral, oblique retinacular, and posterior part of joint capsule
	IP (thumb)	*Ligaments*: collateral and posterior part of joint capsule *Apposition* between distal and proximal phalanges
Extension	CMC (thumb)	*Ligaments*: anterior part of joint capsule *Muscles*: 1st dorsal interosseous, flexor pollicis brevis
	MCP (digits 1–5) PIP and DIP (digits 2–5) IP (thumb)	*Ligaments*: anterior part of joint capsule, palmar ligament
Abduction	CMC and MCP	*Muscles*: 1st dorsal interosseous, adductor pollicis *Fascia and skin* of 1st web space
	MCP (digits 2–5)	*Ligaments*: collateral *Fascia and skin* of web spaces
Adduction	CMC and MCP (thumb)	*Apposition* between thumb and index finger
	MCP (digits 2–5)	*Apposition* between adjacent digits

CMC, carpometacarpal; *DIP*, distal interphalangeal; *IP*, interphalangeal; *MCP*, metacarpophalangeal; *PIP*, proximal interphalangeal.

CLINICAL BOX

Wrist Fractures and Dislocations

Fracture of the distal end of the radius (*Colles fracture*), the most common fracture in people older than 50 years of age, is discussed in the Clinical Box "Fractures of Ulna and Radius." *Fracture of the scaphoid*, relatively common in young adults, is discussed in the Clinical Box "Fractures of Hand."

Anterior dislocation of the lunate is an uncommon but serious injury that usually results from a fall on the dorsiflexed wrist. The lunate is pushed out of its place in the floor of the carpal tunnel toward the palmar surface of the wrist. The displaced lunate may compress the median nerve and lead to *carpal tunnel syndrome* (discussed earlier in this chapter). Because of its poor blood supply, *avascular necrosis of the lunate* may occur. In some cases, excision of the lunate may be required. In *degenerative joint disease of the wrist*, surgical fusion of carpals (*arthrodesis*) may be necessary to relieve the severe pain.

Fracture–separation of the distal radial epiphysis is common in children because of frequent falls in which forces are transmitted from the hand to the radius. In a lateral radiograph of a child's wrist, dorsal displacement of the distal radial epiphysis is obvious (Fig. B3.24). When the epiphysis is placed in its normal position during reduction, the prognosis for normal bone growth is good.

Without knowledge of bone growth and the appearance of bones in radiographic and other diagnostic images at various ages, a displaced epiphyseal plate could be mistaken for a fracture and separation of an epiphysis could be interpreted as a displaced piece of fractured bone. Knowledge of the patient's age and location of epiphyses can prevent these errors.

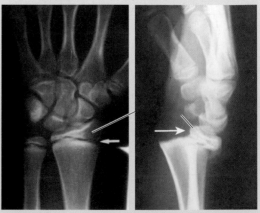

(A) AP view   **(B) Lateral view**
Fracture separation of distal radial epiphysis (*arrow*).

FIGURE B3.24. Dorsal displacement of radial epiphysis.

Skier's Thumb

Skier's thumb (historically, "gamekeeper's thumb") refers to the rupture or chronic laxity of the collateral ligaments of the 1st metacarpophalangeal joint. (Fig. B3.25). The injury results from hyperabduction of the metacarpophalangeal joint of the thumb, which occurs when the thumb is held by a ski pole while the rest of the hand hits the ground or enters the snow. In severe injuries, the head of the metacarpal has an avulsion fracture.

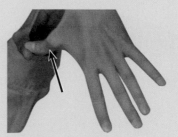

Skier's thumb (*arrow*)

FIGURE B3.25. Skier's thumb.

MEDICAL IMAGING

Upper Limb

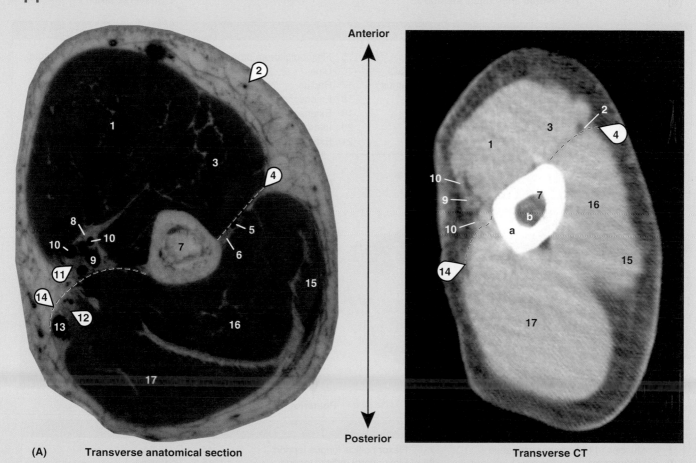

Anterior

Posterior

(A)   Transverse anatomical section

Transverse CT

Key							
1	Biceps brachii	6	Radial nerve	11	Median nerve	16	Medial head of triceps brachii
2	Cephalic vein	7	Humerus	12	Ulnar nerve	17	Long head of triceps brachii
3	Brachialis	8	Musculocutaneous nerve	13	Basilic vein	a	Cortex of humerus
4	Lateral intermuscular septum	9	Brachial artery	14	Medial intermuscular septum	b	Medullary (marrow) cavity
5	Profunda brachii artery	10	Brachial veins	15	Lateral head of triceps brachii		of humerus

FIGURE 3.65. Transverse sections of specimens with correlated transverse MRI or CT scans of left upper limb. A. Arm. *(continued)*

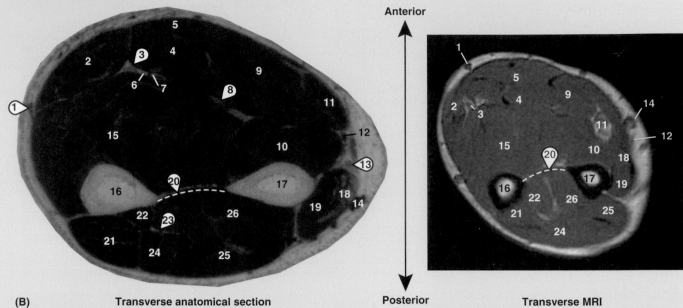

Anterior

Posterior

| (B) | Transverse anatomical section | | Transverse MRI |

Key

1	Basilic vein	8	Median nerve	15	Flexor digitorum profundus	22	Extensor pollicis longus
2	Flexor carpi ulnaris	9	Flexor carpi radialis	16	Ulna	23	Posterior interosseous
3	Ulnar nerve	10	Flexor pollicis longus	17	Radius		vessels and nerve
4	Flexor digitorum superficialis	11	Brachioradialis	18	Extensor carpi radialis longus	24	Extensor digiti minimi
5	Palmaris longus	12	Radial artery	19	Extensor carpi radialis brevis	25	Extensor digitorum
6	Ulnar vein	13	Radial nerve	20	Interosseous membrane	26	Abductor pollicis longus
7	Ulnar artery	14	Cephalic vein	21	Extensor carpi ulnaris		

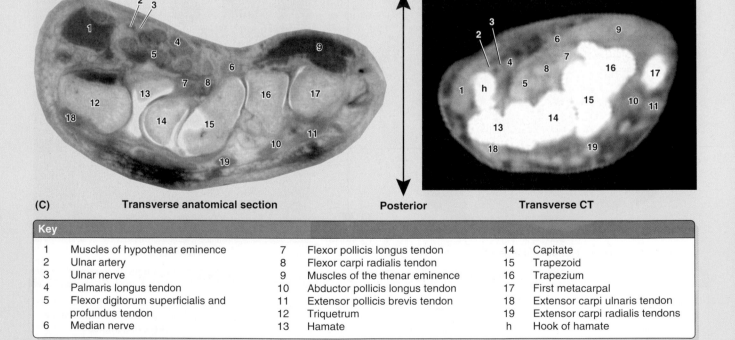

Anterior

Posterior

| (C) | Transverse anatomical section | | Transverse CT |

Key

1	Muscles of hypothenar eminence	7	Flexor pollicis longus tendon	14	Capitate
2	Ulnar artery	8	Flexor carpi radialis tendon	15	Trapezoid
3	Ulnar nerve	9	Muscles of the thenar eminence	16	Trapezium
4	Palmaris longus tendon	10	Abductor pollicis longus tendon	17	First metacarpal
5	Flexor digitorum superficialis and	11	Extensor pollicis brevis tendon	18	Extensor carpi ulnaris tendon
	profundus tendon	12	Triquetrum	19	Extensor carpi radialis tendons
6	Median nerve	13	Hamate	h	Hook of hamate

FIGURE 3.65. *(continued)* **B.** Forearm. **C.** Carpal tunnel.

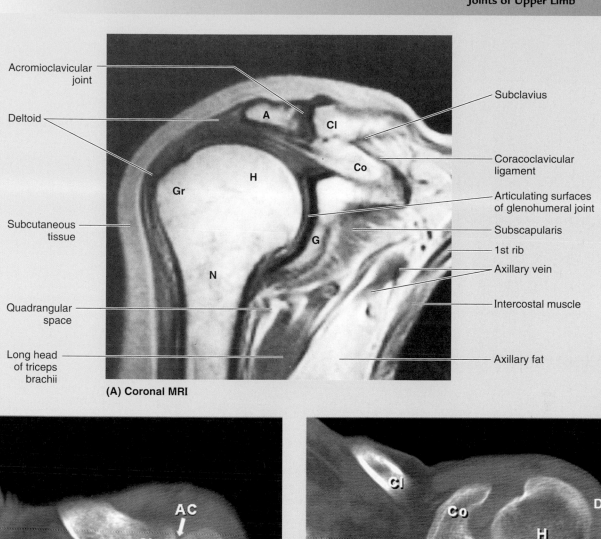

Acromioclavicular joint

Deltoid

Subcutaneous tissue

Quadrangular space

Long head of triceps brachii

Subclavius

Coracoclavicular ligament

Articulating surfaces of glenohumeral joint

Subscapularis

1st rib

Axillary vein

Intercostal muscle

Axillary fat

(A) Coronal MRI

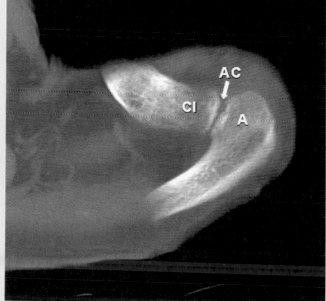

(B) Transverse CT

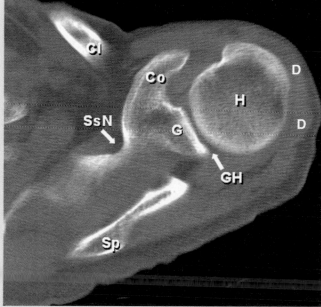

(C) Transverse CT

Key					
A	Acromion	D	Deltoid muscle	H	Head of humerus
AC	Acromioclavicular joint	G	Glenoid cavity (fossa)	N	Surgical neck of humerus
Cl	Clavicle	GH	Glenohumeral joint	Sp	Spine of scapula
Co	Coracoid process	Gr	Greater tubercle of humerus	SsN	Suprascapular notch

FIGURE 3.66. Imaging of glenohumeral and acromioclavicular joints. A. Coronal MRI. The *white* (signal-intense) parts of the identified bones are the fatty matrix of cancellous bone; the *thin black outlines* (absence of signal) of the bones are the compact bones that form their outer surface. **B.** Transverse CT scan through acromioclavicular joint. **C.** Transverse CT scan through glenohumeral joint.

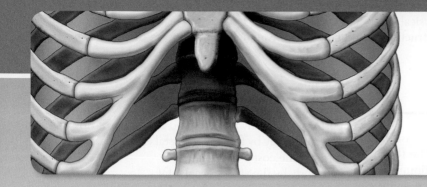

Thorax

CLINICAL BOX KEY

Anatomical
Variations

Diagnostic
Procedures

Life Cycle

Surgical
Procedures

Trauma

Pathology

The **thorax** is the superior part of the trunk between the neck and abdomen. The **thoracic cavity**, surrounded by the thoracic wall, contains the heart, lungs, thymus, distal part of the trachea, and most of the esophagus. To perform a physical examination of the thorax, a working knowledge of its structure and vital organs is required.

THORACIC WALL

The **thoracic wall** consists of skin, fascia, nerves, vessels, muscles, cartilages, and bones. The functions of the thoracic wall include protecting the thoracic and abdominal organs; resisting the negative internal pressures generated by the elastic recoil of the lungs and inspiratory movements; providing attachment for and supporting the weight of the upper limbs; and providing attachment for many of the muscles of the upper limbs, neck, abdomen, and back and the muscles of respiration. The mammary glands of the breasts are located in the subcutaneous tissue overlying the pectoral muscles covering the anterolateral thoracic wall.

Skeleton of Thoracic Wall

The **thoracic skeleton** forms the osteocartilaginous **thoracic cage** (Fig. 4.1). The thoracic skeleton includes 12 pairs of ribs and costal cartilages, 12 thoracic vertebrae and intervertebral (IV) discs, and the sternum. Costal cartilages form the anterior continuation of the ribs, providing a flexible attachment at their articulation with the sternum (Fig. 4.1A). The ribs and their cartilages are separated by **intercostal spaces**, which are occupied by intercostal muscles, vessels, and nerves.

Thoracic Apertures

The thoracic cavity communicates with the neck and upper limb through the **superior thoracic aperture**, the anatomical *thoracic inlet* (Fig. 4.1A). Structures entering and leaving the thoracic cavity through this aperture include the trachea, esophagus, vessels, and nerves. The adult superior thoracic aperture measures approximately 6.5 cm anteroposteriorly and 11 cm transversely. Because of the obliquity of the first pair of ribs, the superior thoracic aperture slopes antero-inferiorly. The superior thoracic aperture is bounded

- posteriorly by the T1 vertebra
- laterally by the first pair of ribs and their costal cartilages
- anteriorly by the superior border of the manubrium

The thoracic cavity communicates with the abdomen through the **inferior thoracic aperture**, the anatomical *thoracic outlet* (Fig. 4.1A). The diaphragm closes the inferior thoracic aperture, separating the thoracic and abdominal cavities almost completely. The inferior thoracic aperture is much larger than the superior thoracic aperture. Structures passing to or from the thorax and abdomen pass through openings in the diaphragm (e.g., the inferior vena cava (IVC) and esophagus) or posterior to it (e.g., aorta).

The inferior thoracic aperture is bounded

- posteriorly by the T12 vertebra
- posterolaterally by the 11th and 12th pairs of ribs
- anterolaterally by the joined costal cartilages of ribs 7–10, forming the costal margin
- anteriorly by the xiphisternal joint

RIBS AND COSTAL CARTILAGES

The **ribs** are curved, flat bones that form most of the thoracic cage (Fig. 4.1). They are remarkably light in weight yet highly resilient. Each rib has a spongy interior containing *bone marrow* (Fig. 4.2), which forms blood cells (hematopoietic tissue). There are three classes of ribs based on attachments (Fig. 4.1):

- **True (vertebrosternal) ribs** (1st to 7th ribs) attach directly to the sternum anteriorly through their own costal cartilages.
- **False (vertebrochondral) ribs** (8th to 10th ribs) have cartilages on their anterior ends that are joined to the cartilage of the rib just superior to them; thus, their connection with the sternum is indirect.
- **Floating (free) ribs** (11th and 12th ribs; sometimes the 10th rib) have rudimentary cartilages on their anterior ends that do not connect even indirectly with the sternum; instead, they end in the posterior abdominal musculature.

Typical ribs (3rd to 9th) have the following parts:

- **Head of rib**, that is wedge-shaped and two facets that are separated by the **crest of the head** (Fig. 4.2A). One facet articulates with the body of the numerically corresponding vertebra, and one facet articulates with that of the superior vertebra.
- **Neck of rib**, that connects the head with the body (shaft) at the level of the tubercle
- **Tubercle of rib** (lump-like enlargement) at the junction of the neck and body. The tubercle has a smooth *articular part* for articulating with the corresponding transverse process of the vertebra (via a synovial joint) and a rough *nonarticular part* for a fibrous attachment to the process via the costotransverse ligament.
- **Body of rib** (shaft) that is thin, flat, and curved along its length, most markedly at the **angle** where the rib begins to turn anterolaterally. The inferior edge has a concavity running along its internal surface, the **costal groove**, that protects the intercostal nerve and vessels (Fig. 4.2).

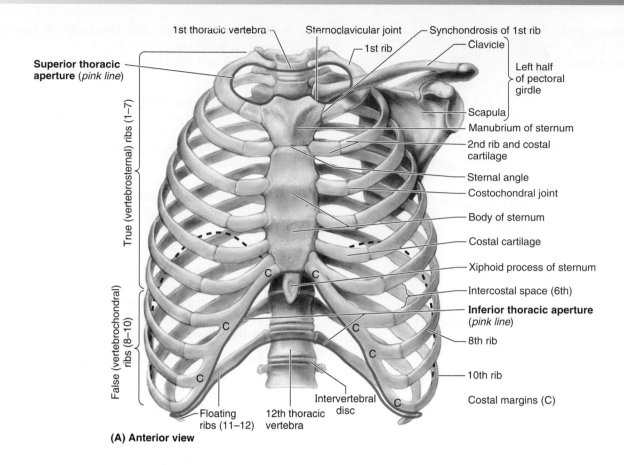

(A) Anterior view

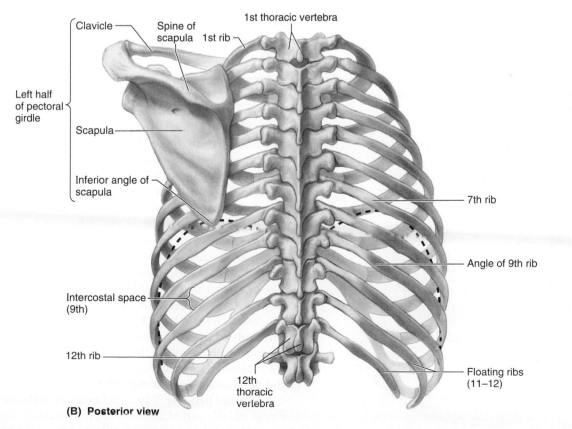

(B) Posterior view

FIGURE 4.1. Thoracic skeleton. The superior and inferior thoracic apertures are outlined in *pink*. The *dotted lines* indicate the position of the diaphragm, which separates the thoracic and abdominal cavities.

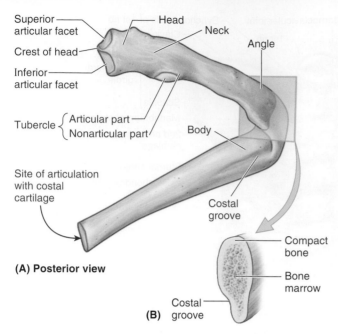

(A) Posterior view

(B) groove

FIGURE 4.2. Typical rib (Right side). A. Features. The 3rd–9th ribs have common characteristics. **B.** Cross section through the midbody of rib.

Atypical ribs (1st, 2nd, and 10th through 12th) are dissimilar (Figs. 4.1 and 4.3):

- The **1st rib** is broad (i.e., its body is widest, and its cross section more nearly horizontal). It is the shortest and most sharply curved of the seven true ribs. It contributes more to the "roof" than to the wall of the thoracic cavity. It has two shallow horizontal grooves crossing its superior surface for the subclavian vessels separated by a **scalene tubercle** and ridge. It articulates only with the T1 vertebra.
- The **2nd rib** is thinner and more typical, except for the formations for attachment of serratus anterior and posterior scalene muscles, and almost twice the length of the 1st rib.

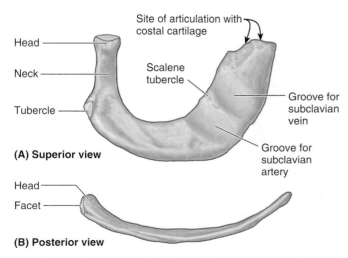

(A) Superior view

(B) Posterior view

FIGURE 4.3. Atypical ribs (Right side). A. 1st rib. **B.** 12th rib.

- The 10th through 12th ribs, like the 1st rib, have only one facet on their heads.
- The **11th** and **12th ribs** are short and have no necks or tubercles.

Costal cartilages prolong the ribs anteriorly and contribute to the elasticity of the thoracic wall. *Intercostal spaces* separate the ribs and their costal cartilages from one another. The spaces and neurovascular structures are named according to the rib forming the superior border of the space; that is, there are 11 intercostal spaces and 11 intercostal nerves. The *subcostal space* is immediately below the 12th rib, and the anterior ramus of spinal nerve T12 is the subcostal nerve.

THORACIC VERTEBRAE

Thoracic vertebrae are typical vertebrae in that they are independent and have bodies, vertebral arches, and seven processes for muscular and articular connections (see Chapter 2). Characteristic features of thoracic vertebrae include the following:

- Bilateral *superior* and *inferior* **costal facets** (demifacets) on their bodies for articulation with the heads of ribs (Fig. 4.4); atypical thoracic vertebrae have a single whole costal facet in place of the demifacets.
- *Costal facets on their transverse processes* for articulation with the tubercles of ribs, except for the inferior two or three thoracic vertebrae
- Long inferiorly slanting spinous processes that overlap the IV disc and vertebra below

STERNUM

The **sternum** is the flat, vertically elongated bone that forms the middle of the anterior part of the thoracic cage. The sternum consists of three parts: manubrium, body, and xiphoid process (Figs. 4.1A and 4.5).

The **manubrium**, the superior part of the sternum, is a roughly trapezoidal bone that lies at the level of the bodies of the T3 and T4 vertebrae. Its thick superior border is indented centrally by the **jugular notch** (suprasternal notch). On each side, a **clavicular notch** articulates with the sternal (medial) end of the clavicle. Just inferior to the latter notch, the costal cartilage of the 1st rib fuses with the lateral border of the manubrium. The manubrium and body of the sternum lie in slightly different planes, forming a projecting **sternal angle** (of Louis). This readily palpable *clinical landmark* is located opposite the second pair of costal cartilages at the level of the IV disc between the T4 and T5 vertebrae (Fig. 4.5B).

The **body of the sternum** (T5–T9 vertebral level) is longer, narrower, and thinner than the manubrium. Its width varies because of the scalloping of its lateral borders by the **costal notches** for articulation with the costal cartilages.

The **xiphoid process** (T10 vertebral level) is the smallest and most variable part of the sternum. It is relatively thin and elongated but varies considerably in form. The process is cartilaginous in young people but more or less ossified in adults

(Continued on page 189)

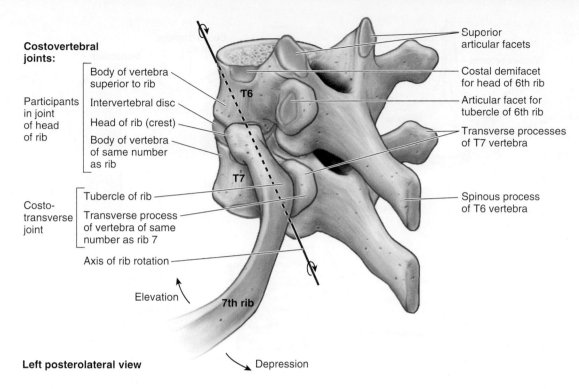

Costovertebral joints:

Participants in joint of head of rib
- Body of vertebra superior to rib
- Intervertebral disc
- Head of rib (crest)
- Body of vertebra of same number as rib

Costo-transverse joint
- Tubercle of rib
- Transverse process of vertebra of same number as rib 7

Axis of rib rotation

Elevation

7th rib

Superior articular facets

Costal demifacet for head of 6th rib

Articular facet for tubercle of 6th rib

Transverse processes of T7 vertebra

Spinous process of T6 vertebra

T6

T7

Left posterolateral view

Depression

FIGURE 4.4. Costovertebral articulations of a typical rib. The costovertebral joints include the joint of the head of the rib, in which the head articulates with two adjacent vertebral bodies and the intervertebral disc between them, and the costotransverse joint, in which the tubercle of the rib articulates with the transverse process of a vertebra.

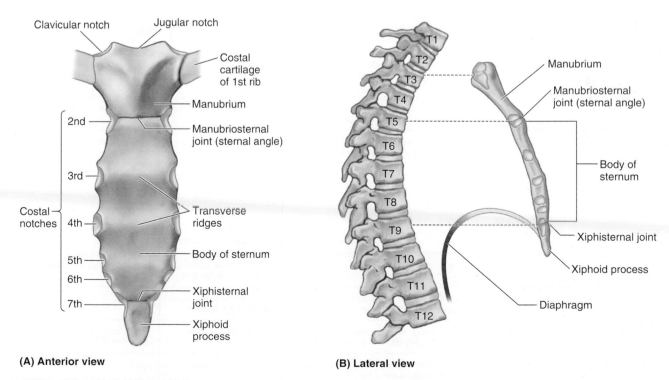

Clavicular notch

Jugular notch

Costal cartilage of 1st rib

Manubrium

Manubriosternal joint (sternal angle)

2nd

3rd

Costal notches   4th

5th

6th

7th

Transverse ridges

Body of sternum

Xiphisternal joint

Xiphoid process

(A) Anterior view

Manubrium

Manubriosternal joint (sternal angle)

Body of sternum

Xiphisternal joint

Xiphoid process

Diaphragm

T1
T2
T3
T4
T5
T6
T7
T8
T9
T10
T11
T12

(B) Lateral view

FIGURE 4.5. Sternum. A. Features. **B.** Relationship of sternum to vertebral column.

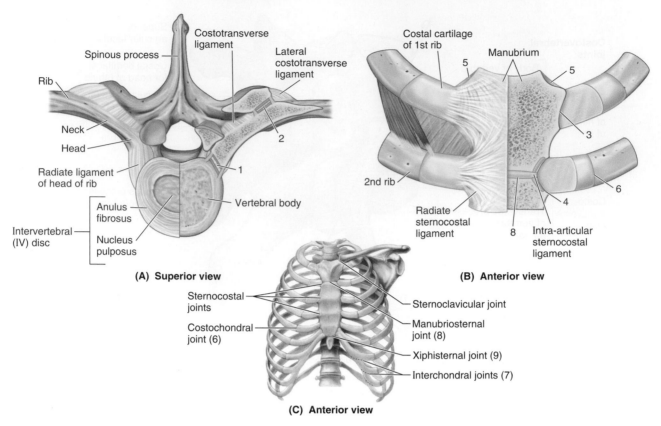

(A) Superior view

(B) Anterior view

(C) Anterior view

FIGURE 4.6. Joints of thoracic wall.

TABLE 4.1. JOINTS OF THORACIC WALL

Joint[a]	Type	Articulations	Ligaments	Comments
Intervertebral	Symphysis (secondary cartilaginous joint)	Adjacent vertebral bodies bound together by intervertebral disc.	Anterior and posterior longitudinal	See Chapter 2
Costovertebral joints of head of ribs (1)	Synovial plane of joint	Head of each rib with superior demifacet or costal facet of corresponding vertebral body and inferior demifacet or costal facet of vertebral body superior to it	Radiate and intra-articular ligaments of head of rib	Heads of 1st, 11th, and 12th ribs (sometimes 10th) articulate only with corresponding vertebral body.
Costotransverse (2)		Articulation of tubercle of rib with transverse process of corresponding vertebra	Lateral and superior costotransverse	11th and 12th ribs do not articulate with transverse process of corresponding vertebrae.
Sternocostal (3, 4)	1st: primary cartilaginous joint 2nd–7th: synovial plane joints	Articulation of 1st costal cartilages with manubrium of sternum Articulation of 2nd–7th pairs of costal cartilages with sternum	Anterior and posterior radiate sternocostal	
Sternoclavicular (5)	Saddle type of synovial joint	Sternal end of clavicle with manubrium and 1st costal cartilage	Anterior and posterior sternoclavicular ligaments; costoclavicular ligament	Joint is divided into two compartments by articular disc.
Costochondral (6)	Primary cartilaginous joint	Articulation of lateral end of costal cartilage with sternal end of rib	Cartilage and bone; bound together by periosteum	Normally, no movement occurs.
Interchondral (7)	Synovial plane joint	Articulation between costal cartilages of 6th–7th, 7th–8th, and 8th–9th ribs	Interchondral ligaments	Articulation between costal cartilages of 9th and 10th ribs is fibrous.
Manubriosternal (8)	Secondary cartilaginous joint (symphysis)	Articulation between manubrium and body of sternum		Often fuse and become synostosis in older people
Xiphisternal (9)	Primary cartilaginous joint (synchondrosis)	Articulation between xiphoid process and body of sternum		

[a]Numbers in parentheses refer to the figures.

older than 40 years of age. In elderly people, the xiphoid process may fuse with the sternal body. The xiphisternal joint (T9 vertebral level) is a midline marker for the superior level of the liver, the central tendon of the diaphragm, and the inferior border of the heart.

Joints of Thoracic Wall

Although movements of the joints of the thoracic wall are frequent (e.g., during respiration), the range of movement at the individual joints is small. Any disturbance that reduces the mobility of these joints interferes with respiration. *Joints of the thoracic wall* occur between the following structures (Fig. 4.6 and Table 4.1):

- Vertebrae (*intervertebral [IV] joints*)
- Ribs and vertebrae (*costovertebral joints*: *joints of the heads of ribs* and the *costotransverse joints*)
- Sternum and costal cartilages (*sternocostal joints*)
- Sternum and clavicle (*sternoclavicular joints*)
- Ribs and costal cartilages (*costochondral joints*)
- Costal cartilages (*interchondral joints*)
- Parts of the sternum (*manubriosternal* and *xiphisternal* joints) in young people; usually, the manubriosternal joint

and sometimes the xiphisternal joint are fused in elderly people.

The IV joints between the bodies of adjacent vertebrae are joined together by longitudinal ligaments and *IV discs* (see Chapter 2).

Movements of Thoracic Wall

Movements of the thoracic wall and diaphragm during inspiration increase the intrathoracic diameters and volume of the thorax. Consequent pressure changes result in air being drawn into the lungs (inspiration) through the nose, mouth, larynx, and trachea. During passive expiration, the diaphragm, intercostal muscles, and other muscles relax, decreasing *intrathoracic volume* and increasing *intrathoracic pressure*, expelling air from the lungs (expiration) through the same passages. The stretched elastic tissue of the lungs recoils, expelling most of the air. Concurrently, *intra-abdominal pressure* decreases and the abdominal viscera are decompressed.

The *vertical dimension* (height) of the central part of the thoracic cavity increases during inspiration as the contracting diaphragm descends, compressing the abdominal viscera (Fig. 4.7A,B). During expiration (Fig. 4.7A,C), the vertical

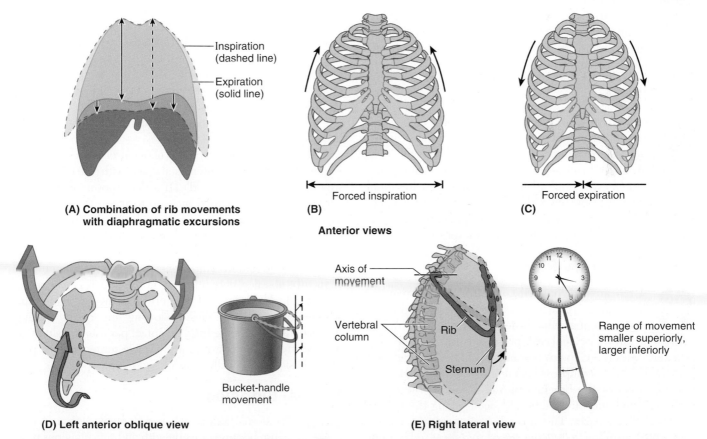

(A) Combination of rib movements with diaphragmatic excursions

Inspiration (dashed line)
Expiration (solid line)

(B) Forced inspiration

(C) Forced expiration

Anterior views

(D) Left anterior oblique view

Bucket-handle movement

Axis of movement

Vertebral column

Rib

Sternum

(E) Right lateral view

Range of movement smaller superiorly, larger inferiorly

FIGURE 4.7. Movements of thoracic wall during respiration. A. The primary movement of inspiration is contraction of the diaphragm, which increases the vertical dimension of the thoracic cavity (*arrows*). **B.** The thorax widens during forced inspiration (*arrows*). **C.** The thorax narrows during forced expiration (*arrows*). **D.** The combination of rib movements (*arrows*) that occur during forced inspiration increase the anteroposterior (AP) and transverse dimensions. The middle parts of the lower ribs move laterally when they are elevated (bucket-handle movement). **E.** When the upper ribs are elevated, the AP dimension of the thorax is increased (pendulum movement).

diameter returns to the neutral position as the elastic recoil of the lungs produces subatmospheric pressure in the pleural cavities, between the lungs and the thoracic wall. As a result of this and the release of resistance to the previously compressed viscera, the domes of the diaphragm ascend, diminishing the vertical dimension. The *anteroposterior (AP) dimension* of the thorax increases considerably when the intercostal muscles contract (Fig. 4.7D,E). Movement of the upper ribs at the costovertebral joints, about an axis passing through the neck of the ribs, causes the anterior ends of the ribs and sternum, especially its inferior end, to move anteriorly and posteriorly like a pendulum (Fig. 4.7E). In addition, the *transverse dimension* of the thorax increases slightly when the intercostal muscles contract, raising the most lateral parts of the ribs, especially the most inferior ones, the "bucket-handle movement" (Fig. 4.7B,D).

CLINICAL BOX

Role of Costal Cartilages

Costal cartilages prolong the ribs anteriorly and contribute to the elasticity of the thoracic wall, preventing many blows from fracturing the sternum and/or ribs. In elderly people, the costal cartilages undergo calcification, making them radiopaque and less resilient.

Rib Fractures

The weakest part of a rib is just anterior to its angle. *Rib fractures* commonly result from direct blows or indirectly from crushing injuries. The middle ribs are most commonly fractured. Direct violence may fracture a rib anywhere, and its broken ends may injure internal organs such as a lung or the spleen.

Flail Chest

Flail chest occurs when a sizable segment of the anterior and/or lateral thoracic wall moves freely because of *multiple rib fractures*. This condition allows the loose segment of the wall to move paradoxically (inward on inspiration and outward on expiration). Flail chest is an extremely painful injury and impairs ventilation, thereby affecting oxygenation of the blood. During treatment, the loose segment may be internally fixed with plates or wires to prevent movement.

Supernumerary Ribs

There are usually 12 ribs on each side, but the number may be increased by the presence of cervical and/or lumbar ribs or decreased by failure of the 12th pair to form. **Cervical ribs** (present in up to 1% of people) articulate with the C7 vertebra and are clinically significant because they may compress spinal nerves C8 and T1 or the inferior trunk of the brachial plexus supplying the upper limb. Tingling and numbness may occur along the medial border of the forearm. They may also compress the subclavian artery, resulting in *ischemic muscle pain* (caused by poor blood supply) in the upper limb. Resection may be required to relieve pressure on these structures, which can be performed through a transaxillary approach (incision in axillary fossa or armpit). **Lumbar ribs** are less common than cervical ribs but have clinical significance in that they may confuse the identification of vertebral levels in diagnostic images.

Thoracotomy, Intercostal Space Incisions, and Rib Excision

The surgical creation of an opening through the thoracic wall to enter a pleural cavity is called a *thoracotomy* (Fig. B4.1). An *anterior thoracotomy* may involve making H-shaped cuts through the perichondrium of one or more costal cartilages and then shelling out segments of costal cartilage to gain entrance to the thoracic cavity. The posterolateral aspects of the 5th–7th intercostal spaces are important sites for *posterior thoracotomy* incisions. In general, a lateral approach is most satisfactory for entry through the thoracic cage. With the patient lying on the contralateral side, the upper limb is fully abducted, placing the forearm beside the patient's head. This elevates and laterally rotates the inferior angle of scapula, allowing access as high as the 4th intercostal space.

Most commonly, rib retraction allows procedures to be performed through a single intercostal space, with care to avoid the superior neurovascular bundle. If wider exposure is required, surgeons use an H-shaped incision to incise the superficial aspect of the periosteum that surrounds the rib, strip the periosteum from the rib, and then remove a wide segment of the rib to gain better access, as might be required to enter the thoracic cavity and remove a lung (*pneumonectomy*), for example. In the rib's absence, entry into the thoracic cavity can be made through the deep aspect of the periosteal sheath, sparing the adjacent intercostal muscles. Following surgery, the missing pieces of ribs regenerate from the intact periosteum, although imperfectly. In many cases, intrathoracic surgery can be performed using a minimally invasive endoscopic approach.

Sternal Biopsies

The sternal body is often used for *bone marrow needle biopsy* because of its breadth and subcutaneous position. The needle first pierces the thin cortical bone and then enters the vascular trabecular (spongy) bone. Sternal biopsy is commonly used to obtain specimens of bone marrow for transplantation and for detection of metastatic cancer.

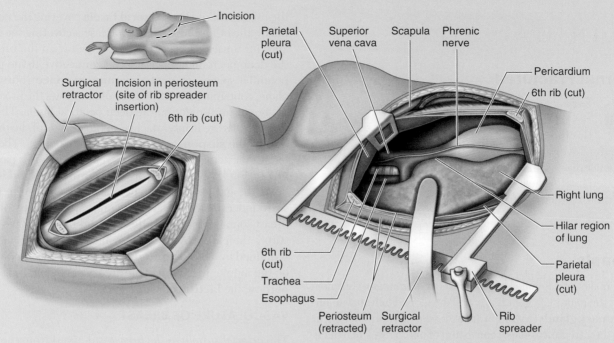

FIGURE B4.1. Thoracotomy.

Surgical Entry into Thorax

To gain wide access to the thoracic cavity for surgical procedures in the mediastinum, the sternum is divided (split) in the median plane (*median sternotomy*) and retracted (e.g., for *coronary artery bypass grafting*). After surgery, the halves of the sternum are reunited and held together with wire sutures. *Lateral thoracotomy* through intercostal spaces provides wide access to the pulmonary cavities. However, *minimally invasive* (or video assisted) *thoracic surgery (thoracoscopy)* allows access to the thorax through small intercostal incisions for many intrathoracic procedures.

Thoracic Outlet Syndrome

When clinicians refer to the superior thoracic aperture as the thoracic "outlet," they are emphasizing the important nerves and arteries that pass through this aperture into the lower neck and upper limb. Hence, various types of *thoracic outlet syndromes* exist, such as the *costoclavicular syndrome*—pallor and coldness of the skin of the upper limb and diminished radial pulse resulting from compression of the subclavian artery between the clavicle and the 1st rib, particularly when the angle between the neck and the shoulder is increased.

Dislocation of Ribs

A rib dislocation (*slipping rib syndrome*) or *dislocation of a sternocostal joint* is the displacement of a costal cartilage from the sternum. This causes severe pain, particularly during deep respiratory movements. The injury produces a lump-like deformity at the dislocation site. Rib dislocations are common in body contact sports, and possible complications are pressure on or damage to nearby nerves, vessels, and muscles.

A rib separation refers to *dislocation of a costochondral junction* between the rib and its costal cartilage. In separations of the 3rd–10th ribs, tearing of the perichondrium and periosteum usually occurs. As a result, the rib may move superiorly, overriding the rib above and causing pain.

Paralysis of Diaphragm

Paralysis of the diaphragm can be detected radiographically by noting its paradoxical movement. Paralysis of half of the diaphragm because of injury to its motor supply from the phrenic nerve does not affect the other half because the domes are separately supplied by the right and left phrenic nerves. Instead of descending on inspiration, the paralyzed dome is pushed superiorly by the abdominal viscera that are being compressed by the active side. The paralyzed dome descends during expiration as it is pushed down by the positive pressure in the lungs (Fig. B4.2).

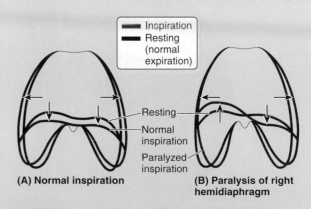

FIGURE B4.2. Normal and paradoxical movements of diaphragm.

(Continued on next page)

CLINICAL BOX

Sternal Fractures

✚ *Sternal fractures* are not common, but crush injuries can occur during traumatic compression of the thoracic wall (e.g., in automobile accidents when the driver's chest is driven into the steering column). The concern in sternal injuries is not primarily for the fracture itself but for the likelihood of heart injury (myocardial contusion, cardiac rupture, tamponade) and/or lung injury.

Breasts

Both males and females have breasts (L. *mammae*), but normally, the mammary glands are well developed only in women. **Mammary glands** in women are accessory to reproduction; in men, they are functionless, consisting of only a few small ducts or cords. The mammary glands are modified sweat glands and therefore have no special capsule or sheath. The contour and volume of the breasts are produced by subcutaneous fat except during pregnancy, when the mammary glands enlarge and new glandular tissue forms. During puberty (8–15 years of age), the female breasts normally grow because of glandular development and increased fat deposition. Breast size and shape result from genetic and nutritional factors.

The roughly circular base of the female breast extends transversely from the lateral border of the sternum to the anterior axillary line and vertically from the 2nd to 6th ribs. A small part of the breast may extend along the inferolateral edge of the pectoralis major muscle toward the axillary fossa, forming an **axillary process** or **tail** (of Spence). Two thirds of

the breast rests on the **pectoral fascia** covering the pectoralis major; the other third rests on the fascia covering the serratus anterior muscle (Figs. 4.8 and 4.9). Between the breast and the deep pectoral fascia is a loose connective tissue plane or potential space—the **retromammary space** (bursa). This plane, containing a small amount of fat, allows the breast some degree of movement on the deep pectoral fascia. The mammary glands are firmly attached to the dermis of the overlying skin by the **suspensory ligaments** (of Cooper). These ligaments, particularly well developed in the superior part of the breast (Fig. 4.8), help support the **mammary gland lobules**.

At the greatest prominence of the breast is the **nipple**, surrounded by a circular pigmented area (the **areola**). The breast contains 15–20 **lobules** of glandular tissue, which constitute the parenchyma of the mammary gland. Each lobule is drained by a **lactiferous duct**, which opens independently on the nipple. Just deep to the areola, each duct has a dilated portion, the **lactiferous sinus** (Fig. 4.8).

VASCULATURE OF BREAST

The *arterial supply of the breast* is derived from the following vessels (Fig. 4.9A):

- **Medial mammary branches of perforating branches** and *anterior intercostal branches of the internal thoracic artery*, originating from the subclavian artery
- **Mammary branches of lateral thoracic** and **thoraco-acromial arteries**, which are branches of the axillary artery
- *Posterior intercostal arteries*, branches of the thoracic aorta that run in the intercostal spaces

The *venous drainage of the breast* (Fig. 4.9B) is mainly to the *axillary vein*, but there is some drainage to the *internal thoracic vein*.

The *lymphatic drainage of the breast* is important because of its role in the metastasis (spread) of cancer cells.

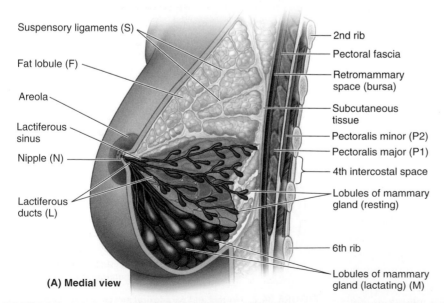

Suspensory ligaments (S)
Fat lobule (F)
Areola
Lactiferous sinus
Nipple (N)
Lactiferous ducts (L)

2nd rib
Pectoral fascia
Retromammary space (bursa)
Subcutaneous tissue
Pectoralis minor (P2)
Pectoralis major (P1)
4th intercostal space
Lobules of mammary gland (resting)
6th rib
Lobules of mammary gland (lactating) (M)

(A) Medial view

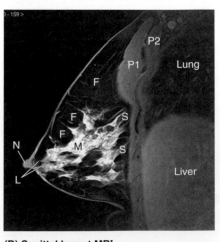

(B) Sagittal breast MRI

FIGURE 4.8. Female breast. A. Sagittal section of female breast and anterior thoracic wall. The *upper part* of the figure demonstrates the fat lobules and suspensory ligaments; the *middle part*, the alveoli of the breast with resting (nonlactating) lobules of the mammary gland; and the *lower part*, lactating lobules of the mammary gland. **B.** Sagittal MRI demonstrating internal structure of breast and posterior relationships.

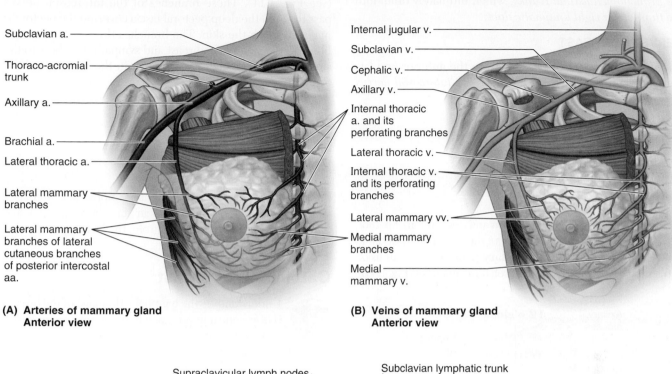

(A) Arteries of mammary gland Anterior view

Subclavian a.
Thoraco-acromial trunk
Axillary a.
Brachial a.
Lateral thoracic a.
Lateral mammary branches
Lateral mammary branches of lateral cutaneous branches of posterior intercostal aa.

Internal thoracic a. and its perforating branches

(B) Veins of mammary gland Anterior view

Internal jugular v.
Subclavian v.
Cephalic v.
Axillary v.
Internal thoracic a. and its perforating branches
Lateral thoracic v.
Internal thoracic v. and its perforating branches
Lateral mammary vv.
Medial mammary branches
Medial mammary v.

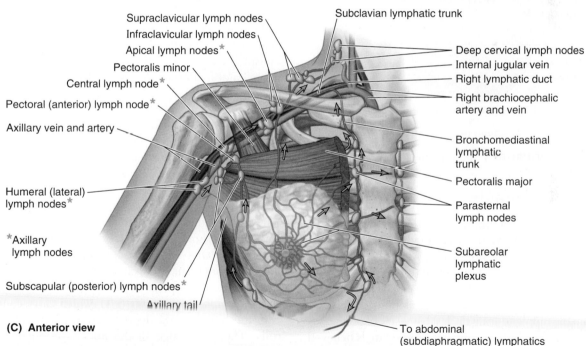

(C) Anterior view

Supraclavicular lymph nodes
Infraclavicular lymph nodes
Apical lymph nodes*
Pectoralis minor
Central lymph node*
Pectoral (anterior) lymph node*
Axillary vein and artery
Humeral (lateral) lymph nodes*
*Axillary lymph nodes
Subscapular (posterior) lymph nodes*
Axillary tail

Subclavian lymphatic trunk
Deep cervical lymph nodes
Internal jugular vein
Right lymphatic duct
Right brachiocephalic artery and vein
Bronchomediastinal lymphatic trunk
Pectoralis major
Parasternal lymph nodes
Subareolar lymphatic plexus
To abdominal (subdiaphragmatic) lymphatics

FIGURE 4.9. Lymphatic drainage and vasculature of breast. A. Arteries. **B.** Veins. **C.** Lymphatic drainage. Axillary lymph nodes are indicated by *asterisks* (*green*). *a.*, artery; *aa.*, arteries; *v.*, vein; *w.*, veins.

Lymph passes from lobules of the gland, nipple, and areola to the **subareolar lymphatic plexus** (Fig. 4.9C). Drainage of lymph then proceeds as follows:

- Most lymph (>75%), especially from the lateral quadrants of the breasts, drains to the *axillary lymph nodes* (that includes the pectoral, humeral, subscapular, central, and apical groups—see *Axillary Lymph Nodes* in Chapter 3, Upper Limb).
- Most of the lymph first drains to the *pectoral (anterior) nodes*. However, some lymph may drain directly to other

axillary nodes or to interpectoral, deltopectoral, supraclavicular, or inferior deep cervical nodes.
- Lymph from the medial breast quadrants drains to the **parasternal lymph nodes** or to the opposite breast.
- Lymph from the inferior breast quadrants may pass deeply to *abdominal lymph nodes* (inferior phrenic nodes).

Lymph from the axillary nodes drains to infraclavicular and supraclavicular nodes and from them to the *subclavian lymphatic trunk*. Lymph from the parasternal nodes enters the *bronchomediastinal trunks*, which ultimately drain into the *thoracic* or *right lymphatic duct*.

NERVES OF BREAST

The *nerves of the breasts* derive from the anterior and lateral cutaneous branches of the *4th to 6th intercostal nerves* (see Fig. 4.11). These branches of the intercostal nerves pass through the deep pectoral fascia covering the pectoralis major to reach the skin. The branches thus convey sensory fibers to the skin of the breast and sympathetic fibers to the smooth muscle of the blood vessels in the breasts and the overlying skin and nipple.

CLINICAL BOX

Breast Quadrants

 For the anatomical location and description of pathology (e.g., cysts and tumors), the breast is divided into four quadrants. The **axillary process** is an extension of the mammary gland of the superolateral quadrant (Fig. B4.3).

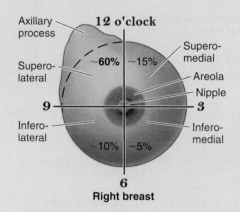

Axillary process
12 o'clock
Superomedial
Superolateral
~60%   ~15%
Areola
Nipple
9   3
Inferolateral
Inferomedial
~10%   ~5%
6
Right breast

FIGURE B4.3. Breast quadrants.

Changes in Breasts

 Changes, such as branching of the lactiferous ducts, occur in the breast tissues throughout the menstrual cycle and during pregnancy. Although mammary glands are prepared for secretion by midpregnancy, they do not produce milk until shortly after the baby is born. *Colostrum*, a creamy white to yellowish premilk fluid, may be secreted from the nipples during the last trimester of pregnancy and during initial episodes of nursing. Colostrum is believed to be especially rich in protein, immune agents, and a growth factor affecting the infant's intestines. In multiparous women (those who have given birth two or more times), the breasts often increase in size and pendulous. The breasts in elderly women are usually small because of the decrease in fat and atrophy of glandular tissue.

Supernumerary Breasts and Nipples

Supernumerary (exceeding two) *breasts* (*polymastia*) or nipples (*polythelia*) may occur superior or inferior to the normal breasts. Usually, supernumerary breasts consist of only a rudimentary nipple and areola. A supernumerary breast may appear anywhere along a line extending from the axilla to the groin, the location of the embryonic *mammary crest* (ridge).

Carcinoma of Breast

Understanding the lymphatic drainage of the breasts is of practical importance in predicting the metastasis (spread) of cancer cells from a *carcinoma of the breast* (breast cancer). Carcinomas of the breast are malignant tumors, usually adenocarcinomas arising from the epithelial cells of the lactiferous ducts in the mammary gland lobules (Fig. B4.4). Metastatic cancer cells that enter a lymphatic vessel usually pass through two or three groups of lymph nodes before entering the venous system. Breast cancer can spread via lymphatics and veins and as well as by direct invasion.

Interference with dermal lymphatics by cancer may cause *lymphedema* (edema, excess fluid in the subcutaneous tissue) in the skin of the breast, which in turn may result in deviation of the nipple and a thickened, leather-like appearance of the skin (Fig. B4.4A). Prominent or "puffy" skin between dimpled pores gives it an orange-peel appearance (*peau d'orange* sign). Larger dimples (fingertip size or bigger) result from cancerous invasion of the glandular tissue and fibrosis (fibrous degeneration), which causes shortening or places traction on the suspensory ligaments. *Subareolar breast cancer* may cause inversion of the nipple by a similar mechanism involving the lactiferous ducts.

Breast cancer typically spreads by means of lymphatic vessels (*lymphogenic metastasis*), which carry cancer cells from the breast to the lymph nodes, chiefly those in the axilla. The cells lodge in the nodes, producing nests of tumor cells (*metastases*). Abundant communications among lymphatic pathways and among axillary, cervical, and parasternal nodes may also cause metastases from the breast to develop in the supraclavicular lymph nodes, the opposite breast, or the abdomen. Because most of the lymphatic drainage of the breast is to the *axillary lymph nodes*, they are the most common site of metastasis from a breast cancer. Enlargement of these palpable nodes suggests the possibility of breast cancer and may be key to early detection. However, the absence of enlarged axillary lymph nodes is no guarantee that metastasis from a breast cancer has not occurred because the malignant cells may have passed to other nodes,

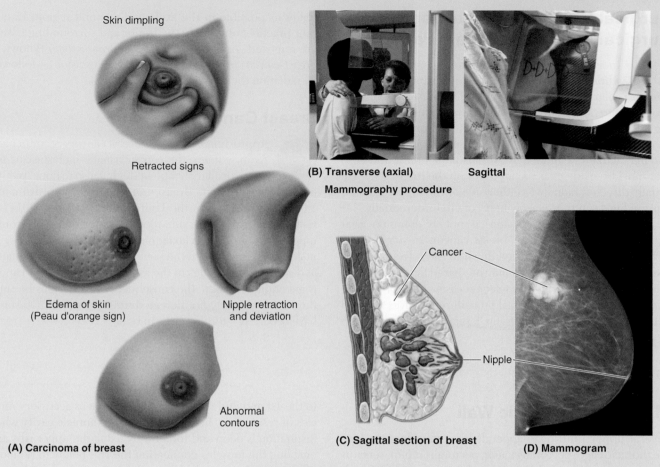

Skin dimpling

Retracted signs

(B) Transverse (axial) **Sagittal**

Mammography procedure

Edema of skin
(Peau d'orange sign)

Nipple retraction
and deviation

Abnormal
contours

Cancer

Nipple

(A) Carcinoma of breast

(C) Sagittal section of breast

(D) Mammogram

FIGURE B4.4. Carcinoma of breast.

such as the infraclavicular and supraclavicular lymph nodes or directly into the circulation. Surgical removal of axillary nodes to which breast cancer has metastasized or damage to the axillary lymph nodes and vessels by radiation therapy for cancer treatment may result in lymphedema in the ipsilateral upper limb, which also drains through the axillary nodes.

The posterior intercostal veins drain into the *azygos/hemi-azygos system of veins* alongside the bodies of the vertebrae and communicate with the internal vertebral venous plexus surrounding the spinal cord. Cancer cells can also spread from the breast by these venous routes to the vertebrae and from there to the cranium and brain. Cancer also spreads by contiguity (invasion of adjacent tissue). When breast cancer cells invade the retromammary space, attach to or invade the pectoral fascia overlying the pectoralis major, or metastasize to the interpectoral nodes, the breast elevates when the muscle contracts. This movement is a clinical sign of advanced cancer of the breast.

Visualizing Breast Structure and Pathology

Examination of the breasts by medical imaging is one of the techniques used to detect breast abnormalities, distinguishing cysts and neoplastic masses from variations in breast architecture. *Mammography* is

radiographic study of the breast, which is flattened to extend the area that can be examined and reduce thickness, making it more uniform for increased visualization (Fig. B4.4B). Mammography is used primarily for screening for problems before they are evident otherwise. Carcinomas often appear as a large, jagged density in the *mammogram* (Fig. B4.4C,D). The skin is thickened over the tumor (*upper two arrows* in Fig. B4.4C), and the nipple is depressed. Figure B4.4D is a *xeromammogram*, in which denser structures (normal stroma, ducts, and the tumor) appear dark. In conventional mammography, denser structures appear light.

Ultrasonography (US) is useful for looking at formations palpated but not clearly observed on a mammogram, especially in women with dense breast tissue, and to gain more specific information about areas of interest in a mammogram or changes detected compared to previous mammograms. Ultrasound is a noninvasive means of distinguishing fluid-filled cysts or abscesses from solid masses. Ultrasound can also be used to guide a biopsy needle or enable fluid aspiration.

Magnetic resonance imaging (MRI) of the breast is performed with specialized machines (*MRI with dedicated breast coils*) to further examine problems detected by mammography or US, to rule out false positive findings, and to plan treatment.

(Continued on next page)

Surgical Incisions of Breast and Surgical Removal of Breast Pathology

 Incisions are placed in the inferior breast quadrants when possible because these quadrants are less vascular than the superior ones. The transition between the thoracic wall and breast is most abrupt inferiorly, producing a line, crease, or deep skin fold—the *inferior cutaneous crease*. Incisions made along this crease will be least evident and may actually be hidden by overlap of the breast. Incisions that must be made near the areola or on the breast itself are usually directed radially to either side of the nipple (Langer tension lines run transversely here) or circumferentially.

Mastectomy (breast excision) is not as common as it once was as a treatment for breast cancer. In *simple mastectomy*, the breast is removed down to the retromammary space. The nipple and areola may be spared, and immediate reconstruction performed in selected cases. *Radical mastectomy*, a more extensive surgical procedure, involves removal of the breast, pectoral muscles, fat, fascia, and as many lymph nodes as possible in the axilla and pectoral region. In current practice, often only the tumor and surrounding tissues are removed—a *lumpectomy* or *quadrantectomy* (known as *breast-conserving surgery*, a wide local excision)—followed by radiation therapy (Goroll, 2014).

Breast Cancer in Men

Approximately 4.5% of breast cancers occur in men. As in women, the cancer usually metastasizes not only to axillary lymph nodes but also to bone, pleura, lung, liver, and skin. Breast cancer affects approximately 1,000 men per year in the United States (Swartz, 2014). A visible and/or palpable subareolar mass or secretion from a nipple may indicate a malignant tumor. Breast cancer in males tends to infiltrate the pectoral fascia, pectoralis major, and apical lymph nodes in the axilla. Although breast cancer is uncommon in men, the consequences are serious because they are frequently not detected until extensive metastases have occurred—for example, in bone.

Muscles of Thoracic Wall

Several upper limb (axio-appendicular) muscles attach to the thoracic cage: pectoralis major, pectoralis minor, serratus anterior anteriorly, and latissimus dorsi posteriorly. In addition, the anterolateral abdominal muscles and some neck and back muscles attach to the thoracic cage. The pectoralis major and minor, the inferior part of the serratus anterior, and the scalene muscles (passing from the cervical vertebrae to the 1st and 2nd ribs) may also function as accessory muscles of respiration, helping expand the thoracic cavity when inspiration is deep and forceful by fixing the upper ribs and enabling the muscles connecting the ribs below to be more effective in elevating the lower ribs during forced inspiration. Muscles of the thoracic wall are illustrated in Figure 4.10 and listed and described in Table 4.2.

Typical intercostal spaces contain three layers of intercostal muscles (Figs. 4.11 and 4.12). The superficial layer is

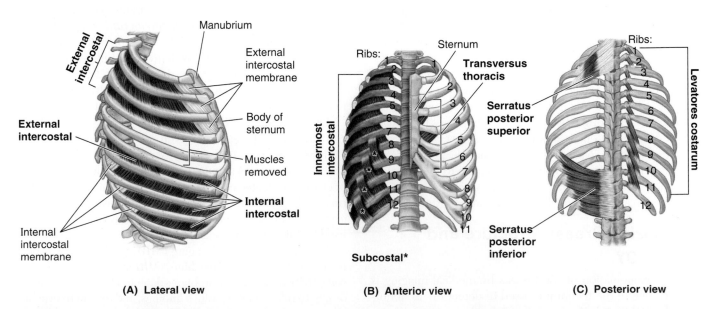

FIGURE 4.10. Muscles of thoracic wall. A. External and internal intercostal muscles. **B.** Innermost intercostals, subcostal, and transversus thoracis muscles. **C.** Serratus posterior superior and inferior and levatores costarum muscles.

TABLE 4.2. MUSCLES OF THORACIC WALL

Muscles	Superior Attachment	Inferior Attachment	Innervation	Main Action
External intercostal	Inferior border of ribs	Superior border of ribs below	Intercostal nerves	During forced inspiration: elevates ribs[a]
Internal intercostal				During forced respiration: Interosseus part depresses ribs; interchondral part elevates ribs[a].
Innermost intercostal				
Transversus thoracis	Posterior surface of lower sternum	Internal surface of costal cartilages 2–6		Weakly depresses ribs
Subcostal	Internal surface of lower ribs near their angles	Superior borders of 2nd or 3rd ribs below		Probably acts in same manner as internal intercostal muscles
Levatores costarum	Transverse processes of C7–T11	Subjacent ribs between tubercle and angle	Posterior rami of C8–T11 nerves	Elevate ribs
Serratus posterior superior	Nuchal ligament, spinous processes of C7–T3 vertebrae	Superior borders of 2nd–4th ribs	2nd–5th intercostal nerves	Elevates ribs[b]
Serratus posterior inferior	Inferior borders of 8th–12th ribs near their angles	Spinous processes of T11–L2 vertebrae	9th–11th intercostal nerves, subcostal (T12) nerve	Depresses ribs[b]

[a]The tonus of the intercostal muscles keeps the intercostal spaces rigid, thereby preventing them from billowing (bulging) out during expiration and from being drawn in during inspiration. The role of individual intercostal muscles and accessory muscles of respiration in moving the ribs is difficult to interpret despite many electromyographic studies.
[b]Action traditionally assigned on the basis of attachments; these muscles appear to be largely proprioceptive in function.

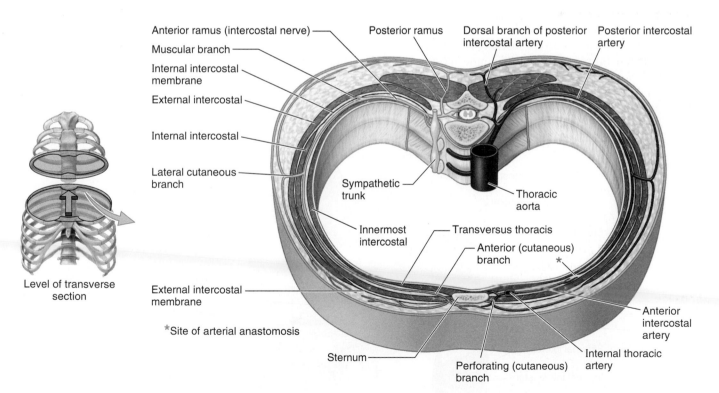

FIGURE 4.11. Intercostal space, transverse section. This section shows nerves (*right side*) and arteries (*left side*).

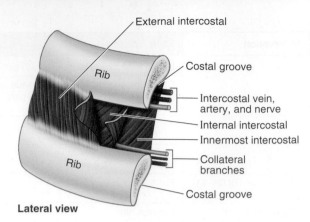

FIGURE 4.12. Contents of typical intercostal space. Remember the structures in the costal groove—from superior to inferior—as *VAN*, for vein, artery, and nerve.

Nerves of Thoracic Wall

The thoracic segments of the spinal cord supply 12 pairs of thoracic spinal nerves to the thoracic wall. As they leave the IV foramina, they divide into anterior and posterior rami (Fig. 4.11). The anterior rami of T1–T11 form the **intercostal nerves** that run along the extent of the intercostal spaces. The anterior rami of the T12 nerves, inferior to the 12th ribs, form the **subcostal nerves**. The posterior rami of the thoracic spinal nerves pass posteriorly immediately lateral to the articular processes of the vertebrae to supply the bones, joints, deep back muscles, and skin of the back in the thoracic region.

Typical intercostal nerves (3rd through 6th) run initially along the posterior aspects of the intercostal spaces between the parietal pleura (serous lining of the thoracic cavity) and the internal intercostal membrane. At first, they run across the internal surface of the internal intercostal membrane and muscle near the middle of the intercostal space. Near the angles of the ribs, the nerves pass between the internal intercostal and innermost intercostal muscles (Figs. 4.12 and 4.13). Here, the nerves pass to and then continue to course within the costal grooves, lying just inferior to the intercostal arteries, which in turn lie inferior to the intercostal veins.

Collateral branches of these nerves arise near the angles of the ribs and run along the superior border of the rib below. The nerves continue anteriorly between the internal and the innermost intercostal muscles, giving branches to these and other muscles and giving rise to **lateral cutaneous branches** approximately at the midaxillary line (Fig. 4.11).

formed by the **external intercostal muscles** (fiber bundles oriented infero-anteriorly), the middle layer is formed by the **internal intercostal muscles** (fiber bundles oriented inferoposteriorly), and the deepest layer is formed by the **innermost intercostal muscles** (similar to internal intercostals but internal to the intercostal neurovasculature). Anteriorly, the fleshy external intercostal muscles are replaced by **external intercostal membranes**; posteriorly, the fleshy internal intercostal muscles are replaced by **internal intercostal membranes**. The innermost intercostal muscles are found only at the most lateral parts of the intercostal spaces.

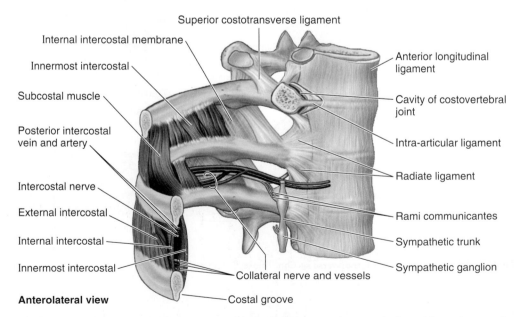

FIGURE 4.13. Posterior part of intercostal space. Note the connection of the intercostal nerve to the sympathetic trunk by rami communicantes (communicating branches).

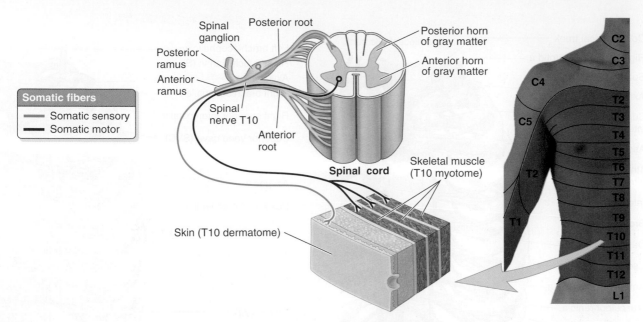

FIGURE 4.14. Dermatomes and myotomes of the trunk. Note the relationship between the area of skin (dermatome) and skeletal muscle (myotome) innervated by a spinal nerve or segment of the spinal cord. The dermatomes of the thorax are shown on the right side of the page.

Anteriorly, the nerves appear on the internal surface of the internal intercostal muscle. Near the sternum, the nerves turn anteriorly, passing between the costal cartilages and entering the subcutaneous tissue as **anterior cutaneous branches**. *Muscular branches* arise all along the course of the intercostal nerves to supply the intercostal, subcostal, transversus thoracis, levatores costarum, and serratus posterior muscles (Table 4.2); and *sensory branches* pass to the parietal pleura.

Atypical intercostal nerves are the 1st and 2nd and 7th through 11th. Intercostal nerves 1 and 2 pass on the internal surfaces of the 1st and 2nd ribs instead of along the inferior margins of the costal grooves. After giving rise to the lateral cutaneous branches, the 7th through 11th intercostal nerves continue to supply the abdominal skin and muscles.

Through the posterior ramus and the lateral and anterior cutaneous branches of the anterior ramus, each spinal nerve supplies a stripe-like area of skin extending from the posterior median line to the anterior median line. These band-like skin areas (**dermatomes**) are each supplied by the *sensory fibers* of a single posterior root through the posterior and anterior rami of its spinal nerve (Fig. 4.14). Because any particular area of skin usually receives innervation from two adjacent nerves, considerable *overlapping of adjacent dermatomes* occurs. Therefore, complete loss of sensation usually does not occur unless two or more intercostal nerves are anesthetized. The muscles supplied by the *motor fibers* of the posterior

and anterior rami of each pair of thoracic spinal nerves constitute a **myotome** (Fig. 4.14) which, in the thoracic region, includes all the muscle of one pair of right and left intercostal spaces and an adjacent portion of the deep back muscles.

Rami communicantes, or communicating branches, connect each intercostal and subcostal nerve to the ipsilateral **sympathetic trunk** (Fig. 4.13). Presynaptic fibers leave the initial portions of the anterior ramus of each thoracic (and upper lumbar) nerve by means of a white ramus communicans and pass to a **sympathetic ganglion**. Postsynaptic fibers distributed to the body wall and limbs pass from the ganglia of the sympathetic trunk via gray rami communicantes to join the anterior ramus of the nearest spinal nerve, including all the intercostal nerves. Sympathetic nerve fibers are distributed through the branches of all spinal nerves (anterior and posterior rami) to reach the blood vessels, sweat glands, and smooth muscle of the body wall and limbs.

Vasculature of Thoracic Wall

The *arteries of the thoracic wall* are derived from the *thoracic aorta* through the posterior intercostal and subcostal arteries, the *subclavian artery* through the internal thoracic and supreme intercostal arteries, and the *axillary artery* through the superior and lateral thoracic arteries (Figs. 4.11 and 4.15A; Table 4.3). Each intercostal space is supplied by three arteries: a large **posterior intercostal artery**

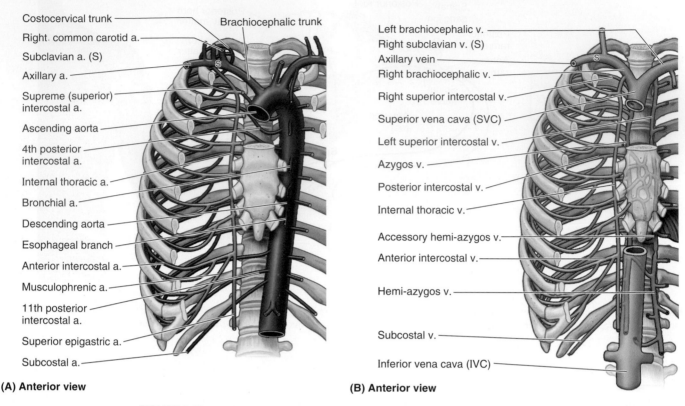

(A) Anterior view

Costocervical trunk
Right common carotid a.
Subclavian a. (S)
Axillary a.
Supreme (superior) intercostal a.
Ascending aorta
4th posterior intercostal a.
Internal thoracic a.
Bronchial a.
Descending aorta
Esophageal branch
Anterior intercostal a.
Musculophrenic a.
11th posterior intercostal a.
Superior epigastric a.
Subcostal a.

Brachiocephalic trunk

(B) Anterior view

Left brachiocephalic v.
Right subclavian v. (S)
Axillary vein
Right brachiocephalic v.
Right superior intercostal v.
Superior vena cava (SVC)
Left superior intercostal v.
Azygos v.
Posterior intercostal v.
Internal thoracic v.
Accessory hemi-azygos v.
Anterior intercostal v.
Hemi-azygos v.
Subcostal v.
Inferior vena cava (IVC)

FIGURE 4.15. Arteries and veins of thoracic wall. A. Arteries. **B.** Veins. *a.*, artery; *v.*, vein.

(and its **collateral branch**) and a small pair of **anterior intercostal arteries**.

The *veins of the thoracic wall* accompany the intercostal arteries and nerves and lie most superior in the costal grooves (Figs. 4.11 and 4.15B). There are 11 **posterior intercostal veins** and one **subcostal vein** on each side.

The posterior intercostal veins anastomose with the anterior intercostal veins, tributaries of the internal thoracic veins. Most posterior intercostal veins end in the azygos/hemi-azygos venous system (discussed later in this chapter), which conveys venous blood to the superior vena cava.

TABLE 4.3. ARTERIAL SUPPLY OF THORACIC WALL

Artery	Origin	Course	Distribution
Posterior intercostals	Supreme intercostal arteries (intercostal spaces 1 and 2) and thoracic aorta (remaining intercostal spaces)	Pass between internal and innermost intercostal muscles	Intercostal muscles and overlying skin, parietal pleura
Anterior intercostals	Internal thoracic arteries (intercostal spaces 1–6) and musculophrenic arteries (intercostal spaces 7–9)		
Internal thoracic	Subclavian artery	Passes inferiorly, lateral to sternum, between costal cartilages and internal intercostal muscles to divide into superior epigastric and musculophrenic arteries	By way of anterior intercostal arteries to intercostal spaces 1–6 and musculophrenic arteries to intercostal spaces 7–9
Subcostal	Thoracic aorta	Courses along interior border of 12th rib	Muscles of anterolateral abdominal wall and overlying skin

CLINICAL BOX

Herpes Zoster Infection

Herpes zoster (*shingles*)—a viral disease of spinal ganglia—is a *skin lesion with a dermatomal distribution*. The *herpes virus* invades a spinal ganglion and is transported along the axon to the skin, where it produces an infection that causes a sharp burning pain in the dermatome supplied by the involved nerve. A few days later, the skin of the dermatome becomes red and vesicular eruptions appear (Fig. B4.5). Vaccination confers protection against herpes zoster and is recommended for adults starting at age 50 years.

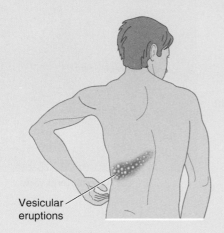

Vesicular eruptions

FIGURE B4.5. **Herpes zoster.**

Dyspnea—Difficult Breathing

When people with respiratory problems such as *asthma* or *emphysema* or with *heart failure* struggle to breathe, they use their accessory respiratory muscles to assist the expansion of their thoracic cavities. The recruitment of the neck muscles (sternocleidomastoid, upper trapezius, and scalene muscles) is visible and particularly striking. They may also lean on a table or their thighs to fix their pectoral girdles (clavicles and scapulae) so the muscles are able to act on their rib attachments and expand the thorax.

Intercostal Nerve Block

Local anesthesia of an intercostal space is produced by injecting a local anesthetic agent around the intercostal nerves. This procedure, an *intercostal nerve block*, is commonly used in patients with rib fractures and sometimes after thoracic surgery. It involves infiltration of the anesthetic around the intercostal nerve and its collateral branches (Fig. B4.6). Because considerable overlap in the innervation of contiguous dermatomes occurs, anesthesia of any particular area of skin usually requires injection of two adjacent nerves. For example, anesthesia for a broken rib requires injection of the anesthetic agent into the region of the intercostal nerves superior and inferior to the rib, proximal to the site of fracture.

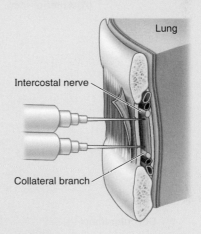

Lung

Intercostal nerve

Collateral branch

FIGURE B4.6. **Intercostal nerve block.**

SURFACE ANATOMY

Thoracic Wall

Several bony landmarks and imaginary vertical lines facilitate anatomical descriptions, identification of thoracic areas, and location of lesions such as a bullet wound:

- **Anterior median (midsternal) line** indicates the intersection of the median plane with the anterior thoracic wall (Fig. SA4.1A).
- **Midclavicular lines** pass through the midpoints of the clavicles, parallel to the anterior median line (Fig. SA4.1A).

- **Anterior axillary line** runs vertically along the anterior axillary fold, which is formed by the border of the pectoralis major as it spans from the thorax to the humerus (arm bone) (Fig. SA4.1B).
- **Midaxillary line** runs from the apex (deepest part) of the axilla, parallel to the anterior axillary line.
- **Posterior axillary line**, also parallel to the anterior axillary line, is drawn vertically along the posterior axillary fold

(Continued on next page)

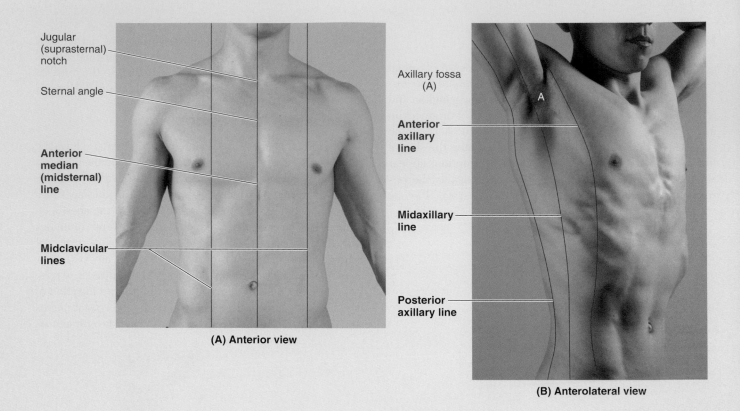

Jugular (suprasternal) notch

Sternal angle

Anterior median (midsternal) line

Midclavicular lines

(A) Anterior view

Axillary fossa (A)

Anterior axillary line

Midaxillary line

Posterior axillary line

(B) Anterolateral view

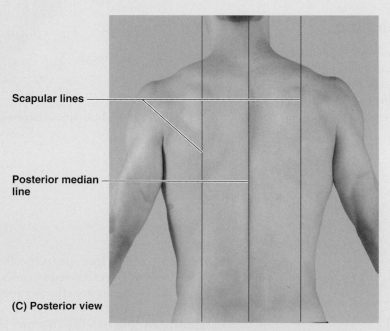

Scapular lines

Posterior median line

(C) Posterior view

FIGURE SA4.1. Vertical lines of thoracic wall.

formed by the latissimus dorsi and teres major muscles as they span from the back to the humerus (Fig. SA4.1B).

- **Posterior median (midvertebral) line** is a vertical line at the intersection of the median plane with the vertebral column (Fig. SA4.1C).
- **Scapular lines** are parallel to the posterior median line and cross the inferior angles of the scapulae (Fig. SA4.1C).

Additional lines (not illustrated) are extrapolated along borders of bony formations—for example, the **parasternal line** (G. *para*, adjacent to).

The **clavicles** lie subcutaneously, forming bony ridges at the junction of the thorax and neck (Fig. SA4.2). They can be palpated easily throughout their length, especially where their medial ends articulate with the manubrium.

The **sternum** also lies subcutaneously in the anterior median line and is palpable throughout its length. The **manubrium of the sternum**

- lies at the level of the bodies of **T3 and T4 vertebrae**
- is anterior to the *arch of the aorta*
- has a **jugular notch** that can be palpated between the prominent sternal ends of the clavicles
- has a *sternal angle* where it articulates with the sternal body at the level of the **T4–T5 intervertebral (IV) disc**

The **sternal angle** is a palpable landmark that lies at the level of the second pair of costal cartilages. The main bronchi pass inferolaterally from the bifurcation of the trachea at the level of the sternal angle. The sternal angle also demarcates

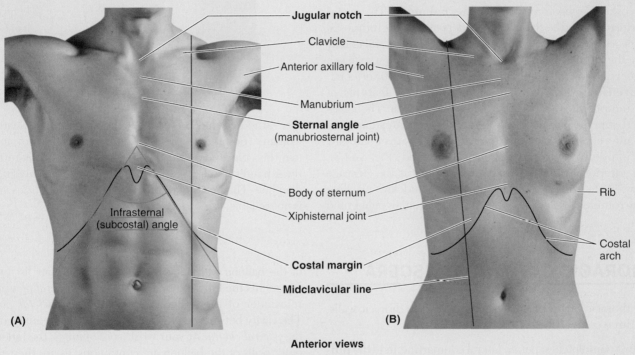

Anterior views

(C) Lateral view

* Transverse thoracic plane

FIGURE SA4.2. Surface features of anterior thoracic wall.

(Continued on next page)

the division between the superior and inferior mediastina and the beginning of the arch of the aorta. The *superior vena cava* (SVC) passes inferiorly deep to the manubrium, projecting as much as a fingerbreadth to the right of this bone.

The 1st rib cannot be palpated because it lies deep to the clavicle; thus, *count the ribs and intercostal spaces anteriorly* by sliding the fingers laterally from the sternal angle onto the 2nd costal cartilage. Start counting with rib 2 and count the ribs and spaces by moving the fingers inferolaterally. The 1st intercostal space is inferior to the 1st rib; likewise, the other spaces lie inferior to the similarly numbered ribs.

The **body of the sternum** lies anterior to the right border of the heart and vertebrae T5–T9. The xiphoid process lies in a slight depression (the **epigastric fossa**) where the converging costal margins form the **infrasternal angle**. The **costal margins**, formed by the medial borders of the 7th–10th costal cartilages, are easily palpable where they extend inferolaterally from the **xiphisternal joint**. This articulation, often seen as a ridge, is at the level of the inferior border of the T9 vertebra.

Breasts are the most prominent surface features of the anterior thoracic wall, especially in women. Their flattened superior surfaces show no sharp demarcation from the anterior surface of the thoracic wall; however, laterally and inferiorly, their borders are well defined (Fig. SA4.3). The anterior median **intermammary cleft** is the cleavage between the breasts. The **nipple** in the midclavicular line is surrounded by a slightly raised and circular pigmented

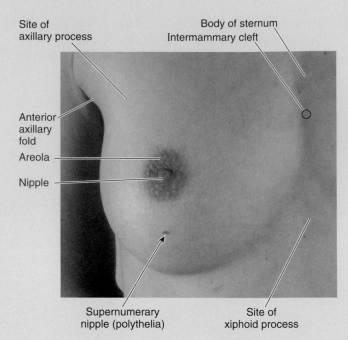

FIGURE SA4.3. Surface anatomy of female breast.

area—the **areola**. The color of the areolas varies with the woman's complexion; they darken during pregnancy and retain this color thereafter. The nipple in men lies anterior to the 4th intercostal space, about 10 cm from the anterior median line. The position of the nipple in women varies and so is not reliable as a surface landmark.

THORACIC CAVITY AND VISCERA

The thoracic cavity, the space enclosed by the thoracic walls, has three compartments (Fig. 4.16A):

- Two completely separate lateral compartments—the **pulmonary cavities**—that contain the lungs and pleurae (lining membranes)
- One central compartment—the **mediastinum**—that contains all other thoracic structures: heart, great vessels, trachea, esophagus, thymus, and lymph nodes

Endothoracic Fascia

The **endothoracic fascia** is a thin fibro-areolar layer between the internal aspect of the thoracic wall and the lining of the pleural cavities (parietal pleura) (Fig. 4.16). The endothoracic fascia provides a cleavage plane, allowing the surgeon to separate the parietal pleura from the thoracic wall, providing access to intrathoracic structures.

Pleurae and Lungs

To visualize the relationship of the pleurae and lungs, push your fist into an underinflated balloon (Fig. 4.16A, *inset*). The part of the balloon wall adjacent to the skin of your fist (which represents the lung) is comparable to the *visceral pleura*; the remainder of the balloon represents the *parietal pleura*. The cavity between the layers of the balloon is analogous to the *pleural cavity*. At your wrist (*root of lung*), the inner and outer walls of the balloon are continuous, as are the visceral and parietal layers of pleura, together forming a *pleural sac*.

PLEURAE

Each lung is invested by and enclosed in a **pleural sac** that consists of two concentric, continuous membranes—the pleurae (Fig. 4.16):

- The **visceral pleura** (pulmonary pleura) covers the lungs and is adherent to all its surfaces, including the surfaces within the horizontal and oblique fissures.
- The **parietal pleura** lines the pulmonary cavities, adhering to the thoracic wall, the mediastinum, and the diaphragm.

The *root of the lung* is enclosed within the area of continuity between the visceral and parietal layers of pleura, the **pleural sleeve**. Inferior to the root of the lung, this continuity between parietal and visceral pleura forms the **pulmonary ligament** extending between the lung and the mediastinum (Fig. 4.17).

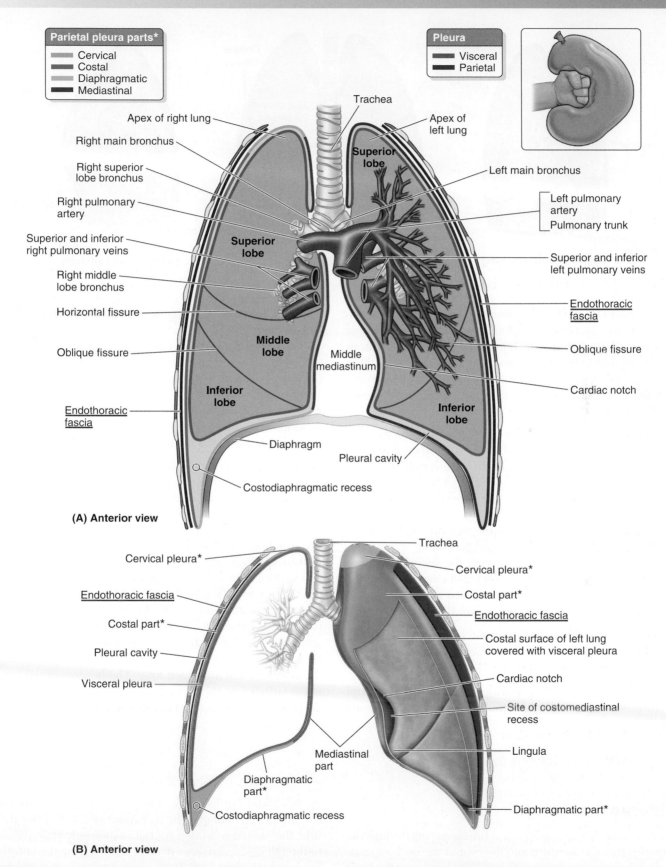

Parietal pleura parts*
- Cervical
- Costal
- Diaphragmatic
- Mediastinal

Pleura
- Visceral
- Parietal

Trachea

Apex of right lung

Apex of left lung

Right main bronchus

Superior lobe

Right superior lobe bronchus

Left main bronchus

Right pulmonary artery

Left pulmonary artery

Pulmonary trunk

Superior and inferior right pulmonary veins

Superior lobe

Superior and inferior left pulmonary veins

Right middle lobe bronchus

Horizontal fissure

Endothoracic fascia

Oblique fissure

Middle lobe

Oblique fissure

Middle mediastinum

Cardiac notch

Inferior lobe

Inferior lobe

Endothoracic fascia

Diaphragm

Pleural cavity

Costodiaphragmatic recess

(A) Anterior view

Trachea

Cervical pleura*

Cervical pleura*

Endothoracic fascia

Costal part*

Costal part*

Endothoracic fascia

Pleural cavity

Costal surface of left lung covered with visceral pleura

Visceral pleura

Cardiac notch

Site of costomediastinal recess

Mediastinal part

Lingula

Diaphragmatic part*

Diaphragmatic part*

Costodiaphragmatic recess

(B) Anterior view

FIGURE 4.16. Lungs and pleurae. A. Lungs and pleural cavity. *Inset*: A fist invaginating an underinflated balloon demonstrates the relationship of the lung (represented by fist) to the walls of the pleural sac (parietal and visceral layers of pleura). The cavity of the pleural sac (pleural cavity) is comparable to the cavity of the balloon. **B.** Parts of parietal pleura and recesses of pleural cavities. *Asterisks* indicate parts of parietal pleura.

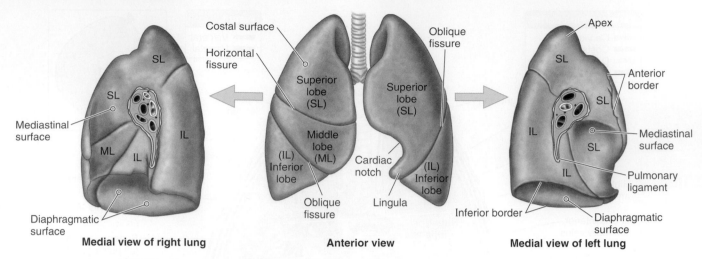

FIGURE 4.17. Lobes and fissures of lungs. The hilum of each lung is centered in the mediastinal surface.

The **pleural cavity**—the potential space between the visceral and the parietal layers of pleura—contains a capillary layer of serous **pleural fluid**, which lubricates the pleural surfaces and allows the layers of pleura to slide smoothly over each other during respiration. Its surface tension also provides the cohesion that keeps the lung surface in contact with the thoracic wall. The *parietal pleura* consists of four parts (Fig. 4.16):

- **Costal part** covers the internal surfaces of the thoracic wall (sternum, ribs, costal cartilages, intercostal muscles and membranes, and sides of thoracic vertebrae) and is separated from the wall by *endothoracic fascia*.
- **Mediastinal part** covers the lateral aspects of the mediastinum.
- **Diaphragmatic part** covers the superior surface of the diaphragm on each side of the mediastinum.
- **Cervical pleura** extends through the superior thoracic aperture into the root of the neck 2–3 cm superior to the medial third of the clavicle to the level of the neck of the 1st rib. It forms a cup-shaped dome over the apex of the lung.

The lines along which the parietal pleura changes direction from one wall of the pleural cavity to another are the **lines of pleural reflection**:

- The **sternal line of pleural reflection** is an abrupt turn of the parietal pleura that occurs where the costal pleura becomes continuous with the mediastinal pleura anteriorly.
- The **costal line of pleural reflection** is also an abrupt turn of the parietal pleura that occurs where the costal pleura becomes continuous with the diaphragmatic pleura inferiorly.
- The **vertebral line of pleural reflection** is a much rounder, gradual reflection where the costal pleura becomes continuous with the mediastinal pleura posteriorly.

The lungs do not completely occupy the pleural cavities during expiration, thus forming areas where two layers of parietal pleura are separated only by pleural fluid. Therefore, the diaphragmatic pleura which covers the periphery of the diaphragm, lies in contact with the lowest part of the costal pleura. The potential pleural spaces here are the **costodiaphragmatic recesses**, the pleural-lined "gutters" that surround the upward convexity of the diaphragm inside the thoracic wall (Fig. 4.16). Similar but smaller pleural recesses are located posterior to the sternum where the costal pleura is in contact with the mediastinal pleura. The potential spaces here are the **costomediastinal recesses** (Fig. 4.16B); the left recess is potentially larger (less occupied) because of the cardiac notch in the left lung. The borders of the lungs move farther into the pleural recesses during deep inspiration and retreat from them during expiration.

SURFACE ANATOMY

Pleurae and Lungs

The cervical pleurae and apices of the lungs pass through the superior thoracic aperture into the root of the neck superior and posterior to the clavicles. The anterior borders of the lungs lie adjacent to the anterior line of reflection of the parietal pleura between the 2nd and 4th costal cartilages (Fig. SA4.4). Here, the margin of the left pleural reflection moves laterally and then inferiorly at the cardiac notch to reach the level of the 6th costal cartilage. The anterior border of the left lung is more deeply indented by its cardiac notch. On the right side, the pleural reflection continues inferiorly from the 4th to the 6th costal cartilage, paralleled closely by the anterior border of the right lung. Both pleural reflections pass laterally and reach the midclavicular line at the level of the 8th costal cartilage, the 10th rib at the midaxillary line, and the 12th rib at the scapular line, proceeding toward the spinous process of the T12 vertebra. Thus, the parietal pleura extends

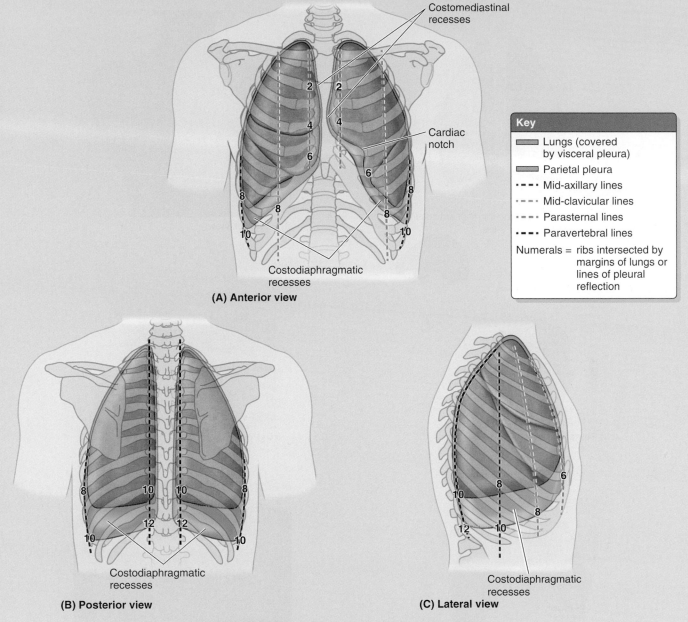

Costomediastinal
recesses

Cardiac
notch

Key
Lungs (covered
by visceral pleura)
Parietal pleura
- - - Mid-axillary lines
- - - Mid-clavicular lines
- - - Parasternal lines
- - - Paravertebral lines
Numerals = ribs intersected by
margins of lungs or
lines of pleural
reflection

Costodiaphragmatic
recesses

(A) Anterior view

Costodiaphragmatic
recesses

(B) Posterior view

Costodiaphragmatic
recesses

(C) Lateral view

FIGURE SA4.4. Surface anatomy of pleurae and lungs.

approximately two ribs inferior to the lung. The *oblique fissure of the lungs* extends from the level of the spinous process of the T2 vertebra posteriorly to the 6th costal cartilage anteriorly, which coincides approximately with the medial border of the scapula when the upper limb is elevated above the head (causing the inferior angle to be rotated laterally). The *horizontal fissure of the right lung* extends from the oblique fissure along the 4th rib and costal cartilage anteriorly.

Auscultation of the lungs (listening to their sounds with a stethoscope) and *percussion of the thorax* (tapping on fingers pressed firmly on the thoracic wall over the lungs to detect sounds in the lungs) are important techniques used during physical examinations (Fig. SA4.5). Percussion helps establish whether the underlying tissues are air filled (*resonant* sound), fluid filled (*dull* sound), or solid (*flat* sound). An awareness

of normal anatomy, particularly the projection of the lungs and the portions that are overlapped by bone (e.g., the scapula) with associated muscles, enables the examiner to know where flat and resonant sounds should be expected. Auscultation assesses airflow through the tracheobronchial tree into the lobes of the lung. The patterns of breath sounds can be characterized by their intensity, pitch, and relative duration throughout inspiration and expiration. The areas of percussion and auscultation of the right and left lungs are outlined in Figure SA4.5. When clinicians refer to "auscultating the base of the lung," they are not usually referring to its diaphragmatic surface or anatomical base. They are usually referring to the inferoposterior part of the inferior lobe.

(Continued on next page)

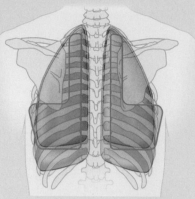

Green: normal area for resonant sound
(A) Posterior view

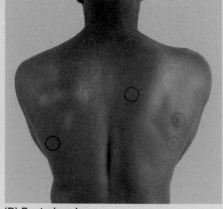

(B) Posterior view

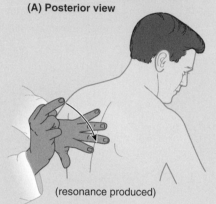

(resonance produced)

(C) Posterior view

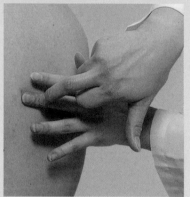

(D) Posterior view

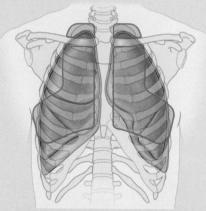

Green: normal area for resonant sound
(E) Anterior view

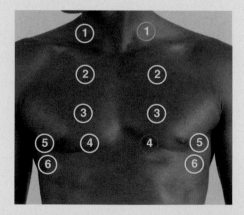

(F) Anterior view

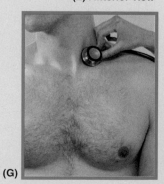

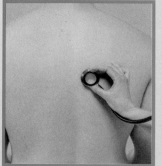

(G)   **Anterior views**   **Posterior views**

FIGURE SA4.5. Auscultation of lungs.

LUGS

The **lungs** are the vital organs of respiration. Their main function is to oxygenate the blood by bringing inspired air into close relation with the venous blood in the pulmonary capillaries. Whereas cadaveric lungs may be shrunken, firm to the touch, and discolored in appearance, healthy lungs in living people are normally light, soft, and spongy. They are also elastic and recoil to about one-third their size when the thoracic cavity is opened.

The **horizontal and oblique fissures** divide the lungs into lobes (see Fig. 4.16). *The right lung has three lobes; the left lung has two.* The right lung is larger and heavier than the left, but it is shorter and wider because the right dome of the diaphragm is higher and the heart and pericardium bulge more to the left. The anterior margin of the right lung is relatively straight, whereas this margin of the left lung has a **cardiac notch**. The cardiac notch primarily indents the antero-inferior aspect of the superior lobe of the left lung. This often creates a thin, tongue-like process of the superior lobe—the **lingula** (Fig. 4.17), which extends below the cardiac notch and slides in and out of the costomediastinal recess during inspiration and expiration. Each lung has the following parts (Figs. 4.17 and 4.18):

- **An apex**: blunt superior end of the lung ascending above the level of the 1st rib into the root of the neck; covered by cervical pleura

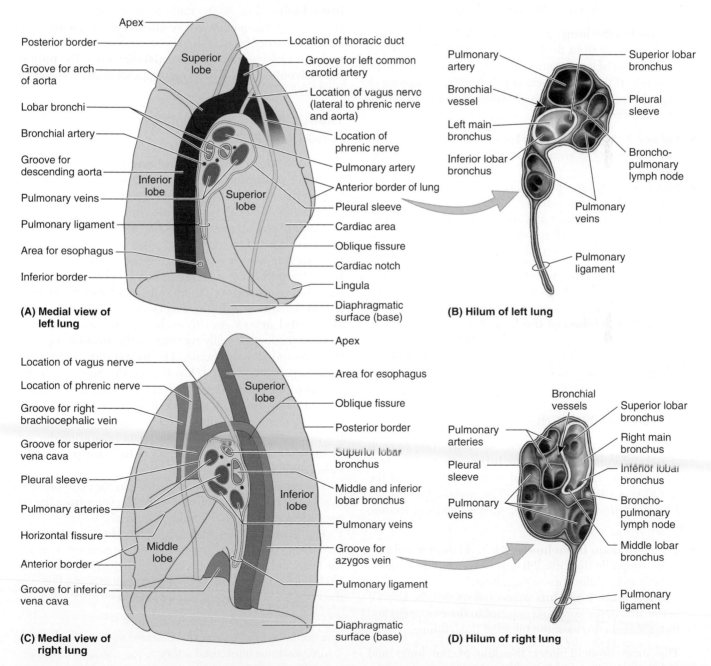

FIGURE 4.18. Mediastinal surfaces and hila of lungs. A. Left lung. **B.** Hilum of left lung. **C.** Right lung. **D.** Hilum of right lung. Impressions are formed in embalmed lungs by contact with adjacent structures (e.g., aorta and superior vena cava).

- **Three surfaces**: *costal surface*, adjacent to the sternum, costal cartilages, and ribs; *mediastinal surface*, including the hilum of the lung and related medially to the mediastinum and posteriorly to the sides of the vertebrae; and *diaphragmatic surface*, resting on the convex dome of the diaphragm
- **Three borders**: *anterior border*, where the costal and mediastinal surfaces meet anteriorly and overlap the heart (the *cardiac notch* indents this border of the left lung); *inferior border*, which circumscribes the diaphragmatic surface of the lung and separates the diaphragmatic surface from the costal and mediastinal surfaces; and *posterior border*, where the costal and mediastinal surfaces meet posteriorly (it is broad and rounded and lies adjacent to the thoracic region of the vertebral column)

The **root of the lung** is composed of the structures entering and emerging from the lung at its hilum (Figs. 4.17 and 4.18). The root of the lung connects the lung with the heart and trachea. If the root is sectioned before the branching of the main bronchus and pulmonary artery, its general arrangement is as follows:

- Pulmonary artery, the most superior structure on the left (the superior lobar bronchus may be most superior on the right)
- Superior and inferior pulmonary veins, located most anterior and inferior, respectively
- Main bronchus, against and approximately in the middle of the posterior boundary, with very small bronchial vessels immediately surrounding it

The root is enclosed within the area of continuity between the parietal and the visceral layers of pleura—the *pleural sleeve* (Fig. 4.18A) or mesopneumonium (mesentery of the lung). The **hilum of the lung** is the area on the medial surface of each lung at which the structures forming the root—the main bronchus, pulmonary vessels, bronchial vessels, lymphatic vessels, and nerves—enter and leave the lung (Fig. 4.19E).

TRACHEA AND BRONCHI

The two **main bronchi** (primary bronchi), one to each lung, pass inferolaterally from the **bifurcation of the trachea**, at the level of the sternal angle, to the hila of the lungs (Figs. 4.19E and 4.20A,B). The walls of the trachea and bronchi are supported by C-shaped rings of hyaline cartilage.

- The **right main bronchus** is wider and shorter and runs more vertically than the left main bronchus as it passes directly to the hilum of the right lung.
- The **left main bronchus** passes inferolaterally, inferior to the arch of the aorta and anterior to the esophagus and thoracic aorta, to reach the hilum of the left lung.

The main bronchi enter the hila of the lungs and branch in a constant fashion within the lungs to form the **bronchial tree**. Each main bronchus divides into **lobar bronchi** (secondary bronchi), two on the left and three on the right, each of which supplies a lobe of the lung. Each lobar bronchus divides into **segmental bronchi** (tertiary bronchi) that supply the bronchopulmonary segments (Fig. 4.19). Each **bronchopulmonary segment** is pyramidal, with its apex directed toward the root of the lung and its base at the pleural surface, and is named according to the segmental bronchus that supplies it.

Each bronchopulmonary segment is supplied independently by a segmental bronchus and a tertiary branch of the pulmonary artery and is drained by intersegmental parts of the pulmonary veins. Beyond the segmental bronchi, there are 20–25 generations of branches that end in **terminal bronchioles** (Fig. 4.20). Each terminal bronchiole gives rise to several generations of **respiratory bronchioles** and each respiratory bronchiole provides 2–11 **alveolar ducts**, each of which gives rise to 5 or 6 **alveolar sacs**. The **pulmonary alveolus** is the basic structural unit of gas exchange in the lung.

VASCULATURE AND NERVES OF LUNGS AND PLEURAE

The large **right** and **left pulmonary arteries** arise from the **pulmonary trunk** at the level of the sternal angle (Fig. 4.21). The pulmonary arteries carry poorly oxygenated (venous) blood to the lungs for oxygenation. The pulmonary arteries pass to the corresponding lung as part of its root. The right artery gives off a lobar artery to the superior lobe before entering the hilum. Within the lungs, the pulmonary arteries descend posterolateral to the main bronchus and divide into lobar and segmental arteries, consecutively. A **lobar artery** goes to each lobe and a **segmental artery** goes to each bronchopulmonary segment of the lung, usually running on the anterior aspect of the corresponding bronchus. The **pulmonary veins** carry well-oxygenated (arterial) blood from the lungs to the left atrium of the heart. Beginning in the pulmonary capillaries, the veins unite into larger and larger vessels. Blood from adjacent bronchopulmonary segments drains into the **intersegmental veins** in the septa separating the segments, then drains to the **superior** or **inferior pulmonary veins** draining each lung.

The veins from the parietal pleura join the systemic veins in adjacent parts of the thoracic wall. The veins from the visceral pleura drain into the pulmonary veins.

The **bronchial arteries** supply blood to the structures comprising the roots of the lungs, the supporting tissues of the lung, and the visceral pleura (Figs. 4.18 and 4.22A). The *left bronchial arteries* arise from the thoracic aorta; however, the *right bronchial artery* may arise from

- a superior posterior intercostal artery
- a common trunk from the thoracic aorta with the right 3rd posterior intercostal artery
- a left superior bronchial artery

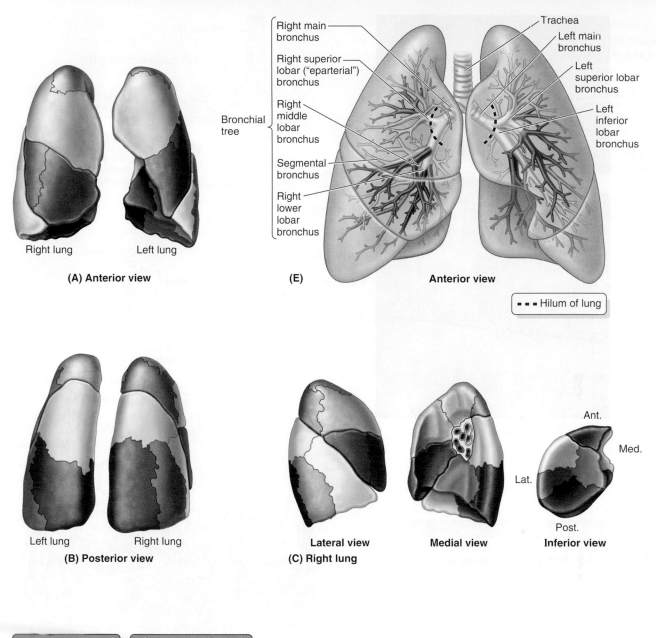

(A) Anterior view

Right main bronchus
Right superior lobar ("eparterial") bronchus
Right middle lobar bronchus
Bronchial tree
Segmental bronchus
Right lower lobar bronchus

Trachea
Left main bronchus
Left superior lobar bronchus
Left inferior lobar bronchus

(E)   Anterior view

▪ ▪ ▪ Hilum of lung

(B) Posterior view

Left lung   Right lung

(C) Right lung

Lateral view   Medial view   Inferior view

Ant.
Med.
Lat.
Post.

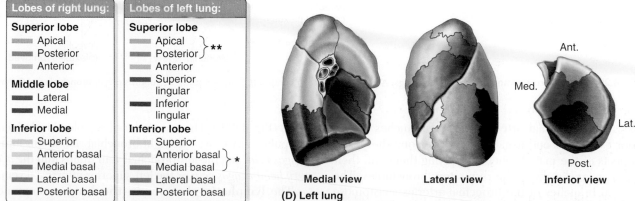

Lobes of right lung:

Superior lobe
▬ Apical
▬ Posterior
▬ Anterior

Middle lobe
▬ Lateral
▬ Medial

Inferior lobe
▬ Superior
▬ Anterior basal
▬ Medial basal
▬ Lateral basal
▬ Posterior basal

Lobes of left lung:

Superior lobe
▬ Apical
▬ Posterior **⁎⁎**
▬ Anterior
▬ Superior lingular
▬ Inferior lingular

Inferior lobe
▬ Superior
▬ Anterior basal
▬ Medial basal **⁎**
▬ Lateral basal
▬ Posterior basal

(D) Left lung

Medial view   Lateral view   Inferior view

Ant.
Med.
Lat.
Post.

⁎⁎ Typically combine into apicoposterior segment
 ⁎ Often combined into anteriomedial basal segment

FIGURE 4.19. Bronchi and bronchopulmonary segments. A–D. The bronchopulmonary segments are demonstrated after injection of different color latex into each tertiary segmental bronchus as shown in **(E)**. *Ant.*, anterior; *Lat.*, lateral; *Med.*, medial; *Post.*, posterior.

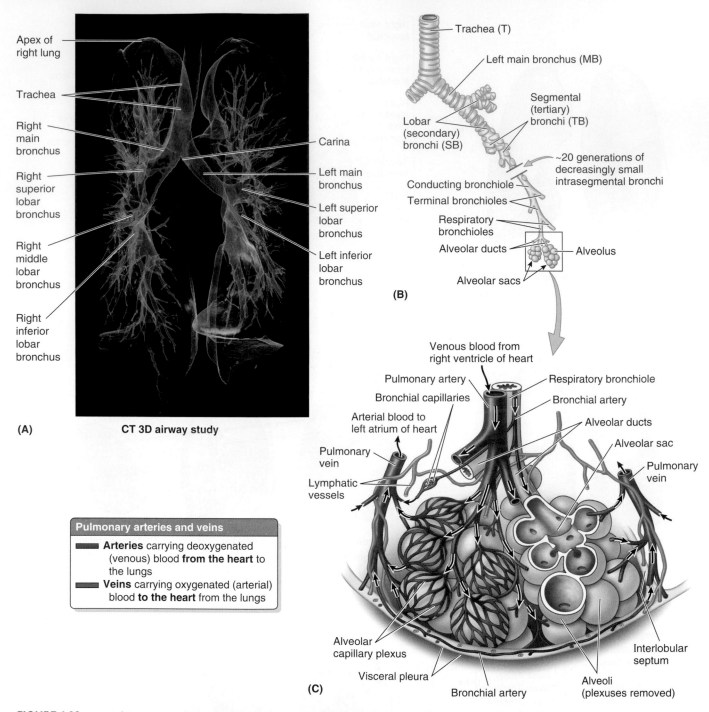

(A)   CT 3D airway study

Apex of right lung

Trachea

Right main bronchus

Right superior lobar bronchus

Right middle lobar bronchus

Right inferior lobar bronchus

Carina

Left main bronchus

Left superior lobar bronchus

Left inferior lobar bronchus

Trachea (T)

Left main bronchus (MB)

Segmental (tertiary) bronchi (TB)

Lobar (secondary) bronchi (SB)

~20 generations of decreasingly small intrasegmental bronchi

Conducting bronchiole

Terminal bronchioles

Respiratory bronchioles

Alveolar ducts

Alveolus

Alveolar sacs

(B)

Venous blood from right ventricle of heart

Pulmonary artery

Bronchial capillaries

Arterial blood to left atrium of heart

Pulmonary vein

Lymphatic vessels

Respiratory bronchiole

Bronchial artery

Alveolar ducts

Alveolar sac

Pulmonary vein

Pulmonary arteries and veins

■ **Arteries** carrying deoxygenated (venous) blood **from the heart** to the lungs

■ **Veins** carrying oxygenated (arterial) blood **to the heart** from the lungs

Alveolar capillary plexus

Visceral pleura

Bronchial artery

Interlobular septum

Alveoli (plexuses removed)

(C)

FIGURE 4.20. Internal structure and organization of lungs. A. 3D computed tomography (CT) airway study. **B.** Subdivisions of bronchial tree. **C.** Alveoli.

The small bronchial arteries provide branches to the superior esophagus and usually then pass along the posterior aspects of the main bronchi, supplying them and their branches as far distally as the respiratory bronchioles. The most distal branches of the bronchial arteries anastomose with branches of the pulmonary arteries in the walls of the bronchioles and in the visceral pleura.

The **bronchial veins** drain only part of the blood supplied to the lungs by the bronchial arteries, primarily, which distributed to or near the more proximal parts of the roots of the lungs (Fig. 4.22B). The remainder of the blood is drained by the pulmonary veins. The right bronchial vein drains into the *azygos vein*, and the left bronchial vein drains into the *accessory hemi-azygos vein* or the left superior intercostal vein.

The **lymphatic plexuses in the lungs** communicate freely (Fig. 4.22C). The **superficial lymphatic plexus** lies deep to the visceral pleura and drains the lung parenchyma (tissue) and visceral pleura. Lymphatic vessels from the plexus drain into the **bronchopulmonary (hilar) lymph nodes** in the hilum of the lung.

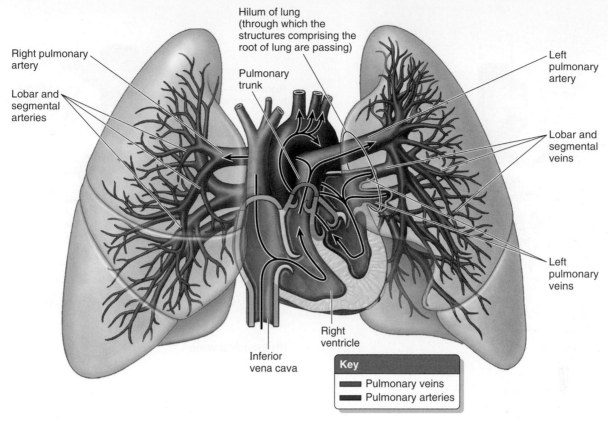

Right pulmonary artery

Lobar and segmental arteries

Hilum of lung (through which the structures comprising the root of lung are passing)

Pulmonary trunk

Left pulmonary artery

Lobar and segmental veins

Left pulmonary veins

Right ventricle

Inferior vena cava

Key
Pulmonary veins
Pulmonary arteries

FIGURE 4.21. Pulmonary circulation. Note that the right pulmonary artery passes under the arch of the aorta to reach the right lung and the left pulmonary artery lies completely to the left of the arch.

The **deep lymphatic plexus** is located in the submucosa of the bronchi and in the peribronchial connective tissue. It is largely concerned with draining structures that form the root of the lung. Lymphatic vessels from this plexus drain into the **pulmonary lymph nodes** located along the lobar bronchi. At the hilum of the lung, they drain into **bronchopulmonary (hilar) lymph nodes** (Fig. 4.22C).

Lymph from the superficial and deep plexuses drains from the bronchopulmonary lymph nodes to the **superior** and **inferior tracheobronchial lymph nodes**, superior and inferior to the bifurcation of the trachea, respectively. Lymph from the tracheobronchial lymph nodes passes to the **right** and **left bronchomediastinal lymph trunks**. These trunks usually terminate on each side at the *venous angles* (junction of the subclavian and internal jugular veins); however, the right bronchomediastinal trunk may first merge with other lymphatic trunks, converging here to form the **right lymphatic duct**. The left bronchomediastinal trunk usually terminates in the *thoracic duct*. The superficial (subpleural) lymphatic plexus drains lymph from the *visceral pleura*. Lymph from the *parietal pleura* drains into the lymph nodes of the thoracic wall (intercostal, parasternal, mediastinal, and phrenic). A few lymphatic vessels from the cervical pleura drain into the axillary lymph nodes.

The **nerves of the lungs and visceral pleura** derive from the pulmonary plexuses located anterior and (mainly) posterior to the roots of the lungs (Fig. 4.22D). These nerve networks contain parasympathetic fibers from the **vagus nerves** (cranial nerve [CN] X) and sympathetic fibers from the sympathetic trunks. *Parasympathetic ganglion cells*—cell bodies of postsynaptic parasympathetic neurons—are in the **pulmonary plexuses** and along the branches of the bronchial tree. The parasympathetic fibers from CN X are motor to the smooth muscle of the bronchial tree (*bronchoconstrictor*), inhibitory to the pulmonary vessels (*vasodilator*), and secretory to the glands of the bronchial tree (*secretomotor*). The visceral afferent fibers of CN X are distributed to the following tissues and structures:

- Bronchial mucosa, where they are probably concerned with tactile sensation for cough reflexes
- Bronchial muscles, possibly involved in stretch reception
- Interalveolar connective tissue, in association with Hering-Breuer reflexes (mechanism that tends to limit respiratory excursions)
- Pulmonary arteries serving pressor receptors (blood pressure) and pulmonary veins serving chemoreceptors (blood gas levels)

Sympathetic ganglion cells—cell bodies of postsynaptic sympathetic neurons—are in the **paravertebral sympathetic ganglia** of the sympathetic trunks. The sympathetic fibers are inhibitory to the bronchial muscle (bronchodilator), motor to the pulmonary vessels (vasoconstrictor), and inhibitory to the alveolar glands of the bronchial tree.

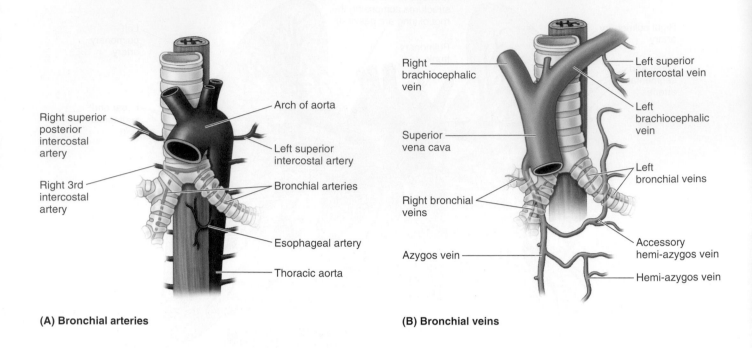

(A) Bronchial arteries

Right superior posterior intercostal artery

Right 3rd intercostal artery

Arch of aorta

Left superior intercostal artery

Bronchial arteries

Esophageal artery

Thoracic aorta

(B) Bronchial veins

Right brachiocephalic vein

Superior vena cava

Right bronchial veins

Azygos vein

Left superior intercostal vein

Left brachiocephalic vein

Left bronchial veins

Accessory hemi-azygos vein

Hemi-azygos vein

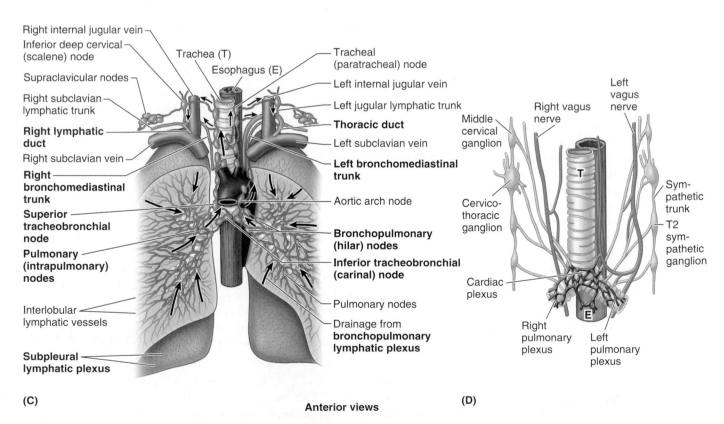

(C)

Right internal jugular vein
Inferior deep cervical (scalene) node
Supraclavicular nodes
Right subclavian lymphatic trunk
Right lymphatic duct
Right subclavian vein
Right bronchomediastinal trunk
Superior tracheobronchial node
Pulmonary (intrapulmonary) nodes
Interlobular lymphatic vessels
Subpleural lymphatic plexus

Trachea (T)
Esophagus (E)

Tracheal (paratracheal) node
Left internal jugular vein
Left jugular lymphatic trunk
Thoracic duct
Left subclavian vein
Left bronchomediastinal trunk
Aortic arch node
Bronchopulmonary (hilar) nodes
Inferior tracheobronchial (carinal) node
Pulmonary nodes
Drainage from **bronchopulmonary lymphatic plexus**

Anterior views

(D)

Middle cervical ganglion
Cervico-thoracic ganglion
Cardiac plexus
Right pulmonary plexus
Right vagus nerve
Left vagus nerve
Sym-pathetic trunk
T2 sym-pathetic ganglion
Left pulmonary plexus

FIGURE 4.22. Vasculature and nerves of lungs and pleurae. A. Bronchial arteries. **B.** Bronchial veins. **C.** Lymphatic drainage. The lymphatic vessels originate from superficial subpleural and deep lymphatic plexuses. *Arrows* indicate the direction of lymph flow. **D.** Innervation. *E,* esophagus; *T,* trachea; *green,* parasympathetic; *purple,* plexus; *yellow,* sympathetic.

CLINICAL BOX

Pulmonary Collapse

 If a sufficient amount of air enters the pleural cavity, the surface tension adhering visceral to parietal pleura (lung to thoracic wall) is broken, and the lung collapses because of its inherent elasticity (elastic recoil). When a lung collapses (*atelectasis*), the pleural cavity—normally a potential space (Fig. B4.7A)—becomes a real space. This reduction in size will be evident radiographically on the affected side by elevation of the diaphragm above its usual levels, intercostal space narrowing (ribs closer together), and displacement of the mediastinum (*mediastinal shift*; most evident via the air-filled trachea within it) toward the affected side. In addition, the collapsed lung will usually appear denser (whiter) and will be surrounded by more radiolucent (blacker) air. One lung may be collapsed after surgery, for example, without collapsing the other because the pleural sacs are separate.

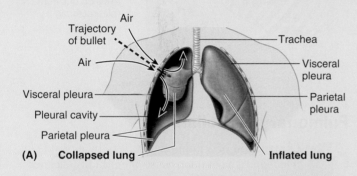

(A) Collapsed lung   Inflated lung

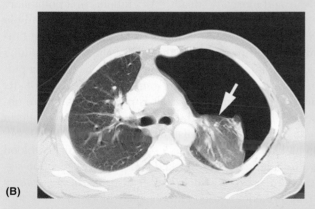

(B)

FIGURE B4.7. A. Pulmonary collapse. **B.** Tension pneumothorax on computed tomography (CT) scan with large left-sided collection of free air and mediastinal shift to the right. Note collapsed lung posteriorly (*arrow*).

Pneumothorax, Hydrothorax, Hemothorax, and Chylothorax

 Entry of air into the pleural cavity—*pneumothorax*—resulting from a penetrating wound of the parietal pleura, tearing of the parietal pleura from a fractured rib, or rupture of a lung from a bullet, for example, results in partial collapse of the lung (Fig. B4.7B). A pneumothorax may also occur as a result of leakage from the lung through an opening in the visceral pleura. The accumulation of a significant amount of fluid in the pleural cavity—*hydrothorax*—may result from *pleural effusion* (escape of fluid into the pleural cavity). With a chest wound, blood may also enter the pleural cavity (*hemothorax*); this condition results more often from injury to a major intercostal vessel than from laceration of a lung. Lymph from a torn thoracic duct may also enter the pleural cavity (*chylothorax*). Chyle is a pale white or yellow lymph fluid in the thoracic duct containing fat absorbed by the intestines (see Chapter 5).

Pleuritis

 During inspiration and expiration, the normally moist, smooth pleurae make no sound detectable by *auscultation* (listening to breath sounds); however, inflammation of the pleurae—*pleuritis* (pleurisy)—makes the lung surfaces rough. The resulting friction (*pleural rub*) may be heard with a stethoscope. Acute pleuritis is marked by sharp, stabbing pain, especially on exertion, such as climbing stairs, when the rate and depth of respiration may be increased even slightly.

Variation in Lobes of Lungs

Occasionally, an extra fissure divides a lung, or a fissure is absent. For example, the left lung sometimes has three lobes and the right lung only two. The most common "accessory" lobe is the *azygos lobe*, which appears in the right lung in approximately 1% of people. In these cases, the azygos vein arches over the apex of the right lung and not over the right hilum, isolating the medial part of the apex as an azygos lobe.

Thoracentesis

 Sometimes, it is necessary to insert a hypodermic needle through an intercostal space into the pleural cavity to obtain a sample of pleural fluid or to remove blood or pus (*thoracentesis*). To avoid damage to the intercostal nerve and vessels, the needle is inserted superior to the rib, high enough to avoid the collateral branches (Fig. B4.8).

Aspiration of Foreign Bodies

Because the right bronchus is wider and shorter and runs more vertically than the left bronchus, *aspirated foreign bodies* are more likely to enter and lodge in it or one of its branches. A potential hazard encountered by dentists is an aspirated foreign body, such as a piece

(Continued on next page)

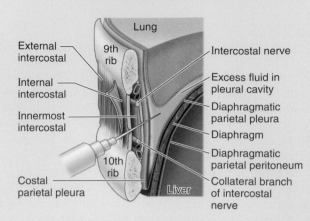

FIGURE B4.8. Technique for midaxillary thoracentesis.

of tooth or filling material. Such objects are also most likely to enter the right main bronchus.

Lung Resections

Knowledge of the anatomy of the bronchopulmonary segments is essential for precise interpretations of diagnostic images of the lungs and for *surgical resection* (removal) of diseased segments. When resecting a bronchopulmonary segment, surgeons follow the interlobar veins to pass between the segments. Bronchial and pulmonary disorders such as tumors or abscesses (collections of pus) often localize in a bronchopulmonary segment, which may be surgically resected. Treatment for lung cancer may include removal of a whole lung (*pneumonectomy*), a lobe (*lobectomy*), or one or more bronchopulmonary segments (*segmentectomy*). Knowledge and understanding of the bronchopulmonary segments and their relationship to the bronchial tree are also essential for planning drainage and clearance techniques used in physical therapy for enhancing drainage from specific areas (e.g., in patients with pneumonia or cystic fibrosis).

Injury to Pleurae

The visceral pleura is insensitive to pain because its innervation is autonomic (motor and visceral afferent). The autonomic nerves reach the visceral pleura in company with the bronchial vessels. The visceral pleura receives no nerves of general sensation.

In contrast, the parietal pleura is sensitive to pain, particularly the costal pleura, because it is richly supplied by branches of the somatic intercostal and phrenic nerves. Irritation of the parietal pleura produces local pain and referred pain to the areas sharing innervation by the same segments of the spinal cord. Irritation of the costal and peripheral parts of the diaphragmatic pleura results in local pain and referred pain along the intercostal nerves to the thoracic and abdominal walls. Irritation of the mediastinal and central diaphragmatic areas of the parietal pleura results in pain

that is referred to the root of the neck and over the shoulder (C3–C5 dermatomes).

Thoracoscopy

Thoracoscopy is a diagnostic and sometimes therapeutic procedure in which the pleural cavity is examined with a thoracoscope (Fig. B4.9). Small incisions are made into the pleural cavity via an intercostal space. In addition to observation, biopsies can be taken and some thoracic conditions can be treated (e.g., disrupting *pleural adhesions* or removing *pleural plaques*, fibrous or calcified thickenings of pleura).

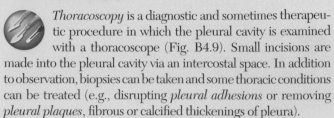
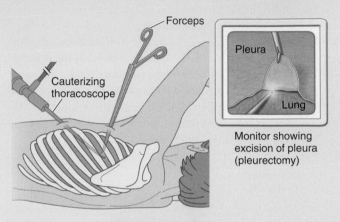

FIGURE B4.9. Pleurectomy.

Pulmonary Embolism

Obstruction of a pulmonary artery by a *blood clot* (*embolus*) is a common cause of morbidity (sickness) and mortality (death). An embolus in a pulmonary artery forms when a blood clot, fat globule, or air bubble travels in the blood to the lungs from a leg vein. The **embolus** passes through the right side of the heart to a lung through a pulmonary artery. The embolus may block a pulmonary artery—*pulmonary embolism*—or one of its branches. The immediate result is partial or complete obstruction of blood flow to the lung. The obstruction results in a sector of lung that is ventilated but not perfused with blood. When a large embolus occludes a pulmonary artery, blood flow through the lung is blocked and blood oxygenation significantly decreases, which may lead to *acute respiratory distress*. A medium-sized embolus may block an artery supplying a bronchopulmonary segment, producing a *pulmonary infarct*, an area of necrotic (dead) lung tissue.

Inhalation of Carbon Particles

Lymph from the lungs carries *phagocytes*, cells possessing the property of ingesting carbon particles from inspired air. In many people, especially cigarette smokers, these particles color the surface of the lungs and associated lymph nodes a mottled gray to black. *Smokers' cough* results from inhalation of irritants in tobacco.

Bronchogenic Carcinoma

Bronchogenic carcinoma is a common type of lung cancer that arises from the epithelium of the bronchial tree. *Lung cancer* is mainly caused by cigarette smoking. Bronchogenic carcinoma usually metastasizes widely because of the arrangement of the lymphatics. The tumor cells probably enter the systemic circulation by invading the wall of a sinusoid or venule in the lung and are transported through the pulmonary veins, left heart, and aorta to all parts of the body, especially the cranium and brain.

Bronchoscopy

When examining the bronchi with a *bronchoscope*—an endoscope for inspecting the interior of the tracheobronchial tree for diagnostic purposes—one can observe a ridge, the **carina**, between the orifices of the main bronchi (Fig. B4.10). The carina is a cartilaginous projection of the last tracheal ring. If the tracheobronchial lymph nodes in the angle between the main bronchi are enlarged because cancer cells have metastasized from a bronchogenic carcinoma, for example, the carina is distorted, widened posteriorly, and immobile.

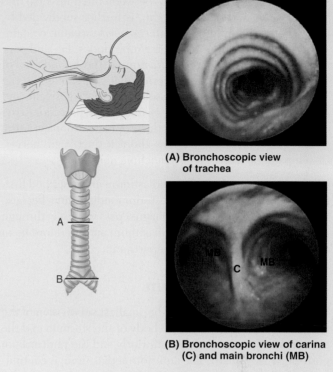

(A) Bronchoscopic view of trachea

(B) Bronchoscopic view of carina (C) and main bronchi (MB)

FIGURE B4.10. Bronchoscopy.

Mediastinum

The **mediastinum**, occupied by the viscera between the pulmonary cavities, is the central compartment of the thoracic cavity (Fig. 4.23). The mediastinum

- is covered on each side by mediastinal pleura and contains all the thoracic viscera and structures, except the lungs
- extends from the superior thoracic aperture to the diaphragm inferiorly and from the sternum and costal cartilages anteriorly to the bodies of the thoracic vertebrae posteriorly
- is a highly mobile region in living persons because it consists primarily of hollow (liquid- or air-filled) visceral structures

The major structures in the mediastinum are also surrounded by blood and lymphatic vessels, lymph nodes, nerves, and fat.

The looseness of the connective tissue and the elasticity of the lungs and parietal pleura on each side of the mediastinum enable it to accommodate movement as well as volume and pressure changes in the thoracic cavity, such as those resulting from movements of the diaphragm, thoracic wall, and tracheobronchial tree during respiration, contraction (beating) of the heart and pulsations of the great arteries, and passage of ingested substances through the esophagus. The connective tissue here becomes more fibrous and rigid with age; hence, the mediastinal structures become less mobile.

The mediastinum is divided into superior and inferior parts for purposes of description:

- The **superior mediastinum** extends between the superior thoracic aperture to the horizontal *transverse thoracic plane* that passes through the sternal angle anteriorly and the IV disc of the T4–T5 vertebrae posteriorly (see Fig. 4.21). The superior mediastinum contains the SVC, brachiocephalic veins, arch of the aorta, thoracic duct, trachea, esophagus, thymus, vagus nerves, left recurrent laryngeal nerve, and phrenic nerves.

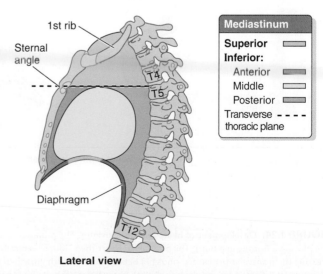

Lateral view

FIGURE 4.23. Subdivisions of mediastinum.

- The **inferior mediastinum**, between the transverse thoracic plane and the diaphragm, is further subdivided by the pericardium into the *anterior mediastinum*, containing remnants of the thymus, lymph nodes, fat, and connective tissue; *middle mediastinum*, the boundaries of which correspond to the pericardial sac, containing the heart, roots of the great vessels, arch of azygos vein, and main bronchi; and *posterior mediastinum*, posterior to the pericardium and containing the esophagus, thoracic aorta, azygos and hemi-azygos veins, thoracic duct, vagus nerves, sympathetic trunks, and splanchnic nerves.

The anterior and middle mediastinum are described first, followed by the superior and posterior mediastinum, because many structures (e.g., the esophagus) pass vertically through the superior and posterior mediastinum and therefore lie in more than one mediastinal compartment.

Anterior Mediastinum

The **anterior mediastinum**, the smallest subdivision of the mediastinum, lies between the body of the sternum and the transversus thoracis muscles anteriorly and the pericardium posteriorly (Fig. 4.23). The anterior mediastinum is continuous with the superior mediastinum at the sternal angle and is limited inferiorly by the diaphragm. The anterior mediastinum consists of *sternopericardial ligaments* (fibrous bands that pass from the pericardium to the sternum), fat, lymphatic vessels, a few lymph nodes, and branches of the internal thoracic vessels. In infants and children, the anterior mediastinum contains the inferior part of the thymus.

Middle Mediastinum

The **middle mediastinum** coincides with the pericardium, containing the heart, ascending aorta, pulmonary trunk, SVC, arch of azygos vein, and main bronchi.

PERICARDIUM

The **pericardium** is a double-walled fibroserous membrane that encloses the heart and the roots of the great vessels, much like the pleura encloses the lungs (Figs. 4.24 and 4.25). A conical **pericardial sac** lies posterior to the body of the sternum and the 2nd–6th costal cartilages at the level of the T5–T8 vertebrae. Its tough external fibrous layer—the **fibrous pericardium**—is continuous with (blends with) the central tendon of the diaphragm (Fig. 4.25A). The internal surface of the fibrous pericardium is lined with a glistening serous membrane, the **parietal layer of serous pericardium**. This layer is reflected onto the heart and great vessels as the **visceral layer of serous pericardium**.

The *pericardial sac* is influenced by movements of the heart and great vessels, sternum, and diaphragm because the fibrous pericardium is

- fused with the tunica adventitia of the great vessels entering and leaving the heart
- attached to the posterior surface of the sternum by sternopericardial ligaments
- fused with the central tendon of the diaphragm

The fibrous pericardium protects the heart against sudden overfilling because it is unyielding and closely related

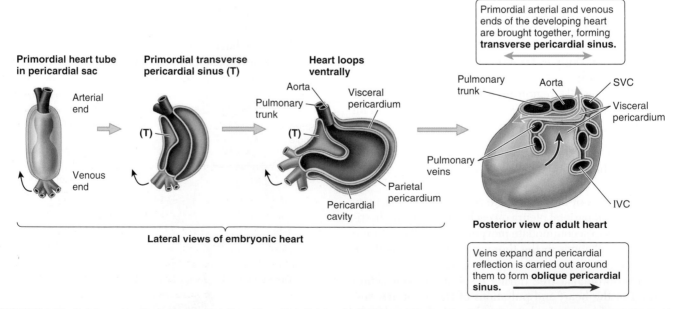

FIGURE 4.24. Development of heart and pericardium. The primordial, longitudinal heart tube invaginates the double-layered pericardial sac (somewhat like placing a hotdog in a bun). The primordial heart then "loops" ventrally, bringing the primordial arterial and venous ends of the heart together and creating the transverse pericardial sinus (*T*) between them. With growth of the embryo, the veins expand and spread apart inferiorly and laterally. The pericardium reflected around them forms the boundaries of the oblique pericardial sinus. *IVC*, inferior vena cava; *SVC*, superior vena cava.

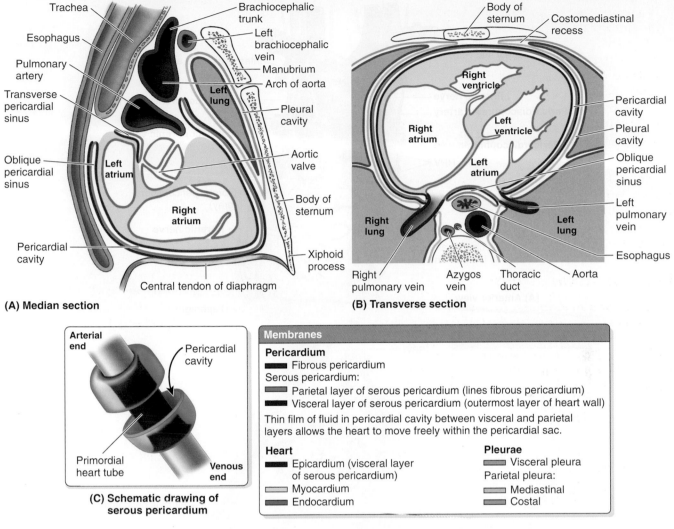

(A) Median section

(B) Transverse section

(C) Schematic drawing of serous pericardium

Membranes

Pericardium
▬ Fibrous pericardium
Serous pericardium:
▬ Parietal layer of serous pericardium (lines fibrous pericardium)
▬ Visceral layer of serous pericardium (outermost layer of heart wall)

Thin film of fluid in pericardial cavity between visceral and parietal layers allows the heart to move freely within the pericardial sac.

Heart
▬ Epicardium (visceral layer of serous pericardium)
▬ Myocardium
▬ Endocardium

Pleurae
▬ Visceral pleura
Parietal pleura:
▬ Mediastinal
▬ Costal

FIGURE 4.25. Layers of pericardium and pericardial cavity.

to the great vessels that pierce it superiorly and posteriorly (Figs. 4.24 and 4.25B). The ascending aorta carries the pericardium superiorly beyond the heart to the level of the sternal angle.

The **pericardial cavity** is the potential space between the opposing layers of the parietal and visceral layers of serous pericardium (Fig. 4.25C). It normally contains a thin film of serous fluid that enables the heart to move and beat in a frictionless environment.

The visceral layer of serous pericardium comprises the *epicardium*, the external layer of the heart wall, and reflects from the heart and great vessels to become continuous with the parietal layer of serous pericardium, where

- the aorta and pulmonary trunk leave the heart; a finger can be inserted through the **transverse pericardial sinus** located posterior to these large vessels and anterior to the SVC (Figs. 4.24, 4.25A, and B4.11).
- the SVC, IVC, and pulmonary veins enter the heart; these vessels are partly covered by serous pericardium, which

forms the **oblique pericardial sinus** (Figs. 4.24 and 4.26), a wide recess posterior to the heart. The oblique sinus can be entered inferiorly and will admit several fingers; however, the fingers cannot pass around any of these vessels because the sinus is a blind recess (cul-de-sac).

These pericardial sinuses form during development of the heart as a consequence of folding of the primordial heart tube (Fig. 4.24). As the heart tube folds, its venous end moves posterosuperiorly so that the venous end of the tube lies adjacent to the arterial end, separated by the transverse pericardial sinus. As these vessels expand and move apart, the pericardium is reflected around them to form the boundaries of the oblique pericardial sinus.

The **arterial supply of the pericardium** is mainly from the **pericardiacophrenic artery** (Fig. 4.26A), a branch of the *internal thoracic artery*, which may accompany or parallel the phrenic nerve to the diaphragm. Smaller contributions of blood to the pericardium come from the *musculophrenic artery*, a terminal branch of the internal thoracic artery; the

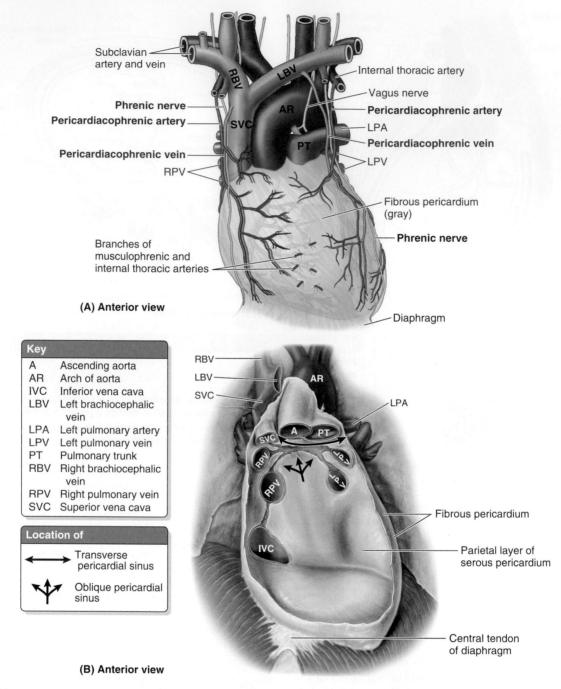

(A) Anterior view

Key

A	Ascending aorta
AR	Arch of aorta
IVC	Inferior vena cava
LBV	Left brachiocephalic vein
LPA	Left pulmonary artery
LPV	Left pulmonary vein
PT	Pulmonary trunk
RBV	Right brachiocephalic vein
RPV	Right pulmonary vein
SVC	Superior vena cava

Location of

⟷ Transverse pericardial sinus

Oblique pericardial sinus

(B) Anterior view

FIGURE 4.26. Pericardium. A. Arterial supply and venous drainage. **B.** Interior of pericardial sac, after removal of the heart, showing the location of the transverse and oblique pericardial sinuses.

bronchial, esophageal, and *superior phrenic arteries* from the thoracic aorta; and the *coronary arteries,* supplying only the visceral layer of serous pericardium (see Fig. 4.15A).

The **venous drainage of the pericardium** is via the (see Fig. 4.15B)

- *pericardiacophrenic veins,* tributaries of the brachiocephalic (or internal thoracic) veins
- variable tributaries of the azygos venous system

The **nerve supply of the pericardium** is from the (Figs. 4.22D and 4.26A)

- *phrenic nerves* (C3–C5)—a primary source of sensory fibers; pain sensations conveyed by these nerves are commonly referred to the skin (C3–C5 dermatomes) of the ipsilateral shoulder region.
- *vagus nerves* (CN X)—function uncertain
- *sympathetic trunks*—vasomotor

CLINICAL BOX

Surgical Significance of Transverse Pericardial Sinus

The transverse pericardial sinus is especially important to cardiac surgeons. After the pericardial sac has been opened anteriorly, a finger can be passed through the transverse pericardial sinus posterior to the aorta and pulmonary trunk (Fig. B4.11). By passing a surgical clamp or placing a ligature around these vessels, inserting the tubes of a bypass machine, and then tightening the ligature, surgeons can stop or divert the circulation of blood in these large arteries while performing cardiac surgery, such as coronary artery bypass grafting. Cardiac surgery is performed while the patient is on a cardiopulmonary bypass machine.

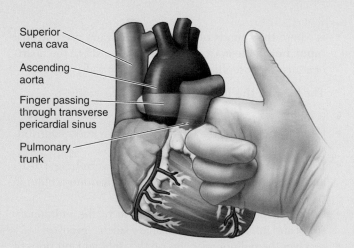

Superior
vena cava

Ascending
aorta

Finger passing
through transverse
pericardial sinus

Pulmonary
trunk

FIGURE B4.11. Transverse pericardial sinus.

Pericarditis and Pericardial Effusion

Inflammation of the pericardium (*pericarditis*) usually causes chest pain. Normally, the layers of serous pericardium make no detectable sound during auscultation. However, pericarditis makes the surfaces rough and the resulting friction, *pericardial friction rub*, sounds like the rustle of silk when listening with a stethoscope. Certain inflammatory diseases may also produce *pericardial*

effusion (passage of fluid from the pericardial capillaries into the pericardial cavity). As a result, the heart becomes compressed (unable to expand and fill fully, *cardiac tamponade*) and is ineffectual.

Cardiac Tamponade

Cardiac tamponade (heart compression) is a potentially lethal condition because the fibrous pericardium is tough and inelastic. Consequently, heart volume is increasingly compromised by the fluid outside the heart but inside the pericardial cavity. When there is a slow increase in the size of the heart, *cardiomegaly*, the pericardium gradually enlarges, allowing the enlargement of the heart to occur without compression. Stab wounds that pierce the heart, causing blood to suddenly enter the pericardial cavity (*hemopericardium*), also produce cardiac tamponade. Hemopericardium may also result from perforation of a weakened area of heart muscle after a heart attack. As blood accumulates, the heart is compressed, and circulation fails.

Pericardiocentesis (drainage of serous fluid from pericardial cavity) is usually necessary to relieve the cardiac tamponade. To remove the excess fluid, a wide-bore needle may be inserted through the left subcostal angle, or 5th or 6th intercostal space near the sternum.

Levels of Viscera in Mediastinum

The level of the viscera relative to the mediastinal subdivisions depends on the position of the person. When a person is lying supine, the level of the viscera relative to the subdivisions of the mediastinum is as shown in Figure B4.12A. Anatomical descriptions traditionally describe the level of the viscera as if the person were supine. However, in the standing position, the levels of the viscera are as shown in Figure B4.12B. This occurs because the soft structures in the mediastinum, the heart and great vessels, and the abdominal viscera supporting them sag inferiorly under the influence of gravity. This movement of mediastinal structures must be considered during physical and radiological examinations.

(Continued on next page)

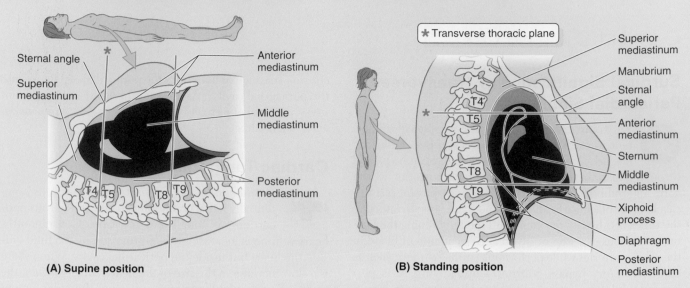

FIGURE B4.12. Position of thoracic viscera in supine and standing positions.

Heart and Great Vessels

The heart, slightly larger than a clenched fist, is a double self-adjusting muscular pump, the parts of which work in unison to propel blood to the body. The right side of the heart receives poorly oxygenated blood from the body through the SVC and IVC and pumps it through the pulmonary trunk to the lungs for oxygenation. The left side of the heart receives well-oxygenated blood from the lungs through the pulmonary veins and pumps it into the aorta for distribution to the body.

The wall of the heart consists of three layers described as follows from superficial to deep (see Fig. 4.24):

- **Epicardium**, a thin external layer (mesothelium) formed by the visceral layer of serous pericardium
- **Myocardium**, a thick middle layer composed of cardiac muscle
- **Endocardium**, a thin internal layer (endothelium and subendothelial connective tissue) or lining membrane of the heart that also covers its valves

ORIENTATION OF HEART

The heart and roots of the great vessels within the pericardial sac are related anteriorly to the sternum, costal cartilages, and the medial ends of the 3rd–5th ribs on the left side. The heart and pericardial sac are situated obliquely, lying about two thirds to the left and one third to the right of the median plane. The heart is shaped like a tipped-over, three-sided pyramid with an apex, base, and four surfaces.

The **apex of the heart** (Figs. 4.27A and 4.28A)

- is directed anteriorly and to the left and is formed by the inferolateral part of the left ventricle
- is located posterior to the left 5th intercostal space in adults, usually 9 cm from the median plane

- is where the sounds of mitral valve closure are maximal (**apex beat**); the apex underlies the site where the heartbeat may be auscultated on the thoracic wall.

The **base of the heart** (Fig. 4.28B)

- is the heart's *posterior aspect*
- is formed mainly by the left atrium, with a lesser contribution by the right atrium
- faces posteriorly toward the bodies of vertebrae T6–T9 and is separated from them by the pericardium, oblique pericardial sinus, esophagus, and aorta
- extends superiorly to the bifurcation of the pulmonary trunk and inferiorly to the coronary sulcus (groove)
- receives the pulmonary veins on the right and left sides of the left atrium and the superior and inferior venae cavae at the superior and inferior ends of the right atrium

The *four surfaces of the heart* are the (Fig. 4.28A,B)

- **anterior (sternocostal) surface**, formed mainly by the right ventricle
- **diaphragmatic (inferior) surface**, formed mainly by the left ventricle and partly by the right ventricle; it is related to the central tendon of the diaphragm.
- **left pulmonary surface**, consists mainly of the left ventricle; it forms the cardiac impression of the left lung.
- **right pulmonary surface**, formed mainly by the right atrium

The heart appears trapezoidal in both anterior and posterior views. The *four borders of the heart* are the (Fig. 4.27)

- **right border** (slightly convex), formed by the right atrium and extending between the SVC and the IVC
- **inferior border** (nearly horizontal), formed mainly by the right ventricle and only slightly by the left ventricle
- **left border** (oblique), formed mainly by the left ventricle and slightly by the left auricle

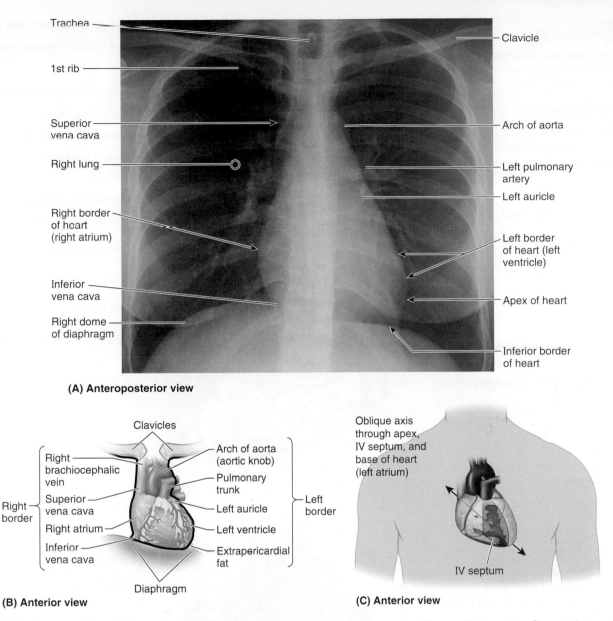

FIGURE 4.27. **Placement of heart in thorax. A.** Radiograph. **B.** Structures forming the margins of the cardiac silhouette. **C.** Orientation of heart. *IV*, interventricular.

- **superior border**, formed by the right and left atria and auricles in an anterior view; the ascending aorta and pulmonary trunk emerge from the superior border, and the SVC enters its right side. Posterior to the aorta and pulmonary trunk and anterior to the SVC, the superior border forms the inferior boundary of the transverse pericardial sinus.

CHAMBERS OF HEART

The heart has four chambers: *right* and *left atria* and *right* and *left ventricles*.

RIGHT ATRIUM

The **right atrium** forms the right border of the heart and receives venous blood from the SVC, IVC, and coronary sinus (Fig. 4.28). The ear-like **right auricle** is a small, conical muscular pouch that projects from the right atrium, increasing the capacity of the atrium as it overlaps the ascending aorta.

The primordial atrium is represented in the adult by the right auricle. The definitive atrium is enlarged by incorporation of most of the embryonic venous sinus (L. *sinus venosus*). The *coronary sinus* lies in the posterior part of the coronary sulcus and receives blood from the cardiac veins. The coronary sinus is also a derivative of the embryonic venous sinus. The part of the venous sinus incorporated into the primordial atrium becomes the smooth-walled **sinus venarum** of the adult right atrium. The separation between the primordial atrium and the sinus venarum is indicated externally by the **sulcus terminalis** (terminal groove) and internally by the **crista terminalis** (terminal crest). The interior of the right atrium has (Figs. 4.29 and 4.30)

- a smooth, thin-walled posterior part (the sinus venarum) on which the SVC, IVC, and coronary sinus open, bringing poorly oxygenated blood into the heart
- a rough, muscular wall composed of **pectinate muscles** (L. *musculi pectinati*)

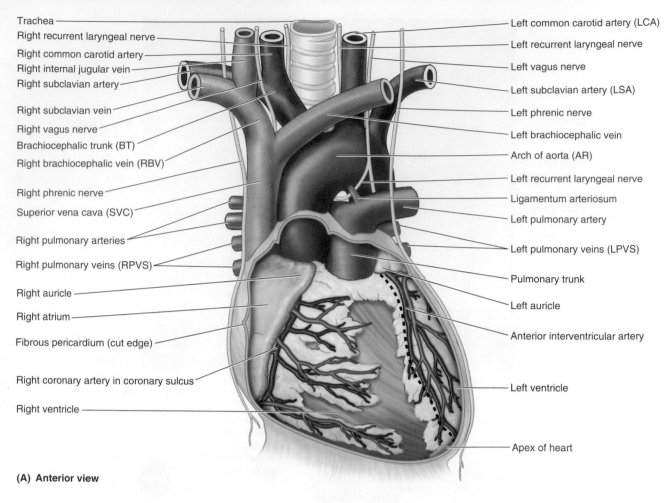

Trachea
Right recurrent laryngeal nerve
Right common carotid artery
Right internal jugular vein
Right subclavian artery
Right subclavian vein
Right vagus nerve
Brachiocephalic trunk (BT)
Right brachiocephalic vein (RBV)
Right phrenic nerve
Superior vena cava (SVC)
Right pulmonary arteries
Right pulmonary veins (RPVS)
Right auricle
Right atrium
Fibrous pericardium (cut edge)
Right coronary artery in coronary sulcus
Right ventricle

Left common carotid artery (LCA)
Left recurrent laryngeal nerve
Left vagus nerve
Left subclavian artery (LSA)
Left phrenic nerve
Left brachiocephalic vein
Arch of aorta (AR)
Left recurrent laryngeal nerve
Ligamentum arteriosum
Left pulmonary artery
Left pulmonary veins (LPVS)
Pulmonary trunk
Left auricle
Anterior interventricular artery
Left ventricle
Apex of heart

(A) Anterior view

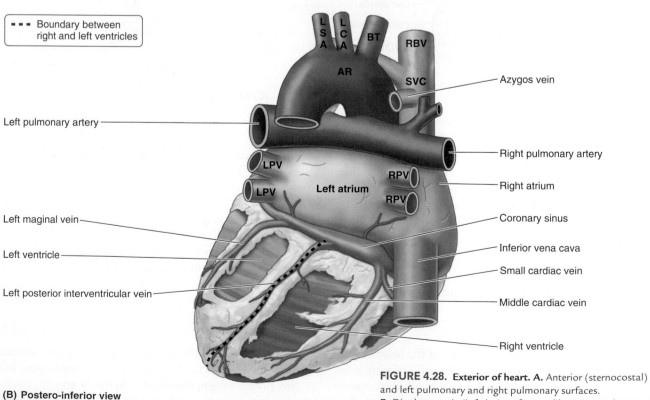

- - - Boundary between
right and left ventricles

LSA LCA BT RBV
AR SVC

Left pulmonary artery

Azygos vein
Right pulmonary artery

LPV
LPV Left atrium RPV RPV

Right atrium

Left maginal vein
Left ventricle
Left posterior interventricular vein

Coronary sinus
Inferior vena cava
Small cardiac vein
Middle cardiac vein
Right ventricle

FIGURE 4.28. Exterior of heart. A. Anterior (sternocostal)
and left pulmonary and right pulmonary surfaces.
B. Diaphragmatic (inferior) surface and base (posterior aspect).

(B) Postero-inferior view

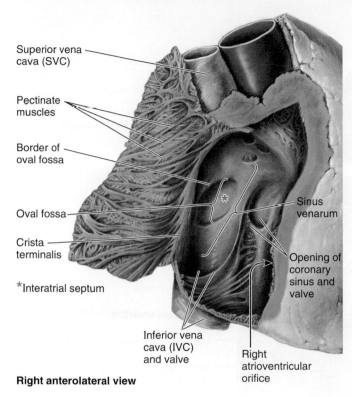

Superior vena cava (SVC)

Pectinate muscles

Border of oval fossa

Oval fossa

Crista terminalis

*Interatrial septum

Sinus venarum

Opening of coronary sinus and valve

Inferior vena cava (IVC) and valve

Right atrioventricular orifice

Right anterolateral view

FIGURE 4.29. Interior of right atrium.

- the **opening of the SVC** into its superior part, at the level of the right 3rd costal cartilage
- the **opening of the IVC** into the inferior part, almost in line with the SVC at approximately the level of the 5th costal cartilage
- the **opening of the coronary sinus** between the right atrioventricular (AV) orifice and the IVC orifice
- a **right AV orifice** through which the right atrium discharges the poorly oxygenated blood into the right ventricle during ventricular relaxation (diastole)
- the **interatrial septum**, separating the atria, has an oval, thumbprint-sized depression, the **oval fossa** (L. *fossa ovalis*), a remnant of the oval foramen and its valve in the fetus

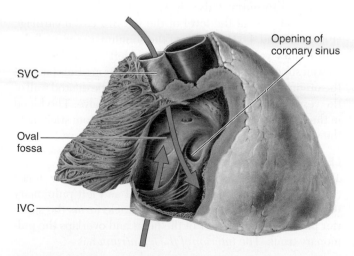

Opening of coronary sinus

SVC

Oval fossa

IVC

FIGURE 4.30. Direction of blood flow in right atrium. *IVC*, inferior vena cava; *SVC*, superior vena cava.

RIGHT VENTRICLE

The right ventricle forms the largest part of the anterior surface of the heart, a small part of the diaphragmatic surface, and almost the entire inferior border of the heart. Superiorly, it tapers into an arterial cone, the **conus arteriosus** (infundibulum), which leads into the pulmonary trunk (Fig. 4.31). The interior of the right ventricle has irregular muscular elevations called **trabeculae carneae**. A thick muscular ridge, the **supraventricular crest**, separates the ridged muscular wall of the inflow part of the chamber from the smooth wall of the conus arteriosus or outflow part of the right ventricle. The inflow part of the right ventricle receives blood from the right atrium through the **right AV (tricuspid) orifice** located posterior to the body of the sternum at the level of the 4th and 5th intercostal spaces (see Fig. SA4.6). The right AV orifice is surrounded by a fibrous ring (part of the fibrous skeleton of heart) that resists the dilation that might otherwise result from blood being forced through it at varying pressures.

The **tricuspid valve** guards the right AV orifice (Figs. 4.31 and 4.32A). The bases of the valve cusps are attached to the fibrous ring around the orifice. **Tendinous cords** (L. *chordae tendineae*) attach to the free edges and ventricular surfaces of the anterior, posterior, and septal cusps—much like the cords attached to a parachute. Because the cords are attached to adjacent sides of two cusps, they prevent separation of the cusps and their inversion when tension is applied to the cords throughout ventricular contraction (*systole*)—that is, the cusps of the tricuspid valve are prevented from prolapsing (being driven into right atrium) as ventricular pressure rises. Thus, regurgitation of blood (backward flow of blood) from the right ventricle into the right atrium is blocked by the valve cusps. The **papillary muscles** form conical projections with their bases attached to the ventricular wall and tendinous cords arising from their apices. There are usually three papillary muscles (anterior, posterior, and septal) in the right ventricle that correspond in name to the cusps of the tricuspid valve. The papillary muscles begin to contract before contraction of the right ventricle, tightening the tendinous cords and drawing the cusps together.

The **interventricular (IV) septum**, composed of membranous and muscular parts, is a strong, obliquely placed partition between the right and the left ventricles (Fig. 4.31), forming part of the walls of each. The superoposterior *membranous part of the IV septum* is thin and is continuous with the fibrous skeleton of the heart. The *muscular part of the IV septum* is thick and bulges into the cavity of the right ventricle because of the higher blood pressure in the left ventricle. The **septomarginal trabecula** (moderator band) is a curved muscular bundle that runs from the inferior part of the IV septum to the base of the anterior papillary muscle. This trabecula is important because it carries part of the *right bundle branches of the AV bundle* of the conducting system of the heart to the anterior papillary muscle (discussed later in this chapter). This "shortcut" across the chamber of the ventricle seems to facilitate conduction time, allowing coordinated contraction of the anterior papillary muscle.

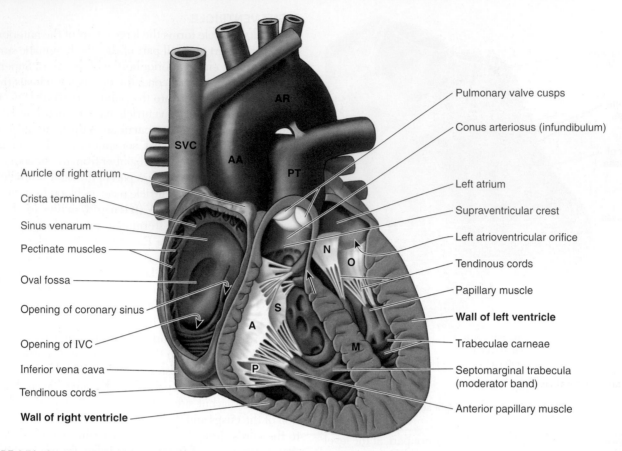

Pulmonary valve cusps

Conus arteriosus (infundibulum)

Left atrium

Supraventricular crest

Left atrioventricular orifice

Tendinous cords

Papillary muscle

Wall of left ventricle

Trabeculae carneae

Septomarginal trabecula (moderator band)

Anterior papillary muscle

Auricle of right atrium

Crista terminalis

Sinus venarum

Pectinate muscles

Oval fossa

Opening of coronary sinus

Opening of IVC

Inferior vena cava

Tendinous cords

Wall of right ventricle

FIGURE 4.31. Interior of heart. Observe the features of each chamber. Note the three cusps of the tricuspid valve—*A*, anterior; *P*, posterior; *S*, septal— and the two cusps of the mitral valve—*N*, anterior; *O*, posterior. *AA*, ascending aorta; *AR*, arch of aorta; *M*, muscular part of interventricular septum; *PT*, pulmonary trunk; *SVC*, superior vena cava; *arrow*, membranous part of interventricular septum.

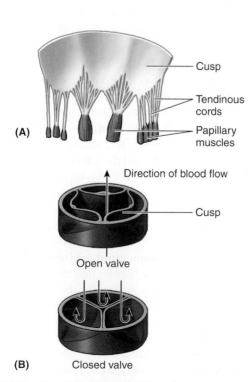

Cusp

Tendinous cords

Papillary muscles

(A)

Direction of blood flow

Cusp

Open valve

Closed valve

(B)

FIGURE 4.32. Tricuspid and pulmonary valves. A. Tricuspid valve spread out. **B.** Blood flow through pulmonary valve.

When the right atrium contracts, blood is forced through the *right AV orifice* into the right ventricle, pushing the cusps of the tricuspid valve aside like curtains. The inflow of blood into the right ventricle (*inflow tract*) enters posteriorly, and the outflow of blood into the pulmonary trunk (*outflow tract*) leaves superiorly and to the left. Consequently, the blood takes a U-shaped path through the right ventricle. The inflow (AV) orifice and outflow (pulmonary) orifice are approximately 2 cm apart.

The **pulmonary valve** is located at the apex of the *conus arteriosus* at the level of the left 3rd costal cartilage (Figs. 4.28 and 4.32B). Each of the semilunar **cusps of the pulmonary valve** (anterior, right, and left) is concave when viewed superiorly. The **pulmonary sinuses** are the spaces at the origin of the pulmonary trunk between the dilated wall of the vessel and each cusp of the pulmonary valve. The blood in the pulmonary sinuses prevents the cusps from sticking to the wall of the pulmonary trunk and failing to close.

LEFT ATRIUM

The **left atrium** forms most of the base of the heart (Fig. 4.33). The pairs of valveless right and left pulmonary veins enter the left atrium. The left auricle forms the superior part of the left border of the heart and overlaps the pulmonary trunk. The *interior of the left atrium* has

- a large smooth-walled part and a small muscular part, the left auricle, that has pectinate muscles in its walls

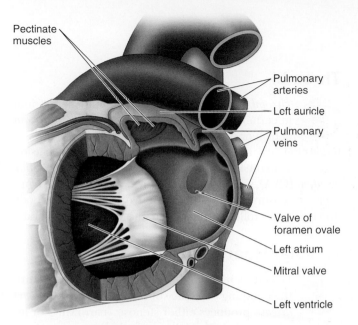

FIGURE 4.33. Interior of left atrium.

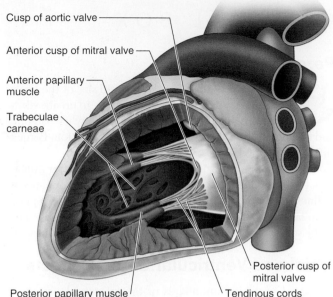

FIGURE 4.34. Interior of left ventricle.

- four pulmonary veins (usually right and left superior and inferior) entering its posterior wall
- a slightly thicker wall than that of the right atrium
- an interatrial septum that slants posteriorly and to the right
- a left AV orifice through which the left atrium discharges the oxygenated blood it receives from the pulmonary veins into the left ventricle during ventricular diastole

The smooth-walled part of the left atrium is formed by absorption of parts of the embryonic pulmonary veins, whereas the rough-walled part, mainly in the auricle, represents the remains of the left part of the primordial atrium.

LEFT VENTRICLE

The **left ventricle** forms the apex of the heart, nearly all of its left (pulmonary) surface and border, and most of the diaphragmatic surface (Figs. 4.31 and 4.34). Because arterial pressure is much higher in the systemic than in the pulmonary circulation, the left ventricle performs more work than the right ventricle.

The *interior of the left ventricle* has (Fig. 4.34):

- a double-leaflet *mitral valve* at the left AV orifice
- walls that are two to three times as thick as that of the right ventricle
- a conical cavity that is longer than that of the right ventricle
- walls that are covered with thick muscular ridges, trabeculae carneae, that are finer but more numerous than those in the right ventricle
- anterior and posterior papillary muscles that are larger than those in the right ventricle
- a smooth-walled, nonmuscular, supero-anterior outflow part, the **aortic vestibule**, leading to the aortic orifice and aortic valve
- an **aortic orifice** that lies in its right posterosuperior part and is surrounded by a fibrous ring to which the right, posterior, and left cusps of the **aortic valve** are attached

The **mitral valve** closing the orifice between the left atrium and left ventricle has two cusps, anterior and posterior (Figs. 4.34 and 4.35). The mitral valve is located posterior to the sternum at the level of the 4th costal cartilage. Each of its cusps receives tendinous cords from more than one papillary muscle. These muscles and their cords support the mitral valve, allowing the cusps to resist the pressure developed during contractions (pumping) of the left ventricle. The tendinous cords become taut, just before and during systole, preventing the cusps from being forced into the left atrium. The **ascending aorta** begins at the aortic orifice.

The aortic valve, obliquely placed, is located posterior to the left side of the sternum at the level of the 3rd intercostal space (see Fig. SA4.6). The **aortic sinuses** are the spaces at the origin of the ascending aorta between the dilated wall of the vessel and each cusp of the aortic (semilunar) valve (see Fig. 4.36). The opening of the right coronary artery is in the *right aortic sinus*; the opening of the left coronary artery is in the *left aortic sinus*; and no artery arises from the *posterior aortic (noncoronary) sinus*.

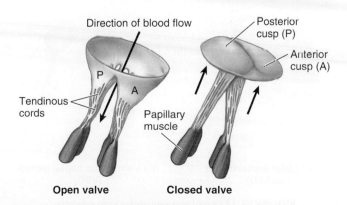

FIGURE 4.35. Mitral valve.

CLINICAL BOX

Percussion of Heart

Percussion defines the density and size of the heart. The classic percussion technique is to create vibration by tapping the chest with a finger while listening and feeling for differences in sound wave conduction. Percussion is performed at the 3rd, 4th, and 5th intercostal spaces from the left anterior axillary line to the right anterior axillary line. Normally, the percussion note changes from resonance to dullness (because of the presence of the heart) approximately 6 cm lateral to the left border of the sternum. The character of the sound changes as different areas of the chest are tapped.

Atrial and Ventricular Septal Defects

Congenital anomalies of the interatrial septum—usually related to incomplete closure of the oval foramen—are *atrial septal defects* or ASDs (Fig. B4.13A). A probe-size patency (defect) appears in the superior part of the oval fossa in 15% to 25% of people. These small ASDs, by themselves, are usually of no clinical significance; however, large ASDs allow oxygenated blood from the lungs to be shunted from the left atrium through the defect into the right atrium, causing enlargement of the right atrium and ventricle and dilation of the pulmonary trunk.

The membranous part of the IV septum develops separately from the muscular part and has a complex embryological origin. Consequently, this part is the common site of *ventricular septal defects* or VSDs (Fig. B4.13B). These congenital anomalies rank first on all lists of cardiac defects. Isolated VSDs account for approximately 25% of all forms of congenital heart disease (Moore et al., 2016). The size of the defect varies from 1–25 mm. A VSD causes a left-to-right shunt of blood through the defect. A large shunt increases pulmonary blood flow, which causes pulmonary disease (*hypertension*, or increased blood pressure) and may cause cardiac failure.

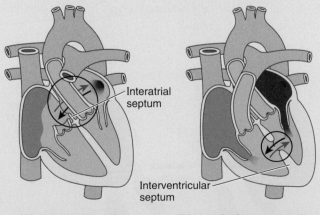

(A) Atrial septal defect (ASD)   Interatrial septum

(B) Ventricular septal defect (VSD)   Interventricular septum

FIGURE B4.13. Atrial and ventricular septal defects.

Thrombi

Thrombi (clots) form on the walls of the left atrium in certain types of heart disease. If these thrombi detach or if pieces break off, they pass into the systemic circulation and occlude peripheral arteries. Occlusion of an artery in the brain results in a stroke or *cerebrovascular accident* (CVA), which may affect, for example, vision, cognition, or sensory or motor function of parts of the body previously controlled by the now-damaged area of the brain.

Valvular Heart Disease

Disorders involving the valves of the heart disturb the pumping efficiency of the heart. *Valvular heart disease* produces either stenosis (narrowing) or insufficiency. *Stenosis* is the failure of a valve to open fully, slowing blood flow from a chamber. *Valvular insufficiency* or *regurgitation*, on the other hand, is failure of the valve to close completely, usually owing to nodule formation on (or scarring and contraction of) the cusps so that the edges do not meet or align. This allows a variable amount of blood (depending on the severity) to flow back into the chamber it was just ejected from. Both stenosis and insufficiency result in an increased workload for the heart. Restriction of high-pressure blood flow (stenosis) and passage of blood through a narrow opening into a larger vessel or chamber (stenosis and regurgitation) produce turbulence. Turbulence sets up eddies (small whirlpools) that produce vibrations that are audible as *murmurs*. Superficial vibratory sensations—*thrills*—may be felt on the skin over an area of turbulence.

Because valvular diseases are mechanical problems, damaged or defective cardiac valves are often replaced surgically in a procedure called *valvuloplasty*. Most commonly, artificial valve prostheses made of synthetic materials are used in these valve replacement procedures, but xenografted valves (valves transplanted from other species, such as pigs) are also used.

A *prolapsed mitral valve* is an insufficient or incompetent valve in which one or both leaflets are enlarged, redundant or "floppy," and extending back into the left atrium during systole. As a result, blood regurgitates into the left atrium when the left ventricle contracts, producing a characteristic murmur.

Aortic valve stenosis is the most frequent valve abnormality and results in *left ventricular hypertrophy*. The great majority of cases of aortic stenosis result from degenerative calcification.

In *pulmonary valve stenosis* (narrowing), the valve cusps are fused, forming a dome with a narrow central opening. In *infundibular pulmonary stenosis*, the conus arteriosus is underdeveloped, producing a restriction of right ventricular outflow. The degree of hypertrophy of the right ventricle is variable.

SURFACE ANATOMY

Heart

The heart and great vessels are approximately in the middle of the thorax, surrounded laterally and posteriorly by the lungs and bounded anteriorly by the sternum and the central part of the thoracic cage (Fig. SA4.5). The *outline of the heart* can be traced on the anterior surface of the thorax by using these guidelines:

- The *superior border* corresponds to a line connecting the inferior border of the 2nd left costal cartilage to the superior border of the 3rd right costal cartilage.
- The *right border* corresponds to a line drawn from the 3rd right costal cartilage to the 6th right costal cartilage; this border is slightly convex to the right.
- The *inferior border* corresponds to a line drawn from the inferior end of the right border to a point in the 5th intercostal space close to the left midclavicular line; the left end of this line corresponds to the location of the apex of the heart and the apex beat.
- The *left border* corresponds to a line connecting the left ends of the lines representing the superior and inferior borders.

- The valves are located posterior to the sternum; however, the sounds produced by them are projected to the **auscultatory areas**: pulmonary, aortic, mitral, and tricuspid (Figs. SA4.6 and 4.7).

The *apex beat* is an impulse that results from the apex being forced against the anterior thoracic wall when the left ventricle contracts. The *location of the apex beat* (mitral area) varies in position; it may be located in the 4th or 5th intercostal spaces, 6–10 cm from the midline of the thorax.

Clinicians' interest in the surface anatomy of the heart and cardiac valves results from their need to listen to individual valve sounds. Blood tends to carry the sound in the direction of its flow. Each area, although overlapping, is situated superficial to the chamber or vessel into which the blood has passed and in a direct line with the valve orifice (Figs. SA4.6 and 4.7). Sounds originating from the valves are best heard at the following locations:

- Aortic valve (A): 2nd right intercostal space to apex of heart
- Pulmonary valve (P): 2nd left intercostal space to left of sternal border
- Tricuspid valve (T): near left sternal border
- Mitral valve (M): apex of heart in 5th intercostal space around midclavicular line

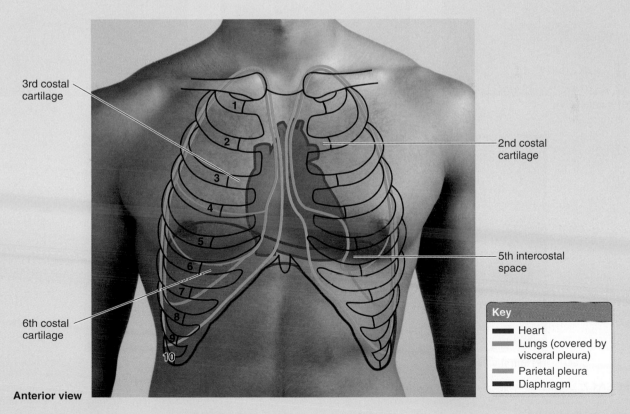

3rd costal cartilage

2nd costal cartilage

5th intercostal space

6th costal cartilage

Key
- Heart
- Lungs (covered by visceral pleura)
- Parietal pleura
- Diaphragm

Anterior view

FIGURE SA4.6. Surface anatomy of the lungs and heart.

(Continued on next page)

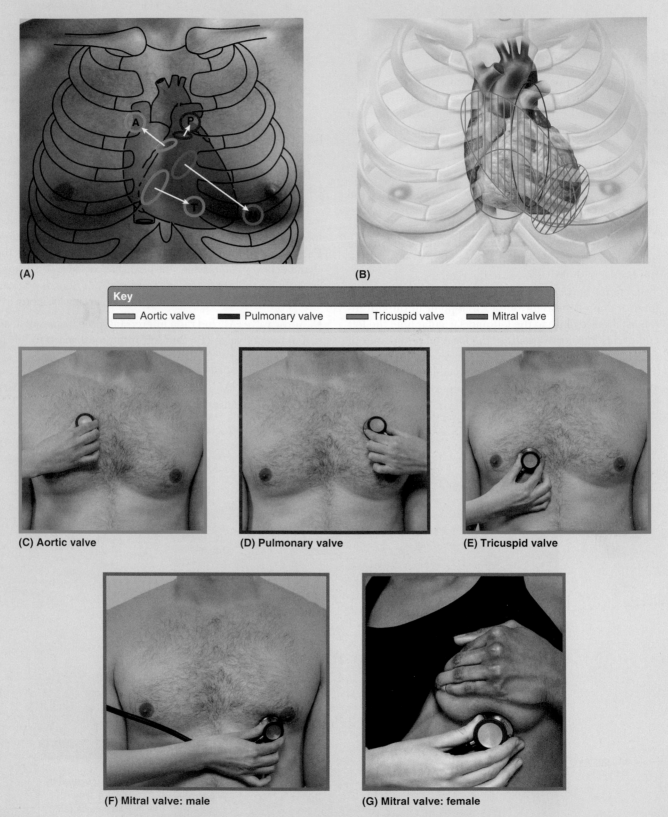

(A)

(B)

Key			
▭ Aortic valve	▬ Pulmonary valve	▬ Tricuspid valve	▬ Mitral valve

(C) Aortic valve

(D) Pulmonary valve

(E) Tricuspid valve

(F) Mitral valve: male

(G) Mitral valve: female

FIGURE SA4.7. Location of valves and areas of auscultation. The location of each valve is indicated by a *colored oval* and the area of auscultation as a *circle* of the same color. Tricuspid valve (*T*) is *green*, mitral valve (*M*) is *purple*, pulmonary valve (*P*) is *pink*, and aortic valve (*A*) is *blue*. The direction of blood flow is indicated by *white arrows*.

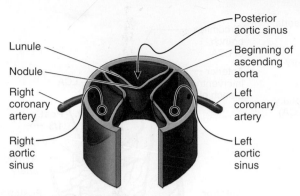

Lunule

Nodule

Right coronary artery

Right aortic sinus

Posterior aortic sinus

Beginning of ascending aorta

Left coronary artery

Left aortic sinus

(A) Anterior view of aortic valve

Backflow of blood due to recoil of elastic aorta (closes valve and causes filling of coronary arteries when myocardium is relaxed)

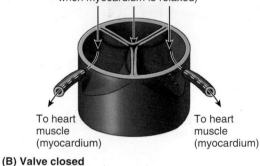

To heart muscle (myocardium)

To heart muscle (myocardium)

(B) Valve closed

FIGURE 4.36. Aortic valve.

ARTERIAL SUPPLY OF HEART

The **coronary arteries** supply the myocardium and epicardium and course just deep to the epicardium, normally embedded in subepicardial fat. The *right* and *left* coronary arteries arise from the corresponding aortic sinuses at the proximal part of the ascending aorta (Figs. 4.36 and 4.37; Table 4.4), just superior to the aortic valve. The endocardium receives oxygen and nutrients directly from the chambers of the heart.

The **right coronary artery (RCA)** arises from the *right aortic sinus* and runs in the coronary sulcus. Near its origin, the RCA usually gives off an ascending **sinu-atrial (SA) nodal branch** (Fig. 4.37A) that supplies the *SA node* (part of the cardiac conducting system). The RCA then descends in the coronary sulcus and gives off the **right marginal branch**, which supplies the right border of the heart as it runs toward (but does not reach) the apex of the heart. After giving off this branch, the RCA turns to the left and continues in the coronary sulcus on the posterior aspect of the heart. At the **crux** (cross) of the heart (Fig. 4.39), the junction of the septa and walls of the four heart chambers, the RCA gives rise to the **AV nodal branch**, which supplies the *AV node* (part of the cardiac conducting system). The RCA then gives off the large posterior IV branch that descends in the posterior IV sulcus toward the apex of the heart (Fig. 4.37). The **posterior IV branch** supplies both ventricles and sends perforating **interventricular septal branches** to the IV septum. The terminal (left ventricular) branch of the

TABLE 4.4. ARTERIAL SUPPLY OF HEART

Artery/Branch	Origin	Course	Distribution	Anastomoses
Right coronary	Right aortic sinus	Follows coronary (AV) sulcus between atria and ventricles	Right atrium, SA and AV nodes, and posterior part of IV septum	Circumflex and anterior IV branches (left coronary artery)
SA nodal	Right coronary artery near its origin (in 60%)	Ascends to SA node	Pulmonary trunk and SA node	
Right marginal	Right coronary artery	Passes to inferior margin of heart and apex	Right ventricle and apex of heart	IV branches
Posterior IV	Right coronary artery (in 67%)	Runs in posterior IV sulcus to apex of heart	Right and left ventricles and posterior third of septum	Anterior IV branches of left coronary artery (at apex)
AV nodal	Right coronary artery near origin of posterior IV artery	Passes to AV node	AV node	
Left coronary	Left aortic sinus	Runs in AV sulcus and gives off anterior IV and circumflex branches	Most of left atrium and ventricle, IV septum, and AV bundles; may supply AV node	Right coronary artery
Anterior IV (LAD)[a]	Left coronary artery	Passes along anterior IV sulcus to apex of heart	Right and left ventricles; anterior two thirds of IV septum	Posterior IV branch of left coronary artery
Circumflex	Left coronary artery	Passes to left in AV sulcus and runs to posterior surface of heart	Left atrium and left ventricle	Right coronary artery
Left marginal	Circumflex branch	Follows left border of heart	Left ventricle	IV branches
Posterior IV	Left coronary artery (in 33%)	Runs in posterior IV sulcus to apex of heart	Right and left posterior third of IV septum	Anterior IV branch of left coronary artery

[a]Clinicians continue to use LAD, the abbreviation for the term "left anterior descending artery."
AV, atrioventricular; *IV*, interventricular; *LAD*, left anterior descending artery; *SA*, sinu-atrial.

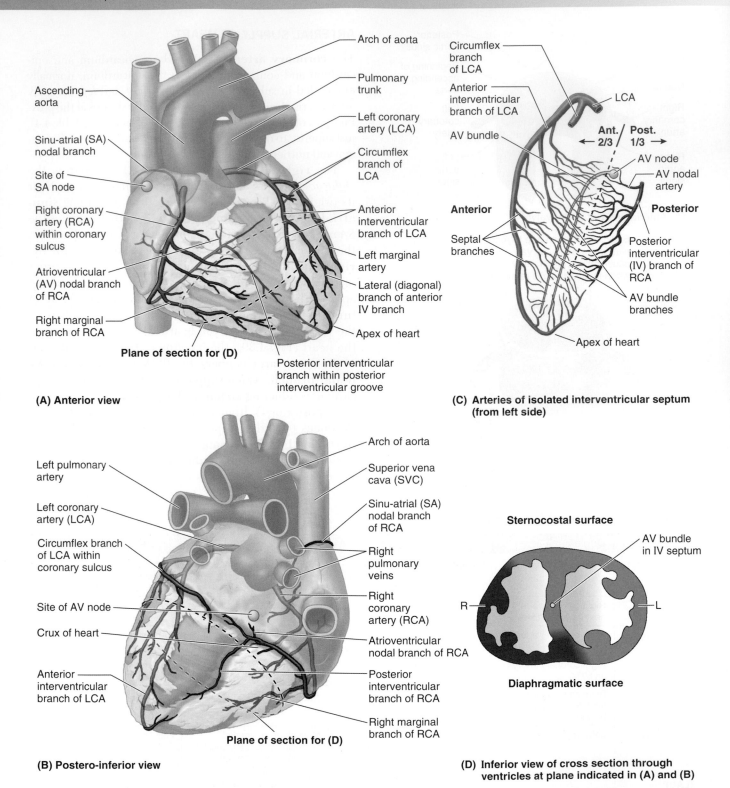

(A) Anterior view

(B) Postero-inferior view

(C) Arteries of isolated interventricular septum (from left side)

(D) Inferior view of cross section through ventricles at plane indicated in (A) and (B)

FIGURE 4.37. Arterial supply of heart. A and B. The most common pattern of distribution of the right coronary artery (RCA) and left coronary artery (LCA). **C.** Arteries of the interventricular septum. **D.** A cross section of the right and left ventricles demonstrates the most common pattern of distribution from the RCA (*red*) and LCA (*orange*).

RCA then continues for a short distance in the coronary sulcus. Typically, the RCA supplies

- the right atrium
- most of the right ventricle
- part of the left ventricle (diaphragmatic surface)
- part of the IV septum (usually the posterior third)
- the SA node (in approximately 60% of people)
- the AV node (in approximately 80% of people)

The **left coronary artery (LCA)** arises from the *left aortic sinus* of the ascending aorta and passes between the left auricle and the left side of the pulmonary trunk in the coronary sulcus. In approximately 40% of people, the SA nodal branch arises from the circumflex branch of the LCA and ascends on the posterior surface of the left atrium to the SA node.

At the left end of the coronary sulcus, located just left of the pulmonary trunk (Fig. 4.37), the LCA divides into two branches: an *anterior IV branch* (left anterior descending [LAD] branch) and a *circumflex branch*. The **anterior IV branch of the LCA** passes along the IV sulcus to the apex of the heart. Here, it turns around the inferior border of the heart and anastomoses with the posterior IV branch of the RCA. The anterior IV branch supplies both ventricles and the IV septum (Fig. 4.37C).

In many people, the anterior IV artery gives rise to a **lateral (diagonal) branch**, which descends on the anterior surface of the heart. The smaller **circumflex branch of the LCA** follows the coronary sulcus around the left border of the heart to the posterior surface of the heart. The **left marginal artery**, a branch of the circumflex branch, follows the left margin of the heart and supplies the left ventricle. The circumflex branch of the LCA terminates in the coronary sulcus on the posterior aspect of the heart before reaching the crux, but in about one third of heart, it continues as the posterior IV branch. Typically, the LCA supplies

- the left atrium
- most of the left ventricle

- part of the right ventricle
- most of the IV septum (usually its anterior two thirds), including the AV bundle of conducting tissue, through its perforating IV septal branches
- the SA node (in approximately 40% of people)

VENOUS DRAINAGE OF HEART

The heart is drained mainly by veins that empty into the coronary sinus and partly by small veins that empty directly into the chambers of the heart. The **coronary sinus**, the main vein of the heart, is a wide venous channel that runs from left to right in the posterior part of the coronary sulcus. The coronary sinus receives the **great cardiac vein** at its left end and the **middle** and **small cardiac veins** at its right end. The **left posterior ventricular vein** and **left marginal vein** also open into the coronary sinus. Small **anterior cardiac veins** from the right ventricular myocardium empty directly into the right atrium (Fig. 4.38). The **smallest cardiac veins** (L. *venae cordis minimae*) are minute vessels that begin in the capillary beds of the myocardium and open directly into the chambers of the heart, chiefly the atria. Although called veins, they are valveless communications with the capillary beds of the myocardium and may carry blood from the heart chambers to the myocardium.

LYMPHATIC DRAINAGE OF HEART

Lymphatic vessels in the myocardium and subendocardial connective tissue pass to the *subepicardial lymphatic plexus*. Vessels from this plexus pass to the coronary sulcus and follow the coronary arteries. A single lymphatic vessel, formed by the union of various vessels from the heart, ascends between the pulmonary trunk and the left atrium and ends in the *inferior tracheobronchial lymph nodes*, usually on the right side (see Fig. 4.22C).

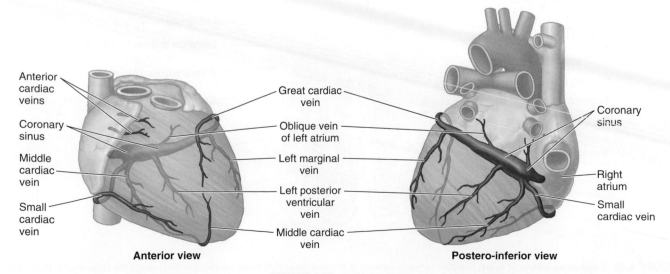

FIGURE 4.38. Cardiac veins.

CONDUCTING SYSTEM OF HEART

The impulse-conducting system, which coordinates the **cardiac cycle**, consists of cardiac muscle cells and highly specialized conducting fibers for initiating impulses and conducting them rapidly through the heart (Fig. 4.39). *Nodal tissue* initiates the heartbeat and coordinates the contractions of the four heart chambers. The **SA node** initiates and regulates the impulses for contraction, giving off an impulse about 70 times per minute in most people. The SA node, *the pacemaker of the heart*, is located anterolaterally just deep to the epicardium at the junction of the SVC and right atrium near the superior end of the sulcus terminalis. The **AV node** is a smaller collection of nodal tissue located in the postero-inferior region of the interatrial septum near the opening of the coronary sinus. The signal generated by the SA node passes through the walls of the right atrium propagated by the cardiac muscle (*myogenic conduction*), which transmits the signal rapidly from the SA node to the AV node. The AV node then distributes the signal to the ventricles through the *AV bundle*. Sympathetic stimulation speeds up conduction, and parasympathetic stimulation slows it down.

The **AV bundle (of His)**, the only bridge of conduction between the atrial and the ventricular myocardium, passes from the AV node through the fibrous skeleton of the heart and along the membranous part of the IV septum. At the junction of the membranous and muscular parts of the septum, the AV bundle divides into **right** and **left bundle branches**. The bundles proceed on each side of the muscular IV septum deep to the endocardium and then ramify into **subendocardial branches** (*Purkinje fibers*), which extend into the walls of the respective ventricles. The subendocardial branches of the right bundle stimulate the muscle of the IV septum, the anterior papillary muscle (through the septomarginal trabecula), and the wall of the right ventricle. The subendocardial branches

of the left bundle stimulate the IV septum, the anterior and posterior papillary muscles, and the wall of the left ventricle.

The following is a brief summary of the conducting system of the heart:

- The SA node initiates an impulse that is conducted to cardiac muscle fibers in the atria, causing them to contract.
- The impulse spreads by myogenic conduction, which transmits the impulse from the SA node to the AV node.
- The signal is distributed from the AV node through the AV bundle and the right and left bundle branches, which pass on each side of the IV septum to supply subendocardial branches to the papillary muscles and the walls of the ventricles.

INNERVATION OF HEART

The heart is supplied by autonomic nerve fibers from superficial and deep **cardiac plexuses** (see Fig. 4.22D). These nerve networks lie anterior to the bifurcation of the trachea and posterior to the ascending aorta. The **sympathetic supply of the heart** is from presynaptic fibers with cell bodies in the intermediolateral cell columns (lateral horns) of the superior five or six thoracic segments of the spinal cord and from postsynaptic sympathetic fibers with cell bodies in the cervical and superior thoracic paravertebral ganglia of the sympathetic trunks. The postsynaptic fibers end in the SA and AV nodes and in relation to the terminations of parasympathetic fibers on the coronary arteries. Sympathetic stimulation of the nodal tissue increases the rate and force of the heart's contractions. Sympathetic stimulation (indirectly) produces dilation of the coronary arteries by inhibiting their constriction. This supplies more oxygen and nutrients to the myocardium during periods of increased activity.

The **parasympathetic supply of the heart** is from presynaptic fibers of the *vagus nerves* (CN X). Postsynaptic parasympathetic cell bodies (intrinsic ganglia) are located

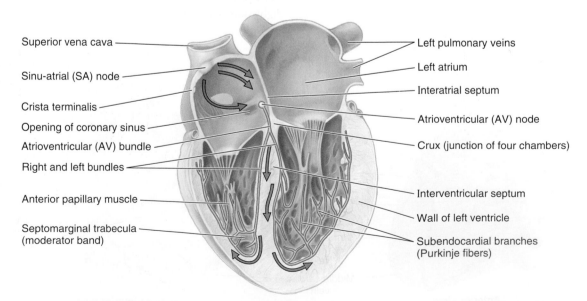

Superior vena cava

Sinu-atrial (SA) node

Crista terminalis

Opening of coronary sinus

Atrioventricular (AV) bundle

Right and left bundles

Anterior papillary muscle

Septomarginal trabecula (moderator band)

Left pulmonary veins

Left atrium

Interatrial septum

Atrioventricular (AV) node

Crux (junction of four chambers)

Interventricular septum

Wall of left ventricle

Subendocardial branches (Purkinje fibers)

FIGURE 4.39. Conducting system of heart. Impulses (*arrows*) initiated at the sinu-atrial (SA) node are propagated through the atrial musculature to the atrioventricular (AV) node and then through the AV bundle and its branches to the myocardium.

near the SA and AV nodes and along the coronary arteries. Parasympathetic stimulation slows the heart rate, reduces the force of the contraction, and constricts the coronary arteries, saving energy between periods of increased demand.

CARDIAC CYCLE

The **cardiac cycle** describes the complete movement of the heart or heartbeat and includes the period from the beginning of one heartbeat to the beginning of the next one.

The synchronous pumping action of the heart's two AV pumps (right and left chambers) constitutes the cardiac cycle.

The atria (the receiving chambers) pump accumulated blood rapidly into the ventricles (the discharging chambers). The right heart (*blue*) is the pump for the pulmonary circuit; the left heart (*red*) is the pump for the systemic circuit (Fig. 4.40). The cycle begins with a period of ventricular elongation and filling (**diastole**) and ends with a period of ventricular shortening and emptying (**systole**). Two *heart sounds*, resulting from valve closures, can be heard with a stethoscope: a *lub* sound as

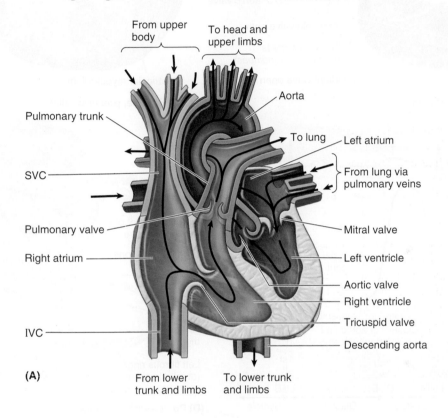

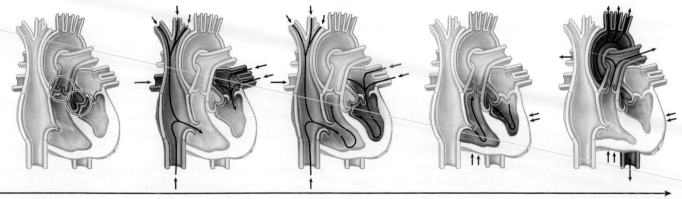

(B) Beginning of diastole upon closure of aortic and pulmonary valves

(C) Opening of atrioventricular valves during early moments of diastole

(D) Atrial contraction during final moments of diastole

(E) Closure of atrioventricular valves (tricuspid and mitral) very soon after systole begins

(F) Opening of aortic and pulmonary valves during systole

Anterior views

FIGURE 4.40. Cardiac cycle. The right heart (*blue side*) is the pump for the pulmonary circuit; the left heart (*red side*) is the pump for the systemic circuit. *IVC*, inferior vena cava; *SVC*, superior vena cava.

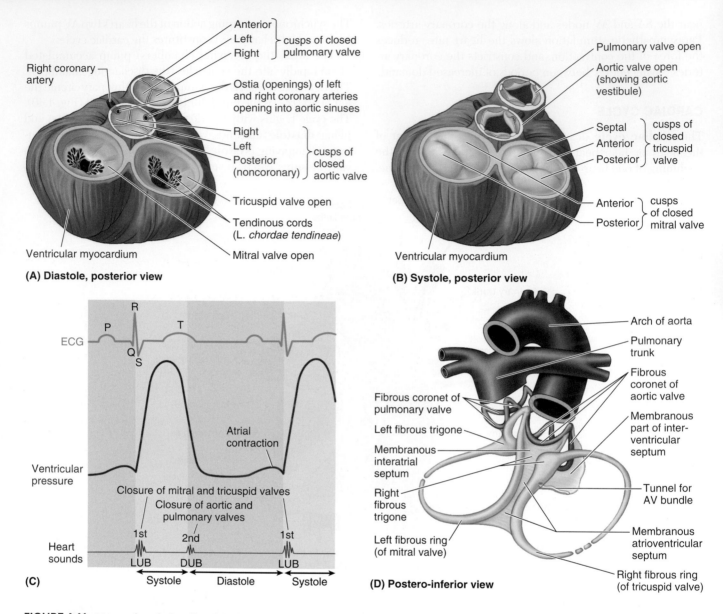

FIGURE 4.41. Heart valves during diastole and systole and outline of cardiac skeleton. A. Ventricular diastole. **B.** Ventricular systole. **C.** Correlation of ventricular pressure, electrocardiogram (*ECG*), and heart sounds. **D.** Cardiac skeleton. *AV,* atrioventricular; *IV,* interventricular.

the blood is transferred from the atria to the ventricles and a *dub* sound as the ventricles contract and expel blood from the heart (Fig. 4.41). The heart sounds are produced by the snapping shut of the one-way valves that normally keep blood from flowing backward during contractions of the heart.

When the ventricles contract, they produce a wringing motion. This motion initially ejects the blood from the ventricles, first narrowing and then shortening the heart, reducing the volume of the ventricular chambers. Continued sequential contraction elongates the heart, followed by widening as the myocardium briefly relaxes, increasing the volume of the chambers to draw blood from the atria.

CARDIAC SKELETON

The muscle fibers are anchored to the **fibrous skeleton of the heart** (Fig. 4.41). The fibrous framework of dense

collagen forms four **fibrous rings**, which surround the orifices of the valves. The right and left **fibrous trigones** connect the rings and the membranous parts of the inter-atrial and IV septa. The fibrous skeleton of the heart

- keeps the orifices of the AV and semilunar valves patent and prevents them from being overly distended by the volume of blood pumping through them
- provides attachments for the leaflets and cusps of the valves
- provides attachment for the myocardium
- forms an electrical "insulator" by separating the myen-terically conducted impulses of the atria and ventricles so that they contract independently and by surrounding and providing passage for the initial part of the AV bundle

CLINICAL BOX

Coronary Artery Disease or Coronary Heart Disease

Coronary artery disease (CAD) is one of the leading causes of death. It has many causes, all of which result in a reduced blood supply to the vital myocardial tissue.

Myocardial Infarction

With sudden occlusion of a major artery by an embolus (G. *embolos*, plug), the region of myocardium supplied by the occluded vessel becomes *infarcted* (rendered virtually bloodless) and undergoes *necrosis* (pathological tissue death). The three most common sites of coronary artery occlusion are (1) the anterior IV (LAD) branch of the LCA (40–50%), (2) the RCA (30–40%), and (3) the circumflex branch of the LCA (15–20%) (Fig. B4.14).

An area of myocardium that has undergone necrosis constitutes a *myocardial infarction* (MI). The most common cause of *ischemic heart disease* is coronary artery insufficiency resulting from atherosclerosis.

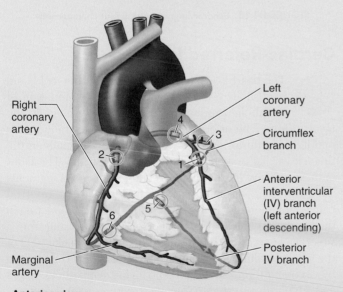

Anterior view

Sites 1–3 account for at least 85% of all occlusions.

FIGURE B4.14. Sites of coronary artery occlusion in order of frequency (1–6).

Coronary Atherosclerosis

The *atherosclerotic process*, characterized by lipid deposits in the intima (lining layer) of the coronary arteries, begins during early adulthood and slowly results in stenosis of the lumina of the arteries (Fig. B4.15). Insufficiency of blood supply to the heart (*myocardial ischemia*) may result in MI.

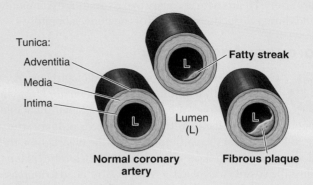

FIGURE B4.15. Atherosclerosis stages of development in a coronary artery.

Coronary Bypass Graft

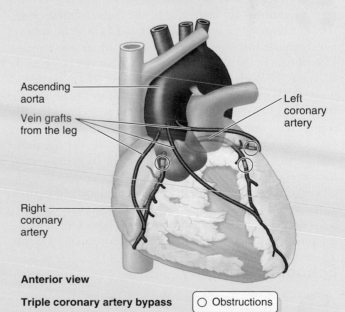

Patients with obstruction of their coronary circulation and severe *angina* (heart pain) may undergo a *coronary bypass graft* operation. A segment of an artery or vein is connected to the ascending aorta or to the proximal part of a coronary artery and then to the coronary artery distal to the stenosis (Fig. B4.16). The great saphenous vein is commonly harvested for coronary bypass surgery because it (1) has a diameter equal to or greater than that of the coronary arteries, (2) can be easily dissected from the lower limb, (3) and offers relatively lengthy portions with a minimum occurrence of valves or branching. Reversal of the implanted segment of vein can negate the effect of a valve if a valved segment must be used. Use of the radial artery in bypass surgery has become increasingly more common. A coronary bypass graft shunts blood from the aorta to a *stenotic*

Anterior view

Triple coronary artery bypass   ○ Obstructions

FIGURE B4.16. Triple coronary artery bypass.

(Continued on next page)

coronary artery to increase the flow distal to the obstruction. Revascularization of the myocardium may also be achieved by surgically anastomosing an internal thoracic artery with a coronary artery. Hearts with coronary bypass grafts are commonly found during dissections in the gross anatomy laboratory.

Coronary Angioplasty

Cardiologists or interventional radiologists use *percutaneous transluminal coronary angioplasty*, in which they pass a catheter with a small inflatable balloon attached to its tip into the obstructed coronary artery (Fig. B4.17). When the catheter reaches the obstruction, the balloon is inflated, flattening the atherosclerotic plaque against the vessel's wall, and the vessel is stretched to increase the size of the lumen, thus improving blood flow. In other cases, *thrombokinase* is injected through the catheter; this enzyme dissolves or reduces the blood clot. After dilation of the vessel, an *intravascular stent* may be introduced to maintain the dilation. These procedures are replacing bypass procedures requiring open surgery at markedly increasing rates.

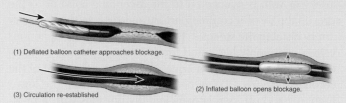

(1) Deflated balloon catheter approaches blockage.

(3) Circulation re-established

(2) Inflated balloon opens blockage.

FIGURE B4.17. Percutaneous transluminal angioplasty.

Variations of Coronary Arteries

Variations in the branching patterns of the coronary arteries are common. In the most common right-dominant pattern, the RCA and LCA share approximately equally in the blood supply to the heart. In approximately 15% of hearts, the LCA is dominant in that the posterior IV branch is a branch of the circumflex artery. There is codominance in about 18% of people, in which branches of both the RCA and LCA reach the crux and give rise to branches that course in or near the posterior IV sulcus. A few people have only a single coronary artery. In other people, the circumflex artery arises from the right aortic sinus. The branches of coronary arteries are considered to be end arteries—ones that supply regions of the myocardium without functional overlap from other large branches. However, anastomoses exist between small branches of the coronary arteries. The potential for development of collateral circulation likely exists in most hearts.

Echocardiography

Echocardiography (ultrasonic cardiography) is a method of graphically recording the position and motion of the heart by the echo obtained from beams of ultrasonic waves directed through the thorax (Fig. B4.18). This technique may detect as little as 20 mL of

fluid in the pericardial cavity, such as that resulting from pericardial effusion. *Doppler echocardiography* is a technique that demonstrates and records the flow of blood through the heart and great vessels by Doppler US, making it especially useful in the diagnosis and analysis of problems with blood flow through the heart, such as septal defects, and in delineating valvular stenosis and regurgitation, especially on the left side of the heart.

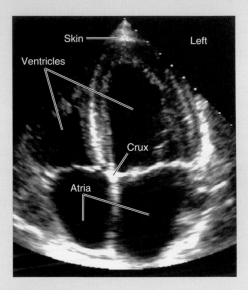

FIGURE B4.18. Echocardiogram. Apical four chamber view.

Cardiac-Referred Pain

The heart is insensitive to touch, cutting, cold, and heat; however, ischemia and the accumulation of metabolic products stimulate pain endings in the myocardium. The afferent pain fibers run centrally in the middle and inferior cervical branches and especially in the thoracic cardiac branches of the sympathetic trunk. The axons of these primary sensory neurons enter spinal cord segments T1–T4 or T5, especially on the left side. Cardiac-referred pain is a phenomenon whereby noxious stimuli originating in the heart are perceived by the person as pain arising from a superficial part of the body—the skin on the medial aspect of the left upper limb, for example. Visceral pain is transmitted by visceral afferent fibers accompanying sympathetic fibers and is typically referred to somatic structures or areas such as the upper limb having afferent fibers with cell bodies in the same spinal ganglion and central processes that enter the spinal cord through the same posterior roots.

Injury to Conducting System of Heart

Damage to the conducting system, often resulting from ischemia caused by *CAD*, produces disturbances of cardiac muscle contraction. Because the anterior IV branch (LAD branch) supplies the AV bundle in most people and because branches of the RCA supply both the SA and the AV nodes, parts of the conducting system of the heart are likely to be affected by their occlusion.

Damage to the AV node or bundle results in a *heart block* because the atrial excitation does not reach the ventricles. As a result, the ventricles begin to contract independently at their own rate (25–30 times per minute), which is slower than the lowest normal rate of 40–45 times per minute.

Damage to one of the bundle branches results in a *bundle branch block*, in which excitation passes along the unaffected branch and causes a normally timed systole of that ventricle only. The impulse then spreads to the other ventricle, producing a late asynchronous contraction.

Superior Mediastinum

The *superior mediastinum* is located superior to the transverse thoracic plane passing through the sternal angle and the junction (IV disc) of vertebrae T4 and T5. From anterior to posterior, **the main contents of the superior mediastinum** are as follows (Figs. 4.42 and 4.43):

- Thymus, a primary lymphoid organ
- Great vessels related to the heart and pericardium
  - Brachiocephalic veins
  - Superior part of SVC
  - Bifurcation of the pulmonary trunk and roots of pulmonary arteries
  - Arch of aorta and roots of its major branches
    - Brachiocephalic trunk
    - Left common carotid artery
    - Left subclavian artery
- Vagus and phrenic nerves
- Cardiac plexus of nerves
- Left recurrent laryngeal nerve
- Trachea
- Esophagus
- Thoracic duct

THYMUS

The **thymus**, a lymphoid organ, is located in the lower part of the neck and the anterior part of the superior mediastinum.

It lies posterior to the manubrium of the sternum and extends into the anterior mediastinum, anterior to the pericardium. After puberty, the thymus undergoes gradual involution and is largely replaced by fat. A rich *arterial supply to the thymus* derives mainly from the anterior intercostal and anterior mediastinal branches of the internal thoracic arteries. The **thymic veins** end in the left brachiocephalic, internal thoracic, and inferior thyroid veins. The **lymphatic vessels of the thymus** end in the parasternal, brachiocephalic, and tracheobronchial lymph nodes (see Fig. 4.22C).

GREAT VESSELS IN MEDIASTINUM

The *brachiocephalic veins* form posterior to the sternoclavicular joints by the union of the internal jugular and subclavian veins (Figs. 4.42 and 4.43A). At the level of the inferior border of the 1st right costal cartilage, the brachiocephalic veins unite to form the SVC. The **left brachiocephalic vein** is more than twice as long as the right brachiocephalic vein because it courses from the left to the right side, passing anterior to the origins (roots) of the three major branches of the arch of the aorta. It shunts blood from the head, neck, and left upper limb to the right atrium. The origin of the **right brachiocephalic vein** is formed by the union of the right internal jugular and subclavian vein, the **right venous angle**, and receives lymph from the right lymphatic duct. Similarly, the origin of the left brachiocephalic vein is formed by union of the left internal jugular and subclavian veins, the **left venous angle**, and receives lymph from the thoracic duct (Fig. 4.42A).

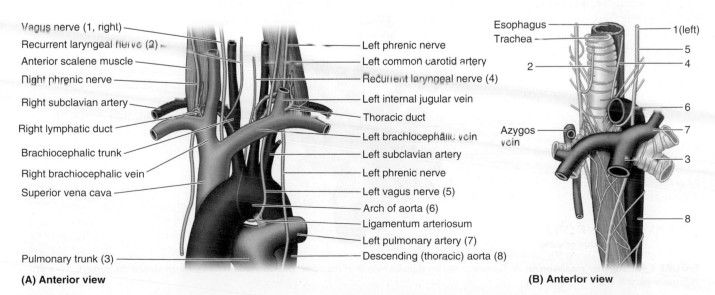

(A) Anterior view

Vagus nerve (1, right)
Recurrent laryngeal nerve (2)
Anterior scalene muscle
Right phrenic nerve
Right subclavian artery
Right lymphatic duct
Brachiocephalic trunk
Right brachiocephalic vein
Superior vena cava
Pulmonary trunk (3)

Left phrenic nerve
Left common carotid artery
Recurrent laryngeal nerve (4)
Left internal jugular vein
Thoracic duct
Left brachiocephalic vein
Left subclavian artery
Left phrenic nerve
Left vagus nerve (5)
Arch of aorta (6)
Ligamentum arteriosum
Left pulmonary artery (7)
Descending (thoracic) aorta (8)

(B) Anterior view

Esophagus
Trachea
1 (left)
5
2
4
6
Azygos vein
7
3
8

FIGURE 4.42. Great vessels and nerves. A. Vessels in the lower neck and superior mediastinum. **B.** Relationships of nerves to trachea, esophagus, and azygos vein.

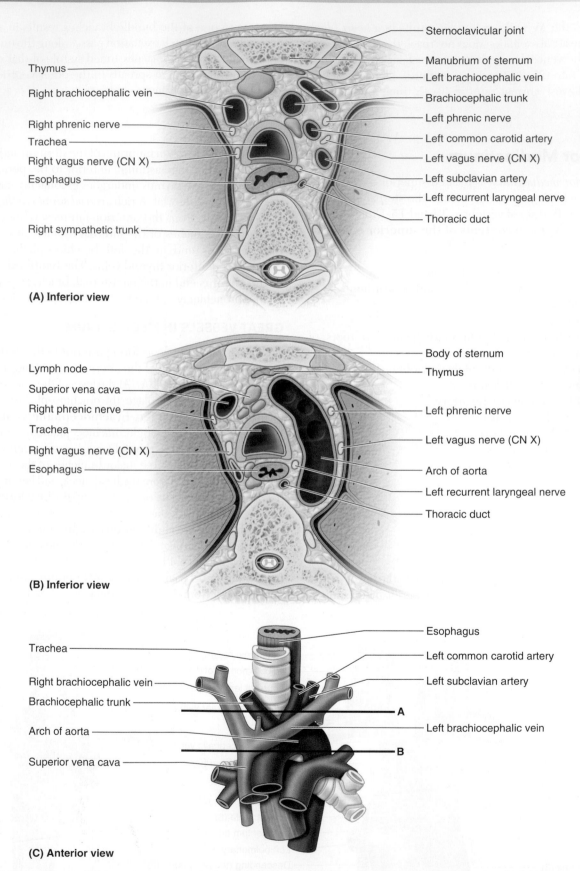

(A) Inferior view

- Sternoclavicular joint
- Manubrium of sternum
- Left brachiocephalic vein
- Brachiocephalic trunk
- Left phrenic nerve
- Left common carotid artery
- Left vagus nerve (CN X)
- Left subclavian artery
- Left recurrent laryngeal nerve
- Thoracic duct

Thymus
Right brachiocephalic vein
Right phrenic nerve
Trachea
Right vagus nerve (CN X)
Esophagus
Right sympathetic trunk

(B) Inferior view

- Body of sternum
- Thymus
- Left phrenic nerve
- Left vagus nerve (CN X)
- Arch of aorta
- Left recurrent laryngeal nerve
- Thoracic duct

Lymph node
Superior vena cava
Right phrenic nerve
Trachea
Right vagus nerve (CN X)
Esophagus

(C) Anterior view

- Esophagus
- Left common carotid artery
- Left subclavian artery
- A
- Left brachiocephalic vein
- B

Trachea
Right brachiocephalic vein
Brachiocephalic trunk
Arch of aorta
Superior vena cava

FIGURE 4.43. Superior mediastinum. A. Transverse section superior to arch of the aorta. **B.** Transverse section through arch of the aorta. **C.** Level of sections in parts A and B.

The SVC returns blood from all structures superior to the diaphragm, except the lungs and heart. It passes inferiorly and ends at the level of the 3rd costal cartilage, where it enters the right atrium. The SVC lies in the right side of the superior mediastinum, anterolateral to the trachea and posterolateral to the ascending aorta (Figs. 4.42 and 4.44A). The *right phrenic nerve* lies between the SVC and the mediastinal pleura. The terminal half of the SVC is in the middle mediastinum, where it is adjacent to the ascending aorta and forms the posterior boundary of the transverse pericardial sinus (see Fig. 4.26B). The **arch of the aorta**, the curved continuation of the ascending aorta, begins posterior to the 2nd right sternocostal joint at the level of the sternal angle and arches superoposteriorly and to the left (Figs. 4.42 and 4.43). The aortic arch ascends anterior to the right pulmonary artery and the bifurcation of the trachea, reaching its apex at the left side of the trachea and esophagus as it passes over the root of the left lung. The arch descends on the left side of the body of the T4 vertebra and ends by becoming the **descending (thoracic) aorta** posterior to the 2nd left sternocostal joint (Fig. 4.44B).

The **ligamentum arteriosum**, the remnant of the fetal ductus arteriosus, passes from the root of the left pulmonary artery to the inferior surface of the arch of the aorta (Fig. 4.42A). The **left recurrent laryngeal nerve** hooks beneath the arch immediately lateral to the ligamentum arteriosum and then ascends between the trachea and esophagus (Fig. 4.42 and Table 4.5). The *branches of the arch of the aorta* are as follows (Figs. 4.42 and 4.43):

- Brachiocephalic trunk
- Left common carotid artery
- Left subclavian artery

The **brachiocephalic trunk**, the first and largest branch of the arch, arises posterior to the manubrium, where it lies anterior to the trachea and posterior to the left brachiocephalic vein. It ascends superolaterally to reach the right side of the trachea and the right sternoclavicular joint, where it divides into the right common carotid and right subclavian arteries. The **left common carotid artery**, the second branch of the aortic arch, arises posterior to the manubrium, slightly posterior and to the left of the brachiocephalic trunk. It ascends anterior to the left subclavian artery and at first anterior to the trachea and then to its left. It enters the neck by passing posterior to the left sternoclavicular joint. The **left subclavian artery**, the third branch of the aortic arch, arises from the posterior part of the arch, just posterior to the left common carotid artery. It ascends lateral to the trachea and the left common carotid artery through the superior mediastinum. The left subclavian artery has no branches in the mediastinum. As it leaves the thorax and enters the root of the neck, it passes posterior to the left sternoclavicular joint and lateral to the left common carotid artery.

NERVES IN SUPERIOR MEDIASTINUM

The vagus nerves (CN X) arise bilaterally from the medulla of the brain, exit the cranium, and descend through the neck posterolateral to the common carotid arteries. Each nerve enters the superior mediastinum posterior to the respective sternoclavicular joint and brachiocephalic vein (Figs. 4.42, 4.43, and 4.45; Table 4.5). The **right vagus nerve** enters the thorax anterior to the right subclavian artery, where it gives rise to the **right recurrent laryngeal nerve**. This posterior branch hooks inferior to the right subclavian artery and ascends between the trachea and the esophagus to supply the

(Continued on page 244)

TABLE 4.5. NERVES OF THORAX

Nerve	Origin	Course	Distribution
Vagus (CN X)	8–10 rootlets from medulla of brainstem	Enters superior mediastinum posterior to sternoclavicular joint and brachiocephalic vein; gives rise to recurrent laryngeal nerve; continues into abdomen	Pulmonary plexus, esophageal plexus, and cardiac plexus
Phrenic	Anterior rami of C3–C5 nerves	Passes through superior thoracic aperture and runs between mediastinal pleura and pericardium	Central portion of diaphragm
Intercostals (1–11)	Anterior rami of T1–T11 nerves	Run in intercostal spaces between internal and innermost layers of intercostal muscles	Muscles in and skin over intercostal space; lower nerves supply muscles and skin of anterolateral abdominal wall.
Subcostal	Anterior ramus of T12 nerve	Follows inferior border of 12th rib and passes into abdominal wall	Abdominal wall and skin of gluteal region
Recurrent laryngeal	Vagus nerve	On right, loops around subclavian artery; on left, loops around arch of aorta and ascends in tracheo-esophageal groove	Intrinsic muscles of larynx (except cricothyroid); sensory inferior to level of vocal folds
Cardiac plexus	Cervical and cardiac branches of vagus nerve and sympathetic trunk	From arch of aorta and posterior surface of heart, fibers extend along coronary arteries and to sinu-atrial node.	Impulses pass to sinu-atrial node; parasympathetic fibers slow rate, reduce force of heartbeat, and constrict coronary arteries; sympathetic fibers have opposite effect.
Pulmonary plexus	Vagus nerve and sympathetic trunk	Forms on root of lung and extends along bronchial subdivisions	Parasympathetic fibers constrict bronchioles; sympathetic fibers dilate them; afferents convey reflexes.
Esophageal plexus	Vagus nerve, sympathetic ganglia, and greater splanchnic nerve	Distal to tracheal bifurcation, vagus, and sympathetic nerves form the plexus around esophagus.	Vagal and sympathetic fibers to smooth muscle and glands of inferior two thirds of esophagus

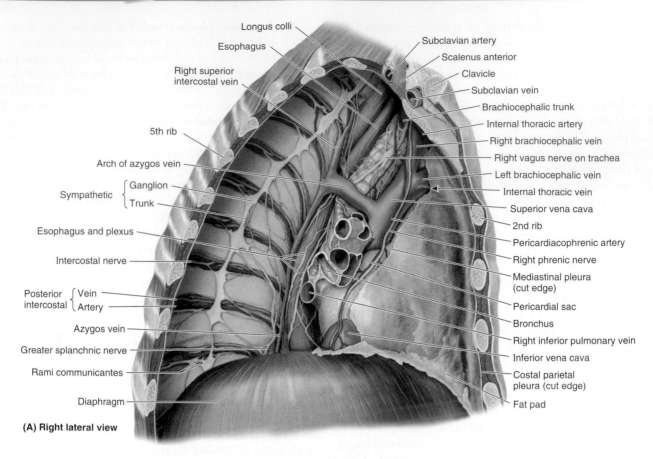

Longus colli
Esophagus
Right superior
intercostal vein
5th rib
Arch of azygos vein
Sympathetic { Ganglion
              Trunk
Esophagus and plexus
Intercostal nerve
Posterior { Vein
intercostal { Artery
Azygos vein
Greater splanchnic nerve
Rami communicantes
Diaphragm

Subclavian artery
Scalenus anterior
Clavicle
Subclavian vein
Brachiocephalic trunk
Internal thoracic artery
Right brachiocephalic vein
Right vagus nerve on trachea
Left brachiocephalic vein
Internal thoracic vein
Superior vena cava
2nd rib
Pericardiacophrenic artery
Right phrenic nerve
Mediastinal pleura
(cut edge)
Pericardial sac
Bronchus
Right inferior pulmonary vein
Inferior vena cava
Costal parietal
pleura (cut edge)
Fat pad

(A) Right lateral view

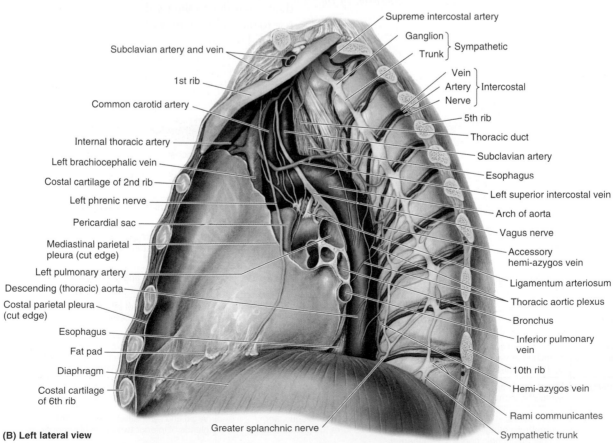

Subclavian artery and vein
1st rib
Common carotid artery
Internal thoracic artery
Left brachiocephalic vein
Costal cartilage of 2nd rib
Left phrenic nerve
Pericardial sac
Mediastinal parietal
pleura (cut edge)
Left pulmonary artery
Descending (thoracic) aorta
Costal parietal pleura
(cut edge)
Esophagus
Fat pad
Diaphragm
Costal cartilage
of 6th rib

Supreme intercostal artery
Ganglion }
Trunk   } Sympathetic
Vein   ⌉
Artery ⎬ Intercostal
Nerve   ⌋
5th rib
Thoracic duct
Subclavian artery
Esophagus
Left superior intercostal vein
Arch of aorta
Vagus nerve
Accessory
hemi-azygos vein
Ligamentum arteriosum
Thoracic aortic plexus
Bronchus
Inferior pulmonary
vein
10th rib
Hemi-azygos vein
Rami communicantes
Sympathetic trunk

Greater splanchnic nerve

(B) Left lateral view

FIGURE 4.44. Right and left sides of mediastinum. A. Right side of mediastinum. **B.** Left side of mediastinum.

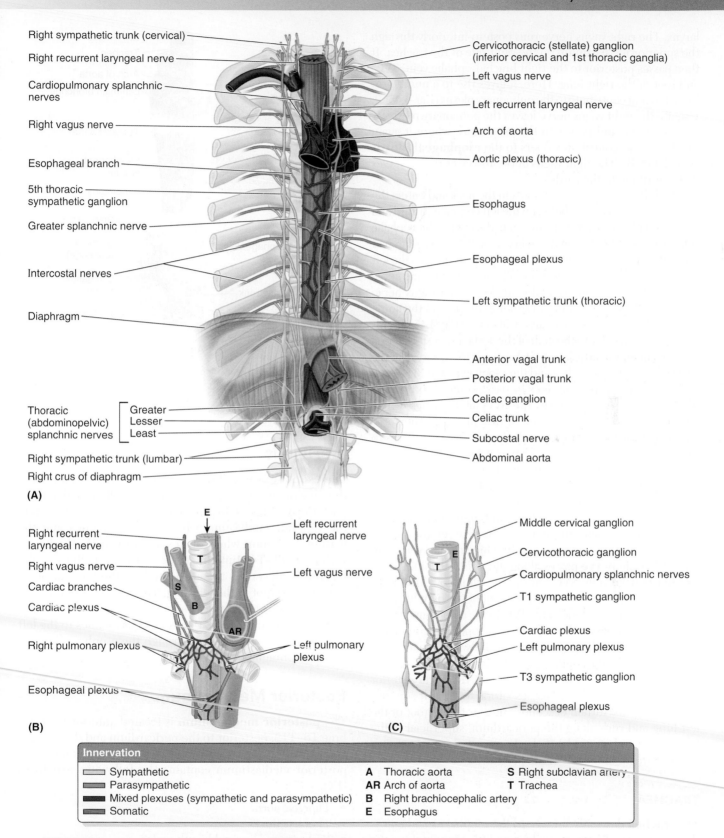

Right sympathetic trunk (cervical)

Right recurrent laryngeal nerve

Cardiopulmonary splanchnic nerves

Right vagus nerve

Esophageal branch

5th thoracic sympathetic ganglion

Greater splanchnic nerve

Intercostal nerves

Diaphragm

Thoracic (abdominopelvic) splanchnic nerves — Greater / Lesser / Least

Right sympathetic trunk (lumbar)

Right crus of diaphragm

(A)

Cervicothoracic (stellate) ganglion (inferior cervical and 1st thoracic ganglia)

Left vagus nerve

Left recurrent laryngeal nerve

Arch of aorta

Aortic plexus (thoracic)

Esophagus

Esophageal plexus

Left sympathetic trunk (thoracic)

Anterior vagal trunk

Posterior vagal trunk

Celiac ganglion

Celiac trunk

Subcostal nerve

Abdominal aorta

Right recurrent laryngeal nerve

Right vagus nerve

Cardiac branches

Cardiac plexus

Right pulmonary plexus

Esophageal plexus

(B)

Left recurrent laryngeal nerve

Left vagus nerve

Left pulmonary plexus

Middle cervical ganglion

Cervicothoracic ganglion

Cardiopulmonary splanchnic nerves

T1 sympathetic ganglion

Cardiac plexus

Left pulmonary plexus

T3 sympathetic ganglion

Esophageal plexus

(C)

Innervation		
▭ Sympathetic	**A** Thoracic aorta	**S** Right subclavian artery
▭ Parasympathetic	**AR** Arch of aorta	**T** Trachea
▭ Mixed plexuses (sympathetic and parasympathetic)	**B** Right brachiocephalic artery	
▭ Somatic	**E** Esophagus	

FIGURE 4.45. **Autonomic nerves in the superior and posterior mediastinum. A.** Overview. **B.** Parasympathetic nerves. **C.** Sympathetic nerves.

larynx. The right vagus nerve runs postero-inferiorly through the superior mediastinum on the right side of the trachea. It then passes posterior to the right brachiocephalic vein, SVC, and root of the right lung. Here, it gives rise to a number of branches that contribute to the pulmonary plexus (Fig. 4.45C). Usually, the right vagus nerve leaves the pulmonary plexus as a single nerve and passes to the esophagus, where it again breaks up and contributes fibers to the **esophageal plexus** (Fig. 4.45A,B). The right vagus nerve also gives rise to nerves that contribute to the cardiac plexus.

The **left vagus nerve** descends in the neck and enters the thorax and mediastinum between the left common carotid and the left subclavian arteries, posterior to the left brachiocephalic vein (see Fig. 4.42). When it reaches the left side of the arch of the aorta, the left vagus nerve diverges posteriorly from the left phrenic nerve. It is separated laterally from the phrenic nerve by the left superior intercostal vein. As the left vagus nerve curves medially at the inferior border of the arch of the aorta, it gives off the left recurrent laryngeal nerve (Fig. 4.45B). This nerve passes inferior to the arch of the aorta just posterolateral to the ligamentum arteriosum and ascends to the larynx in the groove between the trachea and the esophagus (see Fig. 4.42). The left vagus nerve continues to pass posterior to the root of the left lung where it gives rise to many branches, which contribute to the pulmonary and cardiac plexuses. The nerve continues past these plexuses as a single trunk and passes to the esophagus, where it breaks up as it joins fibers from the right vagus in the esophageal plexus (Fig. 4.45B).

The **phrenic nerves** are the sole motor supply to the diaphragm (Fig. 4.44 and Table 4.5); approximately one third of their fibers are sensory to the diaphragm. Each phrenic nerve enters the superior mediastinum between the subclavian artery and the origin of the brachiocephalic vein. The **right phrenic nerve** passes along the right side of the right brachiocephalic vein, SVC, and pericardium over the right atrium. It also passes anterior to the root of the right lung and descends on the right side of the IVC to the diaphragm, which it penetrates or passes through the caval opening (foramen).

The **left phrenic nerve** descends between the left subclavian and the left common carotid arteries (Fig. 4.44B). It crosses the left surface of the arch of the aorta anterior to the left vagus nerve and passes lateral to the left superior intercostal vein. It then descends anterior to the root of the left lung and runs along the pericardium, superficial to the left atrium and ventricle of the heart, where it penetrates the diaphragm to the left of the pericardium.

TRACHEA

The **trachea** descends anterior to the esophagus and enters the superior mediastinum, inclining a little to the right of the median plane (Fig. 4.47C,D). The posterior surface of the trachea is flat where its cartilaginous "rings" are incomplete and where it is related to the esophagus. The trachea ends at the level of the sternal angle by dividing into the right and left main bronchi.

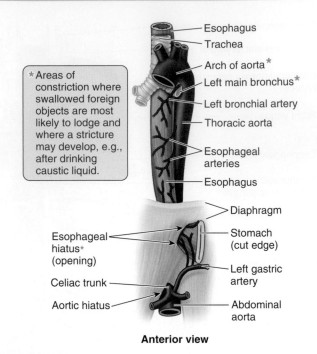

FIGURE 4.46. Esophagus. Blood supply and relationship to surrounding structures.

ESOPHAGUS

The **esophagus** is a fibromuscular tube that extends from the pharynx to the stomach. It is usually flattened antero-posteriorly (Figs. 4.43 and 4.46). The esophagus enters the superior mediastinum between the trachea and the vertebral column, where it lies anterior to the bodies of vertebrae T1–T4. Initially, the esophagus inclines to the left but is moved by the aortic arch to the median plane opposite the root of the left lung. The thoracic duct usually lies on the left side of the esophagus and deep to the aortic arch. Inferior to the arch, the esophagus inclines to the left as it approaches and passes through the esophageal hiatus in the diaphragm.

Posterior Mediastinum

The **posterior mediastinum** is located anterior to vertebrae T5–T12, posterior to the pericardium and diaphragm, and between the parietal pleura of the two lungs. The posterior mediastinum contains the following structures (Fig. 4.47):

- Thoracic aorta
- Thoracic duct
- Posterior mediastinal lymph nodes
- Azygos and hemi-azygos veins
- Esophagus
- Esophageal plexus
- Thoracic sympathetic trunks
- Thoracic splanchnic nerves

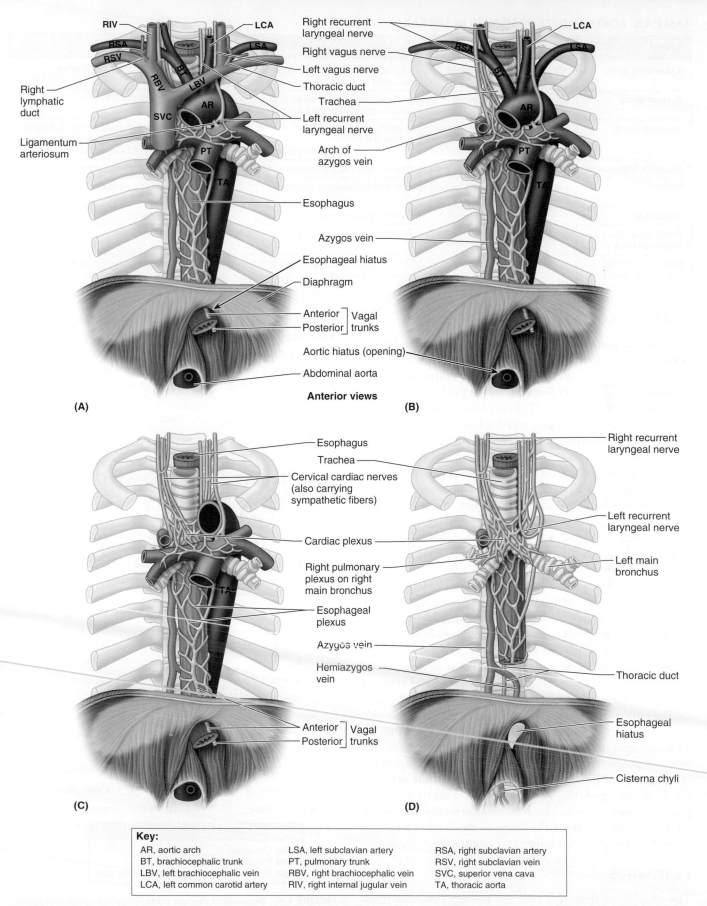

FIGURE 4.47. Structures of superior and posterior mediastinum. A–D. The structures of the mediastinum are revealed by different levels of dissection from anterior to posterior.

Key:
AR, aortic arch
BT, brachiocephalic trunk
LBV, left brachiocephalic vein
LCA, left common carotid artery

LSA, left subclavian artery
PT, pulmonary trunk
RBV, right brachiocephalic vein
RIV, right internal jugular vein

RSA, right subclavian artery
RSV, right subclavian vein
SVC, superior vena cava
TA, thoracic aorta

TABLE 4.6. AORTA AND ITS BRANCHES IN THORAX

Artery	Origin	Course	Branches
Ascending aorta	Aortic orifice of left ventricle	Ascends ~5 cm to sternal angle, where it becomes arch of aorta	Right and left coronary arteries
Arch of aorta	Continuation of ascending aorta	Arches posteriorly on left side of trachea and esophagus and superior to left main bronchus	Brachiocephalic, left common carotid, left subclavian arteries
Thoracic aorta	Continuation of arch of aorta	Descends in posterior mediastinum to left of vertebral column; gradually shifts to right to lie in median plane at aortic hiatus	Posterior intercostal arteries, subcostal, some phrenic arteries, and visceral branches (e.g., esophageal)
Posterior intercostals	Posterior aspect of thoracic aorta	Pass laterally and then anteriorly, parallel to ribs	Lateral and anterior cutaneous branches
Bronchial (one or two branches)	Anterior aspect of aorta or posterior intercostal artery	Run with tracheobronchial tree	Bronchial and peribronchial tissue, visceral pleura
Percardial		Directly to pericardium	To pericardium
Esophageal (four or five branches)	Anterior aspect of thoracic aorta	Run anteriorly to esophagus	To esophagus
Superior phrenic (vary in number)		Arise at aortic hiatus and pass to superior aspect of diaphragm	To diaphragm

THORACIC AORTA

The thoracic aorta, the thoracic part of the descending aorta, is the continuation of the arch of the aorta (Fig. 4.47 and Table 4.6). It begins at the inferior border of the body of T4 vertebra on the left and descends in the posterior mediastinum on the left sides of T5–T12 vertebrae. As it descends, it approaches the median plane and displaces the esophagus to the right. The **thoracic aortic plexus**, an autonomic nerve network, surrounds it (Fig. 4.45A). The thoracic aorta lies posterior to the root of the left lung, the pericardium, and the esophagus. Its name changes to *abdominal aorta* anterior to the inferior border of the T12 vertebra, and it enters the abdomen through the **aortic hiatus** (opening) in the diaphragm (Figs. 4.46 and 4.47). The thoracic duct and azygos vein ascend on the right side of the thoracic aorta and accompany it through this hiatus (Fig. 4.47D).

The *branches of the thoracic aorta* are bronchial, pericardial, posterior intercostals, superior phrenic, esophageal, mediastinal, and subcostal (Fig. 4.48 and Table 4.6). The bronchial arteries consist of one right and two small left vessels. The bronchial arteries supply the trachea, bronchi, lung tissue, and lymph nodes. The **pericardial arteries** send twigs to the pericardium. The **posterior intercostal arteries** (nine pairs) pass into the 3rd through 11th intercostal spaces.

The **superior phrenic arteries** pass to the thoracic side of the diaphragm, where they anastomose with the musculophrenic and pericardiacophrenic branches of the internal thoracic artery. Usually, two **esophageal arteries** supply the middle third of the esophagus. The **mediastinal arteries** are small and supply the lymph nodes and other tissues of the posterior mediastinum. The **subcostal arteries** that course on the abdominal side of the origin of the diaphragm are in series with the posterior intercostal arteries.

ESOPHAGUS

The esophagus descends into the posterior mediastinum from the superior mediastinum, passing posterior and to the right of the arch of the aorta and posterior to the pericardium and left atrium. The esophagus constitutes the primary posterior relationship of the base of the heart. It then deviates to the left and passes through the **esophageal hiatus** in the diaphragm at the level of the T10 vertebra, anterior to the

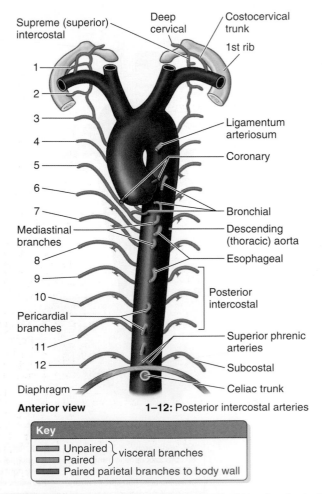

FIGURE 4.48. Branches of thoracic aorta. Unpaired (*green*) and paired (*yellow*) visceral branches; paired parietal branches to body wall (*purple*).

aorta (Figs. 4.46 and 4.47). The esophagus may have three impressions, or "constrictions," in its thoracic part. These may be observed as narrowings of the lumen in oblique chest radiographs that are taken as barium is swallowed.

The esophagus is compressed by three structures: the aortic arch, left main bronchus, and diaphragm. No constrictions are visible in the empty esophagus; however, as it expands during filling, these structures compress its walls.

THORACIC DUCT AND LYMPHATIC TRUNKS

In the posterior mediastinum, the **thoracic duct** lies on the bodies of the inferior seven thoracic vertebrae (Fig. 4.49). The thoracic duct conveys most lymph of the body to the venous system (that from the lower limbs, pelvic cavity, abdominal cavity, left side of thorax, left side of head, neck, and left upper limb). The thoracic duct originates from the **cisterna chyli** in the abdomen and ascends through the aortic hiatus in the diaphragm. The thoracic duct is usually thin-walled and dull white; often, it is beaded because of its numerous valves. It ascends between the thoracic aorta on its left, the azygos vein on its right, the esophagus anteriorly, and the vertebral bodies posteriorly. At the level of the T4–T6 vertebrae, the thoracic duct crosses to the left, posterior to the esophagus, and ascends into the superior mediastinum. The thoracic duct receives branches from the middle and upper intercostal spaces of both sides through several collecting trunks. It also receives branches from posterior mediastinal structures. Near its termination, it often receives the jugular, subclavian, and bronchomediastinal lymphatic trunks. The thoracic duct usually empties into the venous system near the union of the left internal jugular and subclavian veins, the *left venous angle* (Fig. 4.49).

VESSELS AND LYMPH NODES OF POSTERIOR MEDIASTINUM

The thoracic aorta and its branches were discussed previously. The **azygos system of veins**, on each side of the vertebral column, drains the back and thoraco-abdominal walls as well as the mediastinal viscera (Fig. 4.49). The azygos system exhibits much variation not only in its origin but also in its course, tributaries, anastomoses, and termination. The *azygos vein* and its main tributary, the *hemi-azygos vein*, usually arise from "roots" originating from the posterior aspect of the IVC and/or renal vein, respectively, which merge with the ascending lumbar veins.

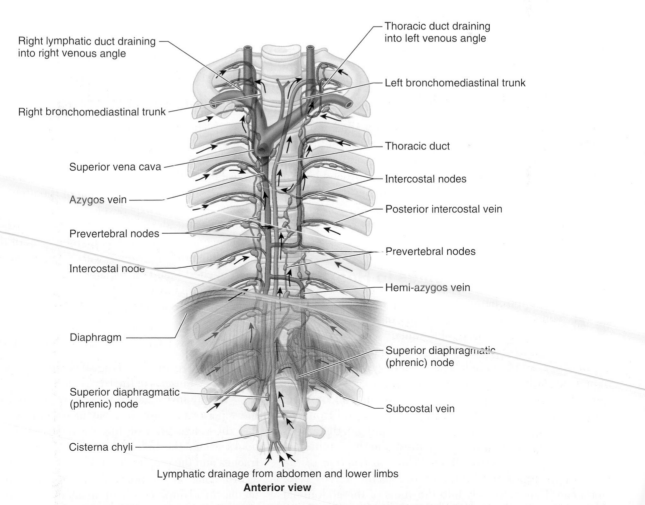

FIGURE 4.49. Posterior mediastinum: lymphatic drainage and azygos system of veins.

The **azygos vein** forms a collateral pathway between the SVC and the IVC and drains blood from the posterior walls of the thorax and abdomen. The azygos vein ascends in the posterior mediastinum, passing close to the right sides of the bodies of the inferior eight thoracic vertebrae. It arches over the superior aspect of the root of the right lung to join the SVC (see Fig. 4.44A). In addition to the posterior intercostal veins, the azygos vein communicates with the vertebral venous plexuses that drain the back, vertebrae, and structures in the vertebral canal (see Chapter 2). The azygos vein also receives the mediastinal, esophageal, and bronchial veins.

The **hemi-azygos vein** ascends on the left side of the vertebral column, posterior to the thoracic aorta as far as T9. Here, it crosses to the right, posterior to the aorta, thoracic duct, and esophagus, and joins the azygos vein.

The **accessory hemi-azygos vein** descends on the left side of the vertebral column from T5 to T8 and then crosses over the T7–T8 vertebrae posterior to the thoracic aorta and thoracic duct to join the azygos vein (see Fig. 4.44B). Sometimes, the accessory hemi-azygos vein joins the hemi-azygos vein and drains with it into the azygos vein.

Posterior mediastinal lymph nodes lie posterior to the pericardium, where they are related to the esophagus and thoracic aorta (Fig. 4.49). There are several nodes posterior to the inferior part of the esophagus and more anterior and lateral to it. The posterior mediastinal lymph nodes receive lymph from the esophagus, the posterior aspect of the pericardium and diaphragm, and the middle posterior intercostal spaces.

NERVES OF POSTERIOR MEDIASTINUM

The sympathetic trunks and their associated ganglia form a major portion of the autonomic nervous system (Fig. 4.49 and Table 4.5). The **thoracic sympathetic trunks** are in continuity with the cervical and lumbar sympathetic trunks. The thoracic sympathetic trunks lie against the heads of the ribs in the superior part of the thorax, the costovertebral joints in the midthoracic level, and the sides of the vertebral bodies in the inferior part of the thorax. The **lower thoracic splanchnic nerves**, also known as *greater, lesser, and least splanchnic nerves*, are part of the *abdominopelvic splanchnic nerves* because they supply viscera inferior to the diaphragm. They consist of presynaptic fibers from the 5th to 12th paravertebral sympathetic ganglia, which pass through the diaphragm and synapse in prevertebral ganglia in the abdomen. They supply sympathetic innervation for most of the abdominal viscera. These splanchnic nerves are discussed further in Chapter 5, Abdomen.

CLINICAL BOX

Laceration of Thoracic Duct

Because the thoracic duct is thin-walled and may be colorless, it may not be easily identified. Consequently, it is vulnerable to inadvertent injury during investigative and/or surgical procedures in the posterior mediastinum. *Laceration of the thoracic duct* results in chyle escaping into the thoracic cavity. Chyle may also enter the pleural cavity, producing *chylothorax*.

Collateral Venous Routes to Heart

The azygos, hemi-azygos, and accessory hemi-azygos veins offer alternate means of venous drainage from the thoracic, abdominal, and back regions when *obstruction of the IVC* occurs. In some people, an accessory azygos vein parallels the main azygos vein on the right side. Other people have no hemi-azygos system of veins. A clinically important variation, although uncommon, is when the azygos system receives all the blood from the IVC, except that from the liver. In these people, the azygos system drains nearly all the blood inferior to the diaphragm, except from the digestive tract. When *obstruction of the SVC* occurs superior to the entrance of the azygos vein, blood can drain inferiorly into the veins of the abdominal wall and return to the right atrium through the IVC and azygos system of veins.

Aneurysm of Ascending Aorta

The distal part of the ascending aorta receives a strong thrust of blood when the left ventricle contracts. Because its wall is not yet reinforced by fibrous pericardium (the fibrous pericardium blends with the aortic adventitia at the beginning of the arch), an *aneurysm* (localized dilation) may develop. An *aortic aneurysm* is evident on a chest film (radiograph of the thorax) or a magnetic resonance angiogram as an enlarged area of the ascending aorta silhouette. Individuals with an aneurysm usually complain of chest pain that radiates to the back. The aneurysm may exert pressure on the trachea, esophagus, and recurrent laryngeal nerve, causing difficulty in breathing and swallowing.

Injury to Recurrent Laryngeal Nerves

The recurrent laryngeal nerves supply all the intrinsic muscles of the larynx, except one. Consequently, any investigative procedure or disease process in the superior mediastinum may involve these nerves and affect the voice. Because the left recurrent laryngeal nerve hooks around the arch of the aorta and ascends between the trachea and the esophagus, it may be involved when there is a bronchial or esophageal carcinoma, enlargement of mediastinal lymph nodes, or an aneurysm of the arch of the aorta. In the latter condition, the nerve may be stretched by the dilated arch of the aorta.

Variations of Great Arteries

The most superior part of the arch of the aorta is usually approximately 2.5 cm inferior to the superior border of the manubrium, but it may be more superior or inferior. Sometimes, the arch curves over the root of the right lung and passes inferiorly on the right side, forming a **right arch of the aorta**. Less frequently, a **double arch of the aorta** or **retro-esophageal right subclavian artery** forms a vascular ring around the esophagus and trachea (Fig. B4.19). If the trachea is compressed enough to affect breathing, surgical division of the vascular ring may be needed.

Variations in the origin of the branches of the arch are fairly common. The usual pattern of branches of the arch of the aorta is present in approximately 65% of people. In approximately 27% of people, the left common carotid artery originates from the brachiocephalic trunk. A brachiocephalic trunk fails to form in approximately 2.5% of people; in these cases, each of the four arteries (right and left common carotid and subclavian arteries) originates independently from the arch of the aorta (Tubbs et al., 2016).

Coarctation of the Aorta

In *coarctation of the aorta*, the arch of the aorta or descending aorta has an abnormal narrowing (*stenosis*) that diminishes the caliber of the aortic lumen, producing an obstruction to blood flow to the inferior part of the body (Fig. B4.20). The most common site for a coarctation is near the ligamentum arteriosum. When the coarctation is inferior to this site (*postductal coarctation*), a good collateral circulation usually develops between the proximal and distal parts of the aorta through the intercostal and internal thoracic arteries.

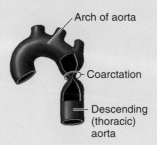

FIGURE B4.20. Coarctation of the aorta.

Age Changes in Thymus

The thymus is a prominent feature of the superior mediastinum during infancy and childhood. In some infants, the thymus may compress the trachea. The thymus plays an important role in the development and maintenance of the immune system. As puberty is reached, the thymus begins to diminish in relative size. By adulthood, it is usually replaced by adipose tissue and is often scarcely recognizable; however, it continues to produce T lymphocytes.

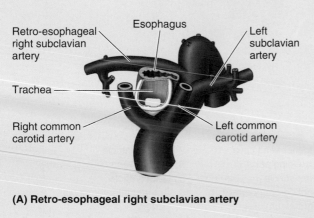

(A) Retro-esophageal right subclavian artery

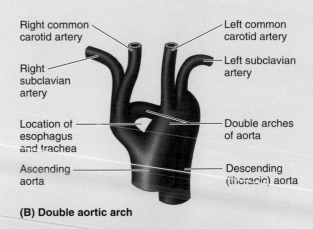

(B) Double aortic arch

FIGURE B4.19. Aortic arch anomalies.

MEDICAL IMAGING

Thorax

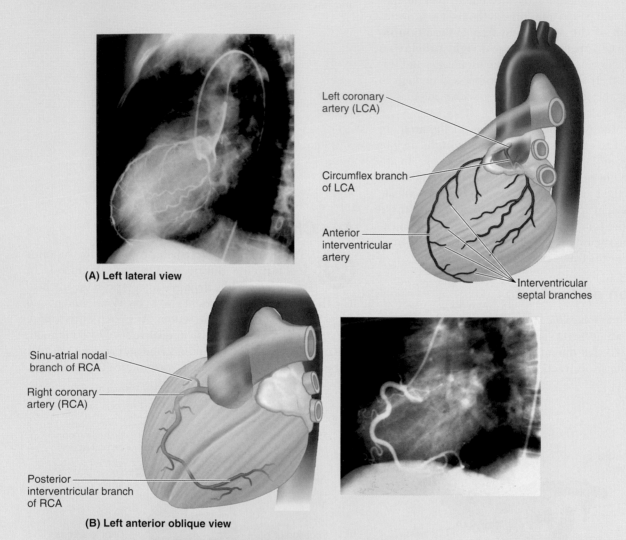

Left coronary artery (LCA)

Circumflex branch of LCA

Anterior interventricular artery

Interventricular septal branches

(A) Left lateral view

Sinu-atrial nodal branch of RCA

Right coronary artery (RCA)

Posterior interventricular branch of RCA

(B) Left anterior oblique view

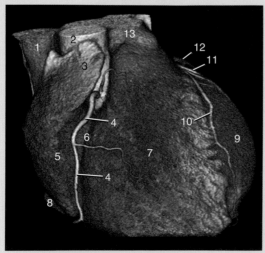

(C) Anterior view

Key	
1. Superior vena cava	7. Right ventricle
2. Ascending aorta	8. Inferior vena cava
3. Right auricle	9. Left ventricle
4. Right coronary artery	10. Anterior interventricular artery
5. Right atrium	
6. Coronary (atrioventricular) sulcus	11. Circumflex branch
	12. Left auricle
	13. Pulmonary trunk

FIGURE 4.50. Imaging of coronary vessels. A and B. Coronary arteriograms. Radiopaque dye has been injected into the left **(A)** and the right **(B)** coronary arteries. **C.** 3D volume reconstruction of heart and coronary vessels.

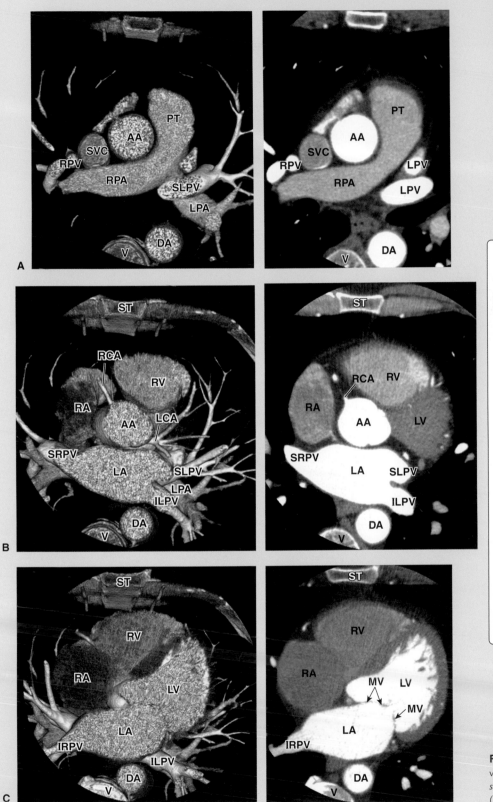

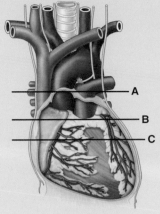

AA	Ascending aorta
DA	Descending aorta
ILPV	Inferior left pulmonary vein
IRPV	Inferior right pulmonary vein
LA	Left atrium
LCA	Left coronary artery
LPA	Left pulmonary artery
LPV	Left pulmonary vein
LV	Left ventricle
MV	Mitral valve
PT	Pulmonary trunk
RA	Right atrium
RCA	Right coronary artery
RPA	Right pulmonary artery
RPV	Right pulmonary vein
RV	Right ventricle
SLPV	Superior left pulmonary vein
SRPV	Superior right pulmonary vein
ST	Sternum
SVC	Superior vena cava
V	Vertebra

FIGURE 4.51. Transverse (axial) 3D volume reconstructions of thorax (on *left side* of page) and CT angiograms of thorax (on *right side* of page).

5

Abdomen

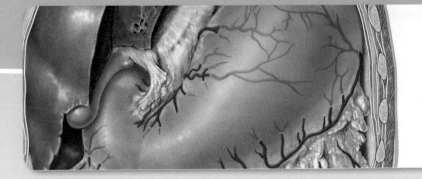

CLINICAL BOX KEY

Anatomical
Variations

Diagnostic
Procedures

Life Cycle

Surgical
Procedures

Trauma

Pathology

The **abdomen** is the part of the trunk between the thorax and the pelvis. The anterolateral wall is musculo-aponeurotic. Posteriorly, the wall includes the lumbar vertebral column and the posterior diaphragm that overlies the thoracic vertebrae and lower ribs (Fig. 5.1A). The abdominal wall encloses the abdominal cavity, containing the peritoneal cavity and housing most of the organs (viscera) of the alimentary system and part of the urogenital system.

ABDOMINAL CAVITY

The **abdominal cavity** is the space bounded by the abdominal walls, diaphragm, and pelvis. The abdominal cavity forms the major part of the **abdominopelvic cavity**—the combined and continuous abdominal and pelvic cavities (Fig. 5.1). The abdominal cavity is

- enclosed anterolaterally by dynamic musculo-aponeurotic abdominal walls
- separated superiorly from the thoracic cavity and posteriorly from the posterior thoracic vertebrae by the diaphragm
- under cover of the thoracic cage superiorly extending to the 4th intercostal space
- continuous inferiorly with the pelvic cavity
- lined with peritoneum, a serous membrane
- the location of most of the digestive organs, spleen, kidneys, and ureters for most of their course

Clinicians subdivide the abdominal cavity into nine regions to locate abdominal organs or pain sites: right and left hypochondriac, right and left lateral (lumbar), right and left inguinal (groin), epigastric, umbilical, and pubic (hypogastric). The nine regions are delineated by four planes (Fig. 5.2A):

- Two horizontal planes:
  - **Subcostal plane**, passing through the inferior border of the 10th costal cartilage on each side
  - **Transtubercular plane**, passing through the iliac tubercles and the body of the L5 vertebra
- Two vertical planes:
  - **Midclavicular planes**, passing from the midpoints of clavicles to the **midinguinal points**, the midpoints of lines joining the *anterior superior iliac spines* and the superior edge of the *pubic symphysis*

For more general clinical descriptions, clinicians use four quadrants of the abdominal cavity: right upper, right lower, left upper, and left lower. The four quadrants are defined by two planes (Fig. 5.2B):

- **Transumbilical plane**, passing through the umbilicus and intervertebral (IV) disc between the L3 and L4 vertebrae
- **Median plane**, passing longitudinally through the body, dividing it into right and left halves

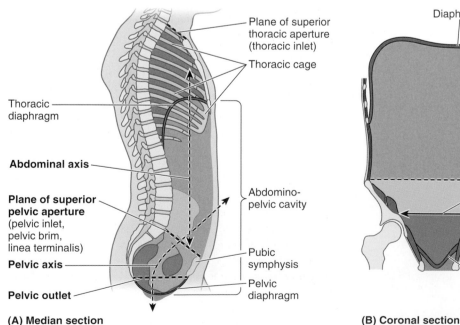

(A) Median section

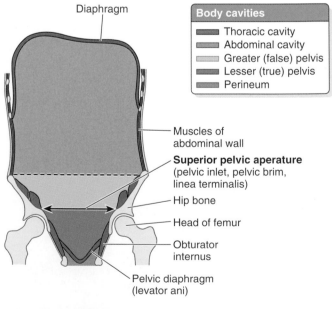

(B) Coronal section

FIGURE 5.1. Abdominopelvic cavity. A. The pelvic inlet (superior pelvic aperture) is the opening into the lesser pelvis. The pelvic outlet (inferior pelvic aperture) is the lower opening of the lesser pelvis. **B.** The plane of the pelvic brim (*double-headed arrow*) separates the greater pelvis (part of the abdominal cavity) from the lesser pelvis (the pelvic cavity).

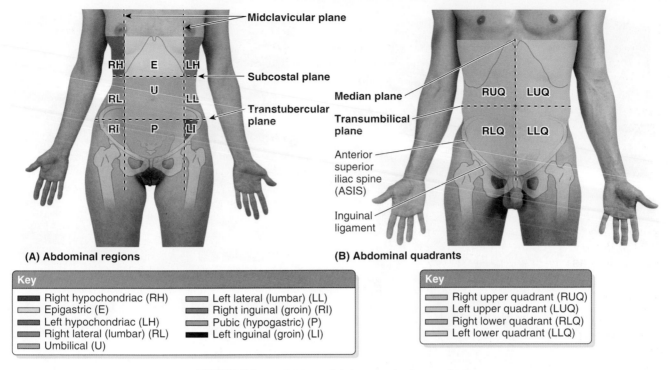

(A) Abdominal regions

Midclavicular plane

RH E LH

Subcostal plane

RL U LL

Transtubercular plane

RI P LI

(B) Abdominal quadrants

Median plane

RUQ LUQ

Transumbilical plane

RLQ LLQ

Anterior superior iliac spine (ASIS)

Inguinal ligament

Key	
Right hypochondriac (RH)	Left lateral (lumbar) (LL)
Epigastric (E)	Right inguinal (groin) (RI)
Left hypochondriac (LH)	Pubic (hypogastric) (P)
Right lateral (lumbar) (RL)	Left inguinal (groin) (LI)
Umbilical (U)	

Key
Right upper quadrant (RUQ)
Left upper quadrant (LUQ)
Right lower quadrant (RLQ)
Left lower quadrant (LLQ)

FIGURE 5.2. Subdivisions of abdomen and reference planes.

ANTEROLATERAL ABDOMINAL WALL

Although the abdominal wall is continuous, it is subdivided for descriptive purposes into the *anterior wall*, *right and left lateral walls*, and *posterior wall*. The boundary between the anterior and the lateral walls is indefinite. Consequently, the combined term **anterolateral abdominal wall**, extending from the thoracic cage to the pelvis, is often used. The anterolateral abdominal wall is bounded superiorly by the thoracic cage (cartilages of the 7th through 10th ribs and the xiphoid process of the sternum) and inferiorly by the inguinal ligament and hip (pelvic) bones (Fig. 5.1). The wall consists of skin, subcutaneous tissue (superficial fascia), muscles and their aponeuroses, deep fascia, extraperitoneal fat, and parietal peritoneum (Fig. 5.3). The skin attaches loosely to the subcutaneous tissue except at the umbilicus, where it adheres firmly.

Fascia of Anterolateral Abdominal Wall

The fascial layers from superficial to deep include the **subcutaneous tissue** (superficial fascia), which lies deep to the skin and contains a variable amount of fat (Fig. 5.3). Inferior to the umbilicus, the subcutaneous tissue is composed of two layers: a **superficial fatty layer** (Camper fascia) and a **deep membranous layer** (Scarpa fascia) (see also Fig. B5.1).

The **investing fascia** (epimysium) covers the external aspects of the three muscle layers of the anterolateral abdominal wall and their aponeuroses.

The **endo-abdominal fascia** is a membranous sheet of varying thickness that lines the internal aspect of the abdominal wall. Although continuous, different parts of this fascia can be named according to the muscle or aponeurosis it is lining—for example, the portion lining the deep surface of the transversus abdominis muscle or aponeurosis is the **transversalis fascia**.

The *parietal peritoneum* lines the abdominal cavity and is located internal to the transversalis fascia. It is separated from the transversalis fascia by a variable amount of **extraperitoneal fat**.

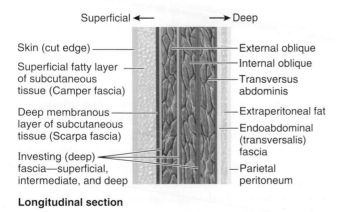

Superficial ← → Deep

Skin (cut edge)

Superficial fatty layer of subcutaneous tissue (Camper fascia)

Deep membranous layer of subcutaneous tissue (Scarpa fascia)

Investing (deep) fascia—superficial, intermediate, and deep

External oblique

Internal oblique

Transversus abdominis

Extraperitoneal fat

Endoabdominal (transversalis) fascia

Parietal peritoneum

Longitudinal section

FIGURE 5.3. Fascia of anterior abdominal wall.

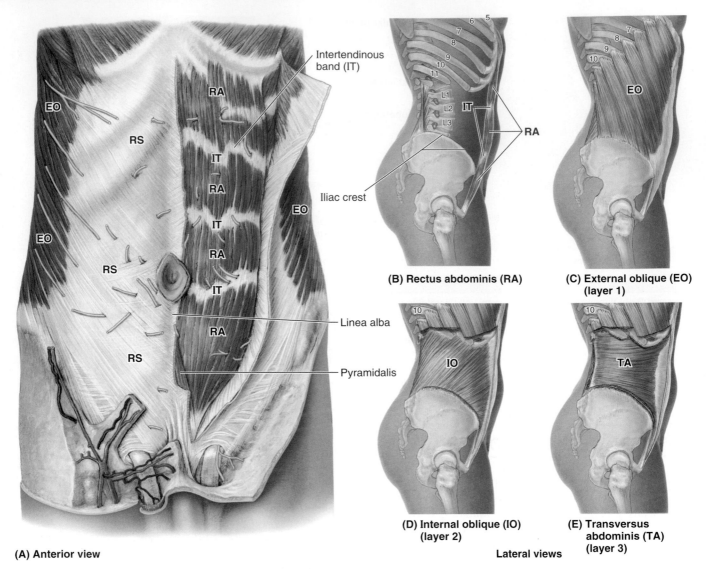

(A) Anterior view

(B) Rectus abdominis (RA)

(C) External oblique (EO) (layer 1)

(D) Internal oblique (IO) (layer 2)

(E) Transversus abdominis (TA) (layer 3)

Lateral views

FIGURE 5.4. Muscles of anterolateral abdominal wall. A. Right side, external oblique (EO) and intact rectus sheath (RS); left side, opened rectus sheath, revealing rectus abdominis (RA) and pyramidalis. **B.** Rectus abdominis. **C.** External oblique. **D.** Internal oblique. **E.** Transversus abdominis.

Muscles of Anterolateral Abdominal Wall

There are five (bilaterally paired) muscles in the anterolateral abdominal wall (Fig. 5.4): three flat muscles and two vertical muscles. Their attachments, nerve supply, and main actions are listed in Table 5.1.

The three flat muscles are as follows:

- **External oblique**, the superficial muscle. Its fibers pass inferomedially and interdigitate with slips of the serratus anterior. The inferior margin is thickened as an undercurving fibrous band that spans between the anterior superior iliac spine and the pubic tubercle as the **inguinal ligament**.
- **Internal oblique**, the intermediate muscle. Its fibers fan out so that its upper fibers are perpendicular and its lower fibers are parallel to those of the external oblique.

- **Transversus abdominis**, the innermost muscle. Its fibers, except for the most inferior ones, run horizontally.

All three flat muscles end anteriorly in a strong sheet-like *aponeurosis*. Between the midclavicular line and the midline, the aponeuroses form the tough *rectus sheath*, enclosing the rectus abdominis. The aponeuroses interweave, forming a midline raphe (G. *rhaphe*, suture, seam)—the **linea alba** (L. white line)—which extends from the xiphoid process to the pubic symphysis. The interweaving is not only between right and left sides but also between superficial, intermediate, and deep layers. For example, the tendinous fibers of the external oblique that decussate at the linea alba, for the most part, become continuous with the tendinous fibers of the contralateral internal oblique, forming a two-bellied muscle sharing a common central tendon. These two muscles work together to flex and rotate the trunk.

TABLE 5.1. PRINCIPAL MUSCLES OF ANTEROLATERAL ABDOMINAL WALL

Muscles	Origin	Insertion	Innervation	Action(s)
External oblique	External surfaces of 5th–12th ribs	Linea alba, pubic tubercle, and anterior half of iliac crest	Thoraco-abdominal and subcostal nerves (anterior rami of T7–T12 spinal nerves)	Compress and support abdominal viscera; flex and rotate trunk
Internal oblique	Thoracolumbar fascia, anterior two thirds of iliac crest, and connective tissue deep to inguinal ligament	Inferior borders of 10th–12th ribs, linea alba, and pubis via conjoint tendon	Thoraco-abdominal nerves (anterior rami of T7–T11), subcostal and first lumbar nerve	
Transversus abdominis	Internal surfaces of 7th–12th costal cartilages, thoracolumbar fascia, iliac crest, and connective tissue deep to inguinal ligament	Linea alba with aponeurosis of internal oblique, pubic crest, and pubis via conjoint tendon		Compresses and supports abdominal viscera
Rectus abdominis	Pubic symphysis and pubic crest	Xiphoid process and 5th–7th costal cartilages	Thoraco-abdominal and subcostal nerves (anterior rami of T7–T12 spinal nerves)	Flexes trunk (lumbar vertebrae) and compresses abdominal viscera[a]; stabilizes and controls tilt of pelvis (antilordosis)

[a]In so doing, these muscles act as antagonists of the diaphragm to produce expiration.

The two vertical muscles are as follows:

- **Rectus abdominis**, a long, broad, strap-like muscle that is mostly enclosed in the *rectus sheath* (Figs. 5.4 and 5.5). The muscle fibers of the rectus do not run the length of the muscle; rather, they run between three or more **tendinous intersections** (Fig. 5.4A), which are typically located at the level of the xiphoid process of the sternum, the umbilicus, and a level halfway between these points. Each intersection is firmly attached to the anterior layer of the rectus sheath.
- **Pyramidalis**, a small triangular muscle (absent in about 20% of people) that lies in the rectus sheath anterior to the inferior part of the rectus abdominis (Fig. 5.4A). It arises from the pubic crest and attaches along the linea alba, which it tenses.

FUNCTIONS AND ACTIONS OF ANTEROLATERAL ABDOMINAL MUSCLES

The muscles of the anterolateral abdominal wall perform the following functions:

- Form a strong expandable support for this region
- Protect the abdominal viscera from injury
- Compress the abdominal viscera to maintain or increase intra-abdominal pressure. Compressing the abdominal viscera and increasing intra-abdominal pressure elevates the relaxed diaphragm to expel air, for example, during respiration, coughing, and voluntary eructation (burping). When the diaphragm contracts during inspiration, the anterolateral abdominal wall expands as the muscles relax to make room for the viscera that are pushed inferiorly.
- Produce the force required for defecation (evacuation of fecal material from the rectum), micturition (urination), vomiting, and parturition (childbirth)
- Produce anterior and lateral flexion and rotation of the trunk and help maintain posture

The **rectus sheath** is formed by the interlaced aponeuroses of the flat abdominal muscles (Fig. 5.5). Superior to the arcuate line (about one third of the distance from the umbilicus to the pubic crest), the rectus abdominis is enveloped by the anterior layer of the rectus sheath, formed by the external oblique aponeurosis and the anterior lamina of the internal oblique aponeurosis, and posterior layer of the rectus sheath, formed by the posterior lamina of the internal oblique aponeurosis and the transversus abdominis aponeurosis (Fig. 5.5A). Inferior to the arcuate line, the aponeuroses of all three muscles, external and internal oblique and transversus abdominis, pass anterior to the rectus abdominis to form the anterior rectus sheath, leaving only the transversalis fascia to cover the rectus abdominis posteriorly (Fig. 5.5B). The **arcuate line** then often demarcates the transition between the posterior rectus sheath covering the superior three quarters of the rectus abdominis proximally and the transversalis fascia covering the inferior quarter (Fig. 5.6).

The *contents of the rectus sheath* are the rectus abdominis and pyramidalis muscles, the anastomosing superior and inferior epigastric arteries and veins, the lymphatic vessels, and the thoraco-abdominal and subcostal nerves (distal portions of the anterior rami of spinal nerves T7–T12), which supply the muscles and overlying skin (Fig. 5.5C).

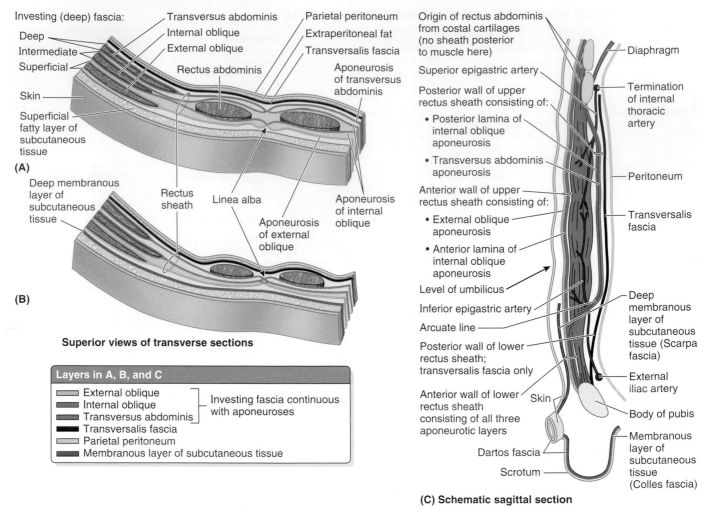

FIGURE 5.5. Structure of anterolateral abdominal wall. A. Transverse section superior to umbilicus. **B.** Transverse section inferior to umbilicus. **C.** Sagittal section of rectus sheath. Planes of sections for **A** and **B** are shown in Figure 5.6.

Internal Surface of Anterolateral Abdominal Wall

The internal surface of the anterolateral abdominal wall is covered with transversalis fascia, a variable amount of extraperitoneal fat, and parietal peritoneum (Figs. 5.3 and 5.5A,B). The infra-umbilical part of this surface of the wall exhibits several peritoneal folds, some of which contain remnants of vessels that carried blood to and from the fetus (Moore et al., 2016).

Five umbilical peritoneal folds—two on each side and one in the median plane—pass toward the umbilicus (Fig. 5.6):

- The **median umbilical fold**, extending from the apex of the urinary bladder to the umbilicus, covers the **median umbilical ligament**, the remnant of the *urachus* that joined the apex of the fetal bladder to the umbilicus.
- Two **medial umbilical folds**, lateral to the median umbilical fold, cover the **medial umbilical ligaments**, formed by the occluded parts of the umbilical arteries.

- Two **lateral umbilical folds**, lateral to the medial umbilical folds, cover the *inferior epigastric vessels* and, therefore, bleed if cut.

The depressions lateral to the umbilical folds are *peritoneal fossae*, some of which are potential sites for a hernia. The location of a hernia in one of these fossae determines how the hernia is classified. The following shallow fossae are located between the umbilical folds (Fig. 5.6):

- **Supravesical fossae** between the median and the medial umbilical folds, formed as the peritoneum reflects from the anterior abdominal wall onto the bladder. The level of the supravesical fossae rises and falls with filling and emptying of the bladder.
- **Medial inguinal fossae** between the medial and the lateral umbilical folds, areas also commonly called **inguinal triangles** (Hesselbach triangles). These are potential sites for *direct* inguinal hernias.
- **Lateral inguinal fossae**, lateral to the lateral umbilical folds; these include the deep inguinal rings and are potential sites for the most common type of inguinal hernia, the *indirect* inguinal hernia.

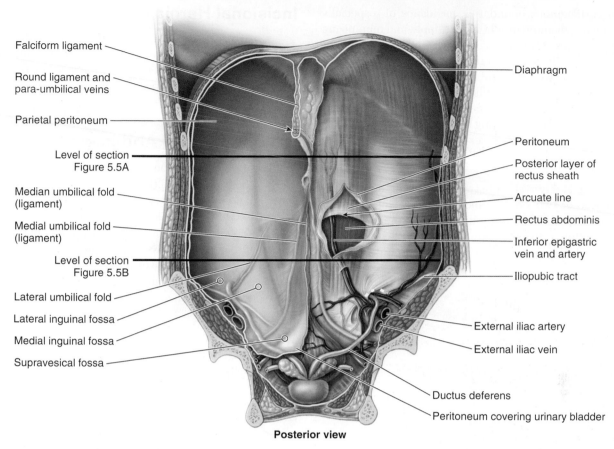

Posterior view

FIGURE 5.6. Posterior aspect of anterolateral abdominal wall showing peritoneal ligaments, folds, and fossae.

CLINICAL BOX

Clinical Significance of Fascia and Fascial Spaces of Abdominal Wall

When closing abdominal skin incisions, surgeons suture the membranous layer of subcutaneous tissue as a separate layer because of its strength. Between the membranous layer and the deep fascia covering the rectus abdominis and external oblique muscles is a potential space where fluid may accumulate (e.g., urine from a ruptured urethra). Although no barriers (other than gravity) prevent fluid from spreading superiorly from this space, it cannot spread inferiorly into the thigh because the membranous layer of subcutaneous tissue attaches to the pubic bone and fuses with the deep fascia of the thigh (fascia lata) along a line inferior and parallel to the inguinal ligament (Fig. B5.1).

Abdominal Surgical Incisions

Surgeons use various incisions to gain access to the abdominal cavity. The incision that allows adequate exposure and, secondarily, the best possible cosmetic effect is chosen. The location of the incision also depends on the type of operation, the location of the organ(s),

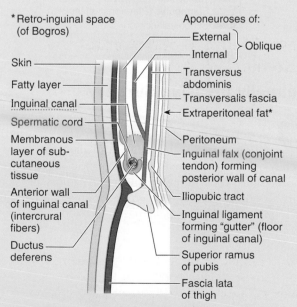

FIGURE B5.1. Schematic sagittal section of inguinal canal.

(Continued on next page)

bony or cartilaginous boundaries, avoidance of (especially motor) nerves, maintenance of blood supply, and minimizing injury to muscles and fascia of the wall while aiming for favorable healing. Instead of transecting muscles, causing irreversible *necrosis* (death) of muscle fibers, the surgeon splits muscles between their fibers. The rectus abdominis is an exception and can be transected because its muscle fibers are short and its nerves entering the lateral part of the rectus sheath can be located and preserved. Cutting a motor nerve paralyzes the muscle fibers supplied by it, thereby weakening the anterolateral abdominal wall. However, because of overlapping areas of innervation between nerves in the abdominal wall, one or two small branches of nerves may be cut without a noticeable loss of motor supply to the muscles or loss of sensation to the skin. Some of the most common abdominal surgical incisions are illustrated in Figure B5.2.

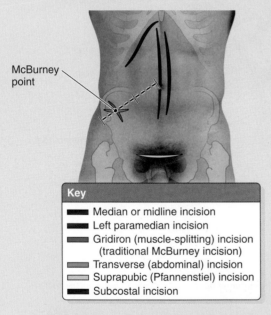

McBurney
point

Key
- Median or midline incision
- Left paramedian incision
- Gridiron (muscle-splitting) incision (traditional McBurney incision)
- Transverse (abdominal) incision
- Suprapubic (Pfannenstiel) incision
- Subcostal incision

FIGURE B5.2. Abdominal surgical incisions.

Minimally Invasive Surgery

 Many abdominopelvic surgical procedures are now performed using an *endoscope*, in which tiny perforations into the abdominal wall allow the entry of remotely operated instruments, replacing the larger conventional incisions. Thus, the potential for nerve injury, incisional hernia, or contamination through the open wound and the time required for healing are minimized.

Incisional Hernia

 If the muscular and aponeurotic layers of the abdomen do not heal properly, a hernia may occur through the defect. An *incisional hernia* is a protrusion of omentum (fold of peritoneum) or an organ through a surgical incision or scar.

Protuberance of Abdomen

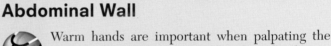 The six common causes of abdominal protrusion begin with the letter F: food, fluid, fat, feces, flatus, and fetus. Eversion of the umbilicus may be a sign of increased intra-abdominal pressure, usually resulting from *ascites* (abnormal accumulation of serous fluid in the peritoneal cavity) or a large mass (e.g., a tumor, a fetus, or an enlarged organ such as the liver).

Excess fat accumulation owing to overnourishment most commonly involves the subcutaneous fatty layer; however, there may also be excessive depositions of extraperitoneal fat.

Palpation of Anterolateral Abdominal Wall

Warm hands are important when palpating the abdominal wall because cold hands make the anterolateral abdominal muscles tense, producing involuntary spasms of the muscles known as *guarding*. Intense guarding, board-like reflexive muscular rigidity that cannot be willfully suppressed, occurs during palpation when an organ (such as the appendix) is inflamed and in itself constitutes a clinically significant sign of *acute abdomen*. The involuntary muscular spasms attempt to protect the inflamed viscera from pressure. The shared segmental nerve supply of the organ and skin and muscles of the wall explains why these spasms occur.

Palpation of abdominal viscera is performed with the patient in the supine position, with thighs and knees semiflexed to enable adequate relaxation of the anterolateral abdominal wall. Otherwise, the deep fascia of the thighs pulls on the membranous layer of abdominal subcutaneous tissue, tensing the abdominal wall. Some people tend to place their hands behind their heads when lying supine, which also tightens the muscles and makes the examination difficult. Placing the upper limbs at the sides and putting a pillow under the person's knees tends to relax the anterolateral abdominal muscles.

SURFACE ANATOMY

Anterolateral Abdominal Wall

The **umbilicus** is where the umbilical cord, from the placenta, entered the fetus and is the reference point for the transumbilical plane (Fig. SA5.1A,B). It indicates the level of the T10 dermatome and is typically at the level of the IV disc between the L3 and L4 vertebrae; however, its position varies with the amount of fat in the person's subcutaneous tissue. The linea alba is a subcutaneous fibrous band extending from the **xiphoid process** to the **pubic symphysis** that is demarcated by a midline vertical skin groove as far inferiorly as the umbilicus (Fig. SA5.1A,B). The pubic symphysis can be felt in the median plane at the inferior end of the linea alba. The bony **iliac crest** at the level of the L4 vertebra can be easily palpated as it extends posteriorly from the **anterior superior iliac spine**.

In an individual with good muscle definition, curved skin grooves, the **semilunar lines** (L. *linae semilunares*) demarcate the lateral borders of the rectus abdominis and rectus sheath. The semilunar lines extend from the inferior costal margin near the 9th costal cartilages to the **pubic tubercles**. Three transverse skin grooves may overlie the **tendinous intersections** of the rectus abdominis (Fig. SA5.1B). The interdigitating bellies of the **serratus anterior** and **external oblique muscles** are also visible. A skin crease, the **inguinal groove**, indicates the site of the inguinal ligament. The groove is located just inferior and parallel to the ligament, marking the division between the anterolateral abdominal wall and the thigh.

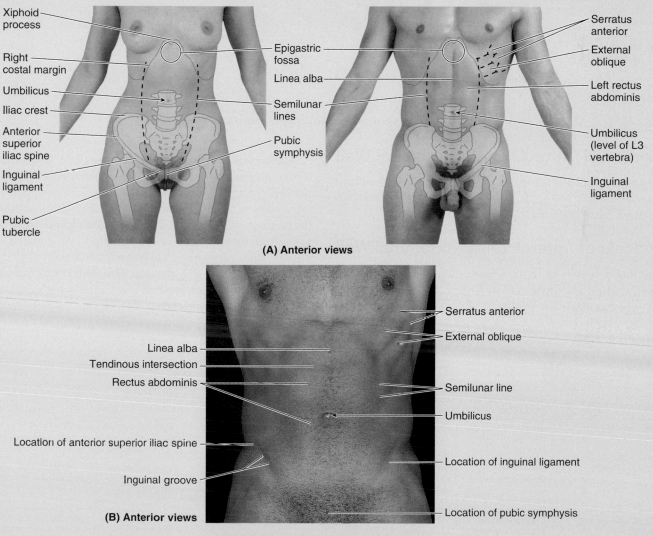

(A) Anterior views

(B) Anterior views

FIGURE SA5.1. Anterolateral abdominal wall.

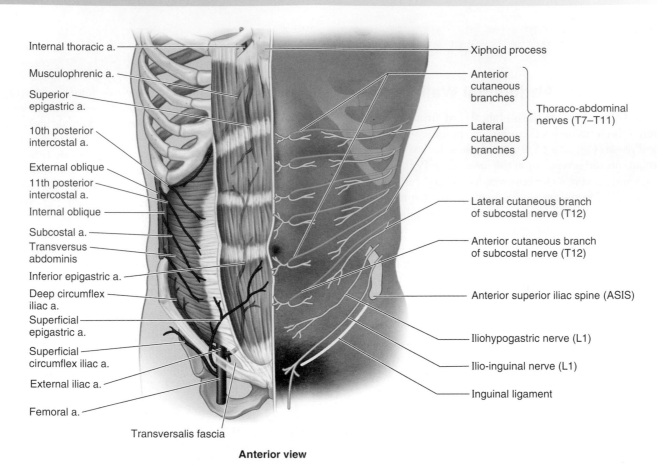

Anterior view

FIGURE 5.7. Arteries and nerves of anterolateral abdominal wall. *a.*, artery.

Nerves of Anterolateral Abdominal Wall

The skin and muscles of the anterolateral abdominal wall are supplied mainly by the nerves illustrated on the left side in Figure 5.7 and listed and described in Table 5.2.

Vessels of Anterolateral Abdominal Wall

The blood vessels of the anterolateral abdominal wall are illustrated on the right side in Figure 5.7 and listed and described in Table 5.3.

The **superior epigastric artery**, the direct continuation of the internal thoracic artery, enters the rectus sheath superiorly through its posterior layer (Fig. 5.5C), supplies the upper part of the rectus abdominis, and anastomoses with the inferior epigastric artery. The **inferior epigastric artery** arises from the external iliac artery deep to the inguinal ligament. It runs superiorly in the transversalis fascia to enter the rectus sheath inferior to the arcuate line. Its branches enter the lower rectus abdominis and anastomose with branches of the superior epigastric artery.

TABLE 5.2. NERVES OF ANTEROLATERAL ABDOMINAL WALL

Nerve	Origin	Course	Distribution
Thoraco-abdominal (T7–T11)	Distal, abdominal parts of lower five intercostal nerves	Run between second and third layers of abdominal muscles; muscular, lateral, and anterior cutaneous branches enter subcutaneous tissue.	Muscles of anterolateral abdominal wall and overlying skin (T7–T9 superior to umbilicus; T10 around umbilicus; T11 immediately below umbilicus)
Subcostal (T12)	Anterior ramus of T12 spinal nerve	Runs along inferior border of 12th rib, then onto subumbilical abdominal wall	Muscles of anterolateral abdominal wall and overlying skin midway between level of umbilicus and iliac crest, inguinal ligament, and pubic crest inferiorly
Iliohypogastric (L1)	Superior terminal branch of anterior ramus of L1 spinal nerve	Pierces transversus abdominis muscle; branches pierce external oblique aponeurosis of most inferior abdominal wall	Skin overlying iliac crest, upper inguinal, and hypogastric regions; internal oblique and transversus abdominis
Ilio-inguinal (L1)	Inferior terminal branch of anterior ramus of L1 spinal nerve	Passes between second and third layers of abdominal muscles, then traverses inguinal canal	Skin of scrotum or labium majus, mons pubis, and adjacent medial aspect of thigh; most inferior internal oblique and transversus abdominis

TABLE 5.3. PRINCIPAL ARTERIES OF ANTEROLATERAL ABDOMINAL WALL

Nerve	Origin	Course	Distribution
Musculophrenic	Internal thoracic artery	Descends along costal margin	Abdominal wall of hypochondriac region, anterolateral, diaphragm
Superior epigastric		Descends in rectus sheath deep to rectus abdominis	Superior rectus abdominis and superior part of anterolateral abdominal wall
10th and 11th posterior intercostal	Abdominal aorta	Arteries continue beyond ribs to descend in abdominal wall between internal oblique and transversus abdominis	Abdominal wall, lateral region
Subcostal			
Inferior epigastric	External iliac artery	Runs superiorly and enters rectus sheath; runs deep to rectus abdominis	Inferior rectus abdominis and medial part of anterolateral abdominal wall
Deep circumflex iliac		Runs on deep aspect of anterior abdominal wall, parallel to inguinal ligament	Iliacus muscle and inferior part of anterolateral abdominal wall
Superficial circumflex iliac	Femoral artery	Runs in superficial fascia along inguinal ligament	Superficial abdominal wall of inguinal region and adjacent anterior thigh
Superficial epigastric		Runs in superficial fascia toward umbilicus	Subcutaneous tissue and skin over pubic and inferior umbilical region

A venous anastomosis between the *superficial epigastric* (femoral) *vein* and the *lateral thoracic* (axillary) *veins*—the **thoraco-epigastric vein**—provides a potential collateral pathway for blood normally draining via the inferior vena cava (IVC) to return to the heart via the superior vena cava when the IVC is blocked.

The **superficial lymphatic vessels** of the abdominal wall accompany the subcutaneous veins; those superior to the umbilicus drain mainly to the **axillary lymph nodes**, whereas those inferior to it drain to the **superficial inguinal lymph nodes** (Fig. 5.8). The **deep lymphatic vessels** accompany the deep veins and drain to the external iliac, common iliac, and lumbar (caval and aortic) lymph nodes.

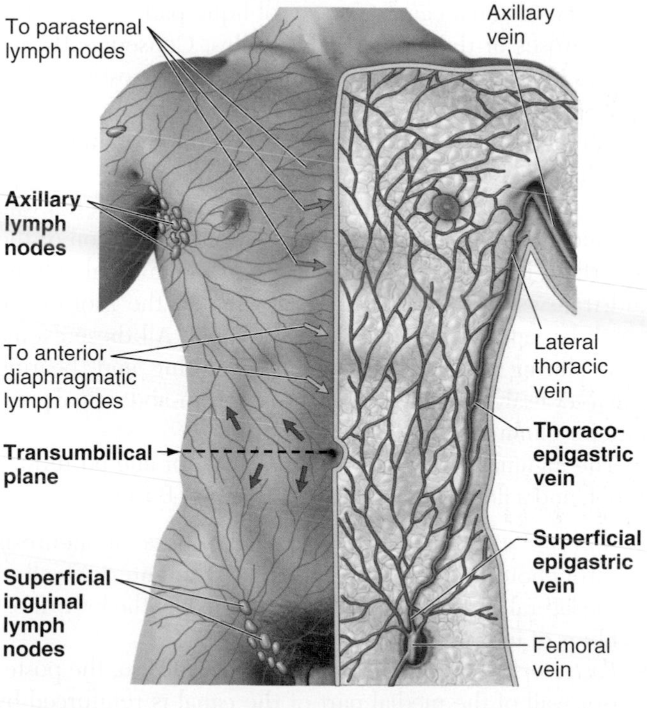

Inguinal Region

The **inguinal region** extends between the anterior superior iliac spine and the pubic tubercle (Fig. 5.9). Anatomically, it is a region where structures exit and enter the abdominal cavity and is, therefore, clinically important because these are potential sites of herniation. Inguinal hernias occur in both sexes, but most (about 86%) occur in males because of the passage of the spermatic cord through the inguinal canal. The migration of the testes from the abdomen into the perineum accounts for many of the structural features of the region (Fig. B5.5). Thus, the testis and scrotum are usually studied in relation to the anterior abdominal wall and inguinal region.

INGUINAL LIGAMENT AND ILIOPUBIC TRACT

The inguinal ligament, the most inferior part of the external oblique aponeurosis, and the *iliopubic tract*, the thickened inferior margin of the transversalis fascia, extend from the anterior superior iliac spine to the pubic tubercle. Most of the fibers of the inguinal ligament insert into the pubic tubercle, but some fibers (Fig. 5.9)

1. attach to the superior ramus of the pubis lateral to the pubic tubercle as the **lacunar ligament** and then continue to run along the pectin pubis as the **pectineal ligament** (of Cooper)
2. arch superiorly to blend with the contralateral external oblique aponeurosis as the **reflected inguinal ligament**

The **iliopubic tract** is a fibrous band that runs parallel and posterior (deep) to the inguinal ligament. It is seen in

FIGURE 5.8. Lymphatics and superficial veins of anterolateral abdominal wall.

Labels: To parasternal lymph nodes; Axillary vein; Axillary lymph nodes; To anterior diaphragmatic lymph nodes; Transumbilical plane; Lateral thoracic vein; **Thoraco-epigastric vein**; **Superficial epigastric vein**; Femoral vein; Superficial inguinal lymph nodes; **Anterior view**

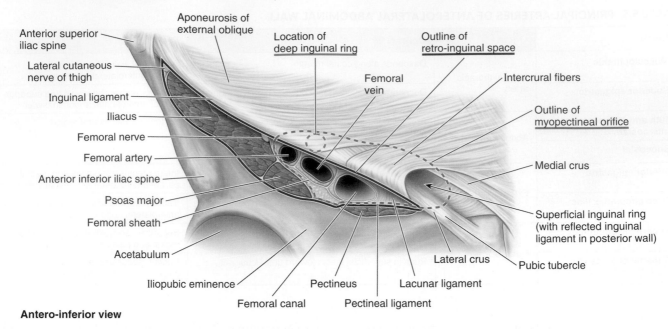

Antero-inferior view

FIGURE 5.9. Inguinal ligament and superficial inguinal ring. Note the lacunar and pectineal ligaments.

place of the inguinal ligament when the inguinal region is viewed from its internal (posterior) aspect, as through an endoscope (Figs. 5.6 and 5.10B). The iliopubic tract reinforces the posterior wall and floor of the inguinal canal as it bridges the structures (hip flexors and much of the neurovascular supply of the lower limb) traversing the **retro-inguinal space** (Fig. 5.9).

INGUINAL CANAL

The **inguinal canal** is formed in relation to the relocation of the gonad (testes or ovary) during fetal development (see clinical box "Relocation of Testes and Ovaries"). The inguinal canal in adults is an approximately 4 cm long, inferomedially directed oblique passage (between the superficial and deep inguinal rings) that runs through the inferior part of the anterior abdominal wall (Fig. 5.10). The inguinal canal lies parallel and just superior to the medial half of the inguinal ligament. The main structure in the inguinal canal is the *spermatic cord* conveying the ductus deferens in males and the vestigial *round ligament of the uterus* in females. The inguinal canal also contains blood and lymphatic vessels and the ilio-inguinal nerve in both sexes. The inguinal canal has an opening at each end (Fig. 5.10):

- The **deep (internal) ring**, the internal entrance to the inguinal canal, is an evagination of the transversalis fascia superior to the middle of the inguinal ligament and lateral to the inferior epigastric vessels.
- The **superficial (external) inguinal ring**, the exit from the inguinal canal, is a slit-like opening in the aponeurosis of the external oblique, superolateral to the pubic

tubercle. The lateral and medial margins of the superficial ring formed by the split in the aponeurosis are the **lateral** and **medial crura** (L. leg-like parts). The **intercrural fibers** form the superolateral margin of the ring (Fig. 5.9).

The deep and superficial inguinal rings do not overlap because the inguinal canal takes an oblique path through the aponeuroses of the abdominal muscles. Consequently, increases in intra-abdominal pressure force the posterior wall of the canal against the anterior wall, closing this passageway and strengthening this potential defect of the abdominal wall. Simultaneous contraction of the external oblique also approximates the anterior wall of the canal to the posterior wall and increases tension on the crura, resisting dilation of the superficial inguinal ring. Contraction of the internal oblique and transversus abdominis muscles makes the roof of the canal descend, which constricts the canal. All these events occur during acts such as sneezing, coughing, and "bearing down" (Valsalva maneuver) to increase intra-abdominal pressure for elimination (e.g., of feces).

The inguinal canal has two walls (anterior and posterior), a roof, and a floor (Fig. 5.10A; see also Fig. B5.1):

- *Anterior wall:* Formed by external oblique aponeurosis throughout the length of the canal; the anterior wall of the lateral part of the canal is reinforced by the lowermost fibers of internal oblique muscle.
- *Posterior wall:* Formed by transversalis fascia; the posterior wall of the medial part of the canal is reinforced by merging of the pubic attachments of the internal oblique and transversus abdominis aponeuroses into a common tendon—the **inguinal falx (conjoint tendon)**.

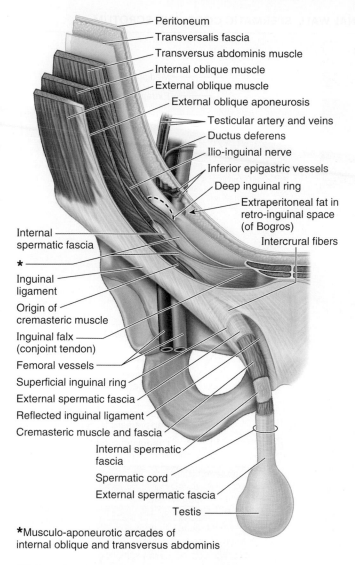

- Peritoneum
- Transversalis fascia
- Transversus abdominis muscle
- Internal oblique muscle
- External oblique muscle
- External oblique aponeurosis
- Testicular artery and veins
- Ductus deferens
- Ilio-inguinal nerve
- Inferior epigastric vessels
- Deep inguinal ring
- Extraperitoneal fat in retro-inguinal space (of Bogros)
- Intercrural fibers

Internal spermatic fascia

*

Inguinal ligament

Origin of cremasteric muscle

Inguinal falx (conjoint tendon)

Femoral vessels

Superficial inguinal ring

External spermatic fascia

Reflected inguinal ligament

Cremasteric muscle and fascia

Internal spermatic fascia

Spermatic cord

External spermatic fascia

Testis

*Musculo-aponeurotic arcades of internal oblique and transversus abdominis

(A) Anterior view

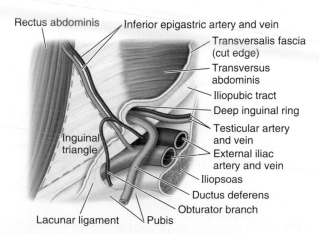

Rectus abdominis

Inferior epigastric artery and vein

Transversalis fascia (cut edge)

Transversus abdominis

Iliopubic tract

Deep inguinal ring

Testicular artery and vein

External iliac artery and vein

Iliopsoas

Ductus deferens

Obturator branch

Inguinal triangle

Lacunar ligament

Pubis

(B) Posterior view of right anterior abdominal wall

FIGURE 5.10. Layers of anterior abdominal wall in inguinal region.
A. Layers of the abdominal wall and the coverings of the spermatic cord and testis derived from them. **B.** Deep inguinal ring.

- *Roof:* Formed laterally by transversalis fascia, centrally by the musculo-aponeurotic arches of internal oblique and transversus abdominis muscles, and medially by the medial crus and intercrural fibers
- *Floor:* Formed laterally by the iliopubic tract (Fig. 5.6), centrally by the superior surface of the "gutter-like" inguinal ligament, and medially by the lacunar ligament (Fig. 5.9)

SPERMATIC CORD

The spermatic cord contains structures running to and from the testis and suspends the testis in the scrotum. The spermatic cord begins at the deep inguinal ring lateral to the inferior epigastric vessels, passes through the inguinal canal, exits at the superficial inguinal ring, and ends in the scrotum at the testis (Fig. 5.10 and Table 5.4). Fascial coverings derived from the anterolateral abdominal wall during the prenatal relocation of the testis include the following:

- **Internal spermatic fascia**: Derived from the transversalis fascia at the deep inguinal ring
- **Cremasteric fascia**: Derived from the fascia of both the superficial and the deep surfaces of the internal oblique muscle
- **External spermatic fascia**: Derived from the external oblique aponeurosis and its investing fascia

The cremasteric fascia contains loops of the **cremaster muscle**, which extend as a continuation of the lowest fascicles of the internal oblique muscle arising from the inguinal ligament. Contraction of the cremaster reflexively draws the testis superiorly in the scrotum, particularly when it is cold; in a warm environment, the cremaster relaxes and the testis descends into the scrotum. Both responses occur in an attempt to maintain the temperature of the testis for *spermatogenesis* (formation of sperms), which requires a constant temperature of approximately one degree cooler than core temperature. The cremaster acts with the **dartos muscle**, a smooth muscle of the fat-free subcutaneous tissue of the scrotum (dartos fascia), which inserts into the skin. The dartos assists in testicular elevation as it produces contraction of the skin of the scrotum. The cremaster is innervated by the **genital branch of the genitofemoral nerve** (L1, L2), a derivative of the lumbar plexus, whereas the dartos receives autonomic innervation.

The round ligament of the uterus in the female receives similar contributions from the layers of the abdominal wall as it traverses the inguinal canal. It is less well-developed and usually is an aggregation of indistinct fibrous strands.

The *constituents of the spermatic cord* are as follows (Fig. 5.11):

- **Ductus deferens** (vas deferens), a muscular tube that conveys sperms from the epididymis to the ejaculatory duct. It courses through the substance of the prostate to open into the prostatic part of the urethra.

TABLE 5.4. CORRESPONDING LAYERS OF ANTERIOR ABDOMINAL WALL, SPERMATIC CORD, AND SCROTUM

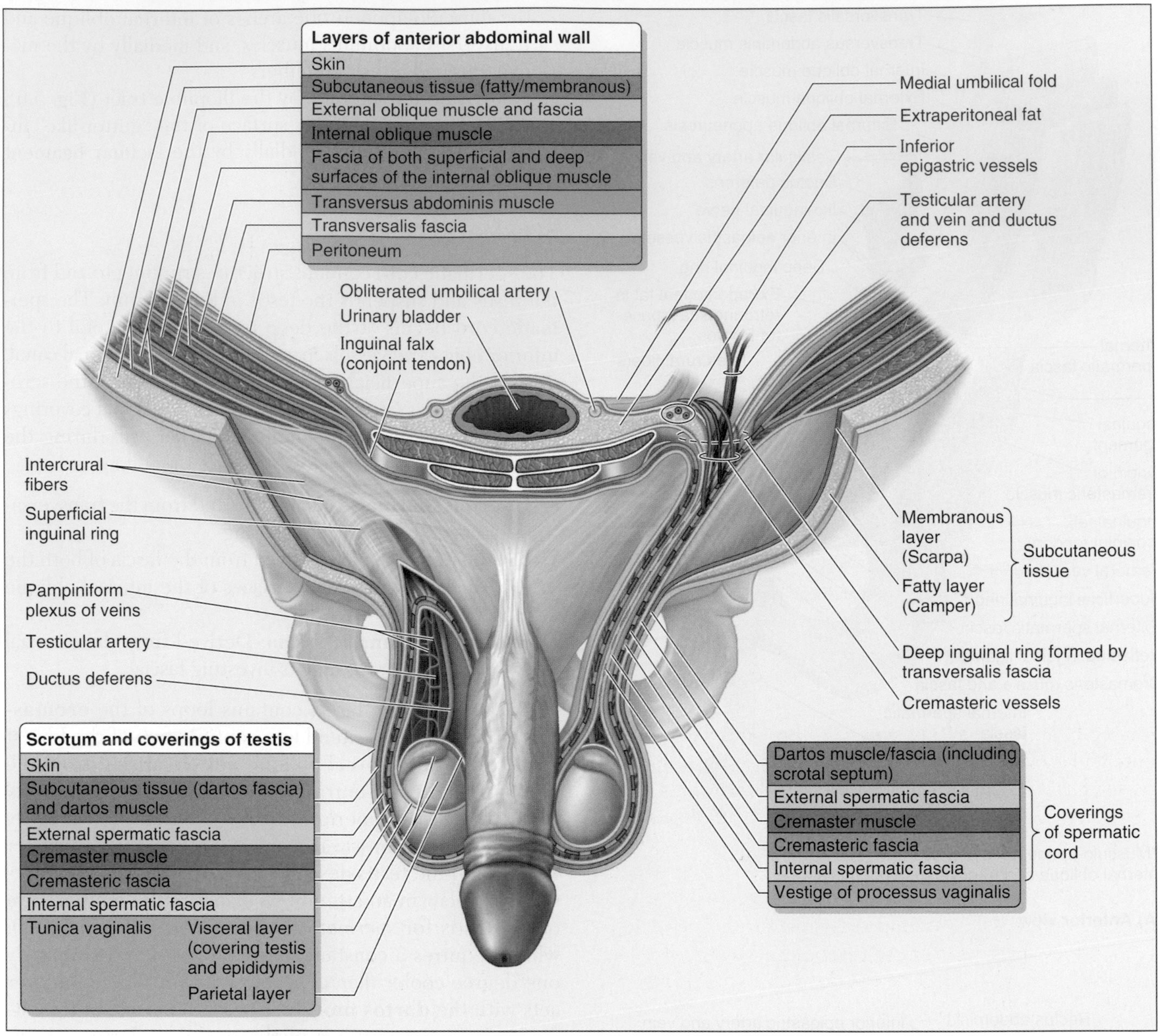

Layers of anterior abdominal wall

Skin
Subcutaneous tissue (fatty/membranous)
External oblique muscle and fascia
Internal oblique muscle
Fascia of both superficial and deep surfaces of the internal oblique muscle
Transversus abdominis muscle
Transversalis fascia
Peritoneum

Obliterated umbilical artery
Urinary bladder
Inguinal falx (conjoint tendon)

Medial umbilical fold
Extraperitoneal fat
Inferior epigastric vessels
Testicular artery and vein and ductus deferens

Intercrural fibers
Superficial inguinal ring
Pampiniform plexus of veins
Testicular artery
Ductus deferens

Membranous layer (Scarpa)
Fatty layer (Camper)
} Subcutaneous tissue

Deep inguinal ring formed by transversalis fascia
Cremasteric vessels

Scrotum and coverings of testis

Skin
Subcutaneous tissue (dartos fascia) and dartos muscle
External spermatic fascia
Cremaster muscle
Cremasteric fascia
Internal spermatic fascia
Tunica vaginalis | Visceral layer (covering testis and epididymis)
| Parietal layer

Dartos muscle/fascia (including scrotal septum)
External spermatic fascia
Cremaster muscle
Cremasteric fascia
Internal spermatic fascia
Vestige of processus vaginalis
} Coverings of spermatic cord

- **Testicular artery** arising from the aorta (vertebral level L2) and supplying the testis and epididymis
- **Artery of ductus deferens** arising from the inferior vesical artery
- **Cremasteric artery** arising from the inferior epigastric artery
- **Pampiniform venous plexus**, a network formed by up to 12 veins that converge superiorly as the right or left testicular veins
- **Sympathetic nerve fibers** on arteries on the ductus deferens
- **Genital branch of genitofemoral nerve** supplying the cremaster muscle

- **Lymphatic vessels** draining the testis and closely associated structures to the lumbar lymph nodes (Fig. 5.12)
- **Vestige of the processus vaginalis**, which may be seen as a fibrous thread in the anterior part of the spermatic cord extending between the abdominal peritoneum and the tunica vaginalis; it may not be detectable.

TESTES

The ovoid testes are suspended in the scrotum by the spermatic cords (Table 5.4). The testes produce sperms (spermatozoa) and hormones, principally testosterone. The sperms are formed in the **seminiferous tubules** that are joined by

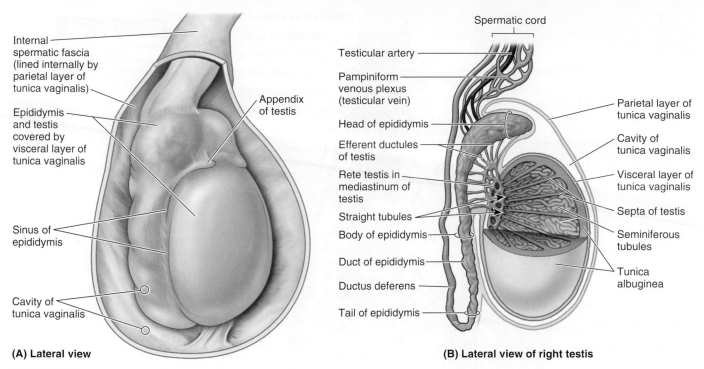

FIGURE 5.11. Structure of testis and epididymis. A. Tunica vaginalis opened. **B.** Contents of the distal spermatic cord, features of the epididymis, and internal structure of the testis.

straight tubules to the **rete testis**. The testes have a tough outer surface, the **tunica albuginea**, that forms a ridge on its internal posterior aspect as the **mediastinum of the testis**. The **tunica vaginalis** is a sequestered peritoneal sac surrounding the testis (Fig. 5.11).

The surface of each testis is covered by the **visceral layer of the tunica vaginalis**, except where the testis attaches to the epididymis and spermatic cord. The visceral layer of the tunica vaginalis—a glistening, transparent serous membrane—is closely applied to the testis, epididymis, and inferior part of the ductus deferens.

The **parietal layer of the tunica vaginalis** lies adjacent to the internal spermatic fascia. The small amount of fluid in the cavity of the tunica vaginalis separates the visceral and parietal layers, allowing the testis to move freely within its side of the scrotum.

The **testicular arteries** arise from the abdominal aorta (at the level of fetal gonadal formation, vertebral level L2) just inferior to the renal arteries (Fig. 5.12). The long, slender testicular arteries indicate the path of prenatal testicular relocation as they pass retroperitoneally (posterior to the peritoneum) in an oblique direction, crossing over the ureters and the inferior parts of the external iliac arteries. They traverse the inguinal canals, becoming part of the spermatic cords to supply the testes.

The **testicular veins** emerging from the testis and epididymis form the pampiniform venous plexus, consisting of 8–12 anastomosing veins lying anterior to the ductus deferens and surrounding the testicular artery in the spermatic cord (Fig. 5.11A). The pampiniform plexus is part of the thermoregulatory system of the testis, helping to keep this gland at a constant temperature. The **left testicular vein** originates as the veins of the pampiniform plexus coalesce; it empties into the left renal vein. The **right testicular vein** has a similar origin and course but enters the IVC.

The *lymphatic drainage of the testis* follows the testicular artery and vein to the **right and left lumbar (caval/aortic)** and **pre-aortic lymph nodes** (Fig. 5.13). The *autonomic nerves of the testis* arise as the **testicular plexus of nerves** on the testicular artery, which contains visceral afferent and sympathetic fibers from the T10 (T11) segment of the spinal cord.

EPIDIDYMIS

The epididymis is an elongated structure on the posterior surface of the testis formed by minute convolutions of the **duct of the epididymis**, so densely compacted that they appear solid (Fig. 5.12). The **efferent ductules** transport newly formed sperms from the rete testis to the epididymis, where they are stored until mature. The rete testis is a network of canals at the termination of the seminiferous tubules.

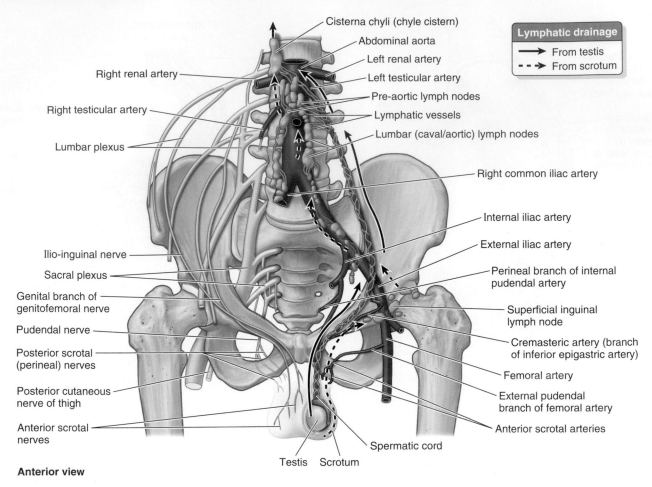

FIGURE 5.12. **Innervation, blood supply, and lymphatic drainage of scrotum, testis, and spermatic cord.** *Arrows,* direction of the flow of lymph to the lymph nodes.

The epididymis has the following parts:

- **Head**: The superior expanded part composed of lobules formed by the coiled ends of 12–14 efferent ductules
- **Body**: The convoluted duct of the epididymis
- **Tail**: Continuous with the ductus deferens, the duct that transports sperms from the epididymis to the ejaculatory duct for expulsion into the prostatic urethra (see Chapter 6)

SCROTUM

The scrotum is a cutaneous sac consisting of two layers: heavily pigmented skin and closely related **dartos fascia** and a fat-free fascial layer, including smooth muscle fibers (dartos muscle) responsible for the rugose (wrinkled) appearance of the scrotum (Table 5.4). Because the *dartos muscle* attaches to the skin, its contraction causes the scrotum to wrinkle when cold, which thickens the integumentary layer while reducing the scrotal surface area. This assists the cremaster in holding the testes closer to the body, thus reducing heat loss.

Scrotal veins accompany the arteries. The *lymphatic vessels* drain into the superficial inguinal lymph nodes.

The *arterial supply of the scrotum* is from the (Fig. 5.12)

- **posterior scrotal branches of the perineal artery**, a branch of the internal pudendal artery
- **anterior scrotal branches of the deep external pudendal artery**, a branch of the femoral artery
- *cremasteric artery*, a branch of the inferior epigastric artery

The *nerves of the scrotum* include the (Fig. 5.12)

- *genital branch of the genitofemoral nerve* (L1, L2) supplying the anterolateral surface
- **anterior scrotal nerves**, branches of the ilio-inguinal nerve (L1) supplying the anterior surface
- **posterior scrotal nerves**, branches of the perineal branch of the **pudendal nerve** (S2–S4) supplying the posterior surface
- **perineal branches of the posterior cutaneous nerve of the thigh** (S2, S3) supplying the inferior surface

CLINICAL BOX

Hydrocele and Hematocele

The presence of excess fluid in a persistent processus vaginalis is a *hydrocele of the testis* (Fig. B5.3A). Certain pathological conditions, such as injury or inflammation of the epididymis, may also produce a *hydrocele of the spermatic cord* (Fig. B5.3B). A *hematocele of the testis* is a collection of blood in the cavity of the tunica vaginalis (Fig. B5.3C).

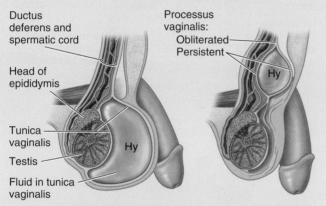

Ductus deferens and spermatic cord

Head of epididymis

Tunica vaginalis

Testis

Fluid in tunica vaginalis

Processus vaginalis:
Obliterated
Persistent

(A) Hydrocele (Hy) of testis   **(B) Hydrocele (Hy) of cord**

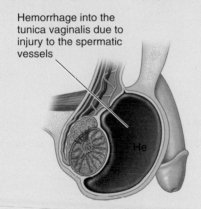

Hemorrhage into the tunica vaginalis due to injury to the spermatic vessels

(C) Hematocele (He) of testis

FIGURE B5.3. Hydrocele and hematocele.

Vasectomy

In male sterilization (*vasectomy*), the *ductus* (*vas*) *deferens* is ligated bilaterally. To perform a vasectomy, the duct is isolated on each side and transected or a small section of it is removed. Spermatozoa can no longer pass to the urethra; they degenerate in the epididymis and proximal end of the ductus deferens. However, the secretions of the *auxiliary genital glands* (seminal glands, bulbo-urethral glands, and prostate) can still be ejaculated. The testis continues to function as an endocrine gland for the production of testosterone.

Palpation of Superficial Inguinal Ring

The superficial inguinal ring (Fig. B5.4A) is palpable superolateral to the pubic tubercle by invaginating the skin of the upper scrotum with the index finger. The examiner's finger follows the spermatic cord superolaterally to the superficial inguinal ring (Fig. B5.4B). If the ring is dilated, it may admit the fingertip without causing pain. With the palmar surface of the finger against the anterior abdominal wall, the deep inguinal ring may be felt as a skin depression superior to the inguinal ligament, 2–4 cm superolateral to the pubic tubercle. Detection of an impulse against the examining finger, when the person coughs, at the superficial ring and a mass at the site of the deep ring suggests an *indirect hernia*. Palpation of a *direct inguinal hernia* is performed by placing the index and/or middle finger over the inguinal triangle (lateral to the superficial ring) and asking the person to cough or strain. If a hernia is present, a forceful impulse is felt against the pad of the finger.

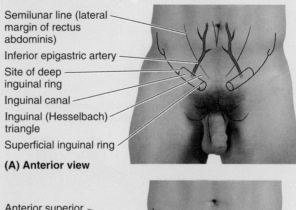

Semilunar line (lateral margin of rectus abdominis)

Inferior epigastric artery

Site of deep inguinal ring

Inguinal canal

Inguinal (Hesselbach) triangle

Superficial inguinal ring

(A) Anterior view

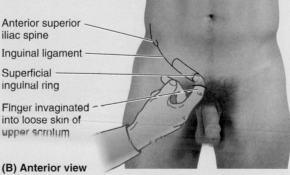

Anterior superior iliac spine

Inguinal ligament

Superficial inguinal ring

Finger invaginated into loose skin of upper scrotum

(B) Anterior view

FIGURE B5.4. Detection of hernias. **A.** The location of superficial and deep inguinal rings. **B.** Palpation of the superficial inguinal ring.

Varicocele

The pampiniform plexus of veins may become *varicose* (dilated) and tortuous. These varicose vessels, usually visible only when a person is standing, often result from defective valves in the testicular vein. The palpable enlargement, which feels like a bundle of worms, usually disappears when the person lies down.

(Continued on next page)

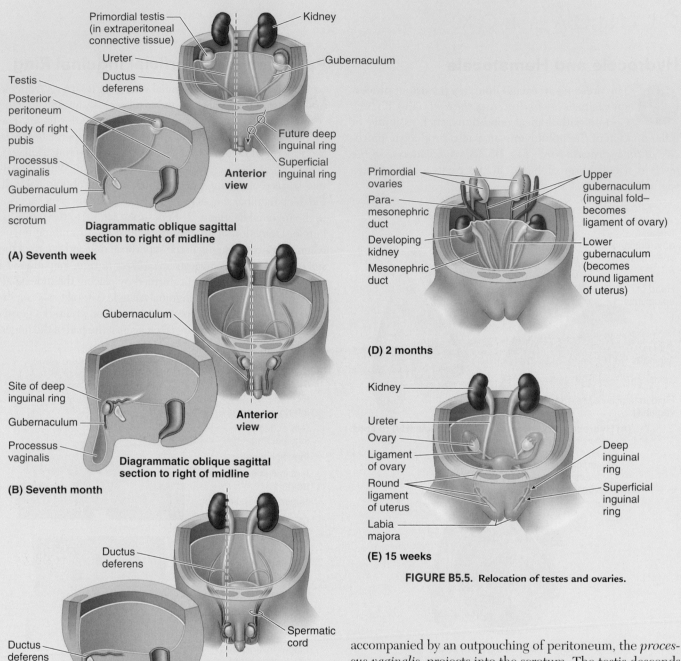

FIGURE B5.5. Relocation of testes and ovaries.

Relocation of Testes and Ovaries

The **fetal testes** relocate from the dorsal abdominal wall in the superior lumbar region to the deep inguinal rings during the 9th–12th fetal weeks (Fig. B5.5A–C). This repositioning probably results from growth of the vertebral column and pelvis. The male *gubernaculum*, attached to the caudal pole of the testis and accompanied by an outpouching of peritoneum, the *processus vaginalis*, projects into the scrotum. The testis descends posterior to the processus vaginalis. The inferior remnant of the processus vaginalis forms the *tunica vaginalis* covering the testis. The ductus deferens, testicular vessels, nerves, and lymphatics accompany the testis. The final location of the testes in the scrotum usually occurs before or shortly after birth.

The **fetal ovaries** also relocate from the dorsal abdominal wall in the superior lumbar region during the 12th week and pass into the lesser pelvis (Fig. B5.5D,E). The female gubernaculum also attaches to the caudal pole of the ovary and projects into the labia majora, attaching en route to the uterus; the part passing from the uterus to the ovary forms the *ovarian ligament*, and the remainder of it becomes the *round ligament of the uterus*. For a complete description of the embryology of the inguinal region, see Moore et al. (2016).

Inguinal Hernias

An *inguinal hernia* is a protrusion of parietal perito-neum and viscera, such as the small intestine, through a normal or abnormal opening from the abdominal cavity. There are two major categories of inguinal hernia: indirect and direct. More than two thirds are indirect hernias. An indirect inguinal hernia can also occur in women, but it is about 20 times more common in males of all ages (Fig. B5.6 and Table B5.1).

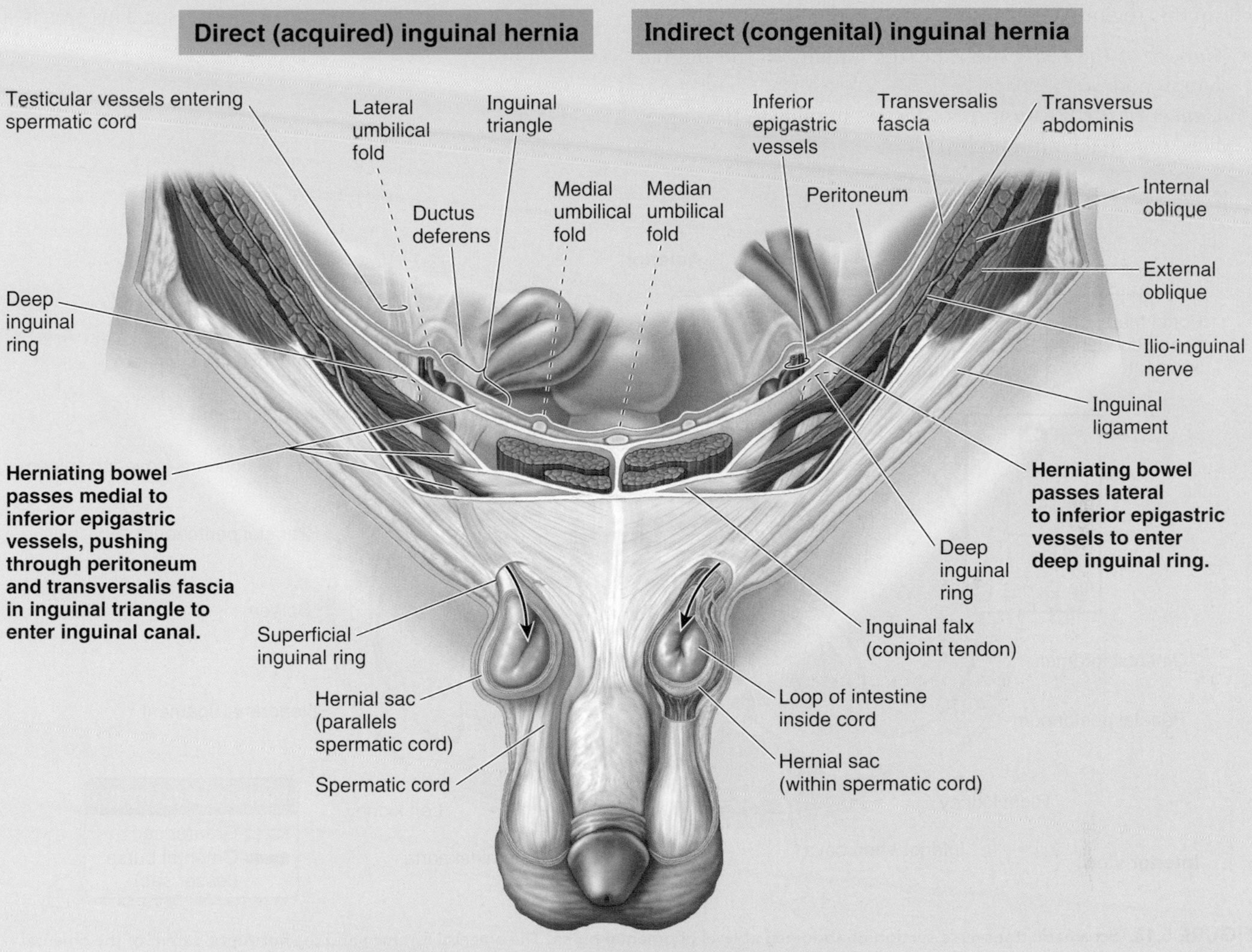

Direct (acquired) inguinal hernia

Indirect (congenital) inguinal hernia

Testicular vessels entering spermatic cord

Lateral umbilical fold

Inguinal triangle

Ductus deferens

Medial umbilical fold

Median umbilical fold

Inferior epigastric vessels

Transversalis fascia

Transversus abdominis

Peritoneum

Internal oblique

Deep inguinal ring

External oblique

Ilio-inguinal nerve

Inguinal ligament

Herniating bowel passes medial to inferior epigastric vessels, pushing through peritoneum and transversalis fascia in inguinal triangle to enter inguinal canal.

Herniating bowel passes lateral to inferior epigastric vessels to enter deep inguinal ring.

Deep inguinal ring

Superficial inguinal ring

Inguinal falx (conjoint tendon)

Hernial sac (parallels spermatic cord)

Loop of intestine inside cord

Spermatic cord

Hernial sac (within spermatic cord)

FIGURE B5.6. Course of direct and indirect inguinal hernias.

TABLE B5.1. CHARACTERISTICS OF INGUINAL HERNIAS

Characteristics	Direct (Acquired)	Indirect (Congenital)
Predisposing factors	Weakness of anterior abdominal wall in inguinal triangle (e.g., owing to distended superficial ring, narrow inguinal falx, or attenuation of aponeurosis in males >40 years of age)	Patency of processus vaginalis (complete or at least of superior part) in younger persons, the great majority of whom are males
Frequency	Less common (one third to one quarter of inguinal hernias)	More common (two thirds to three quarters of inguinal hernias)
Coverings at exit from abdominal cavity	Peritoneum plus transversalis fascia (lies outside inner one or two fascial coverings of cord)	Peritoneum of persistent processus vaginalis plus all three fascial coverings of cord/round ligament
Course	Usually traverses only medial third of inguinal canal, external and parallel to vestige of processus vaginalis	Traverses inguinal canal (entire canal if it is sufficient size) within processus vaginalis
Exit from anterior abdominal wall	Via superficial ring, lateral to cord; rarely enters scrotum	Via superficial ring inside cord, commonly passing into scrotum/labium majus

(Continued on next page)

Testicular Cancer

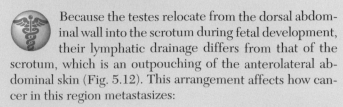

Because the testes relocate from the dorsal abdominal wall into the scrotum during fetal development, their lymphatic drainage differs from that of the scrotum, which is an outpouching of the anterolateral abdominal skin (Fig. 5.12). This arrangement affects how cancer in this region metastasizes:

- *Cancer of the testis* metastasizes initially to the lumbar lymph nodes.
- *Cancer of the scrotum* metastasizes initially to the superficial inguinal lymph nodes.

Cremasteric Reflex

The *cremasteric reflex* is the rapid elevation of the testis on the same side; this reflex is extremely active in children. Contraction of the cremaster muscle—producing the reflex—can be induced by lightly stroking the skin on the medial aspect of the superior part of the thigh with an applicator stick or tongue depressor. This area is supplied by the ilio-inguinal nerve.

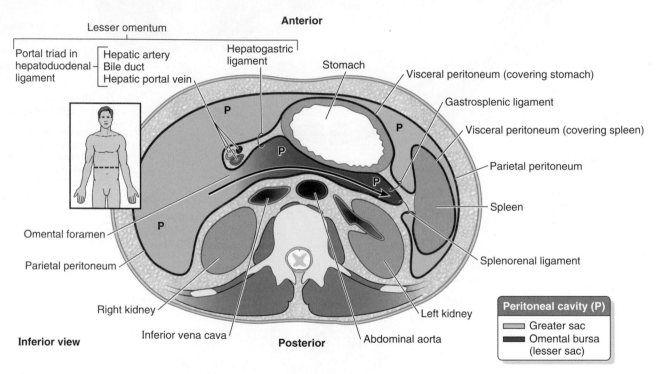

FIGURE 5.13. Schematic transverse section of abdomen at level of omental bursa. The omental foramen and the horizontal extent of the omental bursa (lesser sac) is shown. *Arrow* passes from the greater sac through the omental foramen across the full extent of the omental bursa.

PERITONEUM AND PERITONEAL CAVITY

The **peritoneum** is a glistening, transparent serous membrane that consists of two continuous layers (Fig. 5.13):

- **Parietal peritoneum**, lining the internal surface of the abdominopelvic wall
- **Visceral peritoneum**, investing viscera (organs) such as the spleen and stomach

The peritoneum and viscera are in the abdominopelvic cavity. The relationship of the viscera to the peritoneum is as follows:

- **Intraperitoneal organs** are almost completely covered with visceral peritoneum (e.g., the spleen and stomach); intraperitoneal organs have conceptually, if not literally, invaginated into a closed sac, like pressing your fist into an inflated balloon.
- **Extraperitoneal, retroperitoneal, and subperitoneal organs** are outside the peritoneal cavity—external or posterior to the parietal peritoneum—and are only partially covered with peritoneum (usually on one surface). Organs such as the kidneys are between the parietal peritoneum and the posterior abdominal wall and have parietal peritoneum only on their anterior surfaces, often with a considerable amount of intervening fatty tissue (Fig. 5.13).

The **peritoneal cavity** is within the abdominal cavity and continues into the pelvic cavity. It is a potential space of capillary thinness between the parietal and visceral layers of peritoneum. The peritoneal cavity contains a thin film of **peritoneal fluid** that keeps the peritoneal surfaces moist. *There are no* *organs in the peritoneal cavity.* Peritoneal fluid lubricates the peritoneal surfaces, enabling the viscera to move over each other without friction and allowing the movements of digestion. In addition, the fluid contains leukocytes and antibodies that resist infection. The peritoneal cavity is completely closed in males; however, there is a communication pathway in females to the exterior of the body through the uterine tubes, uterine cavity, and vagina (see Chapter 6). This communication constitutes a potential pathway of infection from the exterior.

Peritoneal Vessels and Nerves

The *parietal peritoneum* is

- served by the same blood and lymphatic vasculature and the same somatic nerve supply as the region of the abdominopelvic wall it lines
- sensitive to pressure, pain, heat, and cold; pain from the parietal peritoneum is generally well localized.

The *visceral peritoneum* is

- served by the same blood and lymphatic vasculature and the same visceral nerve supply as the organs it covers

- insensitive to touch, heat, cold, and laceration; is stimulated primarily by stretching and chemical irritation

Pain from the visceral peritoneum is poorly localized and is referred to the dermatomes of the spinal ganglia providing the sensory fibers. Pain from the foregut derivatives (e.g., pharynx, esophagus, and stomach) is usually experienced in the epigastric region; that from the midgut derivatives (e.g., small intestine, cecum, appendix, and ascending colon), in the umbilical region; and that from the hindgut derivatives (e.g., descending and sigmoid colons), in the pubic region (see Clinical Box "Visceral Referred Pain").

Peritoneal Formations

Various terms are used to describe the parts of the peritoneum that connect organs with other organs or to the abdominal wall and to describe the compartments and recesses that are formed as a consequence (Fig. 5.14). The disposition of peritoneum in adults is easier to visualize when the embryology of the peritoneal cavity and viscera is understood (Moore et al., 2016).

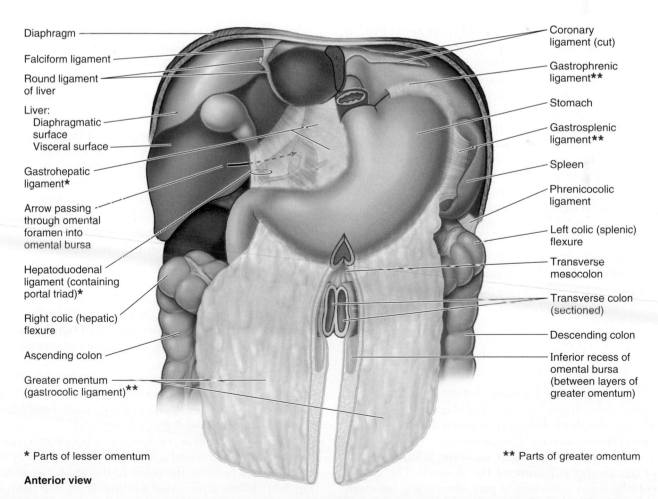

Diaphragm

Falciform ligament

Round ligament of liver

Liver:
 Diaphragmatic surface
 Visceral surface

Gastrohepatic ligament*

Arrow passing through omental foramen into omental bursa

Hepatoduodenal ligament (containing portal triad)*

Right colic (hepatic) flexure

Ascending colon

Greater omentum (gastrocolic ligament)**

Coronary ligament (cut)

Gastrophrenic ligament**

Stomach

Gastrosplenic ligament**

Spleen

Phrenicocolic ligament

Left colic (splenic) flexure

Transverse mesocolon

Transverse colon (sectioned)

Descending colon

Inferior recess of omental bursa (between layers of greater omentum)

* Parts of lesser omentum

** Parts of greater omentum

Anterior view

FIGURE 5.14. Parts of the greater and lesser omentum. The liver and gallbladder have been reflected superiorly. The central part of the greater omentum has been cut out to show its relation to the transverse colon and mesocolon. *Arrow,* site of omental foramen.

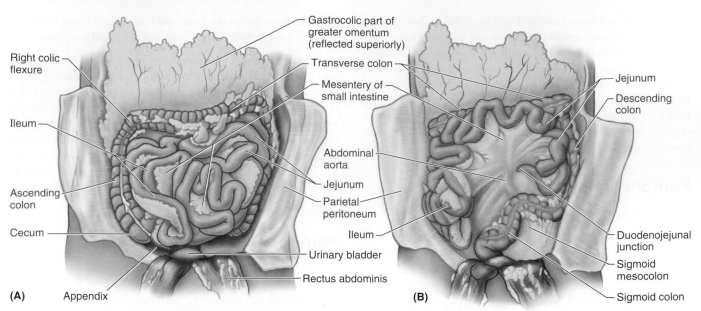

FIGURE 5.15. Greater omentum and mesentery of small intestine. A. The gastrocolic part of the greater omentum has been elevated to reveal the small intestine and ascending and transverse colon. **B.** The small intestine has been retracted superiorly to reveal the mesentery, duodenojejunal junction, sigmoid colon, and sigmoid mesocolon.

A **mesentery** is a double layer of peritoneum that occurs as a result of the invagination of the peritoneum by an organ and constitutes a continuity of the visceral and parietal peritoneum (e.g., *mesentery of small intestine and transverse mesocolon*) (Figs. 5.15 and 5.16). Mesenteries provide a means for neurovascular communication between the organ and the body wall and thus have a core of connective tissue containing blood and lymphatic vessels, nerves, fat, and lymph nodes. Viscera with a mesentery are mobile; the degree of mobility depends on the length of the mesentery.

A **peritoneal ligament** consists of a double layer of peritoneum that connects an organ with another organ or to the abdominal wall. For example, the liver is connected to the anterior abdominal wall by the *falciform ligament* (Fig. 5.14).

An **omentum** is a double-layered extension of peritoneum passing from the stomach and proximal part of the duodenum to adjacent organs. The **greater omentum** extends superiorly, laterally to the left, and inferiorly from the greater curvature of the stomach and the proximal part of the duodenum (Fig. 5.14). The greater omentum has three parts:

1. The **gastrophrenic ligament** between the greater curvature of the stomach and the diaphragm
2. The **gastrosplenic ligament** between the greater curvature of the stomach and the spleen
3. The **gastrocolic ligament** from the inferior portion of the greater curvature of the stomach. The gastrocolic ligament is the largest part, descending anteriorly and inferiorly beyond the transverse colon and then

ascending again posteriorly, fusing with the visceral peritoneum of the transverse colon and the superior layer of its mesentery. The descending and ascending portions of the gastrocolic part of the greater omentum usually fuse together, forming a four-layered fatty "omental apron."

The **lesser omentum** (hepatogastric and hepatoduodenal ligaments) connects the lesser curvature of the stomach and the proximal part of the duodenum to the liver (Fig. 5.14). These ligaments are continuous parts of the lesser omentum and are separated only for descriptive convenience. The stomach is connected to the liver by the **hepatogastric ligament**, the membranous portion of the lesser omentum. The **hepatoduodenal ligament**, the thickened free edge of the lesser omentum, conducts the *portal triad*: portal vein, hepatic artery, and bile duct.

Every organ must have an area that is not covered with visceral peritoneum to allow the entrance and exit of neurovascular structures. Such areas are called **bare areas** and are formed in relation to the attachments of mesenteries, omenta, and ligaments.

A **peritoneal fold** is a reflection of peritoneum that is raised from the body wall by underlying blood vessels, ducts, or obliterated fetal vessels or ducts (e.g., *medial and lateral umbilical folds*) (see Fig. 5.6).

A **peritoneal recess**, or fossa, is a pouch or concavity formed by a peritoneal fold (e.g., *inferior recess of the omental bursa* between the layers of the greater omentum [Fig. 5.14] and the *supravesical* and *umbilical fossae* between the umbilical folds [Fig. 5.6]).

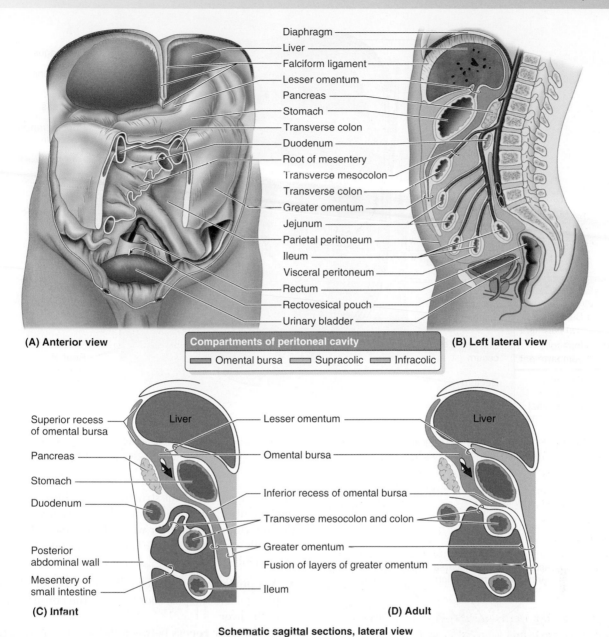

(A) Anterior view

| Compartments of peritoneal cavity |
| Omental bursa   Supracolic   Infracolic |

(B) Left lateral view

Diaphragm
Liver
Falciform ligament
Lesser omentum
Pancreas
Stomach
Transverse colon
Duodenum
Root of mesentery
Transverse mesocolon
Transverse colon
Greater omentum
Jejunum
Parietal peritoneum
Ileum
Visceral peritoneum
Rectum
Rectovesical pouch
Urinary bladder

Superior recess of omental bursa
Pancreas
Stomach
Duodenum
Posterior abdominal wall
Mesentery of small intestine

Liver
Lesser omentum
Omental bursa
Inferior recess of omental bursa
Transverse mesocolon and colon
Greater omentum
Fusion of layers of greater omentum
Ileum

(C) Infant

Liver

(D) Adult

Schematic sagittal sections, lateral view

FIGURE 5.16. Principal peritoneal formations. A. In this opened peritoneal cavity, parts of the greater omentum, transverse colon, and the small intestine and its mesentry have been cut away to reveal deeper structures and the layers of mesenteric structures. **B.** Median section of the abdominopelvic cavity showing the relationships of the peritoneal attachments. **C and D.** Sagittal sections through the inferior recess of the omental bursa showing the formation of the transverse mesocolon and fusion of the layers of the greater omentum in an infant (**C**) and an adult (**D**). The *red arrow* is traversing the omental foramen, connecting the omental bursa and greater peritoneal cavity.

Subdivisions of Peritoneal Cavity

The peritoneal cavity is divided into a greater sac and an omental bursa (Figs. 5.16 and 5.17).

The **greater sac** is the main and larger part of the peritoneal cavity. A surgical incision through the anterolateral abdominal wall enters this sac. The **transverse mesocolon** (mesentery of transverse colon) and the gastrocolic ligament of the greater omentum divide the greater sac into the following (Figs. 5.16–5.18):

- **Supracolic compartment**, containing the stomach, liver, and spleen

- **Infracolic compartment**, containing the small intestine and ascending and descending colon. This compartment lies posterior to the greater omentum and is divided into *right* and *left infracolic spaces* by the mesentery of the small intestine.

Free communication occurs between the supracolic and the infracolic compartments through the **paracolic gutters**, the grooves between the lateral aspect of the ascending or descending colon and the posterolateral abdominal wall, flow less obstructed on the right.

The **omental bursa** (lesser sac), the smaller part of the peritoneal cavity, lies posterior to the stomach, lesser

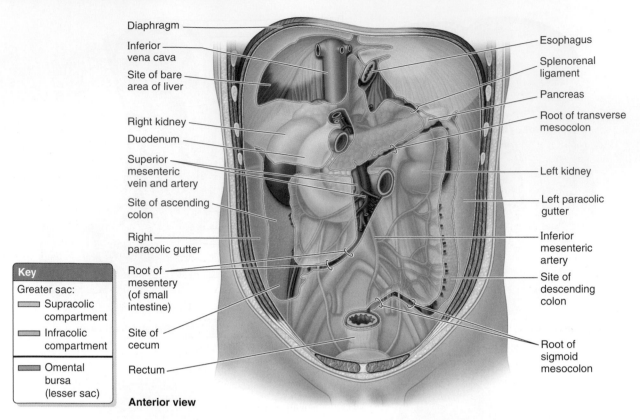

Anterior view

FIGURE 5.17. Posterior wall of peritoneal cavity and roots of peritoneal reflections. The liver and the ascending and descending colon have been mobilized and removed, and the transverse and sigmoid mesocolons and the mesentery of the small intestine have been cut at their roots.

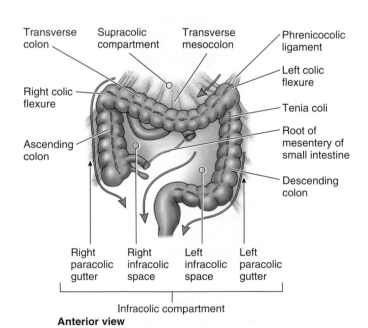

Anterior view

FIGURE 5.18. Supracolic and infracolic compartments of greater sac. The greater omentum has been removed. The infracolic spaces and paracolic gutters determine the flow of ascitic fluid (arrows) when inclined or upright.

omentum, and adjacent structures. This bursa permits free movement of the stomach on adjacent structures because the anterior and posterior walls of the omental bursa slide smoothly over each other. The omental bursa has two recesses (Fig. 5.16):

- A **superior recess**, which is limited superiorly by the diaphragm and the posterior layers of the coronary ligament of the liver
- An **inferior recess** between the superior part of the layers of the greater omentum

Most of the inferior recess of the omental bursa is a potential space sealed off from the main part of the omental bursa posterior to the stomach after adhesion of the anterior and posterior layers of the greater omentum (Fig. 5.16). The omental bursa communicates with the greater sac through the **omental foramen** (epiploic foramen), an opening situated posterior to the free edge of the lesser omentum forming the hepatoduodenal ligament (Figs. 5.4 and 5.14). The boundaries of the omental foramen are as follows:

- *Anteriorly*: The hepatoduodenal ligament (free edge of lesser omentum) containing the portal vein, hepatic artery, and bile duct
- *Posteriorly*: IVC and right crus of diaphragm, covered with parietal peritoneum (They are retroperitoneal.)
- *Superiorly*: The liver, covered with visceral peritoneum
- *Inferiorly*: Superior or first part of the duodenum

CLINICAL BOX

The Peritoneum and Surgical Procedures

Because the peritoneum is well innervated, patients undergoing abdominal surgery experience more pain with large, invasive, open incisions of the peritoneum (*laparotomy*) than they do with small laparoscopic incisions or transvaginal operations. Because of the high incidence of infections such as peritonitis and adhesions after operations in which the peritoneal cavity is opened, efforts are made to remain outside the peritoneal cavity whenever possible (e.g., translumbar approach to the kidneys). When opening the peritoneal cavity is necessary, great effort is made to avoid contamination of the cavity.

Peritonitis and Ascites

When bacterial contamination occurs during laparotomy or when the gut is traumatically penetrated or ruptured as the result of infection and inflammation (e.g., appendicitis), allowing gas, fecal matter, and bacteria to enter the peritoneal cavity, the result is infection and inflammation of the peritoneum—*peritonitis*. Exudation of serum, fibrin, cells, and pus into the peritoneal cavity occurs, accompanied by pain in the overlying skin and an increase in the tone of the anterolateral abdominal muscles. Given the extent of the peritoneal surfaces and the rapid absorption of material, including bacterial toxins, from the peritoneal cavity, when peritonitis becomes *generalized* (widespread in the peritoneal cavity), the condition is dangerous and sometimes lethal. In addition to severe abdominal pain, tenderness, nausea and/or vomiting, fever, and constipation are present.

Excess fluid in the peritoneal cavity is called *ascitic fluid*, clinically called *ascites*. Ascites may also occur as a result of mechanical injury (which may also produce internal bleeding) or other pathological conditions, such as portal hypertension (venous congestion) and widespread metastasis of cancer cells to the abdominal viscera. In all these cases, the peritoneal cavity may be distended with several liters of abnormal fluid, interfering with movements of the viscera.

Rhythmic movements of the anterolateral abdominal wall normally accompany respirations. If the abdomen is drawn in as the chest expands (*paradoxical abdominothoracic rhythm*) and muscle rigidity is present, either peritonitis or pneumonitis (inflammation of the lungs) may be present. Because the intense pain worsens with movement, people with peritonitis commonly lie with their knees flexed to relax their anterolateral abdominal muscles. They also breathe shallowly (and hence more rapidly), reducing the intra-abdominal pressure and pain.

Peritoneal Adhesions and Adhesiotomy

If the peritoneum is damaged, by a stab wound for example, or infected, the peritoneal surfaces become inflamed, making them sticky with *fibrin*. As healing occurs, the fibrin may be replaced with fibrous tissue, forming abnormal attachments between the visceral peritoneum of adjacent viscera or between the visceral peritoneum of a viscus and the parietal peritoneum of the adjacent abdominal wall. *Adhesions* (scar tissue) may also form after an abdominal operation (e.g., owing to a ruptured appendix) and limit the normal movements of the viscera. This tethering may cause chronic pain or emergency complications such as intestinal obstruction when the gut becomes twisted around an adhesion (*volvulus*).

Adhesiotomy refers to the surgical separation of adhesions. Adhesions are often found during dissection of cadavers (e.g., binding of the spleen to the diaphragm).

Abdominal Paracentesis

Treatment of generalized peritonitis includes removal of the ascitic fluid and, in the presence of infection, administration of large doses of antibiotics. Surgical puncture of the peritoneal cavity for the aspiration or drainage of fluid is called *paracentesis*. After injection of a local anesthetic agent, a needle or trocar and a cannula are inserted through the anterolateral abdominal wall into the peritoneal cavity through the linea alba, for example. The needle is inserted superior to the empty urinary bladder and in a location that avoids the inferior epigastric artery.

Functions of Greater Omentum

The greater omentum, large and fat laden, prevents the visceral peritoneum from adhering to the parietal peritoneum. It has considerable mobility and moves around the peritoneal cavity with peristaltic movements of the viscera. It often forms adhesions adjacent to an inflamed organ such as the appendix, sometimes walling it off and thereby protecting other viscera from it.

Spread of Pathological Fluids

Peritoneal recesses are of clinical importance in connection with the spread of pathological fluids such as pus, a product of inflammation. The recesses determine the extent and direction of the spread of fluids that may enter the peritoneal cavity when an organ is diseased or injured.

ABDOMINAL VISCERA

The principal viscera of the abdomen are the esophagus (terminal part), stomach, intestines, spleen, pancreas, liver, gallbladder, kidneys, and suprarenal glands. The esophagus, stomach, and intestine form the **gastrointestinal (GI) tract**. Food passes from the *mouth* and *pharynx* through the *esophagus* to the *stomach*. Digestion mostly occurs in the stomach and *duodenum*. **Peristalsis**, a series of ring-like contraction waves that begin around the middle of the stomach and move slowly toward the pylorus, is responsible for mixing of the masticated (chewed) food mass with gastric juices and for emptying the contents of the stomach into the duodenum.

Absorption of chemical compounds occurs principally in the *small intestine*, consisting of the *duodenum, jejunum,* and *ileum* (Fig. 5.19A). The stomach is continuous with the duodenum, which receives the openings of the ducts from the *pancreas* and *liver* (major glands of digestive tract). Peristalsis also occurs in the jejunum and ileum, although it is not forceful unless an obstruction is present. The *large intestine* consists of the *cecum,* which receives the terminal part of the ileum, *appendix, colon* (*ascending, transverse,* and *descending*), *rectum,* and *anal canal* (which ends at the *anus*). Most reabsorption of water occurs in the ascending colon. Feces (stools) are formed in the descending and sigmoid colon and accumulate in the rectum before defecation.

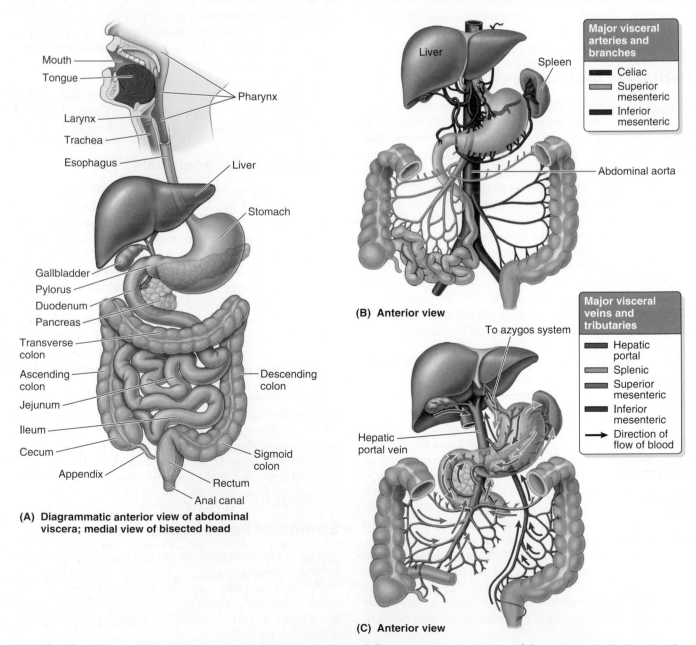

(A) Diagrammatic anterior view of abdominal viscera; medial view of bisected head

(B) Anterior view

(C) Anterior view

FIGURE 5.19. Schematic overview and arterial supply and venous drainage of alimentary system. A. Overview of alimentary system. **B.** Overview of arterial supply. **C.** Overview of portal venous drainage.

The arterial supply to the GI tract, spleen, pancreas, gallbladder, and liver is from the *abdominal aorta* (Fig. 5.19B). The three major branches of the abdominal aorta are the *celiac trunk* and the *superior* and *inferior mesenteric arteries*.

The *hepatic portal vein*, formed by the union of the superior mesenteric and splenic veins (Fig. 5.19C), is the main channel of the *portal venous system*, which collects blood from the abdominal part of the GI tract, pancreas, spleen, and most of the gallbladder and carries it to the liver.

Esophagus

The **esophagus** is a muscular tube, approximately 25 cm (10 inches) long with an average diameter of 2 cm, that extends from the pharynx to the stomach (Figs. 5.19A and 5.20). The esophagus

- follows the vertebral column concavity (thoracic kyphosis)
- passes through the elliptical *esophageal hiatus* in the muscular right crus of the diaphragm, just to the left of the median plane at the level of the T10 vertebra (Fig. 5.20)
- terminates at the **esophagogastric junction**, where ingested matter enters the cardial orifice of the stomach (Fig. 5.21B). It is located to the left of the midline at the level of the 7th left costal cartilage and the T11 vertebra. The esophagus is retroperitoneal during its short abdominal course.

- has circular and external longitudinal layers of muscle. In its superior third, the external layer consists of voluntary striated muscle, the inferior third is composed of smooth muscle, and the middle third is made up of both types of muscle.

The esophagogastric junction is marked internally by the abrupt transition from esophageal to gastric mucosa, referred to clinically as the **Z-line** (Fig. 5.21D). Just superior to this junction, the diaphragmatic musculature forming the esophageal hiatus functions as a physiological **inferior esophageal sphincter** that contracts and relaxes. Radiological studies show that food or liquid may be stopped here momentarily and that the sphincter mechanism is normally efficient in preventing reflux of gastric contents into the esophagus.

The abdominal part of the esophagus has its

- *arterial supply* from the **esophageal branches of the left gastric artery** (Fig. 5.20B), a branch of the *celiac trunk*, and the *left inferior phrenic artery*
- *venous drainage* primarily to the *portal venous system* through the *left gastric vein* (Fig. 5.22B), whereas the proximal thoracic part of the esophagus drains primarily into the *systemic venous system* through the **esophageal veins** entering the *azygos vein* (see Chapter 4). However, the veins of the two parts of the esophagus communicate and provide a clinically important portosystemic anastomosis.
- *lymphatic drainage* into the *left gastric lymph nodes*, which in turn drain mainly to the *celiac lymph nodes* (Fig. 5.20A)
- *innervation* from the *vagal trunks* (becoming anterior and posterior gastric nerves), the *thoracic sympathetic trunks* via the *greater* (abdominopelvic) *splanchnic nerves*, and the *peri-arterial plexus* around the left gastric artery and left inferior phrenic artery (Fig. 5.23B)

Stomach

The **stomach** acts as a food blender and reservoir; its chief function is acidic and mechanical digestion. The *gastric juice* gradually converts a mass of food into a semiliquid mixture, *chyme* (G. juice), which passes into the duodenum.

PARTS AND CURVATURE OF STOMACH

The shape of the stomach is dynamic (changing in shape as it functions) and highly variable from person to person (see Fig. SA5.2B). The stomach has four parts and two curvatures (Fig. 5.21):

- The short **cardia** surrounds the **cardial orifice**, the trumpet-shaped opening of the esophagus into the stomach.
- The **fundus of the stomach** is the dilated superior part of the stomach that is related to the left dome of the

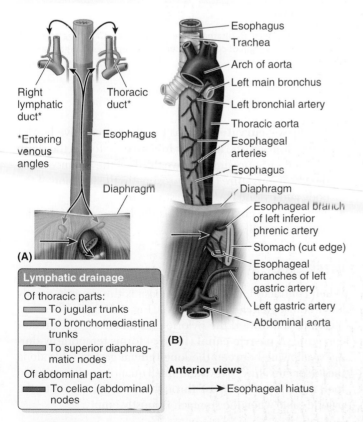

Right lymphatic duct*

Thoracic duct*

*Entering venous angles

Esophagus

Diaphragm

(A)

Esophagus
Trachea
Arch of aorta
Left main bronchus
Left bronchial artery
Thoracic aorta
Esophageal arteries
Esophagus
Diaphragm
Esophageal branch of left inferior phrenic artery
Stomach (cut edge)
Esophageal branches of left gastric artery
Left gastric artery
Abdominal aorta

(B)

Anterior views

⟶ Esophageal hiatus

Lymphatic drainage

Of thoracic parts:
- To jugular trunks
- To bronchomediastinal trunks
- To superior diaphragmatic nodes

Of abdominal part:
- To celiac (abdominal) nodes

FIGURE 5.20. Esophagus. A. Lymphatic drainage. **B.** Arterial supply.

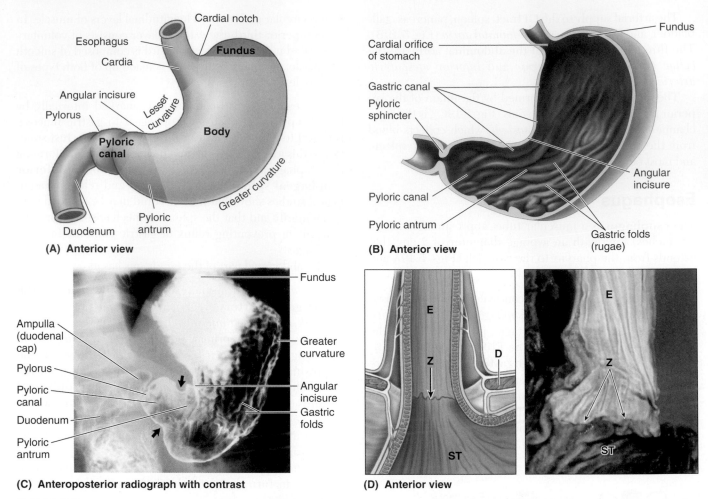

FIGURE 5.21. Esophagus (terminal part), stomach, and proximal duodenum. A. Parts of stomach. **B.** Internal surface of stomach. **C.** Radiograph of stomach and duodenum after barium ingestion. *Arrows*, peristaltic wave. **D.** Illustration and photograph of coronal section of region of esophagogastric junction. *D*, diaphragm; *E*, esophagus; *ST*, stomach; *Z*, esophagogastric junction (Z-line).

diaphragm and is limited inferiorly by the horizontal plane of the cardial orifice. The superior part of the fundus usually reaches the level of the left 5th intercostal space. The **cardial notch** is between the esophagus and the fundus. The fundus may be dilated by gas (especially in the upright position), fluid, food, or any combination of these.

- The **body of the stomach**, the major part of the stomach, lies between the fundus and the pyloric antrum. (Histologists/pathologists often treat the fundus and body as synonyms; hence, the mucosa of the fundus and body is composed of "fundic glands.")
- The **pyloric part of the stomach** is the distal funnel-shaped region; its wide part, the **pyloric antrum**, leads into the **pyloric canal**, its narrow part. The **pylorus**, the distal sphincteric region, is a thickening of the circular layer of smooth muscle, which controls discharge of the stomach contents through the **pyloric orifice** into the duodenum.

- The **lesser curvature** forms the shorter concave border of the stomach; the **angular incisure** (notch) is the sharp indentation approximately two thirds of the distance along the lesser curvature that approximates the junction of the body and pyloric part of the stomach.
- The **greater curvature** forms the longer convex border of the stomach.

INTERIOR OF STOMACH

When contracted, the gastric mucosa is thrown into mostly longitudinal **gastric folds** (rugae) (Fig. 5.21B,C). These are most marked toward the pyloric part and along the greater curvature. A **gastric canal** (furrow) forms temporarily during swallowing between the longitudinal gastric folds along the lesser curvature. Saliva and small quantities of masticated food and other fluids pass through the gastric canal to the pyloric canal when the stomach is mostly empty.

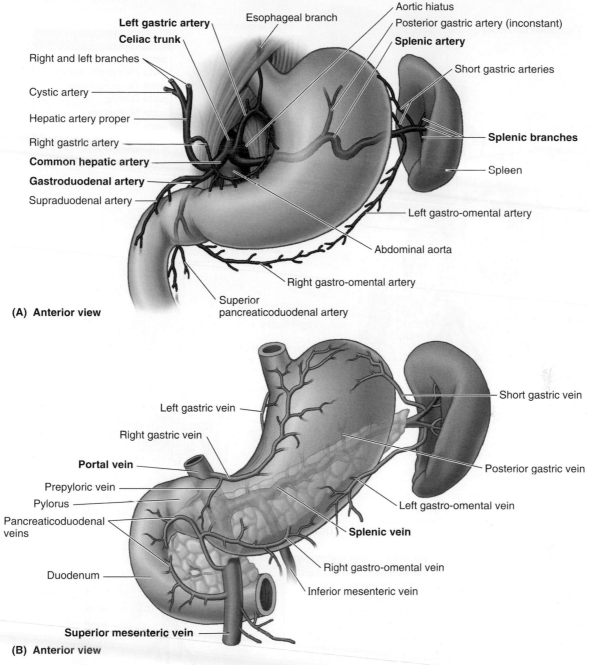

FIGURE 5.22. **Blood vessels of stomach and duodenum. A.** Arterial supply. **B.** Hepatic portal venous drainage.

VASCULATURE AND NERVES OF STOMACH

The stomach has

- a rich *arterial supply*, arising from the celiac trunk and its branches (Fig. 5.22A and Table 5.5). Most of the blood is supplied by anastomoses formed along the lesser curvature by the **right** and **left gastric arteries** and, along the greater curvature, by the **right** and **left gastro-omental artery** (gastro-epiploic artery). The fundus and upper body of stomach receive blood from the short and posterior gastric arteries, branches of the splenic artery.

- **right and left gastric veins** that parallel the arteries and drain directly or indirectly into the hepatic portal venous system (Fig. 5.22B)
- **gastric lymphatic vessels** that drain lymph from the anterior and posterior surfaces of the stomach to the **gastric** and **gastro-omental lymph nodes** located along the lesser and greater curvatures (Fig. 5.23A).

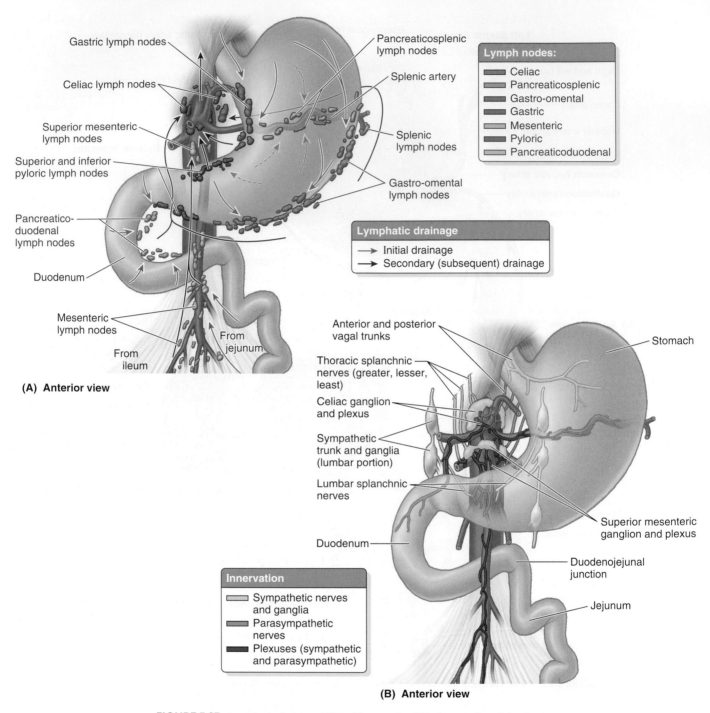

(A) Anterior view

Lymph nodes:
- Celiac
- Pancreaticosplenic
- Gastro-omental
- Gastric
- Mesenteric
- Pyloric
- Pancreaticoduodenal

Lymphatic drainage
→ Initial drainage
→ Secondary (subsequent) drainage

(B) Anterior view

Innervation
- Sympathetic nerves and ganglia
- Parasympathetic nerves
- Plexuses (sympathetic and parasympathetic)

FIGURE 5.23. Lymphatic drainage (A) and innervation (B) of stomach and duodenum.

The efferent vessels from these nodes via the **pancreaticosplenic**, **pyloric**, and **pancreaticoduodenal lymph nodes** accompany the large arteries to the **celiac lymph nodes**.

- *parasympathetic* and *sympathetic innervation*. The *parasympathetic nerve supply* is from the *anterior vagal trunk* (mainly from the left vagus nerve) and the larger *posterior vagal trunk* (mainly from the right vagus nerve) and their branches, which enter the abdomen through the esophageal hiatus (Fig. 5.23B). The *sympathetic nerve supply* is from the T6–T9 segments of the spinal cord, which passes to the *celiac plexus* via the greater splanchnic nerves and is distributed as plexuses around the gastric and gastro-omental arteries. (See Table 1.4 regarding the effects of ANS on GI tract.)

TABLE 5.5. ARTERIAL SUPPLY TO ESOPHAGUS, STOMACH, DUODENUM, LIVER, GALLBLADDER, PANCREAS, AND SPLEEN

Artery[a]	Origin	Course	Distribution
Celiac trunk	Abdominal aorta (T12) just distal to aortic hiatus of diaphragm	After short antero-inferior course, bifurcates into splenic and common hepatic arteries	Esophagus, stomach, duodenum (proximal to bile duct), liver and biliary apparatus, and pancreas
Left gastric	Celiac trunk	Ascends retroperitoneally to esophageal hiatus, giving rise to an esophageal branch; then descending along lesser curvature to anastomose with right gastric artery	Distal portion of esophagus and left portion of lesser curvature of stomach
Splenic		Runs retroperitoneally along superior border of pancreas, then passes between layers of splenorenal ligament to hilum of spleen	Body of pancreas, spleen, and greater curvature of stomach; posterior gastric branch supplies posterior wall and fundus of stomach
Left gastro-omental (gastro-epiploic)	Splenic artery in hilum of spleen	Passes between layers of gastrosplenic ligament to greater curvature of stomach	Left portion of greater curvature of stomach
Short gastric (4 or 5 branches)		Pass between layers of gastrosplenic ligament to fundus of stomach	Fundus of stomach
Hepatic[b]	Celiac trunk	Passes retroperitoneally to reach hepatoduodenal ligament and passes between its layers to porta hepatis; divides into right and left hepatic arteries	Liver, gallbladder, stomach, pancreas, duodenum, and respective lobes of liver
Cystic	Right hepatic artery	Arises within hepatoduodenal ligament	Gallbladder and cystic duct
Right gastric[c]	Hepatic artery	Runs along lesser curvature of stomach	Right portion of lesser curvature of stomach
Gastroduodenal		Descends retroperitoneally posterior to gastroduodenal junction	Stomach, pancreas, first part of duodenum, and distal part of bile duct
Right gastro-omental (gastro-epiploic)[c]	Gastroduodenal artery	Passes between layers of greater omentum to greater curvature of stomach	Right portion of greater curvature of stomach
Anterior and posterior superior pancreaticoduodenal		Descend on head of pancreas	Proximal portion of duodenum and head of pancreas
Anterior and posterior inferior pancreaticoduodenal	Superior mesenteric artery	Ascend on head of pancreas	Distal portion of duodenum and head of pancreas

[a]For anastomoses, see Figure 5.22A.
[b]For descriptive purposes, the hepatic artery is often divided into the common hepatic artery from its origin to the origin of gastroduodenal artery, and the remainder of the vessel is called hepatic artery proper.
[c]Origins are highly variable.

RELATIONS OF STOMACH

The stomach is covered by peritoneum, except where blood vessels run along its curvatures and in a small area posterior to the cardial orifice. The two layers of the lesser omentum separate to extend around the stomach and come together again to leave its greater curvature as the greater omentum.

- *Anteriorly*, the stomach is related to the diaphragm, the left lobe of the liver, and the anterior abdominal wall (Fig. SA5.2A).
- *Posteriorly*, the stomach is related to the omental bursa and pancreas; the posterior surface of the stomach forms most of the anterior wall of the omental bursa (Figs. 5.24 and 5.25).

The **stomach bed** on which the stomach rests when a person is in the supine position is formed by the structures forming the posterior wall of the omental bursa. From superior to inferior, these include the left dome of the diaphragm, spleen, left kidney and suprarenal gland, splenic artery, pancreas, transverse mesocolon, and colon (Fig. 5.24).

Small Intestine

The **small intestine**, consisting of the duodenum, jejunum, and ileum, extends from the pylorus of the stomach to the ileocecal junction where the ileum joins the cecum, the first part of the large intestine.

DUODENUM

The duodenum, the first and shortest (25 cm) part of the small intestine, is also the widest and most fixed part. The duodenum begins at the pylorus and ends at the **duodenojejunal junction**. Whereas the duodenum extends to the right and then to the left, the pylorus and duodenojejunal junctions are both quite close to the midline. Four parts of the duodenum are (Fig. 5.24A) as follows:

- **Superior (first) part**—short (approximately 5 cm), mostly horizontal, and lies anterolateral to the body of L1 vertebra
- **Descending (second) part**—longer (7–10 cm) and runs vertically along the right sides of the L2 and L3 vertebrae, curving around the head of the pancreas; initially, it lies to the right and parallel to the IVC. The *bile duct* and *main*

SURFACE ANATOMY

Stomach

The surface markings of the stomach vary because its size and position change under various circumstances. The surface markings in the supine position include the following (Fig. SA5.2A):

- **Cardial orifice**—usually lies posterior to the *6th left costal cartilage*, 2–4 cm from the median plane at the level of the T10 or T11 vertebra
- **Fundus**—usually lies posterior to the *5th left rib* in the midclavicular plane
- **Greater curvature**—passes inferiorly to the left as far as the *10th left costal cartilage* before turning medially to reach the pyloric antrum

- **Lesser curvature**—passes from the right side of the cardia to the pyloric antrum. The most inferior part of the curvature is marked by the **angular incisure** (Fig. 5.21A), which lies just to the left of the midline.
- **Pyloric part of the stomach**—usually lies at the level of the 9th costal cartilage at the level of the L1 vertebra. The pyloric orifice is approximately 1.25 cm left of the midline.
- **Pylorus**—usually lies on the right side. Its location varies from the L2 to the L4 vertebra.

A heavily built hypersthenic individual with a short thorax and long abdomen is likely to have a stomach that is placed high and more transversely disposed. In people with a slender, asthenic physique, the stomach is low and vertical (Fig. SA5.2B).

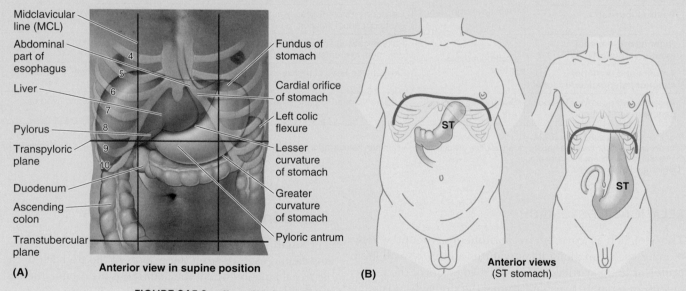

(A) **Anterior view in supine position**

(B) **Anterior views** (ST stomach)

FIGURE SA5.2. Effect of body type (bodily habitus) on disposition and shape of stomach.

pancreatic ducts enter its posteromedial wall via the hepatopancreatic ampulla.

- **Inferior (horizontal or third) part**—6–8 cm long and crosses anterior to the IVC and aorta and posterior to the superior mesenteric artery and vein at the level of the L3 vertebra
- **Ascending (fourth) part**—short (approximately 5 cm) and begins at the left of the L3 vertebra and rises superiorly as far as the superior border of the L2 vertebra, 2–3 cm to the left of the midline. It passes on the left side of the aorta to reach the inferior border of the body of the pancreas. Here, it curves anteriorly to join the jejunum at the

duodenojejunal junction, which takes the form of an acute angle, the **duodenojejunal flexure**. The flexure is supported by the attachment of the **suspensory muscle of the duodenum** (ligament of Treitz).

The suspensory muscle of the duodenum is commonly composed of a slip of skeletal muscle from the diaphragm and a fibromuscular band of smooth muscle from the third and fourth parts of the duodenum. The suspensory muscle passes posterior to the pancreas and splenic vein and anterior to the left renal vein. Its function is not known.

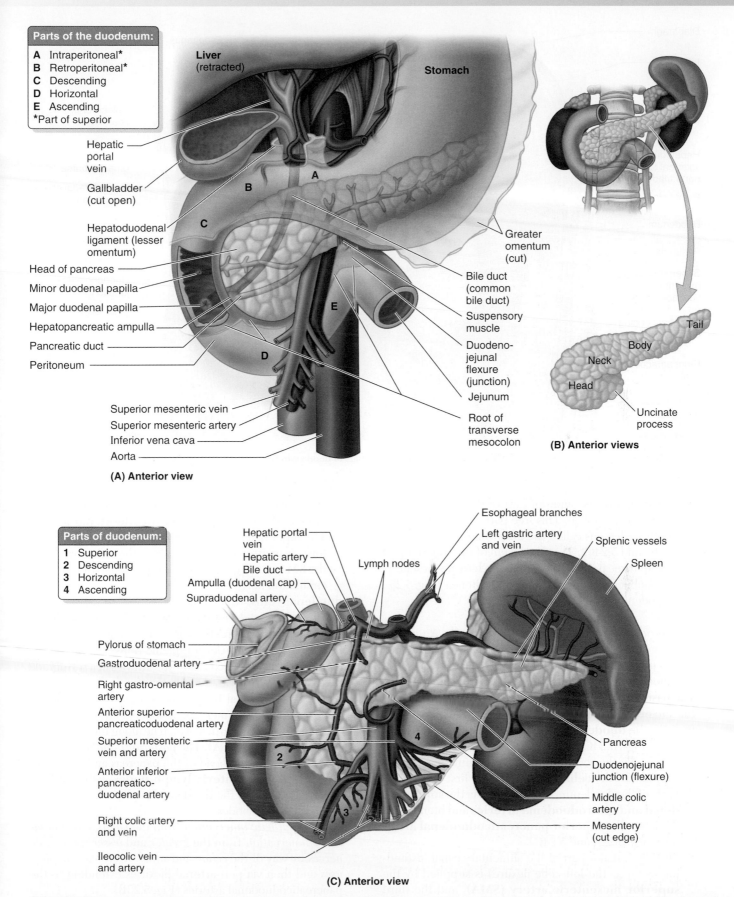

Parts of the duodenum:

A Intraperitoneal*
B Retroperitoneal*
C Descending
D Horizontal
E Ascending
*Part of superior

Liver (retracted)

Stomach

Hepatic portal vein

Gallbladder (cut open)

Hepatoduodenal ligament (lesser omentum)

Head of pancreas

Minor duodenal papilla

Major duodenal papilla

Hepatopancreatic ampulla

Pancreatic duct

Peritoneum

Superior mesenteric vein

Superior mesenteric artery

Inferior vena cava

Aorta

Greater omentum (cut)

Bile duct (common bile duct)

Suspensory muscle

Duodeno-jejunal flexure (junction)

Jejunum

Root of transverse mesocolon

(A) Anterior view

Tail

Body

Neck

Head

Uncinate process

(B) Anterior views

Parts of duodenum:

1 Superior
2 Descending
3 Horizontal
4 Ascending

Hepatic portal vein

Hepatic artery

Bile duct

Ampulla (duodenal cap)

Supraduodenal artery

Esophageal branches

Left gastric artery and vein

Splenic vessels

Spleen

Lymph nodes

Pylorus of stomach

Gastroduodenal artery

Right gastro-omental artery

Anterior superior pancreaticoduodenal artery

Superior mesenteric vein and artery

Anterior inferior pancreatico-duodenal artery

Right colic artery and vein

Ileocolic vein and artery

Pancreas

Duodenojejunal junction (flexure)

Middle colic artery

Mesentery (cut edge)

(C) Anterior view

FIGURE 5.24. Duodenum, spleen, and pancreas. A. Relationships of the duodenum. **B.** Pancreas—relationships and parts. **C.** Vasculature of duodenum and pancreas.

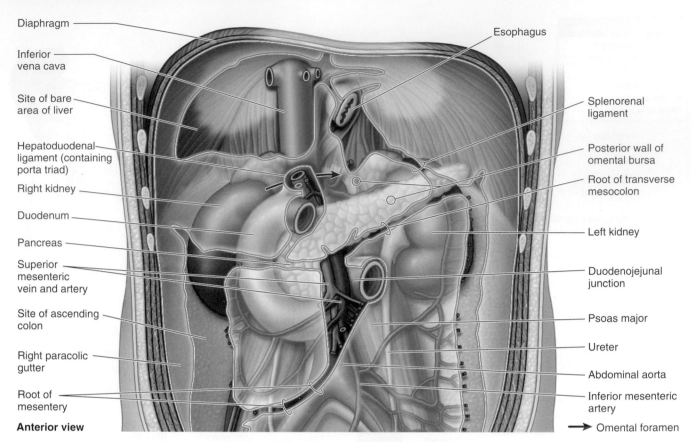

Diaphragm

Inferior vena cava

Site of bare area of liver

Hepatoduodenal ligament (containing porta triad)

Right kidney

Duodenum

Pancreas

Superior mesenteric vein and artery

Site of ascending colon

Right paracolic gutter

Root of mesentery

Anterior view

Esophagus

Splenorenal ligament

Posterior wall of omental bursa

Root of transverse mesocolon

Left kidney

Duodenojejunal junction

Psoas major

Ureter

Abdominal aorta

Inferior mesenteric artery

→ Omental foramen

FIGURE 5.25. Peritoneal relationships of duodenum and pancreas.

The first 2 cm of the superior part of the duodenum has a mesentery and is mobile. This free part—relatively dilated and smooth-walled—is called the **ampulla** or duodenal cap (Figs. 5.21C and 5.24C). The distal 3 cm of the superior part and the other three parts of the duodenum have no mesentery and are immobile because they are retroperitoneal (Fig. 5.25).

The duodenum has

- *an arterial supply* from two different vessels. An important transition in the blood supply of the alimentary tract occurs over the course of the descending (second) part of the duodenum, approximately where the bile duct enters. The basis of this transition is embryological because this is the site of the junction of the foregut and midgut. Consequently, the **duodenal arteries** arise from two different sources (Figs. 5.24C and 5.26 and Table 5.6):
  - Proximally, the abdominal part of the alimentary tract is supplied by the **celiac trunk**, and the first and second parts of the duodenum are supplied via the **supraduodenal and gastroduodenal arteries** and branches of the latter, the **superior pancreaticoduodenal arteries** (Figs. 5.22A and 5.24C).
  - Distally, a major part of the alimentary canal (extending as far as the left colic flexure) is supplied by the **superior mesenteric artery (SMA)**, and the third

and fourth parts of the duodenum are supplied by its branch, the **inferior pancreaticoduodenal artery**. The superior and inferior pancreaticoduodenal arteries form an anastomotic loop between the celiac trunk and the SMA; consequently, there is potential for collateral circulation here.

- **duodenal veins**, which follow the arteries and drain into the *hepatic portal vein* (Figs. 5.19C, 5.22B, and 5.27); some veins drain directly and others indirectly through the superior mesenteric and splenic veins.
- *lymphatic vessels*, which follow the arteries in a retrograde direction. The *anterior lymphatic vessels* drain into the pancreaticoduodenal lymph nodes located along the superior and inferior pancreaticoduodenal arteries and into the **pyloric lymph nodes**, which lie along the gastroduodenal artery (Fig. 5.23A). The *posterior lymphatic vessels* pass posterior to the head of the pancreas and drain into the **superior mesenteric lymph nodes**. Efferent lymphatic vessels from the duodenal lymph nodes drain into the celiac lymph nodes.
- *parasympathetic innervation* from the *vagus* and *sympathetic innervation* from the *greater* and *lesser splanchnic nerves* by way of the *celiac* and *superior mesenteric plexuses* and then via peri-arterial plexuses extending to the pancreaticoduodenal arteries (Fig. 5.23B).

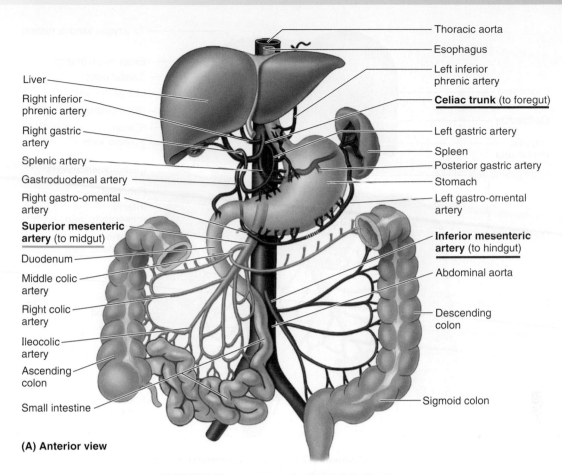

Thoracic aorta

Esophagus

Left inferior phrenic artery

Celiac trunk (to foregut)

Left gastric artery

Spleen

Posterior gastric artery

Stomach

Left gastro-omental artery

Inferior mesenteric artery (to hindgut)

Abdominal aorta

Descending colon

Sigmoid colon

Liver

Right inferior phrenic artery

Right gastric artery

Splenic artery

Gastroduodenal artery

Right gastro-omental artery

Superior mesenteric artery (to midgut)

Duodenum

Middle colic artery

Right colic artery

Ileocolic artery

Ascending colon

Small intestine

(A) Anterior view

FIGURE 5.26. Arterial supply of gastrointestinal tract.

TABLE 5.6. ARTERIAL SUPPLY TO INTESTINES

Artery	Origin	Course	Distribution
Superior mesenteric	Abdominal aorta (L1)	Runs in root of mesentery to ileocecal junction	Part of gastrointestinal tract derived from midgut
Intestinal (*n* = 15–18)		Passes between two layers of mesentery	Jejunum and ileum
Middle colic		Ascends retroperitoneally and passes between layers of transverse mesocolon	Transverse colon
Right colic	Superior mesenteric artery	Passes retroperitoneally to reach ascending colon	Ascending colon
Ileocolic	Terminal branch of superior mesenteric artery	Runs along root of mesentery and divides into ileal and colic branches	Ileum, cecum, and ascending colon
Appendicular	Ileocolic artery	Passes between layers of meso-appendix	Appendix
Inferior mesenteric	Abdominal aorta (L3)	Descends retroperitoneally to left of abdominal aorta	Descending colon
Left colic		Passes retroperitoneally toward left to descending colon	
Sigmoid (*n* = 3 or 4 branches)	Inferior mesenteric artery	Passes retroperitoneally toward left to sigmoid colon	Descending and sigmoid colon
Superior rectal	Terminal branch of inferior mesenteric artery	Descends retroperitoneally to rectum	Proximal part of rectum
Middle rectal	Internal iliac artery	Passes retroperitoneally to rectum	Midpart of rectum
Inferior rectal	Internal pudendal artery	Crosses ischio-anal fossa to reach rectum	Distal part of rectum and anal canal

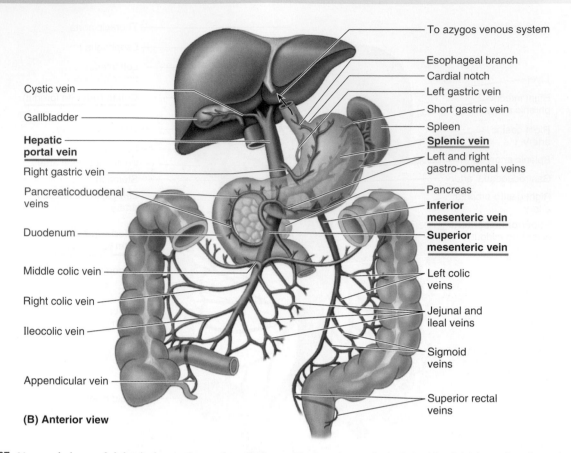

Cystic vein

Gallbladder

**Hepatic
portal vein**

Right gastric vein

Pancreaticoduodenal
veins

Duodenum

Middle colic vein

Right colic vein

Ileocolic vein

Appendicular vein

To azygos venous system

Esophageal branch

Cardial notch

Left gastric vein

Short gastric vein

Spleen

Splenic vein

Left and right
gastro-omental veins

Pancreas

**Inferior
mesenteric vein**

**Superior
mesenteric vein**

Left colic
veins

Jejunal and
ileal veins

Sigmoid
veins

Superior rectal
veins

(B) Anterior view

FIGURE 5.27. **Venous drainage of abdominal part of gastrointestinal tract.** The hepatic portal vein drains blood rich in nutrients but reduced in oxygen from the stomach, intestines, spleen, pancreas, and gallbladder to the liver.

JEJUNUM AND ILEUM

The jejunum begins at the duodenojejunal junction and the ileum ends at the **ileocecal junction**, the union of the terminal ileum and cecum (Fig. 5.28A,B). Together, the jejunum and ileum are 6–7 m long in cadavers; however, tonic contraction makes them substantially shorter in living persons. The jejunum constitutes approximately two fifths of the length; the ileum, the remainder. The terminal ileum usually lies in the pelvis from which it ascends to end in the medial aspect of the cecum. Although no clear line of demarcation between the jejunum and ileum exists, they have distinctive characteristics for most of their lengths (Fig. 5.28C–G and Table 5.7).

The mesentery, a fan-shaped fold of peritoneum, attaches the jejunum and ileum to the posterior abdominal wall. The **root (origin) of the mesentery** (approximately 15 cm long) is directed obliquely, inferiorly, and to the right (Fig. 5.25). It extends from the duodenojejunal junction on the left side of the L2 vertebra to the ileocolic junction and the right sacro-iliac joint. The root of the mesentery crosses (successively) the ascending and horizontal parts of the duodenum, abdominal aorta, IVC, right ureter, right psoas major muscle, and right testicular or ovarian vessels.

The jejunum and ileum have

- *an arterial supply* from the SMA (Figs. 5.19B and 5.26). The SMA runs between the layers of the mesentery and

sends many branches to the jejunum and ileum. The arteries unite to form loops or arches—**arterial arcades**—that give rise to straight arteries, the **vasa recta** (Fig. 5.28C,D).
- *venous drainage* to the **superior mesenteric vein (SMV)** (Fig. 5.27). The SMV lies anterior and to the right of the SMA in the root of the mesentery. The SMV ends posterior to the neck of the pancreas, where it unites with the splenic vein to form the hepatic portal vein.
- specialized *lymphatic vessels*, called **lacteals**, in the intestinal villi that absorb fat and drain into the lymphatic plexuses in the walls of the jejunum and ileum. The lymphatic plexuses drain into lymphatic vessels between the layers of the mesentery and then sequentially through three groups of lymph nodes (see Fig. 5.23A): **juxta-intestinal lymph nodes** (close to the intestinal wall), **mesenteric lymph nodes** (scattered among the arterial arcades), and **central superior nodes** (along the proximal part of the SMA). Efferent lymphatic vessels from these nodes drain into the superior mesenteric lymph nodes. Lymphatic vessels from the terminal ileum follow the ileal branch of the ileocolic artery to the **ileocolic lymph nodes**.
- *sympathetic and parasympathetic innervation*
  - In general, sympathetic stimulation reduces secretion and motility of the intestine and acts as a vasoconstrictor, reducing or stopping digestion and making blood (and energy) available for "fleeing or fighting." Parasympathetic stimulation increases secretion and motility of the

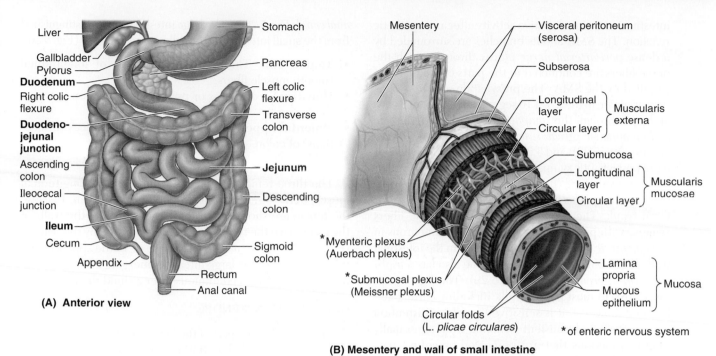

(A) Anterior view

Liver
Gallbladder
Pylorus
Duodenum
Right colic flexure
Duodeno-jejunal junction
Ascending colon
Ileocecal junction
Ileum
Cecum
Appendix
Stomach
Pancreas
Left colic flexure
Transverse colon
Jejunum
Descending colon
Sigmoid colon
Rectum
Anal canal

(B) Mesentery and wall of small intestine

Mesentery
Visceral peritoneum (serosa)
Subserosa
Longitudinal layer
Circular layer
} Muscularis externa
Submucosa
Longitudinal layer
Circular layer
} Muscularis mucosae
*Myenteric plexus (Auerbach plexus)
*Submucosal plexus (Meissner plexus)
Circular folds (L. *plicae circulares*)
Lamina propria
Mucous epithelium
} Mucosa
*of enteric nervous system

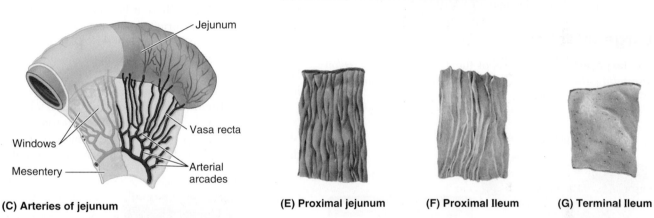

(C) Arteries of jejunum

Jejunum
Windows
Mesentery
Vasa recta
Arterial arcades

(E) Proximal jejunum **(F) Proximal Ileum** **(G) Terminal Ileum**

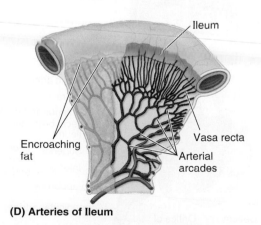

(D) Arteries of Ileum

Ileum
Encroaching fat
Vasa recta
Arterial arcades

FIGURE 5.28. Small intestine. A. Small and large intestine in situ. **B.** Layers of wall of small intestine. **C.** Arteries of jejunum. **D.** Arteries of ilium. **E.** Characteristics of proximal jejunum. **F.** Characteristics of proximal ileum. **G.** Characteristics of terminal ileum.

TABLE 5.7. DISTINGUISHING CHARACTERISTICS OF JEJUNUM AND ILEUM IN LIVING PERSONS

Characteristic	Jejunum	Ileum
Color	Deeper red	Paler pink
Caliber	2–4 cm	2–3 cm
Wall	Thick and heavy	Thin and light
Vascularity	Greater	Less
Vasa recta	Long	Short
Arcades	A few large loops	Many short loops
Fat in mesentery	Less	More
Circular folds (L. *plicae circulares*)	Large, tall, and closely packed	Low and sparse; absent in distal part

intestine, restoring digestive activity after a sympathetic reaction. The SMA and its branches are surrounded by a dense *peri-arterial nerve plexus* through which the nerve fibers are conducted to the parts of the intestine supplied by the SMA. The presynaptic *sympathetic fibers* originate in the T8–T10 segments of the spinal cord and pass through the *sympathetic trunks* and *thoracic abdominopelvic (greater, lesser,* and *least) splanchnic nerves* (Figs. 5.23B and 5.29). They synapse on cell bodies of postsynaptic sympathetic neurons in the *celiac* and *superior mesenteric (prevertebral) ganglia.*

- The *parasympathetic fibers* derive from the posterior vagal trunk. The presynaptic parasympathetic fibers synapse with postsynaptic parasympathetic neurons in the *myenteric* and *submucous plexuses* in the intestinal wall (Fig. 5.28D). The small intestine also has *sensory (visceral afferent) fibers* (Fig. 5.29). The intestine is insensitive to most pain stimuli, including cutting and burning; however, it is sensitive to sudden distention ("gas pains") and transient ischemia from abnormally long contractions that are perceived as **colic** (spasmodic abdominal pains).

Large Intestine

The **large intestine** consists of the *appendix, cecum, colon (ascending, transverse, descending,* and *sigmoid), rectum,* and

anal canal (Fig. 5.28A). The large intestine can be distinguished from the small intestine by the following structures (Fig. 5.30):

- **Teniae coli**: Three thickened bands of longitudinal smooth muscle fibers
- **Haustra**: Sacculations or pouches of the colon between the teniae
- **Omental appendices**: Small, fatty appendices (projections) of colon
- Caliber: The internal diameter is much larger.

The three teniae coli make up most of the longitudinal muscle of the large intestine, except in the rectum. Because the teniae are shorter than the large intestine, the walls of the colon have the typical sacculations formed by the haustra (Fig. 5.30A). The teniae begin at the base of the appendix and run the length of the large intestine, merging at the rectosigmoid junction into a continuous layer around the rectum.

CECUM AND APPENDIX

The **cecum**, the first part of the large intestine, is continuous with the ascending colon. It is a blind intestinal pouch in the right lower quadrant, where it lies in the iliac fossa inferior to the junction of the terminal ileum and cecum. The cecum is usually almost entirely enveloped by peritoneum and can be lifted freely; however, the cecum has no mesentery. The ileum enters the cecum obliquely and partly invaginates into it, forming the **ileal orifice** (Fig. 5.30B).

The **vermiform** (L. worm-like) **appendix**, a blind intestinal diverticulum, extends from the posteromedial aspect of the cecum inferior to the ileocecal junction. The appendix varies in length and has a short triangular mesentery, the

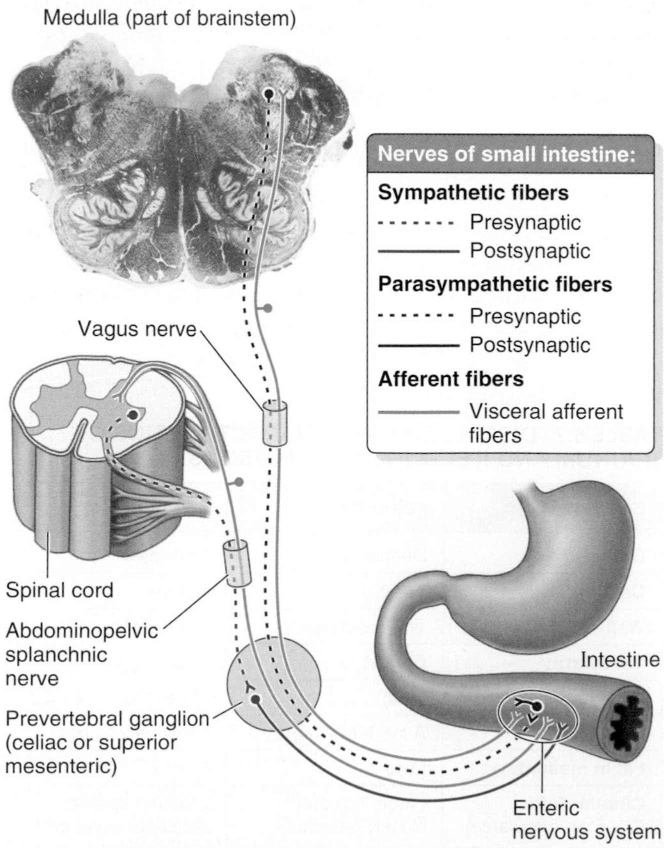

Medulla (part of brainstem)

Nerves of small intestine:

Sympathetic fibers
- - - - - Presynaptic
———— Postsynaptic

Parasympathetic fibers
- - - - - Presynaptic
———— Postsynaptic

Afferent fibers
———— Visceral afferent fibers

Vagus nerve

Spinal cord

Abdominopelvic splanchnic nerve

Prevertebral ganglion (celiac or superior mesenteric)

Intestine

Enteric nervous system

FIGURE 5.29. Innervation of small intestine.

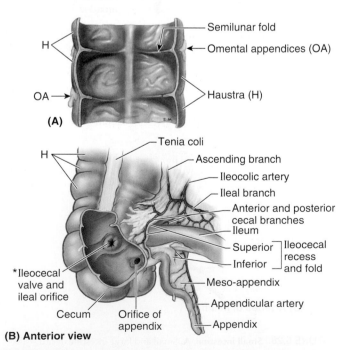

Semilunar fold

Omental appendices (OA)

H

OA →

Haustra (H)

(A)

H

Tenia coli

Ascending branch

Ileocolic artery

Ileal branch

Anterior and posterior cecal branches

Ileum

Superior ⎫ Ileocecal
            ⎬ recess
Inferior ⎭ and fold

*Ileocecal valve and ileal orifice

Meso-appendix

Appendicular artery

Cecum

Orifice of appendix

Appendix

(B) Anterior view

FIGURE 5.30. Characteristics of large intestine, cecum, and appendix. **A.** Features of large intestine. **B.** Blood supply of cecum and appendix. A window has been cut in the wall of the cecum to expose the ileocecal orifice and the orifice of the appendix.

meso-appendix, which derives from the posterior side of the mesentery of the terminal ileum (Fig. 5.30B). The meso-appendix attaches to the cecum and the proximal part of the appendix. The position of the appendix is variable, but it is usually retrocecal (posterior to the cecum). The base of the appendix most often lies deep to a point that is one third of the way along the oblique line joining the right anterior superior iliac spine to the umbilicus (*spino-umbilical* or *McBurney point*).

The cecum is supplied by the **ileocolic artery**, the terminal branch of the SMA. The appendix is supplied by the **appendicular artery**, a branch of the ileocolic artery (Figs. 5.30B and 5.31A and Table 5.6). A tributary of the SMV, the **ileocolic vein**, drains blood from the cecum and appendix (Fig. 5.27). The lymphatic vessels from the cecum and appendix pass to lymph nodes in the meso-appendix and to the ileocolic lymph nodes that lie along the ileocolic artery (Fig. 5.31C). Efferent lymphatic vessels pass to the superior mesenteric lymph nodes. The nerve supply to the cecum and appendix derives from sympathetic and parasympathetic nerves from the superior mesenteric plexus (Fig. 5.31D). The sympathetic nerve fibers originate in the lower thoracic part of the spinal cord (T10–T12), and the parasympathetic nerve fibers derive from the vagus nerves. Afferent nerve fibers from the appendix accompany the sympathetic nerves to the T10 segment of the spinal cord.

COLON

The colon has four parts—ascending, transverse, descending, and sigmoid—that succeed one another in an arch (Fig. 5.31A).

The **ascending colon** passes superiorly on the right side of the abdominal cavity from the cecum, typically in the iliac fossa (greater pelvis), to the right lobe of the liver, where it turns to the left as the **right colic flexure** (hepatic flexure). The ascending colon, narrower than the cecum, lies retroperitoneally along the right side of the posterior abdominal wall. The ascending colon is covered by peritoneum anteriorly and on its sides; however, in approximately 25% of people, it has a short mesentery. The ascending colon is separated from the anterolateral abdominal wall by the greater omentum. A vertical groove, lined with parietal peritoneum (the **right paracolic gutter**), lies lateral to the ascending colon (see Figs. 5.18 and 5.25).

The arterial supply to the ascending colon and right colic flexure is from branches of the SMA—the **ileocolic** and **right colic arteries** (Fig. 5.31A and Table 5.6). Tributaries of the SMV, the **ileocolic** and **right colic veins**, drain blood from the ascending colon (Fig. 5.31B). The lymphatic vessels first pass to the **epicolic and paracolic lymph nodes**, next to the **ileocolic** and intermediate **right colic lymph nodes**, and from then to the **superior mesenteric nodes** (Fig. 5.31C). The *nerves to the ascending colon* derive from the *superior mesenteric plexus* (Fig. 5.31D).

The **transverse colon**, the largest and most mobile part of the large intestine, crosses the abdomen from the right colic flexure to the **left colic flexure** (splenic flexure), where it bends inferiorly to become the descending colon (Fig. 5.31A). The left colic flexure—usually more superior, more acute, and less mobile than the right colic flexure—lies anterior to the inferior part of the left kidney and is attached to the diaphragm by the **phrenicocolic ligament** (see Fig. 5.14). The mesentery of the transverse colon, the *transverse mesocolon*, loops down, so that the central transverse colon is inferior to the level of the iliac crests and is adherent to the posterior wall of the omental bursa. The **root of the transverse mesocolon** lies along the inferior border of the pancreas and is continuous with the parietal peritoneum posteriorly (Fig. 5.25).

The arterial supply of the transverse colon is mainly from the **middle colic artery** (Fig. 5.31A and Table 5.6), a branch of the SMA; however, it may also be supplied to variable degrees by the *right* and *left colic arteries* via anastomoses. *Venous drainage of the transverse colon* is through the *SMV* (Fig. 5.31B). Lymphatic drainage is to the **middle colic lymph nodes**, which in turn drain to the superior mesenteric lymph nodes (Fig. 5.31C). *The nerves of the transverse colon* arise from the superior mesenteric plexus and follow the right and middle colic arteries (Fig. 5.31D). These nerves transmit sympathetic and parasympathetic (vagal) nerve fibers. Some nerves derived from the *inferior mesenteric plexus* may follow anastomoses from the left colic artery.

The **descending colon** passes retroperitoneally from the left colic flexure into the left iliac fossa, where it is continuous with the sigmoid colon. Peritoneum covers the colon anteriorly and laterally and binds it to the posterior abdominal wall. Although retroperitoneal, the inferior descending colon, especially in the iliac fossa, has a short mesentery in approximately 33% of people. As it descends, the colon passes anterior to the lateral border of the left kidney (see Fig. 5.25). As with the ascending colon, a **left paracolic gutter** lies on the lateral side of the descending colon (see Figs. 5.18 and 5.25).

(Continued on page 294)

CLINICAL BOX

Overview of Embryological Rotation of Midgut

The primordial gut consists of the **foregut** (esophagus, stomach, pancreas, duodenum, liver, and biliary ducts), **midgut** (small intestine distal to the bile duct, cecum, appendix, ascending colon, and most of the transverse colon), and **hindgut** (distal transverse colon, descending and sigmoid colon, and rectum). For 4 weeks, the rapidly growing midgut, supplied by the SMA, is herniated into the proximal

(Continued on next page)

part of the umbilical cord (Fig. B5.7A). It is attached to the umbilical vesicle (yolk sac) by the omphalo-enteric duct (yolk stalk). As it returns to the abdominal cavity, the midgut rotates 270 degrees around the axis of the SMA (Fig. B5.7B,C). As the parts of the intestine reach their definitive positions, their mesenteric attachments undergo modifications. Some mesenteries shorten and others disappear (Fig. B5.7D,E). *Malrotation of the midgut* results in several congenital anomalies, such as *volvulus* (twisting) of the intestine (Moore et al, 2016).

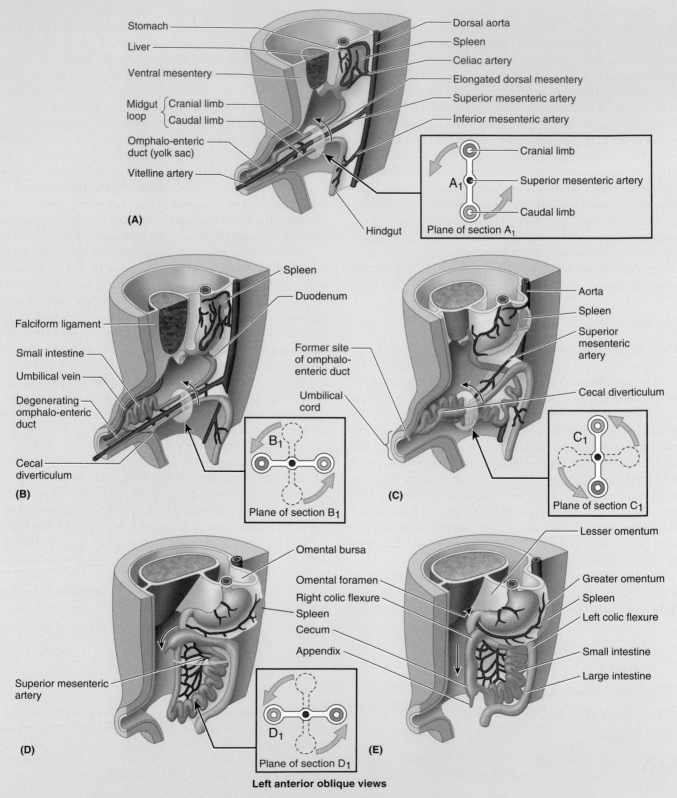

Left anterior oblique views

FIGURE B5.7. Embryological rotation of midgut.

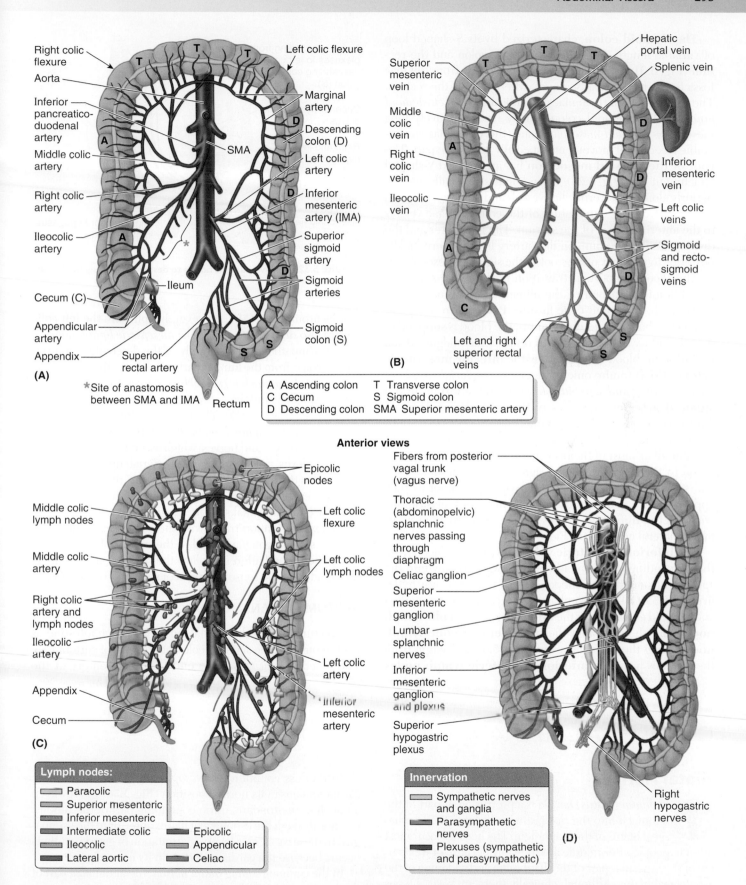

Anterior views

Lymph nodes:	
Paracolic	
Superior mesenteric	
Inferior mesenteric	
Intermediate colic	Epicolic
Ileocolic	Appendicular
Lateral aortic	Celiac

Innervation	
Sympathetic nerves and ganglia	
Parasympathetic nerves	
Plexuses (sympathetic and parasympathetic)	

A	Ascending colon	T	Transverse colon
C	Cecum	S	Sigmoid colon
D	Descending colon	SMA	Superior mesenteric artery

FIGURE 5.31. Large intestine. A. Arterial supply. **B.** Venous drainage. **C.** Lymphatic drainage. **D.** Innervation.

The **sigmoid colon**, characterized by its S-shaped loop of variable length, links the descending colon and the rectum (Fig. 5.31A). The sigmoid colon extends from the iliac fossa to the third sacral segment, where it joins the rectum. The termination of the teniae coli indicates the *rectosigmoid junction*. The sigmoid colon usually has a relatively long mesentery (*sigmoid mesocolon*) and, therefore, has considerable freedom of movement, especially its middle part. The **root of the sigmoid mesocolon** has an inverted V-shaped attachment (see Fig. 5.25), extending first medially and superiorly along the external iliac vessels and then medially and inferiorly from the bifurcation of the common iliac vessels to the anterior aspect of the sacrum. The left ureter and the division of the left common iliac artery lie retroperitoneally posterior to the apex of the root of the sigmoid mesocolon.

The second important transition in the blood supply to the abdominal portion of the alimentary tract occurs approximately at the left colic flexure. Proximal to this point (extending back to mid-duodenum), the blood is supplied to the alimentary tract by the **SMA** (embryonic midgut); distal to this point, blood is supplied by the **inferior mesenteric artery** (IMA) (embryonic hindgut). *The arterial supply of the descending and sigmoid colon is from the **left colic** and **sigmoid arteries**, branches of the IMA (Fig. 5.31A and Table 5.6). The left colic and sigmoid arteries pass to the left, where they divide into ascending and descending branches. Usually, all or most of the branches of the arteries supplying blood to the colon (ileocolic; right, middle, and left colic; and sigmoid arteries) anastomose with each other as they approach the colon, thus forming a continuous anastomotic channel, the **marginal artery**, which may provide important collateral circulation (Fig. 5.31A).

The **inferior mesenteric vein (IMV)** returns blood from the descending and sigmoid colon, flowing usually into the splenic vein and then the hepatic portal vein on its way to the liver (Fig. 5.31B). The lymphatic vessels from the descending and sigmoid colon pass to the **epicolic** and **paracolic lymph nodes** and then through the **intermediate colic lymph nodes** along the left colic artery (Fig. 5.31C). Lymph from these nodes passes to **inferior mesenteric lymph nodes**

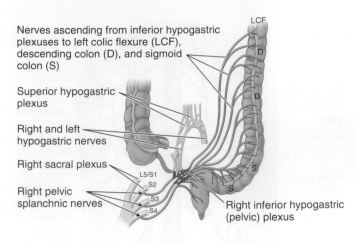

FIGURE 5.32. Parasympathetic nerves to descending and sigmoid colon.

that lie around the IMA; however, lymph from the left colic flexure also drains to the *superior mesenteric lymph nodes*.

The sympathetic nerve supply of the descending and sigmoid colon is from the lumbar part of the sympathetic trunk via *lumbar (abdominopelvic) splanchnic nerves*, the *inferior mesenteric ganglion*, and the *peri-arterial plexuses on the IMA* and its branches (Fig. 5.31D). The parasympathetic nerve supply is from the *pelvic splanchnic nerves* via the inferior hypogastric (pelvic) plexus and nerves, which ascend retroperitoneally from the plexus, independent of the arterial supply (Fig. 5.32). Proximal to the middle of the sigmoid colon, the visceral afferents conveying pain pass retrogradely with sympathetic fibers to thoracolumbar spinal sensory ganglia, whereas those carrying reflex information travel with the parasympathetic fibers to vagal sensory ganglia. Distal to the middle of the sigmoid colon, the visceral afferents follow the parasympathetic fibers retrogradely to the sensory ganglia of spinal nerves S2–S4.

RECTUM AND ANAL CANAL

The rectum, the fixed terminal part of the large intestine, is continuous with the sigmoid colon at the level of vertebra S3. The junction is at the lower end of the mesentery of the

(Continued on page 296)

CLINICAL BOX

Hiatal Hernia

A *hiatal (hiatus) hernia* is a protrusion of part of the stomach into the mediastinum through the esophageal hiatus of the diaphragm. The hernias occur most often in people after middle age, possibly because of weakening of the muscular part of the diaphragm and widening of the esophageal hiatus. Although clinically there are several types of hiatal hernias, the two main types are para-esophageal hiatal hernia and sliding hiatal hernia (Skandalakis et al., 1996).

In the less common *para-esophageal hiatal hernia*, the cardia remains in its normal position (Fig. B5.8A). However, a pouch of peritoneum, often containing part of the fundus, extends through the esophageal hiatus anterior to the esophagus. In these cases, usually no regurgitation of gastric contents occurs because the cardial orifice is in its normal position.

In the common *sliding hiatal hernia*, the abdominal part of the esophagus, the cardia, and parts of the fundus of the stomach slide superiorly through the esophageal hiatus into the thorax, especially when the person lies down or bends over (Fig. B5.8).

Some regurgitation of stomach contents into the esophagus is possible because the clamping action of the right crus of the diaphragm on the inferior end of the esophagus is weak.

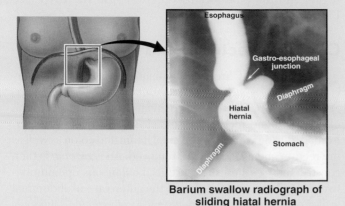

Barium swallow radiograph of sliding hiatal hernia

FIGURE B5.8. Schematic diagram of sliding hiatal hernia.

Carcinoma of Stomach and Gastrectomy

When the body or pyloric part of the stomach contains a malignant tumor, the mass may be palpable. Using *gastroscopy*, physicians can inspect the lining of the air-inflated stomach, enabling them to observe gastric lesions and take biopsies. *Partial gastrectomy* (removal of part of the stomach) may be performed to remove the region of the stomach involved by carcinoma. Because of the anastomoses of the arteries supplying the stomach provide good collateral circulation, one or more arteries may be ligated during this procedure without seriously affecting the blood supply of the remaining part of the stomach.

Partial gastrectomy to remove a carcinoma usually also requires removal of all involved regional lymph nodes. Because cancer frequently occurs in the pyloric region, removal of the *pyloric lymph nodes* as well as the right *gastro-omental lymph nodes* also receiving lymph drainage from this region is especially important. As stomach cancer becomes more advanced, the lymphogenous dissemination of malignant cells involves the *celiac lymph nodes* to which all gastric nodes drain.

Gastric Ulcers, Peptic Ulcers, *Helicobacter pylori*, and Vagotomy

Gastric ulcers are open lesions of the mucosa of the stomach, whereas *peptic ulcers* are lesions of the mucosa of the pyloric canal or, more often, the duodenum. Most ulcers of the stomach and duodenum are associated with an infection of a specific bacterium, *Helicobacter pylori*. It is thought that the high acid level in the stomach and duodenum overwhelms the bicarbonate normally produced by the duodenum and reduces the effectiveness of the mucous

lining, leaving it vulnerable to *H. pylori*. The bacteria erode the protective mucous lining of the stomach, inflaming the mucosa and making it vulnerable to the effects of the gastric acid and digestive enzymes (pepsin) produced by the stomach.

If the ulcer erodes into the gastric arteries, it can cause life-threatening bleeding. Because the secretion of acid by parietal cells of the stomach is largely controlled by the vagus nerves, *vagotomy* (surgical section of the vagus nerves) is performed in some people with chronic or recurring ulcers to reduce the production of acid.

A *posterior gastric ulcer* may erode through the stomach wall into the pancreas, resulting in referred pain to the back. In such cases, *erosion of the splenic artery* results in severe hemorrhage into the peritoneal cavity.

Duodenal (Peptic) Ulcers

Most inflammatory erosions of the duodenal wall, *duodenal ulcers*, are in the posterior wall of the superior part of the duodenum within 3 cm of the pylorus (Fig. B5.9). Occasionally, an ulcer perforates the duodenal wall, permitting its contents to enter the peritoneal cavity and produce *peritonitis*. Because the superior part of the duodenum closely relates to the liver and gallbladder, either of them may adhere to and be ulcerated by a duodenal ulcer. *Erosion of the gastroduodenal artery*, a posterior relation of the superior part of the duodenum, by a duodenal ulcer results in severe hemorrhage into the peritoneal cavity.

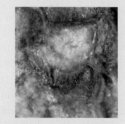

FIGURE B5.9. Duodenal ulcer.

Ileal Diverticulum

An *ileal diverticulum* (of Meckel) is a congenital anomaly that occurs in 1–2% of people. A remnant of the proximal part of the embryonic omphalo-enteric duct (yolk stalk), the diverticulum usually appears as a finger-like pouch 3–6 cm long. It is always on the antimesenteric border of the ileum—the border of the intestine opposite the mesenteric attachment (Fig. B5.10). An ileal diverticulum may become inflamed and produce pain mimicking appendicitis.

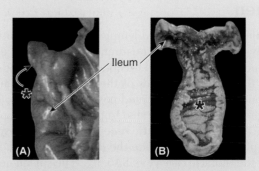

Ileum

(A)   (B)

FIGURE B5.10. Ileal diverticulum(*).

(*Continued on next page*)

Diverticulosis

Diverticulosis is a disorder in which multiple false *diverticula* (external evaginations or outpocketings of the mucosa of the colon) develop along the intestine. It primarily affects middle-aged and elderly people. Diverticulosis is commonly found in the sigmoid colon. Diverticula are subject to infection and rupture, leading to diverticulitis.

Appendicitis

Acute inflammation of the appendix is a common cause of an *acute abdomen* (severe abdominal pain arising suddenly). Digital pressure over the McBurney point produces the maximum abdominal tenderness. The pain of appendicitis usually commences as a vague pain in the peri-umbilical region because afferent pain fibers enter the spinal cord at the T10 level. Later, severe pain in the right lower quadrant results from irritation of the parietal peritoneum lining the posterior abdominal wall.

Appendectomy

Laparoscopic appendectomy has become a standard procedure used to remove the appendix via small incisions. The peritoneal cavity is first inflated with carbon dioxide gas, distending the abdominal wall, to provide viewing and working space. The laparoscope is passed through the incision in the anterolateral abdominal wall (e.g., near or through the umbilicus). One or two other small incisions ("ports") are required for surgical (instrument) access to the appendix and related vessels. An appendectomy may be performed through a transverse or gridiron (muscle-splitting) incision centered at the McBurney point in the right lower quadrant, if indicated.

In unusual cases of *malrotation of the intestine*, or failure of descent of the cecum, the appendix is not in the lower right quadrant (LRQ). When the cecum is high (*subhepatic cecum*), the appendix is in the right hypochondriac region and the pain localizes there, not in the LRQ (see Fig. B5.10).

Colitis, Colectomy, and Ileostomy

Chronic inflammation of the colon (*ulcerative colitis*, *Crohn disease*) is characterized by severe inflammation and ulceration of the colon and rectum. In some cases, a *colectomy* is performed, during which the terminal ileum and colon as well as the rectum and anal canal are removed. An *ileostomy* is then constructed to establish an artificial cutaneous opening between the ileum and the skin of the anterolateral abdominal wall. Following a partial colectomy, a *colostomy* or *sigmoidostomy* is performed to create an artificial cutaneous opening for the terminal part of the colon.

Colonoscopy

The interior surface of the colon can be observed and photographed in a procedure called *colonoscopy*, or *coloscopy*, using a long fiberoptic endoscope (*colonoscope*) inserted into the colon through the anus and rectum. Small instruments can be passed through the colonoscope to perform minor operative procedures, such as biopsies or removal of polyps. Most tumors of the large intestine occur in the rectum; approximately 12% of them appear near the rectosigmoid junction. The interior of the sigmoid colon is observed with a *sigmoidoscope*, a shorter endoscope, in a procedure called *sigmoidoscopy*.

---

sigmoid colon (see Fig. 5.25). The rectum is continuous inferiorly with the anal canal. These parts of the large intestine are described with the pelvis in Chapter 6.

Spleen

The spleen, a mobile ovoid lymphatic organ, lies intraperitoneally in the left upper quadrant. The spleen is entirely surrounded by peritoneum except at the **hilum** (Fig. 5.33), where the splenic branches of the splenic artery and vein enter and leave. It is associated posteriorly with the left 9th through 11th ribs and separated from them by the diaphragm and the **costodiaphragmatic recess**, the cleft-like extension of the pleural cavity between the diaphragm and the lower part of the thoracic cage (Fig. SA5.3B). The spleen normally does not descend inferior to the costal region; it rests on the left colic flexure. The spleen varies considerably in size, weight, and shape; however, it is usually about 12 cm long and 7 cm wide, roughly the size and shape of a clenched fist.

The **diaphragmatic surface of the spleen** is convexly curved to fit the concavity of the diaphragm (Figs. SA5.3 and 5.33). The anterior and superior borders of the spleen are sharp and often notched, whereas its posterior and inferior borders are rounded. The spleen contacts the posterior wall of the stomach and is connected to its greater curvature by the gastrosplenic ligament and to the left kidney by the **splenorenal ligament** (see Fig. 5.13). These ligaments, containing splenic vessels, are attached to the hilum of the spleen on its medial aspect. Except at the hilum, where these peritoneal reflections occur, the spleen is intimately covered with peritoneum. The **hilum of the spleen** is often in contact with the tail of the pancreas and constitutes the left boundary of the omental bursa.

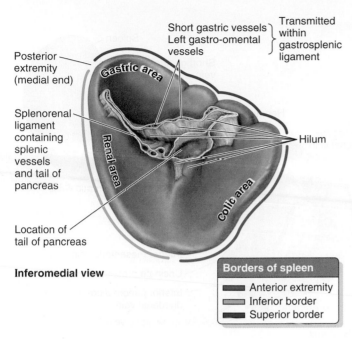

FIGURE 5.33. **Spleen.** Visceral surface.

Labels on figure:
- Posterior extremity (medial end)
- Splenorenal ligament containing splenic vessels and tail of pancreas
- Location of tail of pancreas
- Gastric area
- Renal area
- Colic area
- Short gastric vessels
- Left gastro-omental vessels
- Transmitted within gastrosplenic ligament
- Hilum
- Inferomedial view

Borders of spleen
- Anterior extremity
- Inferior border
- Superior border

The **splenic artery**, the largest branch of the celiac trunk, follows a tortuous course posterior to the omental bursa, anterior to the left kidney, and along the superior border of the pancreas (Fig. 5.34A). Between the layers of the splenorenal ligament, the splenic artery divides into five or more branches that enter the hilum of the spleen, dividing it into two to three vascular segments. The **splenic vein** is formed by several tributaries that emerge from the hilum (Fig. 5.34B). It is joined by the IMV and runs posterior to the body and tail of the pancreas throughout most of its course. The splenic vein unites with the SMV posterior to the neck of the pancreas to form the *hepatic portal vein*.

The *splenic lymphatic vessels* leave the lymph nodes in the hilum and pass along the splenic vessels to the *pancreaticosplenic lymph nodes* (Fig. 5.34C). These nodes relate to the posterior surface and superior border of the pancreas. The *nerves of the spleen* derive from the celiac plexus (Fig. 5.34D). They are distributed mainly along branches of the splenic artery and are vasomotor in function.

SURFACE ANATOMY

Spleen and Pancreas

The **spleen** lies superficially in the left upper abdominal quadrant between the 9th and the 11th ribs (Fig. SA5.3). Its convex, costal surface fits the inferior surface of the diaphragm and the curved bodies of the ribs. In the supine position, the long axis of the spleen is roughly parallel to the long axis of the 10th rib. The spleen is seldom palpable through the anterolateral abdominal wall unless it is enlarged (see Clinical Box "Rupture of Spleen and Splenomegaly").

The **neck of the pancreas** overlies the L1 and L2 vertebrae in the transpyloric plane. Its **head** is to the right and inferior to this plane, and its **body** and **tail** are to the left and superior to this level. Because the pancreas is deep in the abdominal cavity, posterior to the stomach and omental bursa, it is not palpable.

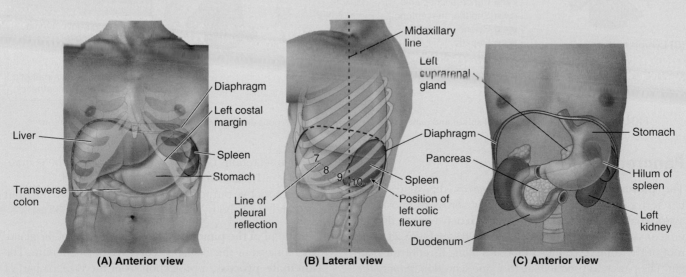

FIGURE SA5.3. Surface anatomy of spleen and pancreas.

Labels on figure:
- (A) Anterior view: Liver, Transverse colon, Diaphragm, Left costal margin, Spleen, Stomach
- (B) Lateral view: Midaxillary line, Diaphragm, Pancreas, Spleen, Position of left colic flexure, Line of pleural reflection, 7, 8, 9, 10
- (C) Anterior view: Left suprarenal gland, Stomach, Hilum of spleen, Left kidney, Duodenum, Pancreas

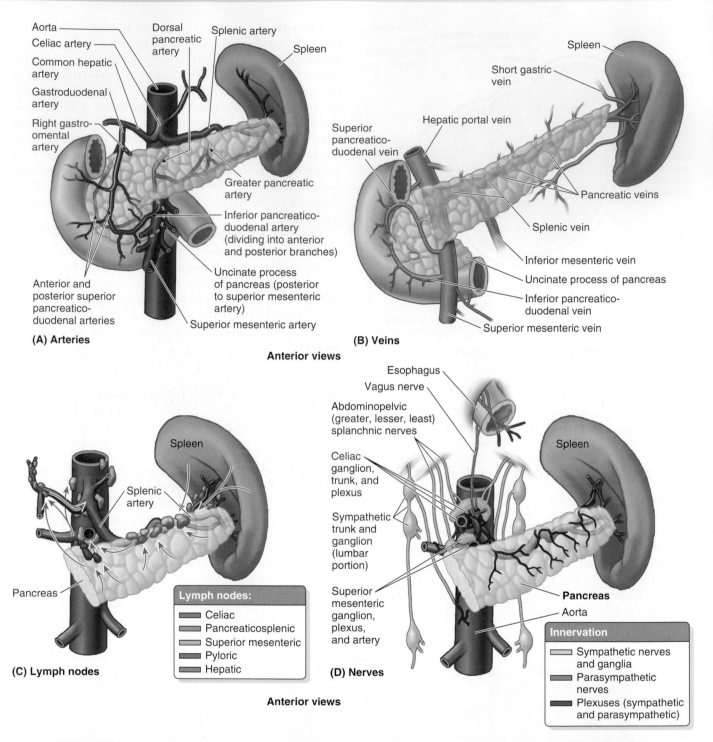

FIGURE 5.34. Neurovasculature of spleen and pancreas. A. Arterial supply. **B.** Venous drainage. **C.** Lymphatic drainage. **D.** Innervation.

Pancreas

The **pancreas**, an elongated accessory digestive gland, lies retroperitoneally and transversely across the posterior abdominal wall, posterior to the stomach between the duodenum on the right and the spleen on the left (see Fig. 5.24). The root of the transverse mesocolon lies along its anterior margin. The pancreas produces an exocrine secretion (*pancreatic juice* from the acinar cells), which enters the duodenum, and endocrine secretions (*glucagon* and *insulin* from the pancreatic islets [of Langerhans]), which enter the blood.

For descriptive purposes, the pancreas is divided into four parts: head, neck, body, and tail (Figs. 5.24 and 5.35).

• The **head of the pancreas**, the expanded part of the gland, is embraced by the C-shaped curve of the duodenum. The **uncinate process**, a projection from the inferior part of the head, extends medially to the left, posterior to the SMA.

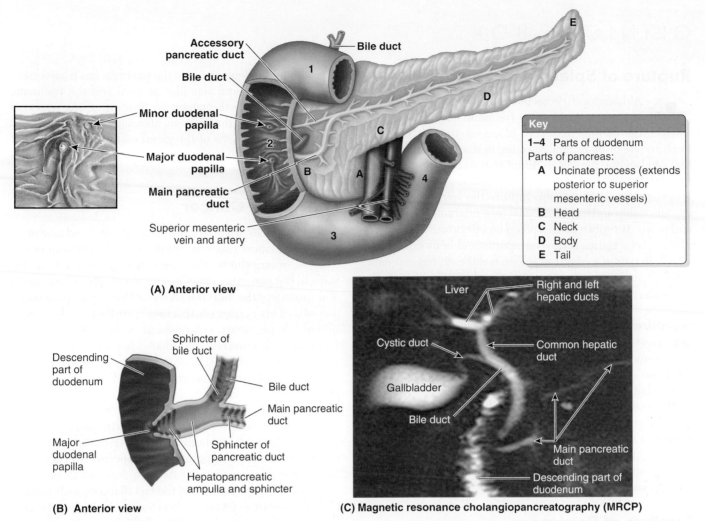

(A) Anterior view

Accessory pancreatic duct
Bile duct
Bile duct
Minor duodenal papilla
Major duodenal papilla
Main pancreatic duct
Superior mesenteric vein and artery

Key

1–4 Parts of duodenum
Parts of pancreas:
  A Uncinate process (extends posterior to superior mesenteric vessels)
  B Head
  C Neck
  D Body
  E Tail

(B) Anterior view

Descending part of duodenum
Sphincter of bile duct
Bile duct
Main pancreatic duct
Sphincter of pancreatic duct
Major duodenal papilla
Hepatopancreatic ampulla and sphincter

(C) Magnetic resonance cholangiopancreatography (MRCP)

Liver
Right and left hepatic ducts
Cystic duct
Common hepatic duct
Gallbladder
Bile duct
Main pancreatic duct
Descending part of duodenum

FIGURE 5.35. Pancreas and biliary system. A. Extrahepatic bile passages and pancreatic ducts. **B.** Sphincters. **C.** Endoscopic retrograde cholangiography and pancreatography (ERCP) reveals the bile and pancreatic ducts. The T tube delivers radiopaque dye into ducts.

- The **neck of the pancreas** is short and overlies the superior mesenteric vessels and origin of the hepatic portal vein, which groove its posterior aspect.
- The **body of the pancreas** continues from the neck and lies to the left of the SMA and SMV, anterior to the splenic vein.
- The **tail of the pancreas** is closely related to the hilum of the spleen and the left colic flexure. The tail is relatively mobile and passes between the layers of the splenorenal ligament with the splenic vessels (see Fig. 5.33).

The **main pancreatic duct** begins in the tail of the pancreas and runs through the parenchyma (substance) of the gland to the head, where it turns inferiorly and merges with the bile duct (Fig. 5.35).

The **bile duct** (common bile duct) crosses the posterosuperior surface of the head of the pancreas or is embedded in its substance. The pancreatic and bile ducts unite to form a short, dilated **hepatopancreatic ampulla** (Fig. 5.35B), which opens into the descending part of the duodenum at the summit of the **major duodenal papilla**. Several smooth-muscle sphincters occur in this area.

The (choledochal) **sphincter of the bile duct**, located around the termination of the bile duct, controls the flow of bile. The **sphincter of the pancreatic duct** (around the terminal part of the pancreatic duct) prevents reflux of bile into the duct, and, the **hepatopancreatic sphincter** (sphincter of Oddi) around the hepatopancreatic ampulla prevents duodenal content from entering the ampulla. The **accessory pancreatic duct** drains the uncinate process and the inferior part of the head of the pancreas and opens into the duodenum at the **minor duodenal papilla** (Fig. 5.35A). Usually, the accessory duct communicates with the main pancreatic duct, but in some people, it is a separate duct.

The **pancreatic arteries** derive mainly from the branches of the splenic artery (Fig. 5.34A and Table 5.5). The *anterior and posterior superior pancreaticoduodenal arteries*, branches of the gastroduodenal artery, and the *anterior and posterior inferior pancreaticoduodenal arteries*, branches of the SMA, supply the head of the pancreas. The **pancreatic veins** are tributaries of the splenic and superior mesenteric parts of the hepatic portal vein; however, most of them empty into the splenic vein (Fig. 5.34B). The *pancreatic lymphatic vessels*

CLINICAL BOX

Rupture of Spleen and Splenomegaly

Although well protected by the 9th through 12th ribs, the spleen is the most frequently injured organ in the abdomen. Severe blows on the left side may fracture one or more ribs, resulting in sharp bone fragments that can lacerate the spleen. Blunt trauma to other regions of the abdomen that cause a sudden, marked increase in intra-abdominal pressure can also rupture the spleen because its capsule is thin and its parenchyma (essential substance) is soft and pulpy. If ruptured, the spleen bleeds profusely. *Rupture of the spleen* causes severe intraperitoneal hemorrhage and shock. Repair of a ruptured spleen is difficult; consequently, *splenectomy* (removal of the spleen) or *subtotal (partial) splenectomy* (removal of one or more segments of the spleen) is often performed to prevent the patient from bleeding to death. Even total splenectomy usually does not produce serious side effects, especially in adults, because most of its functions are assumed by other reticulo-endothelial organs (e.g., liver and bone marrow), but the person will be more susceptible to certain bacterial infections.

When the spleen is diseased, resulting from, for example, granulocytic leukemia (high leukocyte and white blood cell count), it may enlarge to 10 or more times its normal size and weight (*splenomegaly*). Spleen engorgement sometimes accompanies hypertension (high blood pressure). The spleen is not usually palpable in the adult.

Rupture of Pancreas

Pancreatic injury can result from sudden, severe, forceful compression of the abdomen such as the force of impalement on steering wheel in an automobile accident. Because the pancreas lies transversely, the vertebral column acts like an anvil and the traumatic force may rupture the pancreas. *Rupture of the pancreas* frequently tears its duct system, allowing pancreatic juice to enter the parenchyma of the gland and to invade adjacent tissues. Digestion of pancreatic and other tissues by pancreatic juice is very painful.

Pancreatic Cancer

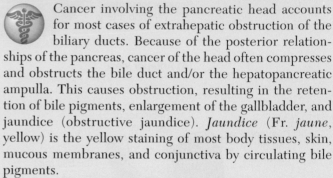

Cancer involving the pancreatic head accounts for most cases of extrahepatic obstruction of the biliary ducts. Because of the posterior relationships of the pancreas, cancer of the head often compresses and obstructs the bile duct and/or the hepatopancreatic ampulla. This causes obstruction, resulting in the retention of bile pigments, enlargement of the gallbladder, and jaundice (obstructive jaundice). *Jaundice* (Fr. *jaune*, yellow) is the yellow staining of most body tissues, skin, mucous membranes, and conjunctiva by circulating bile pigments.

Most people with pancreatic cancer have *ductular adenocarcinoma*. Severe pain in the back is frequently present. Cancer of the neck and body of the pancreas may cause portal or IVC obstruction because the pancreas overlies these large veins. The pancreas's extensive drainage to relatively inaccessible lymph nodes and the fact that pancreatic cancer typically metastasizes to the liver early, via the hepatic portal vein, make surgical resection of the cancerous pancreas nearly futile.

---

follow the blood vessels (Fig. 5.34C). Most of them end in the *pancreaticosplenic nodes* that lie along the splenic artery, but some vessels end in the pyloric lymph nodes. Efferent vessels from these nodes drain to the superior mesenteric lymph nodes or to the celiac lymph nodes via the *hepatic lymph nodes*.

The *nerves of the pancreas* are derived from the *vagus* and *abdominopelvic splanchnic nerves* passing through the diaphragm (Fig. 5.34D). The parasympathetic and sympathetic nerve fibers reach the pancreas by passing along the arteries from the celiac plexus and superior mesenteric plexus. In addition to the sympathetic fibers that pass to blood vessels, sympathetic and parasympathetic fibers are distributed to pancreatic acinar cells and islets. The parasympathetic fibers are secretomotor, but pancreatic secretion is primarily mediated by the hormones, secretin and cholecystokinin formed in the duodenum and proximal intestine. Visceral afferent (pain) fibers accompany the sympathetic fibers.

Liver

The **liver**, the largest internal organ and largest gland in the body, weighs about 1,500 g. The diaphragm separates the liver from the pleura, lungs, pericardium, and heart. With the exception of lipids, every substance absorbed by the alimentary tract is received first by the liver. In addition to its many metabolic activities, the liver stores glycogen and secretes bile.

SURFACES OF LIVER

The liver has a convex **diaphragmatic surface** (anterior, superior, and some posterior) and a relatively flat, concave **visceral surface** (postero-inferior), which are separated anteriorly by the sharp **inferior border** (Fig. 5.36). The diaphragmatic surface is smooth and dome shaped where it is related to the concavity of the inferior surface of the

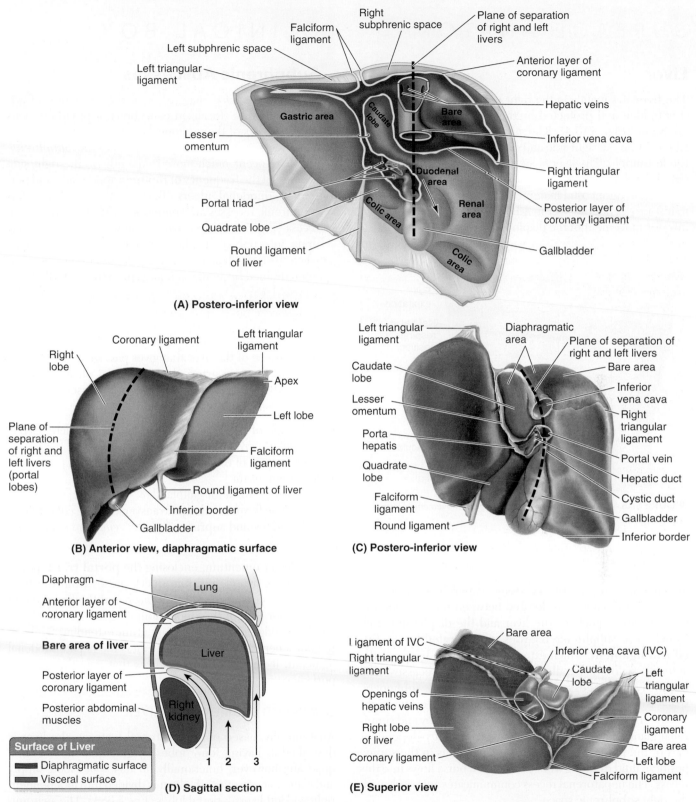

(A) Postero-inferior view

(B) Anterior view, diaphragmatic surface

(C) Postero-inferior view

(D) Sagittal section

(E) Superior view

FIGURE 5.36. Liver and gallbladder. A. Visceral surface of liver. The bare area is demarcated by the reflection of peritoneum from the diaphragm to the liver as the anterior (upper) and posterior (lower) layers of the coronary ligament. These layers meet at the right to form the right triangular ligament and diverge toward the left to enclose the bare area. The anterior layer of the coronary ligament is continuous on the left with the right layer of the falciform ligament, and the posterior layer is continuous with the right layer of the lesser omentum. The left layers of the falciform ligament and lesser omentum meet to form the left triangular ligament. **B.** Diaphragmatic surface of liver. **C.** Visceral surface of liver, portal triad. **D.** Surfaces and recesses. *1,* hepatorenal recess; *2,* subhepatic space; *3,* subphrenic recess. **E.** Superior surface of liver.

SURFACE ANATOMY

Liver

The **liver** lies mainly in the right upper quadrant, where it is hidden and protected by the thoracic cage and diaphragm (Fig. SA5.4). The normal liver lies deep to ribs 7–11 on the right side and crosses the midline toward the left nipple. The liver is located more inferiorly when one is erect because of gravity. Its sharp **inferior border** follows the right costal margin. When the person is asked to inspire deeply, the liver may be palpated because of the inferior movement of the diaphragm and liver.

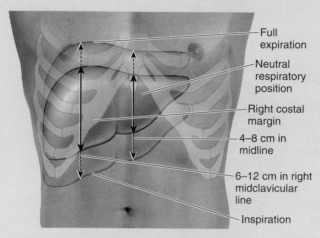

- Full expiration
- Neutral respiratory position
- Right costal margin
- 4–8 cm in midline
- 6–12 cm in right midclavicular line
- Inspiration

FIGURE SA5.4. Vertical dimensions and range of movement of liver.

diaphragm. **Subphrenic recesses**, superior extensions of the peritoneal cavity, are located between the anterior and the superior aspects of the liver and the diaphragm (Fig. 5.36C). The subphrenic recesses are separated by the **falciform ligament**, which extends between the liver and the anterior abdominal wall, into right and left recesses. The **hepatorenal recess** (Morison pouch) of the subhepatic space is a deep recess of the peritoneal cavity on the right side inferior to the liver and anterior to the kidney and suprarenal gland. The *hepatorenal recess* is a gravity-dependent part of the peritoneal cavity when a person is in the supine position; fluid draining from the omental bursa flows into this recess. The hepatorenal recess communicates anteriorly with the right subphrenic space.

The diaphragmatic surface is covered with peritoneum, except posteriorly in the **bare area of the liver**, where it lies in direct contact with the diaphragm (Fig. 5.36A,C,E). The visceral surface of the liver is covered with peritoneum, except at the *bed of the gallbladder* and the *porta hepatis*. The **porta hepatis** is a transverse fissure in the middle

CLINICAL BOX

Subphrenic Abscesses

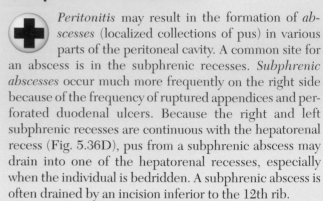 *Peritonitis* may result in the formation of *abscesses* (localized collections of pus) in various parts of the peritoneal cavity. A common site for an abscess is in the subphrenic recesses. *Subphrenic abscesses* occur much more frequently on the right side because of the frequency of ruptured appendices and perforated duodenal ulcers. Because the right and left subphrenic recesses are continuous with the hepatorenal recess (Fig. 5.36D), pus from a subphrenic abscess may drain into one of the hepatorenal recesses, especially when the individual is bedridden. A subphrenic abscess is often drained by an incision inferior to the 12th rib.

visceral surface of the liver that gives passage to the hepatic portal vein, hepatic artery, hepatic nerve plexus, hepatic ducts, and lymphatic vessels (Fig. 5.37). The visceral surface of the liver is related to the

- right side of the anterior aspect of the stomach: *gastric and pyloric areas*
- superior part of the duodenum: *duodenal area*
- lesser omentum
- gallbladder: fossa for gallbladder
- right colic flexure and right transverse colon: *colic area*
- right kidney and suprarenal gland: *renal and suprarenal areas*

The lesser omentum, enclosing the **portal triad** (portal vein, hepatic artery, and bile duct), passes from the liver to the lesser curvature of the stomach and the first 2 cm of the superior part of the duodenum (Fig. 5.37). The thickened free edge of the lesser omentum extending between the porta hepatis and the duodenum is the hepatoduodenal ligament; it encloses the structures that pass through the porta hepatis.

LOBES AND SEGMENTS OF LIVER

Anatomically, based only on external features, the liver is described as having four "lobes": right, left, caudate, and quadrate; however, functionally, in terms of blood supply and glandular secretion, the liver is divided into independent right and left livers—portal lobes (Fig. 5.38A). The anatomical large **right lobe** is separated from the smaller **left lobe** by the falciform ligament and the left sagittal fissure. On the visceral surface, the right and left sagittal fissures and porta hepatis demarcate the **caudate lobe** (posterior and superior) and **quadrate lobe** (anterior and inferior)—both are parts of the right lobe. The **right sagittal fissure** is the continuous

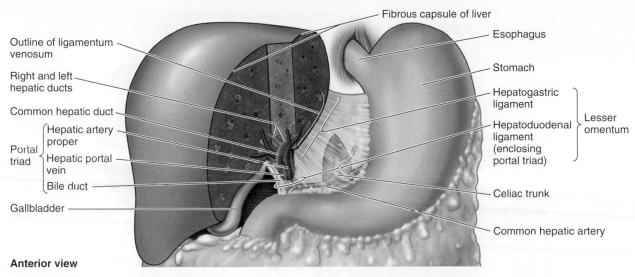

Fibrous capsule of liver

Esophagus

Stomach

Outline of ligamentum venosum

Right and left hepatic ducts

Common hepatic duct

Portal triad
  Hepatic artery proper
  Hepatic portal vein
  Bile duct

Gallbladder

Hepatogastric ligament

Hepatoduodenal ligament (enclosing portal triad)

Lesser omentum

Celiac trunk

Common hepatic artery

Anterior view

FIGURE 5.37. Lesser omentum. The hepatogastric and hepatoduodenal ligaments are shown. The anterior sagittal cut is made in the plane of the fossa for the gallbladder, and the posterior sagittal cut is in the plane of the fissure for the ligamentum venosum. These cuts have been joined by a narrow coronal cut in the plane of the porta hepatis.

groove formed by the fossa for the gallbladder anteriorly and the groove for the IVC posteriorly. The **left sagittal fissure** is the continuous groove formed anteriorly by the **fissure for the round ligament** (L. *ligamentum teres*) and posteriorly by the **fissure for the ligamentum venosum** (Fig. 5.38B). The **round ligament of the liver** is the obliterated remains of the umbilical vein, which carried well-oxygenated blood from the placenta to the fetus. The **ligamentum venosum** is the fibrous remnant of the fetal **ductus venosus**, which shunted blood from the umbilical vein to the IVC, short-circuiting the liver (Moore et al., 2016).

The division between **right** and **left livers** (parts or portal lobes) is the plane of the middle hepatic vein (main portal fissure) approximated by the nearly sagittal plane passing through the **fossa for gallbladder** and the **groove for vena cava** on the visceral surface of the liver and an imaginary line

over the diaphragmatic surface that runs from the fundus of the gallbladder to the IVC (Fig. 5.38). The left liver includes the anatomical caudate lobe and most of the quadrate lobe. The right and left livers are closer in mass than the anatomical lobes, but the right lobe is still somewhat larger. Each portal lobe has its own blood supply from the hepatic artery and hepatic portal vein and its own venous and biliary drainage. The portal lobes of the liver are further subdivided into eight **hepatic segments** (Fig. 5.39). The segmentation is based on the tertiary branches of the right and left hepatic arteries, hepatic portal veins, and hepatic ducts. Each segment is supplied by a tertiary branch of the right or left hepatic artery and hepatic portal vein and drained by a tertiary branch of the right or left hepatic duct. Intersegmental *hepatic veins* pass between and thus further demarcate segments on their way to the IVC.

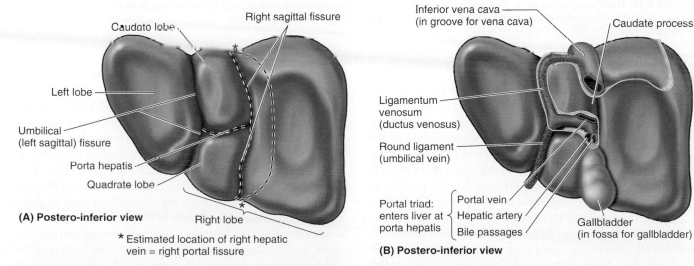

Caudate lobe

Right sagittal fissure

Left lobe

Umbilical (left sagittal) fissure

Porta hepatis

Quadrate lobe

Right lobe

(A) Postero-inferior view

*Estimated location of right hepatic vein = right portal fissure

Inferior vena cava (in groove for vena cava)

Caudate process

Ligamentum venosum (ductus venosus)

Round ligament (umbilical vein)

Portal triad: enters liver at porta hepatis
  Portal vein
  Hepatic artery
  Bile passages

Gallbladder (in fossa for gallbladder)

(B) Postero-inferior view

FIGURE 5.38. Anatomical lobes and fissures of liver, visceral surface. A. Four anatomical lobes. **B.** Structures forming and occupying fissures.

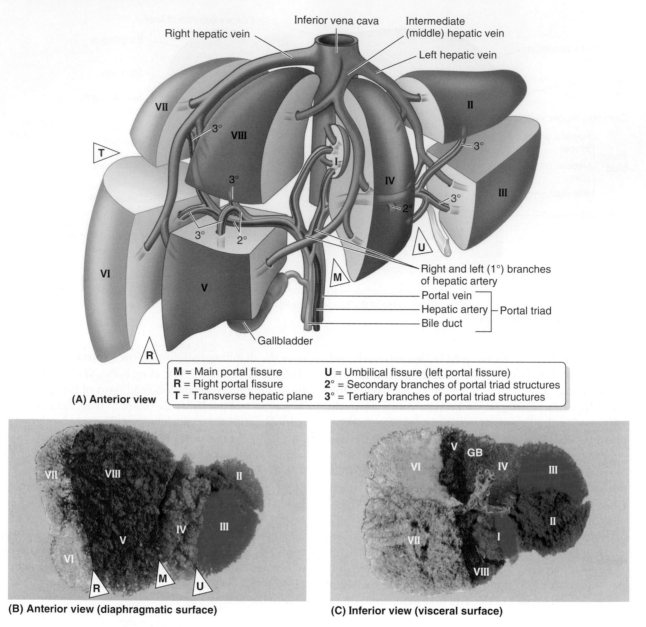

M = Main portal fissure   **U** = Umbilical fissure (left portal fissure)
R = Right portal fissure   **2°** = Secondary branches of portal triad structures
T = Transverse hepatic plane   **3°** = Tertiary branches of portal triad structures

(A) Anterior view

(B) Anterior view (diaphragmatic surface)   **(C) Inferior view (visceral surface)**

FIGURE 5.39.   Hepatic segmentation. A. Each segment (I–VIII) has its own intrasegmental blood supply and biliary drainage. **B and C.** Injection of different colors of latex into the branches of the hepatic portal vein to demonstrate hepatic segments. *GB*, gallbladder.

VASCULATURE AND NERVES OF LIVER

The liver receives blood from two sources (Figs. 5.26, 5.27, and 5.39A): the hepatic portal vein (75–80%) and the hepatic artery (20–25%). The hepatic portal vein carries poorly oxygenated blood from the abdominopelvic portion of the GI tract. The **hepatic artery**, a branch of the celiac trunk, carries well-oxygenated blood from the aorta. At or close to the porta hepatis, the hepatic artery and hepatic portal vein terminate by dividing into **right** and **left branches**, which supply the right and left livers, respectively. Within each lobe, the secondary and tertiary branches of the hepatic portal vein and hepatic artery are consistent enough to form *hepatic segments* (Fig. 5.39). Between the segments are the **right**,

intermediate (middle), and **left hepatic veins**, which drain parts of adjacent segments. The hepatic veins open into the IVC just inferior to the diaphragm (Fig. 5.39A). The attachment of these veins to the IVC helps hold the liver in position.

The liver is a major lymph-producing organ; between one quarter and one half of the lymph received by the thoracic duct comes from the liver. The *lymphatic vessels of the liver* occur as *superficial lymphatics* in the subperitoneal **fibrous capsule of the liver** (Glisson capsule), which form its outer surface, and as *deep lymphatics* in the connective tissue that accompany the ramifications of the portal triad and hepatic veins. Superficial lymphatics from the anterior aspects of the diaphragmatic and visceral surfaces and the deep lymphatic

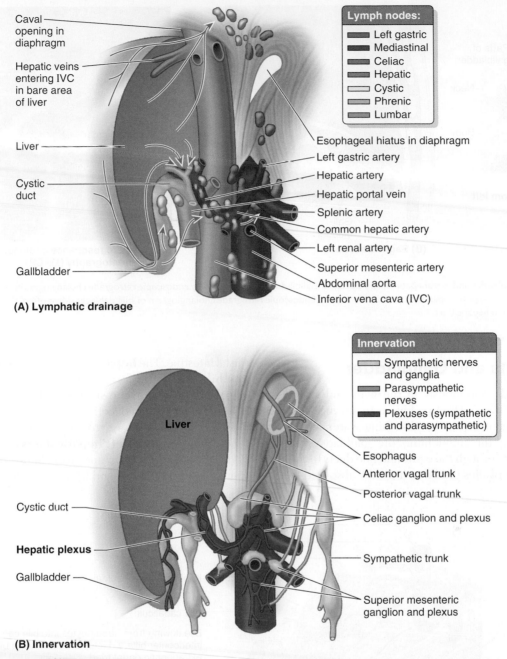

Lymph nodes:
- Left gastric
- Mediastinal
- Celiac
- Hepatic
- Cystic
- Phrenic
- Lumbar

(A) Lymphatic drainage

- Caval opening in diaphragm
- Hepatic veins entering IVC in bare area of liver
- Liver
- Cystic duct
- Gallbladder
- Esophageal hiatus in diaphragm
- Left gastric artery
- Hepatic artery
- Hepatic portal vein
- Splenic artery
- Common hepatic artery
- Left renal artery
- Superior mesenteric artery
- Abdominal aorta
- Inferior vena cava (IVC)

Innervation
- Sympathetic nerves and ganglia
- Parasympathetic nerves
- Plexuses (sympathetic and parasympathetic)

(B) Innervation

- Liver
- Cystic duct
- **Hepatic plexus**
- Gallbladder
- Esophagus
- Anterior vagal trunk
- Posterior vagal trunk
- Celiac ganglion and plexus
- Sympathetic trunk
- Superior mesenteric ganglion and plexus

FIGURE 5.40. Lymphatic drainage and innervation of liver. A. Lymphatic drainage. B. Innervation.

vessels accompanying the interlobular portal triads converge toward the porta hepatis and drain to the **hepatic lymph nodes** scattered along the hepatic vessels and ducts in the lesser omentum (Fig. 5.40A). Efferent lymphatic vessels from these lymph nodes drain into the celiac lymph nodes, which in turn drain into the *cisterna chyli* at the inferior end of the thoracic duct. Superficial lymphatics from the posterior aspects of the diaphragmatic and visceral surfaces of the liver drain toward the bare area of the liver. Here, they drain into **phrenic lymph nodes** or join deep lymphatics that have accompanied the hepatic veins converging on the IVC and then pass with this large vein through the diaphragm to

drain into the **posterior mediastinal lymph nodes**. Efferent vessels from these nodes join the right lymphatic and thoracic ducts. A few lymphatic vessels also drain to the left gastric nodes, along the falciform ligament to the parasternal lymph nodes and along the round ligament of the liver to the lymphatics of the anterior abdominal wall. The *nerves of the liver* derive from the **hepatic nerve plexus** (Fig. 5.40B), the largest derivative of the celiac plexus. The hepatic plexus accompanies the branches of the hepatic artery and hepatic portal vein to the liver. It consists of *sympathetic fibers* from the celiac plexus and *parasympathetic fibers* from the anterior and posterior vagal trunks.

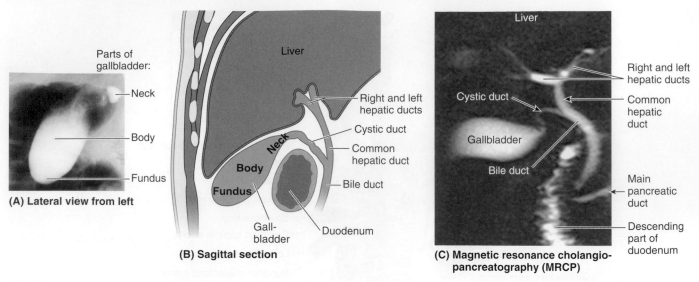

(A) Lateral view from left

Parts of gallbladder:
- Neck
- Body
- Fundus

(B) Sagittal section

Liver
Right and left hepatic ducts
Cystic duct
Common hepatic duct
Bile duct
Neck
Body
Fundus
Gall-bladder
Duodenum

(C) Magnetic resonance cholangio-pancreatography (MRCP)

Liver
Cystic duct
Gallbladder
Bile duct
Right and left hepatic ducts
Common hepatic duct
Main pancreatic duct
Descending part of duodenum

FIGURE 5.41. Gallbladder and extrahepatic biliary ducts. A. Gallbladder demonstrated by endoscopic retrograde cholangiography. **B.** Schematic sagittal section showing relationships to superior part of duodenum. **C.** Endoscopic retrograde cholangiogram of bile passages. Most often, the cystic duct lies anterior to the common hepatic duct.

Biliary Ducts and Gallbladder

Bile is produced continuously in the liver and stored in the gallbladder (Fig. 5.41). In addition to storing bile, the gallbladder concentrates it by absorbing water and salts. When fat enters the duodenum, the gallbladder sends concentrated bile through the cystic and bile ducts to the duodenum. Bile emulsifies the fat so it can be absorbed in the distal intestine. The **hepatocytes** (liver cells) secrete bile into the **bile canaliculi** formed between them (Fig. 5.42). The canaliculi drain into the small **interlobular biliary ducts** and then into large collecting bile ducts of the intrahepatic portal triad, which merge to form the right and left hepatic ducts. The **right** and **left hepatic ducts** drain the right and left livers (portal lobes), respectively. Shortly after leaving the porta hepatis, the right and left hepatic ducts unite to form

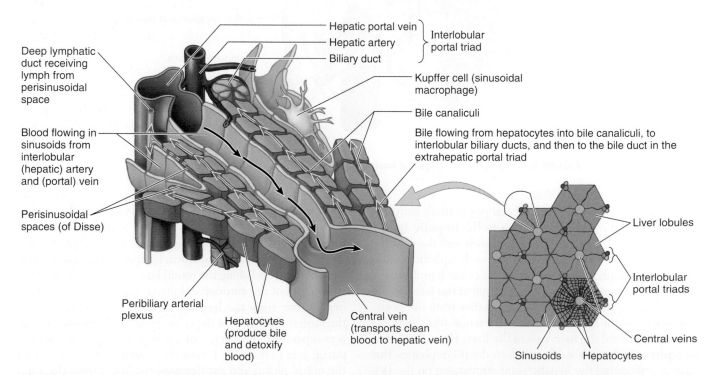

Hepatic portal vein
Hepatic artery
Biliary duct
} Interlobular portal triad

Deep lymphatic duct receiving lymph from perisinusoidal space

Kupffer cell (sinusoidal macrophage)

Bile canaliculi

Blood flowing in sinusoids from interlobular (hepatic) artery and (portal) vein

Bile flowing from hepatocytes into bile canaliculi, to interlobular biliary ducts, and then to the bile duct in the extrahepatic portal triad

Perisinusoidal spaces (of Disse)

Liver lobules

Interlobular portal triads

Peribiliary arterial plexus

Hepatocytes (produce bile and detoxify blood)

Central vein (transports clean blood to hepatic vein)

Central veins

Sinusoids   Hepatocytes

FIGURE 5.42. Flow of blood and bile in the liver. This small part of a liver lobule shows the components of the interlobular portal triad and the positioning of the sinusoids and bile canaliculi. At right, the cut surface of the liver shows the hexagonal pattern of the lobules.

the **common hepatic duct**, which is joined on the right side by the *cystic duct* to form the *bile duct* (Fig. 5.41).

BILE DUCT

The **bile duct** (formerly called the common bile duct) is formed in the free edge of the lesser omentum by the union of the *cystic duct* and *common hepatic duct*. The bile duct descends posterior to the superior part of the duodenum and lies in a groove on the posterior surface of the head of the pancreas. On the left side of the descending part of the duodenum, the bile duct comes into contact with the main pancreatic duct (Figs. 5.35 and 5.43). The two ducts run obliquely through the wall of this part of the duodenum, where they unite to form the *hepatopancreatic ampulla* (ampulla of Vater). The distal end of the ampulla opens into the duodenum through the *major duodenal papilla*. The muscle around the distal end of the bile duct is thickened to form the (choledochal) *sphincter of the bile duct*. When this sphincter contracts, bile cannot enter the ampulla and/or the duodenum; hence, bile backs up and passes along the cystic duct to the *gallbladder* for concentration and storage.

The arteries supplying the bile duct include the following (Figs. 5.37 and 5.44):

- *Posterior superior pancreaticoduodenal artery* and *gastroduodenal artery*, supplying the retroduodenal part of the duct
- *Cystic artery*, supplying the proximal part of the duct
- *Right hepatic artery*, supplying the middle part of the duct

The veins from the proximal part of the bile duct and the hepatic ducts generally enter the liver directly. The **posterior**

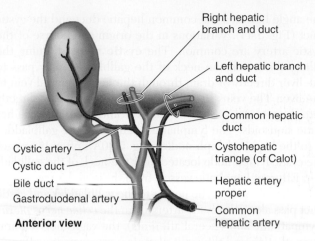

FIGURE 5.44. Blood supply of gallbladder.

superior pancreaticoduodenal vein drains the distal part of the bile duct and empties into the hepatic portal vein or one of its tributaries (see Fig. 5.27). The lymphatic vessels from the bile duct pass to the **cystic lymph node** near the neck of the gallbladder, the **node of the omental foramen**, and the hepatic lymph nodes (Fig. 5.40A). Efferent lymphatic vessels from the bile duct pass to the celiac lymph nodes.

GALLBLADDER

The pear-shaped **gallbladder** (7–10 cm long) lies in the *fossa for gallbladder* on the visceral surface of the liver (Figs. 5.38B and 5.41). Peritoneum completely surrounds the fundus of the gallbladder and binds its body and neck to the liver. The hepatic surface of the gallbladder attaches to the liver by connective tissue of the fibrous capsule of the liver. The gallbladder has three parts (Figs. 5.41 and 5.43):

- The **fundus**, the wide end, projects from the inferior border of the liver and is usually located at the anterior end of the right 9th costal cartilage in the midclavicular line.
- The **body** contacts the visceral surface of the liver, the transverse colon, and the superior part of the duodenum.
- The **neck** is narrow, tapered, and directed toward the porta hepatis.

The neck makes an S-shaped bend and joins the cystic duct. Internally, the mucosa of the neck spirals into a **spiral fold** (spiral "valve"), which *keeps the cystic duct open* so that bile can easily divert into the gallbladder when the distal end of the bile duct is closed by the sphincter of the bile duct and/or the hepatopancreatic sphincter or when bile passes to the duodenum as the gallbladder contracts. The **cystic duct** (approximately 4 cm long) connects the *neck of the gallbladder* to the common hepatic duct. The cystic duct passes between the layers of the lesser omentum, usually parallel to the common hepatic duct, which it joins to form the bile duct.

The **cystic artery**, which supplies the gallbladder and cystic duct, commonly arises from the right hepatic artery in

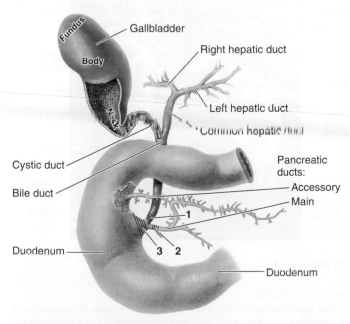

FIGURE 5.43. Extrahepatic bile passages and pancreatic ducts. *1*, sphincter of bile duct; *2*, sphincter of pancreatic duct; *3*, hepatopancreatic sphincter.

the angle between the common hepatic duct and the cystic duct (Fig. 5.44). Variations in the origin and course of the cystic artery are common. The **cystic veins** draining the biliary ducts and the neck of the gallbladder may pass to the liver directly or drain through the hepatic portal vein to the liver. The veins from the fundus and body pass directly into the visceral surface of the liver and drain into the hepatic sinusoids. The lymphatic drainage of the gallbladder is to the hepatic lymph nodes (Fig. 5.40A), often by way of the cystic lymph node located near the neck of the gallbladder. Efferent lymphatic vessels from these nodes pass to the celiac lymph nodes. The nerves to the gallbladder and cystic duct pass along the cystic artery from the *celiac nerve plexus* (sympathetic and visceral afferents), the *vagus nerve* (parasympathetic), and the *right phrenic nerve* (somatic afferent fibers) (Fig. 5.40B). Contraction of the gallbladder is hormonally stimulated.

Hepatic Portal Vein and Portosystemic Anastomoses

The **hepatic portal vein** is the main channel of the **portal venous system** (Fig. 5.45). It collects poorly oxygenated but nutrient-rich blood from the abdominal part of the alimentary tract, including the gallbladder, pancreas, and spleen, and carries it to the liver. Within the liver, its branches are distributed in a segmental pattern and end in noncontractile capillaries, the **venous sinusoids of the liver** (Fig. 5.42).

Portosystemic anastomoses, in which the portal venous system communicates with the systemic venous system, are in the following locations (Fig. 5.45):

- Between the esophageal veins, draining into either the *azygos vein* (systemic system) or the *left gastric vein* (portal system); when dilated, these form *esophageal varices*.
- Between the **rectal veins**, the inferior and middle veins draining into the IVC (systemic system) and the superior rectal vein continuing as the IMV (portal system); when abnormally dilated, these are *hemorrhoids*.
- **Para-umbilical veins** of the anterior abdominal wall (portal system) anastomosing with peri-umbilical **superficial epigastric veins** (systemic system); when dilated, these veins produce *caput medusae*—varicose veins radiating from the umbilicus. These dilated veins were called caput medusae because of their resemblance to the serpents on the head of Medusa, a character in Greek mythology.
- Twigs of *colic veins* (portal system) anastomosing with **retroperitoneal veins** (systemic system)

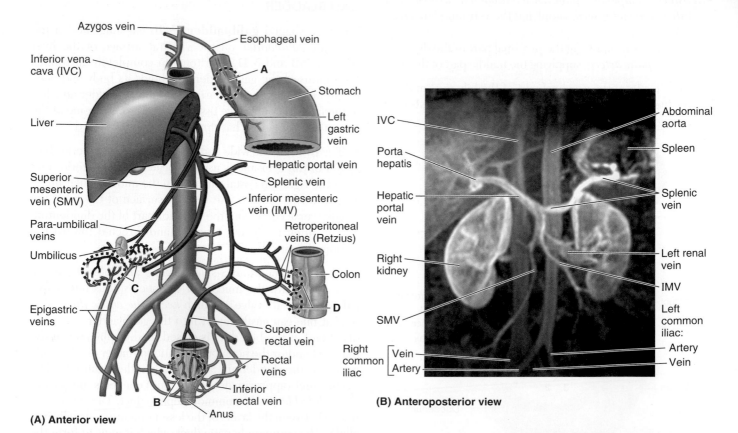

(A) Anterior view

(B) Anteroposterior view

FIGURE 5.45. Hepatic portal venous system. A. Portosystemic anastomoses. These anastomoses provide collateral circulation in cases of obstruction in the liver or hepatic portal vein. *Darker blue,* portal tributaries; *lighter blue,* systemic tributaries; *A,* anastomoses between esophageal veins; *B,* anastomoses between rectal veins; *C,* anastomoses between para-umbilical veins (portal) and small epigastric veins of the anterior abdominal wall; *D,* anastomoses between the twigs of colic veins (portal) and the retroperitoneal veins. **B.** Magnetic resonance (MR) angiogram (portal venogram) demonstrating the tributaries and formation of the portal vein.

CLINICAL BOX

Liver Biopsy

Hepatic tissue may be obtained for diagnostic purposes by *liver biopsy*. The *needle puncture* is commonly made through the right 10th intercostal space in the midaxillary line. Before the physician takes the biopsy, the person is asked to hold his or her breath in full expiration to reduce the costodiaphragmatic recess and to lessen the possibility of damaging the lung and contaminating the pleural cavity.

Rupture of Liver

Although less so than the spleen, the liver is vulnerable to rupture because it is large, fixed in position, and friable. Often, the liver is torn by a fractured rib that perforates the diaphragm. Because of the liver's great vascularity and friability, *liver lacerations* often cause considerable *hemorrhage* and right upper quadrant pain.

Cirrhosis of Liver

In *cirrhosis of the liver*, hepatocytes are destroyed and replaced by fibrous tissue. This tissue surrounds the intrahepatic blood vessels and biliary ducts, making the liver firm and impeding circulation of blood through it. Cirrhosis, the most common of many causes of *portal hypertension*, frequently develops in chronic alcoholics.

Hepatic Lobectomies and Segmentectomy

When it was discovered that the right and left hepatic arteries and ducts, as well as branches of the right and left hepatic portal veins, do not communicate significantly, it became possible to perform *hepatic lobectomies*—removal of the right or left part of the liver—with minimal bleeding. If a severe injury or tumor involves one segment or adjacent segments, it may be possible to resect (remove) only the affected segment(s): *segmentectomy*. The intersegmental hepatic veins serve as guides to the interlobular planes.

Gallstones

Gallstones are concretions (L. *calculi*, pebbles) in the gallbladder cystic duct, hepatic ducts, or bile duct (Figure B5.11). The distal end of the

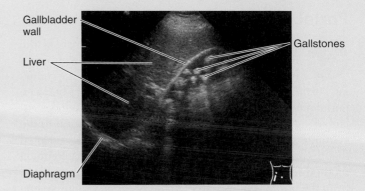

FIGURE B5.11. Longitudinal ultrasonic scan of gallbladder with gallstones.

hepatopancreatic ampulla is the narrowest part of the biliary passages and is the common site for impaction of a gallstone. Gallstones may produce *biliary colic* (pain in the epigastric region). When the gallbladder relaxes, the stone in the cystic duct may pass back into the gallbladder. If a stone blocks the cystic duct, *cholecystitis* (inflammation of the gallbladder) occurs because of bile accumulation, causing enlargement of the gallbladder. Pain develops in the epigastric region and later shifts to the right hypochondriac region at the junction of the 9th costal cartilage and the lateral border of the rectus sheath. Inflammation of the gallbladder may cause pain in the posterior thoracic wall or right shoulder as a result of irritation of the diaphragm. If bile cannot leave the gallbladder, it enters the blood and causes *obstructive jaundice* (see Clinical Box "Pancreatic Cancer" in this chapter).

Cholecystectomy

People with severe *biliary colic* usually have their gallbladders removed. *Laparoscopic cholecystectomy* often replaces the open-incision surgical method. The cystic artery most commonly arises from the right hepatic artery in the **cystohepatic triangle** (Calot triangle). In current clinical use, the cystohepatic triangle is defined inferiorly by the cystic duct, medially by the common hepatic duct, and superiorly by the inferior surface of the liver (Fig. 5.43). Careful dissection of the cystohepatic triangle early during cholecystectomy safeguards these important structures should there be anatomical variations.

Kidneys, Ureters, and Suprarenal Glands

The **kidneys** lie retroperitoneally on the posterior abdominal wall, one on each side of the vertebral column (Figs. 5.44 and 5.46). These urinary organs remove excess water, salts, and wastes of protein metabolism from the blood while returning nutrients and chemicals to the blood. The kidneys convey the waste products from the blood into the urine, which drains through the ureters to the urinary bladder.

CLINICAL BOX

Portal Hypertension

When scarring and fibrosis from cirrhosis of the liver obstruct the hepatic portal vein, pressure rises in the hepatic portal vein and its tributaries, producing *portal hypertension*. At the sites of anastomoses between portal and systemic veins, portal hypertension produces enlarged varicose veins and blood flow from the portal to the systemic system of veins. The veins may become so dilated that their walls rupture, resulting in hemorrhage. A common method for reducing portal hypertension is to divert blood from the portal venous system to the systemic venous system by creating a communication between the portal vein and the IVC or by joining the splenic and left renal veins—a *portacaval anastomosis* or *portosystemic shunt* (Fig. B5.12A).

Bleeding from *esophageal varices* (dilated esophageal veins) at the distal end of the esophagus is often severe and may be fatal (Fig. B5.12B).

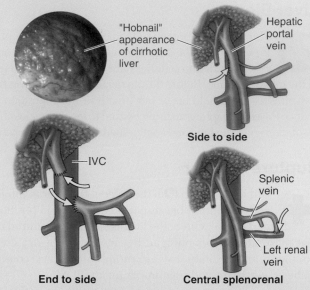

(A) Portosystemic shunts *(yellow arrows)*

"Hobnail" appearance of cirrhotic liver

Hepatic portal vein

Side to side

IVC

End to side

Splenic vein

Left renal vein

Central splenorenal

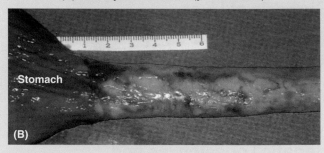

Stomach

(B)

FIGURE B5.12. Portal hypertension. A. Portosystemic shunts. (yellow arrows). **B.** Esophageal varices. Inverted esophagus and stomach. Longitudinal blue venous structures are characteristic of esophageal varices.

The *ureters* run inferiorly from the kidneys, passing over the pelvic brim at the bifurcation of the common iliac arteries. They then run along the lateral wall of the pelvis and enter the *urinary bladder*. The superomedial aspect of each kidney normally contacts a suprarenal gland. A weak fascial septum separates these glands from the kidneys. The *suprarenal glands* function as part of the endocrine system, completely separate in function from the kidneys so they are not attached to each other. They secrete corticosteroids and androgens and make epinephrine and norepinephrine hormones.

RENAL FASCIA AND FAT

Perinephric fat (perirenal fat capsule) surrounds the kidneys and suprarenal glands and is continuous with the fat in the renal sinus (Fig. 5.46). The kidneys, suprarenal glands, and perinephric fat surrounding them are enclosed (except inferiorly) by a membranous layer of **renal fascia**. Inferomedially, the renal fascia is prolonged along the ureters as **peri-ureteric fascia**. External to the renal fascia is the **paranephric fat (pararenal fat body)**, the extraperitoneal

fat of the lumbar region that is most obvious posterior to the kidney. The renal fascia sends collagen bundles through the paranephric fat. Movement of the kidneys occurs during respiration and when changing from supine to erect positions; normal renal mobility is about 3 cm. Superiorly, the renal fascia is continuous with the diaphragmatic fascia on the inferior surface of the diaphragm. Inferiorly, the anterior and posterior layers of renal fascia are loosely united, if attached at all.

KIDNEYS

The kidneys lie on the posterior abdominal wall at the level of the T12–L3 vertebrae. The *right kidney* lies at a slightly lower level than the *left kidney*, probably owing to the presence of the liver (Fig. 5.47). Each kidney has anterior and posterior surfaces, medial and lateral margins, and superior and inferior poles (Fig. 5.48). The lateral margin is convex, and the medial margin is concave where the renal sinus and renal pelvis are located, giving the kidney a somewhat kidney bean–shaped appearance. At the concave medial margin of each kidney is a vertical cleft, the **renal hilum**. The hilum is

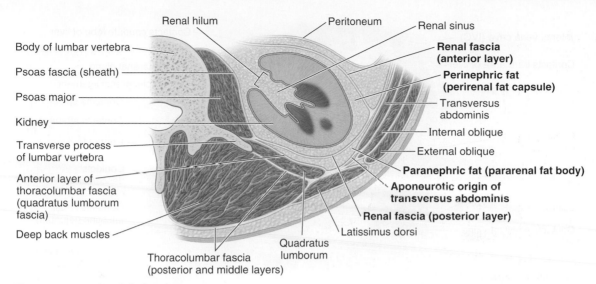

Renal hilum — Peritoneum — Renal sinus

Body of lumbar vertebra —

Psoas fascia (sheath) —

Psoas major —

Kidney —

Transverse process of lumbar vertebra —

Anterior layer of thoracolumbar fascia (quadratus lumborum fascia) —

Deep back muscles —

Renal fascia (anterior layer)

Perinephric fat (perirenal fat capsule)

Transversus abdominis

Internal oblique

External oblique

Paranephric fat (pararenal fat body)

Aponeurotic origin of transversus abdominis

Renal fascia (posterior layer)

Latissimus dorsi

Quadratus lumborum

Thoracolumbar fascia (posterior and middle layers)

Transverse section, inferior view

FIGURE 5.46. Musculofascial relationships of kidneys.

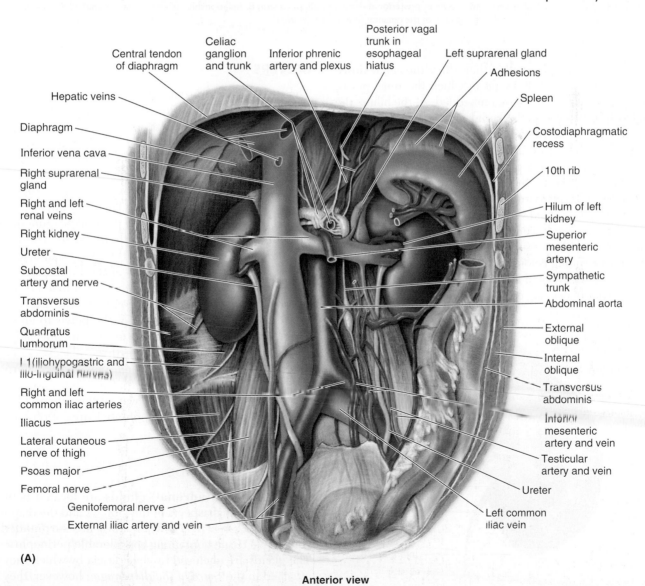

Hepatic veins

Diaphragm

Inferior vena cava

Right suprarenal gland

Right and left renal veins

Right kidney

Ureter

Subcostal artery and nerve

Transversus abdominis

Quadratus lumborum

L1 (iliohypogastric and ilio-inguinal nerves)

Right and left common iliac arteries

Iliacus

Lateral cutaneous nerve of thigh

Psoas major

Femoral nerve

Genitofemoral nerve

External iliac artery and vein

Central tendon of diaphragm

Celiac ganglion and trunk

Inferior phrenic artery and plexus

Posterior vagal trunk in esophageal hiatus

Left suprarenal gland

Adhesions

Spleen

Costodiaphragmatic recess

10th rib

Hilum of left kidney

Superior mesenteric artery

Sympathetic trunk

Abdominal aorta

External oblique

Internal oblique

Transversus abdominis

Inferior mesenteric artery and vein

Testicular artery and vein

Ureter

Left common iliac vein

(A)

Anterior view

FIGURE 5.47. Retroperitoneal viscera and vessels of posterior abdominal wall. **A.** Posterior abdominal wall showing great vessels, kidneys, and suprarenal glands. *(continued)*

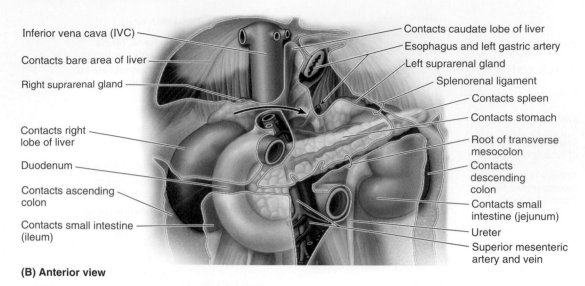

Inferior vena cava (IVC)

Contacts bare area of liver

Right suprarenal gland

Contacts right lobe of liver

Duodenum

Contacts ascending colon

Contacts small intestine (ileum)

Contacts caudate lobe of liver

Esophagus and left gastric artery

Left suprarenal gland

Splenorenal ligament

Contacts spleen

Contacts stomach

Root of transverse mesocolon

Contacts descending colon

Contacts small intestine (jejunum)

Ureter

Superior mesenteric artery and vein

(B) Anterior view

FIGURE 5.47 **Retroperitoneal viscera and vessels of posterior abdominal wall.** *(continued)* **B.** Relationships of kidneys, suprarenal glands, pancreas, and duodenum. The right suprarenal gland is at the level of the omental foramen (*black arrow*).

the entrance to the space within the kidney, the **renal sinus**, which is occupied mostly by fat in which the renal pelvis, calices, vessels, and nerves are embedded. At the hilum, the **renal vein** is anterior to the **renal artery**, which in turn is anterior to the **renal pelvis**.

Superiorly, the kidneys are related to the diaphragm, which separates them from the pleural cavities and the 12th pair of ribs. More inferiorly, the posterior surface of the kidney is related to the quadratus lumborum muscle (Fig. 5.47). The subcostal nerve and vessels and the iliohypogastric and ilio-inguinal nerves descend diagonally across the posterior surfaces of the kidneys (see Fig. SA5.3B). The liver, duodenum, and ascending colon are anterior to the right kidney. The left kidney is related to the stomach, spleen, pancreas, jejunum, and descending colon (Fig. 5.47B).

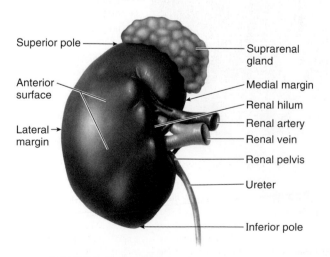

Superior pole

Anterior surface

Lateral margin

Suprarenal gland

Medial margin

Renal hilum

Renal artery

Renal vein

Renal pelvis

Ureter

Inferior pole

FIGURE 5.48. **Right kidney and suprarenal gland.**

URETERS

The **ureters** are muscular ducts with narrow lumina that carry urine from the kidneys to the urinary bladder. The superior expanded end of the ureter, the renal pelvis, is formed through the merging of two or three **major calices** (calyces), each of which was formed by the merging of two or three **minor calices** (Figs. 5.49 and 5.50). Each minor calyx is indented by the apex of the **renal pyramid**—the **renal papilla**. The abdominal parts of the ureters adhere closely to the parietal peritoneum and are retroperitoneal throughout their course. The ureters run inferomedially anterior to the psoas major and the tips of the transverse processes of the lumbar vertebrae (see Fig. SA5.5A) and cross the external iliac artery just beyond the bifurcation of the common iliac artery. They then run along the lateral wall of the pelvis to enter the urinary bladder (Fig. 5.51). The ureters are normally constricted to a variable degree in three places: (1) at the junction of the ureters and renal pelves, (2) where the ureters cross the brim of the pelvic inlet, and (3) during their passage through the wall of the urinary bladder. These constricted areas are potential sites of obstruction by ureteric (kidney) stones.

SUPRARENAL GLANDS

The **suprarenal (adrenal) glands** are located between the superomedial aspects of the kidneys and the diaphragmatic crura (Fig. 5.47), where they are surrounded by connective tissue containing considerable perinephric fat. The glands are enclosed by renal fascia by which they are attached to the *crura of the diaphragm*; however, they are separated from the kidneys by fibrous tissue. The shape

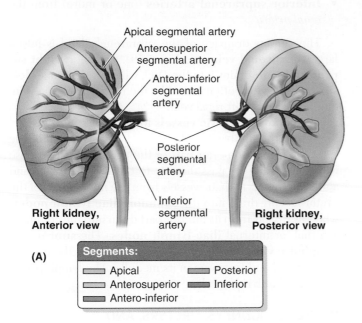

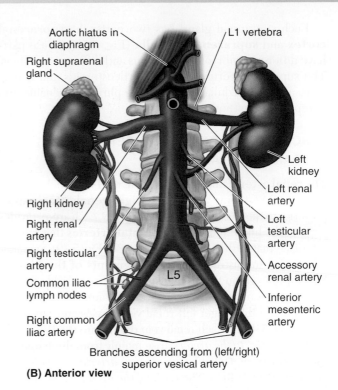

FIGURE 5.49. Blood supply of kidneys and ureters. A. Renal segments and segmental arteries. Only the superior and inferior arteries supply the whole thickness of the kidney. **B.** Blood supply of ureters.

and relations of the suprarenal glands differ on the two sides:

- The pyramid-shaped right gland lies anterior to the diaphragm and makes contact with the IVC anteromedially and the liver anterolaterally.
- The crescent-shaped left gland is related to the spleen, stomach, pancreas, and the left crus of the diaphragm.

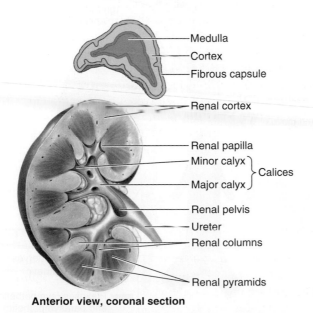

FIGURE 5.50. Internal structure of kidney and suprarenal gland.

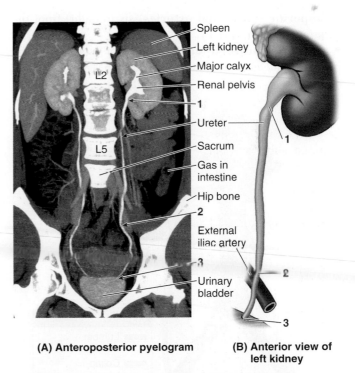

FIGURE 5.51. Normal constrictions of ureters demonstrated by retrograde pyelogram. A. Contrast medium was injected into the ureters from a flexible endoscope (urethroscope) in the bladder. **B.** Sites at which relative constrictions in the ureters normally appear: (*1*) ureteropelvic junction, (*2*) crossing external iliac vessels and/or pelvic brim, and (*3*) as ureter traverses bladder wall.

Each suprarenal gland has two parts: the **suprarenal cortex** and **suprarenal medulla** (Fig. 5.50). These parts have different embryological origins and different functions. The suprarenal cortex secretes corticosteroids and androgens, and the medulla secretes epinephrine (adrenalin) and norepinephrine (noradrenalin).

VASCULATURE OF KIDNEYS, URETERS, AND SUPRARENAL GLANDS

The **renal arteries** arise at the level of the IV disc between the L1 and L2 vertebrae. The longer **right renal artery** passes posterior to the IVC (Fig. 5.47A). Typically, each artery divides close to the hilum into five **segmental arteries** that are end arteries—that is, they do not anastomose (Fig. 5.49). Segmental arteries are distributed to the **segments of the kidney**. Several veins drain the kidney and unite in a variable fashion to form the renal vein. The renal veins lie anterior to the renal arteries, and the longer left renal vein passes anterior to the aorta (Fig. 5.47A). Each renal vein drains into the IVC.

The *arteries to the ureters* arise mainly from three sources: the *renal artery*, *testicular or ovarian arteries*, and *abdominal aorta* (Fig. 5.49A). The *veins of the ureters* drain into the renal and testicular or ovarian veins (Fig. 5.47A).

The endocrine function of the suprarenal glands makes their abundant blood supply necessary. The *suprarenal arteries* arise from three sources:

- **Superior suprarenal arteries** (six to eight) from the *inferior phrenic artery*
- **Middle suprarenal arteries** (one or more) from the *abdominal aorta* near the origin of the SMA

- **Inferior suprarenal arteries** (one or more) from the *renal artery*

The venous drainage of the suprarenal gland is into a large **suprarenal vein** (see Fig. 5.59). The short **right suprarenal vein** drains into the IVC, whereas the longer **left suprarenal vein**, often joined by the inferior phrenic vein, empties into the left renal vein.

The **renal lymphatic vessels** follow the renal veins and drain into the lumbar lymph nodes (Fig. 5.52). Lymphatic vessels from the superior part of the ureter may join those from the kidney or pass directly to the lumbar (caval and aortic) nodes. Lymphatic vessels from the middle part of the ureter usually drain into the **common iliac lymph nodes**, whereas vessels from its inferior part drain into the common, external, or internal **iliac lymph nodes**. The **suprarenal lymphatic vessels** arise from a plexus deep to the capsule of the gland and from one in its medulla. The lymph passes to the lumbar lymph nodes.

NERVES OF KIDNEYS, URETERS, AND SUPRARENAL GLANDS

The nerves to the kidneys and ureters arise from the **renal nerve plexus** and consist of sympathetic and visceral afferent fibers (Fig. 5.53). The renal nerve plexus is supplied by fibers from the abdominopelvic (especially the least) splanchnic nerves. The nerves to the abdominal part of the ureters derive from the renal, abdominal aortic, and superior hypogastric plexuses. Visceral afferent fibers conveying pain sensations follow

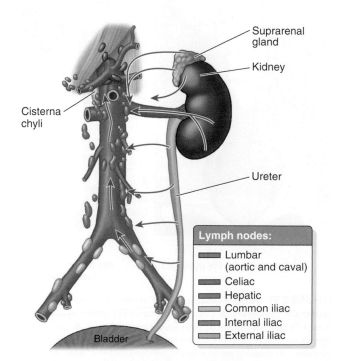

FIGURE 5.52. Lymphatics of kidneys and suprarenal glands. The *arrows* indicate the direction of lymph flow to the lymph nodes.

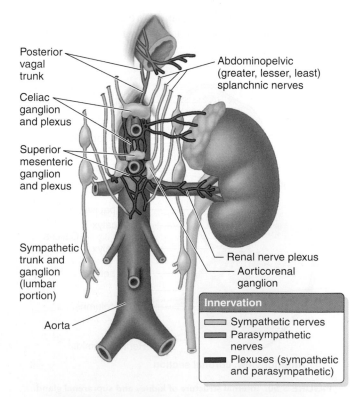

FIGURE 5.53. Innervation of kidneys and suprarenal glands.

SURFACE ANATOMY

Kidneys and Ureters

The **hilum of the left kidney** lies near the level of the trans-pyloric plane, approximately 5 cm from the median plane (Fig. SA5.3). The transpyloric plane passes through the **superior pole of the right kidney**, which is approximately 2.5 cm lower than the left pole. Posteriorly, the superior parts of the kidneys lie deep to the 11th and 12th ribs (Fig. SA5.5A). The levels of the kidneys change during respiration and with changes in posture of 2–3 cm in a vertical direction. The kidneys are generally impalpable. In lean adults, the **inferior pole of the right kidney** is palpable by bimanual examination as a firm, smooth, somewhat rounded mass that descends during inspiration. The left kidney is usually not palpable unless it is enlarged or displaced. The **ureters** occupy a sagittal plane that intersects the tips of the transverse processes of the lumbar vertebrae.

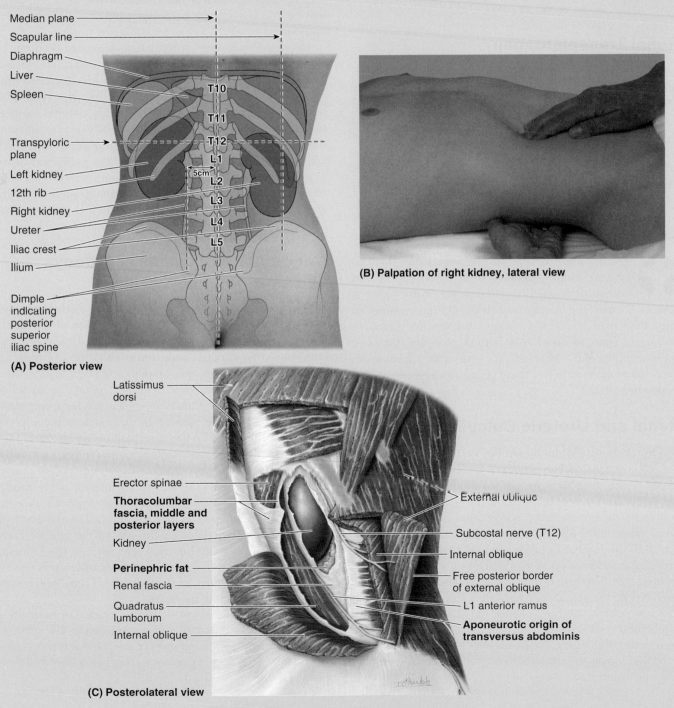

Median plane
Scapular line
Diaphragm
Liver
Spleen
T10
T11
Transpyloric plane
T12
L1
Left kidney
5cm
L2
12th rib
L3
Right kidney
Ureter
L4
Iliac crest
L5
Ilium
Dimple indicating posterior superior iliac spine

(A) Posterior view

(B) Palpation of right kidney, lateral view

Latissimus dorsi
Erector spinae
Thoracolumbar fascia, middle and posterior layers
Kidney
Perinephric fat
Renal fascia
Quadratus lumborum
Internal oblique

External oblique
Subcostal nerve (T12)
Internal oblique
Free posterior border of external oblique
L1 anterior ramus
Aponeurotic origin of transversus abdominis

(C) Posterolateral view

FIGURE SA5.5. Surface anatomy of kidneys and ureters.

CLINICAL BOX

Perinephric Abscess

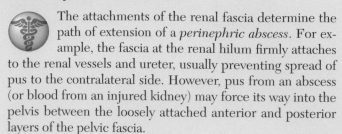

The attachments of the renal fascia determine the path of extension of a *perinephric abscess*. For example, the fascia at the renal hilum firmly attaches to the renal vessels and ureter, usually preventing spread of pus to the contralateral side. However, pus from an abscess (or blood from an injured kidney) may force its way into the pelvis between the loosely attached anterior and posterior layers of the pelvic fascia.

Renal Transplantation

Renal transplantation is now an established operation for the treatment of selected cases of chronic renal failure. The transplanted kidney is placed in the iliac fossa of the greater pelvis (see Chapter 6, Pelvis and Perineum), where it is firmly supported and where only short lengths of renal vessels and ureters are required for implantation. The renal artery and vein are joined to the adjacent external iliac artery and vein, respectively, and the ureter is sutured into the nearby urinary bladder.

Accessory Renal Vessels

During their "ascent" to their final site, the embryonic kidneys receive their blood supply and venous drainage from successively more superior vessels. Usually, the inferior vessels degenerate as superior ones take over the blood supply and venous drainage. Failure of some of these vessels to degenerate results in *accessory (or polar) renal arteries and veins*. Variations in the number and position of these vessels occur in about 25% of people.

Renal and Ureteric Calculi

Excessive distention of the ureter owing to a *renal calculus* (kidney stone) causes severe intermittent pain, *ureteric colic*, as it is gradually forced down the ureter by waves of contraction. The calculus may cause complete or intermittent obstruction of urinary flow. Depending on the level of obstruction, the pain may be referred to the lumbar (loin) or inguinal regions (groin), the proximal anterior aspect of the thigh, or the external genitalia and/or testis. The pain is referred to the cutaneous areas innervated by the spinal cord segments and sensory ganglia, which supply the ureter—mainly T11–L2. Ureteric calculi can be observed and removed with a *nephroscope*. Another technique, *lithotripsy*, focuses a shock wave through the body that breaks the stones into fragments, which then pass with the urine.

Intraperitoneal Injection and Peritoneal Dialysis

The peritoneum is a semipermeable membrane with an extensive surface area, much of which (subdiaphragmatic portions in particular) overlies blood and lymphatic capillary beds. Therefore, fluid injected into the peritoneal cavity is absorbed rapidly. For this reason, anesthetic agents, such as solutions of barbiturate compounds, may be injected into the peritoneal cavity by *intraperitoneal injection*.

In *renal failure*, waste products such as urea accumulate in the blood and tissues and ultimately reach fatal levels. *Peritoneal dialysis* may be performed, in which soluble substances and excess water are removed from the system by transfer across the peritoneum using a dilute sterile solution that is introduced into the peritoneal cavity on one side and then drained from the other side. Diffusible solutes and water are transferred between the blood and the peritoneal cavity as a result of concentration gradients between the two fluid compartments. Peritoneal dialysis is usually employed only temporarily; however, for the long term, it is preferable to use direct blood flow through a renal dialysis machine.

Congenital Anomalies of Kidneys and Ureters

Bifid renal pelvis and ureter are fairly common. These anomalies result from division of the *metanephric diverticulum* (ureteric bud), the primordium of the renal pelvis and ureter. The extent of ureteral duplication depends on the completeness of embryonic division of the metanephric diverticulum. The bifid renal pelvis and/or ureter may be unilateral or bilateral; however, separate openings into the bladder are uncommon. Incomplete division of the metanephric diverticulum results in a bifid ureter; complete division results in a *supernumerary kidney*.

The kidneys are close together in the embryonic pelvis. In approximately 1 in 600 fetuses, the inferior poles (rarely, the superior poles) of the kidneys fuse to form a *horseshoe kidney*. This U-shaped kidney usually lies at the level of the L3–L5 vertebrae because the root of the *inferior mesenteric artery* prevented normal relocation of the kidneys. Horseshoe kidney usually produces no symptoms; however, associated abnormalities of the kidney and renal pelvis may be present, obstructing the ureter.

Sometimes, the embryonic kidney on one or both sides fails to reach the abdomen and lies anterior to the sacrum. Although uncommon, awareness of the possibility of an *ectopic pelvic kidney* should prevent it from being mistaken for a pelvic tumor and removed.

the sympathetic fibers retrograde to spinal ganglia and cord segments T11–L5. The suprarenal glands have a rich nerve supply from the celiac plexus and *abdominopelvic* (greater, lesser, and least) *splanchnic nerves* (Fig. 5.53). The nerves are mainly myelinated presynaptic sympathetic fibers that derive from the lateral horn of the spinal cord and traverse the paravertebral and prevertebral ganglia, without synapse, to be distributed to the chromaffin cells in the suprarenal medulla.

Summary of Innervation of Abdominal Viscera

For autonomic innervation of the abdominal viscera, several different splanchnic nerves and one cranial nerve (the vagus, CN X) deliver presynaptic sympathetic and parasympathetic fibers, respectively, to the abdominal aortic plexus and its associated sympathetic ganglia (Figs. 5.54 and 5.55 and Table 5.8). The peri-arterial extensions of these plexuses deliver postsynaptic sympathetic fibers and the continuations of parasympathetic fibers to the abdominal viscera, where intrinsic parasympathetic ganglia occur.

SYMPATHETIC INNERVATION

The *sympathetic part of the autonomic innervation of the abdominal viscera* consists of the following structures:

- Abdominopelvic splanchnic nerves from the thoracic and abdominal sympathetic trunks
- Prevertebral sympathetic ganglia
- Abdominal aortic plexus and its extensions, the peri-arterial plexuses

The nerve plexuses are mixed, shared with the parasympathetic nervous system and visceral afferent fibers.

The **abdominopelvic splanchnic nerves** convey presynaptic sympathetic fibers to the abdominopelvic cavity. The fibers arise from cell bodies in the IMLs (or lateral horns) of the gray matter of spinal cord segments T5–L2 or L3. The fibers pass successively through the anterior roots, anterior rami, and white communicating branches of thoracic and upper lumbar spinal nerves to reach the sympathetic trunks. They pass through the paravertebral ganglia of the trunks without synapsing to enter the abdominopelvic splanchnic nerves, which convey them to the prevertebral ganglia of the abdominal cavity. The abdominopelvic splanchnic nerves include the following:

- *Lower thoracic splanchnic nerves* (greater, lesser, and least): From the thoracic part of the sympathetic trunks
- *Lumbar splanchnic nerves*: From the lumbar part of the sympathetic trunks

The **lower thoracic splanchnic nerves** are the main source of presynaptic sympathetic fibers serving abdominal viscera. The **greater splanchnic nerve** (from the sympathetic trunk at T5 through T9 or T10 vertebral levels), lesser splanchnic nerve (from T10 and T11 levels), and **least splanchnic nerve** (from the T12 level) are the specific abdominopelvic

splanchnic nerves that arise from the thoracic part of the sympathetic trunks. They pierce the corresponding crus of the diaphragm to convey presynaptic sympathetic fibers to the celiac, superior mesenteric, and aorticorenal (prevertebral) sympathetic ganglia, respectively.

The **lumbar splanchnic nerves** arise from the abdominal part of the sympathetic trunks. Medially, the lumbar sympathetic trunks give off three to four lumbar splanchnic nerves, which pass to the *intermesenteric, inferior mesenteric,* and *superior hypogastric plexuses,* conveying presynaptic sympathetic fibers to the associated prevertebral ganglia of those plexuses.

The cell bodies of postsynaptic sympathetic neurons constitute the major **prevertebral ganglia** that cluster around the roots of the major branches of the abdominal aorta: the **celiac, aorticorenal, superior mesenteric,** and **inferior mesenteric ganglia**. Minor, unnamed prevertebral ganglia occur within the intermesenteric and superior hypogastric plexuses. With the exception of the innervation of the suprarenal medulla, *the synapse between presynaptic and postsynaptic sympathetic neurons occurs in the prevertebral ganglia* (Fig. 5.54B). Postsynaptic sympathetic nerve fibers pass from the prevertebral ganglia to the abdominal viscera by means of the **peri-arterial plexuses** associated with the branches of the abdominal aorta. Sympathetic innervation in the abdomen, as elsewhere, is primarily involved in producing *vasoconstriction.* With regard to the GI tract, it acts to *inhibit* (slow down or stop) *peristalsis.*

PARASYMPATHETIC INNERVATION

The *parasympathetic part of the autonomic innervation of the abdominal viscera* (Figs. 5.54 and 5.55) consists of the following:

- *Anterior and posterior vagal trunks*
- *Pelvic splanchnic nerves*
- *Abdominal* (para-aortic) *autonomic plexuses* and their extensions, the peri-arterial plexuses.
- *Intrinsic* (enteric) *parasympathetic ganglia,* components of intrinsic enteric plexuses of the enteric nervous system.

The nerve plexuses are mixed, shared with the sympathetic nervous system and visceral afferent fibers.

The **anterior** and **posterior vagal trunks** are the continuation of the left and right vagus nerves, that emerge from the esophageal plexus and pass through the esophageal hiatus on the anterior and posterior aspects of the esophagus and stomach (Fig. 5.55). The *vagus nerves convey presynaptic parasympathetic and visceral afferent fibers* (mainly for unconscious sensations associated with reflexes) to the abdominal aortic plexuses and the peri-arterial plexuses, which extend along the branches of the aorta.

The **pelvic splanchnic nerves** are distinct from other splanchnic nerves (Table 5.8) in that they

- have nothing to do with the sympathetic trunks
- derive directly from anterior rami of spinal nerves S2–S4
- convey presynaptic parasympathetic fibers to the inferior hypogastric (pelvic) plexus

(Continued on page 320)

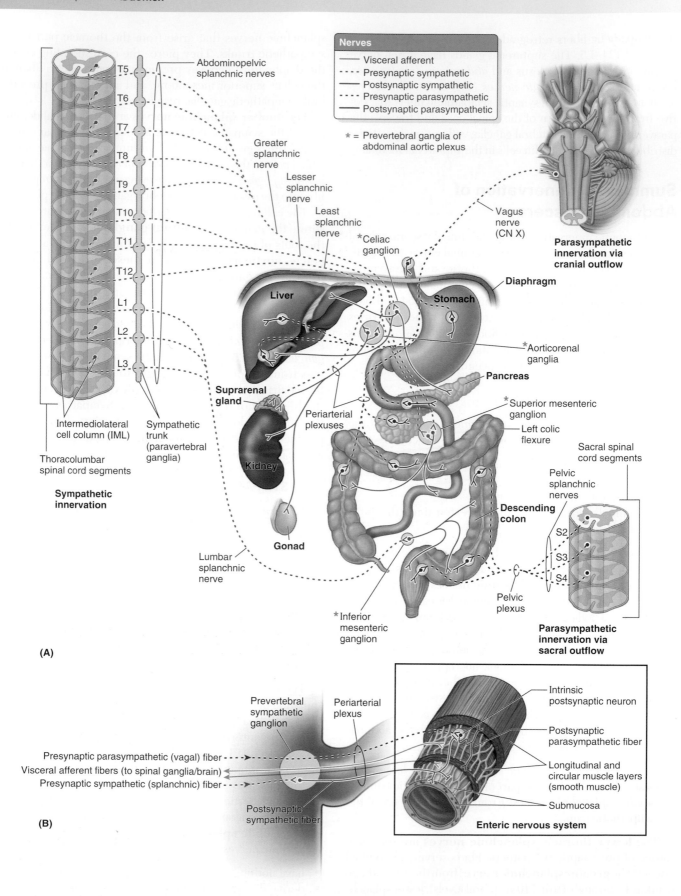

FIGURE 5.54. Autonomic nerves of posterior abdominal wall. A. Origin and distribution of pre- and postsynaptic sympathetic and parasympathetic fibers, and the ganglia involved in supplying abdominal viscera are shown. **B.** The fibers supplying the intrinsic plexuses of abdominal viscera are demonstrated.

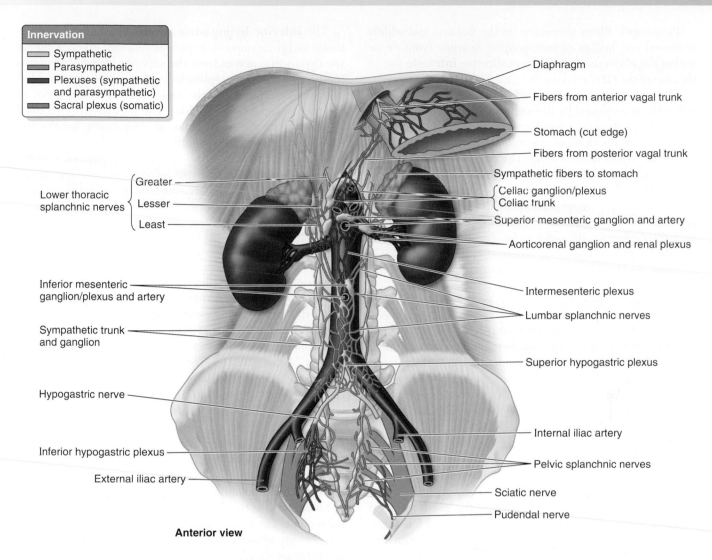

FIGURE 5.55. Splanchnic nerves, nerve plexuses, and sympathetic ganglia in abdomen.

TABLE 5.8. AUTONOMIC INNERVATION OF ABDOMINAL VISCERA (SPLANCHNIC NERVES)

Splanchnic Nerves	Autonomic Fiber Type[a]	System	Origin	Destination
A. Cardiopulmonary (cervical and upper thoracic)	Postsynaptic		Cervical and upper thoracic sympathetic trunk	Thoracic cavity (viscera superior to level of diaphragm)
B. Abdominopelvic			Lower thoracic and abdominopelvic sympathetic trunk:	Abdominopelvic cavity (prevertebral ganglia serving viscera and suprarenal glands inferior to level of diaphragm)

1. Lower thoracic
  a. Greater
  b. Lesser
  c. Least
	Presynaptic	Sympathetic	Thoracic sympathetic trunk:
			a. T5–T9 or T10 level
			b. T10–T11 level
			c. T12 level
				Abdominal prevertebral ganglia:
				a. Celiac ganglia
				b. Aorticorenal ganglia
				c. and

2. Lumbar			Abdominal sympathetic trunk	Other abdominal prevertebral ganglia (superior and inferior mesenteric, and of intermesenteric/hypogastric plexuses)
3. Sacral			Pelvic (sacral) sympathetic trunk	Pelvic prevertebral ganglia
C. Pelvic	Parasympathetic		Anterior rami of S2–S4 spinal nerves	Intrinsic ganglia of descending and sigmoid colon, rectum, and pelvic viscera

[a]Splanchnic nerves also convey visceral afferent fibers, which are not part of the autonomic nervous system.

Presynaptic fibers terminate on the isolated and widely scattered cell bodies of *postsynaptic neurons lying on or within the abdominal viscera*, constituting **intrinsic** (*or, in the case of the GI tract*, enteric) ganglia (Fig. 5.54B).

The presynaptic parasympathetic and visceral afferent reflex fibers conveyed by the vagus nerves extend to intrinsic ganglia of the lower esophagus, stomach, small intestine (including the duodenum), ascending colon, and most of the transverse colon (Fig. 5.54A). The fibers conveyed by the pelvic splanchnic nerves supply the descending and sigmoid parts of the colon, rectum, and pelvic organs. Thus, in terms of the GI tract, *the vagus nerves provide parasympathetic innervation of the smooth muscle and glands of the gut as far as the left colic flexure; the pelvic splanchnic nerves provide the remainder.* Parasympathetic innervation in the abdomen is primarily involved in promotion of *peristalsis* and *secretion*.

EXTRINSIC AUTONOMIC PLEXUSES

The *extrinsic abdominal autonomic plexuses* are nerve networks consisting of both sympathetic and parasympathetic fibers, which surround the abdominal aorta and its major branches (Figs. 5.54 and 5.55). The celiac, superior mesenteric, and inferior mesenteric plexuses are interconnected. The *prevertebral sympathetic ganglia* are scattered among the celiac and mesenteric plexuses.

The *celiac plexus*, surrounding the root of the celiac (arterial) trunk, contains irregular *right* and *left celiac ganglia* (approximately 2 cm long) that unite superior and inferior to the celiac trunk (Figs. 5.54A and 5.55). The *parasympathetic root* of the celiac plexus is a branch of the posterior vagal trunk, which contains fibers from the right and left vagus nerves. The *sympathetic roots* of the plexus are the greater and lesser splanchnic nerves.

The **superior mesenteric plexus** and ganglion or ganglia surround the origin of the SMA. The plexus has one median and two lateral roots. The median root is a branch of the celiac plexus, and the lateral roots arise from the lesser and least splanchnic nerves, sometimes with a contribution from the first lumbar ganglion of the sympathetic trunk.

The **inferior mesenteric plexus** surrounds the IMA and gives offshoots to its branches. It receives a medial root from the intermesenteric plexus and lateral roots from the lumbar ganglia of the sympathetic trunks. An inferior mesenteric ganglion may also appear just inferior to the root of the IMA.

The **intermesenteric plexus** is part of the aortic plexus of nerves between the superior and the inferior mesenteric arteries. It gives rise to renal, testicular or ovarian, and ureteric plexuses.

The **superior hypogastric plexus** is continuous with the intermesenteric plexus and the inferior mesenteric plexus and lies anterior to the inferior part of the abdominal aorta and extends inferiorly across its bifurcation (Table 5.8). **Right** and **left hypogastric nerves** join the superior hypogastric plexus to the inferior hypogastric plexus. The superior hypogastric plexus supplies *ureteric* and *testicular plexuses* and a plexus on each common iliac artery.

The **inferior hypogastric plexuses** are mixed sympathetic and parasympathetic plexuses formed on each side as the hypogastric nerves from the superior hypogastric plexus merge with the pelvic splanchnic nerves. The right and left plexuses are situated on the sides of the rectum, cervix of the uterus, and urinary bladder. The plexuses receive small branches from the superior sacral sympathetic ganglia and the sacral parasympathetic outflow from S2 through S4 sacral spinal nerves (*pelvic [parasympathetic] splanchnic nerves*). Extensions of the inferior hypogastric plexus send autonomic fibers along the blood vessels, which form visceral plexuses on the walls of the pelvic viscera (e.g., *rectal and vesical plexuses*).

INTRINSIC PLEXUSES: THE ENTERIC NERVOUS SYSTEM

Intrinsic ganglionated plexuses of the GI tract, extending from mid-esophagus through internal anal sphincter and along the pancreatobiliary duct system, comprise the **enteric nervous system (ENS)**. The ENS consists of two interconnected plexuses (Fig. 5.54B): (1) the **myenteric plexus** (Auerbach), located between and primarily concerned with motility and vasomotion of the muscular layers of the gut wall (although in the stomach it's also concerned with secretion), and (2) the **submucosal plexus** (Meissner), located in the submucosa of the gut (most prominent in the small intestine, relatively sparse in esophagus and stomach), concerned with the exocrine and endocrine secretion, vasomotion, micromotility, and immune activity (inflammation and immunomodulation) of the mucosa. Vasomotion (control of blood flow) at this level influences water and electrolyte movement. Corresponding plexuses with smaller, sparser ganglia extend to the pancreas, gallbladder, and cystic and major biliary ducts.

The *motor neurons* of these plexuses are intrinsic or enteric ganglia that serve nominally as postsynaptic neurons for the parasympathetic system, but they are much more than relay neurons, merely receiving and passing on efferent impulses sent by presynaptic parasympathetic neurons. They also receive input from postsynaptic sympathetic fibers (making them a third order neuron in that system). They have vast interconnectivity with surrounding efferent neurons, both directly and via *interneurons*, as well as axons terminating on smooth muscle and glands (Fig. 5.56A). *Extrinsic visceral afferent fibers convey long reflex* (hunger, satiety, and nausea) *and pain sensation* to the CNS via vagal (nodose) sensory ganglia, and thoracic, upper lumbar and middle sacral spinal sensory ganglia (Fig. 5.56B). In addition, there are *intrinsic afferent neurons* with cell bodies in the plexuses that *monitor mechanical and chemical conditions in the gut* and communicate with the efferent neurons *providing local (short) reflex circuitry* as well as sending information centrally. Thus, the interconnecting nerve bundles of the plexuses include postsynaptic sympathetic fibers, pre- and postsynaptic parasympathetic fibers, interneuron fibers, and long and short visceral afferent fibers.

These intrinsic neurons and the complex enteric plexuses in which they are enmeshed, integrate and control GI function with remarkable independence, sustaining visceral activities with local reflex mechanisms. *CNS input via the ANS merely modulates the activity of the ENS*, with the parasympathetic system primarily promoting and the sympathetic system primarily inhibiting its motor and secretory activity in response to overall demands placed on the body by environmental and circumstantial factors. With regard to the smooth muscle sphincters, the roles of the sympathetic and parasympathetic systems reverse, with the sympathetic system maintaining tonus and the parasympathetic system inhibiting it. *The ENS can function quite autonomously*, without input from either system; intestine harvested for transplant is not denervated in the usual sense.

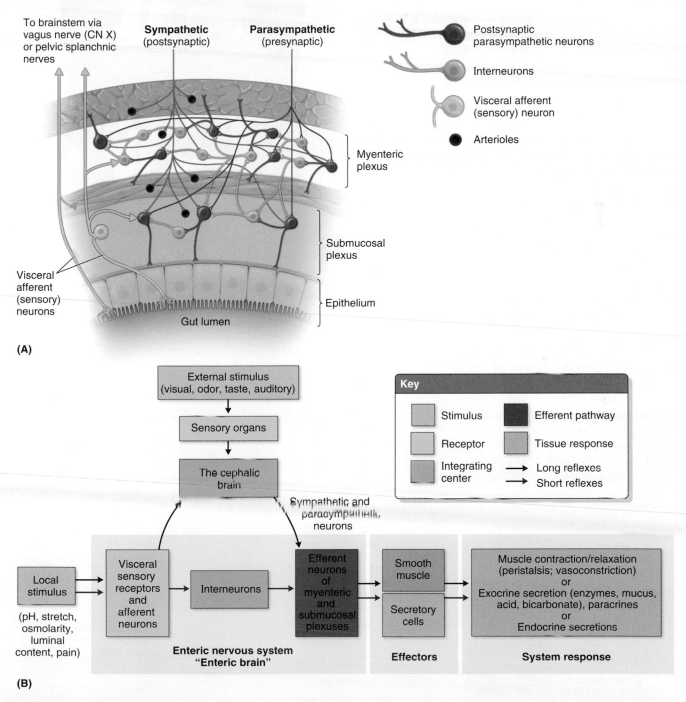

FIGURE 5.56. Enteric nervous system. A. Organization within intestinal wall, schematic illustration. **B.** Flow chart demonstrating long (extrinsic) and short (intrinsic) reflexes involving the enteric nervous system.

The ENS is estimated to include as many as 500 million neurons—more than exist in the entire spinal cord—and employs more than 40 neurotransmitters and neuromodulators, including half the body's dopamine and 95% of all serotonin. The support cells of the intrinsic ENS neurons are more like glial cells (astroglia) of the brain than Schwann cells of the peripheral nervous system. Relatively nonpermeable capillaries associated with the ganglia provide a diffusion barrier resembling the blood–brain barrier of cerebral blood vessels. These facts, combined with complexity and autonomous function, explain why the ENS has come to be considered a "second brain," or at least a third component of the visceral nervous system. Its integrity and appropriate function is vital.

VISCERAL SENSORY INNERVATION

Visceral afferent fibers conveying pain sensations accompany the sympathetic (visceral motor) fibers. The pain impulses pass retrogradely to those of the motor fibers along the splanchnic nerves to the sympathetic trunk, through white communicating branches to the anterior rami of the spinal nerves. Then they pass into the posterior root to the spinal sensory ganglia and spinal cord. Progressively, lower spinal sensory ganglia and spinal cord segments are involved in innervating the abdominal viscera as the tract proceeds caudally. The stomach (foregut) receives innervation from the T6 to T9 levels, small intestine through transverse colon (midgut) from the T8 to T12 levels, and descending colon (hindgut) from the T12 to L2 levels (Fig. 5.57). Starting from the midpoint of the sigmoid colon,

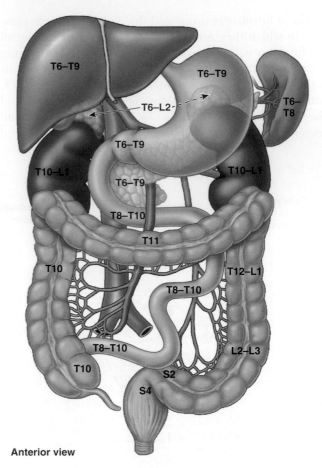

Anterior view

FIGURE 5.57. Segmental innervation of abdominal viscera. Approximate spinal cord segments and spinal sensory ganglia involved in sympathetic and visceral afferent innervation of the abdominal viscera are shown.

CLINICAL BOX

Visceral Referred Pain

Pain arising from a viscus such as the stomach varies from dull to severe. The pain is poorly localized; it radiates to the dermatome level that receives visceral sensory fibers from the organ concerned (Fig. B5.13).

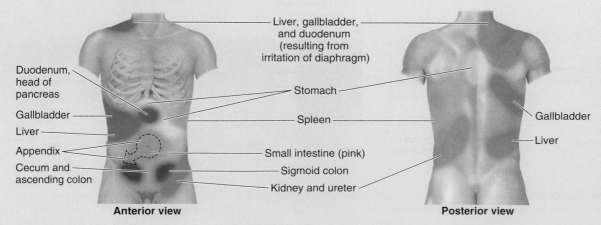

Anterior view **Posterior view**

FIGURE B5.13. Areas of referred pain.

visceral pain fibers run with parasympathetic fibers, the sensory impulses being conducted to S2–S4 sensory ganglia and spinal cord levels. These are the same spinal cord segments involved in the sympathetic innervation of those portions of alimentary tract.

Visceral afferent fibers conveying reflex sensations (that generally do not reach levels of consciousness) accompany the parasympathetic (visceral motor) fibers.

DIAPHRAGM

The **diaphragm** is a dome-shaped, musculotendinous partition separating the thoracic and abdominal cavities. The diaphragm, the chief muscle of inspiration, forms the convex floor of the thoracic cavity and the concave roof of the abdominal cavity (Figs. 5.58 and 5.59). The diaphragm descends during inspiration; however, only its central part moves because its periphery, as the fixed origin of the muscle, attaches to the inferior margin of the thoracic cage and the superior lumbar vertebrae. The diaphragm curves superiorly into **right** and **left domes**; normally, the right dome is higher than the left owing to the presence of the liver (Fig. 5.58). During expiration, the right dome reaches as high as the 5th rib and the left dome ascends to the 5th intercostal space. The level of the domes of the diaphragm varies according to the phase of respiration (inspiration or expiration), posture (e.g., supine or standing), and size and degree of distention of the abdominal viscera.

The muscular part of the diaphragm is situated peripherally with fibers that converge radially on the trifoliate central aponeurotic part, the **central tendon** (Fig. 5.58A,B). This tendon has no bony attachments and is incompletely divided into three leaves, resembling a wide cloverleaf. Although it lies near the center of the diaphragm, the central tendon is closer to the anterior part of the thorax. The superior aspect of the central tendon is fused with the inferior surface of the fibrous pericardium (Fig. 5.58C). The surrounding muscular part of the diaphragm forms a continuous sheet; however, for descriptive purposes it is divided into three parts based on the peripheral attachments (Fig. 5.58A):

- A **sternal part**, consisting of two muscular slips that attach to the posterior aspect of the xiphoid process of the sternum; this part is not always present.
- A **costal part**, consisting of wide muscular slips that attach to the internal surfaces of the inferior six costal cartilages and their adjoining ribs on each side; this part forms the domes of the diaphragm.
- A **lumbar part**, arising from two aponeurotic arches, the *medial* and *lateral arcuate ligaments*, and the three superior lumbar vertebrae; this part forms right and left muscular crura that ascend to the central tendon.

The **crura of the diaphragm** are musculotendinous bundles that arise from the anterior surfaces of the bodies of the superior three lumbar vertebrae, the anterior longitudinal ligament, and the IV discs (Fig. 5.58A). The **right crus**, larger and longer than the left crus, arises from the first

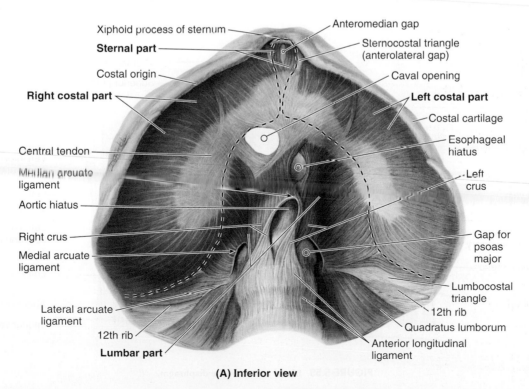

FIGURE 5.58. Attachments, disposition, and features of abdominal aspect of diaphragm. A. Parts of diaphragm. *(continued)*

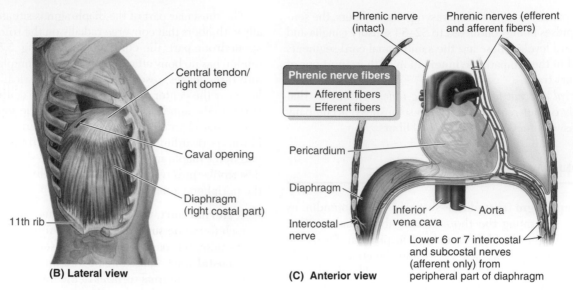

(B) Lateral view

Central tendon/ right dome

Caval opening

Diaphragm (right costal part)

11th rib

Phrenic nerve fibers

— Afferent fibers
— Efferent fibers

Phrenic nerve (intact)

Phrenic nerves (efferent and afferent fibers)

Pericardium

Diaphragm

Intercostal nerve

Inferior vena cava

Aorta

Lower 6 or 7 intercostal and subcostal nerves (afferent only) from peripheral part of diaphragm

(C) Anterior view

FIGURE 5.58. Attachments, disposition, and features of abdominal aspect of diaphragm. *(continued)* **B.** Attachment of right dome of diaphragm. **C.** Innervation of diaphragm.

three or four lumbar vertebrae, whereas the **left crus** arises from only the first two or three. The crura are united by the **median arcuate ligament**, which passes over the anterior surface of the aorta. The diaphragm is also attached on each side to the **medial** and **lateral arcuate ligaments**, which are thickenings of the fascia covering the psoas and quadratus lumborum muscles, respectively.

Diaphragmatic Apertures

The **diaphragmatic apertures** permit structures (e.g., esophagus, vessels, nerves, and lymphatics) to pass between the thorax and the abdomen (Figs. 5.58, 5.59, and 5.60). The three large apertures for the IVC, esophagus, and aorta are the caval opening, esophageal hiatus, and aortic hiatus, respectively.

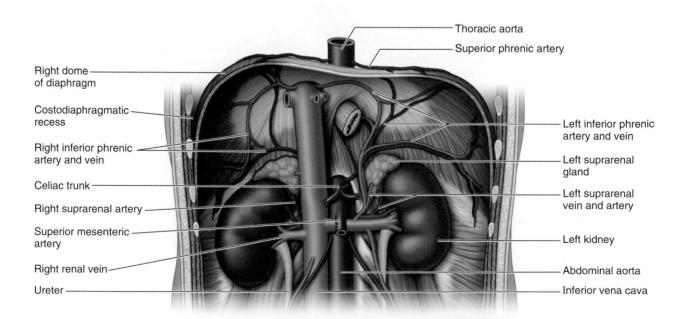

Thoracic aorta

Superior phrenic artery

Right dome of diaphragm

Costodiaphragmatic recess

Right inferior phrenic artery and vein

Celiac trunk

Right suprarenal artery

Superior mesenteric artery

Right renal vein

Ureter

Left inferior phrenic artery and vein

Left suprarenal gland

Left suprarenal vein and artery

Left kidney

Abdominal aorta

Inferior vena cava

FIGURE 5.59. Blood vessels of the diaphragm.

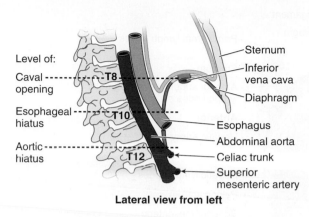

FIGURE 5.60. Diaphragmatic apertures.

CAVAL OPENING

The **caval opening** is an aperture in the central tendon primarily for the IVC. Also passing through the caval opening are terminal branches of the right phrenic nerve and some lymphatic vessels on their way from the liver to the middle phrenic and mediastinal lymph nodes. The caval opening is located to the right of the median plane at the junction of the tendon's right and middle leaves. The most superior of the three diaphragmatic apertures, the caval opening lies at the level of T8 vertebra or the T8/T9 IV disc. The IVC is adherent to the margin of the opening; consequently, when the diaphragm contracts during inspiration, it widens the opening and dilates the IVC. These changes facilitate blood flow to the heart through this large vein.

ESOPHAGEAL HIATUS

The **esophageal hiatus** is an oval aperture for the esophagus in the muscle of the right crus of the diaphragm at the level of the T10 vertebra. The fibers of the right crus decussate (cross one another) inferior to the hiatus, forming a muscular sphincter for the esophagus that constricts it when the diaphragm contracts. In 30% of individuals, a superficial muscular bundle from the left crus contributes to the formation of the right margin of the hiatus. The esophageal hiatus also transmits the anterior and posterior vagal trunks, esophageal branches of the left gastric vessels, and a few lymphatic vessels.

AORTIC HIATUS

The **aortic hiatus** is an opening posterior to the diaphragm. The aortic hiatus transmits the descending aorta, azygos vein, and the thoracic duct. Because the aorta does not pierce the diaphragm, blood flow through it is not affected by the muscle's movements during respiration. The aorta passes between the crura of the diaphragm posterior to the median arcuate ligament, which is at the level of the T12 vertebra (Figs. 5.58A and 5.60).

OTHER APERTURES IN DIAPHRAGM

There is a small opening, the **sternocostal triangle** (foramen), between the sternal and the costal attachments of the diaphragm. This triangle transmits lymphatic vessels from the diaphragmatic surface of the liver and the superior epigastric vessels. The sympathetic trunks pass deep to the medial arcuate ligament. The greater and lesser splanchnic nerves traverse the crura of the diaphragm.

Vasculature and Nerves of Diaphragm

The *arteries of the diaphragm* form a branch-like pattern on both its superior and inferior surfaces. The arteries supplying the superior surface of the diaphragm are the **pericardiacophrenic** and **musculophrenic arteries**, branches of internal thoracic artery, and the **superior phrenic arteries** arising from thoracic aorta (Fig. 5.59). The arteries supplying the inferior surface of the diaphragm are the **inferior phrenic arteries**, which typically are the first branches of the abdominal aorta; however, they may arise from the celiac trunk.

The *veins* draining the superior surface of the diaphragm are the **pericardiacophrenic** and **musculophrenic veins**, which empty into the *internal thoracic veins*, and on the right side, a *superior phrenic vein*, which drains into the IVC. Posteriorly, some veins drain into the *azygos* and *hemi-azygos veins*. The inferior phrenic veins drain blood from the inferior surface of the diaphragm (Fig. 5.59). The **right inferior phrenic vein** usually opens into the IVC, whereas the **left inferior phrenic vein** is usually double, with one branch passing anterior to the esophageal hiatus to end in the IVC and the other, more posterior branch usually joining the left suprarenal vein.

The *lymphatic plexuses* on the thoracic and abdominal surfaces of the diaphragm communicate freely (Fig. 5.61). The anterior and posterior **diaphragmatic lymph nodes** are on the thoracic surface of the diaphragm. Lymph from these nodes drains into the *parasternal*, *posterior mediastinal*, and *phrenic lymph nodes*. Lymph vessels from the abdominal surface of the diaphragm drain into the anterior diaphragmatic, phrenic, and *superior lumbar (caval/aortic) lymph nodes*. Lymphatic vessels are dense on the inferior surface of the diaphragm, constituting the primary means for absorption of peritoneal fluid and substances introduced by intraperitoneal injection.

The entire motor supply to the diaphragm is from the **right** and **left phrenic nerves**, each of which is distributed to half of the diaphragm and arises from the anterior rami of the C3–C5 segments of the spinal cord (Fig. 5.58C). The phrenic nerves also supply sensory fibers (pain and proprioception) to most of the diaphragm. Peripheral parts of the diaphragm receive their sensory nerve supply from the **intercostal nerves** (lower six or seven) and the *subcostal nerves*.

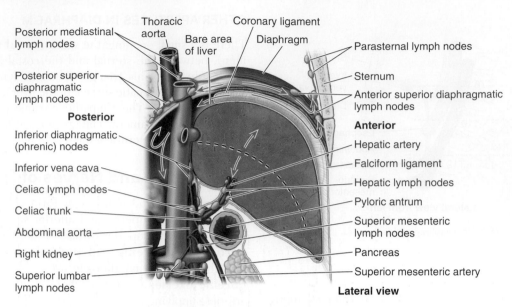

Posterior mediastinal lymph nodes

Thoracic aorta

Bare area of liver

Coronary ligament

Diaphragm

Parasternal lymph nodes

Posterior superior diaphragmatic lymph nodes

Sternum

Anterior superior diaphragmatic lymph nodes

Posterior

Anterior

Inferior diaphragmatic (phrenic) nodes

Hepatic artery

Inferior vena cava

Falciform ligament

Celiac lymph nodes

Hepatic lymph nodes

Celiac trunk

Pyloric antrum

Abdominal aorta

Superior mesenteric lymph nodes

Right kidney

Pancreas

Superior lumbar lymph nodes

Superior mesenteric artery

Lateral view

FIGURE 5.61. Lymphatic drainage of diaphragm.

CLINICAL BOX

Section of a Phrenic Nerve

Section of a phrenic nerve in the neck results in complete paralysis and eventual atrophy of the muscular part of the corresponding half of the diaphragm, except in persons who have an accessory phrenic nerve. *Paralysis of a hemidiaphragm* can be recognized radiographically by its permanent elevation and paradoxical movement.

Referred Pain from Diaphragm

Pain from the diaphragm radiates to two different areas because of the difference in the sensory nerve supply of the diaphragm. Pain resulting from irritation of the diaphragmatic pleura or the diaphragmatic peritoneum is referred to the shoulder region, the area of skin supplied by the C3–C5 segments of the spinal cord. These segments also contribute anterior rami to the phrenic nerves. Irritation of peripheral regions of the diaphragm, innervated by the inferior intercostal nerves, is more localized, being referred to the skin over the costal margins of the anterolateral abdominal wall.

Rupture of Diaphragm and Herniation of Viscera

Rupture of the diaphragm and *herniation of viscera* can result from a sudden large increase in either the intrathoracic or intra-abdominal pressure. The common cause of this injury is severe trauma to the thorax or abdomen during a motor vehicle accident. Most diaphragmatic ruptures are on the left side (95%) because the substantial mass of the liver, intimately associated with the diaphragm on the right side, provides a physical barrier.

A nonmuscular area of variable size called the *lumbocostal triangle* usually occurs between the costal and lumbar parts of the diaphragm. This part of the diaphragm is normally formed only by fusion of the superior and inferior fascias of the diaphragm. When a *traumatic diaphragmatic hernia* occurs, the stomach, small intestine and mesentery, transverse colon, and spleen may herniate through this area into the thorax.

Hiatal or *hiatus hernia*, a protrusion of part of the stomach into the thorax through the esophageal hiatus, was discussed earlier in this chapter. The structures that pass through the esophageal hiatus (vagal trunks, left inferior phrenic vessels, esophageal branches of the left gastric vessels) may be injured in surgical procedures on the esophageal hiatus (e.g., repair of a hiatus hernia).

Congenital Diaphragmatic Hernia

In *congenital diaphragmatic hernia* (CDH), part of the stomach and intestine herniate through a large posterolateral defect (foramen of Bochdalek) in the region of the lumbocostal trigone of the diaphragm. Herniation almost always occurs on the left owing to the presence of the liver on the right. This type of hernia results from the complex development of the diaphragm.

Posterolateral defect of the diaphragm is the only relatively common congenital anomaly of the diaphragm, occurring approximately once in 2,200 newborn infants (Moore et al., 2016). With abdominal viscera in the limited space of the prenatal pulmonary cavity, one lung (usually the left lung) does not have room to develop normally or to inflate after birth. Because of the consequent *pulmonary hypoplasia* (undersized lungs), the mortality rate in these infants is high (approximately 76%).

POSTERIOR ABDOMINAL WALL

The posterior abdominal wall is composed mainly—from deep (posterior) to superficial (anterior)—of the following structures:

- Five lumbar vertebrae and associated IV discs
- Posterior abdominal wall muscles—psoas, quadratus lumborum, iliacus, transversus abdominis, and internal and external oblique muscles
- Lumbar plexus, composed of the anterior rami of lumbar spinal nerves
- Fascia, including thoracolumbar fascia
- Diaphragm, contributing to the superior part of the posterior wall
- Fat, nerves, vessels, and lymph nodes

Fascia of Posterior Abdominal Wall

The posterior abdominal wall is covered with a continuous layer of endo-abdominal fascia, which lies between the parietal peritoneum and the muscles. The fascia lining the posterior abdominal wall is continuous with the transversalis fascia that lines the transversus abdominis muscle (Fig. 5.62).

It is customary to name the fascia according to the structure it covers. The **psoas fascia** covering the psoas major is attached medially to the lumbar vertebrae and pelvic brim. The psoas fascia is thickened superiorly to form the medial arcuate ligament and fuses laterally with the quadratus lumborum and thoracolumbar fascia (Fig. 5.62B). Inferior to the iliac crest, the psoas fascia is continuous with the part of the iliac fascia covering the iliacus muscle.

The **thoracolumbar fascia** is an extensive fascial complex that has anterior, middle, and posterior layers with muscles enclosed between them. It is thin and transparent where it covers thoracic parts of the deep muscles but is thick and strong in the lumbar region. The **posterior and middle layers of thoracolumbar fascia** enclose the vertical deep back muscles (erector spinae). The lumbar part of this posterior layer, extending between the 12th rib and the iliac crest, attaches laterally to the internal oblique and transversus abdominis muscles. The **anterior layer of the thoracolumbar fascia** (quadratus lumborum fascia), covering the quadratus lumborum muscle, attaches to the anterior surfaces of the transverse processes of the lumbar vertebrae, the iliac crest, and the 12th rib and is continuous laterally with the aponeurotic origin of the transversus abdominis muscle.

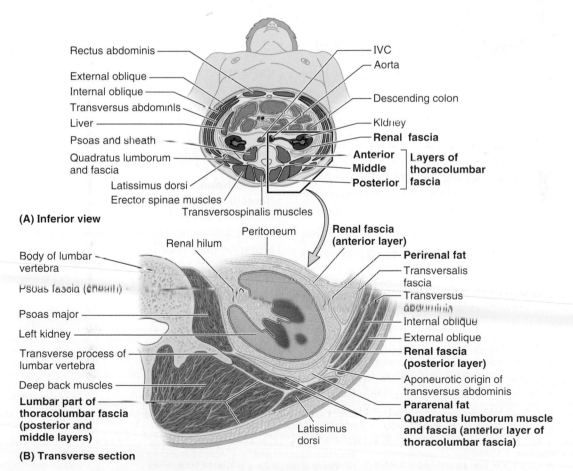

FIGURE 5.62. Fascia of posterior abdominal wall. A. Relationships of fascia and muscle. *IVC,* inferior vena cava. **B.** Layers of thoracolumbar fascia.

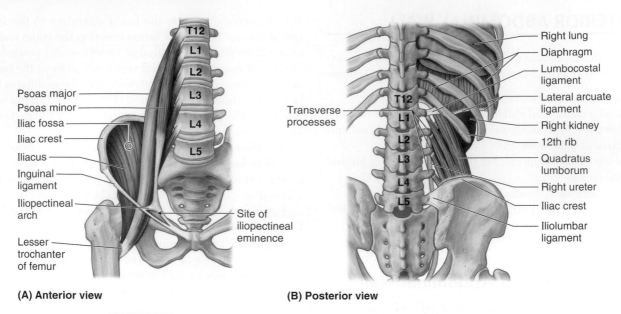

(A) Anterior view   **(B) Posterior view**

FIGURE 5.63. **Muscles of posterior abdominal wall. A.** Iliopsoas. **B.** Quadratus lumborum.

The anterior layer of the thoracolumbar fascia is thickened superiorly to form the lateral arcuate ligaments and is adherent inferiorly to the iliolumbar ligaments (Figs. 5.60 and 5.62).

Muscles of Posterior Abdominal Wall

The main paired muscles in the posterior abdominal wall (Fig. 5.63) are as follows:

- **Psoas major**, passing inferolaterally
- **Iliacus**, lying along the lateral sides of the inferior part of the psoas major; together, the psoas and iliacus form the **iliopsoas**.
- **Quadratus lumborum**, lying adjacent to the transverse processes of the lumbar vertebrae and lateral to the superior parts of the psoas major

The attachments, nerve supply, and main actions of these muscles are summarized in Table 5.9.

Nerves of Posterior Abdominal Wall

There are somatic and autonomic nerves in the posterior abdominal wall. The somatic nerves will be discussed here.

The subcostal nerves, the anterior rami of T12, arise in the thorax, pass posterior to the lateral arcuate ligaments into the abdomen, and run inferolaterally on the anterior surface of the quadratus lumborum muscle (Fig. 5.64) and posterior to the kidneys. They pass through the transversus abdominis and internal oblique muscles to supply the external oblique and skin of the anterolateral abdominal wall.

The **lumbar spinal nerves** pass from the spinal cord through the IV foramina inferior to the corresponding vertebrae, where they divide into posterior and anterior rami. Each ramus contains sensory and motor fibers. The posterior rami pass posteriorly to supply the deep back

TABLE 5.9. MAIN MUSCLES OF POSTERIOR ABDOMINAL WALL

Muscle	Superior Attachments	Inferior Attachment(s)	Innervation	Actions
Psoas major[a]	Transverse processes of lumbar vertebrae; sides of bodies of T12–S1 vertebrae and intervening IV discs	By a strong tendon to lesser trochanter of femur	Lumbar plexus via anterior branches of nerves L2–L4	Acting inferiorly with iliacus, flexes thigh; acting superiorly, flexes vertebral column laterally to balance the trunk; when sitting, acts inferiorly with iliacus to flex trunk
Iliacus[a]	Superior two thirds of iliac fossa, ala of sacrum, and anterior sacro-iliac ligaments	Lesser trochanter of femur and shaft inferior to it and to psoas major tendon	Femoral nerve (L2–L4)	Flexes thigh and stabilizes hip joint; acts with psoas major
Quadratus lumborum	Medial half of inferior border of 12th rib and tips of lumbar transverse processes	Iliolumbar ligament and internal lip of iliac crest	Anterior branches of T12 and L1–L4 nerves	Extends and laterally flexes vertebral column; fixes 12th rib during inspiration

[a]Psoas major and iliacus muscles are often described together as the iliopsoas muscle when flexion of the thigh is discussed (see Chapter 7). The iliopsoas is the chief flexor of the thigh; when thigh is fixed, it is a strong flexor of the trunk (e.g., during sit-ups).
IV, intervertebral.

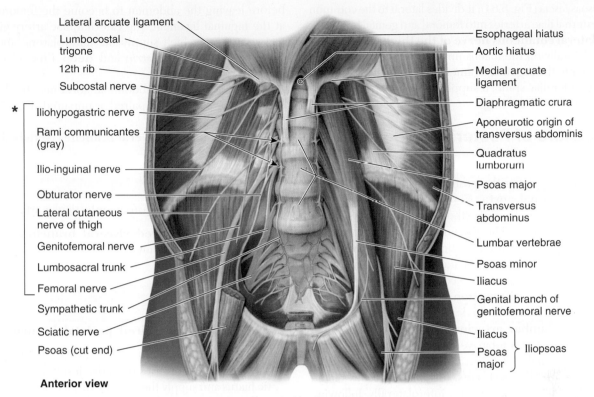

Lateral arcuate ligament
Lumbocostal trigone
12th rib
Subcostal nerve

* Iliohypogastric nerve
Rami communicantes (gray)
Ilio-inguinal nerve
Obturator nerve
Lateral cutaneous nerve of thigh
Genitofemoral nerve
Lumbosacral trunk
Femoral nerve
Sympathetic trunk
Sciatic nerve
Psoas (cut end)

Esophageal hiatus
Aortic hiatus
Medial arcuate ligament
Diaphragmatic crura
Aponeurotic origin of transversus abdominis
Quadratus lumborum
Psoas major
Transversus abdominus
Lumbar vertebrae
Psoas minor
Iliacus
Genital branch of genitofemoral nerve
Iliacus
Psoas major } Iliopsoas

Anterior view

* Lumbar plexus, composed of the anterior rami of lumbar spinal nerves, revealed by the removal of the psoas muscle

FIGURE 5.64. Muscles and nerves of the posterior abdominal wall, lumbosacral plexus.

muscles and skin of the back, whereas the anterior rami pass inferolaterally through the psoas major to supply the skin and muscles of most inferior trunk and lower limb. The proximal parts of the anterior rami of L1–L2 or L3 give rise to *white rami communicantes* that convey presynaptic sympathetic fibers to the lumbar sympathetic trunks. The lumbar sympathetic trunks descend on the anterolateral aspects of the bodies of the lumbar vertebrae in a groove formed by the psoas major (Fig. 5.64).

For the innervation of the abdominal wall and lower limbs, synapses occur in the sympathetic ganglia of the sympathetic trunks. Postsynaptic sympathetic fibers then travel via the *gray communicating branches* to the anterior rami. The anterior rami become the thoraco-abdominal and subcostal nerves, and the lumbar plexus (somatic nerves) and the accompanying postsynaptic sympathetic fibers stimulate vasomotor, sudomotor, and pilomotor action in the distribution of these nerves. The *lumbar splanchnic nerves* that innervate pelvic viscera are described in Chapter 6, Pelvis and Perineum.

The **lumbar plexus of nerves** is in the posterior part of the psoas major, anterior to the lumbar transverse processes (Fig. 5.64). This nerve network is composed of the anterior rami of L1–L4 nerves. All rami receive *gray communicating branches* from the sympathetic trunks. The following nerves

are **branches of the lumbar plexus**; the three largest are listed first:

- The **obturator nerve** (L2–L4) emerges from the medial border of the psoas major and passes through the pelvis to the medial thigh, supplying the adductor muscles.
- The **femoral nerve** (L2–L4) emerges from the lateral border of the psoas major and innervates the iliacus and passes deep to the inguinal ligament to the anterior thigh, supplying the flexors of the hip and extensors of the knee.
- The **lumbosacral trunk** (L4, L5) passes over the ala (wing) of the sacrum and descends into the pelvis to participate in the formation of the sacral plexus along with the anterior rami of the S1–S4 nerves.
- The **ilio-inguinal and iliohypogastric nerves** (L1) arise from the anterior ramus of L1 and enter the abdomen posterior to the medial arcuate ligaments and pass inferolaterally, anterior to the quadratus lumborum. They pierce the transversus abdominis muscles near the anterior superior iliac spines and pass through the internal and external oblique muscles to supply the abdominal muscles and skin of the pubic and inguinal regions.
- The **genitofemoral nerve** (L1, L2) pierces the anterior surface of the psoas major and runs inferiorly on it deep to

the psoas fascia (Fig. 5.64); it divides lateral to the common and external iliac arteries into femoral and genital branches.

- The **lateral cutaneous nerve of the thigh** (L2, L3) runs inferolaterally on the iliacus muscle and enters the thigh posterior to the inguinal ligament, just medial to the anterior superior iliac spine; it supplies the skin on the anterolateral surface of the thigh.

Vasculature of Posterior Abdominal Wall

Most arteries supplying the posterior abdominal wall arise from the **abdominal aorta** (Fig. 5.65); however, the **subcostal arteries** arise from the thoracic aorta and distribute inferior to the 12th rib. The abdominal aorta, approximately 13 cm in length, begins at the aortic hiatus in the diaphragm at the level of the T12 vertebra and ends at the level of the L4 vertebra by dividing into two common iliac arteries. The **level of the aortic bifurcation** is 2–3 cm inferior and to the left of the umbilicus at the level of the iliac crests. Four or five pairs of **lumbar arteries** arise from the abdominal aorta and supply the lumbar vertebrae, back muscles, and posterior abdominal wall.

The **common iliac arteries**, terminal branches of the abdominal aorta, diverge and run inferolaterally, following the medial border of the psoas muscles to the pelvic brim. Here, each common iliac artery divides into the **internal** and **external iliac arteries**. The internal iliac artery enters the pelvis; its course and branches are described in Chapter 6. The external iliac artery follows the iliopsoas muscle. Just before leaving the abdomen to become the femoral artery at the inguinal ligament, the external iliac artery gives rise to two arteries which supply the anterolateral abdominal wall: the *inferior epigastric* and **deep iliac circumflex arteries** (see Fig. 5.7).

The branches of the abdominal aorta may be described as visceral or parietal and paired or unpaired (Figs. 5.65 and 5.66A).

The lateral **paired visceral branches** (vertebral level of origin) are as follows:

- Suprarenal arteries (L1)
- Renal arteries (L1)
- Gonadal arteries, the ovarian or testicular arteries (L2)

The anterior **unpaired visceral branches** (vertebral level of origin) are as follows:

- Celiac trunk (T12)
- SMA (L1)
- IMA (L3)

The posterolateral **paired parietal branches** are as follows:

- Inferior phrenic arteries that arise just inferior to the aortic hiatus and supply the inferior surface of the diaphragm and the suprarenal glands
- Lumbar arteries that pass around the sides of the superior four lumbar vertebrae to supply the posterior abdominal wall

The **unpaired parietal branch** is the **median sacral artery**, which arises from the posterior aspect of the aorta

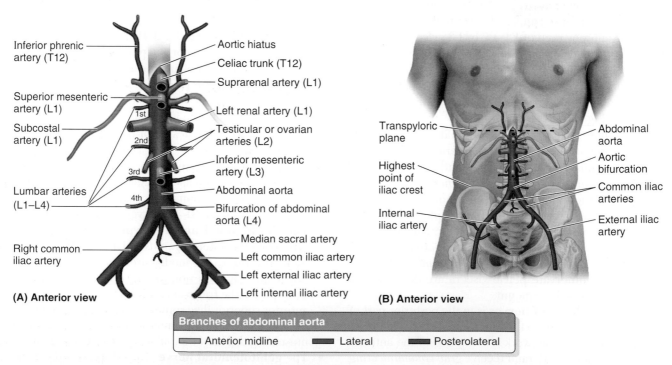

(A) Anterior view

Inferior phrenic artery (T12)
Superior mesenteric artery (L1)
Subcostal artery (L1)
Lumbar arteries (L1–L4)
Right common iliac artery

Aortic hiatus
Celiac trunk (T12)
Suprarenal artery (L1)
Left renal artery (L1)
Testicular or ovarian arteries (L2)
Inferior mesenteric artery (L3)
Abdominal aorta
Bifurcation of abdominal aorta (L4)
Median sacral artery
Left common iliac artery
Left external iliac artery
Left internal iliac artery

(B) Anterior view

Transpyloric plane
Highest point of iliac crest
Internal iliac artery

Abdominal aorta
Aortic bifurcation
Common iliac arteries
External iliac artery

Branches of abdominal aorta

◻ Anterior midline   ◼ Lateral   ◼ Posterolateral

FIGURE 5.65. Branches of abdominal aorta. **A.** Overview. **B.** Surface anatomy.

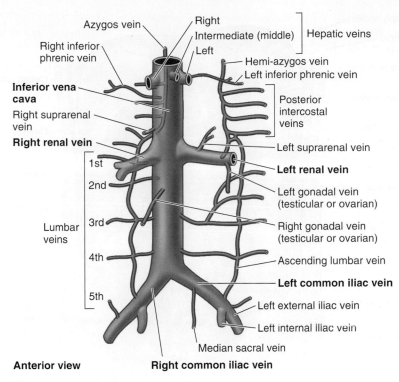

Azygos vein

Right inferior phrenic vein

Right Intermediate (middle) Hepatic veins Left

Inferior vena cava

Right suprarenal vein

Right renal vein

1st

2nd

Lumbar veins 3rd

4th

5th

Anterior view

Hemi-azygos vein

Left inferior phrenic vein

Posterior intercostal veins

Left suprarenal vein

Left renal vein

Left gonadal vein (testicular or ovarian)

Right gonadal vein (testicular or ovarian)

Ascending lumbar vein

Left common iliac vein

Left external iliac vein

Left internal iliac vein

Median sacral vein

Right common iliac vein

FIGURE 5.66. Inferior vena cava and tributaries.

at its bifurcation and descends along the midline into the lesser pelvis.

The *veins of the posterior abdominal wall* are tributaries of the IVC, except for the left testicular or ovarian vein, which enters the left renal vein instead of entering the IVC (Fig. 5.66B). The IVC, the largest vein in the body, has no valves except for a variable, nonfunctional one at its orifice in the right atrium of the heart. The IVC returns poorly oxygenated blood from the lower limbs, most of the back, the abdominal walls, and the abdominopelvic viscera. Blood from the viscera passes through the *portal venous system* and the liver before entering the IVC via the hepatic veins. The IVC begins anterior to the L5 vertebra by the union of the common iliac veins. This union occurs approximately 2.5 cm to the right of the median plane, inferior to the bifurcation of the aorta and posterior to the proximal part of the right *common iliac artery*. The IVC ascends on the right side of the bodies of the L3–L5 vertebrae and on the psoas major muscle to the right of the aorta. The IVC leaves the abdomen by passing through the caval opening in the diaphragm to enter the thorax. The tributaries of the IVC correspond to branches of the aorta:

- Common iliac veins, formed by union of external and internal iliac veins
- Third (L3) and fourth (L4) lumbar veins
- Right testicular or ovarian veins (the left testicular or ovarian veins usually drain into the left renal vein)
- Right and left renal veins

- Ascending lumbar veins. The azygos and hemi-azygos veins arise, in part, from ascending lumbar veins (see Chapter 2, Back). The ascending lumbar and azygos veins connect the IVC and superior vena cava, either directly or indirectly.
- Right suprarenal vein (the left suprarenal vein usually drains into the left renal vein)
- Inferior phrenic veins
- Hepatic veins

Lymphatics of Posterior Abdominal Wall

Lymphatic vessels and lymph nodes lie along the aorta, IVC, and iliac vessels. The common iliac lymph nodes receive lymph from the external and internal iliac lymph nodes. Lymph from the common iliac lymph nodes passes to the lumbar lymph nodes (Fig. 5.67). These nodes receive lymph directly from the posterior abdominal wall, kidneys, ureters, testes or ovaries, uterus, and uterine tubes. They also receive lymph from the descending colon, pelvis, and lower limbs through the inferior mesenteric and common iliac lymph nodes. Efferent lymphatic vessels from the lymph nodes form the right and left **lumbar lymphatic trunks**. Lymphatic vessels from the intestine, liver, spleen, and pancreas pass along the celiac, superior, and inferior mesenteric arteries to the pre-aortic lymph nodes (celiac and superior and inferior mesenteric nodes) scattered around the origins of these arteries from the aorta. Efferent vessels from these nodes

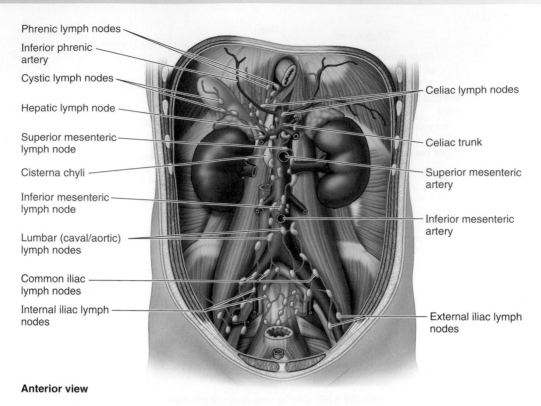

Phrenic lymph nodes
Inferior phrenic artery
Cystic lymph nodes
Hepatic lymph node
Superior mesenteric lymph node
Cisterna chyli
Inferior mesenteric lymph node
Lumbar (caval/aortic) lymph nodes
Common iliac lymph nodes
Internal iliac lymph nodes

Celiac lymph nodes
Celiac trunk
Superior mesenteric artery
Inferior mesenteric artery
External iliac lymph nodes

Anterior view

FIGURE 5.67. Abdominal lymphatic drainage.

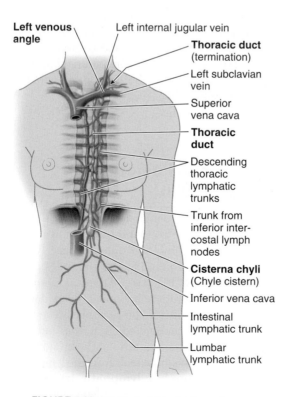

Left venous angle
Left internal jugular vein
Thoracic duct (termination)
Left subclavian vein
Superior vena cava
Thoracic duct
Descending thoracic lymphatic trunks
Trunk from inferior intercostal lymph nodes
Cisterna chyli (Chyle cistern)
Inferior vena cava
Intestinal lymphatic trunk
Lumbar lymphatic trunk

FIGURE 5.68. Abdominal lymphatic trunks.

form the **intestinal lymphatic trunks,** which may be single or multiple and participate in the confluence of lymphatic trunks that gives rise to the thoracic duct.

The **cisterna chyli** is a thin-walled sac at the inferior end of the **thoracic duct,** variable in size and shape, and located anterior to the bodies of the L1 and L2 vertebrae between the right crus of the diaphragm and the aorta (Fig. 5.68). A pair of **descending thoracic lymphatic trunks** carry lymph from the lower six intercostal spaces on each side. More often, there is merely a simple or plexiform convergence of the right and left lumbar lymphatic trunks, the intestinal lymph trunk(s), and a pair of descending thoracic lymphatic trunks. Consequently, essentially all the lymphatic drainage from the lower half of the body (deep lymphatic drainage inferior to the level of the diaphragm and all superficial drainage inferior to the level of the umbilicus) converges in the abdomen to enter the beginning of the thoracic duct. The thoracic duct ascends through the aortic hiatus in the diaphragm into the posterior mediastinum, where it collects more parietal and visceral drainage, particularly from the left upper quadrant of the body, and ultimately ends by entering the venous system at the *junction of the left subclavian and internal jugular veins* (the *left venous angle*).

CLINICAL BOX

Psoas Abscess

 An abscess resulting from tuberculosis in the lumbar region tends to spread from the vertebrae into the psoas sheath, where it produces a *psoas abscess*. As a consequence, the psoas fascia thickens to form a strong stocking-like tube. Pus from the psoas abscess passes inferiorly along the psoas within this fascial tube over the pelvic brim and deep to the inguinal ligament. The pus usually surfaces in the superior part of the thigh. Pus can also reach the psoas sheath by passing from the posterior mediastinum when the thoracic vertebrae are diseased.

Posterior Abdominal Pain

 The iliopsoas muscle has extensive and clinically important relations to the kidneys, ureters, cecum, appendix, sigmoid colon, pancreas, lumbar lymph nodes, and nerves of the posterior abdominal wall. When any of these structures is diseased, movement of the iliopsoas usually causes pain. When intra-abdominal inflammation is suspected, the *iliopsoas test* is performed. The person is asked to lie on the unaffected side and to extend the thigh on the affected side against the resistance of the examiner's hand. Pain resulting from this maneuver is a *positive psoas sign*. An *acutely inflamed appendix*, for example, will produce a positive sign.

Collateral Routes for Abdominopelvic Venous Blood

Three collateral routes, formed by valveless veins of the trunk, are available for venous blood to return to the heart when the IVC is obstructed or ligated:

- The *inferior epigastric veins*, tributaries of the external iliac veins of the inferior caval system, anastomose in the rectus sheath with the *superior epigastric veins*, which drain in sequence through the internal thoracic veins of the superior caval system.
- The *superficial epigastric* or *superficial circumflex iliac veins*, normally tributaries of the great saphenous vein of the inferior caval system, anastomose in the subcutaneous tissues of the anterolateral body wall with one of the tributaries of the axillary vein, commonly the *lateral thoracic vein*. When the IVC is obstructed, this subcutaneous collateral pathway—called the *thoraco-epigastric vein*—becomes particularly conspicuous.
- The *epidural venous plexus* inside the vertebral column (see Chapter 2, Back) communicates with the lumbar veins of the inferior caval system and the tributaries of the azygos system of veins, which is part of the superior caval system.

Abdominal Aortic Aneurysm

 Rupture of an *aneurysm* (localized enlargement) *of the abdominal aorta* causes severe pain in the abdomen or back (Fig. B5.14). If unrecognized, a ruptured aneurysm has a mortality rate of nearly 90% because of heavy blood loss. Surgeons can repair an aneurysm by opening it, inserting a prosthetic graft (such as one made of Dacron), and sewing the wall of the aneurysmal aorta over the graft to protect it. Aneurysms may also be treated by endovascular catheterization procedures.

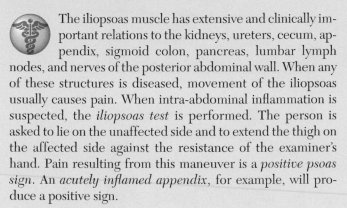
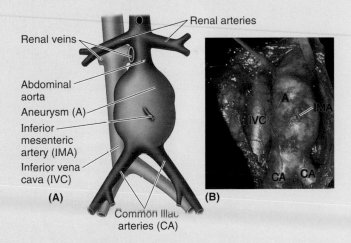

Renal arteries

Renal veins

Abdominal aorta

Aneurysm (A)

Inferior mesenteric artery (IMA)

Inferior vena cava (IVC)

Common iliac arteries (CA)

(A)

(B)

FIGURE B5.14. **Abdominal aortic aneurysm.**

MEDICAL IMAGING

Abdomen

Examples of some of the modalities used in medical imaging of the abdomen follow. Radiographs of the abdomen demonstrate normal and abnormal anatomical relationships, such as those resulting from tumors. Computed tomography (CT) scans (Fig. 5.69), ultrasound (Fig. 5.70), and magnetic resonance imaging (MRI) studies (Fig. 5.71) are also used to examine the abdominal viscera. MRI studies provide better differentiation than CT scans between soft tissues.

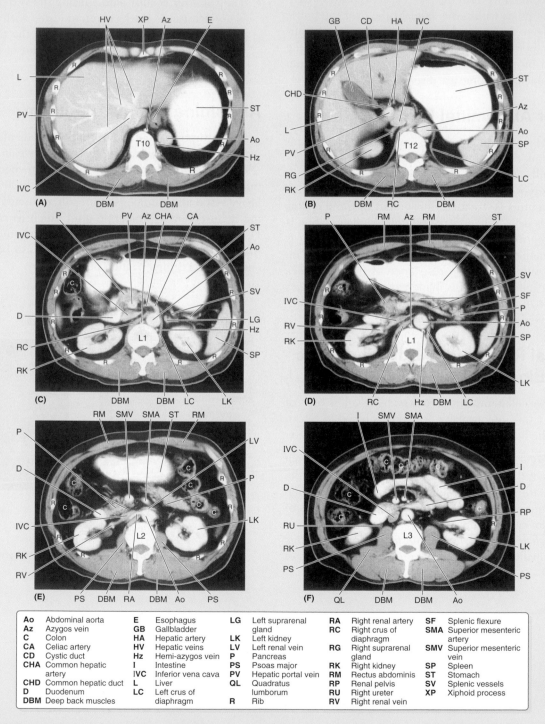

Ao	Abdominal aorta	**E**	Esophagus	**LG**	Left suprarenal gland	**RA**	Right renal artery	**SF**	Splenic flexure
Az	Azygos vein	**GB**	Gallbladder			**RC**	Right crus of diaphragm	**SMA**	Superior mesenteric artery
C	Colon	**HA**	Hepatic artery	**LK**	Left kidney				
CA	Celiac artery	**HV**	Hepatic veins	**LV**	Left renal vein	**RG**	Right suprarenal gland	**SMV**	Superior mesenteric vein
CD	Cystic duct	**Hz**	Hemi-azygos vein	**P**	Pancreas				
CHA	Common hepatic artery	**I**	Intestine	**PS**	Psoas major	**RK**	Right kidney	**SP**	Spleen
		IVC	Inferior vena cava	**PV**	Hepatic portal vein	**RM**	Rectus abdominis	**ST**	Stomach
CHD	Common hepatic duct	**L**	Liver	**QL**	Quadratus lumborum	**RP**	Renal pelvis	**SV**	Splenic vessels
D	Duodenum	**LC**	Left crus of diaphragm			**RU**	Right ureter	**XP**	Xiphoid process
DBM	Deep back muscles			**R**	Rib	**RV**	Right renal vein		

FIGURE 5.69. Transverse (axial) CT images of the abdomen. Progressively inferior levels (**A–F.**) show body wall, viscera, and blood vessels.

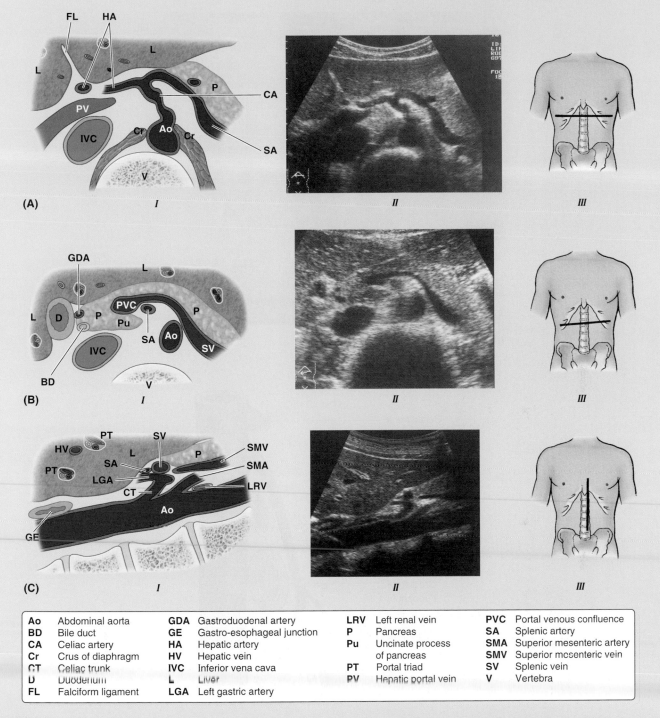

Ao	Abdominal aorta	**GDA**	Gastroduodenal artery	**LRV**	Left renal vein	**PVC**	Portal venous confluence
BD	Bile duct	**GE**	Gastro-esophageal junction	**P**	Pancreas	**SA**	Splenic artery
CA	Celiac artery	**HA**	Hepatic artery	**Pu**	Uncinate process	**SMA**	Superior mesenteric artery
Cr	Crus of diaphragm	**HV**	Hepatic vein		of pancreas	**SMV**	Superior mesenteric vein
CT	Celiac trunk	**IVC**	Inferior vena cava	**PT**	Portal triad	**SV**	Splenic vein
D	Duodenum	**L**	Liver	**PV**	Hepatic portal vein	**V**	Vertebra
FL	Falciform ligament	**LGA**	Left gastric artery				

FIGURE 5.70. Ultrasound scans of abdomen. A. Transverse scan through the celiac trunk. **B.** Transverse scan through the pancreas. **C.** Transverse scan through the aorta.

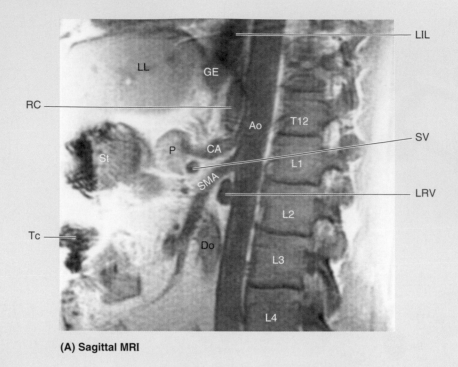

(A) Sagittal MRI

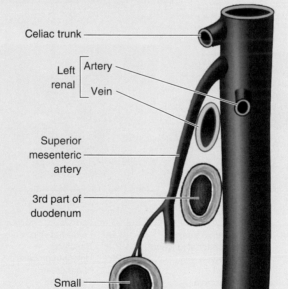

(B) Lateral view (from left)

Key for A					
Ao	Aorta	LL	Left lobe of liver	St	Stomach
CA	Celiac artery	LRV	Left renal vein	SV	Splenic vein
Do	Duodenum	P	Pancreas	Tc	Transverse colon
GE	Gastro-esophageal junction	RC	Right crus		
LIL	Inferior lobe of left lung	SMA	Superior mesenteric artery		

FIGURE 5.71. MRIs of abdomen. A. Sagittal MRI through aorta. **B.** Schematic illustration of relationships of superior mesenteric artery. *(continued)*

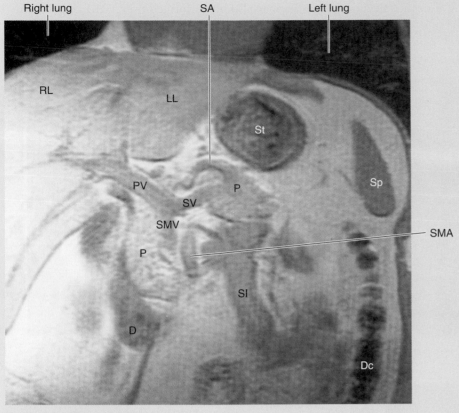

Right lung SA Left lung

RL
LL
St
PV
P
Sp
SV
SMV
SMA
P
SI
D
Dc

(C) Coronal MRI

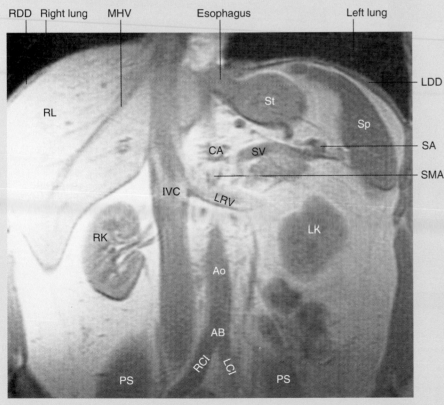

RDD Right lung MHV Esophagus Left lung

RL
St
LDD
Sp
CA SV
SA
IVC
SMA
LRV
RK
LK
Ao
AB
RCI LCI
PS PS

(D) Coronal MRI

Key for C and D	
Ao	Aorta
AB	Aortic bifurcation
CA	Celiac artery
D	Duodenum
Dc	Descending colon
IVC	Inferior vena cava
LCI	Left common iliac artery
LDD	Left dome of diaphragm
LK	Left kidney
LL	Left lobe of liver
LRV	Left renal vein
MHV	Middle hepatic vein
P	Pancreas
PS	Psoas
RCI	Right common iliac artery
RDD	Right dome of diaphragm
RK	Right kidney
PV	Hepatic portal vein
RR	Right lobe of liver
SA	Splenic artery
SI	Small intestine
SMA	Superior mesenteric artery
SMV	Superior mesenteric vein
Sp	Spleen
St	Stomach
SV	Splenic vein

FIGURE 5.71. MRIs of abdomen. *(continued)* **C.** Coronal MRI through portal vein. **D.** Coronal MRI through inferior vena cava.

Abdominal arteriography, radiography after the injection of radiopaque material directly into the bloodstream, detects abnormalities of the abdominal arteries (Fig. 5.72B). *Vessel studies* may also be performed using MRI (Fig. 5.72A). To examine the colon, a barium enema is given after the bowel is cleared of fecal material by a cleansing enema (Fig. 5.72C,D).

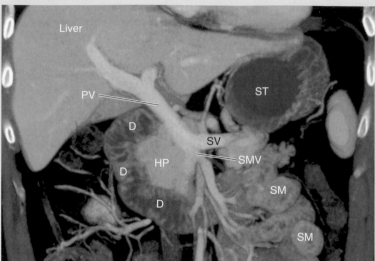

(A) 3-D CT reconstruction, anterior view

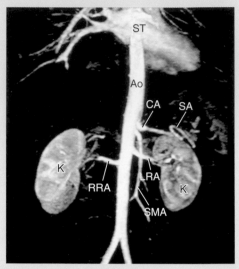

(B) MRI angiogram, anterior view

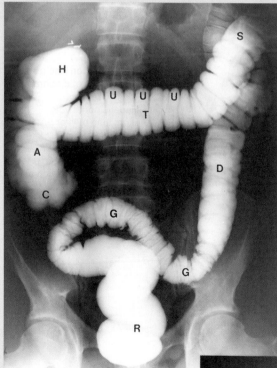

(C) Postero-anterior radiograph

(D) Anterior view

FIGURE 5.72. Other abdominal imaging. A. 3-D reconstruction of abdominal CT scan. *D,* duodenum; *HP,* head of pancreas; *PV,* portal vein; *SM,* small intestine; *SMV,* superior mesenteric vein; *ST,* stomach; *SV,* splenic vein. **B.** Magnetic resonance angiogram (MRA). *Ao,* aorta; *CA,* celiac trunk; *K,* kidney; *LRA,* left renal artery; *RRA,* right renal artery; *SA,* splenic artery; *SMA,* superior mesenteric artery; *ST,* stomach. **C.** Single-contrast radiograph of colon after a barium enema. *Letters* are identified in **D**. **D.** Overview of characteristics of the large intestine.

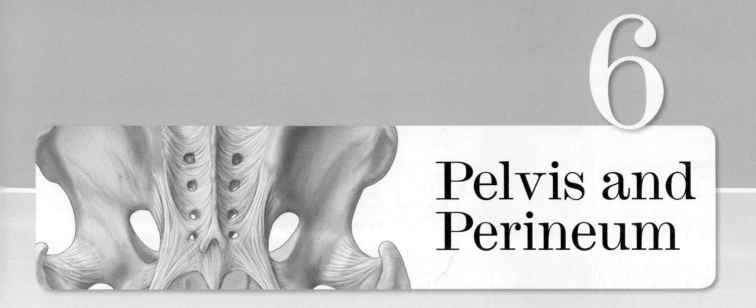

6

Pelvis and Perineum

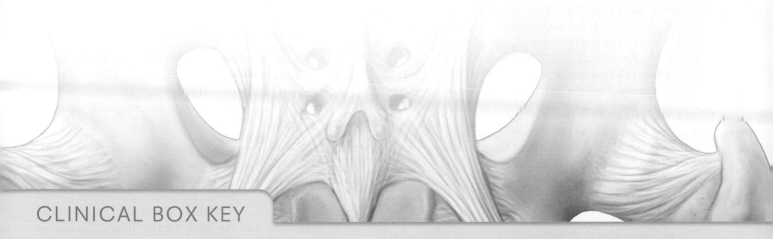

CLINICAL BOX KEY

Anatomical
Variations

Diagnostic
Procedures

Life Cycle

Surgical
Procedures

Trauma

Pathology

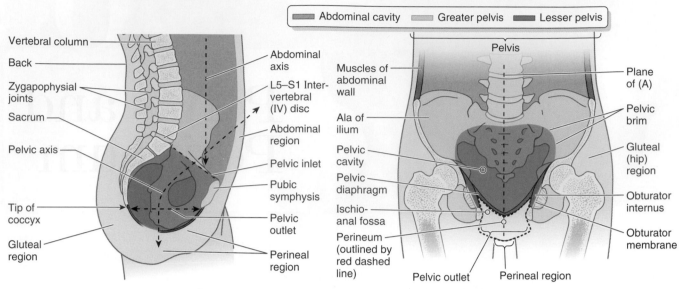

(A) Medial view of left half of bisected lower trunk

(B) Anterior view of posterior half of coronally sectioned lower trunk

FIGURE 6.1. Thoracic and abdominopelvic cavity. A and B. The pelvis is the space within the pelvic girdle, overlapped externally by the abdominal and gluteal (lower limb) regions and the perineum. Thus, the pelvis has no unique external surface area.

The **pelvis** (L. *basin*) is the part of the trunk inferoposterior to the abdomen and is the area of transition between the trunk and the lower limbs (Fig. 6.1). The **pelvic cavity** is a continuation of the abdominal cavity into the pelvis through the *pelvic inlet*. The *perineal region* refers to the area of the trunk between the thighs and the buttocks, extending from the pubis to the coccyx. The *perineum* is a shallow compartment lying deep to this area and inferior to the pelvic diaphragm.

PELVIS

The superior boundary of the *pelvic cavity* is the *pelvic inlet*, the superior pelvic aperture (Figs. 6.1 and 6.2). The pelvis is limited inferiorly by the *pelvic outlet*, which is bounded anteriorly by the *pubic symphysis* (L. *symphysis pubis*) and posteriorly by the *coccyx*.

The **pelvic inlet** (*superior pelvic aperture*) is bounded by the **linea terminalis** of the pelvis, which is formed by the

- superior margin of the pubic symphysis anteriorly
- posterior border of the pubic crest
- pecten pubis, the continuation of the superior ramus of the pubis, which forms a sharp ridge
- **arcuate line** of the ilium
- anterior border of the ala (L. *wing*) of the sacrum
- **sacral promontory**

The **pelvic outlet** (*inferior pelvic aperture*) is bounded by the

- inferior margin of the pubic symphysis anteriorly
- inferior rami of the pubis and ischial tuberosities anterolaterally

- sacrotuberous ligaments posterolaterally (Fig. 6.3B)
- tip of the coccyx posteriorly

Pelvic Girdle

The **pelvic girdle** is a basin-shaped ring of bones that surrounds the pelvic cavity and connects the vertebral column to the two femurs in the thighs. The main functions of the strong pelvic girdle are to (1) transfer the weight of the upper body from the axial to the lower appendicular skeleton for standing and walking, (2) to withstand compression and other forces resulting from its support of body weight, and (3) house and protect the pelvic viscera (including the gravid uterus). In mature individuals, the pelvic girdle is formed by the three bones of the bony pelvis (Fig. 6.2 and Table 6.1):

- Right and left **hip bones**: Two large, irregularly shaped bones, each of which forms at puberty by fusion of three bones—*ilium, ischium,* and *pubis*
- **Sacrum**: Formed by the fusion of five, originally separate, sacral vertebrae

The hip bones are joined at the *pubic symphysis* anteriorly and to the sacrum posteriorly at the **sacro-iliac joints** to form a bony ring, the *pelvic girdle*.

The **ilium** is the superior, flattened, fan-shaped part of the hip bone (Fig. 6.2). The **ala of the ilium** represents the spread of the fan, and the **body of the ilium**, the handle of the fan. The body of the ilium forms the superior part of the **acetabulum**, the cup-shaped depression on the external surface of the hip bone with which the head of the femur articulates. The **iliac crest**, the rim of the ilium, has a curve that follows the contour of the ala between the **anterior** and

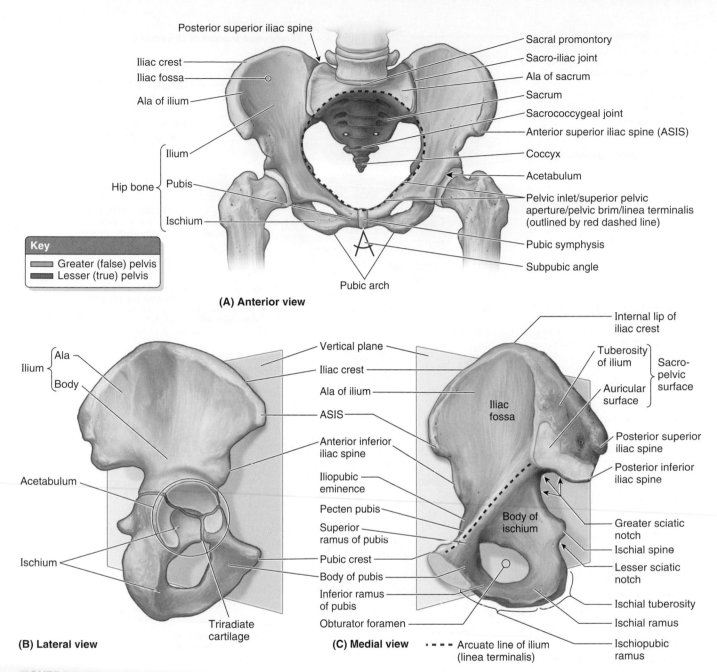

FIGURE 6.2. Bony pelvis. A. ~~Right ilium and pubis.~~ **B** Child's right hip bone. **C.** Adult's right hip bone. In the anatomical position, the anterior superior iliac spine and the anterior aspect of the pubis lie in the same vertical plane.

the **posterior superior iliac spines**. The anterior concave part of the ala forms the **iliac fossa**.

The **ischium** has a body and a ramus (L. *branch*). The **body of the ischium** forms the posterior part of the acetabulum, and the **ramus** forms the posterior part of the inferior boundary of the **obturator foramen**. The large postero-inferior protuberance of the ischium is the **ischial tuberosity** (Fig. 6.2). The small, pointed posterior projection near the junction of the ramus and body is the **ischial spine**.

The **pubis** is an angulated bone that has the **superior pubic ramus**, which forms the anterior part of the

acetabulum, and the **inferior pubic ramus**, which forms the anterior part of the inferior boundary of the *obturator foramen*. The superior pubic ramus has an oblique ridge, the **pecten pubis** (pectineal line of pubis), on its superior aspect. A thickening on the anterior part of the **body of the pubis** is the **pubic crest**, which ends laterally as a swelling—the **pubic tubercle** (Fig. 6.3A).

The **pubic arch** is formed by the **ischiopubic rami** (conjoined inferior rami of the pubis and ischium) of the two sides. These rami meet at the *pubic symphysis*, and their inferior borders define the **subpubic angle** (the distance

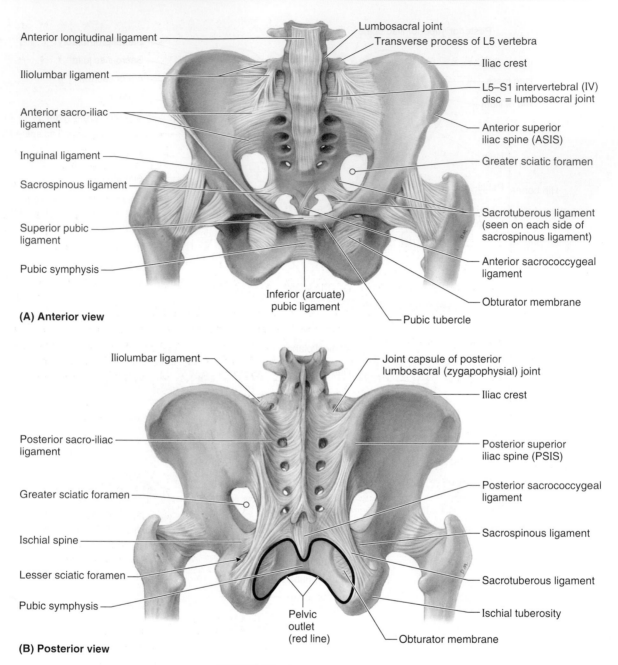

FIGURE 6.3. Ligaments of pelvic girdle.

between the right and the left ischial tuberosities), which can be approximated by the angle between the abducted middle and index fingers for the male, and the angle between the index finger and extended thumb for the female (Fig. 6.4).

The bony pelvis is divided into *greater* (*false*) and *lesser* (*true*) *pelves* by the oblique plane of the *pelvic inlet* (superior pelvic aperture) (Figs. 6.1 and 6.2).

The **greater pelvis** (L. *pelvis major*) is

- superior to the pelvic inlet
- bounded by the abdominal wall anteriorly, the ala of ilium laterally, and the L5 and S1 vertebrae posteriorly

- the location of some abdominal viscera, such as the sigmoid colon and some loops of ileum

The **lesser pelvis** (L. *pelvis minor*) is

- between the *pelvic inlet* and the *pelvic outlet* (Fig. 6.3B)
- the location of the pelvic viscera—urinary bladder and reproductive organs, such as the uterus and ovaries
- bounded by the pelvic surfaces of the hip bones, sacrum, and coccyx
- limited inferiorly by the musculomembranous pelvic diaphragm (levator ani) (Table 6.2 and Fig. 6.1B)

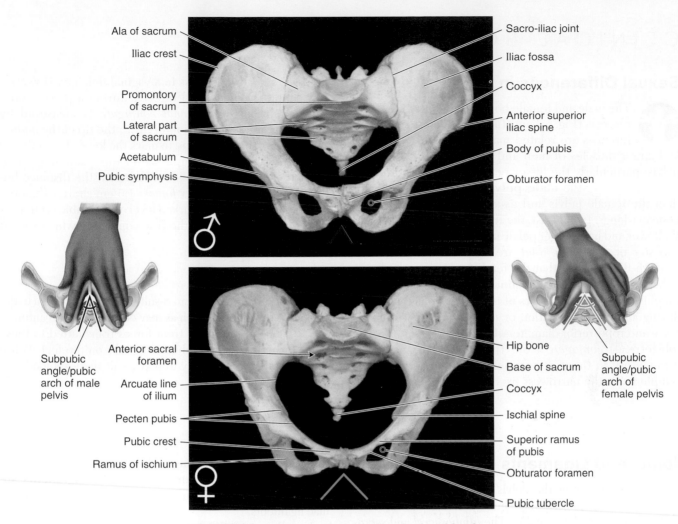

FIGURE 6.4. Comparison of pelvic girdles of male and female.

TABLE 6.1. COMPARISON OF MALE AND FEMALE BONY PELVES

Bony Pelvis	Male (♂)	Female (♀)
General structure	Thick and heavy	Thin and light
Greater pelvis (pelvis major)	Deep	Shallow
Lesser pelvis (pelvis minor)	Narrow and deep	Wide and shallow
Pelvic inlet (superior pelvic aperture)	Heart-shaped	Oval or rounded
Pelvic outlet (inferior pelvic aperture)	Comparatively small	Comparatively large
Pubic arch and subpubic angle (degree)	Narrow (<70 degrees)	Wide (>80 degrees)
Obturator foramen	Round	Oval
Acetabulum	Large	Small

CLINICAL BOX

Sexual Differences in Bony Pelves

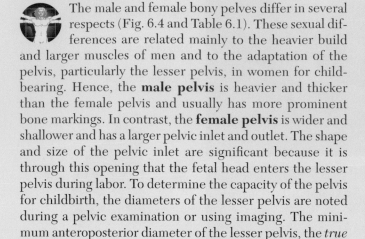

The male and female bony pelves differ in several respects (Fig. 6.4 and Table 6.1). These sexual differences are related mainly to the heavier build and larger muscles of men and to the adaptation of the pelvis, particularly the lesser pelvis, in women for child-bearing. Hence, the **male pelvis** is heavier and thicker than the female pelvis and usually has more prominent bone markings. In contrast, the **female pelvis** is wider and shallower and has a larger pelvic inlet and outlet. The shape and size of the pelvic inlet are significant because it is through this opening that the fetal head enters the lesser pelvis during labor. To determine the capacity of the pelvis for childbirth, the diameters of the lesser pelvis are noted during a pelvic examination or using imaging. The minimum anteroposterior diameter of the lesser pelvis, the *true* (obstetrical) *conjugate* from the middle of the sacral promontory to the posterosuperior margin of the pubic symphysis, is the narrowest fixed distance through which

the baby's head must pass in a vaginal delivery. However, this cannot be measured directly during a pelvic exam. Consequently, the *diagonal conjugate* is measured by palpating the sacral promontory with the tip of the *middle finger*, using the other hand to mark the level of the inferior margin of the pubic symphysis on the examining hand. After the examining hand is withdrawn, the distance between the tip of the *index finger* (1.5 cm shorter than the middle finger) and the marked level of the pubic symphysis is measured to estimate the true conjugate, which should be 11 cm or greater.

Pelvic Fractures

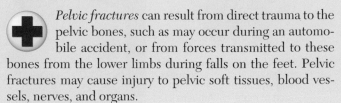

Pelvic fractures can result from direct trauma to the pelvic bones, such as may occur during an automobile accident, or from forces transmitted to these bones from the lower limbs during falls on the feet. Pelvic fractures may cause injury to pelvic soft tissues, blood vessels, nerves, and organs.

Joints and Ligaments of Pelvic Girdle

The primary joints of the pelvis are the *sacro-iliac joints* and the *pubic symphysis*, which link the skeleton of the trunk and the lower limb (Fig. 6.2A). The *lumbosacral* and *sacro-coccygeal* joints are directly related to the pelvic girdle. Strong ligaments support and strengthen these joints (Fig. 6.3).

SACRO-ILIAC JOINTS

The **sacro-iliac joints** are strong, weight-bearing, compound joints consisting of an anterior synovial joint (between the ear-shaped *auricular surfaces* of the sacrum and ilium covered with articular cartilage) and a posterior syndesmosis (between the tuberosities of the same bones) (Figs. 6.2C and 6.5). The articular (auricular) surfaces of the synovial joint have irregular but congruent elevations and depressions that interlock. The sacro-iliac joints differ from most synovial joints in that they have limited mobility, a consequence of their role in transmitting the weight of most of the body to the hip bones.

The sacrum is suspended between the iliac bones and is firmly attached to them by posterior and interosseous sacro-iliac ligaments. The thin **anterior sacro-iliac ligaments** form the anterior part of the fibrous capsule of the synovial joint. The **interosseous sacro-iliac ligaments** occupy an area of about 10 cm² each and are the primary structures involved in transferring the weight of the upper body from the axial skeleton to the two ilia and then to

the femurs during standing and to the ischial tuberosities during sitting. The **posterior sacro-iliac ligaments** are posterior external continuations of the interosseous sacro-iliac ligaments.

Usually, movement is limited to slight gliding and rotary movements, except when subject to considerable force such as occurs after a high jump (or during late pregnancy—see next Clinical Box). Then, the weight of the body is transmitted through the sacrum anterior to the rotation axis, tending to push the superior sacrum inferiorly, thereby causing the inferior sacrum to rotate superiorly. This tendency is resisted by the strong **sacrotuberous** and **sacrospinous ligaments** (Fig. 6.3). These ligaments allow only limited upward movement of the inferior end of the sacrum, thus providing resilience to the sacro-iliac region when the vertebral column sustains sudden weight increases (Fig. 6.5C).

PUBIC SYMPHYSIS

The **pubic symphysis** is a secondary cartilaginous joint that is formed by the union of the bodies of the pubic bones in the median plane (Figs. 6.3 and 6.5D). The fibrocartilaginous **interpubic disc** is generally wider in women than in men. The ligaments joining the pubic bones are thickened superiorly and inferiorly to form the **superior pubic ligament** and the **inferior** (arcuate) **pubic ligament**, respectively. The decussating fibers of tendinous attachments of the rectus abdominis and external oblique muscles also strengthen the pubic symphysis anteriorly.

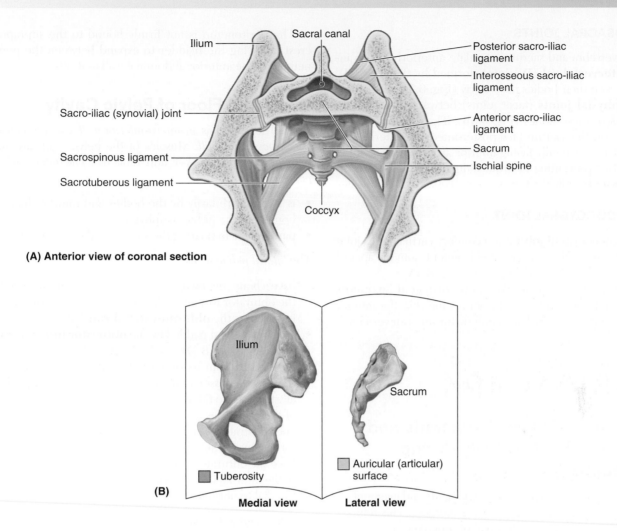

(A) Anterior view of coronal section

Sacral canal

Ilium

Posterior sacro-iliac ligament

Interosseous sacro-iliac ligament

Sacro-iliac (synovial) joint

Anterior sacro-iliac ligament

Sacrum

Sacrospinous ligament

Ischial spine

Sacrotuberous ligament

Coccyx

Ilium

Sacrum

Tuberosity

Auricular (articular) surface

(B)

Medial view   **Lateral view**

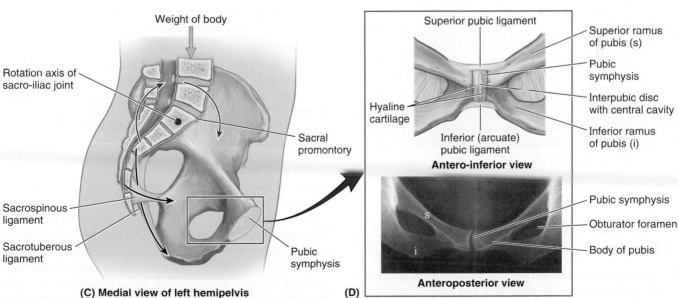

Weight of body

Rotation axis of sacro-iliac joint

Sacral promontory

Sacrospinous ligament

Sacrotuberous ligament

Pubic symphysis

(C) Medial view of left hemipelvis

Superior pubic ligament

Superior ramus of pubis (s)

Pubic symphysis

Interpubic disc with central cavity

Hyaline cartilage

Inferior (arcuate) pubic ligament

Inferior ramus of pubis (i)

Antero-inferior view

Pubic symphysis

Obturator foramen

Body of pubis

Anteroposterior view

(D)

FIGURE 6.5. Sacro-iliac joints and ligaments. A. Posterior half of coronally sectioned pelvis. **B.** Articular surfaces of sacro-iliac joint. **C.** Role of sacrotuberous and sacrospinous ligaments in resisting anterior rotation of pelvis. **D.** Pubic symphysis.

LUMBOSACRAL JOINTS

The L5 vertebra and sacrum articulate anteriorly at the anterior **intervertebral (IV) joint**, formed by the L5–S1 IV disc between their bodies posteriorly (Fig. 6.3A) and at two **zygapophysial joints** (facet joints) between the articular processes of these bones (Fig. 6.3B). The superior articular facets on the sacrum face posteromedially, interlocking with the anterolaterally facing inferior articular facets of the L5 vertebra, preventing L5 from sliding anteriorly. **Iliolumbar ligaments** unite the transverse processes of L5 to the ilia.

SACROCOCCYGEAL JOINT

The **sacrococcygeal joint** is a secondary cartilaginous joint with an IV disc. Fibrocartilage and ligaments join the apex of the sacrum to the base of the coccyx (Fig. 6.3A).

The **anterior** and **posterior sacrococcygeal ligaments** are long strands that reinforce the joint, much like the anterior and posterior longitudinal ligaments do for superior vertebrae.

CLINICAL BOX

Relaxation of Pelvic Ligaments and Increased Joint Mobility during Pregnancy

During pregnancy, the pelvic joints and ligaments relax, and pelvic movements increase. This relaxation during the latter half of pregnancy is caused by the increase in levels of the sex hormones and the presence of the hormone *relaxin*. The sacro-iliac interlocking mechanism is less effective because the relaxation permits greater rotation of the pelvis and contributes to the lordotic posture often assumed during pregnancy with the change in the center of gravity. Relaxation of the sacro-iliac joints and pubic symphysis permits as much as a 10–15% increase in diameters (mostly transverse), facilitating passage of the fetus through the pelvic canal. The coccyx is also allowed to move posteriorly.

Peritoneum and Peritoneal Cavity of Pelvis

The **peritoneum** lining the abdominal cavity continues into the pelvic cavity, reflecting onto the superior aspects of most pelvic viscera (Fig. 6.6 and Table 6.2). Only the uterine tubes—except for their ostia, which are open—are intraperitoneal and suspended by a mesentery. The ovaries, although suspended in the peritoneal cavity by a mesentery, are not covered with peritoneum. The peritoneum creates a number of folds and fossae as it reflects onto most of the pelvic viscera.

The peritoneum is not firmly bound to the suprapubic crest, allowing the bladder to expand between the peritoneum and the anterior abdominal wall as it fills.

Walls and Floor of Pelvic Cavity

The pelvic cavity has an *antero-inferior wall*, two *lateral walls*, and a *posterior wall*. Muscles of the pelvic walls are summarized in Figure 6.7A–E and Table 6.3. The *antero-inferior pelvic wall*

- is formed primarily by the bodies and rami of the pubic bones and the pubic symphysis
- participates in bearing the weight of the urinary bladder

The *lateral pelvic walls*

- have a bony framework formed by the hip bones, including the obturator foramen (Fig. 6.2C); the obturator foramen is closed by the **obturator membrane** (Fig. 6.3).
- are covered and padded by the **obturator internus muscles** (Fig. 6.7A,B,D). Each obturator internus converges posteriorly from its origin within the lesser pelvis, exits through the lesser sciatic foramen, and turns sharply laterally to attach to the femur (Fig. 6.7D). The medial surfaces of these muscles are covered by **obturator fascia**, thickened centrally as a tendinous arch that provides attachment for the levator ani (pelvic diaphragm) (Fig. 6.7B,E).
- have the obturator nerves and vessels and other branches of the internal iliac vessels located on their medial aspects (medial to obturator internus muscles)

The *posterior pelvic wall*

- consists of a bony wall and roof in the midline (formed by the sacrum and coccyx) and musculoligamentous posterolateral walls (formed by the sacro-iliac joints and their associated ligaments and piriformis muscles). Each **piriformis muscle** leaves the lesser pelvis through the *greater sciatic foramen* to attach to the femur (Fig. 6.7A).
- is the site of the nerves forming the **sacral plexus**; the piriformis muscles form a "muscular bed" for this nerve network (Fig. 6.7D,E).

The **pelvic floor** is formed by the bowl- or funnel-shaped **pelvic diaphragm**, which consists of the *levator ani* and *coccygeus* muscles and the fascias (L. *fasciae*) covering the superior and inferior aspects of these muscles (Fig. 6.7A). The **coccygeus muscles** extend from the ischial spines to the pubic bones anteriorly, to the ischial spines posteriorly, and to a thickening in the obturator fascia (**tendinous arch of levator ani**) on each side (Fig. 6.7A,C,E). The **levator ani** consists of three parts, each named according to the attachment of its fibers (Fig. 6.7A,C,E and Table 6.3). The parts of the levator ani are as follows:

- The **puborectalis**, consisting of the thicker, narrower, medial part of the levator ani, which is continuous between the posterior aspects of the right and left pubic bones.

(Continued on page 350)

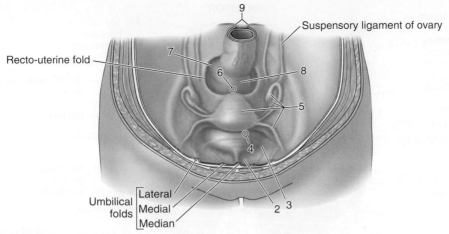

Suspensory ligament of ovary

Recto-uterine fold

Umbilical folds
Lateral
Medial
Median

(A) Anterior view of female

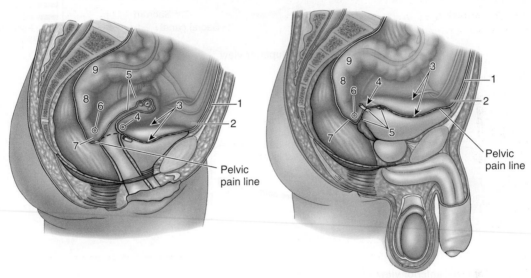

Pelvic pain line

(B) Right lateral view of female

(C) Right lateral view of male

Pelvic pain line

FIGURE 6.6. Pelvic peritoneum.

TABLE 6.2. PERITONEAL REFLECTIONS IN PELVIS

Female (Parts A and B)[a]	Male (Part C)[a]
1 Descends anterior abdominal wall (loose attachment allows insertion of bladder as it fills)	1 Descends anterior abdominal wall (loose attachment allows insertion of bladder as it fills)
2 Reflects onto superior surface of bladder, creating **supravesical fossa**	2 Reflects onto superior surface of bladder, creating **supravesical fossa**
3 Covers convex superior surface of bladder; slopes down sides of bladder to ascend lateral wall of pelvis, creating **paravesical fossae** on each side	3 Covers convex superior surface (roof) of bladder, sloping down sides of roof to ascend lateral wall of pelvis, creating **paravesical fossae** on each side
4 Reflects from bladder to body of uterus, forming **vesico-uterine pouch**	4 Descends posterior surface of bladder as much as 2 cm
5 Covers body and fundus of uterus, posterior fornix of vagina; extends laterally from uterus as double fold of mesentery, the **broad ligament** that engulfs uterine tubes, and round ligaments of uterus, and suspends ovaries	5 Laterally, forms fold over ureters (**ureteric fold**), ductus deferentes, and superior ends of seminal glands
6 Reflects from vagina onto rectum, forming **recto-uterine pouch**[b] (pouch of Douglas)	6 Reflects from bladder and seminal glands onto rectum, forming **rectovesical pouch**[b]
7 Recto-uterine pouch extends laterally and posteriorly to form **pararectal fossae** on each side of rectum.	7 Rectovesical pouch extends laterally and posteriorly to form **pararectal fossae** on each side of rectum.
8 Ascends rectum; from inferior to superior, rectum is subperitoneal and then retroperitoneal.	8 Ascends rectum; from inferior to superior, rectum is subperitoneal and then retroperitoneal.
9 Engulfs sigmoid colon beginning at rectosigmoid junction	9 Engulfs sigmoid colon beginning at rectosigmoid junction

[a]Numbers refer to Figure 6.6.
[b]Low point of peritoneal cavity in erect position.

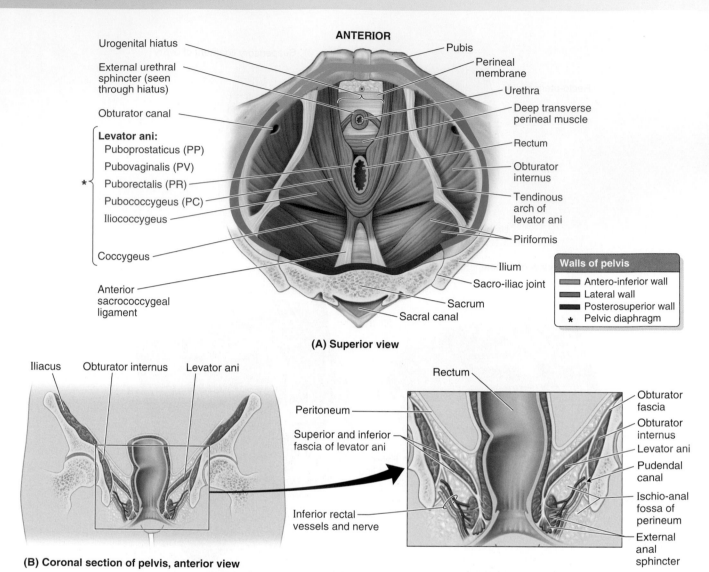

ANTERIOR

Urogenital hiatus

External urethral sphincter (seen through hiatus)

Obturator canal

Levator ani:
  Puboprostaticus (PP)
  Pubovaginalis (PV)
  Puborectalis (PR)
  Pubococcygeus (PC)
  Iliococcygeus

Coccygeus

Anterior sacrococcygeal ligament

Pubis

Perineal membrane

Urethra

Deep transverse perineal muscle

Rectum

Obturator internus

Tendinous arch of levator ani

Piriformis

Ilium

Sacro-iliac joint

Sacrum

Sacral canal

Walls of pelvis
  Antero-inferior wall
  Lateral wall
  Posterosuperior wall
  * Pelvic diaphragm

(A) Superior view

Iliacus   Obturator internus   Levator ani

Peritoneum

Superior and inferior fascia of levator ani

Inferior rectal vessels and nerve

Rectum

Obturator fascia

Obturator internus

Levator ani

Pudendal canal

Ischio-anal fossa of perineum

External anal sphincter

(B) Coronal section of pelvis, anterior view

FIGURE 6.7. **Muscles of pelvic walls and floor. A.** Superior surface of the pelvic diaphragm. **B.** Coronal section of pelvis through levator ani and rectum. *(continued)*

TABLE 6.3. MUSCLES OF PELVIC WALLS AND FLOOR

Muscle	Proximal Attachment	Distal Attachment	Innervation	Main Action
Levator ani (pubococcygeus and iliococcygeus)	Body of pubis, tendinous arch of levator ani, ischial spine	Perineal body, coccyx, anococcygeal ligament, walls of prostate or vagina, rectum, anal canal	Nerve to levator ani (branches of S4), inferior anal (rectal) nerve, coccygeal plexus	Helps support pelvic viscera; resists increases in intra-abdominal pressure
Coccygeus (ischiococcygeus)	Ischial spine	Inferior end of sacrum and coccyx	Branches of S4 and S5 nerves	Forms small part of pelvic diaphragm that supports pelvic viscera; flexes coccyx

(continued)

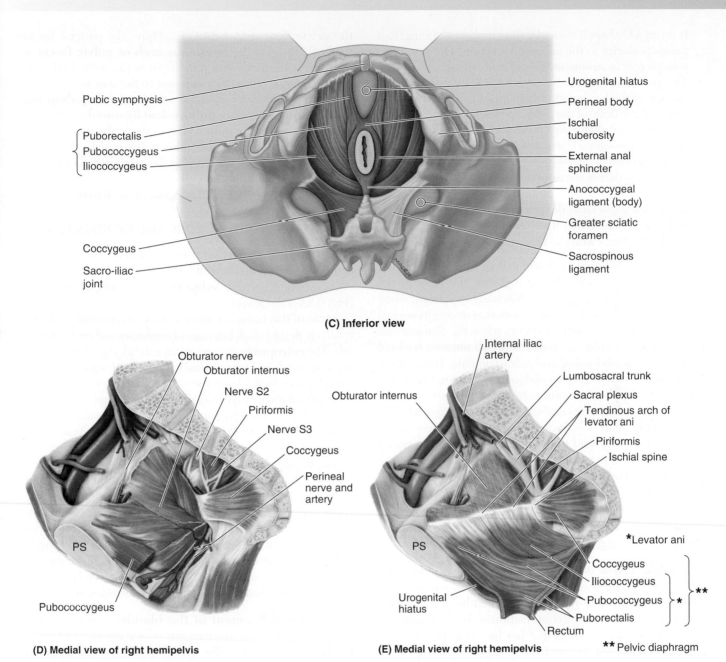

(C) Inferior view

(D) Medial view of right hemipelvis

(E) Medial view of right hemipelvis

FIGURE 6.7 Muscles of pelvic walls and floor. *(continued)* C. Inferior surface of pelvic diaphragm. D. Muscles of lesser pelvis. E. Levator ani added to **D.** *PS,* pubic symphysis.

TABLE 6.3. MUSCLES OF PELVIC WALLS AND FLOOR (continued)

Muscle	Proximal Attachment	Distal Attachment	Innervation	Main Action
Obturator internus	Pelvic surface of ilium and ischium; obturator membrane	Greater trochanter of femur	Nerve to obturator internus (L5, S1, S2)	Laterally rotates hip joint; assists in holding head of femur in acetabulum
Piriformis	Pelvic surface of 2nd–4th sacral segments; superior margin of greater sciatic notch and sacrotuberous ligament		Anterior rami of S1 and S2	Laterally rotates hip joint; abducts hip joint; assists in holding head of femur in acetabulum

It forms a U-shaped muscular sling (puborectal sling) that passes posterior to the anorectal junction. This part plays a major role in maintaining fecal continence.

- The **pubococcygeus**, the wider but thinner intermediate part of the levator ani, which arises from the posterior aspect of the body of the pubis and the anterior part of the tendinous arch and passes posteriorly in a nearly horizontal plane. The lateral fibers attach posteriorly to the coccyx, and the medial fibers merge with those of the contralateral side to form part of the **anococcygeal body** or **ligament**.
- The **iliococcygeus**, the posterolateral part of the levator ani, which arises from the posterior part of the tendinous arch and ischial spine; it is thin and often poorly developed and blends with the anococcygeal body posteriorly.

The levator ani forms a dynamic floor for supporting the abdominopelvic viscera. Acting together, the parts of the levator ani raise the pelvic floor, following its descent when relaxed to allow defecation and urination, restoring its normal position. Further contraction occurs when the thoracic diaphragm and anterolateral abdominal wall muscles contract to compress the abdominal and pelvic contents. Therefore, it can resist the increased intra-abdominal pressure that would otherwise force the abdominopelvic contents (gas, solid and liquid wastes, and the viscera) through the pelvic outlet. This action occurs reflexively during forced expiration, coughing, sneezing, vomiting, and fixation of the trunk during strong movements of the upper limbs, as occurs when lifting a heavy object. The levator ani also has important functions in the voluntary control of urination, fecal continence (via the puborectalis), and support of the uterus.

Pelvic Fascia

The **pelvic fascia** is connective tissue that occupies the space between the membranous peritoneum and the muscular pelvic walls and floor not occupied by pelvic organs (Fig. 6.8). This "layer" is a continuation of the comparatively thin endo-abdominal fascia that lies between the muscular abdominal walls and the peritoneum superiorly.

MEMBRANOUS PELVIC FASCIA: PARIETAL AND VISCERAL

The **parietal pelvic fascia** is a membranous layer of variable thickness that lines the internal (deep or pelvic) aspect of the muscles forming the walls and floor of the pelvis. The parietal pelvic fascia covers the pelvic surfaces of the obturator internus, piriformis, coccygeus, levator ani, and part of the urethral sphincter muscles (Fig. 6.8A–D). The name given to the fascia is derived from the muscle it encloses (e.g., obturator fascia). This layer is continuous superiorly with the transversalis and iliopsoas fascias.

The **visceral pelvic fascia** includes the membranous fascia that directly ensheathes the pelvic organs, forming the adventitial layer of each. The membranous parietal and visceral layers become continuous where the organs penetrate the pelvic floor (Fig. 6.8A,C,E). Here, the parietal fascia thickens, forming the **tendinous arch of pelvic fascia**, a continuous bilateral band running from the pubis to the sacrum along the pelvic floor adjacent to the viscera.

The most anterior part of this tendinous arch (**puboprostatic ligament** in males; **pubovesical ligament** in females) connects the prostate to the pubis in the male or the fundus (base) of the bladder to the pubis in the female. The most posterior part of the band runs as the sacrogenital ligaments from the sacrum around the side of the rectum to attach to the prostate in the male or the vagina in the female.

ENDOPELVIC FASCIA: LOOSE AND CONDENSED

The abundant connective tissue remaining between and continuous with the parietal and visceral membranous layers is extraperitoneal or **subperitoneal endopelvic fascia** (Fig. 6.8A–D).

Some of this fascia is extremely *loose areolar (fatty) tissue*, relatively devoid of all but minor lymphatics and nutrient vessels. The **retropubic** (or *prevesical*, extended posterolaterally as *paravesical*) and **retrorectal** (or *presacral*) **spaces** are *potential spaces* in the loose fatty tissue that accommodate the expansion of the urinary bladder and rectal ampulla as they fill (Fig. 6.8B,D). Other parts of the endopelvic fascia have a fibrous consistency, the *ligamentous fascia*. These parts are often described as "fascial condensations" or pelvic "ligaments."

The **hypogastric sheath** is a thick band of condensed pelvic fascia that gives passage to essentially all the vessels and nerves passing from the lateral wall of the pelvis to the pelvic viscera, along with the ureters and, in the male, the ductus deferens. As it extends medially from the lateral wall, the hypogastric sheath divides into three laminae ("leaflets" or "wings") that pass to or between the pelvic organs, conveying neurovascular structures and providing support. The three laminae of the hypogastric sheath, from anterior to posterior, are

- The **lateral ligament of the bladder**, passing to the bladder, conveying the superior vesical arteries and veins
- The middle lamina in the male, forming the **rectovesical septum** between the posterior surface of the bladder and the prostate anteriorly and the rectum posteriorly (Fig. 6.8D). In the female, the middle lamina is substantial and passes medially to the uterine cervix and vagina as the **transverse cervical** (*cardinal*) **ligament**, also known clinically as the *lateral cervical* or *Mackenrodt ligament* (Fig. 6.8B,E). In its most superior portion, at the base of the broad ligament, the uterine artery runs transversely toward the cervix, whereas the ureters course immediately inferior to them as they pass on each side of the cervix toward the bladder.
- The most posterior lamina passes to the rectum, conveying the middle rectal artery and vein (Fig. 6.8B,D).

The transverse cervical ligament, and the way in which the uterus normally "rests" on top of the bladder, provides the main passive support for the uterus. The bladder, in turn,

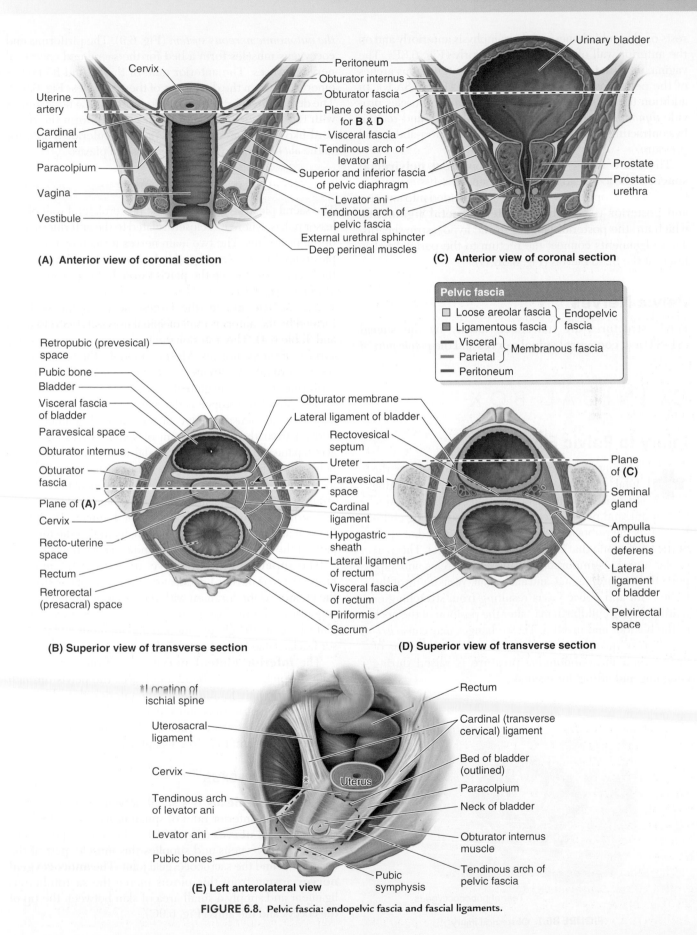

Cervix

Uterine artery

Cardinal ligament

Paracolpium

Vagina

Vestibule

Peritoneum
Obturator internus
Obturator fascia
Plane of section for **B & D**
Visceral fascia
Tendinous arch of levator ani
Superior and inferior fascia of pelvic diaphragm
Levator ani
Tendinous arch of pelvic fascia
External urethral sphincter
Deep perineal muscles

(A) Anterior view of coronal section

Urinary bladder

Prostate
Prostatic urethra

(C) Anterior view of coronal section

Pelvic fascia

☐ Loose areolar fascia ⎫ Endopelvic
■ Ligamentous fascia ⎭ fascia
▬ Visceral ⎫ Membranous fascia
▬ Parietal ⎭
▬ Peritoneum

Retropubic (prevesical) space
Pubic bone
Bladder
Visceral fascia of bladder
Paravesical space
Obturator internus
Obturator fascia
Plane of **(A)**
Cervix
Recto-uterine space
Rectum
Retrorectal (presacral) space

Obturator membrane
Lateral ligament of bladder
Rectovesical septum
Ureter
Paravesical space
Cardinal ligament
Hypogastric sheath
Lateral ligament of rectum
Visceral fascia of rectum
Piriformis
Sacrum

(B) Superior view of transverse section

Plane of **(C)**
Seminal gland
Ampulla of ductus deferens
Lateral ligament of bladder
Pelvirectal space

(D) Superior view of transverse section

*Location of ischial spine
Uterosacral ligament
Cervix
Tendinous arch of levator ani
Levator ani
Pubic bones

Rectum
Cardinal (transverse cervical) ligament
Bed of bladder (outlined)
Paracolpium
Neck of bladder
Obturator internus muscle
Tendinous arch of pelvic fascia

Uterus

Pubic symphysis

(E) Left anterolateral view

FIGURE 6.8. Pelvic fascia: endopelvic fascia and fascial ligaments.

rests on the pubic bones and the symphysis anteriorly and on the anterior wall of the vagina posteriorly (Fig. 6.8E). The vagina, in turn, is suspended between the tendinous arches of the pelvic fascia by the **paracolpium** (Fig. 6.8A,E). In addition to this *passive support*, the perineal muscles provide *dynamic support* for the uterus, bladder, and rectum by contracting during moments of increased intra-abdominal pressure.

There are surgically important potential **pelvirectal spaces** in the loose extraperitoneal connective tissue superior to the pelvic diaphragm. The spaces are divided into anterior and posterior regions by the **lateral rectal ligaments**, which are the posterior laminae of the hypogastric sheaths. These ligaments connect the rectum to the parietal pelvic fascia at the S2–S4 levels (Fig. 6.8B,D).

Pelvic Nerves

Pelvic structures are innervated mainly by the **sacral (S1–S4)** and **coccygeal spinal nerves** and the *pelvic part of the autonomic nervous system* (Fig. 6.9). The piriformis and coccygeus muscles form a bed for the sacral and coccygeal nerve plexuses. The anterior rami of the S2 and S3 nerves emerge between the digitations of these muscles (Fig. 6.9C). The descending part of the anterior ramus of L4 nerve unites with the anterior ramus of the L5 nerve to form the thick, cord-like **lumbosacral trunk**. It passes inferiorly, anterior to the ala of the sacrum, to join the sacral plexus.

SACRAL PLEXUS

The sacral plexus is located on the posterolateral wall of the lesser pelvis, where it is closely related to the anterior surface of the piriformis. The two main nerves formed by the sacral plexus are the *sciatic* and *pudendal nerves*. Most branches of the sacral plexus leave the pelvis through the *greater sciatic foramen* (Fig. 6.9).

The **sciatic nerve**, the largest nerve in the body, is formed by the anterior rami of spinal nerves L4–S3 (Fig. 6.9 and Table 6.4). The anterior rami converge on the anterior surface of the piriformis. Most commonly, the sciatic nerve passes through the *greater sciatic foramen* inferior to the piriformis to enter the gluteal region.

The **pudendal nerve** is the main nerve of the perineum and the chief sensory nerve of the external genitalia. It is derived from the anterior rami of spinal nerves S2–S4. It accompanies the internal pudendal artery and leaves the pelvis through the greater sciatic foramen between the piriformis and the coccygeus muscles. The pudendal nerve hooks around the ischial spine and sacrospinous ligament and enters the perineum through the lesser sciatic foramen. It supplies the skin and muscles of the perineum, including the terminal parts of the reproductive, urinary, and digestive tracts.

The **superior gluteal nerve** arises from the anterior rami of spinal nerves L4–S1 and leaves the pelvis through the greater sciatic foramen with the superior gluteal vessels, superior to the piriformis. It supplies three muscles in the gluteal region: the gluteus medius and minimus and the tensor fasciae latae (see Chapter 7, Lower Limb).

The **inferior gluteal nerve** arises from the anterior rami of spinal nerves L5–S2 and leaves the pelvis through the greater sciatic foramen with the inferior gluteal vessels, inferior to the piriformis and superficial to the sciatic nerve. It breaks up into several branches that supply the overlying gluteus maximus muscle (see Chapter 7).

COCCYGEAL PLEXUS

The **coccygeal plexus** is a small network of nerve fibers formed by the anterior rami of spinal nerves S4 and S5 and the **coccygeal nerves** (Fig. 6.9B). It lies on the pelvic surface of the coccygeus and supplies this muscle, part of the levator ani, and the sacrococcygeal joint. The **anococcygeal nerves** arising from this plexus pierce the sacrotuberous ligament and supply a small area of skin between the tip of the coccyx and the anus (Fig. 6.9C).

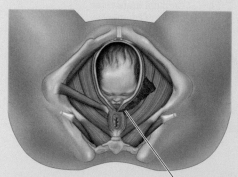

OBTURATOR NERVE

Although it passes through the pelvis, the **obturator nerve** is not a "pelvic nerve" but is rather the primary nerve to the medial thigh. It arises from the lumbar plexus (anterior rami of spinal nerves L2–L4) in the abdomen (greater pelvis) and enters the lesser pelvis (Fig. 6.9C). It runs in the extraperitoneal fat along the lateral wall of the pelvis to the obturator canal, the opening in the obturator membrane, where it exits the pelvis and enters the medial thigh.

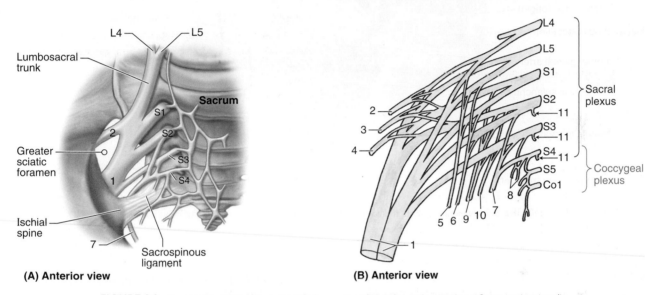

(A) Anterior view

(B) Anterior view

FIGURE 6.9. **Nerves of sacral and coccygeal plexus. A and B.** Schematic overview of nerves. *(continued)*

TABLE 6.4. NERVES OF SACRAL AND COCCYGEAL PLEXUSES

Nerve[a]	Segmental Origin (Anterior Rami)	Distribution
1 Sciatic	L4, L5, S1, S2, S3	Articular branches to hip joint and muscular branches to flexors of knee (hamstring muscles) and all muscles in leg and foot
2 Superior gluteal	L4, L5, S1	Gluteus medius, gluteus minimus, and tensor fasciae latae muscles
3 Inferior gluteal	L5, S1, S2	Gluteus maximus muscle
4 Nerve to piriformis	S1, S2	Piriformis muscle
5 Nerve to quadratus femoris and inferior gemellus	L4, L5, S1	Quadratus femoris and inferior gemellus muscles
6 Nerve to obturator internus and superior gemellus	L5, S1, S2	Obturator internus and superior gemellus muscles
7 Pudendal	S2, S3, S4	Structures in perineum: sensory to genitalia, muscular branches to perineal muscles, sphincter urethrae, and external anal sphincter
8 Nerves to levator ani and coccygeus	S3, S4	Levator ani and coccygeus muscles
9 Posterior femoral cutaneous	S2, S3	Cutaneous branches to buttocks and uppermost medial and posterior surfaces of thigh
10 Perforating cutaneous	S2, S3	Cutaneous branches to medial part of buttocks
11 Pelvic splanchnic	S2, S3, S4	Pelvic viscera via inferior hypogastric and pelvic plexus

[a]Numbers refer to Figure 6.9.

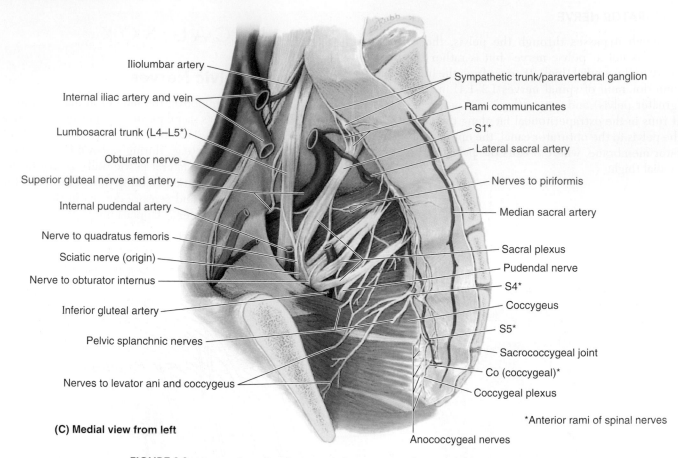

Iliolumbar artery

Internal iliac artery and vein

Lumbosacral trunk (L4–L5*)

Obturator nerve

Superior gluteal nerve and artery

Internal pudendal artery

Nerve to quadratus femoris

Sciatic nerve (origin)

Nerve to obturator internus

Inferior gluteal artery

Pelvic splanchnic nerves

Nerves to levator ani and coccygeus

Sympathetic trunk/paravertebral ganglion

Rami communicantes

S1*

Lateral sacral artery

Nerves to piriformis

Median sacral artery

Sacral plexus

Pudendal nerve

S4*

Coccygeus

S5*

Sacrococcygeal joint

Co (coccygeal)*

Coccygeal plexus

*Anterior rami of spinal nerves

Anococcygeal nerves

(C) Medial view from left

FIGURE 6.9. Nerves of sacral and coccygeal plexus. *(continued)* **C.** Dissection of nerve plexuses.

PELVIC AUTONOMIC NERVES

Autonomic innervation of the pelvic cavity is via four routes: the *sacral sympathetic trunks, hypogastric plexuses, pelvic splanchnic nerves,* and *peri-arterial plexuses.*

The **sacral sympathetic trunks** are the inferior continuations of the lumbar sympathetic trunks (Fig. 6.10). Each sacral trunk usually has four sympathetic ganglia. The sacral trunks descend on the pelvic surface of the sacrum just medial to the pelvic sacral foramina and commonly converge to form the small median **ganglion impar** anterior to the coccyx (Fig. 6.10). The sympathetic trunks descend posterior to the rectum in the extraperitoneal connective tissue and send communicating branches, gray rami communicantes, to each of the anterior rami of the sacral and coccygeal nerves. They also send branches to the median sacral artery and the inferior hypogastric plexus. The primary function of the sacral sympathetic trunks is to provide postsynaptic fibers to the sacral plexus for sympathetic innervation of the lower limb.

The **hypogastric plexuses** (superior and inferior) are networks of sympathetic and visceral afferent nerve fibers. The main part of the **superior hypogastric plexus** lies just inferior to the bifurcation of the aorta and descends

into the pelvis. This plexus is the inferior prolongation of the **intermesenteric plexus** (see Chapter 5, Abdomen), which also receives the L3 and L4 splanchnic nerves. The superior hypogastric plexus enters the pelvis, dividing into **left** and **right hypogastric nerves**, which descend anterior to the sacrum. These nerves descend lateral to the rectum within the *hypogastric sheaths* and then spread as they merge with pelvic splanchnic nerves (parasympathetic) to form the **right** and **left inferior hypogastric plexuses**. Subplexuses of the inferior hypogastric plexuses, **pelvic plexuses**, in both sexes pass to the lateral surfaces of the rectum and to the inferolateral surfaces of the urinary bladder and in males to the prostate and seminal glands (vesicles) and in females to the cervix of the uterus and lateral parts of the fornix of the vagina.

The **pelvic splanchnic nerves** contain presynaptic parasympathetic and visceral afferent fibers derived from the S2–S4 spinal cord segments and visceral afferent fibers from cell bodies in the spinal ganglia of the corresponding spinal nerves (Figs. 6.9B,C and 6.10 and Table 6.4). The pelvic splanchnic nerves merge with the hypogastric nerves to form the inferior hypogastric (and pelvic) plexuses.

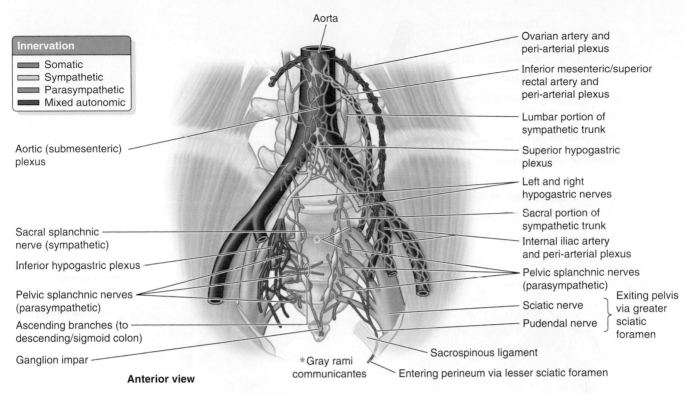

FIGURE 6.10. Autonomic nerves of pelvis.

The **hypogastric/pelvic system of plexuses**, receiving sympathetic fibers via the lumbar splanchnic nerves and parasympathetic fibers via the pelvic splanchnic nerves, innervates the pelvic viscera. The **sympathetic fibers** produce vasomotion, inhibits peristaltic contraction of the rectum, and stimulates contraction of the genital organs during orgasm (producing ejaculation in the male). The **parasympathetic fibers** stimulate contraction of the rectum and bladder for defecation and urination, respectively. Parasympathetic fibers in the prostatic plexus penetrate the pelvic floor to supply the erectile bodies of the external genitalia, producing erection.

The **peri-arterial plexuses** of the superior rectal, ovarian, and internal iliac arteries provide postsynaptic, sympathetic, vasomotor fibers to each of the arteries and its derivative branches.

VISCERAL AFFERENT INNERVATION IN PELVIS

Visceral afferent fibers travel with the autonomic nerve fibers, although the sensory impulses are conducted centrally retrograde to the efferent impulses. In the pelvis, visceral afferent fibers conducting *reflexive sensation* (information that does not reach consciousness) travel with parasympathetic fibers to the spinal sensory ganglia of S2–S4. The route taken by visceral afferent fibers conducting *pain sensation* differs in relationship to an imaginary line, the **pelvic pain line**, that corresponds to the inferior limit of peritoneum

(Fig. 6.6B,C), except in the case of the large intestine, where the pain line occurs midway along the length of the sigmoid colon. Visceral afferent fibers that transmit pain sensations from the viscera *inferior to the pelvic pain line* (structures that do not contact the peritoneum and the distal sigmoid colon and rectum) also travel with parasympathetic fibers to the spinal ganglia of S2–S4. However, visceral afferent fibers conducting pain from the viscera *superior to the pelvic pain line* (structures in contact with the peritoneum, except for the distal sigmoid colon and rectum) follow the sympathetic fibers retrogradely to inferior thoracic and superior lumbar spinal ganglia.

Pelvic Arteries and Veins

Four main arteries enter the lesser pelvis in females, three in males (Fig. 6.11A,D):

- The paired **internal iliac arteries** deliver the most blood to the lesser pelvis. They bifurcate into an *anterior division* and a *posterior division*, providing the visceral branches and parietal branches, respectively.
- The paired *ovarian arteries* (females)
- The **median sacral artery**
- The *superior rectal artery*

The origin, course, and distribution of these arteries and their branches are summarized in Table 6.5.

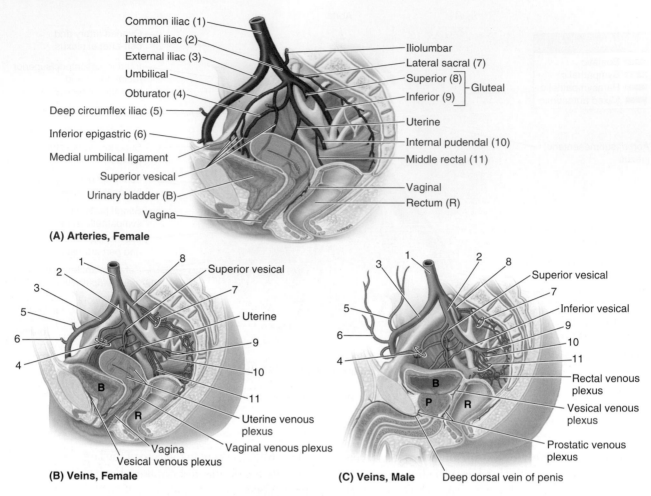

(A) Arteries, Female

Common iliac (1)
Internal iliac (2)
External iliac (3)
Umbilical
Obturator (4)
Deep circumflex iliac (5)
Inferior epigastric (6)
Medial umbilical ligament
Superior vesical
Urinary bladder (B)
Vagina

Iliolumbar
Lateral sacral (7)
Superior (8)
Inferior (9) — Gluteal
Uterine
Internal pudendal (10)
Middle rectal (11)
Vaginal
Rectum (R)

(B) Veins, Female

Superior vesical
Uterine
Uterine venous plexus
Vaginal venous plexus
Vagina
Vesical venous plexus

(C) Veins, Male

Superior vesical
Inferior vesical
Rectal venous plexus
Vesical venous plexus
Prostatic venous plexus
Deep dorsal vein of penis

Veins share names with arteries shown in A and D.

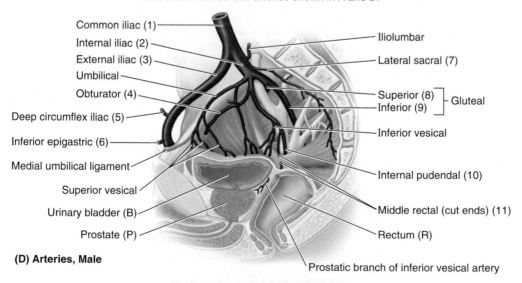

(D) Arteries, Male

Common iliac (1)
Internal iliac (2)
External iliac (3)
Umbilical
Obturator (4)
Deep circumflex iliac (5)
Inferior epigastric (6)
Medial umbilical ligament
Superior vesical
Urinary bladder (B)
Prostate (P)

Iliolumbar
Lateral sacral (7)
Superior (8)
Inferior (9) — Gluteal
Inferior vesical
Internal pudendal (10)
Middle rectal (cut ends) (11)
Rectum (R)
Prostatic branch of inferior vesical artery

Median views of right hemipelves

FIGURE 6.11. Arteries and veins of pelvis.

TABLE 6.5. ARTERIES OF LESSER PELVIS[a]

Artery	Origin	Course	Distribution
Internal iliac (2)	Common iliac artery	Passes over pelvic brim to reach pelvic cavity	Main blood supply to pelvic organs, gluteal muscles, and perineum
Anterior division of internal iliac artery	Internal iliac artery	Passes anteriorly and divides into visceral branches and obturator artery	Pelvic viscera and muscles in medial compartment of thigh
Umbilical	Anterior division of internal iliac artery	Short pelvic course; obliterates after origin of superior vesical artery	Via superior vesical artery
Obturator (4)		Runs antero-inferiorly on lateral pelvic wall	Pelvic muscles, nutrient artery to ilium, and head of femur
Superior vesical artery	Patent part of umbilical artery	Passes to superior aspect of urinary bladder	Superior aspect of urinary bladder; often ductus deferens in male
Artery to ductus deferens	Superior or inferior vesical artery	Runs subperitoneally to ductus deferens	Ductus deferens
Inferior vesical[b]	Anterior division of internal iliac artery	Passes subperitoneally to inferior aspect of male urinary bladder	Urinary bladder and pelvic part of ureter, seminal gland, and prostate in males
Middle rectal (11)		Descends in pelvis to rectum	Seminal gland, prostate, and rectum
Internal pudendal (10)		Leaves pelvis through greater sciatic foramen and enters perineum (ischio-anal fossa) by passing through lesser sciatic foramen	Main artery to perineum, including muscles of anal canal and perineum; skin and urogenital triangle; erectile bodies
Inferior gluteal[c] (9)		Leaves pelvis through greater sciatic foramen inferior to piriformis	Piriformis, coccygeus, levator ani, and gluteal muscles
Uterine		Runs medially on levator ani; crosses ureter to reach base of broad ligament	Pelvic part of ureter, uterus, ligament of uterus, uterine tube, and vagina
Vaginal	Anterior division of internal iliac artery (uterine artery)	At junction of body and cervix of uterus, it descends to vagina	Vagina and branches to inferior part of urinary bladder
Gonadal (testicular and ovarian)	Abdominal aorta	Descends retroperitoneally; testicular artery passes into deep inguinal ring; ovarian artery crosses brim of pelvis and runs medially in suspensory ligament to ovary.	Testis and ovary, respectively
Posterior division of internal iliac artery	Internal iliac artery	Passes posteriorly and gives rise to parietal branches	Pelvic wall and gluteal region
Iliolumbar	Posterior division of internal iliac artery	Ascends anterior to sacro-iliac joint and posterior to common iliac vessels and psoas major	Iliacus, psoas major, quadratus lumborum muscles, and cauda equina in vertebral canal
Lateral sacral (7)		Runs on superficial aspect of piriformis	Piriformis and vertebral canal
Superior gluteal (8)		Leaves pelvis through greater sciatic foramen, superior to piriformis	Gluteal muscles and tensor fasciae latae

[a]Numbers in parentheses refer to Fig. 6.11A & D.
[b]Often arises from uterine artery in females.
[c]Often arises from posterior division of internal iliac artery.

The pelvis is drained by the following:

- Mainly, the **internal iliac veins** and their tributaries
- Superior rectal veins (see portal venous system, Chapter 5, Abdomen)
- Median sacral vein
- Ovarian veins (females)
- Internal vertebral venous plexus (see Chapter 2, Back)

Pelvic venous plexuses are formed by the interjoining of veins in the pelvis (Fig. 6.11B,C). The various plexuses (rectal, vesical, prostatic, uterine, and vaginal) unite and drain mainly into the internal iliac vein, but some drain through the superior rectal vein into the inferior mesenteric vein or through lateral sacral veins into the internal vertebral venous plexus.

Lymph Nodes of Pelvis

The lymph nodes draining pelvic organs are variable in number, size, and location. They are somewhat arbitrarily divided into four primary groups of nodes named for the blood vessels with which they are associated (Fig. 6.12):

- **External iliac lymph nodes** receive lymph mainly from the inguinal lymph nodes; however, they also receive lymph from pelvic viscera, especially the superior parts of the anterior pelvic organs. Whereas most of the lymphatic drainage from the pelvis tends to parallel routes of venous drainage, the lymphatic drainage to the external iliac nodes does not. These nodes drain into the common iliac nodes.
- **Internal iliac lymph nodes** receive drainage from the inferior pelvic viscera, deep perineum, and gluteal region and drain into the common iliac nodes.
- **Sacral lymph nodes**, in the concavity of the sacrum, receive lymph from postero-inferior pelvic viscera and drain either to internal or to common iliac nodes.
- **Common iliac lymph nodes** receive drainage from the three main groups listed above. These nodes begin a common route for drainage from the pelvis that passes next to the lumbar (caval/aortic) nodes.

A smaller group of lymph nodes, **pararectal nodes**, drain primarily to the inferior mesenteric nodes.

Both primary and minor groups of pelvic nodes are highly interconnected, so that many nodes can be removed without disturbing drainage. This also allows cancer to spread in virtually any direction to any pelvic or abdominal viscus. The drainage pattern is not sufficiently predictable to allow the progress of metastatic cancer from pelvic organs to be reliably staged in a manner comparable to that of breast cancer.

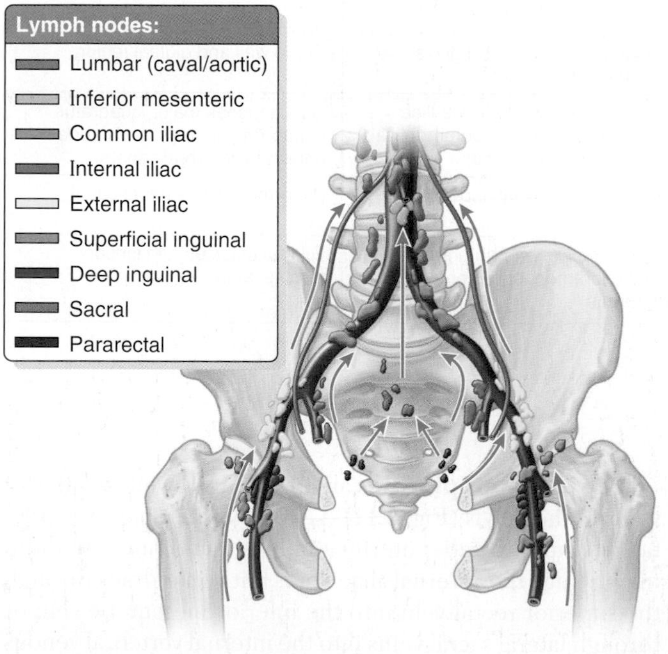

PELVIC VISCERA

The **pelvic viscera** include the caudal parts of the intestinal (rectum) and urinary tracts and the reproductive system (Figs. 6.13–6.15). Although the sigmoid colon and parts of the small bowel extend into the pelvic cavity, they are mobile from their abdominal attachments; therefore, they are abdominal rather than pelvic viscera.

Urinary Organs

The *pelvic urinary organs* include the following (Fig. 6.13):

- *Ureters*, which carry urine from the kidneys
- *Urinary bladder*, which temporarily stores urine
- *Urethra*, which conducts urine from the urinary bladder to the exterior

URETERS

The **ureters** are retroperitoneal muscular tubes that connect the kidneys to the urinary bladder. Urine is transported down the ureters by peristaltic contractions. The ureters run inferiorly from the kidneys, passing over the pelvic brim at the bifurcation of the common iliac arteries (Figs. 6.14 and 6.15). The ureters then run postero-inferiorly on the lateral walls of the pelvis and anterior and parallel to the internal iliac arteries. Opposite the ischial spine, they curve anteromedially, superior to the levator ani, to enter the urinary bladder. The ureters pass inferomedially through the muscular wall of the urinary bladder. This oblique passage through the bladder wall forms a one-way "flap valve";

Lymph nodes:

- Lumbar (caval/aortic)
- Inferior mesenteric
- Common iliac
- Internal iliac
- External iliac
- Superficial inguinal
- Deep inguinal
- Sacral
- Pararectal

FIGURE 6.12. Lymph nodes of pelvis.

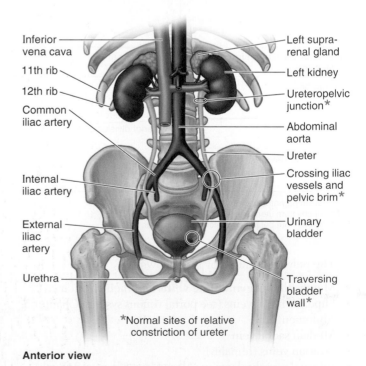

FIGURE 6.13. Urinary organs.

Inferior vena cava
11th rib
12th rib
Common iliac artery
Internal iliac artery
External iliac artery
Urethra

Left suprarenal gland
Left kidney
Ureteropelvic junction*
Abdominal aorta
Ureter
Crossing iliac vessels and pelvic brim*
Urinary bladder
Traversing bladder wall*

*Normal sites of relative constriction of ureter

Anterior view

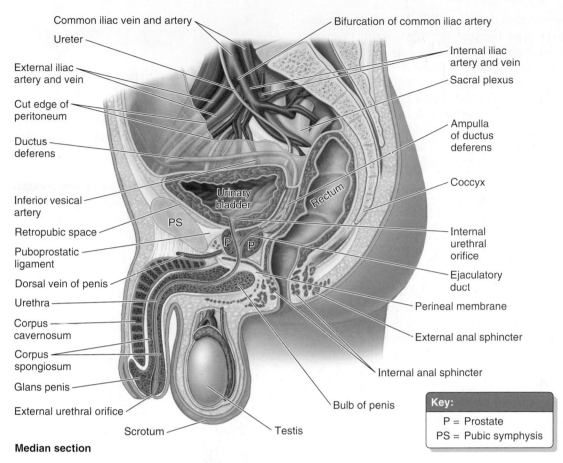

Common iliac vein and artery

Ureter

External iliac
artery and vein

Cut edge of
peritoneum

Ductus
deferens

Inferior vesical
artery

Retropubic space

Puboprostatic
ligament

Dorsal vein of penis

Urethra

Corpus
cavernosum

Corpus
spongiosum

Glans penis

External urethral orifice

Scrotum

Bifurcation of common iliac artery

Internal iliac
artery and vein

Sacral plexus

Ampulla
of ductus
deferens

Coccyx

Internal
urethral
orifice

Ejaculatory
duct

Perineal membrane

External anal sphincter

Internal anal sphincter

Bulb of penis

Testis

Urinary
bladder

Rectum

PS

P   P

Key:
P = Prostate
PS = Pubic symphysis

Median section

FIGURE 6.14. Viscera in hemisected male pelvis. The urinary bladder is distended, as if full.

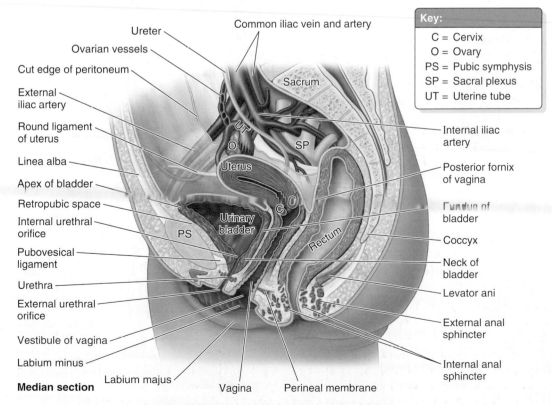

Ureter

Ovarian vessels

Cut edge of peritoneum

External
iliac artery

Round ligament
of uterus

Linea alba

Apex of bladder

Retropubic space

Internal urethral
orifice

Pubovesical
ligament

Urethra

External urethral
orifice

Vestibule of vagina

Labium minus

Labium majus

Common iliac vein and artery

Sacrum

Internal iliac
artery

Posterior fornix
of vagina

Fundus of
bladder

Coccyx

Neck of
bladder

Levator ani

External anal
sphincter

Internal anal
sphincter

Uterus

Urinary
bladder

Rectum

PS

UT

O

SP

C   O

Key:
C = Cervix
O = Ovary
PS = Pubic symphysis
SP = Sacral plexus
UT = Uterine tube

Median section

Vagina   Perineal membrane

FIGURE 6.15.   Viscera in hemisected female pelvis.

the internal pressure of the filling bladder causes the intramural passage to collapse. In males, the only structure that passes between the ureter and the peritoneum is the *ductus deferens*.

The ureter lies posterolateral to the ductus deferens and enters the posterosuperior angle of the bladder (Fig. 6.14; see also Fig. 6.18). In females, the ureter passes medial to the origin of the uterine artery and continues to the level of the ischial spine, where it is crossed superiorly by the uterine artery (Fig. 6.15). The ureter then passes close to the lateral fornix of the vagina and enters the posterosuperior angle of the bladder.

VASCULATURE OF URETERS

Branches of the common and internal iliac arteries supply the pelvic part of the ureters (Fig. 6.16). The most constant arteries supplying this part of the ureters in females are branches of the *uterine arteries*. The sources of similar branches in males are the *inferior vesical arteries*. Veins from the ureters accompany the arteries and have corresponding names. As they course inferiorly, lymph drains sequentially into the lumbar (caval/aortic), common iliac, external iliac, and then internal iliac lymph nodes (Fig. 6.12).

INNERVATION OF URETERS

The *nerves to the ureters* derive from adjacent autonomic plexuses (renal, aortic, superior and inferior hypogastric). The ureters are superior to the pelvic pain line (see Figs. 6.6 and 6.24); therefore, afferent (pain) fibers from the ureters follow sympathetic fibers retrogradely to reach the spinal ganglia and spinal cord segments T11–L1 or L2 (Fig. 6.17).

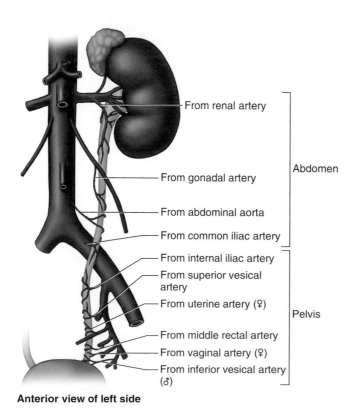

Anterior view of left side

FIGURE 6.16. Blood supply of ureters.

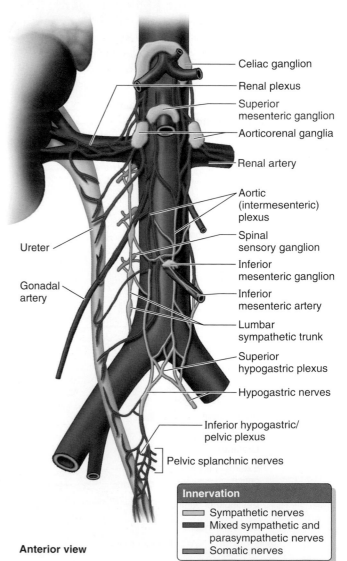

Anterior view

FIGURE 6.17. Innervation of ureters.

URINARY BLADDER

The **urinary bladder**, a hollow viscus (organ) with strong muscular walls, is in the lesser pelvis when empty, its anterior portion directly superior to the pubic bones. It is separated from these bones by the potential *retropubic space* and lies inferior to the peritoneum, where it rests on the pelvic floor (Figs. 6.18–6.20). The bladder is relatively free within the extraperitoneal subcutaneous fatty tissue, except for its neck, which is held firmly by the lateral ligaments of the bladder and the tendinous arch of pelvic fascia, especially the *puboprostatic ligament* in males and the *pubovesical ligament* in females. As the bladder fills, it ascends superiorly into the extraperitoneal fatty tissue of the anterior abdominal wall and enters the greater pelvis. A full bladder may ascend to the level of the umbilicus.

When empty, the bladder is somewhat tetrahedral and externally has an apex, body, fundus, and neck. The four surfaces are a superior surface, two inferolateral surfaces, and a posterior surface (Fig. 6.19). The **apex of the bladder** (anterior end) points toward the superior edge of the pubic symphysis. The **fundus of the bladder (base)** is opposite the apex, formed by the somewhat convex posterior wall. The **body of the bladder** is the part between the apex and the fundus. *In females,* the fundus is closely related to the anterior wall of the vagina; *in males,* it is related to the rectum. The **neck of the bladder** is where the fundus and inferolateral surfaces converge inferiorly.

The **bladder bed** is formed on each side by the pubic bones and the fascia covering the obturator internus and levator ani muscles and posteriorly by the rectum or vagina (Figs. 6.18 and 6.20). The bladder is enveloped by loose connective tissue, the vesical fascia. Only the superior surface is covered by peritoneum.

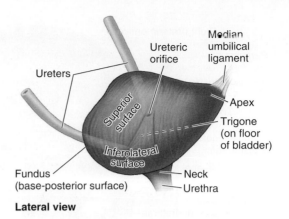

Lateral view

FIGURE 6.19. Surfaces of urinary bladder.

The walls of the bladder are composed chiefly of the **detrusor muscle** (Fig. 6.20A). Toward the neck of the male bladder, its muscle fibers form the involuntary **internal urethral sphincter** (Fig. 6.18). This sphincter contracts during ejaculation to prevent retrograde ejaculation of semen into the bladder. Some fibers run radially and assist in opening the **internal urethral orifice**. In males, the muscle fibers in the neck of the bladder are continuous with the fibromuscular tissue of the prostate, whereas in females, these fibers are continuous with muscle fibers in the wall of the urethra.

The **ureteric orifices** and the internal urethral orifice are at the angles of the **trigone of the bladder** (Fig. 6.20). The ureteric orifices are encircled by loops of detrusor musculature that tighten when the bladder contracts to assist in preventing reflux of urine into the ureters. The **uvula of the bladder** is a slight elevation of the trigone in the internal urethral orifice.

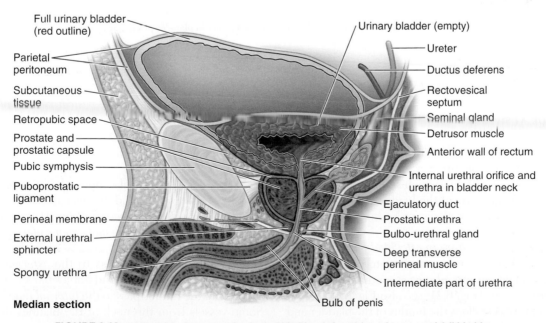

Median section

FIGURE 6.18. Male pelvis demonstrating bed of bladder and position of empty and full bladder.

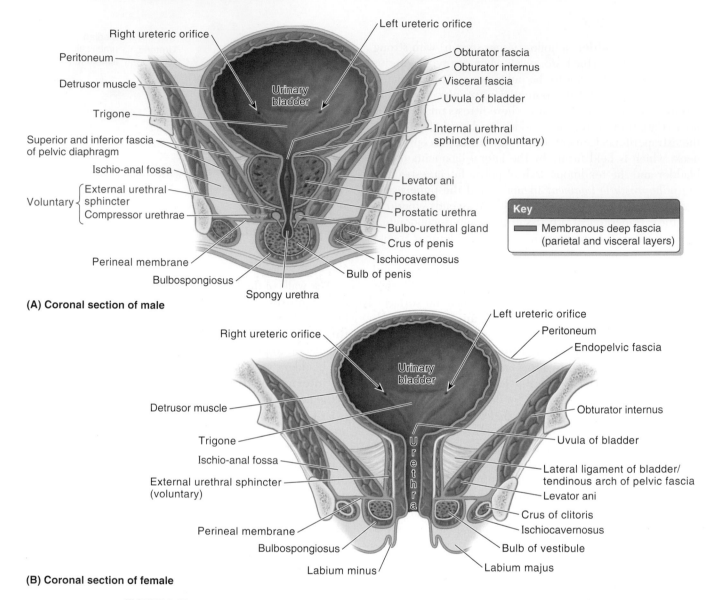

(A) Coronal section of male

(B) Coronal section of female

FIGURE 6.20. Coronal sections of male (A) and female (B) pelves in plane of pelvic portion of urethra.

VASCULATURE OF BLADDER

The main *arteries supplying the bladder* are branches of the *internal iliac arteries* (Fig. 6.11A,D and Table 6.5). The **superior vesical arteries** supply the anterosuperior parts of the bladder. In males, the fundus and neck of the bladder are supplied by the **inferior vesical arteries** (Fig. 6.21). In females, the inferior vesical arteries are replaced by the *vaginal arteries*, which send small branches to the postero-inferior parts of the bladder. The obturator and inferior gluteal arteries also supply small branches to the bladder.

The names of the *veins draining the bladder* correspond to the arteries and are tributaries of the internal iliac veins. In males, the **vesical venous plexus** is continuous with the *prostatic venous plexus* (Fig. 6.21; see also Fig. 6.60C), and the combined plexus envelops the fundus of the bladder and prostate, the seminal glands, the ductus deferentes (plural of

ductus deferens), and the inferior ends of the ureters. The prostatic venous plexus also receives blood from the *deep dorsal vein of the penis*. The *vesical venous plexus* mainly drains through the inferior vesical veins into the internal iliac veins (see Fig. 6.11B,C); however, it may drain through the sacral veins into the *internal vertebral venous plexuses* (see Chapter 2, Back).

In females, the vesical venous plexus envelops the pelvic part of the urethra, and the neck of the bladder receives blood from the *dorsal vein of the clitoris* and communicates with the *vaginal* or *uterovaginal venous plexus* (Fig. 6.11B).

In both sexes, *lymphatic vessels* leave the superior surface of the bladder and pass to the *external iliac lymph nodes* (Figs. 6.22 and 6.23 and Tables 6.6 and 6.7), whereas those from the fundus pass to the *internal iliac lymph nodes*. Some vessels from the neck of the bladder drain into the sacral or common iliac lymph nodes.

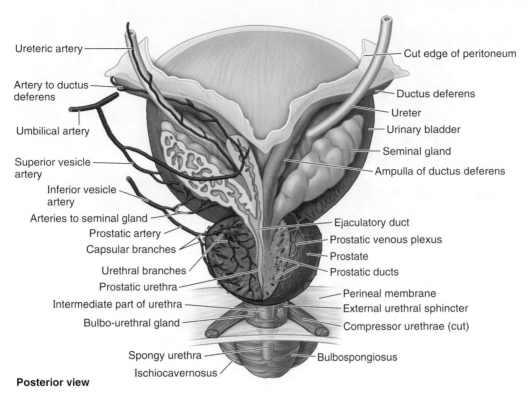

Ureteric artery

Artery to ductus deferens

Umbilical artery

Superior vesicle artery

Inferior vesicle artery

Arteries to seminal gland

Prostatic artery

Capsular branches

Urethral branches

Prostatic urethra

Intermediate part of urethra

Bulbo-urethral gland

Spongy urethra

Ischiocavernosus

Cut edge of peritoneum

Ductus deferens

Ureter

Urinary bladder

Seminal gland

Ampulla of ductus deferens

Ejaculatory duct

Prostatic venous plexus

Prostate

Prostatic ducts

Perineal membrane

External urethral sphincter

Compressor urethrae (cut)

Bulbospongiosus

Posterior view

FIGURE 6.21. Male pelvic genitourinary organs. On the left side, the ampulla of ductus deferens, seminal gland, and prostate have been sectioned to the midline in a coronal plane, and the arterial supply to these structures and the bladder is demonstrated.

CLINICAL BOX

Suprapubic Cystostomy

As the bladder fills, it extends superiorly in the extraperitoneal fatty tissue of the anterior abdominal wall (Fig. 6.18). The bladder then lies adjacent to this wall without the intervention of peritoneum. Consequently, the distended bladder may be punctured (*suprapubic cystostomy*) or approached surgically for the introduction of indwelling catheters or instruments without traversing the peritoneum and entering the peritoneal cavity.

Rupture of Bladder

Because of the superior position of a distended bladder, it may be ruptured by injuries to the inferior part of the anterior abdominal wall or by fractures of the pelvis. The rupture of the superior part of the bladder frequently tears the peritoneum, resulting in passage of urine into the peritoneal cavity. Posterior rupture of the bladder usually results in passage of urine subperitoneally into the perineum.

Cystoscopy

The interior of the bladder and its three orifices can be examined with a *cystoscope*, a lighted tubular endoscope that is inserted through the urethra into the bladder. The cystoscope consists of a light; an observing lens; and various attachments for grasping, removing, cutting, and cauterizing (Fig. B6.2).

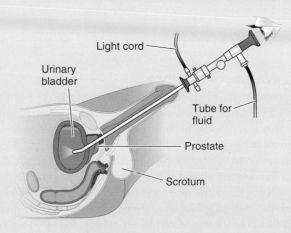

Light cord

Urinary bladder

Tube for fluid

Prostate

Scrotum

FIGURE B6.2. Cystoscopy.

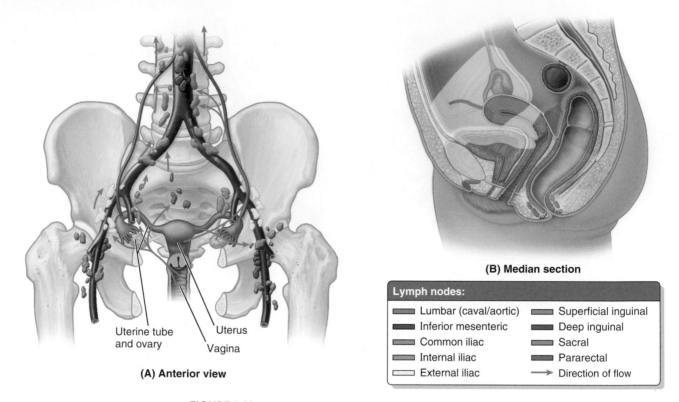

Uterine tube and ovary

Uterus

Vagina

(A) Anterior view

(B) Median section

Lymph nodes:	
Lumbar (caval/aortic)	Superficial inguinal
Inferior mesenteric	Deep inguinal
Common iliac	Sacral
Internal iliac	Pararectal
External iliac	→ Direction of flow

FIGURE 6.22. Lymphatic drainage of female pelvis and perineum.

TABLE 6.6. LYMPHATIC DRAINAGE OF FEMALE PELVIS AND PERINEUM

Lymph Node Group	Typically Drains
Lumbar (along ovarian vessels)	Gonads and associated structures, common iliac nodes (ovary, uterine tube except isthmus and intrauterine parts, fundus of uterus)
Inferior mesenteric	Superiormost rectum, sigmoid colon, descending colon, pararectal nodes
Internal iliac	Inferior pelvic structures, deep perineal structures, sacral nodes (base of bladder, inferior pelvic ureter, anal canal above pectinate line, inferior rectum, middle and upper vagina, cervix, body of uterus)
External iliac	Anterosuperior pelvic structures, deep inguinal nodes (superior bladder, superior pelvic ureter, upper vagina, cervix, lower body of uterus)
Superficial inguinal	Lower limb; superficial drainage of inferolateral quadrant of trunk, including anterior abdominal wall inferior to umbilicus, gluteal region, superficial perineal structures (superolateral uterus near attachment of round ligament, skin of perineum including vulva, ostium of vagina inferior to hymen, prepuce of clitoris, peri-anal skin, anal canal inferior to pectinate line)
Deep inguinal	Glans clitoris, superficial inguinal nodes
Sacral	Postero-inferior pelvic structures, inferior rectum, inferior vagina
Pararectal	Superior rectum

INNERVATION OF BLADDER

Sympathetic fibers to the bladder are conveyed from the T11–L2 or L3 spinal cord levels to the vesical (pelvic) plexuses, primarily through the hypogastric/pelvic plexuses and nerves, whereas parasympathetic fibers from the sacral spinal cord levels are conveyed by the pelvic splanchnic nerves and the inferior hypogastric plexuses (Fig. 6.24). *Parasympathetic fibers* are motor to the detrusor muscle in the bladder wall and inhibitory to the internal sphincter of males. Hence, when the visceral afferent fibers are stimulated by stretching, the detrusor contracts, the internal sphincter relaxes in males, and urine flows into the urethra. Toilet training suppresses this reflex until it is convenient to void. The sympathetic innervation that stimulates ejaculation simultaneously causes contraction of the internal urethral sphincter, preventing reflux of semen into the bladder.

Sensory fibers from the bladder are visceral; reflex afferents and pain afferents (e.g., from overdistention) from the inferior part of the bladder follow the course of the parasympathetic fibers. The superior surface of the bladder is covered with peritoneum and is, therefore, superior to the pain line; thus, pain fibers from the superior part of the bladder follow the sympathetic fibers retrogradely.

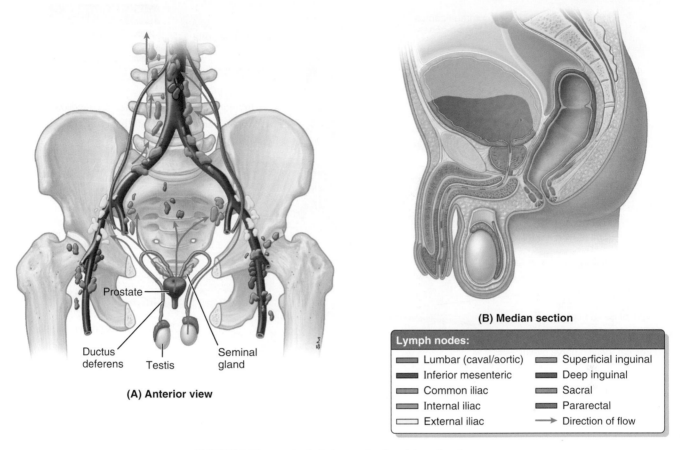

(B) Median section

Lymph nodes:

▬	Lumbar (caval/aortic)	▬	Superficial inguinal
▬	Inferior mesenteric	▬	Deep inguinal
▬	Common iliac	▬	Sacral
▬	Internal iliac	▬	Pararectal
▬	External iliac	→	Direction of flow

(A) Anterior view

Prostate

Ductus deferens   Testis   Seminal gland

FIGURE 6.23. Lymphatic drainage of male pelvis and perineum.

TABLE 6.7. LYMPHATIC DRAINAGE OF MALE PELVIS AND PERINEUM

Lymph Node Group	Typically Drains
Lumbar (near testicular vessels)	Urethra, testis, epididymis
Inferior mesenteric	Superiormost rectum, sigmoid colon, descending colon, pararectal nodes
Internal iliac	External and internal iliac lymph nodes
External iliac	Inferior pelvic structures, deep perineal structures, sacral nodes (prostatic urethra, prostate, base of bladder, inferior pelvic ureter, inferior seminal glands, cavernous bodies, anal canal above pectinate line, inferior rectum)
Superficial inguinal	Lower limb; superficial drainage of inferolateral quadrant of trunk, including anterior abdominal wall inferior to umbilicus, gluteal region, superficial perineal structures (skin of perineum, including skin and prepuce of penis, scrotum, peri-anal skin, anal canal inferior to pectinate line)
Deep inguinal	Glans penis, superficial inguinal nodes, distal spongy urethra
Sacral	Postero-inferior pelvic structures, inferior rectum
Pararectal	Superior rectum

FEMALE URETHRA

The short **female urethra** passes antero-inferiorly from the *internal urethral orifice* of the urinary bladder, posterior, and then inferior to the pubic symphysis to the *external urethral orifice* in the vestibule of the vagina (Fig. 6.20B). The urethra lies anterior to the vagina; its axis is parallel with the vagina. The urethra passes with the vagina through the pelvic diaphragm, external urethral sphincter, and perineal membrane. Urethral glands are present, particularly in its

superior part; the **para-urethral glands** are homologs to the prostate. These glands have a common **para-urethral duct**, which opens (one on each side) near the external urethral orifice. The inferior half of the urethra is in the perineum and is discussed in that section.

VASCULATURE OF FEMALE URETHRA

Blood is supplied by the *internal pudendal* and *vaginal arteries* (see Fig. 6.11A and Table 6.5). The veins follow the

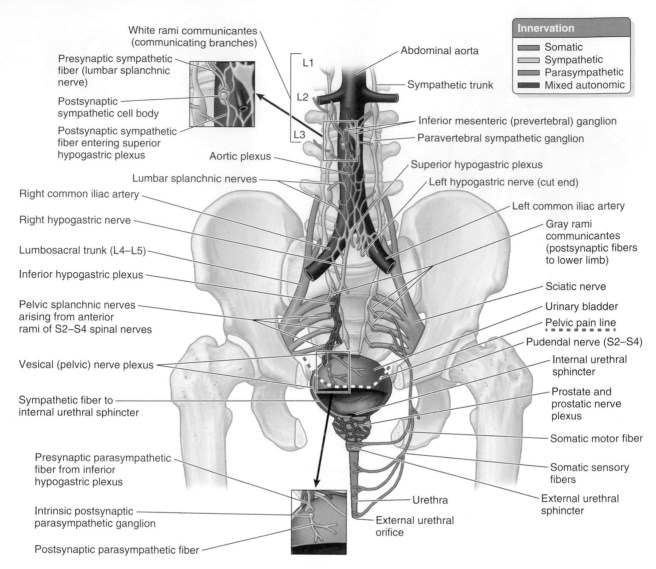

FIGURE 6.24. Innervation of urinary bladder and urethra.

arteries and have similar names. Most lymphatic vessels from the urethra pass to the *sacral* and *internal iliac lymph nodes* (Fig. 6.22 and Table 6.6). A few vessels drain into the *inguinal lymph nodes*.

INNERVATION OF FEMALE URETHRA

The nerves to the urethra arise from the *vesical (nerve) plexus* and the *pudendal nerve* (Fig. 6.24). The pattern is similar to that in the male, given the absence of a prostatic plexus and an internal urethral sphincter. Visceral afferents from most of the urethra run in the *pelvic splanchnic nerves*, but the termination receives somatic afferents from the pudendal nerve.

MALE URETHRA

The **male urethra** is a muscular tube that conveys urine from the *internal urethral orifice* of the urinary bladder to the exterior through the *external urethral orifice* at the tip of the glans penis (Fig. 6.24). The urethra also provides an exit for semen (sperm and glandular secretions). For descriptive purposes, the urethra is divided into four parts: intramural (preprostatic urethra), prostatic, intermediate (membranous), and spongy (penile) (Figs. 6.20A and 6.25; see also Table 6.8).

The *intramural part* is surrounded by an internal urethral sphincter composed of sympathetically innervated smooth muscle (Fig. 6.26). This sphincter prevents semen from entering the bladder during ejaculation (retrograde ejaculation). The prostate surrounds the prostatic urethra. The *intermediate part* is surrounded by the external urethral sphincter, composed of somatically innervated voluntary muscle. The tonic and phasic contraction of this muscle primarily controls urinary continence, but several other muscles may also contribute by compressing the urethra (Fig. 6.26). Stimulation of both sphincters must be inhibited to enable urination.

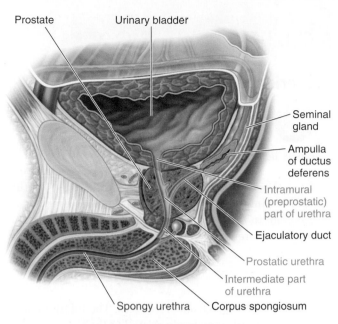

FIGURE 6.25. Parts of male urethra.

VASCULATURE OF MALE URETHRA

The intramural part of the urethra and the prostatic urethra are supplied by the *prostatic branches of the inferior vesical* and *middle rectal arteries* (see Fig. 6.11D and Table 6.5). The intermediate and spongy parts of the urethra are supplied by the *internal pudendal artery*. The veins accompany the arteries and have similar names. The lymphatic vessels drain mainly into the *internal iliac lymph nodes* (Fig. 6.23 and Table 6.7), but some lymph passes to the *external iliac lymph nodes*. Lymphatic vessels from the spongy urethra pass to the *deep inguinal lymph nodes*.

INNERVATION OF MALE URETHRA

The nerves of the male urethra are derived from the *prostatic nerve plexus* (mixed sympathetic, parasympathetic, and visceral afferent fibers) (Fig. 6.24). This plexus is one of the pelvic plexuses (an inferior extension of the vesical plexus) arising as an organ-specific extension of the inferior hypogastric plexus.

TABLE 6.8. PARTS OF MALE URETHRA

Part	Length (cm)	Location/Disposition	Features
Intramural (preprostatic) **part**	0.5–1.5	Extends almost vertically through neck of bladder	Surrounded by internal urethral sphincter; diameter and length vary, depending on whether bladder is filling or emptying.
Prostatic urethra	3.0–4.0	Descends through anterior prostate, forming gentle, anteriorly concave curve; is bounded anteriorly by vertical, trough-like part (rhabdosphincter) of external urethral sphincter	Widest and most dilatable part; features urethral crest with seminal colliculus, flanked by prostatic sinuses into which the prostatic ducts open; ejaculatory ducts open onto colliculus; hence, urinary and reproductive tracts merge in this part.
Intermediate (membranous) **part**	1.0–1.5	Passes through deep perineal pouch, surrounded by circular fibers of external urethral sphincter; penetrates perineal membrane	Narrowest and least distensible part (except for external urethral orifice)
Spongy urethra	~15	Courses through corpus spongiosum; initial widening occurs in bulb of penis; widens again distally as navicular fossa (in the glans penis)	Longest and most mobile part; bulbo-urethral glands open into bulbous part; distally, urethral glands open into small urethral lacunae entering lumen of this part.

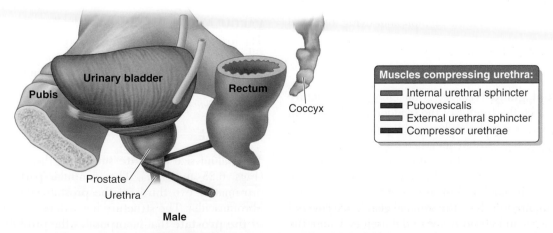

FIGURE 6.26. Compressor muscles of male urethra.

Male Internal Genital Organs

The male internal genital organs include the testes, epididymides (plural of epididymis), ductus deferentes (plural of ductus deferens), seminal glands, ejaculatory ducts, prostate, and bulbo-urethral glands (see Fig. 6.14). The testes and epididymides are described in Chapter 5, Abdomen.

DUCTUS DEFERENS

The **ductus deferens** (vas deferens) is the continuation of the duct of the epididymis (see Chapter 5). The ductus deferens (Figs. 6.14 and 6.21)

- Begins in the tail of the epididymis at the inferior pole of the testis
- Ascends in the spermatic cord
- Passes through the inguinal canal
- Crosses over the external iliac vessels and enters the pelvis
- Passes along the lateral wall of the pelvis where it lies external to the parietal peritoneum
- Ends by joining the duct of the seminal gland to form the *ejaculatory duct*

During the course of the ductus deferens, no other structure intervenes between it and the peritoneum. The ductus crosses superior to the ureter near the posterolateral angle of the bladder, running between the ureter and the peritoneum to reach the fundus of the urinary bladder. Posterior to the bladder, the ductus deferens at first lies superior to the seminal gland, then it descends medial to the ureter and the gland. Here, the ductus deferens enlarges to form the **ampulla of the ductus deferens** before its termination. The ductus then narrows and joins the duct of the seminal gland to form the **ejaculatory duct**.

VASCULATURE OF DUCTUS DEFERENS

The tiny **artery to the ductus deferens** usually arises from a superior (sometimes inferior) vesical artery and accompanies the ductus deferens as far as the testis (Table 6.5). It terminates by anastomosing with the testicular artery, posterior to the testis. The veins accompany the arteries and have similar names. The lymphatic vessels from the ductus deferens drain into the *external iliac lymph nodes* (Fig. 6.23 and Table 6.7).

SEMINAL GLANDS

Each **seminal gland** (vesicle) is an elongated structure that lies between the fundus of the bladder and the rectum (Fig. 6.25). The seminal glands are obliquely placed structures superior to the prostate and do not store sperms. They secrete a thick alkaline fluid that mixes with the sperms as they pass into the ejaculatory ducts and urethra; it is the major constituent (65–75%) of semen (a mixture of secretions). The superior ends of the seminal glands are covered with peritoneum and lie posterior to the ureters, where the peritoneum of the *rectovesical pouch* separates them from

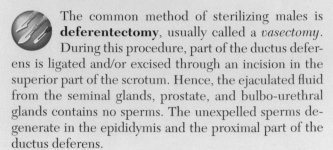

CLINICAL BOX

Sterilization of Males

The common method of sterilizing males is **deferentectomy**, usually called a *vasectomy*. During this procedure, part of the ductus deferens is ligated and/or excised through an incision in the superior part of the scrotum. Hence, the ejaculated fluid from the seminal glands, prostate, and bulbo-urethral glands contains no sperms. The unexpelled sperms degenerate in the epididymis and the proximal part of the ductus deferens.

the rectum (see Fig. 6.6 and Table 6.2). The inferior ends of the seminal glands are closely related to the rectum and are separated from it only by the rectovesical septum.

VASCULATURE OF SEMINAL GLANDS

The arteries to the seminal glands derive from the *inferior vesical* and *middle rectal arteries* (Table 6.5). The veins accompany the arteries and have similar names. The iliac lymph nodes receive lymph from the seminal glands: the *external iliac nodes* from the superior part and the *internal iliac lymph nodes* from the inferior part (Table 6.7).

EJACULATORY DUCTS

Each **ejaculatory duct** is a slender tube that arises by the union of the duct of a seminal gland with the ductus deferens (Figs. 6.21 and 6.25). The ejaculatory ducts arise near the neck of the bladder and run close together as they pass antero-inferiorly through the posterior part of the prostate. The ducts converge to open by slit-like apertures on, or just within, the opening of the prostatic utricle (Figs. 6.27 and 6.28). Prostatic secretions join the seminal fluid in the prostatic urethra after the termination of the ejaculatory ducts.

VASCULATURE OF EJACULATORY DUCTS

The *artery to the ductus deferens*, usually branches off the superior (but frequently inferior) vesical artery, to supply the ejaculatory duct (Table 6.5). The veins join the *prostatic* and *vesical venous plexuses*. The lymphatic vessels drain into the *external iliac lymph nodes* (Table 6.7).

PROSTATE

The walnut-size **prostate** surrounds the *prostatic urethra* (Figs. 6.25 and 6.27). The glandular part makes up approximately two thirds of the prostate; the other third is fibromuscular. The structure has a dense, **fibrous capsule of the prostate** that incorporates the prostatic plexuses of nerves and veins. This is surrounded by the visceral layer of the pelvic fascia, forming a fibrous *prostatic sheath* that is thin anteriorly, continuous anterolaterally with the *puboprostatic ligaments*, and dense posteriorly, continuous with the *rectovesical septum*.

The prostate has (Fig. 6.27B)

- a **base** (superior aspect) that is closely related to the neck of the bladder
- an **apex** (inferior aspect) that is in contact with fascia on the superior aspect of the urethral sphincter and deep perineal muscles
- a muscular *anterior surface* that features mostly transversely oriented muscle fibers forming a vertical trough-like hemisphincter (rhabdosphincter), which is part of the urethral sphincter, separated from the pubic symphysis by retroperitoneal fat in the **retropubic space** (see Fig. 6.18)
- a *posterior surface* that is related to the ampulla of the rectum
- *inferolateral surfaces* that are related to the levator ani

Although not clearly distinct anatomically, the following *lobes and lobules of the prostate* are described (Fig. 6.27A):

- The **isthmus of the prostate** (anterior muscular zone; historically, the anterior lobe) lies anterior to the urethra. It is primarily muscular and represents the superior continuation of the urethral sphincter muscle.
- **Right** and **left lobes** (peripheral zones), each divided in turn into four indistinct *lobules* in two concentric bands, defined by their relationship to the urethra and ejaculatory ducts:

1. A superficial **inferoposterior lobule**, posterior to the urethra and inferior to the ejaculatory ducts, is readily palpable by digital rectal examination.
2. A superficial **inferolateral lobule**, lateral to the urethra, forms the major part of the prostate.
3. A **superomedial lobule** surrounds the ejaculatory duct, deep to the inferoposterior lobule.
4. An **anteromedial lobule**, deep to the inferolateral lobule, is directly lateral to the proximal prostatic urethra.

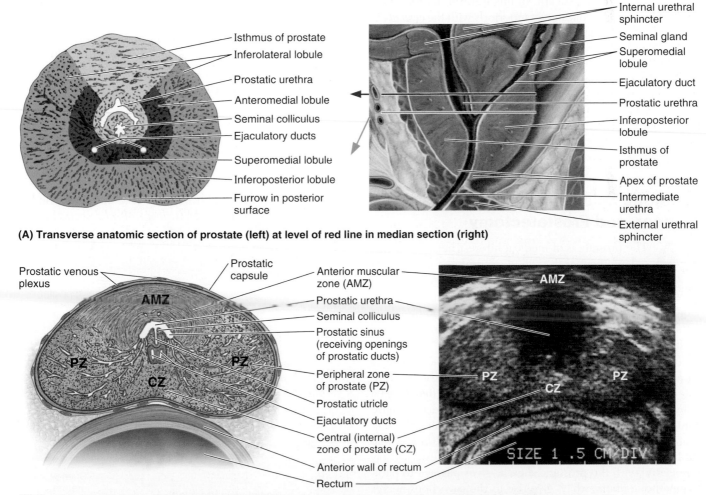

(A) Transverse anatomic section of prostate (left) at level of red line in median section (right)

(B) Graphic interpretation (left) of transverse ultrasound image (right) at level of green line in (A, right).

FIGURE 6.27. Lobules and zones of prostate demonstrated by anatomical section and ultrasonographic imaging.

An embryonic middle (median) lobe gives rise to superomedial and anteromedial lobules. This region tends to undergo hormone-induced hypertrophy in advanced age, forming a **middle lobule** (central zone) considered to be partially responsible for the formation of the *uvula* that may project into the internal urethral orifice (Fig. 6.28).

Urologists and sonographers usually divide the prostate into peripheral and central (internal) zones (Fig. 6.27C,D).

The **prostatic ducts** (20–30) open chiefly into the **prostatic sinuses** that lie on either side of the **seminal colliculus** on the posterior wall of the prostatic urethra (Fig. 6.28). Prostatic fluid provides about 15–30% of the volume of **semen**.

VASCULATURE OF PROSTATE

The *prostatic arteries* are mainly branches of the *internal iliac artery* (Table 6.5), especially the *inferior vesical arteries* and also the *internal pudendal* and *middle rectal arteries*. The *prostatic veins* join to form the **prostatic venous plexus** around the sides and base of the prostate (see Figs. 6.21 and 6.27B). This plexus, between the fibrous capsule of the prostate and the prostatic sheath, drains into the *internal iliac veins*. The plexus is continuous superiorly with the *vesical venous plexus* and communicates posteriorly with the *internal vertebral venous plexus* (see Chapter 2, Back). The lymphatic vessels drain chiefly into the *internal iliac nodes*, but some pass to the *sacral lymph nodes* (Table 6.7).

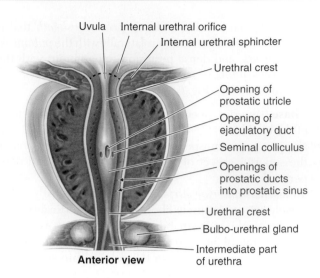

Uvula   Internal urethral orifice
Internal urethral sphincter
Urethral crest
Opening of prostatic utricle
Opening of ejaculatory duct
Seminal colliculus
Openings of prostatic ducts into prostatic sinus
Urethral crest
Bulbo-urethral gland
Intermediate part of urethra
Anterior view

FIGURE 6.28. Posterior wall of prostatic urethra.

BULBO-URETHRAL GLANDS

The two pea-size **bulbo-urethral glands** (Cowper glands) lie posterolateral to the intermediate part of the urethra, largely embedded within the external urethral sphincter (Figs. 6.20A, 6.21, and 6.28). The **ducts of the bulbo-urethral glands** pass through the perineal membrane adjacent to the intermediate urethra and open through minute apertures into the proximal part of the spongy urethra in the bulb of the penis. Their mucus-like secretion enters the urethra during sexual arousal, contributing less than 1% of semen.

CLINICAL BOX

Prostatic Enlargement, Prostatic Cancer, and Prostatectomy

The prostate is of medical interest because benign enlargement or *benign hypertrophy of the prostate* (BHP) is common after middle age. An enlarged prostate projects into the urinary bladder and impedes urination by distorting the prostatic urethra. The middle lobule usually enlarges the most and obstructs the internal urethral orifice.

Prostatic cancer is common in men older than 55 years of age. In most cases, the cancer develops in the posterolateral region. This may be palpated during a digital rectal examination (Fig. B6.3). A malignant prostate feels hard and often irregular. In advanced stages, cancer cells metastasize (spread) to the iliac and sacral lymph nodes and later to distant nodes and bone. The prostatic plexus, closely associated with the prostatic sheath, gives passage to parasympathetic fibers, which give rise to the cavernous nerves that convey the fibers that cause penile erection. A major concern regarding *prostatectomy* is that impotency may be

a consequence. All or part of the prostate, or just the hypertrophied part, is removed (*transurethral resection of the prostate* [TURP]).

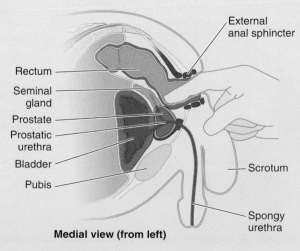

External anal sphincter
Rectum
Seminal gland
Prostate
Prostatic urethra
Bladder
Pubis
Scrotum
Spongy urethra
Medial view (from left)

FIGURE B6.3. Palpation of prostate per rectum.

INNERVATION OF INTERNAL GENITAL ORGANS OF MALE PELVIS

The ductus deferens, seminal glands, ejaculatory ducts, and prostate are richly innervated by *sympathetic nerve fibers* originating from cell bodies in the intermediolateral cell column. They traverse the paravertebral ganglia of the sympathetic trunk to become components of the lumbar (abdominopelvic) splanchnic nerves and the hypogastric and pelvic plexuses (Fig. 6.29). *Presynaptic parasympathetic fibers* from the S2–S4 spinal cord segments traverse the pelvic splanchnic nerves, which also join the inferior hypogastric–pelvic plexuses. Synapses with postsynaptic sympathetic and parasympathetic neurons occur within the plexuses, en route to or near the pelvic viscera. As part of an orgasm, the sympathetic system stimulates contractions of the ductus deferens, and the combined contraction of and secretion from the seminal and prostate glands provide the vehicle (semen) and the expulsive force to discharge the sperms during ejaculation. The function of the pelvic parasympathetic innervation is unclear. However, the parasympathetic fibers in the **prostatic nerve plexus** form the cavernous nerves that pass to the erectile bodies of the penis, which are responsible for producing penile erection.

Female Internal Genital Organs

The female internal genital organs include the uterus, uterine tubes, ovaries, and vagina.

UTERUS

The **uterus** (womb) is a thick-walled, pear-shaped, hollow muscular organ. The nongravid (not pregnant) uterus usually lies in the lesser pelvis, with its body lying on the urinary bladder and its cervix between the urinary bladder and the rectum (see Fig. 6.15). The adult uterus is usually *anteverted* (tipped anterosuperiorly relative to the axis of the vagina) and *anteflexed* (uterine body is flexed or bent anteriorly relative to the cervix) so that its mass lies over the bladder. The position of the uterus changes with the degree of fullness of the bladder and rectum. The uterus is divisible into two main parts (Fig. 6.30):

- The **body of the uterus**, forming the superior two thirds of the structure, includes the **fundus of the uterus**, the rounded part of the body that lies superior to the orifices of the uterine tubes, and the **isthmus of the uterus**, the relatively constricted region of the body (about 1-cm long) just superior to the cervix. The **uterine horns** (L. *cornua*) are the superolateral regions where the uterine tubes enter. The body of the uterus lies between the layers of the broad ligaments and is freely movable.
- The **cervix of the uterus**, the cylindrical, narrow inferior part of the uterus, which has a *supravaginal part* between the isthmus and the vagina and a *vaginal part* that protrudes into the vagina and surrounds the **external**

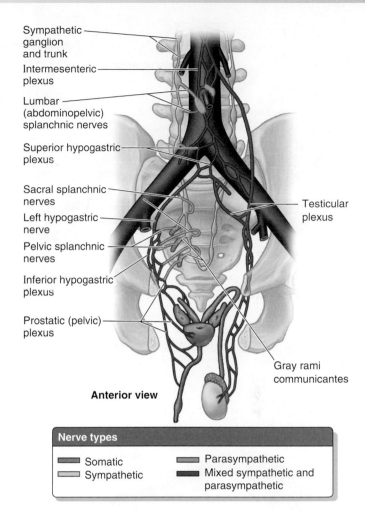

FIGURE 6.29. Autonomic innervation of testis, ductus deferens, prostate, and seminal glands.

os of the uterus. The *supravaginal part of the cervix* is separated from the bladder only anteriorly by loose connective tissue and from the rectum posteriorly by the recto-uterine pouch. The cervix is mostly fibrous, with a small amount of smooth muscle and elastin.

The *wall of the body of the uterus* consists of three layers (Fig. 6.30B):

- **Perimetrium**: The outer serous coat, which consists of peritoneum supported by a thin layer of connective tissue
- **Myometrium**: The middle muscular coat of smooth muscle, which becomes greatly distended during pregnancy; the main branches of the blood vessels and nerves of the uterus are located in this coat.
- **Endometrium**: The inner mucous coat, which firmly adheres to the myometrium and is actively involved in the menstrual cycle, differing in structure with each stage. If conception occurs, the blastocyst implants in this layer; if conception does not occur, the inner surface of the coat is shed during menstruation.

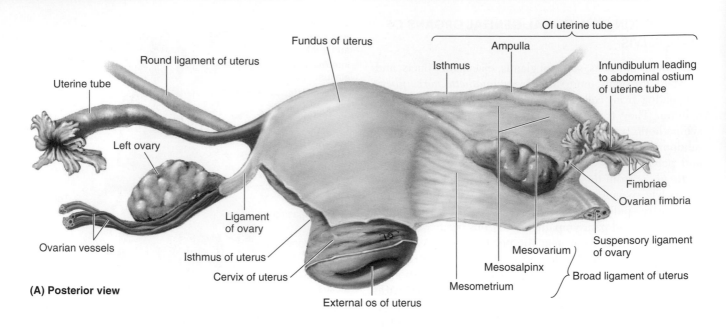

(A) Posterior view

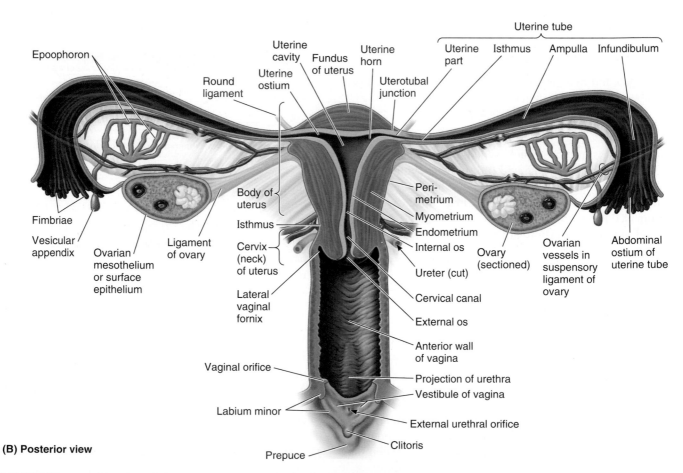

(B) Posterior view

FIGURE 6.30. Internal female genital organs. A. Dissection of isolated specimen. The broad ligament is removed on the left side. **B.** Coronal section demonstrating the internal structure of the female genital organs.

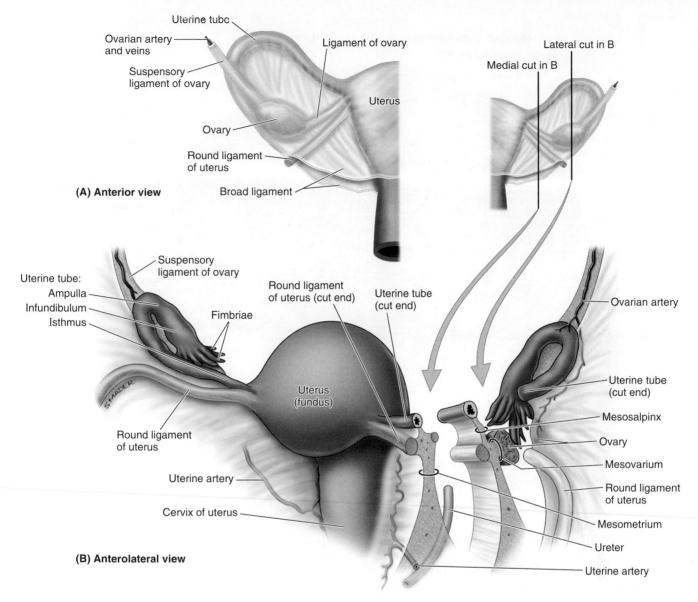

FIGURE 6.31. Uterus, uterine tubes, and broad ligament. A. Relationship of the broad ligament to the ovary and its ligaments. **B.** Sagittal sections showing the mesentry of the uterus (mesometrium), ovary (mesovarium), and uterine tubes (mesosalpinx).

LIGAMENTS OF UTERUS

Externally, the **ligament of the ovary** attaches to the uterus postero-inferior to the uterotubal junction (Fig. 6.31). The **round ligament of the uterus** attaches antero-inferiorly to this junction. These two ligaments are vestiges of the *ovarian gubernaculum* related to the descent of the ovary from its developmental position on the posterior abdominal wall (see Chapter 5, Abdomen).

The **broad ligament of the uterus** is a double layer of peritoneum (mesentery) that extends from the sides of the uterus to the lateral walls and floor of the pelvis. This ligament assists in keeping the uterus relatively centered in the pelvis but mostly contains the ovaries, uterine tubes, and related structures as well as the vasculature that serves them. The two layers of the ligament are continuous with each other at a free edge, which surrounds the uterine tube. Laterally, the ligament is prolonged superiorly over the ovarian vessels as the **suspensory ligament of the ovary** (Fig. 6.31B). Between the layers of the broad ligament on each side of the uterus, the *ligament of the ovary* lies posterosuperiorly and the *round ligament of the uterus* lies antero-inferiorly. The part of the broad ligament by which the ovary is suspended is the **mesovarium** (Fig. 6.31B). The part of the broad ligament forming the mesentery of the uterine tube is the **mesosalpinx**. The major part of the broad ligament serves as a mesentery for the uterus and is the **mesometrium**, which lies inferior to the mesosalpinx and mesovarium.

The principal supports of the uterus are both dynamic and passive. The cervix is the least mobile part of the uterus because of the passive support provided by attached

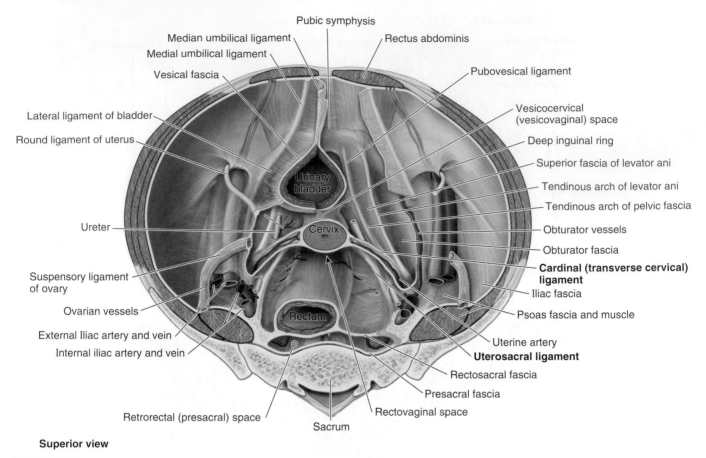

Pubic symphysis
Median umbilical ligament
Medial umbilical ligament
Vesical fascia
Rectus abdominis
Pubovesical ligament
Lateral ligament of bladder
Vesicocervical (vesicovaginal) space
Round ligament of uterus
Deep inguinal ring
Superior fascia of levator ani
Tendinous arch of levator ani
Tendinous arch of pelvic fascia
Ureter
Obturator vessels
Obturator fascia
Cardinal (transverse cervical) ligament
Suspensory ligament of ovary
Iliac fascia
Ovarian vessels
Psoas fascia and muscle
External Iliac artery and vein
Uterine artery
Internal iliac artery and vein
Uterosacral ligament
Rectosacral fascia
Presacral fascia
Retrorectal (presacral) space
Rectovaginal space
Sacrum
Urinary bladder
Cervix
Rectum

Superior view

FIGURE 6.32. Pelvic fascial ligaments. Peritoneum and loose areolar endopelvic fascia have been removed to demonstrate the pelvic fascial ligaments located inferior to the peritoneum but superior to the pelvic floor (pelvic diaphragm).

condensations of endopelvic fascia (ligaments), which may also contain smooth muscle (Figs. 6.8A,B,E and 6.32):

- *Transverse cervical* (cardinal) *ligaments* extend from the cervix and lateral parts of the fornix of the vagina to the lateral walls of the pelvis.
- *Uterosacral ligaments* pass superiorly and slightly posteriorly from the sides of the cervix to the middle of the sacrum (Fig. 6.8E); they are palpable on rectal examination.

Passive support is also provided and the way in which the uterus normally rests on top of the bladder (Fig. 6.33). Dynamic support is provided by the muscles of the pelvic floor (perineal muscles) and supporting/compressive musculature (Fig. 6.34).

RELATIONSHIPS OF UTERUS

Peritoneum covers the body and fundus of uterus anteriorly and superiorly but not the cervix (Figs. 6.6A,C and 6.31 and Table 6.3). The peritoneum is reflected anteriorly from the uterus onto the bladder and posteriorly over the posterior part of the fornix of the vagina onto the rectum. Anteriorly, the uterine body is separated from the urinary bladder by the **vesico-uterine pouch** where the peritoneum is reflected from the uterus onto the posterior margin of the superior

surface of the bladder (Fig. 6.33); the inferior uterine body (isthmus) and cervix lie in direct contact with the bladder without intervening peritoneum. This allows uterine/cervical cancer to invade the urinary bladder. Posteriorly, the uterine body and the supravaginal part of the cervix are separated from the sigmoid colon by a layer of peritoneum and the peritoneal cavity and from the rectum by the *recto-uterine pouch*. Laterally, the uterine artery crosses the ureter superiorly, near the cervix, in the root of the broad ligament (Fig. 6.31).

VASCULATURE OF UTERUS

The *arteries* derive mainly from the *uterine arteries*, with potential collateral supply from the ovarian arteries (Figs. 6.11A and 6.35 and Table 6.5). The *uterine veins* run in the broad ligament, draining the **uterine venous plexus** formed on each side of the uterus and vagina (Fig. 6.35). Veins from this plexus drain into the internal iliac veins.

The *uterine lymphatic vessels* follow three main routes (Fig. 6.22 and Table 6.6):

- Most vessels from the uterine fundus and superior uterine body pass along the ovarian vessels to the lumbar (caval/aortic) lymph nodes, but some vessels pass along the round ligament of the uterus to the *superficial inguinal lymph nodes*.

- Vessels from most of the uterine body pass within the broad ligament to the *external iliac lymph nodes*.
- Vessels from the uterine cervix pass along the uterine vessels, within the transverse cervical ligaments, to the *internal iliac lymph nodes* and along the uterosacral ligaments to the *sacral lymph nodes*.

INNERVATION OF VAGINA AND UTERUS

The innervation of the inferior part of the vagina is somatic, from the *deep perineal nerve*, a branch of the *pudendal nerve*. The innervation of most of the vagina and the entire uterus, however, is visceral. The nerves are derived from the **uterovaginal nerve plexus**, which travels with the uterine artery at the junction of the base of the peritoneal broad ligament and the superior part of the transverse cervical ligament (Fig. 6.36). The uterovaginal plexus is one of the pelvic plexuses that extend to the pelvic viscera from the inferior hypogastric plexus. Sympathetic, parasympathetic, and visceral afferent fibers pass through this plexus. Sympathetic innervation originates in the inferior thoracic spinal cord segments and passes through *lumbar splanchnic nerves* and the intermesenteric–hypogastric–pelvic series of plexuses. Parasympathetic innervation originates in the S2–S4 spinal cord segments and passes through the *pelvic splanchnic nerves* to the inferior hypogastric–uterovaginal plexus. Visceral afferent fibers, carrying pain sensation from the intraperitoneal uterine fundus and body, travel retrogradely with the sympathetic fibers to the lower thoracic from the upper lumbar spinal ganglia; those from the subperitoneal uterine cervix and vagina (inferior to the pelvic pain line) travel with the parasympathetic fibers to the spinal sensory ganglia of S2–S4. All visceral afferent fibers from the uterus and vagina not concerned with pain (those conveying unconscious sensations) also follow the latter route.

UTERINE TUBES

The **uterine tubes** (oviducts, commonly called fallopian tubes) extend laterally from the *uterine horns* and open into the peritoneal cavity near the ovaries (Figs. 6.30 and 6.31). The uterine tubes lie in the *mesosalpinx* in the free edges of

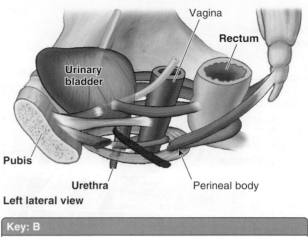
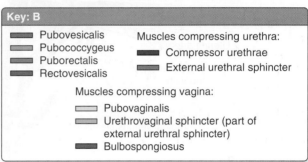

Left lateral view

Key: B

▬ Pubovesicalis	**Muscles compressing urethra:**
▬ Pubococcygeus	▬ Compressor urethrae
▬ Puborectalis	▬ External urethral sphincter
▬ Rectovesicalis	

Muscles compressing vagina:
▭ Pubovaginalis
▬ Urethrovaginal sphincter (part of external urethral sphincter)
▬ Bulbospongiosus

FIGURE 6.34. Supporting and compressive muscles of female pelvis.

the broad ligament. In the "ideal" disposition, the tubes extend posterolaterally to the lateral pelvic walls, where they ascend and arch over the ovaries; however, ultrasound studies demonstrate that the position of the tubes and ovaries is variable (dynamic) in life, and right and left sides are often asymmetrical.

Each uterine tube is divisible into four parts (Fig. 6.31B):

- The **infundibulum** is the funnel-shaped distal end that opens into the peritoneal cavity through the **abdominal ostium**. The finger-like processes of the infundibulum, **fimbriae**, spread over the medial surface of the ovary; one large **ovarian fimbria** is attached to the superior pole of the ovary.
- The **ampulla**, the widest and longest part, begins at the medial end of the infundibulum.
- The **isthmus**, the thick-walled part, enters the uterine horn.
- The **uterine part** is the short intramural segment that passes through the wall of the uterus and opens through the **uterine ostium** into the uterine cavity at the uterine horn.

OVARIES

The almond-shaped **ovaries** are typically located near the attachment of the broad ligament to the lateral pelvic walls, suspended from both by peritoneal folds, the *mesovarium* from the posterosuperior aspect of the broad ligament and the *suspensory ligament of the ovary* from the pelvic wall (Figs. 6.30 and 6.31). The suspensory ligament conveys the ovarian vessels, lymphatics, and nerves to and from the ovary and constitutes the lateral part of the mesovarium. The ovary also attaches to

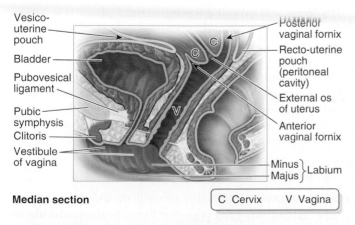

Vesico-uterine pouch	Posterior vaginal fornix
Bladder	Recto-uterine pouch (peritoneal cavity)
Pubovesical ligament	External os of uterus
Pubic symphysis	Anterior vaginal fornix
Clitoris	
Vestibule of vagina	Minus / Majus } Labium

Median section | C Cervix V Vagina

FIGURE 6.33. Parts of uterus and relationship of vagina and uterus.

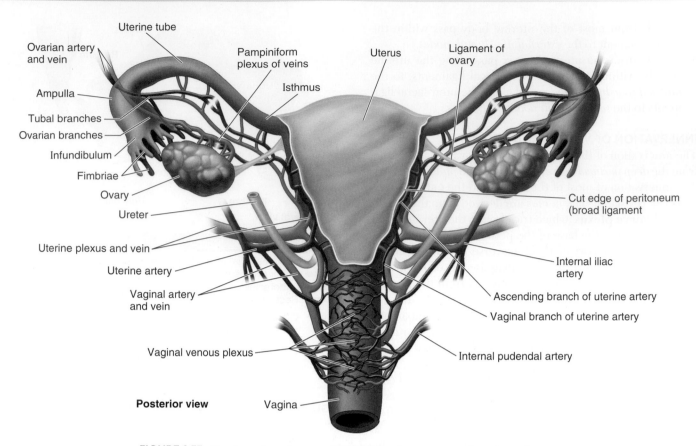

FIGURE 6.35. Blood supply and venous drainage of vagina, uterus, uterine tube, and ovary.

the uterus by the *ligament of ovary*, which runs within the mesovarium. This ligament is a remnant of the superior part of the ovarian gubernaculum of the fetus and connects the proximal (uterine) end of the ovary to the lateral angle of the uterus, just inferior to the entrance of the uterine tube. Because the ovary is suspended in the peritoneal cavity and its surface is not covered by peritoneum, the oocyte expelled at ovulation passes into the peritoneal cavity but is usually trapped by the fimbriae of the uterine tube and carried to the ampulla.

VASCULATURE OF OVARIES AND UTERINE TUBES

The **ovarian arteries** arise from the abdominal aorta and descend along the posterior abdominal wall. At the pelvic brim, they cross over the external iliac vessels and enter the suspensory ligaments (Figs. 6.35). The ovarian artery sends branches through the mesovarium to the ovary and through the mesosalpinx to supply the uterine tube. The **ascending branches of the uterine arteries** (branches of the internal iliac arteries) course along the lateral aspects of the uterus to approach the medial aspects of the ovaries and tubes. The ovarian and ascending uterine arteries terminate by bifurcating into ovarian and tubal branches and anastomose with each other, providing a collateral circulation from abdominal and pelvic sources.

Ovarian veins draining the ovary form a **pampiniform plexus of veins** in the broad ligament near the ovary and uterine tube (Fig. 6.35). The veins of the plexus merge to form

a singular **ovarian vein**, which leaves the lesser pelvis with the ovarian artery. The *right ovarian vein* ascends to enter the *inferior vena cava*; the *left ovarian vein* drains into the *left renal vein*. The tubal veins drain into the *ovarian veins* and *uterine (uterovaginal) venous plexus*. The lymphatic vessels from the ovary join those from the uterine tubes and fundus of the uterus as they ascend to the *right* and *left* (caval/aortic) *lumbar lymph nodes* (see Fig. 6.22 and Table 6.6).

INNERVATION OF OVARIES AND UTERINE TUBES

The nerves descend along the ovarian vessels from the *ovarian plexus* and from the *uterine (pelvic) plexus* (Fig. 6.36). Because the ovaries and uterine tubes are superior to the *pelvic pain line*, the visceral afferent pain fibers ascend retrogradely with the sympathetic fibers of the ovarian plexus and lumbar splanchnic nerves to the cell bodies in the T11–L1 spinal sensory ganglia. Visceral afferent reflex fibers follow parasympathetic fibers retrogradely through the uterine (pelvic) and inferior hypogastric plexuses and pelvic splanchnic nerves to cell bodies in the S2–S4 spinal sensory ganglia.

VAGINA

The **vagina**, a mostly subperitoneal musculomembranous tube, extends from the posterior fornix to the **vestibule of the vagina**, the cleft between the labia minora into which

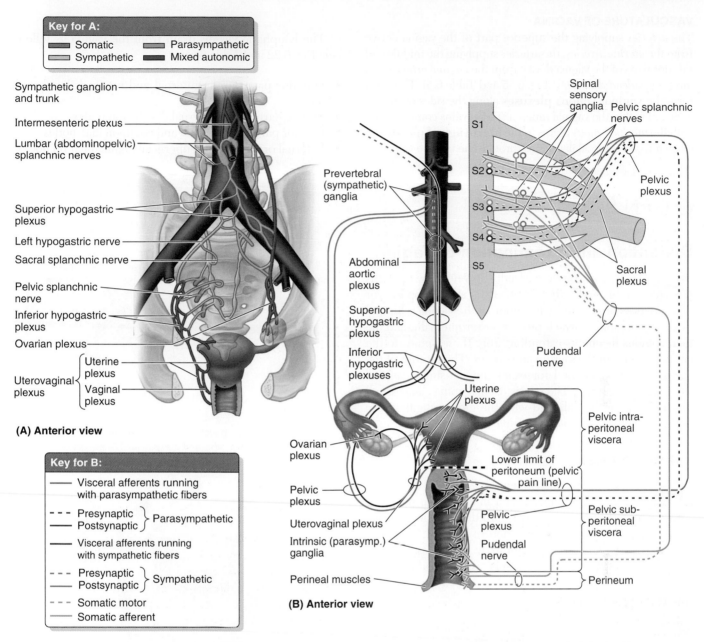

Key for A:

▮ Somatic	▮ Parasympathetic
▮ Sympathetic	▮ Mixed autonomic

Sympathetic ganglion and trunk

Intermesenteric plexus

Lumbar (abdominopelvic) splanchnic nerves

Superior hypogastric plexus

Left hypogastric nerve

Sacral splanchnic nerve

Pelvic splanchnic nerve

Inferior hypogastric plexus

Ovarian plexus

Uterovaginal plexus { Uterine plexus / Vaginal plexus }

(A) Anterior view

Key for B:

— Visceral afferents running with parasympathetic fibers

- - - Presynaptic } Parasympathetic
— Postsynaptic

— Visceral afferents running with sympathetic fibers

- - - Presynaptic } Sympathetic
— Postsynaptic

- - - Somatic motor
— Somatic afferent

Spinal sensory ganglia

Pelvic splanchnic nerves

S1
S2
S3
S4
S5

Prevertebral (sympathetic) ganglia

Abdominal aortic plexus

Superior hypogastric plexus

Inferior hypogastric plexuses

Uterine plexus

Pelvic plexus

Pudendal nerve

Sacral plexus

Ovarian plexus

Lower limit of peritoneum (pelvic pain line)

Pelvic intra-peritoneal viscera

Pelvic plexus

Uterovaginal plexus

Intrinsic (parasymp.) ganglia

Pelvic plexus

Pudendal nerve

Pelvic sub-peritoneal viscera

Perineal muscles

Perineum

(B) Anterior view

FIGURE 6.36. Autonomic innervation of uterus, vagina, and ovaries.

the vagina and urethra open (Fig. 6.30B). The vestibule contains the vaginal and external urethral orifices and the openings of the two greater vestibular glands. The superior end of the vagina surrounds the *cervix* of the uterus. The vagina

- serves as a canal for menstrual fluid
- forms the inferior part of the birth canal
- receives the penis and ejaculate during sexual intercourse
- communicates anteriorly and superiorly with the **cervical canal** and inferiorly with the vestibule. The cervical canal extends from the isthmus of the uterus to the external os (opening) of the uterus.

The vagina is usually collapsed, so its anterior and posterior walls are in contact. The **vaginal fornix**, the recess around the protruding cervix, is usually described as having *anterior, posterior,* and *lateral parts.* The **posterior vaginal fornix** is the deepest part and is closely related to the recto-uterine pouch (Fig. 6.33).

Four muscles compress the vagina and act like sphincters: **pubovaginalis,** *external urethral sphincter,* **urethrovaginal sphincter,** and *bulbospongiosus* (Fig. 6.34). The relations of the vagina are as follows:

- Anteriorly: The fundus of the urinary bladder and urethra
- Laterally: The levator ani, visceral pelvic fascia, and ureters
- Posteriorly (inferior to superior): The anal canal, rectum, and recto-uterine pouch (see Fig. 6.4A)

VASCULATURE OF VAGINA

The *arteries* supplying the superior part of the vagina derive from the *uterine arteries*; the arteries supplying the middle and inferior parts of the vagina derive from the *vaginal arteries* and *internal pudendal arteries* (Fig. 6.35 and Table 6.5). The *veins* form the **vaginal venous plexuses** along the sides of the vagina and within the vaginal mucosa. These veins communicate with the *uterine venous plexus* as the **uterovaginal plexus** and drain into the internal iliac veins through the uterine vein.

The lymphatic vessels drain from the vagina as follows (see Fig. 6.22 and Table 6.6):

- Superior part: To the internal and external iliac lymph nodes
- Middle part: To the internal iliac lymph nodes
- Inferior part: To the sacral and common iliac nodes
- External orifice: To the superficial inguinal lymph nodes

CLINICAL BOX

Distention and Examination of Vagina

The vagina can be markedly distended by the fetus during childbirth, particularly in an anteroposterior direction. Lateral distention of the vagina is limited by the ischial spines, which project posteromedially, and the sacrospinous ligaments extending from these spines to the lateral margins of the sacrum and coccyx. The interior of the vagina can be distended for examination using a *vaginal speculum* (Fig. B6.4). The cervix, ischial spines, and sacral promontory can be palpated with the gloved digits in the vagina and/or rectum (*manual pelvic examination*).

Culdocentesis

An endoscopic instrument (*culdoscope*) can be inserted through an incision made in the posterior part of the vaginal fornix into the peritoneal cavity to drain a pelvic abscess (collection of pus) in the recto-uterine pouch (*culdocentesis*). Similarly, fluid in this part of the perineal cavity (e.g., blood) can be aspirated at this site.

Bicornate Uterus

Incomplete fusion of the embryonic paramesonephric ducts, from which the uterus is formed, results in a variety of congenital anomalies, ranging from the formation of a *unicornuate uterus* (receiving a uterine duct only from the right or left) to duplication in the form of a *bicornuate uterus*, doubled uterine cavities, or a completely doubled uterus (*uterus didelphys*).

Hysterectomy

Hysterectomy (excision of the uterus) is performed through the lower anterior abdominal wall or through the vagina (Fig. B6.5). Because the uterine artery crosses anterior to the ureter near the lateral fornix of the vagina, the ureter is in danger of being inadvertently clamped or severed when the uterine artery is tied off during a hysterectomy. The point of crossing of the artery and the ureter is approximately 2 cm superior to the ischial spine.

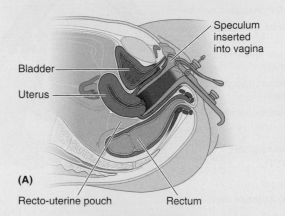

(A)

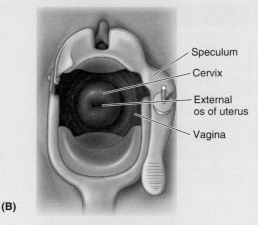

(B)

FIGURE B6.4. Pelvic examination.

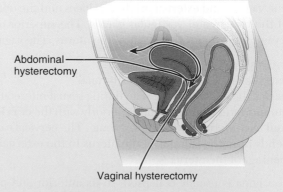

FIGURE B6.5. Routes for hysterectomy.

Cervical Examination and Cytology (Pap Test)

The vagina can be distended with a vaginal speculum to enable inspection of the cervix and obtain a sample of cervical cells for cervical cytology. A spatula is placed on the external os of the uterus (Fig. B6.6) and rotated to scrape cellular material from the vaginal surface of the cervix. This is followed by insertion of a cytobrush into the cervical canal that is used to gather cellular material from the supravaginal cervical mucosa. The cellular material is placed on glass slides for microscopic examination.

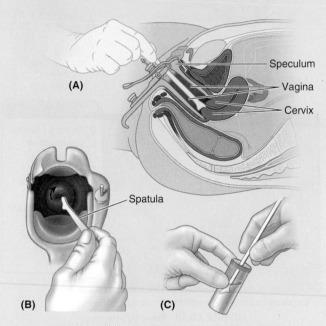

(A)

Speculum

Vagina

Cervix

Spatula

(B)   **(C)**

FIGURE B6.6. Obtaining a sample of cervical cells.

Regional Anesthesia for Childbirth

Several types of regional anesthesia are used to reduce pain during childbirth. *Lumbar epidural* and *low spinal* blocks anesthetize somatic and visceral afferent fibers distributed below waist level, anesthetizing not only the uterus, entire birth canal, and perineum but also the lower limbs (Fig. B6.7A). A *caudal epidural block* is a popular choice for participatory childbirth (Fig. B6.7B). It must be administered in advance of childbirth, which is not possible with precipitous birth. The anesthetic agent is administered using an indwelling catheter in the sacral canal (see Chapter 2, Back), enabling administration of more anesthetic agent for a deeper or prolonged anesthesia if necessary. Within the sacral canal, the anesthesia bathes the S2–S4 spinal nerve roots, including visceral pain fibers from the uterine cervix and upper vagina, and somatic pain fibers of the pudendal nerve. Thus, the birth canal is anesthetized, but the lower limbs are not usually affected. Because visceral pain fibers to the uterine fundus ascend to lower thoracic and upper lumbar spinal levels, they are also not affected and sensations of uterine contraction are still perceived. *Pudendal nerve blocks* (Fig. B6.7C) and local infiltration of the perineum provide only somatic anesthesia of the perineum.

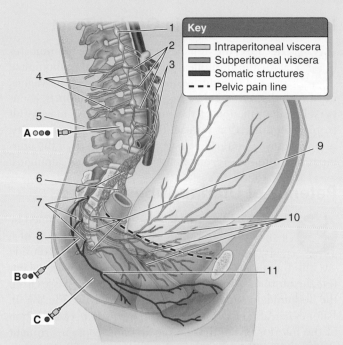

Regional anesthesia for childbirth: sites of injection; (A) spinal block via lumbar puncture, (B) caudal epidural block, (C) pudendal nerve block, (1) sympathetic trunk, (2) lumbar splanchnic nerves, (3) abdominal aortic plexus, (4) spinal ganglia T12–L2(3), (5) L3–L4 level, (6) superior and inferior hypogastric plexuses, (7) spinal ganglia S2–S4, (8) needle tip entering sacral canal, (9) pelvic splanchnic nerves, (10) uterovaginal plexus, (11) pudendal nerve

FIGURE B6.7. Regional anesthesia for childbirth.

Manual Examination of Uterus

The size and disposition of the uterus may be examined by *bimanual palpation* (Fig. B6.8). Two gloved fingers of the examiner's dominant hand are passed superiorly in the vagina, whereas the other hand is pressed inferoposteriorly on the pubic region of the anterior abdominal wall. The size and other characteristics of the uterus can be determined in this way (e.g., whether the anteflexed uterus is in its normal anteverted position).

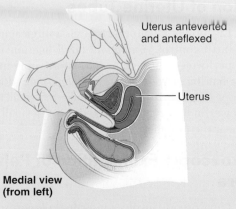

Uterus anteverted and anteflexed

Uterus

Medial view (from left)

FIGURE B6.8. Bimanual palpation of uterus.

(Continued on next page)

Infections of Female Genital Tract

Because the female genital tract communicates with the peritoneal cavity through the abdominal ostia of the uterine tubes, infections of the vagina, uterus, and uterine tubes may result in *peritonitis*. Conversely, inflammation of the tubes (*salpingitis*) may result from infections that spread from the peritoneal cavity. A major cause of infertility in women is blockage of the uterine tubes, often the result of infection that causes salpingitis.

Patency of Uterine Tubes

Patency of the uterine tubes may be determined by a radiographic procedure involving injection of a water-soluble radiopaque material or carbon dioxide gas into the uterus, *hysterosalpingography*. The material enters the uterine tubes and, if the tubes are patent, passes from the abdominal ostium into the peritoneal cavity (Fig. B6.9). Patency can also be determined by *hysteroscopy*, examination of the interior of the tubes using an endoscopic instrument (*hysteroscope*) introduced through the vagina and uterus.

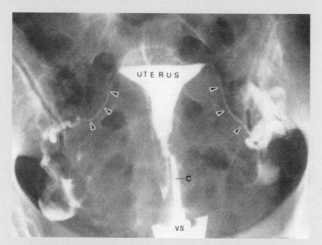

Hysterosalpingogram. *Arrowheads*, uterine tubes; *c*, catheter in the cervical canal; *vs*, vaginal speculum

FIGURE B6.9. Hysterosalpingogram.

Ligation of Uterine Tubes

Ligation of the uterine tubes is a surgical method of birth control. *Abdominal tubal ligation* is usually performed through a short suprapubic incision at the pubic hairline. *Laparoscopic tubal ligation* is done with a laparoscope (Fig. B6.10), which is similar to a small telescope with a powerful light. It is inserted through a small incision, usually near the umbilicus.

Laparoscopic Examination of Pelvic Viscera

Laparoscopy involves inserting a *laparoscope* into the peritoneal cavity through a small incision below the umbilicus (Fig. B6.10). Insufflation of inert gas creates a pneumoperitoneum to provide space to visualize the pelvic organs. Additional openings (ports) can be made to introduce other instruments for manipulation or to enable therapeutic procedures (e.g., ligation of the uterine tubes).

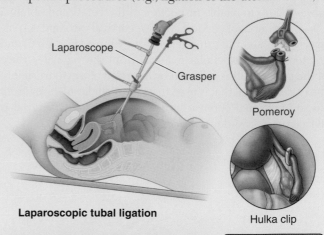

Laparoscope

Grasper

Pomeroy

Hulka clip

Laparoscopic tubal ligation

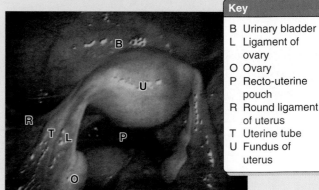

Key

B Urinary bladder
L Ligament of ovary
O Ovary
P Recto-uterine pouch
R Round ligament of uterus
T Uterine tube
U Fundus of uterus

Laparoscopic examination of normal pelvis
FIGURE B6.10. Pelvic laparoscopy.

Ectopic Tubal Pregnancy

Occasionally, a blastocyst fails to reach the uterus and may implant in the mucosa of the uterine tube (most commonly the ampulla), producing an *ectopic tubal pregnancy* (Fig. B6.11). On the right side, the appendix often lies close to the ovary and uterine tube. This close relationship explains why a *ruptured tubal pregnancy* and the resulting peritonitis may be misdiagnosed as acute appendicitis. In both cases, the parietal peritoneum is inflamed in the same general area, and the pain is referred to the right lower quadrant of the abdomen. Tubal rupture and severe hemorrhage constitute a threat to the mother's life and result in death of the embryo. An ectopic pregnancy is not viable and must be removed either surgically or with the use of medications (methotrexate).

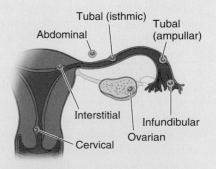

Tubal (isthmic)

Tubal (ampullar)

Abdominal

Interstitial

Infundibular

Ovarian

Cervical

FIGURE B6.11. Ectopic pregnancy.

Rectum

The **rectum** is the pelvic part of the alimentary tract that is continuous proximally with the sigmoid colon and distally with the anal canal (Fig. 6.37A). The **rectosigmoid junction** lies at the level of the S3 vertebra. The rectum follows the curve of the sacrum and coccyx, forming the **sacral flexure of the rectum**. The rectum ends antero-inferior to the tip of the coccyx, where the rectum turns postero-inferiorly and becomes the **anal canal**. The dilated terminal part, the **ampulla of the rectum**, supports and retains the fecal mass before it is expelled during defecation. The rectum is S-shaped in lateral views and has three flexures observable in anterior views as it follows the sacrococcygeal curve (Fig. 6.37B). Its terminal part bends sharply in a posterior

direction, **anorectal flexure**, as it perforates the pelvic diaphragm to become the anal canal (Fig. 6.37A).

The roughly 80-degree anorectal flexure (angle) is an important mechanism for fecal continence and is maintained during the resting state by the tonus of the puborectalis muscle and by its active contraction during peristaltic contractions if defecation is not to occur (Fig. 6.37B). Relaxation of the puborectalis during defecation results in straightening of the anorectal junction. Three sharp **lateral flexures of the rectum** (**superior**, **intermediate**, and **inferior**) are apparent when the rectum is viewed anteriorly (Fig. 6.38A). The flexures are formed in relation to three internal infoldings (**transverse rectal folds**): two on the left and one on the right side (Fig. 6.38B). The folds

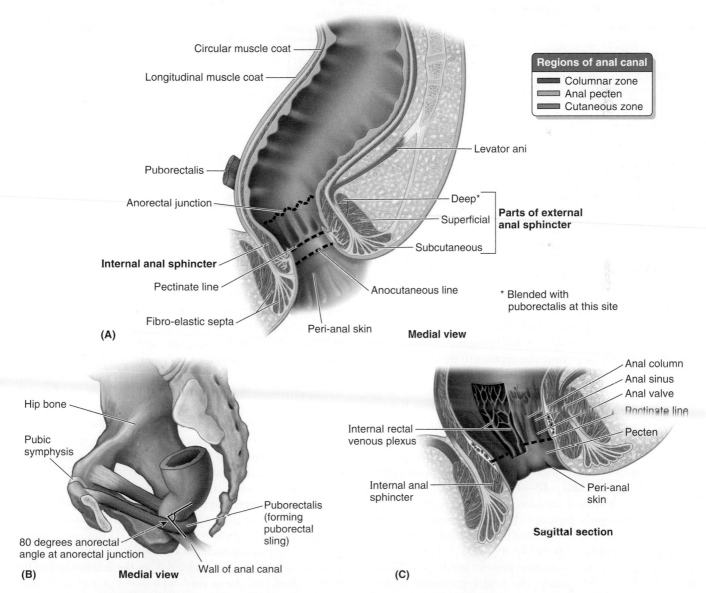

FIGURE 6.37. Rectum and anal canal. A. Musculature and regions of anorectum. **B.** Puborectalis. **C.** Anal canal.

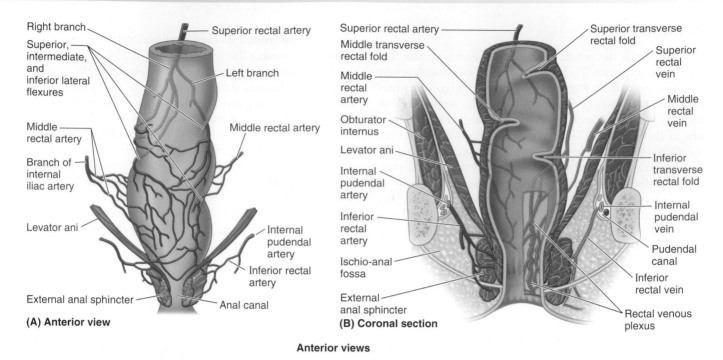

(A) Anterior view

(B) Coronal section

Anterior views

FIGURE 6.38. Vasculature of rectum. A. Overview of arterial supply. **B.** Arterial supply (right side) and venous drainage (left side) of coronally sectioned rectum.

overlie thickened parts of the circular muscle layer of the rectal wall.

Peritoneum covers the anterior and lateral surfaces of the superior third of the rectum (see Fig. 6.6 and Table 6.2), only the anterior surface of the middle third, and no surface of the inferior third because it is subperitoneal. In males, the peritoneum reflects from the rectum to the posterior wall of the bladder, where it forms the floor of the *rectovesical pouch*. In females, the peritoneum reflects from the rectum to the posterior fornix of the vagina, where it forms the floor of the *recto-uterine pouch*. In both sexes, lateral reflections of peritoneum from the upper third of the rectum form *pararectal fossae*, which permit the rectum to distend as it fills with feces.

The rectum rests posteriorly on the inferior three sacral vertebrae and the coccyx, anococcygeal ligament, median sacral vessels, and inferior ends of the sympathetic trunks and sacral plexuses. In males, the rectum is related anteriorly to the fundus of the urinary bladder, terminal parts of the ureters, ductus deferentes, seminal glands, and prostate (see Figs. 6.14 and 6.18). The *rectovesical septum* lies between the fundus of the bladder and the ampulla of the rectum and is closely associated with the seminal glands and prostate. In females, the rectum is related anteriorly to the vagina and is separated from the posterior part of the fornix and cervix by the *recto-uterine pouch* (see Figs. 6.15 and 6.33). Inferior to this pouch, the weak rectovaginal septum separates the superior half of the posterior wall of the vagina from the rectum.

VASCULATURE OF RECTUM

The continuation of the inferior mesenteric artery, the **superior rectal artery**, supplies the proximal part of the rectum. The right and left **middle rectal arteries**, usually arising from the inferior vesical (male) or uterine (female) arteries, supply the middle and inferior parts of the rectum. The **inferior rectal arteries**, arising from the internal pudendal arteries, supply the anorectal junction and anal canal (Fig. 6.38). Blood from the rectum drains via superior, middle, and inferior rectal veins. Because the superior rectal vein drains into the portal venous system and the middle and inferior rectal veins drain into the systemic system, this communication is an important area of *portacaval anastomosis* (see Chapter 5, Abdomen). The submucosal rectal venous plexus surrounds the rectum and communicates with the vesical venous plexus in males and the uterovaginal venous plexus in females. The **rectal venous plexus** consists of two parts: the **internal rectal venous plexus** just deep to the epithelium of the rectum and the **external rectal venous plexus** external to the muscular wall of the rectum (Fig. 6.38B).

Lymphatic vessels from the superior half of the rectum pass to the **pararectal lymph nodes**, located directly on the muscle layer of the rectum (Fig. 6.39), and then ascend to the *inferior mesenteric lymph nodes* either via the *sacral lymph nodes* or by passing through the nodes along the superior rectal vessels. Lymphatic vessels from the inferior half of the rectum drain into the *sacral lymph nodes* or, especially from the distal ampulla, follow the middle rectal vessels to drain into the *internal iliac lymph nodes*.

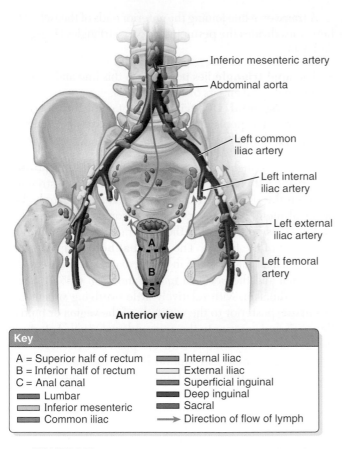

Anterior view

Key

A = Superior half of rectum
B = Inferior half of rectum
C = Anal canal
▬ Lumbar
▬ Inferior mesenteric
▬ Common iliac

▬ Internal iliac
▬ External iliac
▬ Superficial inguinal
▬ Deep inguinal
▬ Sacral
→ Direction of flow of lymph

FIGURE 6.39. Lymphatic drainage of rectum and anal canal.

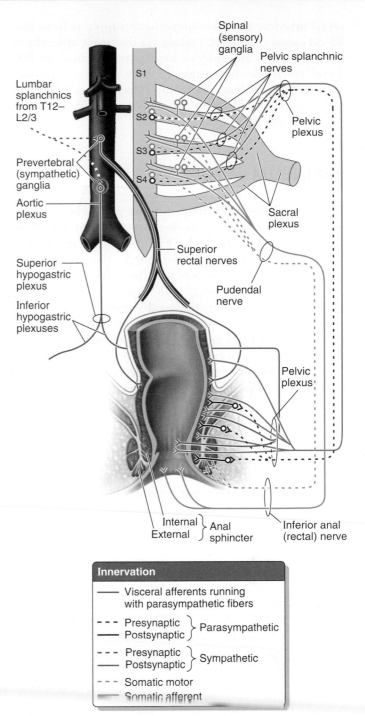

Innervation

— Visceral afferents running
   with parasympathetic fibers

- - - Presynaptic ⎫
—— Postsynaptic ⎬ Parasympathetic

- - - Presynaptic ⎫
—— Postsynaptic ⎬ Sympathetic

- - - Somatic motor
▬▬ Somatic afferent

FIGURE 6.40. **Innervation of rectum and anal canal.** The lumbar and pelvic spinal nerves and hypogastric plexuses have been retracted laterally for clarity.

CLINICAL BOX

Rectal Examination

Many structures related to the antero-inferior part of the rectum may be palpated through its walls (e.g., the prostate and seminal glands in males and the cervix in females). In both sexes, the pelvic surfaces of the sacrum and coccyx may be palpated. The ischial spines and tuberosities may also be palpated. Enlarged internal iliac lymph nodes, pathological thickening of the ureters, and swellings in the ischio-anal fossae (e.g., *ischio-anal abscesses* and abnormal contents in the rectovesical pouch in the male or the recto-uterine pouch in the female) may also be palpated. Tenderness of an *inflamed appendix* may also be detected rectally if it descends into the lesser pelvis (pararectal fossa).

Resection of Rectum

When resecting the rectum in males (e.g., during cancer treatment), the plane of the rectovesical septum (a fascial septum extending superiorly from the perineal body) is located so that the prostate and urethra can be separated from the rectum. In this way, these organs are not often damaged during surgery.

INNERVATION OF RECTUM

The nerve supply to the rectum is from the sympathetic and parasympathetic systems (Fig. 6.40). The *sympathetic supply* is from the lumbar spinal cord, conveyed via the lumbar splanchnic nerves and the hypogastric (pelvic) plexuses and through peri-arterial plexuses on the branches of the inferior mesenteric artery and superior rectal arteries. The *parasympathetic supply* is from the S2–S4 spinal cord level, passing via the pelvic splanchnic nerves (S2–S4) and inferior hypogastric plexuses to the rectal (pelvic) plexus. Because the rectum is inferior (distal) to the pelvic pain line, all visceral afferent fibers follow the parasympathetic fibers retrogradely to the S2–S4 spinal sensory ganglia.

PERINEUM

The term perineum is frequently used to refer to both an external surface area (perineal region) and a shallow "compartment" of the body (Fig. 6.41). The **perineum** (*perineal compartment*) lies inferior to the inferior pelvic aperture and is separated from the pelvic cavity by the pelvic diaphragm (Fig. 6.42). In the anatomical position, the *surface of the perineum* (**perineal region**) is the narrow region between the proximal parts of the thighs. However, when the lower limbs are abducted, the perineal region is a diamond-shaped area extending from the mons pubis anteriorly, the medial surfaces (insides) of the thighs laterally, and the gluteal folds and superior end of the intergluteal (natal) cleft posteriorly (Fig. 6.43).

The osseofibrous structures marking the boundaries of the perineum (perineal compartment) are as follows (Figs. 6.42 and 6.44):

- *Pubic symphysis*, anteriorly
- *Inferior pubic* and *ischial* (*ischiopubic*) *rami*, anterolaterally
- *Ischial tuberosities*, laterally
- *Sacrotuberous ligaments*, posterolaterally
- Inferiormost *sacrum* and *coccyx*, posteriorly

A transverse line joining the anterior ends of the ischial tuberosities divides the perineum into two triangles (Figs. 6.42 and 6.44):

- The **anal triangle** lies posterior to this line and contains the anal canal and its orifice, the anus.
- The **urogenital (UG) triangle**, containing the root of the scrotum and penis in males and the vulva of females, is anterior to this line.

The UG triangle is "closed" by the **perineal membrane** (Fig. 6.45), a thin sheet of tough deep fascia, which stretches between the right and the left sides of the pubic arch. The perineal membrane covers the anterior part of the pelvic outlet and is perforated by the urethra in both sexes and by the vagina of the female. The **perineal body** is an irregular fibromuscular mass located in the median plane between the anal canal and the perineal membrane (Fig. 6.45E). It lies deep to the skin, with relatively little overlying subcutaneous tissue, posterior to the vestibule of the vagina or bulb of the penis and anterior to the anus and anal canal. Anteriorly, the perineal body blends with the posterior border of the perineal membrane and superiorly with the rectovesical or rectovaginal septum. It contains collagenous and elastic fibers and both skeletal and smooth muscle.

The perineal body is the site of convergence of several muscles (Fig. 6.45 and Table 6.9):

- Bulbospongiosus
- External anal sphincter
- Superficial and deep transverse perineal muscles
- Smooth and voluntary slips of muscle from the external urethral sphincter, levator ani, and muscular coats of the rectum

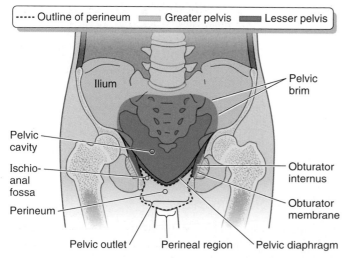

Anterior view of posterior half of coronally sectioned lower trunk

FIGURE 6.41. Perineum and perineal region.

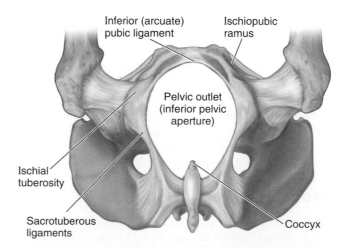

FIGURE 6.42. Osseoligamentous boundaries of perineum.
(Based on Clemente CD. *Anatomy: A Regional Atlas of the Human Body.* 5th ed. Philadelphia, PA: Lippincott Williams & Wilkins; 2006: Fig. 272.1.)

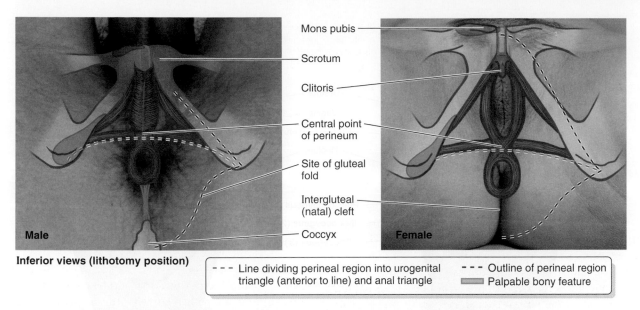

Mons pubis

Scrotum

Clitoris

Central point of perineum

Site of gluteal fold

Intergluteal (natal) cleft

Coccyx

Male

Female

Inferior views (lithotomy position)

--- Line dividing perineal region into urogenital triangle (anterior to line) and anal triangle

--- Outline of perineal region

Palpable bony feature

FIGURE 6.43. Male and female perineal regions. Boundaries and surface features of the perineal region with projections of the osseous boundaries and superficial muscles of perineum. The penis and some of the scrotum are retracted anteriorly.

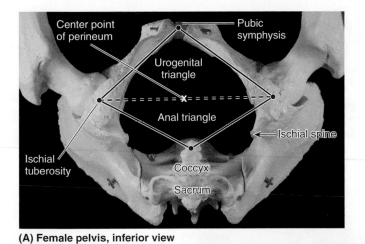

Center point of perineum

Pubic symphysis

Urogenital triangle

Anal triangle

Ischial spine

Ischial tuberosity

Coccyx

Sacrum

(A) Female pelvis, inferior view

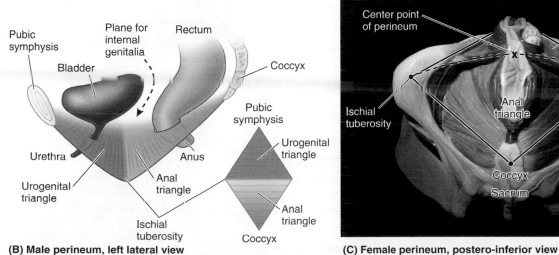

Pubic symphysis

Plane for internal genitalia

Rectum

Bladder

Coccyx

Pubic symphysis

Urogenital triangle

Urethra

Anus

Urogenital triangle

Anal triangle

Anal triangle

Ischial tuberosity

Coccyx

(B) Male perineum, left lateral view

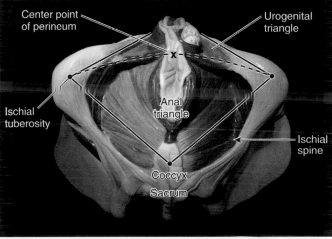

Center point of perineum

Urogenital triangle

Ischial tuberosity

Anal triangle

Ischial spine

Coccyx

Sacrum

(C) Female perineum, postero-inferior view

FIGURE 6.44. Boundaries and disposition of perineum. A. Pelvic girdle demonstrating bony features bounding the perineum. The two triangles comprising the diamond-shaped perineum are superimposed. **B.** The anal and urogenital triangles, schematic illustration. Note that the two triangles do not occupy the same plane. **C.** Musculoskeletal model of perineum.

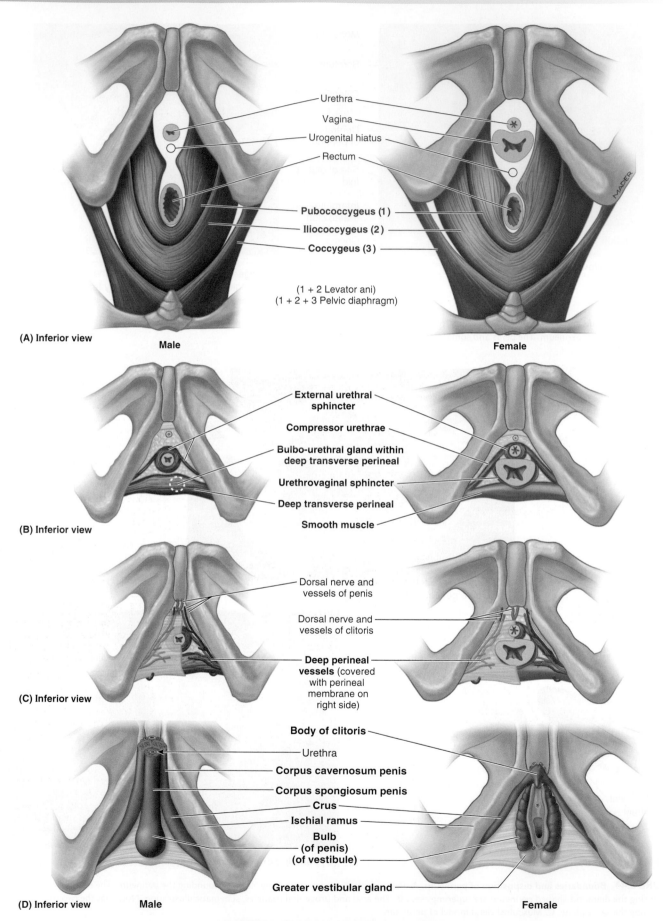

Urethra

Vagina

Urogenital hiatus

Rectum

Pubococcygeus (1)

Iliococcygeus (2)

Coccygeus (3)

(1 + 2 Levator ani)
(1 + 2 + 3 Pelvic diaphragm)

(A) Inferior view   **Male**   **Female**

External urethral sphincter

Compressor urethrae

Bulbo-urethral gland within deep transverse perineal

Urethrovaginal sphincter

Deep transverse perineal

Smooth muscle

(B) Inferior view

Dorsal nerve and vessels of penis

Dorsal nerve and vessels of clitoris

Deep perineal vessels (covered with perineal membrane on right side)

(C) Inferior view

Body of clitoris

Urethra

Corpus cavernosum penis

Corpus spongiosum penis

Crus

Ischial ramus

**Bulb
(of penis)
(of vestibule)**

Greater vestibular gland

(D) Inferior view   **Male**   **Female**

FIGURE 6.45. Muscles of perineum. *(continued)*

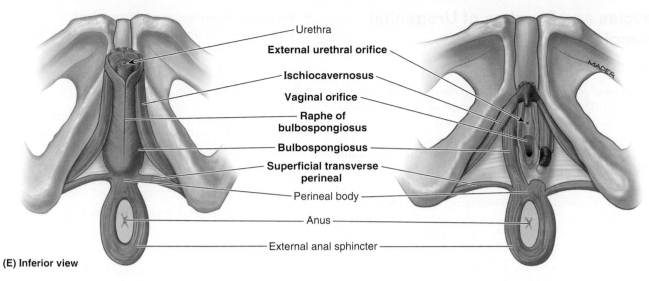

(E) Inferior view

FIGURE 6.45 Muscles of perineum. *(continued)*

TABLE 6.9. MUSCLES OF PERINEUM

Muscle	Origin	Course and Insertion	Innervation	Main Action(s)
External anal sphincter	Skin and fascia surrounding anus and coccyx via anococcygeal ligament	Passes around lateral aspects of anal canal, inserting into perineal body	Inferior anal nerve, branch of pudendal nerve (S2–S4)	Constricts anal canal during peristalsis, resisting defecation; supports and fixes perineal body/pelvic floor
Bulbospongiosus	*Male:* median raphe on ventral surface of bulb of penis and perineal body *Female:* perineal body	*Male:* surrounds lateral aspects of bulb of penis and most proximal part of body of penis, inserting into perineal membrane, dorsal aspect of corpora spongiosum and cavernosa, and fascia of bulb of penis *Female:* passes on each side of lower vagina, enclosing bulb and greater vestibular gland; inserts onto pubic arch and fascia of corpora cavernosa of clitoris	Muscular (deep) branch of perineal nerve, branch of pudendal nerve (S2–S4)	Supports and fixes perineal body/pelvic floor *Male:* compresses bulb of penis to expel last drops of urine/semen; assists erection by compressing outflow via deep perineal vein and by pushing blood from bulb into body of penis *Female:* "sphincter" of vagina; assists in erection of clitoris (and bulb of vestibule); compresses greater vestibular gland
Ischiocavernosus	Internal surface of ischiopubic ramus and ischial tuberosity	Embraces crus of penis or clitoris, inserting onto inferior and medial aspects of crus and to perineal membrane medial to crus		Maintains erection of penis or clitoris by compressing outflow veins and pushing blood from root of penis or clitoris into body
Superficial transverse perineal	Internal surface of ischiopubic ramus and ischial tuberosity; compressor urethrae portion only	Passes along superior posterior border of perineal membrane to perineal body	Muscular (deep) branch of perineal nerve, branch of pudendal nerve (S2–S4); dorsal nerve of penis or clitoris, terminal branch of pudendal nerve (S2–S4)	Support and fix perineal body (pelvic floor) to support abdominopelvic viscera and resist increased intra-abdominal pressure
Deep transverse perineal		Passes along superior posterior border of perineal membrane to perineal body and external anal sphincter		
External urethral sphincter		Surrounds urethra superior to perineal membrane *Male:* also ascends anterior aspect of prostate *Female:* Some fibers also enclose vagina (urethrovaginal sphincter).		Compresses urethra to maintain urinary continence *Female:* Urethrovaginal sphincter portion also compresses vagina.

Fascias and Pouches of Urogenital Triangle

PERINEAL FASCIAS

The *perineal fascias* consists of superficial and deep layers (Fig. 6.46). The **subcutaneous tissue of the perineum**, or *superficial perineal fascia*, consists of a superficial fatty layer and a deep membranous layer. In females, the **fatty layer of subcutaneous tissue of the perineum** makes up the substance of the labia majora and mons pubis and is continuous anteriorly and superiorly with the *fatty layer of subcutaneous tissue of the abdomen (Camper fascia)* (Fig. 6.46A,C). In males, the fatty layer is greatly diminished in the UG triangle and is replaced altogether in the penis and scrotum with smooth (dartos) muscle. It is continuous between the penis or scrotum and the thighs with the fatty layer of subcutaneous tissue of the abdomen (Fig. 6.46B,F). In both sexes, it is continuous posteriorly with the ischio-anal fat pad in the anal region (Fig. 6.46E).

The **membranous layer of subcutaneous tissue of the perineum** (Colles fascia) is attached posteriorly to the posterior margin of the perineal membrane and the perineal body (Fig. 6.46A,B). Laterally, it is attached to the fascia lata (deep fascia) of the superiormost medial aspect of the thigh. Anteriorly, in the male, the membranous layer of subcutaneous tissue is continuous with the dartos fascia of the penis and scrotum; however, on each side of and anterior to the scrotum, the membranous layer becomes continuous with the *membranous layer of subcutaneous tissue of the abdomen (Scarpa fascia)* (Fig. 6.46B,F). In females, the membranous layer passes superior to the fatty layer forming the labia majora and becomes continuous with the membranous layer of the subcutaneous tissue of the abdomen (Fig. 6.46A,C).

The **perineal fascia** (deep perineal, investing, or Gallaudet fascia) intimately invests the ischiocavernosus, bulbospongiosus, and superficial transverse perineal muscles (Fig. 6.46C,D). It is also attached laterally to the ischiopubic rami. Anteriorly, it is fused to the suspensory ligament of the penis or clitoris and is continuous with the deep fascia covering the external oblique muscle of the abdomen and rectus sheath.

SUPERFICIAL PERINEAL POUCH

The **superficial perineal pouch** (compartment) is a potential space between the membranous layer of subcutaneous tissue and the perineal membrane bounded laterally by the ischiopubic rami (Fig. 6.46A–D).

In males, the superficial perineal pouch contains the following structures (Fig. 6.46B,D):

- *Root* (bulb and crura) *of the penis* and associated muscles (*ischiocavernosus* and *bulbospongiosus*)
- Proximal (bulbous) part of the *spongy urethra*

- *Superficial transverse perineal muscles*
- *Deep perineal branches* of the internal pudendal vessels and pudendal nerves

In females, the superficial perineal pouch contains the following structures (Fig. 6.46A,C):

- *Clitoris* and associated muscle (ischiocavernosus)
- *Bulbs of the vestibule* and the surrounding muscle (bulbospongiosus)
- *Greater vestibular glands*
- *Deep perineal branches* of the internal pudendal vessels and pudendal nerves
- *Superficial transverse perineal muscles*

DEEP PERINEAL POUCH

The **deep perineal pouch** (space) is bounded inferiorly by the perineal membrane, superiorly by the inferior fascia of the pelvic diaphragm, and laterally by the inferior portion of the obturator fascia (covering obturator internus muscle). It includes the fat-filled anterior recesses of the ischio-anal fossa (Fig. 6.46C,D). In both sexes, the deep perineal pouch contains part of the urethra centrally, the inferior part of the external urethral sphincter muscle, and the anterior extensions of the ischio-anal fat pads. In males, the deep perineal pouch contains the *intermediate part of the urethra*, *deep transverse perineal muscles*, *bulbo-urethral glands*, and dorsal neurovascular structures of the penis (Fig. 6.46D). In females, it contains the proximal part of the *urethra*, a mass of smooth muscle in place of deep transverse perineal muscles, and the dorsal neurovasculature of the clitoris (Fig. 6.46C).

In the female, deep transverse perineal muscles are mainly smooth muscle. Immediately superior to the posterior half of the perineal membrane, the flat, sheet-like deep transverse perineal muscle, when developed (typically only in males), offers dynamic support for the pelvic viscera. The strong perineal membrane is the inferior boundary (floor) of the deep pouch. The perineal membrane, with the perineal body, is the final passive support of the pelvic viscera.

The *external urethral sphincter* is more tube- and trough-like than disc-like, and in males, only a part of the muscle forms a circular investment (a true sphincter) for the intermediate part of the urethra inferior to the prostate (Fig. 6.47). Its larger, trough-like part extends vertically to the neck of the bladder, displacing the prostate and investing the prostatic urethra anteriorly and anterolaterally only. As the prostate develops from the urethral glands, the posterior and posterolateral muscle atrophies or is displaced by the prostate. Whether this part of the muscle compresses or dilates the prostatic urethra is a matter of some controversy.

In females, the external urethral sphincter is more properly a "UG sphincter," according to Oelrich (1983). Here, too, he described a part forming a true anular sphincter around the urethra, with several additional parts extending from it (Fig. 6.47): a superior part, extending to the neck of the bladder; a subdivision described as extending inferolaterally

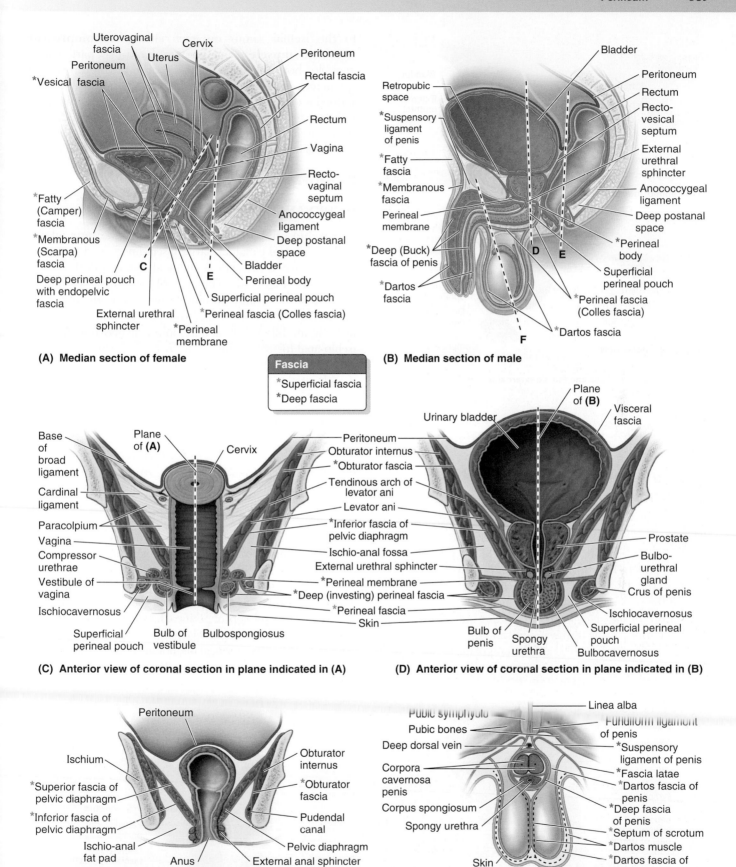

(A) Median section of female

Uterovaginal fascia
Cervix
Uterus
Peritoneum
Peritoneum
Rectal fascia
*Vesical fascia
Rectum
Vagina
Recto-vaginal septum
*Fatty (Camper) fascia
*Membranous (Scarpa) fascia
Deep perineal pouch with endopelvic fascia
External urethral sphincter
*Perineal membrane
Anococcygeal ligament
Deep postanal space
Bladder
Perineal body
Superficial perineal pouch
*Perineal fascia (Colles fascia)
C
E

(B) Median section of male

Bladder
Peritoneum
Rectum
Recto-vesical septum
Retropubic space
*Suspensory ligament of penis
*Fatty fascia
*Membranous fascia
Perineal membrane
*Deep (Buck) fascia of penis
*Dartos fascia
External urethral sphincter
Anococcygeal ligament
Deep postanal space
*Perineal body
Superficial perineal pouch
*Perineal fascia (Colles fascia)
*Dartos fascia
D E
F

Fascia
*Superficial fascia
*Deep fascia

(C) Anterior view of coronal section in plane indicated in (A)

Base of broad ligament
Plane of (A)
Cervix
Cardinal ligament
Paracolpium
Vagina
Compressor urethrae
Vestibule of vagina
Ischiocavernosus
Superficial perineal pouch
Bulb of vestibule
Bulbospongiosus
Peritoneum
Obturator internus
*Obturator fascia
Tendinous arch of levator ani
Levator ani
*Inferior fascia of pelvic diaphragm
Ischio-anal fossa
External urethral sphincter
*Perineal membrane
*Deep (investing) perineal fascia
*Perineal fascia
Skin

(D) Anterior view of coronal section in plane indicated in (B)

Plane of (B)
Visceral fascia
Urinary bladder
Prostate
Bulbo-urethral gland
Crus of penis
Ischiocavernosus
Superficial perineal pouch
Bulbocavernosus
Bulb of penis
Spongy urethra

(E) Anterior view of coronal section in plane indicated in (A) and (B)

Peritoneum
Ischium
*Superior fascia of pelvic diaphragm
*Inferior fascia of pelvic diaphragm
Ischio-anal fat pad
Anus
Obturator internus
*Obturator fascia
Pudendal canal
Pelvic diaphragm
External anal sphincter

(F) Anterior view of coronal section in plane indicated in (B)

Pubic symphysis
Pubic bones
Deep dorsal vein
Corpora cavernosa penis
Corpus spongiosum
Spongy urethra
Skin
Linea alba
Fundiform ligament of penis
*Suspensory ligament of penis
*Fascia latae
*Dartos fascia of penis
*Deep fascia of penis
*Septum of scrotum
*Dartos muscle
*Dartos fascia of scrotum

FIGURE 6.46. Fasciae of perineum.

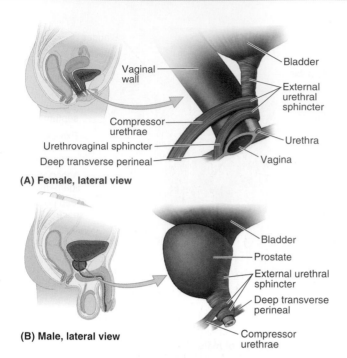

(A) Female, lateral view

- Vaginal wall
- Compressor urethrae
- Urethrovaginal sphincter
- Deep transverse perineal
- Bladder
- External urethral sphincter
- Urethra
- Vagina

(B) Male, lateral view

- Bladder
- Prostate
- External urethral sphincter
- Deep transverse perineal
- Compressor urethrae

FIGURE 6.47. Male and female external urethral sphincters.

to the ischial ramus on each side (the **compressor urethrae muscle**); and another band-like part, which encircles both the vagina and the urethra (urethrovaginal sphincter). In both males and females, the musculature described is oriented perpendicular to the perineal membrane rather than lying in the plane parallel to it. Some dispute the encircling of the urethra in the female, stating that the muscle is not capable of sphincteric action.

Features of Anal Triangle

ISCHIO-ANAL FOSSAE

The **ischio-anal fossae** (ischiorectal fossae) around the wall of the anal canal are large fascia-lined, wedge-shaped spaces between the skin of the anal region and the pelvic diaphragm (Fig. 6.48). The apex of each fossa lies superiorly where the levator ani muscle arises from the obturator fascia. The ischio-anal fossae, wide inferiorly and narrow superiorly, are filled with fat and loose connective tissue. The ischio-anal fossae communicate by means of the *deep post-anal space* over the *anococcygeal ligament* (body), a fibrous mass located between the anal canal and the tip of the coccyx.

CLINICAL BOX

Rupture of Urethra in Males and Extravasation of Urine

Fractures of the pelvic girdle often cause a *rupture of the intermediate part of the urethra*. This results in extravasation of urine and blood into the deep perineal pouch (Fig. B6.12A). The fluid may pass superiorly through the UG hiatus and distribute extraperitoneally around the prostate and bladder.

Rupture of the spongy urethra in the bulb of the penis results in urine passing (extravasating) into the superficial perineal space (Fig. B6.12B). The attachments of the perineal fascia

determine the direction of flow of the extravasated urine. Urine and blood may pass into the loose connective tissue in the scrotum, around the penis, and superiorly, deep to the membranous layer of subcutaneous connective tissue of the inferior anterior abdominal wall. The urine cannot pass far into the thighs because the membranous layer of superficial perineal fascia blends with the fascia lata (deep fascia) enveloping the thigh muscles, just distal to the inguinal ligament. In addition, urine cannot pass posteriorly into the anal triangle because the superficial and deep layers of perineal fascia are continuous with each other around the superficial perineal muscles and with the posterior edge of the perineal membrane between them.

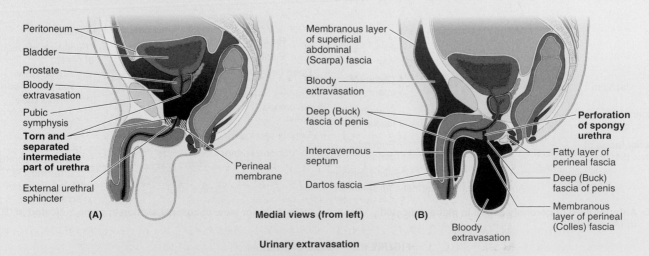

(A)
- Peritoneum
- Bladder
- Prostate
- Bloody extravasation
- Pubic symphysis
- **Torn and separated intermediate part of urethra**
- External urethral sphincter
- Perineal membrane

Medial views (from left)

(B)
- Membranous layer of superficial abdominal (Scarpa) fascia
- Bloody extravasation
- Deep (Buck) fascia of penis
- Intercavernous septum
- Dartos fascia
- **Perforation of spongy urethra**
- Fatty layer of perineal fascia
- Deep (Buck) fascia of penis
- Membranous layer of perineal (Colles) fascia
- Bloody extravasation

Urinary extravasation

FIGURE B6.12. Urinary extravasation.

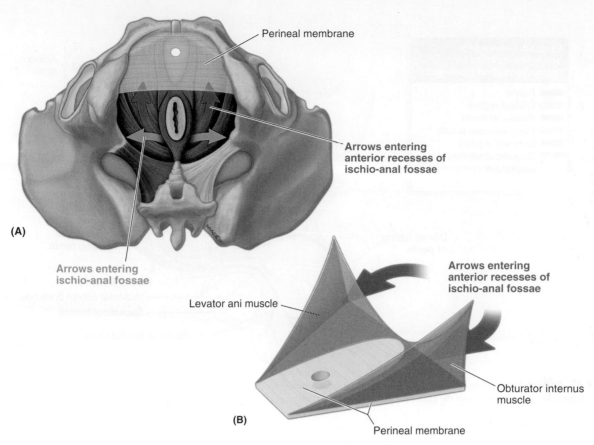

FIGURE 6.48. Pelvic diaphragm and ischio-anal fossae. A. Relationships of pelvic diaphragm, perineal membrane, and ischio-anal fossae. **B.** Schematic illustration of anterior recesses of ischio-anal fossae.

Each ischio-anal fossa is bounded by the following structures: (Fig. 6.48)

- Laterally by the ischium and the inferior part of the obturator internus, covered with obturator fascia
- Medially by the external anal sphincter, with a sloping superior medial wall or roof formed by the levator ani as it descends to blend with the sphincter; both structures surround the anal canal.
- Posteriorly by the sacrotuberous ligament and gluteus maximus
- Anteriorly by the bodies of the pubic bones, inferior to the origin of the puborectalis; these parts of the fossae, extending into the UG triangle superior to the perineal membrane, are known as the **anterior recesses of the ischio-anal fossae**.

The ischio-anal fossae are traversed by tough, fibrous bands and filled with fat, forming the **fat bodies of the ischio-anal fossae**. These bodies support the anal canal but are readily displaced to permit expansion of the anal canal during the passage of feces. The fat bodies are traversed by several neurovascular structures, including the inferior anal/rectal vessels and nerves and two other cutaneous nerves: the perforating branch of S2 and S3 and the perineal branch of the S4 nerve.

PUDENDAL CANAL

The **pudendal canal** (Alcock canal) is essentially a horizontal passageway within the obturator fascia (Figs. 6.48 and 6.49), which covers the medial aspect of the obturator internus muscle and lines the lateral wall of the ischio-anal fossa. The pudendal canal begins at the posterior border of the ischio-anal fossa and runs from the *lesser sciatic notch* adjacent to the ischial spine to the posterior edge of the perineal membrane. The internal pudendal artery and vein, the pudendal nerve, and the nerve to the obturator internus enter this canal at the lesser sciatic notch, inferior to the ischial spine. The internal pudendal vessels supply and drain blood from the perineum; the pudendal nerve innervates most of the same area.

As the artery and nerve enter the canal, they give rise to the **inferior anal** (rectal) **artery** and **nerve** that pass medially to supply the external anal sphincter and peri-anal skin (Fig. 6.49). Toward the distal (anterior) end of the pudendal canal, the artery and nerve both bifurcate, giving rise to the **perineal nerve** and **artery**, which are distributed mostly to the superficial pouch (inferior to the perineal membrane) and to the **dorsal artery** and **nerve of the penis** or **clitoris**, which run in the deep pouch (superior to the membrane) (see Fig. 6.45C).

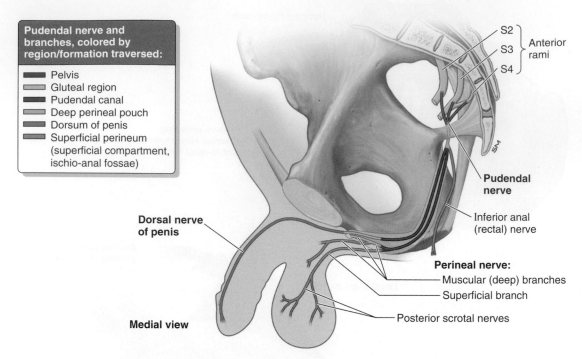

Pudendal nerve and branches, colored by region/formation traversed:

- Pelvis
- Gluteal region
- Pudendal canal
- Deep perineal pouch
- Dorsum of penis
- Superficial perineum (superficial compartment, ischio-anal fossae)

S2 \
S3 } Anterior rami \
S4

Pudendal nerve

Inferior anal (rectal) nerve

Perineal nerve:
Muscular (deep) branches
Superficial branch
Posterior scrotal nerves

Dorsal nerve of penis

Medial view

FIGURE 6.49. Pudendal nerve. The five regions in which the nerve runs are color coded. In females, the superficial perineal nerve gives rise to posterior labial nerves, and the terminal branch of the pudendal nerve is the dorsal nerve of the clitoris.

The perineal nerve has two branches: the **superficial perineal nerves** give rise to *posterior scrotal* or *labial (cutaneous) branches*, and the **deep perineal nerve** supplies the muscles of the deep and superficial perineal pouches, the skin of the vestibule of the vagina, and the mucosa of the inferiormost part of the vagina. The *dorsal nerve of the penis* or *clitoris* is the primary sensory nerve serving the male or female organ, especially the glans.

ANAL CANAL

The anal canal is the terminal part of the large intestine that extends from the superior aspect of the pelvic diaphragm to the **anus**. The canal begins where the ampulla of the rectum abruptly narrows at the level of the U-shaped sling formed by the puborectalis muscle (see Fig. 6.37A,B). The canal ends at the anus, the external outlet of the alimentary tract. The anal canal, surrounded by internal and external anal sphincters, descends postero-inferiorly between the **anococcygeal ligament** and the perineal body. The anal canal is normally collapsed except during passage of feces. Both sphincters must relax before defecation can occur.

The **external anal sphincter** is a large voluntary sphincter that forms a broad band on each side of the inferior two thirds of the anal canal (Fig. 6.50). This sphincter blends superiorly with the puborectalis muscle and is described as having subcutaneous, superficial, and deep parts. The external anal sphincter is supplied mainly by S4 through the inferior anal (rectal) nerve (see Fig. 6.40).

The **internal anal sphincter** is an involuntary sphincter surrounding the superior two thirds of the anal canal (Fig. 6.50). It is a thickening of the circular muscle layer. Its contraction (tonus) is stimulated and maintained by the sympathetic fibers from the superior rectal (peri-arterial) and hypogastric plexuses. It is inhibited (loses its tonic contraction and is allowed to expand passively) by the parasympathetic fibers. This sphincter is tonically contracted most of time to prevent leakage of fluid or flatus; however, it relaxes temporarily in response to distention of the rectal ampulla by feces or gas, requiring voluntary contraction of the puborectalis and the external anal sphincter if defecation or flatulence is not to occur.

INTERIOR OF ANAL CANAL

The superior half of the mucous membrane of the anal canal is characterized by a series of longitudinal ridges called **anal columns** (Figs. 6.37A,C and 6.50). These columns contain the terminal branches of the superior rectal artery and vein. The **anorectal junction**, indicated by the superior ends of the anal columns, is where the rectum joins the anal canal. The inferior ends of these columns are joined by **anal valves**. Superior to the valves are small recesses called **anal sinuses** (see Fig. 6.37C). When compressed by feces, the anal sinuses exude mucus that aids in evacuation of feces from the anal canal. The inferior comb-shaped limit of the anal valves forms an irregular line, the **pectinate line** (Figs. 6.37A,C and 6.51), which indicates the junction of the superior part of the anal canal (visceral; derived from the hindgut) and the inferior part (somatic; derived from

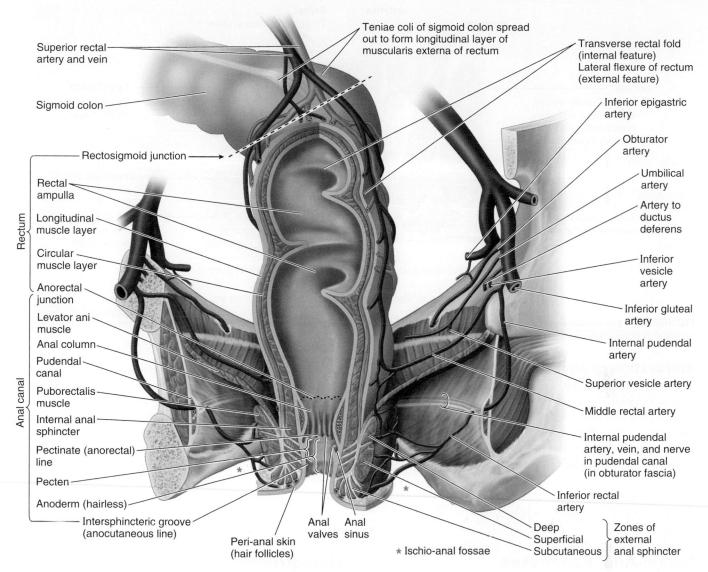

Superior rectal artery and vein

Sigmoid colon

Teniae coli of sigmoid colon spread out to form longitudinal layer of muscularis externa of rectum

Transverse rectal fold (internal feature)
Lateral flexure of rectum (external feature)

Inferior epigastric artery

Obturator artery

Umbilical artery

Artery to ductus deferens

Inferior vesicle artery

Inferior gluteal artery

Internal pudendal artery

Superior vesicle artery

Middle rectal artery

Internal pudendal artery, vein, and nerve in pudendal canal (in obturator fascia)

Inferior rectal artery

Rectosigmoid junction

Rectal ampulla

Longitudinal muscle layer

Circular muscle layer

Anorectal junction

Levator ani muscle

Anal column

Pudendal canal

Puborectalis muscle

Internal anal sphincter

Pectinate (anorectal) line

Pecten

Anoderm (hairless)

Intersphincteric groove (anocutaneous line)

Peri-anal skin (hair follicles)

Anal valves   Anal sinus

* Ischio-anal fossae

Deep
Superficial   } Zones of
Subcutaneous   } external
                          anal sphincter

Rectum

Anal canal

Posterior view of anterior pelvis and perineum

FIGURE 6.50. Rectum and anal canal, levator ani, and ischio-anal fossae.

the embryonic proctodeum). The anal canal superior to the pectinate line differs from the part inferior to the pectinate line in its arterial supply, innervation, and venous and lymphatic drainage. These differences result from their different embryological origins (Moore et al., 2012).

VASCULATURE AND LYMPHATIC DRAINAGE OF ANAL CANAL

The *superior rectal artery* supplies the anal canal superior to the pectinate line (Figs. 6.38 and 6.50). The two *inferior rectal arteries* supply the inferior part of the anal canal as well as the surrounding muscles and peri-anal skin. The *middle rectal arteries* assist with the blood supply to the anal canal by forming anastomoses with the superior and inferior rectal arteries.

The *internal rectal venous plexus* drains in both directions from the level of the pectinate line. Superior to the pectinate line, the internal rectal venous plexus drains chiefly into the *superior rectal vein* (a tributary of the inferior mesenteric vein) and the portal system. Inferior to the pectinate line, the internal rectal venous plexus drains into the *inferior rectal veins* (tributaries of the caval venous system) around the margin of the external anal sphincter (see Fig. 6.38B). The *middle rectal veins* (tributaries of the internal iliac veins) mainly drain the muscularis externa of the rectal ampulla and form anastomoses with the superior and inferior rectal veins. The rectal venous plexuses receive multiple arteriovenous anastomoses from the superior and middle rectal arteries.

Superior to the pectinate line, the lymphatic vessels drain into the *internal iliac lymph nodes* and through them into the common iliac and lumbar lymph nodes (Fig. 6.39 and 6.51). Inferior to the pectinate line, the lymphatic vessels drain into the *superficial inguinal lymph nodes*.

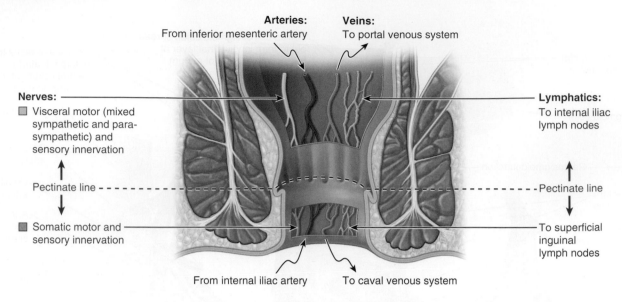

FIGURE 6.51. Innervation and vascular supply of anal canal superior and inferior to pectinate line. The vessels and nerves superior to the pectinate line are visceral; those inferior to the pectinate line are somatic.

INNERVATION OF ANAL CANAL

The nerve supply to the anal canal superior to the pectinate line is visceral innervation from the *inferior hypogastric plexus* (sympathetic, parasympathetic, and visceral afferent fibers) (Figs. 6.40 and 6.51). The superior part of the anal canal is inferior to the pelvic pain line; all visceral afferents travel with the parasympathetic fibers to spinal sensory ganglia S2–S4. Superior to the pectinate line, the anal canal is sensitive only to stretching. The nerve supply of the anal canal inferior to the pectinate line is somatic, derived from the *inferior anal (rectal) nerves*, branches of the pudendal nerve. Therefore, this part of the anal canal is sensitive to pain, touch, and temperature. Somatic efferent fibers stimulate the contraction of the voluntary external anal sphincter.

CLINICAL BOX

Ischio-Anal Abscesses

The ischio-anal fossae are occasionally the sites of infection, which may result in the formation of *ischio-anal abscesses* (Fig. B6.13). These collections of pus are painful. Diagnostic signs of an ischio-anal abscess are fullness and tenderness between the anus and the ischial tuberosity. A peri-anal abscess may rupture spontaneously, opening into the anal canal, rectum, or peri-anal skin.

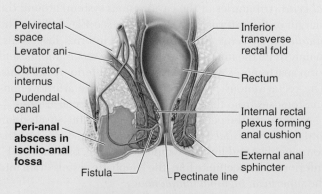

FIGURE B6.13. Ischio-anal abscess.

Hemorrhoids

 Internal hemorrhoids ("piles") are prolapses of the rectal mucosa containing the normally dilated veins of the *internal rectal venous plexus* (Fig. B6.14). They are thought to result from a breakdown of the muscularis mucosae, a smooth muscle layer deep to the mucosa. Internal hemorrhoids that prolapse through the anal canal are often compressed by the contracted sphincters, impeding blood flow. As a result, they tend to strangulate and ulcerate. Owing to the presence of abundant arteriovenous anastomoses, bleeding from internal hemorrhoids is usually bright red.

External hemorrhoids are thrombi (blood clots) in the veins of the *external rectal venous plexus* and are covered by skin (Fig. B6.14). Predisposing factors for hemorrhoids include pregnancy, chronic constipation, and any disorder that impedes venous return, including increased intra-abdominal pressure.

The anastomoses among the superior, middle, and inferior rectal veins form clinically important communications between the portal and the systemic venous systems (Fig. 6.51). The superior rectal vein drains into the inferior

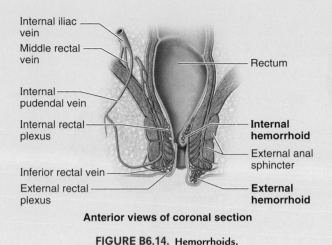

Anterior views of coronal section

FIGURE B6.14. Hemorrhoids.

mesenteric vein, whereas the middle and inferior rectal veins drain through the systemic system into the inferior vena cava. Any abnormal increase in pressure in the valveless portal system or veins of the trunk may cause enlargement of the superior rectal veins, resulting in increase in blood flow or stasis in the internal rectal venous plexus. In *portal hypertension*, the portocaval anastomosis among the superior, middle, and inferior rectal veins, along with portocaval anastomoses elsewhere, may become varicose. It is important to note that the veins of the rectal plexuses *normally* appear varicose (dilated and tortuous) and that internal hemorrhoids occur most commonly in the absence of portal hypertension.

Because visceral afferent nerves supply the anal canal superior to the pectinate line, an incision or a needle insertion in this region is painless. However, the anal canal inferior to the pectinate line is quite sensitive (e.g., to the prick of a hypodermic needle) because it is supplied by the *inferior rectal nerves*, containing somatic sensory fibers.

Male Perineum

The male perineum includes the external genitalia (urethra, scrotum, and penis), perineal muscles, and anal canal.

DISTAL MALE URETHRA

The urethra in the bladder neck (intramural part) and the prostatic urethra, the first two parts of the male urethra, are described with the pelvis (see Fig. 6.25 and Table 6.8). The **intermediate (membranous) part of the urethra** begins at the apex of the prostate and traverses the deep perineal pouch, surrounded by the external urethral sphincter. It then penetrates the perineal membrane, ending as the urethra enters the bulb of the penis (see Fig. 6.18). Posterolateral to this part of the urethra are the small *bulbo-urethral glands* (see Figs. 6.20A and 6.21) and their slender ducts, which open into the proximal part of the spongy urethra.

The **spongy urethra** begins at the distal end of the intermediate part of the urethra and ends at the **external urethral orifice** (Figs. 6.18 and 6.52B,D). The lumen of the spongy urethra is expanded in the bulb of the penis to form the **intrabulbar fossa** and in the glans of the penis to form the **navicular fossa**. On each side, the ducts of the bulbo-urethral glands open into the proximal part of the spongy urethra. There are also many minute openings of the ducts of mucus-secreting **urethral glands** (glands of Littré) into the spongy urethra.

The arterial supply of the intermediate and spongy parts of the urethra is from branches of the *dorsal artery of the penis* (Fig. 6.53B). The veins accompany the arteries and have similar names. Lymphatic vessels from the intermediate part of the urethra drain mainly into the *internal iliac lymph nodes* (Fig. 6.54), whereas most vessels

from the spongy urethra pass to the deep inguinal lymph nodes, but some vessels pass to the external iliac lymph nodes. The innervation of the intermediate part of the urethra is the same as that of the prostatic part (Fig. 6.55). The dorsal nerve of the penis, a branch of the *pudendal nerve*, provides somatic innervation of the spongy part of the urethra.

CLINICAL BOX

Urethral Catheterization

Urethral catheterization is performed to remove urine from a person who is unable to micturate. It is also performed to irrigate the bladder and to obtain an uncontaminated sample of urine. When inserting the catheters and urethral sounds (slightly conical instruments for exploring and dilating a constricted urethra), the curves of the male urethra must be considered.

SCROTUM

The **scrotum** is a cutaneous fibromuscular sac for the testes and associated structures. It is situated postero-inferior to the penis and inferior to the pubic symphysis (Fig. 6.52). The bilateral embryonic formation of the scrotum is indicated by the midline **scrotal raphe** (Fig. 6.52C), which is continuous on the ventral surface of the penis with the **penile raphe** and posteriorly along the median line of the

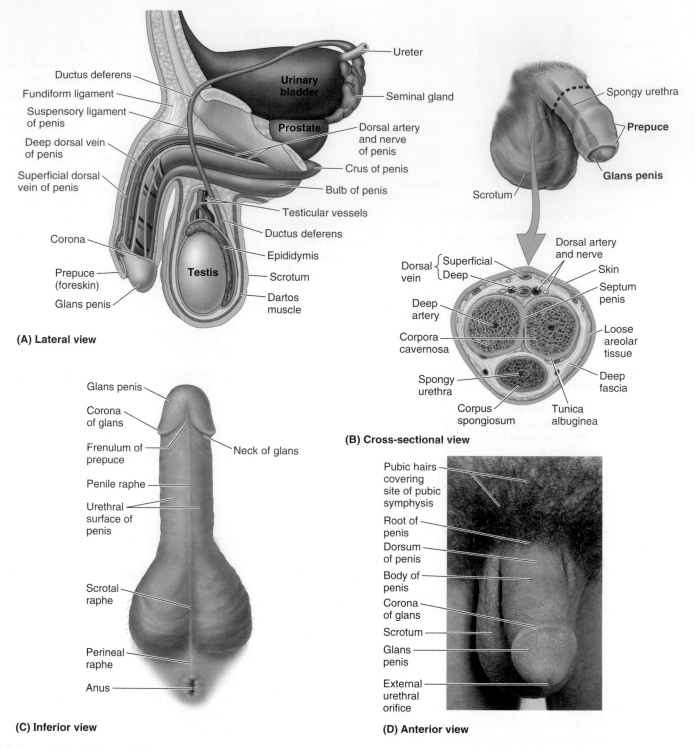

(A) Lateral view

- Ductus deferens
- Fundiform ligament
- Suspensory ligament of penis
- Deep dorsal vein of penis
- Superficial dorsal vein of penis
- Corona
- Prepuce (foreskin)
- Glans penis
- Testis
- Ureter
- Urinary bladder
- Seminal gland
- Prostate
- Dorsal artery and nerve of penis
- Crus of penis
- Bulb of penis
- Testicular vessels
- Ductus deferens
- Epididymis
- Scrotum
- Dartos muscle

(B) Cross-sectional view

- Spongy urethra
- Prepuce
- Glans penis
- Scrotum
- Dorsal vein { Superficial / Deep }
- Deep artery
- Corpora cavernosa
- Spongy urethra
- Corpus spongiosum
- Dorsal artery and nerve
- Skin
- Septum penis
- Loose areolar tissue
- Deep fascia
- Tunica albuginea

(C) Inferior view

- Glans penis
- Corona of glans
- Frenulum of prepuce
- Neck of glans
- Penile raphe
- Urethral surface of penis
- Scrotal raphe
- Perineal raphe
- Anus

(D) Anterior view

- Pubic hairs covering site of pubic symphysis
- Root of penis
- Dorsum of penis
- Body of penis
- Corona of glans
- Scrotum
- Glans penis
- External urethral orifice

FIGURE 6.52. Male urogenital organs. A. Internal structures. **B.** Male external genitalia: uncircumcised penis and scrotum and section through the body of the penis. **C and D.** Surface anatomy of male external genitalia: penis is circumcised.

perineum as the **perineal raphe**. Internally deep to the scrotal raphe, the scrotum is divided into two compartments, one for each testis, by a prolongation of dartos fascia, the **septum of the scrotum**. The contents of the scrotum (testes and epididymides) are described with the abdomen (see Chapter 5).

VASCULATURE OF SCROTUM

The anterior aspect of the scrotum is supplied by the **anterior scrotal arteries**, terminal branches of the **external pudendal arteries** (Fig. 6.55B and Table 6.10), and the posterior aspect is supplied by the **posterior scrotal arteries**, terminal branches of the *internal pudendal arteries*.

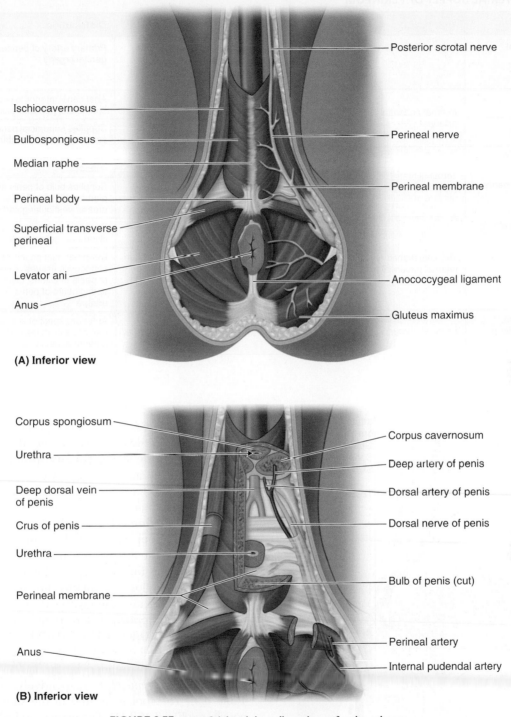

(A) Inferior view

Ischiocavernosus

Bulbospongiosus

Median raphe

Perineal body

Superficial transverse perineal

Levator ani

Anus

Posterior scrotal nerve

Perineal nerve

Perineal membrane

Anococcygeal ligament

Gluteus maximus

(B) Inferior view

Corpus spongiosum

Urethra

Deep dorsal vein of penis

Crus of penis

Urethra

Perineal membrane

Anus

Corpus cavernosum

Deep artery of penis

Dorsal artery of penis

Dorsal nerve of penis

Bulb of penis (cut)

Perineal artery

Internal pudendal artery

FIGURE 6.53. Superficial and deep dissections of male perineum.

The scrotum also receives branches from the cremasteric arteries, branches of inferior epigastric arteries. The *scrotal veins* accompany the arteries and drain primarily to the *external pudendal veins*. Lymphatic vessels from the scrotum drain into the *superficial inguinal lymph nodes* (Fig. 6.54).

INNERVATION OF SCROTUM

The anterior aspect of the scrotum is supplied by the **anterior scrotal nerves** derived from the *ilio-inguinal nerve* and by the *genital branch of the genitofemoral nerve*. The posterior aspect of the scrotum is supplied by **posterior scrotal nerves**, branches of the superficial perineal branches of the pudendal nerve (Fig. 6.56), and by the *perineal branch of the posterior femoral cutaneous nerve*.

PENIS

The penis is the male organ of copulation and the outlet for urine and semen (penile ejaculate, sperms, and a mixture of glandular secretions). The penis consists of a *root*, *body*, and

TABLE 6.10. ARTERIAL SUPPLY OF PERINEUM

Artery	Origin	Course	Distribution
Internal pudendal	Internal iliac artery	Leaves pelvis through greater sciatic foramen; hooks around ischial spine to enter perineum via lesser sciatic foramen; enters pudendal canal	Primary artery of perineum and external genital organs
Inferior rectal	Internal pudendal artery	Arises at entrance to pudendal canal; crosses ischio-anal fossa to anal canal	Anal canal inferior to pectinate line; anal sphincters; peri-anal skin
Perineal		Arises within pudendal canal; passes to superficial perineal pouch (space) on exit	Supplies superficial perineal muscles and scrotum in male; vestibule in female
Posterior scrotal or **labial**	Terminal branch of perineal artery	Runs in subcutaneous tissue of posterior scrotum or labia majora	Skin of scrotum or labia majora and minora
Artery of bulb of penis or **vestibule**		Pierces perineal membrane to reach bulb of penis or vestibule of vagina	Supplies bulb of penis and bulbo-urethral gland in male; bulb of vestibule and greater vestibular gland in female
Deep artery of penis or **clitoris**	Terminal branch of internal pudendal artery	Pierces perineal membrane to run centrally within corpora cavernosa of penis or clitoris	Supplies most erectile tissue of penis or clitoris via helicine arteries
Dorsal artery of penis or **clitoris**		Pierces perineal membrane and passes through suspensory ligament of penis or clitoris to run on dorsum of penis or clitoris, flanked by deep dorsal veins	Deep perineal pouch; skin of penis; connective tissues of erectile tissue of penis or clitoris; distal corpus spongiosum of penis, including spongy urethra
External pudendal, superficial, and deep branches	Femoral artery	Pass medially across thigh to reach scrotum or labia majora (anterior aspect of urogenital triangle)	Anterior aspect of scrotum and skin at root of penis in male; mons pubis and anterior aspect of labia in female

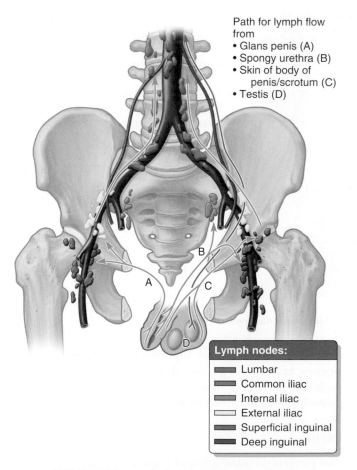

Path for lymph flow from
- Glans penis (A)
- Spongy urethra (B)
- Skin of body of penis/scrotum (C)
- Testis (D)

Lymph nodes:
- Lumbar
- Common iliac
- Internal iliac
- External iliac
- Superficial inguinal
- Deep inguinal

FIGURE 6.54. Lymphatic drainage of male perineum.

glans penis (Fig. 6.52D). It is composed of three cylindrical bodies of erectile cavernous tissue: the paired **corpora cavernosa** and the single **corpus spongiosum** ventrally. (Note that in the anatomical position, the penis is erect; when the penis is flaccid, its dorsum is directed anteriorly.) Each *cavernous body* has a fibrous outer covering or capsule, the **tunica albuginea** (Fig. 6.52B). Superficial to the outer covering is the **deep fascia of the penis** (Buck fascia), the continuation of the deep perineal fascia that forms a membranous covering for the corpora, binding them together. The corpus spongiosum contains the spongy urethra. The corpora cavernosa are fused with each other in the median plane except posteriorly, where they separate to form the *crura of the penis* (Figs. 6.52A and 6.53).

The nonpendulous **root of the penis** consists of the crura and bulb, surrounded by the *ischiocavernosus* and *bulbospongiosus muscles*, respectively (Fig. 6.56 and Table 6.9). The root is located in the superficial perineal pouch (see Fig. 6.46B,D). The **crura** and **bulb of the penis** are the proximal portions of the erectile bodies (corpora). Each crus is attached to the inferior part of the internal surface of the corresponding ischial ramus, anterior to the ischial tuberosity. The bulb of the penis is penetrated by the urethra, continuing from its intermediate part.

The **body of the penis** (usually pendulous) is the free part that is suspended from the pubic symphysis. Except for a few fibers of the bulbospongiosus near the root of the penis and the ischiocavernosus that embrace the crura, the penis has no muscles. Distally, the corpus spongiosum

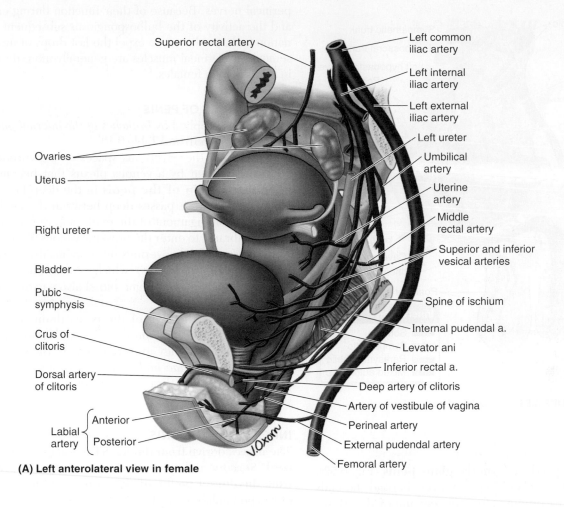

Superior rectal artery

Left common iliac artery

Left internal iliac artery

Left external iliac artery

Left ureter

Umbilical artery

Uterine artery

Middle rectal artery

Superior and inferior vesical arteries

Spine of ischium

Internal pudendal a.

Levator ani

Inferior rectal a.

Deep artery of clitoris

Artery of vestibule of vagina

Perineal artery

External pudendal artery

Femoral artery

Ovaries

Uterus

Right ureter

Bladder

Pubic symphysis

Crus of clitoris

Dorsal artery of clitoris

Labial artery { Anterior / Posterior }

V. Oxorn

(A) Left anterolateral view in female

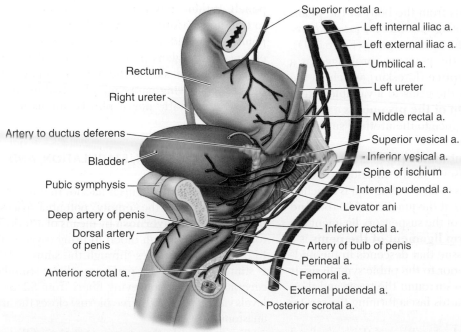

Superior rectal a.

Left internal iliac a.

Left external iliac a.

Umbilical a.

Left ureter

Middle rectal a.

Superior vesical a.

Inferior vesical a.

Spine of ischium

Internal pudendal a.

Levator ani

Inferior rectal a.

Artery of bulb of penis

Perineal a.

Femoral a.

External pudendal a.

Posterior scrotal a.

Rectum

Right ureter

Artery to ductus deferens

Bladder

Pubic symphysis

Deep artery of penis

Dorsal artery of penis

Anterior scrotal a.

(B) Left anterolateral view in male

FIGURE 6.55. Arterial supply of perineum. Superficial and deep dissections of pelvis and perineum. **A.** Female. **B.** Male.

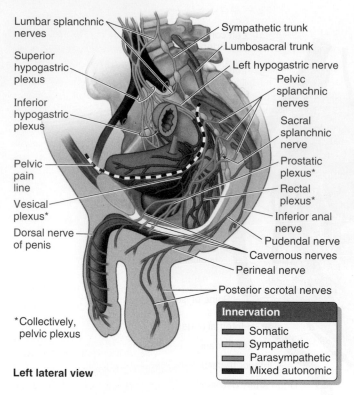

Lumbar splanchnic nerves

Superior hypogastric plexus

Inferior hypogastric plexus

Pelvic pain line

Vesical plexus*

Dorsal nerve of penis

Sympathetic trunk

Lumbosacral trunk

Left hypogastric nerve

Pelvic splanchnic nerves

Sacral splanchnic nerve

Prostatic plexus*

Rectal plexus*

Inferior anal nerve

Pudendal nerve

Cavernous nerves

Perineal nerve

Posterior scrotal nerves

*Collectively, pelvic plexus

Left lateral view

Innervation	
	Somatic
	Sympathetic
	Parasympathetic
	Mixed autonomic

FIGURE 6.56. Innervation of male perineum.

of the penis expands to form the **glans penis** (Fig. 6.52). The margin of the glans (head) projects beyond the ends of the corpora cavernosa to form the **corona of the glans**. The corona overhangs the neck of the glans. The **neck of the glans** separates the glans from the body of the penis. The slit-like opening of the spongy urethra, the *external urethral orifice*, is located near the tip of the glans (Fig. 6.52D). The thin skin and fascia of the penis are prolonged as a double layer of skin, the **prepuce** (foreskin), which, in the uncircumcised penis, covers the glans to a variable extent (Fig. 6.52A,B). The **frenulum of the prepuce** is a median fold that passes from the prepuce to the urethral surface of the glans (Fig. 6.52C).

The **suspensory ligament of the penis** is a condensation of the deep fascia that arises from the anterior surface of the pubic symphysis and splits to form a sling that is attached to the deep fascia of the penis at the junction of its root and body (Fig. 6.52A). The fibers of the suspensory ligament are short and taut. The **fundiform ligament of the penis** is a band of the subcutaneous tissue that descends in the midline from the linea alba superior to the pubic symphysis. It passes inferiorly and splits to surround the penis and then unites and blends with the dartos fascia forming the scrotal septum.

The **superficial perineal muscles** are the superficial transverse perineal, bulbospongiosus, and ischiocavernosus (see Figs. 6.45E and 6.53 and Table 6.9). These muscles are in the superficial perineal pouch and are supplied by the

perineal nerves. Because of their function during erection and the activity of the bulbospongiosus subsequent to urination and ejaculation, to expel the last drops of urine and semen, the perineal muscles are generally more developed in males than in females.

VASCULATURE OF PENIS

The penis is supplied by *branches of the internal pudendal arteries* (Fig. 6.55B and Table 6.10).

Blood from the cavernous spaces of the corpora cavernosa is drained by a venous plexus that becomes the **deep dorsal vein of the penis** in the deep fascia (Fig. 6.52A,B). This vein passes deep between the laminae of the suspensory ligament of the penis, anterior to the perineal membrane, to enter the prostatic venous plexus. Blood from the superficial coverings of the penis drains into the **superficial dorsal vein(s)**, which ends in the *superficial external pudendal vein*. Some blood also passes to the internal pudendal vein.

Lymph from the skin of the penis drains initially to the *superficial inguinal lymph nodes*, and lymph from the glans and distal spongy urethra drains to the *deep inguinal* and *external iliac nodes*. The cavernous bodies and proximal spongy urethra drain to the *internal iliac nodes* (Fig. 6.54).

INNERVATION OF PENIS

The nerves derive from the S2–S4 segments of the spinal cord. Sensory and sympathetic innervation is primarily from the **dorsal nerve of the penis**, a terminal branch of the pudendal nerve (Fig. 6.56), which arises in the pudendal canal and passes anteriorly into the deep perineal pouch. It then runs along the dorsum of the penis lateral to the dorsal artery and supplies the skin and glans. The penis is supplied with a variety of sensory nerve endings, especially the glans penis. Branches of the *ilio-inguinal nerve* supply the skin at the root of the penis. **Cavernous nerves**, conveying parasympathetic fibers independently from the prostatic nerve plexus, innervate the *helicine arteries*.

ERECTION, EMISSION, EJACULATION, AND REMISSION

Most of the time, the penis is flaccid. In this state, most arterial blood bypasses the "empty" potential spaces or **sinuses of the corpora cavernosa** by means of arteriovenous anastomoses. Only enough blood to bring oxygen and nutrition to the tissues circulates through the sinuses. When a male is stimulated erotically, parasympathetic stimulation by the *cavernous nerves* (conveying fibers from S2–S4 spinal cord levels via the prostatic nerve plexus) closes the arteriovenous anastomoses.

Simultaneously, the tonic contraction of the smooth muscle in the fibrous trabeculae and coiled **helicine arteries** (terminal branches of the arteries serving the erectile tissue) is inhibited. As a result, the arteries straighten, enlarging their lumina. Blood flow no longer diverted from the cavernous spaces increases in volume, filling the sinuses of the corpora of the penis. The bulbospongiosus and ischiocavernosus muscles reflexively contract, compressing the veins of the corpora cavernosa, impeding the return of venous blood. **Erection** occurs as the corpora cavernosa and corpus spongiosum become engorged with blood at arterial pressure, causing the erectile bodies to become turgid (enlarged and rigid), elevating the penis.

During **emission**, semen is delivered to the prostatic urethra through the ejaculatory ducts after peristalsis of the ductus deferentes and seminal glands. Prostatic fluid is added to the seminal fluid as the smooth muscle in the prostate contracts. Emission is a sympathetic response (L1–L2 nerves).

During **ejaculation**, semen is expelled from the urethra through the external urethral orifice. Ejaculation results from

- closure of the internal urethral sphincter at the neck of the urinary bladder, a sympathetic response (L1–L2 nerves) preventing retrograde ejaculation into the bladder
- contraction of the urethral muscle, a parasympathetic response (S2–S4 nerves)
- contraction of the bulbospongiosus muscles, from the pudendal nerves (S2–S4)

After ejaculation, the penis gradually returns to a flaccid state (**remission**), resulting from sympathetic stimulation that opens the arteriovenous anastomoses and causes contraction of smooth muscle of the helicine arteries, recoiling them. This reduces blood inflow as the bulbospongiosus and ischiocavernosus muscles relax, allowing more blood to be drained from the cavernous spaces into the deep dorsal vein.

PERINEAL MUSCLES OF MALES

The *superficial perineal muscles* include the **superficial transverse perineal**, **ischiocavernosus**, and **bulbospongiosus** (Fig. 6.53). Details of their attachments, innervation, and actions are provided in Table 6.9. The ischiocavernosus and bulbospongiosus muscles both constrict venous outflow from the erectile bodies to assist erection, simultaneously pushing blood from the penile root into the body. The bulbospongiosus muscle constricts around the bulb of the penis to express the final drops of urine or semen.

Female Perineum

The female perineum includes the female external genitalia, perineal muscles, and anal canal.

FEMALE EXTERNAL GENITALIA

The **female external genitalia** include the mons pubis, labia majora (enclosing the pudendal cleft), labia minora (enclosing the vestibule), clitoris, bulbs of the vestibule, and greater and lesser vestibular glands. The synonymous terms **pudendum** and **vulva** include all these parts. The term *pudendum* is commonly used clinically (Fig. 6.57). The pudendum/vulva serves as sensory and erectile tissue for sexual arousal and intercourse, directs the flow of urine, and prevents entry of foreign material into the UG tract.

MONS PUBIS

The **mons pubis** is the rounded, fatty eminence anterior to the pubic symphysis, pubic tubercle, and superior pubic rami. The amount of fat in the mons increases at puberty and decreases after menopause. After puberty, the mons pubis is covered with coarse pubic hairs (Fig. 6.57A).

CLINICAL BOX

Erectile Dysfunction

Inability to obtain or maintain an erection (*erectile dysfunction* [ED]) may result from several causes. When a lesion of the prostatic plexus or cavernous nerves results in an inability to achieve an erection, a surgically implanted, semirigid, or inflatable penile prosthesis may assume the role of the erectile bodies, providing the rigidity necessary to insert and move the penis within the vagina during intercourse.

ED may occur in the absence of a nerve insult. Central nervous system (hypothalamic) and endocrine (pituitary or testicular) disorders may result in reduced testosterone (male hormone) secretion. Autonomic nerve fibers may fail to stimulate erectile tissues, or blood vessels may be insufficiently responsive to stimulation. In many such cases, erection can be achieved with the assistance of oral medications or injections that increase blood flow into the cavernous sinusoids by causing relaxation of smooth muscle.

Phimosis, Paraphimosis, and Circumcision

An uncircumcised prepuce covers all or most of the glans penis (Fig. 6.52B). The prepuce is usually sufficiently elastic to allow retraction over the glans. In some males, it is tight and cannot be retracted easily (*phimosis*), if at all. Secretions (*smegma*) may accumulate in the preputial sac, located between the glans penis and prepuce, causing irritation.

In some cases, retraction of the prepuce constricts the neck of the glans so that there is interference with the drainage of blood and tissue fluid (*paraphimosis*). The glans may enlarge so much that the prepuce cannot be distracted. *Circumcision*, surgical excision of the prepuce, must be performed.

Circumcision exposes most, or all, of the glans (Fig. 6.52C,D) and is the most common minor surgical operation performed on male infants. Although it is a religious practice in Islam and Judaism, it is often done routinely for nonreligious reasons.

LABIA MAJORA

The **labia majora** are prominent folds of skin that bound the **pudendal cleft**, the slit between the labia majora, and indirectly provide protection for the urethral and vaginal orifices. Each labium majus—largely filled with subcutaneous fat containing smooth muscle and the termination of the round ligament of the uterus—passes inferoposteriorly from the mons pubis toward the anus. The external aspects of the labia in the adult are covered with pigmented skin containing many sebaceous glands and are covered with crisp pubic hair. The internal aspects of the labia are smooth, pink, and hairless. The labia are thicker anteriorly where they join to form the **anterior commissure**. Posteriorly, they merge to form the **posterior commissure**, which usually disappears after the first vaginal birth.

LABIA MINORA

The **labia minora** are folds of fat-free, hairless skin. They have a core of spongy connective tissue containing erectile tissue and many small blood vessels. Although the internal surface of each labium minus consists of thin moist skin, it has the typical pink color of a mucous membrane and contains many sensory nerve endings. The labia minora are enclosed in the pudendal cleft within the labia majora and surround the vestibule into which the external urethral and vaginal orifices open. Anteriorly, the labia minora form two laminae: the medial laminae unite as the **frenulum of the clitoris**, and the lateral laminae unite to form the **prepuce of the clitoris** (Fig 6.57). In young women, especially virgins, the labia minora are connected posteriorly by a small transverse fold, the **frenulum of the labia minora** (fourchette).

CLITORIS

The **clitoris** is an erectile organ located where the labia minora meet anteriorly. The clitoris consists of a **root** and a **body**, which are composed of two crura, two corpora cavernosa, and the **glans of the clitoris**. The glans is covered by the prepuce of the clitoris (Figs. 6.57A and 6.58A). The clitoris is highly sensitive and enlarges on tactile stimulation. The glans is the most highly innervated part of the clitoris.

VESTIBULE

The **vestibule** is the space surrounded by the labia minora, which contains the openings of the urethra, vagina, and ducts of the greater and lesser vestibular glands. The *external urethral orifice* is located postero-inferior to the glans clitoris and anterior to the vaginal orifice. On each side of the external urethral orifice are the openings of the ducts of the para-urethral glands. The size and appearance of the **vaginal orifice** vary with the condition of the **hymen**, a thin fold of mucous membrane within the vaginal orifice surrounding the lumen. After its rupture, only remnants of the hymen, **hymenal caruncles** (tags), are visible (Fig. 6.57A).

BULBS OF VESTIBULE

The **bulbs of the vestibule** are paired masses of elongated erectile tissue that lie along the sides of the vaginal orifice under cover of the bulbospongiosus muscles (Fig. 6.58A). The bulbs are homologous with the bulb of the penis and the corpus spongiosum.

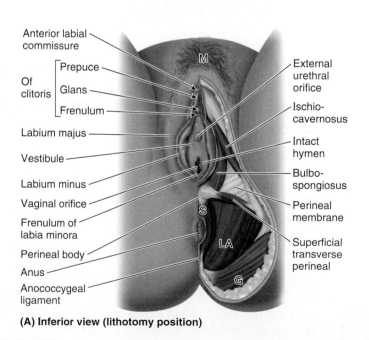

(A) Inferior view (lithotomy position)

Anterior labial commissure · Prepuce · Of clitoris · Glans · Frenulum · Labium majus · Vestibule · Labium minus · Vaginal orifice · Frenulum of labia minora · Perineal body · Anus · Anococcygeal ligament · External urethral orifice · Ischio-cavernosus · Intact hymen · Bulbo-spongiosus · Perineal membrane · Superficial transverse perineal

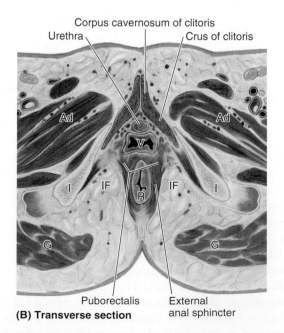

(B) Transverse section

Corpus cavernosum of clitoris · Urethra · Crus of clitoris · Puborectalis · External anal sphincter

FIGURE 6.57. Female perineum. A. Surface anatomy and perineal muscles. **B.** Structures on section. *Ad,* adductor muscles of thigh; *G,* gluteus maximus; *I,* ischium; *IF,* ischio-anal fossa; *LA,* levator ani; *M,* mons pubis; *R,* rectum; *S,* external anal sphincter; *V,* vagina.

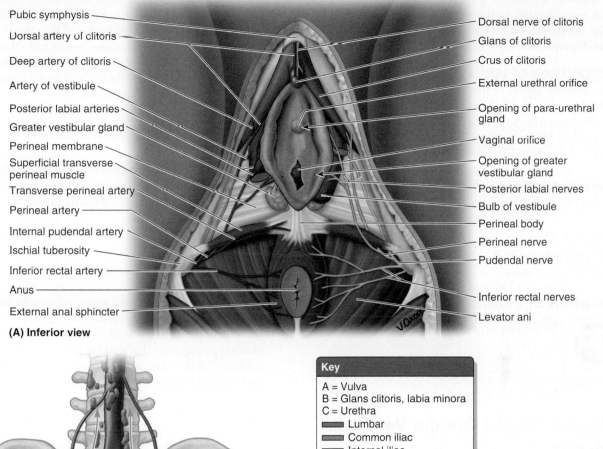

Pubic symphysis
Dorsal artery of clitoris
Deep artery of clitoris
Artery of vestibule
Posterior labial arteries
Greater vestibular gland
Perineal membrane
Superficial transverse perineal muscle
Transverse perineal artery
Perineal artery
Internal pudendal artery
Ischial tuberosity
Inferior rectal artery
Anus
External anal sphincter

Dorsal nerve of clitoris
Glans of clitoris
Crus of clitoris
External urethral orifice
Opening of para-urethral gland
Vaginal orifice
Opening of greater vestibular gland
Posterior labial nerves
Bulb of vestibule
Perineal body
Perineal nerve
Pudendal nerve
Inferior rectal nerves
Levator ani

(A) Inferior view

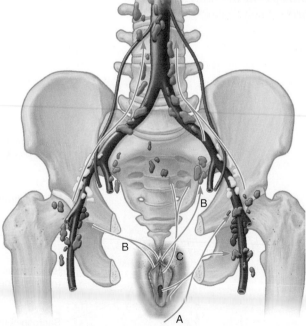

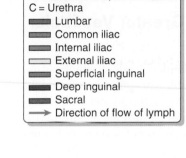

Key

A = Vulva
B = Glans clitoris, labia minora
C = Urethra

Lumbar
Common iliac
Internal iliac
External iliac
Superficial inguinal
Deep inguinal
Sacral
→ Direction of flow of lymph

(B) Anterior view

FIGURE 6.58. Blood supply, innervation, and lymphatic drainage of vulva/pudendum. A. Blood supply and innervation. **B.** Lymphatic drainage.

VESTIBULAR GLANDS

The **greater vestibular glands** (Bartholin glands) are located on each side of the vestibule, posterolateral to the vaginal orifice. These glands are round or oval and are partly overlapped posteriorly by the bulbs of the vestibule and both are partially surrounded by the bulbospongiosus muscles. The slender

ducts of these glands pass deep to the bulbs and open into the vestibule on each side of the vaginal orifice. These glands secrete mucus into the vestibule during sexual arousal. The **lesser vestibular glands** are smaller glands on each side of the vestibule that open into it between the urethral and the vaginal orifices. These glands secrete mucus into the vestibule, which moistens the labia and vestibule (Fig. 6.57A).

VASCULATURE OF VULVA

The *arterial supply to the vulva* is from the *external and internal pudendal arteries* (see Fig. 6.55A and Table 6.10). The *internal pudendal artery* supplies most of the skin, external genitalia, and perineal muscles. The labial arteries are branches of the internal pudendal artery, as are those of the clitoris (Fig. 6.58A). The *labial veins* are tributaries of the *internal pudendal veins* and accompanying veins (L. *venae comitantes*). Venous engorgement during the excitement phase of the sexual response causes an increase in the size and consistency of the clitoris and the bulbs of the vestibule. As a result, the clitoris becomes turgid.

The vulva contains a rich network of *lymphatic vessels* that pass laterally to the *superficial inguinal lymph nodes* (Fig. 6.58B). The glans clitoris and anterior labia minora may also drain to the deep inguinal nodes or internal iliac nodes.

INNERVATION OF VULVA

The anterior aspect of the vulva is supplied by the **anterior labial nerves**, derived from the *ilio-inguinal nerve* and the *genital branch of the genitofemoral nerve*. The posterior aspect is supplied by the *perineal branch of the posterior cutaneous nerve of the thigh* laterally and the *pudendal nerve* centrally. The pudendal nerve is the main nerve of the perineum. Its **posterior labial nerves** supply the labia; *deep* and *muscular branches* supply the orifice of the vagina and superficial perineal

muscles; and the *dorsal nerve of the clitoris* supplies deep perineal muscles and sensation to the clitoris (Fig. 6.58A). The bulb of the vestibule and erectile bodies of the clitoris receive parasympathetic fibers via cavernous nerves from the uterovaginal plexus. Parasympathetic stimulation produces increased vaginal secretion, erection of the clitoris, and engorgement of erectile tissue in the bulbs of the vestibule.

PERINEAL MUSCLES OF FEMALES

The *superficial perineal muscles* include the *superficial transverse perineal, ischiocavernosus*, and *bulbospongiosus* (Fig. 6.57A). Details of the attachments, innervation, and actions of the muscles are provided in Table 6.9.

CLINICAL BOX

Dilation of Female Urethra

 The female urethra is distensible because it contains considerable elastic tissue as well as smooth muscle. It can easily dilate without injury to it. Consequently, the passage of catheters or cystoscopes in females is much easier than it is in males.

Inflammation of Greater Vestibular Glands

 The greater vestibular glands (Bartholin glands) are usually not palpable, except when infected. *Bartholinitis*, inflammation of the greater vestibular glands, may result from a number of pathogenic organisms. Infected glands may enlarge to a diameter of 4–5 cm and impinge on the wall of the rectum.

Pudendal and Ilio-inguinal Nerve Blocks

To relieve the pain experienced during childbirth, *pudendal nerve block anesthesia* may be performed by injecting a local anesthetic agent into the tissues surrounding the pudendal nerve. The injection may be made where the pudendal nerve crosses the lateral aspect of the sacrospinous ligament, near its attachment to the ischial spine. Although a pudendal nerve block anesthetizes most of the perineum, it does not abolish sensation from the anterior part of the perineum that is innervated by the ilio-inguinal nerve. To abolish pain from the anterior part of the perineum, an *ilio-inguinal nerve block* is performed (Fig. B6.15).

Disruption of Perineal Body

 The perineal body is an especially important structure in women because it is the final support of the pelvic viscera. Stretching or tearing of this

attachment of the perineal muscles from the perineal body can occur during childbirth, removing support provided by the pelvic floor. As a result, *prolapse of pelvic viscera*, including prolapse of the bladder (through the urethra), and prolapse of the uterus and/or vagina (through the vaginal orifice) may occur.

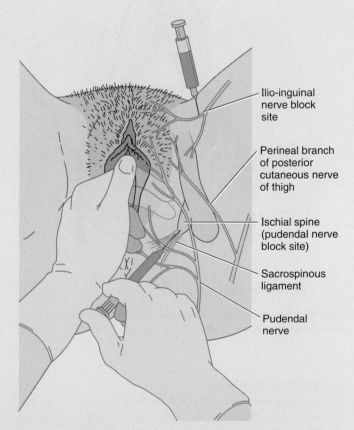

FIGURE B6.15. Pudendal/ilio-inguinal nerve blocks.

Episiotomy

During vaginal surgery and labor, an *episiotomy* (surgical incision of the perineum and inferoposterior vaginal wall) may be made to enlarge the vaginal orifice with the intention of decreasing excessive tearing of the perineum and perineal muscles. Performance of routine prophylactic episiotomy during a vaginal delivery is no longer recommended. Alternative strategies for prevention of tearing include perineal massage and warm compresses applied during the second (pushing) stage of labor. It is generally agreed that episiotomy is indicated when descent of the fetus is arrested or protracted or when instrumentation is necessary (e.g., obstetrical forceps).

MEDICAL IMAGING

Pelvis and Perineum

Magnetic resonance imaging (MRI) provides excellent evaluation of male and female pelvic structures (Figs. 6.59 and 6.60). It also permits the identification of tumors and congenital anomalies.

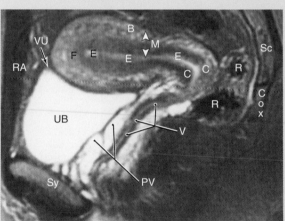

(A) Median section

(C) Transverse section

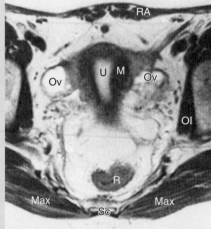

(B) Transverse section

(D) Transverse section

Key	
A	Anus
B	Body of uterus
C	Cervix of uterus
CJ	Ischiopubic ramus
Cox	Coccyx
E	Endometrium
F	Fundus of uterus
IAF	Ischio-anal fossa
IT	Ischial tuberosity
LM	Labium majus
M	Myometrium
Max	Gluteus maximus
OE	Obturator externus
OI	Obturator internus
Ov	Ovary
Pm	Perineal membrane
PR	Puborectalis
PV	Perivaginal veins
R	Rectum (gas)
RA	Rectus abdominis
Sc	Sacrum
SP	Superior ramus of pubis
Sy	Pubic symphysis
U	Uterus
UB	Urinary bladder
V	Vagina
Ve	Vestibule of vagina
VU	Vesico-uterine pouch

FIGURE 6.59. Magnetic resonance imaging (MRI) studies of female pelvis and perineum.

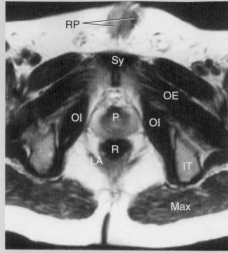

(A) Transverse section

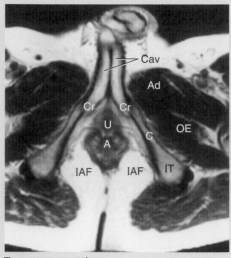

(B) Transverse section

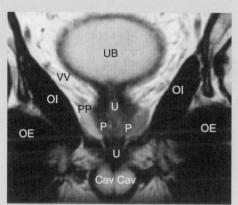

(C) Coronal section - plane of urethra

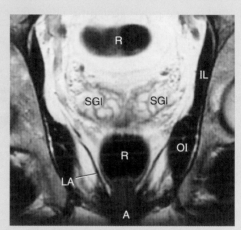

(D) Coronal section - plane of anal canal

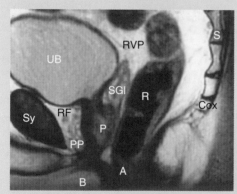

(E) Median section

Key			
A	Anus	P	Prostate
Ad	Adductor muscles	PP	Prostatic venous plexus
B	Bulb of penis	R	Rectum
C	Ischiopubic ramus	RF	Retropubic space
Cav	Corpus cavernosum of penis	RP	Root of penis
Cox	Coccyx	RVP	Rectovesical pouch
Cr	Crus of penis	S	Sacrum
IAF	Ischio-anal fossa	SGI	Seminal gland
IL	Iliacus	Sy	Pubic symphysis
IT	Ischial tuberosity	U	Urethra
LA	Levator ani	UB	Urinary bladder
Max	Gluteus maximus	VV	Vesical venous plexus
OE	Obturator externus		
OI	Obturator internus		

FIGURE 6.60. Magnetic resonance imaging (MRI) studies of male pelvis and perineum.

The female pelvis is commonly examined using ultrasonography. The viscera may be examined by placing a transducer on the lower abdomen, just superior to the pubic symphysis (*1* in Fig. 6.61A). For the nongravid uterus, the full bladder serves as an acoustical "window," conducting transmitted and reflected sound waves to and from the viscera, the uterus retroverted by the full bladder (Fig. 6.61B,E). Currently, viscera are studied most often by means of a slender transducer passed into the vagina (*2* in Fig. 6.61A and Fig. 6.61D). Ultrasonography is the procedure of choice for examining the developing embryo and fetus (Fig. 6.61E,F).

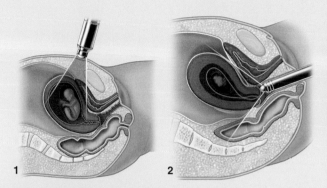

(A) Ultrasound scanning: (1) transabdominal; (2) transvaginal

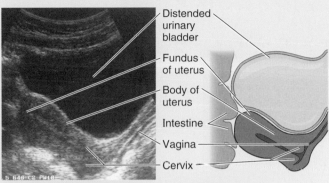

(B) Longitudinal (median) ultrasound image

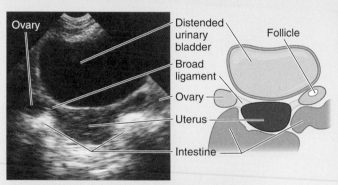

(C) Transverse transabdominal ultrasound (US) image

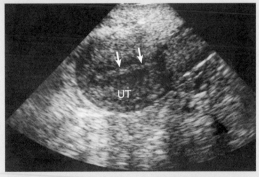

(D) Longitudinal transvaginal US image of nongravid uterus

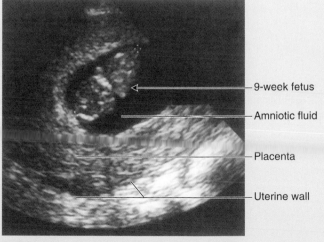

(E) Longitudinal transvaginal US scan of early gravid uterus

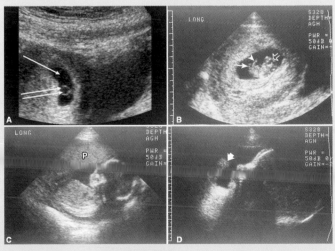

(F) Progressive growth and development of embryo/fetus

FIGURE 6.61. Ultrasonographic (US) studies of the pelvis. A. Placement of the transducer for US scanning of pelvis. **B and C.** Appearance of normal pelvic viscera in transabdominal scans. **D.** Transvaginal US scan of nongravid uterus; *arrows,* endometrium and uterine canal, *UT,* uterus. **E.** Gestational sac in gravid uterus. **F.** US study of embryonic/fetal growth and development. *A,* gestational sac (*single arrow*), embryo (*double arrow*); *B,* limbs (*solid arrows*) and head (*outlined arrow*) are visible; *C,* sagittal section of fetal head, neck, and thorax (*P,* placenta); *D,* profile of face and upper limb (*arrow*).

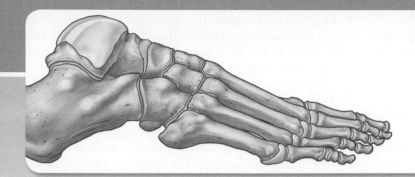

7

Lower Limb

CLINICAL BOX KEY

Anatomical Variations

Diagnostic Procedures

Life Cycle

Surgical Procedures

Trauma

Pathology

The lower limbs (extremities) are specialized for locomotion, supporting body weight, and maintaining balance. The lower limbs are connected to the trunk by the **pelvic girdle**, a bony ring composed of the sacrum and right and left hip bones joined anteriorly at the **pubic symphysis** (L. *symphysis pubis*). The lower limb has six major regions or segments (Fig. 7.1):

1. **Gluteal region** (L. *regio gluealis*) is the transitional zone between the trunk and free lower limbs. It includes the **buttocks** (L. *nates, clunes*) and **hip region** (L. *regio coxae*), which overlie the hip joint and greater trochanter of the femur.
2. **Thigh** or **femoral region** (L. *regio femoris*) includes most of the *femur*, which connects the hip and knee joints.
3. **Knee region** (L. *regio genus*) includes the distal femur, the proximal tibia and fibula, and the *patella* (knee cap) as

well as the joints between these bony structures; the fat-filled hollow posterior to the knee (L. *poples*) is called the *popliteal fossa*.

4. **Leg** or **crural region** (L. *regio cruris*) connects the knee and ankle joints and includes the *tibia* and *fibula*; the **calf** (L. *sura*) of the leg is the posterior prominence. Often, laypersons refer incorrectly to the entire lower limb as "the leg."
5. **Ankle** or **talocrural region** (L. *regio talocruralis*) includes the narrow distal leg and ankle (talocrural) joint.
6. *Foot* or **foot region** (L. *regio pedis*), the distal part of the lower limb, includes the *tarsus*, *metatarsus*, and *phalanges* (toe bones). The superior surface is the **dorsum of the foot**; the inferior, ground-contacting surface is the **plantar region** (**sole**). The **toes** are the **digits of the foot**. As in the hand, digit 1, the **great toe** (L. *hallux*) has only two phalanges, and the other digits have three.

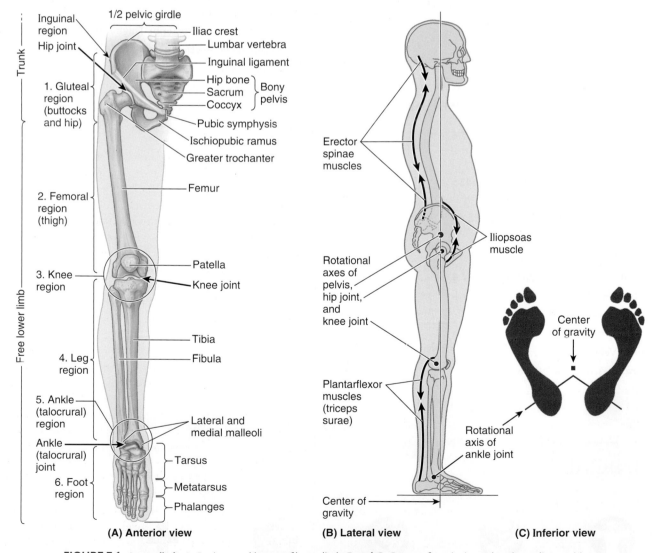

(A) Anterior view   **(B) Lateral view**   **(C) Inferior view**

FIGURE 7.1. Lower limb. A. Regions and bones of lower limb. **B and C.** Center of gravity in a relaxed standing position.

BONES OF LOWER LIMB

Body weight is transferred from the vertebral column through the *sacro-iliac joints* to the pelvic girdle and from the pelvic girdle through the hip joints to the femurs (L. *femora*) and then through the femurs to the knee joints. Weight is then transferred from the knee joint to the ankle joint by the tibia. The fibula does not articulate with the femur and does not bear weight. At the ankle, the weight is transferred to the talus. The talus is the keystone of a longitudinal arch formed by the tarsal and metatarsal bones of each foot, which distribute the weight evenly between the heel and the forefoot when standing. To support the erect bipedal posture better, the femurs are oblique (directed inferomedially) within the thighs so that when standing, the knees are adjacent and are placed directly inferior to the trunk, returning the center of gravity to the vertical lines of the supporting legs and feet (Figs. 7.1 and 7.2A,E). The femurs of females are slightly more oblique than those of males, reflecting the greater width of their pelves.

Hip Bone

Each mature **hip bone** is formed by the fusion of three primary bones: *ilium*, *ischium*, and *pubis* (Fig. 7.3A). At puberty, these bones are still separated by a **triradiate cartilage**. The cartilage disappears and the bones begin to fuse at 15–17 years of age; fusion is complete between 20 and 25 years of age.

The **ilium**, the superior and largest part of the hip bone, contributes to the superior part of the **acetabulum** (Fig. 7.3), the cup-like cavity (socket) on the lateral aspect of the hip bone for articulation with the head of the femur. The ilium consists of a **body**, which joins the pubis and ischium to the acetabulum, and an **ala** (wing), which is bordered superiorly by the **iliac crest**.

The **ischium** forms the postero-inferior part of the acetabulum and hip bone. The ischium consists of a body, where it joins the ilium and superior ramus of the pubis to form the acetabulum. The **ramus of the ischium** joins the inferior ramus of the pubis to form the ischiopubic ramus (Fig. 7.3C).

The **pubis** forms the anterior part of the acetabulum and the anteromedial part of the hip bone. The right pubis has a body that articulates with the left pubis at the *pubic symphysis*. It also has two **rami**, superior and inferior.

To place the hip bone or bony pelvis in the anatomical position (Fig. 7.3B,C), situate it so that the

- *Anterior superior iliac spine* (ASIS) and anterosuperior aspect of the pubis lie in the same coronal (frontal) plane.
- *Symphysial surface of the pubis* is vertical, parallel to the median plane.

- Internal aspect of the body of the pubis faces almost directly superiorly.
- *Acetabulum* faces inferolaterally, with the acetabular notch directed inferiorly.
- Obturator foramen lies inferomedial to the acetabulum.

Femur

The **femur** is the longest and heaviest bone in the body. The femur consists of a **shaft** (body) and superior or proximal and inferior or distal ends (Fig. 7.2). Most of the shaft is smoothly rounded, except for a prominent double-edged ridge on its posterior aspect, the **linea aspera**, which diverges inferiorly. The proximal end of the femur consists of a head, neck, and greater and lesser trochanters. The **head of the femur** is covered with articular cartilage, except for a medially placed depression or pit, the **fovea for the ligament of the head**. The **neck of the femur** is trapezoidal; the narrow end supports the head and its broader base is continuous with the shaft.

Where the neck joins the shaft are two large, blunt elevations, the trochanters. The conical **lesser trochanter**, with its rounded tip, extends medially from the posteromedial part of the junction of the femoral neck and shaft (Fig. 7.2A). The **greater trochanter** is a large, laterally placed mass that projects superomedially where the neck joins the shaft. The **intertrochanteric line** is a roughened ridge running from the greater to the lesser trochanter. A similar but smoother ridge, the **intertrochanteric crest**, joins the trochanters posteriorly (Fig. 7.2B).

The distal end of the femur ends in two spirally curved **femoral condyles (medial and lateral)**. The femoral condyles articulate with the tibial condyles to form the knee joint.

(Continued on page 415)

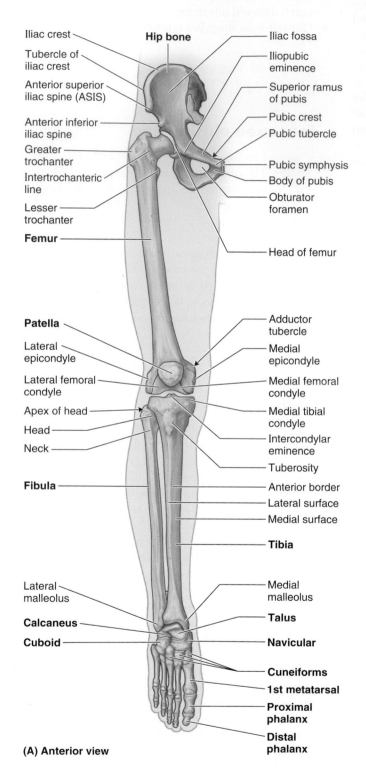

Hip bone

Iliac crest
Tubercle of iliac crest
Anterior superior iliac spine (ASIS)
Anterior inferior iliac spine
Greater trochanter
Intertrochanteric line
Lesser trochanter
Femur

Iliac fossa
Iliopubic eminence
Superior ramus of pubis
Pubic crest
Pubic tubercle
Pubic symphysis
Body of pubis
Obturator foramen

Head of femur

Patella
Lateral epicondyle
Lateral femoral condyle
Apex of head
Head
Neck
Fibula

Adductor tubercle
Medial epicondyle
Medial femoral condyle
Medial tibial condyle
Intercondylar eminence
Tuberosity
Anterior border
Lateral surface
Medial surface
Tibia

Lateral malleolus
Calcaneus
Cuboid

Medial malleolus
Talus
Navicular
Cuneiforms
1st metatarsal
Proximal phalanx
Distal phalanx

(A) Anterior view

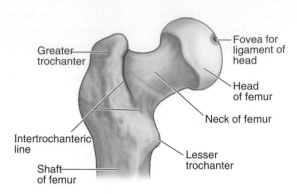

Greater trochanter
Intertrochanteric line
Shaft of femur
Fovea for ligament of head
Head of femur
Neck of femur
Lesser trochanter

(B) Anterior view of proximal femur

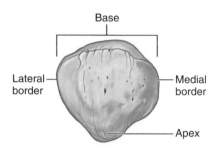

Base
Lateral border
Medial border
Apex

(C) Anterior view of patella

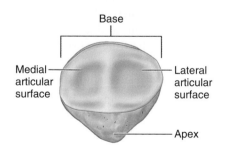

Base
Medial articular surface
Lateral articular surface
Apex

(D) Posterior view of patella

FIGURE 7.2. Bones of lower limb. *(continued)*

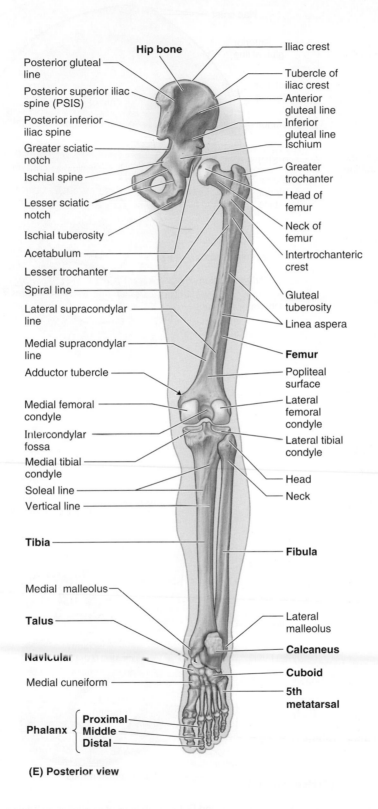

Hip bone

Iliac crest

Posterior gluteal line

Tubercle of iliac crest

Posterior superior iliac spine (PSIS)

Anterior gluteal line

Posterior inferior iliac spine

Inferior gluteal line

Greater sciatic notch

Ischium

Ischial spine

Greater trochanter

Lesser sciatic notch

Head of femur

Ischial tuberosity

Neck of femur

Acetabulum

Intertrochanteric crest

Lesser trochanter

Spiral line

Gluteal tuberosity

Lateral supracondylar line

Linea aspera

Medial supracondylar line

Femur

Adductor tubercle

Popliteal surface

Medial femoral condyle

Lateral femoral condyle

Intercondylar fossa

Lateral tibial condyle

Medial tibial condyle

Soleal line

Head

Vertical line

Neck

Tibia

Fibula

Medial malleolus

Talus

Lateral malleolus

Navicular

Calcaneus

Medial cuneiform

Cuboid

5th metatarsal

Phalanx { **Proximal** **Middle** **Distal**

(E) Posterior view

FIGURE 7.2. Bones of the lower limb. *(continued)*

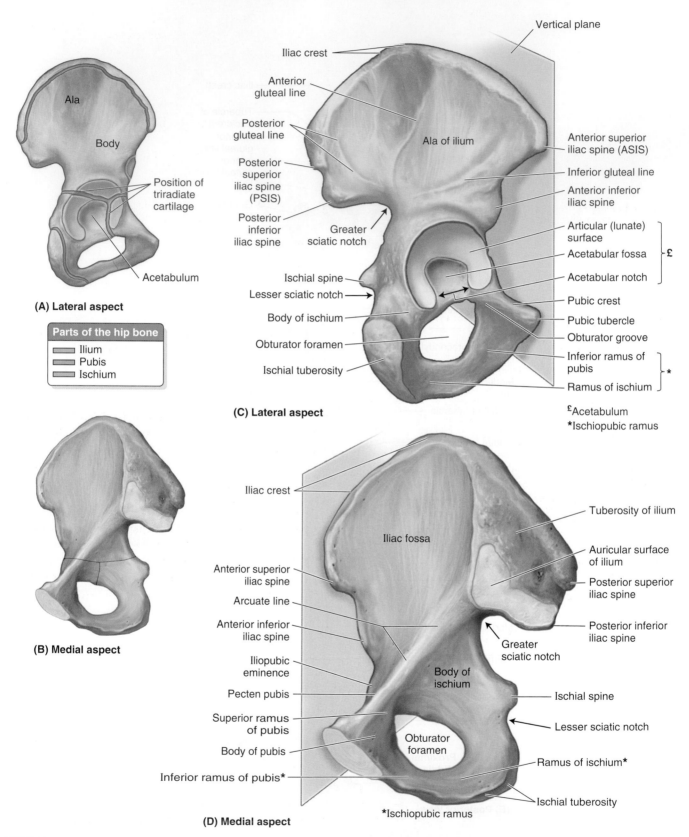

(A) Lateral aspect

Ala

Body

Position of triradiate cartilage

Acetabulum

Parts of the hip bone
- Ilium
- Pubis
- Ischium

(B) Medial aspect

(C) Lateral aspect

Vertical plane

Iliac crest

Anterior gluteal line

Posterior gluteal line

Posterior superior iliac spine (PSIS)

Posterior inferior iliac spine

Ala of ilium

Anterior superior iliac spine (ASIS)

Inferior gluteal line

Anterior inferior iliac spine

Articular (lunate) surface

Acetabular fossa

Acetabular notch

£

Greater sciatic notch

Ischial spine

Lesser sciatic notch

Body of ischium

Obturator foramen

Ischial tuberosity

Pubic crest

Pubic tubercle

Obturator groove

Inferior ramus of pubis

Ramus of ischium

*

£Acetabulum
*Ischiopubic ramus

(D) Medial aspect

Iliac crest

Iliac fossa

Anterior superior iliac spine

Arcuate line

Anterior inferior iliac spine

Iliopubic eminence

Pecten pubis

Superior ramus of pubis

Body of pubis

Inferior ramus of pubis*

Tuberosity of ilium

Auricular surface of ilium

Posterior superior iliac spine

Posterior inferior iliac spine

Greater sciatic notch

Body of ischium

Ischial spine

Lesser sciatic notch

Ramus of ischium*

Ischial tuberosity

Obturator foramen

*Ischiopubic ramus

FIGURE 7.3. Hip bone. A and B. Parts of hip bone of a 13-year-old. **C and D.** Right hip bone of an adult in anatomical position. In this position, the anterior superior iliac spine (ASIS) and the anterior aspect of the pubis lie in the same vertical plane (indicated in *blue*).

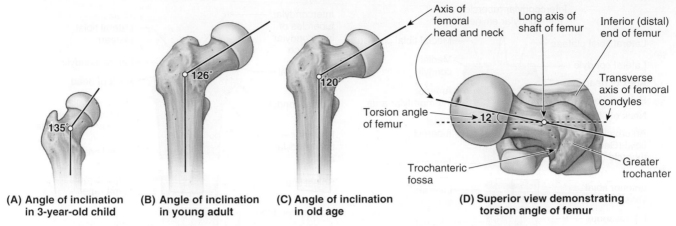

(A) Angle of inclination in 3-year-old child

(B) Angle of inclination in young adult

(C) Angle of inclination in old age

(D) Superior view demonstrating torsion angle of femur

FIGURE 7.4. Angle of inclination and torsion angle of femur.

The proximal femur is bent, making the femur L-shaped, so that the long axis of the head and neck project supero-medially at an angle to that of the obliquely oriented shaft (Fig. 7.4). This obtuse **angle of inclination** in the adult is 115–140 degrees, averaging 126 degrees. The angle is less in females because of the increased width between the ac-etabula and the greater obliquity of the shaft. The angle of inclination allows greater mobility of the femur at the hip joint because it places the head and neck more perpendicular to the acetabulum. This is advantageous for bipedal walking; however, it imposes considerable strain on the neck of the femur. Fractures of the neck may occur in older people as a result of a slight stumble if the neck has been weakened by osteoporosis.

When the femur is viewed superiorly, so that the proxi-mal end is superimposed over the distal end (Fig. 7.4D), it can be seen that the axis of the head and neck of the femur and the transverse axis of the femoral condyles intersect at the long axis of the shaft of the femur, forming the **tor-sion angle**, or **angle of declination**. The mean torsion angle is 7 degrees in males and 12 degrees in females. The torsion angle, combined with the angle of inclination, al-lows rotatory movements of the femoral head within the obliquely placed acetabulum to convert into flexion and extension, abduction and adduction, and rotational move-ments of the thigh.

Patella

The **patella** (knee cap) is a large sesamoid bone that is formed intratendinously after birth. This triangular bone, located anterior to the femoral condyles, articulates with the *patellar surface of the femur* (Fig. 7.2A,C). The subcu-taneous **anterior surface of the patella** is convex; the thick **base** (superior border) slopes infero-anteriorly; the *lateral* and *medial borders* converge inferiorly to form the pointed **apex**; and the **articular** (posterior) **surface of**

the patella is smooth, covered with a thick layer of articu-lar cartilage, and is divided into medial and lateral articular surfaces by a vertical ridge (Fig. 7.2C,D).

Tibia

The large, weight-bearing **tibia** (shin bone) articulates with the femoral condyles superiorly, the talus inferiorly, and the fibula laterally at its proximal and distal ends (Fig. 7.2). The distal end of the tibia is smaller than the proximal end and has facets for articulation with the fibula and talus. The **me-dial malleolus** is an inferiorly directed projection from the medial side of the distal end of the tibia (Fig. 7.5A). The large **nutrient foramen** of the tibia is located on the pos-terior aspect of the proximal third of the bone (Fig. 7.5B). From it, the **nutrient canal** runs inferiorly in the tibia be-fore it opens into the medullary (marrow) cavity. Additional features of the tibia are demonstrated in Figure 7.5.

Fibula

The slender **fibula** lies posterolateral to the tibia and serves mainly for muscle attachment (Figs. 7.2 and 7.5). At its proxi-mal end, the fibula consists of an enlarged **head of fibula** su-perior to a narrow neck (**neck of fibula**). At its distal end, the fibula enlarges to form the **lateral malleolus**, which is more prominent and more posteriorly placed than the medial malle-olus and extends approximately 1 cm farther distally. The fibula is not directly involved in weight bearing; however, its lateral malleolus forms the lateral part of the socket for the trochlea of the talus. The shafts of the tibia and fibula are connected by an **interosseous membrane** throughout most of their lengths.

Bones of the Foot

The **bones of the foot** comprise the tarsus, metatarsus, and phalanges (Figs. 7.1A and 7.6).

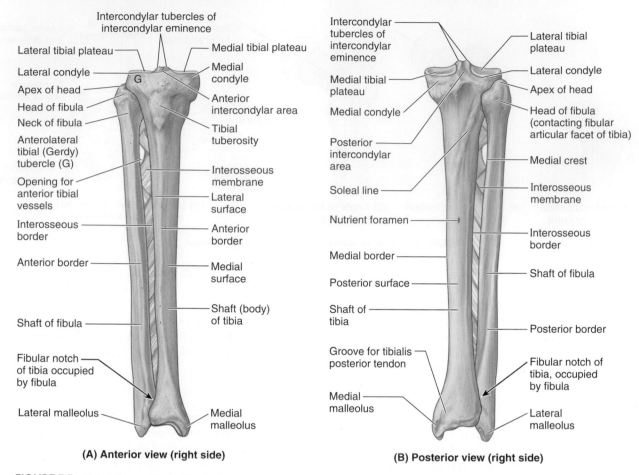

Intercondylar tubercles of intercondylar eminence

Lateral tibial plateau

Lateral condyle

Apex of head

Head of fibula

Neck of fibula

Anterolateral tibial (Gerdy) tubercle (G)

Opening for anterior tibial vessels

Interosseous border

Anterior border

Shaft of fibula

Fibular notch of tibia occupied by fibula

Lateral malleolus

Medial tibial plateau

Medial condyle

Anterior intercondylar area

Tibial tuberosity

Interosseous membrane

Lateral surface

Anterior border

Medial surface

Shaft (body) of tibia

Medial malleolus

(A) Anterior view (right side)

Intercondylar tubercles of intercondylar eminence

Medial tibial plateau

Medial condyle

Posterior intercondylar area

Soleal line

Nutrient foramen

Medial border

Posterior surface

Shaft of tibia

Groove for tibialis posterior tendon

Medial malleolus

Lateral tibial plateau

Lateral condyle

Apex of head

Head of fibula (contacting fibular articular facet of tibia)

Medial crest

Interosseous membrane

Interosseous border

Shaft of fibula

Posterior border

Fibular notch of tibia, occupied by fibula

Lateral malleolus

(B) Posterior view (right side)

FIGURE 7.5. Right tibia and fibula. The shafts are connected by the interosseous membrane composed of strong obliquely oriented fibers.

TARSUS

The **tarsus** consists of seven tarsal bones: talus, calcaneus, cuboid, navicular, and three cuneiforms. Only the talus articulates with the leg bones. The **talus** (ankle bone) has a head, **neck of talus**, and body (Fig. 7.6C). The superior surface of the **body of the talus**, the **trochlea of the talus**, bears the weight of the body transmitted from the tibia, and articulates with the two malleoli (Fig. 7.6A). Most of the surface of the talus is covered with articular cartilage, and thus no muscles or tendons attach to the talus. The talus rests on the anterior two thirds of the calcaneus. The rounded **head of talus** rests partially on the sustentaculum tali of the calcaneus and articulates anteriorly with the navicular (Fig. 7.6B,F)

The **calcaneus** (heel bone) is the largest and strongest bone in the foot. It articulates with the talus superiorly and the cuboid anteriorly (Fig. 7.6A). The calcaneus transmits most of the body weight from the talus to the ground. The **sustentaculum tali** (talar shelf), projecting from the superior border of the medial surface of the calcaneus, supports the head of the talus (Fig. 7.6B,D). The posterior part of the calcaneus has a large prominence,

the **calcaneal tuberosity** (L. *tuber calcanei*), which has *medial* and *lateral processes* on its plantar aspect (Fig. 7.6B). More anteriorly, there is a smaller prominence, the *calcaneal tubercle*.

The **navicular** (L. *little ship*), a flattened, boat-shaped bone, is located between the head of the talus and the cuneiforms. The medial surface of the navicular projects inferiorly as the **tuberosity of navicular**. An overly prominent tuberosity may press against the medial part of the shoe and cause foot pain.

The **cuboid** is the most lateral bone in the distal row of the tarsus. Anterior to the **tuberosity of cuboid** (Fig. 7.6B), on the lateral and plantar surfaces of the bone, is a groove for the tendon of the fibularis longus muscle (Fig. 7.6B,C).

There are three **cuneiforms**: **medial** (first), **intermediate** (second), and **lateral** (third). Each cuneiform (L. *cuneus*, wedge-shaped) articulates with the navicular posteriorly and the base of the appropriate metatarsal anteriorly. In addition, the lateral cuneiform articulates with the cuboid.

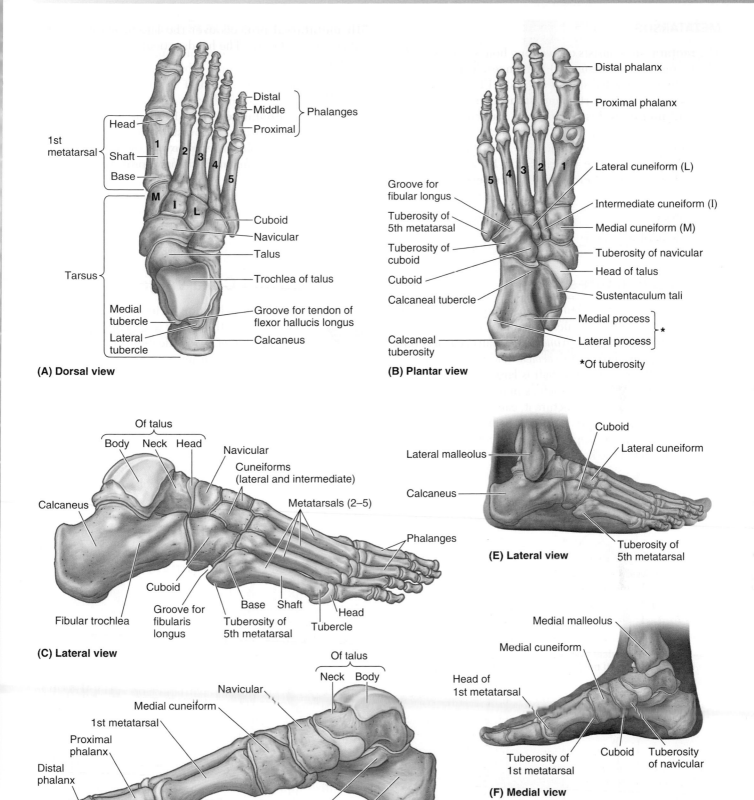

(A) Dorsal view

1st metatarsal
- Head
- Shaft
- Base

Phalanges
- Distal
- Middle
- Proximal

Tarsus
- Cuboid
- Navicular
- Talus
- Trochlea of talus

Medial tubercle
Lateral tubercle

Groove for tendon of flexor hallucis longus
Calcaneus

(B) Plantar view

Distal phalanx
Proximal phalanx

Groove for fibular longus
Tuberosity of 5th metatarsal
Tuberosity of cuboid
Cuboid
Calcaneal tubercle

Lateral cuneiform (L)
Intermediate cuneiform (I)
Medial cuneiform (M)
Tuberosity of navicular
Head of talus
Sustentaculum tali

Calcaneal tuberosity

Medial process
Lateral process

*Of tuberosity

(C) Lateral view

Of talus
Body   Neck   Head

Navicular
Cuneiforms (lateral and intermediate)
Metatarsals (2–5)
Phalanges

Calcaneus

Fibular trochlea
Cuboid
Groove for fibularis longus
Base   Shaft
Tuberosity of 5th metatarsal
Head
Tubercle

(E) Lateral view

Cuboid
Lateral malleolus
Lateral cuneiform
Calcaneus

Tuberosity of 5th metatarsal

(D) Medial view

Of talus
Neck   Body
Navicular
Medial cuneiform
1st metatarsal
Proximal phalanx
Distal phalanx

Sustentaculum tali
Calcaneus

(F) Medial view

Medial malleolus
Medial cuneiform
Head of 1st metatarsal

Tuberosity of 1st metatarsal
Cuboid
Tuberosity of navicular

FIGURE 7.6. Bones of foot. *Blue,* articular cartilage.

Fractures of Foot Bones

Calcaneal fractures occur in people who fall on their heels (e.g., from a ladder). Usually, the bone breaks into several fragments (*comminuted fracture*) that disrupt the subtalar joint, where the talus articulates with the calcaneus (Fig. B7.5A). *Fractures of the talar neck* may occur during severe dorsiflexion of the ankle, for example, when a person is pressing extremely hard on the brake pedal of a car during a head-on collision (Fig. B7.5B). *Metatarsal and phalangeal fractures* are a common injury in endurance athletes and may also occur when a heavy object falls on the foot. Metatarsal fractures are also common in dancers, especially female ballet dancers using the *demi-pointe* technique. The "dancer's fracture" usually occurs when the dancer loses balance, putting the full body weight on the metatarsal and fracturing the bone (Fig. B7.5C).

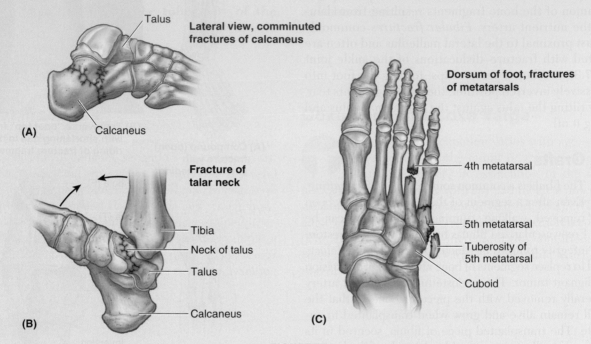

FIGURE B7.5. Fractures of foot bones.

SURFACE ANATOMY

LOWER LIMB BONES

Pelvic Girdle and Femur

When your hands are on your hips, they rest on the **iliac crests**, the curved superior borders of the alae (wings) of the ilium (Fig. SA7.1). The anterior third of the crest is easily palpated because it is subcutaneous. The highest point of the crest is at the level of the intervertebral (IV) disc between the L4 and the L5 vertebrae. The iliac crest ends anteriorly at the pointed **anterior superior iliac spine** (ASIS), which is easy to palpate, especially in thin persons, because it is subcutaneous and often visible (Fig. SA7.1A,B). The ASIS is used as the proximal point for measurement of leg length to the medial malleolus of the tibia. The iliac crest ends posteriorly at the **posterior superior iliac spine** (PSIS), which may be difficult to palpate (Fig. SA7.1C). Its position

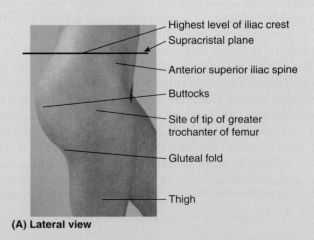

(A) Lateral view

FIGURE SA7.1. Surface projection and palpable features of bones of lower limb. *(continued)*

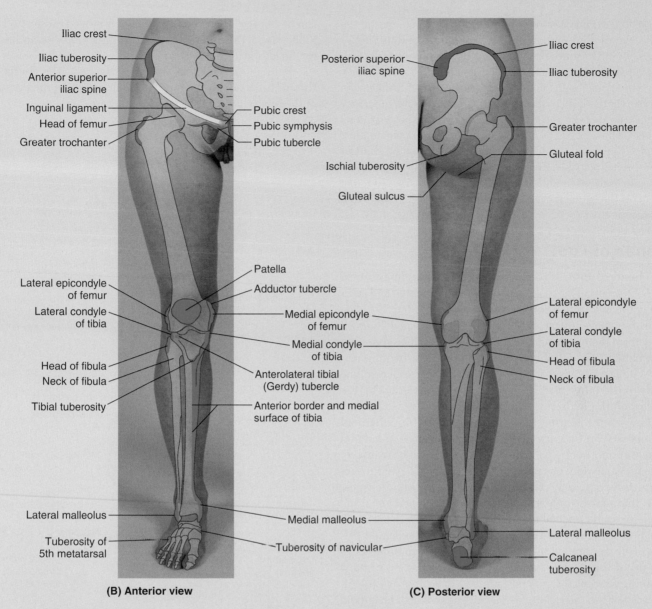

FIGURE SA7.1. Surface projection and palpable features of bones of lower limb. *Green,* palpable features of lower limb bones. *(continued)*

is easy to locate because it lies at the bottom of a skin dimple, approximately 4 cm lateral to the midline, demarcating posteriorly the location of the sacro-iliac joint. The dimple exists because the skin and fascia attach to the PSIS.

The **ischial tuberosity** is easily palpated in the inferior part of the buttocks when the hip joint is flexed. It bears body weight when sitting. The thick gluteus maximus and fat obscure the tuberosity when the hip joint is extended. The **gluteal fold**, a prominent skin fold containing fat, coincides with the inferior border of the gluteus maximus muscle.

The **greater trochanter of the femur** is easily palpable on the lateral side of the hip approximately 10 cm inferior to the iliac crest (Fig. SA7.1B,C). Because it lies close to the skin, the greater trochanter causes discomfort when you lie on your side on a hard surface. In the anatomical position, a line joining the tips of the greater trochanters normally passes through the centers of the femoral heads and pubic tubercles. The shaft of the femur usually is not palpable because it is covered with

large muscles. The **medial and lateral condyles** of the femur are subcutaneous and easily palpated when the knee is flexed or extended. The patellar surface of the femur is where the **patella** slides during flexion and extension of the knee joint. The **lateral and medial margins of the patella** can be palpated when the knee joint is flexed. The **adductor tubercle**, a small prominence of bone, may be felt at the superior part of the medial femoral condyle.

Tibia and Fibula

The **tibial tuberosity**, an oval elevation on the anterior surface of the tibia, is palpable approximately 5 cm distal (inferior) to the apex of the patella to which it is connected by the palpable patellar ligament (Fig. SA7.1B). The subcutaneous **anterior border and medial surface of the tibia** is also easy to palpate.

(Continued on next page)

The skin covering it is freely movable. The prominence at the ankle, the **medial malleolus**, is subcutaneous, and its inferior end is blunt. The **medial** and **lateral tibial condyles** can be palpated anteriorly at the sides of the patellar ligament, especially when the knee joint is flexed. The **head of the fibula** can be palpated at the level of the superior part of the tibial tuberosity because its knob-like head is subcutaneous at the posterolateral aspect of the knee. The **neck of fibula** can be palpated just distal to the fibular head. Only the distal quarter of the **shaft of the fibula** is palpable. Feel your **lateral malleolus**, noting that it is subcutaneous and that its inferior end is sharp. Note that the tip of the lateral malleolus extends farther distally and more posteriorly than does the tip of the medial malleolus.

Bones of Foot

The **head of talus** is palpable anteromedial to the proximal part of the lateral malleolus when the foot is inverted and anterior to the medial malleolus when the foot is everted. Eversion of the foot makes the head of talus more prominent as it moves away from the navicular. The head of talus occupies the space between the sustentaculum tali and the tuberosity of navicular. When the foot is plantarflexed, the superior surface of the **body of the talus** can be palpated on the anterior aspect of the ankle, anterior to the inferior end of the tibia (Fig. SA7.1D).

The weight-bearing **medial process of the calcaneal tuberosity** on the plantar surface of the foot is broad and large but may not be palpable because of the thick overlying skin and subcutaneous tissue (Fig. SA7.1E). The **sustentaculum tali** is the only part of the medial aspect of the calcaneus that may be palpated as a small prominence just distal to the tip of the medial malleolus.

The **tuberosity of the navicular** is easily seen and palpated on the medial aspect of the foot, infero-anterior to the tip of the medial malleolus. Usually, palpation of bony prominences on the plantar surface of the foot is difficult because of the thick skin, fascia, and pads of fat. The cuboid and cuneiforms are difficult to identify individually by palpation. The cuboid can be felt on the lateral aspect of the foot, posterior to the base of the 5th metatarsal. The **medial cuneiform** can be indistinctly palpated between the tuberosity of the navicular and the base of the 1st metatarsal.

The **head of the 1st metatarsal** forms a prominence on the medial aspect of the foot. The **medial** and **lateral sesamoid bones**, located inferior to the head of this metatarsal,

can be felt to slide when the 1st digit is moved passively. The tuberosity of the 5th metatarsal forms a prominent landmark on the lateral aspect of the foot and can be palpated easily at the midpoint of the lateral border of the foot. The **shafts of the metatarsals and phalanges** can be felt on the dorsum of the foot between the extensor tendons.

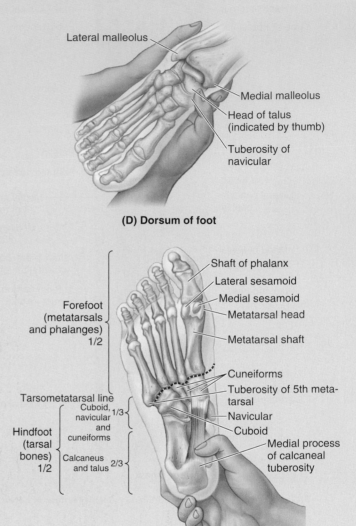

(D) Dorsum of foot

(E) Plantar aspect of foot

FIGURE SA7.1. Surface projection and palpable features of bones of lower limb. *(continued)*

FASCIA, VESSELS, AND NERVES OF LOWER LIMB

Subcutaneous Tissue and Fascia

The **subcutaneous tissue** (superficial fascia) is deep to the skin and consists of loose connective tissue that contains a variable amount of fat, cutaneous nerves, superficial veins, lymphatic vessels, and lymph nodes (Fig. 7.7). The subcutaneous tissue of the hip and thigh is continuous with that of the inferior part of the anterolateral abdominal wall and buttocks. At the knee, the subcutaneous tissue loses its fat anteriorly and laterally, and blends with the deep fascia, but fat is present posteriorly in the popliteal fossa and again distal to the knee in the subcutaneous tissue of the leg.

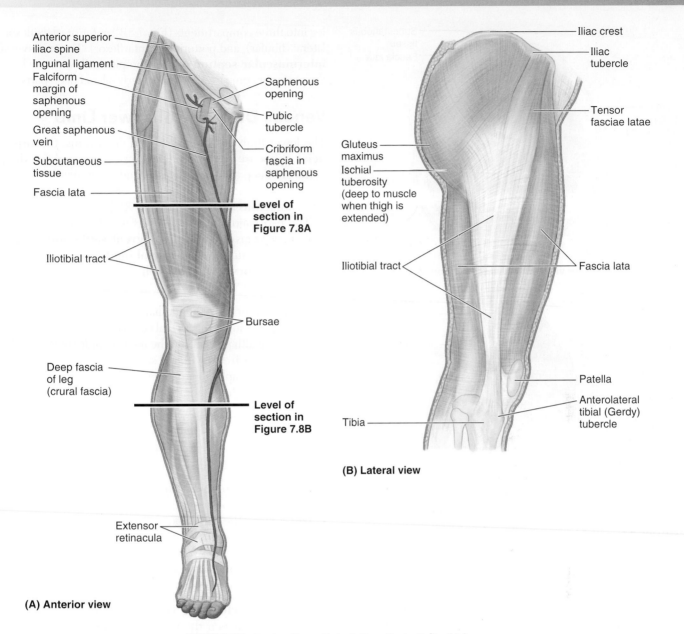

FIGURE 7.7. Fascia of lower limb. A. Deep fascia. **B.** Iliotibial tract.

The **deep fascia** is especially strong, investing the limb like an elastic stocking (Fig. 7.7A). This fascia limits outward extension of contracting muscles, making muscular contraction more efficient in compressing the veins to push blood toward the heart. The *deep fascia of the thigh* is called **fascia lata** (L. *lata*, broad). The fascia lata attaches to and is continuous with

- The inguinal ligament, pubic arch, body of pubis, and pubic tubercle superiorly. The membranous layer of subcutaneous tissue (Scarpa fascia) of the inferior abdominal wall also attaches to the fascia lata just inferior to the inguinal ligament.
- The iliac crest laterally and posteriorly
- The sacrum, coccyx, sacrotuberous ligament, and ischial tuberosity posteriorly
- The superficial aspects of the bones around the knee and the deep fascia of the leg distally

The fascia lata is substantial because it encloses the large thigh muscles, especially laterally where it is thickened to form the **iliotibial tract** (Fig. 7.7B). This broad band of fibers is also the aponeurosis of the *tensor fasciae latae* and *gluteus maximus* muscles. The iliotibial tract extends from the **iliac tubercle** to the **anterolateral tibial tubercle** (Gerdy tubercle) on the lateral condyle of the tibia (Fig. SA7.1).

The thigh muscles are separated into three **fascial compartments**: *anterior*, *medial*, and *posterior*. The walls of these compartments are formed by the fascia lata and three fascial intermuscular septa that arise from the deep aspect of the fascia lata and attach to the linea aspera on the posterior aspect of the femur (Figs. 7.2A,E and 7.8A). The **lateral intermuscular septum** is strong; the other two septa are relatively weak. The iliotibial tract is continuous with the lateral intermuscular septum.

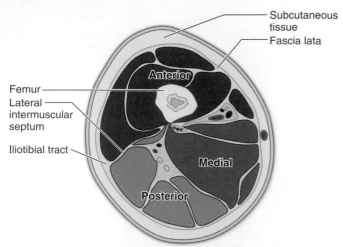

(A) Inferior view of transverse section of thigh

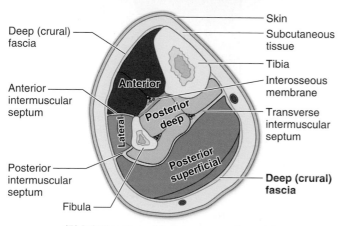

(B) Inferior view of transverse section of leg

FIGURE 7.8. **Fascial compartments. A.** Thigh. **B.** Leg. See Figure 7.7 for level of sections.

The **saphenous opening** is a gap or hiatus in the fascia lata inferior to the medial part of the inguinal ligament (Fig. 7.7A). Its medial margin is smooth, but its superior, lateral, and inferior margins form a sharp edge, the **falciform margin**. The sieve-like **cribriform fascia** (L. *cribrum*, sieve) is a localized membranous layer of subcutaneous tissue over the saphenous opening, enclosing it. The great saphenous vein and some lymphatics pass through the saphenous opening and cribriform fascia to enter the femoral vein and the deep inguinal lymph nodes, respectively.

The **deep fascia of the leg** or **crural fascia** (L. *crus*, leg) is continuous with the fascia lata and attaches to the anterior and medial borders of the tibia, where it is continuous with its periosteum (Fig. 7.7A). The crural fascia is thick in the proximal part of the anterior aspect of the leg, where it forms part of the proximal attachments of the underlying muscles. Although thin in the distal part of the leg, the crural fascia is thickened where it forms the **extensor retinacula. Anterior** and **posterior intermuscular septa** pass from the deep surface of the crural fascia and attach to the corresponding margins of the fibula. The *interosseous membrane* and the *intermuscular septa* divide the

leg into three compartments (Fig. 7.8B): anterior (dorsiflexor), lateral (fibular), and posterior (plantarflexor). The **transverse intermuscular septum** divides the plantarflexor muscles in the posterior compartment into superficial and deep parts.

Venous Drainage of Lower Limb

The lower limb has superficial and deep veins; the **superficial veins** are in the subcutaneous tissue, and the **deep veins** are deep to the deep fascia and accompany the major arteries. Superficial and deep veins have valves, but they are more numerous in deep veins.

The two major *superficial veins* are the great and small saphenous veins (Fig. 7.9). The **great saphenous vein** is formed by the union of the **dorsal digital vein** of the great toe and the **dorsal venous arch** of the foot. The great saphenous vein (Fig. 7.9A,B)

- ascends anterior to the medial malleolus
- passes posterior to the medial condyle of the femur (about a hand's breadth posterior to the medial border of the patella)
- anastomoses freely with the small saphenous vein
- traverses the saphenous opening in the fascia lata (Fig. 7.7A)
- empties into the femoral vein

The **small saphenous vein** arises on the lateral side of the foot from the union of the dorsal digital vein of the 5th digit with the dorsal venous arch (Fig. 7.9A,B). The small saphenous vein (Fig. 7.9D)

- ascends posterior to the lateral malleolus as a continuation of the lateral marginal vein
- passes along the lateral border of the calcaneal tendon
- inclines to the midline of the fibula and penetrates the deep fascia
- ascends between the heads of the gastrocnemius muscle
- empties into the popliteal vein in the popliteal fossa

Abundant **perforating veins** penetrate the deep fascia as they pass between the superficial and deep veins (Fig. 7.9C). They contain *valves* that allow blood to flow only from the superficial to the deep veins. The perforating veins penetrate the deep fascia at oblique angles so that when muscles contract and pressure increases inside the deep fascia, the perforating veins are compressed, preventing blood from flowing from the deep to the superficial veins. This pattern of venous blood flow, from superficial to deep, is important for proper venous return from the limb because it enables muscular contractions to propel blood toward the heart against the pull of gravity (*musculovenous pump*; see Fig. 1.17A in Chapter 1, Overview and Basic Concepts).

The *deep veins* in the lower limb accompany the major arteries and their branches (Fig. 7.10). Instead of occurring as a single vein in the limbs, the deep veins are usually paired, frequently interconnecting **accompanying veins** (L. *venae comitantes*) that flank the artery. They are contained within a vascular sheath with the artery, whose pulsations also help compress and move blood in the veins.

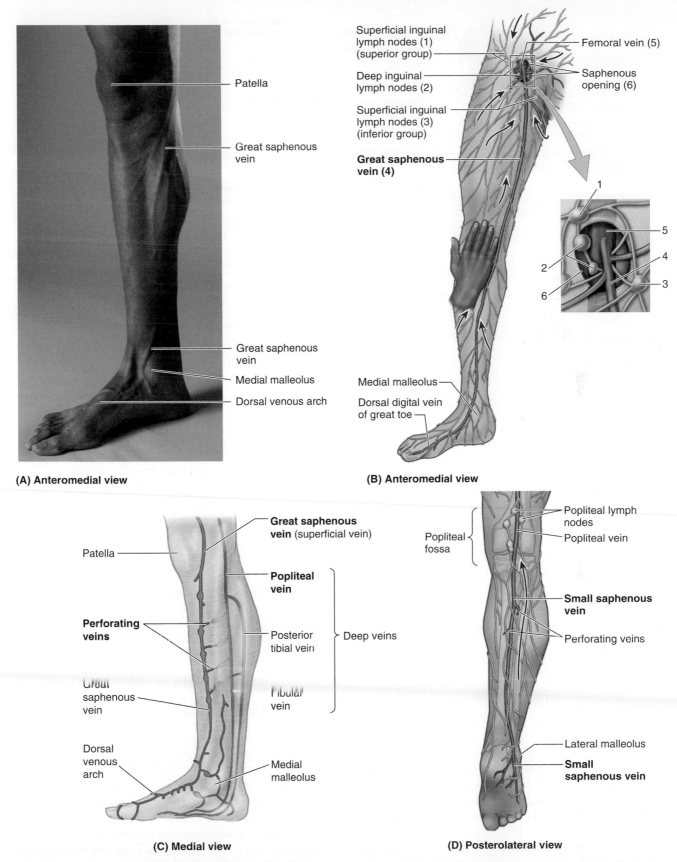

Patella

Great saphenous vein

Great saphenous vein

Medial malleolus

Dorsal venous arch

(A) Anteromedial view

Superficial inguinal lymph nodes (1) (superior group)

Deep inguinal lymph nodes (2)

Superficial inguinal lymph nodes (3) (inferior group)

Great saphenous vein (4)

Femoral vein (5)

Saphenous opening (6)

Medial malleolus

Dorsal digital vein of great toe

(B) Anteromedial view

Great saphenous vein (superficial vein)

Patella

Popliteal vein

Perforating veins

Posterior tibial vein

Great saphenous vein

Fibular vein

Deep veins

Dorsal venous arch

Medial malleolus

(C) Medial view

Popliteal lymph nodes

Popliteal fossa

Popliteal vein

Small saphenous vein

Perforating veins

Lateral malleolus

Small saphenous vein

(D) Posterolateral view

FIGURE 7.9. Superficial venous and lymphatic drainage of lower limb. A. Normal superficial veins distended after exercise. **B.** Great saphenous vein and superficial lymphatic drainage with inset of saphenous opening. *Arrows,* superficial lymphatic drainage to the inguinal nodes. **C.** Perforating veins. **D.** Small saphenous vein and superficial lymphatic drainage (*arrow*) to the popliteal lymph nodes.

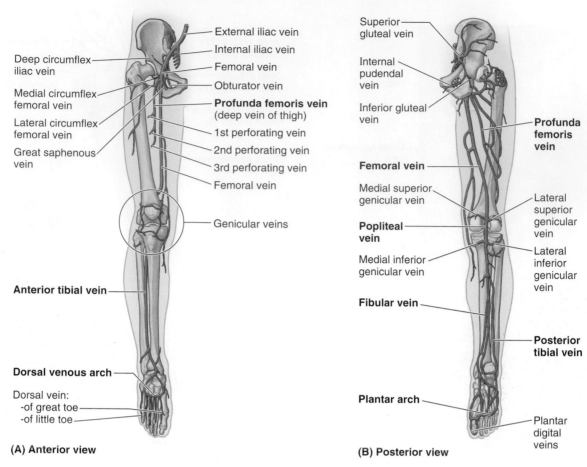

FIGURE 7.10. Deep venous drainage of lower limb.

The deep veins from the leg flow into the popliteal vein posterior to the knee, which becomes the femoral vein in the thigh. The profunda femoris vein joins the terminal portion of the femoral vein. The femoral vein passes deep to the inguinal ligament to become the external iliac vein in the pelvis (Fig. 7.10A).

Arterial Supply of Lower Limb

The *femoral artery* is the main blood supply to the lower limb (Fig. 7.11). The femoral artery is the continuation of the external iliac artery distal to the inguinal ligament. The pulsations of the femoral artery can be palpated inferior to the midpoint of the inguinal ligament (Fig. 7.11C). A major branch, the *profunda femoris artery* (deep artery of thigh) serves the posterior and lateral aspects of the thigh. The femoral artery continues through the anterior compartment of the thigh, passing through the adductor hiatus to the popliteal region, posterior to the knee. The femoral artery becomes the *popliteal artery* at the adductor hiatus. The pulsations of the popliteal artery can be palpated inferiorly in the popliteal fossa, when the knee is in a semiflexed position (Fig. 7.11E). The popliteal artery passes through the fossa and divides into anterior and posterior tibial arteries. The *anterior tibial artery* courses through the anterior compartment of the leg, ending midway

between the malleoli of the ankle, where it becomes the *dorsalis pedis artery* (dorsal artery of foot). The pulsations of the dorsalis pedis artery are palpated over the navicular and medial cuneiform, lateral to the extensor hallucis longus tendon (Fig. 7.11D). The *posterior tibial artery* courses through the deep posterior compartment of the leg and then posterior to the medial malleolus, dividing into medial and lateral plantar arteries to the sole of the foot. Its pulsations can be palpated posterior to the medial malleolus (Fig. 7.11F). The *obturator artery*, usually a branch of the internal iliac artery, supplies the medial compartment of the thigh.

Lymphatic Drainage of Lower Limb

The lower limb has superficial and deep lymphatic vessels. The *superficial lymphatic vessels* converge on and accompany the saphenous veins and their tributaries. The lymphatic vessels accompanying the great saphenous vein end in the **superficial inguinal lymph nodes** (Fig. 7.9B). Most lymph from these nodes passes to the **external iliac lymph nodes**, located along the external iliac vein, but some lymph may also pass to the **deep inguinal lymph nodes**, located on the medial aspect of the femoral vein. The lymphatic vessels accompanying the small saphenous vein enter the **popliteal lymph nodes**, which surround the popliteal

(*Continued on page 429*)

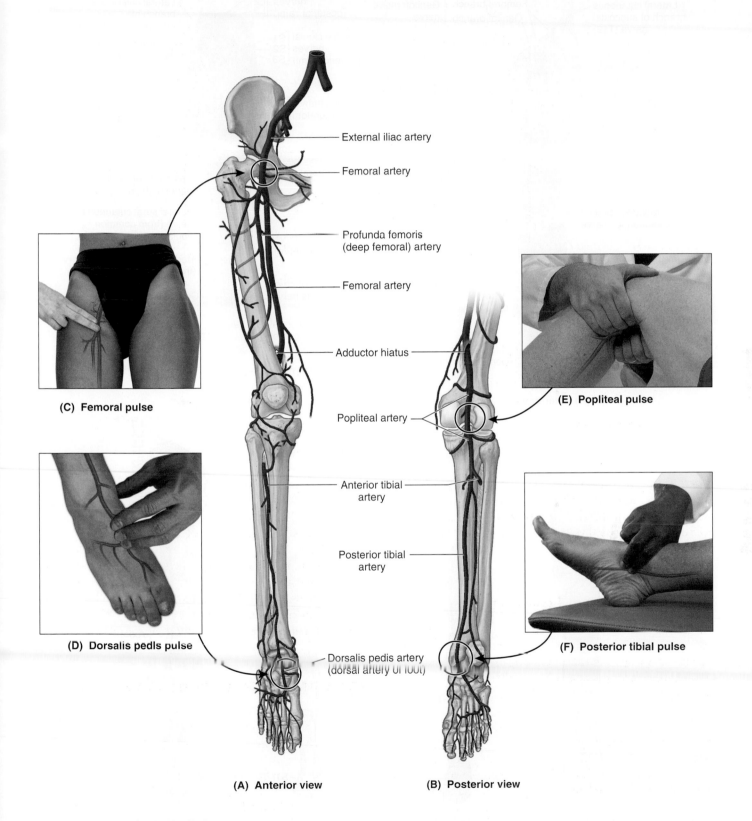

External iliac artery

Femoral artery

Profunda fomoris (deep femoral) artery

Femoral artery

Adductor hiatus

Popliteal artery

Anterior tibial artery

Posterior tibial artery

Dorsalis pedis artery (dorsal artery of foot)

(C) Femoral pulse

(D) Dorsalis pedis pulse

(E) Popliteal pulse

(F) Posterior tibial pulse

(A) Anterior view

(B) Posterior view

FIGURE 7.11. **Arterial supply and pulses of lower limb. A and B.** Overview of arterial supply. **C–F.** Site of palpation of lower limb pulses.

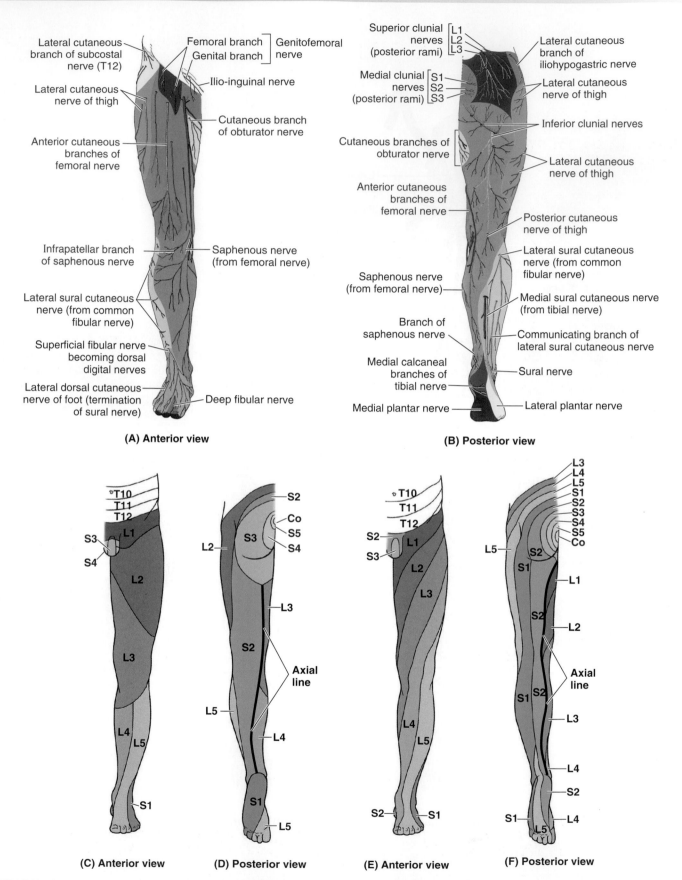

(A) Anterior view

(B) Posterior view

(C) Anterior view   **(D) Posterior view**   **(E) Anterior view**   **(F) Posterior view**

FIGURE 7.12. **Cutaneous innervation of lower limb. A and B.** Peripheral cutaneous nerve distribution. **C–F.** Dermatomes. Two different dermatome maps are frequently used: **C and D,** according to Foerster (1933); **E and F,** according to Keegan and Garrett (1948).

vein in the fat of the popliteal fossa (Fig. 7.9D). The *deep lymphatic vessels* of the leg accompany deep veins and enter the popliteal lymph nodes. Most lymph from these nodes ascends through deep lymphatic vessels to the deep inguinal lymph nodes. Lymph from the deep nodes passes to the external iliac lymph nodes.

Innervation of Lower Limb

CUTANEOUS INNERVATION

Cutaneous nerves in the subcutaneous tissue supply the skin of the lower limb (Fig. 7.12A,B). These nerves, except for some in the proximal part of the limb, are branches of the lumbar and sacral plexuses (see Chapters 2 and 6). The area of skin supplied by cutaneous branches from a single spinal nerve is a *dermatome* (Fig. 7.12C–F). Dermatomes L1–L5 extend as a series of bands from the posterior midline of the trunk into the limbs, passing laterally and inferiorly around the limb to its anterior and medial aspects, reflecting the medial rotation that occurs developmentally. Dermatomes S1 and S2 pass inferiorly down the posterior aspect of the limb, separating near the ankle to pass to the lateral and medial margins of the foot (Fig. 7.12F).

Although simplified into distinct zones in dermatome maps, adjacent dermatomes overlap except at the **axial line**, the line of junction of dermatomes supplied from discontinuous spinal levels.

Two different dermatome maps are commonly used. The pattern according to Foerster (1933) is preferred by many because of its correlation with clinical findings (Fig. 7.12C,D) and that of Keegan and Garrett (1948) by others for its correlation with limb development (Fig. 7.12E,F).

MOTOR INNERVATION

The unilateral embryological muscle mass receiving innervation from a single spinal cord segment or spinal nerve comprises a *myotome*. Lower limb muscles usually receive motor fibers from several spinal cord segments or nerves. Thus, most muscles include more than one myotome, and most often, multiple spinal cord segments are involved in producing the movements. The muscle myotomes are grouped by joint movement to facilitate clinical testing (Fig. 7.13 and Table 7.1).

Somatic motor (general somatic efferent) fibers traveling in the same mixed peripheral nerves that convey sensory fibers to the cutaneous nerves transmit impulses to the lower limb muscles. The somatic motor and somatic sensory innervation of the lower limb is summarized in Figure 7.14.

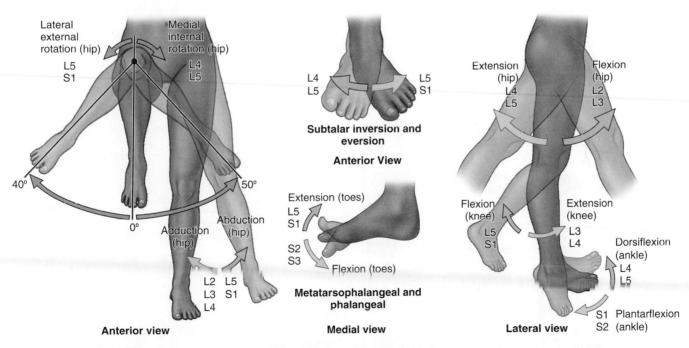

FIGURE 7.13. Myotomes of lower limb.

TABLE 7.1. LOWER LIMB MYOTOMES

Spinal Nerve	Myotome (example)	Spinal Nerve	Myotome (example)
L2	Hip flexors (iliopsoas)	L5	Long toe extensors (extensor hallucis and digitorum longus)
L3	Knee extensors (quadriceps)	S1	Ankle plantarflexors (gastrocnemius/soleus)
L4	Ankle dorsiflexors (tibialis anterior)	S2	Toe abductors (abductor hallucis/plantar interossei)

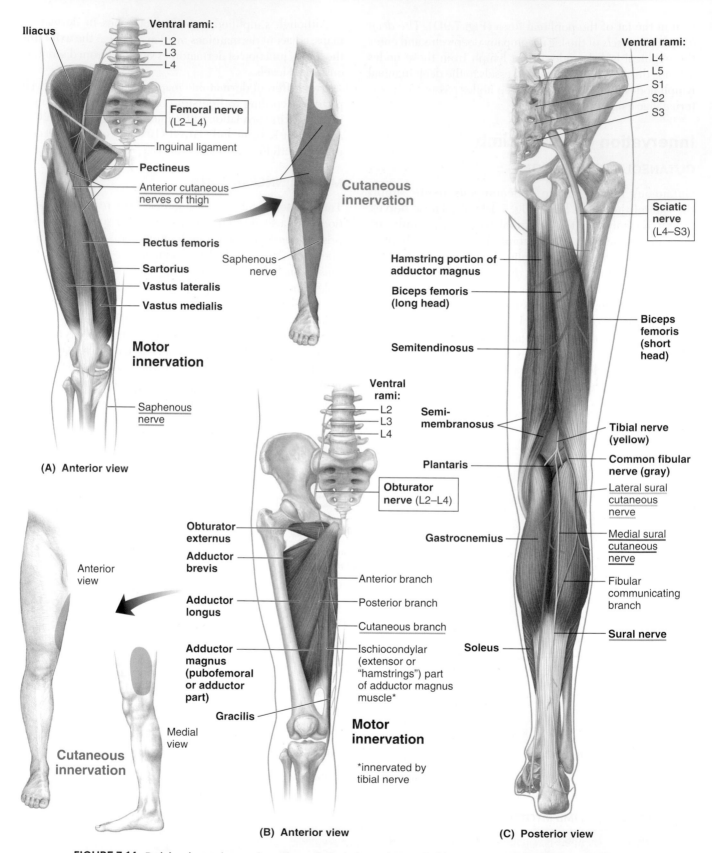

Iliacus

Ventral rami:
- L2
- L3
- L4

Femoral nerve (L2–L4)

Inguinal ligament

Pectineus

Anterior cutaneous nerves of thigh

Cutaneous innervation

Rectus femoris

Sartorius

Vastus lateralis

Vastus medialis

Saphenous nerve

Motor innervation

Saphenous nerve

(A) Anterior view

Anterior view

Medial view

Cutaneous innervation

Obturator externus

Adductor brevis

Adductor longus

Adductor magnus (pubofemoral or adductor part)

Gracilis

Ventral rami:
- L2
- L3
- L4

Obturator nerve (L2–L4)

Anterior branch

Posterior branch

Cutaneous branch

Ischiocondylar (extensor or "hamstrings") part of adductor magnus muscle*

Motor innervation

*innervated by tibial nerve

(B) Anterior view

Ventral rami:
- L4
- L5
- S1
- S2
- S3

Sciatic nerve (L4–S3)

Hamstring portion of adductor magnus

Biceps femoris (long head)

Semitendinosus

Biceps femoris (short head)

Semi-membranosus

Plantaris

Gastrocnemius

Tibial nerve (yellow)

Common fibular nerve (gray)

Lateral sural cutaneous nerve

Medial sural cutaneous nerve

Fibular communicating branch

Sural nerve

Soleus

(C) Posterior view

FIGURE 7.14. **Peripheral nerve innervation of lower limb. A.** Femoral nerve. **B.** Obturator nerve. **C.** Sciatic nerve. *(continued)*

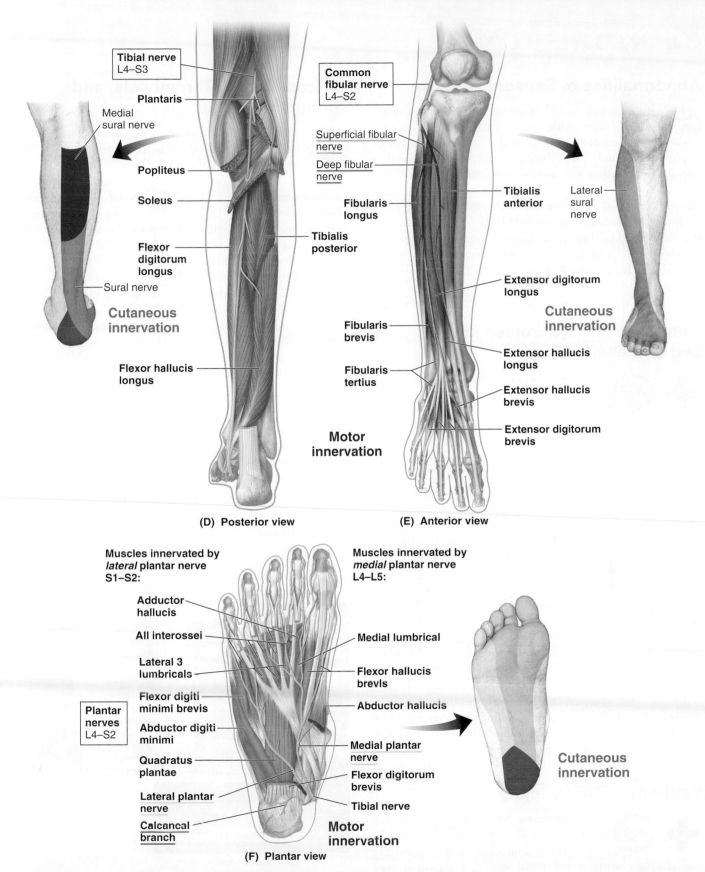

Tibial nerve
L4–S3

Plantaris

Medial sural nerve

Popliteus

Soleus

Flexor digitorum longus

Sural nerve

Cutaneous innervation

Flexor hallucis longus

(D) Posterior view

Common fibular nerve
L4–S2

Superficial fibular nerve

Deep fibular nerve

Fibularis longus

Tibialis posterior

Fibularis brevis

Fibularis tertius

Tibialis anterior

Extensor digitorum longus

Extensor hallucis longus

Extensor hallucis brevis

Extensor digitorum brevis

Lateral sural nerve

Cutaneous innervation

Motor innervation

(E) Anterior view

Muscles innervated by *lateral* plantar nerve S1–S2:

Adductor hallucis

All interossei

Lateral 3 lumbricals

Flexor digiti minimi brevis

Abductor digiti minimi

Quadratus plantae

Lateral plantar nerve

Calcaneal branch

Plantar nerves L4–S2

Muscles innervated by *medial* plantar nerve L4–L5:

Medial lumbrical

Flexor hallucis brevis

Abductor hallucis

Medial plantar nerve

Flexor digitorum brevis

Tibial nerve

Cutaneous innervation

Motor innervation

(F) Plantar view

FIGURE 7.14. Peripheral nerve innervation of lower limb. *(continued)* D. Tibial nerve. E. Common fibular nerve. F. Medial and lateral plantar nerves.

CLINICAL BOX

Abnormalities of Sensory Function

 In the limbs, most cutaneous nerves are multisegmental conveying fibers from more than one segment of the spinal cord. Using a sharp object (a pin or pinwheel), areas lacking sensation are outlined to determine whether the area of numbness matches the dermatome pattern (Fig. 7.12C–F), indicating a segmental (spinal nerve) lesion, or the multisegmental pattern of peripheral cutaneous nerve distribution (Fig. 7.12A,B). Because neighboring dermatomes overlap, the area of numbness resulting from a lesion of a single spinal nerve will be much smaller than indicated by the dermatome map.

Compartment Syndromes in Leg and Fasciotomy

 Increased pressure in a confined anatomical space adversely affects the circulation and threatens the function and viability of tissue within or distal to the space (*compartment syndrome*). The fascial compartments of the lower limbs are generally closed spaces, ending proximally and distally at the joints. Trauma to muscles and/or vessels in the compartments from burns, sustained intense use of muscles, or blunt trauma may produce hemorrhage, edema, and inflammation of the muscles in the compartment. Because the septa and deep fascia of the leg forming the boundaries of the leg compartments are strong, the increased volume consequent to any of these processes increases intracompartmental pressure. The small vessels of muscles and nerves (vasa nervorum) are particularly vulnerable to compression. Structures distal to the compressed area may become ischemic and permanently injured (e.g., muscles with compromised blood supply and/or innervation will not function). Loss of distal leg pulses is an obvious sign of arterial compression, as is lowering of the temperature of tissues distal to the compression. A *fasciotomy* (incision of overlying fascia or a septum) may be performed to relieve the pressure in the compartment(s) concerned.

Saphenous Nerve Injury

 The saphenous nerve accompanies the great saphenous vein in the leg. Should this nerve be injured or caught by a ligature during closure of a surgical wound, the patient may complain of pain, tingling, or numbness (paresthesia) along the medial border of the foot.

Varicose Veins, Thrombosis, and Thrombophlebitis

 Frequently, the great saphenous vein and its tributaries become *varicose* (dilated and/or tortuous so that the cusps of their valves do not close). *Varicose veins* are common in the posteromedial parts of the lower limb and may cause discomfort (Fig. B7.6A). In a healthy vein, the valves allow blood to flow toward the heart while preventing blood flow away from the heart (Fig B7.6B,C). Valves also bear the weight of short columns of blood between two valves. Valves in varicose veins, incompetent due to dilation or rotation, no longer function properly. The resulting reverse flow and the weight of long, unbroken columns of blood produces varicose veins (Fig. B7.6D).

Deep venous thrombosis (DVT) of one or more of the deep veins of the lower limb is characterized by swelling, warmth, and *erythema* (inflammation) and infection. *Venous stasis* (stagnation) is an important cause of thrombus formation. Venous stasis may be caused by the following:

- Incompetent, loose fascia that fails to resist muscle expansion, diminishing the effectiveness of the musculovenous pump
- External pressure on the veins from bedding during prolonged institutional stays or from a tight cast, bandages, or bands of stockings
- Muscular inactivity (e.g., during an overseas flight)

DVT with inflammation around the involved veins (*thrombophlebitis*) may develop. A large thrombus that breaks free from a lower limb vein may travel to a lung, forming a *pulmonary thrombo-embolism* (obstruction of a pulmonary artery). A large embolus may obstruct a main pulmonary artery and may cause death.

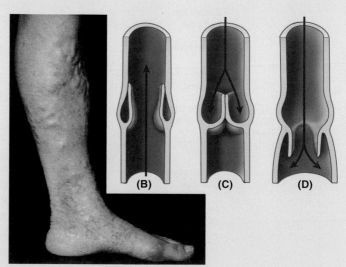

(A) Medial view

FIGURE B7.6. Varicose veins. A. Photograph of varicose veins. **B.** and **C.** Normal **D.** Varicose, incompetent valve

Enlarged Inguinal Lymph Nodes

Lymph nodes enlarge when diseased. *Abrasions* with minor sepsis, caused by pathogenic micro-organisms or their toxins in the blood or other tissues, may produce moderate enlargement of the superficial inguinal lymph nodes (*lymphadenopathy*) in otherwise healthy people. Because these enlarged nodes are located in subcutaneous tissue, they are usually easy to palpate.

When inguinal lymph nodes are enlarged, their entire field of drainage—the trunk inferior to the umbilicus, including the perineum, as well as the entire lower limb—should be examined to determine the cause of their enlargement. In female patients, the relatively remote possibility of metastasis of cancer from the uterus should also be considered because some lymphatic drainage from the uterine fundus may flow along lymphatics accompanying the round ligament of the uterus through the inguinal canal to reach the superficial inguinal lymph nodes.

Regional Nerve Blocks of Lower Limbs

Interruption of the conduction of impulses in peripheral nerves (*nerve block*) may be achieved with *perineural injections* of anesthetics close to the nerves whose conductivity is to be blocked. For example, the femoral nerve (L2–L4) can be blocked 2 cm inferior to the inguinal ligament, approximately a finger's breadth lateral to the femoral artery.

ANTERIOR AND MEDIAL THIGH

The thigh muscles are organized into three compartments—*anterior*, *medial*, and *posterior*—by intermuscular septa (see Fig. 7.8A). Generally, the anterior group is innervated by the femoral nerve, the medial group by the obturator nerve, and the posterior group by the tibial portion of the sciatic nerve.

Anterior Thigh Muscles

The large **anterior compartment of the thigh** contains the **anterior thigh muscles**, *flexors of the hip*, and *extensors of the knee*. The attachments, nerve supply, and main actions of these muscles are summarized in Figs. 7.14A and 7.15 and Table 7.2. The anterior thigh muscles are as follows:

- **Pectineus**, a flat quadrangular muscle, located in the anterior part of the superomedial aspect of the thigh
- **Iliopsoas** (the chief flexor of the hip joint), formed by the merger of two muscles, the psoas major and iliacus. The fleshy parts of the two muscles lie in the posterior abdominal wall and greater pelvis, merging as they enter the thigh by passing deep to the inguinal ligament and attaching to the lesser trochanter of the femur. This muscle is also a postural muscle, active during standing in maintaining normal lumbar lordosis and, indirectly, the compensatory thoracic kyphosis (curvature of vertebral columns).
- **Sartorius**, the tailor's muscle (L. *sart*, a tailor), a long, ribbon-like muscle that is the most superficial muscle in the anterior thigh; it passes obliquely (lateral to medial) across the supero-anterior part of the thigh. It acts across both the hip and knee joints, and when acting bilaterally, the muscles bring the lower limbs into the cross-legged sitting position. None of the actions is strong; therefore, it is mainly a synergist, acting with other thigh muscles that produce these movements.

- **Quadriceps femoris** (L. four-headed femoral muscle), the great extensor of the knee joint that forms the main bulk of the anterior thigh muscles. It covers almost all the anterior aspect and sides of the femur. This muscle has four parts:
  - **Rectus femoris**, the "kicking muscle" (L. *rectus*, straight), which crosses the hip joint and helps the iliopsoas flex this joint. Its ability to extend the knee is compromised during hip flexion.
  - **Vastus lateralis**, the largest component of the quadriceps, located on the lateral aspect of the full length of the thigh
  - **Vastus intermedius**, which lies deep to the rectus femoris between the vastus medialis and the vastus lateralis
  - **Vastus medialis**, which covers the medial aspect of the distal two thirds of the thigh

A small, flat muscle, the **articularis genu** (articular muscle of knee), a derivative of the vastus intermedius (Fig. 7.15D), attaches superiorly to the inferior part of the anterior aspect of the femur and inferiorly to the synovial membrane of the knee joint and the wall of the *suprapatellar bursa*. The muscle pulls the synovial membrane superiorly during extension of the knee, thereby preventing folds of the membrane from being compressed between the femur and the patella within the knee joint.

The tendons of the four parts of the quadriceps unite in the distal part of the thigh to form the **quadriceps tendon**

(Fig. 7.15B). The **patellar ligament** (L. *ligamentum pa-tellae*), the continuation of the quadriceps tendon in which the patella is embedded, is attached to the tibial tuberosity. The vastus medialis and lateralis also attach independently to the patella and form aponeuroses, the **medial** and **lateral patellar retinacula**. The retinacula reinforce the knee joint capsule on each side of the patella en route to their attachment to the *tibial plateau*. The patella provides additional leverage for the quadriceps in placing the tendon more anteriorly, farther from the joint's axis, causing it to approach the tibia from a position of greater mechanical advantage.

Medial Thigh Muscles

The medial thigh muscles—collectively called the **adductor group**—are in the medial compartment of the thigh and are innervated primarily by the obturator nerve (Figs. 7.14B, 7.15 and 7.16, and Table 7.3). The adductor group consists of the following muscles:

- **Adductor longus**: The most anterior muscle in the group

- **Adductor brevis**: Deep (posterior) to the pectineus and adductor longus muscles
- **Adductor magnus**: The largest adductor muscle, composed of adductor and hamstring parts; the parts differ in their attachments, nerve supply, and main actions.
- **Gracilis**: Along, strap-like muscle lying along the medial side of the thigh and knee; it is the only adductor muscle to cross and act at the knee joint as well as the hip joint.
- **Obturator externus**: A deeply placed fan-shaped muscle in the superomedial part of the thigh

The **adductor hiatus** is an opening between the distal aponeurotic attachment of the adductor part of the adductor magnus and the tendon of its hamstring part (Fig. 7.16E). The hiatus transmits the femoral artery and vein from the anterior compartment of the thigh to the popliteal fossa posterior to the knee. The main action of the adductor group of muscles is to adduct the hip joint. They are used to stabilize the stance when standing on both feet, to correct lateral sway of the trunk, and when there is a side-to-side shift. The adductors contribute to flexion of the extended hip joint and to extension of the flexed hip joint when running or against resistance.

TABLE 7.2. ANTERIOR THIGH MUSCLES

Muscle	Proximal Attachment	Distal Attachment	Innervation[a]	Main Action(s)
Pectineus	Superior ramus of pubis	Pectineal line of femur, just inferior to lesser trochanter	Femoral nerve (**L2**, L3); may also receive branch from obturator nerve	Adducts and flexes hip joint; assists with medial rotation of hip joint
Sartorius	Anterior superior iliac spine and superior part of notch inferior to it	Superior part of medial surface of tibia	Femoral nerve (L2, L3)	Flexes, abducts, and laterally rotates hip joint; flexes knee joint
Iliopsoas				
Psoas major[b]	Sides of T12–L5 vertebrae and discs between them; transverse processes of all lumbar vertebrae	Lesser trochanter of femur	Anterior rami of lumbar nerves (**L1**, **L2**, L3)	Acting conjointly in flexing hip joint and in stabilizing this joint; psoas major is also a postural muscle that helps control deviation of the trunk and is active during standing.
Iliacus	Iliac crest, iliac fossa, ala of sacrum, and anterior sacro-iliac ligaments	Tendon of psoas major, lesser trochanter, and femur distal to it	Femoral nerve (**L2**, L3)	
Quadriceps femoris				
Rectus femoris	Anterior inferior iliac spine and ilium superior to acetabulum	Via common tendinous (quadriceps tendon) and independent attachments to base of patella; indirectly via patellar ligament to tibial tuberosity; vastus medialis and lateralis also attach to tibia and patella via aponeuroses (medial and lateral patellar retinacula).	Femoral nerve (L2, **L3**, **L4**)	Extend knee joint; rectus femoris also stabilizes (helps fix in position) hip joint and helps iliopsoas flex hip joint.
Vastus lateralis	Greater trochanter and lateral lip of linea aspera			
Vastus medialis	Intertrochanteric line and medial lip of linea aspera			
Vastus intermedius	Anterior and lateral surfaces of shaft of femur			

[a]The spinal cord segmental innervation is indicated (e.g., "**L1**, **L2**, L3" means that the nerves supplying the psoas major are derived from the first three lumbar segments of the spinal cord). Numbers in boldface (**L1**, **L2**) indicate the main segmental innervation. Damage to one or more of the listed spinal cord segments or to the motor nerve roots arising from them results in paralysis of the muscles concerned.
[b]The psoas minor is a small muscle that attaches proximally to the T12–L1 vertebrae and IV discs and distally to the pectineal line and iliopectineal eminence.

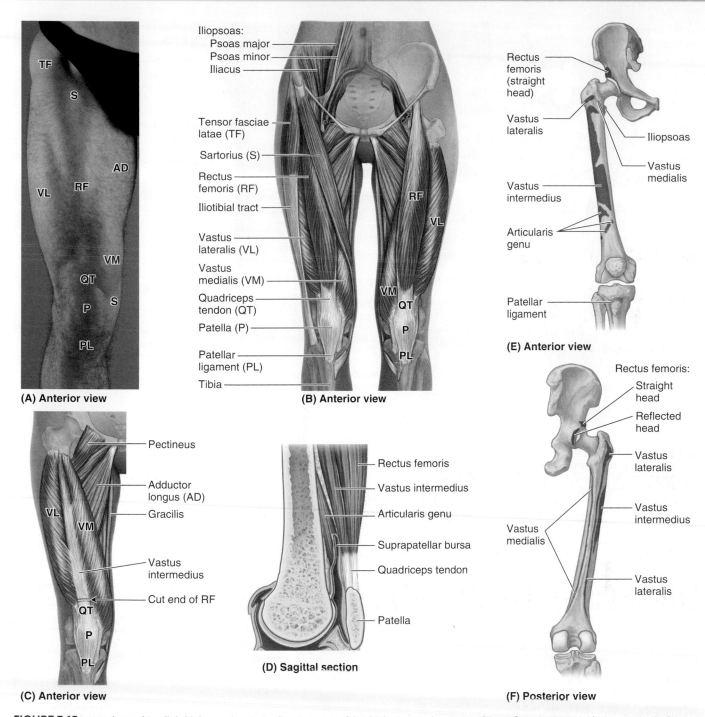

FIGURE 7.15. Anterior and medial thigh muscles. A. Surface anatomy of the thigh. **B.** Muscles. **C.** Quadriceps femoris. **D.** Articularis genu (articular muscle of knee). **E and F.** Muscle attachment sites.

Femoral Triangle and Adductor Canal

The **femoral triangle** is a subfascial space in the anterosuperior third of the thigh (Fig. 7.17). It appears as a triangular depression inferior to the inguinal ligament when the thigh is flexed, abducted, and laterally rotated. The femoral triangle is bounded by the following:

- Superiorly by the inguinal ligament, which forms the base of the femoral triangle

- Medially by the adductor longus
- Laterally by the sartorius; the apex is where the medial border of the sartorius crosses the lateral border of the adductor longus.

The muscular *floor of the femoral triangle* is formed by the iliopsoas laterally and pectineus medially (Fig. 7.17C). The *roof of the femoral triangle* is formed by fascia lata, cribriform fascia, subcutaneous tissue, and skin. Deep to the inguinal

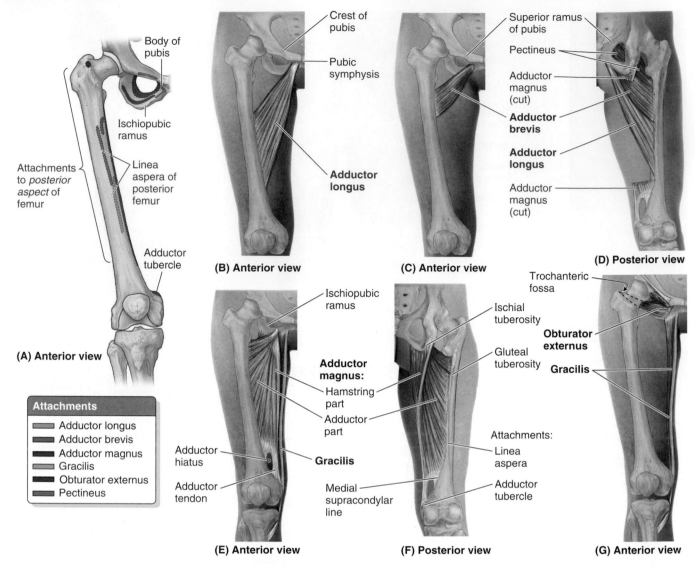

FIGURE 7.16. Medial thigh muscles. A. Muscle attachments. **B.** Adductor longus. **C.** Adductor brevis. **D.** Adductor longus and brevis. **E and F.** Adductor magnus. **G.** Gracilis

TABLE 7.3. MEDIAL THIGH MUSCLES

Muscle[a]	Proximal Attachment[b]	Distal Attachment[b]	Innervation[c]	Main Action(s)
Adductor longus	Body of pubis inferior to pubic crest	Middle third of linea aspera of femur	Obturator nerve (L2, **L3**, L4)	Adducts hip joint
Adductor brevis	Body and inferior ramus of pubis	Pectineal line and proximal part of linea aspera of femur		Adducts hip joint and to some extent flexes it
Adductor magnus	*Adductor part:* inferior ramus of pubis, ramus of ischium *Hamstring part:* ischial tuberosity	*Adductor part:* gluteal tuberosity, linea aspera, medial supracondylar line *Hamstring part:* adductor tubercle of femur	*Adductor part:* obturator nerve (L2, **L3, L4**) *Hamstring part:* tibial part of sciatic nerve (**L4**)	Adducts hip joint; its adductor part also flexes hip joint, and its hamstring part extends it.
Gracilis	Body and inferior ramus of pubis	Superior part of medial surface of tibia	Obturator nerve (**L2**, L3)	Adducts hip joint; flexes knee joint and helps rotate it medially
Obturator externus	Margins of obturator foramen and obturator membrane	Trochanteric fossa of femur	Obturator nerve (L3, **L4**)	Laterally rotates hip joint; pulls head of femur into acetabulum holding pelvis steady

[a]Collectively, the first four muscles listed are the adductors of the thigh, but their actions are more complex (e.g., they act as flexors of the hip joint during flexion of the knee joint and are active during walking).
[b]See Figure 7.16A for muscle attachments.
[c]The spinal cord segmental innervation is indicated (e.g., "L2, **L3**, **L4**" means that the nerves supplying the adductor magnus are derived from the 2nd–4th lumbar segments of the spinal cord). Numbers in boldface (**L3, L4**) indicate the main segmental innervation.

CLINICAL BOX

Hip and Thigh Contusions

Sports broadcasters and trainers refer to a "hip pointer injury," which is a *contusion of the iliac crest*, usually its anterior part. This is one of the most common injuries to the hip region, usually occurring in association with sports, such as football, ice hockey, and volleyball.

Contusions cause bleeding from ruptured capillaries and infiltration of blood into the muscles, tendons, and other soft tissues. The term *hip pointer injury* may also refer to avulsion of the bony site of muscle attachments, for example, of the sartorius or rectus femoris to the anterior superior and inferior iliac spines respectively. However, these injuries should be called *avulsion fractures*.

Another term commonly used is "charley horse," which may refer either to the acute cramping of an individual thigh muscle because of ischemia, nocturnal leg cramps, or to contusion and rupture of blood vessels sufficient to form a *hematoma* (blood clot).

Paralysis of Quadriceps

A person with *paralyzed quadriceps muscles* cannot extend the leg against resistance and usually presses on the distal end of the thigh during walking to prevent inadvertent flexion of the knee joint. Weakness of the vastus medialis or vastus lateralis, resulting from arthritis or trauma to the knee joint, can result in abnormal patellar movement and loss of joint stability.

Chondromalacia Patellae

Chondromalacia patellae (softening of the cartilage; runner's knee) is a common knee injury for marathon runners, but it can also occur in running sports such as tennis or basketball. The aching around or deep to the patella results from *quadriceps imbalance*. Chondromalacia patellae may also result from a blow to the patella or extreme flexion of the knee.

Transplantation of Gracilis

Because the gracilis is a relatively weak member of the adductor group of muscles, it can be removed without noticeable loss of its actions on the leg.

Surgeons often transplant the gracilis, or part of it, with its nerve and blood vessels to replace a damaged muscle in the forearm or to create a replacement for a nonfunctional external anal sphincter, for example.

Patellar Tendon Reflex

Tapping the patellar ligament with a reflex hammer normally elicits the *patellar reflex* ("knee jerk"). This myotatic (deep tendon) reflex is routinely tested during a physical examination by having the person sit with the legs dangling (Fig. B7.7). A firm strike on the ligament with a reflex hammer usually causes the leg to extend. If the reflex is normal, a hand on the person's quadriceps should feel the muscle contract. This tendon reflex tests the integrity of the femoral nerve and the L2–L4 spinal cord segments. *Diminution or absence of the patellar tendon reflex* may result from any lesion that interrupts the innervation of the quadriceps (e.g., peripheral nerve disease).

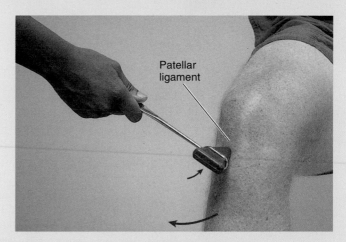

FIGURE B7.7. Patellar tendon reflex.

Groin Pull

Sports broadcasters refer to a "pulled groin" or *groin injury*. These terms refer to a strain, stretching, and probably some tearing of the proximal attachments of the flexor and adductor thigh muscles. The proximal attachments of these muscles are in the inguinal region (groin). Groin injuries usually occur in sports that require quick starts (e.g., sprinting or soccer) or extreme stretching (e.g., gymnastics).

---

ligament, the **retro-inguinal space** is an important passageway connecting the trunk/abdominopelvic cavity to the lower limb. It is created as the inguinal ligament spans the gap between the ASIS and the pubic tubercle (Fig. 7.18). The space is divided into two compartments by the iliopsoas fascia. The lateral compartment is the muscular compartment through which the iliopsoas muscle and femoral nerve pass; the medial compartment allows the passage of the veins, arteries, and lymphatics between the greater pelvis and the femoral triangle.

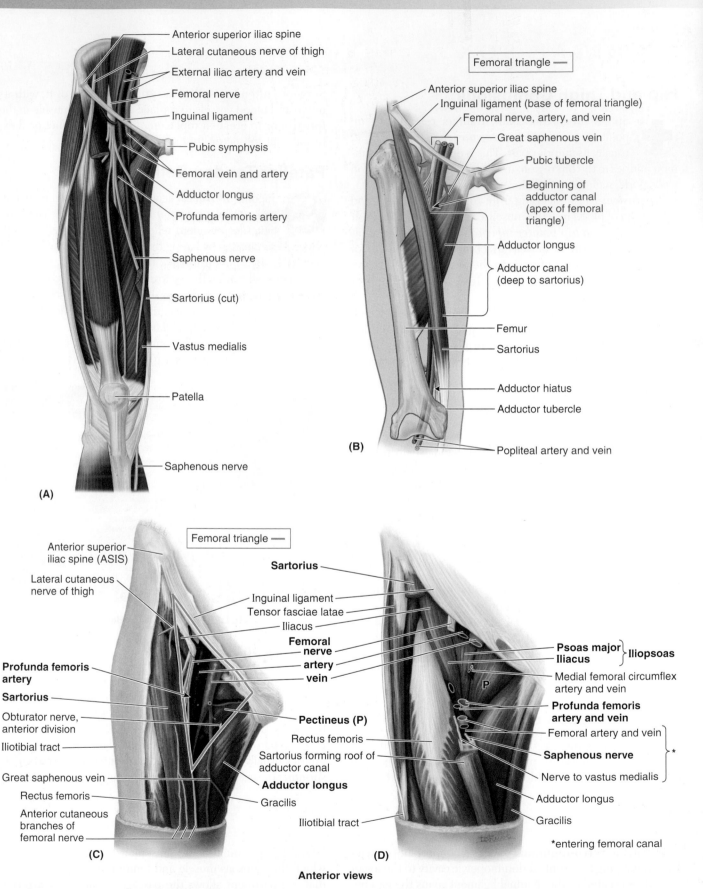

FIGURE 7.17. Nerves and vessels of anterior thigh. A. Overview. **B.** Femoral triangle and adductor canal. **C.** Boundaries and contents of femoral triangle. **D.** Floor of femoral triangle.

The contents of the femoral triangle, from lateral to medial, are as follows (Fig. 7.17):

- Femoral nerve and its (terminal) branches
- Femoral artery and several of its branches
- Femoral vein and its proximal tributaries (e.g., the great saphenous vein and profunda femoris vein)
- Deep inguinal lymph nodes and associated lymphatic vessels

The femoral artery and vein bisect by the femoral triangle and pass to and from the adductor canal at the apex of the triangle (Fig.7.17B). The **adductor canal** (subsartorial canal, Hunter canal) extends from the apex of the femoral triangle, where the sartorius crosses over the adductor longus, to the adductor hiatus in the adductor magnus. It provides an intermuscular passage for the femoral artery and vein, the saphenous nerve, and the nerve to vastus medialis, delivering the femoral vessels to the popliteal fossa where they become popliteal vessels. The adductor canal is bounded anteriorly and laterally by the vastus medialis; posteriorly by the adductor longus and adductor magnus; and medially by the sartorius, which overlies the groove between the above muscles, forming the roof of the canal.

Femoral Nerve

The **femoral nerve** (L2–L4) is the largest branch of the lumbar plexus. The nerve originates in the abdomen within the psoas major and descends posterolaterally through the pelvis to the midpoint of the inguinal ligament. It then passes deep to this ligament (in the muscular compartment of the retro-inguinal space) and enters the femoral triangle, lateral to the femoral vessels (Figs. 7.17 and 7.18). After entering the triangle, the femoral nerve divides into several terminal branches to the anterior thigh muscles (see Fig. 7.14A). It also sends articular branches to the hip and knee joints and provides cutaneous branches to the anteromedial thigh. The terminal cutaneous branch of the femoral nerve, the **saphenous nerve**, descends through the femoral triangle, lateral to the femoral sheath containing the femoral vessels. The saphenous nerve accompanies the femoral artery and vein through the adductor canal and becomes superficial by passing between the sartorius and the gracilis when the femoral vessels transverse the adductor hiatus (Fig. 7.17A). The saphenous nerve runs antero-inferiorly to supply the skin and fascia on the anteromedial aspects of the knee, leg, and foot.

Femoral Sheath

The **femoral sheath** is a funnel-shaped, fascial tube of varying length (usually 3–4 cm) that passes deep to the inguinal ligament and encloses proximal parts of the femoral vessels and creates the femoral canal medial to them (Fig. 7.18). The sheath is formed by an inferior prolongation of the transversalis and iliopsoas fascia from the abdomen/greater pelvis. The femoral sheath does not enclose the femoral nerve. The sheath terminates inferiorly by becoming continuous with the tunica adventitia, the loose connective tissue covering of the femoral vessels. When a long femoral sheath occurs, its medial wall is pierced by the great saphenous vein and

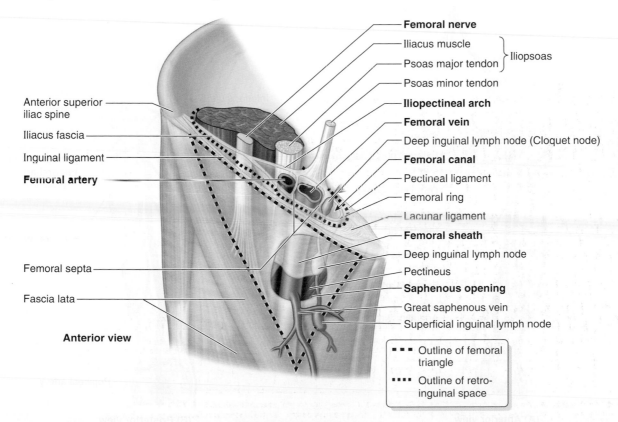

Anterior superior iliac spine
Iliacus fascia
Inguinal ligament
Femoral artery
Femoral septa
Fascia lata

Anterior view

Femoral nerve
Iliacus muscle ⎫
Psoas major tendon ⎬ Iliopsoas
Psoas minor tendon ⎭
Iliopectineal arch
Femoral vein
Deep inguinal lymph node (Cloquet node)
Femoral canal
Pectineal ligament
Femoral ring
Lacunar ligament
Femoral sheath
Deep inguinal lymph node
Pectineus
Saphenous opening
Great saphenous vein
Superficial inguinal lymph node

- - - Outline of femoral triangle
- - - Outline of retro-inguinal space

FIGURE 7.18. Femoral sheath.

lymphatic vessels. The femoral sheath allows the femoral artery and vein to glide deep to the inguinal ligament during movements of the hip joint. The femoral sheath is subdivided into three compartments by vertical septa of extraperitoneal connective tissue that extend from the abdomen along the femoral vessels. The compartments of the femoral sheath are the *lateral compartment* for the femoral artery; *intermediate compartment* for the femoral vein; and *medial compartment*, which constitutes the femoral canal.

The **femoral canal** is the smallest of the three compartments. It is short and conical and lies between the medial wall of the femoral sheath and the femoral vein. The femoral canal

- extends distally to the level of the proximal edge of the saphenous opening
- allows the femoral vein to expand when venous return from the lower limb is increased or when increased intra-abdominal pressure causes a temporary stasis in the vein
- contains loose connective tissue, fat, a few lymphatic vessels, and sometimes a deep inguinal lymph node (Cloquet node)

The base of the femoral canal, formed by the small (approximately 1 cm in diameter) proximal opening at its abdominal end, is the **femoral ring**. The boundaries of the femoral ring are as follows: *laterally*, a **femoral septum** between the femoral canal and the femoral vein; *posteriorly*, the superior ramus of the pubis covered by the pectineal ligament; *medially*, the lacunar ligament; and *anteriorly*, the medial part of the inguinal ligament.

Femoral Artery

The **femoral artery**, the chief artery to the lower limb, is the continuation of the external iliac artery distal to the inguinal ligament (Figs. 7.17 and 7.19). The femoral artery

- enters the femoral triangle deep to the midpoint of the inguinal ligament (midway between the ASIS and the pubic tubercle), lateral to the femoral vein
- lies deep to the fascia lata and descends on the adjacent borders of the iliopsoas and pectineus
- bisects the femoral triangle and exits at its apex to enter the adductor canal, deep to the sartorius
- exits the adductor canal by passing through the adductor hiatus and becoming the *popliteal artery*

The **profunda femoris artery** (deep artery of thigh) is the largest branch of the femoral artery and the chief artery to the thigh. It arises from the femoral artery in the femoral triangle (Figs. 7.17D and 7.19). In the middle third of the thigh, it is separated from the femoral artery and vein by the adductor longus. It gives off three or four **perforating arteries** that wrap around the posterior aspect of the femur and supply the adductor magnus, hamstring, and vastus lateralis muscles.

The *circumflex femoral arteries* are usually branches of the profundus femoris artery, but they may arise from the femoral artery. They encircle the thigh, anastomose with each other and other arteries, and supply the thigh muscles and the proximal

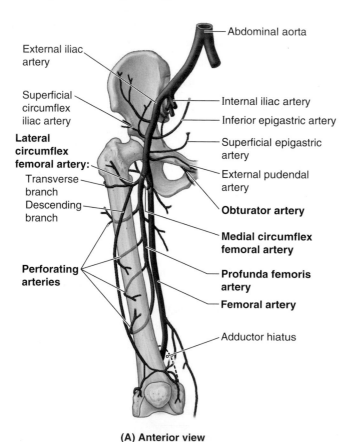

(A) Anterior view

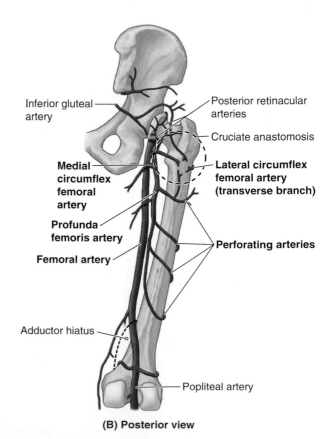

(B) Posterior view

FIGURE 7.19. Arteries of anterior and medial thigh.

end of the femur. The **medial circumflex femoral artery** supplies most of the blood to the head and neck of the femur via its branches, the **posterior retinacular arteries**. It passes deeply between the iliopsoas and pectineus to reach the posterior aspect of the femoral neck, where it runs deep (anterior) to the quadratus femoris. The **lateral circumflex femoral artery** passes laterally across the joint capsule, mainly supplying muscles on the lateral side of the thigh (Fig. 7.19).

Femoral Vein

The **femoral vein** is the continuation of the popliteal vein proximal to the adductor hiatus (Fig. 7.17A). As it ascends through the adductor canal, the femoral vein lies posterolateral and then posterior to the femoral artery (Fig. 7.17B). The femoral vein enters the femoral sheath lateral to the femoral canal and becomes the external iliac vein as it passes posterior to the inguinal ligament. In the inferior part of the femoral triangle, the femoral vein receives the profunda femoris vein, the great saphenous vein, and other tributaries. The *profunda femoris vein (deep vein of thigh)*, formed by the union of three or four perforating veins, enters the femoral vein inferior to the inguinal ligament and inferior to the termination of the great saphenous vein.

Obturator Artery and Nerve

The **obturator artery** usually arises from the internal iliac artery (Fig. 7.19). In approximately 20% of people, an enlarged pubic branch of the inferior epigastric artery either takes the place of the obturator artery (*replaced obturator artery*) or joins it as an *accessory obturator artery*. The obturator artery passes through the obturator foramen, enters the medial compartment of the thigh, and divides into anterior and posterior branches, which straddle the adductor brevis muscle. The obturator artery supplies the obturator externus, pectineus, adductors of thigh, and gracilis. Its posterior branch gives off an acetabular branch that supplies the head of the femur.

The **obturator nerve** (L2–L4) descends along the medial border of the psoas muscle and enters the thigh through the obturator foramen with the obturator artery and vein (Fig. 7.14B). It divides into anterior and posterior branches, which, like the vessels, straddle the adductor brevis. The anterior branch supplies the adductor longus, adductor brevis, gracilis, and pectineus; the posterior branch supplies the obturator externus and adductor magnus.

GLUTEAL REGION AND POSTERIOR THIGH

The *gluteal region* (hip and buttocks) is the prominent area posterior to the pelvis. It is bounded superiorly by the iliac crest, greater trochanter, and ASIS and inferiorly by the *gluteal fold*. The gluteal folds demarcate the inferior border of the buttocks and the superior boundary of the

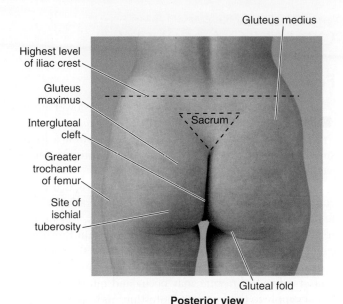

Posterior view

FIGURE 7.20. Surface landmarks of gluteal region.

thigh (Fig. 7.20). The *intergluteal cleft* separates the buttocks from each other.

The parts of the bony pelvis—hip bones, sacrum, and coccyx—are bound together by *gluteal ligaments*. The **sacrotuberous** and **sacrospinous ligaments** convert the sciatic notches in the hip bones into the greater and lesser sciatic foramina (Fig. 7.21). The **greater sciatic foramen** is the passageway for structures entering or leaving the pelvis, whereas the **lesser sciatic foramen** is a passageway for structures entering or leaving the perineum.

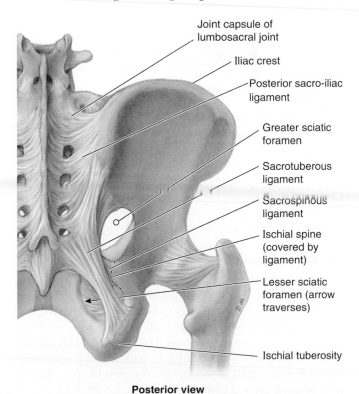

Posterior view

FIGURE 7.21. Sacrotuberous and sacrospinous ligaments.

CLINICAL BOX

Femoral Hernia

The femoral ring is a weak area in the lower anterior abdominal wall that is the site of a *femoral hernia*, a protrusion of abdominal viscera (often a loop of small intestine) through the femoral ring into the femoral canal (Fig. B7.8). A femoral hernia is more common in women than in men (in whom inguinal hernias are more common). The hernial sac displaces the contents of the femoral canal and distends its wall. Initially, the hernia is relatively small because it is contained within the femoral canal, but it can enlarge by passing through the saphenous opening into the subcutaneous tissue of the thigh. *Strangulation of a femoral hernia* may occur and interfere with the blood supply to the herniated intestine, and vascular impairment may result in death of the tissues.

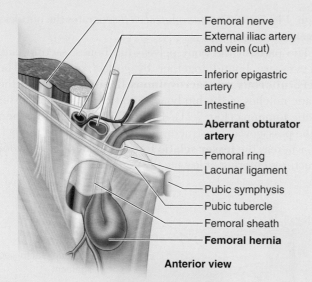

Femoral nerve
External iliac artery and vein (cut)
Inferior epigastric artery
Intestine
Aberrant obturator artery
Femoral ring
Lacunar ligament
Pubic symphysis
Pubic tubercle
Femoral sheath
Femoral hernia

Anterior view

Late stage femoral hernia

FIGURE B7.8. Femoral hernia.

Replaced or Accessory Obturator Artery

An enlarged pubic branch of the inferior epigastric artery either takes the place of the obturator artery (**replaced obturator artery**) or joins it as an **accessory obturator artery** in approximately 20% of people (Fig B7.8). This artery runs close to or across the femoral ring to reach the obturator foramen and could be closely related to a femoral hernia. Consequently, this artery could be involved in a *strangulated femoral hernia*. Surgeons placing staples during endoscopic repair of both inguinal and femoral hernias are vigilant concerning the possible presence of this common arterial variant.

Femoral Pulse and Cannulation of Femoral Artery

The pulse of the femoral artery is usually palpable just inferior to the midpoint of the inguinal ligament (see Fig. 7.11C). Normally, the pulse is strong; however, if the common or external iliac arteries are partially occluded, the pulse may be diminished. The femoral artery may be manually compressed at the midpoint of the inguinal ligament to control arterial bleeding after lower limb trauma (Fig. B7.9).

The femoral artery may be cannulated just inferior to the midpoint of the inguinal ligament (e.g., for cardioangiography—radiography of the heart and great vessels after the introduction of contrast material). For *left cardiac angiography*, a long slender catheter is inserted percutaneously into the femoral artery and passed superiorly in the aorta to the openings of the coronary arteries (see Chapter 4, Thorax).

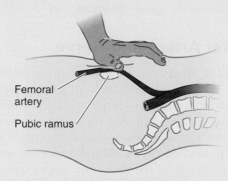

Femoral artery
Pubic ramus

Compression of right femoral artery
Medial view (right side)

FIGURE B7.9. Compression of femoral artery.

Cannulation of Femoral Vein

The femoral vein usually is not palpable; however, its position can be located by feeling the pulsations of the femoral artery, which lies just lateral to it. In thin people, the femoral vein may be close to the surface and may be mistaken for the great saphenous vein. It is thus important to know that the femoral vein has no tributaries at this level, except for the great saphenous vein that joins it approximately 3 cm inferior to the inguinal ligament. To secure blood samples and take pressure recordings from the chambers of the right side of the heart and/or from the pulmonary artery and to perform *right cardiac angiography*, a long, slender catheter is inserted into the femoral vein as it passes through the femoral triangle. Under fluoroscopic control, the catheter is passed superiorly through the external and common iliac veins into the inferior vena cava and right atrium of the heart.

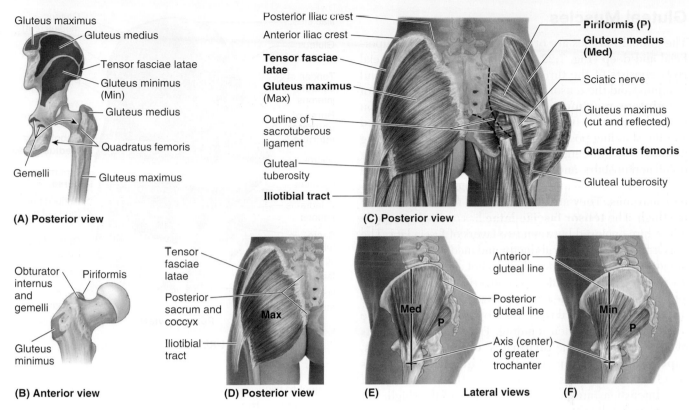

FIGURE 7.22. Gluteal muscles. A and **B.** Muscle attachments. **C.** Overview. **D.** Gluteus maximus. **E.** Gluteus medius. **F.** Gluteus minimus.

TABLE 7.4. MUSCLES OF GLUTEAL REGION

Muscle	Proximal Attachment	Distal Attachment	Innervation[a]	Main Action(s)
Gluteus maximus	Ilium posterior to posterior gluteal line; posterior surface of sacrum and coccyx; and sacrotuberous ligament	Most fibers end in iliotibial tract, which inserts into lateral condyle of tibia; some fibers insert on gluteal tuberosity of femur.	Inferior gluteal nerve (L5, **S1**, **S2**)	Extends hip joint between flexed and standing positions and assists in its lateral rotation; steadies thigh and assists in rising from sitting position
Gluteus medius	External surface of ilium between anterior and posterior gluteal lines	Lateral surface of greater trochanter of femur	Superior gluteal nerve (L4, L5, S1)	Abduct and anterior portions medially rotate hip joint[b]; keep pelvis level when opposite limb is elevated
Gluteus minimus	External surface of ilium between anterior and inferior gluteal lines	Anterior surface of greater trochanter of femur		
Tensor fasciae latae (tensor of fascia lata)	Anterior superior iliac spine; anterior part of iliac crest	Iliotibial tract, which attaches to lateral condyle of tibia (Gerdy tubercle)		Flexes hip joint; acts with gluteus maximus to stabilize the extended knee joint
Piriformis (passes through greater sciatic foramen)	Anterior surface of 2nd–4th sacral segments; superior margin of greater sciatic notch and sacrotuberous ligament	Superior border of greater trochanter of femur	Branches of anterior rami of **S1**, S2	
Obturator internus (passes through lesser sciatic foramen)	Pelvic surface of ilium and ischium; obturator membrane	Medial surface of greater trochanter (trochanteric fossa) of femur[c]	Nerve to obturator internus (L5, **S1**)	Laterally rotate extended hip joint; abduct flexed hip joint; steady femoral head in acetabulum (stabilizes hip joint)
Gemelli, superior and inferior	*Superior*: ischial spine *Inferior*: ischial tuberosity		*Superior*: same as obturator internus *Inferior*: same as quadratus femoris	
Quadratus femoris	Lateral border of ischial tuberosity	Quadrate tubercle on intertrochanteric crest of femur and area inferior to it	Nerve to quadratus femoris (L5, S1)	Laterally rotates hip joint; also pulls femoral head into acetabulum to stabilize hip joint/pelvis

[a]The spinal cord segmental innervation is indicated (e.g., "**S1**, S2" means that the nerves supplying the piriformis are derived from the first two sacral segments of the spinal cord). Numbers in boldface (**S1**) indicate the main segmental innervation.
[b]Also posterior portions laterally rotate the hip joint.
[c]Gemelli muscles blend with the tendon of the obturator internus muscle as it attaches to the greater trochanter of the femur.

Gluteal Muscles

The gluteal muscles are organized into two layers: superficial and deep (Fig. 7.22 and Table 7.4). The superficial layer consists of the three glutei (maximus, medius, and minimus) and the tensor fasciae latae. The main actions of the **gluteus maximus** are extension and lateral rotation of the hip joint. It functions primarily between the flexed and the standing positions, as when rising from the sitting position, straightening from the bending position, walking uphill and upstairs, and running. The **gluteus medius** and **minimus** are fan-shaped muscles that lie deep to the gluteus maximus. They are abductors and medial rotators of the thigh. The **tensor fasciae latae** lies on the lateral side of the hip, enclosed between two layers of fascia lata. The tensor fasciae latae is an abductor and medial rotator of the hip joint; however, it generally does not act independently. To produce flexion, it acts in concert with the iliopsoas and rectus femoris. The tensor fasciae latae also tenses the fascia lata and iliotibial tract, thereby helping stabilize the femur on the tibia when standing. The deep layer consists of smaller muscles: the **piriformis, obturator internus, superior** and **inferior gemelli**, and **quadratus femoris** (Fig. 7.23). These muscles, covered by the inferior half of the gluteus maximus, are lateral rotators of the thigh, but they also stabilize the hip joint, working with the strong ligaments of the hip joint to steady the femoral head in the acetabulum.

Gluteal Bursae

Gluteal bursae, flattened membranous sacs containing a capillary layer of synovial fluid, separate the gluteus maximus from adjacent structures (Fig. 7.24). The bursae are located in areas

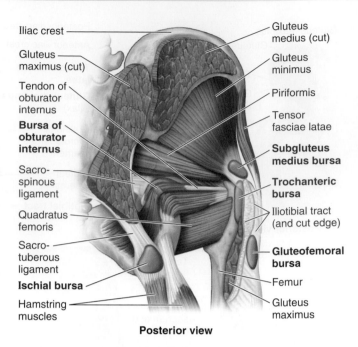

Posterior view

FIGURE 7.24. Gluteal bursae.

subject to friction—for example, between a muscle and a bony prominence—to reduce friction and permit free movement. The bursae associated with the gluteus maximus are as follows:

- The **trochanteric bursa** separates the deep aspect of the gluteus maximus from the greater trochanter of the femur.
- The **ischial bursa** separates the inferior border of the gluteus maximus from the ischial tuberosity.
- The **gluteofemoral bursa** separates the iliotibial tract from the superior part of the proximal attachment of the vastus lateralis.

Posterior Thigh Muscles

Three of the four muscles in the posterior aspect of the thigh are **hamstrings** (Fig. 7.25 and Table 7.5): **semitendinosus, semimembranosus**, and **biceps femoris (long head)**. The hamstring muscles arise from the ischial tuberosity deep to the gluteus maximus, insert on the leg bones, and are innervated by the tibial division of the sciatic nerve (Fig. 7.14C). They span and act on two joints (extension at the hip joint and flexion at the knee joint). Both actions cannot be performed maximally at the same time. A fully flexed knee shortens the hamstrings so they cannot further contract to extend the hip joint. Similarly, a fully extended hip shortens the hamstrings so they cannot act on the knee. When the thighs and legs are fixed, the hamstrings can help extend the trunk at the hip joint. They are active in hip extension under all situations except full flexion of the knee, including maintenance of the standing posture.

The **short head of biceps femoris**, the fourth muscle in the posterior compartment, is not a hamstring because it crosses only the knee joint and is innervated by the fibular division of the sciatic nerve (Fig. 7.14C).

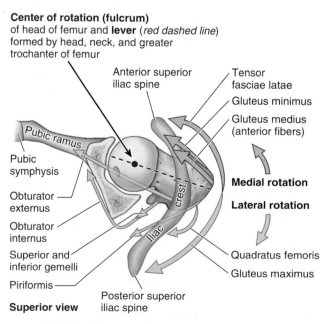

FIGURE 7.23. Medial and lateral rotators of hip joint.

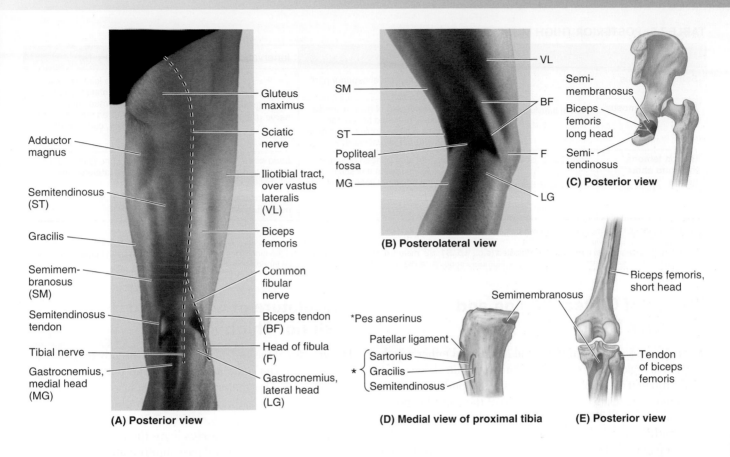

(A) Posterior view

Adductor magnus

Semitendinosus (ST)

Gracilis

Semimem- branosus (SM)

Semitendinosus tendon

Tibial nerve

Gastrocnemius, medial head (MG)

Gluteus maximus

Sciatic nerve

Iliotibial tract, over vastus lateralis (VL)

Biceps femoris

Common fibular nerve

Biceps tendon (BF)

Head of fibula (F)

Gastrocnemius, lateral head (LG)

(B) Posterolateral view

SM

ST

Popliteal fossa

MG

VL

BF

F

LG

(C) Posterior view

Semi- membranosus

Biceps femoris long head

Semi- tendinosus

(D) Medial view of proximal tibia

Semimembranosus

*Pes anserinus

Patellar ligament

{ Sartorius

* { Gracilis

{ Semitendinosus

(E) Posterior view

Biceps femoris, short head

Tendon of biceps femoris

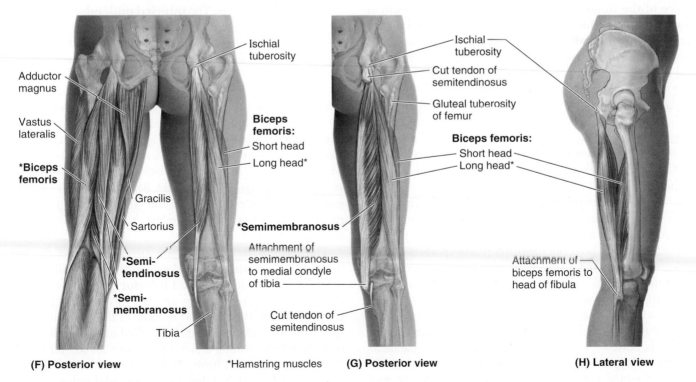

(F) Posterior view   *Hamstring muscles

Adductor magnus

Vastus lateralis

*Biceps femoris

Gracilis

Sartorius

*Semi- tendinosus

*Semi- membranosus

Tibia

Ischial tuberosity

Biceps femoris:
Short head
Long head*

*Semimembranosus

Attachment of semimembranosus to medial condyle of tibia

Cut tendon of semitendinosus

(G) Posterior view

Ischial tuberosity

Cut tendon of semitendinosus

Gluteal tuberosity of femur

Biceps femoris:
Short head
Long head*

(H) Lateral view

Attachment of biceps femoris to head of fibula

FIGURE 7.25. Posterior thigh muscles. A and B. Surface anatomy. **C–E.** Muscle attachments. **F.** Overview. **G.** Semimembranosus and biceps femoris. **H.** Biceps femoris.

TABLE 7.5. POSTERIOR THIGH MUSCLES

Muscle[a]	Proximal Attachment[b]	Distal Attachment[b]	Innervation[c]	Main Action(s)
Semitendinosus	Ischial tuberosity	Medial surface of superior part of tibia	Tibial division of sciatic nerve (**L5**, **S1**, S2)	Extend hip joint; flex knee joint and rotate the leg medially when knee is flexed; when hip and knee are flexed, can extend trunk
Semimembranosus		Posterior part of medial condyle of tibia; reflected attachment forms oblique popliteal ligament (to lateral femoral condyle)		
Biceps femoris, long and short heads	*Long head*: ischial tuberosity *Short head*: linea aspera and lateral supracondylar line of femur	Lateral side of head of fibula; tendon is split at this site by fibular collateral ligament of knee	*Long head*: tibial division of sciatic nerve (L5, **S1**, S2) *Short head*: common fibular division of sciatic nerve (L5, **S1**, S2)	Flexes knee joint and rotates it laterally when knee is flexed; extends hip joint (e.g., when starting to walk)

[a]Collectively, these three muscles are known as hamstrings.
[b]See Figure 7.25C–E for muscle attachments.
[c]The spinal cord segmental innervation is indicated (e.g., "L5, **S1**, S2" means that the nerves supplying the biceps femoris are derived from the 5th lumbar segment and first two sacral segments of the spinal cord). Numbers in boldface (**S1**) indicate the main segmental innervation.

Nerves of Gluteal Region and Posterior Thigh

Several nerves arise from the sacral plexus and either supply the gluteal region (e.g., superior and inferior gluteal nerves) or pass through it to supply the perineum (e.g., pudendal nerve) and thigh (e.g., sciatic nerve). The skin of the gluteal region is richly innervated by the superficial gluteal nerves: the **superior, middle,** and **inferior clunial nerves** (see Fig. 7.12B). The deep gluteal nerves are the **sciatic, posterior cutaneous nerve of the thigh, superior gluteal** and **inferior gluteal nerves, nerve to the quadratus femoris, pudendal nerve,** and **nerve to the obturator internus** (Fig. 7.26B,C and Table 7.6). All of these nerves are branches of the sacral plexus and leave the pelvis through the greater sciatic foramen. Except for the superior gluteal nerve, and sometimes the common fibular nerve (root of the sciatic nerve), they all emerge inferior to the piriformis muscle (Fig. 7.26B–D). The pudendal nerve supplies no structures in the gluteal region; it exits the region via the lesser sciatic foramen to supply structures in the perineum.

The **sciatic nerve** is the largest nerve in the body and is the continuation of the main part of the sacral plexus (Fig. 7.26B,C). The sciatic nerve runs inferolaterally under cover of the gluteus maximus, midway between the greater trochanter and the ischial tuberosity. It descends from the gluteal region into the posterior thigh, where it lies posterior to the adductor magnus and deep (anterior) to the long head of the biceps femoris. The sciatic nerve is really two nerves loosely bound together in the same connective tissue sheath: the *tibial nerve* and the *common fibular (peroneal) nerve* (Figs. 7.14C and 7.26B). The two nerves separate in the inferior third of the thigh; however, in 12% of people, the nerves separate as they leave the pelvis. In these cases, the tibial nerve passes inferior to the piriformis, and the common fibular nerve pierces this muscle or passes superior to it (Fig. 7.26D). The sciatic nerve supplies no structures in the gluteal region; it innervates the posterior thigh muscles, all leg and foot muscles, and the skin of most of the leg and foot. It also supplies articular branches to lower limb joints inferior to the hip.

Vasculature of Gluteal and Posterior Thigh Regions

The *arteries of the gluteal region* arise, directly or indirectly, from the *internal iliac arteries*, but the patterns of origin are variable. The major gluteal branches of the internal iliac artery are the superior and inferior gluteal arteries and the internal pudendal artery (Fig. 7.26A,B). The **superior** and **inferior gluteal arteries** leave the pelvis through the greater sciatic foramen and pass superior and inferior to the piriformis, respectively. The **internal pudendal artery** enters the gluteal region through the greater sciatic foramen inferior to the piriformis and enters the perineum through the lesser sciatic foramen (Fig. 7.26B). It does not supply the buttocks. The posterior compartment of the thigh has no major artery exclusive to the compartment; it receives blood from the inferior gluteal, medial circumflex femoral, and perforating and popliteal arteries. The *profunda femoris artery* (deep artery of thigh) is the chief artery of the thigh, giving off *perforating arteries* which pierce the adductor magnus to enter the posterior compartment and supply the hamstrings (Fig. 7.26A,B). A continuous anastomotic chain thus extends from the gluteal to the popliteal region, which gives rise to branches to the muscles and to the sciatic nerve.

The *veins of the gluteal region* are tributaries of the internal iliac veins that drain blood from the gluteal region (Fig. 7.27). The **superior** and **inferior gluteal veins** accompany the corresponding arteries through the greater sciatic foramen, superior and inferior to the piriformis, respectively. They communicate with tributaries of the femoral vein, thereby providing an alternate route for the return of blood from the lower limb if the femoral vein is occluded or has to be ligated. The **internal pudendal veins** accompany the internal pudendal arteries and join to enter the internal iliac vein. The pudendal veins drain blood from the perineum. *Perforating veins* accompany the arteries of the same name to drain blood from the posterior compartment of the thigh into the *profunda femoris vein*. They also

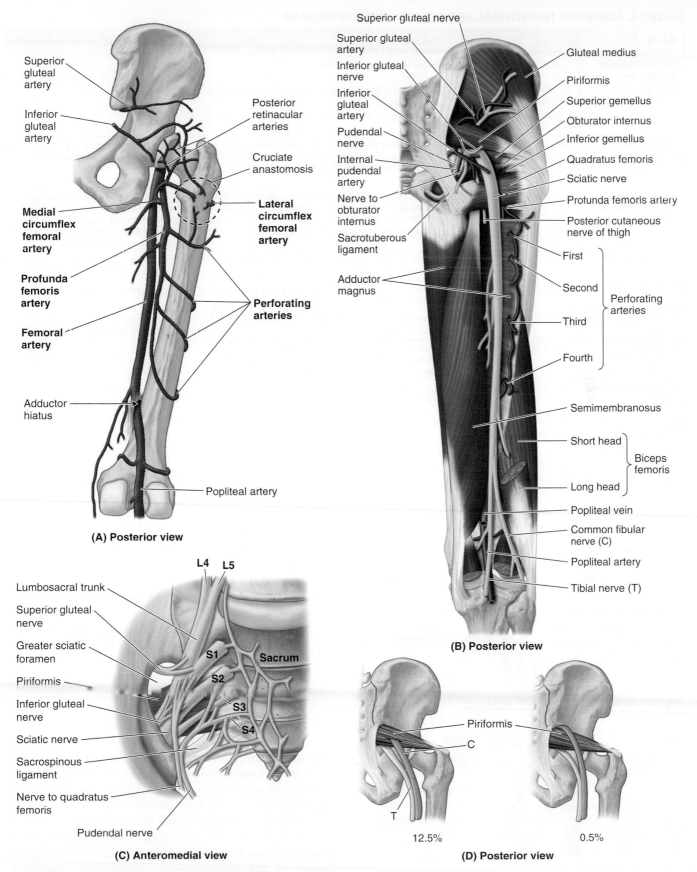

(A) Posterior view

Superior gluteal artery

Inferior gluteal artery

Posterior retinacular arteries

Cruciate anastomosis

Medial circumflex femoral artery

Profunda femoris artery

Femoral artery

Adductor hiatus

Lateral circumflex femoral artery

Perforating arteries

Popliteal artery

(B) Posterior view

Superior gluteal nerve

Superior gluteal artery

Inferior gluteal nerve

Inferior gluteal artery

Pudendal nerve

Internal pudendal artery

Nerve to obturator internus

Sacrotuberous ligament

Adductor magnus

Gluteal medius

Piriformis

Superior gemellus

Obturator internus

Inferior gemellus

Quadratus femoris

Sciatic nerve

Protunda femoris artery

Posterior cutaneous nerve of thigh

First

Second

Third

Fourth

Perforating arteries

Semimembranosus

Short head

Long head

Biceps femoris

Popliteal vein

Common fibular nerve (C)

Popliteal artery

Tibial nerve (T)

(C) Anteromedial view

Lumbosacral trunk

Superior gluteal nerve

Greater sciatic foramen

Piriformis

Inferior gluteal nerve

Sciatic nerve

Sacrospinous ligament

Nerve to quadratus femoris

Pudendal nerve

L4 L5

S1

S2

S3

S4

Sacrum

(D) Posterior view

Piriformis

C

T

12.5%

0.5%

FIGURE 7.26. Nerves and vasculature of gluteal region and posterior thigh. A. Arteries. **B.** Course of arteries and nerves in posterior thigh. **C.** Formation of sciatic nerve in pelvis. **D.** Anomalous relationships of sciatic nerve to piriformis.

TABLE 7.6. NERVES OF THE GLUTEAL AND POSTERIOR THIGH REGIONS

Nerve	Origin	Course	Distribution[a]
Clunial: superior, middle, and inferior	*Superior*: posterior rami of L1–L3 nerves *Middle*: posterior rami of S1–S3 nerves *Inferior*: posterior cutaneous nerve of thigh	*Superior*: crosses iliac crest *Middle*: exits through posterior sacral foramina and enter gluteal region *Inferior*: curves around inferior border of gluteus maximus	Supplies skin of gluteal region (buttocks) as far as greater trochanter
Sciatic	Sacral plexus (L4–S3)	Leaves pelvis through greater sciatic foramen inferior to piriformis; enters gluteal region; descends deep to biceps femoris; bifurcates into tibial and common fibular nerves at apex of popliteal fossa	Supplies no muscles in gluteal region; supplies all muscles in posterior compartment of thigh
Posterior cutaneous nerve of thigh	Sacral plexus (S1–S3)	Leaves pelvis through greater sciatic foramen inferior to piriformis; runs deep to gluteus maximus; emerges from its inferior border; descends in posterior thigh deep to fascia lata giving rise to cutaneous branches	Supplies skin of buttocks through inferior clunial branches and skin over posterior aspect of thigh and calf; lateral perineum, upper medial thigh via perineal branch
Superior gluteal	Sacral plexus (L4–S1)	Leaves pelvis through greater sciatic foramen superior to piriformis; runs between gluteus medius and minimus	Innervates gluteus medius, gluteus minimus, and tensor fasciae latae
Inferior gluteal	Sacral plexus (L5–S2)	Leaves pelvis through greater sciatic foramen inferior to piriformis; divides into several branches	Innervates gluteus maximus
Nerve to quadratus femoris	Sacral plexus (L4, L5–S1)	Leaves pelvis through greater sciatic foramen deep to sciatic nerve	Innervates hip joint, inferior gemellus, and quadratus femoris
Pudendal	Sacral plexus (S2–S4)	Enters gluteal region through greater sciatic foramen inferior to piriformis; descends posterior to sacrospinous ligament; enters perineum through lesser sciatic foramen	Supplies most innervation to the perineum; supplies no structures in gluteal region
Nerve to obturator internus	Sacral plexus (L5–S2)	Enters gluteal region through greater sciatic foramen inferior to piriformis; descends posterior to ischial spine; enters lesser sciatic foramen; passes to obturator internus	Supplies superior gemellus and obturator internus

[a]See Figure 7.12 for cutaneous innervation of the lower limb.

communicate inferiorly with the popliteal vein and superiorly with the inferior gluteal vein.

Lymph from the deep tissues of the gluteal region follows the gluteal vessels to the **gluteal lymph nodes** and from them to the internal, external, and common **iliac lymph nodes** and then to the **lumbar (caval and aortic) lymph nodes** (Fig. 7.27). *Lymph from superficial tissues* of the gluteal region enters the *superficial inguinal lymph nodes.* The superficial inguinal nodes send efferent lymphatic vessels to the **external iliac nodes**.

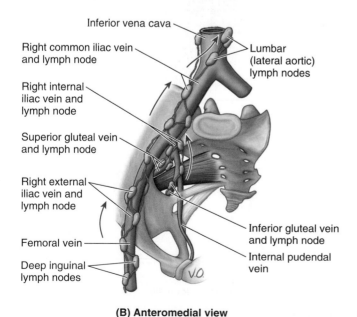

Inferior vena cava
Right common iliac vein and lymph node
Lumbar (lateral aortic) lymph nodes
Right internal iliac vein and lymph node
Superior gluteal vein and lymph node
Right external iliac vein and lymph node
Femoral vein
Deep inguinal lymph nodes
Inferior gluteal vein and lymph node
Internal pudendal vein

(B) Anteromedial view

FIGURE 7.27. Veins and lymphatics of gluteal and posterior thigh regions.

Trochanteric and Ischial Bursitis

Diffuse deep pain in the lateral thigh region, especially during stair climbing or rising from a seated position, may be caused by *trochanteric bursitis*. It is characterized by point tenderness over the greater trochanter; however, the pain often radiates along the iliotibial tract. A commonly overlooked diagnosis that clinically mimics trochanteric bursitis is a tear of the insertion of gluteus medius tendon on the trochanter. *Ischial bursitis* results from excessive friction between the ischial bursae and the ischial tuberosities (e.g., as from cycling). Because the tuberosities bear the body weight during sitting, these pressure points may lead to *pressure sores* in debilitated people, particularly paraplegic persons.

Injury to Superior Gluteal Nerve

Injury to the superior gluteal nerve, for example during hip replacement surgery depending on the surgical approach, results in a disabling *gluteus medius limp* to compensate for weakened abduction of the thigh by the gluteus medius and minimus. Also, a *gluteal gait*, a compensatory list of the body to the weakened side, may be present. Medial rotation of the thigh is also severely impaired.

When a person is asked to stand on one leg, the gluteus medius and minimus normally contract as soon as the contralateral foot leaves the floor, preventing tipping of the pelvis to the unsupported side (Fig. B7.10A). When a person with a lesion of the superior gluteal nerve is asked to stand on one leg, the pelvis descends on the unsupported side (Fig. B7.10B), indicating that the gluteus medius on the contralateral side is weak or nonfunctional. This is referred to clinically as a *positive Trendelenburg test*.

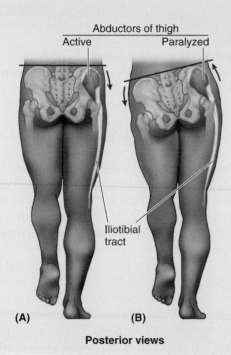

Posterior views

FIGURE B7.10. Trendelenburg test.

When the pelvis descends on the unsupported side, the lower limb becomes, in effect, too long and does not clear the ground when the foot is brought forward in the swing phase of walking. To compensate, the individual leans away from the unsupported side, raising the pelvis to allow adequate room for the foot to clear the ground as it swings forward. This results in a characteristic "waddling" or gluteal gait. Other ways to compensate are to lift the foot higher as it is brought forward or to swing the foot outward.

Hamstring Injuries

Hamstrings strains (pulled and/or torn hamstrings) are common in people who run and/or kick hard (e.g., quick-start sports such as sprinting, baseball, and soccer). The muscular exertion required to excel in these sports may tear part of the proximal attachment of the hamstrings to the ischial tuberosity.

Injury to Sciatic Nerve

Incomplete section of the sciatic nerve (e.g., from a stab wound) may involve the inferior gluteal and/or the posterior femoral cutaneous nerves. Recovery from a sciatic lesion is slow and usually incomplete. With respect to the sciatic nerve, the buttocks has a side of safety (its lateral side) and a side of danger (its medial side). Wounds or surgery on the medial side may injure the sciatic nerve and its branches to the hamstrings. Paralysis of these muscles results in impairment of thigh extension and leg flexion. A pain in the buttocks may possibly result from compression of the sciatic nerve by the piriformis muscle (*piriformis syndrome*).

Intragluteal Injections

The gluteal region is a common site for *intramuscular injection* of drugs because the gluteal muscles are thick and large, providing a large area for venous absorption of drugs. Injections into the buttocks are safe only in the superolateral quadrant of the buttocks (Fig. B7.11). Complications of improper technique include nerve injury, hematoma, and abscess formation.

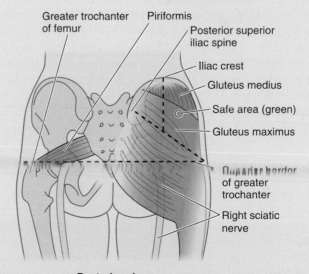

Posterior view
Safe area (green) for intragluteal injections

FIGURE B7.11. Intragluteal injections.

POPLITEAL FOSSA

The **popliteal fossa** is a mostly fat-filled, diamond-shaped space posterior to the knee joint (Fig. 7.28). All important vessels and nerves from the thigh to the leg pass through this fossa.

The popliteal fossa is bounded by the following:

- Biceps femoris superolaterally
- Semimembranosus superomedially, medial to which is the semitendinosus tendon
- Lateral and medial heads of the gastrocnemius, inferolaterally and inferomedially, respectively
- Skin and popliteal fascia posteriorly (roof)
- Popliteal surface of the femur, posterior capsule of the knee joint, and the popliteus fascia covering the popliteus muscle (floor)

The contents of the popliteal fossa include the following (Fig. 7.28B):

- Termination of the small saphenous vein
- Popliteal artery and vein and their branches and tributaries
- Tibial and common fibular nerves
- Posterior cutaneous nerve of the thigh
- Popliteal lymph nodes and lymphatic vessels
- Fat

Fascia of Popliteal Fossa

The *subcutaneous tissue* overlying the fossa contains fat, the small saphenous vein (unless it has penetrated the deep fascia at a more inferior level), and three cutaneous nerves: the terminal branch(es) of the *posterior cutaneous nerve of the thigh* and the *medial and lateral sural cutaneous nerves* (Fig. 7.28A). The **popliteal fascia** is a strong sheet of deep fascia that forms a protective covering for neurovascular structures passing from the thigh through the popliteal fossa to the leg. The popliteal fascia is continuous with the *fascia lata* superiorly and the *deep (crural) fascia of the leg* inferiorly.

Neurovasculature of Popliteal Fossa

The **popliteal artery**, the continuation of the femoral artery, begins where the femoral artery passes through the adductor hiatus (Figs. 7.19B and 7.28B). The popliteal artery passes through the popliteal fossa and ends at the inferior border of the popliteus by dividing into the **anterior** and **posterior tibial arteries** (Fig. 7.28D). The deepest structure in the popliteal fossa, the popliteal artery, runs close to the joint capsule of the knee joint. Five genicular branches of the popliteal artery supply the joint capsule and ligaments of the knee joint. The genicular arteries are the **superior lateral**, **superior medial**, **middle**, **inferior lateral**, and **inferior medial genicular arteries** (Fig. 7.28D). They participate in the formation of the **genicular anastomosis** (L. *genu*, knee), a peri-articular arterial anastomosis around the knee that provides collateral circulation capable of maintaining blood supply to the leg during full knee flexion. Other contributors to the anastomosis are also shown in Figure 7.28D. The muscular branches of the popliteal artery supply the hamstring, gastrocnemius, soleus, and plantaris muscles. The superior muscular branches of the popliteal artery have clinically important anastomoses with the terminal part of the profunda femoris artery and gluteal arteries.

The **popliteal vein** is formed at the inferior border of the popliteus as a continuation of the *posterior tibial veins*. Throughout its course, the vein lies superficial to and in the same fibrous sheath as the popliteal artery (Fig. 7.28B). Superiorly, the popliteal vein becomes the *femoral vein* as it traverses the adductor hiatus. The small saphenous vein passes from the posterior aspect of the lateral malleolus to the popliteal fossa, where it pierces the deep popliteal fascia and enters the popliteal vein (Fig. 7.28A).

The **superficial popliteal lymph nodes** are usually small and lie in the subcutaneous tissue. The **deep popliteal lymph nodes** surround the vessels and receive lymph from the joint capsule of the knee and the lymphatic vessels that accompany the deep veins of the leg (see Fig. 7.9D). Lymphatic vessels from the popliteal lymph nodes follow the femoral vessels to the *deep inguinal lymph nodes*.

The *sciatic nerve* usually ends at the superior angle of the popliteal fossa by dividing into the tibial and common fibular nerves (Fig. 7.28A–C). The **tibial nerve**—the medial, larger terminal branch of the sciatic nerve—is the most superficial of the three main central components of the popliteal fossa (nerve, vein, and artery). The tibial nerve bisects the fossa as it passes from its superior to its inferior angle. While in the fossa, the tibial nerve gives branches to the soleus, gastrocnemius, plantaris, and popliteus muscles. A **medial sural cutaneous nerve** is also derived from the tibial nerve in the popliteal fossa (Fig. 7.28A,C). It is joined by the **sural communicating branch of the common fibular nerve** at a highly variable level to form the **sural nerve**. This nerve supplies skin on the posterior and lateral aspects of the leg and lateral side of the foot. The **lateral sural cutaneous nerve** is a branch of the common fibular nerve that supplies the skin of the lateral aspect of the leg.

The **common fibular nerve** (Fig. 7.28A–C)—the lateral, smaller terminal branch of the sciatic nerve—usually begins at the superior angle of the popliteal fossa and follows closely the medial border of the biceps femoris and its tendon along the superolateral boundary of the popliteal fossa. The common fibular nerve leaves the fossa by passing superficial to the lateral head of the gastrocnemius and winding around the fibular neck, where it is vulnerable to injury.

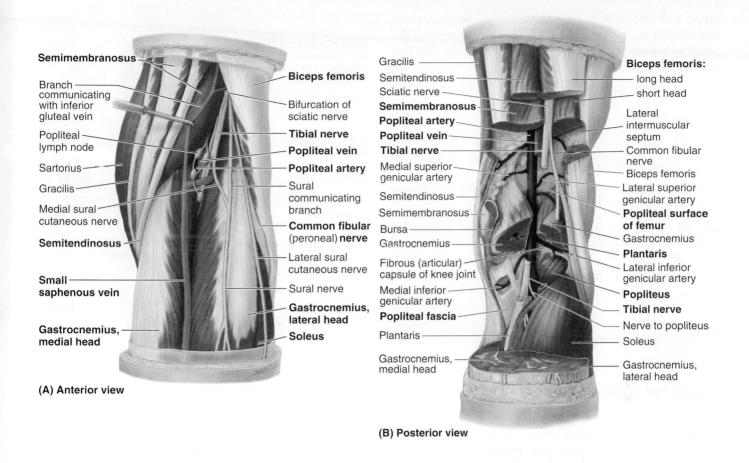

Semimembranosus

Branch communicating with inferior gluteal vein

Popliteal lymph node

Sartorius

Gracilis

Medial sural cutaneous nerve

Semitendinosus

Small saphenous vein

Gastrocnemius, medial head

Biceps femoris

Bifurcation of sciatic nerve

Tibial nerve

Popliteal vein

Popliteal artery

Sural communicating branch

Common fibular (peroneal) nerve

Lateral sural cutaneous nerve

Sural nerve

Gastrocnemius, lateral head

Soleus

(A) Anterior view

Gracilis

Semitendinosus

Sciatic nerve

Semimembranosus

Popliteal artery

Popliteal vein

Tibial nerve

Medial superior genicular artery

Semitendinosus

Semimembranosus

Bursa

Gastrocnemius

Fibrous (articular) capsule of knee joint

Medial inferior genicular artery

Popliteal fascia

Plantaris

Gastrocnemius, medial head

Biceps femoris:

long head

short head

Lateral intermuscular septum

Common fibular nerve

Biceps femoris

Lateral superior genicular artery

Popliteal surface of femur

Gastrocnemius

Plantaris

Lateral inferior genicular artery

Popliteus

Tibial nerve

Nerve to popliteus

Soleus

Gastrocnemius, lateral head

(B) Posterior view

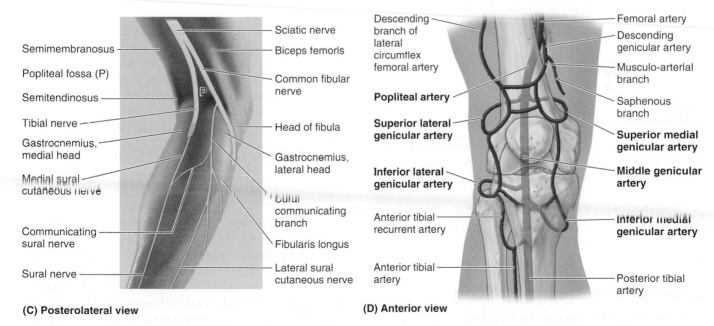

Semimembranosus

Popliteal fossa (P)

Semitendinosus

Tibial nerve

Gastrocnemius, medial head

Medial sural cutaneous nerve

Communicating sural nerve

Sural nerve

Sciatic nerve

Biceps femoris

Common fibular nerve

Head of fibula

Gastrocnemius, lateral head

Sural communicating branch

Fibularis longus

Lateral sural cutaneous nerve

(C) Posterolateral view

Descending branch of lateral circumflex femoral artery

Popliteal artery

Superior lateral genicular artery

Inferior lateral genicular artery

Anterior tibial recurrent artery

Anterior tibial artery

Femoral artery

Descending genicular artery

Musculo-arterial branch

Saphenous branch

Superior medial genicular artery

Middle genicular artery

Inferior medial genicular artery

Posterior tibial artery

(D) Anterior view

FIGURE 7.28. Popliteal fossa. A. Superficial dissection. **B.** Deep dissection. **C.** Surface anatomy of major peripheral and cutaneous nerves. **D.** Genicular peri-articular arterial anastomosis.

Here, it divides into its terminal branches, the superficial and deep fibular nerves. The most inferior branches of the *posterior cutaneous nerve of the thigh* supply the skin that overlies the popliteal fossa.

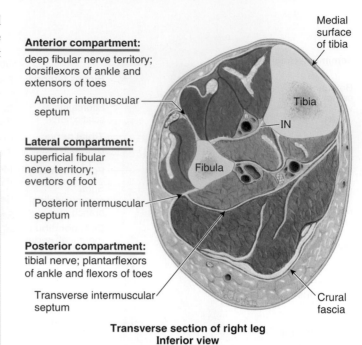

Anterior compartment:
deep fibular nerve territory; dorsiflexors of ankle and extensors of toes

Anterior intermuscular septum

Lateral compartment:
superficial fibular nerve territory; evertors of foot

Posterior intermuscular septum

Posterior compartment:
tibial nerve; plantarflexors of ankle and flexors of toes

Transverse intermuscular septum

Medial surface of tibia

Tibia

IN

Fibula

Crural fascia

Transverse section of right leg
Inferior view

FIGURE 7.29. **Compartments of leg.** *IN*, interosseous membrane.

CLINICAL BOX

Popliteal Pulse

Because the popliteal artery is deep in the popliteal fossa, it may be difficult to feel the *popliteal pulse*. To palpate this pulse, the person is placed in the prone position with the knee flexed to relax the popliteal fascia and hamstrings (see Fig. 7.11E). The pulsations are best felt in the inferior part of the fossa. Weakening or loss of the popliteal pulse is a sign of femoral artery obstruction. The popliteal artery is vulnerable in knee dislocations; downstream pulses should be tested if dislocation has occurred.

Popliteal Aneurysm

A *popliteal aneurysm* (abnormal dilation of all or part of the popliteal artery) usually causes edema (swelling) and pain in the popliteal fossa. If the femoral artery has to be ligated, usually, blood can bypass the occlusion through the genicular anastomosis and reach the popliteal artery distal to the ligation. Gradual ligation may be necessary.

LEG

The *leg* contains the *tibia* and *fibula*, bones that connect the knee and ankle. The tibia, the weight-bearing bone, is larger and stronger than the non–weight-bearing fibula. The leg bones are connected by the *interosseous membrane*. The leg is divided into three compartments—anterior, lateral, and posterior—which are formed by the anterior and posterior *intermuscular septa*, the *interosseous membrane*, and the two leg bones (Fig. 7.29).

Anterior Compartment of Leg

The **anterior compartment**, or dorsiflexor (*extensor*) compartment, is located anterior to the *interosseous membrane*, between the lateral surface of the tibial shaft and the medial surface of the fibular shaft (Figs. 7.29, 7.30, and 7.31 and Table 7.7). The anterior compartment is bounded anteriorly by the deep fascia of the leg and skin. Inferiorly, two

band-like thickenings of the deep fascia form retinacula that bind the tendons of the anterior compartment muscles, preventing them from bow-stringing anteriorly during dorsiflexion of the ankle joint. The **superior extensor retinaculum** is a strong, broad band of deep fascia (Fig. 7.30B) passing from the fibula to the tibia, proximal to the malleoli. The **inferior extensor retinaculum**, a Y-shaped band of deep fascia, attaches laterally to the anterosuperior surface of the calcaneus and medially to the medial malleolus and medial cuneiform. It forms a strong loop around the tendons of the fibularis tertius and extensor digitorum longus muscles. The four muscles in the anterior compartment are as follows (see Fig. 7.27):

- Tibialis anterior
- Extensor digitorum longus
- Extensor hallucis longus
- Fibularis tertius

These muscles are mainly dorsiflexors of the ankle joint and extensors of the toes (Table 7.7).

The **deep fibular (peroneal) nerve**, one of the two terminal branches of the common fibular nerve, is the nerve of the anterior compartment (Fig. 7.30C). The deep fibular nerve arises between the fibularis longus muscle and the neck of the fibula. After entering the compartment, the nerve accompanies the anterior tibial artery. Also see the overview of the innervation of the leg (see Figs. 7.14E and 7.36A and Table 7.9).

The **anterior tibial artery** supplies structures in the anterior compartment (Fig. 7.30C). The smaller terminal branch of the popliteal artery, the anterior tibial artery, begins

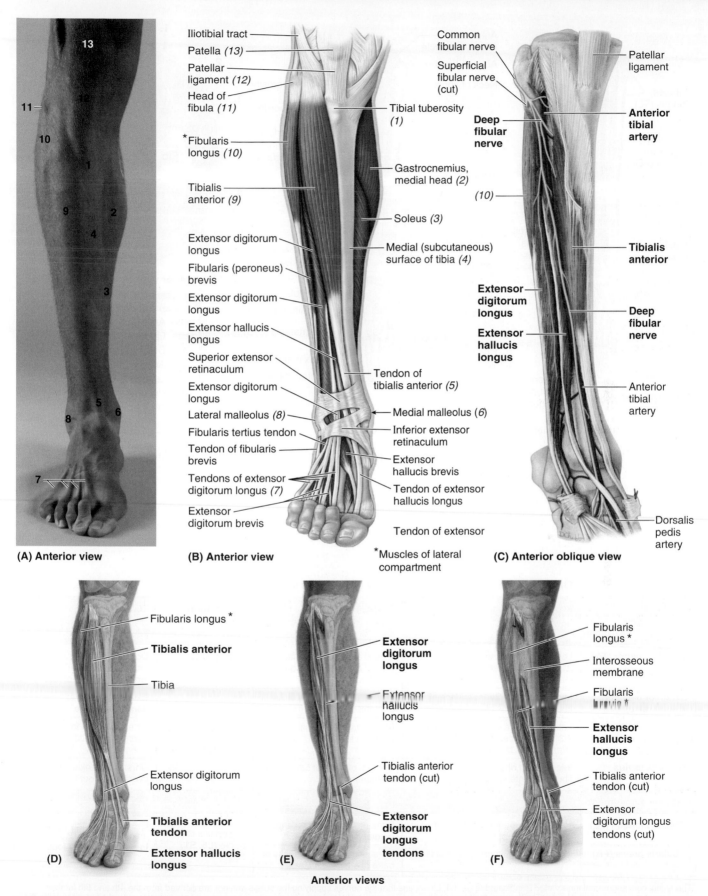

(A) Anterior view

Iliotibial tract
Patella (13)
Patellar ligament (12)
Head of fibula (11)
*Fibularis longus (10)
Tibialis anterior (9)
Extensor digitorum longus
Fibularis (peroneus) brevis
Extensor digitorum longus
Extensor hallucis longus
Superior extensor retinaculum
Extensor digitorum longus
Lateral malleolus (8)
Fibularis tertius tendon
Tendon of fibularis brevis
Tendons of extensor digitorum longus (7)
Extensor digitorum brevis

Common fibular nerve
Superficial fibular nerve (cut)
Tibial tuberosity (1)
Gastrocnemius, medial head (2)
Soleus (3)
Medial (subcutaneous) surface of tibia (4)
Tendon of tibialis anterior (5)
Medial malleolus (6)
Inferior extensor retinaculum
Extensor hallucis brevis
Tendon of extensor hallucis longus
Tendon of extensor

(B) Anterior view

*Muscles of lateral compartment

Patellar ligament
Deep fibular nerve
Anterior tibial artery
(10)
Tibialis anterior
Extensor digitorum longus
Extensor hallucis longus
Deep fibular nerve
Anterior tibial artery
Dorsalis pedis artery

(C) Anterior oblique view

Fibularis longus *
Tibialis anterior
Tibia
Extensor digitorum longus
Tibialis anterior tendon
Extensor hallucis longus

(D)

Extensor digitorum longus
Extensor hallucis longus
Tibialis anterior tendon (cut)
Extensor digitorum longus tendons

(E)

Fibularis longus *
Interosseous membrane
Fibularis brevis *
Extensor hallucis longus
Tibialis anterior tendon (cut)
Extensor digitorum longus tendons (cut)

(F)

Anterior views

FIGURE 7.30. Anterior compartment of leg and dorsum of foot. A. Surface anatomy. The *numbers* are defined in part B. **B.** Overview. **C.** Nerves and vessels. The muscles have been separated to display these structures. **D.** Tibialis anterior. **E.** Extensor digitorum longus. **F.** Extensor hallucis longus and fibularis tertius.

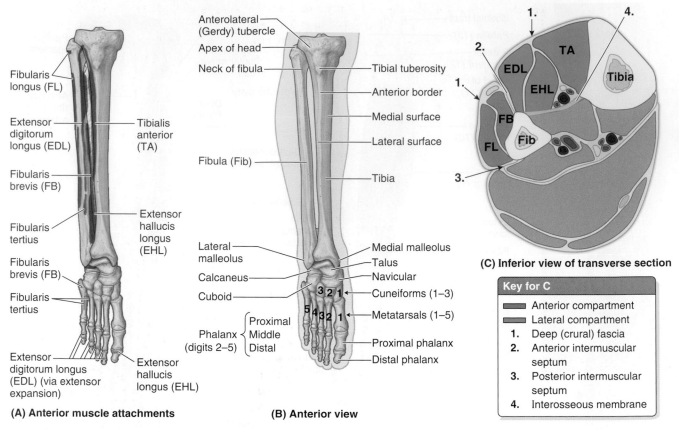

(A) Anterior muscle attachments

Fibularis longus (FL)

Extensor digitorum longus (EDL)

Tibialis anterior (TA)

Fibularis brevis (FB)

Fibularis tertius

Extensor hallucis longus (EHL)

Fibularis brevis (FB)

Fibularis tertius

Extensor digitorum longus (EDL) (via extensor expansion)

Extensor hallucis longus (EHL)

(B) Anterior view

Anterolateral (Gerdy) tubercle

Apex of head

Neck of fibula

Tibial tuberosity

Anterior border

Medial surface

Lateral surface

Fibula (Fib)

Tibia

Lateral malleolus

Medial malleolus

Calcaneus

Talus

Cuboid

Navicular

3 2 1 — Cuneiforms (1–3)

5 4 3 2 1 — Metatarsals (1–5)

Phalanx (digits 2–5) { Proximal / Middle / Distal }

Proximal phalanx

Distal phalanx

(C) Inferior view of transverse section

Key for C
Anterior compartment
Lateral compartment
1. Deep (crural) fascia
2. Anterior intermuscular septum
3. Posterior intermuscular septum
4. Interosseous membrane

FIGURE 7.31. **Anterior and lateral compartment of leg. A.** Muscle attachments. **B.** Bones. **C.** Contents, transverse section.

TABLE 7.7. MUSCLES OF THE ANTERIOR AND LATERAL COMPARTMENTS OF THE LEG

Muscle	Proximal Attachment	Distal Attachment	Innervation[a]	Main Action(s)
Anterior compartment				
Tibialis anterior (TA)	Lateral condyle and superior 2/3 of lateral surface of tibia and interosseous membrane	Medial and inferior surfaces of medial cuneiform and base of 1st metatarsal	Deep fibular nerve (**L4**, L5)	Dorsiflexes ankle; inverts foot; supports medial longitudinal arch of foot
Extensor hallucis longus (EHL)	Middle part of anterior surface of fibula and interosseous membrane	Dorsal aspect of base of distal phalanx of great toe (hallux)	Deep fibular nerve (L5, S1)	Extends great toe; dorsiflexes ankle
Extensor digitorum longus (EDL)	Lateral condyle of tibia and superior 2/3 of anterior surface of fibula and interosseous membrane	Middle and distal phalanges of lateral four digits		Extends lateral four digits; dorsiflexes ankle
Fibularis tertius	Inferior third of anterior surface of fibula and interosseous membrane	Dorsum of base of 5th metatarsal		Dorsiflexes ankle; aids in eversion of foot
Lateral compartment				
Fibularis longus (FL)	Head and superior two thirds of lateral surface of fibula	Base of 1st metatarsal and medial cuneiform	Superficial fibular nerve (**L5**, **S1**, S2)	Evert foot; weakly plantarflex ankle. FL supports transverse arch of foot.
Fibularis brevis (FB)	Middle part of lateral surface of fibula	Dorsal surface of tuberosity of base of 5th metatarsal		

[a]The spinal cord segmental innervation is indicated (e.g., "**L4**, L5" means that the nerves supplying the tibialis anterior are derived from the 4th and 5th lumbar segments of the spinal cord). Numbers in boldface (**L4**) indicate the main segmental innervation. Damage to one or more of the listed spinal cord segments or to the motor nerve roots arising from them results in paralysis of the muscles concerned.

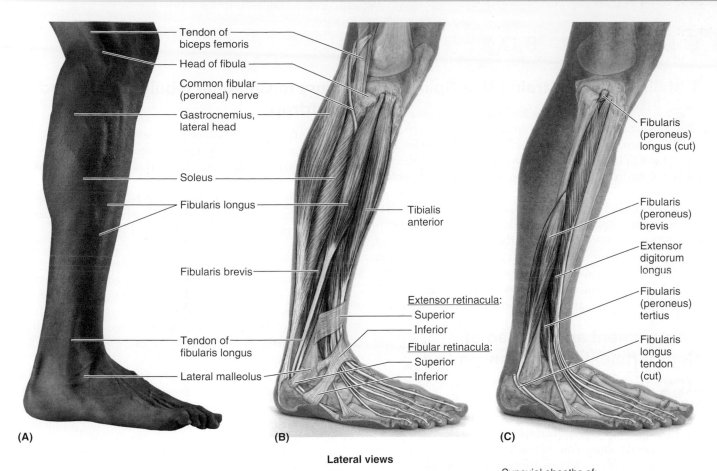

Lateral views

FIGURE 7.32. Lateral compartment of leg and lateral aspect of foot.
A. Surface anatomy. **B.** Overview. **C.** Fibularis (peroneus) longus and brevis. **D.** Retinacula and synovial sheaths of the tendons (*purple*).

at the inferior border of the popliteus muscle (Fig. 7.28D). It passes anteriorly through a gap in the superior part of the interosseous membrane and descends on the anterior surface of this membrane between the tibialis anterior and the extensor digitorum longus. It ends at the ankle joint, midway between the malleoli (Fig. 7.30C), where it becomes the *dorsalis pedis artery* (dorsal artery of foot). Also see the overview of the blood supply of the leg (see Figs. 7.11 and 7.37 and Table 7.10).

Lateral Compartment of Leg

The **lateral compartment**, or *evertor compartment*, is bounded by the lateral surface of the fibula, the anterior and posterior intermuscular septa, and the deep fascia of the leg (Figs. 7.31C and 7.32 and Table 7.7). The lateral compartment contains two muscles—the **fibularis longus** and **brevis**—that pass posterior to the lateral malleolus (Fig. 7.32).

The **superficial fibular nerve**, the nerve in the lateral compartment, is a terminal branch of the common fibular nerve. After supplying the two muscles, it continues as a cutaneous nerve, supplying the skin on the distal part of the

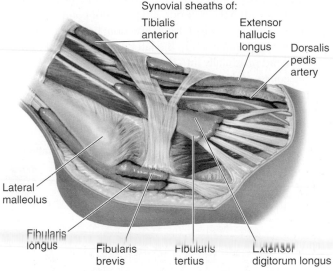

(D) Anterolateral view

anterior surface of the leg and nearly all the dorsum of the foot (see Figs. 7.14E, 7.37A, and 7.44A).

The lateral compartment of the leg does not have an artery coursing through it. The muscles are supplied proximally by perforating branches of the anterior tibial artery and distally by perforating branches of the **fibular artery** (see Fig. 7.38). These perforating arteries have accompanying veins (L. *venae comitantes*).

CLINICAL BOX

Tibialis Anterior Strain (Shin Splints)

Shin splints—edema and pain in the area of the distal two thirds of the tibia—result from repetitive microtrauma of the tibialis anterior, which causes small tears in the periosteum covering the shaft of the tibia and/or of fleshy attachments to the overlying deep fascia of the leg. Shin splints are a mild form of the *anterior compartment syndrome*.

Shin splints commonly occur during traumatic injury or athletic overexertion of muscles in the anterior compartment, especially tibialis anterior. Muscles in the anterior compartment swell from sudden overuse, and the edema and muscle–tendon inflammation reduce the blood flow to the muscles. The swollen muscles are painful and tender to pressure.

Containment and Spread of Compartmental Infections in Leg

The fascial compartments of the lower limbs are generally closed spaces, ending proximally and distally at the joints. Because the septa and deep fascia of the leg forming the boundaries of the leg compartments are strong, the increased volume consequent to infection with *suppuration* (formation of pus) increases intracompartmental pressure. Inflammation within the anterior and posterior compartments of the leg spreads chiefly in a distal direction; however, a *purulent* (pus-forming) *infection* in the lateral compartment of the leg can ascend proximally into the popliteal fossa, presumably along the course of the common fibular nerve. *Fasciotomy* may be necessary to relieve compartmental pressure and *débride* (remove by scraping) pockets of infection.

Injury to Common Fibular Nerve and Footdrop

Because of its superficial and lateral position, *the common fibular nerve is the nerve most often injured in the lower limb*. It winds subcutaneously around the fibular neck, leaving it vulnerable to direct trauma. This nerve may also be severed during fracture of the fibular neck or severely stretched when the knee joint is injured or dislocated.

Severance of the common fibular nerve results in flaccid paralysis of all muscles in the anterior and lateral compartments of the leg (dorsiflexors of ankle and evertors of foot). The loss of dorsiflexion of the ankle causes *footdrop*, which is exacerbated by unopposed inversion of the foot. This has the effect of making the limb "too long": The toes do not clear the ground during the swing phase of walking (Fig. B7.12A).

There are several other conditions that may result in a lower limb that is "too long" functionally—for example, pelvic tilt and spastic paralysis or contraction of the soleus. There are at least three means of compensating for this problem:

1. A *waddling gait*, in which the individual leans to the side opposite the long limb, "hiking" the hip (Fig. B7.12B)
2. A *swing-out gait*, in which the long limb is swung out laterally (abducted) to allow the toes to clear the ground (Fig. B7.12C)
3. A high-stepping *steppage gait*, in which extra flexion is employed at the hip and knee to raise the foot as high as necessary to keep the toes from hitting the ground (Fig. B7.12D)

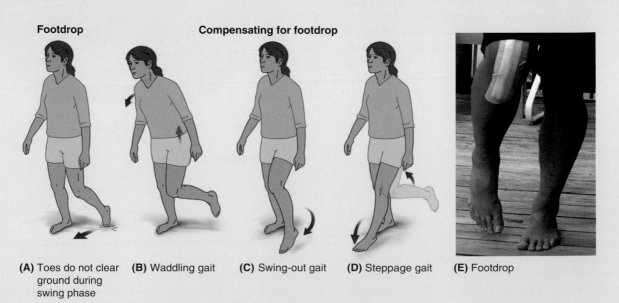

Footdrop   **Compensating for footdrop**

(A) Toes do not clear ground during swing phase   **(B)** Waddling gait   **(C)** Swing-out gait   **(D)** Steppage gait   **(E)** Footdrop

FIGURE B7.12. Footdrop and compensating gait patterns.

Because the dropped foot makes it difficult to make the heel strike the ground first as in a normal gait, a steppage gait is commonly employed in the case of flaccid paralysis (Fig. B7.12E). Sometimes, an extra "kick" is added as the free limb swings forward in an attempt to flip the forefoot upward just before setting the foot down.

The braking action normally produced by eccentric contraction of the dorsiflexors is also lost in flaccid paralysis footdrop. Therefore, the foot is not lowered to the ground in a controlled manner after heel strike; instead, the foot slaps the ground suddenly, producing a distinctive *clop* and greatly increasing the shock both received by the forefoot and transmitted up the tibia to the knee. Individuals with a common fibular nerve injury may also experience a variable loss of sensation on the anterolateral aspect of the leg and the dorsum of the foot.

Deep Fibular Nerve Entrapment

Excessive use of muscles supplied by the deep fibular nerve (e.g., during skiing, running, and dancing) may result in muscle injury and edema in the anterior compartment. This may entrap (cause compression of) the deep fibular nerve or its vasa nervorum and result in pain in the anterior compartment.

Compression of the nerve by tight-fitting ski boots, for example, may occur where the nerve passes deep to the inferior extensor retinaculum and the extensor hallucis brevis. Pain occurs in the dorsum of the foot and usually radiates to the web space between the first and second toes. Because ski boots are a common cause of this type of nerve entrapment, this condition has been called the "ski boot syndrome"; however, the syndrome also occurs in soccer players and runners and can also result from tight shoes.

Superficial Fibular Nerve Entrapment

Chronic ankle sprains may produce recurrent stretching of the superficial fibular nerve, which may cause pain along the lateral side of the leg and the dorsum of the ankle and foot. Numbness and *paresthesia* (pain, numbness or tingling) may be present and increase with activity.

Palpation of Dorsalis Pedis Pulse

The *dorsalis pedis pulse* is evaluated during a physical examination of the peripheral vascular system. Dorsalis pedis pulses may be palpated with the feet slightly dorsiflexed. The pulses are usually easy to palpate because the dorsal arteries are subcutaneous and pass along a line from the extensor retinaculum to a point just lateral to the extensor hallucis longus tendons (Fig. B7.13; see also Fig. 7.11D). A diminished or absent dorsalis pedis pulse usually suggests vascular insufficiency resulting from arterial disease. The five P signs of acute arterial occlusion are *pain, pallor, paresthesia, paralysis,* and *pulselessness*. Some healthy adults (and even children) have *congenitally nonpalpable dorsalis pedis pulses*; the variation is usually bilateral. In these cases, the dorsalis pedis artery is replaced by an extended perforating fibular artery of smaller caliber than the typical dorsalis pedis artery but running in the same location.

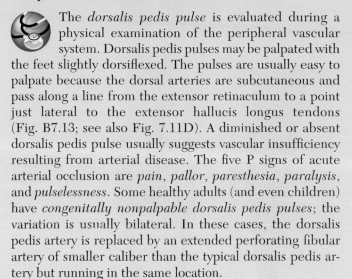

Extensor hallucis longus

Dorsalis pedis artery

Tibialis anterior tendon

FIGURE B7.13. Dorsalis pedis pulse.

Posterior Compartment of Leg

The **posterior compartment**, or plantarflexor compartment, is the largest of the three leg compartments. The posterior compartment and the *calf muscles* within it are divided into superficial and deep subcompartments/muscle groups by the *transverse intermuscular septum* (see Fig. 7.29). The tibial nerve and posterior tibial and fibular vessels supply both parts of the posterior compartment but run in the deep part, just deep (anterior) to the transverse intermuscular septum.

SUPERFICIAL MUSCLE GROUP

The superficial group of plantarflexors, including the *gastrocnemius, soleus,* and *plantaris,* forms a powerful muscular mass in the calf (Figs. 7.33 and 7.34 and Table 7.8).

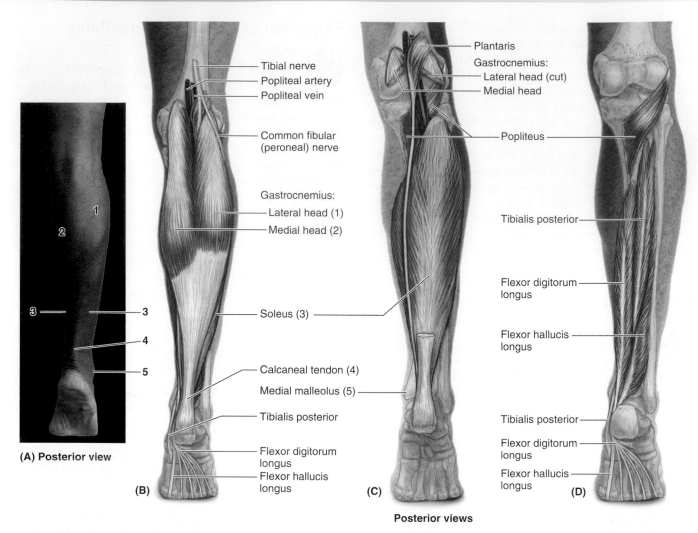

(A) Posterior view

Plantaris
Gastrocnemius:
— Lateral head (cut)
— Medial head

Tibial nerve
Popliteal artery
Popliteal vein

Common fibular
(peroneal) nerve

Popliteus

Gastrocnemius:
— Lateral head (1)
— Medial head (2)

Tibialis posterior

Soleus (3)

Flexor digitorum
longus

Calcaneal tendon (4)

Flexor hallucis
longus

Medial malleolus (5)

Tibialis posterior

Tibialis posterior

Flexor digitorum
longus

Flexor digitorum
longus
Flexor hallucis
longus

Flexor hallucis
longus

(B)   (C)   (D)

Posterior views

FIGURE 7.33. **Muscles of posterior compartment of leg. A.** Surface anatomy. *Numbers* are identified in part **B. B.** Gastrocnemius. **C.** Soleus and plantaris. **D.** Muscles of deep compartment.

The two-headed gastrocnemius and the soleus share a common tendon, the **calcaneal tendon** (L. *tendo calcaneus*, Achilles tendon), which attaches to the calcaneus. Collectively, these two muscles form the three-headed **triceps surae** (L. *sura*, calf). The triceps surae elevates the heel and thus depresses the forefoot, generating as much as 93% of the plantarflexion force.

The calcaneal tendon typically spirals a quarter turn (90 degrees) during its descent, so that the gastrocnemius fibers attach laterally and the soleal fibers attach medially. This arrangement is thought to be significant to the tendon's elastic ability to absorb energy (shock) and recoil, releasing the energy as part of the propulsive force it exerts. Although they share a common tendon, the two muscles of the triceps surae are capable of acting alone and often do so: "You stroll with the soleus but win the long jump with the gastrocnemius."

To test the triceps surae, the foot is plantarflexed against resistance (e.g., by "standing on the toes," in which case body weight [gravity] provides resistance). If normal, the calcaneal tendon and triceps surae can be seen and palpated.

A *subcutaneous calcaneal bursa*, located between the skin and the calcaneal tendon, allows the skin to move over the taut tendon. A deep *bursa of the calcaneal tendon* (retrocalcaneal bursa), located between the tendon and the calcaneal tuberosity, allows the tendon to glide over the bone.

The **gastrocnemius** is the most superficial muscle in the posterior compartment and forms the proximal, most prominent part of the *calf* (Fig. 7.33A,B and Table 7.8). It is a fusiform, two-headed, two-joint muscle with a medial head that is slightly larger and extends more distally than the lateral head. The heads form the inferolateral and inferomedial boundaries of the popliteal fossa and then merge at the inferior angle of the fossa.

The gastrocnemius crosses and is capable of acting on both the knee and the ankle joints; however, it cannot exert its full power on both joints at the same time. It functions most effectively when the knee is extended and is maximally

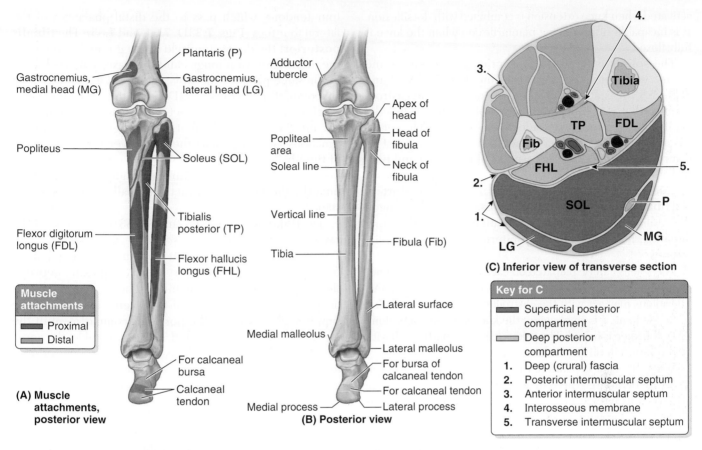

FIGURE 7.34. Posterior compartment of leg. A. Muscle attachments. **B.** Bones. **C.** Contents, transverse section.

TABLE 7.8. MUSCLES OF POSTERIOR COMPARTMENT OF LEG

Muscle	Proximal Attachment	Distal Attachment	Innervation[a]	Main Action(s)
Superficial muscle group				
Gastrocnemius: **Lateral head** **Medial head**	*Lateral head*: lateral aspect of lateral condyle of femur *Medial head*: popliteal surface of femur, superior to medial condyle	Posterior surface of calcaneus via calcaneal tendon	Tibial nerve (S1, S2)	Plantarflexes ankle when knee is extended, raises heel during walking, and flexes knee joint
Soleus	Posterior aspect of head of fibula, superior quarter of posterior surface of fibula, soleal line, and medial border of tibia			Plantarflexes ankle; steadies leg on foot
Plantaris	Inferior end of lateral supracondylar line of femur and oblique popliteal ligament			Weakly assists gastrocnemius in plantarflexing ankle; function is probably mainly proprioceptive.
Deep muscle group				
Popliteus	Lateral surface of lateral condyle of femur and lateral meniscus (intra-articular; within cavity of knee joint)	Posterior surface of tibia, superior to soleal line	Tibial nerve (L4, L5, S1)	Weakly flexes knee and unlocks it by laterally rotating femur on fixed tibia, may also medially rotate tibia of unplanted limb
Flexor hallucis longus	Inferior two thirds of posterior surface of fibula and inferior part of interosseous membrane	Base of distal phalanx of great toe (hallux)	Tibial nerve (**S2**, S3)	Flexes great toe at all joints; weakly plantarflexes ankle; supports medial longitudinal arch of foot
Flexor digitorum longus	Medial part of posterior surface of tibia inferior to soleal line and by a broad tendon to fibula	Bases of distal phalanges of lateral four digits		Flexes lateral four digits; plantarflexes ankle; supports longitudinal arches of foot
Tibialis posterior	Interosseous membrane, posterior surface of tibia inferior to soleal line, and posterior surface of fibula	Primarily to tuberosity of navicular; also to cuneiforms, cuboid, and bases of 2nd–4th metatarsals	Tibial nerve (L4, L5)	Plantarflexes ankle; inverts foot; supports medial longitudinal arch of foot

[a]The spinal cord segmental innervation is indicated (e.g., "S2, S3" means that the nerves supplying the flexor hallucis longus are derived from the 2nd and 3rd sacral segments of the spinal cord). Numbers in boldface (**S2**) indicate the main segmental innervation.

activated when knee extension is combined with dorsiflexion. It is incapable of producing plantarflexion when the knee is fully flexed.

The **soleus** is located deep to the gastrocnemius and is the "workhorse" of plantarflexion (Fig. 7.33A–C and Table 7.8). It is a large muscle, broader than the gastrocnemius, that is named for its resemblance to a sole—the flat fish that reclines on its side on the sea floor. The soleus has a continuous proximal attachment in the shape of an inverted U to the posterior aspects of the fibula and tibia and a tendinous arch between them, the **tendinous arch of soleus** (L. *arcus tendineus soleus*). The popliteal artery and tibial nerve exit the popliteal fossa by passing through this arch, the popliteal artery simultaneously bifurcating into its terminal branches, the anterior and posterior tibial arteries.

The soleus can be palpated on each side of the gastrocnemius when the individual is standing on tiptoes (Fig. 7.33A). The soleus may act with the gastrocnemius in plantarflexing the ankle joint; it cannot act on the knee joint and acts alone when the knee is flexed. The soleus has many parts, each with fiber bundles in different directions.

When the foot is planted, the soleus pulls posteriorly on the bones of the leg. This is important to standing because the line of gravity passes anterior to the leg's bony axis. The soleus is thus an antigravity muscle (the predominant plantarflexor for standing and strolling), which contracts antagonistically but cooperatively (alternately) with the dorsiflexor muscles of the leg to maintain balance.

The **plantaris** is a small muscle with a short (small finger-sized) belly, a long tendon, and a high density of muscle spindles (Fig. 7.33C and Table 7.8). This vestigial muscle is absent in 5–10% of people. Because of its minor motor role, the plantaris tendon can be removed for grafting (e.g., during reconstructive surgery of the tendons of the hand) without causing disability.

DEEP MUSCLE GROUP

Four muscles make up the deep group in the posterior compartment (Figs. 7.33D and 7.34 and Table 7.8):

- Popliteus
- Flexor digitorum longus
- Flexor hallucis longus
- Tibialis posterior

The **popliteus** is a thin, triangular muscle in the floor of the popliteal fossa (Fig. 7.33C,D). The popliteus acts to unlock the fully extended knee joint, whereas the other muscles act on the ankle and foot joints. The **flexor hallucis longus** is the powerful flexor of all the joints of the great toe. The **flexor digitorum longus** is smaller than the flexor hallucis longus, even though it moves four digits. It passes diagonally into the sole of the foot, superficial to the tendon of the flexor hallucis longus, and divides into

four tendons, which pass to the distal phalanges of the lateral four toes (Figs. 7.33D, 7.34, and 7.35). The **tibialis posterior**, the deepest muscle in the group, lies between the flexor digitorum longus and the flexor hallucis longus in the same plane as the tibia and fibula within the deep subcompartment (Figs. 7.33D, 7.34, and 7.35). When the foot is off the ground, it can act synergistically with the tibialis anterior to invert the foot, their otherwise antagonistic functions canceling each other. However, the primary role of the tibialis posterior is to support or maintain (fix) the medial longitudinal arch during weight bearing; consequently, the muscle contracts statically throughout the stance phase of gait.

The **tibial nerve** (L4, L5, and S1–S3) is the larger of the two terminal branches of the *sciatic nerve* (Figs. 7.35A and 7.36B; see also Fig. 7.14C). It runs through the popliteal fossa with the popliteal artery and vein passing between the heads of the gastrocnemius. These structures pass deep to the tendinous arch of the soleus. The tibial nerve supplies all muscles in the posterior compartment of the leg (Fig. 7.14C, Tables 7.8 and 7.9). At the ankle, the nerve lies between the flexor hallucis longus and the flexor digitorum longus. Postero-inferior to the medial malleolus, the tibial nerve divides into the medial and lateral plantar nerves. A branch of the tibial nerve, the *medial sural cutaneous nerve*, usually unites with the *sural communicating branch of the common fibular nerve* to form the *sural nerve* (Figs. 7.14C and 7.28A,C and Table 7.9). The sural nerve supplies the skin of the lateral and posterior part of the inferior third of the leg and the lateral side of the foot. Articular branches of the tibial nerve supply the knee joint, and medial calcaneal branches supply the skin of the heel.

The **posterior tibial artery** (Figs. 7.35A and 7.37A and Table 7.10), the larger terminal branch of the popliteal artery, provides the blood supply to the posterior compartment of the leg and to the foot. It begins at the distal border of the popliteus and passes deep to the tendinous arch of the soleus. After giving off the fibular artery, its largest branch, the posterior tibial artery passes inferomedially on the posterior surface of the tibialis posterior. During its descent, it is accompanied by the tibial nerve and veins. The posterior tibial artery runs posterior to the medial malleolus (Fig. 7.37A). Deep to the flexor retinaculum and the origin of the abductor hallucis, the posterior tibial artery divides into *medial* and *lateral plantar arteries*, the arteries of the sole of the foot.

The *fibular artery* arises inferior to the distal border of the popliteus and the tendinous arch of soleus (Figs. 7.35A and 7.37A). It descends obliquely toward the fibula and then passes along its medial side, usually within the flexor hallucis longus. The fibular artery gives muscular branches to the muscles in the posterior and lateral compartments of the leg. It also gives rise to the *nutrient artery of the fibula*. The *perforating branch of the fibular artery* pierces the interosseous membrane and passes to the dorsum of the foot.

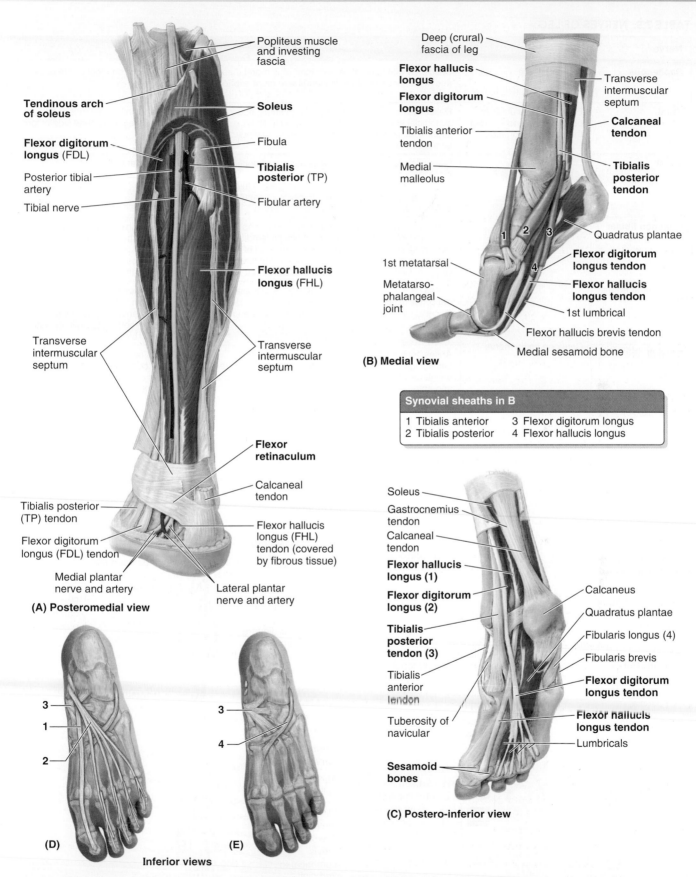

Popliteus muscle and investing fascia

Tendinous arch of soleus

Soleus

Fibula

Flexor digitorum longus (FDL)

Posterior tibial artery

Tibial nerve

Tibialis posterior (TP)

Fibular artery

Flexor hallucis longus (FHL)

Transverse intermuscular septum

Transverse intermuscular septum

Flexor retinaculum

Calcaneal tendon

Tibialis posterior (TP) tendon

Flexor digitorum longus (FDL) tendon

Medial plantar nerve and artery

Flexor hallucis longus (FHL) tendon (covered by fibrous tissue)

Lateral plantar nerve and artery

(A) Posteromedial view

Deep (crural) fascia of leg

Flexor hallucis longus

Flexor digitorum longus

Tibialis anterior tendon

Medial malleolus

Transverse intermuscular septum

Calcaneal tendon

Tibialis posterior tendon

Quadratus plantae

1st metatarsal

Metatarso-phalangeal joint

Flexor digitorum longus tendon

Flexor hallucis longus tendon

1st lumbrical

Flexor hallucis brevis tendon

Medial sesamoid bone

(B) Medial view

Synovial sheaths in B	
1 Tibialis anterior	3 Flexor digitorum longus
2 Tibialis posterior	4 Flexor hallucis longus

Soleus

Gastrocnemius tendon

Calcaneal tendon

Flexor hallucis longus (1)

Flexor digitorum longus (2)

Tibialis posterior tendon (3)

Tibialis anterior tendon

Tuberosity of navicular

Sesamoid bones

Calcaneus

Quadratus plantae

Fibularis longus (4)

Fibularis brevis

Flexor digitorum longus tendon

Flexor hallucis longus tendon

Lumbricals

(C) Postero-inferior view

(D)

(E)

Inferior views

FIGURE 7.35. Nerves, vessels, and tendon sheaths of posterior leg. A. Vessels and nerves are exposed by removal of most of soleus muscle. **B.** Structures passing posterior to medial malleolus. Synovial sheaths of the tendons are *purple*; each is named in key. **C.** Relationships of tendons of deep posterior compartment muscles posterior to medial malleolus and in sole of foot. **D and E.** Muscle attachments in sole of foot. Number labels in parts D and E pertain to the numbers in part C.

TABLE 7.9. NERVES OF LEG

Nerve	Origin	Course	Distribution
Saphenous	Femoral nerve	Descends with femoral vessels through femoral triangle and adductor canal; then descends with great saphenous vein	Supplies skin on medial side of leg and foot
Sural	Formed by the union of cutaneous branches from the tibial and common fibular nerves	Descends between heads of gastrocnemius; becomes superficial at middle of leg; descends with small saphenous vein; passes inferior to lateral malleolus to lateral side of foot	Supplies skin on posterior and lateral aspects of leg and lateral side of foot
Tibial	Sciatic nerve	Forms as sciatic nerve bifurcates at apex of popliteal fossa; descends through popliteal fossa and lies on popliteus; runs inferiorly on tibialis posterior with posterior tibial vessels; terminates beneath flexor retinaculum by dividing into medial and lateral plantar nerves	Supplies plantarflexor muscles of posterior compartment of leg and knee joint
Common fibular	Sciatic nerve	Forms as sciatic nerve bifurcates at apex of popliteal fossa and follows medial border of biceps femoris and its tendon; passes over posterior aspect of head of fibula; then winds around neck of fibula deep to fibularis longus, where it divides into deep and superficial fibular nerves	Supplies skin on lateral part of posterior aspect of leg via its branch, the lateral sural cutaneous nerve; also supplies knee joint via its articular branch
Superficial fibular	Common fibular nerve	Arises between fibularis longus and neck of fibula; descends in lateral compartment of leg; pierces deep fascia at distal third of leg to become subcutaneous	Supplies fibular muscles of lateral compartment of leg and skin on distal third of anterior surface of leg and dorsum of foot, except skin of first interdigital cleft
Deep fibular	Common fibular nerve	Arises between fibularis longus and neck of fibula; passes through extensor digitorum longus and descends on interosseous membrane; crosses distal end of tibia and enters dorsum of foot	Supplies dorsiflexor muscles of anterior compartment of leg, extensor muscles on dorsum of foot, and skin of first interdigital cleft; sends articular branches to joints it crosses

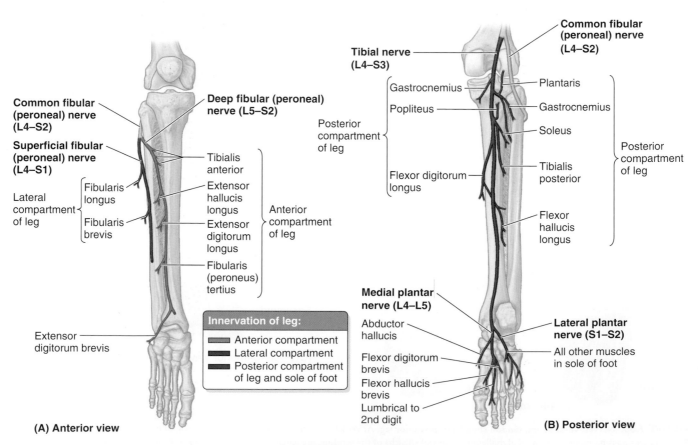

FIGURE 7.36. Nerves of leg.

TABLE 7.10. ARTERIES OF LEG

Artery	Origin	Course	Distribution
Popliteal	Continuation of femoral artery at adductor hiatus in adductor magnus	Passes through popliteal fossa to leg; ends at inferior border of popliteus muscle by dividing into anterior and posterior tibial arteries	Superior, middle, and inferior genicular arteries to knee; muscular branches to hamstrings and muscles of superficial posterior compartment of leg
Anterior tibial	Popliteal artery	Passes into anterior compartment through gap in superior part of interosseous membrane; descends on this membrane between tibialis anterior and extensor digitorum longus	Anterior compartment of leg
Dorsalis pedis (dorsal artery of foot)	Continuation of anterior tibial artery distal to inferior extensor retinaculum	Descends anteromedially to first interosseous space; divides into deep plantar and 1st dorsal metatarsal arteries	Muscles on dorsum of foot; pierces first dorsal interosseous muscle as deep plantar artery to contribute to formation of plantar arch
Posterior tibial	Popliteal	Passes through posterior compartment of leg; terminates distal to flexor retinaculum by dividing into medial and lateral plantar arteries	Posterior and lateral compartments of leg; circumflex fibular branch joins anastomoses around knee; nutrient artery passes to tibia
Fibular	Posterior tibial	Descends in posterior compartment adjacent to posterior intermuscular septum	Posterior compartment of leg: perforating branches supply lateral compartment of leg

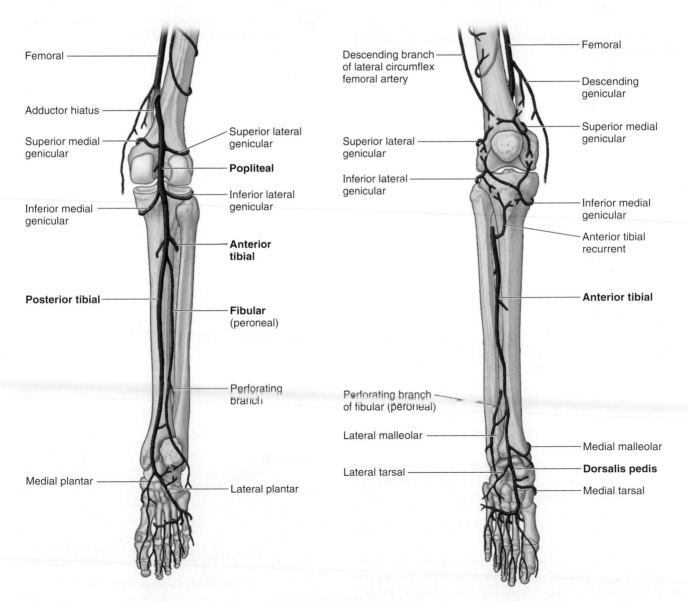

FIGURE 7.37. Arteries of leg.

CLINICAL BOX

Gastrocnemius Strain

Gastrocnemius strain (tennis leg) is a painful calf injury resulting from partial tearing of the medial belly of the gastrocnemius at or near its musculotendinous junction. It is caused by overstretching the muscle by concomitant full extension of the knee and dorsiflexion of the ankle joint.

Posterior Tibial Pulse

The *posterior tibial pulse* can usually be palpated between the posterior surface of the medial malleolus and the medial border of the calcaneal tendon (Fig. B7.14; see also Fig 7.11F). Because the posterior tibial artery passes deep to the flexor retinaculum, it is important when palpating this pulse to have the person relax the retinaculum by inverting the foot. Failure to do this may lead to the erroneous conclusion that the pulse is absent.

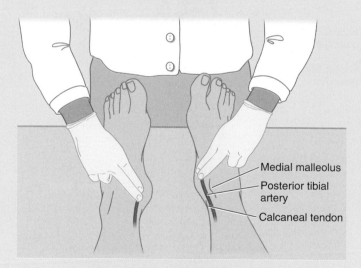

Medial malleolus
Posterior tibial artery
Calcaneal tendon

FIGURE B7.14. Posterior tibial artery pulse.

Both posterior tibial arteries are examined simultaneously for equality of force. Palpation of the posterior tibial pulses is essential for examining patients with occlusive *peripheral arterial disease*. Although posterior tibial pulses are absent in approximately 15% of normal young people, absence of posterior tibial pulses is a sign of occlusive peripheral arterial disease in people older than 60 years of age. For example, *intermittent claudication*, characterized by leg pain and cramps, develops during walking and disappears after rest. These conditions result from ischemia of the leg muscles caused by narrowing or occlusion of the leg arteries.

Injury to Tibial Nerve

Injury to the tibial nerve is uncommon because of its protected position in the popliteal fossa; however, the nerve may be injured by deep lacerations in the fossa. *Posterior dislocation of the knee joint* may also damage the tibial nerve. *Severance of the tibial nerve* produces paralysis of the flexor muscles in the leg and the intrinsic muscles in the sole of the foot. People with a tibial nerve injury are unable to plantarflex their ankle or flex their toes. Loss of sensation also occurs on the sole of the foot.

Absence of Plantarflexion

If the muscles of the calf are paralyzed, the calcaneal tendon is ruptured, or normal push-off is painful, a much less effective and efficient push-off (from the midfoot) can still be accomplished by the actions of the gluteus maximus and hamstrings in extending the thigh at the hip joint and the quadriceps in extending the knee. Because push-off from the forefoot is not possible (in fact, the ankle will be passively dorsiflexed as the body's weight moves anterior to the foot), those attempting to walk in the absence of plantarflexion often rotate the foot as far laterally (externally) as possible during the stance phase to disable passive dorsiflexion and allow a more effective push-off through hip and knee extension exerted at the midfoot.

Calcaneal Tendon Reflex

The *ankle (jerk) reflex* is elicited by striking the calcaneal tendon briskly with a reflex hammer while the person's legs are dangling over the side of the examining table (Fig. B7.15). This tendon reflex tests the S1 and S2 nerve roots. If the S1 nerve root is cut or compressed, the ankle reflex is virtually absent.

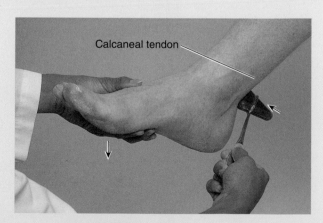

Calcaneal tendon

FIGURE B7.15. Calcaneal tendon reflex.

Inflammation and Rupture of Calcaneal Tendon

Inflammation of the calcaneal tendon constitutes 9–18% of running injuries. Microscopic tears of collagen fibers in the tendon, particularly just superior to its attachment to the calcaneus, result in *tendinitis*, which causes pain during walking.

Calcaneal tendon rupture is often sustained by people with a history of calcaneal tendinitis. After complete rupture of the tendon, passive dorsiflexion is excessive, and the person cannot plantarflex against resistance.

Calcaneal Bursitis

Calcaneal bursitis (Achilles bursitis) results from inflammation of the bursa of the calcaneal tendon located between the calcaneal tendon and the superior part of the posterior surface of the calcaneal tuberosity. Calcaneal bursitis causes pain posterior to the heel and occurs commonly during long-distance running, basketball, and tennis.

FOOT

The **foot**, distal to the ankle, provides a platform for supporting the weight of the body when standing and has an important role in locomotion. The skeleton of the foot consists of 7 tarsals, 5 metatarsals, and 14 phalanges (Fig. 7.38). The foot and its bones may be considered in terms of three anatomical and functional parts:

- The **hindfoot**: talus and calcaneus
- The **midfoot**: navicular, cuboid, and cuneiforms
- The **forefoot**: metatarsals and phalanges

The regions of the foot include the following:

- The *plantar region* (sole): the part contacting the ground
- The *dorsum of the foot* (dorsal region of the foot): the part directed superiorly
- The **heel region** (heel): the sole underlying the calcaneus
- The **ball of the foot**: the sole underlying the sesamoid bones and heads of the medial two metatarsals

The great toe (L. *hallux*) is also called the **first toe** (L. *digitus primus*); the **little toe** (L. *digitus minimus*) is also called the **fifth toe** (L. *digitus quintus*).

Deep Fascia of Foot

Over the lateral and posterior aspects, the deep fascia of the foot is continuous with the **plantar fascia**, the deep fascia of the sole, which has a thick central part, the **plantar aponeurosis**, and weaker medial and lateral parts (Figs. 7.39

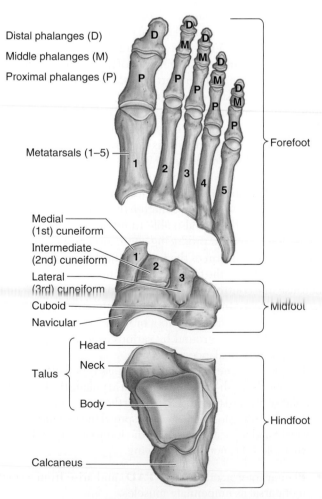

Distal phalanges (D)
Middle phalanges (M)
Proximal phalanges (P)

Metatarsals (1–5)

Forefoot

Medial (1st) cuneiform
Intermediate (2nd) cuneiform
Lateral (3rd) cuneiform
Cuboid
Navicular

Midfoot

Head
Neck
Talus
Body
Hindfoot

Calcaneus

Dorsum of foot, superior view

FIGURE 7.38. Parts of foot.

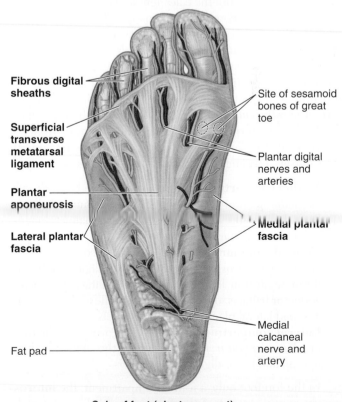

Fibrous digital sheaths

Superficial transverse metatarsal ligament

Plantar aponeurosis

Lateral plantar fascia

Fat pad

Site of sesamoid bones of great toe

Plantar digital nerves and arteries

Medial plantar fascia

Medial calcaneal nerve and artery

Sole of foot (plantar aspect), inferior view

FIGURE 7.39. Plantar aponeurosis.

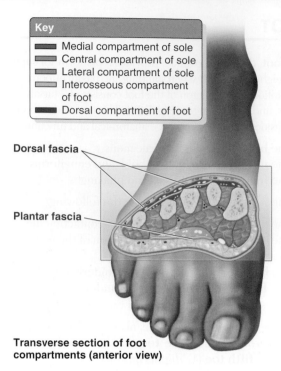

Key
- Medial compartment of sole
- Central compartment of sole
- Lateral compartment of sole
- Interosseous compartment of foot
- Dorsal compartment of foot

Dorsal fascia

Plantar fascia

Transverse section of foot compartments (anterior view)

FIGURE 7.40. Fascia and compartments of foot.

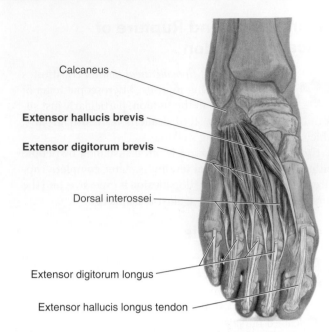

Calcaneus

Extensor hallucis brevis

Extensor digitorum brevis

Dorsal interossei

Extensor digitorum longus

Extensor hallucis longus tendon

FIGURE 7.41. Extensor digitorum brevis and extensor hallucis brevis.

and 7.40). The plantar fascia holds parts of the foot together, helps protect the sole from injury, and passively supports the longitudinal arches of the foot. The plantar aponeurosis arises posteriorly from the calcaneus and distally divides into five bands that become continuous with the fibrous digital sheaths that enclose the flexor tendons that pass to the toes. Inferior to the heads of the metatarsals, the aponeurosis is reinforced by transverse fibers forming the *superficial transverse metatarsal ligament*. The deep fascia is thin on the dorsum of the foot, where it is continuous with the *inferior extensor retinaculum* (Fig. 7.40). In the forefoot and midfoot, vertical intermuscular septa extend superiorly from the margins of the plantar aponeurosis toward the 1st and 5th metatarsals, forming three *compartments of the sole*:

- **Medial compartment of the sole**, covered superficially by *medial plantar fascia*, contains the abductor hallucis, flexor hallucis brevis, tendon of the flexor hallucis longus, and medial plantar nerve and vessels.
- **Central compartment of the sole**, covered by the *plantar aponeurosis*, contains the flexor digitorum brevis, flexor digitorum longus, quadratus plantae, lumbricals, adductor hallucis, distal part of tendon flexor hallucis longus, and lateral plantar nerve and vessels.
- **Lateral compartment of the sole**, covered by the thinner *lateral plantar fascia*, contains the abductor digiti minimi and flexor digiti minimi brevis.

In the forefoot only, a fourth compartment, the **interosseous compartment of the foot**, contains the metatarsals, the dorsal and plantar interosseous muscles, and the deep plantar and metatarsal vessels.

Muscles of Foot

Of the 20 individual muscles of the foot, 14 are located on the plantar aspect, 2 are on the dorsal aspect, and 4 are intermediate in position. (Figs. 7.41 and 7.42).

From the plantar aspect, muscles of the sole are arranged in four layers within four compartments. The muscles of the sole of the foot are illustrated in Figure 7.42, and their attachments, innervation, and actions are described in Table 7.11. Despite their compartmental and layered arrangement, the **plantar muscles** function primarily as a group during the support phase of stance to maintain the arches of the foot.

The muscles become most active in the later portion of the movement to stabilize the foot for propulsion (push-off), a time when forces also tend to flatten the foot's transverse arch. Concurrently, they are also able to refine further the efforts of the long muscles, producing supination and pronation in enabling the platform of the foot to adjust to uneven ground.

The muscles of the foot are of little importance individually because fine control of the individual toes is not important to most people. Rather than producing actual movement, they are most active in fixing the foot or in increasing the pressure applied against the ground by various aspects of the sole or toes to maintain balance.

Despite its name, the adductor hallucis is probably most active during the push-off phase of stance in pulling the lateral four metatarsals toward the great toe, fixing the transverse arch of the foot, and resisting forces that would spread the metatarsal heads as weight and force are applied to the forefoot (see Table 7.13).

In Table 7.11, note the following:

- **P**lantar interossei **AD**duct (**PAD**) and arise from a single metatarsal as unipennate muscles.
- **D**orsal interossei **AB**duct (**DAB**) and arise from two metatarsals as bipennate muscles.

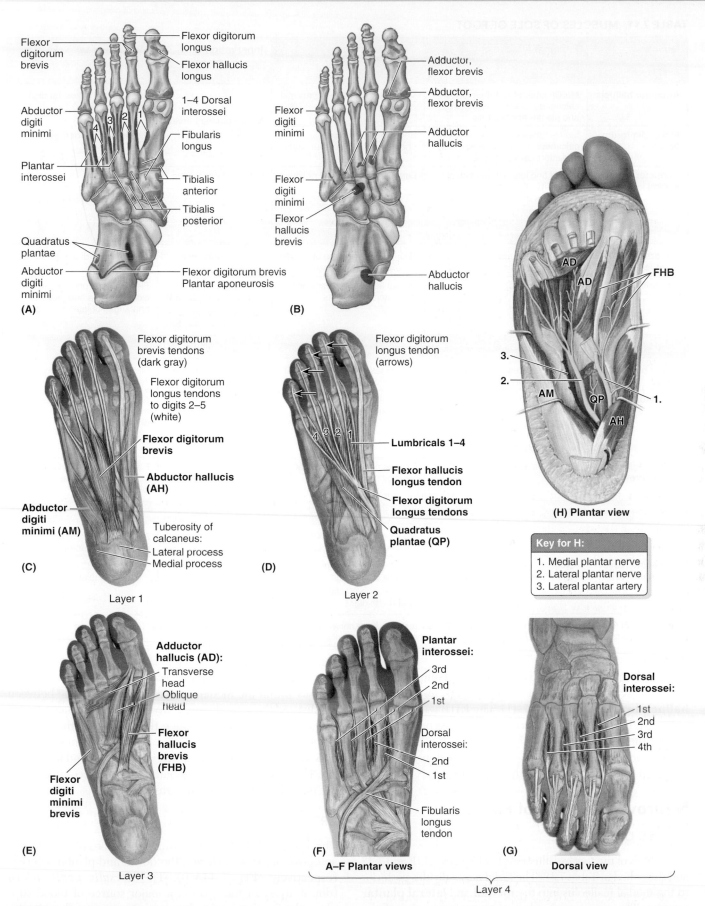

FIGURE 7.42. Muscles of sole of foot. A and B. Muscle attachments. **C.** First layer. **D.** Second layer. **E.** Third layer. **F and G.** Fourth layer. **H.** Medial and lateral plantar nerves.

TABLE 7.11. MUSCLES OF SOLE OF FOOT

Muscle	Proximal Attachment	Distal Attachment	Innervation[a]	Main Action(s)
First layer				
Abductor hallucis	Medial tubercle of tuberosity of calcaneus, flexor retinaculum, and plantar aponeurosis	Medial side of base of proximal phalanx of 1st digit	Medial plantar nerve (S2, **S3**)	Abducts and flexes 1st digit (great toe, hallux)
Flexor digitorum brevis	Medial tubercle of tuberosity of calcaneus, plantar aponeurosis, and intermuscular septa	Both sides of middle phalanges of lateral four digits		Flexes lateral four digits
Abductor digiti minimi	Medial and lateral tubercles of tuberosity of calcaneus, plantar aponeurosis, and intermuscular septa	Lateral side of base of proximal phalanx of 5th digit	Lateral plantar nerve (S2, **S3**)	Abducts and flexes 5th digit
Second layer				
Quadratus plantae	Medial surface and lateral margin of plantar surface of calcaneus	Posterolateral margin of tendon of flexor digitorum longus	Lateral plantar nerve (S2, **S3**)	Assists flexor digitorum longus in flexing lateral four digits
Lumbricals	Tendons of flexor digitorum longus	Medial aspect of expansion over lateral four digits	*Medial one*: medial plantar nerve (S2, **S3**) *Lateral three*: lateral plantar nerve (S2, **S3**)	Flex proximal phalanges; extend middle and distal phalanges of lateral four digits
Third layer				
Flexor hallucis brevis	Plantar surfaces of cuboid and lateral cuneiform	Both sides of base of proximal phalanx of 1st digit	Medial plantar nerve (S2, **S3**)	Flexes proximal phalanx of 1st digit
Adductor hallucis	*Oblique head*: bases of metatarsals 2–4 *Transverse head*: plantar ligaments of 3rd–5th metatarsophalangeal joints	Tendons of both heads attach to lateral side of base of proximal phalanx of 1st digit.	Deep branch of lateral plantar nerve (S2, **S3**)	Adducts 1st digit; assists in maintaining transverse arch of foot
Flexor digiti minimi brevis	Base of 5th metatarsal	Base of proximal phalanx of 5th digit	Superficial branch of lateral plantar nerve (S2, **S3**)	Flexes proximal phalanx of 5th digit, thereby assisting with its flexion
Fourth layer				
Plantar interossei (three muscles)	Bases and medial sides of metatarsals 3–5	Medial sides of bases of proximal phalanges of 3rd–5th digits	Lateral plantar nerve (S2, **S3**)	Adduct digits (3–5) and flex metatarsophalangeal joints
Dorsal interossei (four muscles)	Adjacent sides of metatarsals 1–5	*First*: medial side of proximal phalanx of 2nd digit *Second to fourth*: lateral sides of 2nd–4th digits		Abduct digits (2–4) and flex metatarsophalangeal joints

[a]The spinal cord segmental innervation is indicated (e.g., "S2, **S3**" means that the nerves supplying the abductor hallucis are derived from the 2nd and 3rd sacral segments of the spinal cord). Numbers in boldface (**S3**) indicate the main segmental innervation.

Two closely connected muscles on the dorsum of the foot are the **extensor digitorum brevis (EDB)** and **extensor hallucis brevis (EHB)** (Fig. 7.41). The EHB is actually part of the EDB. These muscles form a fleshy mass on the lateral part of the dorsum of the foot, anterior to the lateral malleolus, and aid the extensor digitorum and extensor hallucis longus in extending digits one through four.

Neurovasculature of Foot

NERVES OF FOOT

The nerves of the foot are illustrated in Figures 7.14F and 7.43 and described in Table 7.12. The *tibial nerve* divides posterior to the medial malleolus into the **medial** and **lateral plantar nerves**. These nerves supply the intrinsic muscles of the foot, except for the EDB and EHB on the dorsum, which are supplied by the *deep fibular nerve*. The medial plantar nerve courses within the medial compartment of the sole between the first and the second muscle layers. Initially, the lateral plantar nerve runs laterally between the muscles of the first and second layers of plantar muscles. Their deep branches then pass medially between the muscles of the third and fourth layers. The medial and lateral plantar nerves are accompanied by the medial and lateral plantar arteries and veins.

ARTERIES OF FOOT

The arteries of the foot are terminal branches of the anterior and posterior *tibial arteries*, the dorsal and plantar arteries, respectively (Fig. 7.44A,B). The *dorsalis pedis artery* (dorsal artery of foot), often a major source of blood supply to the forefoot, is the direct continuation of the anterior tibial artery. The dorsalis pedis artery begins midway between

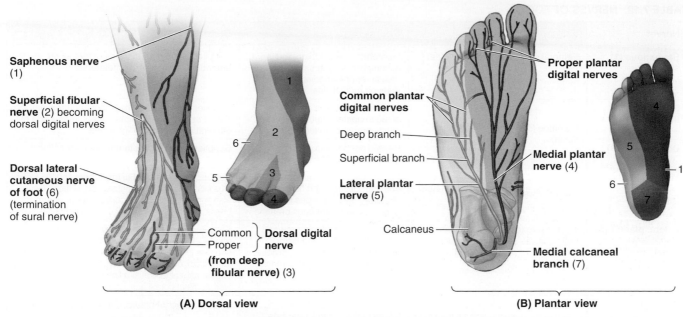

FIGURE 7.43. Cutaneous innervation of foot.

the malleoli (at the ankle joint) and runs anteromedially, deep to the inferior extensor retinaculum between the extensor hallucis longus and the extensor digitorum longus tendons on the dorsum of the foot. This artery gives off the **lateral tarsal artery** and then passes distally to the first interosseous space, where it gives off the **arcuate artery** and then divides into the **1st dorsal metatarsal artery** and a **deep plantar artery** (Fig. 7.44A). The deep plantar artery passes deeply between the heads of the first dorsal interosseous muscle to enter the sole of the foot, where it joins the lateral plantar artery to form the **deep plantar arch** (Fig. 7.44B). The arcuate artery gives off the **2nd, 3rd,** and **4th dorsal metatarsal arteries,** which

run to the clefts of the toes, where each of them divides into two **dorsal digital arteries** (Fig. 7.44A).

The sole of the foot has prolific blood supply from the posterior tibial artery, which divides deep to the flexor retinaculum. The terminal branches pass deep to the abductor hallucis as the *medial* and *lateral plantar arteries,* which accompany similarly named nerves. The **medial plantar artery** supplies the muscles of the great toe and the skin on the medial side of the sole and has digital branches that accompany digital branches of the medial plantar nerve.

Initially, the **lateral plantar artery** and nerve course laterally between the muscles of the first and second layers of

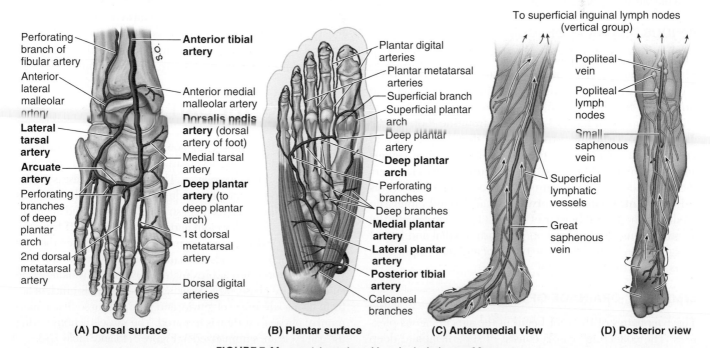

FIGURE 7.44. Arterial supply and lymphatic drainage of foot.

TABLE 7.12. NERVES OF FOOT

Nerve[a]	Origin	Course	Distribution[a]
Saphenous (1)	Femoral nerve (in femoral triangle)	Descends through thigh and leg; accompanies great saphenous vein anterior to medial malleolus; ends on medial side of foot	Supplies skin on medial side of foot as far anteriorly as head of 1st metatarsal
Superficial fibular (2)	Common fibular nerve (at neck of fibula)	Pierces deep fascia in distal third of leg to become cutaneous; then sends branches to foot and digits	Supplies skin on dorsum of foot and proximal dorsal aspects of all digits, except lateral side of 5th digit and first interdigital cleft
Deep fibular (3)		Passes deep to extensor retinaculum to enter dorsum of foot	Supplies extensor digitorum/extensor hallucis brevis and skin of first interdigital cleft
Medial plantar (4)	Tibial nerve (posterior to medial malleolus, as larger terminal branch)	Passes distally in foot between abductor hallucis and flexor digitorum brevis; divides into muscular and cutaneous branches	Supplies plantar aspect of medial foot and 3½ digits, plus sides and distal dorsal aspects of those digits; also supplies abductor hallucis, flexor digitorum brevis, flexor hallucis brevis, and first lumbrical
Lateral plantar (5)	Tibial nerve (posterior to medial malleolus, as smaller terminal branch)	Passes laterally in foot between quadratus plantae and flexor digitorum brevis muscles; divides into superficial and deep branches	Supplies quadratus plantae, abductor digiti minimi, and flexor digiti minimi brevis; deep branch supplies plantar and dorsal interossei, lateral three lumbricals, and adductor hallucis; supplies skin on plantar aspect lateral to a line splitting 4th digit as well as distal dorsal aspect of lateral 1½ toes
Sural (6)	Formed in popliteal fossa or calf by the union of cutaneous branches from the tibial and common fibular nerves	Passes posterior and inferior to lateral malleolus to lateral side of foot	Lateral aspect of hindfoot, midfoot, and 5th digit
Calcaneal branches (7)	Tibial and sural nerves (posterior to malleoli)	Pass from distal part of posterior aspect of leg to skin on heel	Skin of heel

[a]Numbers refer to Figure 7.43.

plantar muscles. The deep plantar arch begins opposite the base of the 5th metatarsal as the continuation of the *lateral plantar artery*, coursing between the third and the fourth muscle layers (Fig. 7.44B). The arch is completed medially by union with the *deep plantar artery*, a branch of the dorsal artery of the foot. As it crosses the foot, the deep plantar arch gives rise to four **plantar metatarsal arteries**; three **perforating branches**; and many branches to the skin, fascia, and muscles in the sole. The plantar digital arteries arise from the plantar metatarsal arteries near the base of the proximal phalanx, supplying adjacent digits.

VENOUS DRAINAGE OF FOOT

There are both superficial and deep veins in the foot. The *deep veins* consist of inter-anastomosing paired veins accompanying all the arteries internal to the deep fascia. The *superficial veins* are subcutaneous, are unaccompanied by arteries, and drain most of the blood from the foot. Dorsal digital veins continue proximally as **dorsal metatarsal veins**, which join to form the subcutaneous dorsal venous arch, proximal to which a **dorsal venous network** covers the remainder of the dorsum of the foot. Superficial veins from a **plantar venous network** drain around either the medial or the lateral border of the foot to converge with the dorsal venous arch and network to form medial and lateral marginal veins, which become the *great* and *small saphenous veins*, respectively (Fig. 7.44C,D).

LYMPHATIC DRAINAGE OF FOOT

The lymphatics of the foot begin in the subcutaneous plexuses. The collecting vessels consist of superficial and deep lymphatic vessels, which follow the superficial veins and major vascular bundles, respectively. Superficial lymphatic vessels are most numerous in the sole. The *medial superficial lymphatic vessels* leave the foot medially along the *great saphenous vein* and accompany it to the *superficial inguinal lymph nodes* (Fig. 7.44C), located along the vein's termination, and then to the *deep inguinal lymph nodes*. The *lateral superficial lymphatic vessels* drain the lateral side of the foot and accompany the *small saphenous vein* to the popliteal fossa, where they enter the *popliteal lymph nodes* (Fig. 7.44D). The *deep lymphatic vessels* from the foot also drain into the popliteal lymph nodes. Lymphatic vessels from them follow the femoral vessels to the deep inguinal lymph nodes. All lymph from the lower limb then passes to the iliac lymph nodes.

WALKING: THE GAIT CYCLE

Locomotion is a complex function. The movements of the lower limb during walking on a level surface may be divided into alternating swing and stance phases. The **gait cycle** consists of one cycle of swing and stance by one limb. The **stance phase** begins with **heel strike** when the heel strikes the ground and begins to assume the body's full weight and ends with **push-off** from the forefoot. The **swing phase** begins after push-off, when the toes leave the ground, and ends when the heel strikes the ground. The swing phase occupies approximately 40% of the walking cycle and the stance phase, 60%. Walking is a remarkably efficient activity, taking advantage of gravity and momentum so that a minimum of physical exertion is required. The muscle actions during the gait cycle are summarized in Figure 7.45 and Table 7.13.

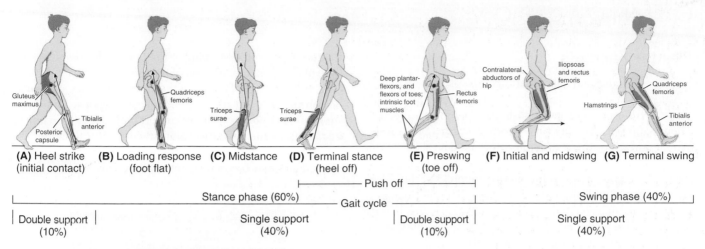

FIGURE 7.45. Gait cycle. Eight phases are typically described, two of which have been combined in **F**.

TABLE 7.13. MUSCLE ACTION DURING GAIT CYCLE

Phase of Gait		Mechanical Goals	Active Muscle Groups
STANCE PHASE	Heel strike (initial contact)	Lower forefoot to ground	Ankle dorsiflexors (eccentric contraction)
		Continue deceleration (reverse forward swing)	Hip extensors
		Preserve longitudinal arch of foot	Intrinsic muscles of foot
			Long tendons of foot
	Loading response (flat foot)	Accept weight	Knee extensors
		Decelerate mass	Ankle plantarflexors
		Stabilize pelvis	Hip abductors
		Preserve longitudinal arch of foot	Intrinsic muscles of foot
			Long tendons of foot
	Midstance	Stabilize knee	Knee extensors
		Control dorsiflexion (preserve momentum)	Ankle plantarflexors (eccentric contraction)
		Stabilize pelvis	Hip abductors
		Preserve longitudinal arch of foot	Intrinsic muscles of foot
	Terminal stance (heel off)	Accelerate mass	Ankle plantarflexors (concentric contraction)
		Stabilize pelvis	Hip abductors
		Preserve arches of foot; fix forefoot	Intrinsic muscles of foot
	Preswing (toe off)	Accelerate mass	Long flexors of digits
		Preserve arches of foot; fix forefoot	Intrinsic muscles of foot
			Long tendons of foot
		Decelerate thigh; prepare for swing	Flexor of hip (eccentric contraction)
SWING PHASE	Initial swing	Accelerate thigh, vary cadence	Flexor of hip (concentric contraction)
		Clear foot	Ankle dorsiflexors
	Midswing	Clear foot	Ankle dorsiflexors
	Terminal swing	Decelerate thigh	Hip extensors (eccentric contraction)
		Decelerate leg	Knee flexors (eccentric contraction)
		Position foot	Ankle dorsiflexors
		Extend knee to place foot (control stride); prepare for contact	Knee extensors

Modified from Rose J, Gamble JG. *Human Walking*. 2nd ed. Baltimore, MD: Lippincott Williams & Wilkins; 1994.

CLINICAL BOX

Plantar Fasciitis

 Straining and inflammation of the plantar aponeurosis, a condition called *plantar fasciitis*, may result from running and high-impact aerobics, especially when inappropriate footwear is worn. It causes pain on the plantar surface of the heel and on the medial aspect of the foot. Point tenderness is located at the proximal attachment of the plantar aponeurosis to the medial tubercle of the calcaneus and on the medial surface of this bone. The pain increases with passive extension of the great toe and may be further exacerbated by dorsiflexion of the ankle and/or weight bearing. A *calcaneal spur* (abnormal bony process) protruding from the medial tubercle has long been associated with plantar fasciitis and pain on the medial side of the foot when walking; however, many asymptomatic patients are found to have such spurs.

Sural Nerve Grafts

 Pieces of the sural nerve are often used for *nerve grafts* in procedures such as repairing nerve defects resulting from wounds. The surgeon is usually able to locate this nerve in relation to the small saphenous vein.

Medial Plantar Nerve Entrapment

 Compressive irritation of the medial plantar nerve as it passes deep to the flexor retinaculum or curves deep to the abductor hallucis may cause aching, burning, numbness, and tingling (paresthesia) on the medial side of the sole and in the region of the navicular tuberosity. Medial plantar nerve compression may occur during repetitive eversion of the foot (e.g., during gymnastics and running). Because of its frequency in runners, these symptoms have been called "jogger's foot."

Plantar Reflex

The *plantar reflex* (L4, L5, S1, and S2 nerve roots) is a myotatic (deep tendon) reflex. The lateral aspect of the sole is stroked with a blunt object, such as a tongue depressor, beginning at the heel and crossing to the base of the great toe. Flexion of the toes is a normal response (Fig. B7.16). Slight fanning of the lateral four toes and dorsiflexion of the great toe is an abnormal response (*Babinski sign*), indicating brain injury or cerebral disease, except in infants. Because the corticospinal tracts (motor function) are not fully developed in newborns, a Babinski sign is usually elicited and may be present until children are 4 years of age.

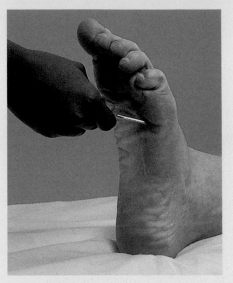

FIGURE B7.16. Plantar reflex.

Contusion of Extensor Digitorum Brevis

Clinically, knowing the location of the belly of the EDB is important for distinguishing this muscle from abnormal edema. Contusion and tearing of the muscle fibers and associated blood vessels result in a *hematoma*, producing edema anteromedial to the lateral malleolus. Most people who have not seen this inflamed muscle assume they have a severely sprained ankle.

JOINTS OF LOWER LIMB

The joints of the lower limb include the articulations of the pelvic girdle (lumbosacral joints, sacro-iliac joints, and pubic symphysis), which are discussed in Chapter 6. The remaining joints of the lower limb are the hip joint, knee joint, tibiofibular joints, ankle joint, and foot joints.

Hip Joint

The **hip joint** forms the connection between the lower limb and the pelvic girdle. It is a strong, stable multiaxial ball and socket type of synovial joint. The femoral head is the ball, and the acetabulum is the socket (Fig. 7.46). This joint is designed for stability over a wide range of movement. During standing, the weight of the upper body is transferred through the hip bones to the heads of the femurs.

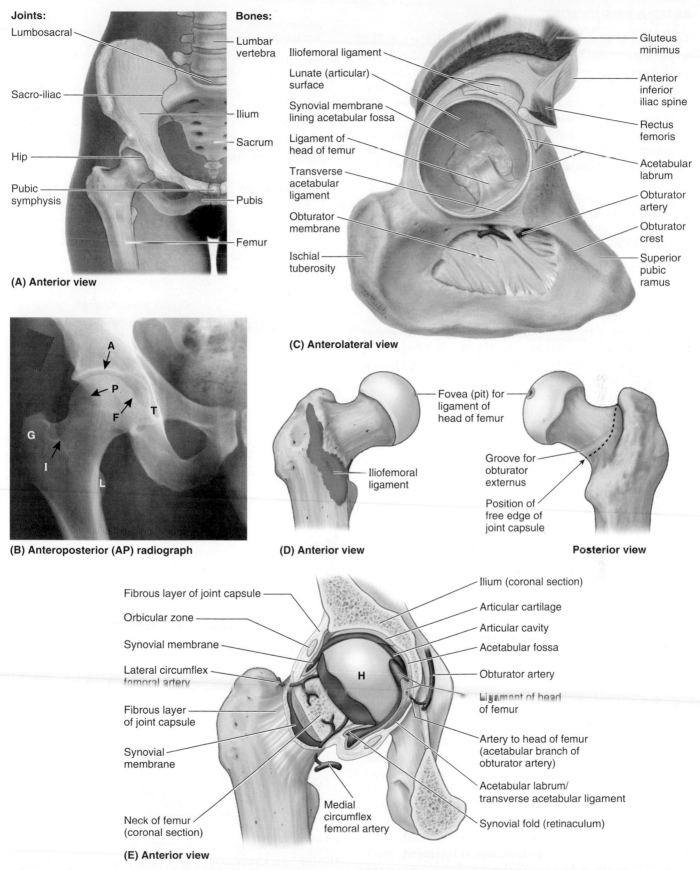

FIGURE 7.46. Articular surfaces and blood supply of hip joint. A. Joints and bones of pelvic girdle and hip. **B.** Radiograph of hip joint. *A,* roof of acetabulum; *F,* fovea (pit) for the ligament of the head of femur; *G,* greater trochanter; *I,* intertrochanteric crest; *L,* lesser trochanter; *P,* posterior rim of acetabulum; *T,* "teardrop" appearance caused by superimposition (*H*) of structures at the inferior margin of the acetabulum. **C.** Acetabular region of hip bone. **D.** Bony features of proximal femur. **E.** Blood supply of head and neck of femur. A section of bone has been removed from the femoral neck.

ARTICULAR SURFACES

The round head of the femur articulates with the cup-like acetabulum of the hip bone. The head is covered with **articular cartilage**, except for the pit or *fovea for the ligament of the head of femur* (Fig. 7.46D). The rim of the acetabulum consists of a semilunar articular part covered with articular cartilage, the **lunate surface of the acetabulum**. Because the depth of the acetabulum is increased by the fibrocartilaginous **acetabular labrum** (L. *labrum*, lip) and the **transverse acetabular ligament** (bridging the *acetabular notch*), more than half of the head fits within the acetabulum (Fig. 7.46A–C). Centrally, a deep nonarticular part, the **acetabular fossa**, is formed mainly by the ischium.

JOINT CAPSULE

The external **fibrous layer of the joint capsule** attaches proximally on the hip bone to the bony rim of the acetabulum and the transverse acetabular ligament. Distally, it attaches to the femoral neck only anteriorly at the intertrochanteric line and at the root of the greater trochanter (Fig. 7.46E). Posteriorly, the fibrous layer has an arched border that crosses the neck proximal to the intertrochanteric crest but is not attached to it. The joint capsule covers approximately the proximal two thirds of the neck of the femur posteriorly. A protrusion of the **synovial membrane** beneath and beyond the free posterior margin of the joint capsule onto the femoral neck forms a bursa for the obturator externus tendon (Fig. 7.47B).

Most fibers of the fibrous layer take a spiral course from the hip bone to the intertrochanteric line; some deep fibers, most marked in the posterior part of capsule, wind circularly around the neck, forming an **orbicular zone** (Fig. 7.47B). Thick parts of the fibrous layer form the ligaments of the hip joint, which pass in a spiral fashion from the pelvis to the femur. Extension winds the spiraling ligaments and fibers more tightly, constricting the capsule and drawing the femoral head tightly into the acetabulum, increasing stability.

The hip joint is reinforced (Fig. 7.47) by the following structures:

- *Anteriorly and superiorly* by the strong Y-shaped **iliofemoral ligament** (Bigelow ligament), which attaches to the anterior inferior iliac spine and acetabular rim proximally and the intertrochanteric line distally. The iliofemoral ligament prevents hyperextension of the hip joint during standing by screwing the femoral head into the acetabulum.
- *Inferiorly and anteriorly* by the **pubofemoral ligament**, which arises from the obturator crest of the pubic bone and passes laterally and inferiorly to merge with the fibrous layer of the joint capsule. This ligament blends with the medial part of the iliofemoral ligament and tightens during extension and abduction of the hip joint. The pubofemoral ligament resists excessive abduction of the hip joint.
- *Posteriorly* by the weak **ischiofemoral ligament**, which arises from the ischial part of the acetabular rim and spirals superolaterally to the neck of the femur, medial to the base of the greater trochanter

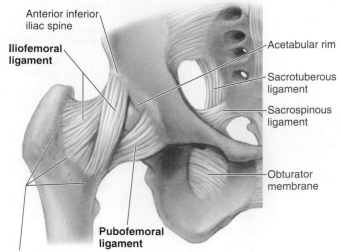

(A) Anterior view

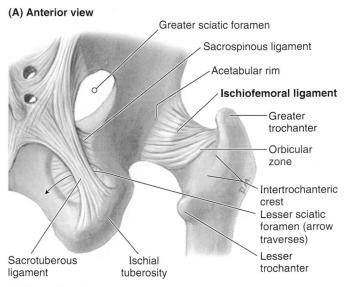

(B) Posterior view

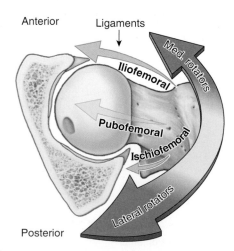

(C) Superior view

FIGURE 7.47. Ligaments of hip joint. A. Iliofemoral and pubofemoral ligaments. **B.** Ischiofemoral ligament. **C.** Transverse section through right hip joint demonstrates the reciprocal pull of the medial (*Med*) and lateral rotators (*reddish brown arrows*) and the intrinsic ligaments of the hip joint. Relative strengths are indicated by arrow width.

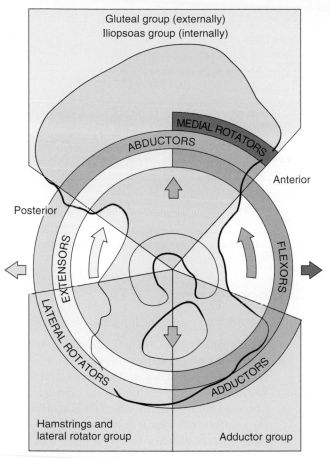

Functional groups of muscles acting at hip joint

Flexors
Iliopsoas
Sartorius
Tensor fasciae latae
Rectus femoris tendon
Pectineus
Adductor longus
Adductor brevis
Adductor magnus—anterior part
Gracilis

Adductors
Pectineus
Adductor longus
Adductor brevis
Adductor magnus
Obturator externus
Gracilis

Lateral rotators
Obturator externus and internus
Piriformis
Gemelli
Quadratus femoris
Gluteus maximus
(Gluteus medius and minimus)

Extensors
Gluteus maximus
Hamstrings:
Semitendinosus
Semimembranosus
Long head, biceps femoris
Adductor magnus—posterior part

Abductors
Gluteus medius
Gluteus minimus
Tensor fasciae latae

Medial rotators
Gluteus medius ⎫ Anterior parts
Gluteus minimus ⎭
Tensor fasciae latae

Diagrammatic lateral view

Circular zones:
The zones represent the position of origin of functional groups relative to center of femoral head in acetabulum (point of rotation). Pull is applied on the femur (femoral trochanters or shaft) from these positions.

Colored arrows:
The curved arrows show the direction of rotation of femoral head and neck caused by activity of extensors and flexors. The short arrows indicate the direction of movement of the femoral neck and greater trochanter caused by activity of the lateral/medial rotators and abductors/adductors.

FIGURE 7.48. Relative positions of muscles producing movements of hip joint.

Both muscles (medial and lateral rotators of the thigh) and ligaments pull the femoral head medially into the acetabulum, increasing stability. They are reciprocally balanced when doing so (Fig. 7.47C).

The **synovial membrane of the hip joint** lines the fibrous layer as well as any intracapsular bony surfaces not lined with articular cartilage (Fig. 7.46E). Thus, where the fibrous layer attaches to the femur, the synovial membrane reflects proximally along the femoral neck to the edge of the femoral head. The **synovial folds** (retinacula), which reflect superiorly along the femoral neck as longitudinal bands, contain subsynovial **retinacular arteries** (branches of the medial and a few from the lateral femoral circumflex artery), which supply the head and neck of the femur.

The **ligament of head of femur**, primarily a synovial fold conducting a blood vessel, is weak and of little importance in strengthening the hip joint (Fig. 7.46C,E). Its wide end attaches to the margins of the acetabular notch and the *transverse acetabular ligament*; its narrow end attaches to the femur at the *fovea for the ligament of the head of femur*. Usually, the ligament contains a small artery to the head of the femur. A fat pad in the acetabular fossa fills the part of the fossa that is not occupied by the ligament of the head of femur. Both the ligament and the fat pad are covered with synovial membrane.

HIP MOVEMENTS

Hip movements are flexion–extension, abduction–adduction, medial–lateral rotation, and circumduction (Fig. 7.48 and Table 7.14). Movements of the trunk at the hip joints are also important, such as those occurring when a person

TABLE 7.14. STRUCTURES LIMITING MOVEMENTS OF HIP JOINT

Movement	Limiting Structures
Flexion	Soft tissue apposition Tension of joint capsule posteriorly Tension of gluteus maximus
Extension	*Ligaments*: iliofemoral, ischiofemoral, and pubofemoral Tension of iliopsoas
Abduction	*Ligaments*: pubofemoral, ischiofemoral, and inferior band of iliofemoral Tension of hip adductors
Adduction	Soft tissue apposition (thighs) Tension of iliotibial band, superior joint capsule, superior band of iliofemoral ligament, and hip abductors (especially when contralateral hip joint is abducted or flexed)
Internal rotation	*Ligaments*: ischiofemoral and posterior joint capsule Tension of external rotators of hip joint
External rotation	*Ligaments*: iliofemoral, pubofemoral, and anterior joint capsule

Modified from Clarkson HM. *Musculoskeletal Assessment. Joint Range of Motion and Manual of Muscle Strength*. 2nd ed. Baltimore, MD: Lippincott Williams & Wilkins; 2000.

lifts the trunk from the supine position during sit-ups or keeps the pelvis level when one foot is off the ground. The degree of flexion and extension possible at the hip joint depends on the position of the knee. If the knee is flexed, relaxing the hamstrings, the thigh can be actively flexed until it almost reaches the anterior abdominal wall. Not all this movement occurs at the hip joint; some results from flexion of the vertebral column. During extension of the hip joint, the fibrous layer of the joint capsule, especially the iliofemoral ligament, is taut; therefore, the hip can usually be extended only slightly beyond the vertical except by movement of the bony pelvis (flexion of the lumbar vertebrae). Abduction of the hip joint is usually somewhat freer than adduction. Lateral rotation is much more powerful than medial rotation.

BLOOD SUPPLY

The arteries supplying the hip joint are as follows (Fig. 7.49):

- *Medial and lateral circumflex femoral arteries*, which are usually branches of the *profunda femoris artery* but are occasionally branches of the femoral artery. The main blood supply is from the retinacular arteries arising as branches from the circumflex femoral arteries (especially the *medial circumflex femoral artery*).
- *Artery to the head of femur*, a branch of the obturator artery that traverses the ligament of the head

NERVE SUPPLY

The *Hilton law* states that the nerve supplying the muscles extending directly across and acting at a given joint also

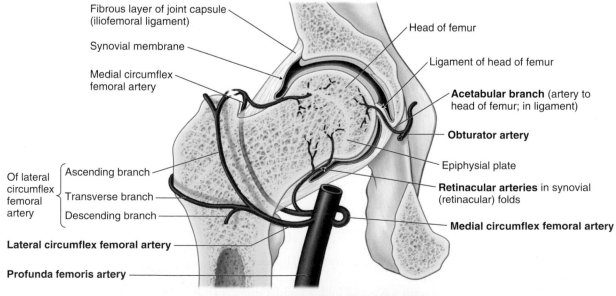

Anterior view of coronally sectioned hip joint

FIGURE 7.49. Blood supply of hip joint.

CLINICAL BOX

Fractures of Femoral Neck (Hip Fractures)

Fracture of the neck of the femur often disrupts the blood supply to the head of the femur. The medial circumflex femoral artery supplies most of the blood to the head and neck of the femur. Its retinacular arteries often are torn when the femoral neck is fractured or the hip joint is dislocated. In some cases, the blood supplied to the femoral head through the artery to the ligament of the femoral head may be the only remaining source of blood to the proximal fragment. This artery is frequently inadequate for maintaining the femoral head; consequently, the fragment may undergo *avascular necrosis* (AVN— also called osteonecrosis), the result of deficient blood supply. These fractures are especially common in individuals older than 60 years of age, especially in women because their femoral necks are often weak and brittle as a result of *osteoporosis*.

Surgical Hip Replacement

The hip joint is subject to severe traumatic injury and degenerative disease. *Osteoarthritis of the hip joint*, characterized by pain, edema, limitation of motion, and erosion of articular cartilage, is a common cause of disability. During *hip replacement*, a metal prosthesis anchored to the person's femur replaces the femoral head and neck and the acetabulum is often lined with a metal/plastic socket (Fig. B7.17).

Dislocation of Hip Joint

Congenital dislocation of the hip joint is common, occurring in approximately 1.5 per 1,000 live births; it affects more girls and is bilateral in approximately half the cases. Dislocation occurs when the femoral head is not properly located in the acetabulum. The affected limb appears (and functions as if) shorter because the dislocated femoral head is more superior than on the normal side, resulting in a positive *Trendelenburg sign* (hip appears to drop to one side during walking). Inability to abduct the thigh is characteristic of congenital dislocation.

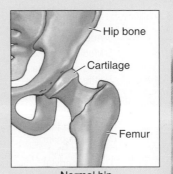
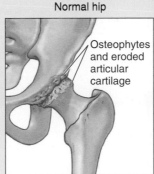
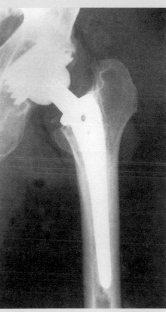

Hip bone
Cartilage
Femur
Normal hip

Osteophytes and eroded articular cartilage

(A) Hip with moderate arthritis   **(B)** Hip prosthesis

FIGURE B7.17. Surgical hip replacement.

Acquired dislocation of the hip joint is uncommon because this joint is so strong and stable. Nevertheless, dislocation may occur during an automobile accident when the hip is flexed, adducted, and medially rotated, the usual position of the lower limb when a person is riding in a car. Posterior dislocations are most common. The fibrous layer of the joint capsule ruptures inferiorly and posteriorly, allowing the femoral head to pass through the tear in the capsule and over the posterior margin of the acetabulum onto the lateral surface of the ilium, shortening and medially rotating the affected limb. Because of the close relationship of the sciatic nerve to the hip joint, it may be injured (stretched and/or compressed) during posterior dislocation or fracture–dislocation of the hip joint.

innervate the joint. Therefore, the nerve supply of the hip joint is from the

- *femoral nerve* or its muscular branches, anteriorly
- *obturator nerve*, inferiorly
- *superior gluteal nerve*, superiorly
- *nerve to quadratus femoris*, posteriorly

Knee Joint

The knee is primarily a hinge type of synovial joint, allowing flexion and extension; however, the hinge movements are combined with gliding and rolling and with rotation about a vertical axis. Although the knee joint is well constructed, its function is commonly impaired when it is hyperextended (e.g., in body contact sports such as hockey).

ARTICULAR SURFACES

The **articular surfaces of the knee joint** are characterized by their large size and incongruent shapes (Fig. 7.50). The knee joint consists of three articulations:

- Two **femorotibial articulations (lateral** and **medial)** between the lateral and the medial femoral and tibial condyles

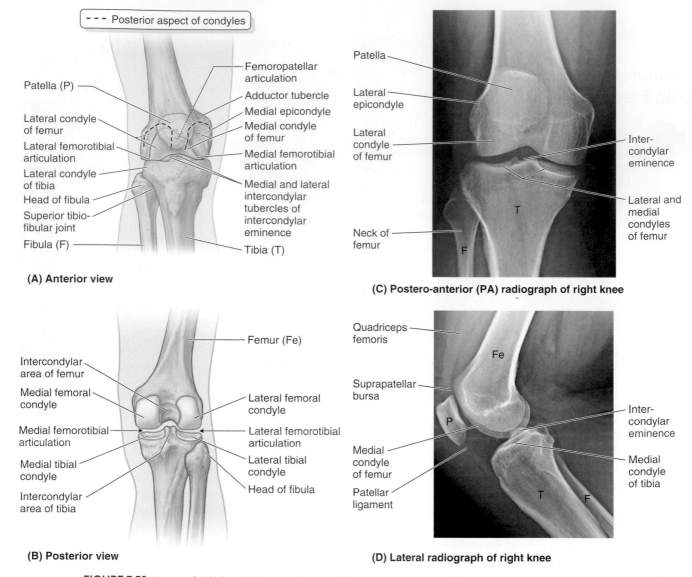

FIGURE 7.50. **Bones of right knee joint. A and B.** Bony features. **C and D.** Radiographs. *Letters* are defined in **A** and **B**.

- One intermediate **femoropatellar articulation** between the patella and the femur

The fibula is not involved in the knee joint. The stability of the knee joint depends on the

- strength and actions of surrounding muscles and their tendons
- ligaments connecting the femur and tibia

Of these supports, the muscles are most important; therefore, many sport injuries are preventable through appropriate conditioning and training. The most important muscle in stabilizing the knee joint is the large *quadriceps femoris*, particularly the inferior fibers of the vastus medialis and lateralis.

JOINT CAPSULE

The **joint capsule** consists of an external *fibrous layer* (fibrous capsule) and an internal *synovial membrane* that lines all internal surfaces of the articular cavity not covered with articular cartilage.

The fibrous layer has a few thickened parts that make up intrinsic ligaments but, for the most part, it is thin posteriorly and laterally. The fibrous layer attaches to the femur superiorly (Fig. 7.50), just proximal to the articular margins of the condyles. Posteriorly, it encloses the condyles and the *intercondylar fossa* (Fig. 7.51A). The fibrous layer has an opening posterior to the lateral tibial condyle to allow the popliteus tendon to pass out of the joint capsule to attach to the tibia (see Fig. 7.33D). Inferiorly, the fibrous layer attaches to the margin of the articular surface of the tibia (tibial plateau), except where the popliteus tendon crosses the bone. The quadriceps tendon, patella, and patellar ligament serve as a capsule anteriorly—that is, the fibrous layer is continuous with the lateral and medial margins of these structures (Fig. 7.51).

The extensive *synovial membrane* lines the internal aspect of the fibrous capsule and attaches to the periphery of the patella

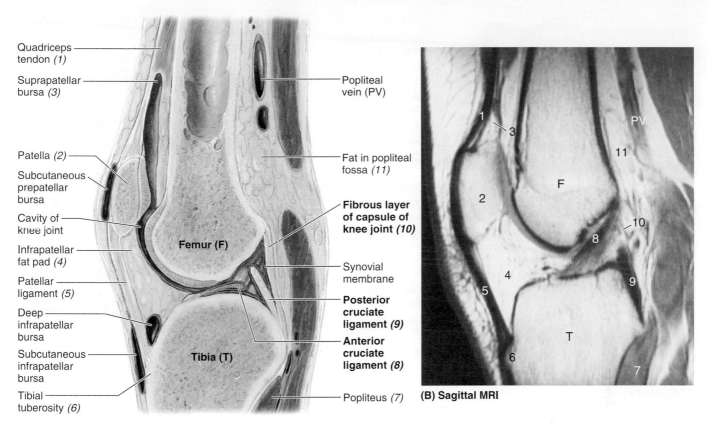

(A) Sagittal section

Labels on left (A):
- Quadriceps tendon (1)
- Suprapatellar bursa (3)
- Patella (2)
- Subcutaneous prepatellar bursa
- Cavity of knee joint
- Infrapatellar fat pad (4)
- Patellar ligament (5)
- Deep infrapatellar bursa
- Subcutaneous infrapatellar bursa
- Tibial tuberosity (6)
- Femur (F)
- Tibia (T)

Labels on right (A):
- Popliteal vein (PV)
- Fat in popliteal fossa (11)
- Fibrous layer of capsule of knee joint (10)
- Synovial membrane
- Posterior cruciate ligament (9)
- Anterior cruciate ligament (8)
- Popliteus (7)

(B) Sagittal MRI

FIGURE 7.51. Joint capsule and bursae around the knee joint. A. Sagittal section. **B.** Sagittal magnetic resonance imaging (MRI). The numbers are defined in part **A**.

and the edges of the *menisci*. It lines the fibrous layer laterally and medially, but centrally, it becomes separated from the fibrous layer. The synovial membrane reflects from the posterior aspect of the joint anteriorly into the intercondylar region, covering the cruciate ligaments and the **infrapatellar fat pad**, so they are excluded from the articular cavity (Fig. 7.51). This creates a median infrapatellar synovial fold, a vertical fold of synovial membrane that approaches the posterior aspect of the patella. Thus, it almost subdivides the articular cavity into right and left femorotibial articular cavities. Fat-filled lateral and medial alar folds of synovial membrane extend into the joint from the infrapatellar fold. More reflections or plicae have been identified with arthroscopy. If these plicae become inflamed, they can cause pain on movement and may be arthroscopically removed.

Superior to the patella, the knee joint cavity extends deep to the vastus intermedius as the *suprapatellar bursa*. The synovial membrane of the joint capsule is continuous with the synovial lining of this bursa (Fig. 7.51). Muscle slips deep to the vastus intermedius form the *articularis genu muscle* (articular muscle of the knee), which attaches to the synovial membrane and retracts the suprapatellar bursa during extension of the knee.

LIGAMENTS

The joint capsule is strengthened by four capsular (intrinsic) ligaments, the patellar, tibial collateral, oblique popliteal, and

arcuate popliteal ligaments and one extracapsular ligament, the fibular collateral ligament (Fig. 7.52).

The patellar ligament, the distal part of the quadriceps tendon, is a strong, thick, fibrous band passing from the apex and adjoining margins of the patella to the tibial tuberosity. Laterally, it receives the *medial* and *lateral patellar retinacula*, aponeurotic expansions of the vastus medialis and lateralis, and overlying deep fascia. The retinacula play an important role in maintaining alignment of the patella relative to the patellar articular surface of the femur.

The *collateral ligaments of the knee* are taut when the knee is fully extended; however, as flexion proceeds, they become increasingly slack, permitting rotation at the knee.

The **fibular** or **lateral collateral ligament** (LCL), rounded and cord-like, is strong. It extends inferiorly from the lateral epicondyle of femur to the lateral surface of the head of the fibula (Fig. 7.52). The tendon of the popliteus passes deep to the LCL, separating it from the lateral meniscus. The tendon of the biceps femoris is split into two parts by this ligament.

The **tibial** or **medial collateral ligament** (MCL) is a strong flat band that extends from the medial epicondyle of the femur to the medial condyle and superior part of the medial surface of the tibia. At its midpoint, the deep fibers of the LCL are firmly attached to the medial meniscus (Fig. 7.52).

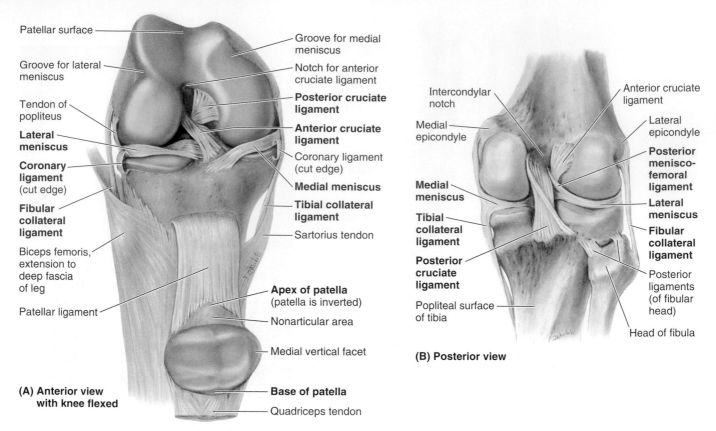

Patellar surface

Groove for lateral meniscus

Tendon of popliteus

Lateral meniscus

Coronary ligament (cut edge)

Fibular collateral ligament

Biceps femoris, extension to deep fascia of leg

Patellar ligament

(A) Anterior view with knee flexed

Groove for medial meniscus

Notch for anterior cruciate ligament

Posterior cruciate ligament

Anterior cruciate ligament

Coronary ligament (cut edge)

Medial meniscus

Tibial collateral ligament

Sartorius tendon

Apex of patella (patella is inverted)

Nonarticular area

Medial vertical facet

Base of patella

Quadriceps tendon

Intercondylar notch

Medial epicondyle

Medial meniscus

Tibial collateral ligament

Posterior cruciate ligament

Popliteal surface of tibia

(B) Posterior view

Anterior cruciate ligament

Lateral epicondyle

Posterior menisco-femoral ligament

Lateral meniscus

Fibular collateral ligament

Posterior ligaments (of fibular head)

Head of fibula

FIGURE 7.52. **Relations and ligaments of knee joint. A.** Anterior view of flexed knee with quadriceps tendon cut and reflected inferiorly. **B.** Posterior view.

The **oblique popliteal ligament** is a reflected expansion of the tendon of the semimembranosus that strengthens the joint capsule posteriorly. It arises posterior to the medial tibial condyle and passes superolaterally to attach to the central part of the posterior aspect of the joint capsule.

The **arcuate popliteal ligament** arises from the posterior aspect of the fibular head, passes superomedially over the tendon of the popliteus, and spreads over the posterior surface of the knee joint.

The *intra-articular structures* within the knee joint consist of the cruciate ligaments and menisci. The popliteus tendon is also intra-articular during part of its course.

The **cruciate ligaments** (L. *crux*, cross) join the femur and tibia, crisscrossing within the joint capsule but outside the articular cavity (Figs. 7.52 and 7.53). The cruciate ligaments cross each other obliquely like the letter *X*. During medial rotation of the tibia on the femur, the cruciate ligaments wind around each other; thus, the amount of medial rotation possible is limited to about 10 degrees. Because they become unwound during lateral rotation, nearly 60 degrees of lateral rotation is possible when the knee is flexed more than 90 degrees. The crossing-over point of the cruciate ligaments serves as the pivot for rotatory movements at the knee. Because of their oblique orientation, in every position, one cruciate ligament, or parts of one or both ligaments, are tense.

The **anterior cruciate ligament** (ACL), the weaker of the two cruciate ligaments, arises from the anterior intercondylar area of the tibia, just posterior to the attachment of the medial meniscus (Fig. 7.53). It extends superiorly, posteriorly, and laterally to attach to the posterior part of the medial side of the lateral condyle of the femur. The ACL limits posterior rolling of the femoral condyles on the tibial plateau during flexion, converting it to spin. It also prevents posterior displacement of the femur on the tibia and hyperextension of the knee joint. When the joint is flexed to a right angle, the tibia cannot be pulled anteriorly because it is held by the ACL. The ACL has a relatively poor blood supply.

The **posterior cruciate ligament** (PCL), the stronger of the two cruciate ligaments, arises from the posterior intercondylar area of the tibia (Fig. 7.53). The PCL passes superiorly and anteriorly on the medial side of the ACL to attach to the anterior part of the lateral surface of the medial condyle of the femur. The PCL limits anterior rolling of the femur on the tibial plateau during extension, converting it to spin. It also prevents anterior displacement of the femur on the tibia or posterior displacement of the tibia on the femur and helps prevent hyperflexion of the knee joint. In the weight-bearing flexed knee, the PCL is the main stabilizing factor for the femur (e.g., when walking downhill).

The **menisci of the knee joint** are crescentic plates of fibrocartilage on the articular surface of the tibia that deepen the surface and play a role in shock absorption (Fig. 7.53C,D). The menisci are thicker at their external margins and taper to thin, unattached edges in the interior of the joint. Wedge-shaped

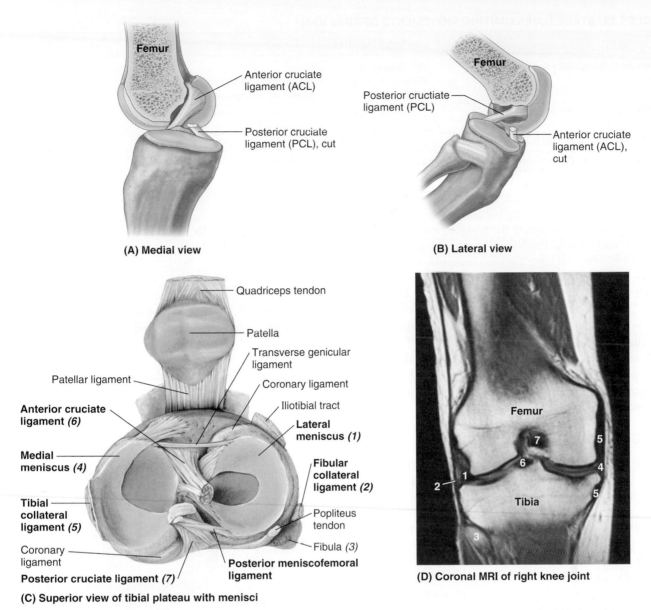

(A) Medial view

Femur

Anterior cruciate
ligament (ACL)

Posterior cruciate
ligament (PCL), cut

(B) Lateral view

Femur

Posterior cruciate
ligament (PCL)

Anterior cruciate
ligament (ACL),
cut

(C) Superior view of tibial plateau with menisci

Quadriceps tendon

Patella

Transverse genicular
ligament

Patellar ligament

Coronary ligament

Iliotibial tract

**Anterior cruciate
ligament (6)**

**Lateral
meniscus (1)**

**Medial
meniscus (4)**

**Fibular
collateral
ligament (2)**

**Tibial
collateral
ligament (5)**

Popliteus
tendon

Coronary
ligament

Fibula (3)

Posterior cruciate ligament (7)

**Posterior meniscofemoral
ligament**

(D) Coronal MRI of right knee joint

Femur

Tibia

FIGURE 7.53. Cruciate ligaments and menisci of knee joint. A. Anterior cruciate ligament. **B.** Posterior cruciate ligament. In **A and B**, the femur has been sectioned longitudinally and the near half has been removed with the proximal part of the corresponding cruciate ligament. **C.** Attachments to tibial plateau. The quadriceps tendon is cut, and the patella reflected anteriorly. **D.** The numbers on this MRI image of the right knee are defined in part **C**.

in transverse section, the menisci are firmly attached at their ends to the *intercondylar area of the tibia*. Their external margins attach to the fibrous layer of the capsule of the knee joint.

The **coronary ligaments** are capsular fibers that attach the margins of the menisci to the tibial condyles. A slender fibrous band, the **transverse ligament of knee**, joins the anterior edges of the menisci (Fig. 7.53C), allowing them to move together during knee movements. The **medial meniscus** is C-shaped and broader posteriorly than anteriorly. Its anterior end (horn) attaches to the anterior intercondylar area of the tibia, anterior to the attachment of the ACL. Its posterior end attaches to the posterior intercondylar area, anterior to the attachment of the PCL. The medial meniscus firmly adheres to the deep surface of the tibial collateral ligament. The **lateral**

meniscus is nearly circular and is smaller and more freely movable than the medial meniscus. The tendon of the popliteus separates the lateral meniscus from the fibular collateral ligament. A strong tendinous slip, the **posterior meniscofemoral ligament**, joins the lateral meniscus to the PCL and the medial femoral condyle (Fig. 7.52B).

MOVEMENTS OF KNEE JOINT

Flexion and extension are the main knee movements; some rotation occurs when the knee is flexed (Table 7.15). When the leg is fully extended with the foot on the ground, the knee passively "locks" because of medial rotation of the femur on the planted tibia. This position makes the lower limb a solid

TABLE 7.15. STRUCTURES LIMITING MOVEMENTS OF KNEE JOINT

Movement	Limiting Structures
Flexion (femoropatellar and femorotibial)	Soft tissue apposition posteriorly Tension of vastus lateralis, medialis, and intermedius Tension of rectus femoris (especially with hip joint extended)
Extension (femoropatellar and femorotibial)	*Ligaments*: anterior cruciate and posterior cruciate, fibular and tibial collateral, posterior joint capsule, and oblique popliteal ligament
Internal rotation (femorotibial with knee flexed)	*Ligaments*: anterior cruciate and posterior cruciate
External rotation (femorotibial with knee flexed)	*Ligaments*: fibular and tibial collateral

Modified from Clarkson HM. *Musculoskeletal Assessment. Joint Range of Motion and Manual of Muscle Strength.* 2nd ed. Baltimore, MD: Lippincott Williams & Wilkins; 2000.

column and more adapted for weight bearing. When the knee is "locked," the thigh and leg muscles can relax briefly without making the knee joint too unstable. To "unlock" the knee, the popliteus contracts, rotating the femur laterally about 5 degrees on the tibial plateau so that flexion of the knee can occur. The menisci must be able to move on the tibial plateau as the points of contact between the femur and the tibia change.

Three paired facets (superior, middle, and inferior) on the posterior surface of the patella articulate with the patellar surface of the femur successively during flexion and extension of the knee (Fig. 7.54).

BURSAE AROUND KNEE

There are at least 12 bursae around the knee joint because most tendons run parallel to the bones and pull lengthwise across the joint during knee movements (Fig. 7.55 and Table 7.16). The **subcutaneous prepatellar** and **infrapatellar bursae** are located at the convex surface of the joint, allowing the skin to be able to move freely during knee movements. Four bursae communicate with the articular cavity of the knee joint: **suprapatellar bursa** (deep to the distal quadriceps), *popliteus bursa, anserine bursa,* and *gastrocnemius bursa.*

ARTERIES AND NERVES OF KNEE JOINT

The genicular branches that form the peri-articular genicular anastomosis around the knee are from the femoral, popliteal, anterior and posterior recurrent branches of the anterior tibial, and circumflex fibular arteries (see Fig. 7.28D). The middle genicular branches of the popliteal artery penetrate the fibrous layer of the joint capsule and supply the cruciate ligaments, synovial membrane, and peripheral margins of the menisci.

The *nerves of the knee joint* are articular branches from the femoral, tibial, and common fibular nerves and the obturator and saphenous nerves.

Tibiofibular Joints

The tibia and fibula are connected by two joints: the *superior tibiofibular joint* and the *tibiofibular syndesmosis* (inferior

tibiofibular joint). In addition, an *interosseous membrane* joins the shafts of the two bones (Fig. 7.56). Movement at the proximal joint is impossible without movement at the distal one. The fibers of the interosseous membrane and all ligaments of tibiofibular articulations run inferiorly from the tibia to the fibula, resisting the downward pull placed on the fibula by most muscles attached to it. However, they allow slight upward movement of the fibula during dorsiflexion of the ankle.

The **superior tibiofibular joint** is a plane type of synovial joint between the flat facet on the fibular head and

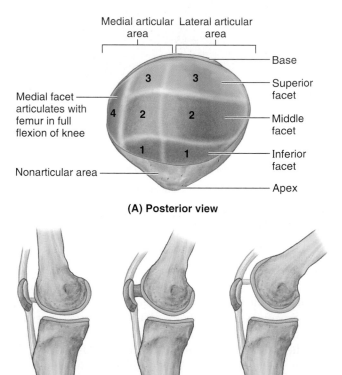

(A) Posterior view

(B) Medial view

Extension (1) Slight flexion (2) Flexion (3)

FIGURE 7.54. Femoropatellar articulation. A. Articular surfaces of patella. **B.** Articulation of patella with femur during flexion and extension of knee.

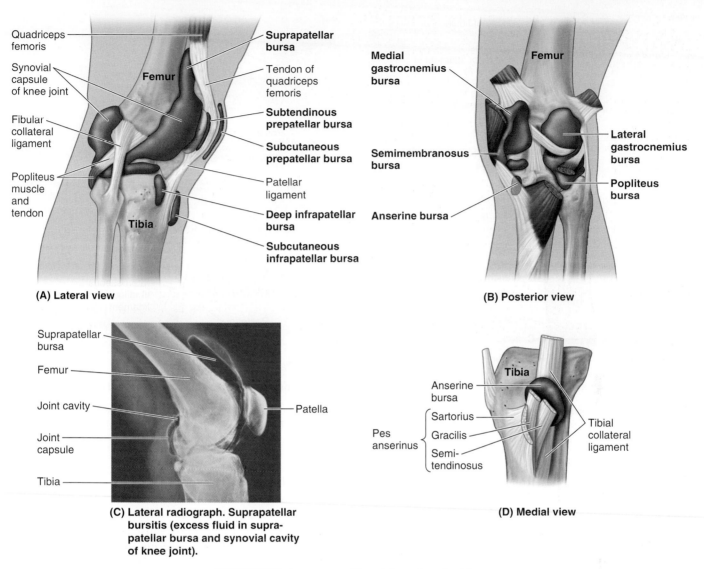

FIGURE 7.55. Bursae around knee joint and proximal leg.

TABLE 7.16. BURSAE AROUND KNEE JOINT

Bursae	Locations	Comments
Suprapatellar	Between femur and tendon of quadriceps femoris	Held in position by articularis genu muscle; communicates freely with synovial cavity of knee joint
Popliteus	Between tendon of popliteus and lateral condyle of tibia	Opens into synovial cavity of knee joint inferior to lateral meniscus
Anserine	Separates tendons of sartorius, gracilis, and semitendinosus from tibia and tibial collateral ligament	Area where tendons of these muscles attach to tibia; resembles a goose's foot (L. *pes*, foot; L. *anserinus*, goose)
Gastrocnemius	Lie deep to proximal attachment of tendon of medial and lateral heads of gastrocnemius	Extensions of synovial cavity of knee joint
Semimembranosus	Between medial head of gastrocnemius and semimembranosus tendon	Related to distal attachment of semimembranosus
Subcutaneous prepatellar	Between skin and anterior surface of patella	Allows free movement of skin over patella during movements of leg
Subcutaneous infrapatellar	Between skin and tibial tuberosity	Helps knee withstand pressure when kneeling
Deep infrapatellar	Between patellar ligament and anterior surface of tibia	Separated from knee joint by infrapatellar fat pad

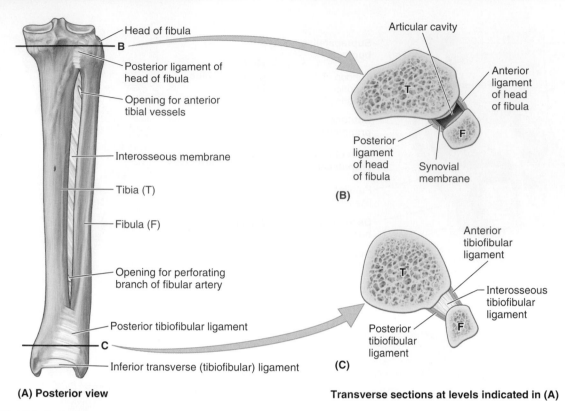

(A) Posterior view

Transverse sections at levels indicated in (A)

FIGURE 7.56. Tibiofibular joints. A. Superior tibiofibular joint and tibiofibular syndesmosis, posterior view. The level of the transverse sections shown in parts **B** and **C** are identified. **B.** Transverse section through the superior tibiofibular joint. **C.** Transverse section through the tibiofibular syndesmosis.

a similar facet located posterolaterally on the lateral tibial condyle. The tense joint capsule surrounds the joint and attaches to the margins of the articular surfaces of the fibula and tibia. The joint capsule is strengthened by **anterior** and **posterior ligaments of head of fibula** (Fig. 7.56B). The synovial membrane lines the fibrous capsule. Slight gliding movements occur during dorsiflexion of the ankle.

The **tibiofibular syndesmosis** is a compound fibrous joint (Fig. 7.56C). The integrity of this articulation is essential for stability of the ankle joint because it keeps the lateral malleolus firmly against the lateral surface of the talus. The strong **interosseous tibiofibular ligament** is continuous superiorly with the interosseous membrane and forms the principal connection between the distal

(continued on page 488)

CLINICAL BOX

Genu Varum and Genu Valgum

The femur is placed diagonally within the thigh, whereas the tibia is almost vertical within the leg, creating a **Q-angle** at the knee between the long axes of the bones. The Q-angle is assessed by drawing a line from the ASIS to the middle of the patella and extrapolating a second (vertical) line through the middle of the patella and tibial tuberosity (Fig. B7.18A). The Q-angle is typically greater in adult females owing to their wider pelves. A medial angulation of the leg in relation to the thigh, in which the femur is abnormally vertical and the Q-angle is small,

is a deformity called *genu varum* (bowleg) that causes unequal weight distribution (Fig. B7.18B). Excess pressure is placed on the medial aspect of the knee joint, which results in *arthrosis* (destruction of knee cartilage). A lateral angulation of the leg (Fig. B7.18C) in relation to the thigh (exaggeration of knee angle) is *genu valgum* (knock-knee). Consequently, in genu valgum, excess stress is placed on the lateral structures of the knee. The patella, normally pulled laterally by the tendon of the vastus lateralis, is pulled even farther laterally when the leg is extended in the presence of genu varum so that its articulation with the femur is abnormal.

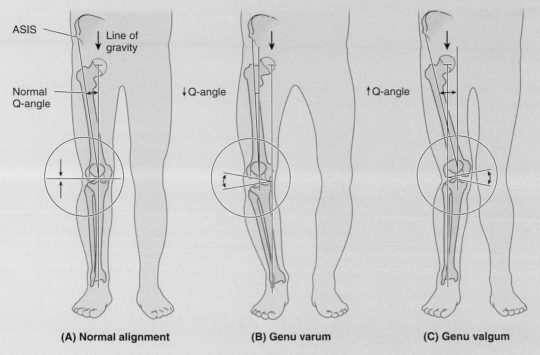

(A) Normal alignment   **(B) Genu varum**   **(C) Genu valgum**

FIGURE B7.18. **Alignment of lower limb bones.** Normal alignment (**A**), genu varum (**B**), and genu valgum (**C**) are shown. *ASIS*, anterior superior iliac spine.

Patellofemoral Syndrome

Pain deep to the patella often results from excessive running, especially downhill; hence, this type of pain is often called "runner's knee." The pain results from repetitive microtrauma caused by abnormal tracking of the patella relative to the patellar surface of the femur, a condition known as the *patellofemoral syndrome*. This syndrome may also result from a direct blow to the patella and from *osteoarthritis of the patellofemoral compartment* (degenerative wear and tear of articular cartilages). In some cases, strengthening of the vastus medialis corrects *patellofemoral dysfunction*. This muscle tends to prevent lateral dislocation of the patella resulting from the Q-angle because the vastus medialis attaches to and pulls on the medial border of the patella. Hence, weakness of the vastus medialis predisposes the individual to patellofemoral dysfunction and patellar dislocation.

Patellar Dislocation

When the patella is dislocated, it nearly always dislocates laterally. Patellar dislocation is more common in women, presumably because of their greater Q-angle, which, in addition to representing the oblique placement of the femur relative to the tibia, represents the angle of pull of the quadriceps relative to the axis of the patella and tibia (the term *Q-angle* was actually coined in reference to the angle of pull of the quadriceps). The tendency toward lateral dislocation is normally counterbalanced by the medial, more horizontal pull of the powerful vastus medialis. In addition, the more anterior projection of the lateral femoral condyle and deeper slope for the larger lateral patellar facet provide a mechanical deterrent to lateral dislocation. An imbalance of the lateral pull and the mechanisms resisting it results in abnormal tracking of the patella within the patellar groove and chronic patellar pain, even if actual dislocation does not occur.

Popliteal Cysts

Popliteal cysts (Baker cysts) are abnormal fluid-filled sacs of synovial membrane in the region of the popliteal fossa. A popliteal cyst is almost always a complication of chronic knee joint effusion. The cyst may be a herniation of the gastrocnemius or semimembranosus bursa through the fibrous layer of the joint capsule into the popliteal fossa, communicating with the synovial cavity of the knee joint by a narrow stalk. Synovial fluid may also escape from the knee joint (*synovial effusion*) or a bursa around the knee and collect in the popliteal fossa. Here, it forms a new synovial-lined sac, or popliteal cyst. In adults, popliteal cysts can be large, extending as far as the midcalf, and may interfere with knee movements.

(Continued on next page)

Knee Joint Injuries

 Knee joint injuries are common because the knee is a low-placed, mobile, weight-bearing joint and its stability depends almost entirely on its associated ligaments and muscles. The most common knee injuries in contact sports are ligament sprains, which occur when the foot is fixed on the ground. If a force is applied against the knee when the foot cannot move, ligament injuries are likely to occur. The MCL and LCL are tightly stretched when the leg is extended, preventing disruption of the sides of the joint. The firm attachment of the MCL to the medial meniscus is of clinical significance because tearing of this ligament frequently results in concomitant tearing of the medial meniscus. The injury is frequently caused by a blow to the lateral side of the extended knee or excessive lateral twisting of the flexed knee, which disrupts the MCL and concomitantly tears and/or detaches the medial meniscus from the joint capsule. This injury is common in athletes who twist their flexed knees while running (e.g., in football and soccer). The ACL, which serves as a pivot for rotatory movements of the knee, is taut during flexion and may also tear subsequent to the rupture of the MCL (Fig. B7.19A). *ACL rupture*, one of the most common knee injuries in skiing accidents, for example, causes the free tibia to slide anteriorly under the femur, a sign known as the *anterior drawer sign* (see Fig. 7.19B). Although strong, *PCL rupture* may occur when a person lands on the tibial tuberosity when the knee is flexed. PCL ruptures usually occur in conjunction with tibial or fibular ligament tears. The *posterior drawer sign*, in which the free tibia slides posteriorly under the fixed femur, occurs as a result of PCL rupture (Fig. B7.19C).

Arthroscopy of Knee Joint

 Arthroscopy is an endoscopic examination that allows visualization of the interior of the knee joint cavity with minimal disruption of tissue (Fig. B7.19D). The arthroscope and one (or more) additional cannula(e) are inserted through tiny incisions known as portals. The second cannula is for passage of specialized tools (e.g., manipulative probes or forceps) or equipment for trimming, shaping, or removing damaged tissue. This technique allows removal of torn menisci and loose bodies in the joint, such as bone chips, and débridement (the excision of devitalized articular cartilaginous material in advanced cases of arthritis). Ligament repair or replacement may also be performed using an arthroscope.

Knee Replacement

 If a person's knee is diseased (e.g., from osteoarthritis), an artificial knee joint may be inserted (*total knee replacement arthroplasty*) (Fig. B7.19E). The artificial knee joint consists of plastic and metal components that are cemented to the femoral and tibial bone ends after removal of the defective areas.

Bursitis in Knee Region

Prepatellar bursitis ("housemaid's knee") is usually a friction bursitis caused by friction between the skin and the patella. If the inflammation is chronic,

(Continued on page 488)

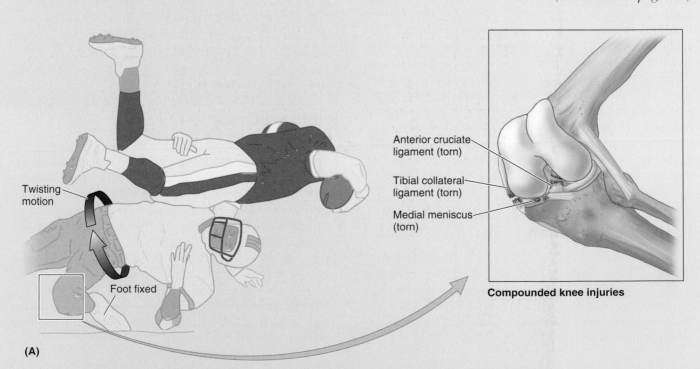

Twisting motion

Foot fixed

(A)

Anterior cruciate ligament (torn)

Tibial collateral ligament (torn)

Medial meniscus (torn)

Compounded knee injuries

FIGURE B7.19. Knee joint injuries, arthroscopy, and knee replacement. *(continued)*

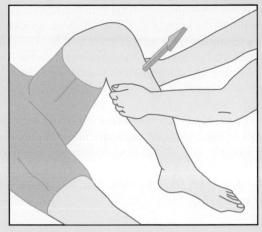

(B) Anterior drawer sign (ACL)

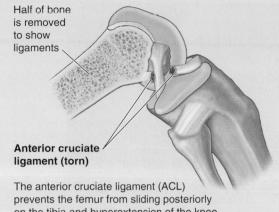

Half of bone is removed to show ligaments

Anterior cruciate ligament (torn)

The anterior cruciate ligament (ACL) prevents the femur from sliding posteriorly on the tibia and hyperextension of the knee and limits medial rotation of the femur when the foot is on the ground, and the leg is flexed.

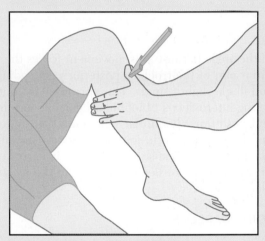

(C) Posterior drawer sign (PCL)

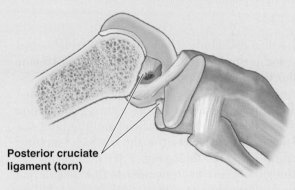

Posterior cruciate ligament (torn)

The posterior cruciate ligament (PCL) prevents the femur from sliding anteriorly on the tibia, particularly when the knee is flexed.

(D) Arthroscopy

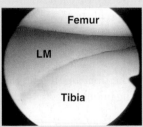

Femur

LM

Tibia

Normal lateral meniscus of the knee

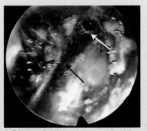

ACL graft *(black arrow)* with femoral anchoring screw visible *(white arrow)*

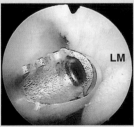

LM

Trimming of a torn lateral meniscus *(LM)*

(E) Osteoarthritis

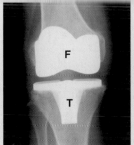

F

T

(F) Total knee replacement arthroplasty

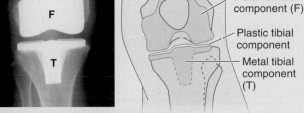

Metal femoral component (F)

Plastic tibial component

Metal tibial component (T)

FIGURE B7.19. *(continued)*

(Continued on next page)

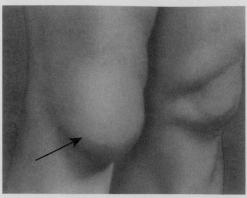

Prepatellar bursitis (arrow)

FIGURE B7.20. Prepatellar bursitis.

the bursa becomes distended with fluid and forms a swelling anterior to the knee (Fig. B7.20). *Subcutaneous infrapatellar bursitis* results from excessive friction between the skin and the tibial tuberosity; the edema occurs over the proximal end of the tibia. *Deep infrapatellar bursitis* results in edema between the patellar ligament and the tibia, superior to the tibial tuberosity.

The suprapatellar bursa communicates with the articular cavity of the knee joint; consequently, abrasions or penetrating wounds (e.g., a stab wound) superior to the patella may result in *suprapatellar bursitis* caused by bacteria entering the bursa from the torn skin. The infection may spread to the knee joint.

ends of the tibia and fibula. The joint is also strengthened anteriorly and posteriorly by the **anterior** and **posterior tibiofibular ligaments**. The distal, deep continuation of the posterior inferior tibiofibular ligament, the **inferior transverse (tibiofibular) ligament**, forms a strong connection between the medial and lateral malleoli and the posterior "wall" of the malleolar mortise for the trochlea (superior articular surface) of the talus (Fig. 7.57B). Slight movement of the joint occurs to accommodate the talus during dorsiflexion of the ankle.

Ankle Joint

The **ankle joint (talocrural articulation)** is a hinge-type synovial joint that is located between the distal ends of the tibia and fibula and the superior part of the talus (Fig. 7.57).

ARTICULAR SURFACES

The distal ends of the tibia and fibula (along with the inferior transverse part of the posterior tibiofibular ligament) form a *malleolar mortise* (deep socket) into which the pulley-shaped *trochlea of the talus* fits (Fig. 7.57B). The trochlea (L. pulley) is the rounded superior articular surface of the talus. The medial surface of the lateral malleolus articulates with the lateral surface of the talus. The tibia articulates with the talus in two places:

- Its inferior surface forms the roof of the malleolar mortise, transferring the body's weight to the talus.
- Its medial malleolus articulates with the medial surface of the talus.

The malleoli grip the talus tightly as it rocks in the mortise during movements of the ankle joint. The grip of the malleoli on the trochlea is strongest during dorsiflexion

of the ankle because this movement forces the wider, anterior part of the trochlea posteriorly, spreading the tibia and fibula slightly apart. This spreading is limited by the strong interosseous tibiofibular ligament and the anterior and posterior tibiofibular ligaments that unite the tibia and fibula. The ankle joint is relatively unstable during plantarflexion because the trochlea is narrower posteriorly and therefore lies loosely within the mortise during plantarflexion.

JOINT CAPSULE

The joint capsule is thin anteriorly and posteriorly but is supported on each side by strong collateral ligaments (Fig. 7.58). The fibrous layer of the capsule is attached superiorly to the borders of the articular surfaces of the tibia and malleoli and inferiorly to the talus. The synovial membrane lining the fibrous layer of the joint capsule extends superiorly between the tibia and the fibula as far as the interosseous tibiofibular ligament.

LIGAMENTS

The ankle joint is reinforced laterally by the **lateral ligament of the ankle**, which consists of three separate ligaments (Fig. 7.58A,C):

- **Anterior talofibular ligament**, a flat, weak band that extends anteromedially from the lateral malleolus to the neck of the talus
- **Posterior talofibular ligament**, a thick, fairly strong band that runs horizontally medially and slightly posteriorly from the malleolar fossa of the fibula to the lateral tubercle of the talus
- **Calcaneofibular ligament**, a round cord that passes postero-inferiorly from the tip of the lateral malleolus to the lateral surface of the calcaneus

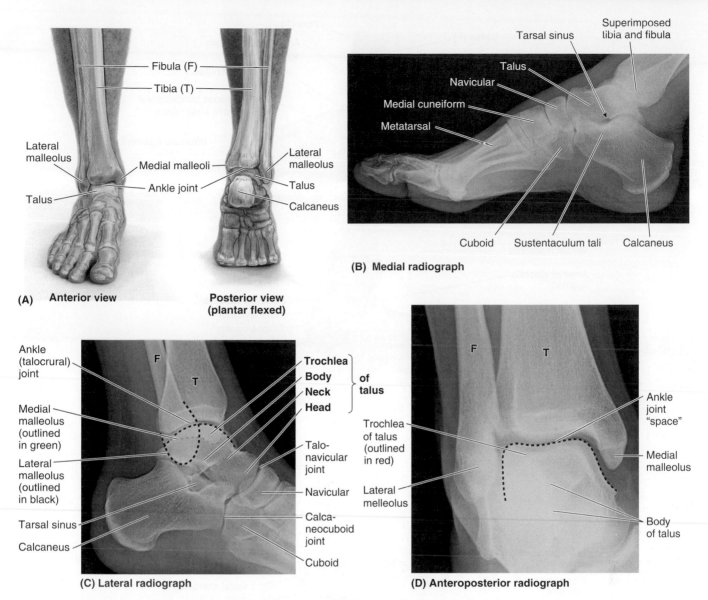

FIGURE 7.57. Bones of leg and ankle joint. A. Bones in situ. **B.** Medial radiograph. **C.** Lateral radiograph. **D.** Anteroposterior radiograph.

The joint capsule of the ankle joint is reinforced medially by the large, strong **medial ligament of the ankle** (deltoid ligament) that attaches proximally to the medial malleolus and fans out from it to attach distally to the talus, calcaneus, and navicular via four adjacent and continuous parts (Fig. 7.58B): the **tibionavicular part**, the **tibiocalcaneal part**, and the **anterior** and **posterior tibiotalar parts**. The medial ligament stabilizes the ankle joint during eversion of the foot and prevents subluxation (partial dislocation) of the ankle joint.

MOVEMENTS

The main movements of the ankle joint are **dorsiflexion** (elevating forefoot and toes) and **plantarflexion** (depressing forefoot and toes). When the ankle joint is plantarflexed, some "wobble" (small amounts of abduction, adduction, inversion, and eversion) is possible in this unstable position. Structures limiting movements of the ankle joint are outlined in Table 7.17.

- *Dorsiflexion of the ankle* is produced by muscles in the anterior compartment of the leg. Dorsiflexion is usually limited by passive resistance of the triceps surae to stretching and by tension in the medial and lateral ligaments.
- *Plantarflexion of ankle* is produced by muscles in the posterior and lateral compartments of the leg.

ARTERIES AND NERVES

The arteries are derived from malleolar branches of the fibular and anterior and posterior tibial arteries. The nerves are derived from the tibial nerve and deep fibular nerve.

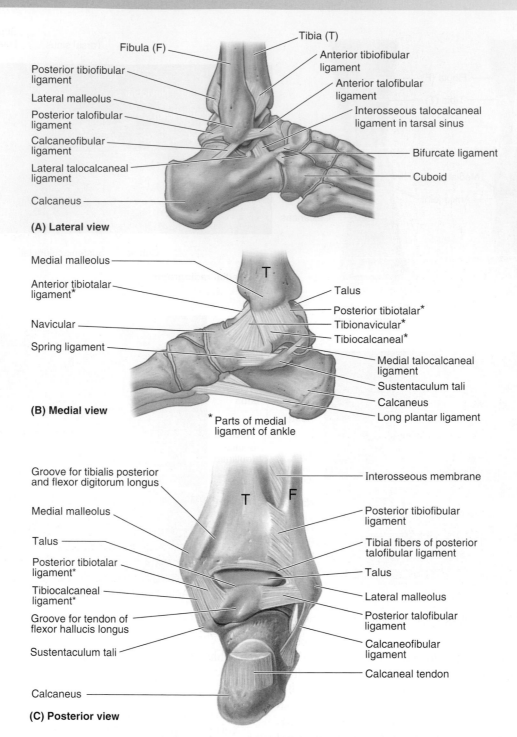

FIGURE 7.58. Ligaments of ankle and talocalcaneal joints.

TABLE 7.17. STRUCTURES LIMITING MOVEMENTS OF ANKLE JOINT

Movement	Limiting Structures
Plantarflexion	*Ligaments*: anterior talofibular, anterior part of medial ligament of ankle, anterior joint capsule
	Contact of talus with tibia
	Tension of dorsiflexors of ankle
Dorsiflexion	*Ligaments*: medial ligament of ankle, calcaneofibular, posterior talofibular, posterior joint capsule
	Contact of talus with tibia
	Tension of plantarflexors of ankle

Modified from Clarkson HM. *Musculoskeletal Assessment. Joint Range of Motion and Manual of Muscle Strength.* 2nd ed. Baltimore, MD: Lippincott Williams & Wilkins; 2000.

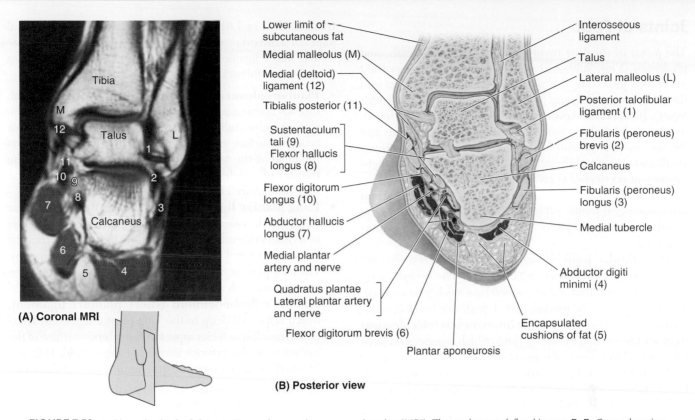

(A) Coronal MRI

Lower limit of subcutaneous fat
Medial malleolus (M)
Medial (deltoid) ligament (12)
Tibialis posterior (11)
Sustentaculum tali (9)
Flexor hallucis longus (8)
Flexor digitorum longus (10)
Abductor hallucis longus (7)
Medial plantar artery and nerve
Quadratus plantae Lateral plantar artery and nerve
Flexor digitorum brevis (6)
Plantar aponeurosis

Interosseous ligament
Talus
Lateral malleolus (L)
Posterior talofibular ligament (1)
Fibularis (peroneus) brevis (2)
Calcaneus
Fibularis (peroneus) longus (3)
Medial tubercle
Abductor digiti minimi (4)
Encapsulated cushions of fat (5)

(B) Posterior view

FIGURE 7.59. Ankle and subtalar joints. A. Coronal magnetic resonance imaging (MRI). The *numbers* are defined in part **B. B.** Coronal section.

CLINICAL BOX

Tibial Nerve Entrapment

Entrapment and compression of the tibial nerve (*tarsal tunnel syndrome*) occurs when there is edema and tightness in the ankle involving the synovial sheaths of the tendons of muscles in the posterior compartment of the leg. The area involved is from the medial malleolus to the calcaneus. The heel pain results from compression of the tibial nerve by the flexor retinaculum.

Ankle Sprains

The ankle is the most frequently injured major joint in the body. *Ankle sprains* (torn fibers of ligaments) are most common. A sprained ankle is nearly always an *inversion injury*, involving twisting of the weight-bearing plantarflexed foot. The anterior talofibular ligament (part of the lateral ligament) is most commonly torn during ankle sprains, either partially or completely, resulting in instability of the ankle joint. The calcaneofibular ligament may also be torn.

Pott Fracture–Dislocation of Ankle

A *Pott fracture–dislocation of the ankle* occurs when the foot is forcibly everted. This action pulls on the extremely strong medial ligament, often tearing off the medial malleolus (Fig. B7.21). The talus then moves laterally, shearing off the lateral malleolus or, more commonly, breaking the fibula superior to the tibiofibular syndesmosis. If the tibia is carried anteriorly, the posterior margin of the distal end of the tibia is also sheared off by the talus.

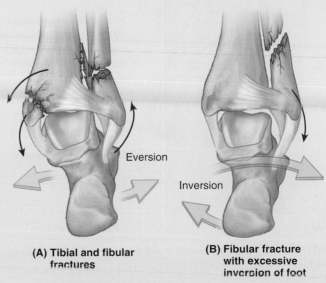

Eversion

Inversion

(A) Tibial and fibular fractures

(B) Fibular fracture with excessive inversion of foot

Posterior views

FIGURE B7.21. Fracture–dislocations of ankle joint.

Joints of Foot

The *joints of the foot* involve the tarsals, metatarsals, and phalanges (Figs. 7.59–7.61 and Table 7.19). The important intertarsal joints are the *subtalar (talocalcaneal) joint* and the *transverse tarsal joint (calcaneocuboid and talonavicular joints)*. Inversion and eversion of the foot are the main movements involving these joints. The other intertarsal joints and the *tarsometatarsal* and *intermetatarsal joints* are relatively small and are so tightly joined by ligaments that only slight movement occurs between them. In the foot, flexion and extension occurs in the forefoot at the metatarsophalangeal and interphalangeal joints. All of the foot bones proximal to the metatarsophalangeal joints are united by dorsal and plantar ligaments.

The **subtalar joint** occurs where the talus rests on and articulates with the calcaneus (Fig. 7.59). The subtalar joint is a synovial joint that is surrounded by a weak joint capsule, which is supported by medial, lateral, posterior, and interosseous *talocalcaneal ligaments*. The **interosseous talocalcaneal ligament** lies within the *tarsal sinus*, which separates the subtalar and calcaneonavicular joints and is especially strong.

The **transverse tarsal joint** is a compound joint formed by the **talonavicular part of the talocalcaneonavicular** and the **calcaneocuboid joints**—two separate joints aligned

transversely (Fig. 7.61). Transection across the transverse tarsal joint is a standard method for *surgical amputation of the foot*.

The major ligaments of the plantar aspect of the foot are (Fig. 7.60) as follows:

- **Plantar calcaneonavicular (spring) ligament**, which extends across and fills a wedge-shaped gap between the sustentaculum tali and the inferior margin of the posterior articular surface of the navicular. This ligament supports the head of the talus and plays an important role in the transfer of weight from the talus and in maintaining the longitudinal arch of the foot.
- **Long plantar ligament**, which passes from the plantar surface of the calcaneus to the groove on the cuboid. Some of its fibers extend to the bases of the metatarsals, thereby forming a tunnel for the tendon of the fibularis longus. The long plantar ligament is important in maintaining the longitudinal arch of the foot.
- **Plantar calcaneocuboid (short plantar) ligament**, which is located deep to the long plantar ligament. It extends from the anterior aspect of the inferior surface of the calcaneus to the inferior surface of the cuboid. It is also involved in maintaining the longitudinal arch of the foot.

The structures limiting movements of the feet and toes are summarized in Table 7.18.

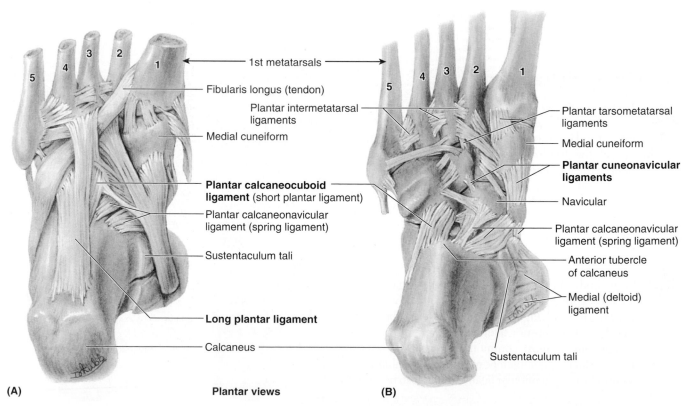

FIGURE 7.60. Plantar ligaments. A and B. Sequential stages of dissection of the sole of the right foot showing the attachments of the ligaments and tendons of the long invertor and evertor muscles.

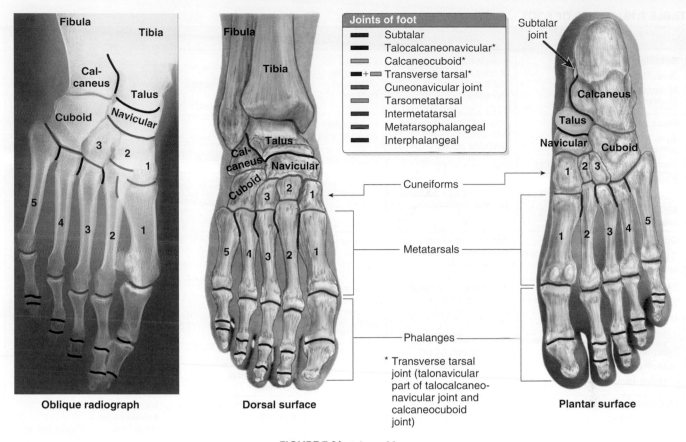

FIGURE 7.61. Joints of foot.

TABLE 7.18. STRUCTURES LIMITING MOVEMENTS OF FOOT AND TOES

Movement	Joint	Limiting Structures
Inversion	Subtalar, transverse tarsal	*Ligaments*: lateral ligament of ankle, talocalcaneal ligament, lateral joint capsule Tension of evertor muscles of ankle
Eversion	Subtalar, transverse tarsal	*Ligaments*: medial ligament of ankle, medial talocalcaneal ligament, medial joint capsule Tension of tibialis posterior, flexor hallucis longus, flexor digitorum longus Contact of talus with calcaneus
Flexion	MTP, PIP, DIP	*MTP*: tension of posterior joint capsule, extensor muscles, and collateral ligaments Tension of tibialis posterior, flexor hallucis longus, flexor digitorum longus Contact of talus with calcaneus
Extension	MTP, PIP, DIP	*MTP*: tension of plantar joint capsule, plantar ligaments, and flexor muscles *PIP*: tension in plantar joint capsule *DIP*: ligaments and plantar joint capsule
Abduction	MTP	*Ligaments*: collateral ligaments, medial joint capsule Tension of adductor muscles Skin between web spaces
Adduction	MTP	Apposition of toes

DIP, distal interphalangeal (toes 2–5); *MTP*, metatarsophalangeal; *PIP*, proximal interphalangeal.
Modified from Clarkson HM. *Musculoskeletal Assessment. Joint Range of Motion and Manual of Muscle Strength*. 2nd ed. Baltimore, MD: Lippincott Williams & Wilkins; 2000.

TABLE 7.19. JOINTS OF FOOT

Joint	Articulating Surfaces	Joint Capsule	Ligaments	Blood Supply	Nerve Supply
Subtalar (talocalcaneal) *Type:* Plane synovial *Movements:* Inversion and eversion of foot	Inferior surface of body of talus articulates with superior surface of calcaneus.	Attached to margins of articular surfaces	Medial, lateral, and posterior talocalcaneal ligaments and interosseous talocalcaneal ligament	Posterior tibial and fibular arteries	*Plantar aspect:* medial or lateral plantar nerve *Dorsal aspect:* deep fibular nerve
Talocalcaneonavicular *Type:* Synovial joint; talonavicular part is ball-and-socket type. *Movements:* Gliding and rotatory	Head of talus articulates with calcaneus and navicular bones.	Incompletely encloses joint	Plantar calcaneonavicular (spring) ligament supports head of talus.	Anterior tibial artery via lateral tarsal artery, a branch of dorsalis pedis artery	
Calcaneocuboid *Type:* Plane synovial *Movements:* Inversion and eversion of foot; circumduction	Anterior end of calcaneus articulates with posterior surface of cuboid.	Encloses joint	Dorsal and plantar calcaneocuboid and long plantar ligaments		
Cuneonavicular joint *Type:* Plane synovial *Movements:* Little	Anterior end of navicular articulates with bases of cuneiform bones.	Common capsule encloses joints	Dorsal and plantar cuneonavicular ligaments		
Tarsometatarsal *Type:* Plane synovial *Movements:* Gliding or sliding	Anterior ends of tarsal bones articulate with bases of metatarsal bones.	Separate joint capsules enclose each joint	Dorsal, plantar, and interosseous tarsometatarsal ligaments		Deep fibular; medial and lateral plantar nerves; sural nerve
Intermetatarsal *Type:* Plane synovial *Movements:* Little	Bases of metatarsal bones articulate with each other.	Separate joint capsules enclose each joint	Dorsal, plantar, and interosseous tarsometatarsal ligaments	Lateral metatarsal artery, (a branch of dorsalis pedis artery of foot)	Digital nerves
Metatarsophalangeal *Type:* Condyloid synovial *Movements:* Flexion, extension, and some abduction, adduction, and circumduction	Heads of metatarsal bones articulate with bases of proximal phalanges.		Collateral and plantar ligaments		
Interphalangeal *Type:* Hinge synovial *Movements:* Flexion and extension	Head of one phalanx articulates with base of one distal to it.		Collateral and plantar ligaments	Digital branches of plantar arch	

Arches of Foot

The foot is composed of numerous bones connected by ligaments that provide considerable flexibility which allow it to deform with each ground contact, thereby absorbing much of the shock. Furthermore, the tarsal and metatarsal bones are arranged in longitudinal and transverse arches passively supported and actively restrained by flexible tendons that add to the weight-bearing capabilities and resiliency of the foot (Fig. 7.62). The arches distribute weight over the foot (*pedal platform*), acting not only as shock absorbers but also as springboards for propelling it during walking, running, and jumping. The resilient arches add to the foot's ability to adapt to changes in surface contour. The weight of the body is transmitted to the talus from the tibia. Then it is transmitted posteriorly to the calcaneus and anteriorly to the "ball of the foot" (the sesamoid bones of the 1st metatarsal and the head of the 2nd metatarsal), and that weight/pressure is shared laterally with the heads of the 3rd through 5th metatarsals as necessary for balance and comfort (Fig. 7.62A). Between these weight-bearing points are the relatively elastic arches of the foot, which become slightly flattened by the body weight during standing, but they normally resume their curvature (recoil) when body weight is removed.

The **longitudinal arch of the foot** is composed of medial and lateral parts (Fig. 7.62B). Functionally, both parts act as a unit, with the transverse arch spreading the weight in all directions. The **medial longitudinal arch** is higher and more important than the lateral longitudinal arch. The medial longitudinal arch is composed of the calcaneus, talus, navicular, three cuneiforms, and three metatarsals. *The talar head is the keystone of the medial longitudinal arch.* The tibialis anterior and posterior via their tendinous attachments help support the medial longitudinal arch (Fig. 7.62C). The fibularis longus tendon, passing from lateral to medial, also helps support this arch. The **lateral longitudinal arch** is much flatter than the medial longitudinal arch and rests on the ground during standing. It is composed of the calcaneus, cuboid, and lateral two metatarsals.

The **transverse arch of the foot** runs from side to side. It is formed by the cuboid, cuneiforms, and bases of the metatarsals. The medial and lateral parts of the longitudinal arch serve as pillars for the transverse arch. The tendon of the fibularis longus and tibialis posterior, crossing the sole of the foot obliquely, helps maintain the curvature of the transverse arch.

The integrity of the bony arches of the foot is maintained by both passive factors and dynamic supports (Fig. 7.62C). The passive factors include the shape of the united bones and the four successive layers of fibrous tissue: plantar aponeurosis, long plantar ligament, plantar calcaneocuboid (short plantar) ligament, and calcaneonavicular (spring) ligament.

The dynamic supports include the active (reflexive) bracing action of the intrinsic muscles of the foot and the active and tonic contraction of the muscles with long tendons extending into the foot (flexor hallucis longus and flexor digitorum longus for the longitudinal arch and fibularis longus and tibialis anterior for the transverse arch). Of these factors, the plantar ligaments and plantar aponeurosis bear the greatest stress and are most important in maintaining the arches.

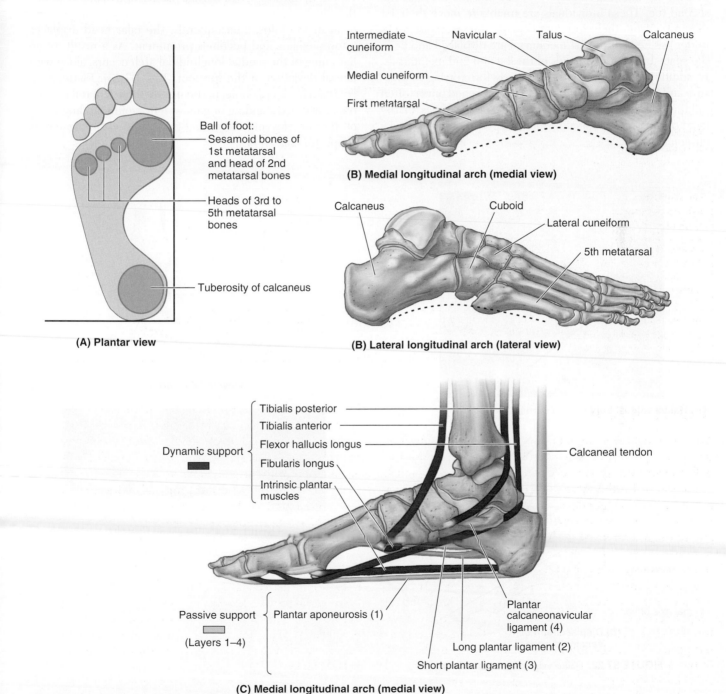

FIGURE 7.62. **Arches of foot. A.** Weight-bearing areas of foot. **B.** Medial longitudinal arch and lateral longitudinal arch. **C.** Passive and dynamic supports of foot. There are four layers of passive support (1–4).

CLINICAL BOX

Hallux Valgus

Hallux valgus is a foot deformity caused by degenerative joint disease; it is characterized by lateral deviation of the great toe (L. *hallux*). In some people, the deviation is so great that the first toe overlaps the second toe. These individuals are unable to move their 1st digit away from their 2nd digit because the sesamoid bones under the head of the 1st metatarsal are displaced and lie in the space between the heads of the 1st and 2nd metatarsals. In addition, a subcutaneous bursa may form owing to pressure and friction against the shoe. The thickened bursa (often inflamed and tender) and/or *reactive hyperostosis* of the head of the 1st metatarsal results in a protuberance called a *bunion* (Fig. B7.22).

Pes Planus (Flatfeet)

Acquired flatfeet ("fallen arches") are likely to be secondary to dysfunction of the tibialis posterior owing to trauma, degeneration with age, or denervation. In the absence of normal passive or dynamic support, the plantar calcaneonavicular ligament fails to support the head of the talus. Consequently, the talar head displaces inferomedially and becomes prominent. As a result, some flattening of the medial longitudinal arch occurs, along with lateral deviation of the forefoot (Fig. B7.23). Flatfeet are common in older people, particularly if they undertake much unaccustomed standing or gain weight rapidly, adding stress on the muscles and increasing the strain on the ligaments supporting the arches.

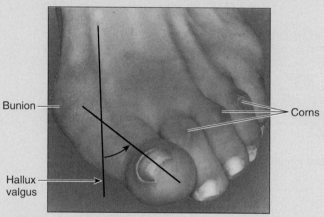

Bunion

Corns

Hallux valgus

(A) Hallux valgus, bunion, and corns

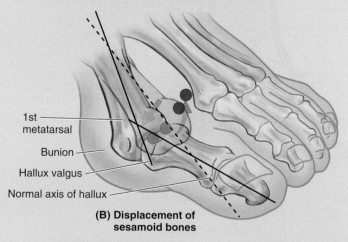

1st metatarsal

Bunion

Hallux valgus

Normal axis of hallux

(B) Displacement of sesamoid bones

FIGURE B7.22. Hallux valgus, bunions, and corns.

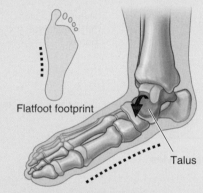

Flatfoot footprint

Talus

View of fallen arch

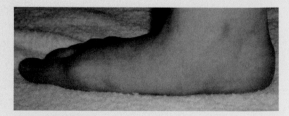

FIGURE B7.23. Pes planus (flatfeet).

MEDICAL IMAGING

Lower Limb

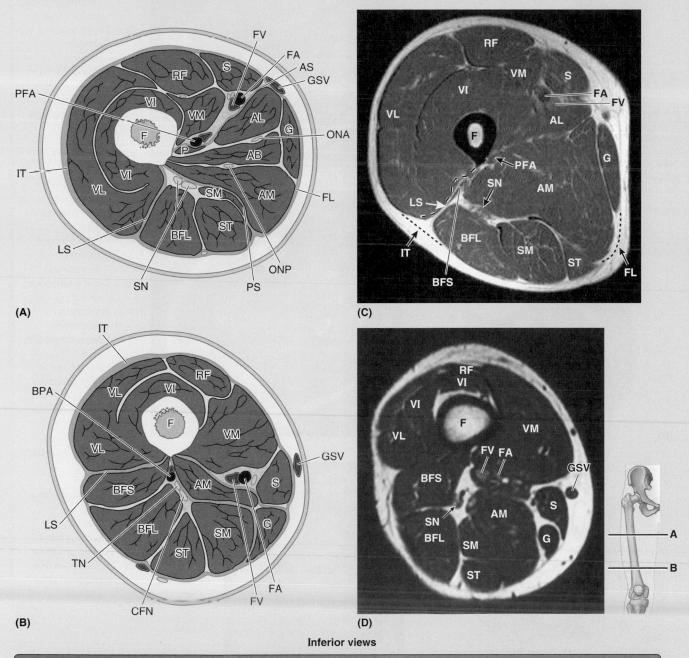

Inferior views

Key					
AB	Adductor brevis	FL	Fascia lata	RF	Rectus femoris
AL	Adductor longus	FV	Femoral vein	S	Sartorius
AM	Adductor magnus	G	Gracilis	SM	Semimembranosus
AS	Anteromedial intermuscular septum	GSV	Great saphenous vein	SN	Sciatic nerve
		IT	Iliotibial tract	ST	Semltendinosus
BFL	Long head of biceps femoris	LS	Lateral intermuscular septum	TN	Tibial nerve
BFS	Short head of biceps femoris	ONA	Anterior branch of obturator nerve	VI	Vastus intermedius
BPA	Branch of profunda femoris artery	ONP	Posterior branch of obturator nerve	VL	Vastus lateralis
CFN	Common fibular nerve	P	Pectineus	VM	Vastus medialis
F	Femur	PFA	Profunda femoris artery		
FA	Femoral artery	PS	Posteromedial intermuscular septum		

FIGURE 7.63. Transverse sections (A and B) and magnetic resonance imaging (MRI) scans (C and D) of thigh.

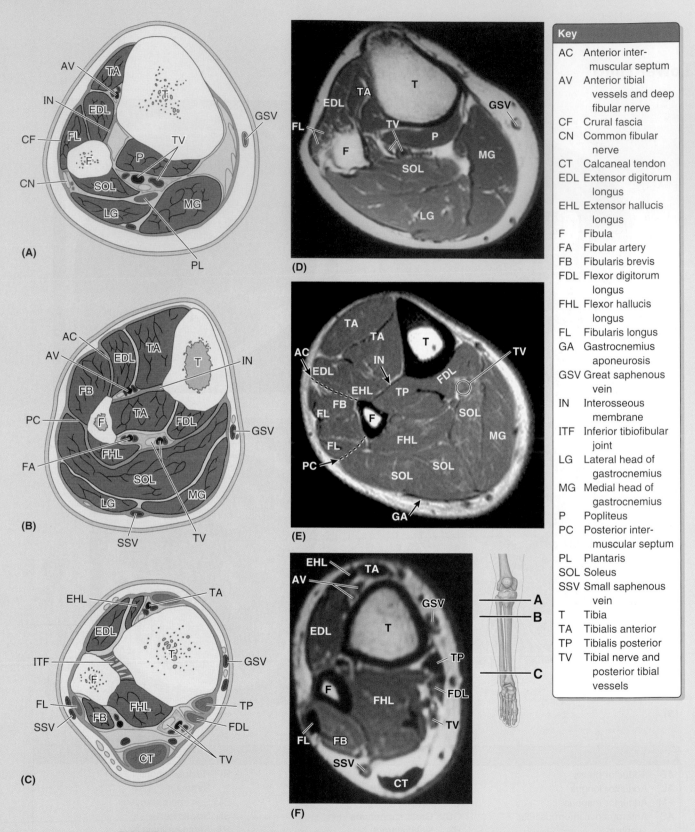

FIGURE 7.64. Transverse sections (A–C) and magnetic resonance imaging (MRI) scans (D–F) of leg.

Key

AC	Anterior inter-muscular septum
AV	Anterior tibial vessels and deep fibular nerve
CF	Crural fascia
CN	Common fibular nerve
CT	Calcaneal tendon
EDL	Extensor digitorum longus
EHL	Extensor hallucis longus
F	Fibula
FA	Fibular artery
FB	Fibularis brevis
FDL	Flexor digitorum longus
FHL	Flexor hallucis longus
FL	Fibularis longus
GA	Gastrocnemius aponeurosis
GSV	Great saphenous vein
IN	Interosseous membrane
ITF	Inferior tibiofibular joint
LG	Lateral head of gastrocnemius
MG	Medial head of gastrocnemius
P	Popliteus
PC	Posterior inter-muscular septum
PL	Plantaris
SOL	Soleus
SSV	Small saphenous vein
T	Tibia
TA	Tibialis anterior
TP	Tibialis posterior
TV	Tibial nerve and posterior tibial vessels

Head

CLINICAL BOX KEY

Anatomical
Variations

Diagnostic
Procedures

Life Cycle

Surgical
Procedures

Trauma

Pathology

The **head** consists of the *brain*, its protective coverings, and the *ears* and *face*. The **cranium** (skull) is the skeleton of the head (Fig. 8.1). Learning the features of the cranium provides an important framework to facilitate the understanding of the head region.

In the *anatomical position*, the cranium is oriented so that the inferior margin of the orbit (orbital cavity) and the superior margin of the external acoustic meatus of both sides lie in the same horizontal plane (Fig. 8.1B). This standard craniometric reference is the **orbitomeatal plane** (Frankfort horizontal plane).

CRANIUM

The *cranium* consists of two parts, structurally and functionally: the neurocranium and viscerocranium (Fig. 8.1). The **neurocranium** (cranial vault) is the bony case of the brain and its membranous coverings, the cranial meninges. It also contains the proximal parts of the cranial nerves (CNs) and the vasculature of the brain. The neurocranium has a dome-like roof, the **calvaria** (skullcap), and a floor or **cranial base** (basicranium). The neurocranium is formed by eight bones: four singular bones centered on the midline (*frontal, ethmoid, sphenoid,* and *occipital*) and two sets of bones occurring as bilateral pairs (*temporal* and *parietal*). Most calvarial bones are united by fibrous interlocking *sutures*; however, during childhood, some bones (sphenoid and occipital) are united by hyaline cartilage (*synchondroses*).

The **viscerocranium** (facial skeleton) is made up of the facial bones that mainly develop in the mesenchyme of the embryonic pharyngeal arches (Moore et al., 2016). The viscerocranium forms the anterior part of the cranium and consists of bones surrounding the mouth, nose, and most of the orbits (Fig. 8.1A). Fifteen irregular bones form the cranial base: three singular bones lying in the midline (*mandible, ethmoid,* and *vomer*) and six paired bones occurring bilaterally (*maxilla; inferior nasal concha* [turbinate], *zygomatic, palatine, nasal,* and *lacrimal bones*).

Facial Aspect of Cranium

Features of the anterior or **facial (frontal) aspect of the cranium** are the frontal and zygomatic bones, orbits, nasal region, maxillae, and mandible (Fig. 8.1A).

The **frontal bone** forms the skeleton of the forehead, articulating inferiorly with the nasal and zygomatic bones. It also articulates with the lacrimal, ethmoid, and sphenoid bones and forms the roof of the orbit and part of the floor of the anterior part of the cranial cavity. The intersection of the frontal and nasal bones is the **nasion** (L. *nasus*, nose). The **supra-orbital margin** of the frontal bone, the angular boundary between the squamous (flat) and orbital parts, has either a **supra-orbital foramen** or **notch**. Just superior to the supra-orbital margin is a ridge, the **superciliary arch**. In some crania of adults, a remnant of the developmental frontal suture, the **metopic suture**, is visible in the midline of the **glabella**, the smooth area between the superciliary arches.

The **zygomatic bones**, forming the prominences of the cheeks, lie on the inferolateral sides of the orbits and rest on the maxillae (Fig. 8.1A,B). A small **zygomaticofacial foramen** pierces the lateral aspect of each bone. Inferior to the nasal bones is the **piriform** (pear-shaped) **aperture**, the anterior nasal opening of the cranium. The bony **nasal septum** can be observed, dividing the nasal cavity into right and left parts. On the lateral wall of each nasal cavity are curved bony plates, the **nasal conchae** (the middle and inferior nasal conchae are shown in Fig. 8.1A).

The **maxillae** form the upper jaw and are united at the **intermaxillary suture** in the median plane. Their **alveolar processes** include the tooth sockets (alveoli) and constitute the supporting bone for the **maxillary teeth**. The maxillae surround most of the piriform aperture and form the infra-orbital margins medially. They have a broad connection with the zygomatic bones laterally and have an **infra-orbital foramen** inferior to each orbit.

The **mandible** is the U-shaped bone forming the lower jaw; it has an **alveolar part** that supports the **mandibular teeth**. It consists of a horizontal part, the **body**, and a vertical part, the **ramus**. Inferior to the second premolar teeth are **mental foramina** (Fig. 8.1B). Forming the prominence of the chin is the **mental protuberance**, a triangular elevation of bone inferior to the **mandibular symphysis**, the region where the halves of the infantile mandible fuse (Fig. 8.1A,B).

The bones of the orbit are illustrated and described later (see Fig. 8.22). Openings within the orbits are the **superior and inferior orbital fissures** and **optic canals**.

Lateral Aspect of Cranium

The **lateral aspect of the cranium** is formed by both the neurocranium and viscerocranium (Fig. 8.1B). The main features of the neurocranial part are the *temporal fossa*, which is bounded superiorly and posteriorly by **superior** and **inferior temporal lines**, anteriorly by the frontal and zygomatic bones, and inferiorly by the **zygomatic arch** that is formed by the union of the **temporal process of the zygomatic bone** and the **zygomatic process of the temporal bone**. The *infratemporal fossa* is an irregular space inferior and deep to the zygomatic arch and the mandible and posterior to the maxilla.

In the anterior part of the temporal fossa, superior to the midpoint of the zygomatic arch, is the **pterion** (G. *pteron*, wing). It is usually indicated by a roughly H-shaped formation of sutures that unite the frontal, parietal, sphenoid (greater wing), and temporal bones.

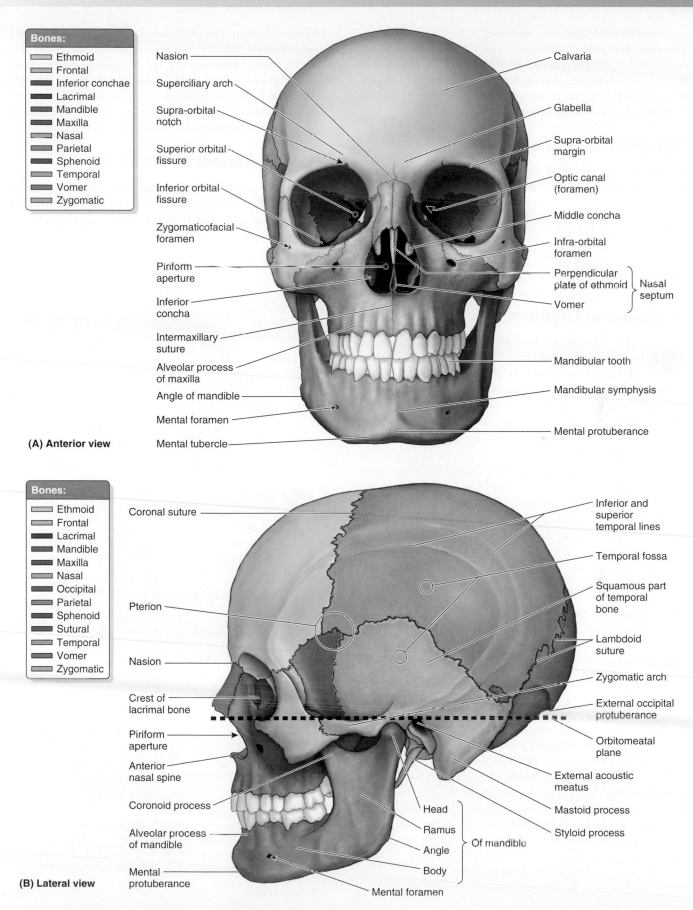

Bones:
- Ethmoid
- Frontal
- Inferior conchae
- Lacrimal
- Mandible
- Maxilla
- Nasal
- Parietal
- Sphenoid
- Temporal
- Vomer
- Zygomatic

Nasion
Superciliary arch
Supra-orbital notch
Superior orbital fissure
Inferior orbital fissure
Zygomaticofacial foramen
Piriform aperture
Inferior concha
Intermaxillary suture
Alveolar process of maxilla
Angle of mandible
Mental foramen
Mental tubercle

Calvaria
Glabella
Supra-orbital margin
Optic canal (foramen)
Middle concha
Infra-orbital foramen
Perpendicular plate of ethmoid } Nasal septum
Vomer
Mandibular tooth
Mandibular symphysis
Mental protuberance

(A) Anterior view

Bones:
- Ethmoid
- Frontal
- Lacrimal
- Mandible
- Maxilla
- Nasal
- Occipital
- Parietal
- Sphenoid
- Sutural
- Temporal
- Vomer
- Zygomatic

Coronal suture
Pterion
Nasion
Crest of lacrimal bone
Piriform aperture
Anterior nasal spine
Coronoid process
Alveolar process of mandible
Mental protuberance

Inferior and superior temporal lines
Temporal fossa
Squamous part of temporal bone
Lambdoid suture
Zygomatic arch
External occipital protuberance
Orbitomeatal plane
External acoustic meatus
Mastoid process
Styloid process

Head
Ramus
Angle   } Of mandible
Body

Mental foramen

(B) Lateral view

FIGURE 8.1. A and B. Adult cranium (skull). In **B**, the pterion is the area of junction of four bones within the temporal fossa.

The **external acoustic opening** is the entrance to the **external acoustic meatus** (ear canal), which leads to the tympanic membrane (eardrum). The **mastoid process** of the temporal bone lies postero-inferior to the external acoustic meatus (Fig. 8.1B). Anteromedial to the mastoid process is the slender **styloid process** of the temporal bone.

Occipital Aspect of Cranium

The posterior or **occipital aspect of the cranium** is formed by the rounded posterior aspect of the head or **occiput** (L. back of head; Fig. 8.2A). The occipital bone, parts of the parietal bones, and mastoid parts of the temporal bones form this

part of the cranium. The **external occipital protuberance** is usually an easily palpable elevation in the median plane. The **superior nuchal line**, marking the superior limit of the neck, extends laterally from each side of this protuberance; the **inferior nuchal line** is less distinct. In the center of the occiput, the **lambda** indicates the junction of the sagittal and lambdoid sutures. The lambda can sometimes be felt as a depression.

Superior Aspect of Cranium

The **superior aspect of the cranium**, usually somewhat oval in form, broadens posterolaterally at the **parietal eminences** (Fig. 8.2B). The four bones forming the *calvaria*,

CLINICAL BOX

Fractures of Cranium

The convexity of the calvaria (skullcap) distributes and thereby minimizes the effects of a blow to it. However, hard blows to the head in thin areas are likely to produce *depressed fractures* in which a fragment of bone is depressed inward, compressing and/or injuring the brain (Fig. B8.1A). In *comminuted fractures*, the bone is broken into several pieces. *Linear calvarial fractures*, the most frequent type, usually occur at the point of impact, but fracture lines often radiate away from it in two or more directions. If the area of the calvaria is thick at the site of impact, the bone usually bends inward without fracturing; however, a fracture may occur some distance from the site of direct trauma where the calvaria is thinner. In a *contre-coup (counterblow) fracture*, the fracture occurs on the opposite side of the cranium rather than at the point of impact.

Basilar fractures involve the bones forming the cranial base (e.g., occipital bone around the foramen magnum, temporal and/or sphenoid bones, or the roof of the orbit). As a result of the fracture, cerebrospinal fluid (CSF) may leak into the nose (CSF rhinorrhea) and ear (CSF otorrhea), and CN and blood vessel injury may occur, depending on the site of the fracture.

Fracture of the pterion can be life-threatening because it overlies the frontal (anterior) branches of the middle meningeal vessels, which lie in grooves on the internal aspect of the lateral wall of the calvaria (Fig. B8.1B). A hard blow to the side of the head may fracture the thin bones forming the pterion, rupturing the frontal branches deep to the pterion. The resulting *epidural hematoma* exerts pressure on the underlying cerebral cortex. Untreated *middle meningeal artery hemorrhage* may cause death in a few hours.

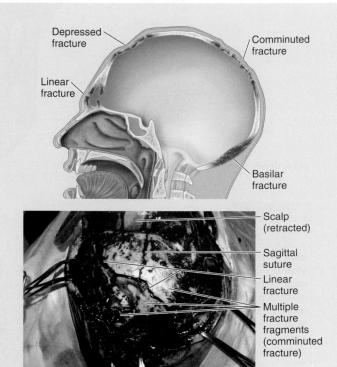

(A) Posterosuperior view

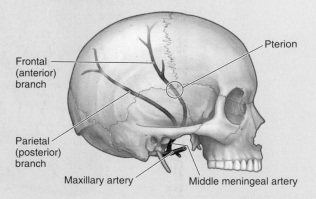

(B) Lateral view

FIGURE B8.1. Fractures of cranium.

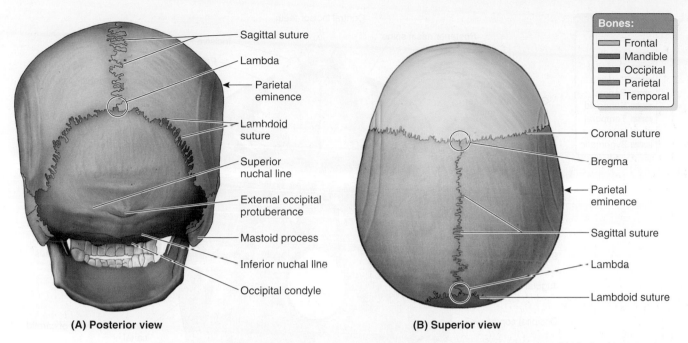

FIGURE 8.2. Adult cranium (skull). A. Occiput. **B.** Features of calvaria (skullcap).

the dome-like roof of the neurocranium, are visible from this aspect: the frontal bone anteriorly, the right and left parietal bones laterally, and the occipital bone posteriorly. The **coronal suture** unites the frontal and parietal bones, the **sagittal suture** unites the right and left parietal bones, and the **lambdoid suture** unites the occipital bone with the right and left parietal and temporal bones. The **bregma** is the landmark formed by the intersection of the sagittal and coronal sutures. The **vertex**, the superiormost point of the cranium, is near the midpoint of the sagittal suture (Fig 8.2A).

External Surface of Cranial Base

The **external aspect of the cranial base** (basicranium) features the **alveolar arch of the maxillae** (the free border of the alveolar processes surrounding and supporting the maxillary teeth); the **palatine processes** of the maxillae; and the palatine, sphenoid, vomer, temporal, and occipital bones (Fig. 8.3A). The *hard palate* (bony palate) is formed by the **palatine processes of the maxillae** anteriorly and the **horizontal plates of the palatine bones** posteriorly. Posterior to the central incisor teeth is the *incisive fossa*. Posterolaterally are the **greater** and **lesser palatine foramina**. The posterior edge of the palate forms the inferior boundary of the **choanae** (posterior nasal apertures), which are separated from each other by the **vomer**. The vomer is a thin, flat bone that forms a part of the bony nasal septum (Fig. 8.1A). Wedged between the frontal, temporal,

and occipital bones is the **sphenoid bone**, which consists of a body and three pairs of processes: the **greater** and **lesser wings** and the pterygoid processes (Fig. 8.3A,D). The **pterygoid processes**, consisting of **medial** and **lateral pterygoid plates**, extend inferiorly on each side of the sphenoid from the junction of the body and greater wings (Fig. 8.3A). The opening of the bony part of the pharyngotympanic (auditory) tube and the *sulcus (groove) for the cartilaginous part of the tube* lies medial to the **spine of the sphenoid**, inferior to the junction of the greater wing of the sphenoid, and the **petrous** (L. rock-like) part of the temporal bone. Depressions in the **squamous** (L. flat) **part of the temporal bone**, called the **mandibular fossae**, accommodate the heads of the mandible when the mouth is closed.

The cranial base is formed posteriorly by the **occipital bone**, which articulates with the sphenoid anteriorly. The parts of the occipital bone encircle the large **foramen magnum**. On each side of the foramen are two large protuberances, the **occipital condyles**, by which the cranium articulates with the vertebral column (Fig. 8.3A). The large fissure between the occipital bone and the petrous part of the temporal bone is the **jugular foramen**. The internal carotid artery enters the carotid canal at the **external opening of the carotid canal** just anterior to the jugular foramen. The palpable **mastoid processes** provide for muscle attachments. The **stylomastoid foramen** lies between the mastoid and styloid processes.

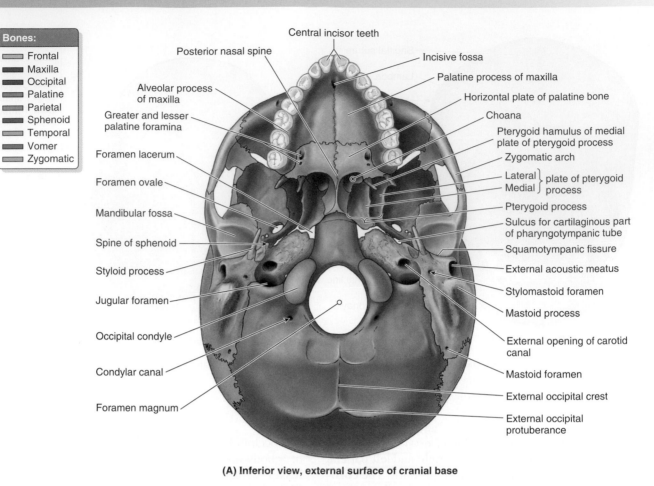

Bones:

- Frontal
- Maxilla
- Occipital
- Palatine
- Parietal
- Sphenoid
- Temporal
- Vomer
- Zygomatic

(A) Inferior view, external surface of cranial base

FIGURE 8.3. Cranial base. A. Features of external surface. *(continued)*

TABLE 8.1. FORAMINA/APERTURES OF CRANIAL FOSSAE AND CONTENTS

Foramina/Apertures	Contents
Anterior cranial fossa	
Foramen cecum	Nasal emissary vein (1% of population; in danger of injury during surgery)
Cribriform foramina in cribriform plate	Axons of olfactory cells in olfactory epithelium that form olfactory nerves (CN I)
Anterior and posterior ethmoidal foramina	Vessels and nerves with same names as foramina
Middle cranial fossa	
Optic canals	Optic nerves (CN II) and ophthalmic arteries
Superior orbital fissure	Ophthalmic veins; ophthalmic nerve (CN V_1); CN III, IV, and VI; and sympathetic fibers
Foramen rotundum	Maxillary nerve (CN V_2)
Foramen ovale	Mandibular nerve (CN V_3) and accessory meningeal artery
Foramen spinosum	Middle meningeal artery and vein and meningeal branch of CN V_3
Foramen lacerum[a]	Internal carotid artery and its accompanying sympathetic and venous plexuses
Groove or hiatus of greater petrosal nerve	Greater petrosal nerve and petrosal branch of middle meningeal artery
Posterior cranial fossa	
Foramen magnum	Medulla and meninges, vertebral arteries, CN XI, dural veins, anterior and posterior spinal arteries
Jugular foramen	CN IX, X, and XI; superior bulb of internal jugular vein; inferior petrosal and sigmoid sinuses; and meningeal branches of ascending pharyngeal and occipital arteries
Hypoglossal canal	Hypoglossal nerve (CN XII)
Condylar canal	Emissary vein that passes from sigmoid sinus to vertebral veins in neck
Mastoid foramen	Mastoid emissary vein from sigmoid sinus and meningeal branch of occipital artery

[a]Structures actually pass horizontally across (rather than vertically through) the area of the foramen lacerum, an artifact of dry skulls, which is closed by cartilage in life.

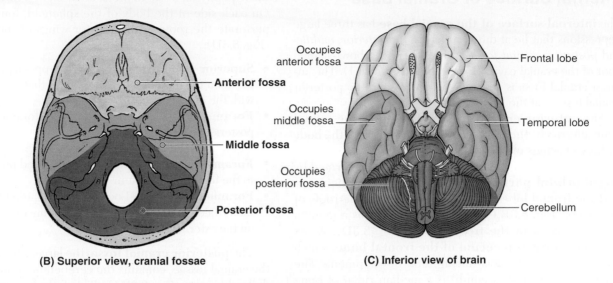

(B) Superior view, cranial fossae

Occupies anterior fossa
Occupies middle fossa
Occupies posterior fossa

Frontal lobe
Temporal lobe
Cerebellum

(C) Inferior view of brain

Anterior fossa
Middle fossa
Posterior fossa

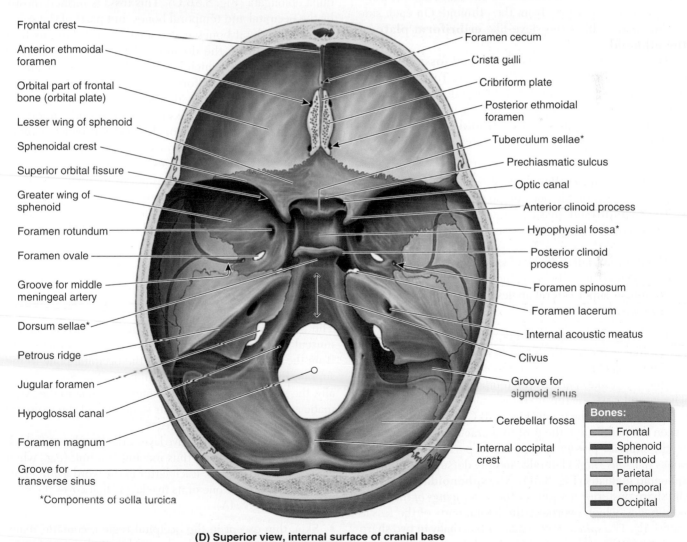

Frontal crest
Anterior ethmoidal foramen
Orbital part of frontal bone (orbital plate)
Lesser wing of sphenoid
Sphenoidal crest
Superior orbital fissure
Greater wing of sphenoid
Foramen rotundum
Foramen ovale
Groove for middle meningeal artery
Dorsum sellae*
Petrous ridge
Jugular foramen
Hypoglossal canal
Foramen magnum
Groove for transverse sinus

Foramen cecum
Crista galli
Cribriform plate
Posterior ethmoidal foramen
Tuberculum sellae*
Prechiasmatic sulcus
Optic canal
Anterior clinoid process
Hypophysial fossa*
Posterior clinoid process
Foramen spinosum
Foramen lacerum
Internal acoustic meatus
Clivus
Groove for sigmoid sinus
Cerebellar fossa
Internal occipital crest

*Components of sella turcica

Bones:
Frontal
Sphenoid
Ethmoid
Parietal
Temporal
Occipital

(D) Superior view, internal surface of cranial base

FIGURE 8.3. Cranial base. *(continued)* **B.** Cranial fossae of internal surface of cranial base. **C.** Lobes and cerebellum of brain related to cranial fossae. **D.** Features of internal surface.

Internal Surface of Cranial Base

The **internal surface of the cranial base** has three large depressions that lie at different levels: the *anterior*, *middle*, and *posterior cranial fossae*, which form the bowl-shaped floor of the cranial cavity (Fig. 8.3B and Table 8.1). The anterior cranial fossa is at the highest level, and the posterior cranial fossa is at the lowest level.

The **anterior cranial fossa** is formed by the frontal bone anteriorly, the ethmoid bone centrally, and the body and lesser wings of the sphenoid posteriorly (Fig. 8.3D). The greater part of the anterior cranial fossa is formed by ridged **orbital parts of the frontal bone**, which support the frontal lobes of the brain and form the roofs of the orbits (Fig. 8.3B,C). The **frontal crest** is a median bony extension of the frontal bone (Fig. 8.3D). At its base is the **foramen cecum of the frontal bone**, which gives passage to vessels during fetal development. The **crista galli** (L. cock's comb) is a median ridge of bone that projects superiorly from the ethmoid. On each side of the crista galli is the sieve-like **cribriform plate of the ethmoid**.

The butterfly-shaped **middle cranial fossa** has a *central part* composed of the *sella turcica* (Turkish saddle) on the body of the sphenoid and large depressed *lateral parts* on each side. The **sella turcica** is surrounded by the **anterior** and **posterior clinoid processes** (*clinoid* means "bedpost"). The sella turcica is composed of three parts:

- The **tuberculum sellae** (horn of saddle), the slight elevation anteriorly on the body of the sphenoid
- The **hypophysial fossa** (pituitary fossa), a saddle-like depression for the pituitary gland (L. *hypophysis*) in the middle
- The **dorsum sellae** (back of saddle) posteriorly, formed by a square plate of bone on the body of the sphenoid. Its prominent superolateral angles are the *posterior clinoid processes*.

The bones forming the larger, lateral parts of the middle cranial fossa are the greater wings of the sphenoid, squamous (flat) parts of the temporal bones laterally, and petrous (rock-like) parts of the temporal bones posteriorly. The lateral parts of the middle cranial fossa support the temporal lobes of the brain (Fig. 8.3B,C). The boundary between the middle and the posterior cranial fossae is formed by the *superior border of the petrous part of the temporal bones (petrous ridge)* laterally and the dorsum sellae of the sphenoid medially (Fig. 8.3D). The **sphenoidal crests** are the sharp posterior margins of the *lesser wings of the sphenoid bones*, which overhang the lateral parts of the fossae anteriorly. The *sphenoidal crests* end medially in two sharp bony projections: the *anterior clinoid processes*. The **prechiasmatic sulcus** extends between the right and the left optic canals. The **foramen lacerum** lies posterolateral to

the hypophysial fossa. In life, it is closed by a cartilage plate. On each side of the body of the sphenoid, four foramina perforate the roots of the greater wings of the sphenoid (Fig. 8.3D):

- **Superior orbital fissure**: A teardrop-shaped opening between the greater and lesser wings that communicates with the orbit
- **Foramen rotundum**: A circular foramen located posterior to the larger medial end of the superior orbital fissure
- **Foramen ovale**: An oval foramen located posterolateral to the foramen rotundum
- **Foramen spinosum**: Located posterolateral to the foramen ovale, opening anterior to the spine of the sphenoid on the external surface (Fig. 8.3A)

The **posterior cranial fossa**, the largest and deepest of the cranial fossae, contains the cerebellum, pons, and medulla oblongata (Fig. 8.3B,C). This fossa is formed mostly by the occipital and temporal bones, but parts of the sphenoid and parietal bones make smaller contributions to it (Fig. 8.3D). From the dorsum sellae, there is a marked incline, the **clivus**, which leads to the foramen magnum. Posterior to this large foramen, the **internal occipital crest** is a landmark that divides the posterior part of the fossae into two **cerebellar fossae**; the crest ends superiorly in the **internal occipital protuberance**. Broad grooves in this fossa are formed by the *transverse* and *sigmoid sinuses*. At the base of the petrous ridges of the temporal bones are the **jugular foramina**. Anterosuperior to the jugular foramen is the **internal acoustic meatus**. The **hypoglossal canals** lie superior to the anterolateral margin of the foramen magnum, passing through the bases of the occipital condyles.

SCALP

The **scalp** consists of skin, subcutaneous tissue, and a musculo-aponeurotic layer that cover the neurocranium from the superior nuchal lines on the occipital bone to the supra-orbital margins of the frontal bone (Fig. 8.1A). Laterally, the scalp extends over the temporal fascia to the zygomatic arches. The neurovascular structures of the scalp are discussed with those of the face.

The scalp is composed of five layers, the first three of which are connected intimately, thus moving as a unit (e.g., when wrinkling the forehead). Each letter in the word *scalp* serves as a memory key for one of its five layers that cover the neurocranium (Fig. 8.4A):

- **S**kin, thin except in the occipital region, contains many sweat and sebaceous glands and hair follicles; it has an abundant arterial supply and good venous and lymphatic drainage.

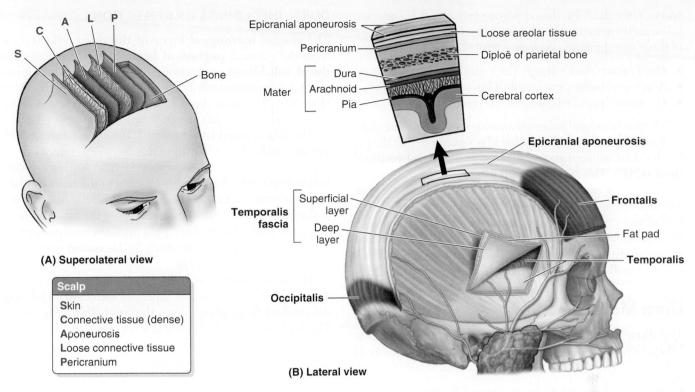

(A) Superolateral view

Scalp
Skin
Connective tissue (dense)
Aponeurosis
Loose connective tissue
Pericranium

(B) Lateral view

FIGURE 8.4. Scalp. A. Layers of scalp. **B.** Epicranial aponeurosis and layers of scalp, cranium, and meninges.

- Connective tissue, forming the thick, dense, richly vascularized subcutaneous layer, is well supplied with cutaneous nerves.
- Aponeurosis (**epicranial aponeurosis**), a strong tendinous sheet that covers the calvaria, serves as the broad intermediate tendon of the frontal and occipital bellies of the occipitofrontalis muscle and the superior auricular muscle (Fig. 8.4B); collectively, these structures form the musculo-aponeurotic *epicranius*.
- Loose connective tissue, a sponge-like layer, has potential spaces that may distend with fluid as a result of injury or infection (Fig. 8.4A); this layer allows free movement of the **scalp proper** (the first three layers) over the underlying calvaria.
- Pericranium, a dense layer of connective tissue, forms the external periosteum of the neurocranium; it is firmly attached but can be stripped fairly easily from the calvaria of living people, except where the pericranium is continuous with the fibrous tissue uniting the cranial sutures.

CRANIAL MENINGES

The **cranial meninges** are coverings of the brain that lie immediately internal to the cranium. The meninges protect and enclose the brain in a fluid-filled cavity, the subarachnoid

CLINICAL BOX

Scalp Injuries and Infections

The loose connective tissue layer is the *danger area of the scalp* because pus or blood spreads easily in it. Infection in this layer can also pass into the cranial cavity through emissary veins, which pass through the calvaria and reach intracranial structures such as the meninges. An infection cannot pass into the neck because the occipital belly of the occipitofrontalis muscle attaches to the occipital bone and mastoid parts of the temporal bones. Neither can the infection spread laterally beyond the zygomatic arches because the epicranial aponeurosis is continuous with the temporal fascia that attaches to these arches. An infection or fluid (e.g., pus or blood) can enter the eyelids and the root of the nose because the frontal belly of the occipitofrontalis muscle inserts into the skin and subcutaneous tissue and does not attach to the bone. Consequently, "black eyes" can result from an injury to the scalp or forehead. *Ecchymoses*, or purple patches, develop as a result of extravasation of blood into the subcutaneous tissue and skin of the eyelids and surrounding regions.

space. They also form the supporting framework for arteries, veins, and venous sinuses. The cranial meninges are composed of three membranous connective tissue layers (Fig. 8.5):

- *Dura mater* (dura): Tough, thick external fibrous layer
- *Arachnoid mater* (arachnoid): Thin intermediate layer
- *Pia mater* (pia): Delicate internal vascular layer

The arachnoid and pia are continuous membranes that make up the **leptomeninx**. The arachnoid is separated from the pia by the subarachnoid space, which contains **cerebrospinal fluid (CSF)**. This is a clear liquid similar in constitution to blood; it provides nutrients but has less protein and a different ion concentration. CSF is formed predominantly by the *choroid plexuses* within the four ventricles of the brain. CSF leaves the ventricular system of the brain and enters the subarachnoid space, where it cushions and nourishes the brain and presses the arachnoid to the inner surface of the dura (see Fig. 8.12).

Dura Mater

The **dura mater** (dura), a two-layered membrane that is adherent to the internal surface of the cranium, consists of (Figs. 8.5 and 8.6)

- an external *periosteal layer*, formed by the periosteum covering the internal surface of the calvaria
- an internal *meningeal layer*, a strong fibrous membrane that is continuous at the foramen magnum with the dura covering the spinal cord

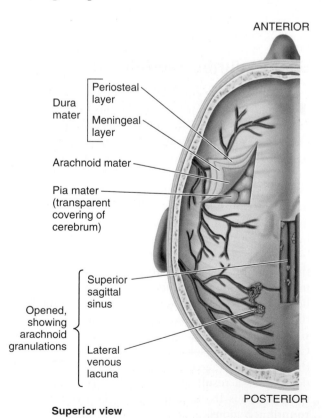

FIGURE 8.5. Cranial meninges.

DURAL INFOLDINGS OR REFLECTIONS

The internal **meningeal layer of the dura** reflects away from the external **periosteal layer of the dura** to form **dural infoldings** (reflections), which divide the cranial cavity into compartments and support parts of the brain (Fig. 8.6). The four dural infoldings are the *falx cerebri*, *tentorium cerebelli*, *falx cerebelli*, and *diaphragma sellae*.

The **falx cerebri** (cerebral falx), the largest dural infolding, is a sickle-shaped partition that lies in the **longitudinal cerebral fissure**, which separates the right and left cerebral hemispheres. The falx cerebri attaches in the median plane to the internal surface of the calvaria from the *frontal crest* of the frontal bone and the crista galli of the ethmoid bone anteriorly to the internal occipital protuberance posteriorly. The falx cerebri ends posteriorly by becoming continuous with the tentorium cerebelli.

The **tentorium cerebelli** (cerebellar tentorium) is a wide crescentic septum that separates the occipital lobes of the cerebral hemispheres from the cerebellum (Fig. 8.6A). The tentorium cerebelli attaches anteriorly to the clinoid processes of the sphenoid bone, anterolaterally to the petrous part of the temporal bone, and posterolaterally to the internal surface of the occipital bone and part of the parietal bone. The falx cerebri attaches to the tentorium cerebelli in the midline and holds it up, giving it a tent-like appearance (L. *tentorium*, tent). The concave anteromedial border of the tentorium cerebelli is free, leaving a gap called the **tentorial notch** through which the brainstem extends from the posterior into the middle cranial fossa. The tentorium cerebelli divides the cranial cavity into *supratentorial* and *infratentorial compartments* (Fig. 8.7).

The **falx cerebelli** (cerebellar falx) is a vertical dural infolding that lies inferior to the tentorium cerebelli in the posterior part of the posterior cranial fossa (Fig. 8.7A). It partially separates the cerebellar hemispheres.

The **diaphragma sellae** (sellar diaphragm), the smallest dural infolding, is a circular extension of dura that is suspended between the clinoid processes, forming a partial roof over the hypophysial fossa. The diaphragma sellae covers the pituitary gland in this fossa and has an aperture for passage of the infundibulum (pituitary stalk) and hypophysial veins (Figs. 8.7B and 8.8).

DURAL VENOUS SINUSES

The **dural venous sinuses** are endothelial-lined spaces between the periosteal and meningeal layers of the dura (Fig. 8.6A). They largely form along attachments of dural infoldings and centrally on the cranial base. Large veins from the surface of the brain and from the diploë empty into these sinuses, and most of the blood from the brain and diploë ultimately drains through them into the internal jugular veins (IJVs).

The **superior sagittal sinus** lies in the convex attached (superior) border of the falx cerebri (Fig. 8.6A,B). It begins at the crista galli and ends near the internal occipital protuberance at the **confluence of sinuses**. The superior sagittal

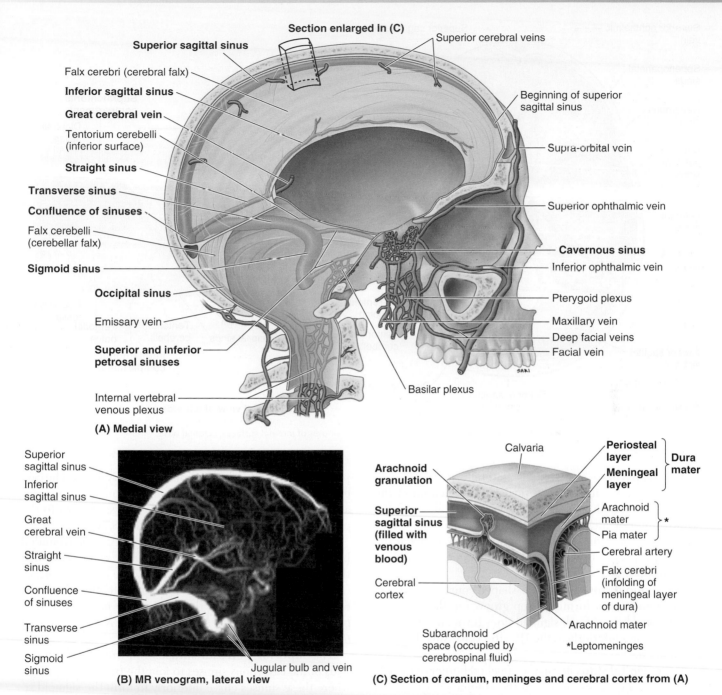

Section enlarged In (C) — Superior cerebral veins

Superior sagittal sinus
Falx cerebri (cerebral falx)
Inferior sagittal sinus
Great cerebral vein
Tentorium cerebelli (inferior surface)
Straight sinus
Transverse sinus
Confluence of sinuses
Falx cerebelli (cerebellar falx)
Sigmoid sinus
Occipital sinus
Emissary vein
Superior and inferior petrosal sinuses
Internal vertebral venous plexus

(A) Medial view

Beginning of superior sagittal sinus
Supra-orbital vein
Superior ophthalmic vein
Cavernous sinus
Inferior ophthalmic vein
Pterygoid plexus
Maxillary vein
Deep facial veins
Facial vein
Basilar plexus

Superior sagittal sinus
Inferior sagittal sinus
Great cerebral vein
Straight sinus
Confluence of sinuses
Transverse sinus
Sigmoid sinus
Jugular bulb and vein
(B) MR venogram, lateral view

Calvaria
Arachnoid granulation
Superior sagittal sinus (filled with venous blood)
Cerebral cortex
Subarachnoid space (occupied by cerebrospinal fluid)
Periosteal layer / **Meningeal layer** — **Dura mater**
Arachnoid mater / Pia mater — *
Cerebral artery
Falx cerebri (infolding of meningeal layer of dura)
Arachnoid mater
*Leptomeninges
(C) Section of cranium, meninges and cerebral cortex from (A)

FIGURE 8.6. Dural infoldings (reflections) and dural venous sinuses. A. Venous sinuses of the dura mater and their communications are shown. **B.** Maximum intensity projection (MIP) displays magnetic resonance (MR) venogram of dural venous sinuses and cerebral veins. **C.** Relationship of dural venous sinuses and meninges.

sinus receives the superior cerebral veins and communicates on each side through slit-like openings with the **lateral venous lacunae**, lateral expansions of the superior sagittal sinus (Fig. 8.5).

Arachnoid granulations (collections of arachnoid villi) are tufted prolongations of the arachnoid that protrude through the meningeal layer of the dura mater into the dural venous sinuses and lateral venous lacunae. The arachnoid granulations transfer CSF to the venous system (Figs. 8.5 and 8.6C).

The **inferior sagittal sinus**, much smaller than the superior sagittal sinus, runs in the inferior, free concave border of the falx cerebri and ends in the straight sinus (Figs. 8.6A,B and 8.7B).

The **straight sinus** is formed by the union of the inferior sagittal sinus with the great cerebral vein. It runs inferoposteriorly along the line of attachment of the falx cerebri to the tentorium cerebelli to join the *confluence of sinuses* (Fig. 8.7A,B).

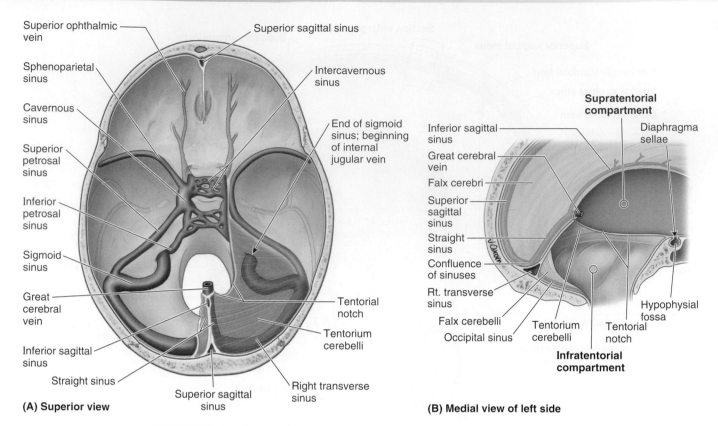

(A) Superior view

(B) Medial view of left side

FIGURE 8.7. Dural venous sinuses. Dural venous sinuses of internal surface of cranial base.

The **transverse sinuses** pass laterally from the *confluence of sinuses* in the posterior attached margin of the tentorium cerebelli, grooving the occipital bones and the postero-inferior angles of the parietal bones (Fig. 8.7A). The transverse sinuses leave the tentorium cerebelli at the posterior aspect of the petrous temporal bone and become sigmoid sinuses.

The **sigmoid sinuses** follow S-shaped courses in the posterior cranial fossa, forming deep grooves in the temporal and occipital bones. Each sigmoid sinus turns anteriorly and then continues inferiorly as the IJV after traversing the jugular foramen.

The **occipital sinus** lies in the attached border of the falx cerebelli and ends superiorly in the confluence of sinuses (Fig. 8.7B). The occipital sinus communicates inferiorly with the internal vertebral venous plexus.

The **cavernous sinus** is located bilaterally on each side of the sella turcica on the body of the sphenoid bone (Figs. 8.6A and 8.7A). The cavernous sinus consists of a venous plexus of thin-walled veins that extend from the superior orbital fissure anteriorly to the apex of the petrous part of the temporal bone posteriorly. The cavernous sinus receives blood from the superior and inferior ophthalmic veins, superficial middle cerebral vein, and sphenoparietal sinus. The venous channels in the cavernous sinuses communicate with each other through **intercavernous sinuses** anterior and

posterior to the infundibulum of the pituitary gland. The cavernous sinuses drain postero-inferiorly through the *superior* and *inferior petrosal sinuses* and via emissary veins to the *pterygoid venous plexuses* (Figs. 8.6A and 8.7B).

The **internal carotid artery** (Fig. 8.8A,B), surrounded by the *carotid plexus of sympathetic nerves*, courses through the cavernous sinus and is crossed by the *abducent nerve* (*CN VI*). From superior to inferior, the lateral wall of each cavernous sinus contains the *oculomotor nerve* (*CN III*), *trochlear nerve* (*CN IV*), and *CN V_1 and CN V_2 divisions of the trigeminal nerve.*

The **superior petrosal sinuses** run from the posterior ends of the cavernous sinuses to join the transverse sinuses, where these sinuses curve inferiorly to form the sigmoid sinuses (Fig. 8.7A). Each superior petrosal sinus lies in the anterolateral attached margin of the tentorium cerebelli, which attaches to the superior border of the petrous part of the temporal bone.

The **inferior petrosal sinuses** commence at the posterior end of the cavernous sinus and drain the cavernous sinuses directly into the origins of the IJVs. The *basilar plexus* connects the inferior petrosal sinuses and communicates inferiorly with the internal vertebral venous plexus (Fig. 8.6A). **Emissary veins** connect the dural venous sinuses with veins outside the cranium (Fig. 8.6A). The size and number of emissary veins vary.

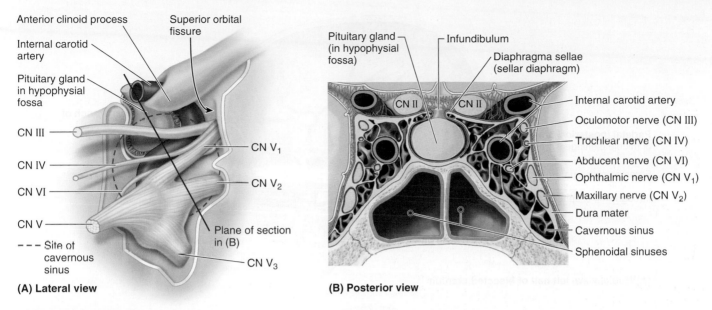

FIGURE 8.8. Cavernous sinus. A. Relationships of the oculomotor, trochlear, trigeminal, and abducent nerves to the internal carotid artery. **B.** Coronal section through cavernous sinus.

CLINICAL BOX

Occlusion of Cerebral Veins and Dural Venous Sinuses

Occlusion of cerebral veins and *dural venous sinuses* may result from thrombi (clots), thrombophlebitis (venous inflammation), or tumors. The facial veins make clinically important connections with the cavernous sinus through the superior ophthalmic veins (Fig. 8.6A). Blood from the medial angle of the eye, nose, and lips usually drains inferiorly into the facial vein. However, because the facial vein has no valves, blood may pass superiorly to the superior ophthalmic vein and enter the cavernous sinus. In people with *thrombophlebitis of the facial vein*, pieces of an infected thrombus may extend into the cavernous sinus, producing *thrombophlebitis of the cavernous sinus*.

Metastasis of Tumor Cells to Dural Sinuses

The basilar and occipital sinuses communicate through the foramen magnum with the internal vertebral venous plexuses (Fig. 8.6D). Because these venous channels are valveless, compression of the thorax, abdomen, or pelvis, as occurs during heavy coughing and straining, may force venous blood from these regions into the internal vertebral venous system and subsequently into the dural venous sinuses. As a result, pus in abscesses and tumor cells in these regions may spread (*metastasize*) to the vertebrae and brain.

Fractures of Cranial Base

In fractures of the cranial base, the internal carotid artery may be torn, producing an *arteriovenous fistula* within the cavernous sinus. Arterial blood rushes into the cavernous sinus, enlarging it and forcing retrograde blood into its venous tributaries, especially the ophthalmic veins. As a result, the eyeball protrudes (*exophthalmos*) and the conjunctiva becomes engorged (*chemosis*). The protruding eyeball pulsates in synchrony with the radial pulse, a phenomenon known as *pulsating exophthalmos*. Because CNs III, IV, V_1, V_2, and VI lie in or close to the lateral wall of the cavernous sinus, they may also be affected when the sinus is injured (Fig. 8.8A,B).

A blow to the head can detach the periosteal layer of dura from the calvaria without fracturing the cranial bones. However, in the cranial base, the two dural layers are firmly attached and difficult to separate from the bones. Consequently, a fracture of the cranial base usually tears the dura and results in leakage of CSF.

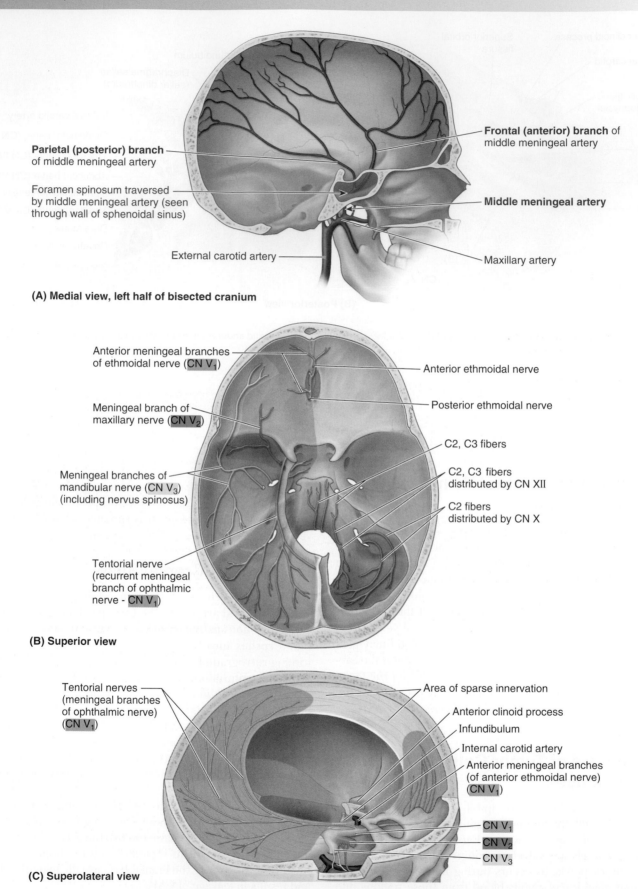

Parietal (posterior) branch
of middle meningeal artery

Foramen spinosum traversed
by middle meningeal artery (seen
through wall of sphenoidal sinus)

External carotid artery

Frontal (anterior) branch of
middle meningeal artery

Middle meningeal artery

Maxillary artery

(A) Medial view, left half of bisected cranium

Anterior meningeal branches
of ethmoidal nerve (CN V₁)

Meningeal branch of
maxillary nerve (CN V₂)

Meningeal branches of
mandibular nerve (CN V₃)
(including nervus spinosus)

Tentorial nerve
(recurrent meningeal
branch of ophthalmic
nerve - CN V₁)

Anterior ethmoidal nerve

Posterior ethmoidal nerve

C2, C3 fibers

C2, C3 fibers
distributed by CN XII

C2 fibers
distributed by CN X

(B) Superior view

Tentorial nerves
(meningeal branches
of ophthalmic nerve)
(CN V₁)

Area of sparse innervation

Anterior clinoid process

Infundibulum

Internal carotid artery

Anterior meningeal branches
(of anterior ethmoidal nerve)
(CN V₁)

CN V₁

CN V₂

CN V₃

(C) Superolateral view

FIGURE 8.9. Middle meningeal artery and innervation of dura mater. A. Middle meningeal artery. **B.** Innervation. The right side of the calvaria and brain is removed, and CN V is dissected. **C.** The internal aspect of the cranial base illustrating the innervation of the dura by cranial and spinal nerves.

VASCULATURE AND NERVE SUPPLY OF DURA MATER

The **arteries of the dura** supply more blood to the calvaria than to the dura. The largest of these vessels, the **middle meningeal artery** (Figs. 8.9A and 8.10A), is a branch of the maxillary artery, a terminal branch of the external carotid artery. The middle meningeal artery enters the middle cranial fossa through the *foramen spinosum*, runs laterally in the fossa, and turns supero-anteriorly on the greater wing of the sphenoid, where it divides into frontal and parietal branches. The **anterior (frontal) branch** runs superiorly to cross the pterion where it sends branches to the anterior calvaria. The **posterior (parietal) branch** runs posterosuperiorly and ramifies over the posterior aspect of the calvaria. The **veins of the dura** accompany the meningeal arteries (Fig. 8.10A).

The **innervation of the dura** is largely by the three divisions of CN V (Fig. 8.9B,C). Sensory branches are also conveyed from the vagus (CN X) and hypoglossal (CN XII) nerves, but the fibers probably are peripheral branches from sensory ganglia of the superior three cervical nerves. The sensory (pain) endings are more numerous in the dura along each side of the superior sagittal sinus and where arteries and veins course in the dura. They are more abundant in the tentorium cerebelli than they are in the floor of the cranium. Pain arising from the dura is generally referred, perceived as a headache arising in cutaneous or mucosal regions supplied by the involved cervical nerve or division of the trigeminal nerve.

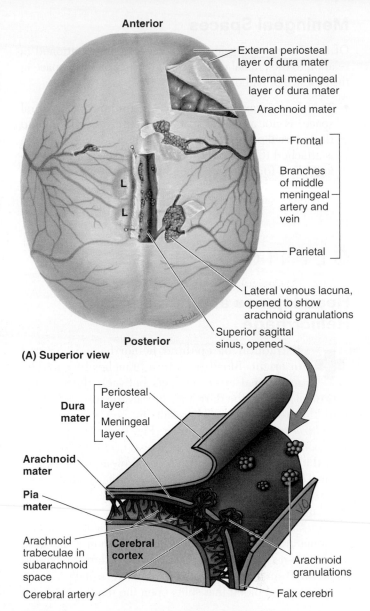

(A) Superior view

Labels (A):
External periosteal layer of dura mater
Internal meningeal layer of dura mater
Arachnoid mater
Frontal
Branches of middle meningeal artery and vein
Parietal
Lateral venous lacuna, opened to show arachnoid granulations
Superior sagittal sinus, opened
Anterior
Posterior

(B) Coronal section, opened superior sagittal sinus

Labels (B):
Dura mater — Periosteal layer, Meningeal layer
Arachnoid mater
Pia mater
Arachnoid trabeculae in subarachnoid space
Cerebral artery
Cerebral cortex
Arachnoid granulations
Falx cerebri

FIGURE 8.10. Layers, formations, and relations of cranial meninges. **A.** Meningeal layers in situ and branches of middle meningeal vessels. **B.** Superior sagittal sinus opened to demonstrate arachnoid granulations.

Arachnoid Mater and Pia Mater

The *arachnoid mater* and *pia mater* (*leptomeninx*) develop from a single layer of mesenchyme surrounding the embryonic brain. CSF-filled spaces form within this layer and coalesce to form the *subarachnoid space* (Fig. 8.10A,B). Web-like **arachnoid trabeculae** pass between the arachnoid and pia. The avascular arachnoid mater, although closely applied to the meningeal layer of the dura, is held against the inner surface of the dura by the pressure of the CSF. The *pia mater* is a thin membrane that is highly vascularized by a network of fine blood vessels and adheres to the surface of the brain and follows its contours (Figs. 8.6B and 8.10). Where cerebral arteries penetrate the cerebral cortex, the pia follows them for a short distance, forming a pial coat and a peri-arterial space.

Meningeal Spaces

Of the three meningeal "spaces" commonly mentioned in relation to the cranial meninges, only one exists as a space in the absence of pathology:

- The **dura–cranium interface** (extradural or epidural space) is not a natural space between the cranium and the external periosteal layer of the dura because the dura is attached to the bones. It becomes a space only pathologically—for example, when blood from torn meningeal vessels pushes the periosteum away from the cranium and accumulates.
- The **dura–arachnoid junction** or interface (subdural space) is likewise not a natural space between the dura and the arachnoid. A space may develop in the dural border cell layer as the result of trauma, such as after a blow to the head (Haines, 2013).
- The **subarachnoid space**, between the arachnoid and pia, is a real space that contains CSF, trabecular cells, cerebral arteries, and bridging superior cerebral veins that drain into the superior sagittal sinus (Fig. 8.10B).

CLINICAL BOX

Head Injuries and Intracranial Hemorrhage

Extradural or epidural hemorrhage is arterial in origin. Blood from torn branches of a middle meningeal artery collects between the external periosteal layer of the dura and the calvaria, usually after a hard blow to the head. This results in the formation of an *extradural* or *epidural hematoma* (Fig. B8.2A). Typically, a brief *concussion* (loss of consciousness) occurs followed by a lucid interval of some hours. Later, drowsiness and coma occur. The brain is compressed as the blood mass increases, necessitating evacuation of the blood and occlusion of the bleeding vessels.

A *dural border hematoma* classically is called a subdural hematoma; however, this term is a misnomer because there is no naturally occurring space at the dura–arachnoid junction. Hematomas at this junction are usually caused by extravasated blood that splits open the dural border cell layer (Fig. B8.2B). The blood does not collect within a preexisting space but rather creates a space at the dura–arachnoid junction (Haines, 2013). Dural border hemorrhage usually follows a blow to the head that jerks the brain inside the cranium and injures it. The precipitating trauma may be trivial or forgotten, but a hematoma may develop over many weeks from venous bleeding. Dural border hemorrhage is typically venous in origin and commonly results from tearing of a superior cerebral vein bridging in as it enters the superior sagittal sinus.

Subarachnoid hemorrhage is an extravasation (escape) of blood, usually arterial, into the subarachnoid space (Fig. B8.2C). Most subarachnoid hemorrhages result from *rupture of a saccular aneurysm* (sac-like dilation on an artery). Some subarachnoid hemorrhages are associated with head trauma involving cranial fractures and cerebral lacerations. Bleeding into the subarachnoid space results in meningeal irritation, a severe headache, stiff neck, and often loss of consciousness.

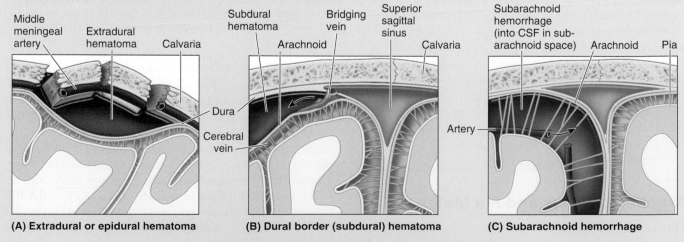

(A) Extradural or epidural hematoma

(B) Dural border (subdural) hematoma

(C) Subarachnoid hemorrhage

FIGURE B8.2. Intracranial hemorrhage. *CSF*, cerebrospinal fluid.

BRAIN

The following is a brief discussion of the parts of the brain, vasculature, and ventricular system because the brain is usually studied in neuroscience courses. The brain is composed of the *cerebrum*, *cerebellum*, and *brainstem* (midbrain, pons, and medulla oblongata) (Fig. 8.11A,B). Of the 12 CNs, *11 CNs arise from the brain* (Fig. 8.11C). They have motor, parasympathetic, and/or sensory functions. Generally, these nerves are surrounded by a dural sheath as they leave the cranium; the dural sheath becomes continuous with the connective tissue of the epineurium. For a summary of the CNs, see Chapter 10, Review of Cranial Nerves.

Parts of Brain

When the calvaria and dura mater are removed, **gyri** (folds), **sulci** (grooves), and **fissures** (clefts) of the cerebral cortex are visible through the delicate arachnoid–pia layer. The parts of brain are as follows (Fig. 8.11A,B):

- The **cerebrum** includes the **cerebral hemispheres**, which form the largest part of the brain and are separated by a longitudinal fissure into which the falx cerebri extends. Each hemisphere is divided into four lobes: frontal, parietal, temporal, and occipital. The frontal lobes occupy the anterior cranial fossa, the temporal lobes occupy the lateral parts of the middle cranial fossae, and the occipital lobes extend posteriorly over the tentorium cerebelli (see Fig. 8.3B).
- The **diencephalon** is composed of the epithalamus, thalamus, and hypothalamus and forms the central core of the brain (Fig. 8.11B).
- The **midbrain**, the rostral part of the brainstem, lies at the junction of the middle and posterior cranial fossae. CN III and IV are associated with the midbrain.

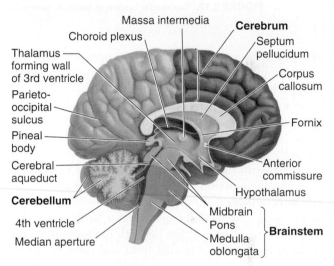

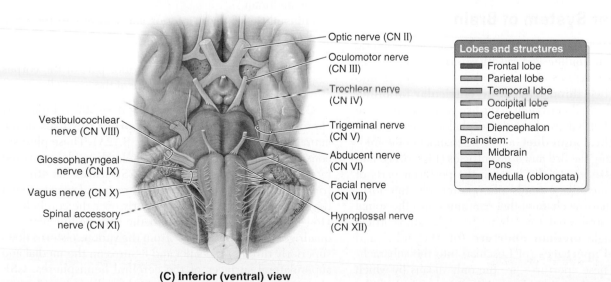

FIGURE 8.11.   Structure of brain. A. Right cerebral hemisphere, cerebellum, and brainstem. **B.** Parts of brain identified on median section. **C.** Brainstem and cranial nerves.

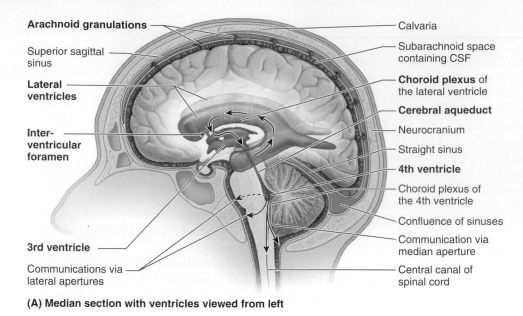

Arachnoid granulations

Superior sagittal sinus

Lateral ventricles

Inter-ventricular foramen

3rd ventricle

Communications via lateral apertures

Calvaria

Subarachnoid space containing CSF

Choroid plexus of the lateral ventricle

Cerebral aqueduct

Neurocranium

Straight sinus

4th ventricle

Choroid plexus of the 4th ventricle

Confluence of sinuses

Communication via median aperture

Central canal of spinal cord

(A) Median section with ventricles viewed from left

FIGURE 8.12. **Ventricular system of brain. A.** Ventricles. *Arrows,* direction of cerebrospinal fluid (*CSF*) flow. *(continued)*

- The **pons**, the part of the brainstem between the midbrain rostrally and the medulla oblongata caudally, lies in the anterior part of the posterior cranial fossa. CN V is associated with the pons.
- The **medulla oblongata (medulla)**, the most caudal part of the brainstem, is continuous with the spinal cord and lies in the posterior cranial fossa. CNs IX, X, and XII are associated with the medulla, whereas CN VI to VIII are located at the junction of the pons and medulla.
- The **cerebellum** is the large brain mass lying posterior to the pons and medulla and inferior to the posterior part of the cerebrum. It lies beneath the tentorium cerebelli in the posterior cranial fossa and consists of two hemispheres united by a narrow middle part, the **vermis**.

Ventricular System of Brain

The ventricular system of the brain consists of two lateral ventricles and the midline 3rd and 4th ventricles (Fig. 8.12A). The **lateral ventricles** (1st and 2nd ventricles) open into the third ventricle through the **interventricular foramina** (of Monro). The **3rd ventricle**, a slit-like cavity between the right and the left halves of the diencephalon, is continuous with the **cerebral aqueduct**, a narrow channel in the midbrain connecting the 3rd and 4th ventricles (Figs. 8.11B and 8.12B). The **4th ventricle**, lying in the posterior parts of the pons and medulla, extends inferoposteriorly. Inferiorly, it tapers to a narrow channel that continues into the spinal cord as the central canal. CSF drains from the 4th ventricle through a single **median aperture** (of Magendie) and paired **lateral apertures** (of Luschka) into the subarachnoid space. These apertures are the only means by which CSF enters the subarachnoid space. If they are blocked, the ventricles distend, producing compression of the cerebral

hemispheres. At certain areas, mainly at the base of the brain, the arachnoid and pia mater are widely separated by large pools (cisterns) of CSF (Fig. 8.12B). Major **subarachnoid cisterns** include the

- **cerebellomedullary cistern**, the largest of the cisterns, located between the cerebellum and the medulla; receives CSF from the apertures of the 4th ventricle; divided into the **posterior cerebellomedullary cistern** (L. *cisterna magna*) and the **lateral cerebellomedullary cistern**
- **pontocerebellar cistern** (pontine cistern), an extensive space ventral to the pons and continuous inferiorly with the spinal subarachnoid space
- **interpeduncular cistern** (basal cistern), located in the interpeduncular fossa between the cerebral peduncles of the midbrain
- **chiasmatic cistern**, inferior and anterior to the optic chiasm
- **quadrigeminal cistern** (cistern of the great cerebral vein), located between the posterior part of the corpus callosum and the superior surface of the cerebellum

CSF is secreted (at the rate of 400–500 mL/day) by choroidal epithelial cells of the **choroid plexuses** in the lateral, 3rd, and 4th ventricles (Fig. 8.12A). These plexuses consist of vascular fringes of pia (tela choroidea) covered by cuboidal epithelial cells. Some CSF leaves the 4th ventricle to pass inferiorly into the subarachnoid space around the spinal cord and posterosuperiorly over the cerebellum. However, most CSF flows into the interpeduncular and quadrigeminal cisterns. CSF from the various cisterns flows superiorly through the sulci and fissures on the medial and superolateral surfaces of the cerebral hemispheres. CSF also passes into the extensions of the subarachnoid space around the CNs.

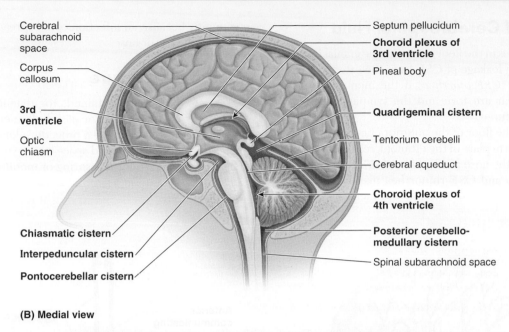

Cerebral subarachnoid space — Corpus callosum — 3rd ventricle — Optic chiasm — Chiasmatic cistern — Interpeduncular cistern — Pontocerebellar cistern

Septum pellucidum — Choroid plexus of 3rd ventricle — Pineal body — Quadrigeminal cistern — Tentorium cerebelli — Cerebral aqueduct — Choroid plexus of 4th ventricle — Posterior cerebellomedullary cistern — Spinal subarachnoid space

(B) Medial view

FIGURE 8.12. Ventricular system of brain. *(continued)* **B.** Subarachnoid cisterns.

The main site of CSF absorption into the venous system is through the arachnoid granulations, protrusions of arachnoid villi into the walls of dural venous sinuses, especially the superior sagittal sinus and its lateral venous lacunae (Figs. 8.10 and 8.12A). Along with the meninges and calvaria, CSF protects the brain by providing a cushion against blows to the head. The CSF in the subarachnoid space provides the buoyancy that prevents the weight of the brain from compressing the CN roots and blood vessels against the internal surface of the cranium.

CLINICAL BOX

Cerebral Injuries

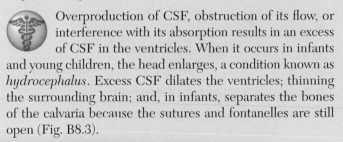

Cerebral contusion (bruising) results from brain trauma in which the pia is stripped from the injured surface of the brain and may be torn, allowing blood to enter the subarachnoid space. The bruising results from the sudden impact of the moving brain against the stationary cranium or from the suddenly moving cranium against the stationary brain. Cerebral contusion may result in an extended loss of consciousness.

Cerebral lacerations are often associated with depressed cranial fractures or gunshot wounds. Lacerations result in rupture of blood vessels and bleeding into the brain and subarachnoid space, causing increased intracranial pressure and cerebral compression. *Cerebral compression* may be produced by the following:

- Intracranial collections of blood
- Obstruction of CSF circulation or absorption
- Intracranial tumors or abscesses
- Brain swelling caused by *brain edema*, an increase in brain volume resulting from an increase in water and sodium content

Hydrocephalus

Overproduction of CSF, obstruction of its flow, or interference with its absorption results in an excess of CSF in the ventricles. When it occurs in infants and young children, the head enlarges, a condition known as *hydrocephalus*. Excess CSF dilates the ventricles; thinning the surrounding brain; and, in infants, separates the bones of the calvaria because the sutures and fontanelles are still open (Fig. B8.3).

FIGURE B8.3. Hydrocephalus.

(Continued on next page)

Leakage of Cerebrospinal Fluid

Fractures in the floor of the middle cranial fossa may result in leakage of CSF from the external acoustic meatus (*CSF otorrhea*) if the meninges superior to the middle ear are torn and the tympanic membrane (eardrum) is ruptured.

Fractures in the floor of the anterior cranial fossa may involve the cribriform plate of the ethmoid, resulting in leakage of CSF through the nose (*CSF rhinorrhea*).

CSF otorrhea and CSF rhinorrhea may be primary indications of a cranial base fracture and increase the risk of *meningitis* because an infection could spread to the meninges from the ear or nose.

Cisternal Puncture

CSF may be obtained, for diagnostic purposes, from the posterior cerebellomedullary cistern (Fig. 8.12B), using a procedure known as *cisternal puncture*. The subarachnoid space or the ventricular system may also be entered for measuring or monitoring CSF pressure, injecting antibiotics, or administering contrast media for radiography.

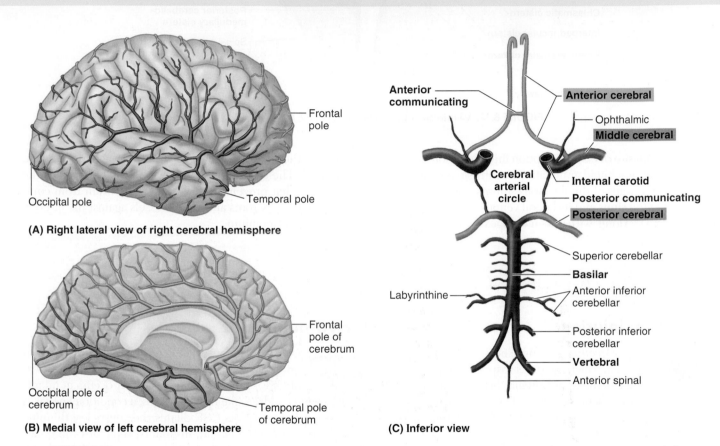

FIGURE 8.13. **Arterial supply of cerebrum. A.** Lateral surface of cerebrum. **B.** Medial surface of cerebrum. **C.** Schematic overview. *(continued)*

TABLE 8.2. ARTERIAL SUPPLY OF CEREBRAL HEMISPHERES

Artery	Origin	Distribution
Internal carotid	Common carotid artery at superior border of thyroid cartilage	Gives branches to walls of cavernous sinus, pituitary gland, and trigeminal ganglion; provides primary supply to brain
Anterior cerebral	Internal carotid artery	Cerebral hemispheres, except for occipital lobes
Anterior communicating	Anterior cerebral artery	Cerebral arterial circle (of Willis)
Middle cerebral	Continuation of internal carotid artery distal to anterior cerebral artery	Most of lateral surface of cerebral hemispheres
Vertebral	Subclavian artery	Cranial meninges and cerebellum
Basilar	Formed by union of vertebral arteries	Brainstem, cerebellum, and cerebrum
Posterior cerebral	Terminal branch of basilar artery	Interior aspect of cerebral hemisphere and occipital lobe
Posterior communicating	Posterior cerebral artery	Optic tract, cerebral peduncle, internal capsule, and thalamus

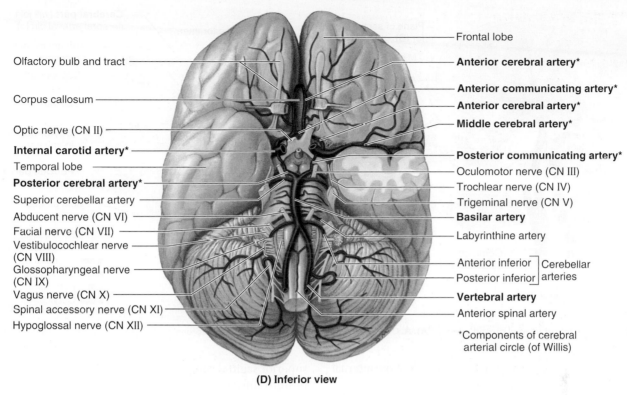

Frontal lobe

Anterior cerebral artery*

Anterior communicating artery*

Anterior cerebral artery*

Middle cerebral artery*

Posterior communicating artery*

Oculomotor nerve (CN III)

Trochlear nerve (CN IV)

Trigeminal nerve (CN V)

Basilar artery

Labyrinthine artery

Anterior inferior ⎤ Cerebellar
Posterior inferior ⎦ arteries

Vertebral artery

Anterior spinal artery

*Components of cerebral
arterial circle (of Willis)

Olfactory bulb and tract

Corpus callosum

Optic nerve (CN II)

Internal carotid artery*

Temporal lobe

Posterior cerebral artery*

Superior cerebellar artery

Abducent nerve (CN VI)

Facial nerve (CN VII)

Vestibulocochlear nerve
(CN VIII)

Glossopharyngeal nerve
(CN IX)

Vagus nerve (CN X)

Spinal accessory nerve (CN XI)

Hypoglossal nerve (CN XII)

(D) Inferior view

FIGURE 8.13. Arterial supply of cerebrum. *(continued)* **D.** Cerebral arterial circle and cranial nerves.

Vasculature of Brain

Although it accounts for only about 2.5% of body weight, the brain receives about one sixth of the cardiac output and one fifth of the oxygen consumed by the body at rest. The blood supply to the brain is from the internal carotid and vertebral arteries (Figs. 8.13 and Fig. 8.14 and Table 8.2).

The **internal carotid arteries** arise in the neck from the common carotid arteries and enter the cranial cavity with the carotid plexus of sympathetic nerves through the carotid canals. The intracranial course of the internal carotid artery is shown in Figure 8.14. The cervical part of this artery ascends to the entrance to the carotid canal in the petrous temporal bone. The petrous part of the artery turns horizontally and medially in the carotid canal to emerge superior to the foramen lacerum and enters the cranial cavity. The cavernous part of the artery runs on the lateral side of the sphenoid in the carotid groove as it traverses the cavernous sinuses. Inferior to the anterior clinoid process, the artery makes a 180-degree turn to join the cerebral arterial circle. The internal carotid arteries course anteriorly through the cavernous sinuses, with the abducent nerves (CN VI) and in close proximity to the oculomotor (CN III) and trochlear (CN IV) nerves. The terminal branches of the internal carotids are the **anterior** and **middle cerebral arteries** (Fig. 8.13C,D and Table 8.2).

The **vertebral arteries** begin in the root of the neck as branches of the first part of the subclavian arteries, pass through the transverse foramina of the first six cervical vertebrae, and perforate the dura and arachnoid to pass through the foramen magnum (Fig. 8.14A). The intracranial parts of the vertebral arteries unite at the caudal border of the pons to form the **basilar artery**. The basilar artery runs through the pontocerebellar cistern (Fig. 8.12B) to the superior border of the pons, where it ends by dividing into the two **posterior cerebral arteries**.

In addition to supplying branches to deeper parts of the brain, the cortical branches of each cerebral artery supply a surface and a pole of the cerebrum as follows:

- *Anterior cerebral arteries* supply most of the medial and superior surfaces and the frontal pole.
- *Middle cerebral arteries* supply the lateral surface and temporal pole.
- *Posterior cerebral arteries* supply the inferior surface and occipital pole.

The **cerebral arterial circle** (of Willis) at the base of the brain is an important anastomosis between the four arteries (two vertebral and two internal carotid arteries) that supply the brain (Fig. 8.13C,D). The arterial circle is formed by the *posterior cerebral, posterior communicating, internal carotid, anterior cerebral,* and *anterior communicating arteries.* The various components of the cerebral arterial circle give numerous small branches to the brain. Variations in the origin and size of the vessels forming the cerebral arterial circle are common (e.g., the posterior communicating arteries may be absent, or there may be two anterior communicating arteries). In approximately one in three people,

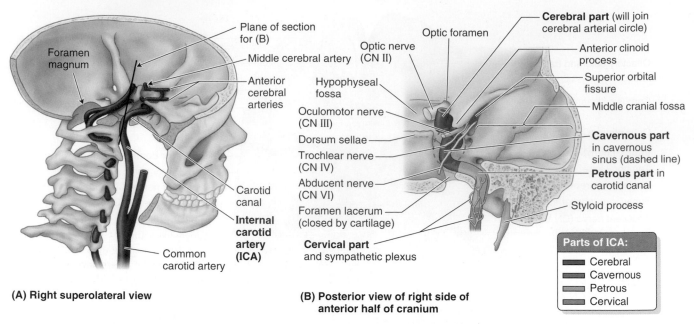

(A) Right superolateral view

(B) Posterior view of right side of anterior half of cranium

FIGURE 8.14. **Course of internal carotid artery (ICA). A.** Overview. **B.** Parts of internal carotid artery. Coronal section that intersects the carotid canal.

one posterior cerebral artery is a major branch of the internal carotid artery.

The thin-walled, valveless **cerebral veins** draining the brain pierce the arachnoid and meningeal layer of dura to end in the nearest dural venous sinuses. The sinuses drain for the most part into the IJVs. The superior cerebral veins on the superolateral surface of the brain drain into the superior sagittal sinus (see Fig. 8.6A); cerebral veins on the postero-inferior aspect drain into the straight, transverse, and superior petrosal sinuses. The **great cerebral vein** (of Galen), a single midline vein, is formed inside the brain by the union of two internal cerebral veins and ends by merging with the inferior sagittal sinus to form the straight sinus (see Figs. 8.6A and 8.7B).

CLINICAL BOX

Strokes

An *ischemic stroke* is the sudden development of neurological deficits that are related to impaired cerebral blood flow. The most common causes of strokes are *spontaneous cerebrovascular accidents* such as *cerebral embolism, cerebral thrombosis, cerebral hemorrhage,* and *subarachnoid hemorrhage* (Esenwa et al., 2016). The cerebral arterial circle is an important means of collateral circulation in the event of gradual obstruction of one of the major arteries forming the circle. Sudden occlusion, even if only partial, results in neurological deficits. In elderly persons, the anastomoses are often inadequate when a large artery (e.g., internal carotid) is occluded, even if the occlusion is gradual (in which case, function is impaired at least to some degree).

Hemorrhagic stroke follows the rupture of an artery or a *saccular aneurysm,* a sac-like dilation on a weak part of the arterial wall. The most common type of saccular aneurysm is a *berry aneurysm,* occurring in the vessels of or near the cerebral arterial circle and the medium arteries at the base of the brain (Fig. B8.4). In time, especially in people with

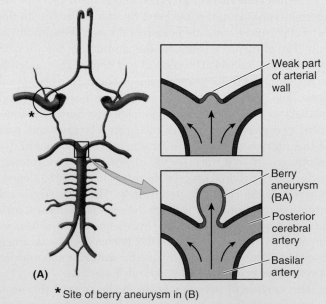

(A)

*** Site of berry aneurysm in (B)

FIGURE B8.4. Berry aneurysm (BA). *(continued)*

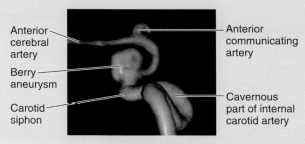

Anterior cerebral artery

Berry aneurysm

Carotid siphon

Anterior communicating artery

Cavernous part of internal carotid artery

(B) 3D CT reconstruction of berry aneurysm

FIGURE B8.4. Berry aneurysm (BA). *(continued)*

hypertension (high blood pressure), the weak part of the arterial wall expands and may rupture, allowing blood to enter the subarachnoid space.

Transient Ischemic Attack

Transient ischemic attack (TIA) refers to neurological symptoms resulting from ischemia (deficient blood supply) of the brain. The symptoms of a TIA may be ambiguous: staggering, dizziness, light-headedness, fainting, and paresthesias (e.g., tingling in a limb). Most TIAs last a few minutes, but some persist for up to an hour. Individuals with TIAs are at increased risk for myocardial infarction and ischemic stroke (Marshall, 2016). Magnetic resonance imaging (MRI) is used to differentiate between a TIA and a *completed stroke* (infraction of brain tissue).

FACE

The **face** is the anterior aspect of the head from the forehead to the chin and from one ear to the other. The basic shape of the face is determined by the underlying bones, the facial muscles, and the subcutaneous tissue. The skin of the face is thin, pliable, and firmly attached to the underlying cartilages of the external ear and nose.

CLINICAL BOX

Facial Injuries

Because the face does not have a distinct layer of deep fascia and the subcutaneous tissue is loose between the attachments of facial muscles, *facial lacerations* tend to gape (part widely). Consequently, the skin must be sutured carefully to prevent scarring. The looseness of the subcutaneous tissue also enables fluid and blood to accumulate in the loose connective tissue after bruising of the face. *Facial inflammation* causes considerable swelling.

Muscles of Face

The **facial muscles (muscles of facial expression)** are in the subcutaneous tissue of the anterior and posterior scalp, face, and neck (Fig. 8.15 and Table 8.3). Most of these muscles attach to bone or fascia and produce their effects by pulling the skin. They move the skin and change facial expressions to convey mood. The *muscles of facial expression* also surround the orifices of the mouth, eyes, and nose and act as sphincters and dilators that close and open the orifices.

The **orbicularis oris** is the sphincter of the mouth and is the first of a series of sphincters associated with the alimentary (digestive) tract. The **buccinator** (L. trumpeter), active in smiling, also keeps the cheek taut, thereby preventing it from folding and being injured during chewing. The orbicularis oris and buccinator work with the tongue to keep food between the teeth during *mastication* (chewing). The buccinator is also active during sucking, whistling, and blowing (e.g., when playing a wind instrument).

The **orbicularis oculi** closes the eyelids and assists the flow of lacrimal fluid (tears). It has three parts: the *palpebral part*, which gently closes the eyelids; the *lacrimal part*, which passes posterior to the lacrimal sac, aiding drainage of tears; and the *orbital part*, which tightly closes the eyelids to protect the eyeballs against glare and dust.

Nerves of Face

Cutaneous (sensory) innervation of the face is provided primarily by the *trigeminal nerve* (CN V; Fig. 8.16), whereas the motor innervation to the muscles of facial expression is provided by the *facial nerve* (CN VII; Fig. 8.15B) and the motor innervation to the muscles of mastication by the *mandibular nerve*, the motor root of the trigeminal nerve.

The cutaneous nerves of the neck overlap those of the face (Fig. 8.16B). Cutaneous branches of the cervical nerves from the *cervical plexus* extend over the ear, the posterior aspect of the neck and scalp. The *great auricular nerve* innervates the inferior aspect of the auricle and much of the area overlying the angle of the mandible.

The **trigeminal nerve** (CN V) is the sensory nerve for the face and the motor nerve for the muscles of mastication and several small muscles (Fig. 8.16 and Table 8.4). Three large groups of peripheral processes from nerve cell bodies of the **trigeminal ganglion**—the large sensory ganglion of CN V—form the *ophthalmic nerve* (CN V_1), the *maxillary nerve* (CN V_2), and the sensory component of the *mandibular nerve* (CN V_3). These nerves are named according to their main regions of termination: the eye, maxilla, and mandible, respectively. The first two divisions (CN V_1 and CN V_2) are wholly sensory. CN V_3 is largely sensory but also receives motor fibers (axons) from the motor root

(continued on page 524)

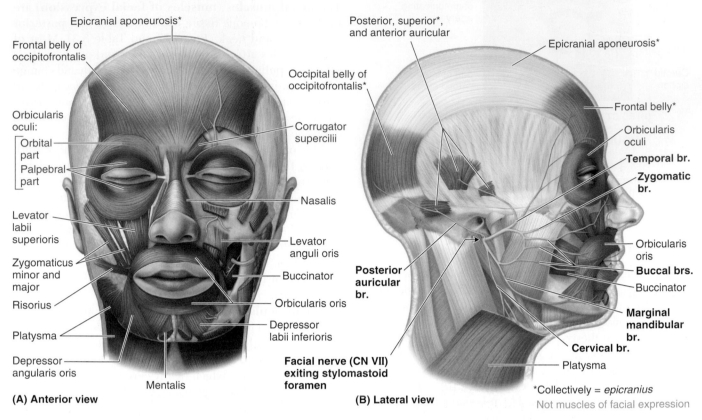

FIGURE 8.15. **Muscles of face and scalp. A.** Muscles of facial expression. **B.** Innervation, branches of facial nerve (CN VII) *br*, branch.

TABLE 8.3. MAJOR FUNCTIONAL MUSCLES OF FACE AND SCALP

Muscle[a]	Origin	Insertion	Main Action(s)
Occipitofrontalis			
Frontal belly	Epicranial aponeurosis	Skin and subcutaneous tissue of eyebrows and forehead	Elevates eyebrows and wrinkles skin of forehead; protracts scalp (indicating surprise or curiosity)
Occipital belly	Lateral two thirds of superior nuchal line	Epicranial aponeurosis	Retracts scalp; increasing effectiveness of frontal belly
Orbicularis oculi (orbital sphincter)	Medial orbital margin; medial palpebral ligament; lacrimal bone	Skin around margin of orbit; superior and inferior tarsi (tarsal plates)	Closes eyelids: palpebral part does so gently; orbital part tightly (winking)
Orbicularis oris (oral sphincter)	Medial maxilla and mandible; deep surface of peri-oral skin; angle of mouth	Mucous membrane of lips	Tonus closes mouth; phasic contraction compresses and protrudes lips (kissing) or resists distention (when blowing).
Buccinator (cheek muscle)	Mandible, alveolar processes of maxilla and mandible, pterygomandibular raphe	Angle of mouth (modiolus); orbicularis oris	Presses cheek against molar teeth; works with tongue to keep food between occlusal surfaces and out of oral vestibule; resists distention (when blowing)
Platysma	Subcutaneous tissue of infraclavicular and supraclavicular regions	Base of mandible; skin of cheek and lower lip; angle of mouth; orbicularis oris	Depresses mandible (against resistance); tenses skin of inferior face and neck (conveying tension and stress)

[a]All facial muscles are innervated by the facial nerve (CN VII) via its posterior auricular branch or via the temporal, zygomatic, buccal, marginal mandibular, or cervical branches of the parotid plexus.

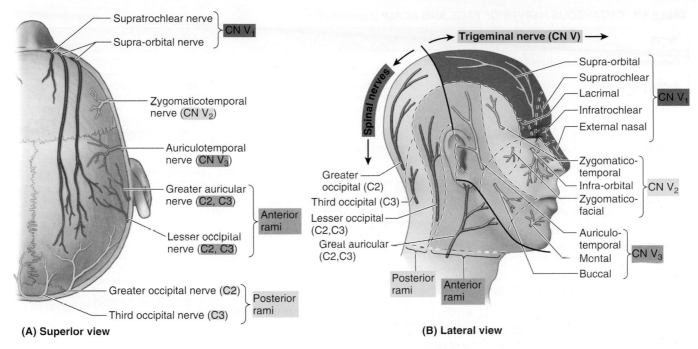

FIGURE 8.16. Cutaneous nerves of face and scalp.

TABLE 8.4. CUTANEOUS NERVES OF FACE AND SCALP

Nerve	Origin	Course	Distribution
Cutaneous nerves derived from ophthalmic nerve (CN V₁)			
Supra-orbital	Branch from bifurcation of *frontal nerve*, approximately in middle of orbital roof	Continues anteriorly along roof of orbit, emerging via supra-orbital notch or foramen; ascends forehead, breaking into branches	Mucosa of *frontal sinus*; skin and conjunctiva of middle of *superior eyelid*; skin and pericranium of *anterolateral forehead and scalp* to vertex
Supratrochlear	Branch from bifurcation of *frontal nerve*, approximately in middle of orbital roof	Continues anteromedially along roof of orbit, passing lateral to trochlea and ascending forehead	Skin and conjunctiva of medial aspect of *superior eyelid*; skin and pericranium of *anteromedial forehead*
Lacrimal	Branch of *CN V₁* proximal to superior	Runs superolaterally through orbit, receiving secretomotor fibers via a communicating branch from the zygomaticotemporal nerve	*Lacrimal gland* (secretomotor fibers); small area of skin and conjunctiva of *lateral part of superior eyelid*
Infratrochlear	Terminal branch (with anterior ethmoidal nerve) of *nasociliary nerve*	Follows medial wall of orbit, passing inferior to trochlea	Skin lateral to *root of nose*; skin and conjunctiva of *eyelids adjacent to medial canthus*, *lacrimal sac*, and *lacrimal caruncle*
External nasal	Terminal branch of *anterior ethmoidal nerve*	Emerges from nasal cavity by passing between nasal bone and lateral nasal cartilage	Skin of nasal *ala*, *vestibule, and dorsum of nose*, including *apex*
Cutaneous nerves derived from maxillary nerve (CN V₂)			
Infra-orbital	Continuation of *CN V₂* distal to its entrance into the orbit via the inferior orbital fissure	Traverses infra-orbital groove and canal in orbital floor, giving rise to superior alveolar branches; then emerges via infra-orbital foramen, immediately dividing into inferior palpebral, internal and external nasal, and superior labial branches	Mucosa of *maxillary sinus*; premolar, canine, and incisor *maxillary teeth*; skin and conjunctiva of *inferior eyelid*; skin of *cheek*, *lateral nose*, and antero-inferior *nasal septum*; skin and oral mucosa of *upper lip*
Zygomaticofacial	Smaller terminal branch (with zygomaticotemporal nerve) of *zygomatic nerve*	Traverses zygomaticofacial canal in zygomatic bone at inferolateral angle of orbit	Skin on prominence of *cheek*
Zygomaticotemporal	Larger terminal branch (with zygomaticofacial nerve) of *zygomatic nerve*	Sends communicating branch to lacrimal nerve in orbit; then passes to temporal fossa via zygomaticotemporal canal in zygomatic bone	Skin overlying *anterior part of temporal fossa*

(continued)

TABLE 8.4. CUTANEOUS NERVES OF FACE AND SCALP (continued)

Nerve	Origin	Course	Distribution
Cutaneous nerves derived from mandibular nerve (CN V₃)			
Auriculotemporal	In infratemporal fossa via two roots from *posterior trunk of CN V₃* that encircle middle meningeal artery	Passes posteriorly deep to ramus of mandible and superior deep part of parotid gland, emerging posterior to temporomandibular joint	Skin anterior to auricle and posterior two thirds of *temporal region*; skin of tragus and adjacent helix of *auricle*; skin of roof of *external acoustic meatus*; and skin of superior *tympanic membrane*
Buccal	In infratemporal fossa as sensory branch of *anterior trunk of CN V₃*	Passes between two parts of lateral pterygoid muscle, emerging anteriorly from cover of ramus of mandible and masseter, uniting with buccal branches of facial nerve	Skin and oral mucosa of *cheek* (overlying and deep to anterior part of buccinator); *buccal gingiva* (gums) adjacent to second and third molars
Mental	Terminal branch of *inferior alveolar nerve* (branch of V₃)	Emerges from mandibular canal via mental foramen in anterolateral aspect of body of mandible	Skin of *chin*; oral mucosa of *lower lip*
Cutaneous nerves derived from anterior rami of cervical spinal nerves			
Great auricular	Spinal nerves C2 and C3 via cervical plexus	Ascends vertically across sternocleidomastoid, posterior to external jugular vein	Skin overlying angle of mandible and inferior lobe of auricle; parotid sheath
Lesser occipital		Follows posterior border of sternocleidomastoid; then ascends posterior to auricle	Scalp posterior to auricle
Cutaneous nerves derived from posterior rami of cervical spinal nerves			
Greater occipital nerve	As medial branch of posterior ramus of spinal nerve C2	Emerges between axis and obliquus capitis inferior; then pierces trapezius	Scalp of occipital region
Third occipital nerve	As lateral branch of posterior ramus of spinal nerve C3	Pierces trapezius	Scalp of lower occipital and suboccipital regions

of CN V. The major cutaneous branches of the trigeminal nerve (Table 8.4) are as follows:

- Ophthalmic nerve (CN V₁): Lacrimal, supra-orbital, supratrochlear, infratrochlear, and external nasal nerves
- Maxillary nerve (CN V₂): Infra-orbital, zygomaticotemporal, and zygomaticofacial nerves
- Mandibular (CN V₃): Auriculotemporal, buccal, and mental nerves

The **motor nerves of the face** are the *facial nerve* (CN VII) to the muscles of facial expression and the mandibular nerve (CN V₃) to the muscles of mastication (masseter, temporal, medial, and lateral pterygoids). These nerves also supply some more deeply placed muscles (described later in this chapter in relation to the mouth, middle ear, and neck). The **facial nerve** (CN VII) emerges from the cranium via the *stylomastoid foramen* (Fig. 8.15B and Tables 8.1 and 8.3). Its extracranial branches (temporal, zygomatic, buccal, marginal mandibular, cervical, and posterior auricular nerves) supply the superficial muscle of the neck and chin (platysma), muscles of facial expression, muscle of the cheek (buccinator), muscles of the ear (auricular), and muscles of the scalp (occipital and frontal bellies of occipitofrontal muscle).

Innervation of the scalp anterior to the auricles is by branches of all three divisions of the **trigeminal nerve** (CN V₁, CN V₂, CN V₃) (Fig. 8.16B and Table 8.4). Posterior to the auricles, innervation of the scalp is by spinal cutaneous nerves (C2 and C3).

Superficial Vasculature of Face and Scalp

The face is richly supplied by superficial arteries and drained by external veins, as is evident in blushing and blanching (becoming pale). The terminal branches of both arteries and veins anastomose freely, including anastomoses across the midline with contralateral partners. Most arteries supplying the face are branches of the *external carotid arteries* (Fig. 8.17 and Table 8.5). Most external facial veins are drained by veins that accompany the arteries of the face. As with most superficial veins, they are subject to many variations and have abundant anastomoses that allow drainage to occur by alternate routes during periods of temporary compression. The alternate routes include both superficial pathways and deep drainage.

The **facial artery** provides the major arterial supply to the superficial face (Figs. 8.17B and 8.18 and Table 8.5).

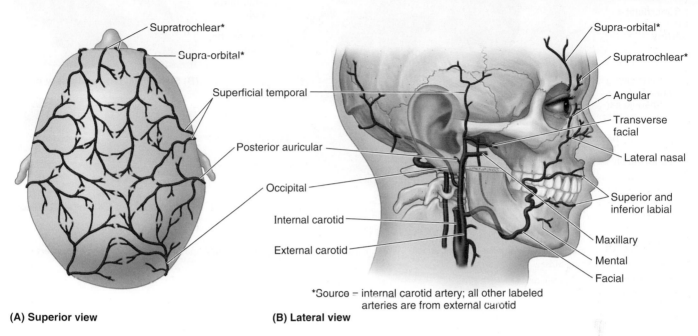

(A) Superior view   **(B) Lateral view**

*Source – internal carotid artery; all other labeled arteries are from external carotid

FIGURE 8.17. Superficial arteries of face and scalp.

TABLE 8.5. SUPERFICIAL ARTERIES OF FACE AND SCALP

Artery	Origin	Course	Distribution
Facial	External carotid artery	Ascends deep to submandibular gland; winds around inferior border of mandible and enters face	Muscles of facial expression and face
Inferior labial	Facial artery near angle of mouth	Runs medially in lower lip	Lower lip
Superior labial		Runs medially in upper lip	Upper lip and ala (side) and septum of nose
Lateral nasal	Facial artery as it ascends alongside nose	Passes to ala of nose	Skin on ala and dorsum of nose
Angular	Terminal branch of facial artery	Passes to medial angle (canthus) of eye	Superior part of cheek and inferior eyelid
Occipital	External carotid artery	Passes medial to posterior belly of digastric and mastoid process; accompanies occipital nerve in occipital region	Scalp of back of head, as far as vertex
Posterior auricular		Passes posteriorly, deep to parotid gland, along styloid process between mastoid process and ear	Auricle and scalp posterior to auricle
Superficial temporal	Smaller terminal branch of external carotid artery	Ascends anterior to ear to temporal region and ends in scalp	Facial muscles and skin of frontal and temporal regions
Transverse facial	Superficial temporal artery within parotid gland	Crosses face superficial to masseter and inferior to zygomatic arch	Parotid gland and duct, muscles and skin of face
Mental	Terminal branch of inferior alveolar artery	Emerges from mental foramen and passes to chin	Facial muscles and skin of chin
Supra-orbital	Terminal branch of ophthalmic artery, a branch of internal carotid artery	Passes superiorly from supra-orbital foramen	Muscle and skin of forehead and scalp
Supratrochlear		Passes superiorly from supratrochlear notch	Muscles and skin of scalp

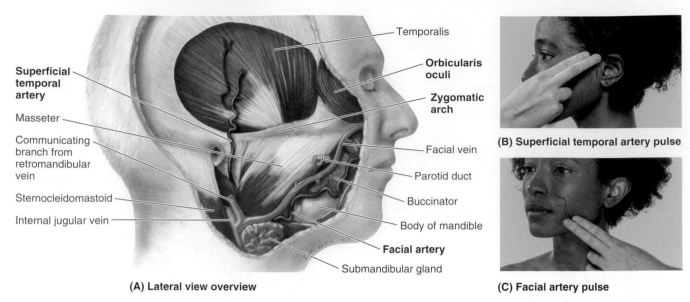

Superficial temporal artery

Masseter

Communicating branch from retromandibular vein

Sternocleidomastoid

Internal jugular vein

Temporalis

Orbicularis oculi

Zygomatic arch

Facial vein

Parotid duct

Buccinator

Body of mandible

Facial artery

Submandibular gland

(A) Lateral view overview

(B) Superficial temporal artery pulse

(C) Facial artery pulse

FIGURE 8.18. **Vasculature and arterial pulses of face. A.** Vasculature. Parotid gland has been removed. **B.** Palpation of superficial temporal artery pulse. **C.** Palpation of facial artery pulse.

It arises from the external carotid artery and winds its way to the inferior border of the mandible, just anterior to the masseter. It then courses over the face to the medial angle (canthus) of the eye. The facial artery sends branches to the upper and lower lips (**superior** and **inferior labial arteries**). The facial artery also sends branches to the side of the nose (**lateral nasal artery**) and then terminates as the **angular artery**, which supplies the medial angle of the eye.

The **superficial temporal artery** is the smaller terminal branch of the external carotid artery; the other branch is the *maxillary artery.* The superficial temporal artery emerges on the face between the temporomandibular joint (TMJ) and the auricle and ends in the scalp by dividing into **frontal** and **parietal branches** (Fig. 8.18). The **transverse facial artery** arises from the superficial temporal artery within the parotid gland and crosses the face superficial to the masseter. It divides into numerous branches that supply the parotid gland and duct, the masseter, and the skin of the face. It anastomoses with branches of the facial artery.

The **arteries of the scalp** course within the subcutaneous connective tissue layer between the skin and the epicranial aponeurosis. They anastomose freely with one another. The arterial walls are firmly attached to the dense connective tissue in which they are embedded, limiting their ability to constrict when cut. Consequently, bleeding from scalp wounds is profuse. The arterial supply is from the *external carotid arteries* through the **occipital, posterior auricular**, and **superficial temporal arteries** and from the *internal carotid arteries* by way of the **supratrochlear**

CLINICAL BOX

Pulses of Arteries of Face

The pulses of the superficial temporal and facial arteries can be used for taking the pulse (Fig. 8.18B). For example, anesthesiologists at the head of the operating table often take the *temporal artery pulse* anterior to the auricle as the artery crosses the zygomatic arch to supply the scalp. The *facial artery pulse* can be palpated where the facial artery crosses the inferior border of the mandible immediately anterior to the masseter (Fig. 8.18C).

Compression of Facial Artery

The facial artery can be occluded by pressure against the mandible where the vessel crosses it. Because of the numerous anastomoses between the branches of the facial artery and other arteries of the face, *compression of the facial artery* on one side does not stop all bleeding from a lacerated facial artery or one of its branches. In lacerations of the lip, pressure must be applied on both sides of the cut to stop the bleeding. In general, facial wounds bleed freely but heal quickly.

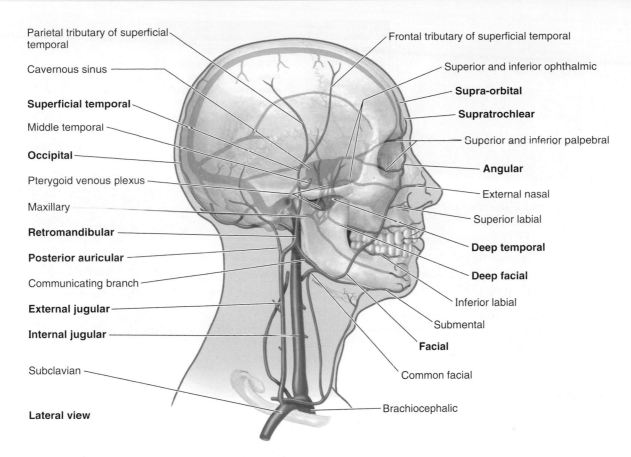

Parietal tributary of superficial temporal

Cavernous sinus

Superficial temporal

Middle temporal

Occipital

Pterygoid venous plexus

Maxillary

Retromandibular

Posterior auricular

Communicating branch

External jugular

Internal jugular

Subclavian

Lateral view

Frontal tributary of superficial temporal

Superior and inferior ophthalmic

Supra-orbital

Supratrochlear

Superior and inferior palpebral

Angular

External nasal

Superior labial

Deep temporal

Deep facial

Inferior labial

Submental

Facial

Common facial

Brachiocephalic

FIGURE 8.19. Veins of face and scalp.

and **supra-orbital arteries** (Fig. 8.17A and Table 8.5). Arteries of the scalp supply little blood to the cranium, which is supplied primarily by the middle meningeal artery.

The **facial vein** provides the primary superficial venous drainage of the face (Figs. 8.18 and 8.19). It begins at the medial angle of the eye as the **angular vein**. Among the tributaries of the facial vein is the **deep facial vein**, which drains the *pterygoid venous plexus* of the infratemporal fossa (Fig. 8.19). Inferior to the margin of the mandible, the facial vein is joined by the anterior branch of the retromandibular vein. The facial vein drains directly or indirectly into the **internal jugular vein** (Fig. 8.19). At the medial angle of the eye, the facial vein communicates with the *superior ophthalmic vein*, which drains into the *cavernous sinus*.

The **superficial temporal vein** drains the forehead and scalp and receives tributaries from the veins of the temple and face. Near the auricle, the superficial temporal vein enters the parotid gland (Fig. 8.18). The **retromandibular vein**, formed by the union of the superficial temporal vein and the maxillary vein, is a deep vein that descends within the parotid gland, superficial to the external carotid artery and deep to the facial nerve (Fig. 8.19). The retromandibular

vein divides into an *anterior branch*, which unites with the facial vein, and a *posterior branch*, which joins the posterior auricular vein to form the **external jugular vein** (EJV). The EJV crosses the superficial surface of the sternocleidomastoid muscle to enter the subclavian vein in the root of the neck.

Venous drainage of the superficial parts of the scalp is through the accompanying veins of the scalp arteries, the **supra-orbital** and **supratrochlear veins**, which descend to unite at the medial angle of the eye to form the angular vein, which becomes the facial vein at the inferior margin of the orbit. The superficial temporal veins and **posterior auricular veins** drain the scalp anterior and posterior to the auricles, respectively. The **occipital veins** drain the occipital region of the scalp. Venous drainage of deep parts of the scalp in the temporal region is through **deep temporal veins**, which are tributaries of the pterygoid venous plexus.

There are no lymph nodes in the scalp or face except for the parotid/buccal region. Lymph from the scalp, face, and neck drains into the *superficial ring (pericervical collar) of lymph nodes*—the *submental, submandibular, parotid, mastoid,* and *occipital* located at the junction of

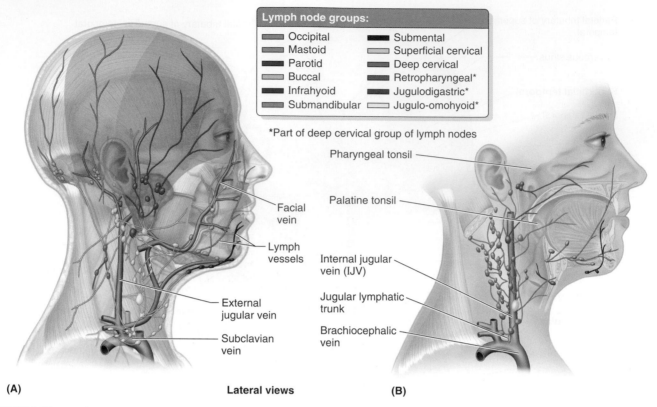

Lymph node groups:
- Occipital
- Mastoid
- Parotid
- Buccal
- Infrahyoid
- Submandibular
- Submental
- Superficial cervical
- Deep cervical
- Retropharyngeal*
- Jugulodigastric*
- Jugulo-omohyoid*

*Part of deep cervical group of lymph nodes

Pharyngeal tonsil

Palatine tonsil

Facial vein

Lymph vessels

Internal jugular vein (IJV)

Jugular lymphatic trunk

External jugular vein

Brachiocephalic vein

Subclavian vein

(A)   Lateral views   (B)

FIGURE 8.20. Lymphatic drainage of face and scalp. A. Superficial drainage. **B.** Deep drainage. All lymphatic vessels from the head and neck ultimately drain into the deep cervical nodes, either directly or indirectly.

the head and neck (Fig. 8.20). Lymph from the superficial ring of nodes drains into the **deep cervical lymph nodes** along the IJV. Lymph from these nodes passes to the jugular lymphatic trunk, which joins the thoracic duct on the left side and the IJV or brachiocephalic vein on the right side. A summary of the lymphatic drainage of the face follows:

- Lymph from the lateral part of the face and scalp drains to the superficial **parotid lymph nodes**.
- Lymph from the deep parotid nodes drains to the deep cervical lymph nodes.
- Lymph from the upper lip and lateral parts of the lower lip drains into the **submandibular lymph nodes**.
- Lymph from the chin and central part of the lower lip drains into the **submental lymph nodes**.

Parotid Gland

The **parotid gland** is the largest of three paired salivary glands. It is enclosed within a tough fascial capsule, the **parotid sheath**, derived from the investing layer of deep cervical fascia. The parotid gland has an irregular shape because the area it occupies, the **parotid bed**, is antero-inferior to

CLINICAL BOX

Squamous Cell Carcinoma of Lip

Squamous cell carcinoma (cancer) of the lip usually involves the lower lip (Fig. B8.5). Overexposure to sunshine and irritation from pipe smoking over many years are contributing factors. Cancer cells from the central part of the lower lip, the floor of the mouth, and apex of the tongue spread to the submental lymph nodes, whereas cancer cells from lateral parts of the lower lip drain to the submandibular lymph nodes.

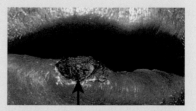

FIGURE B8.5. Squamous cell carcinoma of lower lip.

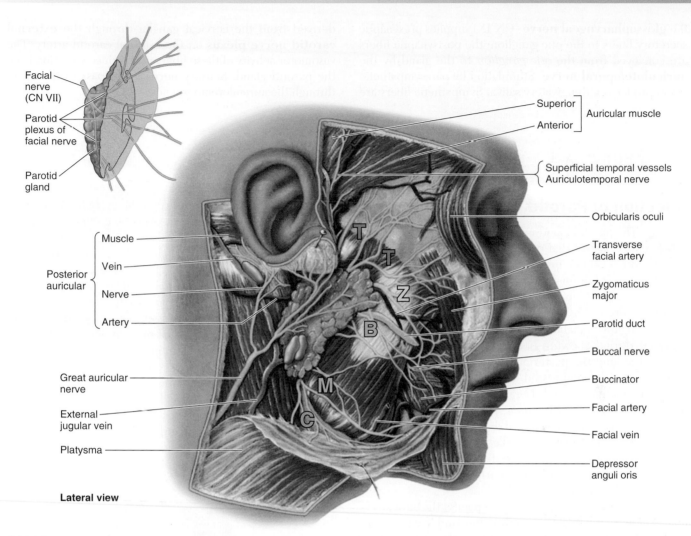

Facial nerve (CN VII)

Parotid plexus of facial nerve

Parotid gland

Superior
Anterior } Auricular muscle

Superficial temporal vessels
Auriculotemporal nerve

Orbicularis oculi

Posterior auricular {
Muscle
Vein
Nerve
Artery

Transverse facial artery

Zygomaticus major

Parotid duct

Buccal nerve

Buccinator

Facial artery

Facial vein

Depressor anguli oris

Great auricular nerve

External jugular vein

Platysma

Lateral view

FIGURE 8.21. Relationships of parotid gland. *Inset*, parotid plexus of facial nerve; the parotid gland has been sectioned in the coronal plane. Branches of facial nerve: *B*, buccal; *C*, cervical; *M*, marginal mandibular; *T*, temporal; *Z*, zygomatic.

the external acoustic meatus, where it is wedged between the ramus of the mandible and the mastoid process (Fig. 8.21). The inferiorly directed apex of the parotid gland is posterior to the angle of the mandible, and its base is related to the zygomatic arch. The **parotid duct** passes horizontally from the anterior edge of the gland. At the anterior border of the masseter, the duct turns medially, pierces the buccinator, and enters the oral cavity through a small orifice opposite the second maxillary molar tooth. Embedded within the substance of the parotid gland, from superficial to deep, are the *parotid plexus of the facial nerve* (CN VII) and its branches, the *retromandibular vein* and the *external carotid artery*. On the parotid sheath and within the gland are *parotid lymph nodes*.

The **great auricular nerve** (C2 and C3), a branch of the cervical plexus, provides sensory innervation to the parotid sheath and overlying skin (see Fig. 8.16B and Table 8.4) and then passes superior to it with the superficial temporal vessels (Fig. 8.18). The parasympathetic component of

CLINICAL BOX

Trigeminal Neuralgia

Trigeminal neuralgia (tic douloureux) is a sensory disorder of the sensory root of *CN V* characterized by sudden attacks of excruciating, lightning like jabs of facial pain. A *paroxysm* (sudden sharp pain) can last for 15 minutes or more. The maxillary nerve (CN V$_2$) is most frequently involved; then the mandibular nerve (CN V$_3$); and, least frequently, the ophthalmic nerve (CN V$_1$). The pain often is initiated by touching a sensitive *trigger zone* of the skin.

The cause of trigeminal neuralgia is unknown; however, some investigators believe that most affected people have an anomalous blood vessel that compresses the sensory root of CN V. When the aberrant artery is moved away from the root, the symptoms usually disappear. Other researchers believe the condition is caused by pathological processes affecting neurons of the trigeminal ganglion. In some cases, it is necessary to section the sensory root for relief of trigeminal neuralgia.

the **glossopharyngeal nerve** (CN IX) supplies presynaptic secretory fibers to the otic ganglion; the postsynaptic fibers are conveyed from the *otic ganglion* to the gland by the **auriculotemporal nerve**. Stimulation by parasympathetic fibers produces a thin, watery saliva. Sympathetic fibers are derived from the cervical ganglia through the **external carotid nerve plexus** on the external carotid artery. The vasomotor activity of these fibers may reduce secretion from the parotid gland. Sensory nerve fibers pass to the gland through the auriculotemporal nerve.

CLINICAL BOX

Infection of Parotid Gland

The parotid gland may become infected by pathologic agents that pass through the bloodstream, as occurs in *mumps*, an acute communicable viral disease. Infection of the gland causes inflammation (*parotiditis*) of the gland. Severe pain occurs because the tough parotid sheath, innervated by the great auricular nerve (Fig. 8.21), becomes tightly stretched by swelling. The pain may be aggravated during chewing because the enlarged gland is wrapped around the posterior border of the ramus of the mandible and is compressed against the mastoid process when the mouth is opened. The mumps virus also may cause *inflammation of the parotid duct*, producing redness of the *parotid papilla*, where the parotid duct opens into the mouth opposite the second maxillary molar tooth. Because the pain produced by mumps may be confused with a toothache, redness of the papilla is often an early sign that the disease involves the gland and not a tooth.

Parotid gland disease often causes pain in the auricle, external acoustic meatus, temporal region, and TMJ because the auriculotemporal nerve, from which the parotid gland receives sensory fibers, also supplies sensory fibers to the skin over the temporal fossa and auricle.

Lesions of Trigeminal Nerve

Lesions of the entire trigeminal nerve cause widespread anesthesia involving the following areas:

- Corresponding anterior half of the scalp
- Face, except for an area overlying the angle of the mandible
- Cornea and conjunctiva
- Mucous membranes of the nose and paranasal sinuses, mouth, and anterior part of the tongue
- Paralysis of the muscles of mastication also occurs.

Bell Palsy

Injury to the facial nerve (CN VII) or its branches produces paralysis of some or all the facial muscles on the affected side (*Bell palsy*). The affected areas sag and facial expression is distorted (Fig. B8.6). The loss of tonus of the orbicularis oculi causes the inferior eyelid to evert (fall away from the surface of the eyeball). As a result, the lacrimal fluid is not spread over the cornea, preventing adequate lubrication, hydration, and flushing of the cornea. This makes the cornea vulnerable to ulceration. If the injury weakens or paralyzes the buccinator and orbicularis oris, food will accumulate in the oral vestibule during chewing, usually requiring continual removal with a finger. When the sphincters or dilators of the mouth are affected, displacement of the mouth (drooping of the corner) is produced by gravity and contraction of unopposed contralateral facial muscles, resulting in food and saliva dribbling out of the side of the mouth. Weakened lip muscles affect speech. Affected people cannot whistle or blow a wind instrument effectively. They frequently dab their eyes and mouth with a handkerchief to wipe the fluid (tears and saliva), which runs from the drooping eyelid and mouth.

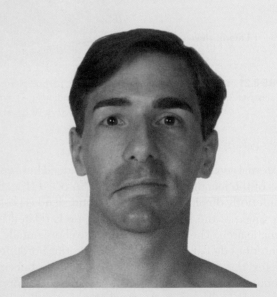

FIGURE B8.6. Bell palsy.

Parotidectomy

About 80% of salivary gland tumors occur in the parotid glands. Surgical excision of the parotid gland (*parotidectomy*) is often performed as part of the treatment. Because the parotid plexus of CN VII is embedded in the parotid gland, the plexus and its branches are in jeopardy during surgery. An important step in parotidectomy is the identification and preservation of the facial nerve. Preoperative computerized tomography (CT) or MRI is used for surgical planning to establish the relationship of a parotid tumor to the expected location of CN VII (not visible on CT or MRI) adjacent to the retromandibular vein (which is visible on the images).

ORBITS

The **orbits** are pyramidal, bony cavities in the facial skeleton with their bases (*orbital openings*) directed anterolaterally and their apices, posteromedially (Fig. 8.22; see also Fig. 8.33D). The orbits contain and protect the *eyeballs* and their muscles, nerves, and vessels together with most of the lacrimal apparatus. All space in the orbits not occupied by structures is filled with **orbital fat**.

The orbit has a base, four walls, and an apex:

- The **superior wall** (roof) is approximately horizontal and is formed mainly by the *orbital part of the frontal bone*, which separates the orbital cavity from the anterior cranial fossa. Near the apex of the orbit, the superior wall is formed by the *lesser wing of the sphenoid*. Anterolaterally, the lacrimal gland occupies the **fossa for the lacrimal gland** (lacrimal fossa) in the orbital part of the frontal bone.
- The **medial wall** is formed by the **ethmoid bone**, along with contributions from the frontal, lacrimal, and sphenoid bones. Anteriorly, the medial wall is indented by the

lacrimal groove and **fossa for the lacrimal sac**. The bone forming the medial wall is paper-thin, and the ethmoid air cells are often visible through the bone of a dried cranium.
- The **lateral wall** is formed by the **frontal process of the zygomatic bone** and the *greater wing of the sphenoid*. This is the strongest and thickest wall, which is important because it is most exposed and vulnerable to direct trauma. Its posterior part separates the orbit from the temporal lobes of the brain and middle cranial fossae.
- The **inferior wall** (floor) is formed mainly by the *maxilla* and partly by the *zygomatic* and *palatine bones*. The thin inferior wall is shared by the orbit superiorly and the maxillary sinus inferiorly. It slants inferiorly from the apex to the inferior orbital margin. The inferior wall is demarcated from the lateral wall by the inferior orbital fissure.
- The **apex** of the orbit is at the optic canal in the lesser wing of the sphenoid, just medial to the superior orbital fissure.

The bones forming the orbit are lined with **periorbita** (periosteum). The periorbita is continuous with the following structures:

- Periosteal layer of dura at the optic canal and superior orbital fissure
- Periosteum covering the external surface of the cranium (pericranium) at the orbital margins and through the inferior orbital fissure
- Orbital septa at the orbital margins
- Fascial sheaths of the extra-ocular muscles
- Orbital fascia that forms the fascial sheath of the eyeball

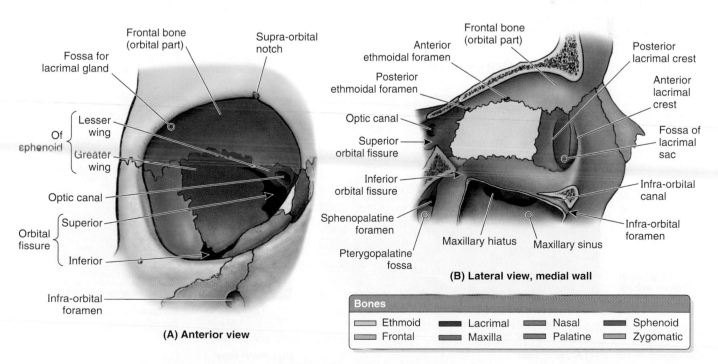

(A) Anterior view

(B) Lateral view, medial wall

Bones			
Ethmoid	Lacrimal	Nasal	Sphenoid
Frontal	Maxilla	Palatine	Zygomatic

FIGURE 8.22. Bones of right orbit.

CLINICAL BOX

Fractures of Orbit

When blows are powerful enough and the impact is directly on the bony rim of the orbit, the resulting fractures usually occur at the sutures between the bones forming the orbital margin. Because of the thinness of the medial and inferior walls of the orbit, a blow to the eye may fracture the orbital walls while the margin remains intact. Indirect traumatic injury that displaces the orbital walls is called a *"blowout" fracture*. Fractures of the medial wall may involve the ethmoidal and sphenoidal sinuses, whereas fractures in the inferior wall may involve the maxillary sinus and may entrap the inferior rectus muscle, limiting upward gaze. Although the superior wall is stronger than the medial and inferior walls, it is thin enough to be translucent and may be readily penetrated. Thus, a sharp object may pass through it into the frontal lobe of the brain. Orbital fractures often result in intra-orbital bleeding, which exerts pressure on the eyeball, causing *exophthalmos* (protrusion of the eyeball).

Orbital Tumors

Because of the closeness of the optic nerve to the sphenoidal and posterior ethmoidal sinuses, a malignant tumor in these sinuses may erode the thin bony walls of the orbit and compress the optic nerve and orbital contents. *Tumors in the orbit* produce *exophthalmos*. A tumor in the middle cranial fossa may enter the orbital cavity through the superior orbital fissure.

Eyelids and Lacrimal Apparatus

The eyelids and lacrimal fluid, secreted by the lacrimal glands, protect the cornea and eyeball from injury and irritation.

EYELIDS

When closed, the **eyelids** (L. *palpebrae*) cover the eyeball anteriorly, thereby protecting it from injury and excessive light (Fig. 8.24). They also keep the cornea moist by spreading the lacrimal fluid. The eyelids are movable folds that are covered externally by thin skin and internally by a transparent mucous membrane, the **palpebral conjunctiva**. The palpebral conjunctiva is reflected onto the eyeball, where it is continuous with the **bulbar conjunctiva** (Figs. 8.23 and 8.24A). The bulbar conjunctiva is loose and wrinkled over the sclera and contains small blood vessels. The bulbar conjunctiva is adherent to the periphery of the cornea. The lines of reflection of the palpebral conjunctiva onto the eyeball form deep recesses, the **superior** and **inferior conjunctival fornices**. The **conjunctival sac** is the space bound by the palpebral and bulbar conjunctivae. This sac is a specialized form of mucosal "bursa" that enables the eyelids to move freely over the surface of the eyeball as they open and close.

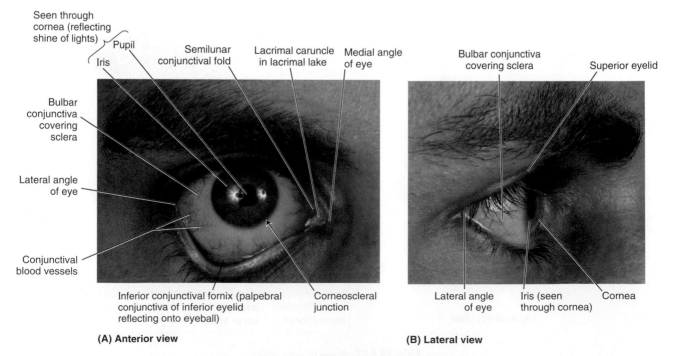

(A) Anterior view

(B) Lateral view

FIGURE 8.23. Surface anatomy of eyeballs and eyelids.

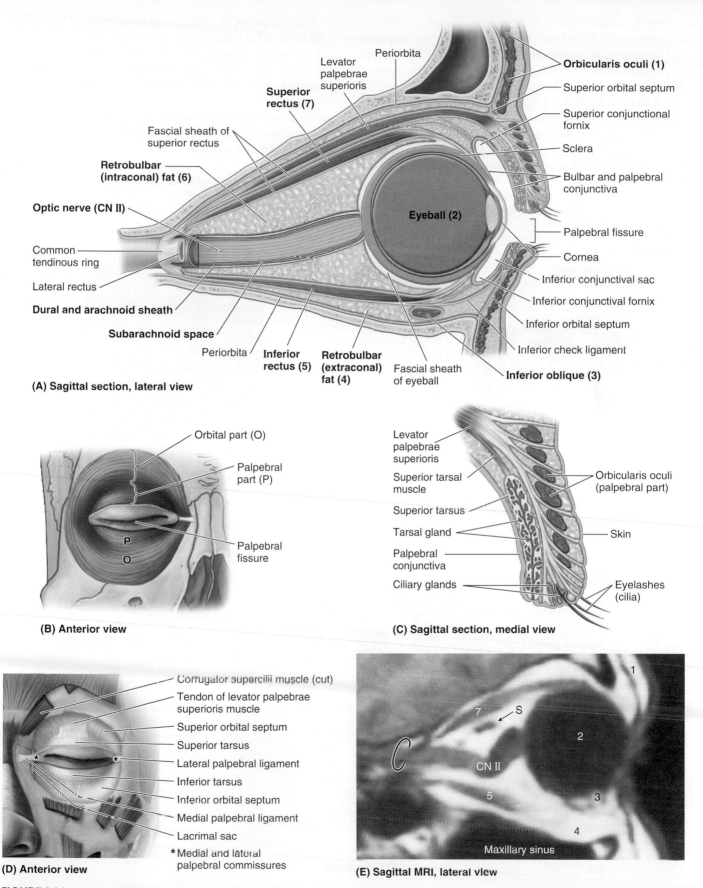

(A) Sagittal section, lateral view

Periorbita
Levator palpebrae superioris
Superior rectus (7)
Fascial sheath of superior rectus
Retrobulbar (intraconal) fat (6)
Optic nerve (CN II)
Common tendinous ring
Lateral rectus
Dural and arachnoid sheath
Subarachnoid space
Periorbita
Inferior rectus (5)
Retrobulbar (extraconal) fat (4)
Fascial sheath of eyeball
Inferior oblique (3)
Orbicularis oculi (1)
Superior orbital septum
Superior conjunctional fornix
Sclera
Bulbar and palpebral conjunctiva
Eyeball (2)
Palpebral fissure
Cornea
Inferior conjunctival sac
Inferior conjunctival fornix
Inferior orbital septum
Inferior check ligament

(B) Anterior view

Orbital part (O)
Palpebral part (P)
Palpebral fissure
P
O

(C) Sagittal section, medial view

Levator palpebrae superioris
Superior tarsal muscle
Superior tarsus
Tarsal gland
Palpebral conjunctiva
Ciliary glands
Orbicularis oculi (palpebral part)
Skin
Eyelashes (cilia)

(D) Anterior view

Corrugator supercilii muscle (cut)
Tendon of levator palpebrae superioris muscle
Superior orbital septum
Superior tarsus
Lateral palpebral ligament
Inferior tarsus
Inferior orbital septum
Medial palpebral ligament
Lacrimal sac
*Medial and lateral palpebral commissures

(E) Sagittal MRI, lateral view

1
7   S
2
CN II
5   3
4
Maxillary sinus
C

FIGURE 8.24. Orbit, eyeball, and eyelids. A. Contents of orbit. The numbers are identified in part **E. B.** Parts of orbicularis oculi. **C.** Superior eyelid. **D.** Skeleton of eyelids and orbital septum. **E.** Sagittal magnetic resonance imaging (MRI) of orbit. *S,* superior ophthalmic vein.

The superior (upper) and inferior (lower) eyelids are strengthened by dense bands of connective tissue, the **superior** and **inferior tarsi** (singular **tarsus**; Fig. 8.24C,D). Fibers of the palpebral portion of the orbicularis oculi are in the subcutaneous tissue superficial to these tarsi and deep to the skin of the eyelid (Fig 8.24A,C). Embedded in the tarsi are **tarsal glands**, the lipid secretion, which lubricates the edges of the eyelids and prevents them from sticking together when they close (Fig. 8.24C). This secretion also forms a barrier that lacrimal fluid does not cross when produced in normal amounts. When production is excessive, it spills over the barrier onto the cheeks as tears.

The **eyelashes** (L. *cilia*) are in the margins of the eyelids. The large sebaceous glands associated with the eyelashes are the **ciliary glands**. The junctions of the superior and inferior eyelids make up the **medial** and **lateral palpebral commissures**, defining the **angles of the eyes** (Fig. 8.23). Thus, each eye has medial and lateral angles, or *canthi*.

In the **medial angle of the eye**, there is a reddish shallow reservoir of tears, the **lacrimal lake**. Within the lake is the **lacrimal caruncle**, a small mound of moist modified skin (Figs. 8.23A and 8.25A,B). Lateral to the caruncle is a **semilunar conjunctival fold**, which slightly overlaps

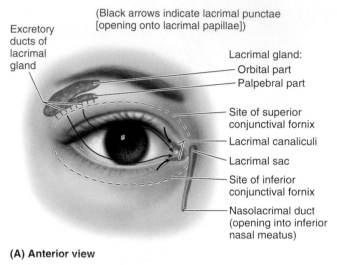

(Black arrows indicate lacrimal punctae [opening onto lacrimal papillae])

Excretory ducts of lacrimal gland

Lacrimal gland:
— Orbital part
— Palpebral part

Site of superior conjunctival fornix

Lacrimal canaliculi

Lacrimal sac

Site of inferior conjunctival fornix

Nasolacrimal duct (opening into inferior nasal meatus)

(A) Anterior view

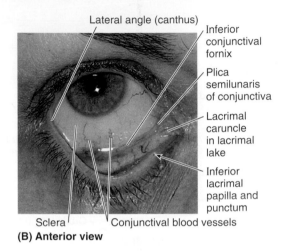

Lateral angle (canthus)

Inferior conjunctival fornix

Plica semilunaris of conjunctiva

Lacrimal caruncle in lacrimal lake

Inferior lacrimal papilla and punctum

Sclera   Conjunctival blood vessels

(B) Anterior view

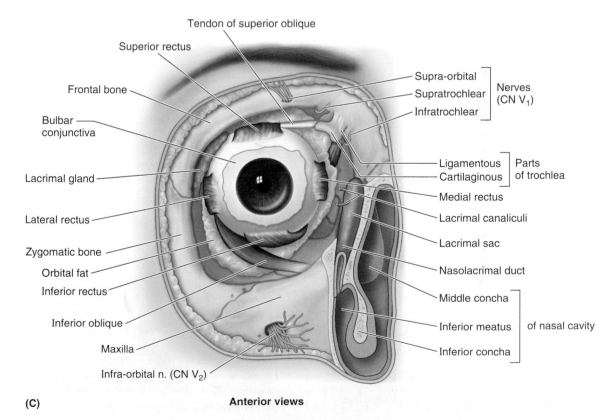

Tendon of superior oblique

Superior rectus

Frontal bone

Bulbar conjunctiva

Lacrimal gland

Lateral rectus

Zygomatic bone

Orbital fat

Inferior rectus

Inferior oblique

Maxilla

Infra-orbital n. (CN V₂)

Supra-orbital
Supratrochlear } Nerves (CN V₁)
Infratrochlear

Ligamentous } Parts
Cartilaginous } of trochlea

Medial rectus

Lacrimal canaliculi

Lacrimal sac

Nasolacrimal duct

Middle concha

Inferior meatus } of nasal cavity

Inferior concha

(C)   **Anterior views**

FIGURE 8.25. Lacrimal apparatus. A. Surface anatomy of the lacrimal apparatus. **B.** Surface anatomy of the eye, with the inferior eyelid retracted. **C.** Dissection of the anterior orbit and nose *n.*, nerve.

the eyeball. When the edges of the eyelids are everted, a minute circular opening, the **lacrimal punctum**, is visible at its medial end on the summit of a small elevation, the **lacrimal papilla** (Fig. 8.25B).

Between the nose and the medial angle of the eye is the **medial palpebral ligament**, which connects the tarsi to the medial margin of the orbit. The orbicularis oculi muscle originates and inserts onto this ligament (Fig. 8.24D). A similar **lateral palpebral ligament** attaches the tarsi to the lateral margin of the orbit. The **orbital septum**, a weak membrane, spans from the tarsi to the margins of the orbit, where it becomes continuous with the periosteum (Fig. 8.24D). It keeps the orbital fat contained and can limit the spread of infection to and from the orbit.

LACRIMAL APPARATUS

The lacrimal apparatus consists of the following (Fig. 8.25):

- *Lacrimal glands* secrete lacrimal fluid (tears).
- **Lacrimal ducts** convey lacrimal fluid from the lacrimal glands to the conjunctival sac.
- **Lacrimal canaliculi** (L. small canals), each commencing at a *lacrimal punctum* (opening) on the *lacrimal papilla* near the medial angle of the eye (Fig. 8.25B), convey the lacrimal fluid from the *lacrimal lake* to the *lacrimal sac*, the dilated superior part of the nasolacrimal duct (Fig. 8.25A).
- **Nasolacrimal duct** conveys the lacrimal fluid to the nasal cavity.

The almond-shaped **lacrimal gland** lies in the *fossa for the lacrimal gland* in the superolateral part of each orbit. The production of lacrimal fluid is stimulated by parasympathetic impulses from CN VII. It is secreted through 8–12 **excretory ducts**, which open into the *superior conjunctival fornix* of the conjunctival sac (Fig. 8.25A). The fluid flows inferiorly within the sac under the influence of gravity. When the cornea becomes dry, the eyelid blinks. The eyelids come together in a lateral to medial sequence, pushing a film of fluid medially over the cornea. The lacrimal fluid

containing foreign material such as dust is pushed toward the medial angle of the eye, accumulating in the *lacrimal lake* from which it drains by capillary action through the *lacrimal puncta* and *lacrimal canaliculi* to the *lacrimal sac*. From this sac, the lacrimal fluid passes to the nasal cavity through the *nasolacrimal duct* (Fig. 8.25C). Here, the fluid flows posteriorly to the nasopharynx and is swallowed.

The *nerve supply of the lacrimal gland* is both sympathetic and parasympathetic. The presynaptic parasympathetic secretomotor fibers are conveyed from the facial nerve by the *greater petrosal nerve* and then by the *nerve of the pterygoid canal* to the *pterygopalatine ganglion*, where they synapse with the cell body of the postsynaptic fiber (see Fig. 8.64D). Vasoconstrictive, postsynaptic sympathetic fibers—brought from the *superior cervical ganglion* by the *internal carotid plexus* and deep petrosal nerve—join the parasympathetic fibers to form the nerve of the pterygoid canal and traverse the pterygopalatine ganglion (see Fig. 8.64E). Branches of the *zygomatic nerve* (from the maxillary nerve) then bring both types of fibers to the lacrimal branch of the ophthalmic nerve (CN V$_1$), by which they enter the gland.

Eyeball

The **eyeball** contains the optical apparatus of the visual system. It occupies most of the anterior portion of the orbit, suspended by six extrinsic muscles that control its movements, and a fascial *suspensory apparatus*. It measures approximately 25 mm in diameter. All anatomical structures within the eyeball have a circular or spherical arrangement.

The *eyeball proper* has three layers (coats or tunics); however, there is an additional connective tissue layer that surrounds the eyeball, supporting it within the orbit. The connective tissue layer is composed posteriorly of the **fascial sheath of the eyeball** (bulbar fascia or Tenon capsule), which forms the actual socket for the eyeball, and anteriorly of bulbar conjunctiva (Fig. 8.24A).

The fascial sheath is the most substantial portion of the suspensory apparatus. A very loose connective tissue layer,

CLINICAL BOX

Injury to Nerves Supplying Eyelids

Because it supplies somatic motor innervation to the levator palpebrae superioris and sympathetic innervation to the superior tarsal muscle, a lesion of the oculomotor nerve (CN III) causes paralysis of the muscle, and the superior eyelid droops (*ptosis*). Damage to the facial nerve (CN VII) involves paralysis of the orbicularis oculi, preventing the eyelids from closing fully. Normal rapid protective blinking of the eye is also lost. The loss of tonus of the muscle in the lower eyelid causes the lid to fall away (*evert*) from the surface of the eye. This leads to drying of the cornea and leaves it unprotected from dust and small particles. Thus, irritation of the unprotected eyeball results in excessive but inefficient *lacrimation* (tear formation).

Inflammation of Palpebral Glands

 Any of the glands in the eyelid may become inflamed and swollen from infection or obstruction of their ducts. If the ducts of the ciliary glands become obstructed, a painful red *suppurative* (pus-producing) swelling, a *sty*, develops on the eyelid. Cysts of the sebaceous glands of the eyelids, called *chalazia*, may also form.

the **episcleral space** (a potential space), lies between the fascial sheath and the outer layer of the eyeball, facilitating movements of the eyeball within the fascial sheath.

The three layers of the eyeball are as follows (Fig. 8.26):

1. *Fibrous layer* (outer coat), consisting of the *sclera* and *cornea*
2. *Vascular layer* (middle coat), consisting of the *choroid*, *ciliary body*, and *iris*
3. *Inner layer* (inner coat), consisting of the *retina*, which has both *optic* and *nonvisual parts*

FIBROUS LAYER OF EYEBALL

The **fibrous layer of the eyeball** is the external fibrous skeleton of the eyeball, providing shape and resistance. The **sclera** is the tough opaque part of the fibrous layer (coat) of the eyeball, covering the posterior five sixths of the eyeball (Fig. 8.26A) and providing attachment for both the extrinsic (extra-ocular) and the intrinsic muscles of the eye. The anterior part of the sclera is visible through the transparent bulbar conjunctiva as "the white of the eye" (Fig. 8.24B).

The **cornea** is the transparent part of the fibrous layer covering the anterior one sixth of the eyeball. The convexity of the cornea is greater than that of the sclera (Figs. 8.26A and 8.27), and so it appears to protrude from the eyeball when viewed laterally.

The two parts of the fibrous coat differ primarily in terms of the regularity of the arrangement of the collagen fibers of which they are composed and the degree of hydration of each. Whereas the sclera is relatively avascular, the cornea is completely avascular, receiving its nourishment from capillary beds around its periphery and fluids on its external and internal surfaces, the *lacrimal fluid* and *aqueous humor*, respectively (Fig. 8.27). Lacrimal fluid also provides oxygen absorbed from the air.

The cornea is highly sensitive to touch; its innervation is provided by the ophthalmic nerve (CN V_1). Even very small foreign bodies (e.g., dust particles) elicit blinking, flow of tears, and sometimes severe pain. Drying of the corneal surface may cause ulceration.

The **limbus** of the cornea is the angle formed by the intersecting curvatures of sclera and cornea at the **corneoscleral junction** (Figs. 8.26A and 8.27). The junction is a 1-mm wide, gray, and translucent circle, including numerous capillary loops involved in nourishing the avascular cornea.

VASCULAR LAYER OF EYEBALL

The **vascular layer of the eyeball** (also called the **uvea** or uveal tract) consists of the choroid, ciliary body, and iris (Fig. 8.26B).

The **choroid**, a dark reddish-brown layer between the sclera and the retina, forms the largest part of the vascular layer of the eyeball and lines most of the sclera (Fig. 8.27B). Within this pigmented and dense vascular bed, larger vessels are located externally (near the sclera). The finest vessels (the **capillary lamina of the choroid** or *choriocapillaris*, an extensive capillary bed) are innermost, adjacent to the avascular light-sensitive layer of the retina, which it supplies with oxygen and nutrients. Engorged with blood in life (it has the highest perfusion rate per gram of tissue of all vascular beds of the body), this layer is responsible for the "red eye" reflection that occurs in flash photography. The choroid attaches firmly to the pigment layer of the retina, but it can easily be stripped from the sclera. The choroid is continuous anteriorly with the ciliary body.

The **ciliary body** is a ring-like thickening of the layer posterior to the corneoscleral junction that is muscular as well as vascular (Figs. 8.26B and 8.27B). It connects the choroid with the circumference of the iris. The ciliary body provides

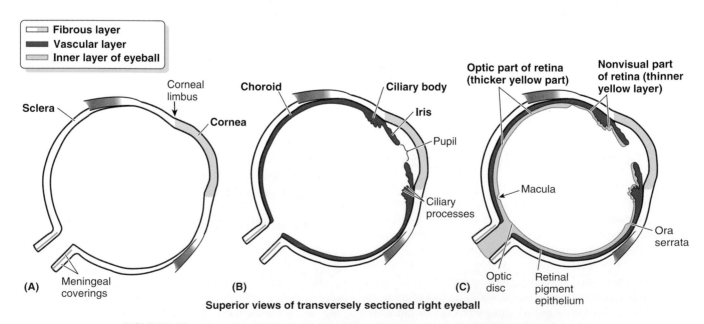

FIGURE 8.26. Layers of eyeball. A. Outer fibrous layer. **B.** Middle vascular layer. **C.** Inner layer (retina).

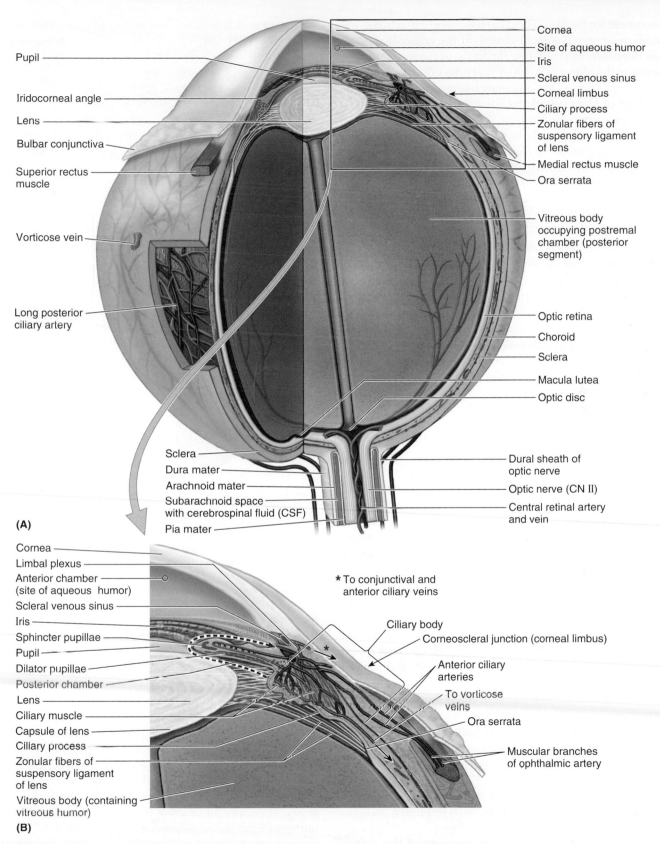

Pupil

Iridocorneal angle

Lens

Bulbar conjunctiva

Superior rectus
muscle

Vorticose vein

Long posterior
ciliary artery

Cornea

Site of aqueous humor

Iris

Scleral venous sinus

Corneal limbus

Ciliary process

Zonular fibers of
suspensory ligament
of lens

Medial rectus muscle

Ora serrata

Vitreous body
occupying postremal
chamber (posterior
segment)

Optic retina

Choroid

Sclera

Macula lutea

Optic disc

Sclera

Dura mater

Arachnoid mater

Subarachnoid space
with cerebrospinal fluid (CSF)

Pia mater

Dural sheath of
optic nerve

Optic nerve (CN II)

Central retinal artery
and vein

(A)

Cornea

Limbal plexus

Anterior chamber
(site of aqueous humor)

Scleral venous sinus

Iris

Sphincter pupillae

Pupil

Dilator pupillae

Posterior chamber

Lens

Ciliary muscle

Capsule of lens

Cillary process

Zonular fibers of
suspensory ligament
of lens

Vitreous body (containing
vitreous humor)

*** To conjunctival and
anterior ciliary veins**

Ciliary body

Corneoscleral junction (corneal limbus)

Anterior ciliary
arteries

To vorticose
veins

Ora serrata

Muscular branches
of ophthalmic artery

(B)

FIGURE 8.27. Eyeball with quarter section removed. A. Structure of eyeball. The inner aspect of the optic part of the retina is supplied by the central retinal artery, whereas the outer, light-sensitive aspect is nourished by the capillary lamina of the choroid. The branches of the central artery are end arteries that do not anastomose with each other or any other vessel. **B.** Structures of ciliary region. The ciliary body is both muscular and vascular, as is the iris, the latter including two muscles: the sphincter pupillae and dilator pupillae. Venous blood from this region and the aqueous humor in the anterior chamber drain into the scleral venous sinus.

attachment for the lens. The contraction and relaxation of the circularly arranged smooth muscle of the ciliary body controls thickness, and therefore the focus, of the lens. Folds on the internal surface of the ciliary body, the **ciliary processes**, secrete *aqueous humor*. Aqueous humor fills the **anterior segment of the eyeball**, the interior of the eyeball anterior to the lens, suspensory ligament, and ciliary body (Fig. 8.27B).

The **iris**, which literally lies on the anterior surface of the lens, is a thin contractile diaphragm with a central aperture, the **pupil**, for transmitting light (Figs. 8.26B and 8.27). When awake, the size of the pupil varies continually to regulate the amount of light entering the eye (Fig. 8.28). Two involuntary muscles control the size of the pupil: The parasympathetically stimulated, circularly arranged **sphincter pupillae** decreases its diameter (constrict or contracts the pupil, *pupillary miosis*) and the sympathetically stimulated, radially arranged **dilator pupillae** increases its diameter (dilates the pupil). The nature of the pupillary responses is paradoxical: Sympathetic responses usually occur immediately, yet it may take up to 20 minutes for the pupil to dilate in response to low lighting, as in a darkened theater. Parasympathetic responses are typically slower than sympathetic responses, yet parasympathetically stimulated pupillary constriction is normally instantaneous. Abnormal sustained pupillary dilation (*mydriasis*) may occur in certain diseases or as a result of trauma or the use of certain drugs.

INNER LAYER OF EYEBALL

The inner layer of the eyeball is the **retina** (Figs. 8.26C and 8.27). It consists grossly of two functional parts with distinct locations: the optic and nonvisual parts. The **optic part of the retina** is sensitive to visual light rays and has two layers: a neural layer and pigmented layer. The **neural layer** is light receptive. The **pigmented layer** consists of a single layer of cells that reinforces the light-absorbing property of the choroid in reducing the scattering of light in the eyeball. The **nonvisual retina** is an anterior continuation of the pigmented layer and a layer of supporting cells. The nonvisual retina extends over the ciliary body (**ciliary part** of the retina) and the posterior surface of the iris (**iridial part** of the retina) to the pupillary margin.

Clinically, the internal aspect of the posterior part of the eyeball, where light entering the eyeball is focused, is referred to as the **fundus of the eyeball** (ocular fundus). The retina of the fundus includes a distinctive circular area, the **optic disc** (optic papilla), where the sensory fibers and vessels conveyed by the optic nerve (CN II) enter and radiate to the eyeball (Figs. 8.26C, 8.27A, and 8.29). Because it contains no photoreceptors, the optic disc is insensitive to light. Hence, it is commonly called the *blind spot*.

Just lateral to the optic disc is the **macula of the retina** or **macula lutea** (L. yellow spot). The yellow color of the macula is apparent only when the retina is examined with red-free light. The macula lutea is a small oval area of the retina with special photoreceptor cones that is specialized

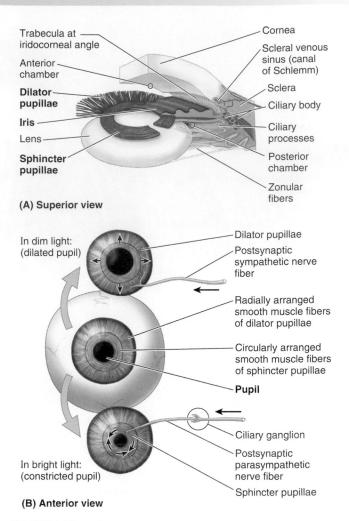

(A) Superior view

(B) Anterior view

FIGURE 8.28. Structure and function of the iris. A. Iris dissected, in situ. The iris separates the anterior and posterior chambers of the anterior segment of the eyeball as it bounds the pupil. **B.** Dilation and constriction of pupil. In dim light, sympathetic fibers stimulate dilation of the pupil. In bright light, parasympathetic fibers stimulate constriction of the pupil.

for acuity of vision. It is not normally observed with an *ophthalmoscope* (a device for viewing the interior of the eyeball through the pupil). At the center of the macula lutea is a depression, the **fovea centralis** (L. central pit), the area of most acute vision. The fovea is approximately 1.5 mm in diameter; its center, the **foveola**, does not have the capillary network visible elsewhere deep to the retina.

The optic part of the retina terminates anteriorly along the **ora serrata** (L. serrated edge), the irregular posterior border of the ciliary body (Figs. 8.26C and 8.27B). Except for the cones and rods of its neural layer, the retina is supplied by the **central retinal artery**, a branch of the ophthalmic artery. The cones and rods of the outer neural layer receive nutrients from the *capillary lamina of the choroid*, or choriocapillaris (discussed in "Vasculature of Orbit," later in this chapter). Its inner surface has the finest vessels of the choroid, against which the retina is pressed. A corresponding system of retinal veins unites to form the **central retinal vein** (Fig. 8.27A).

REFRACTIVE MEDIA AND COMPARTMENTS OF EYEBALL

On their way to the retina, light waves pass through the refractive media of the eyeball: cornea, aqueous humor, lens, and vitreous humor (Fig. 8.27). The *cornea* is the primary refractory medium of the eyeball—that is, it bends light to the greatest degree, focusing an inverted image on the light-sensitive retina, especially that of the *optic fundus*.

The **aqueous humor** (often shortened clinically to "aqueous") occupies the *anterior segment of the eyeball* (Fig. 8.27B). The anterior segment is subdivided by the iris and pupil. The **anterior chamber of the eye** is the space between the cornea anteriorly and the iris/pupil posteriorly. The **posterior chamber of the eye** is between the iris/pupil anteriorly and the lens and ciliary body posteriorly. Aqueous humor is produced in the posterior chamber by the ciliary processes of the ciliary body. This clear watery solution provides nutrients for the avascular cornea and lens. After passing through the pupil into the anterior chamber, the aqueous humor drains through a trabecular meshwork at the **iridocorneal angle** into the *scleral venous sinus* (L. *sinus venosus sclerae*, canal of Schlemm) (Fig. 8.28A). The humor is removed by the **limbal plexus**, a network of scleral veins close to the limbus, which drain in turn into both tributaries of the *vorticose* and the *anterior ciliary veins* (Fig. 8.27B). Intra-ocular pressure (IOP) is a balance between production and outflow of aqueous humor.

The **lens** is posterior to the iris and anterior to the vitreous humor of the vitreous body (Figs. 8.27 and 8.28A). It is a transparent, biconvex structure enclosed in a capsule. The highly elastic **capsule of the lens** is anchored by **zonular fibers** (collectively constituting the **suspensory ligament of the lens**) to the encircling ciliary processes. Although most

Branches of retinal vessels (arterioles and venules)

Macula of retina   Optic disc

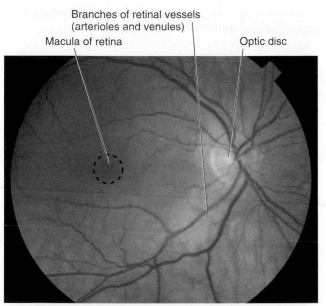

(A) Ophthalmoscopic view

Branches of retinal vessels (arterioles and venules)

Macula of retina   Optic disc

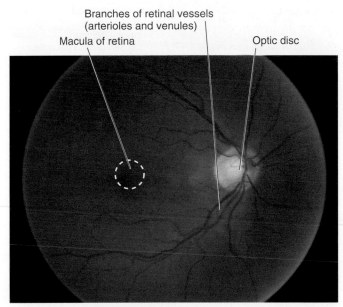

(B) Digital retinal photography

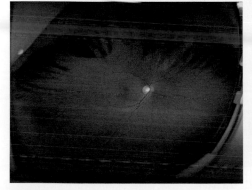

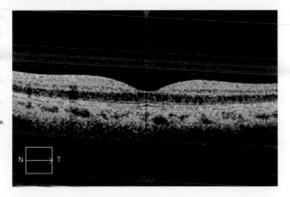

Blue line: plane of section of (through macula)

(C) Optical coherence tomography

FIGURE 8.29. Views of the retina. A. Right ocular fundus, ophthalmoscopic view. **B.** Right ocular fundus, digital retinal photography. Retinal venules (wider) and retinal arterioles (narrower) radiate from the center of the oval optic disc. The dark area lateral to the disc is the macula. Branches of retinal vessels extend toward this area but do not reach its center, the fovea centralis—the area of most acute vision. **C.** Macular thickness, optical coherence tomography.

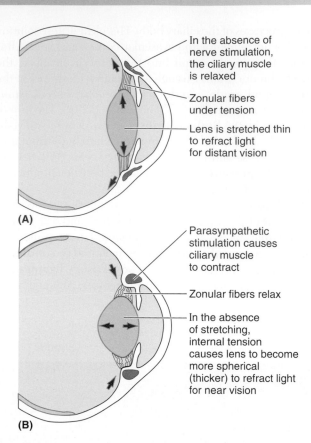

(A)

In the absence of nerve stimulation, the ciliary muscle is relaxed

Zonular fibers under tension

Lens is stretched thin to refract light for distant vision

(B)

Parasympathetic stimulation causes ciliary muscle to contract

Zonular fibers relax

In the absence of stretching, internal tension causes lens to become more spherical (thicker) to refract light for near vision

FIGURE 8.30. Changing lens shape for distant and near vision (accommodation). A. Distant vision. **B.** Near vision.

refraction is produced by the cornea, the convexity of the lens, particularly its anterior surface, constantly varies to fine-tune the focus of near or distant objects on the retina (Fig. 8.30). The isolated unattached lens assumes a nearly spherical shape. In other words, in the absence of external attachment and stretching, it becomes nearly round.

The **ciliary muscle** of the ciliary body changes the shape of the lens. In the absence of nerve stimulation, the diameter of the relaxed muscular ring is larger. The lens suspended within the ring is under tension as its periphery is stretched, causing it to be thinner (less convex). The less convex lens brings more distant objects into focus (far vision). Parasympathetic stimulation via the oculomotor nerve (CN III) causes sphincter-like contraction of the ciliary muscle. The ring becomes smaller, and tension on the lens is reduced. The relaxed lens thickens (becomes more convex), bringing near objects into focus (near vision). The active process of changing the shape of the lens for near vision is called **accommodation**. The thickness of the lens increases with aging so that the ability to accommodate typically becomes restricted after age 40 years.

The **vitreous humor** is a watery fluid enclosed in the meshes of the **vitreous body**, a transparent jelly-like substance in the posterior four fifths of the eyeball posterior to the lens (*posterior segment of the eyeball*, also called the *postremal* or *vitreous chamber*) (Fig. 8.27A). In addition to transmitting light, the vitreous humor holds the retina in place and supports the lens.

CLINICAL BOX

Ophthalmoscopy

Physicians view the fundus (inner surface of the posterior part) of the eye with an *ophthalmoscope*. The retinal arteries and veins radiate over the fundus from the optic disc. The pale, oval optic disc appears on the medial side, with retinal vessels radiating from its center in the ophthalmoscopic view of the retina (Fig. 8.29). Pulsation of the retinal arteries is usually visible. Centrally, at the posterior pole of the eyeball, the macula lutea appears darker than the reddish hue of surrounding areas of the retina.

Detachment of Retina

The layers of the developing retina are separated in the embryo by an intraretinal space. During the early fetal period, the embryonic layers fuse, obliterating this space. Although the pigment cell layer becomes firmly fixed to the choroid, its attachment to the neural layer is not firm. Consequently, detachment of the retina may follow a blow to the eye. A *detached retina* usually results from seepage of fluid between the neural and pigmented layers of the retina, perhaps days or even weeks after trauma to the eye (Fig. B8.7). People with a retinal detachment may complain of flashes of light or specks floating in front of their eye.

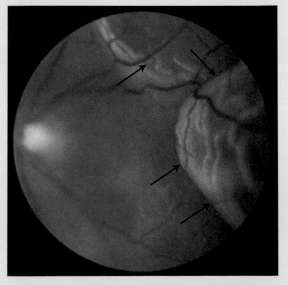

FIGURE B8.7. Detached retina. *Arrows*, edges of wrinkled, detached portions of retina.

Papilledema

 An increase in CSF pressure slows venous return from the retina, causing *edema of the retina* (fluid accumulation). The edema is viewed during ophthalmoscopy as swelling of the optic disc, a condition called *papilledema*.

Presbyopia and Cataracts

 As people age, their lenses become harder and more flattened. These changes gradually reduce the focusing power of the lenses, a condition known as *presbyopia* (G. *presbyos*, old). Some people also develop cataracts, a loss of transparency (cloudiness) of the lens from areas of opaqueness. *Cataract extraction* combined with an *intra-ocular lens implant* has become a common operation. An extracapsular cataract extraction involves removing the lens but leaving the capsule of the lens intact to receive a synthetic intra-ocular lens (Fig. B8.8A,B). Intracapsular lens extraction involves removing the lens and lens capsule and implanting a synthetic intra-ocular lens in the anterior chamber (Fig. B8.8C).

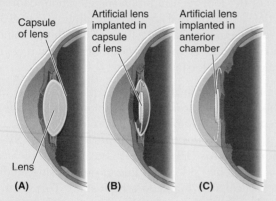

FIGURE B8.8. Cataract extraction with intra-ocular lens implant.

Glaucoma

Outflow of aqueous humor through the scleral venous sinus into the blood circulation must occur at the same rate at which the aqueous is produced. If the outflow decreases significantly because the outflow pathway is blocked, *intra-ocular pressure* (IOP) builds up in the anterior and posterior chambers of the eye, a condition called *glaucoma* (Fig. B8.9). Blindness can result from compression of the inner

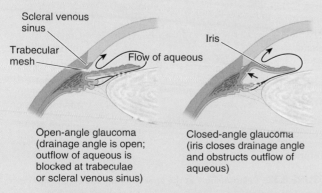

FIGURE B8.9. Open- versus closed-angle glaucoma.

layer of the eyeball (retina) and the retinal arteries if aqueous humor production is not reduced to maintain normal IOP.

Corneal Ulcers and Transplants

 Damage to the sensory innervation of the cornea from CN V₁ leaves the cornea vulnerable to injury by foreign particles. People with scarred or opaque corneas may receive *corneal transplants* from donors. Corneal implants of nonreactive plastic material are also used.

Development of Retina

 The retina and optic nerve develop from the **optic cup**, an outgrowth of the embryonic forebrain, the **optic vesicle** (Fig. B8.10A). As it evaginates from the forebrain (Fig. B8.10B), the optic vesicle carries the developing meninges with it. Hence the optic nerve is invested with cranial meninges and an extension of the subarachnoid space (Fig. B8.10C). The central artery and vein of the retina cross the subarachnoid space and run within the distal part of the optic nerve. The pigment cell layer of the retina develops from the outer layer of the optic cup, and the neural layer develops from the inner layer of the cup.

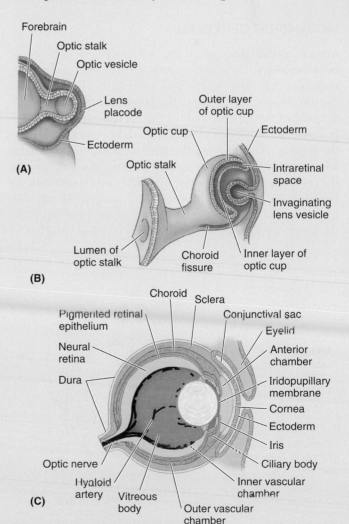

FIGURE B8.10. Development of retina.

Extra-ocular Muscles of Orbit

The **extra-ocular muscles of the orbit** are the *levator palpebrae superioris*, four *recti* (*superior, inferior, medial,* and *lateral*), and two *obliques* (*superior* and *inferior*). These muscles work together to move the superior eyelids and eyeballs (Figs. 8.31–8.33 and Table 8.6).

LEVATOR PALPEBRAE SUPERIORIS

The **levator palpebrae superioris** broadens into a wide bilaminar aponeurosis as it approaches its distal attachments. The superficial lamina attaches to the skin of the superior eyelid and the deep lamina to the superior tarsus (Fig. 8.24B). This muscle is opposed most of the time by gravity and is the antagonist of the superior half of the orbicularis oculi, the sphincter of the palpebral fissure. The deep lamina of the distal (palpebral) part of the muscle includes smooth muscle fibers, the **superior tarsal muscle**, that produce additional widening of the palpebral fissure, especially during a sympathetic response (e.g., fright). However, they seem to function continuously (in the absence of a sympathetic response per se) because an interruption of the sympathetic supply produces a constant *ptosis*—drooping of the upper eyelid.

MOVEMENTS OF EYEBALL

Movements of the eyeball can be described in terms of rotations around three *axes—vertical, transverse,* and *anteroposterior* (AP) (Fig. 8.31)—and are described according to the direction of movement of the pupil from the primary position or of the superior pole of the eyeball from the neutral position. Rotation of the eyeball around the vertical axis moves the pupil medially (toward the midline, **adduction**) or laterally (away from the midline, **abduction**). Rotation around the transverse axis moves the pupil superiorly (**elevation**) or inferiorly (**depression**). Movements around the AP axis (corresponding to the axis of gaze in the primary position) move the superior pole of the eyeball medially (**medial rotation**, or

intorsion) or laterally (**lateral rotation,** or *extorsion*). These rotational movements accommodate moderate changes in the tilt of the head. Absence of these movements resulting from nerve lesions contributes to double vision.

Movements may occur around the three axes simultaneously, requiring three terms to describe the direction of movement from the primary position (e.g., the pupil is elevated, adducted, and medially rotated).

RECTI AND OBLIQUE MUSCLES

The four **recti muscles** (L. *rectus,* straight) run anteriorly to the eyeball, arising from a fibrous cuff, the **common tendinous ring**, that surrounds the optic canal and part of the superior orbital fissure at the apex of the orbit (Figs. 8.32 and 8.33A,B and Table 8.6). Structures that enter the orbit through this canal and the adjacent part of the fissure lie initially within the cone of recti. The four recti are named for their individual positions relative to the eyeball. Because they mainly run anteriorly to attach to the superior, inferior, medial, and lateral aspects of the eyeball anterior to its equator, the primary actions of the four recti in producing elevation, depression, adduction, and abduction are relatively intuitive (Fig. 8.34).

Several factors make the actions of the obliques and the secondary actions of the superior and inferior recti more challenging to understand:

- The *apex of the orbit* is medially placed relative to the orbit, so that the *axis of the orbit* does not coincide with the *optical axis* (Fig. 8.33D). Therefore, *when the eye is in the primary position,* the **superior rectus (SR)** and **inferior rectus (IR)** muscles also approach the eyeball from its medial side, their line of pull passing medial to the vertical axis (Fig. 8.33A, right side). This gives both muscles a secondary action of *adduction.* The SR and IR also extend laterally, passing superior and inferior to the AP axis, respectively, giving the SR a secondary action of *medial rotation* and the IR a secondary action of *lateral rotation* (Fig. 8.33A, left side).

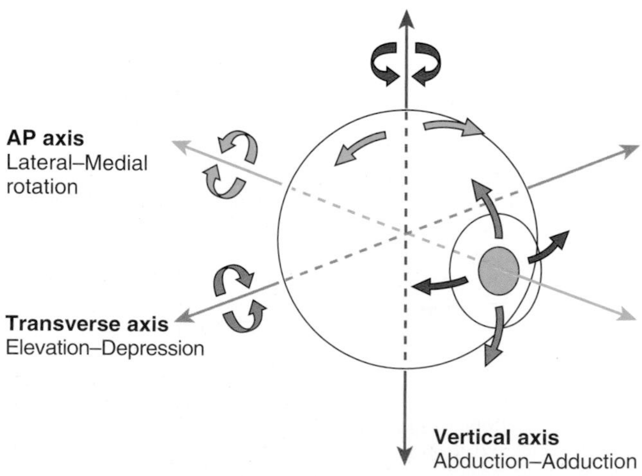

AP axis
Lateral–Medial rotation

Transverse axis
Elevation–Depression

Vertical axis
Abduction–Adduction

FIGURE 8.31. Axes around which movements of eyeball occur.

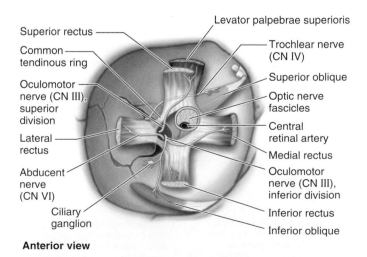

Anterior view

FIGURE 8.32. Relationships at the apex of orbit. The eyeball has been excised (enucleated).

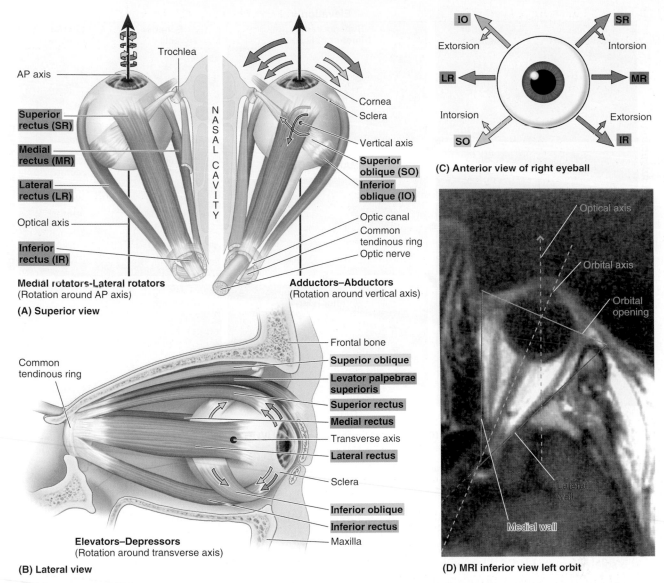

FIGURE 8.33. Extra-ocular muscles and their movements. A. Medial–lateral rotators (left eye) and adductors–abductors (right eye). *Arrows* indicate movements of the eyeball around the anteroposterior (AP) axis on the left and around the vertical axis on the right. **B.** Elevators–depressors. *Arrows* indicate movements of the eyeball around the transverse axis. **C.** Unilateral diagram of extra-ocular muscle actions, starting from the primary position. For movements in any of the six cardinal directions (*large arrows*), the indicated muscle is the prime mover. Movements in directions between *large arrows* require synergistic actions by the adjacent muscles. *Small arrows* indicate muscles producing rotational movements around the AP axis. **D.** Orbital and optical axes.

TABLE 8.6. MUSCLES OF ORBIT

Muscle	Origin	Insertion	Innervation	Main Action(s)[a]
Levator palpebrae superioris	Lesser wing of sphenoid bone, superior and anterior to optic canal	Superior tarsus and skin of superior eyelid	Oculomotor nerve; deep layer (superior tarsal muscle) supplied by sympathetic fibers	Elevates superior eyelid
Superior oblique (SO)	Body of sphenoid bone	Tendon passes through trochlea to insert into sclera, deep to SR	Trochlear nerve (CN IV)	Abducts, depresses, and rotates eyeball medially (intorsion)
Inferior oblique (IO)	Anterior part of floor of orbit	Sclera deep to lateral rectus muscle	Oculomotor nerve (CN III)	Abducts, elevates, and rotates eyeball laterally (extorsion)
Superior rectus (SR)	Common tendinous ring	Sclera just posterior to corneoscleral junction	Oculomotor nerve (CN III)	Elevates, adducts, and rotates eyeball medially (intorsion)
Inferior rectus (IR)				Depresses, adducts, and rotates eyeball laterally (extorsion)
Medial rectus (MR)				Adducts eyeball
Lateral rectus (LR)			Abducent nerve (CN VI)	Abducts eyeball

[a]It is essential to appreciate that all muscles are continuously involved in eyeball movements; thus, the individual actions are not usually tested clinically.

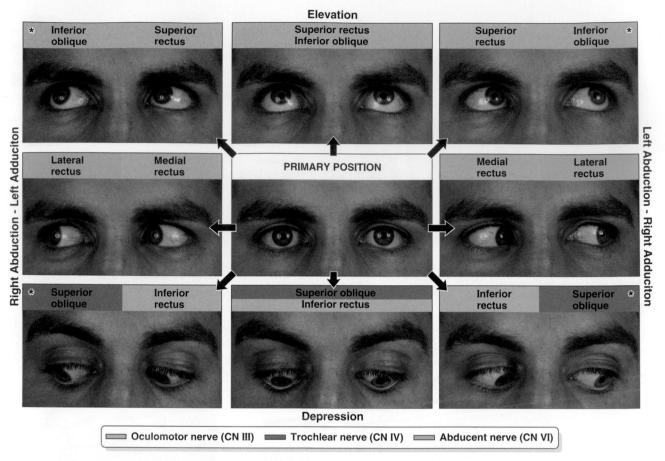

FIGURE 8.34. Anatomical movements of extra-ocular muscles (single movements directly from primary position). Note that results in outer corner frames (asterisks) are opposite to those in Figure 8.35A showing clinical testing of the eye muscles.

- *If the gaze is first directed laterally* (abducted by the **lateral rectus [LR]**) so that the line of gaze coincides with plane of the IR and SR, *the SR produces elevation only* (and is solely responsible for the movement) (Fig. 8.35B), and *the IR produces depression only* (and is likewise solely responsible) (Fig. 8.35C). During a physical examination, the physician directs the patient to follow his or her finger laterally (testing the LR and abducent nerve [CN VI]), then superiorly and inferiorly to isolate and test the function of the SR and IR and the integrity of the oculomotor nerve (CN III) that supplies both (Fig. 8.35A)

- The **inferior oblique** (**IO**) is the only muscle to originate from the anterior part of the orbit (immediately lateral to the lacrimal fossa) (Fig. 8.32). The **superior oblique** (**SO**) originates from the apex region like the rectus muscles (but superomedial to the common tendinous ring); however, its tendon traverses the *trochlea* just inside the superomedial orbital rim, redirecting its line of pull (Fig. 8.33A). Thus, the inserting tendons of the oblique muscles lie in the same oblique vertical plane. When the inserting tendons are viewed anteriorly (see Fig. 8.25C) or superiorly (Fig. 8.33A,B) with the eyeball in the primary position, it can be seen that the tendons of the oblique muscles pass mainly laterally to insert on the lateral half of the eyeball, posterior to its equator. Because they pass

inferior and superior to the AP axis as they pass laterally, the IO is the primary lateral rotator, and the SO the primary medial rotator, of the eye (Fig. 8.33A, left side).

- However, in the primary position, the obliques also pass posteriorly across the transverse axis (Fig. 8.33B) and posterior to the vertical axis (Fig. 8.33A, right side), giving the SO a secondary function as a depressor, the IO a secondary function as an elevator, and both muscles a secondary function as abductors.

- If the gaze is first directed medially (adducted by the **medial rectus [MR]**) so that the line of gaze coincides with plane of the inserting tendons of the SO and IO, the SO produces depression only (and is solely responsible for the movement) (Fig. 8.35D), and the IO produces elevation only (and is likewise solely responsible) (Fig. 8.35E). During a physical examination, the physician directs the patient to follow his or her finger medially (testing the MR and oculomotor nerve), then inferiorly and superiorly to isolate and test the functions of the SO and IO and the integrity of the trochlear nerve (CN IV) supplying the SO and of the inferior division of the oculomotor nerve (CN III) supplying the IO (Fig. 8.35A). *In practice, the main action of the*

- *SO is depression of the pupil in the adducted position* (e.g., directing the gaze down the page when the

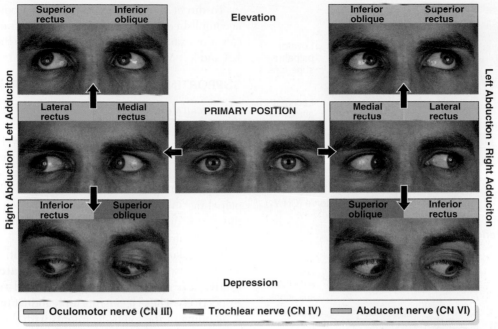

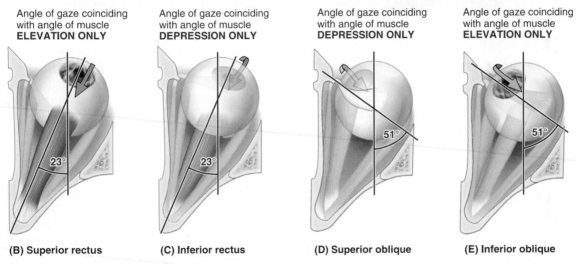

FIGURE 8.35. Clinical testing of extra-ocular muscles. Right eye is shown. **A.** Two-movement sequences (elevation or depression following left or right gaze). Following movements of the examiner's finger, the pupil is moved in an extended H pattern to isolate and test individual extra-ocular muscles and the integrity of their nerves. **B and C.** When the eye is initially abducted by lateral rectus (LR), only the rectus muscles can produce elevation and depression. **D and E.** When the eye is adducted by medial rectus (MR), only the oblique muscles can produce these movements.

gaze of both eyes is directed medially [*converged*] for reading).

- *IO is elevation of the pupil in the adducted position* (e.g., directing the gaze up the page during **convergence** for reading).

Although the actions produced by the extra-ocular muscles have been considered individually, all motions require the action of several muscles in the same eye, assisting each other as synergists or opposing each other as antagonists. Muscles that are synergistic for one action may be antagonistic for another. For example, no single muscle can act to elevate the pupil directly from the primary position (Fig. 8.33C). The two elevators (SR and IO) act as synergists to do so (Fig. 8.35). However, these muscles are antagonistic as rotators and so neutralize each other so that no rotation occurs as they work together to elevate the pupil.

Similarly, no single muscle can act to depress the pupil directly from the primary position. The two depressors, the SO and IR, both produce depression when acting alone and also produce opposing actions in terms of adduction–abduction and medial–lateral rotation. However, when the SO and IR act simultaneously, their synergistic actions depress the pupil as their antagonistic actions neutralize each other; therefore, pure depression results (Fig. 8.35).

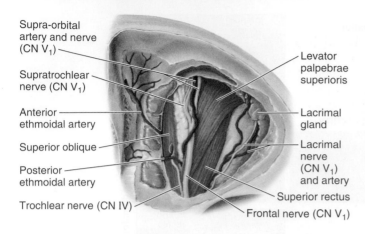

Supra-orbital artery and nerve (CN V₁)

Supratrochlear nerve (CN V₁)

Anterior ethmoidal artery

Superior oblique

Posterior ethmoidal artery

Trochlear nerve (CN IV)

Levator palpebrae superioris

Lacrimal gland

Lacrimal nerve (CN V₁) and artery

Superior rectus

Frontal nerve (CN V₁)

(A) Superior view

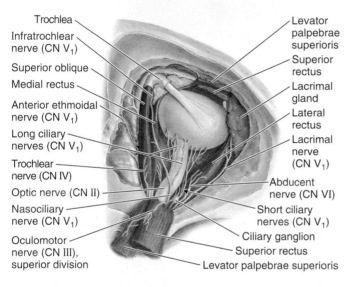

Trochlea

Infratrochlear nerve (CN V₁)

Superior oblique

Medial rectus

Anterior ethmoidal nerve (CN V₁)

Long ciliary nerves (CN V₁)

Trochlear nerve (CN IV)

Optic nerve (CN II)

Nasociliary nerve (CN V₁)

Oculomotor nerve (CN III), superior division

Levator palpebrae superioris

Superior rectus

Lacrimal gland

Lateral rectus

Lacrimal nerve (CN V₁)

Abducent nerve (CN VI)

Short ciliary nerves (CN V₁)

Ciliary ganglion

Superior rectus

Levator palpebrae superioris

(B) Superior view

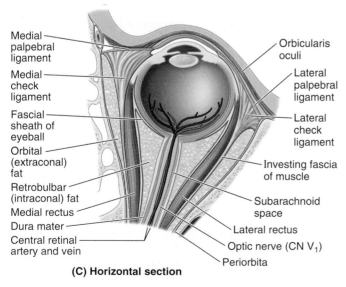

Medial palpebral ligament

Medial check ligament

Fascial sheath of eyeball

Orbital (extraconal) fat

Retrobulbar (intraconal) fat

Medial rectus

Dura mater

Central retinal artery and vein

Orbicularis oculi

Lateral palpebral ligament

Lateral check ligament

Investing fascia of muscle

Subarachnoid space

Lateral rectus

Optic nerve (CN V₁)

Periorbita

(C) Horizontal section

FIGURE 8.36. Dissections of orbit. A. Superficial dissection of the right orbit. **B.** Deep dissection of left orbit. **C.** Fascial sheath of eyeball and check ligaments.

To direct the gaze, coordination of both eyes must be accomplished by the paired action of contralateral *yoke muscles*. For example, in directing the gaze to the right, the right LR and left MR act as yoke muscles.

SUPPORTING APPARATUS OF EYEBALL

The *fascial sheath of the eyeball* envelops the eyeball, extending posteriorly from the conjunctival fornices to the optic nerve, forming the actual socket for the eyeball (Fig. 8.36C). The cup-like fascial sheath is pierced by the tendons of the extra-ocular muscles and is reflected onto each of them as a tubular *muscle sheath*. The muscle sheaths of the levator palpebrae superioris and SR muscles are fused; thus, when the gaze is directed superiorly, the superior eyelid is further elevated out of the line of vision.

Triangular expansions from the sheaths of the medial and LR muscles, called the **medial** and **lateral check ligaments**, are attached to the lacrimal and zygomatic bones, respectively. These ligaments limit abduction and adduction. A blending of the check ligaments with the fascia of the IR and IO muscles forms a hammock-like sling, the *suspensory ligament of the eyeball*. A similar check ligament from the fascial sheath of the IR retracts the inferior eyelid when the gaze is directed downward. Collectively, the check ligaments act with the oblique muscles and the **retrobulbar fat** to resist the posterior pull on the eyeball produced by the rectus muscles. In starvation or diseases that reduce the retrobulbar fat, the eyeball is retracted into the orbit (*enophthalmos*).

Nerves of Orbit

The large **optic nerves** (**CN II**; Fig. 8.36B) are purely sensory nerves that transmit impulses generated by optical stimuli and develop as paired anterior extensions of the forebrain. Throughout their course in the orbit, the optic nerves are surrounded by extensions of the *cranial meninges* and *subarachnoid space*, the latter occupied by a thin layer of CSF (Fig. 8.38A, *inset*). The intra-orbital extensions of the cranial dura and arachnoid mater constitute the **optic sheath**, which becomes continuous anteriorly with the fascial sheath of the eyeball and the sclera. A layer of pia mater covers the surface of the optic nerve within the sheath. They exit the orbits via the optic canals.

In addition to the optic nerves, the nerves of the orbit include those that enter through the *superior orbital fissure* and supply the ocular muscles (Figs. 8.35 and 8.37A,B): **oculomotor** (CN III), **trochlear** (CN IV), and **abducent** (CN VI) nerves. A memory device for the innervation of the extra-ocular muscles moving the eyeball is similar to a chemical formula: $LR_6SO_4AO_3$ (**l**ateral **r**ectus, CN **VI**; **s**uperior **o**blique, CN **IV**; **a**ll **o**thers, CN **III**). The trochlear and abducent nerves pass directly to the single muscle supplied by each nerve. The oculomotor nerve divides into a superior branch supplying SR and levator palpebrae superioris and an inferior branch supplying the medial and IR and IO and carrying presynaptic parasympathetic fibers to the ciliary ganglion.

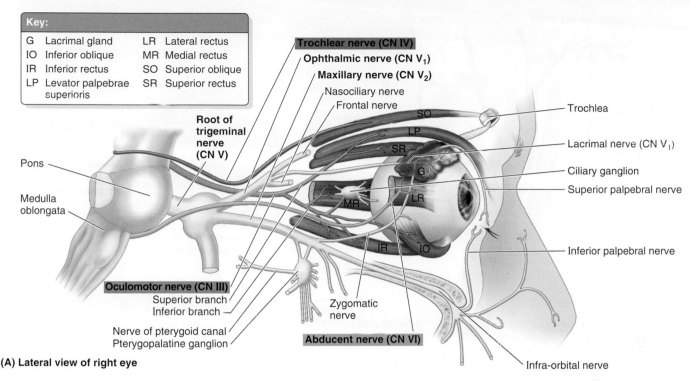

(A) Lateral view of right eye

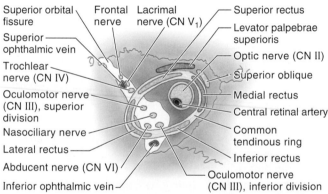

(B) Anterior view

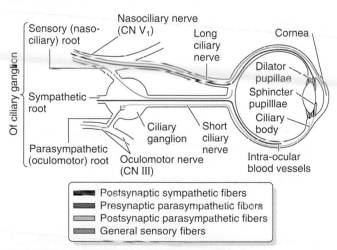

- Postsynaptic sympathetic fibers
- Presynaptic parasympathetic fibers
- Postsynaptic parasympathetic fibers
- General sensory fibers

(C) Diagram of non-visual innervation of eyeball

FIGURE 8.37. Nerves of orbit. A. Overview. **B.** Relationships at apex of orbit. **C.** Distribution of nerve fibers to ciliary ganglion and eyeball.

The three branches of the **ophthalmic nerve (CN V₁)** that pass through the superior orbital fissure and supply structures in the orbit are as follows (Figs. 8.36A and 8.37A,B):

- The **lacrimal nerve**, which arises in the lateral wall of the cavernous sinus and passes to the lacrimal gland, giving sensory branches to the conjunctiva and skin of the superior eyelid; its distal part also carries secretomotor fibers conveyed to it from the zygomatic nerve (CN V₂).
- The **frontal nerve**, which enters the orbit through the superior orbital fissure and divides into the supra-orbital and supratrochlear nerves, providing sensory innervation to the superior eyelid, scalp, and forehead
- The **nasociliary nerve**, the sensory nerve to the eyeball, which also supplies several branches to the orbit, face, paranasal sinuses, nasal cavity, and anterior cranial fossa. The **infratrochlear nerve**, a terminal branch of the nasociliary nerve, supplies the eyelids, conjunctiva, skin of the nose, and lacrimal sac. The anterior and posterior **ethmoidal nerves**, also branches of the nasociliary nerve, supply the mucous membrane of the sphenoidal and ethmoidal sinuses and the nasal cavities and dura mater of the anterior cranial fossa. The *long ciliary nerves* are branches of the nasociliary nerve (CN V₁). The *short ciliary nerves* are branches of the ciliary ganglion (Figs. 8.36B and 8.37C).

The **ciliary ganglion** is a small group of postsynaptic parasympathetic nerve cell bodies associated with CN V₁. It is located between the optic nerve (CN II) and the LR toward the posterior limit of the orbit. This ganglion receives nerve fibers from three sources:

- Sensory fibers from CN V₁ via the nasociliary nerve
- Presynaptic parasympathetic fibers from CN III

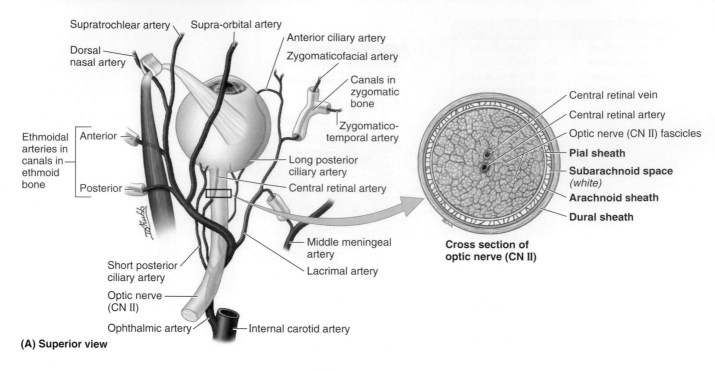

(A) Superior view

Supratrochlear artery
Supra-orbital artery
Anterior ciliary artery
Zygomaticofacial artery
Dorsal nasal artery
Canals in zygomatic bone
Zygomatico-temporal artery
Ethmoidal arteries in canals in ethmoid bone — Anterior / Posterior
Long posterior ciliary artery
Central retinal artery
Middle meningeal artery
Lacrimal artery
Short posterior ciliary artery
Optic nerve (CN II)
Ophthalmic artery
Internal carotid artery

Central retinal vein
Central retinal artery
Optic nerve (CN II) fascicles
Pial sheath
Subarachnoid space *(white)*
Arachnoid sheath
Dural sheath

Cross section of optic nerve (CN II)

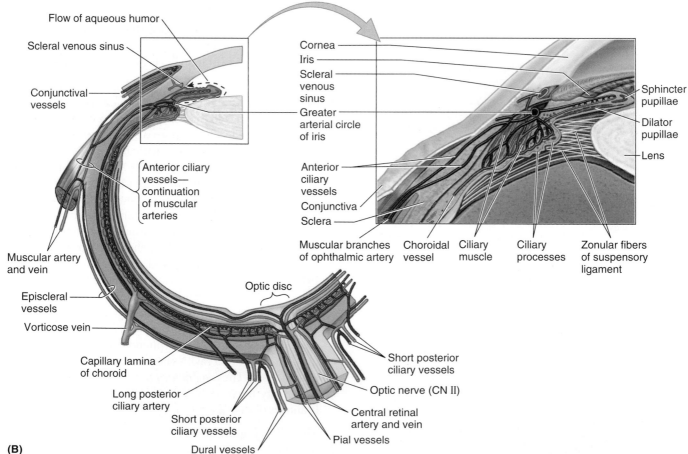

(B)

Flow of aqueous humor
Scleral venous sinus
Conjunctival vessels
Anterior ciliary vessels— continuation of muscular arteries
Muscular artery and vein
Episcleral vessels
Vorticose vein
Capillary lamina of choroid
Long posterior ciliary artery
Short posterior ciliary vessels
Dural vessels
Pial vessels
Central retinal artery and vein
Optic nerve (CN II)
Short posterior ciliary vessels
Optic disc

Cornea
Iris
Scleral venous sinus
Greater arterial circle of iris
Anterior ciliary vessels
Conjunctiva
Sclera
Muscular branches of ophthalmic artery
Choroidal vessel
Ciliary muscle
Ciliary processes
Zonular fibers of suspensory ligament
Sphincter pupillae
Dilator pupillae
Lens

FIGURE 8.38. Arteries of orbit and eyeball. A. Branches of ophthalmic artery. *Inset,* cross section of optic nerve (CN II). **B.** Partial horizontal section of right eyeball. The artery supplying the inner part of the retina (central retinal artery) and the choroid, which in turn nourishes the outer nonvascular layer of the retina, are shown. The vorticose vein (one of four or five) drains venous blood from the choroid into the posterior ciliary and ophthalmic veins. The scleral venous sinus returns the aqueous humor, secreted into the anterior chamber by the ciliary processes, to the venous circulation.

TABLE 8.7. ARTERIES OF ORBIT

Artery	Origin	Course and Distribution
Ophthalmic	Internal carotid artery	Traverses optic canal to reach orbital cavity
Central retinal artery	Ophthalmic artery	Pierces dural sheath and runs in optic nerve to eyeball; branches in center of optic disc; supplies optic retina (except cones and rods)
Supra-orbital		Passes superiorly and posteriorly from supra-orbital foramen to supply forehead and scalp
Supratrochlear		Passes from supra-orbital margin to forehead and scalp
Lacrimal		Passes along superior border of lateral rectus muscle to supply lacrimal gland, conjunctiva, and eyelids
Dorsal nasal		Courses along dorsal aspect of nose and supplies its surface
Short posterior ciliaries		Pierces sclera at periphery of optic nerve to supply choroid, which in turn supplies cones and rods of optic retina
Long posterior ciliaries		Pierces sclera to supply ciliary body and iris
Posterior ethmoidal		Passes through posterior ethmoidal foramen to posterior ethmoidal cells
Anterior ethmoidal		Passes through anterior ethmoidal foramen to supply anterior and middle ethmoidal cells, frontal sinus, nasal cavity, and skin on dorsum of nose
Anterior ciliary	Muscular branches of ophthalmic and infra-orbital arteries	Pierces sclera at attachments of rectus muscles and forms network in iris and ciliary body
Infra-orbital	Third part of maxillary artery	Passes along infra-orbital groove and foramen to face

- Postsynaptic sympathetic fibers from the internal carotid plexus

The **short ciliary nerves** arise from the ciliary ganglion and carry postsynaptic parasympathetic fibers originating in the ciliary ganglion, afferent fibers from the nasociliary nerve, and postsynaptic sympathetic fibers that pass through the ganglion to the iris and cornea. The **long ciliary nerves**, which pass to the eyeball, bypassing the ciliary ganglion, convey postsynaptic sympathetic fibers to the dilator pupillae and afferent fibers from the iris and cornea.

Vasculature of Orbit

The *arteries of the orbit* are mainly from the **ophthalmic artery**, a branch of the internal carotid artery (Fig. 8.38A and Table 8.7). The **infra-orbital artery**, from the external carotid artery, also contributes to the supply of the orbital floor and adjacent structures. The central retinal artery, a branch of the ophthalmic artery arising inferior to the optic nerve, pierces the dural sheath of the optic nerve and runs within the nerve to the eyeball, emerging at the optic disc (Figs. 8.36C and 8.38B). Branches of this artery spread over the internal surface of the retina. The terminal branches (arterioles) of the central retinal artery are *end arteries*, which provide the only blood supply to the internal aspect of the retina.

The external aspect of the retina is also supplied by the **capillary lamina of the choroid** (Fig. 8.38B). Of the eight or so posterior ciliary arteries (also branches of the ophthalmic artery), six **short posterior ciliary arteries** directly supply the choroid, which nourishes the outer nonvascular layer of the retina. Two **long posterior ciliary arteries**, one on each side of the eyeball, pass between the sclera and the choroid to anastomose with the **anterior ciliary arteries**

(continuations of the **muscular branches of the ophthalmic artery** supplying the rectus muscles) to supply the ciliary plexus (Figs. 8.27 and 8.38B).

Venous drainage of the orbit is through the **superior** and **inferior ophthalmic veins**, which pass through the superior orbital fissure and enter the cavernous sinus (Fig. 8.39). The inferior ophthalmic vein also drains to the pterygoid venous plexus. The central retinal vein usually enters the cavernous sinus directly, but it may join one of the ophthalmic veins (Fig. 8.36C). The **vorticose veins** from the vascular layer of the eyeball drain primarily to the inferior ophthalmic vein (Figs. 8.27A, 8.38B, and 8.39). The **scleral venous sinus** is a vascular structure encircling the anterior chamber of the eyeball through which the aqueous humor is returned to the blood circulation (see Fig. 8.27B).

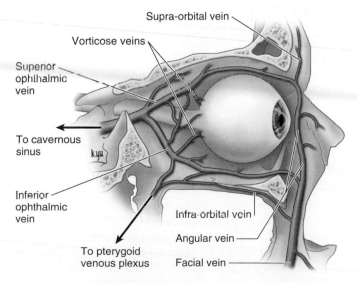

FIGURE 8.39. Ophthalmic veins.

CLINICAL BOX

Blockage of Central Retinal Artery

 Because terminal branches of the central retinal artery are end arteries, obstruction of them by an embolus results in instant and total blindness. Blockage of the artery is usually unilateral and occurs in older people.

Blockage of Central Retinal Vein

Because the central retinal vein enters the cavernous sinus, *thrombophlebitis* of this sinus may result in passage of a thrombus to the central retinal vein and produce a blockage in one of the small retinal veins. Occlusion of a branch of the central retinal vein usually results in slow, painless loss of vision.

Subconjunctival Hemorrhages

Subconjunctival hemorrhages are manifested by bright or dark red patches deep to and within the bulbar conjunctiva. A blow to the eye, excessively hard blowing of the nose, and paroxysms of coughing or violent sneezing can cause hemorrhages resulting from rupture of small conjunctival capillaries.

Pupillary Light Reflex

The *pupillary light reflex* is tested using a penlight during a neurological examination. This reflex, involving CN II (afferent limb) and CN III (efferent limb), is the rapid constriction of the pupil in response to light. When light enters one eye, both pupils constrict because each retina sends fibers into the optic tracts of both sides. The sphincter pupillae muscle is innervated by parasympathetic fibers; consequently, interruption of these fibers causes dilation of the pupil because of the unopposed action of the sympathetically innervated dilator pupillae muscle. The first sign of *compression of the oculomotor nerve* is ipsilateral slowness of the pupillary response to light.

Corneal Reflex

During a neurological examination, the examiner touches the cornea with a wisp of cotton to evoke a *corneal reflex*. A normal (positive) response is a blink. Absence of a response suggests a lesion of CN V$_1$; a lesion of CN VII (the motor nerve to the orbicularis oculi) may also impair this reflex. The examiner must be certain to touch the cornea (not just the sclera) to evoke the reflex. The presence of a contact lens may hamper or abolish the ability to evoke this reflex.

Paralysis of Extra-Ocular Muscles/ Palsies of Orbital Nerves

 One or more extra-ocular muscles may be paralyzed by disease in the brainstem or by a head injury, resulting in *diplopia* (double vision). Paralysis of a muscle is apparent by the limitation of movement of the eyeball in the field of action of the muscle and by the production of two images when one attempts to use the muscle.

Oculomotor Nerve Palsy

Complete *oculomotor nerve palsy* affects most of the ocular muscles, the levator palpebrae superioris, and the sphincter pupillae. The superior eyelid droops and cannot be raised voluntarily because of the unopposed activity of the orbicularis oculi (supplied by the facial nerve) (Fig. B8.11A). The pupil is also fully dilated and nonreactive because of the unopposed dilator pupillae. The pupil is fully abducted and depressed ("down and out") because of the unopposed activity of the lateral rectus and superior oblique, respectively.

Abducent Nerve Palsy

When the abducent nerve (CN VI) supplying only the lateral rectus is paralyzed, the individual cannot abduct the pupil on the affected side (Fig. B8.11B). The pupil is fully adducted by the unopposed pull of the medial rectus.

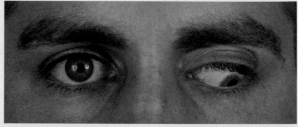

Right eye   **Left eye:** Downward and outward gaze, dilated pupil, eyelid drooping (ptosis)

(A) Left oculomotor (CN III) nerve paralysis

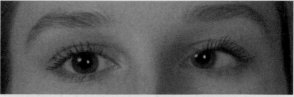

Right eye: Does not abduct   **Left eye**

← Direction of gaze

(B) Right abducent (CN VI) nerve paralysis

FIGURE B8.11. Oculomotor and abducent nerve palsy.

TEMPORAL REGION

The **temporal region** includes the temporal and infratemporal fossae—superior and inferior to the zygomatic arch, respectively (Fig. 8.40).

Temporal Fossa

The **temporal fossa** (Fig. 8.40A,B), where most of the temporalis muscle is located, is bounded by the following:

- Posteriorly and superiorly by the superior and inferior temporal lines
- Anteriorly by the frontal and zygomatic bones

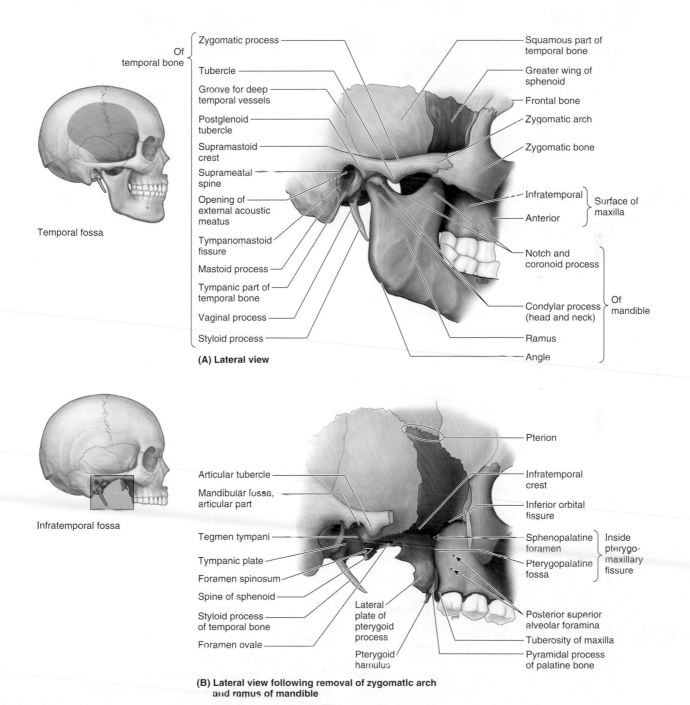

Of temporal bone
- Zygomatic process
- Tubercle
- Groove for deep temporal vessels
- Postglenoid tubercle
- Supramastoid crest
- Suprameatal spine
- Opening of external acoustic meatus
- Tympanomastoid fissure
- Mastoid process
- Tympanic part of temporal bone
- Vaginal process
- Styloid process

Temporal fossa

Squamous part of temporal bone
Greater wing of sphenoid
Frontal bone
Zygomatic arch
Zygomatic bone
Infratemporal — Surface of maxilla
Anterior
Notch and coronoid process
Condylar process (head and neck) — Of mandible
Ramus
Angle

(A) Lateral view

Infratemporal fossa

- Articular tubercle
- Mandibular fossa, articular part
- Tegmen tympani
- Tympanic plate
- Foramen spinosum
- Spine of sphenoid
- Styloid process of temporal bone
- Foramen ovale
- Lateral plate of pterygoid process
- Pterygoid hamulus

Pterion
Infratemporal crest
Inferior orbital fissure
Sphenopalatine foramen — Inside pterygomaxillary fissure
Pterygopalatine fossa
Posterior superior alveolar foramina
Tuberosity of maxilla
Pyramidal process of palatine bone

(B) Lateral view following removal of zygomatic arch and ramus of mandible

FIGURE 8.40. Bony boundaries of temporal and infratemporal fossae. A. The lateral wall of the infratemporal fossa is formed by the ramus of the mandible. The space is deep to the zygomatic arch and is traversed by the temporalis muscle and the deep temporal nerves and vessels. Through this interval, the temporal fossa communicates with the infratemporal fossa. **B.** Infratemporal fossa. This fossa communicates with the pterygopalatine fossa through the pterygomaxillary fissure.

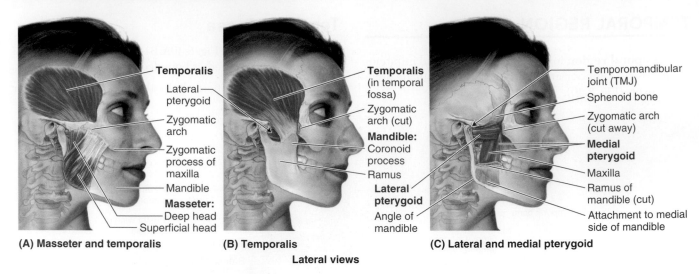

(A) Masseter and temporalis   **(B) Temporalis**   **(C) Lateral and medial pterygoid**

Lateral views

FIGURE 8.41. **Muscles of mastication. A.** Temporalis and masseter muscles. **B.** Temporalis muscle. **C.** Lateral and medial pterygoid muscles.

TABLE 8.8. MUSCLES OF MASTICATION ACTING ON THE MANDIBLE AT THE TEMPOROMANDIBULAR JOINT (TMJ)

Muscle	Proximal Attachment	Distal Attachment	Innervation		Action on Mandible
Temporalis	Triangular muscle with broad attachment to floor of temporal fossa and deep surface of temporalis fascia	Narrow attachment to tip and medial surface of coronoid process and anterior border of ramus of mandible		Via deep temporal nerves	Elevates mandible, closing jaws; posterior, more horizontal fibers are retractors of mandible.
Masseter	Quadrate muscle attaching to inferior border and medial surface of maxillary process of zygomatic bone and the zygomatic arch	Angle and lateral surface of ramus of mandible		Via masseteric nerve	Elevates mandible; superficial fibers make limited contribution to protrusion of mandible.
Lateral pterygoid	Triangular two-headed muscle from (1) infratemporal surface and crest of greater wing of sphenoid and (2) lateral surface of lateral pterygoid plate	Superior head attaches primarily to joint capsule and articular disc of temporomandibular joint (TMJ); inferior head attaches primarily to pterygoid fovea on anteromedial aspect of neck of condyloid process of mandible.	Anterior trunk of mandibular nerve (CN V$_3$)	Via nerves to lateral pterygoid	Acting bilaterally, protracts mandible and depresses chin; acting unilaterally, swings jaw toward contralateral side; alternate unilateral contraction produces larger lateral chewing movements.
Medial pterygoid	Quadrangular two-headed muscle from (1) medial surface of lateral pterygoid plate and pyramidal process of palatine bone and (2) tuberosity of maxilla	Medial surface of ramus of mandible, inferior to mandibular foramen; in essence, a "mirror image" of the ipsilateral masseter, the two muscles flanking the ramus		Via nerve to medial pterygoid	Acts synergistically with masseter to elevate mandible; contributes to protrusion; alternate unilateral activity produces smaller grinding movements.

- Laterally by the zygomatic arch
- Inferiorly by the infratemporal crest

The *floor of the temporal fossa* is formed by parts of the four bones (frontal, parietal, temporal, and greater wing of the sphenoid) that form the pterion. The fan-shaped *temporalis muscle* arises from the bony floor and the overlying **temporalis fascia**, which forms the *roof of the temporal fossa* (Fig. 8.41 and Table 8.8). The temporalis fascia extends from the *superior temporal line* to the zygomatic arch. When the powerful masseter, attached to the inferior border of the arch, contracts and exerts a strong downward pull on the arch, the temporalis fascia provides resistance.

Infratemporal Fossa

The **infratemporal fossa** is an irregularly shaped space deep and inferior to the zygomatic arch, deep to the ramus of the mandible, and posterior to the maxilla. The *boundaries of the fossa* are as follows (Fig. 8.40B):

- Laterally: Ramus of the mandible
- Medially: Lateral pterygoid plate
- Anteriorly: Posterior aspect of the maxilla
- Posteriorly: Tympanic plate and the mastoid and styloid processes of the temporal bone
- Superiorly: Inferior surface of the greater wing of the sphenoid bone

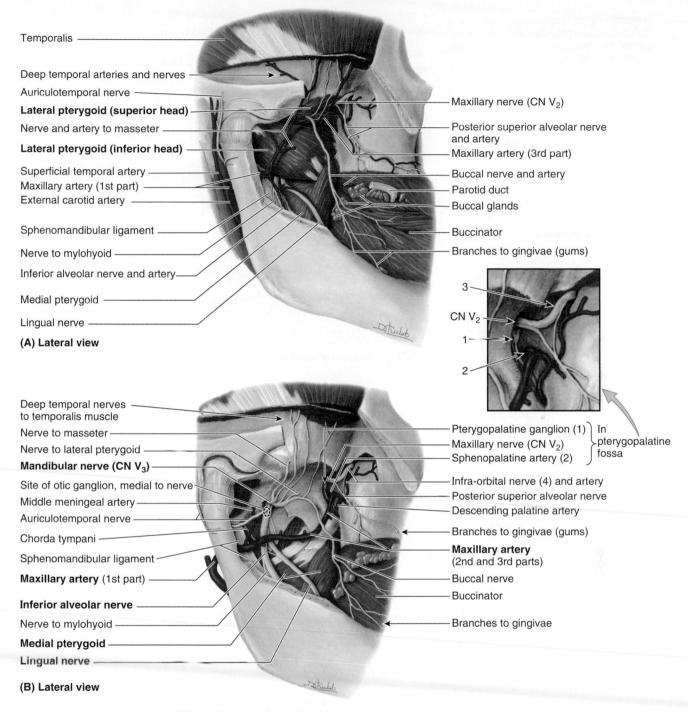

Temporalis

Deep temporal arteries and nerves

Auriculotemporal nerve

Lateral pterygoid (superior head)

Nerve and artery to masseter

Lateral pterygoid (inferior head)

Superficial temporal artery

Maxillary artery (1st part)

External carotid artery

Sphenomandibular ligament

Nerve to mylohyoid

Inferior alveolar nerve and artery

Medial pterygoid

Lingual nerve

(A) Lateral view

Maxillary nerve (CN V$_2$)

Posterior superior alveolar nerve and artery

Maxillary artery (3rd part)

Buccal nerve and artery

Parotid duct

Buccal glands

Buccinator

Branches to gingivae (gums)

3

CN V$_2$

1

2

Deep temporal nerves to temporalis muscle

Nerve to masseter

Nerve to lateral pterygoid

Mandibular nerve (CN V$_3$)

Site of otic ganglion, medial to nerve

Middle meningeal artery

Auriculotemporal nerve

Chorda tympani

Sphenomandibular ligament

Maxillary artery (1st part)

Inferior alveolar nerve

Nerve to mylohyoid

Medial pterygoid

Lingual nerve

(B) Lateral view

Pterygopalatine ganglion (1)   In
Maxillary nerve (CN V$_2$)   pterygopalatine
Sphenopalatine artery (2)   fossa

Infra-orbital nerve (4) and artery

Posterior superior alveolar nerve

Descending palatine artery

Branches to gingivae (gums)

Maxillary artery (2nd and 3rd parts)

Buccal nerve

Buccinator

Branches to gingivae

FIGURE 8.42. Dissections of right infratemporal region. A. Superficial. **B.** Deep.

• Inferiorly: Where the medial pterygoid muscle attaches to the mandible near its angle (Table 8.8)

The *contents of the infratemporal fossa* are as follows (Fig. 8.42):

• Inferior part of the temporalis muscle
• Lateral and medial pterygoid muscles
• Maxillary artery
• Pterygoid venous plexus

• Mandibular, inferior alveolar, lingual, buccal, and chorda tympani nerves and the otic ganglion

The **temporalis muscle** has a broad proximal attachment to the floor of the temporal fossa and is attached distally to the tip and medial surface of the coronoid process and anterior border of the ramus of the mandible (Fig. 8.41A,B and Table 8.8). It elevates the mandible (closes the lower jaw); its posterior fibers retrude (retract) the protruded mandible.

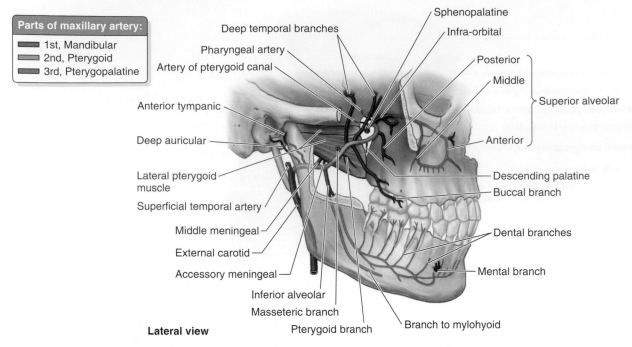

Parts of maxillary artery:
▬ 1st, Mandibular
▬ 2nd, Pterygoid
▬ 3rd, Pterygopalatine

Deep temporal branches
Pharyngeal artery
Artery of pterygoid canal
Anterior tympanic
Deep auricular
Lateral pterygoid muscle
Superficial temporal artery
Middle meningeal
External carotid
Accessory meningeal
Inferior alveolar
Masseteric branch
Pterygoid branch

Sphenopalatine
Infra-orbital
Posterior
Middle
} Superior alveolar
Anterior
Descending palatine
Buccal branch
Dental branches
Mental branch
Branch to mylohyoid

Lateral view

FIGURE 8.43. Branches of maxillary artery.

The two-headed **lateral pterygoid muscle** passes posteriorly. Its superior head attaches to the joint capsule and disc of the TMJ, and the inferior head attaches primarily to the pterygoid fovea at the condylar process of the mandible.

The **medial pterygoid muscle** lies on the medial aspect of the ramus of the mandible. Its two heads embrace the inferior head of the lateral pterygoid and then unite (Fig. 8.42A). The medial pterygoid passes inferoposteriorly and attaches to the medial surface of the mandible near its angle. The attachments, nerve supply, and actions of the pterygoid muscles are described in Table 8.8.

The **maxillary artery**, the larger of the two terminal branches of the external carotid artery, is the major artery to the deep face. It arises posterior to the neck of the mandible, courses anteriorly deep to the neck of the mandibular condyle, and then passes superficial or deep to the lateral pterygoid (Figs. 8.43 and 8.44A). The artery passes medially from the infratemporal fossa through the *pterygomaxillary fissure* to enter the *pterygopalatine fossa* (Fig. 8.40B). The maxillary artery is thus divided into three parts by its relation to the lateral pterygoid muscle (Fig. 8.43).

Branches of the first, or retromandibular, part of the maxillary artery are as follows:

- *Deep auricular artery*, supplying the external acoustic meatus
- *Anterior tympanic artery*, supplying the tympanic membrane
- *Middle meningeal artery*, supplying the dura and calvaria
- *Accessory meningeal arteries*, supplying the cranial cavity

- *Inferior alveolar artery*, which supplies the mandible, gingivae (gums), teeth, and floor of the mouth

Branches of the second, or pterygoid part, of the maxillary artery are as follows:

- *Deep temporal arteries*, anterior and posterior, which ascend to supply the temporalis muscle
- *Pterygoid arteries*, which supply the pterygoid muscles
- *Masseteric artery*, which passes laterally through the mandibular notch to supply the masseter muscle
- *Buccal artery*, which supplies the buccinator muscle and mucosa of the cheek

Branches of the third, or pterygopalatine, part of the maxillary artery are the

- *Posterior superior alveolar artery*, supplying the maxillary molar and premolar teeth, the buccal gingiva, and the lining of the maxillary sinus
- *Infra-orbital artery*, supplying the inferior eyelid, lacrimal sac, infra-orbital region of the face, side of the nose, and the upper lip
- *Descending palatine artery*, supplying the mucous membrane and glands of the palate (roof of the mouth) and palatine gingiva
- *Artery of pterygoid canal*, supplying the superior part of the pharynx, the pharyngotympanic (auditory) tube, and the tympanic cavity
- *Pharyngeal artery*, supplying the roof of the pharynx, the sphenoidal sinus, and the inferior part of the pharyngotympanic tube

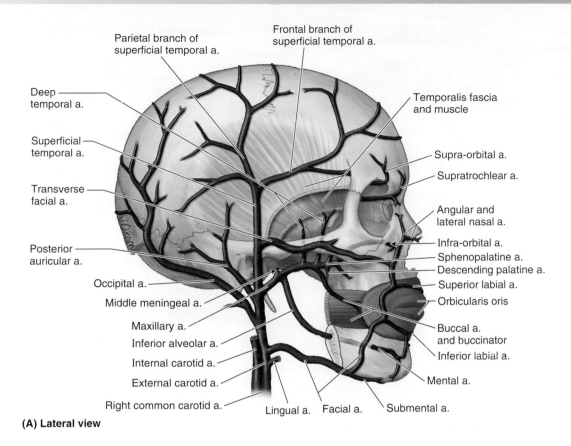

(A) Lateral view

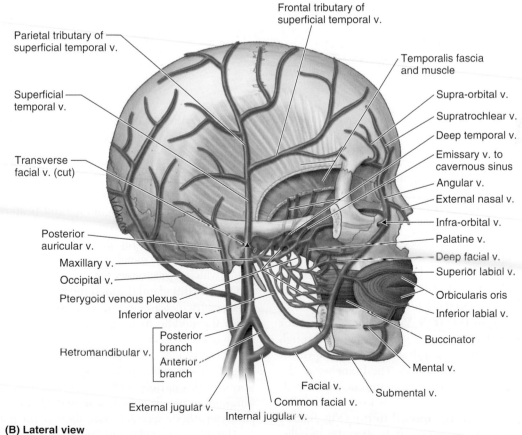

(B) Lateral view

FIGURE 8.44. Vasculature of head. A. Branches of external carotid artery. *a.*, artery. **B.** Venous drainage of face, scalp, and infratemporal fossa. *v.*, vein.

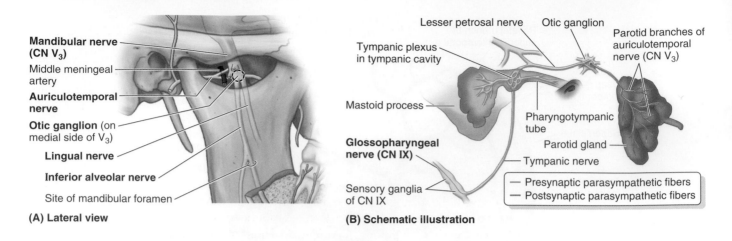

Mandibular nerve (CN V₃)

Middle meningeal artery

Auriculotemporal nerve

Otic ganglion (on medial side of V₃)

Lingual nerve

Inferior alveolar nerve

Site of mandibular foramen

(A) Lateral view

Lesser petrosal nerve   Otic ganglion

Tympanic plexus in tympanic cavity

Parotid branches of auriculotemporal nerve (CN V₃)

Mastoid process

Pharyngotympanic tube

Glossopharyngeal nerve (CN IX)

Parotid gland

Tympanic nerve

Sensory ganglia of CN IX

— Presynaptic parasympathetic fibers
— Postsynaptic parasympathetic fibers

(B) Schematic illustration

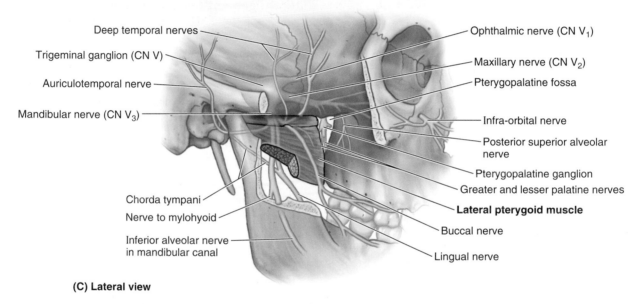

Deep temporal nerves

Trigeminal ganglion (CN V)

Auriculotemporal nerve

Mandibular nerve (CN V₃)

Chorda tympani

Nerve to mylohyoid

Inferior alveolar nerve in mandibular canal

Ophthalmic nerve (CN V₁)

Maxillary nerve (CN V₂)

Pterygopalatine fossa

Infra-orbital nerve

Posterior superior alveolar nerve

Pterygopalatine ganglion

Greater and lesser palatine nerves

Lateral pterygoid muscle

Buccal nerve

Lingual nerve

(C) Lateral view

FIGURE 8.45. Nerves of infratemporal fossa. A. At foramen ovale. **B.** Innervation of parotid gland. **C.** Overview. Relationship of nerves to lateral pterygoid is shown.

- *Sphenopalatine artery*, the termination of the maxillary artery, which supplies the nasal cavity (lateral nasal wall, the nasal septum, and the adjacent paranasal sinuses)

The **pterygoid venous plexus** occupies most of the infratemporal fossa (Fig. 8.44B). It is located partly between the temporalis and pterygoid muscles. The plexus drains anteriorly to the facial vein via the deep facial vein but mainly drains posteriorly via the maxillary and then the retromandibular veins.

The **mandibular nerve (CN V₃)** receives the motor root of the trigeminal nerve (CN V) and descends through the foramen ovale to enter the infratemporal fossa, dividing into anterior and posterior trunks. The branches of the large posterior trunk are the auriculotemporal, inferior alveolar, and lingual nerves (Figs. 8.42 and 8.45A). The smaller anterior trunk gives rise to the **buccal nerve** (Fig. 8.45C) and branches to the four muscles of mastication (temporalis, masseter, and medial and lateral pterygoids) but not the buccinator, which is supplied by the facial nerve (CN VII).

The **otic ganglion** (parasympathetic) is in the infratemporal fossa (Fig. 8.45A,B), just inferior to the foramen ovale, medial to the mandibular nerve, and posterior to the lateral pterygoid muscle. Presynaptic parasympathetic fibers, derived mainly from the glossopharyngeal nerve (CN IX), synapse in the otic ganglion. Postsynaptic parasympathetic fibers, which are secretory to the parotid gland, pass from the ganglion to this gland through the auriculotemporal nerve.

The auriculotemporal nerve arises via two roots that encircle the middle meningeal artery and then unite into a single trunk (Figs. 8.42 and 8.45A,C). The trunk divides into numerous branches, the largest of which passes posteriorly, medial to the neck of the mandible and supplies sensory fibers to the auricle and temporal region. The auriculotemporal nerve also sends articular fibers to the TMJ and parasympathetic secretomotor fibers to the parotid gland.

The inferior alveolar and lingual nerves descend between the lateral and medial pterygoid muscles. The **inferior alveolar nerve** enters the mandibular foramen and passes through the mandibular canal, forming the **inferior dental plexus**, which sends branches to all mandibular teeth on that side. The *nerve to mylohyoid*, a small branch of the inferior alveolar nerve, is given off just before the nerve enters the mandibular foramen (Fig. 8.45C). A branch of the inferior dental plexus, the **mental nerve**, passes through the mental foramen and supplies the skin and mucous membrane of the lower lip, the skin of the chin, and the vestibular gingiva of the mandibular incisor teeth (see Fig. 8.52A).

The **lingual nerve** lies anterior to the inferior alveolar nerve (Figs. 8.42 and 8.52). It is sensory to the anterior two thirds of the tongue, the floor of the mouth, and the lingual gingivae. It enters the mouth between the medial pterygoid and the ramus of the mandible and passes anteriorly under cover of the oral mucosa, just medial and inferior to the third molar tooth.

The **chorda tympani nerve**, a branch of CN VII (Fig. 8.45C), carries taste fibers from the anterior two thirds of the tongue and presynaptic parasympathetic secretomotor fibers for the submandibular and sublingual salivary glands. The chorda tympani joins the lingual nerve in the infratemporal fossa.

TEMPOROMANDIBULAR JOINT

The **temporomandibular joint** is a modified hinge type of synovial joint permitting movement in three planes. The articular surfaces involved are the **head of the mandible**, the **articular tubercle** of the temporal bone, and the *mandibular fossa* (Fig. 8.46). The articular surfaces of the TMJ are covered by fibrocartilage rather than hyaline cartilage as in a typical synovial joint. An **articular disc** divides the joint cavity into two separate synovial compartments. The **joint capsule** of the TMJ is loose. The fibrous layer of the capsule attaches to the margins of the articular area on the temporal bone and around the neck of the mandible. The thick part of the joint capsule forms the intrinsic **lateral ligament** (temporomandibular ligament), which strengthens the TMJ laterally and, with the **postglenoid tubercle**, acts to prevent posterior dislocation of the joint (Fig. 8.46A,C).

Two extrinsic ligaments and the lateral ligament connect the mandible to the cranium. The **stylomandibular ligament**, a thickening of the fibrous capsule of the parotid gland, runs from the styloid process to the angle of the mandible

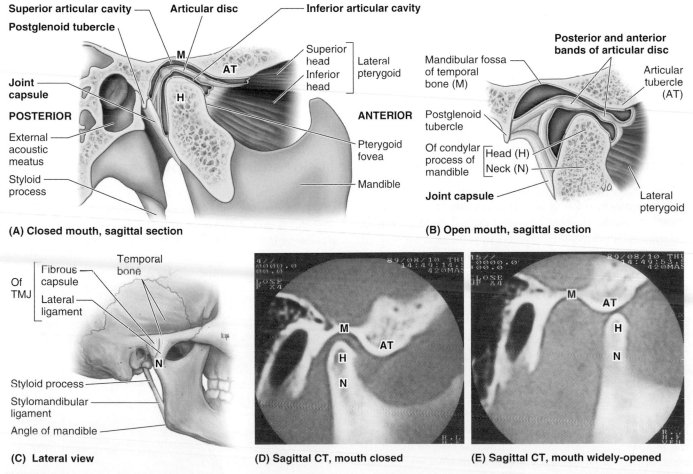

FIGURE 8.46. Temporomandibular joint (TMJ). Anatomical and computerized tomography (CT) images of the TMJ in the closed-mouth **(A, C, and D)** and open-mouth **(B and E)** positions. **C.** Lateral and stylomandibular ligaments of TMJ.

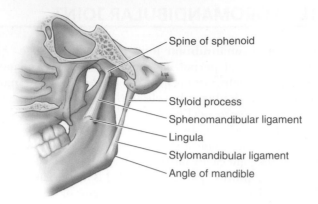

FIGURE 8.47. Sphenomandibular and stylomandibular ligaments.

- Spine of sphenoid
- Styloid process
- Sphenomandibular ligament
- Lingula
- Stylomandibular ligament
- Angle of mandible

TABLE 8.9. MOVEMENTS AT THE TEMPOROMANDIBULAR JOINT

Movements of Mandible	Muscles
Elevation (close mouth)	Temporalis, masseter, and medial pterygoid
Depression (open mouth)	Lateral pterygoid and suprahyoid and infrahyoid muscles[a]
Protrusion (protrude chin)	Lateral pterygoid, masseter, and medial pterygoid[b]
Retrusion (retrude chin)	Temporalis (posterior oblique and near horizontal fibers) and masseter
Lateral movements (grinding and chewing)	Temporalis of same side, pterygoids of opposite side, and masseter

[a]The prime mover is normally gravity; these muscles are mainly active against resistance.
[b]The lateral pterygoid is the prime mover here, with minor secondary roles played by the masseter and medial pterygoid.

(Figs. 8.46 and 8.47). It does not contribute significantly to the strength of the TMJ. The **sphenomandibular ligament** runs from the spine of the sphenoid to the lingula of the mandible (Fig. 8.47) and is the primary passive support and "swing rope" of the mandible.

To enable more than a small amount of depression of the mandible—that is, to open the mouth wider than just separating the upper and lower teeth—the head of the mandible and articular disc must move anteriorly on the articular surface until the head lies inferior to the articular tubercle (Fig. 8.46B,E), a movement referred to as *translation* by dentists. If this anterior gliding occurs unilaterally, the head

of the mandible on the retracted side rotates (pivots) on the inferior surface of the articular disc, permitting simple side-to-side chewing or grinding movements over a small range. During protrusion and retrusion of the mandible, the mandibular head and articular disc slide anteriorly and posteriorly on the articular surface of the temporal bone, with both sides moving together. TMJ movements are produced chiefly by the muscles of mastication. The attachments, nerve supply, and actions of these muscles are described in Tables 8.8 and 8.9.

CLINICAL BOX

Mandibular Nerve Block

To perform a mandibular nerve block, an anesthetic agent is injected near the mandibular nerve where it enters the infratemporal fossa. This block usually anesthetizes the auriculotemporal, inferior alveolar, lingual, and buccal branches of the mandibular nerve.

Inferior Alveolar Nerve Block

An alveolar nerve block—commonly used by dentists when repairing mandibular teeth—anesthetizes the inferior alveolar nerve, a branch of CN V₃. The anesthetic agent is injected around the mandibular foramen, the opening into the mandibular canal on the medial aspect of the ramus of the mandible. This canal gives passage to the inferior alveolar nerve, artery, and vein. When this nerve block is successful, all mandibular teeth are anesthetized to the median plane. The skin and mucous membrane of the lower lip, the labial alveolar mucosa and gingiva, and the skin of the chin are also anesthetized because they are supplied by the mental branch of this nerve.

Dislocation of Temporomandibular Joint

During yawning or taking a large bite, excessive contraction of the lateral pterygoids may cause the heads of the mandibles to dislocate anteriorly, by passing anterior to the articular tubercles (Fig. B8.12).

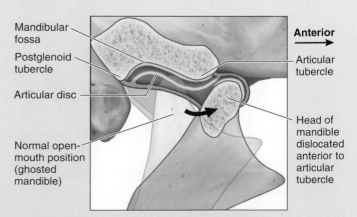

- Mandibular fossa
- Postglenoid tubercle
- Articular disc
- Normal open-mouth position (ghosted mandible)
- **Anterior**
- Articular tubercle
- Head of mandible dislocated anterior to articular tubercle

FIGURE B8.12. Dislocation of temporomandibular joint (TMJ).

In this position, the mandible remains depressed and the person may not be able to close the mouth. Most commonly, a sideways blow to the chin when the mouth is open dislocates the TMJ on the side that received the blow. *Fracture(s) of the mandible* may be accompanied by dislocation of the TMJ. Because of the close relationship of the facial and auriculotemporal nerves to the TMJ, care must be taken during surgical procedures to preserve both the branches of the facial nerve overlying it and the articular branches of the auriculotemporal nerve that enter the posterior part of the joint. Injury to articular branches of the auriculotemporal nerve supplying the TMJ—associated

with traumatic dislocation and rupture of the joint capsule and lateral ligament—leads to laxity and instability of the TMJ.

Arthritis of Temporomandibular Joint

The TMJ may become inflamed from *degenerative arthritis*. Abnormal function of the TMJ may result in structural problems, such as dental occlusion and joint clicking (*crepitus*). The clicking is thought to result from delayed anterior disc movements during mandibular depression and elevation.

ORAL REGION

The **oral region** includes the oral cavity, teeth, gingivae (gums), tongue, palate, and the region of the palatine tonsils. The oral cavity is where food is ingested and prepared for digestion in the stomach and small intestine. When food is chewed, the teeth and saliva from the salivary glands facilitate the formation of a manageable *food bolus* (L. lump).

Oral Cavity

The **oral cavity** (mouth) consists of two parts: the *oral vestibule* and the *oral cavity proper* (Fig. 8.48). The oral vestibule communicates with the exterior through the mouth. The size of the **oral fissure** (opening) is controlled by muscles such as the orbicularis oris (the sphincter of the oral fissure).

The **oral cavity proper** is the space posterior and medial to the upper and lower **dental arches** (maxillary and mandibular alveolar arches and the teeth they bear). The oral cavity is limited laterally and anteriorly by the dental arches. The **roof of the oral cavity proper** is formed by the palate. Posteriorly, the oral cavity communicates with the **oropharynx**, the oral part of the pharynx. When the mouth is closed and at rest, the oral cavity is fully occupied by the tongue.

Oral Vestibule

The **oral vestibule** is the slit-like space between the lips and cheeks superficially and the teeth and gingivae deeply. The **lips**, the mobile, fleshy muscular folds surrounding the mouth, contain the orbicularis oris and superior and inferior labial muscles, vessels, and nerves. They are covered externally by skin and internally by mucous membrane. The upper lip has a vertical groove, the **philtrum** (Fig. 8.49). As the skin of the lips approaches the mouth, it changes color abruptly to red; this red margin of the lips is the **vermillion border**, a transitional zone between the skin and mucous membrane. The skin of the **transitional zone** (commonly considered as the lip by itself) is hairless and so thin that it is bright red or darker brown because of the underlying capillary bed. The upper lip is supplied by superior labial arteries of the *facial* and *infra-orbital arteries*.

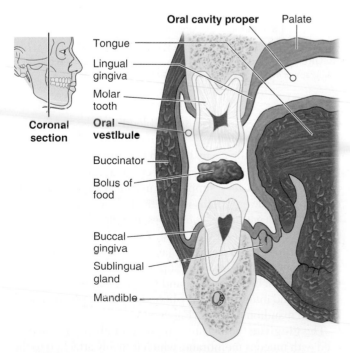

FIGURE 8.48. Oral cavity. The orientation drawing shows the site of the coronal section.

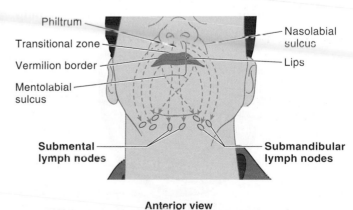

Anterior view

FIGURE 8.49. Surface anatomy and lymphatic drainage of cheeks, lips, and chin.

The lower lip is supplied by inferior labial arteries of the *facial* and **mental arteries**. The upper lip is supplied by the superior labial branches of the **infra-orbital nerves** (CN V₂), and the lower lip is supplied by the inferior labial branches of the mental nerves (CN V₃) (see Fig. 8.52A).

Lymph from the upper lip and lateral parts of the lower lip passes primarily to the submandibular lymph nodes (Fig. 8.49), whereas lymph from the medial part of the lower lip passes initially to the submental lymph nodes.

The **cheeks** (L. *buccae*) include the lateral distensible walls of the oral cavity and the facial prominences over the zygomatic bones. The cheeks have essentially the same structure as the lips, with which they are continuous. The principal muscles of the cheeks are the buccinators (Fig. 8.48). The lips and cheeks function as an oral sphincter that pushes food from the oral vestibule into the oral cavity proper. The tongue and buccinators work together to keep the food between the occlusal surfaces of the molar teeth during chewing. The **labial** and **buccal glands** are small mucous glands between the mucous membrane and the underlying orbicularis oris and buccinator muscles (see Fig. 8.42B).

Teeth and Gingivae

The **teeth** are hard conical structures set in the **dental alveoli** (tooth sockets) of the upper and lower jaws that are used in mastication (chewing) and assisting in articulation (speech). Children have 20 **deciduous (primary) teeth**. The first tooth usually erupts at 6–8 months of age and the last tooth by 20–24 months of age. Eruption of the **permanent (secondary) teeth**, normally 16 in each jaw (3 molars, 2 premolars, 1 canine, and 2 incisors on each side), usually is complete by the midteens (Fig. 8.50), except for the third molars (wisdom teeth), which usually erupt during the late teens or early 20s.

A tooth has a crown, neck, and root. Each type of tooth has a characteristic appearance (Figs. 8.51 and 8.52). The **crown** projects from the gingiva. The **neck of the tooth** is the part of the tooth between the crown and the root. The **root(s) of the tooth** is/are fixed in the alveolus by the fibrous *periodontium* (periodontal membrane). Most of the tooth is composed of **dentine** (L. *dentinium*), which is covered by **enamel** over the crown and **cement** (L. *cementum*) over the root. The **pulp**

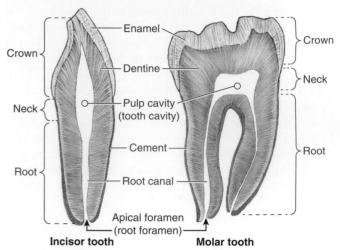
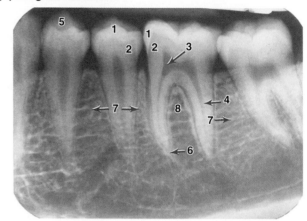

(A) Longitudinal section

Incisor tooth   Molar tooth

Crown — Enamel — Crown
Neck — Dentine — Neck
Neck — Pulp cavity (tooth cavity)
Root — Cement — Root
Root — Root canal
Apical foramen (root foramen)

(B) Lateral radiograph

1 Enamel	2 Dentine	3 Pulp cavity
4 Root canal	5 Buccal cusp	6 Root apex
7 Interalveolar septa (alveolar bone)		
8 Interradicular septum (alveolar bone)		

FIGURE 8.51. Parts of a tooth. A. An incisor and molar tooth. **B.** Bite-wing radiograph of maxillary premolar and molar teeth.

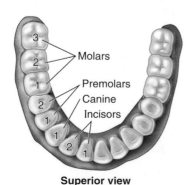

Superior view
Mandibular (lower) dental arch

Molars
Premolars
Canine
Incisors

FIGURE 8.50. Adult mandibular dentition.

cavity contains connective tissue, blood vessels, and nerves. The **root canal** (pulp canal) transmits the nerves and vessels to and from the pulp cavity through the **apical foramen**.

The **superior** and **inferior alveolar arteries**, branches of the maxillary artery, supply the maxillary (upper) and the mandibular (lower) teeth, respectively (see Figs. 8.43 and 8.44A). **Veins** with the same names and distribution accompany the arteries (see Fig. 8.44B). **Lymphatic vessels** from the teeth and gingivae pass mainly to the submandibular lymph nodes (Fig. 8.49). The superior and inferior **alveolar nerves**, branches of CN V₂ and CN V₃, respectively, form superior and inferior **dental plexuses** that supply the maxillary and mandibular teeth (Fig. 8.52A).

The **gingivae** (gums) are composed of fibrous tissue covered with mucous membrane, which is firmly attached to the alveolar processes of the mandible and maxilla and the necks of the teeth. The **buccal gingivae** of the mandibular molar

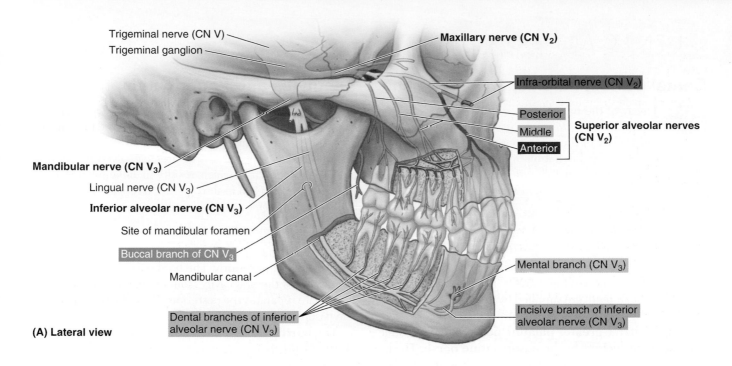

(A) Lateral view

- Trigeminal nerve (CN V)
- Trigeminal ganglion
- **Maxillary nerve (CN V₂)**
- Infra-orbital nerve (CN V₂)
- Posterior
- Middle
- Anterior
- **Superior alveolar nerves (CN V₂)**
- **Mandibular nerve (CN V₃)**
- Lingual nerve (CN V₃)
- **Inferior alveolar nerve (CN V₃)**
- Site of mandibular foramen
- Buccal branch of CN V₃
- Mandibular canal
- Dental branches of inferior alveolar nerve (CN V₃)
- Mental branch (CN V₃)
- Incisive branch of inferior alveolar nerve (CN V₃)

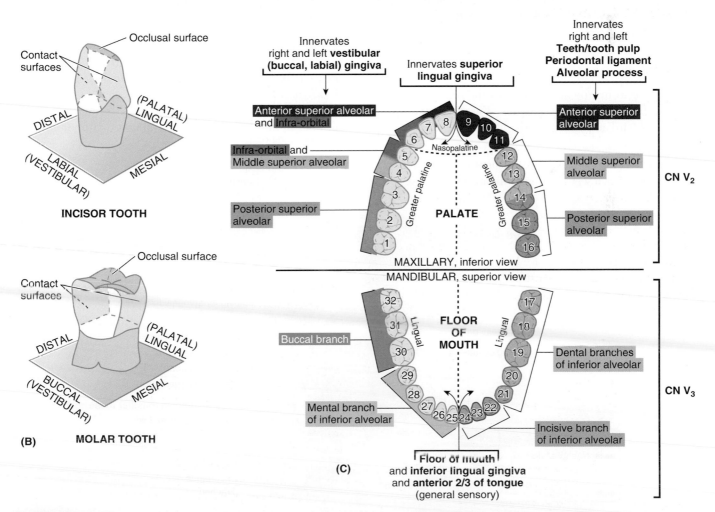

FIGURE 8.52. Innervation of teeth and gingiva. A. Superior and inferior alveolar nerves. **B.** Surfaces of an incisor and molar tooth. **C.** Innervation of the mouth and teeth.

CLINICAL BOX

Dental Caries, Pulpitis, and Toothache

Decay of the hard tissues of a tooth results in the formation of *dental caries* (cavities). Invasion of the pulp cavity of the tooth by a carious lesion (cavity) results in infection and irritation of the tissues in the cavity. This condition causes an inflammatory process (*pulpitis*). Because the pulp cavity is a rigid space, the swollen pulpal tissues cause pain (*toothache*).

Gingivitis and Periodontitis

Improper oral hygiene results in food deposits in tooth and gingival crevices, which may cause inflammation of the gingivae (*gingivitis*). If untreated, the disease spreads to other supporting structures (e.g., alveolar bone), producing *periodontitis*, which results in inflammation of the gingivae. It may result in absorption of alveolar bone and *gingival recession* that exposes the sensitive cement of the teeth.

teeth (Fig. 8.48) are supplied by the buccal nerve, a branch of the mandibular nerve (Fig. 8.52C). The **lingual gingivae** of all mandibular teeth are supplied by the lingual nerve. The **palatine gingivae** of the maxillary premolar and molar teeth are supplied by the **greater palatine nerve** and the palatine gingivae of the incisors by the **nasopalatine nerve**. The labial and buccal aspects of the maxillary gingivae are supplied by the anterior, middle, and posterior **superior alveolar nerves** (Fig. 8.52A).

Palate

The **palate** forms the arched roof of the oral cavity proper and the floor of the nasal cavities (Fig. 8.53). The palate consists of hard and soft parts: the hard palate anteriorly and

the soft palate posteriorly. The hard palate separates the anterior part of the oral cavity from the nasal cavities, and the soft palate separates the posterior part of the oral cavity from the nasopharynx superior to it.

The **hard palate** is the anterior vaulted (concave) two thirds of the palate; this space is filled with the tongue when it is at rest. The hard palate (covered by a mucous membrane) is formed by the palatine processes of the maxillae and the horizontal plates of the palatine bones (Fig. 8.54A). Three foramina open on the oral aspect of the hard palate: the incisive fossa and the greater and lesser palatine foramina. The **incisive fossa** is a slight depression posterior to the central incisor teeth. The nasopalatine nerves pass from the nose through a variable number of incisive canals and foramina that open into the

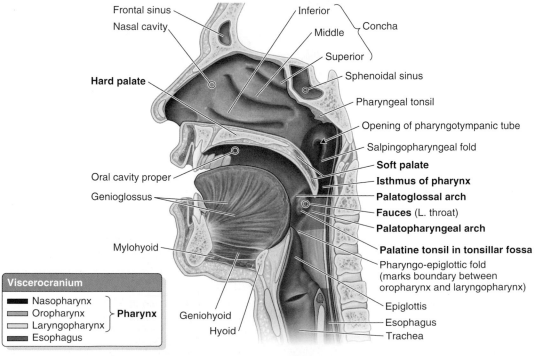

Medial view of right half of viscerocranium

FIGURE 8.53. Palate, nasal and oral cavities, and pharynx.

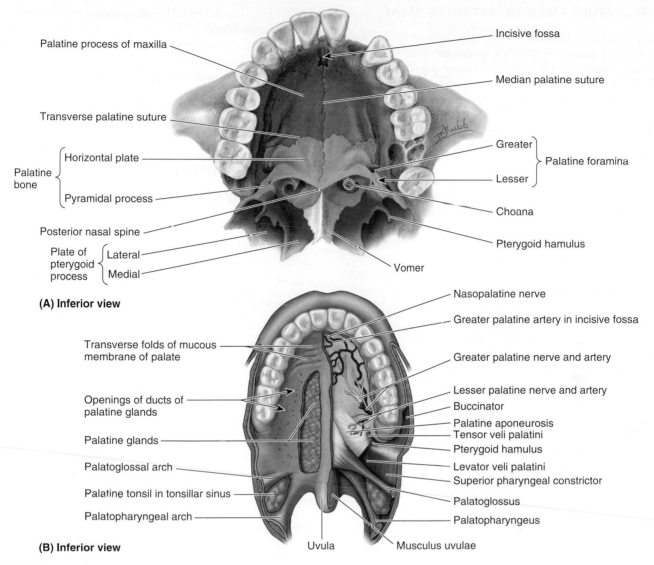

(A) Inferior view

Palatine process of maxilla

Transverse palatine suture

Palatine bone { Horizontal plate

Pyramidal process

Posterior nasal spine

Plate of pterygoid process { Lateral / Medial

Incisive fossa

Median palatine suture

Greater } Palatine foramina
Lesser }

Choana

Pterygoid hamulus

Vomer

(B) Inferior view

Transverse folds of mucous membrane of palate

Openings of ducts of palatine glands

Palatine glands

Palatoglossal arch

Palatine tonsil in tonsillar sinus

Palatopharyngeal arch

Uvula

Nasopalatine nerve

Greater palatine artery in incisive fossa

Greater palatine nerve and artery

Lesser palatine nerve and artery

Buccinator

Palatine aponeurosis

Tensor veli palatini

Pterygoid hamulus

Levator veli palatini

Superior pharyngeal constrictor

Palatoglossus

Palatopharyngeus

Musculus uvulae

FIGURE 8.54. Palate. A. The bones forming the hard palate. **B.** Part of the right side has been dissected to show the palatine glands. The left side has been dissected to show the muscles of the soft palate and palatine arteries and nerves.

incisive fossa (Fig. 8.54A,B). Medial to the third molar tooth, the **greater palatine foramen** pierces the lateral border of the bony palate. The *greater palatine vessels and nerve* emerge from this foramen and run anteriorly on the palate. The lesser palatine foramina transmit the *lesser palatine nerves and vessels* to the soft palate and adjacent structures.

The **soft palate** is the movable posterior third of the palate, which is suspended from the posterior border of the hard palate (Figs. 8.54B and 8.55A). The soft palate extends postero-inferiorly as a curved free margin from which hangs a conical process, the **uvula**. The soft palate is strengthened by the **palatine aponeurosis**, formed by the expanded tendon of the **tensor veli palatini**. The aponeurosis, attached to the posterior margin of the hard palate, is thick anteriorly and thin posteriorly. The anterior part of the soft palate is

formed mainly by the palatine aponeurosis, whereas its posterior part is muscular.

When one swallows, the soft palate is initially tensed to allow the tongue to press against it, squeezing the bolus of food to the back of the oral cavity proper. The soft palate is then elevated posteriorly and superiorly against the wall of the pharynx, thereby preventing passage of food into the nasal cavity. Laterally, the soft palate is continuous with the wall of the pharynx and is joined to the tongue and pharynx by the **palatoglossal** and **palatopharyngeal arches** (Figs. 8.54B and 8.55A), respectively. The **palatine tonsils**, usually called "the tonsils," are masses of lymphoid tissue, one on each side of the oropharynx (Fig. 8.55A). Each tonsil lies in a *tonsillar sinus* (fossa) bounded by the palatoglossal and palatopharyngeal arches and the tongue.

VASCULATURE AND INNERVATION OF PALATE

The palate has a rich blood supply, chiefly from the right and left **greater palatine arteries**, branches of the descending palatine arteries (Fig. 8.54B). The **lesser palatine artery**, a smaller branch of the descending palatine artery, enters the palate through the **lesser palatine foramen** and anastomoses with the ascending palatine artery, a branch of the facial artery. **Venous drainage of the palate**, corresponding to and accompanying the branches of the maxillary artery, involves tributaries of the pterygoid venous plexus (see Fig. 8.44B).

The *sensory nerve fibers from the palate* pass through the **pterygopalatine ganglion** and are considered branches of the maxillary nerve. The greater palatine nerve supplies the gingivae, mucous membrane, and glands of most of the hard palate (Fig. 8.54B). The nasopalatine nerve supplies the mucous membrane of the anterior part of the hard palate. The **lesser palatine nerves** supply the soft palate. The palatine nerves accompany the arteries through the greater and lesser palatine foramina, respectively. Except for the tensor veli palatini supplied by CN V₃, all muscles of the soft palate are supplied through the *pharyngeal plexus of nerves* (see Chapter 8) derived from pharyngeal branches of the vagus nerve (CN X).

MUSCLES OF SOFT PALATE

The **muscles of the soft palate** arise from the cranial base and descend to the palate (Figs. 8.54B and 8.55B). The soft palate may be elevated so that it is in contact with the posterior wall of the pharynx, sealing off the oral passage from the

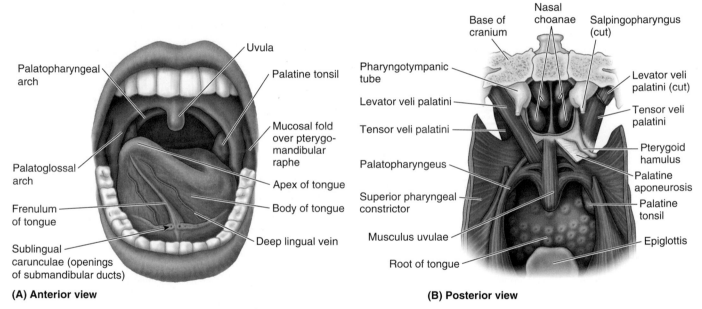

(A) Anterior view   **(B) Posterior view**

FIGURE 8.55. Soft palate. A. Surface anatomy of oral cavity and soft palate. **B.** Dissection of the soft palate shows the muscles and their relationship to the posterior part of the tongue.

TABLE 8.10. MUSCLES OF SOFT PALATE

Muscle	Origin	Insertion	Innervation	Main Action(s)
Tensor veli palatini	Scaphoid fossa at root of posterior border of medial pterygoid plate, spine of sphenoid bone, and cartilage of pharyngotympanic tube	Palatine aponeurosis (Fig. 8.54B)	Nerve to medial pterygoid (a branch of CN V₃) via otic ganglion	Tenses soft palate and opens mouth of pharyngotympanic tube during swallowing and yawning
Levator veli palatini	Cartilage of pharyngotympanic tube and petrous part of temporal bone		Pharyngeal branch of CN X via pharyngeal plexus	Elevates soft palate during swallowing and yawning
Palatoglossus	Palatine aponeurosis	Side of tongue		Elevates posterior part of tongue and draws soft palate onto tongue
Palatopharyngeus	Hard palate and palatine aponeurosis	Lateral wall of pharynx		Tenses soft palate and pulls walls of pharynx superiorly, anteriorly, and medially during swallowing
Musculus uvulae	Posterior nasal spine and palatine aponeurosis	Mucosa of uvula		Shortens uvula and pulls it superiorly

nasopharynx (e.g., when swallowing or breathing through the mouth). The soft palate can also be drawn inferiorly so that it is in contact with the posterior part of the tongue, sealing off the oral cavity from the nasal passage (e.g., when breathing exclusively through the nose, even with the mouth open). For attachments, nerve supply, and actions of the five muscles of the soft palate, see Figure 8.55B and Table 8.10. The muscles of the soft palate are as follows:

- The **levator veli palatini** (lifter of soft palate) is a cylindrical muscle that runs infero-anteriorly, spreading out in the soft palate where it attaches to the superior surface of the palatine aponeurosis.
- The tensor veli palatini (tensor of soft palate) is a muscle with a triangular belly that passes inferiorly; the tendon formed at its apex hooks around the **pterygoid hamulus**—the hook-shaped inferior projection of the medial pterygoid plate—before spreading out as the *palatine aponeurosis*.
- The **palatoglossus** is a slender slip of muscle that is covered with a mucous membrane; it forms the *palatoglossal arch*. Unlike the other muscles ending in *-glossus*, the palatoglossus is a palatine muscle (in function and innervation) rather than a tongue muscle.
- The **palatopharyngeus** is a thin, flat muscle also covered with a mucous membrane; it forms the *palatopharyngeal arch* and blends inferiorly with the longitudinal muscle of the pharynx.
- The **musculus uvulae** inserts into the mucosa of the uvula.

Tongue

The **tongue** (L. *lingua*; G. *glossa*) is a mobile muscular organ that can assume a variety of shapes and positions.

The tongue is partly in the oral cavity proper and partly in the oropharynx (Fig. 8.53). At rest, it occupies most of the oral cavity proper. The tongue—mainly composed of muscles and covered by mucous membrane—assists with mastication (chewing), taste, deglutition (swallowing), articulation (speech), and oral cleansing. The tongue has a root, a body, an apex, a curved dorsal surface, and an inferior surface (Fig. 8.56A). A V-shaped groove, the **terminal sulcus** (L. *sulcus terminalis*) of the tongue (Fig. 8.56B), marks the separation between the *anterior (presulcal) part* and the *posterior (postsulcal) part*.

The **root of the tongue** is the posterior third that rests on the floor of the mouth. The anterior two thirds of the tongue form the **body of the tongue**. The pointed anterior part of the body is the **apex (tip) of the tongue**. The body and apex are extremely mobile. The **dorsum (dorsal surface) of the tongue** is the posterosuperior surface of the tongue, which includes the **terminal sulcus**. At the apex of this groove is the **foramen cecum** (Fig. 8.56B), a small pit that is the nonfunctional remnant of the proximal part of the embryonic thyroglossal duct from which the thyroid gland developed. The mucous membrane on the anterior part of the tongue is rough because of the presence of numerous **lingual papillae**:

- **Vallate papillae** are large and flat-topped; they lie directly anterior to the terminal sulcus and are surrounded by deep moat-like trenches, the walls of which are studded by *taste buds*; the ducts of serous *lingual glands* (of von Ebner) open into these trenches.
- **Foliate papillae** are small lateral folds of lingual mucosa; they are poorly developed in humans.
- **Filiform papillae** are long, numerous, thread-like, and scaly; they contain afferent nerve endings that are sensitive to touch.

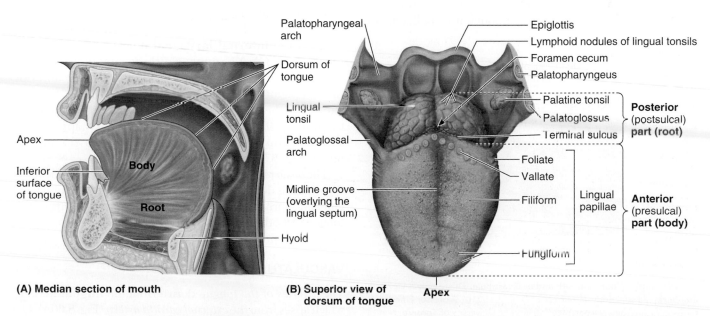

(A) Median section of mouth

Palatopharyngeal arch
Dorsum of tongue
Lingual tonsil
Palatoglossal arch
Midline groove (overlying the lingual septum)
Hyoid
Apex
Inferior surface of tongue
Body
Root

(B) Superior view of dorsum of tongue

Epiglottis
Lymphoid nodules of lingual tonsils
Foramen cecum
Palatopharyngeus
Palatine tonsil
Palatoglossus
Terminal sulcus
Foliate
Vallate
Filiform
Funglform
Lingual papillae
Apex

Posterior (postsulcal) part (root)
Anterior (presulcal) part (body)

FIGURE 8.56. Tongue. A. Parts. **B.** Features of dorsum of tongue.

- **Fungiform papillae** are mushroom-shaped and appear as pink or red spots; they are scattered among the filiform papillae but are most numerous at the apex and sides (margins) of the tongue.

The vallate, foliate, and most of the fungiform papillae contain taste receptors in the **taste buds**. A few taste buds are also in the epithelium covering the oral surface of the soft palate, the posterior wall of the oropharynx, and the epiglottis.

The mucous membrane of the dorsum of the tongue is thin over the anterior part of the tongue and is closely attached to the underlying muscle (Fig. 8.56A). A depression on the dorsal surface, the **midline groove of the tongue** (median sulcus of tongue), divides the tongue into right and left halves (Fig. 8.56B). It also indicates the site of fusion of the embryonic distal tongue buds.

The *root of the tongue* lies within the oropharynx, posterior to the *terminal sulcus* and the *palatoglossal arches* (Fig. 8.56B). Its mucous membrane is thick and freely movable. It has no lingual papillae, but the underlying **lymphoid nodules**, known collectively as the lingual tonsil, give this part of the tongue its cobblestone appearance.

The **inferior surface of the tongue** is covered with a thin, transparent mucous membrane through which one can see the underlying **deep lingual veins**. With the tongue raised, the **lingual frenulum** (Fig. 8.57), a large midline fold of mucosa that passes from the gingiva covering the lingual aspect of the anterior alveolar ridge to the postero-inferior surface of the tongue, can be seen. The frenulum connects the tongue to the floor of the mouth while allowing the anterior part of the tongue to move freely. At the base of the frenulum are the *openings of the submandibular ducts* from the submandibular salivary glands.

MUSCLES OF TONGUE

The tongue is essentially a mass of muscles that is mostly covered by mucous membrane. Although it is traditional to do so, providing descriptions of the actions of tongue muscles by ascribing a single action to a specific muscle greatly oversimplifies the actions of the tongue and is misleading. The muscles of the tongue do not act in isolation, and some muscles perform multiple actions with parts of one muscle capable of acting independently, producing different—even antagonistic—actions. *In general, however, extrinsic muscles alter the position of the tongue and intrinsic muscles alter its shape* (Fig. 8.58 and Table 8.11).

The four intrinsic and four extrinsic muscles in each half of the tongue are separated by a fibrous **lingual septum**, which extends vertically from the midline groove of the tongue (Fig. 8.58C). The **intrinsic muscles of the tongue** (superior and inferior longitudinal, transverse, and vertical) are confined to the tongue and are not attached to bone. The **extrinsic muscles of the tongue** (genioglossus, hyoglossus, styloglossus, and palatoglossus) originate from bony formations outside the tongue and attach to it.

INNERVATION OF TONGUE

All the muscles of the tongue are supplied by CN XII, the **hypoglossal nerve** (Fig. 8.59), except for the palatoglossus (actually a palatine muscle supplied by the *pharyngeal plexus*, the plexus of nerves that includes motor branches of CN X). For general sensation (touch and temperature), the mucosa of the anterior two thirds of the tongue is supplied by the lingual nerve, a branch of CN V₃ (Fig. 8.59B). For special sensation (taste), this part of the tongue, except for the vallate papillae, is supplied through the chorda tympani nerve, a branch of CN VII. The nerve joins the lingual nerve and runs anteriorly in its sheath (Fig. 8.59A).

The mucous membrane of the posterior third of the tongue and the vallate papillae are supplied by the lingual branch of the glossopharyngeal nerve (CN IX) for both general and special sensation (taste) (Fig. 8.59B). Twigs of the **internal laryngeal nerve**, a branch of the vagus nerve (CN X), supply mostly general but some special sensation to a small area of the tongue just anterior to the epiglottis. These mostly sensory nerves also carry parasympathetic secretomotor fibers to serous glands in the tongue. These nerve fibers probably synapse in the **submandibular ganglion** suspended from the lingual nerve (Fig. 8.59A).

The *basic taste sensations* are *sweet, salty, sour,* and *bitter.* Sweetness is detected at the apex, saltiness at the lateral margin, and sourness and bitterness at the posterior part of the tongue. All other "tastes" expressed by gourmets are olfactory (smell and aroma).

VASCULATURE OF TONGUE

The *arteries of the tongue* derive from the **lingual artery**, which arises from the *external carotid artery* (Fig. 8.60A). On entering the tongue, the lingual artery passes deep (medial)

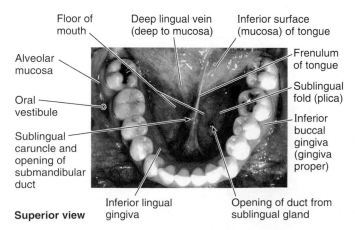

Floor of mouth
Deep lingual vein (deep to mucosa)
Inferior surface (mucosa) of tongue
Alveolar mucosa
Frenulum of tongue
Oral vestibule
Sublingual fold (plica)
Sublingual caruncle and opening of submandibular duct
Inferior buccal gingiva (gingiva proper)
Inferior lingual gingiva
Opening of duct from sublingual gland
Superior view

FIGURE 8.57. Floor of mouth and oral vestibule. The tongue is elevated and retracted superiorly. (Courtesy of Dr. B. Liebgott, Professor Emeritus, Division of Anatomy, Department of Surgery, University of Toronto, Ontario, Canada.)

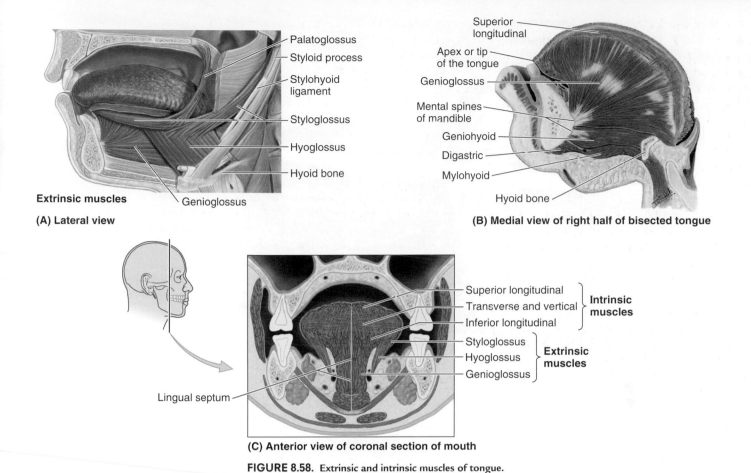

Palatoglossus
Styloid process
Stylohyoid ligament
Styloglossus
Hyoglossus
Hyoid bone

Extrinsic muscles
Genioglossus

(A) Lateral view

Superior longitudinal
Apex or tip of the tongue
Genioglossus
Mental spines of mandible
Geniohyoid
Digastric
Mylohyoid
Hyoid bone

(B) Medial view of right half of bisected tongue

Superior longitudinal
Transverse and vertical   } **Intrinsic muscles**
Inferior longitudinal
Styloglossus
Hyoglossus   } **Extrinsic muscles**
Genioglossus

Lingual septum

(C) Anterior view of coronal section of mouth

FIGURE 8.58. Extrinsic and intrinsic muscles of tongue.

TABLE 8.11. MUSCLES OF TONGUE

Muscle	Shape and Position	Proximal Attachment	Distal Attachment	Main Action(s)
Extrinsic muscles of tongue				
Genioglossus	Fan-shaped muscle; constitutes bulk of tongue	Via a short tendon from superior part of mental spine of mandible	Entire dorsum of tongue; inferior most and posterior most fibers attach to body of hyoid.	Bilateral activity depresses tongue, especially central part, creating a longitudinal furrow; posterior part pulls tongue anteriorly for protrusion[a]; most anterior part retracts apex of protruded tongue; unilateral contraction deviates ("wags") tongue to contralateral side
Hyoglossus	Thin, quadrilateral muscle	Body and greater horn of hyoid	Inferior aspects of lateral part of tongue	Depresses tongue, especially pulling its sides inferiorly; helps shorten (retrude) tongue
Styloglossus	Short triangular muscle	Anterior border of distal styloid process; stylohyoid ligament	Margins of tongue posteriorly, interdigitating with hyoglossus	Retrudes tongue and curls (elevates) its sides, working with genioglossus to form a central trough during swallowing
Palatoglossus	Narrow crescent-shaped palatine muscle; forms posterior column of isthmus of fauces	Palatine aponeurosis of soft palate	Enters posterolateral tongue transversely, blending with intrinsic transverse muscles	Capable of elevating posterior tongue or depressing soft palate; most commonly acts to constrict isthmus of fauces (L. the throat)
Intrinsic muscles of tongue				
Superior longitudinal	Thin layer deep to mucous membrane of dorsum of tongue	Submucosal fibrous layer and median fibrous septum	Margins of tongue and mucous membrane	Curls tongue longitudinally upward, elevating apex and sides of tongue; shortens (retrudes) tongue
Inferior longitudinal	Narrow band close to inferior surface of tongue	Root of tongue and body of hyoid	Apex of tongue	Curls tongue longitudinally downward, depressing apex; shortens (retrudes) tongue
Transverse	Deep to superior longitudinal muscle	Median fibrous septum	Fibrous tissue at lateral lingual margins	Narrows and elongates (protrudes) tongue[a]
Vertical	Fibers intersect transverse muscle	Submucosal fibrous layer of dorsum of tongue	Inferior surface of borders of tongue	Flattens and broadens tongue[a]

[a]The transverse and vertical intrinsic muscles act simultaneously as posterior genioglossus pulls root anteriorly to protrude tongue.

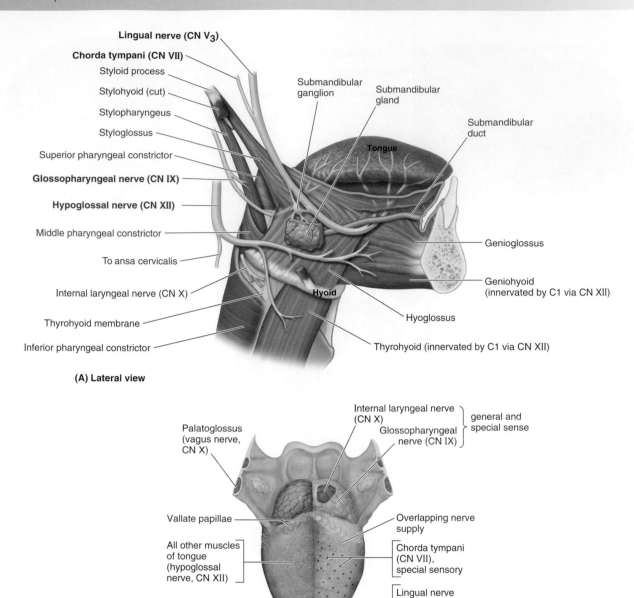

Lingual nerve (CN V₃)
Chorda tympani (CN VII)
Styloid process
Stylohyoid (cut)
Stylopharyngeus
Styloglossus
Superior pharyngeal constrictor
Glossopharyngeal nerve (CN IX)
Hypoglossal nerve (CN XII)
Middle pharyngeal constrictor
To ansa cervicalis
Internal laryngeal nerve (CN X)
Thyrohyoid membrane
Inferior pharyngeal constrictor

Submandibular ganglion
Submandibular gland
Submandibular duct
Tongue
Genioglossus
Geniohyoid (innervated by C1 via CN XII)
Hyoglossus
Hyoid
Thyrohyoid (innervated by C1 via CN XII)

(A) Lateral view

Internal laryngeal nerve (CN X)
Glossopharyngeal nerve (CN IX)
} general and special sense
Palatoglossus (vagus nerve, CN X)
Vallate papillae
All other muscles of tongue (hypoglossal nerve, CN XII)
Overlapping nerve supply
Chorda tympani (CN VII), special sensory
Lingual nerve (CN V₃), general sensory

MOTOR NERVES   **SENSORY NERVES**

(B) Superior view

FIGURE 8.59. Innervation of tongue. A. Course of lingual and hypoglossal nerves. **B.** Overview of sensory and motor innervation.

to the hyoglossus muscle. The *main branches of the lingual artery* are as follows:

- **Dorsal lingual arteries**, which supply the posterior part, the root of the tongue, and send a tonsillar branch to the palatine tonsil
- **Deep lingual artery**, which supplies the anterior part of the tongue; the dorsal and deep arteries communicate with each other near the apex of the tongue.
- **Sublingual artery**, which supplies the sublingual gland and the floor of the mouth

The **veins of the tongue** are as follows (Fig. 8.60B):

- *Dorsal lingual veins*, which accompany the lingual artery

- *Deep lingual veins*, which begin at the apex of the tongue and run posteriorly beside the lingual frenulum to join the *sublingual vein*

All lingual veins terminate, directly or indirectly, in the IJV.

Lymphatic drainage of the tongue takes the following routes (Fig. 8.60C,D):

- Lymph from the posterior third drains to the *superior deep cervical lymph nodes* on both sides.
- Lymph from the *medial part of the anterior two thirds* drains to the *inferior deep cervical lymph nodes*.
- Lymph from *lateral parts of the anterior two thirds* drains to the *submandibular lymph nodes*.

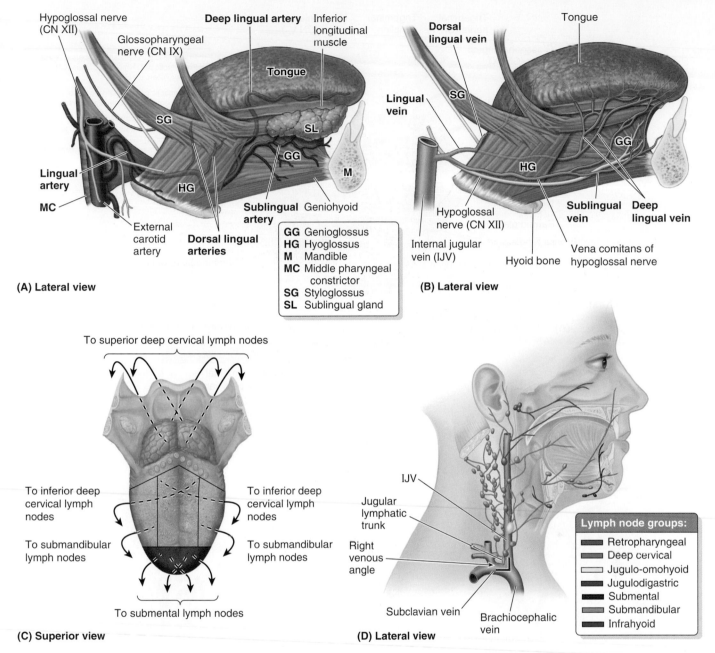

FIGURE 8.60. Blood supply and lymphatic drainage of tongue. A. Arterial supply. **B.** Venous drainage. **C and D.** Lymphatic drainage. *IJV*, internal jugular vein.

- Lymph from the *apex of the tongue and frenulum* drains to the *submental lymph nodes.*
- Lymph from the *posterior third* and the area near the midline groove drains bilaterally.

Salivary Glands

The **salivary glands** include the parotid, submandibular, and sublingual glands (Fig. 8.61A). **Saliva**, the clear, tasteless, odorless viscid fluid secreted by these glands and the mucous glands of the oral cavity serves the following functions:

- Keeps the mucous membrane of the mouth moist
- Lubricates the food during mastication

- Begins digestion of starches
- Serves as an intrinsic "mouthwash"
- Plays a significant role in the prevention of tooth decay and in the ability to taste

In addition to the three major salivary glands, small *accessory salivary glands* are scattered over the palate, lips, cheeks, tonsils, and tongue.

The **parotid glands** are the largest of the major salivary glands (Fig. 8.61A). Each parotid gland has an irregular shape because it occupies the gap between the ramus of the mandible and the styloid and mastoid processes of the temporal bone. The purely serous secretion of the gland passes through the parotid duct and empties into the vestibule of

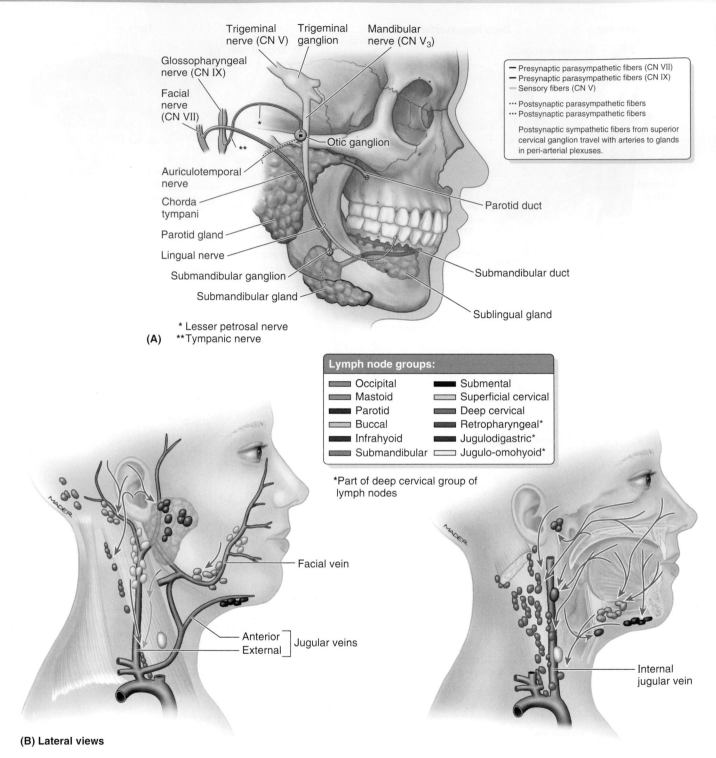

Trigeminal nerve (CN V)

Trigeminal ganglion

Mandibular nerve (CN V₃)

Glossopharyngeal nerve (CN IX)

Facial nerve (CN VII)

Otic ganglion

Auriculotemporal nerve

Chorda tympani

Parotid gland

Lingual nerve

Submandibular ganglion

Submandibular gland

Parotid duct

Submandibular duct

Sublingual gland

— Presynaptic parasympathetic fibers (CN VII)
— Presynaptic parasympathetic fibers (CN IX)
— Sensory fibers (CN V)
··· Postsynaptic parasympathetic fibers
··· Postsynaptic parasympathetic fibers

Postsynaptic sympathetic fibers from superior cervical ganglion travel with arteries to glands in peri-arterial plexuses.

* Lesser petrosal nerve
(A)   **Tympanic nerve

Lymph node groups:

▬ Occipital	▬ Submental
▬ Mastoid	▬ Superficial cervical
▬ Parotid	▬ Deep cervical
▬ Buccal	▬ Retropharyngeal*
▬ Infrahyoid	▬ Jugulodigastric*
▬ Submandibular	▬ Jugulo-omohyoid*

*Part of deep cervical group of lymph nodes

Facial vein

Anterior ⎱ Jugular veins
External ⎰

Internal jugular vein

(B) Lateral views

FIGURE 8.61. Salivary glands. A. Location and innervation. **B.** Lymphatic drainage of face and glands.

the oral cavity opposite the second maxillary molar tooth. In addition to its digestive function, the secretion washes food particles into the mouth proper. The *arterial supply of the parotid gland* and its duct is from branches of the *external carotid* and *superficial temporal arteries* (see Fig. 8.44A).

The *veins from the parotid gland* drain into the *retromandibular veins* (see Fig. 8.44B). The *lymphatic vessels from the parotid gland* end in the *superficial and deep cervical lymph nodes* (Fig. 8.61B). The parotid gland was discussed earlier in this chapter, when its innervation was described.

The **submandibular glands** lie along the body of the mandible, partly superior and partly inferior to the posterior half of the mandible, and partly superficial and partly deep to the mylohyoid muscle (Fig. 8.61A). The **submandibular duct** arises from the part of the gland that lies between the mylohyoid and hyoglossus muscles. Passing from lateral to medial, the lingual nerve loops under the duct as it runs anteriorly to open via one to three orifices on a small, fleshy *sublingual papilla* on each side of the lingual frenulum (Fig. 8.59B). The orifices of the submandibular ducts are visible, and saliva often sprays from it when the tongue is elevated and retracted.

The *arterial supply of the submandibular glands* is from the **submental arteries** (see Fig. 8.44A). The veins accompany the arteries. The submandibular gland is supplied by presynaptic parasympathetic secretomotor fibers conveyed from the facial nerve to the lingual nerve by the *chorda tympani nerve* (Fig. 8.61A), which synapse with postsynaptic neurons in the *submandibular ganglion*. The latter fibers accompany arteries to reach the gland, along with vasoconstrictive postsynaptic sympathetic fibers from the superior cervical ganglion. The lymphatic vessels of the submandibular gland drain into the *deep cervical lymph nodes*, particularly the *jugulo-omohyoid lymph node* (Fig. 8.61B).

The **sublingual glands** are the smallest and most deeply situated (Fig. 8.61A). Each gland lies in the floor of the mouth between the mandible and the genioglossus muscle. The glands from each side unite to form a horseshoe-shaped mass around the lingual frenulum. Numerous small **sublingual ducts** open into the floor of the mouth alongside the lingual folds.

The *arterial supply of the sublingual glands* is from the *sublingual* and submental *arteries*—branches of the lingual and facial arteries, respectively (Figs. 8.44A and 8.60A). The *innervation of the sublingual glands* is the same as that described for the submandibular gland.

CLINICAL BOX

Imaging of Salivary Glands

The parotid and submandibular salivary glands may be examined radiographically after the injection of a contrast medium into their ducts. This special type of radiograph (*sialogram*) demonstrates the salivary ducts and some secretory units. *Salivary duct calculi* (*stones*) are visible on CT. Salivary gland tumors are evaluated with CT or MRI.

Gag Reflex

One may touch the anterior part of the tongue without feeling discomfort; however, when the posterior tongue or mouth is touched, one usually gags. CN IX and CN X are responsible for the muscular contraction of each side of the oropharynx. Glossopharyngeal branches (CN IX) provide the afferent limb of the *gag reflex*.

Paralysis of Genioglossus

When the genioglossus is paralyzed, the tongue mass has a tendency to shift posteriorly, obstructing the airway and presenting the risk of suffocation. Total relaxation of the genioglossus muscles occurs during general anesthesia; therefore, the tongue of an anesthetized patient must be prevented from relapsing by inserting an airway.

Injury to Hypoglossal Nerve

Trauma, such as a fractured mandible, may injure the hypoglossal nerve (CN XII), resulting in paralysis and eventual atrophy of one side of the tongue. The tongue deviates to the paralyzed side during protrusion because of the (unopposed) action of the unaffected genioglossus on the other side (see also Chapter 10, Fig. B10.7).

Sublingual Absorption of Drugs

For quick *transmucosal absorption* of a drug—for instance, when nitroglycerin is used as a vasodilator in *angina pectoris* (chest pain)—the pill (or spray) is put under the tongue, where the thin mucosa allows the absorbed drug to enter the deep lingual veins (see Fig. 8.57) in less than a minute.

Lingual Carcinoma

Malignant tumors in the posterior part of the tongue metastasize to the superior deep cervical lymph nodes on both sides. In contrast, tumors in the apex and anterolateral parts usually do not metastasize to the inferior deep cervical nodes until late in the disease. Because the deep nodes are closely related to the IJVs, metastases from the carcinoma may spread to the submental and submandibular regions and along the IJVs into the neck.

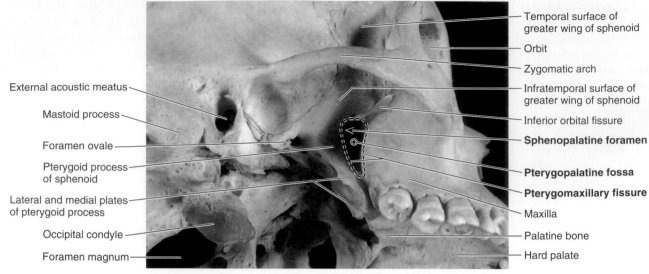

(A) Inferolateral and slightly posterior view, looking into infratemporal and pterygopalatine fossae

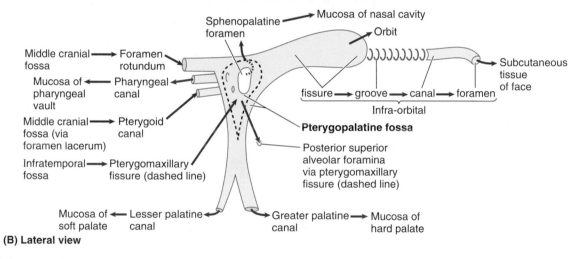

(B) Lateral view

FIGURE 8.62. Pterygopalatine fossa—communications and contents. The pterygopalatine fossa communicates with most compartments of the deep face via many passages (foramina, fissures, and canals). **A.** Photograph. **B.** Schematic illustration.

PTERYGOPALATINE FOSSA

The **pterygopalatine fossa** is a small conical space inferior to the apex of the orbit. It lies between the pterygoid process of the sphenoid posteriorly and the posterior aspect of the maxilla anteriorly (Fig. 8.62A). The fragile perpendicular plate of the palatine bone forms its medial wall. The incomplete *roof of the pterygopalatine fossa* is formed by the *greater wing of the sphenoid*. The *floor of the pterygopalatine fossa* is formed by the *pyramidal process of the palatine bone*. Its superior, larger end opens into the *inferior orbital fissure*; its inferior end is closed except for the palatine foramina. The pterygopalatine fossa communicates (Fig. 8.62B)

- laterally with the *infratemporal fossa* through the **pterygomaxillary fissure**
- medially with the *nasal cavity* through the **sphenopalatine foramen**

- anterosuperiorly with the *orbit* through the *inferior orbital fissure*
- posterosuperiorly with the *middle cranial fossa* through the *foramen rotundum* and *pterygoid canal*

The contents of the pterygopalatine fossa are the

- maxillary nerve (CN V$_2$), with which are associated the nerve of the pterygoid canal and the pterygopalatine ganglion (Figs. 8.63 and 8.64B)
- terminal (third) part of the maxillary artery and its branches (Figs. 8.43 and 8.64A) with accompanying veins draining to the pterygoid venous plexus

The **maxillary nerve (CN V$_2$)** enters the pterygopalatine fossa posterosuperiorly through the foramen rotundum and runs anterolaterally in the fossa (Figs. 8.63 and 8.64). Within the fossa, the maxillary nerve gives off the *zygomatic nerve*, which divides into the *zygomaticofacial* and

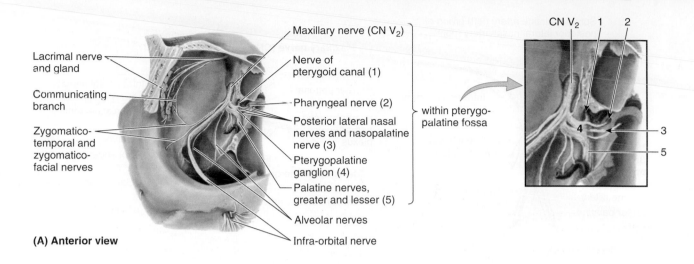

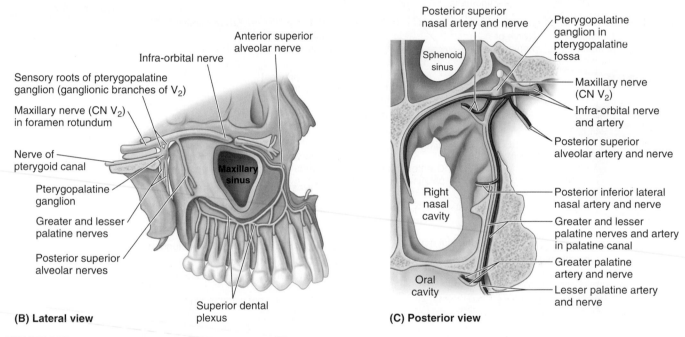

FIGURE 8.63. Nerves of pterygopalatine fossa. A. The fossa is viewed through the floor of the orbit to show the maxillary nerve (CN V₂) and its branches. **B.** The fossa is viewed laterally. Part of the lateral wall of the maxillary sinus has been removed. **C.** In this coronal section, the nasopalatine and greater and lesser palatine nerves can be seen.

zygomaticotemporal nerves (Fig. 8.63A). These nerves emerge from the zygomatic bone through the cranial foramina of the same name and supply the lateral region of the cheek and the temple. The *zygomaticotemporal nerve* also gives rise to a communicating branch, which conveys parasympathetic secretomotor fibers to the lacrimal gland by way of the lacrimal nerve from CN V₁.

While in the pterygopalatine fossa, the maxillary nerve also gives off the two *pterygopalatine nerves*, which suspend the parasympathetic *pterygopalatine ganglion* in the superior part of the pterygopalatine fossa (Fig. 8.63A,B). The pterygopalatine nerves convey general sensory fibers of the maxillary nerve, which pass through the pterygopalatine

ganglion without synapsing, and supply the nose, palate, tonsil, and gingivae (Fig. 8.64B,E). The maxillary nerve leaves the pterygopalatine fossa through the inferior orbital fissure, after which it is known as the *infra-orbital nerve*.

The *parasympathetic fibers to the pterygopalatine ganglion* come from the facial nerve by way of its first branch, the *greater petrosal nerve* (Fig. 8.64C). This nerve joins the *deep petrosal nerve* as it traverses the foramen lacerum region to form the **nerve of the pterygoid canal**. This nerve passes anteriorly through the pterygoid canal to the pterygopalatine fossa. The parasympathetic fibers of the greater petrosal nerve synapse in the pterygopalatine ganglion (Fig. 8.64D).

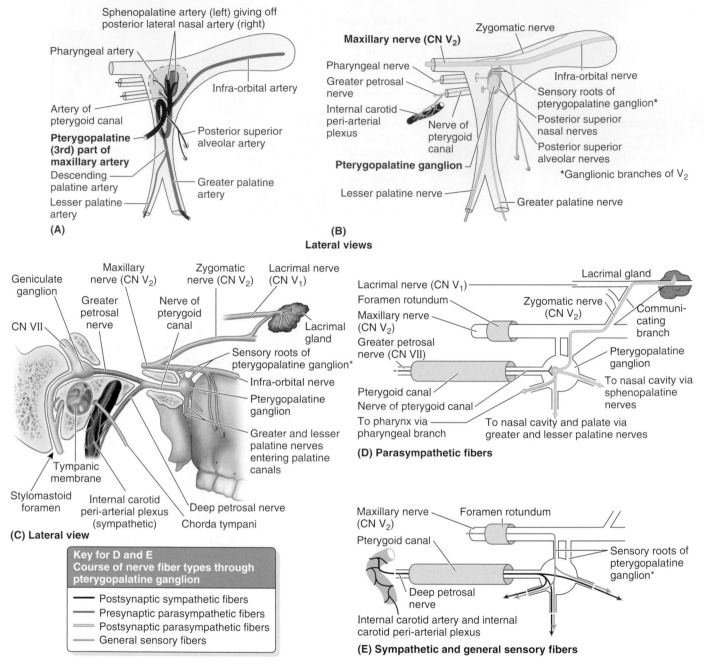

FIGURE 8.64. **Schematic illustrations of arteries and nerves of pterygopalatine fossa. A.** Pterygopalatine part of the maxillary artery. **B.** Pterygopalatine part of the maxillary nerve. **C.** Pterygopalatine ganglion in situ. **D.** Course of parasympathetic fibers. **E.** Course of sympathetic fibers.

The **deep petrosal nerve** is a sympathetic nerve that arises from the *sympathetic plexus on the internal carotid artery* (Fig. 8.64C,E). It conveys postsynaptic fibers from nerve cell bodies in the superior cervical sympathetic ganglion. Thus, these fibers do not synapse in the pterygopalatine ganglion; they pass directly to join the branches of the ganglion (maxillary nerve). The postsynaptic parasympathetic and sympathetic fibers pass to the lacrimal gland and the glands of the nasal cavity, palate, and superior pharynx (Fig. 8.63C).

The maxillary artery, a terminal branch of the external carotid artery, passes anteriorly and traverses the infratemporal

fossa. It passes over the lateral pterygoid muscle and enters the pterygopalatine fossa. The **pterygopalatine part of the maxillary artery**, its third part, passes through the *pterygomaxillary fissure* and enters the pterygopalatine fossa (Fig. 8.64A). The artery gives rise to branches that accompany all the nerves in the fossa with the same names. The *branches of the third, or pterygopalatine, part of the maxillary artery* are as follows (Fig. 8.64B):

- Posterior superior alveolar artery
- Descending palatine artery, which divides into greater and lesser palatine arteries

- Artery of the pterygoid canal
- Sphenopalatine artery, which divides into posterior lateral nasal branches to the lateral wall of the nasal cavity and its associated paranasal sinuses and the posterior septal branches (Fig. 8.63C)
- Infra-orbital artery, which gives rise to the anterior superior alveolar artery and terminates as branches to the inferior eyelid, nose, and upper lip

NOSE

The **nose** is the part of the respiratory tract superior to the hard palate; it contains the organ of smell. It includes the external nose and nasal cavities, which are divided into right and left cavities by the *nasal septum* (Fig. 8.65A). Each nasal cavity is divisible into an *olfactory area* and a *respiratory area*. The functions of the nose and nasal cavities are as follows:

- Olfaction (smelling)
- Respiration (breathing)
- Filtration of dust
- Humidification of inspired air
- Reception and elimination of secretions from the nasal mucosa, paranasal sinuses, and nasolacrimal ducts

External Nose

The **external nose** varies considerably in size and shape, mainly because of differences in the nasal cartilages. The **dorsum of the nose** extends from its superior angle, the **root of the nose** (Fig. 8.65A), to the **apex** (tip) **of the nose**. The inferior surface of the nose is pierced by two piriform (L. pear-shaped) openings, the **nares** (nostrils, anterior nasal apertures), which are bound laterally by the **alae** (wings) of the nose and separated from each other by the nasal septum. The external nose consists of bony and cartilaginous parts (Fig. 8.65B).

The **bony part of the nose** consists of the following structures:

- Nasal bones
- Frontal processes of the maxillae
- Nasal part of the frontal bone and its nasal spine
- Bony part of the nasal septum

The **cartilaginous part of the nose** consists of five main cartilages: two **lateral cartilages**, two **alar cartilages**, and a **septal cartilage**. The U-shaped alar cartilages are free and movable; they dilate or constrict the nares when the muscles acting on the nose contract.

Nasal Cavities

The nasal cavities, entered through the nares (Fig. 8.65A), open posteriorly into the nasopharynx through the *choanae*.

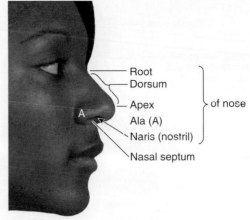

(A) Lateral view

Root
Dorsum
Apex
Ala (A)
Naris (nostril)
Nasal septum
} of nose

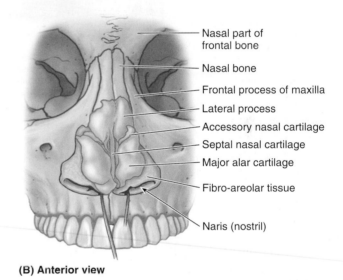

Nasal part of frontal bone
Nasal bone
Frontal process of maxilla
Lateral process
Accessory nasal cartilage
Septal nasal cartilage
Major alar cartilage
Fibro-areolar tissue
Naris (nostril)

(B) Anterior view

FIGURE 8.65. External nose. A. Surface anatomy of nose. **B.** Nasal bones and cartilages. The cartilages are retracted inferiorly.

Mucosa lines the nasal cavities, except the *vestibule of the nose*, which is lined with skin (Fig. 8.66). The **nasal mucosa** is firmly bound to the periosteum and perichondrium of the supporting bones and cartilages of the nose (Fig. 8.67A). The mucosa is continuous with the lining of all the chambers with which the nasal cavities communicate: the nasopharynx posteriorly, the paranasal sinuses superiorly and laterally, and the lacrimal sac and conjunctiva superiorly. The inferior two thirds of the nasal mucosa is the **respiratory area**, and the superior one third is the **olfactory area** (Fig. 8.67B). Air passing over the respiratory area is warmed and moistened before it passes through the rest of the upper respiratory tract to the lungs. The *olfactory area* is specialized mucosa containing the peripheral organ of smell; sniffing draws air to the area. The central processes of the olfactory receptor neurons in the olfactory epithelium unite to form nerve bundles that pass through the cribriform plate (collectively constituting the **olfactory nerve (CN 1)** [Fig. 8.67B]), and enter the **olfactory bulb** (see also Chapter 9, Neck, Fig. 9.5).

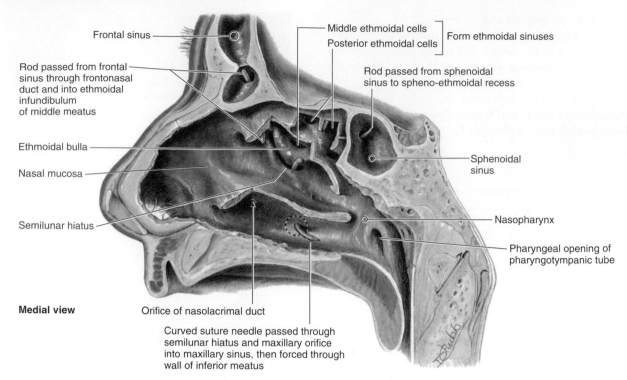

Frontal sinus

Rod passed from frontal sinus through frontonasal duct and into ethmoidal infundibulum of middle meatus

Ethmoidal bulla

Nasal mucosa

Semilunar hiatus

Medial view

Orifice of nasolacrimal duct

Curved suture needle passed through semilunar hiatus and maxillary orifice into maxillary sinus, then forced through wall of inferior meatus

Middle ethmoidal cells
Posterior ethmoidal cells
} Form ethmoidal sinuses

Rod passed from sphenoidal sinus to spheno-ethmoidal recess

Sphenoidal sinus

Nasopharynx

Pharyngeal opening of pharyngotympanic tube

FIGURE 8.66. **Features and openings of lateral wall of nose.** Parts of the conchae have been removed to show the openings of sinuses and other structures.

The *boundaries of the nasal cavity* (Fig. 8.67A) are as follows:

- The *roof of the nasal cavity* is curved and narrow, except at the posterior end.
- The *floor of the nasal cavity* is wider than the roof and is formed by the *hard palate*.
- The *medial wall of the nasal cavity* is formed by the nasal septum, the main components of which are the *perpendicular plate of the ethmoid*, *vomer*, *septal cartilage*, and the *nasal crests of the maxillary and palatine bones*.
- The *lateral wall of the nasal cavity* is uneven because of the nasal conchae (**superior**, **middle**, and **inferior**), three elevations that project inferiorly like scrolls. The conchae curve inferomedially, each forming a roof and partial medial wall for a **meatus**, or recess.

The nasal conchae (L. *shells*) divide the nasal cavity into four air passages (Figs. 8.66 and 8.67A): spheno-ethmoidal recess, superior nasal meatus, middle nasal meatus, and inferior nasal meatus. The **spheno-ethmoidal recess**, lying superoposterior to the superior concha, receives the *opening of the sphenoidal sinus*. The **superior nasal meatus** is a narrow passage between the superior and middle nasal conchae (parts of the ethmoid bone) into which the posterior ethmoidal sinuses open by one or more orifices. The **middle nasal meatus** is longer and deeper than the superior one. The anterosuperior part of this passage leads into the *ethmoidal infundibulum*, an opening through which it communicates with the frontal sinus, via the **frontonasal duct**. The **semilunar hiatus** (L. *hiatus semilunaris*) is a semicircular groove into which the frontonasal duct opens. The **ethmoidal bulla** (L. bubble), a rounded

elevation located superior to the semilunar hiatus, is visible when the middle concha is removed. The bulla is formed by *middle ethmoidal cells*, which constitute the *ethmoidal sinuses* (Fig. 8.66). The *maxillary sinus* also opens into the posterior end of the semilunar hiatus. The **inferior nasal meatus** is a horizontal passage, inferolateral to the inferior nasal concha (an independent, paired bone). The **nasolacrimal duct** from the lacrimal sac opens into the anterior part of this meatus.

The *arterial supply of the medial and lateral walls of the nasal cavity* is from branches of the **sphenopalatine artery**, **anterior and posterior ethmoidal arteries**, **greater palatine artery**, **superior labial artery**, and the **lateral nasal branches of the facial artery** (Figs. 8.63C and 8.67C). On the anterior part of the nasal septum is an area rich in capillaries (*Kiesselbach area*) where all five arteries supplying the septum anastomose. This area is often where profuse bleeding from the nose occurs. A rich *plexus of veins* drains deep to the nasal mucosa into the sphenopalatine, facial, and ophthalmic veins.

The *nerve supply of the postero-inferior half to two thirds of the nasal mucosa* is chiefly from CN V$_2$ by way of the nasopalatine nerve to the nasal septum and posterior lateral nasal branches of the greater palatine nerve to the lateral wall (Fig. 8.67B). The anterosuperior part of the nasal mucosa (both the septum and lateral wall) is supplied by the **anterior ethmoidal nerves**, branches of CN V$_1$.

Paranasal Sinuses

The **paranasal sinuses** are air-filled extensions of the respiratory part of the nasal cavity into the following cranial bones: frontal, ethmoid, sphenoid, and maxilla (Fig. 8.68). They are named according to the bones in which they are located.

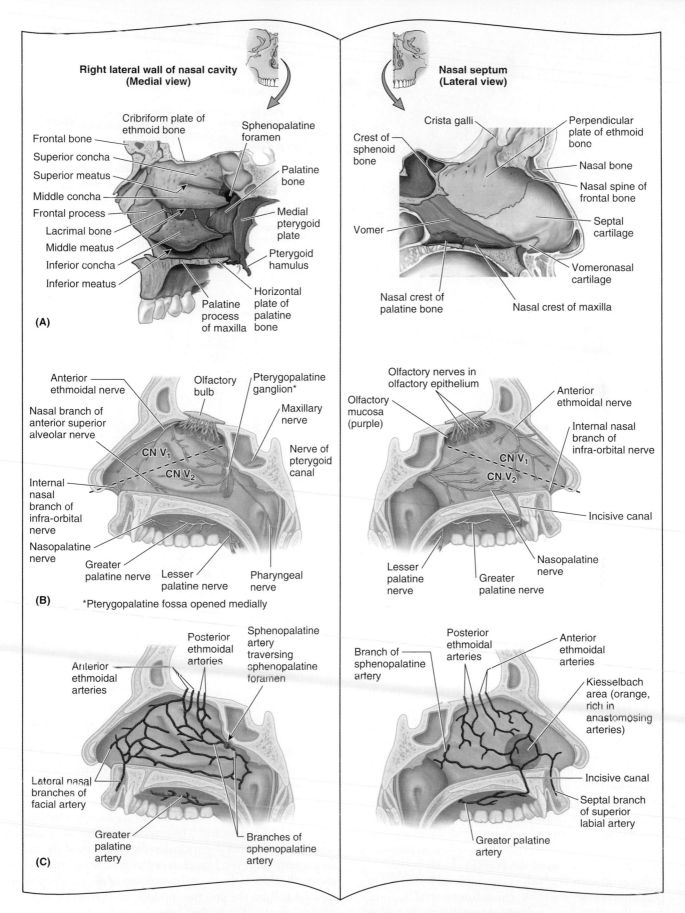

FIGURE 8.67. Bones, nerves, and arteries of lateral wall of nose and nasal septum.

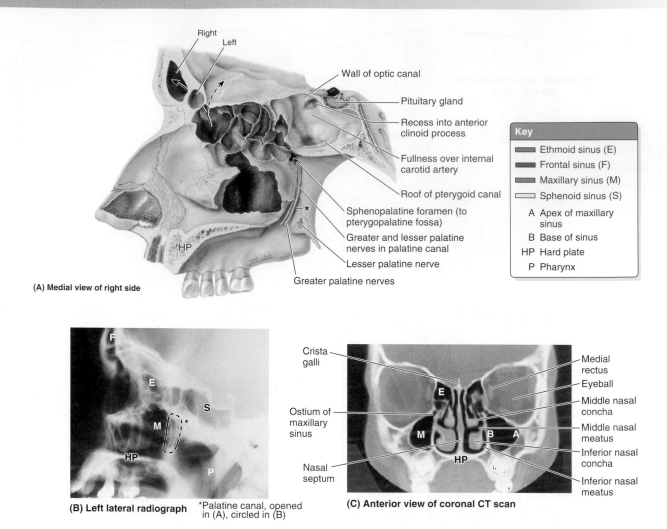

FIGURE 8.68. Paranasal sinuses. A. Paranasal sinuses on the right side have been opened from a nasal approach and color coded. **B.** Lateral radiograph. **C.** Coronal CT scan.

The **frontal sinuses** are located between the outer and inner tables of the frontal bone, posterior to the superciliary arches and the root of the nose. Each sinus drains through a frontonasal duct into the *ethmoidal infundibulum*, which opens into the *semilunar hiatus* of the middle meatus (Fig. 8.66). The frontal sinuses are innervated by branches of the *supra-orbital nerves* (CN V$_1$).

The **ethmoidal cells (sinuses)** include several cavities that are located in the lateral mass of the ethmoid bone between the nasal cavity and the orbit. The **anterior ethmoidal cells** drain directly or indirectly into the middle meatus through the infundibulum (Fig. 8.66). The **middle ethmoidal cells** open directly into the middle meatus. The **posterior ethmoidal cells**, which form the ethmoidal bulla, open directly into the superior meatus. The ethmoidal sinuses are supplied by the anterior and posterior ethmoidal branches of the *nasociliary nerves* (CN V$_1$).

The **sphenoidal sinuses**, unevenly divided and separated by a bony septum, occupy the body of the sphenoid bone; they may extend into the wings of this bone in elderly people. Because of these sinuses, the body of the sphenoid is fragile. Only thin plates of bone separate the sinuses from several important structures: the optic nerves and optic chiasm, the pituitary gland,

the internal carotid arteries, and the cavernous sinuses. The *posterior ethmoidal artery* and *nerve* supply the sphenoidal sinuses.

The **maxillary sinuses** are the largest of the paranasal sinuses (Fig. 8.68). These large pyramidal cavities occupy the bodies of the maxillae. The apex of the maxillary sinus extends laterally and often into the zygomatic bone. The *base of the maxillary sinus* forms the inferior part of the lateral wall of the nasal cavity. The *roof of the maxillary sinus* is formed by the floor of the orbit. The *floor of the maxillary sinus* is formed by the alveolar part of the maxilla. The roots of the maxillary teeth, particularly the first two molars, often produce conical elevations in the floor of the maxillary sinus. Each sinus drains by an opening, the **maxillary ostium** (Figs. 8.66 and 8.68), into the middle meatus of the nasal cavity by way of the semilunar hiatus. Because of the superior location of this opening, it is impossible for the sinus to drain when the head is erect until the sinus is full. The *arterial supply of the maxillary sinus* is mainly from superior alveolar branches of the *maxillary artery*; however, branches of the *greater palatine artery* supply the floor of the sinus. *Innervation of the maxillary sinus mucosa* is from the anterior, middle, and posterior *superior alveolar nerves* (Fig. 8.63B), branches of CN V$_2$.

CLINICAL BOX

Nasal Fractures

 Because of the prominence of the nose, *fractures of the nasal bones* are common facial fractures in automobile accidents and sports (unless face guards are worn). Fractures usually result in deformation of the nose, particularly when a lateral force is applied by someone's elbow, for example. *Epistaxis* (nosebleed) usually occurs. In severe fractures, disruption of the bones and cartilages results in displacement of the nose. When the injury results from a direct blow, the cribriform plate of the ethmoid bone may also fracture, often accompanied by *CSF rhinorrhea* (leaking of CSF through the nose).

Deviation of Nasal Septum

 The nasal septum is usually *deviated* (bent) to one side or the other (Fig. B8.13). This could be the result of a birth injury, but more often, the deviation results during adolescence and adulthood from trauma (e.g., during a fist fight). Sometimes, the deviation is so severe that the nasal septum is in contact with the lateral wall of the nasal cavity and often obstructs breathing or exacerbates snoring. The deviation can be corrected surgically.

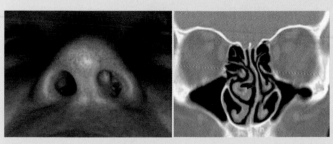

Nasal septum deviated to left side

FIGURE B8.13. Deviation of nasal septum.

Rhinitis

 The nasal mucosa becomes swollen and inflamed (*rhinitis*) during severe upper respiratory infections and allergic reactions (e.g., hay fever). Swelling of the mucosa occurs readily because of its vascularity and glandular nature. Infections of the nasal cavities may spread to the following structures:

- Anterior cranial fossa through the cribriform plate
- Nasopharynx and retropharyngeal soft tissues
- Middle ear through the *pharyngotympanic tube* (auditory tube), which connects the tympanic cavity and nasopharynx
- Paranasal sinuses
- Lacrimal apparatus and conjunctiva

Epistaxis

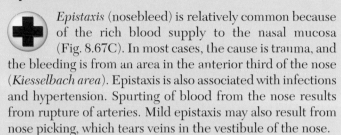 *Epistaxis* (nosebleed) is relatively common because of the rich blood supply to the nasal mucosa (Fig. 8.67C). In most cases, the cause is trauma, and the bleeding is from an area in the anterior third of the nose (*Kiesselbach area*). Epistaxis is also associated with infections and hypertension. Spurting of blood from the nose results from rupture of arteries. Mild epistaxis may also result from nose picking, which tears veins in the vestibule of the nose.

Sinusitis

Because the paranasal sinuses are continuous with the nasal cavities through apertures that open into them, infection may spread from the nasal cavities, producing inflammation and swelling of the mucosa of the sinuses (*sinusitis*) and local pain. Sometimes, several sinuses are inflamed (*pansinusitis*), and the swelling of the mucosa may block one or more openings of the sinuses into the nasal cavities.

Infection of Ethmoidal Cells

If nasal drainage is blocked, infections of the ethmoidal cells may break through the fragile medial wall of the orbit. Severe infections from this source may cause blindness because some posterior ethmoidal cells lie close to the optic canal, which gives passage to the optic nerve and ophthalmic artery. Spread of infection from these cells could also affect the dural nerve sheath of the optic nerve, causing *optic neuritis* (inflammation of optic nerve).

Infection of Maxillary Sinuses

The maxillary sinuses are the most commonly infected, probably because their ostia are commonly small and are located high on their superomedial walls. When the mucous membrane of the sinus is congested, the maxillary ostia are often obstructed. Because of the high location of the ostia, when the head is erect, it is impossible for the sinuses to drain until they are full. Because the ostia of the right and left sinuses lie on the medial sides (i.e., are directed toward each other), when lying on one's side, only the upper sinus (e.g., the right sinus if lying on the left side) drains. A cold or allergy involving both sinuses can result in nights of rolling from side-to-side in an attempt to keep the sinuses drained. A maxillary sinus can be *cannulated* (intubated) and drained by passing a cannula from the nares through the maxillary ostium into the sinus.

(Continued on next page)

Relationship of Teeth to Maxillary Sinus

The close proximity of the three maxillary molar teeth to the floor of the maxillary sinus poses potentially serious problems. During removal of a molar tooth, *a fracture of a root* may occur. If proper retrieval methods are not used, a piece of the root may be driven superiorly into the maxillary sinus. A communication may be created between the oral cavity and the maxillary sinus as a result, and an infection may occur.

EAR

The **ear** is divided into *external, middle,* and *internal parts* (Fig. 8.69A). The external and middle parts are mainly concerned with the transference of sound to the internal ear, which contains the organ for equilibrium (the condition of being evenly balanced) as well as for hearing. The *tympanic membrane* (eardrum) separates the external ear from the middle ear (Fig. 8.69A). The *pharyngotympanic (auditory) tube* joins the middle ear to the nasopharynx.

External Ear

The **external ear** is composed of the *auricle* (pinna), which collects sound, and the *external acoustic meatus (canal),* which conducts sound to the tympanic membrane (Fig. 8.69A).

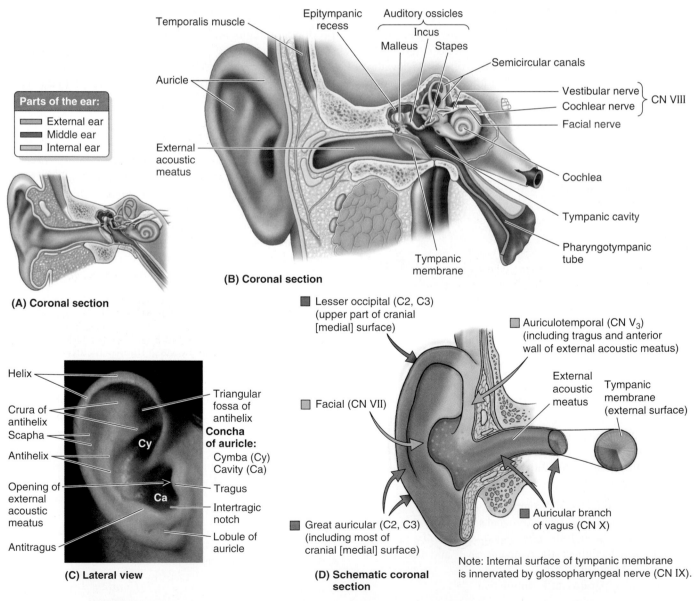

FIGURE 8.69. Ear. A and B. The external, middle, and internal ear are detailed. **C.** Surface anatomy. **D.** Innervation.

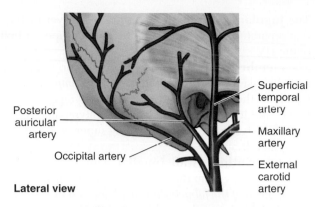

Lateral view

FIGURE 8.70. Arterial supply of auricle.

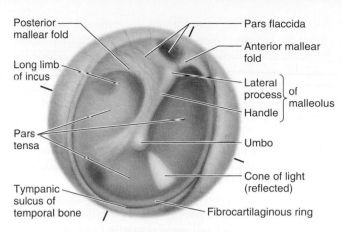

(A) Otoscopic view of right tympanic membrane

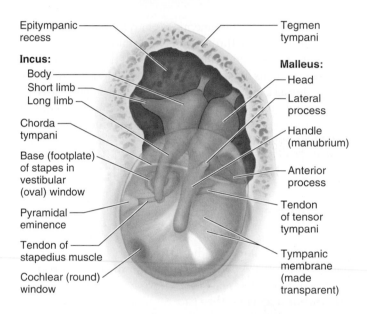

(B) Ossicles of ear seen through tympanic membrane

FIGURE 8.72. Tympanic membrane and lateral approach to tympanic cavity. A. Otoscopic view of the right tympanic membrane. The *cone of light* is a reflection of the light of the otoscope. **B.** The tympanic membrane has been rendered semitransparent, and the lateral wall of the epitympanic recess has been removed to demonstrate the ossicles of the ear in situ.

The **auricle** (L. *auricula*) is composed of elastic cartilage covered by thin skin. The auricle has several depressions and elevations. The **concha** is the deepest depression, and the elevated margin of the auricle is the **helix** (Fig. 8.69C). The noncartilaginous **lobule** (earlobe) consists of fibrous tissue, fat, and blood vessels. It is easily pierced for taking small blood samples and inserting earrings. The **tragus** is a tongue-like projection overlapping the opening of the external acoustic meatus. The arterial supply to the auricle is derived mainly from the *posterior auricular* and *superficial temporal arteries* (Fig. 8.70). The main **nerves to the skin of the auricle** are the *great auricular* and *auriculotemporal nerves* (Fig. 8.69D), with minor contributions from the facial (CN VII) and vagus (CN X) nerves.

Lymphatic drainage from the lateral surface of the superior half of the auricle is to the *superficial parotid lymph nodes*. Lymph from the cranial surface of the superior half of the auricle drains to the *mastoid* and *deep cervical lymph nodes* (Fig. 8.71). Lymph from the remainder of the auricle, including the lobule, drains to the *superficial cervical lymph nodes*.

The external acoustic meatus is a canal that leads from the auricle to the tympanic membrane, a distance of 2–3 cm in adults (Fig. 8.69A). The lateral third of this slightly S-shaped canal is cartilaginous and lined with skin, which is continuous with the skin of the auricle. Its medial two thirds is bony and lined with thin skin that is continuous with the external layer of the tympanic membrane. The ceruminous and sebaceous glands produce *cerumen* (earwax).

The **tympanic membrane**, approximately 1 cm in diameter, is a thin, oval, semitransparent membrane at the medial end of the external acoustic meatus (Fig. 8.72). It forms a partition between the meatus and the *tympanic cavity* of the middle ear. The elastic lamina propria of the tympanic membrane is covered with thin skin externally and the mucous membrane of the middle ear internally.

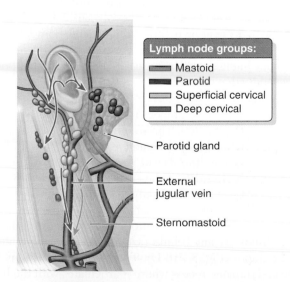

Lymph node groups:
- Mastoid
- Parotid
- Superficial cervical
- Deep cervical

— Parotid gland

— External jugular vein

— Sternomastoid

FIGURE 8.71. Lymphatic drainage of auricle.

Viewed through an otoscope (an instrument used for examining the tympanic membrane), the tympanic membrane is normally translucent and pearly gray. It has a concavity toward the external acoustic meatus with a shallow, cone-like central depression, the peak of which is the **umbo** (Fig. 8.72). The handle of the malleus (one of the small ear bones, or auditory ossicles, of the middle ear) is usually visible near the umbo. From the umbo at the inferior end of the handle of the malleus, a bright **cone of light** is reflected from the otoscope's illuminator. This light reflex is visible, radiating antero-inferiorly in a healthy ear. Superior to the attachment of the lateral process of the malleus, the membrane is thin and is called the **flaccid part** (L. *pars flaccida*). Its lamina propria lacks the radial and circular elastic fibers present in the remainder of the tympanic membrane, called the **tense part** (L. *pars tensa*).

The tympanic membrane moves in response to air vibrations that pass to it through the external acoustic meatus. Vibrations of the membrane are transmitted by the **auditory ossicles** (malleus, incus, and stapes) through the middle ear to the internal ear (Fig. 8.73). The external surface of the tympanic membrane is supplied mainly by the *auriculotemporal nerve*, a branch of CN V$_3$ (Fig. 8.69D). Some innervation is supplied by a small *auricular branch of the vagus nerve* (CN X). The internal surface of the tympanic membrane is supplied by the *glossopharyngeal nerve* (CN IX).

Middle Ear

The *cavity of the middle ear*, or **tympanic cavity**, is the narrow air-filled chamber in the petrous part of the temporal bone. The cavity has two parts: the **tympanic cavity proper**, the space directly internal to the tympanic membrane, and the **epitympanic recess**, the space superior to the membrane (Fig. 8.73A,B). The tympanic cavity is connected anteromedially with the nasopharynx by the **pharyngotympanic tube** and posterosuperiorly with the **mastoid antrum**. The tympanic cavity is lined with mucous membrane, which is continuous with the lining of the pharyngotympanic tube, mastoid cells, and mastoid antrum.

The contents of the middle ear are as follows:

- Auditory ossicles: malleus, incus, and stapes
- Tendons of the stapedius and tensor tympani muscles
- Chorda tympani nerve, a branch of CN VII
- Tympanic plexus of nerves

WALLS OF TYMPANIC CAVITY

The middle ear, shaped like a lozenge or red blood cell with concave sides, has six walls (Fig. 8.73):

- The **tegmental wall (roof)** is formed by a thin plate of temporal bone, the *tegmen tympani*, which separates the tympanic cavity from the dura mater on the floor of the middle cranial fossa.

- The **jugular wall (floor)** is formed by a layer of bone that separates the tympanic cavity from the superior bulb of the IJV.
- The **membranous wall (lateral wall)** is formed mostly by the peaked convexity of the *tympanic membrane*. The handle of the malleus is attached to the tympanic membrane, and its head extends into the epitympanic recess, part of the tympanic cavity extending superior to the tympanic membrane.
- The **labyrinthine wall (medial wall)** separates the tympanic cavity from the internal ear. It also features the *promontory of the labyrinthine wall*, formed by the initial part (basal turn) of the cochlea, and the *oval* and *round windows*.
- The **carotid wall (anterior wall)** separates the tympanic cavity from the carotid canal, which contains the internal carotid artery; superiorly, it has the **opening of the pharyngotympanic tube** and the **canal for the tensor tympani muscle**.
- The **mastoid wall (posterior wall)** has an opening superiorly, the **aditus** (L. access) to the mastoid antrum, connecting the tympanic epitympanic recess to the mastoid cells; the canal for the facial nerve descends between the posterior wall and the antrum, medial to the aditus. The tendon of the **stapedius muscle** emerges from the apex of the **pyramidal eminence** (*pyramid*), a hollow, bony cone enclosing the stapedius muscle.

The mastoid antrum is a cavity in the mastoid process of the temporal bone into which the mastoid cells open (Fig. 8.74). The antrum and mastoid cells are lined by mucous membrane, which is continuous with the lining of the middle ear.

AUDITORY OSSICLES

The auditory ossicles (malleus, incus, and stapes) form a mobile chain of small bones across the tympanic cavity from the tympanic membrane to the **oval window** (L. *fenestra vestibuli*), an oval opening on the labyrinthine wall of the tympanic cavity leading to the vestibule of the bony labyrinth (Fig. 8.73B). The ossicles are covered with the mucous membrane lining the tympanic cavity, but unlike other bones of the body, they are not directly covered with a layer of periosteum.

The **malleus** (L. hammer) is attached to the tympanic membrane (Fig. 8.73C). The rounded **head** of the malleus lies superiorly in the epitympanic recess. The **neck** of the malleus lies against the flaccid part of the tympanic membrane, and the **handle** of the malleus is embedded in the tense part of the tympanic membrane with its tip at the umbo. The head of the malleus articulates with the incus; the tendon of the tensor tympani inserts into the handle of the malleus.

The **incus** (L. anvil) links (articulates with) the malleus and the stapes (Fig. 8.73B,D). The **body of the incus** lies in the epitympanic recess where it articulates with the head

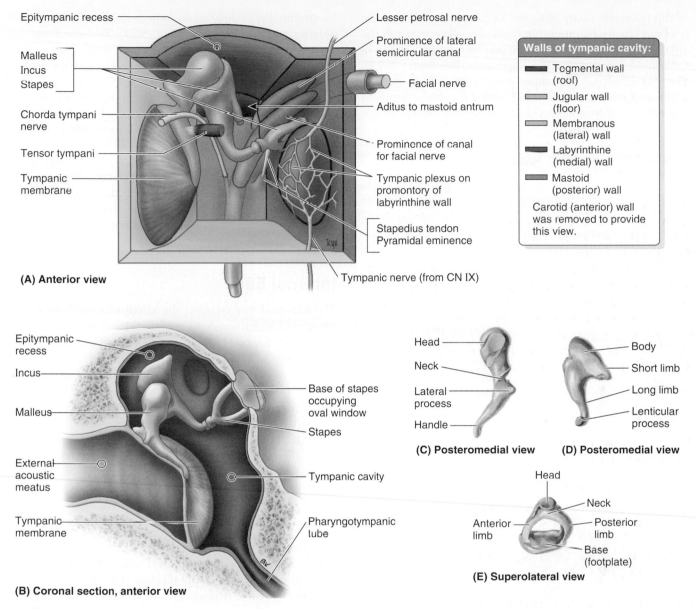

(A) Anterior view

Epitympanic recess

Malleus
Incus
Stapes

Chorda tympani nerve

Tensor tympani

Tympanic membrane

Lesser petrosal nerve

Prominence of lateral semicircular canal

Facial nerve

Aditus to mastoid antrum

Prominence of canal for facial nerve

Tympanic plexus on promontory of labyrinthine wall

Stapedius tendon
Pyramidal eminence

Tympanic nerve (from CN IX)

Walls of tympanic cavity:

- Tegmental wall (roof)
- Jugular wall (floor)
- Membranous (lateral) wall
- Labyrinthine (medial) wall
- Mastoid (posterior) wall

Carotid (anterior) wall was removed to provide this view.

(B) Coronal section, anterior view

Epitympanic recess

Incus

Malleus

External acoustic meatus

Tympanic membrane

Base of stapes occupying oval window

Stapes

Tympanic cavity

Pharyngotympanic tube

(C) Posteromedial view

Head

Neck

Lateral process

Handle

(D) Posteromedial view

Body

Short limb

Long limb

Lenticular process

(E) Superolateral view

Head

Neck

Anterior limb

Posterior limb

Base (footplate)

FIGURE 8.73. Auditory ossicles. A. Walls of the right tympanic cavity. **B.** Ossicles in situ. **C–E.** Features of malleus (**C**), incus (**D**), and stapes (**E**).

of the malleus. The **long limb** lies parallel to the handle of the malleus, and its inferior end articulates with the stapes by way of the **lenticular process**. The **short limb** is connected by a ligament to the posterior wall of the tympanic cavity.

The **stapes** (L. stirrup) is the smallest ossicle (Fig. 8.73E). The **base** (footplate) **of the stapes** is attached to the margins of the oval window on the labyrinthine wall. The base is considerably smaller than the tympanic membrane; as a result, the vibratory force of the stapes is increased approximately 10 times over that of the tympanic membrane. Consequently, the auditory ossicles increase the force but decrease the amplitude of the vibrations transmitted from the tympanic membrane.

Two muscles dampen or resist movements of the auditory ossicles; one also dampens movements (vibrations) of the tympanic membrane. The **tensor tympani** is a short muscle that arises from the superior surface of the cartilaginous part of the pharyngotympanic tube, the greater wing of the sphenoid, and the petrous part of temporal bone (Fig. 8.73A). The tensor tympani inserts into the handle of the malleus. The tensor tympani, supplied by CN V_3, pulls the handle of the malleus medially, tensing the tympanic membrane and reducing the amplitude of vibrations. This action tends to prevent damage to the internal ear when one is exposed to loud sounds. The **stapedius** is a tiny muscle (the body's smallest) inside the *pyramidal eminence*, a hollow, cone-shaped prominence on the posterior wall of the tympanic cavity (Fig. 8.73A). Its tendon enters the tympanic cavity by emerging from a small foramen in the apex of the pyramidal eminence and inserts on the neck of the stapes. The nerve to the stapedius arises from CN VII. The stapedius pulls the stapes posteriorly and tilts its base in the *oval window*, thereby tightening the anular ligament and reducing the oscillatory range. It also prevents excessive movement of the stapes.

PHARYNGOTYMPANIC TUBE

The **pharyngotympanic tube** (auditory tube, Eustachian tube) connects the tympanic cavity to the nasopharynx (Fig. 8.74), where it opens posterior to the inferior nasal meatus. The posterolateral third of the tube is bony and the remainder is cartilaginous. The pharyngotympanic tube is lined by mucous membrane, which is continuous posteriorly with the lining of the tympanic cavity and anteriorly with the lining of the nasopharynx. The function of the pharyngotympanic tube is to equalize pressure in the middle ear with the atmospheric pressure, thereby allowing free movement of the tympanic membrane. By allowing air to enter and leave the tympanic cavity, this tube balances the pressure on both sides of the membrane. Because the walls of the cartilaginous part of the tube are normally in apposition, the tube must be actively opened. It is opened by the expanding girth of the belly of the *levator veli palatini* as it contracts

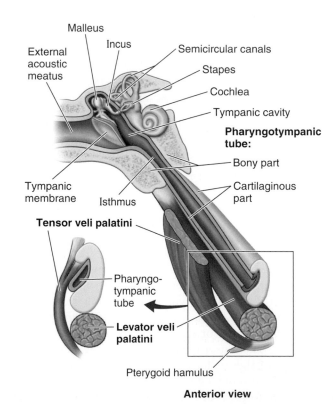

FIGURE 8.74. Right pharyngotympanic tube. The tube is open throughout its length by removing its membranous wall and the lateral part of its bony wall.

longitudinally, pushing against one wall while the *tensor veli palatini* pulls on the other (Fig. 8.74). Because these are muscles of the soft palate, equalizing pressure "popping the eardrums" is commonly associated with activities such as yawning and swallowing.

The *arteries of the pharyngotympanic tube* are derived from the *ascending pharyngeal artery*, a branch of the external carotid artery, the *middle meningeal artery*, and the *artery of the pterygoid canal*, branches of the maxillary artery (see Fig. 8.43). The *veins of the pharyngotympanic tube* drain into the *pterygoid venous plexus*. The *nerves of the pharyngotympanic tube* arise from the *tympanic plexus* (Fig. 8.73A), which is formed by fibers of CN IX. The anterior part of the tube also receives nerve fibers from the *pterygopalatine ganglion*.

Internal Ear

The **internal ear** contains the **vestibulocochlear organ** concerned with the reception of sound and the maintenance of balance. Embedded in the petrous part of the temporal bone (Figs. 8.75 and 8.76A), the internal ear consists of the sacs and ducts of the *membranous labyrinth*. The *membranous labyrinth*, containing *endolymph*, is suspended within the perilymph-filled *bony labyrinth* by delicate filaments similar to the filaments of the arachnoid mater that traverse the subarachnoid space and the spiral ligament. These fluids are involved in stimulating the end organs for balance and hearing, respectively, and providing ionic differentials for the sensory organs.

BONY LABYRINTH

The **bony labyrinth** is a series of cavities (cochlea, vestibule, and semicircular canals) contained within the otic capsule of the petrous part of the temporal bone (Figs. 8.75 and 8.76B). The **otic capsule** is made of bone that is denser than the remainder of the petrous temporal bone and can be isolated from it using a dental drill. The otic capsule is often erroneously illustrated and identified as being the bony labyrinth. However, the bony labyrinth is the *fluid-filled space* that is surrounded by the otic capsule; it is most accurately represented by a cast of the otic capsule after removal of the surrounding bone (Fig. 8.76C).

The **cochlea** is the shell-shaped cavity of the bony labyrinth that contains the **cochlear duct**, the part of the internal ear concerned with hearing (Figs. 8.75 and 8.76B). The **spiral canal** of the cochlea begins at the vestibule and makes 2.5 turns around a bony core, the **modiolus** (Fig. 8.77). The modiolus contains canals for blood vessels and for the distribution of the peripheral fibers of the cochlear nerve. The large basal turn of the cochlea features the round window, closed by the secondary tympanic membrane, and produces the *promontory of the labyrinthine wall* of the tympanic cavity. At the basal turn, the bony labyrinth communicates

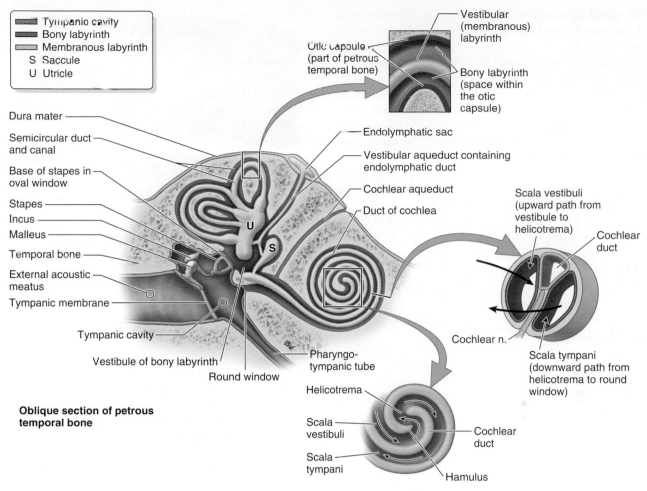

FIGURE 8.75. Internal ear. Schematic illustration of bony and membranous labyrinth in situ. *n.*, nerve.

with the subarachnoid space superior to the jugular foramen through the **cochlear aqueduct** (Fig. 8.75). The **vestibule of the bony labyrinth** is a small oval chamber (approximately 5 mm long) that contains the **utricle** and **saccule** and parts of the balancing apparatus (vestibular labyrinth). The vestibule features the *oval window* on its lateral wall, occupied by the base of the stapes. The vestibule communicates with the bony cochlea anteriorly, the semicircular canals posteriorly, and the posterior cranial fossa by the **vestibular aqueduct**. The aqueduct extends to the posterior surface of the petrous part of the temporal bone, where it opens posterolateral to the *internal acoustic meatus*. The vestibular aqueduct transmits the **endolymphatic duct** and two small blood vessels.

The **semicircular canals** (anterior, posterior, and lateral) lie posterosuperior to the vestibule, into which they open. They occupy three planes in space and are set at right angles to each other (Figs. 8.75 and 8.76). Each semicircular canal forms about two thirds of a circle and is about 1.5 mm in diameter, except at one end where there is a swelling, the **bony ampulla**. The canals have only

five openings into the vestibule because the anterior and posterior canals share a common limb. Lodged within the canals are the *semicircular ducts* of the membranous labyrinth (Fig. 8.76C,D).

MEMBRANOUS LABYRINTH

The **membranous labyrinth** consists of a series of communicating sacs and ducts that are suspended in the bony labyrinth (Figs. 8.75 and 8.76C,D). The membranous labyrinth contains **endolymph**, a watery fluid similar in composition to intracellular fluid, thus differing in composition from the surrounding **perilymph**, which is like extracellular fluid, and fills the remainder of the bony labyrinth. The membranous labyrinth is composed of two divisions, the *vestibular labyrinth* and the *cochlear labyrinth*, and consists of more parts than does the bony labyrinth:

- **Vestibular labyrinth**: Utricle and saccule, two small communicating sacs in the vestibule of the bony labyrinth and three **semicircular ducts** in the semicircular canals
- **Cochlear labyrinth**: Cochlear duct in the cochlea

The **spiral ligament**, a spiral thickening of the cochlear canal, secures the cochlear duct to the spiral canal of the cochlea (Fig. 8.77)

The semicircular ducts open into the *utricle* through five openings, reflecting the way the surrounding semicircular canals open into the vestibule. The utricle communicates with the saccule through the **utriculosaccular duct**, from which the *endolymphatic duct* arises (Fig. 8.75). The *saccule* is continuous with the cochlear duct through the **ductus reunions**, a uniting duct (Fig. 8.76B). The utricle and saccule have specialized areas of sensory organs sensitive to gravitational pull and linear acceleration called *maculae*. The **macula of the utricle** (L. *macula utriculi*) is in the floor of the utricle, parallel to the base of the cranium (Fig. 8.76D), whereas the **macula of the saccule** (L. *macula sacculi*) is vertically placed on the medial wall of the saccule. The **hair cells in the maculae** are innervated by fibers of the **vestibular division of the vestibulocochlear nerve** (CN VIII). The cell bodies of the sensory neurons are in the **vestibular ganglia**, which are in the internal acoustic meatus (Fig. 8.78).

The *endolymphatic duct* traverses the vestibular aqueduct and emerges through the bone of the posterior cranial fossa, where it expands into a blind pouch called the **endolymphatic sac**. It is located under the dura on the posterior surface of the petrous part of the temporal bone (Fig. 8.76A,D). The endolymphatic sac is a reservoir for accommodating volume and pressure changes in the excess endolymph formed by the blood capillaries in the membranous labyrinth.

Each semicircular duct has an **ampulla** at one end containing a sensory organ, the **ampullary crest** (L. *crista ampullaris*) (Figs. 8.76 and 8.78). The crests are sensors for recording movements of the endolymph in the ampulla, resulting from rotation and rotational acceleration of the head in the plane of the duct. The hair cells of the crest, like those of the maculae, stimulate primary sensory neurons whose cell bodies are in the *vestibular ganglia*.

The cochlear duct is a spiral, blind tube, closed at one end and triangular in cross section (Fig. 8.75). The duct is firmly suspended across the cochlear canal between the *spiral ligament* on the external wall of the cochlear canal and the **osseous spiral lamina** of the modiolus (Fig. 8.77). Spanning the spiral canal in this manner, the endolymph-filled cochlear duct divides the perilymph-filled spiral canal into two channels that communicate at the apex of the cochlea via the **helicotrema** (Fig. 8.75).

Waves of hydraulic pressure created in the perilymph of the vestibule by the vibrations of the base of the stapes ascend to the apex of the cochlea by one channel, the **scala vestibuli** (Fig. 8.79). The pressure waves then pass through the helicotrema and then descend back to the basal turn of the cochlea by the other channel, the **scala tympani**. There, the pressure waves again become vibrations, this time of the *secondary tympanic membrane*, which occupies the round window. Here, the energy initially received by the (primary)

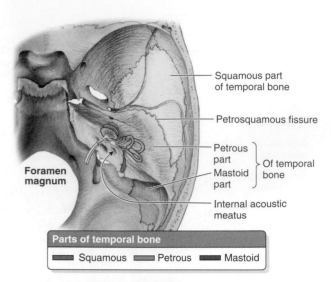

(A) Superior view of internal surface of cranial base

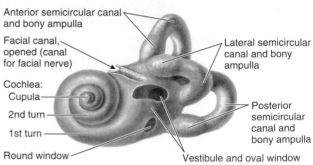

(B) Anterolateral view of left otic capsule

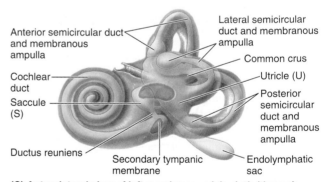

(C) Anterolateral view of left membranous labyrinth (through transparent otic capsule)

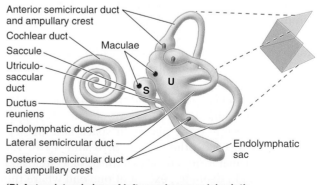

(D) Anterolateral view of left membranous labyrinth

FIGURE 8.76. Bony and membranous labyrinth of internal ear.

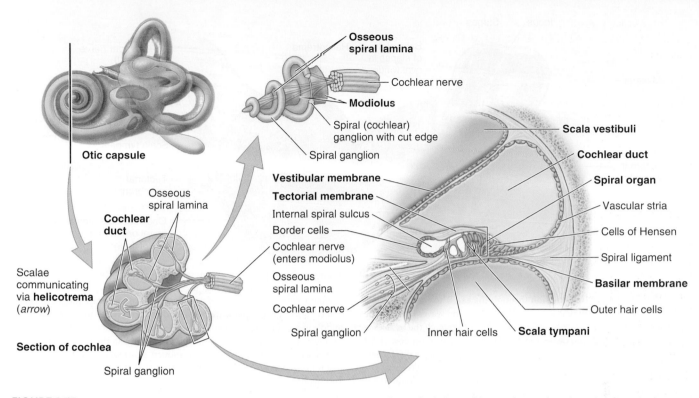

FIGURE 8.77. Structure of cochlea. The cochlea has been sectioned along the axis about which the cochlea winds (see the orientation figure in the upper left). An isolated, cone-like, bony core of the cochlea, the modiolus, is shown after the turns of the cochlea are removed, leaving only the spiral lamina winding around it like the thread of a screw. Details of the area enclosed in the *rectangle* are also shown.

tympanic membrane is finally dissipated into the air of the tympanic cavity.

The roof of the cochlear duct is formed by the **vestibular membrane** (Fig. 8.77). The floor of the duct is formed by part of the duct, the **basilar membrane**, plus the outer edge of the osseous spiral lamina. The receptor for auditory stimuli is the **spiral organ** (of Corti), situated on the basilar membrane. It is overlaid by the gelatinous **tectorial membrane**. The spiral organ contains hair cells, the tips of which are embedded in the tectorial membrane. The spiral organ is stimulated by deformation of the cochlear duct induced by hydraulic pressure waves in the perilymph, which ascend and descend in the surrounding scala vestibuli and tympani (Fig. 8.79). Impulses are conducted centrally by the **cochlear division of the vestibulocochlear nerve (CN VIII)**.

INTERNAL ACOUSTIC MEATUS

The **internal acoustic meatus** is a narrow canal that runs laterally from the posterior cranial fossa for approximately 1 cm within the petrous part of the temporal bone (Fig. 8.76A).

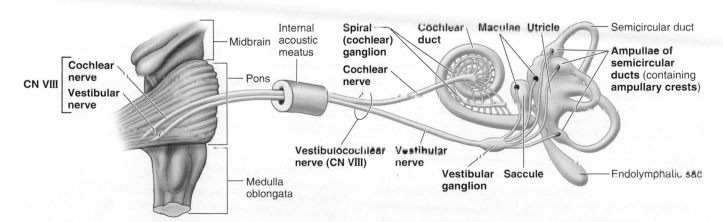

FIGURE 8.78. Vestibulocochlear nerve (CN VIII).

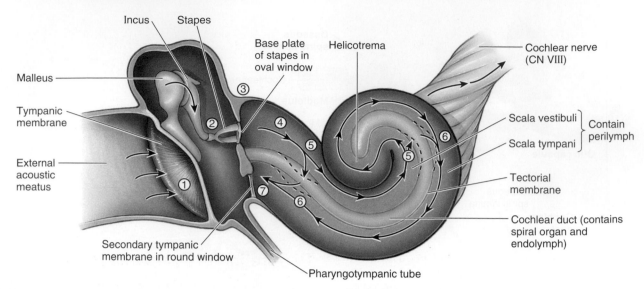

Incus   Stapes

Base plate   Helicotrema
of stapes in
oval window

Malleus

③

Tympanic
membrane

② ④ ⑥

Cochlear nerve
(CN VIII)

Scala vestibuli } Contain
Scala tympani } perilymph

⑤ ⑤

External
acoustic
meatus

① ⑦ ⑥

Tectorial
membrane

Secondary tympanic
membrane in round window

Cochlear duct (contains
spiral organ and
endolymph)

Pharyngotympanic tube

FIGURE 8.79. Sound transmission through the ear. The cochlea is depicted schematically as if consisting of a single coil to demonstrate the transmission of sound stimuli through the ear. *1.* Sound waves entering the external ear strike the tympanic membrane, causing it to vibrate. *2.* Vibrations initiated at the tympanic membrane are transmitted through the ossicles of the middle ear and their articulations. *3.* The base of the stapes vibrates with increased strength and decreased amplitude in the oval window. *4.* Vibrations of the base of the stapes create pressure waves in the perilymph of the scala vestibuli. *5.* Pressure waves in the scala vestibuli cause displacement of the basilar membrane of the cochlear duct. Short waves (high pitch) cause displacement near the oval window; longer waves (low pitch) cause more distant displacement, nearer the helicotrema at the apex of the cochlea. Movement of the basilar membrane bends the hair cells of the spiral organ. Neurotransmitter is released, stimulating action potentials conveyed by the cochlear nerve to the brain. *6.* Vibrations are transferred across the cochlear duct to the perilymph of the scala tympani. *7.* Pressure waves in the perilymph are dissipated (dampened) by the secondary tympanic membrane at the round window into the air of the tympanic cavity.

The meatus aligns with the external acoustic meatus. The internal acoustic meatus is closed laterally by a thin, perforated plate of bone that separates it from the internal ear. The facial nerve (CN VII), the vestibulocochlear nerve (CN VIII), and blood vessels pass through small openings in this plate of bone. The vestibulocochlear nerve divides near the lateral end of the internal acoustic meatus into two divisions: a **cochlear nerve** and a **vestibular nerve** (Fig. 8.78). Sound transmission through the ear is summarized in Figure 8.79.

CLINICAL BOX

External Ear Injury

Bleeding within the auricle resulting from trauma may produce an *auricular hematoma*. A localized collection of blood forms between the perichondrium and the auricular cartilage, causing distortion of the contours of the auricle. As the hematoma enlarges, it compromises the blood supply to the cartilage. If untreated (e.g., by aspiration of blood), *fibrosis* (formation of fibrous tissue) develops in the overlying skin, forming a deformed auricle (e.g., the cauliflower or boxer's ear of some professional fighters).

Otoscopic Examination

Examination of the external acoustic meatus and tympanic membrane begins by straightening the meatus. In adults, the helix is grasped and pulled posterosuperiorly (up, out, and back). These movements reduce the curvature of the external acoustic meatus, facilitating insertion of the *otoscope* (Fig. B8.14A).

The meatus is relatively short in infants; therefore, extra care must be exercised to prevent injury to the tympanic membrane. The meatus is straightened in infants by pulling the auricle inferoposteriorly (down and back). The examination also provides a clue to tenderness, which can indicate inflammation of the auricle and/or the meatus.

The tympanic membrane is normally translucent and pearly gray (Fig. B8.14B). The handle of the malleus is usually visible near the center of the membrane (the umbo). From the inferior end of the handle, a bright *cone of light* is reflected from the otoscope's illuminator. This *light reflex* is visible radiating antero-inferiorly in the healthy ear.

Acute Otitis Externa

Otitis externa is an inflammation of the external acoustic meatus. The infection often develops in swimmers who do not dry their meatus after swimming and/or use ear drops, but it may also be the result of a

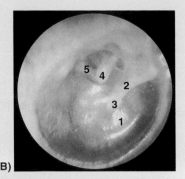

FIGURE B8.14. Otoscopic examination. *1*, cone of light; *2*, handle of malleus; *3*, umbo; *4*, long limb of incus; *5*, posterior limb of stapes.

bacterial infection of the skin lining the meatus. The affected individual complains of itching and pain in the external ear. Pulling the auricle or applying pressure on the tragus increases the pain.

Otitis Media

An earache and a bulging red tympanic membrane may indicate pus or fluid in the middle ear, a sign of *otitis media* (Fig. B8.15A). Infection of the middle ear is often secondary to upper respiratory infections.

Inflammation and swelling of the mucous membrane lining the tympanic cavity may cause partial or complete blockage of the pharyngotympanic tube. The tympanic membrane becomes red and bulges, and the person may complain of "ear popping." An amber-colored bloody fluid may be observed through the tympanic membrane. If untreated, otitis media may produce impaired hearing as the result of scarring of the auditory ossicles, limiting their ability to move in response to sound.

Perforation of Tympanic Membrane

Perforation of the tympanic membrane ("ruptured eardrum") may result from otitis media and is one of several causes of middle ear deafness. Perforation may also result from foreign bodies in the external acoustic meatus, trauma, or excessive pressure (e.g., during scuba diving).

Minor ruptures of the tympanic membrane often heal spontaneously. Large ruptures usually require surgical repair. Because the superior half of the tympanic membrane is much more vascular than the inferior half, incisions to release pus from a middle ear abscess (*myringotomy*), for example, are made postero-inferiorly through the membrane (Fig. B8.15B). This incision also avoids injury to the chorda tympani nerve and auditory ossicles. In persons with chronic middle ear infections, myringotomy may be followed by insertion of *tympanostomy* or *pressure-equalization* (*PE*) *tubes* in the incision to enable drainage of effusion and ventilation of pressure (Fig. B8.15C).

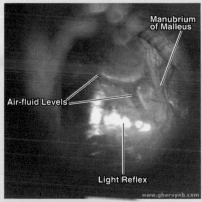

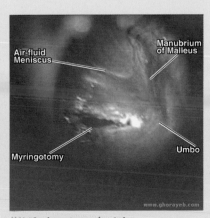

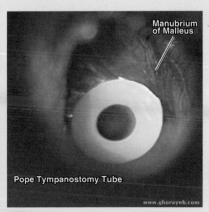

(A) Otitis media   **(B) Myringotomy incision**   **(C) Tympanostomy tube inserted**

FIGURE B8.15. Otitis media, myringotomy, and tympanostomy.

MEDICAL IMAGING

Head

Radiography, although replaced by CT and/or MRI in most cases, is sometimes used for cranial examinations. Because crania vary considerably in shape, one must examine radiographs carefully for abnormalities (Fig. 8.80A,B). To visualize the arteries of the brain, a radiopaque contrast medium is injected into the carotid or vertebral artery and radiographs are taken, producing *arteriograms* (Fig. 8.80C). This type of radiograph is used for detecting cerebral aneurysms and arteriovenous malformations.

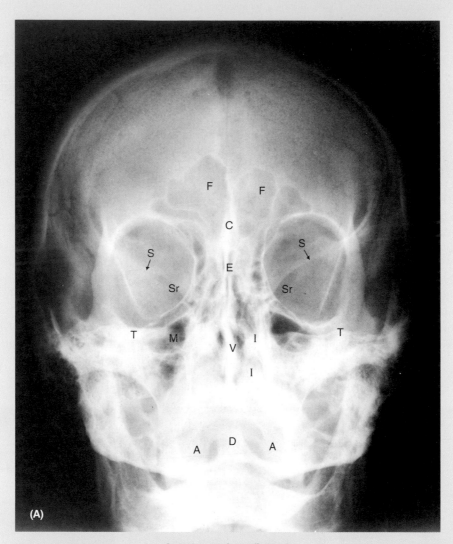

Anteroposterior radiograph

FIGURE 8.80. Radiographs of cranium (skull). A. The lateral masses of the atlas (*A*) and the dens of the axis (*D*) are superimposed on the facial skeleton (viscerocranium). Also identified are crista galli (*C*), nasal septum formed by the perpendicular plate of the ethmoid (*E*), and the vomer (*V*); frontal sinus (*F*); inferior and middle conchae (*I*) of lateral wall of the nasal cavity; maxillary sinus (*M*); lesser wings of sphenoid (*S*); superior orbital fissure (*Sr*); and superior surface of petrous part of temporal bone (*T*). *(continued)*

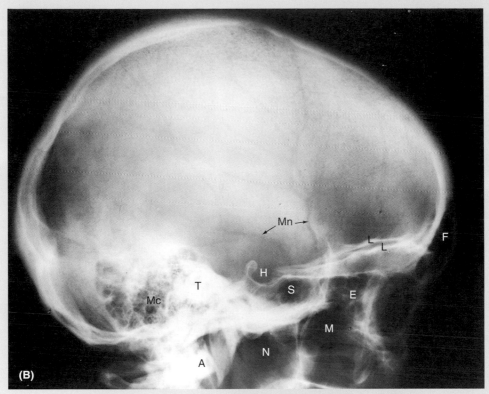

Lateral radiograph

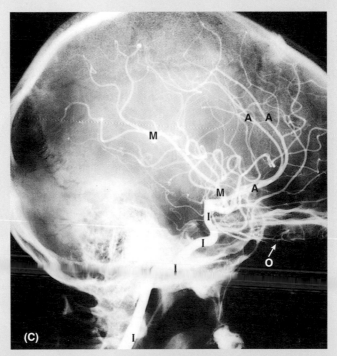

Lateral arteriogram

FIGURE 8.80. Radiographs of cranium (skull). *(continued)* **B.** Identified are anterior arch of the atlas (*A*); paranasal sinuses: ethmoidal (*E*), frontal (*F*), maxillary (*M*), sphenoidal (*S*), and mastoid cells (*Mc*); hypophysial fossa (*H*) for the pituitary gland; bony grooves for the branches of the middle meningeal vessels (*Mn*); nasopharynx (*N*); and the petrous part of the temporal bone (*T*). The right and left orbital parts of the frontal bone are not superimposed; thus, the floor of the anterior cranial fossa appears as two lines (*L*). **C.** Vertebrobasilar arteriogram. Identified are the anterior cerebral artery (*A*), internal carotid artery (*I*), middle cerebral artery (*M*), and ophthalmic artery (*O*).

MRI is slower (longer acquisition time) and more expensive than CT but shows much more detail in the soft tissues than does CT (Fig. 8.81). MRI is the gold standard for detecting and delineating intracranial and spinal lesions because it provides good soft tissue contrast of normal and pathological structures. It also permits multiplanar capability, which provides three-dimensional information and relationships that are not so readily available with CT. MRI can also demonstrate blood and CSF flow. Magnetic resonance angiography (MRA) is useful for determining the patency of vessels of the cerebral arterial circle.

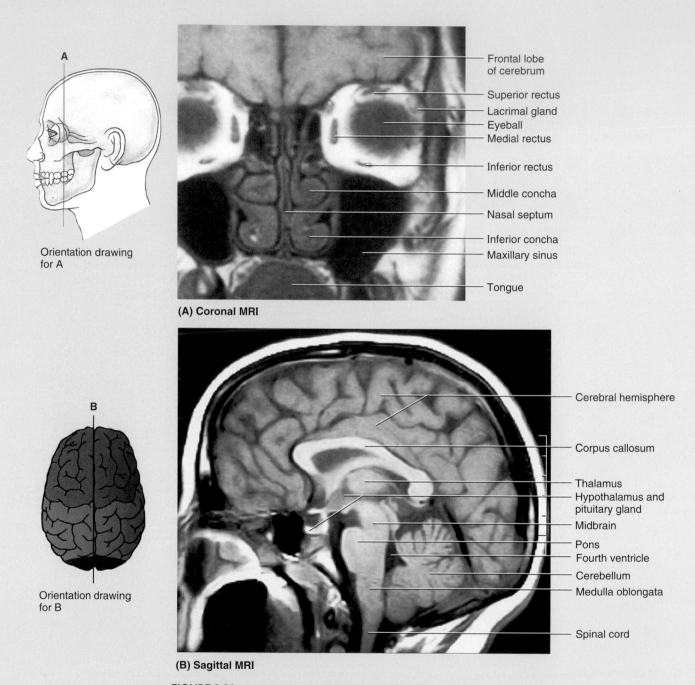

A

Orientation drawing
for A

(A) Coronal MRI

- Frontal lobe of cerebrum
- Superior rectus
- Lacrimal gland
- Eyeball
- Medial rectus
- Inferior rectus
- Middle concha
- Nasal septum
- Inferior concha
- Maxillary sinus
- Tongue

B

Orientation drawing
for B

(B) Sagittal MRI

- Cerebral hemisphere
- Corpus callosum
- Thalamus
- Hypothalamus and pituitary gland
- Midbrain
- Pons
- Fourth ventricle
- Cerebellum
- Medulla oblongata
- Spinal cord

FIGURE 8.81. Magnetic resonance imaging (MRI) studies of head. *(continued)*

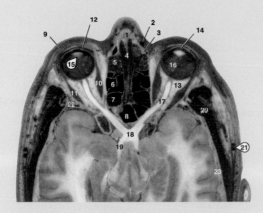

(C) Transverse section of cadaveric head

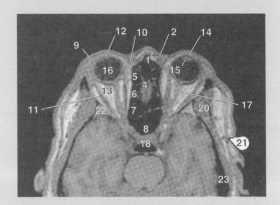

(D) Transverse (axial) MRI scan

Key							
1	Nasal bones	7	Posterior ethmoidal air cell	13	Retrobulbar fat	19	Optic tract
2	Angular artery	8	Sphenoidal sinus	14	Anterior chamber	20	Temporalis muscle
3	Frontal process of maxilla	9	Orbicularis oculi muscle	15	Lens	21	Superficial temporal vessels
4	Nasal septum	10	Medial rectus muscle	16	Vitreous body	22	Greater wing of sphenoid
5	Anterior ethmoidal air cell	11	Lateral rectus muscle	17	Optic nerve	23	Squamous portion of temporal bone
6	Middle ethmoidal air cell	12	Cornea	18	Optic chiasm		

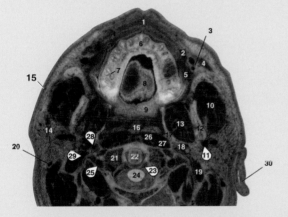

(E) Transverse section of cadaveric head

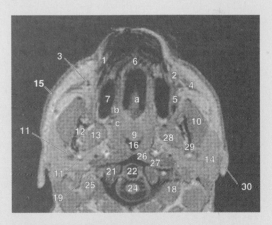

(F) Transverse (axial) MRI scan

Key							
1	Orbicularis oris muscle	12	Ramus of mandible	23	Transverse ligament of atlas		
2	Levator anguli oris muscle	13	Lateral pterygoid muscle	24	Spinal cord		
3	Facial artery and vein	14	Parotid gland	25	Vertebral artery in foramina transversaria		
4	Zygomaticus major muscle	15	Skin	26	Longus colli muscle		
5	Buccinator muscle	16	Region of pharyngeal tubercle	27	Longus capitis muscle		
6	Maxilla	17	Sphenoid bone	28	Internal carotid artery		
7	Alveolar process of maxilla	18	Stylohyoid ligament and muscle	29	Internal jugular vein		
8	Dorsum of tongue	19	Posterior belly of digastric muscle	30	Inferior portion of helix of auricle		
9	Soft palate	20	Occipital artery	a	Hard palate		
10	Masseter muscle	21	First cervical vertebra (Atlas)	b	Palatoglossus muscle		
11	Retromandibular vein	22	Dens (Axis)	c	Palatopharyngeus muscle		

FIGURE 8.81. Magnetic resonance imaging (MRI) studies of head. *(continued)*

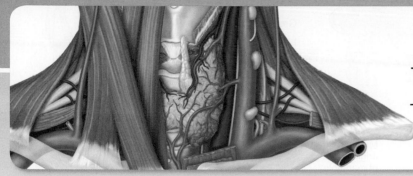

9

Neck

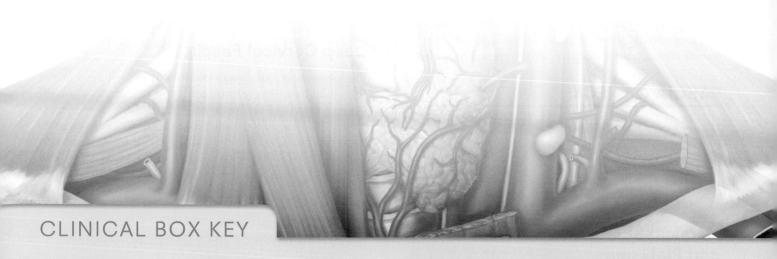

CLINICAL BOX KEY

Anatomical
Variations

Diagnostic
Procedures

Life Cycle

Surgical
Procedures

Trauma

Pathology

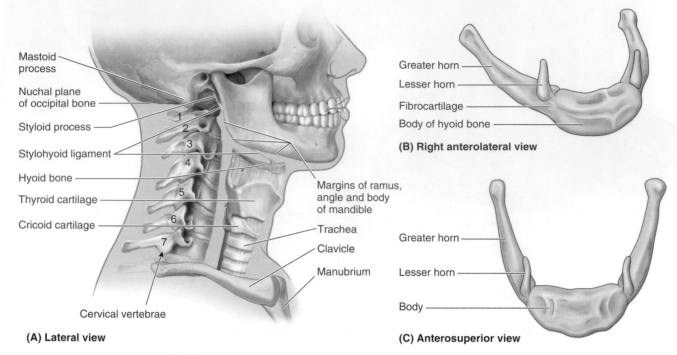

(A) Lateral view

Mastoid process
Nuchal plane of occipital bone
Styloid process
Stylohyoid ligament
Hyoid bone
Thyroid cartilage
Cricoid cartilage

Cervical vertebrae

Margins of ramus, angle and body of mandible
Trachea
Clavicle
Manubrium

1
2
3
4
5
6
7

(B) Right anterolateral view

Greater horn
Lesser horn
Fibrocartilage
Body of hyoid bone

(C) Anterosuperior view

Greater horn
Lesser horn
Body

FIGURE 9.1. Bones and cartilages of neck. A. Overview. **B and C.** Features of hyoid bone.

The **neck** (L. *collum, cervix*) joins the head to the trunk and limbs and serves as a major conduit for structures passing between them. In addition, several important organs with unique functions are located here: the larynx, thyroid, and parathyroid glands, for example.

The *skeleton of the neck* is formed by the cervical vertebrae (C1–C7), **hyoid bone** (usually referred to as the *hyoid*), manubrium of the sternum, and clavicles (Fig. 9.1A). The mobile hyoid lies in the anterior part of the neck at the level of the C3 vertebra in the angle between the mandible and thyroid cartilage. The hyoid does not articulate with any other bone and functionally serves as an attachment for anterior neck muscles and a prop to keep the airway open (Fig. 9.1B,C).

FASCIA OF NECK

Structures in the neck are surrounded by a layer of fatty subcutaneous tissue (superficial fascia) and are compartmentalized by layers of deep cervical fascia. The fascial planes determine the direction in which an infection in the neck may spread.

Cervical Subcutaneous Tissue and Platysma

The **subcutaneous tissue of the neck** (superficial cervical fascia) is a layer of connective tissue that lies between the dermis of the skin and the investing layer of deep cervical fascia (Fig. 9.2). It contains cutaneous nerves, blood and lymphatic vessels, superficial lymph nodes, and variable amounts of fat; anterolaterally, it contains the platysma.

The **platysma**, a muscle of facial expression, arises in subcutaneous tissue covering the superior parts of the deltoid and pectoralis major muscles and sweeps superomedially over the clavicle to the inferior border of the mandible (Fig. 9.2B). It is a broad thin sheet of muscle.

Deep Cervical Fascia

The **deep cervical fascia** consists of three fascial layers (Fig. 9.2): *investing, pretracheal,* and *prevertebral,* which support the viscera, muscles, vessels, and deep lymph nodes. The fascial layers provide the slipperiness that allows structures in the neck to move and pass over one another without difficulty (e.g., when swallowing and turning the head and neck). The fascial layers also form *natural cleavage planes,* allowing separation of tissues during surgery.

INVESTING LAYER OF DEEP CERVICAL FASCIA

The **investing layer of deep cervical fascia**, the most superficial deep fascial layer, surrounds the entire neck deep to the skin and subcutaneous tissue (Fig. 9.2). At the

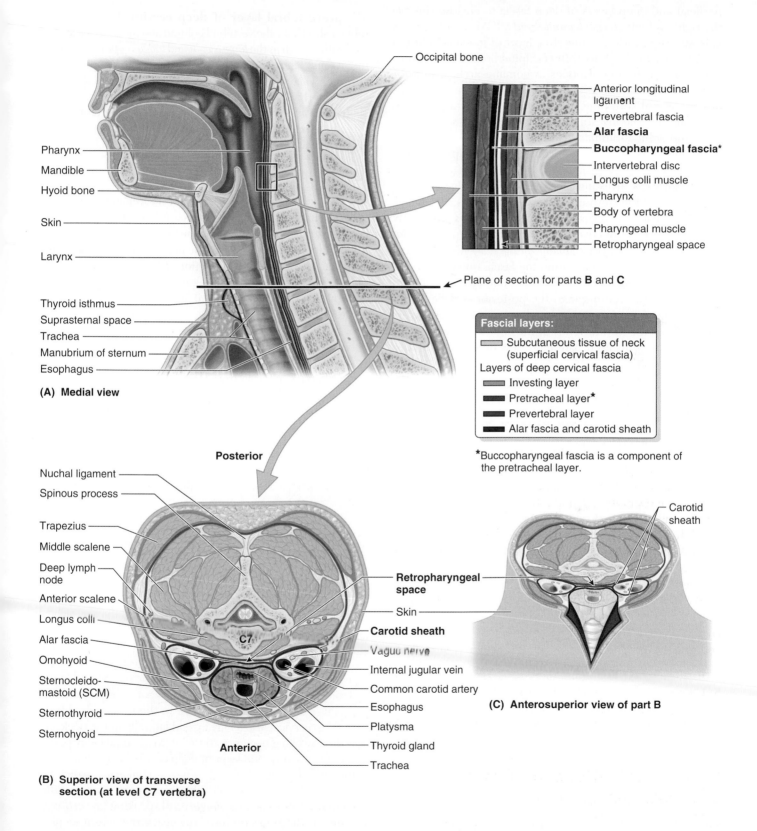

Occipital bone

Anterior longitudinal ligament
Prevertebral fascia
Alar fascia
Buccopharyngeal fascia*
Intervertebral disc
Longus colli muscle
Pharynx
Body of vertebra
Pharyngeal muscle
Retropharyngeal space

Pharynx
Mandible
Hyoid bone

Skin

Larynx

Thyroid isthmus
Suprasternal space
Trachea
Manubrium of sternum
Esophagus

(A) Medial view

Plane of section for parts **B** and **C**

Fascial layers:
Subcutaneous tissue of neck (superficial cervical fascia)
Layers of deep cervical fascia
Investing layer
Pretracheal layer*
Prevertebral layer
Alar fascia and carotid sheath

*Buccopharyngeal fascia is a component of the pretracheal layer.

Posterior

Nuchal ligament
Spinous process

Trapezius

Middle scalene

Deep lymph node

Anterior scalene

Longus colli

Alar fascia

Omohyoid

Sternocleido-mastoid (SCM)

Sternothyroid

Sternohyoid

C7

Retropharyngeal space

Skin

Carotid sheath

Vagus nerve

Internal jugular vein

Common carotid artery

Esophagus

Platysma

Thyroid gland

Trachea

Anterior

(B) Superior view of transverse section (at level C7 vertebra)

Carotid sheath

(C) Anterosuperior view of part B

FIGURE 9.2. Cervical fascia. A. Fascia of the retropharyngeal region. **B.** Cross-section of the neck at the level of the thyroid gland. **C.** Fascial compartments of the neck demonstrating an anterior midline approach to the thyroid gland.

"four corners" of the neck, the investing layer splits into superficial and deep layers of deep fascia to enclose (invest) the right and left *sternocleidomastoid* (SCM) and *trapezius muscles*. Superiorly, the investing layer of fascia attaches to the superior nuchal line of the occipital bone, mastoid processes of the temporal bones, zygomatic arches, inferior border of the mandible, hyoid bone, and spinous processes of the cervical vertebrae. Just inferior to its attachment to the mandible, the investing layer of deep fascia splits to enclose the submandibular gland (see Fig. 9.6A). Posterior to the mandible, it splits to form the fibrous capsule of the parotid gland.

Inferiorly, the investing layer of deep fascia attaches to the manubrium of the sternum, clavicles, acromions, and spines of the scapulae. The investing layer is continuous posteriorly with the periosteum covering the C7 spinous process and the nuchal ligament (L. *ligamentum nuchae*) (Fig. 9.2B,C). Just superior to the manubrium, the fascia remains divided into two layers that enclose the SCM; one layer attaches to the anterior and the other to the posterior surface of the manubrium. A *suprasternal space* lies between these layers and encloses the inferior ends of the anterior jugular veins, the jugular venous arch, fat, and a few deep lymph nodes (Fig. 9.2A).

PRETRACHEAL LAYER OF DEEP CERVICAL FASCIA

The thin **pretracheal layer of deep cervical fascia** is limited to the anterior part of the neck (Fig. 9.2). It extends inferiorly from the hyoid bone into the thorax, where it blends with the fibrous pericardium covering the heart. The pretracheal layer includes a thin *muscular part*, which encloses the infrahyoid muscles, and a *visceral part*, which encloses the thyroid gland, trachea, and esophagus. The pretracheal layer is continuous posterosuperiorly with the *buccopharyngeal fascia* and blends laterally with the *carotid sheaths*.

The **carotid sheath** is a tubular fascial investment that extends from the cranial base to the root of the neck. This sheath blends anteriorly with the investing and pretracheal layers of fascia and posteriorly with the prevertebral layer of deep cervical fascia. The carotid sheath contains the following structures (Fig. 9.2B,C):

- Common and internal carotid arteries
- Internal jugular vein (IJV)
- Vagus nerve (cranial nerve [CN] X)
- Deep cervical lymph nodes (some)
- Carotid sinus nerve
- Sympathetic nerve fibers (carotid peri-arterial plexuses)

The carotid sheath and pretracheal fascia communicate with the mediastinum of the thorax inferiorly and the cranial cavity superiorly. These communications represent potential pathways for the spread of infection and extravasated blood.

PREVERTEBRAL LAYER OF DEEP CERVICAL FASCIA

The **prevertebral layer of deep cervical fascia** forms a tubular sheath for the vertebral column and the muscles associated with it, such as the *longus colli* and *longus capitis* anteriorly, the *scalenes* laterally, and the *deep cervical muscles* posteriorly (Fig. 9.2). This layer of fascia is fixed to the cranial base superiorly and inferiorly and fuses with the *anterior longitudinal ligament* centrally at approximately T3 vertebra. The prevertebral layer extends laterally as the *axillary sheath* (see Chapter 3, Upper Limb), which surrounds the axillary vessels and brachial plexus.

RETROPHARYNGEAL SPACE

The **retropharyngeal space** permits movement of the pharynx, esophagus, larynx, and trachea relative to the vertebral column during swallowing. It is the largest and most clinically important interfascial space in the neck because it is the major pathway for the spread of infection (Fig. 9.2A). It is a potential space that consists of loose connective tissue between the visceral part of the prevertebral layer of deep cervical fascia and the *buccopharyngeal fascia*. Inferiorly, the buccopharyngeal fascia is continuous with the pretracheal layer of deep cervical fascia. The *alar fascia* crosses the retropharyngeal space. This thin layer is attached along the midline of the buccopharyngeal fascia from the cranium to the level of the C7 vertebra and extends laterally to blend with the carotid sheath. The retropharyngeal space is closed superiorly by the base of the cranium and on each side by the carotid sheath.

CLINICAL BOX

Spread of Infection in Neck

The investing layer of deep cervical fascia helps prevent the spread of *abscesses* (a collection of pus). If an infection occurs between the investing layer of deep cervical fascia and the muscular part of the pretracheal fascia surrounding the infrahyoid muscles, the infection usually does not spread beyond the superior edge of the manubrium. If, however, the infection occurs between the investing fascia and the visceral part of the pretracheal fascia, it can spread into the thoracic cavity anterior to the pericardium.

Pus from an abscess posterior to the prevertebral layer of deep cervical fascia may extend laterally in the neck and form a swelling posterior to the SCM. The pus may perforate the prevertebral layer of deep cervical fascia and enter the retropharyngeal space, producing a bulge in the pharynx (*retropharyngeal abscess*). This swelling may cause difficulty in swallowing (*dysphagia*) and speaking (*dysarthria*). Similarly, air from a ruptured trachea, bronchus, or esophagus (*pneumomediastinum*) may pass superiorly in the neck.

SUPERFICIAL STRUCTURES OF NECK: CERVICAL REGIONS

The platysma is a broad, thin sheet of muscle in the subcutaneous tissue of the neck. It covers the anterolateral aspect of the neck and like other muscles of facial expression, it is innervated by the facial nerve (CN VII). The attachments, innervation, and actions are summarized in Figure 9.3 and Table 9.1.

The neck is divided into regions. The four major regions are the SCM region, posterior cervical region, lateral cervical region, and anterior cervical region. Each region can be further subdivided into triangles. The boundaries and contents of each region are summarized in Figure 9.4 and Tables 9.2 and 9.3.

The **sternocleidomastoid (SCM) muscle**, defining the SCM region, visibly divides each side of the neck into *anterior* and *lateral cervical regions*. The SCM has two heads: the rounded tendon of the **sternal head** and thicker **clavicular head**. The two heads are separated inferiorly by a space, the **lesser supraclavicular fossa**. The attachments, innervation, and actions of the SCM are summarized in Figure 9.3 and Table 9.1.

The **descending part of trapezius** is the major landmark of the posterior region (Fig. 9.4). The suboccipital region is deep to the superior part of this region. See Extrinsic Back Muscles in Chapter 2, Back.

Lateral Cervical Region

MUSCLES IN LATERAL CERVICAL REGION

The floor of the lateral cervical region is formed by prevertebral fascia (Fig. 9.5A,C) overlying four muscles (Fig. 9.5D): splenius capitis, levator scapulae, middle scalene (L. *scalenus medius*), and posterior scalene (L. *scalenus posterior*). Sometimes, part of the inferior, part of the anterior scalene (L. *scalenus anterior*) appears in the inferomedial angle of the lateral cervical region.

NERVES OF LATERAL CERVICAL REGION

The **spinal accessory nerve (CN XI)** passes deep to the SCM, supplying it before it enters the lateral cervical region at or inferior to the junction of the superior and middle thirds of the posterior border of the SCM (Fig. 9.5A,C,D). It passes postero-inferiorly, within or deep to the investing layer of deep cervical fascia, running on the levator scapulae from which it is separated by the prevertebral layer of fascia. CN XI disappears deep to the anterior border of the trapezius at the junction of its superior two thirds with its inferior one third, then enters the muscle.

The **roots of brachial plexus** (anterior rami of C5–C8 and T1) appear between the anterior and middle scalene muscles (Fig. 9.5D,E). Five rami unite to form the *three trunks* (*superior, middle, and inferior*) of the *brachial plexus* (Fig. 9.5E), which descend inferolaterally through the lateral cervical region. The plexus then passes between the 1st rib, clavicle, and superior border of the scapula (the *cervico-axillary canal*) to enter the axilla, providing innervation for most of the upper limb.

The **suprascapular nerve**, which arises from the superior trunk of the brachial plexus, runs across the lateral cervical region to supply the supraspinatus and infraspinatus muscles on the posterior aspect of the scapula (Fig. 9.5E). It also sends articular branches to the glenohumeral joint.

The anterior rami of C1–C4 make up the roots of the **cervical plexus**, forming a series of nerve loops. The plexus lies anteromedial to the levator scapulae and middle scalene

---

CLINICAL BOX

Congenital Torticollis

Torticollis is a contraction of the cervical muscles that produces twisting of the neck and slanting of the head (Fig. B9.1). The most common type of *congenital torticollis* (wry neck) results from a fibrous tissue tumor (L. *fibromatosis colli*) that develops in the SCM before or shortly after birth. Occasionally, the SCM is injured when an infant's head is pulled excessively during a difficult birth, tearing its fibers (*muscular torticollis*). This tearing results in a hematoma that may develop into a fibrous mass entrapping a branch of the spinal accessory nerve (CN XI), thus denervating part of the SCM. Surgical release of a partially fibrotic SCM from its distal attachments to the manubrium and clavicle may be necessary.

Cervical dystonia (abnormal tonicity of the cervical muscles), commonly known as *spasmodic torticollis*, usually begins in adulthood. It may involve any bilateral combination of lateral neck muscles, especially the SCM and trapezius.

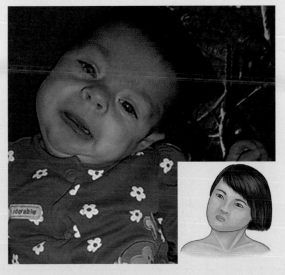

FIGURE B9.1. Congenital torticollis.

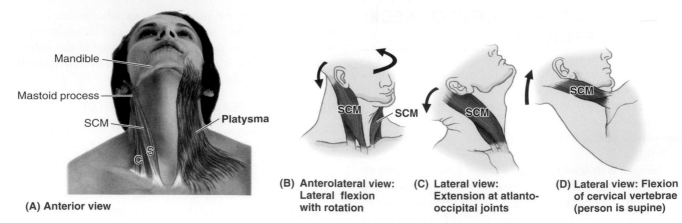

Mandible

Mastoid process

SCM

Platysma

C S

(A) Anterior view

SCM   SCM

(B) Anterolateral view:
Lateral flexion
with rotation

SCM

(C) Lateral view:
Extension at atlanto-
occipital joints

SCM

(D) Lateral view: Flexion
of cervical vertebrae
(person is supine)

FIGURE 9.3. Platysma and sternocleidomastoid. A. Overview. **B–D.** Actions of sternocleidomastoid. *SCM*, sternocleidomastoid: *C*, clavicular head, *S*, sternal head.

muscle and deep to the SCM. The superficial branches of the plexus that initially pass posteriorly are cutaneous branches (Fig. 9.5C). The deep branches passing anteromedially are motor branches, including the roots of the phrenic nerve and the **ansa cervicalis** (Figs. 9.5E and 9.6A,B).

Cutaneous branches of the cervical plexus emerge around the middle of the posterior border of the SCM, often called the **nerve point of the neck**, and supply the skin of the anterolateral neck, superolateral thoracic wall, and the scalp between the auricle and the external occipital protuberance (Fig. 9.5C). Close to their origin, the roots of the cervical plexus receive communicating branches (L. *rami communicantes*), most of which descend from the *superior cervical ganglion* in the superior part of the neck.

The branches of the cervical plexus arising from the nerve loop between the anterior rami of C2 and C3 are as follows (Fig. 9.5A–D):

- **Lesser occipital nerve** (C2), supplying the skin of the neck and scalp posterosuperior to the auricle
- **Great auricular nerve** (C2 and C3), ascending vertically across the SCM onto the parotid gland, where it divides

and supplies the skin and sheath over the gland, the posterior aspect of the auricle, and the area of skin overlying the angle of the mandible to the mastoid process
- **Transverse cervical nerve** (C2 and C3), supplying the skin covering the anterior cervical region; the nerve curves around the middle of the posterior border of the SCM and passes anteriorly and horizontally across it, deep to the external jugular vein (EJV) and platysma.

Branches of the cervical plexus arising from the loop formed between the anterior rami of C3 and C4 are the **supraclavicular nerves** (C3 and C4), which emerge as a common trunk under cover of the SCM and send small branches to the skin of the neck and cross the clavicle to supply the skin over the shoulder (Fig. 9.4B,C). Deep motor branches include branches arising from the anterior rami of cervical nerves supplying the rhomboids (dorsal scapular nerve, C4 and C5), serratus anterior (long thoracic nerve, C5–C7), and nearby prevertebral muscles (Fig. 9.5D).

The **phrenic nerves** originate chiefly from the 4th cervical nerve (C4) but receive contributions from the C3 and C5 nerves. The phrenic nerves contain motor, sensory, and

TABLE 9.1. CUTANEOUS AND SUPERFICIAL MUSCLES OF NECK

Muscle[a]	Superior Attachment	Inferior Attachment	Innervation	Main Action(s)
Platysma	Inferior border of mandible, skin, and subcutaneous tissues of lower face	Fascia covering superior parts of pectoralis major and deltoid muscles	Cervical branch of facial nerve (CN VII)	Draws corners of mouth inferiorly and widens it as in expressions of sadness and fright; draws skin of neck superiorly when teeth are clenched, indicating tension
Sternocleidomastoid (SCM)	Lateral surface of mastoid process of temporal bone and lateral half of superior nuchal line	*Sternal head:* Anterior surface of manubrium of sternum *Clavicular head:* Superior surface of medial third of clavicle	Spinal accessory nerve (CN XI; motor), C2 and C3 nerves (pain and proprioception)	*Unilateral contraction:* Tilts head to same side (i.e., laterally flexes neck) and rotates it so face is turned superiorly toward opposite side *Bilateral contraction:* (1) extends neck at atlanto-occipital joints, (2) flexes cervical vertebrae so that chin approaches manubrium, or (3) extends superior cervical vertebrae while flexing inferior vertebrae so chin is thrust forward with head kept level With cervical vertebrae fixed, may elevate manubrium and medial end of clavicles, assisting pump-handle action of deep respiration

[a]Trapezius. See pg. 115.

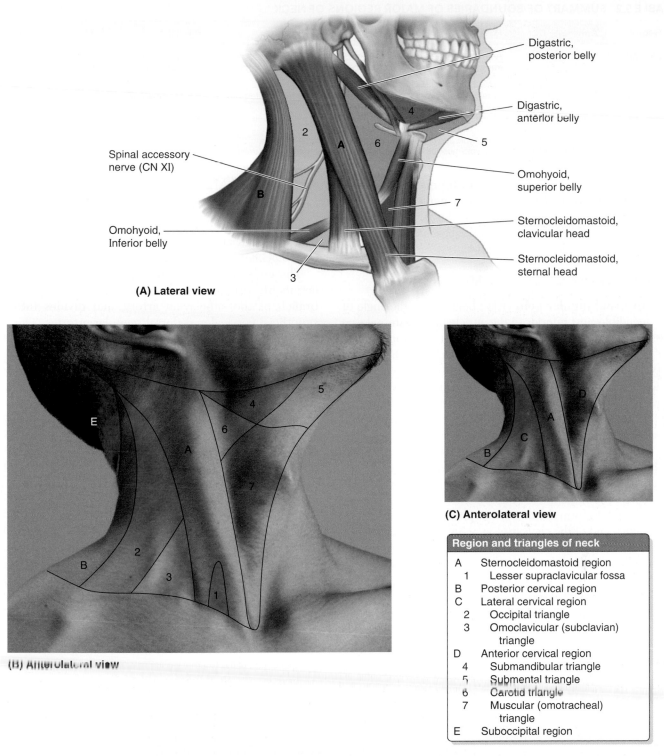

Digastric, posterior belly

Digastric, anterior belly

Spinal accessory nerve (CN XI)

Omohyoid, superior belly

Omohyoid, Inferior belly

Sternocleidomastoid, clavicular head

Sternocleidomastoid, sternal head

(A) Lateral view

(B) Anterolateral view

(C) Anterolateral view

Region and triangles of neck	
A	Sternocleidomastoid region
1	Lesser supraclavicular fossa
B	Posterior cervical region
C	Lateral cervical region
2	Occipital triangle
3	Omoclavicular (subclavian) triangle
D	Anterior cervical region
4	Submandibular triangle
5	Submental triangle
6	Carotid triangle
7	Muscular (omotracheal) triangle
E	Suboccipital region

FIGURE 9.4. Regions and triangles of neck. A. Boundaries. **B.** Triangles. **C.** Regions.

sympathetic nerve fibers. These nerves provide the sole motor supply to the diaphragm as well as sensation to its central part. In the thorax, the nerves supply the mediastinal pleura and the pericardium. Receiving variable communicating fibers in the neck and fibers from the cervical sympathetic ganglia or their branches, each phrenic nerve forms on the anterior scalene muscle at the level of the superior border of the thyroid cartilage (Fig. 9.5E).

The phrenic nerves lie anterior to the subclavian arteries and posterior to the subclavian veins as they enter the thorax (Fig. 9.5E). The contribution from C5 to the phrenic nerve may derive from an **accessory phrenic nerve**, frequently a

TABLE 9.2. SUMMARY OF BOUNDARIES OF MAJOR REGIONS OF NECK

Region	Anterior	Posterior	Superior	Inferior	Roof	Floor
Lateral[a]	Posterior border of SCM	Anterior border of trapezius	Merging of SCM and trapezius	Clavicle (between) SCM and trapezius)	Investing layer of deep cervical fascia; platysma	Muscles covered by prevertebral layer of deep cervical fascia
Anterior[b]	Median line of neck	Anterior border of SCM	Inferior border of mandible	Superior sternum	Subcutaneous tissue; platysma	Pharynx, larynx, thyroid gland

SCM, sternocleidomastoid.
[a]Further subdivided by the inferior belly of omohyoid into occipital (2) and omoclavicular (3) triangles.
[b]Further subdivided by the digastric and omohyoid muscles into submandibular (4), submental (5), carotid (6), and muscular (7) triangles.

branch of the nerve to the subclavius. If present, the accessory phrenic nerve lies lateral to the main nerve and descends posterior and sometimes anterior to the subclavian vein. The accessory phrenic nerve joins the phrenic nerve either in the root of the neck or in the thorax.

VEINS IN LATERAL CERVICAL REGION

The **external jugular vein** (EJV) begins near the angle of the mandible (just inferior to the auricle of the external ear) by the union of the posterior division of the *retromandibular vein* with the *posterior auricular vein* (Fig. 9.5A). The EJV crosses the SCM obliquely, deep to the platysma, and then pierces the investing layer of deep cervical fascia, which forms the roof of this region, at the posterior border of the SCM (Fig. 9.5C). The EJV descends to the inferior part of the lateral cervical region and terminates in the subclavian vein.

The major venous channel draining the upper limb, the **subclavian vein**, courses through the inferior part of the lateral cervical region, passing anterior to the anterior scalene muscle and phrenic nerve (Fig. 9.5E). The subclavian vein joins the IJV to form the **brachiocephalic vein** posterior to the medial end of the clavicle (Fig. 9.5A,E). Just superior to the clavicle, the EJV receives the *cervicodorsal (transverse cervical)*, *suprascapular*, and *anterior jugular veins*.

ARTERIES IN LATERAL CERVICAL REGION

The arteries in the lateral cervical region are the cervicodorsal trunk and suprascapular artery, the third part of the subclavian artery, and part of the occipital artery (Fig. 9.5C,E).

The **cervicodorsal trunk** (transverse cervical artery) commonly originates from the *thyrocervical trunk*, a branch of the subclavian artery, and divides into the superficial cervical and dorsal scapular arteries. The cervicodorsal trunk runs superficially and laterally across the phrenic nerve and anterior scalene muscle, 2–3 cm superior to the clavicle. It then crosses (passes through) the *trunks of the brachial plexus*, supplying branches to their *vasa nervorum* (blood vessels of nerves) and passing deep to the trapezius (Fig. 9.5E). The superficial cervical artery accompanies CN XI along the anterior (deep) surface of the trapezius. The dorsal scapular artery runs anterior to the insertions of the rhomboid muscles, accompanying the dorsal scapular nerve. The dorsal scapular artery may arise independently, directly from the subclavian artery with no trunk formed.

The **suprascapular artery**, arising from the cervicodorsal trunk, or directly from the subclavian artery, passes inferolaterally across the anterior scalene muscle and

(Continued on page 605)

TABLE 9.3. SUMMARY OF CONTENTS OF REGIONS/TRIANGLES OF NECK[a]

Region	Main Contents and Underlying Structures
Sternocleidomastoid region (**A**) Lesser supraclavicular fossa (*1*)	Sternocleidomastoid (SCM) muscle; superior part of the external jugular vein; greater auricular nerve; transverse cervical nerve Inferior part of internal jugular vein
Posterior cervical region (**B**)	Descending part of trapezius muscle; cutaneous branches of posterior rami of cervical spinal nerves; suboccipital region (**E**) lies deep to superior part of this region

Lateral cervical region (posterior triangle of neck) (**C**)
  Occipital triangle (*2*)
  Omoclavicular (subclavian) triangle (*3*)
	Part of external jugular vein; posterior branches of cervical plexus of nerves; spinal accessory nerve; trunks of brachial plexus; cervicodorsal trunk; cervical lymph nodes
	Subclavian artery (third part), part of subclavian vein (sometimes); suprascapular artery; supraclavicular lymph nodes

Anterior cervical region (anterior triangle of neck) (**D**)
  Submandibular (digastric) triangle (*4*)
  Submental triangle (*5*)
  Carotid triangle (*6*)
	Submandibular gland almost fills triangle; submandibular lymph nodes; hypoglossal nerve; mylohyoid nerve; parts of facial artery and vein
	Submental lymph nodes and small veins that unite to form anterior jugular vein
	Common carotid artery and its branches; internal jugular vein and its tributaries; vagus nerve; external carotid artery and some of its branches; hypoglossal nerve and superior root of ansa cervicalis; spinal accessory nerve; thyroid gland, larynx; pharynx; deep cervical lymph nodes; branches of cervical plexus

Muscular (omotracheal) triangle (*7*)	Sternothyroid and sternohyoid muscles; thyroid and parathyroid glands

[a]Letter and numbers in parentheses refer to regions/triangles demonstrated in Figure 9.4.

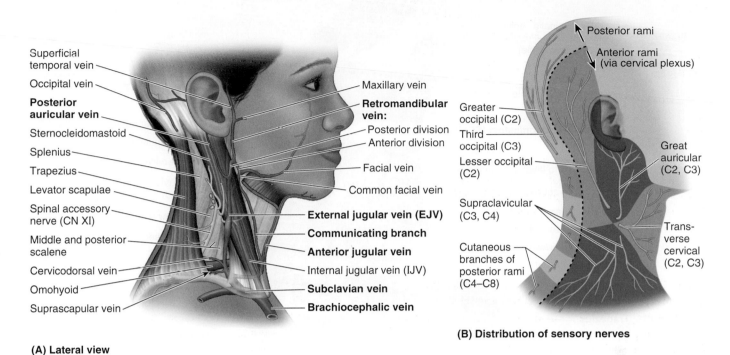

(A) Lateral view

Superficial temporal vein
Occipital vein
Posterior auricular vein
Sternocleidomastoid
Splenius
Trapezius
Levator scapulae
Spinal accessory nerve (CN XI)
Middle and posterior scalene
Cervicodorsal vein
Omohyoid
Suprascapular vein

Maxillary vein
Retromandibular vein:
Posterior division
Anterior division
Facial vein
Common facial vein
External jugular vein (EJV)
Communicating branch
Anterior jugular vein
Internal jugular vein (IJV)
Subclavian vein
Brachiocephalic vein

(B) Distribution of sensory nerves

Posterior rami
Anterior rami (via cervical plexus)
Greater occipital (C2)
Third occipital (C3)
Lesser occipital (C2)
Great auricular (C2, C3)
Supraclavicular (C3, C4)
Cutaneous branches of posterior rami (C4–C8)
Transverse cervical (C2, C3)

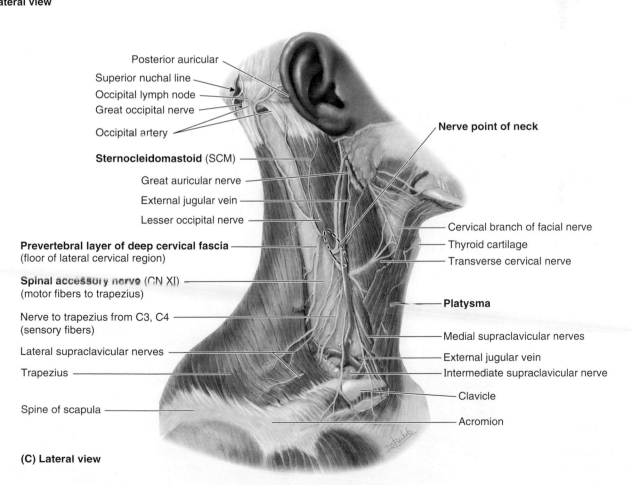

(C) Lateral view

Posterior auricular
Superior nuchal line
Occipital lymph node
Great occipital nerve
Occipital artery
Sternocleidomastoid (SCM)
Great auricular nerve
External jugular vein
Lesser occipital nerve
Prevertebral layer of deep cervical fascia (floor of lateral cervical region)
Spinal accessory nerve (CN XI) (motor fibers to trapezius)
Nerve to trapezius from C3, C4 (sensory fibers)
Lateral supraclavicular nerves
Trapezius
Spine of scapula

Nerve point of neck
Cervical branch of facial nerve
Thyroid cartilage
Transverse cervical nerve
Platysma
Medial supraclavicular nerves
External jugular vein
Intermediate supraclavicular nerve
Clavicle
Acromion

FIGURE 9.5. Lateral cervical region. A. Superficial veins of neck. **B.** Distribution of sensory nerves. **C.** Superficial dissection. *(continued)*

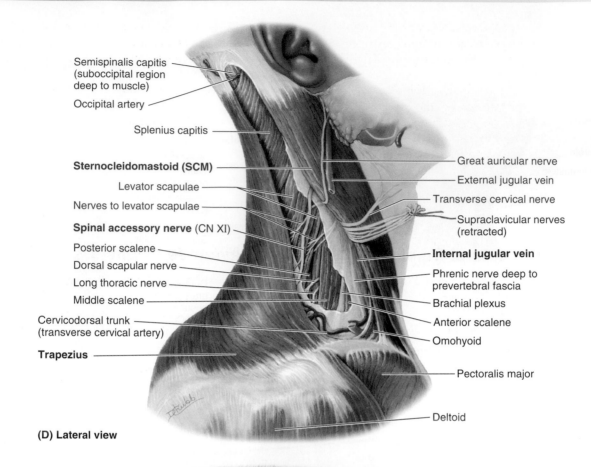

Semispinalis capitis
(suboccipital region
deep to muscle)

Occipital artery

Splenius capitis

Sternocleidomastoid (SCM)

Levator scapulae

Nerves to levator scapulae

Spinal accessory nerve (CN XI)

Posterior scalene

Dorsal scapular nerve

Long thoracic nerve

Middle scalene

Cervicodorsal trunk
(transverse cervical artery)

Trapezius

Great auricular nerve

External jugular vein

Transverse cervical nerve

Supraclavicular nerves
(retracted)

Internal jugular vein

Phrenic nerve deep to
prevertebral fascia

Brachial plexus

Anterior scalene

Omohyoid

Pectoralis major

Deltoid

(D) Lateral view

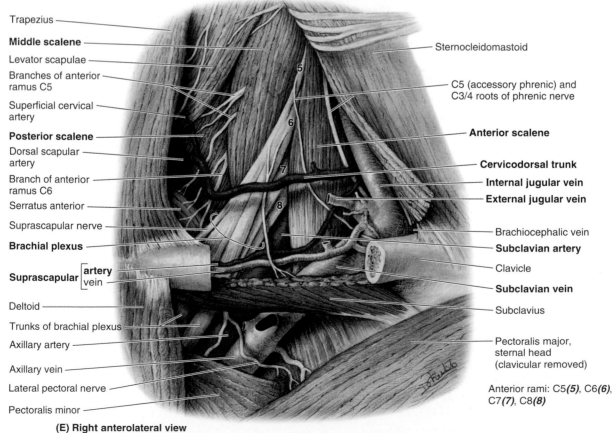

Trapezius

Middle scalene

Levator scapulae

Branches of anterior
ramus C5

Superficial cervical
artery

Posterior scalene

Dorsal scapular
artery

Branch of anterior
ramus C6

Serratus anterior

Suprascapular nerve

Brachial plexus

Suprascapular ⌈artery
           ⌊vein

Deltoid

Trunks of brachial plexus

Axillary artery

Axillary vein

Lateral pectoral nerve

Pectoralis minor

Sternocleidomastoid

C5 (accessory phrenic) and
C3/4 roots of phrenic nerve

Anterior scalene

Cervicodorsal trunk

Internal jugular vein

External jugular vein

Brachiocephalic vein

Subclavian artery

Clavicle

Subclavian vein

Subclavius

Pectoralis major,
sternal head
(clavicular removed)

Anterior rami: C5*(5)*, C6*(6)*,
C7*(7)*, C8*(8)*

(E) Right anterolateral view

FIGURE 9.5. Lateral cervical region. *(continued)* **D.** Deep dissection. **E.** A deeper dissection of the inferior part of the lateral cervical region.

phrenic nerve. It crosses the subclavian artery (third part) and the cords of the brachial plexus. It then passes posterior to the clavicle to supply muscles on the posterior aspect of the scapula (Fig. 9.5E).

The **occipital artery**, a branch of the external carotid artery (Fig. 9.5C), crosses the apex of the lateral cervical region, ascending to supply the posterior half of the scalp.

The **third part of the subclavian artery** supplies blood to the upper limb. It begins approximately a finger's breadth superior to the clavicle, opposite the lateral border of the anterior scalene muscle. It lies posterosuperior to the subclavian vein in the inferior part of the lateral cervical region (Fig. 9.5E). The pulsations of the artery can be felt via deep pressure in the omoclavicular triangle just superior to the clavicle (Fig. 9.3). The artery is in contact with the 1st rib as it passes posterior to the anterior scalene muscle; consequently, compression of the artery against this rib can control bleeding in the upper limb.

CLINICAL BOX

Nerve Blocks in Lateral Cervical Region

Regional anesthesia is often used for surgical procedures in the neck region or upper limb. In a *cervical plexus block*, an anesthetic agent is injected at several points along the posterior border of the SCM, mainly at the junction of its superior and middle thirds, the *nerve point of the neck* (Fig. B9.2). For anesthesia of the upper limb, the anesthetic agent in a *supraclavicular brachial plexus block* is injected around the supraclavicular part of the brachial plexus. The main injection site is superior to the midpoint of the clavicle.

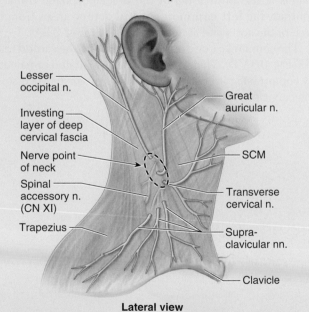

Lateral view

FIGURE B9.2. Nerve point of neck. *n.*, nerve; *nn.*, nerves; *SCM*, sternocleidomastoid.

Severance of Phrenic Nerve and Phrenic Nerve Block

Severance of a phrenic nerve results in paralysis of the corresponding half of the diaphragm. A phrenic nerve block produces a short period of paralysis of the diaphragm on one side (e.g., for a lung operation). The anesthetic agent is injected around the nerve where it lies on the anterior surface of the anterior scalene muscle.

Subclavian Vein Puncture

The right or left subclavian vein often provides a point of entry into the venous system for *central line placement* (Fig. B9.3). Central lines are inserted to administer *parenteral* (venous nutritional) *fluids* and medications and to measure *central venous pressure*. The pleura and/or the subclavian artery are at risk of puncture during this procedure. Alternative sites of central venous line placement are the IJV and femoral vein.

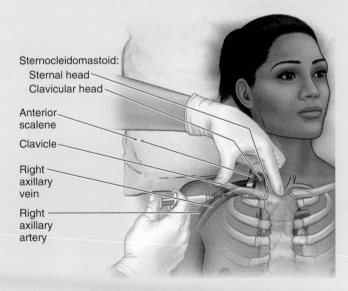

FIGURE B9.3. Subclavian vein puncture.

Prominence of External Jugular Vein

The EJV may serve as an "internal barometer." When venous pressure is in the normal range, the EJV is usually visible superior to the clavicle for only a short distance. However, when venous pressure rises (e.g., as in heart failure) the vein is prominent throughout its course along the side of the neck. Consequently, routine observation for distention of the EJVs during physical examinations may reveal diagnostic signs of heart failure, obstruction of the superior vena cava, enlarged supraclavicular lymph nodes, or increased intrathoracic pressure.

Anterior Cervical Region

MUSCLES IN ANTERIOR CERVICAL REGION

In the anterolateral part of the neck, the *hyoid bone* provides attachments for the suprahyoid muscles superior to it and the infrahyoid muscles inferior to it (Figs. 9.6–9.8). These **hyoid muscles** steady or move the hyoid and larynx. The attachments, innervation, and main actions of the suprahyoid and infrahyoid muscles are presented in Table 9.4.

The **suprahyoid muscles** are superior to the hyoid bone and connect it to the cranium. This group includes the mylohyoid, geniohyoid, stylohyoid, and digastric muscles. The group constitutes the substance of the floor of the mouth, supporting the hyoid bone in providing a base from which the tongue functions and in elevating the hyoid and larynx in relation to swallowing and tone production. Each **digastric muscle** has anterior and posterior bellies joined by an **intermediate tendon** that descends toward the hyoid bone. A **fibrous sling** allows the intermediate tendon to slide anteriorly and posteriorly as it connects this tendon to the body and greater horn of the hyoid bone (Fig. 9.8A,B).

The **infrahyoid muscles** (strap muscles) are inferior to the hyoid bone. These four muscles anchor the hyoid, sternum, clavicle, and scapula and depress the hyoid and larynx during swallowing and speaking (Fig. 9.7 and Table 9.4). They also work with the suprahyoid muscles to steady the hyoid, providing a firm base for the tongue. The infrahyoid group of muscles is arranged in two planes: a *superficial plane* made up of the sternohyoid and omohyoid and a *deep plane* composed of the sternothyroid and thyrohyoid. The **omohyoid** has two bellies united by an intermediate tendon that is connected to the clavicle by a fascial sling (Fig. 9.7C). The **sternothyroid** is wider than the **sternohyoid**, under which it lies. The sternothyroid covers the lateral lobe of the thyroid gland, attaching to the oblique line of the lamina of the thyroid cartilage immediately superior to the gland, limiting superior expansion of an enlarged thyroid gland. The **thyrohyoid**, running superiorly from the oblique line of the thyroid cartilage to the hyoid, appears to be a continuation of the sternothyroid muscle.

ARTERIES IN ANTERIOR CERVICAL REGION

The anterior cervical region contains the **carotid system of arteries**, consisting of the common carotid artery and its terminal branches, the internal and external carotid arteries (Figs. 9.8A and 9.9C). This region also contains the IJV and its tributaries and the anterior jugular veins. The common carotid artery and one of its terminal branches, the *external carotid artery*, are the main arterial vessels in the carotid triangle.

Each **common carotid artery** ascends within the *carotid sheath* with the IJV and vagus nerve to the level of the superior border of the thyroid cartilage. Here, each common carotid artery terminates by dividing into the internal and external carotid arteries. The **right common carotid artery** begins at the bifurcation of the brachiocephalic trunk. In contrast, the **left common carotid artery** arises from the arch of the aorta and ascends in the neck (Fig. 9.9A).

The common carotid arteries ascend into the carotid triangle (Fig. 9.8A,B). Their pulse can be auscultated or palpated by compressing it lightly against the transverse processes of the cervical vertebrae.

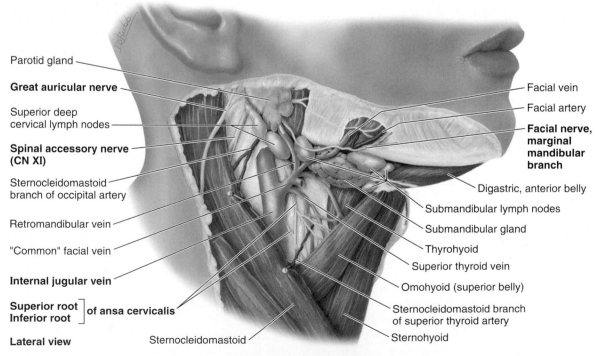

Parotid gland
Great auricular nerve
Superior deep cervical lymph nodes
Spinal accessory nerve (CN XI)
Sternocleidomastoid branch of occipital artery
Retromandibular vein
"Common" facial vein
Internal jugular vein
Superior root ⎤ of ansa cervicalis
Inferior root ⎦
Lateral view
Sternocleidomastoid

Facial vein
Facial artery
Facial nerve, marginal mandibular branch
Digastric, anterior belly
Submandibular lymph nodes
Submandibular gland
Thyrohyoid
Superior thyroid vein
Omohyoid (superior belly)
Sternocleidomastoid branch of superior thyroid artery
Sternohyoid

FIGURE 9.6. Superficial dissection of anterior cervical region.

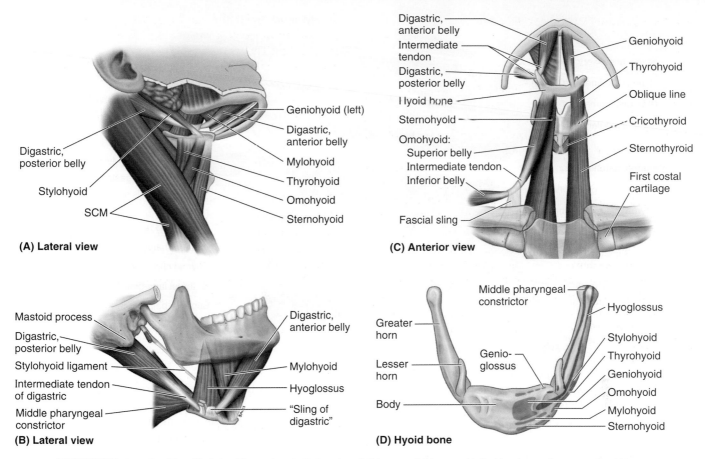

FIGURE 9.7. Suprahyoid and infrahyoid muscles. A–C. Overview. *SCM,* sternocleidomastoid. **D.** Muscle attachments to hyoid bone.

TABLE 9.4. MUSCLES OF ANTERIOR CERVICAL REGION (EXTRINSIC MUSCLES OF LARYNX)

Muscle	Origin	Insertion	Innervation	Main Action(s)
Suprahyoid muscles				
Mylohyoid	Mylohyoid line of mandible	Mylohyoid raphe and body of hyoid bone	Nerve to mylohyoid, a branch of inferior alveolar nerve (from mandibular nerve, CN V$_3$)	Elevates hyoid bone, floor of mouth, and tongue during swallowing and speaking
Geniohyoid	Inferior mental spine of mandible	Body of hyoid bone	C1 via hypoglossal nerve (CN XII)	Pulls hyoid bone anterosuperiorly; shortens floor of mouth; widens pharynx
Stylohyoid	Styloid process of temporal bone		Stylohyoid branch of facial nerve (CN VII)	Elevates and retracts hyoid bone, thus elongating floor of mouth
Digastric	*Anterior belly:* digastric fossa of mandible *Posterior belly:* mastoid notch of temporal bone	Intermediate tendon to body and greater horn of hyoid bone	*Anterior belly:* nerve to mylohyoid, a branch of inferior alveolar nerve *Posterior belly:* digastric branch of facial nerve (CN VII)	Working with infrahyoid muscles, depresses mandible against resistance; elevates and steadies hyoid bone during swallowing and speaking
Infrahyoid muscles				
Sternohyoid	Manubrium of sternum and medial end of clavicle	Body of hyoid bone	C1–C3 by a branch of ansa cervicalis	Depresses hyoid bone after elevation during swallowing
Omohyoid	Superior border of scapula near suprascapular notch	Inferior border of hyoid bone		Depresses, retracts, and steadies hyoid bone
Sternothyroid	Posterior surface of manubrium of sternum	Oblique line of thyroid cartilage	C2 and C3 by a branch of ansa cervicalis	Depresses hyoid bone and larynx
Thyrohyoid	Oblique line of thyroid cartilage	Inferior border of body and greater horn of hyoid bone	C1 via hypoglossal nerve	Depresses hyoid bone and elevates larynx

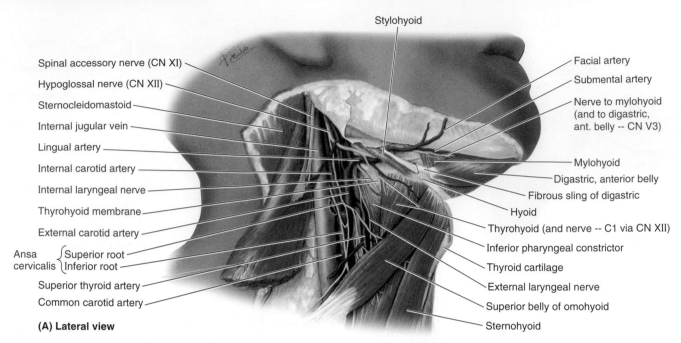

(A) Lateral view

FIGURE 9.8. **Anterior cervical region and suprahyoid region. A.** Deep dissection. *(continued)*

At the bifurcation of the common carotid artery into external and internal carotid arteries, there is a slight dilation of the proximal part of the internal carotid artery—the **carotid sinus** (Fig. 9.9C). Innervated principally by the glossopharyngeal nerve (CN IX) via its **carotid branch**, as well as the vagus nerve, the carotid sinus is a *baroreceptor* (pressoreceptor) stimulated by increases in arterial blood pressure.

The **carotid body**, an ovoid mass of tissue, lies on the medial (deep) side of the bifurcation of the common carotid artery in close relation to the carotid sinus (Fig. 9.9C). Supplied mainly by the carotid branch of CN IX and by CN X, the carotid body is a *chemoreceptor* that monitors the level of oxygen in the blood (PO_2). It is stimulated by low levels of oxygen and initiates a reflex that increases the rate and depth of respiration, cardiac rate, and blood pressure.

The **internal carotid arteries**, the direct continuation of the common carotid arteries, have no branches in the neck. They enter the cranium through the *carotid canals* and become the main arteries of the brain and structures in the orbits.

The **external carotid arteries** supply most structures external to the cranium; the orbit and part of the forehead and scalp supplied by the supra-orbital artery are the major exceptions (Figs. 9.8A,B and 9.9C). Each external carotid artery runs posterosuperiorly to the region between the neck of the mandible and the lobule of the auricle, where it is embedded in the parotid gland. Here, it divides into two terminal branches: the *maxillary* and *superficial temporal arteries* (Fig. 9.9C). Before these terminal branches, six arteries arise from the external carotid artery (Figs. 9.8A,B and 9.9C):

- **Ascending pharyngeal artery** arises as the first or second branch of the external carotid artery and is its only medial branch; ascends on the pharynx and sends branches to the pharynx, prevertebral muscles, middle ear, and cranial meninges.

- **Occipital artery** arises from the posterior aspect of the external carotid artery, superior to the origin of the facial artery; passes posteriorly, immediately medial and parallel to the attachment of the posterior belly of the digastric muscle, ending in the posterior part of the scalp. During its course, it passes superficial to the internal carotid artery and CN IX to CN XI.

- **Posterior auricular artery**, a small posterior branch of the external carotid artery, ascends posteriorly between the external acoustic meatus and the mastoid process and contributes to the blood supply of adjacent muscles, parotid gland, facial nerve, structures in the temporal bone, auricle, and scalp.

- **Superior thyroid artery**, the most inferior of the three anterior branches of the external carotid artery, runs anteroinferiorly deep to the infrahyoid muscles to reach the thyroid gland. In addition to supplying this gland, it gives off branches to the infrahyoid muscles and the SCM and gives rise to the *superior laryngeal artery*, supplying the larynx.

- **Lingual artery** also arises from the anterior aspect of the external carotid artery, where it lies on the middle constrictor muscle of the pharynx (see Fig. 8.60A). It passes deep to CN XII, the stylohyoid muscle, and the posterior belly of the digastric muscle and disappears deep to the hyoglossus muscle. The lingual artery gives *dorsal lingual arteries* to the posterior tongue and then bifurcates into the *deep lingual* and *sublingual arteries*.

- **Facial artery** also arises anteriorly from the external carotid artery, either in common with the lingual artery or immediately superior to it. After giving rise to the *ascending palatine artery* and a *tonsillar branch*, it passes *superiorly* under cover of the digastric and stylohyoid

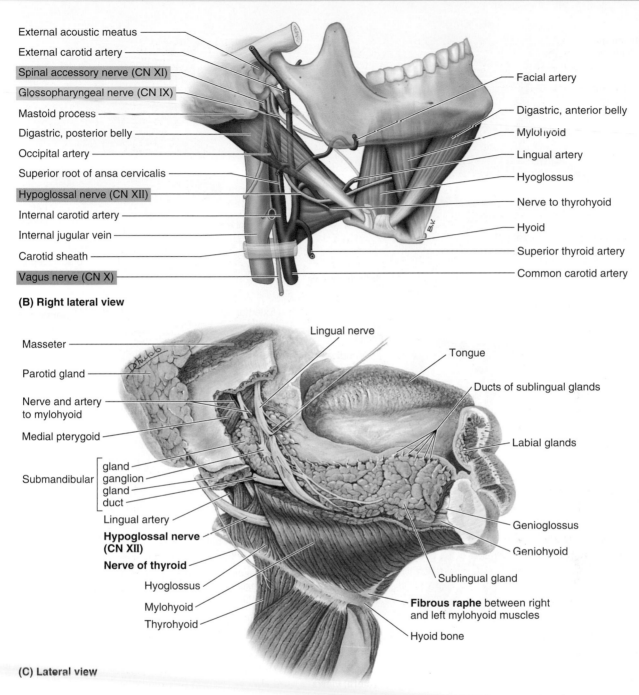

External acoustic meatus
External carotid artery
Spinal accessory nerve (CN XI)
Glossopharyngeal nerve (CN IX)
Mastoid process
Digastric, posterior belly
Occipital artery
Superior root of ansa cervicalis
Hypoglossal nerve (CN XII)
Internal carotid artery
Internal jugular vein
Carotid sheath
Vagus nerve (CN X)

Facial artery
Digastric, anterior belly
Mylohyoid
Lingual artery
Hyoglossus
Nerve to thyrohyoid
Hyoid
Superior thyroid artery
Common carotid artery

(B) Right lateral view

Masseter
Parotid gland
Nerve and artery to mylohyoid
Medial pterygoid
Submandibular { gland ganglion gland duct }
Lingual artery
Hypoglossal nerve (CN XII)
Nerve of thyroid
Hyoglossus
Mylohyoid
Thyrohyoid

Lingual nerve
Tongue
Ducts of sublingual glands
Labial glands
Genioglossus
Geniohyoid
Sublingual gland
Fibrous raphe between right and left mylohyoid muscles
Hyoid bone

(C) Lateral view

FIGURE 9.8. Anterior cervical region and suprahyoid region. *(continued)* **B.** Relationships of the nerves and vessels to the suprahyoid muscles. **C.** Dissection of suprahyoid region. The right half of the mandible and the superior half of the mylohyoid muscle have been removed.

muscles and the angle of the mandible. It supplies the submandibular gland and then gives rise to the *submental artery* to the floor of the mouth, hooking around the middle of the inferior border of the mandible (where its pulse can be palpated) to enter the face.

VEINS IN ANTERIOR CERVICAL REGION

Most veins in the anterior cervical region are tributaries of the **internal jugular vein (IJV)**, usually the largest vein in the neck (Figs. 9.6, 9.8A,B, and 9.9B). The *IJV* drains blood

from the brain, anterior face, cervical viscera, and deep muscles of the neck. The IJV commences at the jugular foramen in the posterior cranial fossa as the direct continuation of the sigmoid sinus (see Chapter 8, Head). From the dilation at its origin, the **superior bulb of the IJV** (Fig. 9.9D), the vein runs inferiorly through the neck in the *carotid sheath* with the internal carotid artery superior to the carotid bifurcation and the common carotid artery and CN X inferiorly (Fig. 9.8B). The vein lies laterally within the sheath, with the nerve located posteriorly. The *cervical sympathetic trunk* lies posterior to the carotid sheath, embedded in the prevertebral

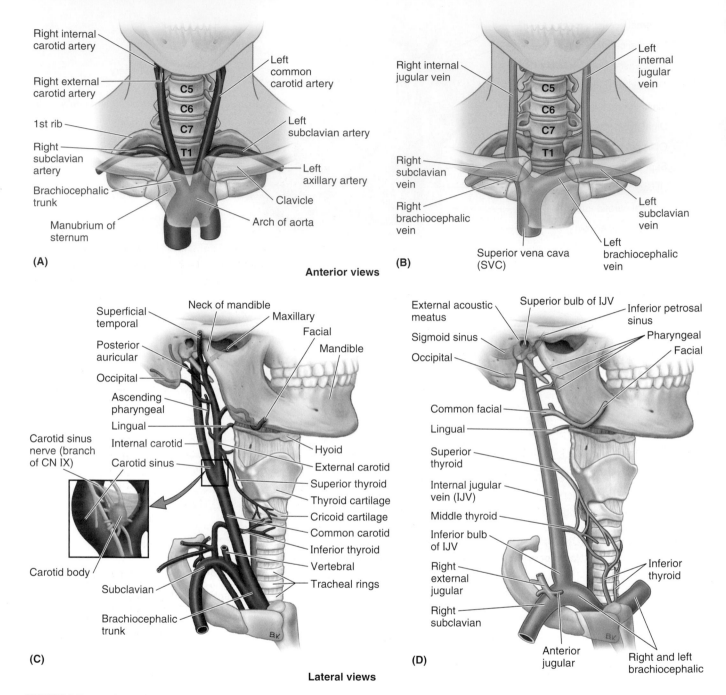

(A)

Anterior views

(B)

(C)

(D)

Lateral views

FIGURE 9.9. Arteries and veins in neck. A. Subclavian and carotid arteries. **B.** Internal jugular and subclavian veins. **C.** Branches of the subclavian and external carotid arteries. **D.** Tributaries of internal jugular vein.

layer of deep cervical fascia. The IJV leaves the anterior cervical region by passing deep to the SCM.

Posterior to the sternal end of the clavicle, the IJV unites with the subclavian vein to form the *brachiocephalic vein*. The inferior end of the IJV dilates to form the **inferior bulb of the IJV** (Fig. 9.9D). This bulb has a bicuspid valve that permits blood to flow toward the heart while preventing backflow into the vein. The tributaries of the IJV are the inferior petrosal sinus and the facial, lingual, pharyngeal, and superior and middle thyroid veins.

NERVES IN ANTERIOR CERVICAL REGION

The transverse cervical nerve (C2 and C3) supplies the skin covering the anterior cervical region (Fig. 9.8A). The **hypoglossal nerve** (CN XII), the motor nerve of the tongue, enters the submandibular triangle deep to the posterior belly of the digastric muscle to supply the muscles of the tongue (Fig. 9.8A–C). Branches of the **glossopharyngeal** and **vagus nerves** (CNs IX and X) are located in the submandibular and carotid triangles (Fig. 9.8B).

CLINICAL BOX

Ligation of External Carotid Artery

Sometimes, *ligation of an external carotid artery* is necessary to control bleeding from one of its relatively inaccessible branches. This procedure decreases blood flow through the artery and its branches but does not eliminate it. Blood flows in a retrograde (backward) direction into the artery from the external carotid artery on the other side through communications between its branches (e.g., those in the face and scalp) and across the midline. When the external carotid or subclavian arteries are ligated, the descending branch of the occipital artery provides the main collateral circulation, anastomosing with the vertebral and deep cervical arteries.

Surgical Dissection of Carotid Triangle

The carotid triangle provides an important surgical approach to the carotid system of arteries, the IJV, the vagus and hypoglossal nerves, and the cervical sympathetic trunk. Damage or compression of the vagus and/or recurrent laryngeal nerves during surgical dissection of the triangle may produce an alteration in the voice because these nerves supply laryngeal muscles.

Carotid Occlusion and Endarterectomy

Atherosclerotic thickening of the intima of the internal carotid artery may obstruct blood flow. Symptoms resulting from this obstruction depend on the degree of obstruction and the amount of collateral blood flow to the brain from other arteries. A partial occlusion may cause a *transient ischemic attack* (TIA), a sudden focal loss of neurological function (e.g., dizziness and disorientation) that disappears within 24 hours. Arterial occlusion may also cause a *stroke*.

Carotid occlusion, causing *stenosis* (narrowing), can be relieved by opening the artery at its origin and stripping off the atherosclerotic plaque with the intima. This procedure is called *carotid endarterectomy*. Because of the relations of the internal carotid artery, there is risk of cranial nerve injury during the procedure involving one or more of the following nerves: CN IX, CN X (or its branch, the superior laryngeal nerve), CN XI, or CN XII.

There is growing use of *carotid angioplasty and stenting*, similar to the procedure described for coronary angioplasty.

Carotid Pulse

The *carotid pulse* ("neck pulse") is easily felt by palpating the common carotid artery in the side of the neck, where it lies in a groove between the trachea and infrahyoid muscles. It is usually easily palpated just deep to the anterior border of the SCM at the level of the superior border of the thyroid cartilage. It is routinely checked during *cardiopulmonary resuscitation* (CPR). Absence of a carotid pulse indicates *cardiac arrest*.

Internal Jugular Pulse

Pulsations of the IJV can provide information about heart activity corresponding to *electrocardiogram* (*ECG*) *recordings* and right atrial pressure. The vein's pulsations are transmitted through the surrounding tissues and may be observed deep to the SCM superior to the medial end of the clavicle. Because there are no valves in the brachiocephalic vein or the superior vena cava, a wave of contraction passes up these vessels to the IJV. The pulsations are especially visible when the person's head is inferior to the feet (the *Trendelenburg position*). The internal jugular pulse increases considerably in conditions such as mitral valve disease, which increases pressure in the pulmonary circulation and the right side of the heart.

Internal Jugular Vein Puncture

A needle and catheter may be inserted into the IJV for diagnostic or therapeutic purposes. The right IJV is preferable because it is usually larger and straighter. During this procedure, the clinician palpates the common carotid artery and inserts the needle into the IJV just lateral to it at a 30-degree angle, aiming at the apex of the triangle between the sternal and clavicular heads of the SCM. The needle is then directed inferolaterally toward the ipsilateral nipple (Fig. B9.4).

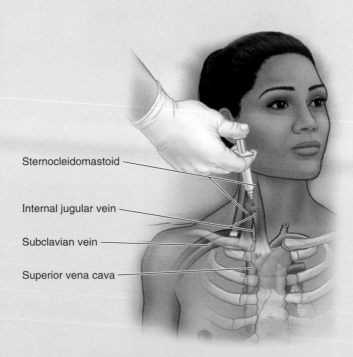

Sternocleidomastoid

Internal jugular vein

Subclavian vein

Superior vena cava

FIGURE B9.4. Internal jugular vein puncture.

SURFACE ANATOMY

Cervical Regions and Triangles of Neck

The **skin of the neck** is thin and pliable. The subcutaneous connective tissue contains the platysma, a thin sheet of striated muscle that ascends to the face (Figs. SA9.1A and 9.4A).

The **SCM** is the key muscular landmark of the neck. It defines the **SCM region** and divides the neck into anterior and lateral cervical regions (Fig. SA9.1C). This muscle is easy to observe and palpate throughout its length as it passes superolaterally from the clavicle and manubrium to the mastoid process of the temporal bone. The SCM can be made to stand out by asking the person to rotate the face toward the contralateral side and elevate the chin.

The **EJV** runs vertically across the SCM toward the angle of the mandible (Fig. SA9.1C). This vein may be prominent, especially if distended, and can be visualized by asking the person to take a deep breath (*Valsalva maneuver*). The **jugular notch** in the manubrium is the fossa between the sternal heads of the SCM. The **lesser supraclavicular fossa**, between the sternal and clavicular heads of the SCM, overlies the inferior end of the IJV. Deep to the superior half of the SCM is the cervical plexus, and deep to the inferior half of the SCM are the IJV, common carotid artery, and vagus nerve in the carotid sheath.

The **anterior border of the trapezius** defines the posterior cervical region. It may be observed and palpated when the shoulders are shrugged against resistance (Fig. SA9.1B).

Just inferior to the belly of the omohyoid is the **greater supraclavicular fossa** (Fig. SA9.1D), the depression overlying the omoclavicular triangle. The **subclavian arterial pulsations** can be palpated here in most people.

The **occipital triangle** contains the **spinal accessory nerve** (CN XI). Because of its vulnerability and frequency of iatrogenic injury (damage resulting from medical treatment), it is important to be able to estimate the location of the nerve (Fig. SA9.1B). Its course can be approximated by

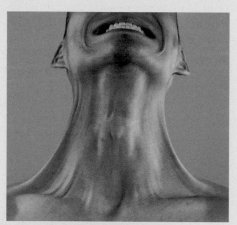

(A) Anterior view

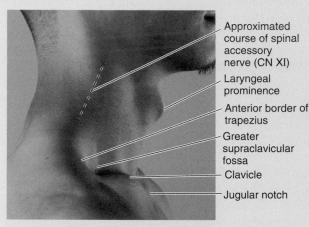

Approximated course of spinal accessory nerve (CN XI)

Laryngeal prominence

Anterior border of trapezius

Greater supraclavicular fossa

Clavicle

Jugular notch

(B) Lateral view

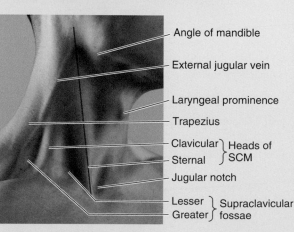

Angle of mandible

External jugular vein

Laryngeal prominence

Trapezius

Clavicular ⎱ Heads of
Sternal ⎰ SCM

Jugular notch

Lesser ⎱ Supraclavicular
Greater ⎰ fossae

(C) Anterolateral view

(D) Palpation of submandibular lymph nodes

FIGURE SA9.1. Surface anatomy of cervical regions. *SCM*, sternocleidomastoid.

a line that intersects the junction of the superior and middle thirds of the posterior border of the SCM and the junction of the middle and lower thirds of the anterior border of the trapezius.

The **submandibular gland** nearly fills the submandibular triangle (Figs. 9.6 and 9.8C). It is palpable as a soft mass inferior to the body of the mandible, especially when the tongue is pushed against the maxillary incisor teeth. The **submandibular lymph nodes** lie superficial to the gland and, if enlarged, can be palpated by moving the fingers from the **angle of the mandible** along its inferior border (Fig. SA9.1D). If this is continued until the examiner's fingers meet under the chin, enlarged **submental lymph nodes** can be palpated in the **submental triangle**.

The carotid arterial system is located in the **carotid triangle**. The **carotid sheath** can be mapped out by a line joining the **sternoclavicular (SC) joint** to a point midway between the **mastoid process** and the angle of the mandible (Fig. SA9.1C). The **carotid pulse** can be palpated by placing the index and 3rd fingers on the thyroid cartilage and pointing them posterolaterally between the trachea and SCM. The pulse is palpable just medial to the SCM.

DEEP STRUCTURES OF NECK

The **deep structures of the neck** are the prevertebral muscles, located posterior to the cervical viscera and anterolateral to the vertebral column, and structures located on the cervical side of the superior thoracic aperture, the root of the neck (Fig. 9.10).

Prevertebral Muscles

The **anterior** and **lateral vertebral muscles** or **prevertebral muscles**, consisting of the longus colli and capitis and rectus capitis anterior and the anterior scalene muscles, lie directly posterior to the retropharyngeal space (see Fig. 9.2). The *lateral vertebral muscles*, consisting of the rectus capitis

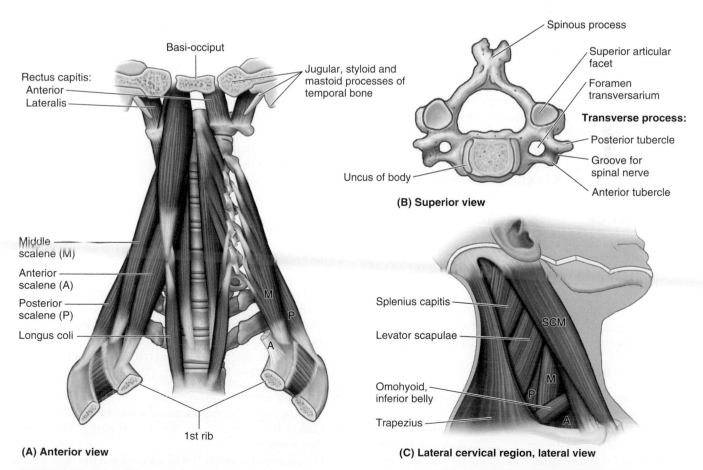

FIGURE 9.10. Prevertebral muscles. A. Overview. **B.** Muscle attachments to cervical vertebra. **C.** In lateral cervical region. *SCM*, sternocleidomastoid.

(A) Anterior view

(C) Lateral cervical region, lateral view

(B) Superior view

TABLE 9.5. PREVERTEBRAL MUSCLES

Muscle	Superior Attachment	Inferior Attachment	Innervation	Main Action(s)
Anterior vertebral muscles				
Longus colli	Anterior tubercle of C1 vertebra (atlas); bodies of C1–C3 and transverse processes of C3–C6 vertebrae	Bodies of C5–T3 vertebrae; transverse processes of C3–C5 vertebrae	Anterior rami of C2–C6 spinal nerves	Flexes neck (anterior [or lateral] bending of cervical vertebrae C2–C7)
Longus capitis	Basilar part of occipital bone	Anterior tubercles of C3–C6 transverse processes	Anterior rami of C1–C3 spinal nerves	Flexion of head on neck (anterior [or lateral] bending of the head relative to the vertebral column at the atlanto-occipital joints)
Rectus capitis anterior	Base of cranium, just anterior to occipital condyle	Anterior surface of lateral mass of atlas (C1 vertebra)	Branches from loop between C1 and C2 spinal nerves	
Anterior scalene	Anterior tubercles of transverse processes of C3–C6 vertebrae	1st rib	Cervical spinal nerves C4–C6	Flexes neck laterally; elevates 1st rib during forced inspiration[a]
Lateral vertebral muscles				
Rectus capitis lateralis	Jugular process of occipital bone	Transverse process of atlas (C1 vertebra)	Branches from loop between C1 and C2 spinal nerves	Flexes head and helps stabilize it[b]
Splenius capitis	Inferior half of nuchal ligament and spinous processes of superior six thoracic vertebrae	Lateral aspect of mastoid process and lateral third of superior nuchal line	Posterior rami of middle cervical spinal nerves	Laterally flexes and rotates head and neck to same side; acting bilaterally, extends head and neck[c]
Levator scapulae	Posterior tubercles of transverse processes of C1–C4 vertebrae	Superior part of medial border of scapula	Dorsal scapular nerve C5 and cervical spinal nerves C3 and C4	Elevates scapula and tilts glenoid cavity inferiorly by rotating scapula
Middle scalene	Posterior tubercles of transverse processes of C4–C7 vertebrae	Superior surface of 1st rib; posterior to groove for subclavian artery	Anterior rami of cervical spinal nerves	Flexes neck laterally; elevates 1st rib during forced inspiration[a]
Posterior scalene	Posterior tubercles of transverse processes of C4–C6 vertebrae	External border of 2nd rib	Anterior rami of cervical spinal nerves C7 and C8	Flexes neck laterally; elevates 2nd rib during forced inspiration[a]

[a]Flexion of neck = anterior (or lateral) bending of cervical vertebrae C2–C7.
[b]Flexion of head = anterior (or lateral) bending of the head relative to the vertebral column at the atlanto-occipital joints.
[c]Rotation of the head occurs at the atlanto-axial joints.

lateralis, splenius capitis, levator scapulae, and middle and posterior scalene muscles, lie posterior to the neurovascular plane of the cervical and brachial plexuses and subclavian artery, except the rectus capitis lateralis, which lies in the floor of the lateral cervical region. The prevertebral muscles are illustrated in Figure 9.10A,C and described in Table 9.5.

Root of Neck

The **root of the neck** is the junctional area between the thorax and neck (Fig. 9.11C). The inferior boundary of the root is formed laterally by the first pair of ribs and their costal cartilages, anteriorly by the manubrium of the sternum, and posteriorly by the body of the T1 vertebra. Only the neurovascular elements of the root of the neck are described here; the visceral structures are discussed later in this chapter.

ARTERIES IN ROOT OF NECK

The **brachiocephalic trunk**, covered anteriorly by the sternohyoid and sternothyroid muscles, is the largest branch of the arch of the aorta. It arises in the midline, posterior to the manubrium, and passes superolaterally to the right. It divides into the right common carotid and right subclavian arteries posterior to the right SC joint (Fig. 9.11A–D).

The **subclavian arteries** supply the upper limbs and send branches to the neck and brain. The **right subclavian artery** arises from the brachiocephalic trunk, and the **left subclavian artery** arises from the arch of the aorta (Fig. 9.11A–D). Their courses in the neck begin posterior to the respective SC joints as they ascend through the superior thoracic aperture. The arteries arch superolaterally, extending between their origin and the medial margin of the anterior scalene muscle. As the arteries begin to descend, they travel deep to the middle of the clavicles and cross the superior surface of the 1st rib. At the outer margin of the 1st rib, their name changes to the axillary arteries.

For purposes of description, the anterior scalene muscle divides each subclavian artery into three parts: The first part is medial to the muscle, the second is posterior to it, and the third is lateral to it (Fig. 9.11A,C). The cervical pleurae, covering the apices of the lungs, and sympathetic trunk lie posterior to the arteries (Fig. 9.11C). The **branches of the subclavian artery** are as follows (Fig. 9.11A–C):

- *Vertebral artery, internal thoracic artery*, and *thyrocervical trunk* from the first part of the subclavian artery
- *Costocervical trunk* from the second part of the subclavian artery
- *Dorsal scapular artery*, often arising from the third part of the subclavian artery

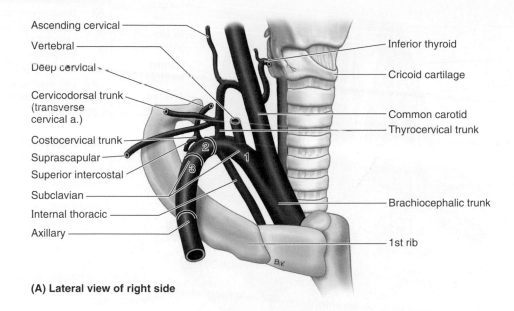

Ascending cervical

Vertebral

Deep cervical

Cervicodorsal trunk
(transverse
cervical a.)

Costocervical trunk

Suprascapular

Superior intercostal

Subclavian

Internal thoracic

Axillary

Inferior thyroid

Cricoid cartilage

Common carotid
Thyrocervical trunk

Brachiocephalic trunk

1st rib

(A) Lateral view of right side

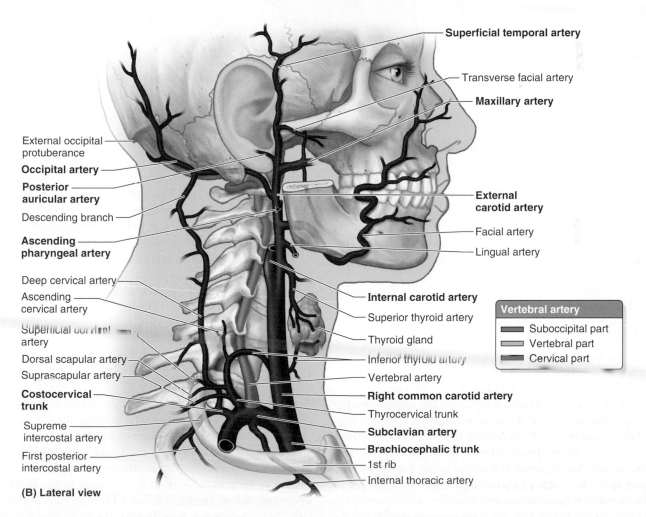

External occipital
protuberance

Occipital artery

**Posterior
auricular artery**

Descending branch

**Ascending
pharyngeal artery**

Deep cervical artery

Ascending
cervical artery

Superficial cervical
artery

Dorsal scapular artery

Suprascapular artery

**Costocervical
trunk**

Supreme
intercostal artery

First posterior
intercostal artery

Superficial temporal artery

Transverse facial artery

Maxillary artery

**External
carotid artery**

Facial artery

Lingual artery

Internal carotid artery

Superior thyroid artery

Thyroid gland

Inferior thyroid artery

Vertebral artery

Right common carotid artery

Thyrocervical trunk

Subclavian artery

Brachiocephalic trunk

1st rib

Internal thoracic artery

Vertebral artery	
	Suboccipital part
	Vertebral part
	Cervical part

(B) Lateral view

FIGURE 9.11. Root of neck and prevertebral region. A. Branches of subclavian artery. The subclavian artery is divided into three parts by the anterior scalene muscle: (*1*) medial, (*2*) posterior, and (*3*) lateral. *a.,* artery. **B.** Overview of the arteries of the head and neck. *(continued)*

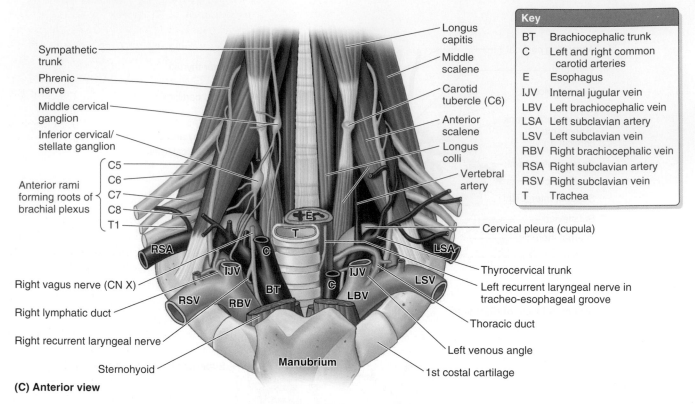

Sympathetic trunk
Phrenic nerve
Middle cervical ganglion
Inferior cervical/ stellate ganglion
Anterior rami forming roots of brachial plexus
— C5
— C6
— C7
— C8
— T1
RSA
Right vagus nerve (CN X)
Right lymphatic duct
Right recurrent laryngeal nerve
Sternohyoid
IJV
RSV
RBV
BT
C
T
E
C
Manubrium

Longus capitis
Middle scalene
Carotid tubercle (C6)
Anterior scalene
Longus colli
Vertebral artery
Cervical pleura (cupula)
Thyrocervical trunk
Left recurrent laryngeal nerve in tracheo-esophageal groove
Thoracic duct
Left venous angle
1st costal cartilage
LSA
IJV
LSV
LBV

Key

BT	Brachiocephalic trunk
C	Left and right common carotid arteries
E	Esophagus
IJV	Internal jugular vein
LBV	Left brachiocephalic vein
LSA	Left subclavian artery
LSV	Left subclavian vein
RBV	Right brachiocephalic vein
RSA	Right subclavian artery
RSV	Right subclavian vein
T	Trachea

(C) Anterior view

FIGURE 9.11. Root of neck and prevertebral region. *(continued)* **C.** Dissection of the root of the neck.

The **cervical part of the vertebral artery** arises from the first part of the subclavian artery and ascends in the pyramidal space formed between the scalene and longus muscles (Fig. 9.10A). The artery then passes through the foramina of the transverse processes of vertebrae C1–C6. This **vertebral part of the vertebral artery** may enter a foramen more superior than the C6 vertebra. The **suboccipital part of the vertebral artery** courses in a groove on the posterior arch of the atlas before it enters the cranial cavity through the foramen magnum, demarcating the beginning of the **cranial part of the vertebral artery**.

The **internal thoracic artery** arises from the anteroinferior aspect of the subclavian artery and passes inferomedially into the thorax (Fig. 9.11A–C). The internal thoracic artery has no branches in the neck; its thoracic distribution is described in Chapter 1.

The **thyrocervical trunk** arises from the anterosuperior aspect of the first part of the subclavian artery, near the medial border of the anterior scalene muscle. It has two lateral branches: the suprascapular artery, supplying muscles on the posterior scapula, and the cervicodorsal trunk (Fig. 9.11A–C). Arising from the cervicodorsal trunk are the *dorsal scapular* and *superficial cervical arteries*, sending branches to muscles in the lateral cervical region, the trapezius, and medial scapular muscles. The terminal branches of the thyrocervical trunk are the *inferior thyroid artery*, the primary visceral artery of the neck, and the ascending cervical artery, supplying lateral muscles of the upper neck.

The **costocervical trunk** arises posteriorly from the second part of the subclavian artery (posterior to the anterior scalene muscle on the right side and usually just medial to this muscle on the left side). The trunk passes posterosuperiorly and divides into the superior intercostal and deep cervical arteries, which supply the first two intercostal spaces and the posterior deep cervical muscles, respectively (Fig. 9.11A,B).

The **dorsal scapular artery** often arises from the cervicodorsal trunk, but it may be an independent branch of the second or third part of the subclavian artery. It runs deep to supply the levator scapulae and rhomboid muscles, supplying both and participating in the arterial anastomoses around the scapula (see Chapter 3).

VEINS IN ROOT OF NECK

Two large veins terminate in the root of the neck: the EJV, draining blood received mostly from the scalp and face, and the variable **anterior jugular vein (AJV)** (see Fig. 9.5A). The AJV typically arises near the hyoid bone from the confluence of superficial submandibular veins. At the root of the neck, the vein turns laterally, posterior to the SCM, and opens into the termination of the EJV or into the subclavian vein. Superior to the manubrium, the right and left AJVs commonly unite across the midline to form the **jugular venous arch** in the suprasternal space.

The subclavian vein, the continuation of the axillary vein, begins at the lateral border of the 1st rib and ends

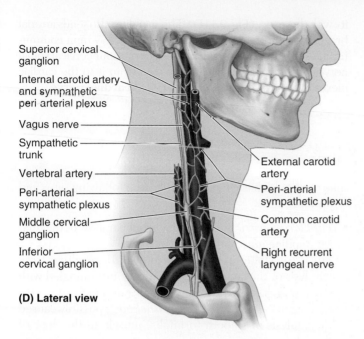

Superior cervical ganglion

Internal carotid artery and sympathetic peri arterial plexus

Vagus nerve

Sympathetic trunk

Vertebral artery

Peri-arterial sympathetic plexus

Middle cervical ganglion

Inferior cervical ganglion

External carotid artery

Peri-arterial sympathetic plexus

Common carotid artery

Right recurrent laryngeal nerve

(D) Lateral view

FIGURE 9.11.   Root of neck and prevertebral region. *(continued)* **D.** Cervical sympathetic trunk and peri-arterial plexuses.

when it unites with the IJV posterior to the medial end of the clavicle to form the brachiocephalic vein (Fig. 9.11C). This union is commonly referred to as the **venous angle** and is the site where the *thoracic duct* (left side) and the *right lymphatic trunk* (right side) drain lymph collected throughout the body into the venous circulation. Throughout its course, the IJV is enclosed by the *carotid sheath* (Fig. 9.8B).

NERVES IN ROOT OF NECK

There are three pairs of major nerves in the root of the neck: (1) the vagus nerves, (2) the phrenic nerves (described earlier in this chapter with the cervical plexus), and (3) the sympathetic trunks (Fig. 9.11D).

VAGUS NERVES (CN X)

After their exit from the jugular foramen (see Fig. 8.3A,D), each vagus nerve passes inferiorly in the neck within the posterior part of the carotid sheath in the angle between the IJV and the common carotid artery (see Figs. 9.2B and 9.8B). The right vagus nerve passes anterior to the first part of the subclavian artery and posterior to the brachiocephalic vein and SC joint to enter the thorax (Fig. 9.11C,D). The left vagus nerve descends between the left common carotid and the left subclavian arteries and posterior to the SC joint to enter the thorax.

The **recurrent laryngeal nerves** arise from the vagus nerves in the inferior part of the neck. The nerves of the two sides have essentially the same distribution; however, they arise and recur (loop around) different structures and at different levels on the two sides. The **right recurrent laryngeal nerve** loops inferior to the right subclavian artery (Fig. 9.11C), and the **left recurrent laryngeal nerve**

loops inferior to the arch of the aorta (see Fig. 9.13B). After looping, both recurrent laryngeal nerves ascend superiorly to the posteromedial aspect of the thyroid gland, where they ascend in the **tracheo-esophageal groove** (see Fig. 9.13A), supplying both the trachea and esophagus and all the intrinsic muscles of the larynx except the cricothyroid.

The **cardiac branches of CN X** originate in the neck as well as in the thorax and convey presynaptic parasympathetic and visceral afferent fibers to the cardiac plexus of nerves.

SYMPATHETIC TRUNKS

The **cervical portion of the sympathetic trunks** lies anterolateral to the vertebral column, extending superiorly to the level of the C1 vertebra or the cranial base (Fig. 9.11C,D). The sympathetic trunks receive no white rami communicantes (communicating branches) in the neck. The cervical portion of the trunks contains three **cervical sympathetic ganglia**: superior, middle, and inferior. These ganglia receive presynaptic fibers conveyed to the sympathetic trunk by the superior thoracic spinal nerves and their associated white rami communicantes, which then ascend through the sympathetic trunk to the ganglia. After synapsing with the postsynaptic neuron in the cervical sympathetic ganglia, postsynaptic neurons send fibers to the following structures:

- Cervical spinal nerves via *gray* rami communicantes
- Thoracic viscera via *cardiopulmonary splanchnic nerves*
- Head and viscera of the neck via *cephalic arterial branches*, which accompany arteries (especially the vertebral and internal and external carotid arteries) as the *sympathetic peri-arterial plexuses*

The **inferior cervical ganglion** usually fuses with the first thoracic ganglion to form the **cervicothoracic ganglion (stellate ganglion)**. This star-shaped (L. *stella*, a star) ganglion lies anterior to the transverse process of the C7 vertebra, just superior to the neck of the 1st rib on each side and posterior to the origin of the vertebral artery. Some postsynaptic fibers from the ganglion pass via gray rami communicantes to the anterior rami of the C7 and C8 spinal nerves. Other fibers pass to the heart via the **inferior cervical cardiac nerve** (a cardiopulmonary splanchnic nerve), which passes along the trachea to the deep cardiac plexus. Other fibers pass via arterial branches to contribute to the sympathetic peri-arterial nerve plexus around the vertebral artery running into the cranial cavity.

The **middle cervical ganglion**, usually small and occasionally absent, lies on the anterior aspect of the inferior thyroid artery at the level of the cricoid cartilage and the transverse process of the C6 vertebra, just anterior to the vertebral artery. Postsynaptic fibers pass from the ganglion via gray rami communicantes to the anterior rami of the C5 and C6 spinal nerves, via a **middle cervical cardiac nerve** (cardiopulmonary splanchnic nerve) to the heart and via arterial branches to form peri-arterial plexuses to the thyroid gland.

The **superior cervical ganglion** is at the level of the C1 and C2 vertebrae. Because of its large size, it forms a good landmark for locating the sympathetic trunk. Postsynaptic fibers pass from it by means of cephalic arterial branches, forming the internal carotid sympathetic plexus that enters the cranial cavity with the artery (Fig. 9.11D). This ganglion also sends arterial branches to the external carotid artery and gray rami communicantes to the anterior rami of the superior four cervical spinal nerves. Other postsynaptic fibers pass from it to the cardiac plexus of nerves via a **superior cervical cardiac nerve**.

CLINICAL BOX

Cervicothoracic Ganglion Block

 Anesthetic injected around the cervicothoracic ganglion blocks transmission of stimuli through the cervical and superior thoracic ganglia. This ganglion block may relieve vascular spasms involving the brain and upper limb. It is also useful when deciding if *surgical resection* (removal) of the ganglion would be beneficial to a person with *excess vasoconstriction of the ipsilateral limb*.

Lesion of Cervical Sympathetic Trunk

A *lesion of a sympathetic trunk in the neck* results in a sympathetic disturbance called *Horner syndrome*, which is characterized by the following signs and symptoms:

- *Pupillary constriction*, resulting from paralysis of the dilator pupillae muscle
- *Ptosis* (drooping of the superior eyelid), resulting from paralysis of the smooth (tarsal) muscle intermingled with striated muscle of the levator palpebrae superioris
- Sinking in of the eyeball (*enophthalmos*), possibly caused by paralysis of smooth (orbitalis) muscle in the floor of the orbit
- Vasodilation and absence of sweating on the face and neck (*anhydrosis*), caused by a lack of sympathetic (vasoconstrictive) nerve supply to the blood vessels and sweat glands

VISCERA OF NECK

The cervical viscera (organs) are organized in three layers, named for their primary function (Fig. 9.12). Superficial to deep, they are the *endocrine layer* (thyroid and parathyroid glands), the *respiratory layer* (larynx and trachea), and the *alimentary layer* (pharynx and esophagus).

Endocrine Layer of Cervical Viscera

The viscera of **the endocrine layer** are part of the body's endocrine system of ductless, hormone-secreting glands. The *thyroid gland* produces *thyroid hormone*, which controls the rate of metabolism, and *calcitonin*, a hormone controlling calcium metabolism. The *parathyroid glands* produce

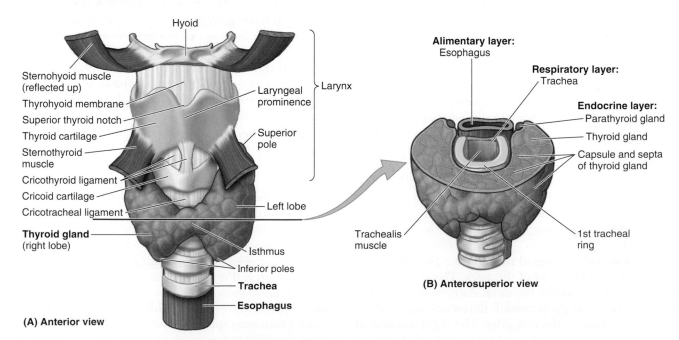

FIGURE 9.12. Functional layers of cervical viscera.

parathormone (PTH), which controls the metabolism of phosphorus and calcium in the blood.

THYROID GLAND

The **thyroid gland** is located anteriorly in the neck. It lies deep to the sternothyroid and sternohyoid muscles from the level of the C5–T1 vertebrae (see Fig. 9.2A,B). It consists primarily of **right** and **left lobes**, anterolateral to the larynx and trachea. A relatively thin **isthmus** unites the lobes over the trachea, usually anterior to the second and third tracheal rings (Fig. 9.12). The thyroid gland is surrounded by a thin **fibrous capsule**, which sends septa deeply into the gland. Dense connective tissue attaches the fibrous capsule to the cricoid cartilage and superior tracheal rings. External to the capsule is a loose *fascial sheath* formed by the visceral portion of the pretracheal layer of deep cervical fascia.

The rich *blood supply of the thyroid gland* is from the paired *superior* and *inferior thyroid arteries* (Figs. 9.13 and 9.14). These vessels lie between the fibrous capsule and the loose fascial sheath. Usually, the first branches of the external carotid artery, the **superior thyroid arteries**, descend to the superior poles of the gland, pierce the pretracheal layer of deep cervical fascia, and divide into anterior and posterior branches. The **inferior thyroid arteries**, the largest branches of the thyrocervical trunks, arising from the subclavian arteries, run superomedially posterior to the carotid sheaths to reach the posterior aspect of the thyroid gland. The right and left superior and inferior thyroid arteries anastomose extensively within the gland, ensuring its supply while providing potential collateral circulation between the subclavian and the external carotid arteries.

In approximately 10% of people, a **thyroid ima artery** (L. *arteria thyroidea ima*) arises from the brachiocephalic trunk; the arch of the aorta; or from the right common carotid, subclavian, or internal thoracic arteries (Fig. 9.13B). This small artery ascends on the anterior surface of the trachea, which it supplies, and continues to the isthmus of the thyroid gland. The possible presence of this artery must be considered when performing procedures in the midline of the neck inferior to the isthmus because it is a potential source of bleeding.

Three pairs of thyroid veins usually drain the **thyroid plexus of veins** on the anterior surface of the thyroid gland and trachea (Fig. 9.13B). The **superior thyroid veins** accompany the superior thyroid arteries and drain the superior poles of the gland. The **middle thyroid veins** drain the middle of the lobes, and the **inferior thyroid veins** drain the inferior poles. The superior and middle thyroid veins drain into the IJVs, and the inferior thyroid veins drain into the brachiocephalic veins posterior to the manubrium.

The *lymphatic vessels of the thyroid gland* communicate with a capsular network of lymphatic vessels. From this network, the vessels pass initially to **prelaryngeal**,

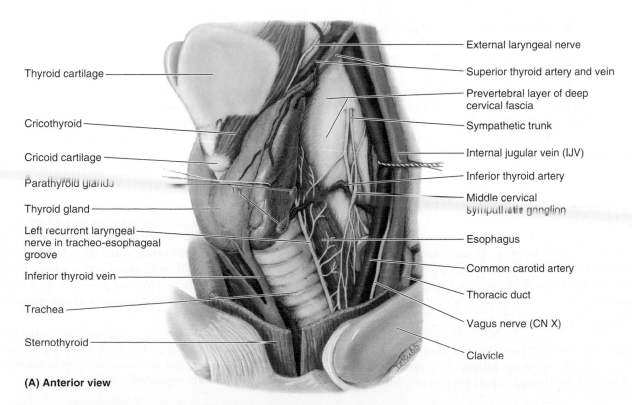

Thyroid cartilage
Cricothyroid
Cricoid cartilage
Parathyroid glands
Thyroid gland
Left recurrent laryngeal nerve in tracheo-esophageal groove
Inferior thyroid vein
Trachea
Sternothyroid

External laryngeal nerve
Superior thyroid artery and vein
Prevertebral layer of deep cervical fascia
Sympathetic trunk
Internal jugular vein (IJV)
Inferior thyroid artery
Middle cervical sympathetic ganglion
Esophagus
Common carotid artery
Thoracic duct
Vagus nerve (CN X)
Clavicle

(A) Anterior view

FIGURE 9.13. Thyroid and parathyroid glands and larynx. A. Dissection of left side of root of neck. *(continued)*

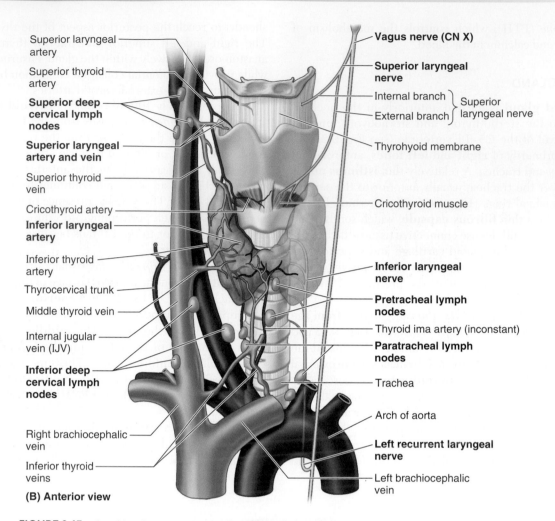

Superior laryngeal
artery

Superior thyroid
artery

**Superior deep
cervical lymph
nodes**

**Superior laryngeal
artery and vein**

Superior thyroid
vein

Cricothyroid artery

**Inferior laryngeal
artery**

Inferior thyroid
artery

Thyrocervical trunk

Middle thyroid vein

Internal jugular
vein (IJV)

**Inferior deep
cervical lymph
nodes**

Right brachiocephalic
vein

Inferior thyroid
veins

(B) Anterior view

Vagus nerve (CN X)

**Superior laryngeal
nerve**

Internal branch ⎤ Superior
                ⎬ laryngeal
External branch ⎦ nerve

Thyrohyoid membrane

Cricothyroid muscle

**Inferior laryngeal
nerve**

**Pretracheal lymph
nodes**

Thyroid ima artery (inconstant)

**Paratracheal lymph
nodes**

Trachea

Arch of aorta

**Left recurrent laryngeal
nerve**

Left brachiocephalic
vein

FIGURE 9.13. **Thyroid and parathyroid glands and larynx.** *(continued)* **B.** Vessels, nerves, and lymph nodes of larynx.

pretracheal, and **paratracheal lymph nodes**, which drain in turn to the **superior** and **inferior deep cervical nodes** (Fig. 9.14B). Inferior to the thyroid gland, the lymphatic vessels pass directly to the *inferior deep cervical lymph nodes*. Some lymphatic vessels may drain into *brachiocephalic lymph nodes* or the *thoracic duct*.

The *nerves of the thyroid gland* are derived from the superior, middle, and inferior *cervical sympathetic ganglia* (Fig. 9.13A). They reach the gland through the *cardiac and superior and inferior thyroid peri-arterial plexuses* that accompany the thyroid arteries. These fibers are vasomotor, causing constriction of blood vessels. Endocrine secretion from the thyroid gland is hormonally regulated by the pituitary gland.

PARATHYROID GLANDS

The small, flattened oval **parathyroid glands** lie external to the fibrous capsule on the medial half of the posterior surface of each lobe of the thyroid gland (Fig. 9.14A). Most people

have four parathyroid glands. Approximately 5% of people have more; some have only two glands. The two **superior parathyroid glands** are usually at the level of the inferior border of the cricoid cartilage. The **inferior parathyroid glands** are usually near the inferior poles of the thyroid gland, but they may lie in a variety of positions.

The *inferior thyroid arteries* supply both the superior and the inferior parathyroid glands; however, these glands may also receive branches from the superior thyroid arteries, the thyroid ima artery, or the laryngeal, tracheal, and esophageal arteries. The **parathyroid veins** drain into the *thyroid plexus of veins* of the thyroid gland and trachea. The *lymphatic vessels from the parathyroid glands* drain with those of the thyroid gland into the deep cervical and paratracheal lymph nodes (Fig. 9.14B).

The *nerves of the parathyroid glands* are derived from *thyroid branches of the cervical sympathetic ganglia*. The nerves are vasomotor but not secretomotor because these glands are hormonally regulated.

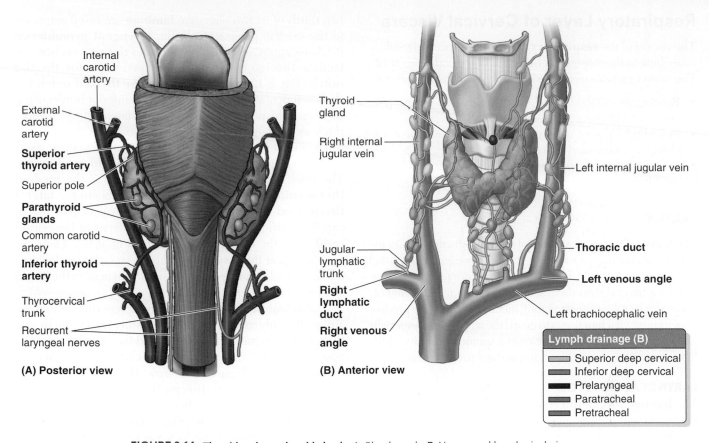

Internal carotid artery

External carotid artery

Superior thyroid artery

Superior pole

Parathyroid glands

Common carotid artery

Inferior thyroid artery

Thyrocervical trunk

Recurrent laryngeal nerves

(A) Posterior view

Thyroid gland

Right internal jugular vein

Left internal jugular vein

Jugular lymphatic trunk

Right lymphatic duct

Right venous angle

Thoracic duct

Left venous angle

Left brachiocephalic vein

(B) Anterior view

Lymph drainage (B)	
	Superior deep cervical
	Inferior deep cervical
	Prelaryngeal
	Paratracheal
	Pretracheal

FIGURE 9.14. Thyroid and parathyroid glands. A. Blood supply. **B.** Venous and lymphatic drainage.

CLINICAL BOX

Thyroidectomy

During a *thyroidectomy* (e.g., excision of a malignant thyroid gland), the parathyroid glands are in danger of being inadvertently damaged or removed. These glands are safe during *subtotal thyroidectomy* because the most posterior part of the thyroid gland usually is preserved. Variability in the position of the parathyroid glands, especially the inferior ones, puts them in danger of being removed during surgery on the thyroid gland. If the parathyroid glands are inadvertently removed during surgery, the patient suffers from *tetany*, a severe convulsive disorder. The generalized convulsive muscle spasms result from a fall in blood calcium levels. Hormone replacement therapy is required.

Accessory Thyroid Tissue

Accessory thyroid tissue may develop in the neck lateral to the thyroid cartilage (Fig. B9.5); usually, the tissue lies on the thyrohyoid muscle. A **pyramidal lobe**, an extension of thyroid tissue from the superior aspect of the isthmus, and its connective tissue continuation may also contain thyroid tissue. Accessory thyroid tissue, like that of a pyramidal lobe, originates from remnants of the

thyroglossal duct—a transitory endodermal tube extending from the posterior tongue region of the embryo carrying the thyroid-forming tissue at its descending distal end. Although the accessory tissue may be functional, it is usually too small to maintain normal function if the thyroid gland is removed.

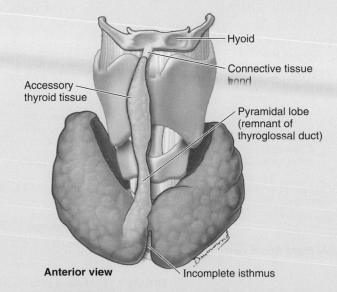

Hyoid

Connective tissue band

Accessory thyroid tissue

Pyramidal lobe (remnant of thyroglossal duct)

Anterior view

Incomplete isthmus

FIGURE B9.5. Accessory thyroid glandular tissue.

Respiratory Layer of Cervical Viscera

The viscera of the **respiratory layer**, the *larynx* and *trachea*, contribute to the respiratory functions of the body (Fig. 9.12). The main functions of the respiratory viscera are as follows:

- Routing air and food into the respiratory tract and esophagus, respectively
- Providing a patent airway and an active valve for it, enabling it to be sealed off temporarily
- Producing tone for the mouth (tongue, teeth, and lips) to modify into voice

LARYNX

The **larynx**, the complex organ of voice production, lies in the anterior part of the neck at the level of the bodies of the C3–C6 vertebrae (see Fig. 9.1). It connects the inferior part of the pharynx (oropharynx) with the trachea. Although most commonly known for its role as the phonating mechanism for voice production, its most vital function is to guard the air passages, especially during swallowing, when it serves as a sphincter or valve of the lower respiratory tract, thus maintaining a patent airway.

LARYNGEAL SKELETON

The **laryngeal skeleton** consists of nine cartilages joined by ligaments and membranes (Fig. 9.15). Three cartilages are single (thyroid, cricoid, and epiglottic) and three are paired (arytenoid, corniculate, and cuneiform).

The **thyroid cartilage** is the largest of the cartilages. Its superior border lies opposite the C4 vertebra. The inferior two thirds of its two plate-like **laminae** are fused anteriorly in the median plane to form the **laryngeal prominence** ("Adam's apple" of males). Superior to this prominence, the laminae diverge to form the V-shaped **superior thyroid notch** (Fig. 9.12A). The small **inferior thyroid notch** is a shallow indentation in the middle of the inferior border of the cartilage. The posterior border of each lamina projects superiorly as the **superior horn** and inferiorly as the **inferior horn** (Fig. 9.15A). The superior border and superior horns attach to the hyoid bone by the **thyrohyoid membrane**. The thick median part of this membrane is the **median thyrohyoid ligament**, and its lateral parts are the **lateral thyrohyoid ligaments**. The inferior horns of the thyroid cartilages articulate with the lateral surfaces of the cricoid cartilage at the **cricothyroid joints** (Fig. 9.15). The main movements at these synovial joints are rotation and gliding of the thyroid cartilage, which result in changes in the length and tension of the vocal folds.

The **cricoid cartilage** forms a complete ring around the airway, the only cartilage of the respiratory tract to do so. It is shaped like a signet ring with its band facing anteriorly. The posterior (signet) part of the cricoid cartilage is the *lamina*; the anterior (band) part is the *arch*. The cricoid cartilage is smaller but thicker and stronger than the thyroid cartilage. The cricoid cartilage is attached to the inferior margin of the thyroid cartilage by the **median cricothyroid ligament** and to the first tracheal ring by the **cricotracheal ligament** (Fig. 9.15). Where the larynx is closest to the skin and most accessible, the median cricothyroid ligament may be felt as a soft spot during palpation inferior to the thyroid cartilage.

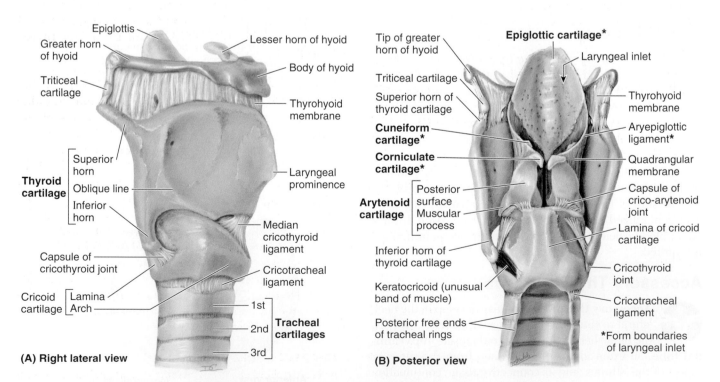

FIGURE 9.15. Skeleton of larynx and associated ligaments and membranes.

SURFACE ANATOMY

Larynx

The U-shaped hyoid bone lies superior to the thyroid cartilage at the level of the C4 and C5 vertebrae (Fig. SA9.2). The laryngeal prominence is produced by the fused *laminae of the thyroid cartilage*, which meet in the median plane. The cricoid cartilage can be felt inferior to the laryngeal prominence. It lies at the level of the C6 vertebra. The cartilaginous **tracheal rings** are palpable in the inferior part of the neck. The 2nd through 4th rings cannot be felt because the **isthmus of the thyroid**, connecting its right and left lobes, covers them. The first tracheal ring is just superior to the isthmus.

Key			
C	Cricoid cartilage	**RL**	Right lobe of thyroid gland
H	Hyoid	**S**	Isthmus
IP	Inferior pole of lobe	**SP**	Superior pole of lobe
LL	Left lobe of thyroid gland	**T**	Thyroid cartilage
P	Laryngeal prominence	✴	Tracheal rings

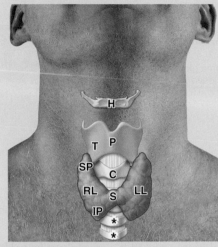

Anterior view

FIGURE SA9.2. Surface anatomy of larynx and thyroid gland.

The **arytenoid cartilages** are paired, three-sided pyramidal cartilages that articulate with lateral parts of the superior border of the cricoid cartilage lamina. Each cartilage has an apex superiorly, a vocal process anteriorly, and a large muscular process that projects laterally from its base (Fig. 9.15B). The **apex** of each arytenoid cartilage bears the corniculate cartilage and attaches to the aryepiglottic fold. The **vocal process** provides the posterior attachment for the vocal ligament (see Figs. 9.17 and 9.18A), and the **muscular process** serves as a lever to which the posterior and lateral crico-arytenoid muscles are attached.

The **crico-arytenoid joints**, located between the bases of the arytenoid cartilages and the superolateral surfaces of the lamina of the cricoid cartilage, permit the arytenoid cartilages to slide toward or away from one another, to tilt anteriorly and posteriorly, and to rotate. These movements are important in approximating, tensing, and relaxing the vocal folds. The elastic **vocal ligaments** extend from the junction of the laminae of the thyroid cartilage anteriorly to the vocal process of the arytenoid cartilage posteriorly (Figs. 9.16 and 9.17). The vocal ligaments form the submucosal skeleton of the vocal folds. The vocal ligaments are the thickened, free superior border of the **conus elasticus** or **cricovocal membrane** (Fig. 9.18A). The parts of the cricovocal membrane, extending laterally between the vocal folds and the superior border of the cricoid, are the **lateral cricothyroid ligaments**. The fibro-elastic conus elasticus blends anteriorly with the *median cricothyroid ligament*. The *conus elasticus* and overlying mucosa close the tracheal inlet, except for the central **rima glottidis** (aperture between vocal folds).

The **epiglottic cartilage**, consisting of elastic cartilage, gives flexibility to the **epiglottis** (Figs. 9.17 and 9.18A). It is a heart-shaped cartilage covered with mucous membrane. Situated posterior to the root of the tongue and the hyoid bone and anterior to the **laryngeal inlet**, the epiglottic cartilage forms the superior part of the anterior wall and the superior margin of the inlet. Its broad superior end is free; its tapered inferior end, the **stalk of the epiglottis**, is attached to the angle formed by the thyroid laminae and the **thyro-epiglottic ligament** (Fig. 9.18A).

The **hyo-epiglottic ligament** attaches the anterior surface of the epiglottic cartilage to the hyoid bone. A thin submucosal sheet of connective tissue, the **quadrangular membrane**, extends between the lateral aspects of the arytenoid and epiglottic cartilages (Fig. 9.17). Its free inferior margin constitutes the **vestibular ligament**, which is covered loosely by mucosa to form the **vestibular fold** (Figs. 9.16 and 9.17). This fold lies superior to the vocal fold and extends from the thyroid cartilage to the arytenoid cartilage. The free superior margin of the quadrangular membrane forms the **aryepiglottic ligament**, which is covered with mucosa to form the **aryepiglottic fold**.

The **corniculate and cuneiform cartilages** appear as small nodules in the posterior part of the aryepiglottic folds (Figs. 9.15 and 9.17). The *corniculate cartilages* attach to the apices of the arytenoid cartilages; the *cuneiform cartilages* do not directly attach to other cartilages.

INTERIOR OF THE LARYNX

The **laryngeal cavity** extends from the *laryngeal inlet*, through which it communicates with the *laryngopharynx*, to the level

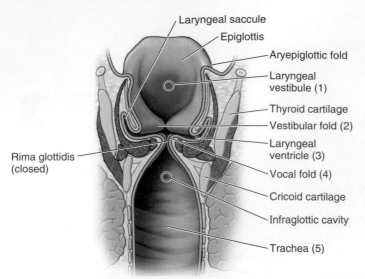

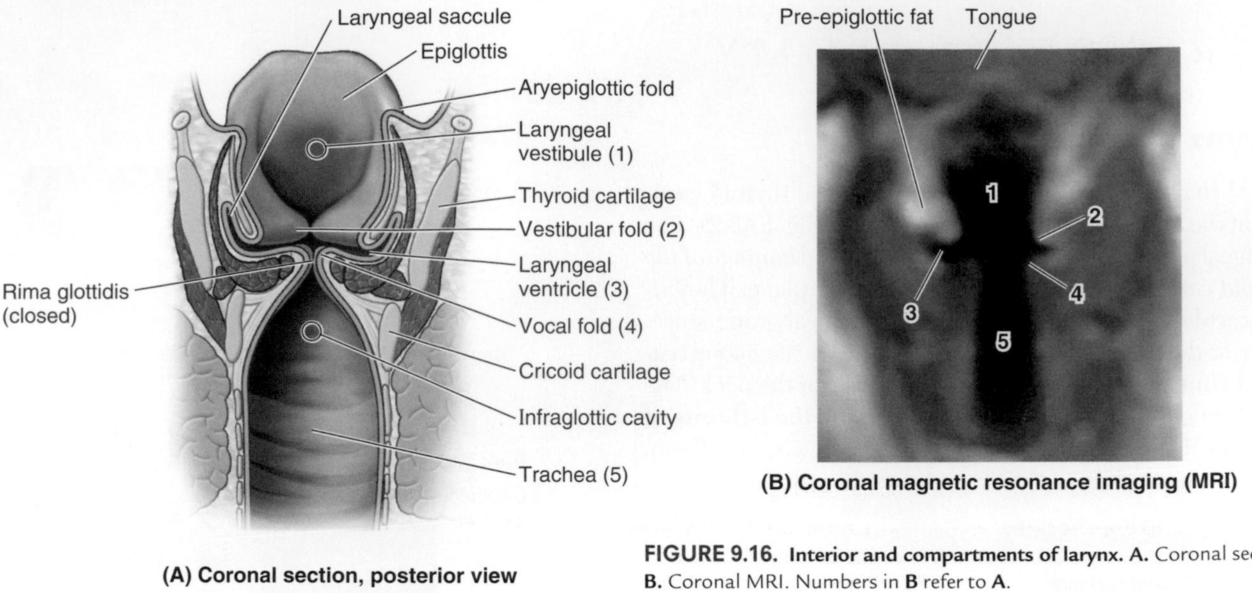

(A) Coronal section, posterior view

Labels on image A:
Laryngeal saccule
Epiglottis
Aryepiglottic fold
Laryngeal vestibule (1)
Thyroid cartilage
Vestibular fold (2)
Laryngeal ventricle (3)
Vocal fold (4)
Cricoid cartilage
Infraglottic cavity
Trachea (5)
Rima glottidis (closed)

(B) Coronal magnetic resonance imaging (MRI)

Labels on image B:
Pre-epiglottic fat   Tongue

FIGURE 9.16. **Interior and compartments of larynx. A.** Coronal section. **B.** Coronal MRI. Numbers in **B** refer to **A**.

of the inferior border of the cricoid cartilage. Here, the laryngeal cavity is continuous with the lumen of the trachea. The laryngeal cavity includes the following structures (Fig. 9.16):

- **Laryngeal vestibule**, between the laryngeal inlet and vestibular folds
- **Middle part of laryngeal cavity**, the central cavity (airway) between the vestibular and vocal folds

- **Laryngeal ventricle**, recesses extending laterally from the middle part of the laryngeal cavity between vestibular and vocal folds. The **laryngeal saccule** is a blind pocket opening into each ventricle that is lined with mucosal glands.
- **Infraglottic cavity**, the inferior cavity of the larynx between the vocal folds and the inferior border of the cricoid cartilage, where it is continuous with the lumen of the trachea

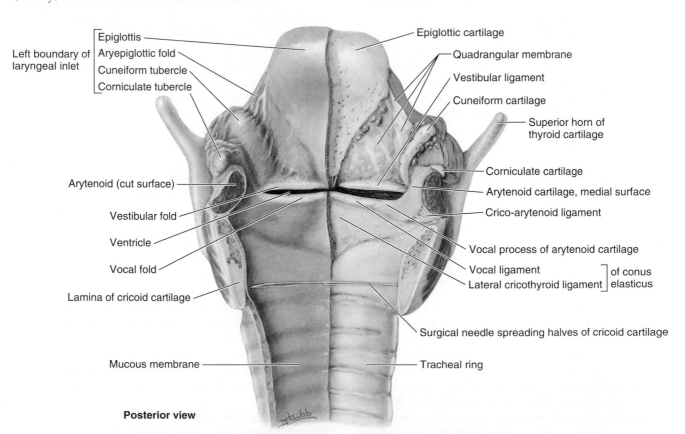

Labels on image:
Left boundary of laryngeal inlet
Epiglottis
Aryepiglottic fold
Cuneiform tubercle
Corniculate tubercle
Epiglottic cartilage
Quadrangular membrane
Vestibular ligament
Cuneiform cartilage
Superior horn of thyroid cartilage
Corniculate cartilage
Arytenoid (cut surface)
Arytenoid cartilage, medial surface
Vestibular fold
Crico-arytenoid ligament
Ventricle
Vocal process of arytenoid cartilage
Vocal fold
Vocal ligament ⎤ of conus
Lamina of cricoid cartilage
Lateral cricothyroid ligament ⎦ elasticus
Surgical needle spreading halves of cricoid cartilage
Mucous membrane
Tracheal ring

Posterior view

FIGURE 9.17. **Interior of larynx.** The posterior wall of the larynx is split in the median plane, and the two sides are separated.

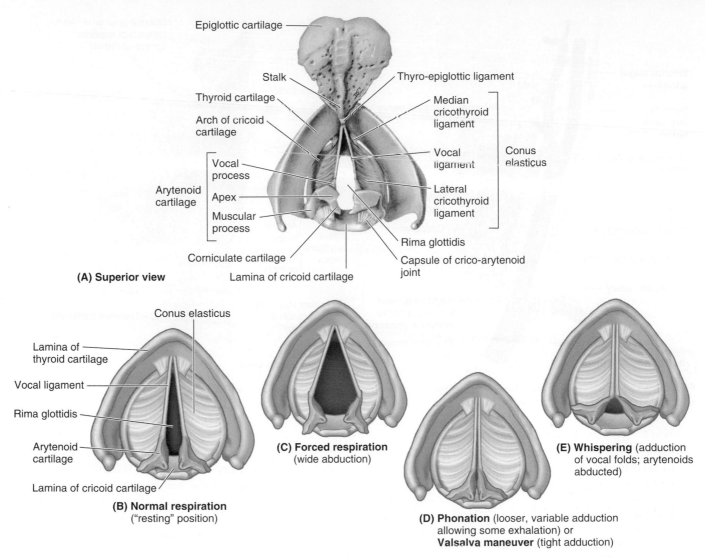

FIGURE 9.18. Rima glottidis. A. Conus elasticus. **B–E.** Variation in shape of rima glottidis. The shape of the rima glottidis varies according to the position of the vocal folds.

The **vocal folds** (true vocal cords) control sound production. The apex of each wedge-shaped fold projects medially into the laryngeal cavity (Figs. 9.16–9.18). Each fold contains a

- *Vocal ligament*, consisting of thickened elastic tissue that is the medial free edge of the conus elasticus
- **Vocalis muscle**, composed of exceptionally fine muscle fibers immediately lateral to and terminating at intervals relative to the length of the vocal ligaments (Table 9.6)

The *vocal folds* are the source of sounds (tone) that come from the larynx. The vocal folds produce audible vibrations when their free margins are closely (but not tightly) apposed during phonation, and air is forcibly expired intermittently. The vocal folds also serve as the main inspiratory sphincter of the larynx when they are tightly closed. Complete adduction of the folds forms an effective sphincter that prevents entry of air.

The **glottis** (vocal apparatus of the larynx) makes up the vocal folds and processes, together with the *rima glottidis*. The shape of the rima (L. slit) varies according to the position of the vocal folds. During ordinary breathing, the rima is narrow and wedge shaped (Fig. 9.18B); during forced respiration it is wide and kite-shaped (Fig. 9.18C). The rima glottidis is slit-like when the vocal folds are closely approximated during phonation (Fig. 9.18D). Variation in the tension and length of the vocal folds, in the width of the rima glottidis, and in the intensity of the expiratory effort produces changes in the pitch of the voice. The lower range of pitch of the voice of postpubertal males results from the increased laryngeal prominence resulting in greater length of the vocal folds.

The vestibular folds (false vocal cords), extending between the posterior aspect of the laryngeal prominence and arytenoid cartilages (Figs. 9.16 and 9.17), play little or no part in voice production. They are protective in function. They consist of two thick folds of mucous membrane enclosing the

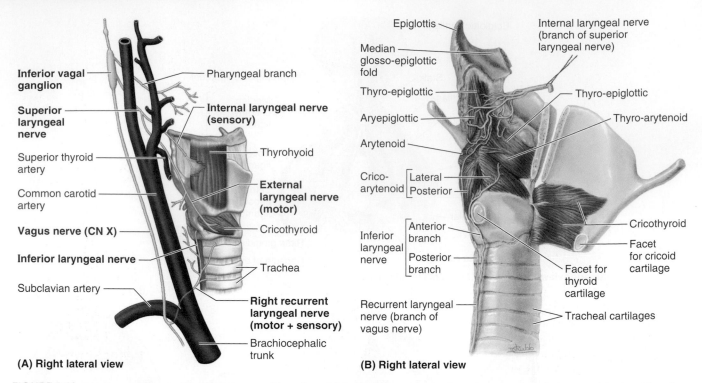

FIGURE 9.19. **Muscles and nerves of larynx. A.** Laryngeal branches of right vagus nerve. **B.** Muscles and nerves. The right lamina of the thyroid cartilage is turned anteriorly (like opening a book).

vestibular ligaments. The space between these ligaments is the *rima vestibuli.* The lateral recesses between the vocal and the vestibular folds are the laryngeal ventricles.

MUSCLES OF LARYNX

The laryngeal muscles are divided into extrinsic and intrinsic groups:

- The **extrinsic laryngeal muscles** move the entire larynx (Table 9.4). The *infrahyoid muscles* are depressors of the hyoid bone and larynx, whereas the *suprahyoid* and *stylopharyngeus muscles* are elevators of the hyoid and larynx.
- The **intrinsic laryngeal muscles** move the laryngeal parts, making alterations in the length and tension of the vocal folds and in the size and shape of the rima glottidis. All but one of the intrinsic muscles of the larynx are supplied by the *recurrent laryngeal nerve* (Fig. 9.19), a branch of CN X. The cricothyroid muscle is supplied by the external laryngeal nerve, one of the two terminal branches of the *superior laryngeal nerve* (Fig. 9.19). The actions of the intrinsic laryngeal muscles are illustrated in Figure 9.20 and described in Table 9.6.

VESSELS OF THE LARYNX

The *laryngeal arteries,* branches of the superior and inferior thyroid arteries, supply the larynx (Fig. 9.13B). The superior laryngeal artery accompanies the internal branch of the superior laryngeal nerve through the thyrohyoid membrane and branches to supply the internal surface of

the larynx. The inferior laryngeal artery, a branch of the inferior thyroid artery, accompanies the inferior laryngeal nerve (terminal part of the recurrent laryngeal nerve) and supplies the mucous membrane and muscles in the inferior part of the larynx.

The *laryngeal veins* accompany the laryngeal arteries (Fig. 9.13B). The **superior laryngeal vein** usually joins the superior thyroid vein and through it drains into the IJV. The **inferior laryngeal vein** joins the inferior thyroid vein or the thyroid plexus of veins on the anterior aspect of the trachea, which empties into the left brachiocephalic vein.

The *lymphatic vessels of the larynx* superior to the vocal folds accompany the superior laryngeal artery through the thyrohyoid membrane and drain into the *superior deep cervical lymph nodes* (Fig. 9.14B). The lymphatic vessels inferior to the vocal folds drain into the *pretracheal* or *paratracheal lymph nodes,* which drain into the inferior deep cervical lymph nodes.

NERVES OF THE LARYNX

The *nerves of the larynx* are the *superior and inferior laryngeal branches of the vagus nerve* (Fig. 9.19). The superior laryngeal nerve arises from the inferior vagal ganglion and divides into two terminal branches within the carotid sheath: the *internal laryngeal nerve* (sensory and autonomic) and the *external laryngeal nerve* (motor).

The **internal laryngeal nerve,** the larger terminal branch of the superior laryngeal nerve, pierces the thyrohyoid membrane with the superior laryngeal artery, supplying

(Continued on page 628)

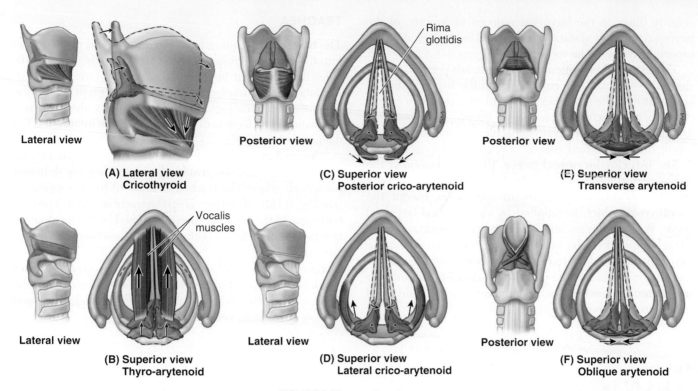

Lateral view

(A) Lateral view
Cricothyroid

Posterior view

Rima glottidis

(C) Superior view
Posterior crico-arytenoid

Posterior view

(E) Superior view
Transverse arytenoid

Vocalis muscles

Lateral view

(B) Superior view
Thyro-arytenoid

Lateral view

(D) Superior view
Lateral crico-arytenoid

Posterior view

(F) Superior view
Oblique arytenoid

FIGURE 9.20. Muscles of larynx.

TABLE 9.6. MUSCLES OF THE LARYNX

Muscle	Origin	Insertion	Innervation	Main Action(s)
Cricothyroid	Anterolateral part of cricoid cartilage	Inferior margin and inferior horn of thyroid cartilage	External laryngeal nerve (from CN X)	Stretches and tenses vocal ligament
Thyro-arytenoid[a]	Lower half of posterior aspect of angle of thyroid laminae and cricothyroid ligament	Anterolateral arytenoid surface	Inferior laryngeal nerve (terminal part of recurrent laryngeal nerve, from CN X)	Relaxes vocal ligament
Posterior crico-arytenoid	Posterior surface of lamina of cricoid cartilage	Vocal process of arytenoid cartilage		Abducts vocal folds
Lateral crico-arytenoid	Arch of cricoid cartilage			Adducts vocal folds (interligamentous portion)
Transverse and oblique arytenoids[b]	One arytenoid cartilage	Contralateral arytenoid cartilage		Adduct arytenoid cartilages (adducting intercartilaginous portion of vocal folds, closing posterior rima glottidis)
Vocalis[c]	Lateral surface of vocal process of arytenoid cartilage	Ipsilateral vocal ligament		Relaxes posterior vocal ligament while maintaining (or increasing) tension of anterior part

[a]Superior fibers of the thyro-arytenoid muscles pass into the aryepiglottic fold, and some of them reach the epiglottic cartilage; these fibers constitute the thyro-epiglottic muscle, which widens the laryngeal inlet.
[b]Some fibers of the oblique arytenoid muscles continue as aryepiglottic muscles.
[c]This slender muscle slip lies medial to and is composed of fibers finer than those of the thyro-arytenoid muscle.

sensory fibers to the laryngeal mucous membrane of the laryngeal vestibule and middle laryngeal cavity, including the superior surface of the vocal folds.

The **external laryngeal nerve** descends posterior to the sternothyroid muscle in company with the superior thyroid artery. At first, the nerve lies on the inferior constrictor muscle of the pharynx; it then pierces the muscle, contributing to its innervation (with the pharyngeal plexus), and continues to supply the cricothyroid muscle.

The **inferior laryngeal nerve**, the continuation of the *recurrent laryngeal nerve* (a branch of the vagus nerve), supplies all intrinsic muscles of the larynx except the cricothyroid, which is supplied by the external laryngeal nerve. It also supplies sensory fibers to the mucosa of the infraglottic cavity. The inferior laryngeal nerve enters the larynx by passing deep to the inferior border of the inferior constrictor muscle of the pharynx. It divides into anterior and posterior branches that accompany the inferior laryngeal artery into the larynx.

TRACHEA

The **trachea**, extending from the inferior end of the larynx into the thorax, terminates at the sternal angle, where it divides into the right and left main bronchi (see Fig. 9.13). Deviation of the trachea from the midline often signals the presence of a pathological process. In adults, the trachea is approximately 2.5 cm in diameter, whereas in infants, it is the diameter of a pencil.

The trachea is a fibrocartilaginous tube, supported by incomplete cartilaginous **tracheal rings**. They are deficient posteriorly where the trachea is adjacent to the esophagus (see Fig. 9.12B). The rings keep the trachea patent. The posterior gap in the tracheal rings is spanned by the involuntary **trachealis muscle**, smooth muscle connecting the ends of the tracheal rings.

Lateral to the trachea are the common carotid arteries and lobes of thyroid gland (see Fig. 9.13B). Inferior to the isthmus of the thyroid gland are the jugular venous arch and the inferior thyroid veins.

CLINICAL BOX

Injury to Laryngeal Nerves

The inferior laryngeal nerves are vulnerable to injury during *thyroidectomy* and other surgical operations in the anterior triangles of the neck. Because the inferior laryngeal nerve innervates the muscles moving the vocal fold, injury results in *paralysis of the vocal fold*. The voice is initially poor because the paralyzed fold cannot adduct to meet the normal vocal fold. When bilateral paralysis of the vocal folds occurs, the voice is almost absent because the vocal folds are motionless in a position that is slightly narrower than the usual neutral respiratory position. They cannot be adducted for phonation, nor can they be abducted for increased respiration, resulting in *stridor* (high-pitched, noisy respiration) often accompanied by anxiety. Injury to the external branch of the superior laryngeal nerve results in a voice that is monotonous in character because the paralyzed cricothyroid muscle supplied by it is unable to vary the length and tension of the vocal fold.

Hoarseness is the most common symptom of serious disorders of the larynx, such as carcinoma of the vocal folds.

Fractures of Laryngeal Skeleton

Laryngeal fractures may result from blows received in sports, such as kick boxing and hockey, or from compression by a shoulder strap during an automobile accident. Laryngeal fractures produce submucous hemorrhage and edema, respiratory obstruction, hoarseness, and sometimes a temporary inability to speak. The thyroid, cricoid, and most of the arytenoid cartilages often ossify as age advances, commencing at approximately 25 years of age in the thyroid cartilage.

Aspiration of Foreign Bodies

A foreign object, such as a piece of steak, may accidentally *aspirate* through the laryngeal inlet into the vestibule of the larynx, where it becomes trapped superior to the vestibular folds. When a foreign object enters the vestibule, the laryngeal muscles go into spasm, tensing the vocal folds. The rima glottidis closes and no air enters the trachea. *Asphyxiation* occurs, and the person will die in approximately 5 minutes from lack of oxygen if the obstruction is not removed. Emergency therapy must be given to open the airway.

The procedure used depends on the condition of the patient, the facilities available, and the experience of the person giving first aid. Because the lungs still contain air, sudden compression of the abdomen (*Heimlich maneuver*) causes the diaphragm to elevate and compress the lungs, expelling air from the trachea into the larynx (Fig. B9.6). This maneuver usually dislodges the food or other material from the larynx.

FIGURE B9.6. Heimlich maneuver.

Tracheostomy

A transverse incision through the skin of the neck and anterior wall of the trachea (*tracheostomy*) establishes an airway in patients with upper airway obstruction or respiratory failure. The infrahyoid muscles are retracted laterally, and the isthmus of the thyroid gland is either divided or retracted superiorly. An opening is made in the trachea between the 1st and 2nd tracheal rings or through the 2nd through 4th rings. A *tracheostomy tube* is then inserted into the trachea and secured (Fig. B9.7). To avoid complications during a tracheostomy, the following anatomical relationships are important:

- The inferior thyroid veins arise from a venous plexus on the thyroid gland and descend anterior to the trachea.
- A small thyroid ima artery is present in approximately 10% of people; it ascends from the brachiocephalic trunk or the arch of the aorta to the isthmus of the thyroid gland.
- The left brachiocephalic vein, jugular venous arch, and pleurae may be encountered, particularly in infants and children.
- The thymus covers the inferior part of the trachea in infants and children.
- The trachea is small, mobile, and soft in infants, making it easy to cut through its posterior wall and damage the esophagus.

Laryngoscopy

Laryngoscopy is the procedure used to examine the interior of the larynx. The larynx may be examined visually by *indirect laryngoscopy* using a laryngeal mirror, or it may be viewed by *direct laryngoscopy* using a tubular endoscopic instrument, a *laryngoscope*. The vestibular and vocal folds can be observed (Fig. B9.8).

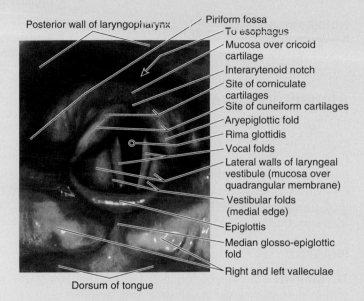

FIGURE B9.8. Laryngoscopic examination.

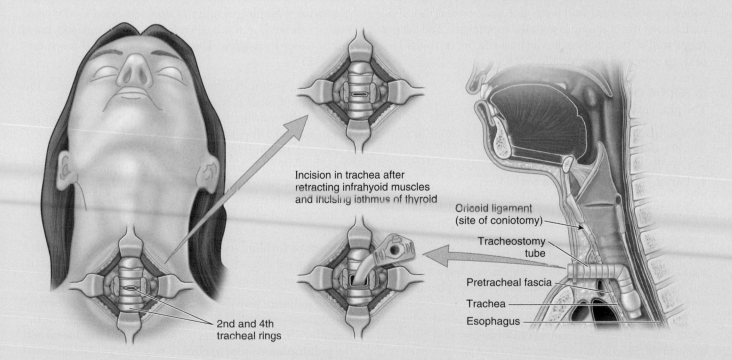

FIGURE B9.7. Tracheostomy.

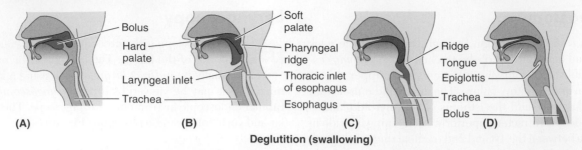

FIGURE 9.21. Swallowing. A. The bolus of food is squeezed to the back of the oral cavity by pushing the tongue against the palate. **B.** The nasopharynx is sealed off and the larynx is elevated, enlarging the pharynx to receive food. **C.** The pharyngeal sphincters contract sequentially, squeezing food into the esophagus. **D.** The bolus of food moves down the esophagus by peristaltic contractions.

Alimentary Layer of Cervical Viscera

The viscera of **the alimentary layer** take part in the digestive functions of the body. Although the *pharynx* conducts air to the larynx, trachea, and lungs, its constrictor muscles direct (and the epiglottis deflects) food to the esophagus (Fig. 9.21). The *esophagus*, also involved in food propulsion, is the beginning of the *alimentary canal* (digestive tract).

PHARYNX

The **pharynx** is the superior expanded part of the alimentary system posterior to the nasal, oral, and laryngeal cavities (Fig. 9.22A). The pharynx extends from the cranial base to the inferior border of the cricoid cartilage anteriorly and the inferior border of C6 vertebra posteriorly. The pharynx is widest opposite the hyoid bone and narrowest at its inferior end, where it is continuous with the esophagus. The flat posterior wall of the pharynx lies against the prevertebral layer of deep cervical fascia (see Fig. 9.2A).

INTERIOR OF PHARYNX

The pharynx is divided into three parts:

- *Nasopharynx*, posterior to the nose and superior to the soft palate
- *Oropharynx*, posterior to the mouth
- *Laryngopharynx*, posterior to the larynx

The **nasopharynx**, the posterior extension of the nasal cavities, has a respiratory function (Fig. 9.22). The nasal cavities open into the nasopharynx through two **choanae** (paired openings between the nasal cavity and nasopharynx). The roof and posterior wall of the nasopharynx form a continuous surface that lies inferior to the body of the sphenoid bone and the basilar part of the occipital bone.

The **pharyngeal tonsils** (commonly called adenoids when enlarged) are concentrations of aggregated lymphoid tissue in the mucous membrane of the roof and posterior wall of the nasopharynx (Fig. 9.22B).

Extending inferiorly from the medial end of the pharyngotympanic tube (auditory tube) is a vertical fold of mucous membrane, the **salpingopharyngeal fold** (Fig. 9.23B).

It covers the salpingopharyngeus muscle (Fig. 9.22C), which opens the pharyngeal orifice of the pharyngotympanic tube during swallowing. The collection of lymphoid tissue in the submucosa of the pharynx near the pharyngeal orifice of the pharyngotympanic tube is the **tubal tonsil** (Fig. 9.23C). Posterior to the **torus** (elevation) **of the pharyngotympanic tube** and the salpingopharyngeal fold is a slit-like lateral extension of the pharynx, the **pharyngeal recess**, which extends laterally and posteriorly (Fig. 9.22B).

The **oropharynx** has a digestive function. It is bounded by the soft palate superiorly, the base of the tongue inferiorly, and the palatoglossal and palatopharyngeal arches laterally (Figs. 9.22 and 9.23). It extends from the soft palate to the superior border of the epiglottis.

Deglutition (swallowing) is the process that transfers a food bolus (masticated morsel) from the mouth through the pharynx and esophagus into the stomach. Solid food is masticated (chewed) and mixed with saliva to form a soft bolus that is easier to swallow. Deglutition occurs in three stages (Fig. 9.21):

- *Stage 1*: Voluntary; the bolus is compressed against the palate and pushed from the mouth into the oropharynx, mainly by coordinated movements of the muscles of the tongue and soft palate.
- *Stage 2*: Involuntary and rapid; the soft palate is elevated, sealing off the nasopharynx from the oropharynx and laryngopharynx. The pharynx widens and shortens to receive the bolus of food as the suprahyoid muscles and longitudinal pharyngeal muscles contract, elevating the larynx.
- *Stage 3*: Involuntary; sequential contraction of all three pharyngeal constrictor muscles forces the food bolus inferiorly into the esophagus.

The **palatine tonsils** are concentrated collections of lymphoid tissue on each side of the oropharynx that lie in the **tonsillar sinus**. The sinus is between the **palatoglossal** and the **palatopharyngeal arches** (Fig. 9.23). The tonsillar bed is formed by the superior constrictor of the pharynx and the thin sheet of **pharyngobasilar fascia** (Fig. 9.22C). This fascia blends with the periosteum of the cranial base and defines the limits of the pharyngeal wall superiorly.

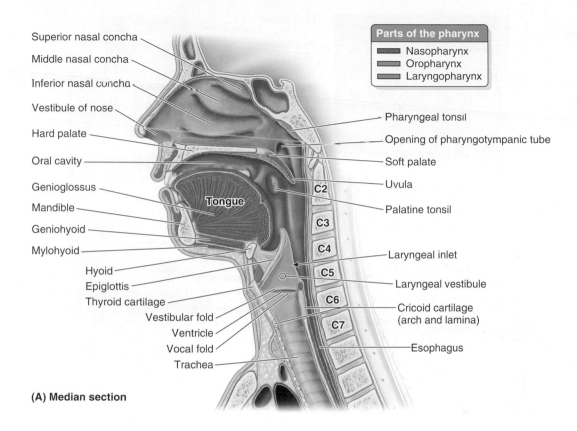

Superior nasal concha
Middle nasal concha
Inferior nasal concha
Vestibule of nose
Hard palate
Oral cavity
Genioglossus
Mandible
Geniohyoid
Mylohyoid
Hyoid
Epiglottis
Thyroid cartilage
Vestibular fold
Ventricle
Vocal fold
Trachea

Tongue

C2
C3
C4
C5
C6
C7

Parts of the pharynx
Nasopharynx
Oropharynx
Laryngopharynx

Pharyngeal tonsil
Opening of pharyngotympanic tube
Soft palate
Uvula
Palatine tonsil
Laryngeal inlet
Laryngeal vestibule
Cricoid cartilage (arch and lamina)
Esophagus

(A) Median section

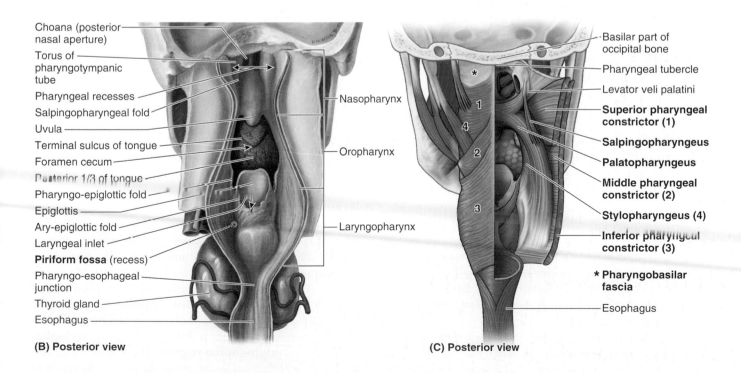

Choana (posterior nasal aperture)
Torus of pharyngotympanic tube
Pharyngeal recesses
Salpingopharyngeal fold
Uvula
Terminal sulcus of tongue
Foramen cecum
Posterior 1/3 of tongue
Pharyngo-epiglottic fold
Epiglottis
Ary-epiglottic fold
Laryngeal inlet
Piriform fossa (recess)
Pharyngo-esophageal junction
Thyroid gland
Esophagus

Nasopharynx
Oropharynx
Laryngopharynx

(B) Posterior view

Basilar part of occipital bone
Pharyngeal tubercle
Levator veli palatini
Superior pharyngeal constrictor (1)
Salpingopharyngeus
Palatopharyngeus
Middle pharyngeal constrictor (2)
Stylopharyngeus (4)
Inferior pharyngeal constrictor (3)
* Pharyngobasilar fascia
Esophagus

(C) Posterior view

FIGURE 9.22. Nasopharynx, oropharynx, and laryngopharynx. A. Parts of pharynx. **B.** Anterior wall of pharynx. The posterior wall has been incised along the midline and spread apart. **C.** Muscles. The posterior wall of the pharynx has been incised in the midline and reflected laterally, and the mucous membrane has been removed from the right side.

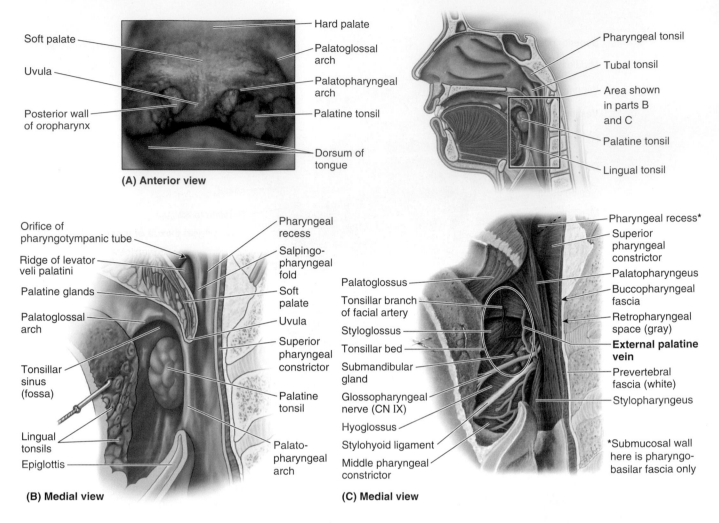

FIGURE 9.23. Oral cavity and tonsils. A. Structures of the oral cavity in an adult male whose mouth is wide open with the tongue protruded. **B.** Internal aspect of the lateral wall of the pharynx showing the palatine tonsil and its relationship to surrounding structures. **C.** Deep dissection of the tonsillar bed.

The **laryngopharynx** (hypopharynx) lies posterior to the larynx, extending from the superior border of the epiglottis and the pharyngo-epiglottic folds to the inferior border of the cricoid cartilage, where it narrows and becomes continuous with the esophagus (Fig. 9.22). Posteriorly, the laryngopharynx is related to the bodies of the C4–C6 vertebrae. Its posterior and lateral walls are formed by the *middle* and *inferior pharyngeal constrictor muscles*. Internally, the wall is formed by the *palatopharyngeus* and *stylopharyngeus muscles* (Fig. 9.22C). The laryngopharynx communicates with the larynx through the *laryngeal inlet* on its anterior wall (Fig. 9.22A).

The **piriform fossa** (recess) is a small depression of the laryngopharyngeal cavity on each side of the inlet (Fig. 9.22B). This mucosa-lined fossa is separated from the laryngeal inlet by the *aryepiglottic fold*. Laterally, the piriform fossa is bounded by the medial surfaces of the thyroid cartilage and the *thyrohyoid membrane*. Branches of the internal laryngeal and recurrent laryngeal nerves lie deep to the mucous membrane of the piriform fossa.

PHARYNGEAL MUSCLES

The wall of the pharynx has a muscular layer composed entirely of voluntary muscle arranged mainly into an external circular and an internal longitudinal layer. In most of the **alimentary canal**, the muscular layer consists of smooth muscle. The external layer consists of three **pharyngeal constrictor muscles**: **superior**, **middle**, and **inferior** (Figs. 9.24 and 9.25). The internal, mainly longitudinal, layer of muscles consists of the **palatopharyngeus**, **stylopharyngeus**, and **salpingopharyngeus**. These muscles elevate the larynx and shorten the pharynx during swallowing and speaking. The attachments, nerve supply, and actions of the pharyngeal muscles are described in Table 9.7.

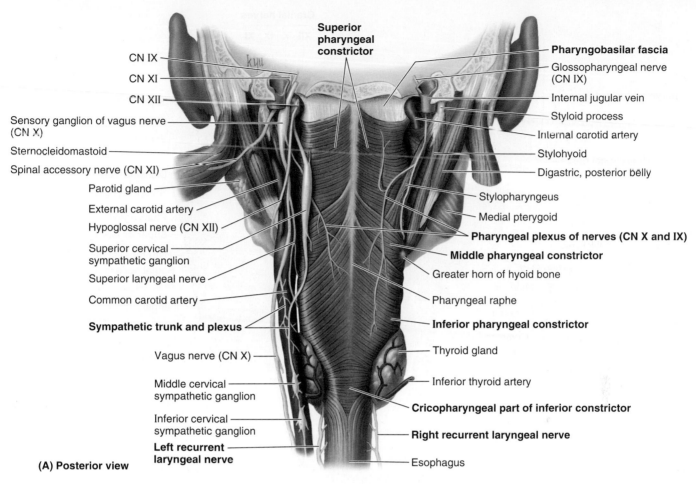

FIGURE 9.24. Pharynx and cranial nerves. A. Overview. *(continued)*

The pharyngeal constrictors have a strong internal fascial lining, the *pharyngobasilar fascia*, and a thin external fascial lining, the *buccopharyngeal fascia*. The pharyngeal constrictors contract involuntarily so that contraction takes place sequentially from the superior to the inferior end of the pharynx, propelling food into the esophagus. All three constrictors are supplied by the *pharyngeal plexus of nerves* that lies on the lateral wall of the pharynx, mainly on the middle constrictor (Fig. 9.24A). The overlapping of the constrictor muscles leaves four gaps in the musculature for structures to enter or leave the pharynx (Fig. 9.25A and Table 9.7):

1. Superior to the superior constrictor, the levator veli palatini, pharyngotympanic tube, and ascending palatine artery pass through the *gap between the superior constrictor and the cranium*. It is here that the pharyngobasilar fascia blends with the buccopharyngeal fascia to form, with the mucous membrane, the thin wall of the pharyngeal recess (Fig. 9.24B).

2. A *gap between the superior and middle pharyngeal constrictors* forms a passageway that allows the stylopharyngeus, glossopharyngeal nerve, and stylohyoid ligament to pass to the internal aspect of the pharyngeal wall.

3. A *gap between the middle and inferior pharyngeal constrictors* allows the internal laryngeal nerve and superior laryngeal artery and vein to pass to the larynx.

4. A *gap inferior to the inferior pharyngeal constrictor* allows the recurrent laryngeal nerve and inferior laryngeal artery to pass superiorly into the larynx.

VESSELS OF PHARYNX

The **tonsillar artery**, a branch of the facial artery (Fig. 9.23C), passes through the superior constrictor muscle and enters the inferior pole of the tonsil. The tonsil also receives arterial twigs from the ascending palatine, lingual, descending palatine, and ascending pharyngeal arteries. The large **external palatine vein** (*paratonsillar vein*) descends from the soft palate and passes close to the lateral surface of the tonsil before it enters the pharyngeal venous plexus.

The **tonsillar lymphatic vessels** pass laterally and inferiorly to the lymph nodes near the angle of the mandible and the **jugulodigastric node** (Fig. 9.26B). The jugulodigastric node is referred to as the *tonsillar node* because of its frequent enlargement when the tonsil is inflamed (*tonsillitis*).

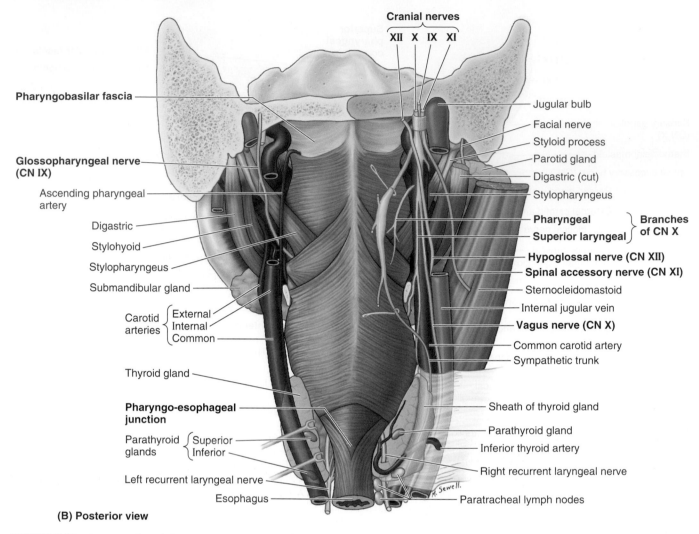

Cranial nerves
XII X IX XI

Pharyngobasilar fascia

Glossopharyngeal nerve (CN IX)

Ascending pharyngeal artery

Digastric

Stylohyoid

Stylopharyngeus

Submandibular gland

Carotid arteries { External / Internal / Common }

Thyroid gland

Pharyngo-esophageal junction

Parathyroid glands { Superior / Inferior }

Left recurrent laryngeal nerve

Esophagus

Jugular bulb

Facial nerve

Styloid process

Parotid gland

Digastric (cut)

Stylopharyngeus

Pharyngeal } Branches of CN X
Superior laryngeal }

Hypoglossal nerve (CN XII)

Spinal accessory nerve (CN XI)

Sternocleidomastoid

Internal jugular vein

Vagus nerve (CN X)

Common carotid artery

Sympathetic trunk

Sheath of thyroid gland

Parathyroid gland

Inferior thyroid artery

Right recurrent laryngeal nerve

Paratracheal lymph nodes

(B) Posterior view

FIGURE 9.24. **Pharynx and cranial nerves.** *(continued)* **B.** Relationships of vessels and nerves. In both **A** and **B**, a large wedge of occipital bone (including the foramen magnum) and the articulated cervical vertebrae have been separated from the remainder (anterior portion) of the head and cervical viscera at the retropharyngeal space and removed.

The palatine, lingual, and pharyngeal tonsils form the pharyngeal **tonsillar ring** (of Waldeyer), an incomplete circular band of lymphoid tissue around the superior part of the pharynx. The antero-inferior part of the ring is formed by the **lingual tonsil**, a collection of lymphoid tissue aggregations in the posterior part of the tongue (Fig. 9.23). Lateral parts of the ring are formed by the *palatine* and *tubal tonsils*, and posterior and superior parts are formed by the *pharyngeal tonsil*.

PHARYNGEAL NERVES

The *nerve supply to the pharynx* (motor and most of sensory) derives from the **pharyngeal plexus of nerves** (Fig. 9.24). Motor fibers in the plexus are derived from the vagus nerve (CN X) via its pharyngeal branch(es). They supply all the muscles of the pharynx and soft palate, except the stylopharyngeus (supplied by CN IX) and the tensor veli palatini (supplied by CN V$_3$). The inferior pharyngeal constrictor also receives some motor fibers from the external and recurrent laryngeal branches of the vagus. Sensory fibers in

the plexus are derived from CN IX. They supply most of the mucosa of all three parts of the pharynx. The sensory nerve supply of the mucous membrane of the anterior and superior nasopharynx is mainly from the maxillary nerve (CN V$_2$). The *tonsillar nerves* are derived from the **tonsillar plexus of nerves**, formed by branches of CN IX and CN X, and the pharyngeal plexus of nerves.

ESOPHAGUS

The **esophagus** is a muscular tube that extends from the laryngopharynx at the **pharyngo-esophageal junction** to the stomach at the cardial orifice (Fig. 9.22A). The esophagus consists of striated (voluntary) muscle in its upper third, smooth (involuntary) muscle in its lower third, and a mixture of striated and smooth muscle in between. Its first part, the **cervical esophagus**, begins at the inferior border of the cricoid cartilage (the level of C6 vertebra) in the median plane.

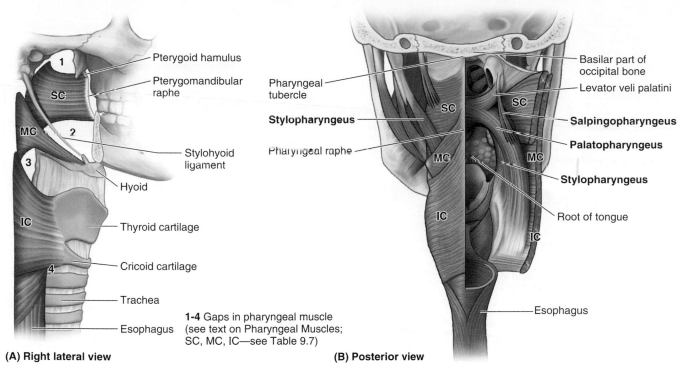

Pterygoid hamulus
Pterygomandibular raphe
Stylohyoid ligament
Hyoid
Thyroid cartilage
Cricoid cartilage
Trachea
Esophagus

Pharyngeal tubercle
Stylopharyngeus
Pharyngeal raphe

Basilar part of occipital bone
Levator veli palatini
Salpingopharyngeus
Palatopharyngeus
Stylopharyngeus
Root of tongue
Esophagus

1-4 Gaps in pharyngeal muscle (see text on Pharyngeal Muscles; SC, MC, IC—see Table 9.7)

(A) Right lateral view   **(B) Posterior view**

FIGURE 9.25. Muscles of pharynx.

TABLE 9.7. MUSCLES OF PHARYNX

Muscle	Origin	Insertion	Innervation	Main Action(s)
External layer				
Superior pharyngeal constrictor (SC)	Pterygoid hamulus, pterygomandibular raphe; posterior end of mylohyoid line of mandible and side of tongue	Pharyngeal tubercle on basilar part of occipital bone	Pharyngeal branch of vagus (CN X) and pharyngeal plexus	Constrict walls of pharynx during swallowing
Middle pharyngeal constrictor (MC)	Stylohyoid ligament and greater and lesser horns of hyoid bone	(Median) pharyngeal raphe	Pharyngeal branch of vagus (CN X) and pharyngeal plexus, plus branches of external and recurrent laryngeal nerves of vagus	
Inferior pharyngeal constrictor (IC)	Oblique line of thyroid cartilage and side of cricoid cartilage	Cricopharyngeal part encircles pharyngo-esophageal junction without forming a raphe.		
Internal layer				
Palatopharyngeus	Hard palate and palatine aponeurosis	Posterior border of lamina of thyroid cartilage and side of pharynx and esophagus	Pharyngeal branch of vagus (CN X) and pharyngeal plexus	Elevate (shorten and widen) pharynx and larynx during swallowing and speaking
Salpingopharyngeus	Cartilaginous part of pharyngotympanic tube	Blends with palatopharyngeus		
Stylopharyngeus	Styloid process of temporal bone	Posterior and superior borders of thyroid cartilage with palatopharyngeus	Glossopharyngeal nerve (CN IX)	

Externally, the pharyngo-esophageal junction appears as a constriction produced by the **cricopharyngeal part of the inferior pharyngeal constrictor muscle** (the superior esophageal sphincter). The cervical esophagus lies between the trachea and cervical vertebral bodies and is in contact with the cervical pleura at the root of the neck (see Fig. 9.11C).

The thoracic duct adheres to the left side of the esophagus and lies between the pleura and the esophagus.

The *arteries of the cervical esophagus* are branches of the *inferior thyroid arteries* (see Fig. 9.13A). Each artery gives off ascending and descending branches that anastomose with each other and across the midline. The *veins* are

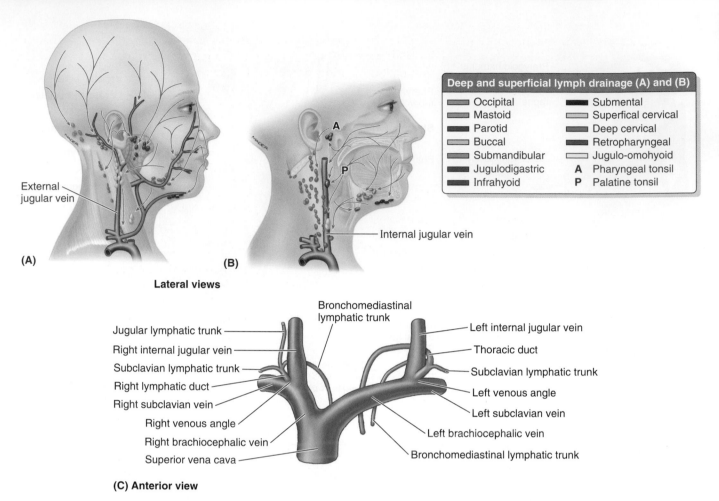

Deep and superficial lymph drainage (A) and (B)

Occipital	Submental
Mastoid	Superfical cervical
Parotid	Deep cervical
Buccal	Retropharyngeal
Submandibular	Jugulo-omohyoid
Jugulodigastric	**A** Pharyngeal tonsil
Infrahyoid	**P** Palatine tonsil

External jugular vein

Internal jugular vein

(A)   (B)

Lateral views

Bronchomediastinal lymphatic trunk

Jugular lymphatic trunk

Right internal jugular vein

Subclavian lymphatic trunk

Right lymphatic duct

Right subclavian vein

Right venous angle

Right brachiocephalic vein

Superior vena cava

Left internal jugular vein

Thoracic duct

Subclavian lymphatic trunk

Left venous angle

Left subclavian vein

Left brachiocephalic vein

Bronchomediastinal lymphatic trunk

(C) Anterior view

FIGURE 9.26. Lymphatic drainage of head and neck. A. Superficial lymph nodes. **B.** Deep lymph nodes. **C.** Termination of thoracic and right lymphatic ducts.

tributaries of the *inferior thyroid veins. Lymphatic vessels of the cervical esophagus* drain into the *paratracheal lymph nodes* and *inferior deep cervical lymph nodes* (Figs. 9.13B and 9.26B).

The *nerve supply of the esophagus* is somatic motor and sensory to the superior half and parasympathetic (vagal), sympathetic, and visceral sensory to the inferior half. The cervical esophagus receives the somatic fibers via branches from the *recurrent laryngeal nerves* and vasomotor fibers from the *cervical sympathetic trunks* through the plexus around the inferior thyroid artery (see Fig. 9.13A).

LYMPHATICS IN NECK

Most superficial tissues of the neck are drained by lymphatic vessels that enter the *superficial cervical lymph nodes*, which are located along the course of the EJV (Fig. 9.26A). Lymph from these nodes drains into *inferior deep cervical lymph nodes* (Fig. 9.26B,C). The specific group of inferior deep

cervical nodes involved here descends across the lateral cervical region with the spinal accessory nerve (CN XI). Most lymph from the 6th to 8th nodes then drains into the *supraclavicular lymph nodes*, which accompany the transverse cervical artery. The main group of deep cervical nodes forms a chain along the IJV, mostly under cover of the SCM.

Other deep cervical nodes include the prelaryngeal, pretracheal, paratracheal, and retropharyngeal nodes (Fig. 9.26C). Efferent lymphatic vessels from the deep cervical nodes join to form the **jugular lymphatic trunks**, which usually join the thoracic duct on the left side. On the right side, the vessels enter the junction of the internal jugular and subclavian veins (*right venous angle*) directly or via a short right lymphatic duct.

The **thoracic duct** passes through the superior thoracic aperture along the left border of the esophagus. It arches laterally in the root of the neck, posterior to the carotid sheath and anterior to the sympathetic trunk and vertebral and subclavian arteries (see Fig. 9.11C). This duct enters the left brachiocephalic vein at the junction of the subclavian and

IJVs (*left venous angle*) (Fig. 9.26C). The duct drains lymph from the entire body, except the upper right quarter (right side of the head and neck, the right upper limb, and the upper right quarter of the thorax) which drains through the *right lymphatic duct* (see Fig. 1.17 in "Overview and Basic Concepts"). The left jugular, subclavian, and bronchomediastinal lymphatic trunks usually unite to form the thoracic duct, which enters the left venous angle. Often, however, these lymphatic trunks enter the venous system independently in the region of the right venous angle.

CLINICAL BOX

Radical Neck Dissections

Radical neck dissections are performed when cancer invades the lymphatics. During the procedure, the deep cervical lymph nodes and the tissues around them are removed as completely as possible. Although major arteries, the brachial plexus, CN X, and the phrenic nerve are preserved, most cutaneous branches of the cervical plexus are removed. The aim of the dissection is to remove all tissue that contains lymph nodes in one piece. The deep cervical lymph nodes, particularly those located along the transverse cervical artery, may be involved in the spread of cancer from the thorax and abdomen. Because their enlargement may give the first clue to cancer in these regions, they are often referred to as the *cervical sentinel lymph nodes*.

Adenoiditis

Inflammation of the pharyngeal tonsils (adenoids) is called *adenoiditis*. This condition can obstruct the passage of air from the nasal cavities through the choanae into the nasopharynx, making mouth breathing necessary. Infection from the enlarged pharyngeal tonsils may also spread to the tubal tonsils, causing swelling and closure of the pharyngotympanic tubes. Impairment of hearing may result from nasal obstruction and blockage of the pharyngotympanic tubes. Infection spreading from the nasopharynx to the middle ear causes *otitis media* (middle ear infection), which may produce temporary or permanent hearing loss.

Foreign Bodies in Laryngopharynx

Foreign bodies entering the pharynx may become lodged in the piriform fossae. If the object (e.g., a chicken bone) is sharp, it may pierce the mucous membrane and injure the internal laryngeal nerve. The superior laryngeal nerve and its internal laryngeal branch are also vulnerable to injury if the instrument used to remove the foreign body accidentally pierces the mucous membrane. Injury to these nerves may result in anesthesia of the laryngeal mucous membrane as far inferiorly as the vocal folds. Young children swallow various objects, most of which reach the stomach and subsequently pass through the alimentary tract without difficulty. In some cases, the foreign body stops at the inferior end of the laryngopharynx, its narrowest part. A medical image such as a radiograph or a computed tomography (CT) scan will reveal the presence of a radiopaque foreign body. Foreign bodies in the pharynx are often removed under direct vision through a *pharyngoscope*.

Tonsillectomy

Tonsillectomy (removal of the palatine tonsil) is performed by dissecting the tonsil from the tonsillar sinus or by a guillotine or snare operation. Each procedure involves removal of the tonsil and the fascial sheet covering the tonsillar sinus. Because of the rich blood supply of the tonsil, bleeding commonly arises from the large external palatine vein or less commonly from the tonsillar artery or other arterial twigs (Fig. 9.23C). The glossopharyngeal nerve accompanies the tonsillar artery on the lateral wall of the pharynx and is vulnerable to injury because this wall is thin. The internal carotid artery is especially vulnerable when it is tortuous as it lies directly lateral to the tonsil (Fig. B9.9).

(Continued on next page)

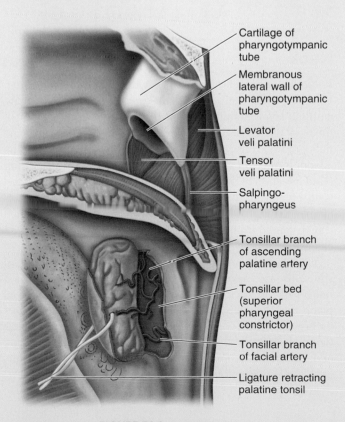

Cartilage of pharyngotympanic tube

Membranous lateral wall of pharyngotympanic tube

Levator veli palatini

Tensor veli palatini

Salpingo-pharyngeus

Tonsillar branch of ascending palatine artery

Tonsillar bed (superior pharyngeal constrictor)

Tonsillar branch of facial artery

Ligature retracting palatine tonsil

FIGURE B9.9. Tonsillectomy.

Zones of Penetrating Trauma

Three zones are common *clinical guides to the seriousness of neck penetrating trauma* (Fig. B9.10). The zones give physicians an understanding of structures that are at risk with penetrating neck injuries.

- *Zone I* includes the root of neck extending from the clavicles and manubrium to the inferior border of the cricoid cartilage. Structures at risk are the cervical pleurae, apices of lungs, thyroid and parathyroid glands, trachea, esophagus, common carotid arteries, jugular veins, and the cervical region of the vertebral column.
- *Zone II* extends from the cricoid cartilage to the angles of the mandible. Structures at risk are the superior poles of the thyroid gland, thyroid and cricoid cartilages, larynx, laryngopharynx, carotid arteries, jugular veins, esophagus, and cervical region of the vertebral column.
- *Zone III* occurs superiorly from the angles of the mandible. Structures at risk are the salivary glands, oral and nasal cavities, oropharynx, and nasopharynx.

Injuries in zones I and III obstruct the airway and have the greatest risk for *morbidity* (complications after surgical procedures and other treatments) and *mortality* (a fatal outcome) because injured structures are difficult to visualize and repair and vascular damage is difficult to control. Injuries in zone II are most common; however, morbidity and mortality are lower because physicians can control vascular damage by direct pressure and surgeons can visualize and treat injured structures more easily than they can in zones I and III.

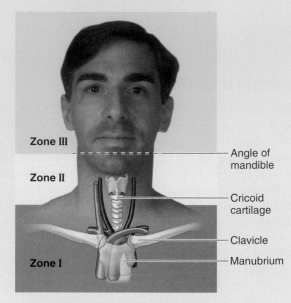

FIGURE B9.10. Zones of penetrating neck trauma.

MEDICAL IMAGING

Neck

Radiography has limited and specific uses in neck imaging. Upright radiography of the sinuses can be used to evaluate air–fluid levels in purulent sinusitis. Soft tissue radiography of the neck (different radiographic technique than cervical spine radiography) (Fig. 9.27) is used to look for enlargement of the adenoids and to examine the contour of the airway in croup (viral infection of the subglottic trachea). In cases of suspected acute epiglottitis (life-threatening bacterial infection of the epiglottis), the rapid identification of an enlarged epiglottis, which can be gained from a single lateral soft tissue neck radiograph, can lead to lifesaving protection of a compromised airway.

CT scans are used to diagnose inflammatory paranasal sinus disease, severe facial fractures, and cross-sectional images of the neck (Fig. 9.28A). CT is acquired in the axial plane, and the dataset can then be used to reconstruct images in the sagittal and coronal planes. CT scans are superior to radiographs because they reveal radiodensity differences among and within soft tissues (e.g., in salivary glands). CT angiograms enable reconstruction of the arteries in 3-D (Fig. 9.29).

Magnetic resonance imaging (MRI) systems construct images of transverse, sagittal, and coronal sections of the neck and have the advantage of using no radiation (Fig. 9.28B). MRI studies of the neck are superior to CT studies for showing detail in soft tissues, but they provide little information about bones. *Ultrasonography (US)* is also a useful imaging technique for studying soft tissues of the neck. US provides images of many abnormal conditions noninvasively, at relatively low cost, and with minimal discomfort. It is useful for distinguishing solid from cystic masses, for example, which may be difficult to determine during physical examination. US is the major imaging modality used to evaluate morphologic changes in the thyroid gland (functional thyroid disease is evaluated by nuclear medicine procedures and with laboratory studies). Vascular imaging of arteries and veins of the neck is possible using intravascular ultrasonography (Fig. 9.30A,B). The images are produced by placing the transducer within the blood vessel. *Doppler ultrasound techniques* help evaluate blood flow through a vessel (e.g., for detecting stenosis [narrowing] of a carotid artery).

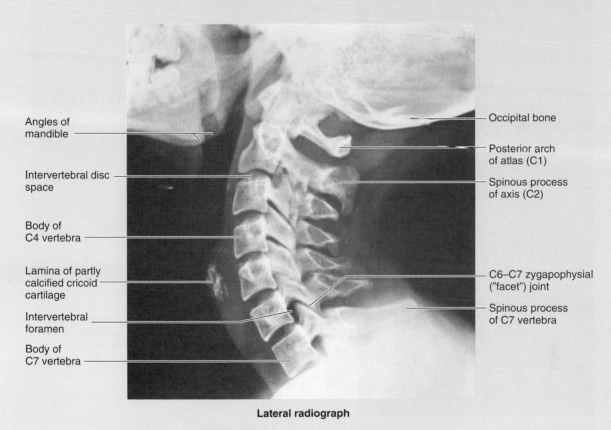

Lateral radiograph

FIGURE 9.27. Cervical region of vertebral column.

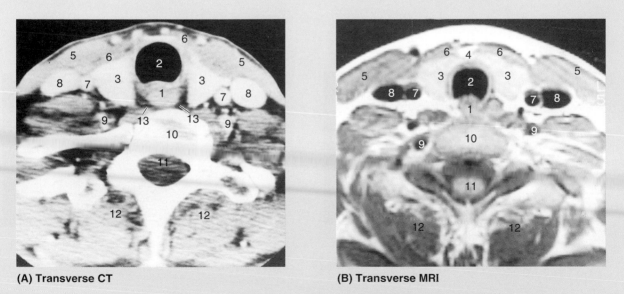

(A) Transverse CT

(B) Transverse MRI

FIGURE 9.28. Scans of neck through thyroid gland. Transverse studies via CT **(A)** and MRI **(B)** revealing the structures of the neck. *1*, esophagus; *2*, trachea; *3*, lobes of thyroid gland; *4*, thyroid isthmus; *5*, sternocleidomastoid (SCM); *6*, sternohyoid; *7*, common carotid artery; *8*, internal jugular vein (IJV); *9*, vertebral artery; *10*, vertebral body; *11*, spinal cord in cerebrospinal fluid; *12*, deep muscles of the back; *13*, retropharyngeal space.

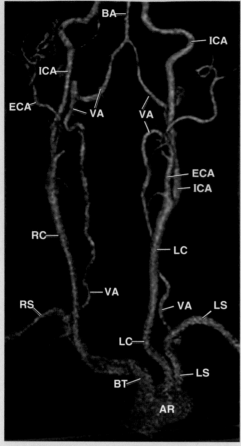

(A) Anterior View

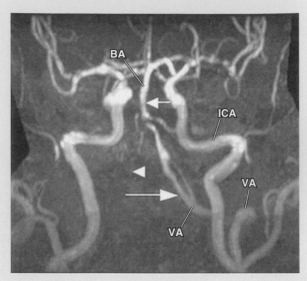

(B) Anterior View

Key for A and B			
AR	Arch of aorta	LC	Left common carotid artery
BA	Basilar artery	LS	Left subclavian artery
BT	Brachiocephalic trunk	RC	Right common carotid artery
ECA	External carotid artery	RS	Right subclavian artery
ICA	Internal carotid artery	VA	Vertebral artery

FIGURE 9.29. Arteries of head and neck. A. CT angiogram of arteries of head and neck. **B.** Frontal magnetic resonance angiogram showing a severe stenosis at the junction of the left vertebral artery (*long arrow*) and basilar artery (*short arrow*). The right distal vertebral artery is also occluded (*arrowhead*).

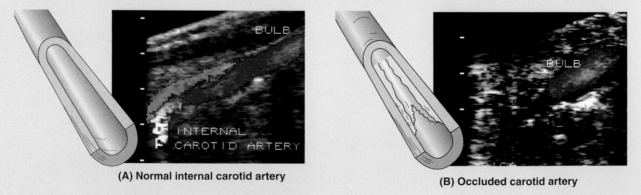

(A) Normal internal carotid artery

(B) Occluded carotid artery

FIGURE 9.30. Doppler color flow study of internal carotid artery. A. Normal. **B.** Occluded artery.

10
Review of Cranial Nerves

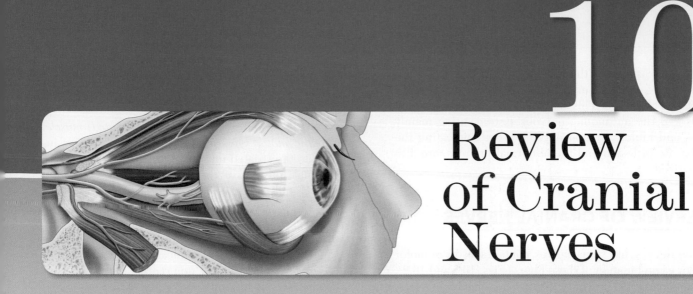

CLINICAL BOX KEY

Anatomical Variations

Diagnostic Procedures

Life Cycle

Surgical Procedures

Trauma

Pathology

The regional aspects of the cranial nerves are described in the preceding chapters, especially those for the head, neck, thorax, and abdomen. This chapter summarizes the cranial nerves and the autonomic nervous system, using mainly figures and tables. Cranial nerve injuries, indicating the type or site of lesion and the abnormal findings, are also summarized.

OVERVIEW OF CRANIAL NERVES

Cranial nerves, like spinal nerves, contain sensory or motor fibers or a combination of these fibers (Figs. 10.1 and 10.2). Cranial nerves innervate muscles or glands or carry impulses from sensory receptors. They are called cranial nerves because they exit from the cranial cavity via foramina or fissures in the cranium and are covered by tubular sheaths derived from the cranial meninges (Fig. 10.3). There are 12 pairs of cranial nerves, which are numbered I to XII, from rostral to caudal, according to their attachments to the brain (Fig. 10.1 and Table 10.1). Their names reflect their general distribution or function.

Cranial nerves carry one or more of the following five main functional components (Fig. 10.2):

- **Motor (efferent) fibers**
  1. *Motor fibers innervating voluntary (striated) muscle.* *Somatic motor* (general somatic efferent) axons innervate the striated muscles in the orbit, tongue, and external muscles of the neck (sternocleidomastoid [SCM] and trapezius) as well as striated muscles of the face, palate, pharynx, and larynx. The muscles of the face, palate, pharynx, and larynx are derived from the pharyngeal arches, and their somatic motor innervation can be referred to more specifically as *branchial motor.*
  2. *Motor fibers involved in innervating glands and involuntary (smooth) muscle* (e.g., in viscera and blood vessels). These include *visceral motor* (general visceral efferent) axons that constitute the cranial outflow of the parasympathetic division of the autonomic nervous system. The presynaptic (preganglionic) fibers that emerge from the brain synapse outside the central nervous system (CNS) in a parasympathetic ganglion. The postsynaptic (postganglionic) fibers continue to innervate glands and smooth muscle throughout the body.

(Continued on page 648)

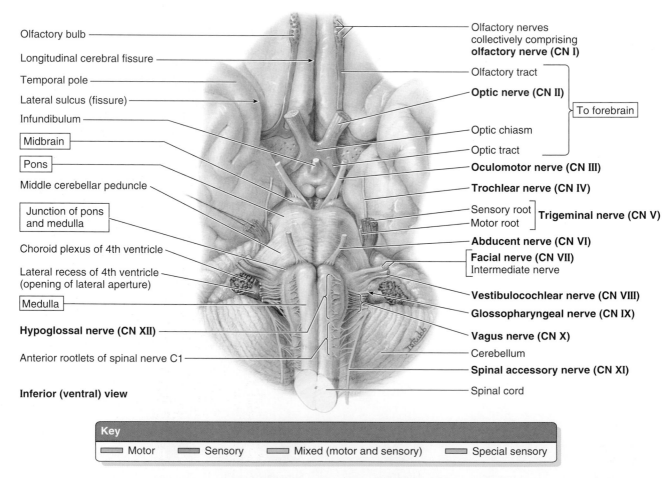

Olfactory bulb
Longitudinal cerebral fissure
Temporal pole
Lateral sulcus (fissure)
Infundibulum
Midbrain
Pons
Middle cerebellar peduncle
Junction of pons and medulla
Choroid plexus of 4th ventricle
Lateral recess of 4th ventricle (opening of lateral aperture)
Medulla
Hypoglossal nerve (CN XII)
Anterior rootlets of spinal nerve C1

Inferior (ventral) view

Olfactory nerves collectively comprising **olfactory nerve (CN I)**
Olfactory tract
Optic nerve (CN II)
To forebrain
Optic chiasm
Optic tract
Oculomotor nerve (CN III)
Trochlear nerve (CN IV)
Sensory root | **Trigeminal nerve (CN V)**
Motor root
Abducent nerve (CN VI)
Facial nerve (CN VII)
Intermediate nerve
Vestibulocochlear nerve (CN VIII)
Glossopharyngeal nerve (CN IX)
Vagus nerve (CN X)
Cerebellum
Spinal accessory nerve (CN XI)
Spinal cord

Key

| ☐ Motor | ☐ Sensory | ☐ Mixed (motor and sensory) | ☐ Special sensory |

FIGURE 10.1. Superficial origin of cranial nerves.

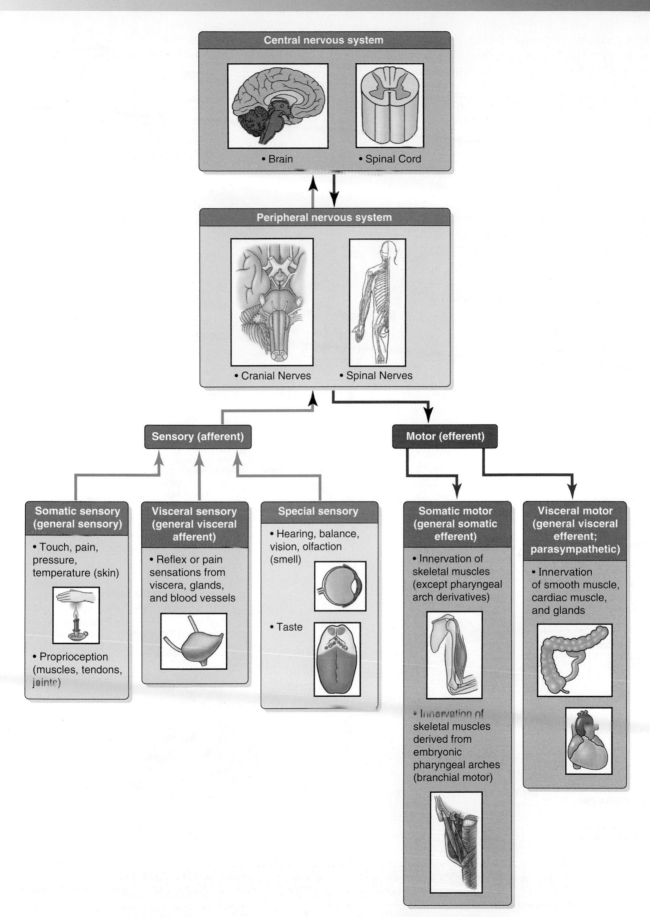

FIGURE 10.2. Overview of sensory and motor components of cranial and spinal nerves.

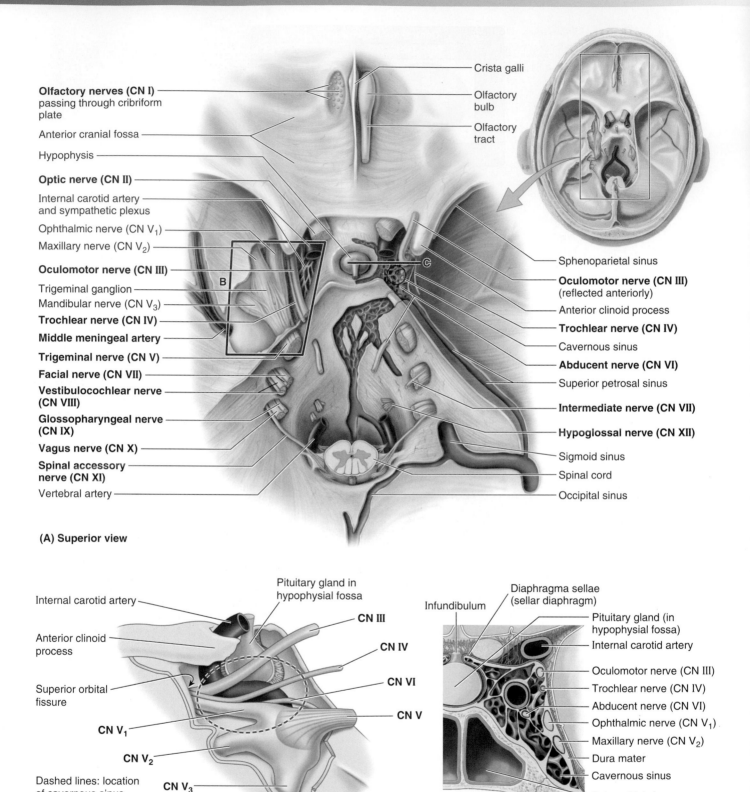

(A) Superior view

(B) Lateral view

**(C) Posterior view of coronal section
of cavernous sinus**

FIGURE 10.3. Cranial nerves in relation to internal aspect of cranial base. A. The tentorium cerebelli has been removed and the venous sinuses have been opened on the right side. The dural roof of the trigeminal cave has been removed on the left side, and CN V, CN III, and CN IV have been dissected from the lateral wall of cavernous sinus. **B.** Lateral view of area outlined in red in **(A)** demonstrating nerves related to cavernous sinus. **C.** Structures and relationships within the cavernous sinus and its lateral wall.

TABLE 10.1. SUMMARY OF CRANIAL NERVES

Trochlear — CN IV
Motor: superior oblique muscle of eye

Abducent — CN VI
Motor: lateral rectus muscle of eye

Oculomotor — CN III
Motor: ciliary muscles, sphincter pupillae, all extrinsic muscles of eye except those listed for CN IV and VI

Optic — CN II
Sensory: vision

Cranial nerve fibers
— Efferent (motor)
— Afferent (sensory)

Facial — CN VII
Primary root
Motor: muscles of facial expression

Olfactory — CN I
Sensory: smell

Trigeminal — CN V
Sensory root
Sensory: face, sinuses, teeth

Trigeminal — CN V
Motor root
Motor: muscles of mastication

Facial — CN VII
Intermediate nerve
Motor: submandibular, sublingual, lacrimal glands
Sensory: taste to anterior two thirds of tongue, soft palate

Vestibulocochlear — CN VIII
Vestibular nerve, sensory: orientation, motion
Cochlear nerve, sensory: hearing

Hypoglossal — CN XII
Motor: all intrinsic and extrinsic muscles of tongue (excluding palatoglossus— a palatine muscle)

Spinal accessory — CN XI
Motor: sternocleidomastoid and trapezius muscles

Vagus — CN X
Motor: palate, pharynx, larynx, trachea, bronchial tree, heart, GI tract to left colic flexure
Sensory: pharynx, larynx: reflex sensory from tracheo-bronchial tree, lungs, heart, GI tract to left colic flexure

Glossopharyngeal — CN IX
Motor: stylopharyngeus muscle, parotid gland
Sensory: taste: posterior third of tongue; general sensation: pharynx, tonsillar sinus, pharyngotympanic tube, middle ear cavity

CN I
CN II
CN III
CN IV
CN VI
CN V₁
CN V₂
CN V
CN V₃
CN V₃
CN VII
CN VII
CN VIII
CN IX
CN X
CN XI
CN XII

TABLE 10.1. SUMMARY OF CRANIAL NERVES

Nerve	Components	Location of Nerve Cell Bodies	Cranial Exit	Main Action(s)
Olfactory (CN I)	Special sensory (olfaction)	Olfactory epithelium (olfactory cells)	Foramina in cribriform plate of ethmoid bone	Smell from nasal mucosa of roof of each nasal cavity and superior sides of nasal septum and superior concha
Optic (CN II)	Special sensory (vision)	Retina (ganglion cells)	Optic canal	Vision from retina
Oculomotor (CN III)	Somatic motor	Midbrain (nucleus of oculomotor nerve)	Superior orbital fissure	Motor to superior rectus, inferior rectus, medial rectus, inferior oblique, and levator palpebrae superioris muscles; raises superior eyelid; turns eyeball superiorly, inferiorly, and medially
	Visceral motor	Presynaptic: midbrain (Edinger-Westphal nucleus) Postsynaptic: ciliary ganglion		Parasympathetic innervation to sphincter pupillae and ciliary muscle; constricts pupil and accommodates lens of eye
Trochlear (CN IV)	Somatic motor	Midbrain (nucleus of trochlear nerve)		Motor to superior oblique to assist in turning eye inferolaterally (or inferiorly when adducted)
Trigeminal (CN V)				
Ophthalmic (CN V$_1$)	Somatic (general) sensory	Trigeminal ganglion Synapse: sensory nucleus of trigeminal nerve	Superior orbital fissure	Sensation from cornea, skin of forehead, scalp, eyelids, nose, and mucosa of nasal cavity and paranasal sinuses
Maxillary (CN V$_2$)		Trigeminal ganglion Synapse: sensory nucleus of trigeminal nerve	Foramen rotundum	Sensation from skin of face over maxilla, including upper lip, maxillary teeth, mucosa of nose, maxillary sinuses, and palate
Mandibular (CN V$_3$)		Trigeminal ganglion Synapse: sensory nucleus of trigeminal nerve	Foramen ovale	Sensation from skin over mandible, including lower lip, side of head, mandibular teeth, temporomandibular joint, mucosa of mouth, and anterior two thirds of tongue
	Somatic (branchial) motor	Pons (motor nucleus of trigeminal nerve)		Motor to muscles of mastication, mylohyoid, anterior belly of digastric, tensor veli palatini, and tensor tympani
Abducent (CN VI)	Somatic motor	Pons (nucleus of abducent nerve)	Superior orbital fissure	Motor to lateral rectus to turn eye laterally
Facial (CN VII)	Somatic (branchial) motor	Pons (motor nucleus of facial nerve)	Internal acoustic meatus; facial canal; stylomastoid foramen	Motor to muscles of facial expression and scalp; also supplies stapedius of middle ear, stylohyoid, and posterior belly of digastric
	Special sensory (taste)	Geniculate ganglion Synapse: nuclei of solitary tract		Taste from anterior two thirds of tongue and palate
	Somatic (general) sensory	Geniculate ganglion Synapse: sensory nucleus of trigeminal nerve		Sensation from skin of external acoustic meatus
	Visceral motor	Presynaptic: pons (superior salivatory nucleus) Postsynaptic: pterygopalatine ganglion, submandibular ganglion		Parasympathetic innervation to submandibular and sublingual salivary glands, lacrimal gland, and glands of nose and palate
Vestibulocochlear (CN VIII)				
Vestibular	Special sensory (balance)	Vestibular ganglion Synapse: vestibular nuclei	Internal acoustic meatus	Vestibular sensation from semicircular ducts, utricle, and saccule related to position and movement of head
Cochlear	Special sensory (hearing)	Spiral ganglion Synapse: cochlear nuclei		Hearing from spiral organ
Glossopharyngeal (CN IX)	Somatic (branchial) motor	Medulla (nucleus ambiguus)	Jugular foramen	Motor to stylopharyngeus to assist with swallowing
	Visceral motor	Presynaptic: medulla (inferior salivatory nucleus) Postsynaptic: otic ganglion		Parasympathetic innervation to parotid gland
	Special sensory (taste)	Sensory ganglion (nuclei of solitary tract)		Taste from posterior third of tongue
	Somatic (general) sensory	Sensory ganglion Synapse: sensory nucleus of CN V		External ear, pharynx, middle ear
	Visceral sensory	Sensory ganglion (nuclei of solitary tract)		Carotid body and sinus

(continued)

TABLE 10.1. SUMMARY OF CRANIAL NERVES (continued)

Nerve	Components	Location of Nerve Cell Bodies	Cranial Exit	Main Action(s)
Vagus (CN X)	Somatic (branchial) motor	Medulla (nucleus ambiguus)	Jugular foramen	Motor to constrictor muscles of pharynx (except stylopharyngeus), intrinsic muscles of larynx, muscles of palate (except tensor veli palatini), and striated muscle in superior two thirds of esophagus
	Visceral motor	Presynaptic: medulla Postsynaptic: neurons in, on, or near viscera		Parasympathetic innervation to smooth muscle of trachea, bronchi, digestive tract, and cardiac muscle of heart
	Visceral sensory	Inferior ganglion Synapse: nuclei of solitary tract		Visceral sensation from base of tongue, pharynx, larynx, trachea, bronchi, heart, esophagus, stomach, and intestine
	Special sensory (taste)	Inferior ganglion Synapse: nuclei of solitary tract		Taste from epiglottis and palate
	Somatic (general) sensory	Superior ganglion Synapse: sensory nucleus of trigeminal nerve		Sensation from auricle, external acoustic meatus, and dura mater of posterior cranial fossa
Spinal accessory (CN XI)	Somatic motor	Cervical spinal cord		Motor to sternocleidomastoid and trapezius
Hypoglossal (CN XII)	Somatic motor	Medulla (nucleus of CN XII)	Hypoglossal canal	Motor to intrinsic and extrinsic muscles of tongue (except palatoglossus)

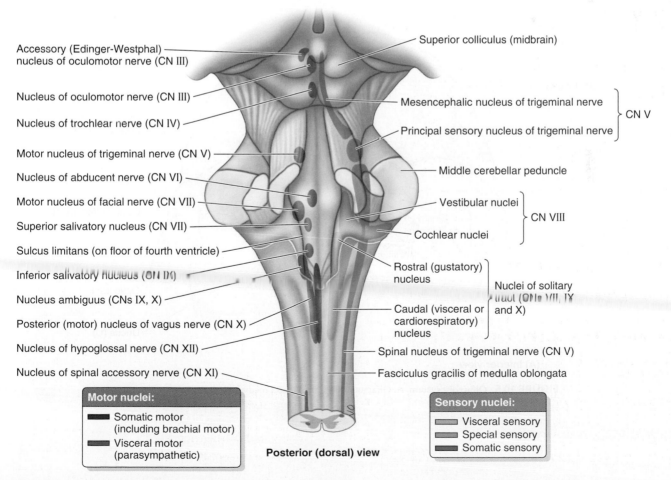

FIGURE 10.4. Cranial nerve nuclei. The motor nuclei are shown on the left side of the brainstem and the sensory nuclei on the right side. The sensory and motor nuclei are all paired—that is, located in both the right and left sides of the brainstem.

- **Sensory (afferent) fibers**
  3. *Fibers conveying sensation from the viscera.* These are visceral sensory (general visceral afferent) fibers, such as those conveying reflex and pain information from the carotid body and sinus, pharynx, larynx, trachea, bronchi, lungs, heart, and gastrointestinal tract.
  4. *Fibers transmitting general sensation (e.g., touch, pressure, heat, cold) from the skin and mucous membranes.* These are somatic (general) sensory fibers, which are carried mainly by CN V and also by CN VII, CN IX, and CN X.
  5. *Fibers transmitting unique sensations.* These are *special sensory* fibers conveying taste and smell and those serving the special senses of vision, hearing, and balance.

The fibers of cranial nerves connect centrally to **cranial nerve nuclei**, groups of neurons in which sensory or afferent fibers terminate and from which motor or efferent fibers originate (Fig. 10.4). Except for CN I and CN II, which involve extensions of the forebrain, and CN XI with nuclei in the C1–C3 segments of the spinal cord, the nuclei of the cranial nerves are located in or adjacent to the brainstem.

Nuclei of similar functional components are generally aligned into functional columns in the brainstem region.

OLFACTORY NERVE (CN I)

The **olfactory nerves (CN I)** convey the sense of smell (Fig. 10.5). The cell bodies of the **olfactory receptor neurons** are located in the olfactory part of the nasal mucosa, or olfactory area, in the roof of the nasal cavity and along the nasal septum and medial wall of the superior nasal concha (Fig. 10.5B). The central processes of the bipolar olfactory neurons are collected into bundles to form approximately 20 olfactory nerves on each side that together form the right or left olfactory nerve (Fig. 10.5C). The fibers pass through tiny foramina in the **cribriform plate** of the ethmoid bone, surrounded by sleeves of dura and arachnoid, and enter the **olfactory bulb** in the anterior cranial fossa. The olfactory nerve fibers synapse with **mitral cells** in the olfactory bulb. The axons of these cells form the **olfactory tract**, which conveys the impulses to the brain (Fig. 10.5A,C). The olfactory bulbs and tracts are technically anterior extensions of the forebrain.

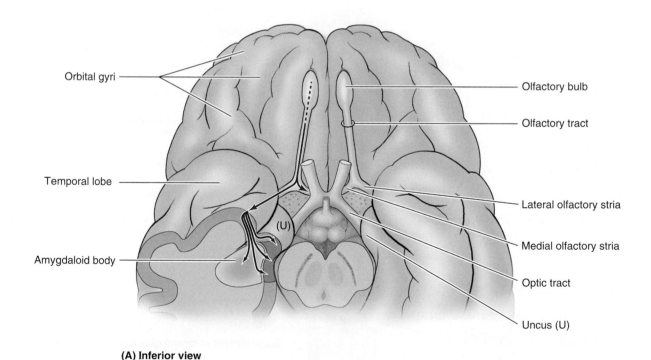

(A) Inferior view

FIGURE 10.5. Olfactory system. A. Olfactory bulbs, tracts, and medial and lateral striae. *(continued)*

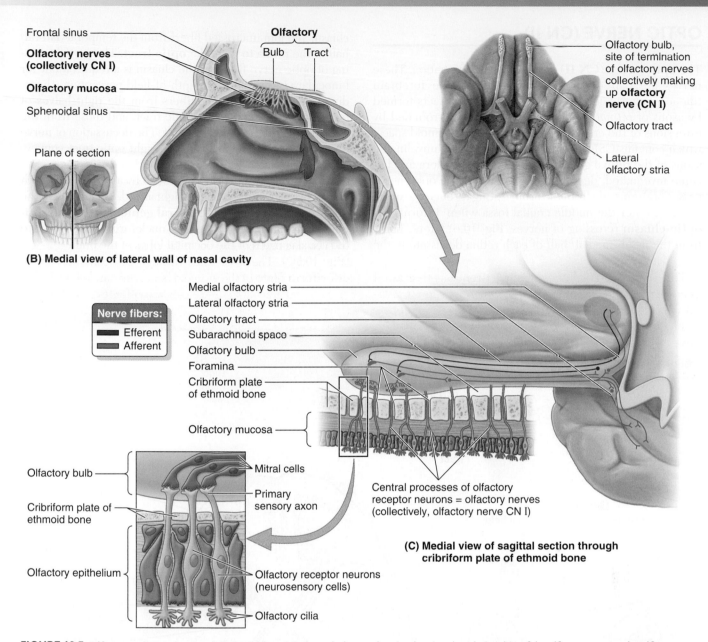

FIGURE 10.5 Olfactory system. *(continued)* **B.** Sagittal section through the nasal cavity showing the relationship of the olfactory area to the olfactory bulb. **C.** Bodies of the olfactory receptor neurons are in the olfactory epithelium. These bundles of axons are collectively called the olfactory nerve (CN I).

CLINICAL BOX

Anosmia—Loss of Smell

Loss or decrease in olfaction usually occurs with aging. This may also occur due to excessive smoking and cocaine use. The chief complaint of most people with *anosmia* is the loss or alteration of taste; however, clinical studies reveal that in all but a few people, the dysfunction is in the olfactory system (Simpson & Sweazey, 2013). Transient olfactory impairment occurs as a result of *viral* or *allergic rhinitis* (inflammation of the nasal mucous membrane).

Injury to the nasal mucosa, olfactory nerve fibers, olfactory bulbs, or olfactory tracts may also impair smell. In severe head injuries, the olfactory bulbs may be torn away from the olfactory nerves, or some olfactory nerve fibers may be torn as they pass through a *fractured cribriform plate*. If all the nerve bundles on one side are torn, a complete loss of smell occurs on that side; consequently, anosmia may be a clue to a *fracture of the cranial base* and *cerebrospinal fluid* (CSF) *rhinorrhea*, a leakage of the fluid through the nose from the subarachnoid space. *Olfaction disorders* are also linked with psychiatric illnesses (e.g., schizophrenia) and epilepsy. These patients may experience distortion of smell (*parosmia*) or perceive an odor when there is none present (*olfactory hallucination*).

OPTIC NERVE (CN II)

The **optic nerve** (**CN II**) conveys visual information. These nerves are paired, anterior extensions of the forebrain (diencephalon) and are therefore CNS fiber tracts formed by axons of *retinal ganglion cells*. CN II is surrounded by extensions of the cranial meninges and subarachnoid space, which contains CSF. CN II begins where the unmyelinated axons of the retinal ganglion cells pierce the sclera and become myelinated, deep to the **optic disc**. The optic nerve passes posteromedially in the orbit, exiting through the **optic canal** to enter the middle cranial fossa where it forms the **optic chiasm** (crossing of nerves; Fig. 10.6). Here, fibers from the nasal (medial) half of each retina decussate in the chiasm and join uncrossed fibers from the temporal (lateral) half of the retina to form the **optic tract**. The partial crossing of optic nerve fibers in the chiasm is a requirement for binocular vision, allowing depth-of-field perception (three-dimensional vision). Thus, fibers from the right halves of both retinas form the right optic tract, and those from the left halves form the left optic tract. The decussation of nerve fibers in the chiasm results in the right optic tract conveying impulses from the left visual field and vice versa. The combined **visual field** is what is seen by a person with both eyes wide open and looking straight ahead. Most fibers in the optic tracts terminate in the **lateral geniculate bodies (nuclei)** of the thalamus. From these nuclei, axons are relayed to the visual cortices of the occipital lobes of the brain.

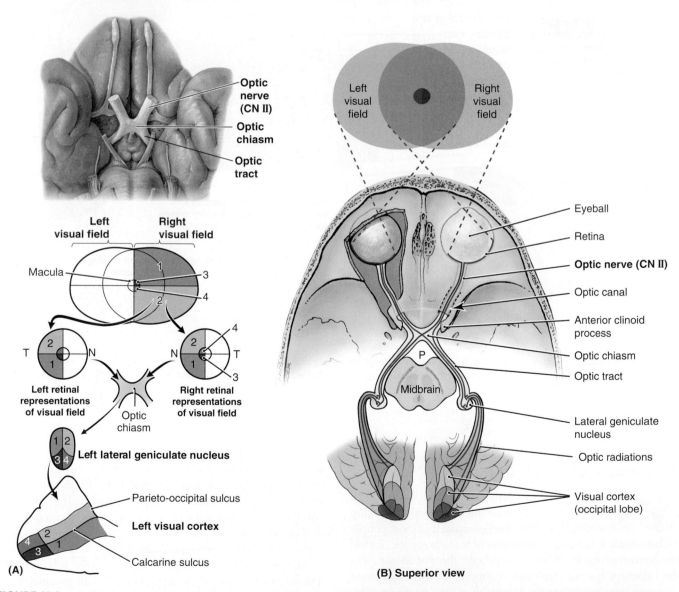

FIGURE 10.6. Visual system. A. Right visual field representation on retinas, left lateral geniculate body, and left visual cortex. Areas corresponding to: (*1*) upper general, (*2*) lower general, (*3*) upper macular and (*4*) lower macular portions of right visual field. *N*, nasal; *T*, temporal (aspects of visual fields). **B.** Overview of visual pathway. *P*, location of pituitary gland.

CLINICAL BOX

Visual Field Defects

Visual field defects may result from a large number of neurological diseases. It is clinically important to be able to link the defect to a likely location of the lesion (Fig. B10.1).

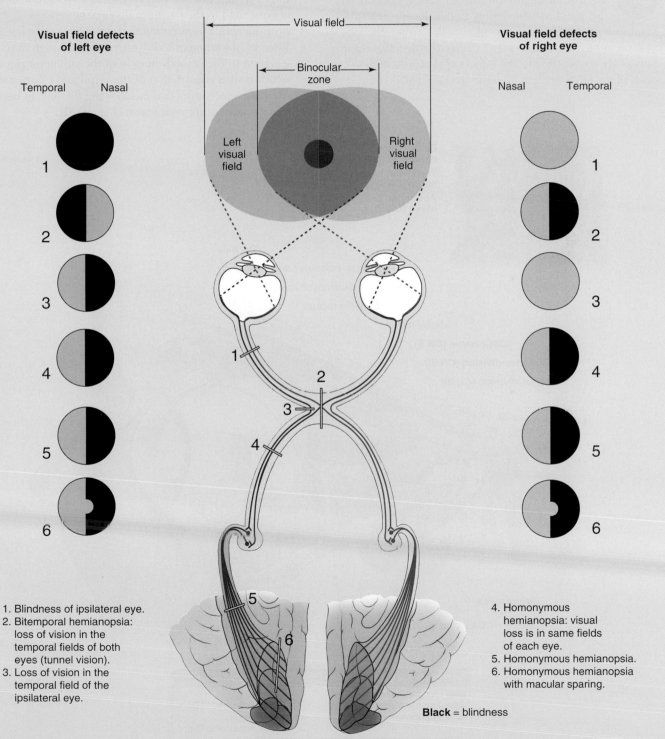

1. Blindness of ipsilateral eye.
2. Bitemporal hemianopsia: loss of vision in the temporal fields of both eyes (tunnel vision).
3. Loss of vision in the temporal field of the ipsilateral eye.

4. Homonymous hemianopsia: visual loss is in same fields of each eye.
5. Homonymous hemianopsia.
6. Homonymous hemianopsia with macular sparing.

Black = blindness

FIGURE B10.1. Visual field defects.

Demyelinating Diseases and the Optic Nerve

Because the optic nerves are actually CNS tracts, the myelin sheath that surrounds the fibers from the point at which they penetrate the sclera is formed by oligodendrocytes (glial cells) rather than by neurolemma (Schwann cells). Consequently, the optic nerves are susceptible to the effects *of demyelinating diseases of the CNS*, such as *multiple sclerosis* (MS).

NERVES TO EXTRA-OCULAR MUSCLES

Oculomotor Nerve (CN III)

The **oculomotor nerve** (**CN III**) provides the following (Figs. 10.7 and 10.8):

- *Somatic motor innervation* to four of the six extra-ocular muscles (*superior*, *medial*, and *inferior rectus* and *inferior oblique*) and to the *levator palpebrae superioris*
- *Proprioceptive innervation* to the previous muscles
- *Visceral (parasympathetic) innervation* through the ciliary ganglion to the smooth muscle of the sphincter pupillae, which causes constriction of the pupil and ciliary muscle to produce accommodation (allowing the lens to become more rounded) for near vision (Fig. 10.8B)

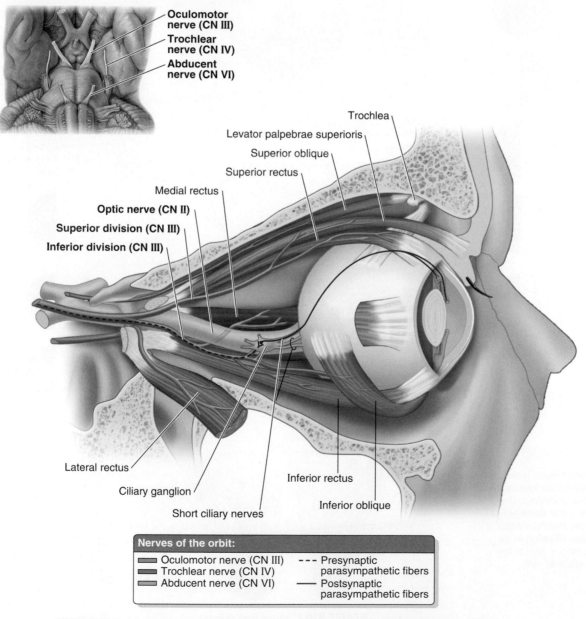

Nerves of the orbit:

▬ Oculomotor nerve (CN III)	- - - Presynaptic parasympathetic fibers
▬ Trochlear nerve (CN IV)	
▬ Abducent nerve (CN VI)	—— Postsynaptic parasympathetic fibers

FIGURE 10.7. Distribution of oculomotor (CN III), trochlear (CN IV), and abducent (CN VI) nerves.

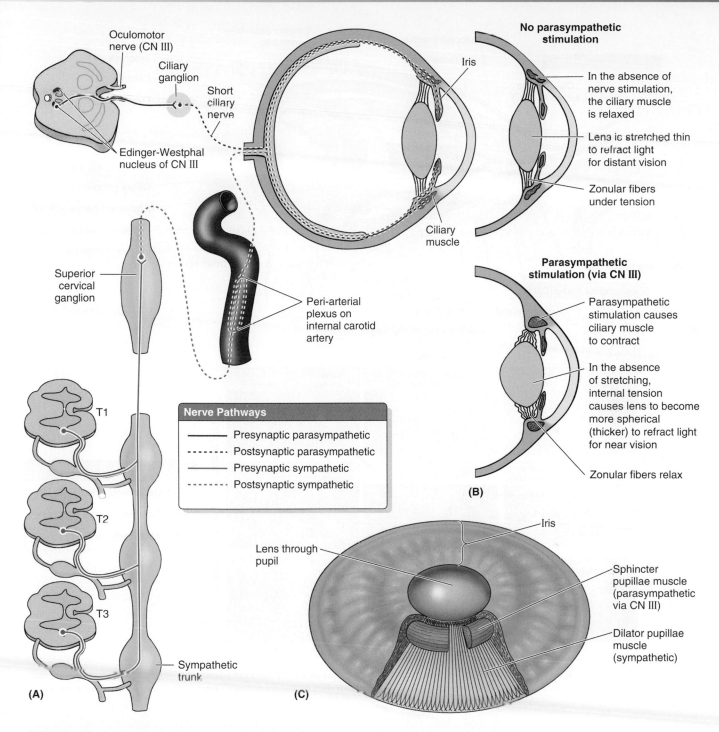

No parasympathetic stimulation

In the absence of nerve stimulation, the ciliary muscle is relaxed

Lens is stretched thin to refract light for distant vision

Zonular fibers under tension

Iris

Ciliary muscle

Oculomotor nerve (CN III)

Ciliary ganglion

Short ciliary nerve

Edinger-Westphal nucleus of CN III

Superior cervical ganglion

Peri-arterial plexus on internal carotid artery

Nerve Pathways

—— Presynaptic parasympathetic
- - - - Postsynaptic parasympathetic
—— Presynaptic sympathetic
- - - - Postsynaptic sympathetic

T1

T2

T3

Sympathetic trunk

(A)

Parasympathetic stimulation (via CN III)

Parasympathetic stimulation causes ciliary muscle to contract

In the absence of stretching, internal tension causes lens to become more spherical (thicker) to refract light for near vision

Zonular fibers relax

(B)

Lens through pupil

Iris

Sphincter pupillae muscle (parasympathetic via CN III)

Dilator pupillae muscle (sympathetic)

(C)

FIGURE 10.8. **Autonomic innervation of intra-ocular muscles. A.** Overview of nerve pathway. **B.** Function of ciliary muscle. **C.** Iris and muscles of iris.

CN III is the main motor nerve to the ocular and extra-ocular muscles. It emerges from the midbrain, pierces the dura, and runs through the roof and lateral wall of the *cavernous sinus*. CN III leaves the cranial cavity and enters the orbit through the *superior orbital fissure*. Within this fissure, CN III divides into a **superior division**, which supplies the superior rectus and levator palpebrae superioris,

and an **inferior division**, which supplies the inferior and medial rectus and inferior oblique (Figs. 10.7 and 10.9). The inferior division also carries presynaptic parasympathetic (visceral efferent) fibers to the **ciliary ganglion**, where they synapse. Postsynaptic fibers from this ganglion pass to the eyeball in the *short ciliary nerves* to innervate the ciliary muscle and the sphincter pupillae (Fig. 10.8C).

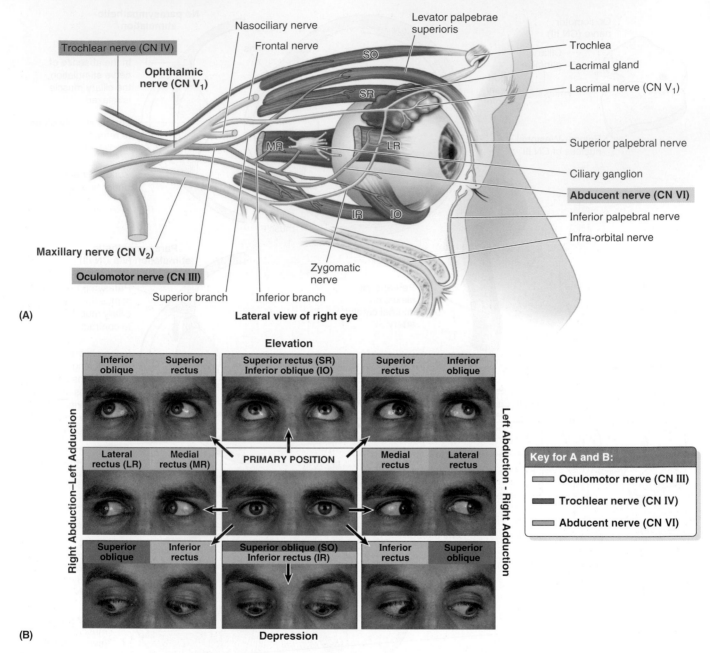

FIGURE 10.9. Innervation of extra-ocular muscles. A. Schematic overview. **B.** Anatomical movements of extra-ocular muscles. Single movements start from the center (rest or primary position). See Fig. 8.35A for sequential movements used for clinical testing of extra-ocular muscles and cranial nerves.

Trochlear Nerve (CN IV)

The **trochlear nerve** (**CN IV**) provides somatic motor and proprioceptive innervation to the contralateral *superior oblique*. The trochlear nerve, the smallest cranial nerve, arises from the nucleus of the trochlear nerve and crosses the midline prior to emerging inferior to the inferior colliculus of the posterior surface of the midbrain. It then passes anteriorly around the brainstem and pierces the dura mater at the margin of the tentorium cerebelli to course anteriorly in the lateral wall of the cavernous sinus. The nerve continues along the wall of the sinus to pass through the superior orbital fissure into the orbit, where it supplies one extra-ocular muscle, the superior oblique (Figs. 10.7 and 10.9).

Abducent Nerve (CN VI)

The **abducent nerve** (**CN VI**) provides somatic motor to and proprioceptive information from one extra-ocular muscle (*lateral rectus*). The abducent nerve emerges from the brainstem between the pons and the medulla and traverses the pontine cistern of the subarachnoid space. It then pierces the dura and runs the longest intracranial course within the cranial cavity of all the cranial nerves. During its intracranial course, it bends sharply over the crest of the petrous part of the temporal bone and then courses through the cavernous sinus, surrounded by venous blood such as the internal carotid artery. CN VI then enters the orbit through the superior orbital fissure and runs anteriorly to supply the lateral rectus, which abducts the eye (Figs. 10.7 and 10.9).

The innervation and movements of the extra-ocular muscles from the rest (primary) position are summarized in Figure 10.9B.

CLINICAL BOX

OCULAR PALSIES

The oculomotor (CN III), trochlear (CN IV), and abducent (CN VI) nerves may be compressed and the muscles they supply completely paralyzed intra- and extracranially by different mechanisms, including neurological diseases, tumors, or aneurysms.

Oculomotor Nerve (CN III)

Complete CN III Palsy

 Characteristic signs of a complete lesion of CN III are (Fig. B10.2) as follows for the ipsilateral eye:

- *Ptosis* (drooping) of the superior eyelid, caused by paralysis of the levator palpebrae superioris
- Eyeball (pupil) abducted and directed slightly inferiorly (down and out) because of unopposed actions of the lateral rectus and superior oblique
- No *pupillary (light) reflex* (constriction of the pupil in response to bright light)
- *Dilation of pupil*, resulting from the interruption of parasympathetic fibers to the sphincter pupillae, leaving the dilator pupillae unopposed
- *No accommodation of the lens* (adjustment to increase convexity for near vision) because of paralysis of the ciliary muscle

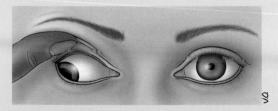

Right eye: Downward and outward directed pupil, dilated pupil, eyelid manually elevated due to ptosis.   Left: Normal

Gaze directed anteriorly

FIGURE B10.2. Oculomotor nerve (CN III) lesion.

Partial CN III Palsy

Rapidly increasing intracranial pressure (e.g., resulting from an acute extradural or subdural hematoma) often compresses CN III against the petrous part of the temporal bone.

Because the parasympathetic fibers in CN III are superficial, they are affected first (*internal ophthalmoplegia*). *External ophthalmoplegia* results from selective damage of the somatic motor fibers.

An aneurysm of a posterior cerebral or superior cerebellar artery may exert pressure on CN III as it passes between these vessels. Because CN III lies in the lateral wall of the cavernous sinus, injuries, infections, or tumors may also affect this nerve.

Trochlear Nerve (CN IV)

 CN IV is rarely injured in isolation. The characteristic sign of trochlear nerve injury is *diplopia* (double vision) when looking down (e.g., when descending stairs). Diplopia occurs because the superior oblique normally assists the inferior rectus in depressing the pupil (directing the gaze downward) and is the only muscle to do so from the adducted position.

Abducent Nerve (CN VI)

 Because CN VI has a long intracranial course, it is often stretched when intracranial pressure increases, partly because of the sharp bend it makes over the crest of the petrous part of the temporal bone after entering the dura. A *space-occupying lesion* such as a *brain tumor* may compress CN VI, causing *paralysis of the lateral rectus muscle*. A *complete lesion of CN VI* causes medial deviation of the affected eye—that is, it is fully adducted at rest and does not fully abduct owing to the unopposed action of the medial rectus, leaving the person unable to abduct the pupil (Fig. B10.3).

Right: Normal   Left eye: Does not abduct

Direction of gaze ⟶

FIGURE B10.3. Abducent nerve (CN VI) lesion.

TRIGEMINAL NERVE (CN V)

The **trigeminal nerve (CN V)** emerges from the lateral aspect of the pons by a large *sensory root* and a small *motor root* (see Fig. 10.1). CN V is the principal general sensory nerve for the head (face, teeth, mouth, nasal cavity, and dura of the cranial cavity) (Fig. 10.10). The **sensory root of CN V** is composed mainly of the central processes of neurons in the **trigeminal ganglion** (Fig. 10.10B). The peripheral processes of the ganglionic neurons that form three nerves or divisions are the **ophthalmic nerve (CN V_1)**, **maxillary nerve (CN V_2)**, and sensory component of the **mandibular nerve (CN V_3)**. For a summary of CN V, see Figure 10.10 and Table 10.2. The fibers of the **motor root of CN V** are distributed exclusively via the mandibular nerve (CN V_3) to the muscles of mastication, mylohyoid, anterior belly of the digastric, tensor veli palatini, and tensor tympani.

TABLE 10.2. SUMMARY OF DIVISIONS OF TRIGEMINAL NERVE (CN V)

Divisions/Distributions	Branches

Ophthalmic nerve (CN V_1)
Somatic sensory only
Passes through superior orbital fissure
Supplies cornea, upper conjunctiva, mucosa of anterosuperior nasal cavity, frontal and ethmoidal sinuses, anterior and supratentorial dura mater, skin of dorsum of external nose, superior eyelid, forehead, and scalp
	Tentorial nerve (a meningeal branch)
	Lacrimal nerve
	  Communicating branch from zygomatic n.
	Frontal nerve
	  Supra-orbital nerve
	  Supratrochlear nerve
	Nasociliary nerve
	  Sensory root of ciliary ganglion
	  Short ciliary nerves
	  Long ciliary nerves
	  Anterior and posterior ethmoidal nerves
	  Infratrochlear nerves

Maxillary nerve (CN V_2)
Somatic sensory only
Passes through foramen rotundum
Supplies dura mater of anterior part of middle cranial fossa; conjunctiva of inferior eyelid; mucosa of postero-inferior nasal cavity, maxillary sinus, palate, and anterior part of superior oral vestibule; maxillary teeth; and skin of lateral external nose, inferior eyelid, anterior cheek, and upper lip
	Meningeal branch
	Zygomatic nerve
	  Zygomaticofacial branch
	  Zygomaticotemporal branch
	  Communicating branch to lacrimal nerve
	Ganglionic branches to (sensory root of) pterygopalatine ganglion
	Posterior superior alveolar branches
	Infra-orbital nerve
	  Anterior and middle superior alveolar branches
	  Superior labial branches
	  Inferior palpebral branches
	  External nasal branches
	Greater palatine nerves
	  Posterior inferior lateral nasal nerves
	Lesser palatine nerves
	Posterior superior lateral nasal branches
	Nasopalatine nerve
	Pharyngeal nerve

Mandibular nerve (CN V_3)
Somatic sensory and somatic (branchial) motor
Passes through the foramen ovale
Supplies sensory innervation to mucosa of anterior two thirds of tongue, floor of mouth, and posterior and anterior inferior oral vestibule; mandibular teeth; and skin of lower lip, buccal, parotid, and temporal regions of face; and external ear (auricle, upper external auditory meatus, and tympanic membrane)
Supplies motor innervation to muscles of mastication, mylohyoid, anterior belly of digastric, tensor tympani, and tensor veli palatini
	Somatic sensory branches
	  Meningeal branch (nervus spinosum)
	  Buccal nerve
	  Auriculotemporal nerve
	  Lingual nerve
	  Inferior alveolar nerve
	  Inferior dental plexus
	  Mental nerve
	Somatic (branchial) motor branches to:
	  Masseter
	  Temporalis
	  Medial and lateral pterygoids
	  Mylohyoid
	  Anterior belly of digastric
	  Tensor tympani
	  Tensor veli palatini

Somatic sensory CN V_1

Somatic sensory CN V_2

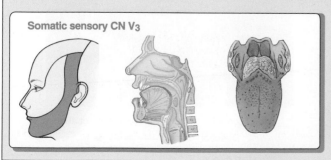

Somatic sensory CN V_3

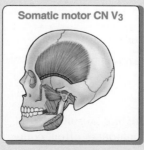

Somatic motor CN V_3

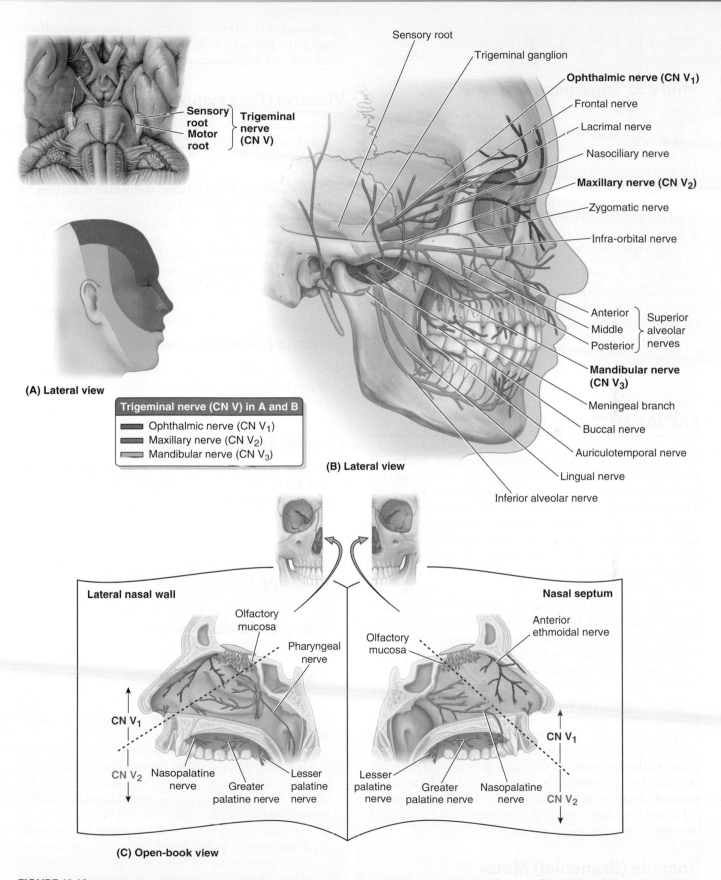

(A) Lateral view

Trigeminal nerve (CN V) in A and B

- Ophthalmic nerve (CN V₁)
- Maxillary nerve (CN V₂)
- Mandibular nerve (CN V₃)

Sensory root

Trigeminal ganglion

Ophthalmic nerve (CN V₁)

Frontal nerve

Lacrimal nerve

Nasociliary nerve

Maxillary nerve (CN V₂)

Zygomatic nerve

Infra-orbital nerve

Anterior
Middle
Posterior } Superior alveolar nerves

Mandibular nerve (CN V₃)

Meningeal branch

Buccal nerve

Auriculotemporal nerve

Lingual nerve

Inferior alveolar nerve

Sensory root
Motor root } Trigeminal nerve (CN V)

(B) Lateral view

Lateral nasal wall

Olfactory mucosa

Pharyngeal nerve

CN V₁

CN V₂   Nasopalatine nerve

Greater palatine nerve

Lesser palatine nerve

Nasal septum

Anterior ethmoidal nerve

Olfactory mucosa

CN V₁

CN V₂

Lesser palatine nerve

Greater palatine nerve

Nasopalatine nerve

(C) Open-book view

FIGURE 10.10. Distribution of trigeminal nerve (CN V). A. Cutaneous (sensory) distribution of the three divisions of the trigeminal nerve. **B.** Branches of the ophthalmic (CN V₁), maxillary (CN V₂), and mandibular divisions (CN V₃). **C.** CN V₁ and CN V₂ innervation of the palate and lateral wall and septum of the nasal cavity.

CLINICAL BOX

Injury to Trigeminal Nerve

CN V may be injured by trauma, tumors, aneurysms, or meningeal infections, causing the following signs and symptoms:

- *Paralysis of the muscles of mastication*, producing deviation of the mandible toward the side of the lesion
- *Loss of general sensation* (the ability to appreciate soft tactile, thermal, or painful sensations in the face)
- *Loss of the corneal reflex* (blinking in response to the cornea being touched) and the *sneezing reflex*

Trigeminal neuralgia (tic douloureux), the principal disease affecting the sensory root of CN V, produces excruciating, episodic pain that is usually restricted to the areas supplied by the maxillary and/or mandibular divisions of CN V.

FACIAL NERVE (CN VII)

The **facial nerve (CN VII)** emerges from the junction of the pons and medulla as two divisions: the *motor root* and the *intermediate nerve* (L. *nervus intermedius*) (see Fig. 10.1). The larger **motor root** (facial nerve proper) innervates the muscles of facial expression, and the smaller **intermediate nerve** carries taste, presynaptic parasympathetic, and somatic sensory fibers (Fig. 10.11). During its course, CN VII traverses the posterior cranial fossa, internal acoustic meatus, facial canal, stylomastoid foramen of the temporal bone, and parotid gland. After traversing the internal acoustic meatus, the nerve proceeds a short distance anteriorly within the temporal bone and then turns abruptly posteriorly to course along the medial wall of the tympanic cavity. The sharp bend is the **geniculum of the facial nerve** (Fig. 10.11A), the site of the **geniculate ganglion** (sensory ganglion of CN VII). Within the facial canal, CN VII gives rise to the greater petrosal nerve, the nerve to the stapedius, and the chorda tympani nerve. After running the longest intra-osseous course of any cranial nerve, CN VII emerges from the cranium via the *stylomastoid foramen*; gives off the posterior auricular branch; enters the parotid gland; and forms the **parotid plexus**, which gives rise to the following five terminal motor branches: temporal, zygomatic, buccal, marginal mandibular, and cervical.

Somatic (Branchial) Motor

As the nerve of the 2nd pharyngeal arch, the facial nerve supplies the striated muscle derived from its mesoderm, mainly the muscles of facial expression and auricular muscles. It also supplies the posterior belly of the digastric muscle and the stylohyoid and stapedius muscles.

Visceral (Parasympathetic) Motor

The parasympathetic distribution of the facial nerve is detailed in Figure 10.12. CN VII provides presynaptic parasympathetic fibers to the **pterygopalatine ganglion** for innervation of the lacrimal, nasal, pharyngeal, and palatine glands, and to the **submandibular ganglion** for innervation of the sublingual and submandibular salivary glands. The main features of parasympathetic ganglia associated with the facial nerve and other cranial nerves are summarized at the end of the chapter in Table 10.4. Parasympathetic fibers synapse in these ganglia, whereas sympathetic and other fibers pass through them without synapse.

Somatic (General) Sensory

Some fibers from the geniculate ganglion supply a small area of skin on both aspects of the auricle and in the region of the external acoustic meatus (Fig. 10.11).

Special Sensory (Taste)

Fibers carried by the chorda tympani join the **lingual nerve** of (CN V$_3$) to convey taste sensation from the anterior two thirds of the tongue and soft palate (Fig. 10.11).

CLINICAL BOX

Injury to Facial Nerve

A *proximal lesion of CN VII* near its origin or near the geniculate ganglion is accompanied by loss of motor, gustatory (taste), and autonomic functions. The *motor paralysis of facial muscles* involves upper and lower parts of the face on the ipsilateral (same) side (*Bell palsy*).

A *central lesion of CN VII* (lesion of the CNS) results in paralysis of muscles of the inferior face on the contralateral side. However, forehead wrinkling is not visibly impaired because it is innervated bilaterally. Lesions between the geniculate ganglion and the origin of the chorda tympani produce the same effects as that resulting from injury near the ganglion, except that lacrimal secretion is not affected. Because it passes through the facial canal, CN VII is vulnerable to compression when a viral infection produces inflammation of the nerve (*viral neuritis*).

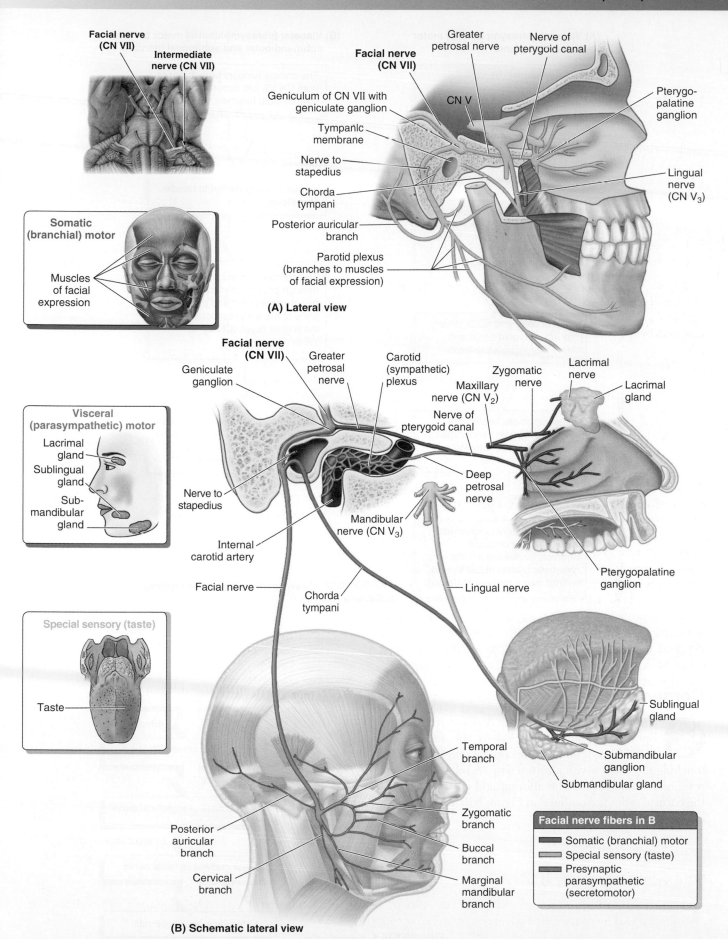

(A) Lateral view

(B) Schematic lateral view

Facial nerve
(CN VII)
Intermediate
nerve (CN VII)

Somatic
(branchial) motor

Muscles
of facial
expression

Visceral
(parasympathetic) motor
Lacrimal
gland
Sublingual
gland
Sub-
mandibular
gland

Special sensory (taste)

Taste

Facial nerve
(CN VII)
Greater
petrosal nerve
Nerve of
pterygoid canal
Geniculum of CN VII with
geniculate ganglion
CN V
Pterygo-
palatine
ganglion
Tympanic
membrane
Nerve to
stapedius
Chorda
tympani
Lingual
nerve
(CN V₃)
Posterior auricular
branch
Parotid plexus
(branches to muscles
of facial expression)

Facial nerve
(CN VII)
Geniculate
ganglion
Greater
petrosal
nerve
Carotid
(sympathetic)
plexus
Zygomatic
nerve
Maxillary
nerve (CN V₂)
Lacrimal
nerve
Lacrimal
gland
Nerve of
pterygoid canal
Nerve to
stapedius
Deep
petrosal
nerve
Internal
carotid artery
Mandibular
nerve (CN V₃)
Facial nerve
Chorda
tympani
Lingual nerve
Pterygopalatine
ganglion
Sublingual
gland
Submandibular
ganglion
Submandibular gland
Temporal
branch
Posterior
auricular
branch
Zygomatic
branch
Buccal
branch
Cervical
branch
Marginal
mandibular
branch

Facial nerve fibers in B
Somatic (branchial) motor
Special sensory (taste)
Presynaptic
parasympathetic
(secretomotor)

FIGURE 10.11. **Distribution of facial nerve (CN VII). A.** Facial nerve in situ; intra-osseous course and branches. **B.** Regional distribution of facial nerve.

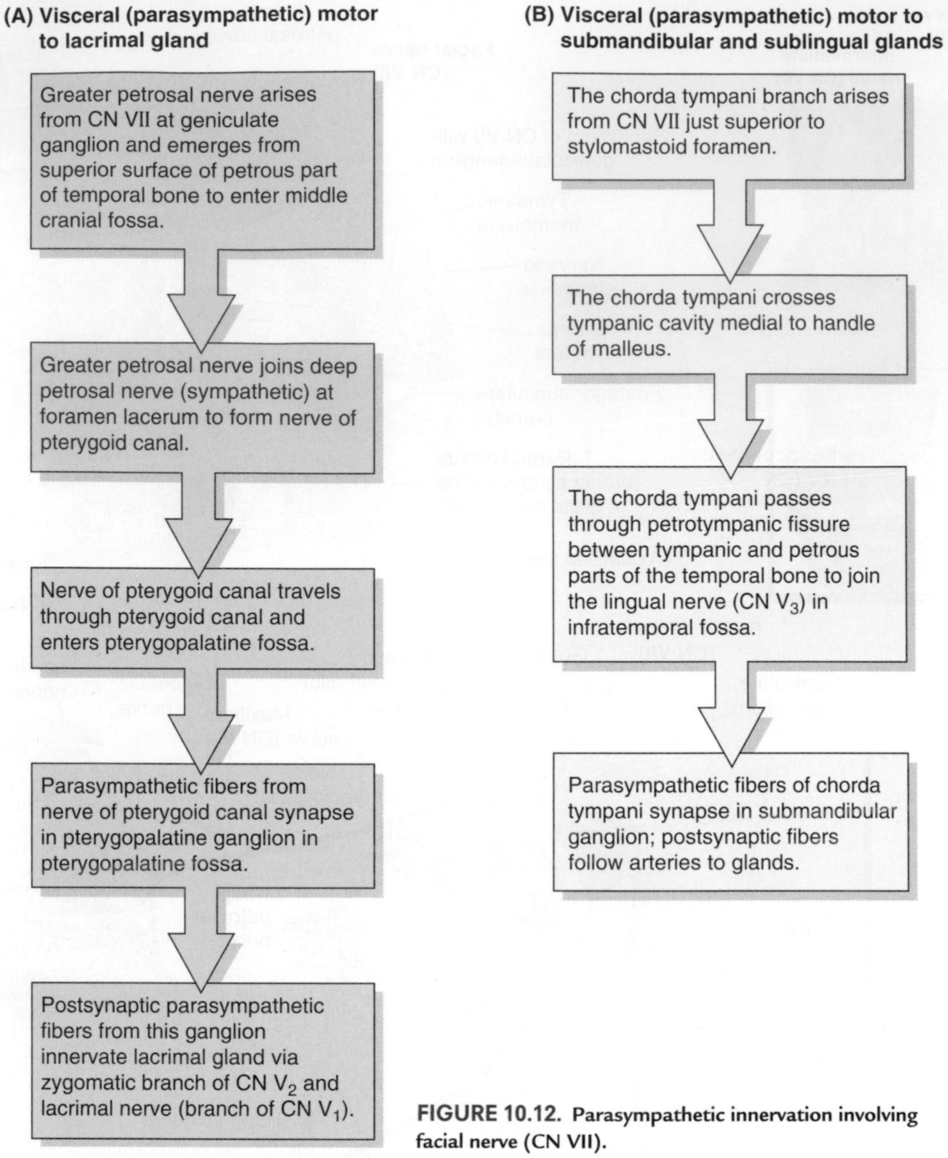

(A) Visceral (parasympathetic) motor to lacrimal gland

Greater petrosal nerve arises from CN VII at geniculate ganglion and emerges from superior surface of petrous part of temporal bone to enter middle cranial fossa.

↓

Greater petrosal nerve joins deep petrosal nerve (sympathetic) at foramen lacerum to form nerve of pterygoid canal.

↓

Nerve of pterygoid canal travels through pterygoid canal and enters pterygopalatine fossa.

↓

Parasympathetic fibers from nerve of pterygoid canal synapse in pterygopalatine ganglion in pterygopalatine fossa.

↓

Postsynaptic parasympathetic fibers from this ganglion innervate lacrimal gland via zygomatic branch of CN V$_2$ and lacrimal nerve (branch of CN V$_1$).

(B) Visceral (parasympathetic) motor to submandibular and sublingual glands

The chorda tympani branch arises from CN VII just superior to stylomastoid foramen.

↓

The chorda tympani crosses tympanic cavity medial to handle of malleus.

↓

The chorda tympani passes through petrotympanic fissure between tympanic and petrous parts of the temporal bone to join the lingual nerve (CN V$_3$) in infratemporal fossa.

↓

Parasympathetic fibers of chorda tympani synapse in submandibular ganglion; postsynaptic fibers follow arteries to glands.

FIGURE 10.12. Parasympathetic innervation involving facial nerve (CN VII).

CLINICAL BOX

Corneal Reflex

Loss of the *corneal reflex* may occur if either the ophthalmic nerve (CN V$_1$) or the facial nerve (CN VII) is lesioned. The corneal reflex is tested by touching the cornea with a cotton wisp. A bilateral blinking response should result. The afferent and efferent limbs of the corneal reflex are outlined in Figure B10.4.

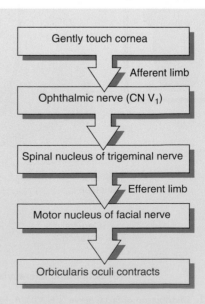

Gently touch cornea

↓ Afferent limb

Ophthalmic nerve (CN V$_1$)

↓

Spinal nucleus of trigeminal nerve

↓ Efferent limb

Motor nucleus of facial nerve

↓

Orbicularis oculi contracts

FIGURE B10.4. Corneal reflex.

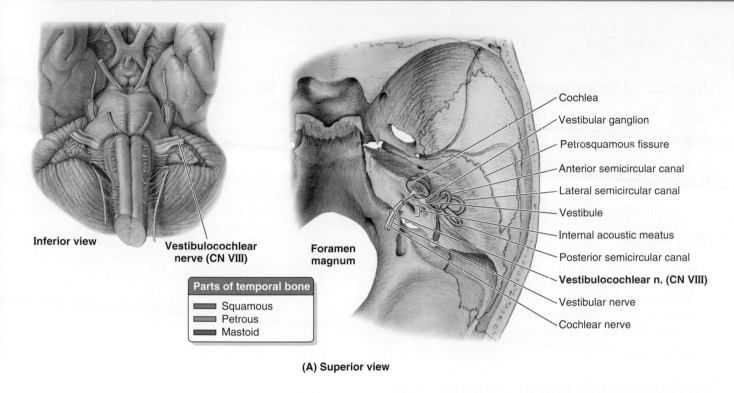

Inferior view

Vestibulocochlear nerve (CN VIII)

Foramen magnum

Cochlea
Vestibular ganglion
Petrosquamous fissure
Anterior semicircular canal
Lateral semicircular canal
Vestibule
Internal acoustic meatus
Posterior semicircular canal
Vestibulocochlear n. (CN VIII)
Vestibular nerve
Cochlear nerve

Parts of temporal bone	
	Squamous
	Petrous
	Mastoid

(A) Superior view

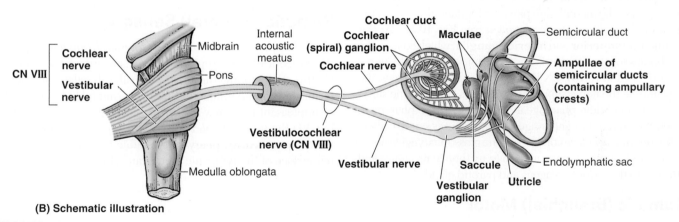

CN VIII
Cochlear nerve
Vestibular nerve
Midbrain
Pons
Medulla oblongata

Internal acoustic meatus

Vestibulocochlear nerve (CN VIII)

Cochlear duct
Cochlear (spiral) ganglion
Cochlear nerve
Maculae

Semicircular duct

Ampullae of semicircular ducts (containing ampullary crests)

Vestibular nerve
Vestibular ganglion
Saccule
Utricle
Endolymphatic sac

(B) Schematic illustration

FIGURE 10.13 Distribution of vestibulocochlear nerve (CN VIII). A. Internal surface of cranial base showing the location of the bony labyrinth of the internal ear within the temporal bone and the internal acoustic meatus for CN VIII. B. Schematic overview.

VESTIBULOCOCHLEAR NERVE (CN VIII)

The **vestibulocochlear nerve** (**CN VIII**) is a special sensory nerve of hearing and equilibrium. This nerve emerges from the junction of the pons and medulla of the brainstem and enters the *internal acoustic meatus* (see Fig. 10.1). Here, it separates into the *vestibular* and *cochlear nerves* (Fig. 10.13):

- The **vestibular nerve** is concerned with *equilibrium*. It is composed of the central processes of bipolar neurons in the **vestibular ganglion**; the peripheral processes of the neurons extend to the **maculae of the utricle** and **saccule** (sensitive to linear acceleration relative to the position of the head) and to the **ampullae of semicircular ducts** (sensitive to rotational acceleration).

- The **cochlear nerve** is concerned with *hearing*. It is composed of the central processes of bipolar neurons in the **spiral ganglion**; the peripheral processes of the neurons extend to the *spiral organ*.

CLINICAL BOX

Injuries of Vestibulocochlear Nerve

Although the vestibular and cochlear nerves are essentially independent, peripheral lesions often produce concurrent clinical effects because of their close relationship. Hence, *lesions of CN VIII* may cause *tinnitus* (ringing or buzzing of the ears), *vertigo* (dizziness, loss of balance), and impairment or loss of hearing. Central lesions may involve either the cochlear or vestibular divisions of CN VIII.

Deafness

There are two kinds of deafness: *conductive deafness*, involving the external or middle ear (e.g., *otitis media*, inflammation in the middle ear), and *sensorineural deafness*, which results from disease in the cochlea or in the pathway from the cochlea to the brain.

Acoustic Neuroma

An *acoustic neuroma* is a benign tumor of the **neurolemma** (Schwann cells). The tumor begins in the vestibular nerve and may develop within the internal acoustic meatus or at the cerebellopontine angle. The tumor usually presents initially with CN VIII dysfunction (i.e., unilateral hearing loss and *vestibular ataxia*—loss of balance and coordination). As the tumor grows, it may involve CN VII (especially within the internal acoustic meatus) or CN V, resulting in *facial palsy* and/or trigeminal sensory loss. Further progression of the tumor may compress CN IX, the cerebellum, and the brainstem.

GLOSSOPHARYNGEAL NERVE (CN IX)

The **glossopharyngeal nerve** (**CN IX**) emerges from the lateral aspect of the medulla and passes anterolaterally to leave the cranium through the *jugular foramen* (see Fig. 10.16). At this foramen are **superior** and **inferior ganglia**, which contain the cell bodies for the afferent (sensory) components of the nerve (Figs. 10.14 and 10.15). CN IX follows the stylopharyngeus, the only muscle the nerve supplies, as it passes between the superior and the middle pharyngeal constrictor of the pharynx to reach the oropharynx and tongue. It contributes sensory fibers to the *pharyngeal plexus* of nerves. The glossopharyngeal nerve is afferent from the tongue and pharynx (hence its name) and efferent to the stylopharyngeus and parotid gland.

Somatic (Branchial) Motor

Motor fibers pass to one muscle, the *stylopharyngeus*, derived from the 3rd pharyngeal arch.

Visceral (Parasympathetic) Motor

Following a circuitous route involving the **tympanic nerve**, **tympanic plexus**, and **lesser petrosal nerve**, presynaptic parasympathetic fibers are provided to the **otic ganglion** for innervation of the parotid gland (Figs. 10.14 and 10.15).

Somatic (General) Sensory

The *pharyngeal*, *tonsillar*, and *lingual branches* supply the mucosa of the oropharynx and isthmus of the fauces (L. throat), including the palatine tonsil, soft palate, and posterior third of the tongue. Stimuli determined to be unusual or unpleasant here may evoke the gag reflex or even vomiting. Via the *tympanic plexus*, CN IX supplies the mucosa of the tympanic cavity, pharyngotympanic tube, and the internal surface of the tympanic membrane (Fig. 10.14B).

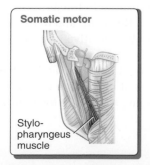

Somatic motor

Stylopharyngeus muscle

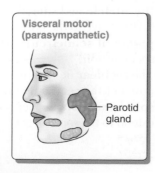

Visceral motor (parasympathetic)

Parotid gland

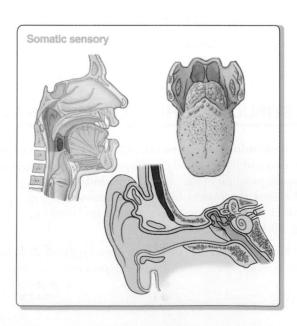

Somatic sensory

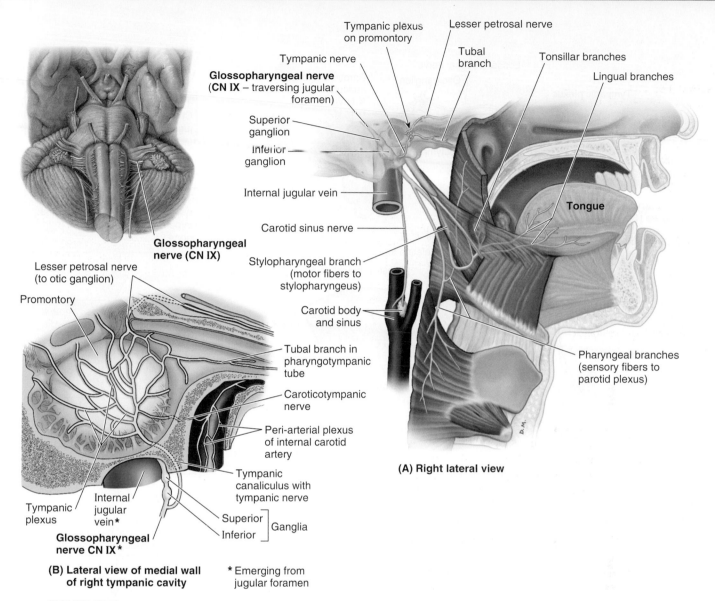

FIGURE 10.14. Distribution of glossopharyngeal nerve (CN IX). A. Pharynx. **B.** Middle ear (tympanic cavity and pharyngotympanic tube).

Special Sensory (Taste)

Taste fibers are conveyed from the posterior third of the tongue to the sensory ganglia.

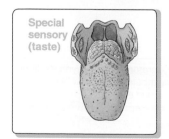

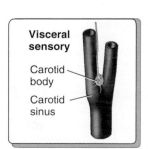

Visceral Sensory

The carotid sinus nerve supplies the **carotid sinus**, a *baro- (presso-) receptor* sensitive to changes in blood pressure, and the **carotid body**, a *chemoreceptor* sensitive to blood gas (oxygen and carbon dioxide) levels (Figs. 10.14 and 10.15).

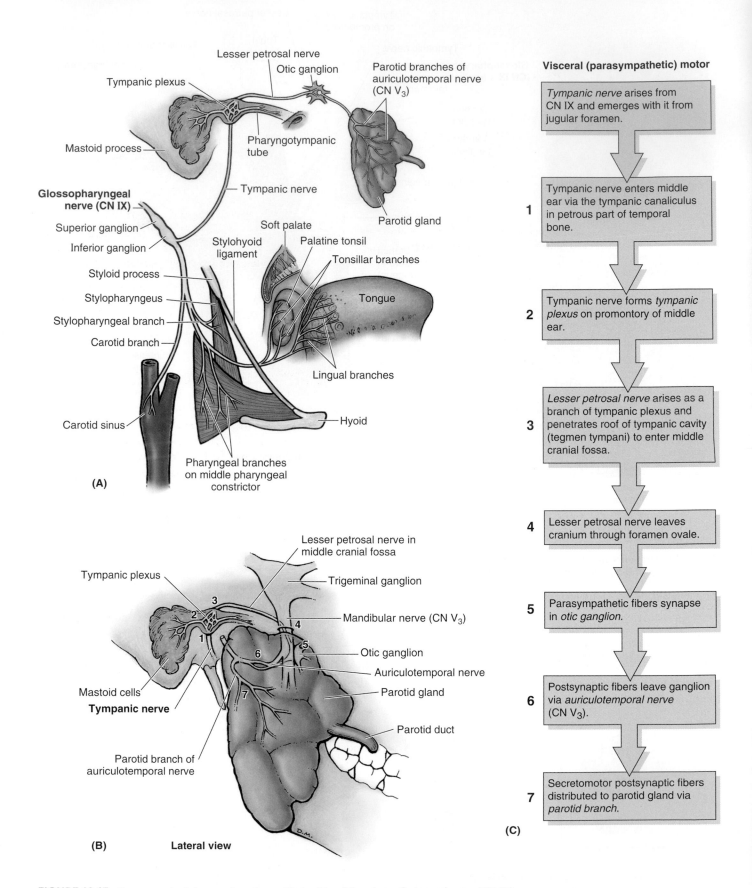

Visceral (parasympathetic) motor

Tympanic nerve arises from CN IX and emerges with it from jugular foramen.

1 Tympanic nerve enters middle ear via the tympanic canaliculus in petrous part of temporal bone.

2 Tympanic nerve forms *tympanic plexus* on promontory of middle ear.

3 *Lesser petrosal nerve* arises as a branch of tympanic plexus and penetrates roof of tympanic cavity (tegmen tympani) to enter middle cranial fossa.

4 Lesser petrosal nerve leaves cranium through foramen ovale.

5 Parasympathetic fibers synapse in *otic ganglion*.

6 Postsynaptic fibers leave ganglion via *auriculotemporal nerve* (CN V₃).

7 Secretomotor postsynaptic fibers distributed to parotid gland via *parotid branch*.

(C)

FIGURE 10.15. Parasympathetic innervation of parotid gland involving glossopharyngeal nerve (CN IX).

CLINICAL BOX

Lesions of Glossopharyngeal Nerve

Isolated lesions of CN IX or its nuclei are uncommon. Injuries of CN IX resulting from infection or tumors are usually accompanied by signs of involvement of adjacent nerves. Because CN IX, CN X, and CN XI pass through the jugular foramen, tumors in this region produce multiple cranial nerve palsies—the *jugular foramen syndrome*.

An isolated lesion would result in absence of taste on the posterior third of the tongue, changes in swallowing, absent *gag reflex* on the side of the lesion, and *palatal deviation* toward the unaffected side (Fig. B10.5). The afferent (sensory) limb of the gag reflex is via the glossopharyngeal nerve (CN IX) and the efferent (motor) limb is via the vagus nerve (CN X). The gag reflex is absent in about 37% of normal individuals (Davies et al., 1995).

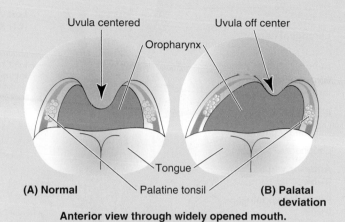

Anterior view through widely opened mouth.

FIGURE B10.5. Gag reflex. In **(B)**, note that the palate and posterior wall of the pharynx deviate to the left side when the gag reflex is elicited. This is due to a right CN IX/CN X lesion and is called the "curtain sign."

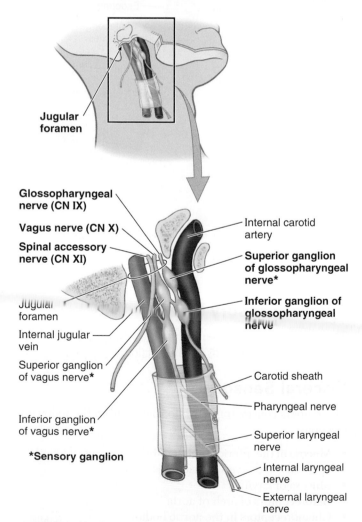

FIGURE 10.16. Relationship of structures traversing jugular foramen. CN IX, CN X, and CN XI are in numerical order, anterior to the internal jugular vein.

VAGUS NERVE (CN X)

The **vagus nerve** (**CN X**) arises by a series of rootlets from the lateral aspect of the medulla that merge and leave the cranium through the jugular foramen positioned between CN IX and CN XI (Fig. 10.16). What was formerly called "the cranial root of the accessory nerve" is actually a part of CN X (see Fig. 10.1). CN X has a **superior ganglion** in the jugular foramen that is mainly concerned with the general sensory component of the nerve. Inferior to the foramen is an **inferior ganglion** (nodose ganglion) concerned with the visceral sensory components of the nerve. In the region of the superior ganglion are connections to CN IX and the superior cervical (sympathetic) ganglion. CN X continues inferiorly in the carotid sheath to the root of the neck, supplying branches to the palate, pharynx, and larynx (Fig. 10.17 and Table 10.3).

The course that CN X takes in the thorax differs on the two sides (see Table 10.3). CN X supplies branches to the heart, bronchi, and lungs. The vagi join the *esophageal plexus* surrounding the esophagus, which is formed by branches of the vagi and sympathetic trunks. This plexus follows the esophagus through the diaphragm into the abdomen, where the **anterior** and **posterior vagal trunks** break up into branches that innervate the esophagus, stomach, and intestinal tract as far as the left colic flexure (Fig. 10.17).

Somatic (Branchial) Motor

Fibers from the *nucleus ambiguus* supply the following muscles:

- Pharyngeal muscles, except stylopharyngeus, via the pharyngeal plexus (with sensory fibers of the glossopharyngeal nerve)
- Muscles of the soft palate
- All muscles of the larynx

(continues on p. 667)

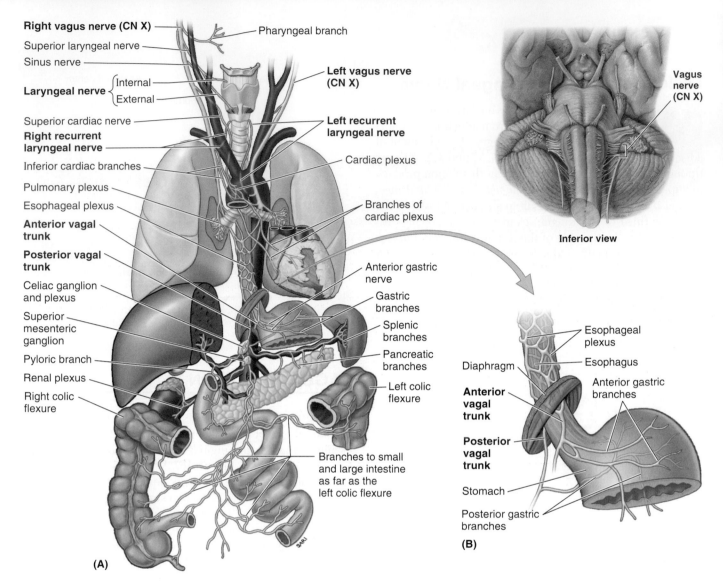

FIGURE 10.17. Distribution of vagus nerves (CN X). A. Course of nerves in neck, thorax, and abdomen. **B.** Anterior and posterior vagal trunks.

Visceral (Parasympathetic) Motor

Fibers from the *posterior (dorsal) nucleus of the vagus nerve* supply the thoracic and abdominal viscera to the left colic (splenic) flexure.

Somatic (General) Sensory

These fibers convey sensory information from the following locations:

- Dura mater of posterior cranial fossa
- Skin posterior to the ear
- External acoustic meatus

Special Sensory (Taste)

These fibers carry the sense of taste from the root of the tongue and the taste buds on the epiglottis.

Visceral Sensory

These fibers convey sensory information from the following locations:

- Mucosa of the inferior pharynx at the esophageal junction, epiglottis, and aryepiglottic folds
- Mucosa of larynx
- Baroreceptors of arch of aorta
- Chemoreceptors in the aortic bodies
- Thoracic and abdominal viscera

TABLE 10.3. SUMMARY OF VAGUS NERVE (CN X)

Divisions (Parts)	Branches
Cranial Vagi arise by a series of rootlets from medulla (includes traditional cranial root of CN XI).	Meningeal branch to dura mater (sensory; actual fibers of C2 spinal ganglion neurons that "hitch a ride" with vagus nerve) Auricular branch
Cervical Exit cranium/enter neck through jugular foramen; right and left vagus nerves enter carotid sheaths and continue to root of neck.	Pharyngeal branches to pharyngeal plexus (motor) Cervical cardiac branches (parasympathetic, visceral afferent) Superior laryngeal nerve (mixed), internal (sensory) and external (motor) branches Right recurrent laryngeal nerve (mixed)
Thoracic Vagi enter thorax through superior thoracic aperture; left vagus contributes to anterior esophageal plexus; right vagus to posterior plexus; form anterior and posterior vagal trunks	Left recurrent laryngeal nerve (mixed); all distal branches convey parasympathetic and visceral afferent fibers for reflex stimuli: Thoracic cardiac branches Pulmonary branches Esophageal plexus
Abdominal Anterior and posterior vagal trunks enter abdomen through esophageal hiatus in diaphragm.	Esophageal branches Gastric branches Hepatic branches Celiac branches (from posterior trunk) Pyloric branch (from anterior trunk) Renal branches Intestinal branches (to left colic flexure)

CLINICAL BOX

Lesions of Vagus Nerve

Isolated lesions of CN X are uncommon. Injury to the pharyngeal branches of CN X results in *dysphagia* (difficulty in swallowing). *Lesions of the superior laryngeal nerve* produce *anesthesia* of the superior part of the larynx and *paralysis of the cricothyroid muscle*. The voice is weak and tires easily. *Injury of a recurrent laryngeal nerve* may be caused by *aneurysms of the arch of the aorta* and may occur during neck operations. Injury of the recurrent laryngeal nerve causes *hoarseness* and *dysphonia* (difficulty in speaking) because of paralysis of the vocal folds (cords). Paralysis of both recurrent laryngeal nerves causes *aphonia* (loss of voice) and *inspiratory stridor* (a harsh, high-pitched respiratory sound). Because of its longer course, lesions of the left recurrent laryngeal nerve are more common than those of the right. *Proximal lesions of CN X* also affect the pharyngeal and superior laryngeal nerves, causing difficulty in swallowing and speaking. *Tachycardia* (accelerated heartbeat) and *cardiac arrhythmia* (irregular heartbeat) may occur.

SPINAL ACCESSORY NERVE (CN XI)

The **spinal accessory nerve** (**CN XI**) is somatic motor to the SCM and trapezius muscles (Fig. 10.18). The traditional "cranial root" of CN XI is actually a part of CN X (Lachman et al., 2002). CN XI emerges as a series of rootlets from the first five or six cervical segments of the spinal cord. It joins CN X temporarily as they pass through the jugular foramen, separating again as they exit (Fig. 10.16).

CN XI descends along the internal carotid artery, penetrates and innervates the SCM, and emerges from the muscle near the middle of its posterior border. It crosses the lateral cervical region (posterior triangle) and passes deep to the superior border of the trapezius to innervate it. Branches of the cervical plexus conveying sensory fibers from spinal nerves C2–C4 join the spinal accessory nerve in the lateral cervical region or within the muscles, providing them with pain and proprioceptive fibers.

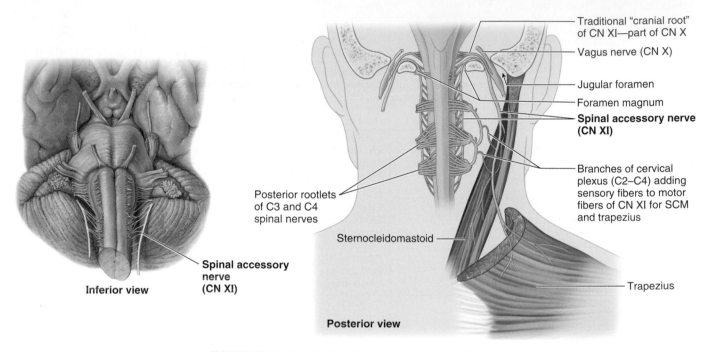

FIGURE 10.18. Distribution of spinal accessory nerve (CN XI).

CLINICAL BOX

Injury to Spinal Accessory Nerve

Because of its nearly subcutaneous passage through the lateral cervical region, CN XI is susceptible to injury during surgical procedures, such as *lymph node biopsy, cannulation of the internal jugular vein,* and *carotid endarterectomy* (surgical removal of sclerotic plaque from bifurcation of common carotid artery). *Lesions of CN XI produce atrophy of the trapezius with consequent weakness in elevating (shrugging) of the shoulder and impairment of rotary movements of the neck and chin to the opposite side as a result of weakness of the SCM* (Fig. B10.6).

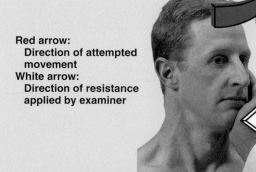

Red arrow:
   Direction of attempted movement
White arrow:
   Direction of resistance applied by examiner

(A) Test for sternocleidomastoid function

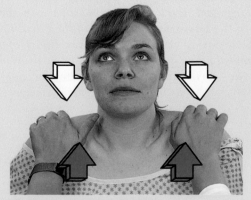

(B) Test for trapezius function

FIGURE B10.6. Muscle tests for sternocleidomastoid and trapezius.

HYPOGLOSSAL NERVE (CN XII)

The **hypoglossal nerve** (**CN XII**) is somatic motor to intrinsic and extrinsic muscles of the tongue (styloglossus, hyoglossus, and genioglossus). The hypoglossal nerve arises as a purely motor nerve by several rootlets from the medulla (see Fig. 10.1) and leaves the cranium through the *hypoglossal canal* (Fig. 10.19). After exiting the cranial cavity, the nerve is joined by a branch or branches of the cervical plexus, conveying general somatic motor fibers from C1 and C2 spinal nerves and general somatic sensory fibers from the spinal ganglion of C2. These spinal nerve fibers "hitch a ride" with CN XII to reach the hyoid muscles, with some of the sensory fibers passing retrograde along it to reach the dura mater of the posterior cranial fossa. CN XII passes inferiorly medial to the angle of the mandible and then curves anteriorly to enter the tongue.

CN XII ends in many branches that supply all the extrinsic muscles of the tongue, except the palatoglossus (which is a palatine muscle). CN XII has the following branches:

- A **meningeal branch** that returns to the cranium through the hypoglossal canal and innervates the dura mater on the floor and posterior wall of the posterior cranial fossa. The nerve fibers conveyed are from the sensory spinal ganglion of spinal nerve C2, not from CN XII.
- The **superior root of the ansa cervicalis** that branches from CN XII to supply the infrahyoid muscles (sternohyoid, sternothyroid, and omohyoid). This branch conveys only fibers from the cervical plexus (loop between the anterior rami of C1 and C2) that joined the nerve outside the cranial cavity. Some fibers reach the thyrohyoid muscle.
- Terminal **lingual branches** that supply the styloglossus, hyoglossus, genioglossus, and intrinsic muscles of the tongue

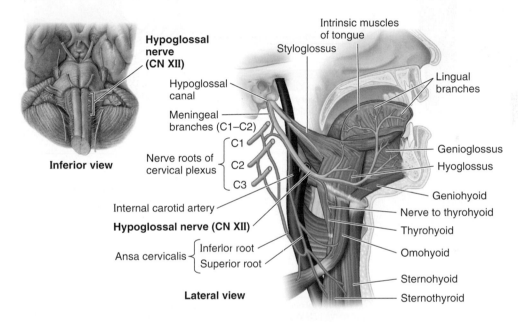

Inferior view

Hypoglossal nerve (CN XII)
Hypoglossal canal
Meningeal branches (C1–C2)
Nerve roots of cervical plexus — C1, C2, C3
Internal carotid artery
Hypoglossal nerve (CN XII)
Ansa cervicalis — Inferior root, Superior root
Lateral view

Intrinsic muscles of tongue
Styloglossus
Lingual branches
Genioglossus
Hyoglossus
Geniohyoid
Nerve to thyrohyoid
Thyrohyoid
Omohyoid
Sternohyoid
Sternothyroid

FIGURE 10.19. Distribution of hypoglossal nerve (CN XII).

CLINICAL BOX

Injury to Hypoglossal Nerve

Injury to CN XII paralyzes the ipsilateral half of the tongue, resulting in *dysarthria*. After some time, the tongue *atrophies*, making it appear shrunken and wrinkled. When the tongue is protruded, its apex deviates toward the paralyzed side due to lack of muscle contraction on that side (Fig. B10.7).

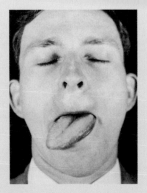

FIGURE B10.7. Hypoglossal (CN XII) nerve lesion.

TABLE 10.4. SUMMARY OF CRANIAL PARASYMPATHETIC GANGLIA AND RELATED VISCERAL MOTOR AND SENSORY FIBER DISTRIBUTION.

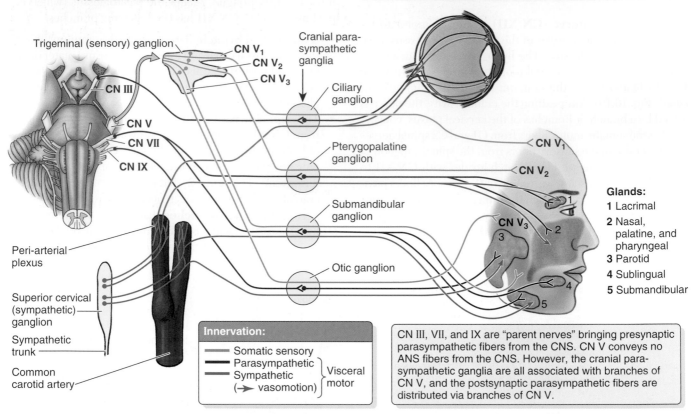

CN III, VII, and IX are "parent nerves" bringing presynaptic parasympathetic fibers from the CNS. CN V conveys no ANS fibers from the CNS. However, the cranial parasympathetic ganglia are all associated with branches of CN V, and the postsynaptic parasympathetic fibers are distributed via branches of CN V.

Innervation:
— Somatic sensory
— Parasympathetic } Visceral motor
— Sympathetic (→ vasomotion)

Glands:
1 Lacrimal
2 Nasal, palatine, and pharyngeal
3 Parotid
4 Sublingual
5 Submandibular

Parasympathetic Ganglion	Location	Parasympathetic Root	Sympathetic Root[a]	Main Distribution
Ciliary	Between optic nerve and lateral rectus, close to apex of orbit	Inferior branch of oculomotor nerve (CN III)	Postsynaptic fibers from superior cervical ganglion branch from peri-arterial plexus on internal carotid artery in cavernous sinus	Parasympathetic postsynaptic fibers from ciliary ganglion pass to ciliary muscle and sphincter pupillae of iris; sympathetic postganglionic fibers from superior cervical ganglion pass to dilator pupillae and blood vessels of eye.
Pterygopalatine	In pterygopalatine fossa, where it is suspended by ganglionic branches of maxillary nerve (sensory roots of pterygopalatine ganglion); just anterior to opening of pterygoid canal and inferior to CN V₂	Greater petrosal nerve from facial nerve (CN VII) via nerve of pterygoid canal	Deep petrosal nerve, a branch of peri-arterial plexus on internal carotid artery that is a continuation of postsynaptic fibers of cervical sympathetic trunk; fibers from superior cervical ganglion pass through pterygopalatine ganglion and enter branches of CN V₂.	Parasympathetic postsynaptic (secretomotor) fibers from pterygopalatine ganglion innervate lacrimal gland via zygomatic branch of CN V₂; sympathetic postsynaptic fibers from superior cervical ganglion accompany branches of pterygopalatine nerve that are distributed to blood vessels of nasal cavity, palate, and superior parts of pharynx.
Otic	Between tensor veli palatini and mandibular nerve (CN V₃); lies inferior to foramen ovale of sphenoid bone	Tympanic nerve from glossopharyngeal nerve (CN IX); from tympanic plexus, tympanic nerve continues as lesser petrosal nerve.	Fibers from superior cervical ganglion come from peri-arterial plexus on middle meningeal artery.	Parasympathetic postsynaptic fibers from otic ganglion are distributed to parotid gland via auriculotemporal nerve (branch of CN V₃); sympathetic postsynaptic fibers from superior cervical ganglion pass to parotid gland and supply its blood vessels.
Submandibular	Suspended from lingual nerve by two ganglionic branches (sensory roots); lies on surface of hyoglossus muscle inferior to submandibular duct	Parasympathetic fibers join facial nerve (CN VII) and leave it in its chorda tympani branch, which unites with lingual nerve.	Sympathetic fibers from superior cervical ganglion via peri-arterial plexus on facial artery	Parasympathetic postsynaptic (secretomotor) fibers from submandibular ganglion are distributed to sublingual and submandibular glands; sympathetic fibers supply sublingual and submandibular glands.

[a]Sympathetic fibers traverse ganglia en route to blood vessels and dilator pupillae muscle, but do not synapse in the cranial parasympathetic ganglia

APPENDIX A | References

CHAPTER 1 Overview and Basic Concepts

Terminologia Anatomica: International Anatomical Terminology. Federative Committee on Anatomical Terminology; 1998.

Moore KL, Persaud TVN, Torchia MG. *The Developing Human: Clinically Oriented Embryology.* 10th ed. Philadelphia, PA: Elsevier/Saunders; 2016.

Swartz MH. *Textbook of Physical Diagnosis: History and Examination.* 7th ed. Philadelphia, PA: Elsevier/Saunders; 2014.

Wilson-Pauwels L, Stewart PA, Akesson E. *Autonomic Nerves: Basic Science, Clinical Aspects, Case Studies.* Ontario, Canada: Decker; 2010.

CHAPTER 2 Back

Bogduk N. *Clinical and Radiological Anatomy of the Lumbar Spine and Sacrum.* 5th ed. London, United Kingdom: Churchill Livingstone; 2012.

Haines DE, ed. *Fundamental Neuroscience for Basic and Clinical Applications.* 4th ed. Philadelphia, PA: Elsevier/Saunders; 2013.

Moore KL, Persaud TVN, Torchia MG. *The Developing Human: Clinically Oriented Embryology.* 10th ed. Philadelphia, PA: Elsevier/Saunders; 2016.

CHAPTER 3 Upper Limb

Foerster O. The dermatomes in man. *Brain.* 1933;56:1–39.

Keegan JJ, Garrett FD. The segmental distribution of the cutaneous nerves in the limbs of man. *Anat Rec.* 1948;102:409–437.

CHAPTER 4 Thorax

Goroll AH, Mulley AG. *Primary Care Medicine: Office Evaluation and Management of the Adult Patient.* 7th ed. Philadelphia, PA: Wolters Kluwer; 2014.

Moore KL, Persaud TVN, Torchia MG. *The Developing Human: Clinically Oriented Embryology.* 10th ed. Philadelphia, PA: Elsevier/Saunders; 2016.

Swartz MH. *Textbook of Physical Diagnosis: History and Examination.* 7th ed. Philadelphia, PA: Elsevier/Saunders; 2014.

Tubbs RS, Shoja MM, Loukas M. *Bergman's Comprehensive Encyclopedia of Human Anatomic Variation.* 3rd ed. Hoboken, NJ: John Wiley & Sons; 2016. https://onlinelibrary.wiley.com/doi/book/10.1002/9781118430309.

CHAPTER 5 Abdomen

Moore KL, Persaud TVN, Torchia MG. *The Developing Human: Clinically Oriented Embryology.* 10th ed. Philadelphia, PA: Elsevier/Saunders; 2016.

Skandalakis LJ, Gadacz TR, Mansberger AR Jr, et al. *Modern Hernia Repair: The Embryological and Anatomical Basis of Surgery.* New York, NY: Parthenon; 1996.

CHAPTER 6 Pelvis and Perineum

Moore KL, Persaud TVN, Torchia MG. *The Developing Human: Clinically Oriented Embryology.* 10th ed. Philadelphia, PA: Elsevier/Saunders; 2016.

Oelrich TM. The striated urogenital sphincter muscle in the female. *Anat Rec.* 1983;205:223–232.

CHAPTER 7 Lower Limb

Foerster O. The dermatomes in man. *Brain.* 1933;56:1–39.

Keegan JJ, Garrett FD. The segmental distribution of the cutaneous nerves in the limbs of man. *Anat Rec.* 1948;102:409–437.

CHAPTER 8 Head

Esenwa CC, Czeisler BM, Mayer SA. Acute ischemic stroke. In: Louis ED, Mayer SA, Rowland LP, eds. *Merritt's Neurology.* 13th ed. Philadelphia, PA: Wolters Kluwer; 2016:237–384.

Haines DE, ed. *Fundamental Neuroscience for Basic and Clinical Applications.* 4th ed. Philadelphia, PA: Elsevier/Saunders; 2013.

Marshall RS. Transient ischemic attack. In: Louis ED, Mayer SA, Rowland LP, eds. *Merritt's Neurology.* 13th ed. Philadelphia, PA: Wolters Kluwer; 2016:285–287.

Moore KL, Persaud TVN, Torchia MG. *The Developing Human: Clinically Oriented Embryology.* 10th ed. Philadelphia, PA: Elsevier/Saunders; 2016.

CHAPTER 10 Cranial Nerves

Davies AE, Kidd D, Stone SP, et al. Pharyngeal sensation and gag reflex in healthy subjects. *Lancet.* 1995;345:487–488.

Lachman N, Acland RD, Rosse C. Anatomical evidence for the absence of a morphologically distinct cranial root of the accessory nerve in man. *Clin Anat.* 2002;15:4–10.

Simpson KL, Sweazey RD. Olfaction and taste. In: Haines DE, ed. *Fundamental Neuroscience for Basic and Clinical Applications.* 4th ed. New York, NY: Saunders; 2013.

Index

Note: Page numbers in **boldface** indicate the primary entry for the term, which appears in boldface or as a heading on the page indicated and is followed by a definition or explanation; page numbers in *italics* denote figures; and those followed by "t" denote tables.

A

Abdomen, 253–338
  acute, 260, 296
  definition of, **254**
  fat accumulation on, 260
  hernias of, 269, 271, 294–295, 326
  inguinal region, 254, 255, **263**–272
  medical imaging of, 334–338, *334–338*
  minimally invasive surgery, 260
  paracentesis of, 277
  planes of, 254, *255*
  protuberance of, 260
  quadrants of, 254, *255*
  regions of, 254, *255*
  surgical incisions, 259–260, *260*
  transverse section of, *272*
  vessel studies of, 338, *338*
  viscera of, *278*, **278**–323 (*See also*
      individual organs of abdomen)
    innervation of, 317–323
    overview of, 278–279
    palpation of, 260
    parasympathetic innervation of,
        317–320, *318–319*, 319t
    sympathetic innervation of, 317,
        318–319, 319t
    visceral sensory innervation of, *322*,
        322–323
  walls of (*See* Abdominal wall)
Abdominal region, *340*
Abdominal tubal ligation, 380
Abdominal wall
  anterior, 255
  anterolateral, 254, **255**–263
    arteries of, 262, *262*, 263t
    boundaries of, 255
    fascia of, 255, *255*, 258, *258*, 259
    internal surface of, *255*, 258, *258*–259
    lymphatic drainage of, 263, *263*
    muscles of, *256*, 256–257, 257t
      flat, 256
      functions and actions of, 257, 257t
      vertical, 257

nerves of, 262, *262*, 262t
  palpation of, 260
  surface anatomy of, 261, *261*
  veins of, 263, *263*
lateral, 255
posterior, 255, **327**–332
  composition of, 327
  fascia of, *327*, 327–328
  lymphatic drainage of, 331–332, *332*
  muscles of, *328*, 328, 328t
  nerves of, 328–330, *329*
  pain, 333
  vasculature of, 311–312, 330–331,
      330–331
  viscera of, *311–312*
Abduction, *6–7*
  of digits of foot, 466, 493t
  of digits of hand, 152, 153t, 154, 177t
  of joints of upper limb, *109*
  of pupil of eyeball, *542*, 542–544, *543*
  structures limiting,
    of foot and toes, 493t
    of hand joints, 177t
    of hip joint, 476t
    of shoulder joint/region, 165t
    of wrist and carpal joints, 175t
  of thigh at hip joint, *475*, 475–476,
      476t
  of thumb, 151, *151*, 178t
  of vocal folds, *625*
  of wrist (carpal joints), 175t
Abrasions, 433
Abscess
  ischio-anal, 394, *394*
  neck (cervical), 598
  perinephric, 316
  psoas, 333
  retropharyngeal, 598
  subphrenic, 302
Absorption
  sublingual, 571
  transmucosal, 571
Accommodation, in vision, **540**, 655

Acetabulum, **340**, *341*, 411, *413*, *414*,
    472–474
  comparison of male and female, 343t
  labrum of, *473*, **474**
  lunate surface of, *473*, **474**
  rim of, 474, *474*
Acetylcholine, 33
Acoustic meatus
  external, *501*, **502**, *504*, 580, *580*, 581,
      583, *585*, 588
  internal, *505*, **506**, 585, **587–588**
Acoustic neuroma, 662
Acromial end, of clavicle, **93**, *93*, *94*, 101
Acromion of scapula, **93**, *93*, *94*, *96*, 101,
    101, 163, *163*, *166*
Adduction, *6–7*
  of digits of foot, 466, 493t
  of digits of hand, 152, 153t, 154, 177t
  of joints of upper limb, 109
  of pupil of eyeball, 542, *542*, *543*, 544–545
  structures limiting,
    of foot and toes, 493t
    of hand joints, 177t
    of hip joint, 476t
    of shoulder joint/region, 165t
    of wrist and carpal joints, 175t
  of thigh at hip joint, *475*, 475–476, 476t
  of thumb, 151, *151*, 178t
  of vocal folds, *625*, *625*
  of wrist (carpal joints), 175t
Adenocarcinoma, ductular, 000
Adenoid, 637
Adenoiditis, 637
Adhesions
  peritoneal, 277
  pleural, 216
Adhesiotomy, 277
Adhesive capsulitis, 168
Aditus, **582**, *583*
Aging
  of intervertebral (I–V) discs, 61, 68
  of vertebrae, 61, 68, *68*
  of zygapophysial joints, 68